Muscles of Chordates

Development, Homologies, and Evolution

Rui Diogo
Janine M. Ziermann
Julia Molnar
Natalia Siomava
Virginia Abdala

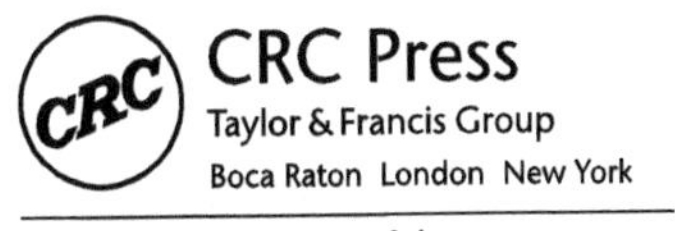

CRC Press is an imprint of the
Taylor & Francis Group, an **informa** business

CRC Press
Taylor & Francis Group
6000 Broken Sound Parkway NW, Suite 300
Boca Raton, FL 33487-2742

Printed on acid-free paper

International Standard Book Number-13: 978-1-138-57116-7 (Paperback)
International Standard Book Number-13: 978-1-138-57123-5 (Hardback)

Library of Congress Cataloging-in-Publication Data

Names: Diogo, Rui, author.
Title: Muscles of chordates : development, homologies, and evolution / Rui Diogo, Janine M. Ziermann, Julia Molnar, Natalia Siomava, and Virginia Abdala.
Description: Boca Raton : Taylor & Francis, 2018. | Includes bibliographical references and index.
Identifiers: LCCN 2017049446 | ISBN 9781138571167 (paperback : alk. paper)
Subjects: LCSH: Chordata--Anatomy. | Muscles--Anatomy.
Classification: LCC QL605 .D56 2018 | DDC 596--dc23
LC record available at https://lccn.loc.gov/2017049446

Visit the Taylor & Francis Web site at
http://www.taylorandfrancis.com

and the CRC Press Web site at
http://www.crcpress.com

Printed and bound in the United States of America by Sheridan

Contents

Preface

In 2010, two of us (Diogo and Abdala) published the book *Muscles of Vertebrates*, which had a wide impact within the scientific community, as well as in courses of zoology and comparative anatomy across the globe. A major reason for that impact was that before the publication of that book, there had been no attempt to combine, in a single book, information about the head, neck, and pectoral appendage muscles of all major extant vertebrate groups. Because of that impact, many scientists as well as teachers and students have demanded from us an even more complete book that (a) also includes muscles of the pelvic appendages as well as of the median appendages; (b) embraces even more taxa, not only the other extant chordates, but also more subgroups within each of the major vertebrate clades; (c) reflects the large amount of data that has been obtained in experimental evolutionary developmental biology (evo–devo) on chordate muscle development, including the strong links between the heart and head muscles; and (d) combines all these items in order to discuss broader issues linking the study of muscles and their implications for macroevolution, the links between phylogeny and ontogeny, homology and serial homology, regeneration, and evolutionary medicine. This book is the answer to those demands, as it compiles the information available on the evolution, development, and homologies of all skeletal muscles of all major extant groups of chordates. The chordates are a fascinating group of animals that includes about 70,000 living species that have an outstanding anatomical, ecological, and behavioral diversity, including forms living in fresh and seawaters, forests, deserts and the arctic, and flying high in the skies. This book will thus have a crucial impact in fields such as evo–devo, developmental biology, evolutionary biology, comparative anatomy, ecomorphology, functional anatomy, zoology, and biological anthropology, because it also pays special attention to the configuration, evolution and variations of the skeletal muscles of humans. Moreover, it is written and illustrated in a way that makes it useful for not only scientists working in these and other fields but also teachers and students related to any of these fields or simply interested in knowing more about the development, comparative anatomy, and evolution of chordates in general or about the origin and evolutionary history of the structures of our own body in particular.

Rui Diogo
Washington, DC

About the Authors

Rui Diogo is an associate professor at the Howard University College of Medicine and a resource faculty at the Center for the Advanced Study of Hominid Paleobiology of George Washington University. He was one of the youngest researchers to be nominated as fellow of the American Association of Anatomists, and he won several prestigious awards, being the only researcher selected for first and second places for best article of the year in the top anatomical journal two times in just three years (2013/2015). In addition to being the single author or coauthor of more than 100 papers in top journals, such as *Nature*, and of numerous book chapters, he is the coeditor of five books and the sole or first author of 13 books covering subjects as diverse as fish evolution, chordate development, human medicine and pathology, and the links between evolution and behavioral ecology. One of these books was adopted at medical schools worldwide, *Learning and Understanding Human Anatomy and Pathology: An Evolutionary and Developmental Guide for Medical Students*, and another one has been often listed as one of the best 10 books on evolutionary biology in 2017, *Evolution Driven by Organismal Behavior: A Unifying View of Life, Function, Form, Mismatches, and Trends.*

Janine M. Ziermann is an assistant professor at the Howard University College of Medicine. She received her PhD in Germany studying the evolution and development of head muscles in larval amphibians. This was followed by a postdoctoral in Netherlands and one in the United States to further study vertebrates. Her current research focuses on the evolution and development of the cardiopharyngeal field, which gives rise to head, neck, and heart musculature. Additionally, she aims to use the knowledge from her research to better understand congenital defects, which often affect both head and heart structures. She won several awards, including the American Association of Anatomists (AAA) and the Keith and Marion Moore Young Anatomist's Publication Award (YAPA). She single authored or coauthored more than 30 papers in top journals, such as *Nature*, book chapters, commentaries, and books.

Julia Molnar is an assistant professor at New York Institute of Technology, College of Osteopathic Medicine. She received a prestigious postdoctoral fellowship from the American Association of Anatomists and many illustration awards, including the Lazendorf Award, for paleontological illustration. She has an extensive publication record that spans the fields of biomechanics, comparative anatomy, and paleontology. Her scientific illustrations and animations have been featured on numerous news websites, including PBS, National Geographic, and the History Channel, and at paleontology museums around the world.

Natalia Siomava won a prestigious stipend from DAAD (German Academic Exchange Service) to study in Germany where, at the age of 27, she obtained her PhD degree in developmental and evolutionary biology. She then moved to the United States as a young research fellow at Howard University College of Medicine, where she developed her skills in vertebrate comparative anatomy. She has experience working in leading researcher groups in Europe and the United States, and her works are used as a basis for lab manuals for students. She is a member of the American Association of Anatomists and a volunteer in several projects aimed to help scientists save the biodiversity of life on earth.

Virginia Abdala is an associate professor at the Universidad Nacional de Tucumán and researcher at the Consejo Nacional de Investigaciones Científicas y Técnicas, Argentina. In addition to being the single author or coauthor of more than 85 papers and of numerous book chapters, she is the academic editor of two prestigious international journals known worldwide. She is also the coauthor of two books and coeditor of another one, which is the first one produced to be used in courses of vertebrate comparative morphology in Argentina.

Acknowledgments

We want to thank above all the curators and staff of the numerous collections and all the institutions that kindly provided the specimens we dissected, as well as all the authors that worked on other specimens and reported them in the publications that were reviewed and compiled by us. In particular, we would like to thank our coauthors who agreed to share portions of our joint publications in this book, including Borja Esteve-Altava, Peter Johnston, Elena Voronezhskaya, Fedor Shkil, Raul E. Diaz, Tautis Skorka, and Grant Dagliyan. We also want to thank all the numerous researchers, teachers, and students with whom we discussed vertebrate anatomy, functional morphology, development, phylogeny, paleontology, and evolution as well as on any other subjects addressed in the present book. Also, thanks to all those who have been involved in administering the various grants and other awards that we have received and that are related in one way or another with this book, without which this work would really not have been possible. We also want to thank all our colleagues, friends, and families for their kind support and encouragement.

Thanks to *all* of you!

1 Introduction

Chordates are characterized by possession of a notochord, pharyngeal slits, and a hollow dorsal nerve during at least some of their developmental stages, and they represent over 550 million years of evolution. About 70,000 living species comprise this fascinating and ecologically, behaviorally, and anatomically diverse group of animals. A cladogram showing the relationships of those main extant chordate clades to which we refer in the present volume is shown in Figure 1.1. As can be seen in that cladogram, the phylogenetically most basal extant chordate clade is the Cephalochordata, which includes lancelets, also known as amphioxus, and consists of about 30 living species. The Olfactores thus includes both the vertebrates and the tunicates (also known as urochordates), which comprises more than 2150 living species that mainly live in shallow ocean waters, including sea squirts (ascidians), sea porks, sea livers, and sea tulips. More than 66,000 species of vertebrates—chordates characterized by features such as backbones and spinal columns—have been described so far. Vertebrates originated about 525 million years ago during the Cambrian explosion and include the extant clades Cyclostomata (hagfishes and lampreys) and Gnathostomata, which is in turn subdivided into chondrichthyans (holocephalans and elasmobranchs) and osteichthyans.

The Osteichthyes is a highly speciose group of animals, divided into two extant clades: the Sarcopterygii (lobe-finned fishes and tetrapods) and the Actinopterygii (ray-finned fishes). The Polypteridae (included in the Cladistia) are commonly considered to be the phylogenetically most basal extant actinopterygian taxon. The Chondrostei (including the Acipenseridae and Polyodontidae) is usually considered the sister-group of a clade including the Lepisosteidae (included in the Ginglymodi) and the Amiidae (included in Halecomorphi) plus the Teleostei. Within the Teleostei, four main living clades are usually recognized: the Elopomorpha, Osteoglossomorpha, Otocephala (Clupeomorpha + Ostariophysi), and Euteleostei. Authors continue to debate whether Halecomorphi is the sister-group of teleosts or of Ginglymodi; in the latter case, the Ginglymodi and Halecomorphi would be included in the clade Holostei. However, we do not consider the data published since *Muscles of Vertebrates* (Diogo and Abdala 2010) was written to be conclusive enough to contradict the more traditional Halecomorphi–Teleostei sister-group relationship followed in that book. In fact, in a paper published just 2 months ago that specifically addressed this topic, the authors concluded that at least concerning cytogenetic data the Amiidae are more similar to teleosts than to any non-teleostean actinopterygians and that there are actually "striking differences" between the Amiidae and the Lepisosteidae (Majtanova et al. 2017). Therefore, in the present book, we follow the Halecomorphi–Teleostei sister-group relationship. Be that as it may, the broader ideas presented in this book—for instance, regarding muscle homologies and macroevolution—would not be significantly changed if we followed the alternative phylogenetic hypothesis. These groups are very closely related clades of just a specific subgroup of fishes (actinopterygians), in a book that also includes sarcopterygian fishes, chondrichthyan fishes, tetrapods, and cyclostomes as well as nonvertebrate chordates.

The Sarcopterygii includes two groups of extant fishes, the coelacanths (Actinistia) and lungfishes (Dipnoi), and the Tetrapoda. Within tetrapods, Amphibia is the sister-group of Amniota, which includes the Mammalia and the Reptilia (note: when we use the term *reptiles*, we refer to the group including lepidosaurs, birds, crocodylians, and turtles, which, despite some controversy, continues to be considered a monophyletic taxon by most taxonomists: see, e.g., Gauthier et al. 1988; Kardong 2002; Dawkins 2004; Diogo 2007; Conrad 2008). The Amphibia include three main extant groups: caecilians (Gymnophiona or Caecilia), frogs (Anura or Salientia), and salamanders (Caudata or Urodela), the two latter groups being possibly more closely related to each other than to the caecilians (see, e.g., Carroll 2007). As noted just above, the Reptilia include four main extant groups: turtles (Testudines), lepidosaurs (Lepidosauria), crocodylians (Crocodylia), and birds (Aves). The Lepidosauria comprises the Rhynchocephalia, which includes a single extant genus, *Sphenodon*, and the Squamata, which according to Conrad (2008) includes amphisbaenians, mosasaurs, snakes, and "lizards" (as explained by this author, "lizards" do not form a monophyletic group, because some "lizards" are more closely related to taxa such as snakes than to other "lizards": see Conrad 2008 for more details on the interrelationships of squamates). At the time when *Muscles of Vertebrates* (Diogo and Abdala 2010) was written, it was often thought that the Lepidosauria was more closely related to the Crocodylia and Aves (i.e., to the Archosauria) than was the Testudines, and the former three clades were usually included in the clade Diapsida (see, e.g., Gauthier et al. 1988; Dilkes 1999; Kardong 2002; Meers 2003; Dawkins 2004; Conrad 2008). That idea was thus followed in the book *Muscles of Vertebrates*. However, recent molecular studies have consistently suggested that turtles are instead the sister-group of archosaurs and therefore that lepidosaurs are the extant sister-group of all other extant reptiles. Although many morphologists did not follow this new classification, we consider the molecular data supporting it to be strong (see, e.g., Hedges 2012). Therefore, we follow here this new classification and thus group turtles and archosaurs in the clade Archosauromorpha, which is the sister-group of the clade Lepidosauromorpha.

The Mammalia includes the Monotremata and Theria, which in turn is subdivided into marsupials and placentals. Within the latter, the Primates (including modern humans),

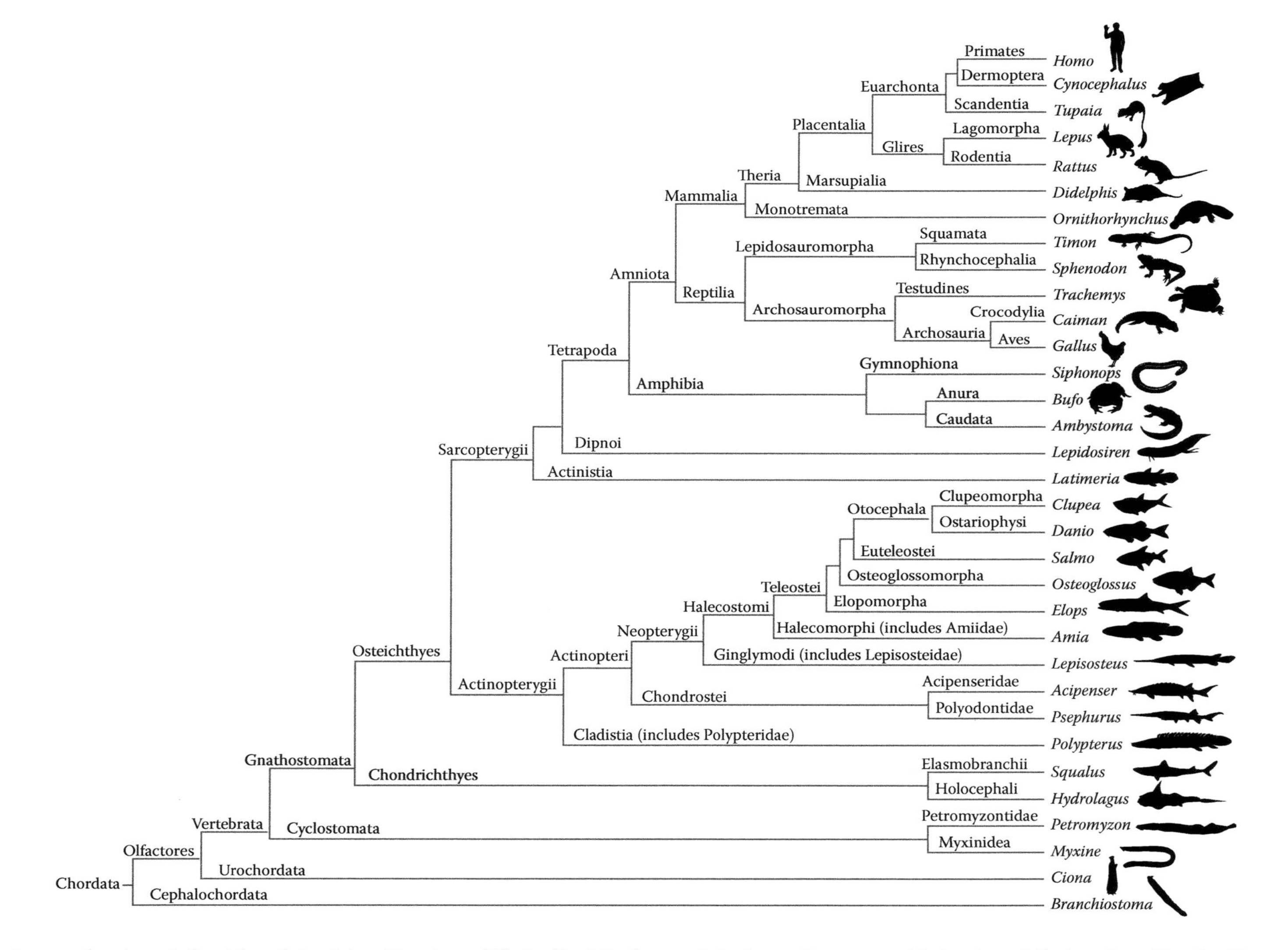

FIGURE 1.1 Cladogram showing relationships of chordates. (Courtesy of PhylopPic, http://www.phylopic.org. Images provided under public domain or Creative Commons—Attribution-ShareAlike 3.0 Unported license [http://creativecommons.org/licenses/by-sa/3.0/]. Artist credits: Rebecca Groom, Sarah Werning, Matt Reinbold (modified by T. Michael Keesey), Andrew A. Farke, Gareth Monger, Database Center for Life Science, and Mali'o Kodis [photograph by Hans Hillewaert]).

Dermoptera (including colugos or "flying lemurs"), and Scandentia (including tree shrews) are included in the clade Euarchonta, which is the sister-group of the clade Glires including rodents (e.g., mice and rats) and Lagomorpha (e.g., rabbits). In *Muscles of Vertebrates* (Diogo and Abdala 2010), dermopterans, tree shrews, and primates were placed in an unresolved trichotomy because the relationships between these three groups were unresolved (some authors grouped colugos with tree shrews, others grouped tree shrews with primates, and still others grouped colugos with primates: see, e.g., Sargis 2002a,b, 2004; Dawkins 2004; Marivaux et al. 2006; Janeka et al. 2007; Silcox et al. 2007; Diogo 2009). However, since that book was written, both molecular studies and morphological studies, including our own phylogenetic studies based on muscle data, have strongly supported a Dermopteran–Primates sister-group relationship, which is therefore followed in the present volume (see, e.g., reviews of Diogo and Wood 2011).

Several other studies have provided information on the musculature of the chordates, but most of them concentrated on a single taxon or on a specific subgroup of muscles. Moreover, the few more inclusive comparative analyses that were based on dissections were published at least half a century ago or even earlier (e.g., Humphry 1872a,b; Edgeworth 1902, 1911, 1923, 1926a,b,c, 1928, 1935; Luther 1913, 1914; Huber 1930a,b, 1931; Brock 1938; Kesteven 1942–1945). Furthermore, none of those works covered in detail all the skeletal muscles of all major extant groups of chordates. Also, the authors of those works did not have access to crucial information that is now available about, for example, the coelacanth *Latimeria chalumnae* (discovered only in 1938), the important role played by neural crest cells in the development and patterning of the head muscles of vertebrates, or the molecular and other types of evidence that has been accumulated about the phylogenetic interrelationships of chordates (e.g., Millot and Anthony 1958; Jarvik 1963, 1980; Alexander 1973; Le Lièvre and Le Douarin 1975; Anthony 1980; Lauder 1980b; Rosen et al. 1981; Noden 1983a, 1984, 1986; Hatta et al. 1990, 1991; Adamicka and Ahnelt 1992; Couly et al. 1992; Miyake et al. 1992; Köntges and Lumsden 1996; Pough et al. 1996; Schilling and Kimmel 1997; Kardong and Zalisko 1998; McGonnell 2001; Olsson et al. 2001; Hunter and Prince 2002; Kardong 2002; West-Eberhard 2003; Diogo 2004a,b, 2007, 2008a,b; Ericsson and Olsson 2004; Ericsson et al. 2004; Carroll et al. 2005; Kisia and Onyango 2005; Thorsen and Hale 2005; Noden and Schneider 2006; Diogo and Abdala 2007; Diogo and Wood 2011, 2012a; Dutel et al. 2015).

This is therefore the first book that compiles the available information, obtained from our own dissections of thousands of specimens and from a detailed literature review, for all skeletal muscles of chordates, including the muscles of amphioxus and tunicates and the muscles of the head and paired and median appendages of vertebrates. As emphasized in our previous works (reviewed in Diogo and Abdala 2010), one of the major problems researchers face when they compare the muscles of a certain chordate taxon with those of other taxa is the use of different names to designate the same muscle in the members of different clades and even of the same clade. In order to reconcile the different nomenclatures, we use a unifying nomenclature for all skeletal muscles of chordates that takes into account all of the data compiled for this book. In fact, we are fully aware of the new, ambitious, and necessary ontological projects that are now being developed in different biological disciplines. Such ontologies are extremely important and are becoming increasingly popular, because they provide a vocabulary for representing and communicating knowledge about a certain topic and a set of relationships that hold among the terms in that vocabulary. Therefore, we hope that the information provided here will stimulate researchers to develop a detailed ontology of the skeletal muscles of chordates, as well as to undertake future studies about the evolution, homologies, and development of these muscles and of other vertebrate anatomical structures in general. We sincerely hope that this volume will further contribute to the revival of the field of vertebrate chordate myology, which was too often neglected in the late twentieth century. Fortunately, this field is becoming more and more crucial again due to the rise of evolutionary developmental biology, as it is key to understanding the development and evolution of chordates as a whole, as well as the evolutionary history, anatomical variations, ontogeny, and pathologies of the skeletal muscles of humans in particular.

2 Methodology

BIOLOGICAL MATERIAL

The general phylogenetic framework for the comparisons provided in the present work is set out in Figure 1.1 (see also text of Chapter 1). The specimens we dissected are from the Colección Mamíferos Lillo of the Universidad Nacional de Tucumán (CML), the Primate Foundation of Arizona (PFA), the Department of Anatomy (GWU-ANA) and the Department of Anthropology (GWU-ANT) of the George Washington University, the Department of Anatomy of Howard University (HU-ANA), the Smithsonian Institution's National Museum of Natural History (USNM), the Department of Anatomy of Valladolid University (VU), the Cincinnati Museum of Natural History (CMNH), the San Diego Zoo (SDZ), the Canadian Museum of Nature (CMN), the Cleveland Metroparks Zoo (CMZ), the Yerkes National Primate Research Center (YNPRC), the Duke Lemur Center (DLC), the Museo Nacional de Ciencias Naturales de Madrid (MNCN), the Centro Nacional Patagónico de Argentina (CONICET), the Macquarie University of Australia (MU), the herpetological collection of Diamante-CONICET-Argentina (DIAMR), the Fundación Miguel Lillo of Argentina (FML), the San Diego State University (SDSU), the Laboratory of Functional and Evolutionary Morphology of the University of Liège (LFEM), the American Museum of Natural History (AMNH), the Academy of Natural Sciences of Philadelphia (ANSP), the Chinese Academy of Sciences at Wuhan (CASW), the California Academy of Sciences (CAS), the Field Museum of Natural History (FMNH), the Illinois Natural History Survey (INHS), the Museum National d'Histoire Naturelle de Paris (MNHN), the Musée Royal de l'Afrique Centrale (MRAC), the Université Nationale du Bénin (UNB), the collection of Anthony Herrel (AH), the herpetological collection of the Hebrew University of Jerusalem–Israel (HUJ), the Museo de Zoologia of the San Pablo University–Brasil (MZUSP), the Tupinambis Project Tucumán–Argentina (PT), the personal collection of Richard Thomas in Puerto Rico University (RT), the Antwerp Zoo (ANZ), the Center for Regenerative Therapies Dresden (CRTD), the Peabody Museum of Natural History of Yale University (YPM), the Reptile Breeding Facility at La Sierra University (LSU), California State University Northridge (CSUN), the Institüt für Evolution und Ökologie, Universität Tübingen (IEOUT), the University of Auckland, New Zealand (UANZ), the Mount Desert Island Biological Laboratory (MDIBL), the University of Alabama Ichthyological Collection (UAIC), the Warm Springs National Fish Hatchery (WSNFH), the Fish and Wildlife Service (FWS), the Hammond Bay Biological Station (HBBS), the Ward's Natural Science (WNS), donated by Ed Gilland at Howard University (HUG), donated by Lionel Christiaen New York (NYC), the Carolina Biological Support (CBS), Hazen and Alburg (HA), donated by Richard Elinson at Duquesne University Pittsburgh (DUP), and the Wisconsin Department of Natural Resources (WDNR). The list of specimens we examined is given below; the number of specimens dissected is followed by an abbreviation that refers to the state of the specimen (alc, alcohol fixed; fre, fresh; for, formalin embalmed; cands, trypsin-cleared and alizarin-stained; GFP, muscles shown with green fluorescent protein; ant, antibody staining of muscles). In our dissections, other than their color, there were no notable differences regarding the attachments, overall configuration, and general appearance of the muscles of fresh, alcohol fixed, and formalin embalmed specimens.

NON-SARCOPTERYGIAN TAXA—Non-actinopterygian chordates: *Branchiostoma floridae*, CBS, 3 (alc). *Ciona intestinalis*: NYC, 2 (alc). *Hydrolagus colliei*: WNS, 3 (alc). *Leucoraja erinacea*: HBBS, 3 (alc). *Mustelus laevis*: HUG, 1 (alc). *Myxine glutinosa*, MDIBL, 2 (alc). *Petromyzon marinus*, MDIBL, 3 (fre). *Squalus acanthias*: MDIBL, 3 (fre). **Non-teleostean actinopterygians**: *Acipenser brevirostum*: ANSP 178482, 1 (alc). *Acipenser fulvescens*: WSNFH, FWS and WDNR, 1 (fre). *Acipenser sturio*: MNCN 152172, 3 (alc). *Amia calva*: MNCN 35961, 1 (alc), 1 (cands); 1 (alc). *Lepisosteus oculatus*: uncatalogued, 1 (alc). *Lepisosteus osseus*: ANSP 107961, 2 (alc); ANSP 172630, 1 (alc); MNCN 246557, 1 (cands). *Lepisosteus platyrhincus*: AMNH 74789, 2 (alc). *Polyodon spathula*: UAIC 3536.06, 2 (alc). *Polypterus bichir*: MNCN 1579, 7 (alc), 1 (cands). *Polypterus delhizi*: UANZ (alc), 1. *Polypterus senegalus*: HU-ANA (fre), 3. *Psephurus gladius*: CASW, uncatalogued, 1 (alc). **Clupeomorpha**: *Denticeps clupeoides*: MRAC 76-032-P-1, 2 (alc). *Engraulis encrasicolus*: MNCN 68048, 2 (alc); MNCN 65097, 8 (alc); MNCN 1099, 3 (alc). *Engraulis* sp: MNCN 48896, 3 (alc). *Ethmalosa fimbriata*: MNCN 48865, 3 (alc). *Ilisha fuerthii*: MNCN 49338, 8 (alc). *Thryssa setirostris*: MNCN 49294, 2 (alc). **Elopomorpha:** *Albula vulpes*: MNCN 52124, 2 (alc). *Anguilla anguilla*: MNCN 41049, 3 (alc). *Elops lacerta*: LFEM, 2 (alc). *Elops saurus*: MNCN 48752, 2 (alc). *Conger conger*: MNCN 1530, 5 (alc). *Eurypharynx pelecanoides*: AMNH 44315, 1 (alc); AMNH 44344, 1 (alc). *Megalops cyprinoides*: MNCN 48858, 3 (alc). *Notacanthus bonaparte*: MNCN 107324, 3 (alc). **Euteleostei**: *Alepocephalus rostratus*: MNCN 108199, 2 (alc). *Argentina brucei*: USNM 239005, 2 (alc). *Argentina sphyraena*: MNCN 001134, 12 (alc); MNCN 78530, 5 (alc). *Astronesthes niger*: MNCN 1102, 1 (alc). *Aulopus filamentosus*: MNCN 1170, 6 (alc). *Bathylagus euryops*: MNCN 124597, 1 (alc). *Bathylagus longirostris*: USNM 384823, 2 (alc). *Bathylagus tenuis*: MNHN 2005-1978, 2 (alc). *Chlorophthalmus agassizi*: MNCN 1193, 3 (alc); MNCN 1182, 5 (alc). *Coregonus lavaretus*: MNCN 75424, 1 (alc). *Coregonus tugun*: MNCN 75422,

2 (alc). *Esox lucius*: MNCN 197706, 5 (alc). *Galaxias maculatus*: USNM 344889, 2 (alc). *Osmerus eperlanus*: MNCN 193795, 11 (alc). *Osmerus mordax*: USNM 32565, 2 (alc). *Plecoglossus altivelis*: MNCN 192036, 1 (alc). *Retropinna retropinna*: AMNH 30890, 1 (alc). *Salmo trutta*: MNCN 136179, 2 (alc), 1 (cands); MNCN 16373, 2 (alc); MNCN 40685, 2 (alc). *Salmo* sp: MNCN 48863, 2 (alc). *Searsia koefoedi*: USNM 206896, 2 (alc). *Stokellia anisodon*: AMNH 31037, 1 (alc). *Stomias boa*: MNCN 74444, 8 (alc); MNCN 74456, 4 (alc). *Thymallus thymallus*: MNCN 115147, 1 (alc); MNCN 114992, 1 (alc). *Umbra limi*: MNCN 35672, 2 (alc); 36072, 2 (alc). *Umbra krameri*: MNCN 36659, 3 (alc). *Xenodermichthys copei*: MNCN 78950, 2 (alc); MNCN 1584, 2 (alc); USNM 215527, 2 (alc). **Ostariophysi**: *Bagrus bajad*: LFEM, 1 (alc), 1 (cands). *Bagrus docmak*: MRAC 86-07-P-512, 1 (alc). *Barbus barbus*: LFEM, 1 (cands). *Barbus guiraonis*: MNCN 245730, 3 (alc). *Brachyhypopomus brevirostris*: LFEM, 2 (alc). *Brachyhypopomus* sp: INHS 89761, 2 (alc). *Brycon guatemalensis*: MNCN 180536, 3 (alc). *Brycon henni*: CAS 39499, 1 (alc). *Callichthys callichthys*: USNM 226210, 2 (alc). *Catostomus commersonii*: MNCN 36124, 10 (alc). *Citharinus* sp.: 86-016-P-72, 3 (alc). *Cetopsis coecutiens*: USNM 265628, 2 (alc). *Chanos chanos*: USNM 347536, 1 (alc), LFEM, 1 (alc). *Chrysichthys auratus*: UNB, 2 (alc). *Chrysichthys nigrodigitatus*: LFEM, 1 (cands). *Cobitis paludica*: MNCN 248076, 7 (alc). *Cromeria nilotica*: MRAC P.141098, 2 (alc). *Danio rerio*: MNCN, 10 (alc); 5 (alc). *Diplomystes chilensis*: LFEM, 3 (alc). *Distichodus notospilus*: MRAC A0-048-P-2630, 3 (alc). *Gonorynchus gonorynchus*: LFEM, 2 (alc). *Gonorynchus greyi*: FMNH 103977, 1 (alc). *Grasseichthys gabonensis*: MRAC 73-002-P-264, 3 (alc). *Gymnotus carapo*: INHS 35493, 2 (alc). MNCN 115675, 2 (alc). *Kneria wittei*: MRAC P-33512, 2 (alc). *Nematogenys inermis*: USNM 084346, 2 (alc). *Opsariichthys uncirostris*: MNCN 56668, 3 (alc). *Parakneria abbreviata*: MRAC 99-090-P-703, 3 (alc). *Phractolaemus ansorgii*: MRAC P.137982, 3 (alc). *Pimelodus blochii*: LFEM, 2 (alc), 1 (cands). *Silurus aristotelis*: LFEM, 2 (alc). *Silurus glanis*: LFEM, 2 (alc). *Sternopygus macrurus*: CAS 48241, 1 (alc); INHS 62059, 2 (alc). *Trichomycterus areolatus*: LFEM, 2 (alc). *Xenocharax spilurus*: MRAC A0-048-P-2539, 3 (alc). **Osteoglossomorpha**: *Hiodon tergisus*: MNCN 36019, 3 (alc). *Mormyrus niloticus*: LFEM, 1 (alc). *Mormyrus tapirus*: MNCN 80593, 3 (alc); MNCN 85283, 1 (alc). *Pantodon buchholzi*: MNCN 73493, 4 (alc). *Xenomystus nigri*: MNCN 227824, 25 (alc).

SARCOPTERYGII—Amphibia: *Ambystoma mexicanum*: MNCN, uncatalogued, 2 (alc); CRTD, uncatalogued, >200 (fre+GFP: with nonregenerated and with regenerated limbs). *Ambystoma ordinarium*: MNCN, uncatalogued, 2 (alc). *Ambystoma texanum*: FML 03402, 1 (alc). *Aspidoscelis uniparens*: LSU(fre), 3. *Bufo arenarum*: FML 01352-1, 3 (alc). *Chtonerpethon indistinctum*: JC, uncatalogued, 1 (alc). *Eleutherodactylus coqui*: DUP, 2 (alc), several embryos and juveniles (alc). *Leptodactylus fuscus*: FML, uncatalogued, 2 (alc). *Litoria caerulea*: DIAM 0313, 1 (alc). *Phyllomedusa sauvagi*: FML 04899, 2 (alc), and DIAM 0337, 1 (alc). *Rana pipiens*: HA, 1 (alc). *Siphonops paulensis*: FML, uncatalogued, 1 (alc). *Siphonops* sp.: DB, uncatalogued, 2 (alc). *Telmatobius laticeps*: FML 3960, 1 (alc). *Xenopus laevis*: DUP, 2 adult (alc), several embryos and larvae (alc). **Aves**: *Cairina moschata*: FML w/d, 1 (alc). *Coturnyx coturnyx*: FML w/d, 2 (alc). *Gallus domesticus*: FML w/d, 3 (alc). *Nothura* (alc). FML w/d 1 (alc). *Pitangus sulphuratus*: FML w/d, 1 (alc). *Thraupis sayaca*: FML w/d, 1 (alc). **Cladistia**: *Latimeria chalumnae*: IEOUT SZ 10378, 1. **Crocodylia**: *Caiman latirostris*: FML w/d, 1 (alc), and CCyTTP w/d, 4 (alc). **Dipnoi**: *Lepidosiren paradoxa*: CONICET, uncatalogued, 1 (alc). *Neoceratodus forsteri*: MU, uncatalogued, 2 (alc); JVM-I-105-2, 2 (for). **Lepidosauria**: *Ameiva ameiva*: FML 03637, 4 (alc). *Amphisbaena alba*: FML uncatalogued, 2 (alc). *Anisolepis longicauda*: UNNEC no number, 1 (alc). *Basiliscus vittatus*: SDSU 02097, 1 (alc). *Bogertia lutzae*: MZU(ALC) 54747, 1 (alc). *Briba brasiliana*: MZU(ALC) 73851, 1 (alc). *Callopistes maculatus*: MZU(ALC) 58107, 1 (alc). *Calyptommatus leiolepis*: MZU(ALC) 71339, 1 (alc). *Chalcides chalcides*: FML 03712, 1 (alc). *Chamaeleo calyptratus*: LSU 3 (fre). *Cnemidophorus ocellifer*: FML 03389, 2 (alc); FML 03409, 4 (alc); without data, 1 (alc); FML 17606, 1 (alc). *Cordylus tropidosternon*: AH no number, 1 (alc). *Crocodilurus lacertinus*: MZU(ALC) 12622, 1 (alc). *Dicrodon guttulatum*: FML 02017, 1 (alc). *Diplolaemus bibroni*: MACN 35850, 1 (alc). *Dracaena paraguayensis*: MZU(ALC) 52369, 1 (alc). *Echinosaura horrida*: MZU(ALC) 54452, 1 (alc). *Enyalius iheringii*: MZU(ALC) 74901, 1 (alc). *Garthia gaudichaudii*: MZU(ALC) 45329, 1 (alc). *Garthia penai*: MZU(ALC) 60937, 1 (alc). *Gekko vittatus*: AH no number, 2 (alc). *Gerrohsaurus major*: AH no number, 1 (alc). *Gymnodactylus geckoides*: MZ(ALC) 48128, 1 (alc). *Hemidactylus garnoti*: AH no number, 2 (alc). *Hemidactylus mabouia*: FML 02142, 1 (alc)., and FML 02421, 1 (alc). *Homonota fasciata*: FML 02137, 1 (alc)., and FML 00915, 2 (alc). *Leiosaurus paronae*: MACN 4386, 1 (alc). *Liolaemus cuyanus*: FML 02021, 7 (alc). *Mabuya frenata*: FML 00277, 1 (alc)., and FML 01713, 1 (alc). *Microlophus theresioides*: FML 03674, 1 (alc). *Phelsuma madagascariensis*: AH no number, 2 (alc). *Phyllodactylus gerrophygus*: FML 01563, 2 (alc). *Phyllopezus pollicaris*: FML 02913, 2 (alc). *Phymaturus* (alc): FML 13834-13844, 3 (alc). *Phymaturus punae*: FML 2942, 4 (alc). *Podarcis sicula*: FML 03714, 1 (alc). *Polychrus acutirostris*: MZU(ALC) 48151, 1 (alc). MZU(ALC) 08605, 1 (alc). *Pristidactylus achalensis*: MACN 32779, 1 (alc). *Proctoporus guentheri*: FML 02010, 1 (alc). *Teius teyous*: FML 00290, 2 (alc). *Stenocercus caducus*: FML 00260, 1 (alc)., and FML 00901, 1 (alc). *Thecadactylus rapicauda*: MZU(ALC) 11476, 1 (alc). *Trioceros melleri*: CSUN(alc), 1. *Tropidurus etheridgei*: FML 03562, 2 (alc). *Tropidurus hygomi*: FML 08796, 1 (alc). *Tropidurus oreadicus*: FML 08771, 1 (alc). *Tropidurus (alc)inulosus*: FML 00129, 2 (alc)., and FML 03559, 2 (alc). *Tupinambis rufescens*: PT 0084, 1 (alc)., PT 0085, 1 (alc)., FML 06412, 1 (alc), FML 06425, 1 (alc)., and FML 07420, 1 (alc). *Vanzoia klugei*: MZU(ALC) 59130, 1 (alc). *Varanus* (alc): AH no number, 1 (alc). *Xantusia* (alc): AH no number 1, 1 (alc). *Zonosaurus* (alc): AH no number, 1 (alc). **Mammalia**: *Aotus nancymaae*: GWUANT AN1, 1 (fre). *Callithrix jacchus*: GWUANT

CJ1, 1 (fre). *Cercopithecus diana*: GWUANT CD1, 1 (fre). *Colobus guereza*: GWUANT CG1, 1 (fre). *Cynocephalus volans*: USNM, 144941, 1 (alc); USNM, uncatalogued, 1 (alc). *Didelphis albiventris*: CML 5971, 1 (alc). *Didelphis virginiana*: HUDV1-5, 5 (alc). *Gorilla gorilla*: CMS GG1, 1 (fre); VU GG1, 1 (fre). *Homo sapiens*: GWU-ANA, 1-16, 16 (for). *Hylobates gabriellae*: VU HG1, 1 (fre). *Hylobates lar*: HU-ANA, H01, 1 (for). *Lepilemur ruficaudatus*: HU-ANA, L01, 1 (for). *Lemur catta*: GWUANT LC1, 1 (fre). *Leptailurus serval*: VU, 1 (fre). *Loris tardigradus*: SDZ LT53090, 1 (fre). *Lutreolina crassicaudata*: CML 4114, 1 (alc). *Macaca fascicularis*: VU MF1, 1 (fre). *Macaca mulatta*: HU-ANA, M01, 1 (for); YNPRC, M1-9, 9 (for). *Macaca silenus*: VU MS1, 1 (fre). *Monodelphis dimidiata*: CML 4118, 1 (alc). *Nycticebus coucang*: SDZ NC41235, 1 (fre); SDZ NC43129, 1 (fre). *Nycticebus pygmaeus*: VU NP1, 1 (fre); VU NP2, 1 (fre); SDZ NP40684, 1 (fre); SDZ NP51791, 1 (fre). *Otolemur garnettii*: DLC, OG1-10, 10 (for). *Otolemur crassicaudatus*: DLC, OC1-12, 12 (for). *Ornithorhynchus anatinus*: USNM, 13678, 1 (alc); USNM, uncatalogued, 1 (alc). *Pan paniscus*: ANZ, 7 (fre). *Pan troglodytes*: PFA, 1016, 1 (fre); PFA, 1009, 1 (fre); PFA, 1051, 1 (alc); HU-ANA, C104, 1 (for); GWU-ANT, 01, 1 (for); GWU-ANT, 02, 1 (for); YNPRC, C1-2, 2 (for); CMZ, C1-2, 2 (for). *Panthera tigris*: VU, 1 (fre). *Papio anubis*: GWUANT PA1, 1 (fre). *Pithecia pithecia*: VU PP1, 1 (fre); GWUANT PP1, 1 (fre). *Pongo pygmaeus*: HU-ANA, O01, 1 (for); GWU-ANT, 01, 1 (for). *Propithecus verreauxi*: GWUANT PV1, 1 (fre); GWUANT PV2, 1 (fre). *Rattus norvegicus*: USNM, uncatalogued, 2 (alc). *Saimiri sciureus*: GWUANT SC1, 1 (fre). *Tarsius syrichta*: CMNH M-3135, 1 (alc). *Thylamys venustus*: CML 5586, 1 (alc). *Tupaia sp.*: UNSM, 87244, 1 (alc), USNM, uncatalogued, 1 (alc). **Testudines**: *Cuora amboinensis*: YPM R 14443m 1 (alc). *Cuora galbinifrons*: YPM R 12735, 1 (alc). *Geochelone chilensis*: DIAMR-038, 2 (alc); DIAMR-039, 2 (alc); DIAMR-040, 1 (alc); FML 16879, 1 (alc); FML 16880, 1 (alc); FML16595, 1 (alc); FML 00005, 1 (alc); FML 16978, 1 (alc). *Glyptemys insculpta*: YPM R 5952, 1 (alc). *Mauremys caspica rivulata*: YPM R 16233-36, 2 (alc). *Phrynops hilarii*: DIAMR-044, 1 (alc); DIAMR-042, 1 (alc); DIAMR-041, 1 (alc); DIAMR-043, 1 (alc); DIAMR-037, 1 (alc); DIAMR-005, 1 (alc); DIAMR-006, 1 (alc); DIAMR-007, 1 (alc). *Podocnemys unifilis*: DIAMR-078, 6 (alc). *Rhinoclemmys pulcherrima*: AH uncatalogued, 1 (alc). *Sacalia bealei*: YPM R 14670-71, 2 (alc). *Terrapene carolina*: YPM R 13624, 1 (alc). YPM R 13622, 1 (alc). *Testudo graeca*: HUJ-R 22843, 2 (alc); HUJ-R 22845, 2 (alc). *Trachemys scripta*: RT uncatalogued, 2 (alc).

NOMENCLATURE

The myological nomenclature used in the present work essentially follows that used in the book "Muscles of Vertebrates" (Diogo and Abdala 2010), with a few exceptions that will be mentioned in the text and tables provided in the following chapters. Regarding the pectoral and forelimb musculature, we recognize five main groups of muscles: the axial muscles of the pectoral girdle, the appendicular muscles of the pectoral girdle and arm, the appendicular muscles of the ventral forearm, the appendicular muscles of the hand, and the appendicular muscles of the dorsal forearm. Regarding the pelvic and hind limb musculature, we also recognize five main groups of muscles: the axial muscles of the pelvic girdle, the appendicular muscles of the pelvic girdle and thigh, the appendicular muscles of the ventral leg, the appendicular muscles of the foot, and the appendicular muscles of the dorsal leg. The appendicular musculature of the pectoral girdle, arm, forearm, and hand and of the pelvic girdle, thigh, leg, and foot derives mainly from the adductor and abductor muscle masses of the pectoral fin of phylogenetically basal fishes and essentially corresponds to the "abaxial musculature" *sensu* Shearman and Burke (2009). The axial pectoral girdle musculature and the axial pelvic girdle musculature are derived from the postcranial axial musculature, and, together with most of the remaining epaxial and hypaxial muscles of the body (with the exception of, e.g., various muscles of the pectoral girdle and hind limb), form the "primaxial musculature" *sensu* Shearman and Burke (2009). As explained by these authors, the muscles of the vertebrate body are classically described as epaxial or hypaxial according to the innervation by either the dorsal or ventral rami of the spinal nerves, respectively, while the terms *abaxial musculature* and *primaxial musculature* reflect embryonic criteria that are used to distinguish domains relative to embryonic patterning. The "primaxial" domain is composed of somitic cells that develop within somite-derived connective tissue, and the "abaxial" domain includes muscle and bone that originates from somites but then mixes with, and develops within, lateral plate-derived connective tissue.

Concerning the head and neck musculature, the main groups of muscles recognized here correspond to those proposed by Edgeworth (1902–1935): external ocular, mandibular, hyoid, branchial, epibranchial, and hypobranchial. Edgeworth (1935) viewed the development of these muscles in the light of developmental pathways leading from presumptive premyogenic condensations to different states in each cranial arch (see Figure 2.1; the condensations of the first and second arches corresponding respectively to Edgeworth's "mandibular and hyoid muscle plates" and those of the more posterior, "branchial" arches corresponding to his "branchial muscle plates"). According to him, these developmental pathways involve the migration of premyogenic cells, differentiation of myofibers, directional growth of myofibers, and possibly interactions with surrounding structures. These events occur in very specific locations, e.g., dorsal, medial, or ventral areas of each cranial arch, as shown in Figure 2.1: for instance, the mandibular muscle plate gives rise dorsally to the premyogenic condensation constrictor dorsalis, medially to the premyogenic condensation adductor mandibulae, and ventrally to the intermandibularis (no description of a ventral mandibular premyogenic condensation was given by Edgeworth); the hyoid condensation usually gives rise to dorsomedial and ventral derivatives; the hypobranchial condensation gives rise to the "geniohyoideus" and to the "rectus cervicis" (as noted by Miyake et al. [1992], it is not clear if Edgeworth's

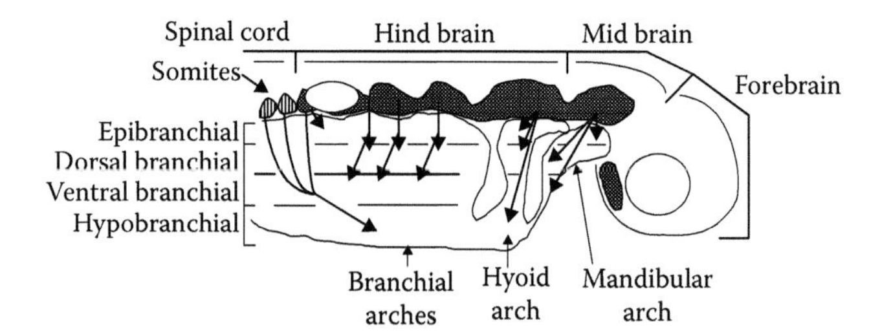

FIGURE 2.1 Schematic presentation of embryonic origin of cranial muscles in gnathostomes based on Edgeworth's works (e.g., Edgeworth 1902, 1911, 1923, 1926a,b,c, 1928, 1935). Premyogenic cells originate from the paraxial mesoderm (hatched areas) and several somites (areas with vertical bars). Large arrows indicate a contribution of cells in segments of the mesoderm to the muscle formation of different cranial arches. For more details, see text. (The nomenclature of the structures illustrated basically follows that of Miyake et al. [1992].) (Modified from Miyake, T. et al., *J Morphol*, 212, 213–256, 1992.)

geniohyoideus and rectus cervicis represent separate premyogenic condensations or later states of muscle development).

According to Edgeworth (1935), although exceptions may occur (see the following), the mandibular muscles are generally innervated by the fifth cranial nerve (CNV); the hyoid muscles, by CNVII; and the branchial muscles, by CNIX and CNX. Diogo et al. (2008a) divided the branchial muscles *sensu lato* (that is, all the branchial muscles *sensu* Edgeworth 1935) into three main groups. The first comprises the "true" branchial muscles, which are subdivided into (a) the branchial muscles *sensu stricto* that are directly associated with the movements of the branchial arches and are usually innervated by the glossopharyngeal nerve (CNIX); (b) the protractor pectoralis and its derivatives, which are instead mainly associated with the pectoral girdle and are often innervated by the spinal accessory nerve (CNXI) but are said to be innervated by CNX in phylogenetically plesiomorphic gnathostomes such as chondrichthyans (Edgeworth 1935). The second group consists of the pharyngeal muscles, which are only present as independent structures in extant mammals. They are considered to be derived from branchial arches 4–6, and they are usually innervated by the vagus nerve (CNX). As will been seen in the following chapters, the mammalian stylopharyngeus is considered to be derived from the third arch and is primarily innervated by the glossopharyngeal nerve; thus, it is grouped with the true branchial muscles rather than with the pharyngeal muscles. The third group is made up of the laryngeal muscles, which are considered to be derived from branchial arches 4–6 and are usually innervated by the vagus nerve (CNX). Regarding the epibranchial and hypobranchial muscles, according to Edgeworth these are "developed from the anterior myotomes of the body" and thus "are intrusive elements of the head"; they "retain a spinal innervation" and "do not receive any branches from the Vth, VIIth, IXth and Xth nerves" (Edgeworth 1935: 189). It is worth mentioning that in addition to the mandibular, hyoid, branchial, hypobranchial, and epibranchial musculature, Edgeworth (1935: 5) referred to a primitive "premandibular arch" in "which passed the IIIrd nerve." This third cranial nerve, together with CNIV and CNVI—which, according to Edgeworth (1935: 5), are "not segmental nerves; they innervate muscles of varied segmental origin and are, phylogenetically, of later development than are the other cranial nerves"—innervate the external ocular muscles of most extant vertebrates. These external ocular muscles will not be discussed in the present volume.

Some of the hypotheses defended by Edgeworth have been contradicted by recent studies (e.g., certain phylogenetic hypotheses that he used to formulate his theories: see the following chapters). However, many of his conclusions have been corroborated by more recent developmental and genetic studies. For instance, Miyake et al. (1992) published a paper that reexamined, discussed, and supported some of the general ideas proposed by Edgeworth (1935). For example, they noted that "Noden (1983a, 1984, 1986) elegantly demonstrated with quail-chick chimeras that cranial muscles are embryologically of somitic origin, and not, as commonly thought, of lateral plate origin, and in doing so corroborated the nearly forgotten work of Edgeworth" (Miyake et al. 1992: 214). They also pointed out that molecular developmental studies such as Hatta et al. (1990, 1991) "have corroborated one of Edgeworth's findings: the existence of one pre-myogenic condensation (the constrictor dorsalis) in the cranial region of teleost fish" (Miyake et al. 1992: 214). The existence of this and other condensations (e.g., the hyoid condensation) has received further support in developmental studies published in recent years (e.g., Knight et al. 2008; Kundrat et al. 2009). For instance, in the zebrafish, engrailed immunoreactivity is only detected in the levator arcus palatini + dilatator operculi muscles; i.e., in the two muscles that are derived from the dorsal portion of the mandibular muscle plate (constrictor dorsalis *sensu* Edgeworth 1935). Remarkably, in mammals such as the mouse, engrailed immunoreactivity is detected in mandibular muscles that are very likely derived from a more ventral ("adductor mandibulae") portion of that plate; i.e., in the masseter, temporalis, pterygoideus medialis, and/or pterygoideus lateralis. Also interestingly, authors such as Tzahor (2009) have shown that among members of the same species, muscles from the same arch (e.g., from the mandibular arch) might originate from different types of cells. For instance, the mandibular "adductor mandibulae complex" and its derivatives (e.g., masseter) derive from cranial paraxial mesoderm, while the more ventral mandibular muscle intermandibularis and its derivatives (e.g., mylohyoideus) originate from medial splanchnic mesoderm.

As stated by Miyake et al. (1992) and more recently by Diogo and Abdala (2010), Edgeworth's (1935) division of the head and neck muscles in external ocular, mandibular, hyoid, branchial, epibranchial, and hypobranchial muscles continues to be widely used by both comparative anatomists and developmental biologists. For instance, Edgeworth's schematic is similar to that proposed in Mallatt's anatomical studies (e.g., Mallatt 1997; the differences between the two schematics are mainly nomenclatural ones, for example, the "hyoidean and mandibular superficial constrictors" *sensu* Edgeworth correspond to the "hyoidean and mandibular interbranchial muscles" *sensu* Mallatt—see Table 2 of Mallatt [1997] and the following chapters), as well as to the schematics used in numerous recent developmental and molecular works, such as those by Holland et al. (1993, 2008), Kuratani et al. (2002, 2004), Trainor et al. (2003), Kuratani (2004, 2005a,b, 2008a), Kusakabe and Kuratani (2005), Olsson et al. (2005), Kuratani and Ota (2008a), and Kuratani and Schilling (2008). However, as expected, some researchers prefer to catalog the head and neck muscles into groups that do not fully correspond to those proposed by Edgeworth (1935). For instance, Noden and Francis-West (2006) refer to three main types of head and neck muscles (Figure 2.2): the "extraocular" muscles, which correspond to Edgeworth's extraocular muscles, the "branchial" muscles, which correspond to the mandibular, the hyoid, and most of the branchial muscles *sensu* Edgeworth and the "laryngoglossal" muscles, which include not only the hypobranchial muscles but also part of the branchial muscles *sensu* Edgeworth (namely, the laryngeal muscles *sensu* Diogo and Abdala 2010]). A main advantage of recognizing these three groups is to stress that at least in vertebrate taxa such as salamanders, chickens, and mice, laryngeal muscles such as the dilatator laryngis and constrictor laryngis receive a contribution from somitic myogenic cells (e.g., Noden 1983a; Noden et al. 1999; Yamane 2005; Piekarski and Olsson 2007), as do the hypobranchial muscles *sensu* Edgeworth (see preceding text and the following chapters). That is, the main difference between the "branchial" and "laryngoglossal" groups *sensu* Noden and Francis-West (2006) is that unlike the former, the latter receives a contribution from these somitic cells. However, developmental studies have shown that some of the "branchial" muscles *sensu* Noden and Francis-West (2006), including some true (nonlaryngeal) branchial muscles *sensu* Diogo et al. (2008a,b), such as the protractor pectoralis and the levatores arcuum branchialium of salamanders, the trapezius of chickens and mice and even possibly some hyoid muscles such as the urodelan interhyoideus, do also receive a contribution of somitic myogenic cells (see, e.g., Piekarski and Olsson 2007; NB: Edgeworth 1935 included the protractor pectoralis and its derivatives—which include the trapezius of amniotes—in the branchial musculature, but he was already aware of the controversy concerning the body vs. head origin of this muscle). Moreover, while it might seem appropriate to designate the laryngeal and hypobranchial muscles of derived vertebrate clades such as birds as "laryngoglossal" muscles, it would be less suitable to use the name *laryngoglossal* to designate the hypobranchial muscles of taxa such as lampreys or sharks, because the latter muscles are not functionally associated with a larynx or with a tongue in those taxa (see the following chapters below). That is why authors that usually work with non-osteichthyan clades often prefer to follow the names that Edgeworth (1935) used to designate the main groups of head and neck muscles; i.e., external ocular, mandibular, hyoid, branchial, hypobranchial, and epibranchial (see, e.g., Holland et al. 1993; Kuratani et al. 2002, 2004; Kuratani 2004, 2005a,b, 2008a; Kusakabe and Kuratani 2005; Olsson et al. 2005; Holland et al. 2008; Kuratani and Ota 2008a; Kuratani and Schilling 2008; see also the following chapters). As one of the main goals of this volume is to propose a unifying nomenclature for muscles of the Chordata as a whole; we will also use these names throughout the book.

One major advantage of using and expanding the nomenclature proposed by Diogo and Abdala (2010) is that it combines, and thus creates a bridge between, names that are normally used in human anatomy and names that are more typically used in works dealing with other chordate taxa, including not only bony fishes but also phylogenetically more plesiomorphic vertebrates such as agnathans, elasmobranchs, and holocephalans. For instance, coracomandibularis, intermandibularis, and interhyoideus are names that are often used in the literature to designate the muscles of nonosteichthyan vertebrates. As some of these muscles are directly homologous to muscles that are present in osteichthyans and particularly in phylogenetically plesiomorphic sarcopterygian and actinopterygian groups such as cladistians, actinistians, and dipnoans, it makes sense to use these names in the descriptions of the latter groups. At the same time, this nomenclature allows us to keep almost all the names that are currently used to designate the muscles of humans (see, e.g., *Terminologia Anatomica* 1998) and takes into account major nomenclatural reviews that have been performed for other groups of tetrapods (e.g., *Nomina Anatomica Avium*: Baumel et al. [1979]). Maintaining the stability of the names used in human anatomy is an important aspect of our nomenclature, because these names have been employed for centuries in thousands of publications dealing with human anatomy and medicine and by thousands of teachers, physicians, and practitioners. As one of main goals of using this unifying nomenclature is to avoid the confusion created by using different names to designate the same muscles in distinct vertebrate groups; some of the names that we use to designate the muscles of certain taxa do not correspond to the names that are more usually used in the literature for those taxa. So, using the muscles of dipnoans as an example, the adductor mandibulae A3, the adductor mandibulae A2, the adductor mandibulae A2-PVM, the protractor pectoralis, the coracomandibularis, and the sternohyoideus *sensu* in this book correspond, respectively, to the "adductor mandibulae anterior," to the "more anterior/lateral part of the adductor mandibulae posterior," to the "more posterior/mesial part of the adductor mandibulae posterior," to the "cucullaris," to the "geniothoracicus," and to the "rectus cervicus" *sensu* Miyake et al. (1992) and Bemis and Lauder (1986) (see the following chapters). When we cite works that use a nomenclature that differs from that proposed here, the respective synonymy is given in the tables provided throughout the book. The muscles listed in these tables are those that are usually present in adults of the respective taxa; we do not list all the muscles that occasionally appear as variations (e.g., although a few adult modern humans have a platysma cervicale, in the vast majority of them, this muscle is absent). The terms *anterior*, *posterior*, *dorsal*, and *ventral* are used as they relate to pronograde tetrapods (e.g., in mammals the eye, and thus the muscle orbicularis oculi, is usually anterior to the ear, and thus to the muscle auricularis superior, and dorsal to the mandible, and thus to the muscle orbicularis oris: see the following chapters). Although the identification

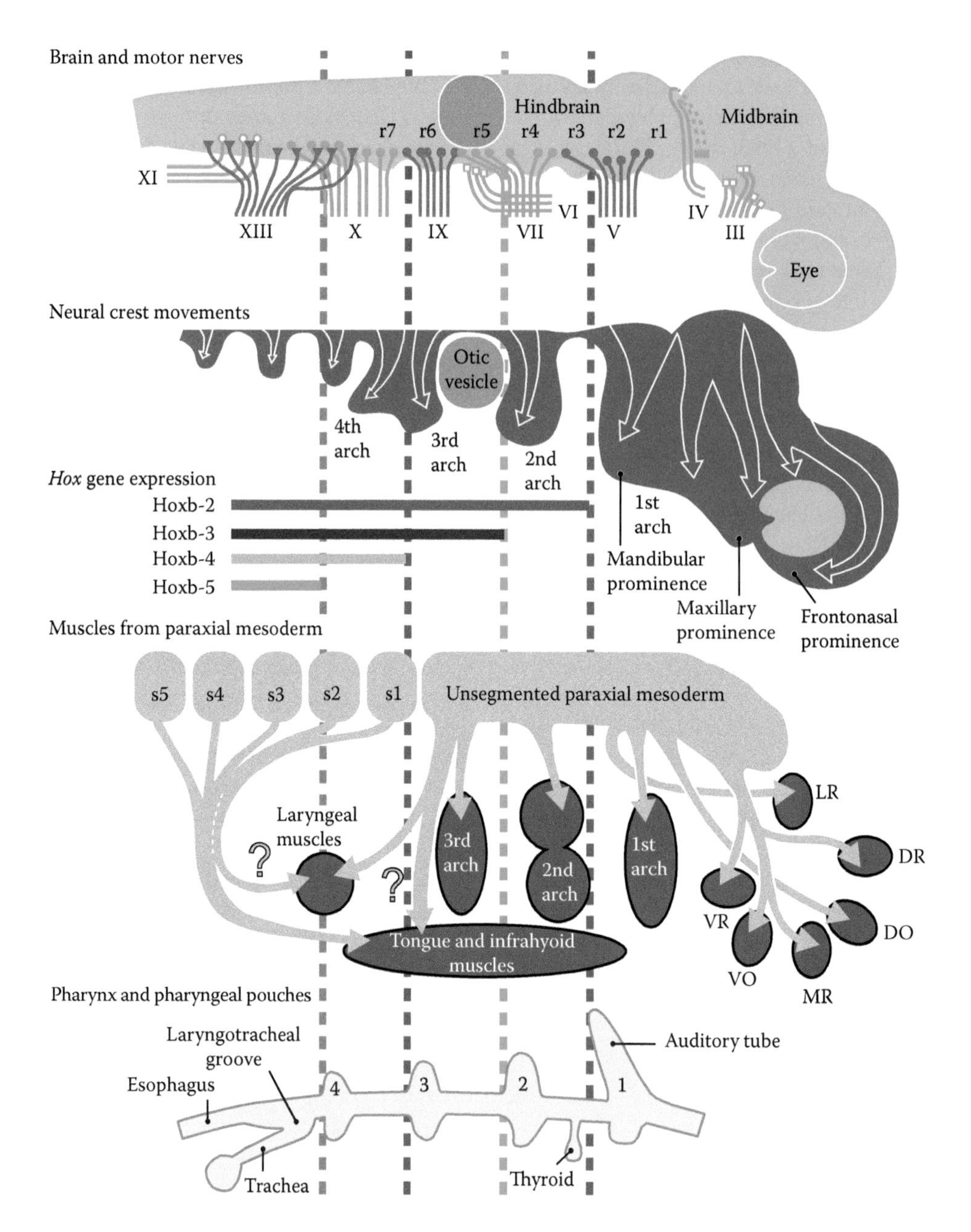

FIGURE 2.2 General diagram showing the developmental origins of the head and neck muscles in amniotes. (The nomenclature of the structures illustrated basically follows that of Diogo et al. [2016b].) (Modified from Diogo, R. et al., *Taylor & Francis*, 2016b. With permission). It is remarkable that the use of these new techniques has confirmed a great part of Edgeworth's hypotheses (e.g., Edgeworth 1902, 1911, 1923, 1926a,b,c, 1928, 1935) about the origins and homologies of the vertebrate head and neck muscles, for instance, that the "adductor mandibulae complex" ("mandibular adductors"), the pterygomandibularis ("pterygoideus"), and the intermandibularis derive from the first arch (mandibular muscles *sensu* Edgeworth [1935]); that the masseter and temporalis of mammals correspond to part of the adductor mandibulae complex of non-mammalian such as birds; that the levator hyoideus ("columella") and the depressor mandibulae ("mandibular depressors") derive from the second arch (hyoid muscles *sensu* Edgeworth [1935]); that the mammalian stapedius ("stapedial") corresponds to the levator hyoideus of non-mammalian groups such as birds; that part of the "digastricus" of mammals (i.e., the digastricus posterior) derives from the depressor mandibulae of non-mammalian groups such as birds; that the hyobranchialis ("branchiomandibularis") derives from the third arch, i.e., that it is a branchial muscle *sensu* Edgeworth (1935); that the intrinsic and extrinsic tongue muscles are mainly derived from somites and anteriorly migrate during the ontogeny in order to make part of the craniofacial musculature, i.e., that they are hypobranchial muscles *sensu* Edgeworth (1935; but see text). It should be noted that some authors, such as Noden and Francis-West (2006), argue that the laryngeal muscles are also hypobranchial muscles *sensu* Edgeworth; that is, they do not consider these muscles as part of the branchial musculate as did Edgeworth and as supported by most current developmental studies (see text).

of separate muscles is obviously somewhat subjective, we followed as strictly as possible Edgeworth's (1935) criteria for analyzing the evidence acquired by others and ourselves, including, for instance, the degree of separation of muscular fibers, the differences in function, orientation and attachments of these fibers, and the innervation of the various myological structures being investigated (see Diogo and Abdala [2010] for a review on this subject).

PHYLOGENY AND HOMOLOGY

The definition of homology and its use in systematics and comparative anatomy has been discussed by several authors (e.g., Patterson 1988; de Pinna 1991; Agnarsson and Coddington 2007). The simplest meaning of homology is equivalence of parts (e.g., de Pinna 1991). In the present work, we follow the phylogenetic definition of homology, as proposed by Patterson (1988): homology is equal to synapomorphy. Therefore, following de Pinna (1991), we recognize two main types of muscular homology. "Primary homology" hypotheses are conjectures or hypotheses about the common origin of muscular characters that are established after a careful analysis of criteria such as function, topology, and ontogeny (i.e., after the so-called test of similarity). In this volume, we follow the same methodology that we have employed and carefully explained in previous works (e.g., Diogo and Abdala 2010) and thus take into account all the lines of evidence obtained from our dissections and gleaned from the literature in order to formulate such primary homology hypotheses (e.g., the innervation of the muscles; their relationships with other muscular structures; their relationships with hard tissues; the configuration/orientation of their fibers; their development; their function; and the configuration or absence/presence of the muscles in embryos of model organisms that were previously the subject of genetic manipulations, e.g., the knockdown of certain Hox genes or the induction of C-met mutations).

We use multiple lines of evidence because, as noted by Edgeworth (1935), no single one is infallible. For instance, although the innervation of a muscle generally remains constant and corresponds to its segment of origin (e.g., Luther 1913, 1914; Edgeworth 1935; Kesteven 1942–1945; Köntges and Lumsden, 1996), there are cases in which the same muscle may have different innervations in different taxa. One of the examples provided by Edgeworth (1935: 221) to illustrate this concerns the intermandibularis of extant dipnoans, which "is innervated by the vth and viith nerves though wholly of mandibular origin." Also, there are cases in which the same muscle may originate from different regions and/or segments of the body in different taxa. An example provided by Edgeworth (1935: 220) concerns the branchial muscle protractor pectoralis (which he often called "cucullaris"), which "has diverse origins in *Ornithorhynchus*, *Talusia* and *Sus*; in the first-named it is developed from the 3rd [arch], in the second from the 2nd and in the last from the 1st branchial muscle-plate. These changes are secondary to the non-development of the branchial muscle-plates, from behind forwards. The muscles are homologous and have a constant primary innervation from the xith nerve." According to Edgeworth (1935: 224), there are also cases in which "an old structure may be lost" (e.g., the branchiomandibularis is lost in extant ginglymodians and teleosts), in which "new muscles may be developed" (e.g., the glossal muscles of tetrapods), and in which "an old structure or group of structures may be transformed" (e.g., the levator hyoideus "is transformed, either partially or wholly, into a Depressor mandibulae"). The occurrence of such phenomena thus raises further difficulties for comparative analyses among different clades. There are also cases in which "similar secondary developments occur in separate genera or phyla," i.e., cases of convergence and parallelism (see, e.g., Diogo 2004a, 2005a for discussions on these two concepts; see also the following chapters).

Following de Pinna (1991), the primary homology hypotheses must pass the second, or "hard," test of homology, the test of phylogenetic conjunction and congruence (agreement in supporting the same phylogenetic relationships), before they can be considered "secondary homology" hypotheses. For example, if muscle A of taxon X and muscle B of taxon Y have similar innervations, functions, topologies, and development but the phylogenetic data available strongly support the idea that muscles A and B were the result of convergent evolution (i.e., that they were independently acquired and do not correspond to a structure that was present in the last common ancestor of A and B), then the phylogenetic criterion has preponderance over the other criteria. The phylogenetic framework that we use in the present work is shown in Figure 1.1. Following the methodology explained earlier, if the data provided by some lines of evidence (e.g., innervation, function, and relationships with other muscular and hard structures) indicate that muscles C and D might be homologous (primary homology hypothesis), but muscle C is only present in monotremes, and muscle D, in modern humans among mammals, then we would conclude that muscles C and D are likely not homologous (i.e., the primary homology hypothesis did not pass the test of phylogenetic conjunction and congruence). Therefore, the hypotheses of homology that are shown in the tables provided in the present work are hypotheses that are phylogenetically congruent with the scenario shown in the cladogram of Figure 1.1; i.e., they are secondary homology hypotheses *sensu* de Pinna (1991).

3 Non-Vertebrate Chordates and the Origin of the Muscles of Vertebrates

The origin and evolution of chordates and vertebrates, particularly the origin of the vertebrate head, has fascinated researchers for centuries (e.g., Gill 1895; Minot 1897; Gregory 1935; Holland and Holland 1998, 2001; Holland et al. 2008; Koop et al. 2014; Ziermann et al. 2014; Diogo et al. 2015b). The origin, development, and comparative anatomy of hard (e.g., skeleton) and soft tissues (e.g., muscles, nervous system, and cardiovascular system) are crucial pieces of information for this investigation. Moreover, the findings that urochordates (e.g., tunicates, *Ciona*) and not cephalochordates (e.g., amphioxus, *Branchiostoma*) are the closest sister-group of vertebrates (Figure 3.1) (Delsuc et al. 2006) has dramatically changed our understanding of the origin and evolution of both chordates and vertebrates. Cephalochordates are the sister taxon of Olfactores (= urochordates + vertebrates; Figure 3.1), and amphioxus (lancelet) is therefore one of the best models to analyze chordate and vertebrate evolution (Koop and Holland 2008). The adult amphioxus has morphological features that are more easily compared with features found in vertebrates than the adult tunicate, and their genome sequence has more archetypal characters of ancestral chordates preserved than either tunicates or vertebrates (e.g., Garcia-Fernàndez and Holland 1994; Shimeld and Holland 2000; Putnam et al. 2008; Candiani 2012). For instance, amphioxus has segmented muscles and pharyngeal gill slits, a dorsal notochord, and a hollow nerve cord (Shimeld and Holland 2000). However, other vertebrate characters such as the presence of a cartilaginous or bony skeleton are absent in amphioxus (Shimeld and Holland 2000).

The discovery that both branchiomeric muscles and myocardium are derived from a cardiopharyngeal field was a crucial contribution to our understanding of the evolution of chordate muscles (Diogo et al. 2015b). Contributions from myogenic progenitors to cardiac and branchiomeric derivatives were experimentally shown to be present in the sea squirt *Ciona* (tunicates, urochordates; e.g., Stolfi et al. 2010; Razy-Krajka et al. 2014; Kaplan et al. 2015), in chickens (e.g., Tirosh-Finkel et al. 2006), and in mice (e.g., Tzahor 2009; Lescroart et al. 2010, 2015). As urochordates are the closest sister taxon of vertebrates, the cardiopharyngeal field must have been present in at least the last common ancestor (LCA) of urochordates + vertebrates (Diogo et al. 2015b), i.e., in the LCA of Olfactores (Figure 3.1). In the amphioxus larvae, a structure that is sometimes called a "heart" (a contractile vessel) (Willey 1894) lies posterior to the first three gill slits (Holland et al. 2003; Simões-Costa et al. 2005). However, there are doubts about whether this heart is related to the heart of other chordates and to the head muscles because it consists of a coelomic epithelium (myoepithelium) opposed to the gut (walled off from basal lamina: Holland et al. 2003). In fact, in the most recent review on the subject, Diogo et al. (2015b) argued that amphioxus does not have a heart that is homologous with that of urochordates + vertebrates. Hence their suggestion that although cephalochordates likely have branchiomeric muscles as vertebrates and urochordates do (Figures 3.2 through 3.5), they do not have a true cardiopharyngeal field such as that present in the Olfactores.

The discovery of the cardiopharyngeal field also revealed genetic mechanisms that are conserved in vertebrates and seem to have been present in the LCA of Olfactores (Figures 3.1 and 3.4). This complex gene network was extensively studied in *Ciona intestinalis* and mice (Lescroart et al. 2015; reviewed by Diogo et al. 2015b). In short, it shows that Olfactores had common pan-cardiopharyngeal (mesodermal) progenitors that produce the first heart field (FHF) (left ventricle, atria) and the Tbx1-positive cardiopharyngeal progenitors (Figure 3.4). The latter field differentiates into the cardiopharyngeal mesoderm, and in mice, the anterior part of progenitor cells activate *Lhx2*, self-renew and produce the second heart field (SHF)-derived right ventricle and outflow tract and the first (mandibular, muscle of mastication) and second (hyoid, muscles of facial expression) arch branchiomeric muscles (Diogo et al. 2015b). Other crucial genes involved in the differentiation of cephalic muscles are *Islet 1*, *Nkx2-5*, and *Mesp1* (the respective genes in *Ciona*: *Islet*, *Mesp*, *Nkx4*, *Tbx1/10*) (for more details, see Diogo et al. [2015b] and Lescroart et al. [2015]). The majority of cephalic muscles in vertebrates are embryologically derived from muscle plates from the mandibular, hyoid, and branchial arches (i.e., the branchiomeric muscles), while some originate from anterior somatic myotomes (i.e., the epibranchial and hypobranchial muscles) and some (e.g., pharyngeal muscles) from the mesoderm surrounding the pharynx or esophagus (Edgeworth 1935). Recent studies have shown that the pharyngeal muscles and at least part of the esophageal muscles are developmentally closely related to the muscles derived from the branchial arches (Gopalakrishnan et al. 2015).

These new insights from the developmental processes in vertebrates, tunicates, and amphioxus place our understanding of the comparative anatomy and evolution of chordates in a broader, more informed context and thus help us infer the ancestral states for chordates. Accordingly, based on our recent comparative studies and literature reviews about the cephalic muscles in a wide range of chordates, we inferred that the cephalic muscles present in the LCA of extant vertebrates were probably (Table 3.1) (a) mandibular muscles: an undifferentiated intermandibularis muscle sheet, labial muscles, some other mandibular muscles; (b) at least one hyoid muscle (at least some constrictores hyoidei); (c) at least some branchial muscles

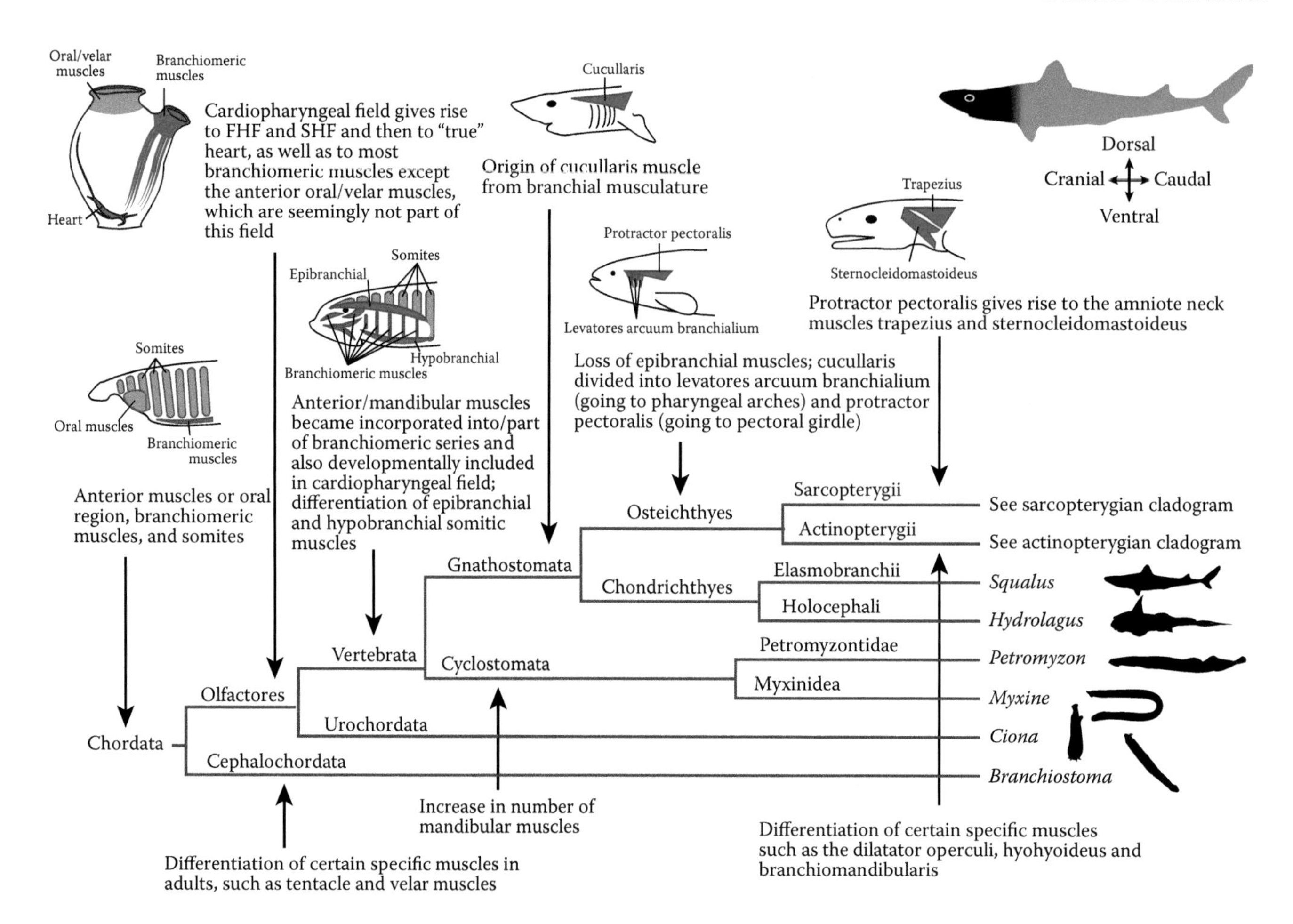

FIGURE 3.1 Some of the major features defining the Chordata and some of its key subgroups, according to our own data and review of the literature. These include, among others, the following: (a) In Chordata: somites and branchiomeric muscles. (b) In Olfactores: placodes, neural crest-like cells, and cardiopharyngeal field (CPF; NB: although within nonvertebrate chordates, conclusive evidence for these features was only reported in urochordates, some of them may have been already present in the LCA of extant chordates: see text) giving rise to first heart field and second heart field and to branchiomeric muscles (possibly not all of them; i.e., the inclusion of oral/velar muscles into CPF might have occurred during vertebrate evolution: see text). (c) In Vertebrata: skull, cardiac chambers, and differentiation of epibranchial and hypobranchial somitic muscles. (d) In Gnathostomata: jaws and differentiation between hypaxial and epaxial somitic musculature, paired appendages and fin muscles, and origin of the branchiomeric muscle cucullaris. (e) In Osteichthyes: loss of epibranchial muscles, cucullaris divided into levatores arcuum branchialium (going to pharyngeal arches) and protractor pectoralis (going to pectoral girdle), an exaptation that later allowed the emergence of the tetrapod neck. (f) Within sarcopterygians, the protractor pectoralis gave rise to the amniote neck muscles trapezius and sternocleidomastoideus.

(at least some constrictores branchiales); (d and e) undifferentiated epibranchial and hypobranchial muscle sheets (Ziermann et al. 2014). Therefore, the first of the following sections will focus in more detail on these new findings (see Figures 3.2, 3.3, and 3.5). Then, in the subsequent sections, we will combine these observations and comparisons with recent developmental and comparative data in order to address broader evolutionary and anatomical questions and to pave the way for the next chapters about the muscles of vertebrates.

CIONA INTESTINALIS AND *BRANCHIOSTOMA FLORIDAE* AS EXAMPLES OF UROCHORDATES AND CEPHALOCHORDATES

In addition to the numerous vertebrates, including cyclostomes, that we have dissected in the past (see Chapter 2), we recently dissected adult amphioxus (*Branchiostoma floridae*, cephalochordates) and adult *Ciona intestinalis* (Tunicata, Ascidiacea, urochordates) specimens. The myological terminology used in the following text and in Figures 3.2 and 3.3 follows that of Moreno and Rocha (2008) for *Ciona* and Willey (1894) for amphioxus, unless explicitly stated.

The adult morphology of *Ciona* was described in some detail by Moreno and Rocha (2008), so here we merely summarize new findings and/or key structures for comparison with amphioxus and with vertebrates. The adult *C. intestinalis* is sessile, and the individual is surrounded by a dense translucent tunica (Figure 3.2A). The body of *Ciona* can be divided in a thorax (pharynx/branchial basket) and an abdomen that contains the digestive tract, the heart, and the gonads (Figure 3.2B). The orientation of transverse and longitudinal muscle (LoM) fibers is visible on the tunica (compare

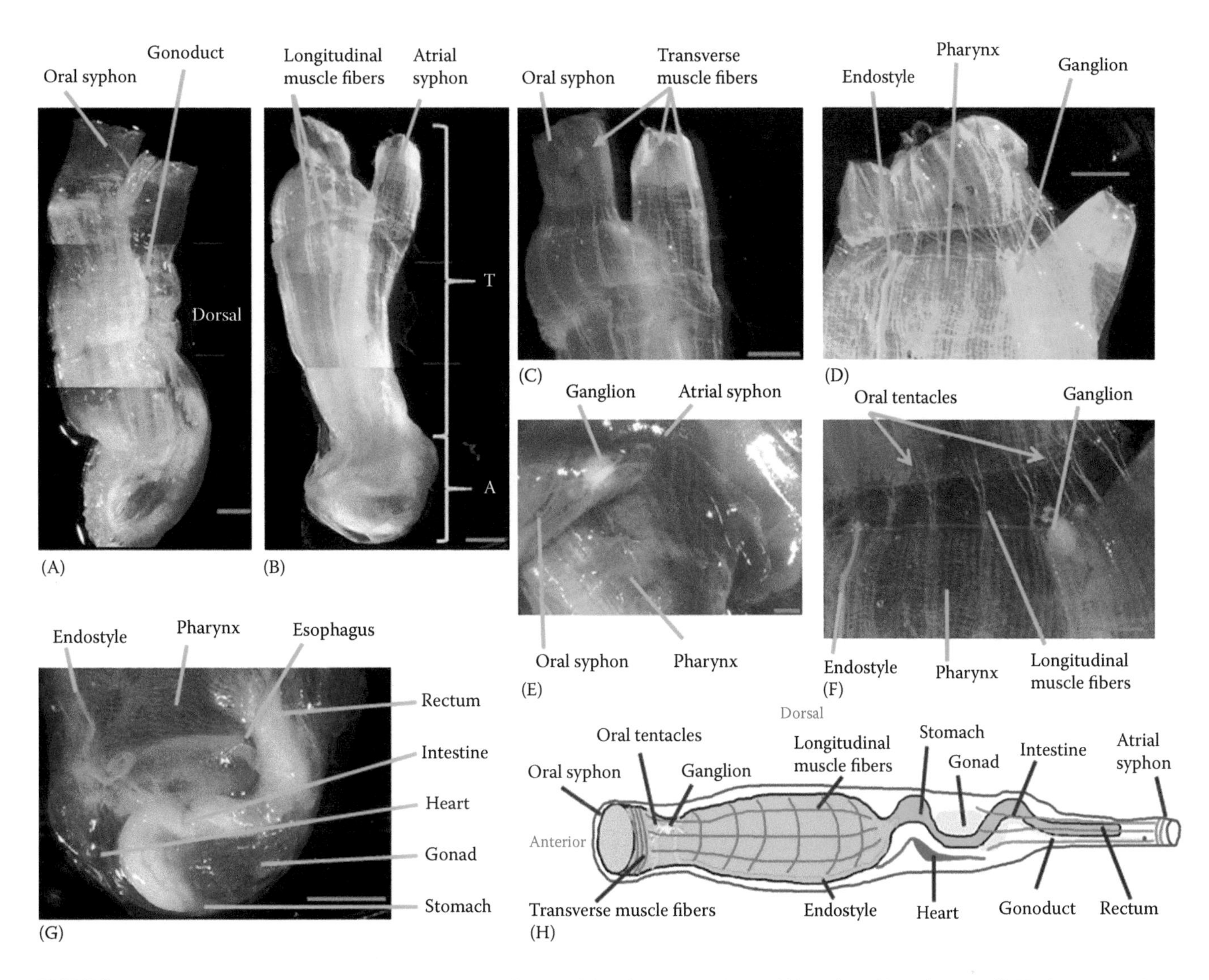

FIGURE 3.2 Adult *C. intestinalis* (based on new dissections and data from the authors). (A) In situ with tunica. (B–G) Tunica removed and specimen stained with alcian blue. (C) Atrial and oral syphons with dense transverse fibers. (D) Oral syphon laterally opened. The ventral endostyle is clearly visible beginning at the anterior end of the pharynx. (E) Oral syphon flexed ventrolaterally to see the dorsal ganglion (nerve center). (F) Red light added to increase contrast. The longitudinal fibers clearly end just anterior to the oral ring. The nerve fibers extending from the ganglion toward the oral syphon and the oral tentacles at the oral ring. (G) Abdominal region with the ventral heart and the rectum in the extended cylindrical tube that is bended dorsally and ends in the atrial syphon just next to the oral syphon (see (A)). (H) Theoretical schematic of an adult *Ciona* if one would unfold it in the abdominal region (see (G)). Scale bar in (E) = 1 mm; all other = 5 mm.

Figure 3.2A and B). The oral syphon, or inflow opening, is the larger opening, with a lobed margin (Figure 3.2A through C). The atrial syphon is a cylindrical extension of the body, as indicated by the termination of the gonoduct and the rectum well before they reach the syphon (Figure 3.2A). Through the atrial syphon (Figure 3.2), gametes and feces leave the body. Both syphons have a dense area of transverse muscle fibers (Figure 3.2A through D). In addition, *Ciona* has transverse and longitudinal body muscles. The transverse muscles lying in the thoracic region are less dense than those of the syphons (Figure 3.2C). The longitudinal fibers run from the oral syphon to the extremity of the abdomen (Figure 3.2B) and are parallel to the endostyle (Figure 3.2D and F). However, they do not extend through the whole oral syphon but seem to start just superior to the oral ring with the oral tentacles (Figure 3.2F). Posterior (inferior) to the oral ring is a "second ring," where the pharynx starts (Figure 3.2F). Remarkably, the "second ring" of *Ciona* lies in a region that seems to topologically correspond to the region displaying a buccal ring with tentacles/cirri and a velum in amphioxus.

The adult body of *Ciona* includes the large pharynx that ends at an esophagus dorsally to the heart (Figure 3.2G). Here, the peculiar morphology of adult sea squirts becomes obvious. The oral syphon is the anterior end followed by the pharynx (branchial basket) that ends in the esophagus, which itself ends at the (poorly defined) stomach (Figure 3.2G). With the ganglion (neural complex) located dorsally at the base of the oral syphon (Figure 3.2E), the anterior/posterior, dorsal/ventral, and left/right orientation of the animal is defined (Figure 3.2A). However, the intestine curves at the bottom of the animal and the rectum continues dorsally and anteriorly in the extended cylindrical tube leading to the atrial syphon

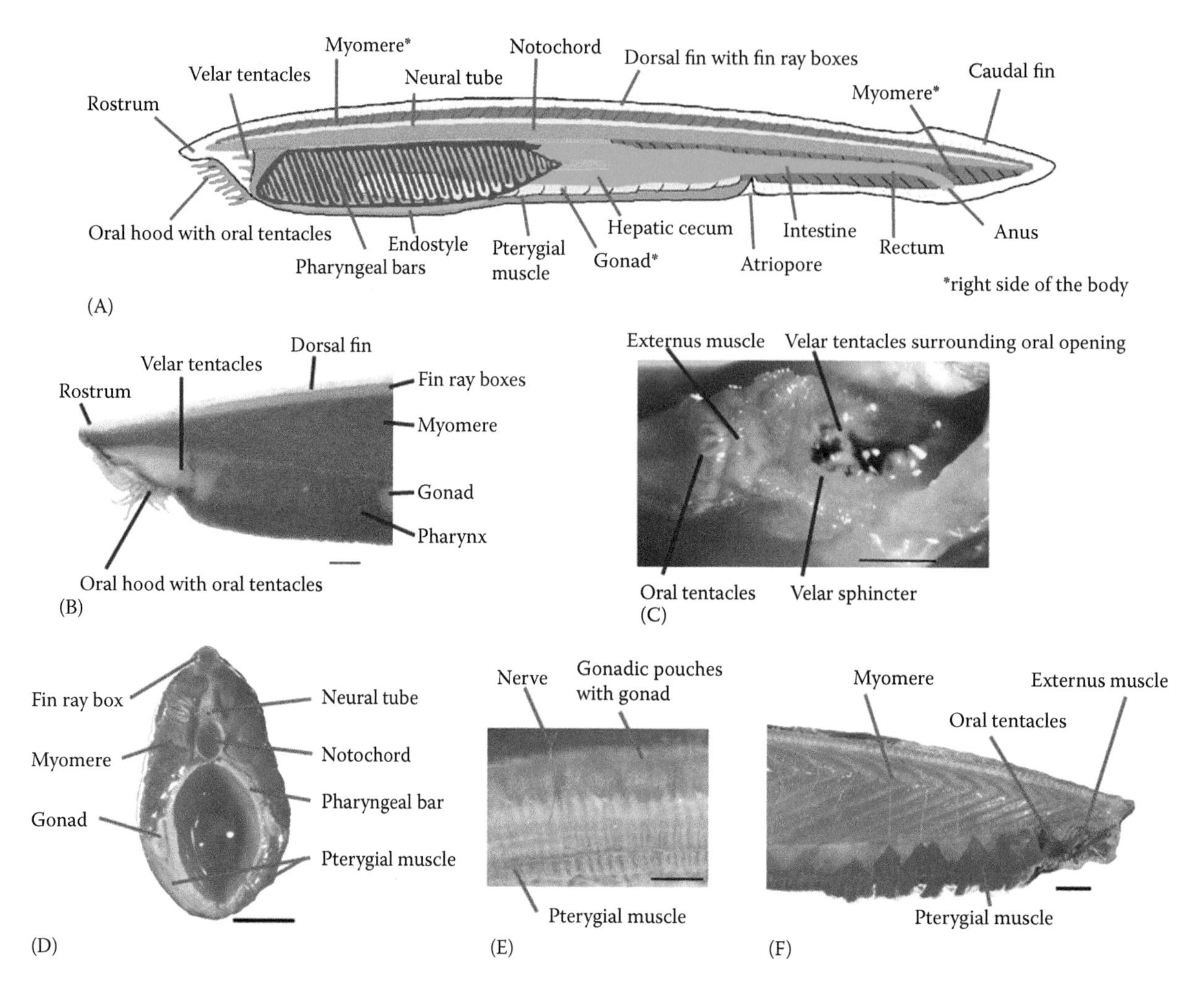

FIGURE 3.3 Adult amphioxus (*B. floridae*) (based on new dissections and data from the authors). (A) Figure of an adult amphioxus. Myomeres and gonads not shown (*) on the left side of the body to show underlying structures. (B–E) Specimen stained with alcian blue and Lugol's solution. (B) View of the anterior region. (C) Ventral view, anterior to the left; oral and velar region. Oral hood with tentacles reflected anteriorly. Scale bar = 0.5 mm. (D) Transverse section through pharyngeal region. (E) Ventrolateral view, anterior to the left. Nerve redrawn to increase visibility. (F) Right lateral view of the anterior body region. Pterygial muscle cut in midline and reflected laterally showing clear striation. (B–F): scale bar = 1 mm.

(Figure 3.2G). The gonads lie between the curled intestine, and the gonoduct has a similar path as the rectum but extends further up (Figure 3.2A and G). That is, when the animal is "unfolded" at its abdominal area (Figure 3.2H), its body plan does not seem extremely different from that of other adult chordates (Figure 3.2H). The ganglion (neural center) lies anterodorsally; the heart lies ventral to the stomach. The heart is ventral to the pharynx in gnathostome fishes, and its caudal position in *Ciona* might be related to the enlarged pharynx, which is likely a feature associated with filtration feeding.

The adult features of amphioxus were described in some detail by Willey (1894). However, due to the new phylogenetic, morphological, genetic, and developmental insights into chordates and cephalochordates and the controversial interpretations about—and lack of detailed studies of—their cephalic musculature, it is crucial to take a fresh, comprehensive look at those muscles and related structures in amphioxus adults. Amphioxus is an elongated animal with a dorsal neural tube that extends far anterior into the cephalic region and posterior into the tail (Figure 3.3A). The notochord lies ventral to the neural tube (Figure 3.3B) and spans the entire length of the body. The anteroventral mouth is surrounded by oral tentacles (buccal cirri) and ends in a large pharynx (Figure 3.3A and B). The entrance to the pharynx is surrounded by velar tentacles (Figure 3.3A through C). An endostyle spans the entire ventral length of the pharynx (Figure 3.3A). The esophagus connects to the intestine; the hepatic caecum is situated shortly thereafter. The intestine (rectum) ends in an anus, ventral and anterior to the short tail. The atrium is the most ventral organ, terminating *via* an atrial opening in the atrial syphon that contains atrial sphincter muscles, well anterior to the anal opening (Figure 3.3A).

Segmented muscles (myomeres, myotomes) cover the dorsal body in its entire length and extend into the anterior tip dorsal to the buccal cavity (Figure 3.3B). The myotomes are LoMs used for locomotion and stretch from the notochord down to just cover the gonads (Figure 3.3D). The cross-striated pterygial muscle (subatrial or transverse muscle *sensu*

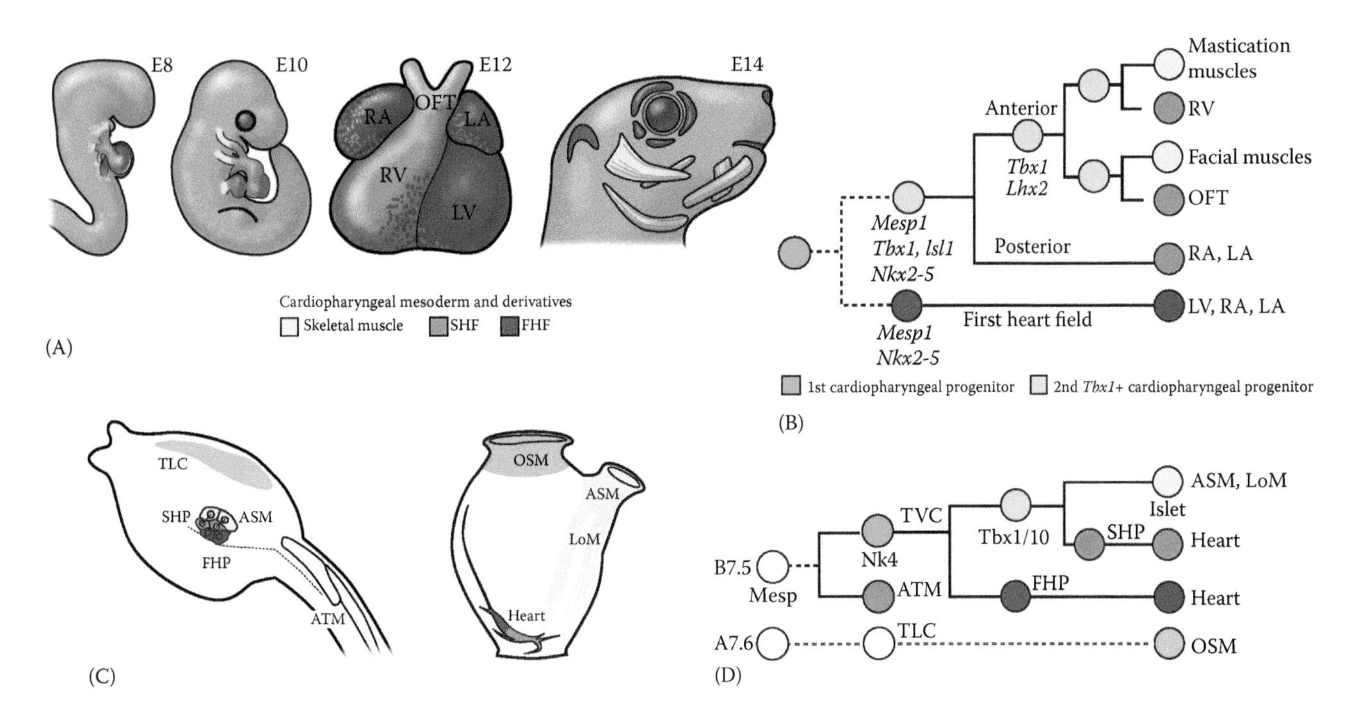

FIGURE 3.4 Evolutionary conserved cardiopharyngeal ontogenetic motif. (Modified from Diogo, R. et al., *Nature*, 520, 466–473, 2015. With permission.) (A) Mouse embryos at embryonic days (E) E8 and E10, the four-chambered mouse heart at E12, and the mouse head at E14. *Red*, FHF-derived regions of heart (left ventricle [LV] and atria); *orange*, SHF-derived regions of heart (right ventricle [RV], left atrium [LA], right atrium [RA], and outflow tract [OFT]); *yellow*, branchiomeric skeletal muscles; *purple*, extraocular muscles. (B) Cell lineage tree depicting the origins of cardiac compartments and branchiomeric muscles in mice. All cells derive from common pan-cardiopharyngeal progenitors (*dark green*) that produce the FHF, precursors of the LV and atria (RA, LA), and the second, *Tbx1+*, cardiopharyngeal progenitors (*light green*). Broken lines indicate that the early FHF/SHF progenitor remains to be identified in mice. In anterior cardiopharyngeal mesoderm, progenitor cells activate *Lhx2* and self-renew and produce the SHF-derived RV and OFT and first and second arch branchiomeric muscles (including muscles of mastication and facial expression). (C) Cardiopharyngeal precursors in *C. intestinalis* hatching larva (*left*) and their derivatives in the metamorphosed juvenile (*right*). The FHF (*red*) and SHF (*orange*) heart precursors contribute to the heart (*red–orange mix*), while ASM precursors (*yellow*) form atrial siphon and LoMs (*yellow*). OSMs (*right: blue*) derive from a heterogenous larval population of trunk lateral cells (TLCs) (*left: blue*). Cardiopharyngeal mesoderm is bilaterally symmetrical around the midline (*dotted line*). (D) Cell lineage tree depicting clonal relationship and gene activities deployed in *Ciona* cardiopharyngeal precursors. All cells derive from *Mesp+* B7.5 blastomeres, which produce anterior tail muscles (*gray disks*) (see also left panel in (C)) and trunk ventral cells (TVCs) (*dark green disk*). The latter pan-cardiopharyngeal progenitors express *Nk4* and divide asymetrically to produce the first heart precursors (*red disk*) and second TVCs, the *Tbx1/10+* second cardiopharyngeal progenitors (second TVC) (*light green disk*). The latter again divide asymmetrically to produce second heart precursors (*orange disk*) and the precursors of ASM and LoM (*yellow disk*), which upregulate *Islet*. The OSM arise from A7.6-derived TLCs (*light blue disk*).

Willey [1894]) lies ventrally (Figure 3.3A, D, and E), covering the floor of the atrium and extending anteriorly to end in the velar sphincter (Figure 3.3C) and posteriorly where it seems to form the atrial sphincter muscle. The pterygial muscle is divided by a median longitudinal septum into two halves that are further divided by thin transverse septa into a series of compartments that are not segmentally arranged (Figure 3.3E). The velar sphincter seems to be the only velar muscle. The muscles of the oral hood include one externus muscle and one internus muscle (Figure 3.3C). The outer muscle (externus) lies at the base of the cirri, and the fibers of one side interlace ventrally with those of the other side. The inner muscle (internus) lies between two consecutive cirri and is composed of multiple tiny muscles, each lying at the base between two oral cirri.

The central nervous system (CNS) (neural tube) is a long tube above the notochord, and both structures extend far anteriorly in amphioxus (Figure 3.3A). The dorsal nerve roots divide into dorsal and ventral rami that run externally to the myotomes. The dorsal rami extend from the dorsal region over the myomeres to the ventral region where they split into cutaneous branchesm and branches that turn medially around the metapleural fold to innervate the pterygial muscle (Figure 3.3E).

EVOLUTION AND HOMOLOGY OF CHORDATE MUSCLES BASED ON DEVELOPMENTAL AND ANATOMICAL STUDIES

Amphioxus and lamprey larvae (ammocoete; cyclostomes) are filter feeders, and both have a functional endostyle (Holland et al. 2008). However, in larval amphioxus, the pharynx has an asymmetric development, a feature that was likely not present in the LCA of chordates (Stokes and Holland 1995;

Presley et al. 1996). Almost the entire muscular system of amphioxus is composed of striated muscle fibers. Smooth muscles are found in the postpharyngeal gut, excluding the gut diverticulum (Holmes 1953). The striated muscles can be divided into the parietal muscles, which are the myotomes, and the visceral and splanchnic muscles (Willey 1894). The visceral muscles include the pterygial (transverse or subatrial) muscle, the muscles of the oral hood and cirri, and the velar and anal sphincter muscles. Another striated muscle was described by Holmes (1953) under the confusing name "trapezius," lying dorsal to the pharynx at so-called funnels and lateroventrally to the notochord (NB: we could not identify a muscle that follows this description). This muscle does not seem to be homologous to the trapezius/cucullaris of gnathostomes because cyclostomes and in urochordates lack such a muscle. All striated muscles of amphioxus are composed of flat lamelliform plates, which, in cyclostomes, are found in connection with the lateral muscles only (Willey 1894).

The fibers of oral and velar muscles of amphioxus closely resemble those fibers found in the walls of the heart of the "higher" vertebrates (Willey 1894). This resemblance may be relevant in the context of the discovery of strong links between the head and branchiomeric muscles in urochordates and vertebrates (cardiopharyngeal field) (see Diogo et al. 2015b). In amphioxus larvae, the muscle fibers on the peritoneum on the pharyngeal floor are functionally related to the closing of the gill slits (Yasui et al. 2014). In contrast, the myoepithelial cells in the branchial muscles of Hemichordata (the extant sister-group of chordates) derive from LoMs (Cameron 2002) and give elasticity to the lacunae in the tongue bar and the blood vessels (Yasui et al. 2014). The anterior myotomes of amphioxus that overlap the region of the oral hood might correspond to the supraocularis of lampreys, which is also an anterior extension of the parietal (epibranchial) somitic musculature only separated from the trunk somitic musculature by a septum (connective tissue that also separates myotomes) (Ziermann et al. 2014; Diogo and Ziermann 2015a). This condition is similar to that of the nasalis muscle in hagfishes (anterior extension of somitic muscle parietalis in the study by Ziermann et al. [2014]).

Both larval and adult amphioxus have orobranchial muscles that are developmentally and anatomically similar to the vertebrate branchiomeric musculature (Diogo et al. 2015b) (Figure 3.5). Yasui et al. (2014) suggested that the amphioxus larval orobranchial muscles might be anatomically more similar to the branchiomeric muscles of adult vertebrates than the adult oral, velar, and pterygial muscles of amphioxus. The authors described five distinct larval orobranchial muscles and stated that these muscles disappear during metamorphosis and are topologically replaced by the adult oral, velar, and pterygial muscles. However, despite the observed apoptosis in the larval pharyngeal region of amphioxus (Willey 1894; Yasui et al. 2014), it is not clear whether all larval pharyngeal muscles are absent in the adult. A detailed developmental study of the transformations that occur in the gill slits/pharyngeal arches region during metamorphosis would be needed to resolve the developmental origin of the adult pharyngeal muscles of amphioxus.

The adult amphioxus has two oral muscles, the externus and internus, related to the oral tentacles, and one velar sphincter muscle (Figure 3.5). The oral muscles appear to develop without segmental patterning (Yasui et al. 2014). According to Willey (1894), these oral muscles of amphioxus relate to the cirri and do not resemble any vertebrate muscles. The pterygial muscle of amphioxus is a branchiomeric muscle *sensu* Diogo et al. (2015b) (Table 3.1), extending anteriorly and posteriorly to form the velar and atrial sphincter, respectively (Holmes, 1953). In fact, almost the same configuration is seen in *Ciona* (urochordates): the muscles related to both the oral and atrial sphincters (siphons) express *Tbx1* and seem to correspond to vertebrate branchiomeric muscles (e.g., Stolfi et al. 2010; Sambasivan et al. 2011; Diogo et al. 2015b). Moreover, several authors have noted that the pterygial muscle of amphioxus develops ventrally in the pharynx of this animal and is innervated by peripheral nerves that are similar to the nerves of the branchiomeric muscles of vertebrates (Fritzsch and Northcutt 1993; Yasui et al. 2014). The musculature of the adult amphioxus velar sphincter might correspond to the transversus oris of adult hagfish and/or to the annularis of adult lamprey (Table 3.1). However, the transversus oris of hagfish and the annularis of lampreys do not seem to be homologous to each other, although both are part of the nasal muscle group of mandibular muscles in cyclostomes (see Chapters 4 and 5). The amphioxus oral internus muscle consists of multiple small muscles associated with the base between two oral tentacles; this might indicate a relationship with the large number of cephalic muscles in cyclostomes (lingual, dental, and velar muscles; see Chapters 4 and 5; NB: the nasal muscles in cyclostomes might be better explained by the splitting of myotomal structures in the head).

The atrium of amphioxus shares some similarities with the atrium of urochordates (e.g. of *Ciona*) (Figure 3.5); cyclostomes do not have an atrium. In amphioxus, the innervation (sensory and motor) of the atrial region, including the pterygial muscle that covers this region ventrally, is by dorsal nerve roots (Holmes 1953; Bone 1960). The motor axons also control the lateral ciliary tracts of the pharyngeal bars (Bone 1960). The atrial nervous system includes connected neurons on the visceral and parietal borders of the atrium (Holmes 1953). The atrial epithelium arises from the invagination of larval ectoderm and the majority of neurons of the atrial nervous system lie in the epithelium suggesting that this nervous system is associated with an ectodermal layer (Holmes 1953). Therefore, this atrial nervous system does not seem to be homologous with the sympathetic systems in craniate vertebrates (cyclostomes + gnathostomes), suggesting that at least some of the features of the visceral nervous systems of amphioxus and vertebrates might be nonhomologous (Holmes 1953; Bone 1960).

The atrial cavity in amphioxus seems to play an important role in filter feeding (Dennell 1950) and develops during the larval period (development of first gill slit on the right side until metamorphosis) (Willey 1894). The contraction of the pterygial muscle reduces the atrial cavity and water is expelled through the atriopore (Willey 1894; Dennell 1950).

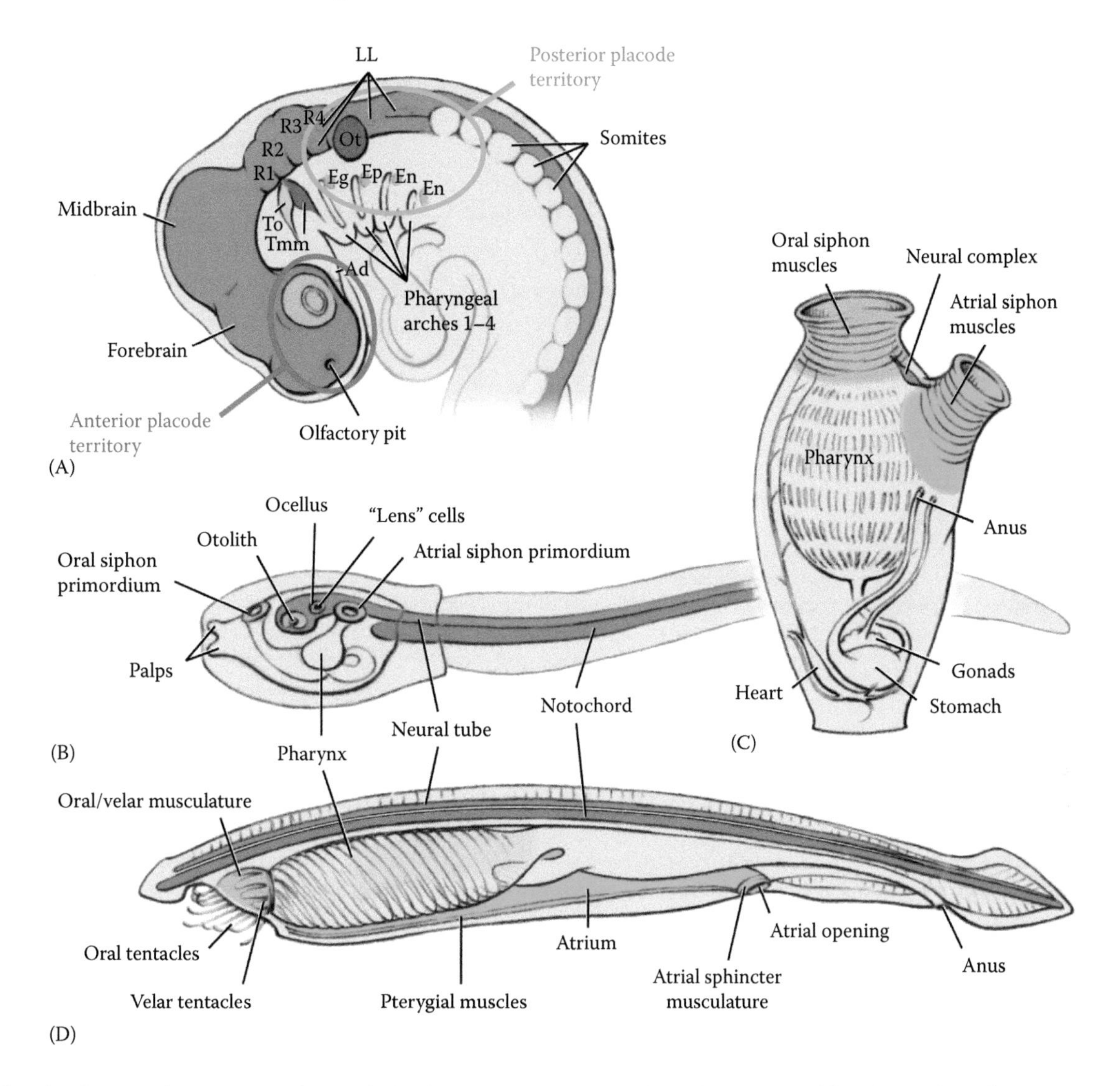

FIGURE 3.5 Comparative anatomy of cephalochordates, urochordates, and vertebrates. (Modified from Diogo, R. et al., *Nature*, 520, 466–473, 2015. With permission.) (A) Location of ectodermal placodes in vertebrate head according to Graham and Shimeld's (2013) hypothesis (*anterior to the left*): olfactory placode/pit (*red*) at tip of forebrain; lens placodes (*orange*) form posteriorly as part of eye; adenohypophyseal placode (Ad; *yellow*) lies ventrally to forebrain; trigeminal placodes form alongside anterior hindbrain at the levels of rhombomeres 1 and 2 (R1, R2), the anterior one being the ophthalmic placode (To; *light blue*) and the posterior one the maxillomandibular placode (Tmm; *purple*); otic placode (Ot; *brown*) forms opposite the central domain of hindbrain; lateral line placodes (LL; *pink*) form anteriorly and posteriorly to otic placode; epibranchial placodes (*green*)—geniculate (Eg), petrosal (Ep) and nodose (En)—form as part of pharyngeal series. Forebrain, midbrain, R1 (R2, R3, R4) (*dark blue*)—rhombomere 1 (2, 3, 4) and somites. (B) Urochordate tadpole larva (*anterior to the left*): notochord in red and two siphon primordia (*green*; *orange*), with putative relationships to the anterior and posterior placode territories shown in (A). (C) Adult urochordate showing siphon primordia after metamorphosis. (D) Adult cephalochordate showing the hypotheses of urochordate–cephalochordate muscle homology proposed in the present review. (Modified from Willey, A., *Amphioxus and the Ancestry of the Vertebrates*, MacMillan and Co., New York, 1894; Sambasivan, R. et al., *Development*, 138, 2401–2415, 2011; Graham, A., and Shimeld, S. M., *J Anat*, 222, 32–40, 2013.)

This mechanism depends on closing the atriopore, which is not just a perforation of the atrium floor (Dennell 1950). Between the external aperture of the atriopore and the atrial cavity lies a short chamber (atrial siphon) (Dennell 1950). This asymmetrical atrial siphon is ventrally cut off from the main atrium cavity. Anteriorly, the false floor, disconnecting the siphon from the atrium, extends horizontally and freely reaches into the atrium as two processes (Dennell 1950). Posteriorly, the siphon opens to the exterior; the floor of the atrium and the siphon walls are muscular, and a contraction of the latter results in the occlusion of the cavity and closure of the atriopore (Dennell 1950).

In order to clean the oral (buccal) cirri after feeding, the cirri are spasmodically flexed into and out of the oral hood cavity accompanied by an expulsion of water from the hood, removing the external particles (Dennell 1950). The inward movement of the cirri results from the contraction of the oral (labial) muscles together with the powerful movements of the atrium floor *via* contractions of the pterygial muscle (Dennell 1950). The latter movement changes the volume of

TABLE 3.1
Muscles Inferred to be Present in the LCA of Extant Chordates and the LCA of Extant Vertebrates

LCA Extant Chordata	*B. floridae* (Amphioxus; Cephalochordata)	LCA Extant Olfactores	*C. intestinalis* (Sea Squirt; Tunicata or Urochordata)	LCA Extant Vertebrata
Anterior Muscles				**Mandibular Muscles**
Muscles of oral region [*The circular oral siphon muscles in urochordates, which are not part of the cardiopharyngeal field, could correspond to oral/velar muscles of amphioxus and of vertebrates such as cyclostomes and thus to the first arch (mandibular) muscles of gnathostomes, which would thus be included as part of the branchiomeric muscle series only during vertebrate evolution]	External and internal oral tentacle muscles, velar sphincter muscle	Muscles of oral region	Oral syphon muscles (corresponding to oral/velar muscles in cyclostomes and/or mandibular muscles in gnathostomes?*)	Undifferentiated; intermandibularis; muscle sheet Other mandibular muscles (e.g., at least one labial muscle and probably at least one velar and/or dorsal mandibular muscle; see Ziermann et al. [2014] for discussion)
Branchiomeric Muscles (NB: mandibular muscles only became incorporated into/part of branchiomeric series in vertebrates*)				
Muscles corresponding to atrial sphincter and/or pterygial muscle of cephalochordates/urochordates + some transverse and longitudinal body muscles (genetic/morphogenetic program for differentiation into branchiomeric versus body/somitic muscles was probably not as sharply defined as in the LCA of vertebrates)	Pterygial (subatrial) muscle, atrial sphincter muscle (also "trapezius muscle" *sensu* Holmes [1953], which very likely does not correspond to trapezius muscle of amniotes?)	Branchiomeric muscle derivatives from cardiopharyngeal field	Atrial syphon muscles and some associated muscles (e.g., some transverse/longitudinal body muscles)	*Hyoid*: Constrictores hyoidei (at least one) *True branchial*: Constrictores branchiales (one or more) and adductores branchiales and/or interbranchiales? (see Ziermann et al. [2014], for discussion) *Other branchial muscles*:
No clear differentiation of epibranchial and/or hypobranchial muscle groups derived from somitic musculature				Epibranchial and hypobranchial musculature

the atrium and, with it, the direction of water streaming in and out through the pharynx and the oral hood rather than through the atriopore (Dennell 1950). This functional association between the oral and pterygial musculatures in amphioxus is fascinating, because it might suggest that there is also a developmental link between these muscles. Therefore, although we do not include the oral muscles of amphioxus among the branchiomeric series (i.e., as mandibular muscles) present in the LCA of chordates (Table 3.1), these muscles might have been incorporated into the cardiopharyngeal field later in vertebrate evolution (see the following).

The somatic muscles of amphioxus, which are restricted to the myotomes (Yasui et al. 2014), have a peculiar mode of innervation: muscle tails (extensions of the muscles) take their innervation from the ventral surface of the nerve cord (Holmes 1953; Flood 1966), while other muscles are innervated by peripheral nerves from the dorsal roots (Holmes 1953; Yasui et al. 2014). In vertebrates, branchial motor neurons are located dorsally to somatic motor neurons, although their division is not as distinct as that of the postcranial spinal nerves (Yasui et al. 2014). In cephalochordates (amphioxus), the peripheral nerves from the CNS show a metameric pattern, as seen in vertebrates, and do not directly innervate the mouth and gills. Instead, they extend into the metapleural folds, where they anastomose and form the oral nerve ring (Kaji et al. 2001, 2009) or metapleural longitudinal nerves (Yasui et al. 2014) before innervating the oral and branchial targets. In gnathostomes, the hypobranchial (somitic) muscle precursors migrate together with the hypoglossal along the boundary of the pharynx and the body cavity to reach the oral floor (Oisi et al. 2015). During the development of lampreys, muscle precursors arise from rostral somites, migrate caudally and ventrally along the caudal end of the pharynx at the interface to the rostral part of the body wall, and turn rostrally reaching the pharyngeal wall (Marinelli and Strenger 1954). Therefore, typically both the myotomal muscle precursors and the hypoglossal nerve do not migrate into pharyngeal arches (Mackenzie et al. 1998). Branchiomeric nerves (cranial nerves) are associated with pharyngeal arches and are characterized by their lateral positions, while spinal nerves are associated with somites, and their dorsal roots are more medially, medial to the dermomyotome (Oisi et al. 2015). At the head–trunk transition area in gnathostomes, the relationship between the vagus nerve (cranial) and the hypoglossal nerve (spinal) is reversed, and the latter nerve is more lateral (Kuratani 1997). This pattern is also found in the lamprey, but greatly modified in the hagfish (Oisi et al. 2015); still, it was likely present in the LCA of gnathostomes and cyclostomes (i.e. in the LCA of extant vertebrates).

The CNS in adult amphioxus; i.e., the neural tube, was described in detail by Willey (1894). It ends anteriorly behind the anterior end of the notochord. Also anteriorly, a pair of nerves project from the sides of the nerve tube, followed by a pair that arises more dorsally—also called cranial nerves—that lie in front of the first myotomes, have no ventral roots, and seem to be only sensory. They do not innervate any muscles, are only found in the snout, and have peripheral ganglionic enlargements. All following spinal nerve pairs are not symmetrically arranged but alternate with one another, similar to the alternation of myotomes. This asymmetrical alternation becomes more pronounced posteriorly. Behind the second pair of nerves ascend dorsal and ventral nerve roots, arising dorsally and ventrally from the neural tube, respectively. The dorsal roots are compact nerves from collected nerve fibers, while the ventral fibers emerge separately in loose bundles from the neural tube. Each body segment has one pair of dorsal roots and one pair of ventral root bundles, and both types of roots are completely independent of each other, in contrast to vertebrates in which the dorsal and ventral roots coalesce (Willey 1894). The first dorsal spinal nerve pair (i.e., the third pair in total) passes from the neural tube to the skin through the septum that separates the first and second myotomes. All following dorsal roots show this pattern (second dorsal spinal nerve pair passes through the septum between second and third myotomes, and so on). The dorsal roots divide into the ramus dorsalis and ramus ventralis shortly after leaving the neural tube, and both rami run externally to the muscles in the subepidermic cutis (Holmes 1953). The corresponding branches of spinal nerves in vertebrates lie medially to the muscles during the first part of their course—i.e., between the muscle and the notochord. Those cranial nerves of vertebrates resemble the dorsal roots of amphioxus in the sense that they are external to the somites of the head.

The ramus dorsalis divides into smaller nerves innervating the skin of the back, while the ventral ramus divides into several cutaneous nerves and a visceral branch turning medially below the myotomes and passing between the myotomes and the pterygial muscle (Willey 1894; Holmes 1953). The dorsal spinal nerves of amphioxus are therefore mixed nerves, i.e., sensory and motor. The ventral spinal nerves are entirely motor nerves and, after leaving the neural tube (spinal cord), they fan out and innervate the myotomes. Interestingly, in vertebrates, the ventral roots are motor and the dorsal roots are sensory (Kaji et al. 2009), before they exchange fibers in the spinal nerve, which is then mixed and gives rise to a mixed dorsal and a mixed ventral primary ramus. The visceral neurons are different in amphioxus compared to other vertebrates (craniate animals *sensu* Holmes [1953]). The descending visceral branch of each segment in the atrial region of amphioxus runs over the pterygial muscle and is often described as a branching transverse nerve. This transverse nerve passes between the atrial floor epithelium and the pterygial muscle fibers and provides the motor innervation for these fibers. Experiments by Holmes (1953) showed that the motor nerves from the dorsal root induce the contractions of the atrial floor; the motor division of the descending visceral ramus comes straight from the cord to the pterygial muscle. In summary, the pterygial muscle, the velar muscles, and the oral hood muscles are innervated by the visceral branches of the dorsal nerves, as is the atrial sphincter (i.e., the caudal fibers of the pterygial muscle surrounding the anus) in all probability. This relationship is the basis for our statement earlier that with respect to their innervation, the muscles in the atrial region in amphioxus probably correspond to the branchiomeric muscles in vertebrates (see also Gans 1989 and Diogo and Ziermann 2015a).

The innervation of the larval mouth of amphioxus dramatically differs from the innervation of the adult oral region. It seems that the larval oral nerve ring plays a crucial role in patterning the nervous system in the oral region but is not homologous with any structures in vertebrates (Kaji et al. 2009) and is not the precursor of the inner oral hood nerve plexus and the velar nerve ring as described by Kaji et al. (2001). The visceral branches from the dorsal spinal nerves that innervate the oral hood in the adult amphioxus arise from the branches of the third to seventh dorsal nerves (Willey 1894). One set of those branches courses beneath to the outer surface of the oral hood and forms frequent anastomoses which gave this network the term *outer plexus* (Willey 1894; Kaji et al. 2009). The other set is below the inner surface of the oral hood and is the inner plexus. Both plexi are distinct from each other, besides the fact that the nerves have a common origin from the dorsal roots (Willey 1894). The outer plexus continues up into the individual buccal cirri, and the inner plexus seems to end at the base of the buccal cirri (Willey 1894). The inner plexus of both sides of the oral hood is exclusively formed by nerves arising from the left third and fourth nerves (Willey 1894; Kaji et al. 2001). The innervation of the velum is by the fourth, fifth, and sometimes sixth dorsal nerves of the left site only (Willey 1894; Kaji et al. 2009). This asymmetry seems to be related to the peculiar development of amphioxus.

The dorsal spinal nerves of amphioxus have some characteristics typical of the cranial nerves of vertebrates, but the walls of the gill slits are innervated in amphioxus by spinal nerves, while they are innervated by cranial nerves in vertebrates (Willey 1894). The cerebral vesicle is a widening of the central canal in the region of the cranial nerves and is not divided into ventricles. The cerebral vesicle opens in young amphioxus by an aperture called the neuropore into the base of the olfactory pit. The neuropore closes in later stages and is only indicated by a groove at the base of a stalk connecting the olfactory pit with the roof of the brain (Willey 1894). Behind the cerebral vesicle, the central canal widens into a dorsal portion that is independent of the ventral tube. The region of the nerve tube over which the dorsal portion extends was compared to the medulla oblongata of craniate vertebrates. During the development of the CNS of vertebrates, there might be a stage that is comparable to the adult condition in amphioxus (Willey 1894). However, in vertebrates, the anterior portion of the medullary tube enlarges and divides into fore-, mid-, and hindbrain.

RECENT FINDINGS ON THE "NEW HEAD HYPOTHESIS" AND THE ORIGIN OF VERTEBRATES

Data obtained since Gans and Northcutt's (1983) paper on the new head hypothesis can be divided into three major categories: those that support parts of the new head hypothesis, those that revive earlier ideas, and those that present new, and often surprising, scenarios. As an example of the first category, paleontological studies support the idea that a head skeleton composed of cartilage and calcified tissues derived from neural crest and sclerotomal mesoderm is an ancestral vertebrate feature (e.g., Valentine 2004). However, these studies also revealed specific evolutionary changes that markedly differ from previous assumptions; for example, the first gnathostomes (e.g., placoderms) probably possessed not only calcified endochondral bones, but also dermal bones (e.g., maxillary) similar to those found only in extant bony fish (osteichthyans) (Zhu et al. 2013). Gans and Northcutt's hypothesis that the evolution of chordates and early vertebrates relates to a shift from filter feeding to suction feeding, and thus to a more active mode of predation has also been supported in recent decades (Northcutt 2005), but the specific phenotypic changes involved are still heatedly debated. For instance, Mallatt's (2008) neoclassical hypothesis for the origin of the vertebrate jaw is more conservative in assuming that the upper lip and its muscles in sharks are homologous with those of lampreys, while Kuratani et al.'s (2013) heterotopic hypothesis assumes that the upper lip was lost in gnathostomes and acquired de novo in some gnathostome groups, such as sharks.

The second category of data includes a surprising revival of ideas defended by classical authors such as Goodrich, Garstang, Gegenbaur, Edgeworth, and even Darwin. Some of these ideas were widely accepted in the late nineteenth and/or early twentieth centuries, but they were largely abandoned during the second half of the twentieth century and were therefore not incorporated in the new head hypothesis. They include (a) the sister-group relationship between urochordates and vertebrates (Delsuc et al. 2006), previously advanced by authors such as Garstang (1928) and Darwin (1871); (b) Gegenbaur's (1878) hypothesis that the pectoral appendage (girdle + fin) originated as an integral part of the head (Gillis et al. 2009); and (c) Edgeworth's (1935) hypothesis that at least part of the esophageal musculature and the cucullaris derivatives (e.g., trapezius) derive from the branchiomeric musculature and/or follow a head program (e.g., Piotrowski and Nüsslein-Volhard 2000; Diogo and Abdala 2010; Sambasivan et al. 2011; Minchin et al. 2013, but see, e.g., Minchin et al. 2013), which was supported by the recent clonal studies of Lescroart et al. (2015) (see Figure 3.6).

The third category comprises new and mostly unexpected scenarios. For instance, contrary to what was usually accepted at the time of the writing of the new head hypothesis, cranial neural crest cells, while giving rise to numerous skeletal elements of the head and serving as precursors for connective tissue and tendons, do *not* form muscles (Noden 1983b, 1986; Noden and Francis-West 2006). Instead, the mesoderm-derived muscle progenitors fuse together to form myofibers within cranial neural crest-derived connective tissue in a precisely coordinated manner. Muscles of a certain arch are usually associated with connective tissue and, through this tissue also with skeletal elements, of the same arch (Köntges and Lumsden 1996).

As mentioned earlier, another remarkable discovery was that of the cardiopharyngeal field (Figures 3.4 through 3.6; reviewed by Diogo et al. 2015b). Strikingly, the results of these analyses suggest that some branchiomeric muscles are more closely related to certain heart muscles than to other branchiomeric muscles, contradicting the view, which has been long accepted, that the branchiomeric muscles mainly constitute a single anatomical and developmental unit (e.g., Edgeworth 1935; Figure 3.6). Furthermore, these studies suggest that the first (mandibular) arch might well have not been part of the original series of branchial arches (Miyashita and Diogo 2016). For instance, the most rostral branchial arch of basal chordates such as cephalochordates and "prevertebrate" fossils such as *Haikouella* is thought to correspond to the second (hyoid) branchial arch of vertebrates (Mallatt 2008). According to this idea, the first arch was incorporated into the branchial arches only in more derived chordates, which might explain, for instance, why it is the only arch in vertebrates in which *Hox* genes are not expressed and do not pattern arch formation (Mallatt 2008; Miyashita and Diogo 2016; see below).

It was recently shown that *Hox1* is essential for the anterior–posterior (AP) axial identity of the endostyle in the urochordate *C. intestinalis* (Yoshida et al. 2017). That is remarkable because *Hox1* represses *Otx* expression. If *Hox1* is knocked out, the posterior end of the endostyle is transformed to an anterior identity because of ectopic expression of *Otx* and the atrial siphon and gill slits are lost (Sasakura et al. 2012; Yoshida et al. 2017). In turn, the overexpression of *Hox1* can repress the anterior endostyle identity. A change in regional identity of the endoderm causes a disruption of the body wall muscle formation implying that the endostyle, a major part of the pharyngeal endoderm, is essential for coordinated pharyngeal development (Yoshida et al. 2017). Experiments by Yoshida et al. (2017) suggest that the identity of the anterior and posterior endostyles is by default the one expression *Otx*, i.e., anterior. Furthermore, retinoic acid (RA) receptor and RAsignaling from larval endoderm and muscle induce *Hox1* expression in the posterior endostyle, and RA synthesis is required to maintain *Hox1* expression (Yoshida et al. 2017). The posterior endodermal identity and posterior RA synthesis is needed for the elongation of the body wall muscles toward the posterior end in *C. intestinalis*. In chordates, *Otx* and *Hox1* transcription factors are expressed in the embryonal pharyngeal endoderm. During mouse development, *Otx2* expression is observed in the first arch endoderm (Ang et al. 1994), and *Hox1a* and *Hox1b* expressions, in the caudal

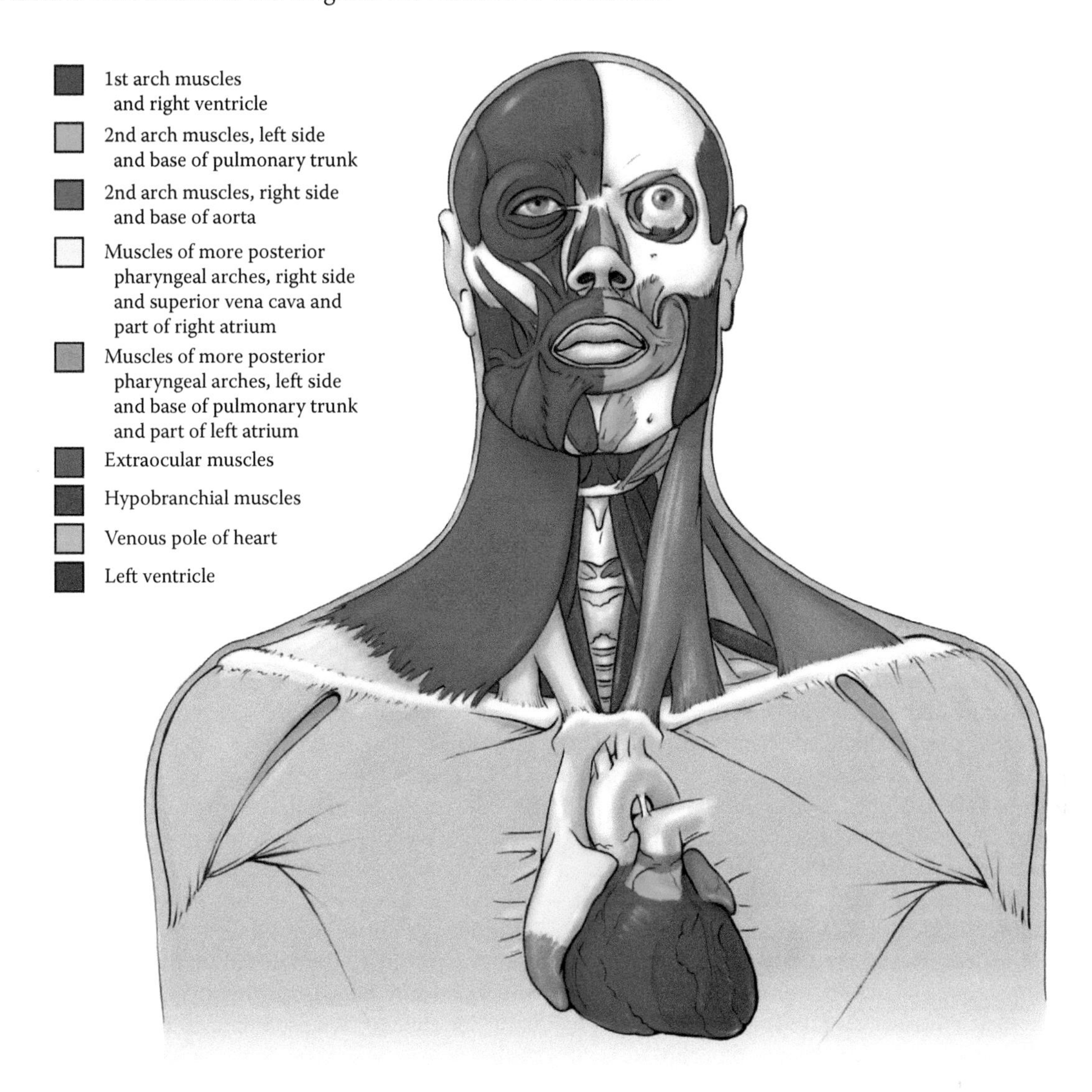

FIGURE 3.6 Striking heterogeneity of the human head musculature. (Modified from Diogo, R. et al., *Nature*, 520, 466–473, 2015. With permission; modified to include new data provided by Lescroart, F. et al., *PNAS*, 112, 1446–1451, 2015.) The head musculature includes at least seven different muscle groups, all arising from the cardiopharyngeal field and being branchiomeric, except the hypobranchial, and perhaps (but not likely) the extraocular, muscles. On the left side of the body (right part of figure) the facial expression muscles were removed to show the masticatory muscles. The seven groups are (a) *first/mandibular arch muscles*, including cells clonally related to the right ventricle (*purple*) and apparently to the extraocular muscles (see below); (b) *left second/hyoid arch muscles*, with cells related to myocardium at the base of the pulmonary trunk (*green*); (c) *right second/hyoid arch muscles*, related to myocardium at the base of the aorta (*red*); (d) *left muscles of the most posterior pharyngeal arches*, including muscles of the pharynx and larynx and the cucullaris-derived neck muscles trapezius and sternocleidomastoideus, which are related to the base of the pulmonary trunk and part of the left atrium (orange); (e) *right muscles of the most posterior pharyngeal arches*, including muscles of the pharynx and larynx and the cucullaris-derived neck muscles trapezius and sternocleidomastoideus, which are related to the superior vena cava and part of the right atrium (*yellow*); (f) *extraocular muscles* (*pink*), which are often not considered to be branchiomeric, but that according to classical embryologic studies and recent retrospective clonal analyses in mice contain cells related to those of branchiomeric mandibular muscles; and (g) *hypobranchial muscles*, including tongue and infrahyoid muscles that derive from somites and migrate into the head and neck (*dark gray*) (to show that, they are not part of the colored cardiopharyngeal field). The venous pole of the heart is shown in blue, and the left ventricle, derived from the first heart field, in brown.

pharynx which are dependent on RA (Wendling et al. 2000; Niederreither et al. 2003). The cephalochordate amphioxus was shown to express *Hox1* in the endoderm, repressing *Otx* expression and, under RA control, determining the posterior limit of the pharynx (Schubert et al. 2005). All this evidence points toward a genetic mechanism present in the LCA of chordates needed for the proper AP axis specification of the early pharyngeal endoderm, which, in turn, is needed for the proper formation of pharyngeal muscles (Yoshida et al. 2017).

Within their new head hypothesis, Gans and Northcutt (1983) argued that one of the main differences between vertebrates and invertebrates is that vertebrates possess complex sense organs and associated cranial ganglia, while invertebrates have poorly specialized sense organs and no neurogenic placodes. However, studies performed in the last three decades, particularly on urochordates, strongly contradict this scenario. In addition to the discovery of a cardiopharyngeal field in urochordates (Figures 3.4 and 3.5; reviewed by Diogo et al. 2015b), the results of these studies have shown that urochordates also have placodes and neural crest-like cells, as recently summarized by Graham and Shimeld (2013) and Hall and Gillis (2013). The points made by the latter authors are briefly summarized below and in Figure 3.5.

DEVELOPMENT AND EVOLUTION OF CHORDATE MUSCLES AND THE ORIGIN OF HEAD MUSCLES OF VERTEBRATES

In Table 3 and Figure 1 of their study, Gans and Northcutt (1983) suggested that branchiomeric muscles were acquired at the origin of vertebrates. However, recent works, as well as some older and unfortunately often ignored studies, clearly show that branchiomeric muscles related to the pharynx were present in the LCA of chordates (Figure 3.1; Table 3.1). More than 100 years ago, it was reported that larvae of the giant cephalochordate *Epigonichthys* develop complex orobranchial musculature, but almost no investigation of the orobranchial musculature of this clade has been completed since (reviewed by Yasui et al. [2014]). As noted earlier, in the cephalochordate amphioxus, the larval mouth and unpaired primary gills develop five groups of orobranchial muscles as the larval mouth enlarges posteriorly, the oral musculature developing without segmental patterning (Yasui et al. 2014). During metamorphosis, the orobranchial musculature is said to completely disappear (but see preceding text) and the adult oral, velar, and pterygial (= subatrial or transverse) muscles (Figure 3.2) develop. Yasui et al. (2014) suggested that the cephalochordate orobranchial muscles are probably a larval adaptation to prevent harmful intake, but they noted that the larval orobranchial muscles are perhaps more similar anatomically to the vertebrate branchiomeric muscles than are adult cephalochordate oral, velar, and pterygial muscles. They also noted that vestigial muscles transiently appear with secondary gill formation, suggesting an ancestral state of bilateral muscular gills and a segmental pattern of branchiomeric muscles in chordates. Six years after Gans and Northcutt's (1983) study, Gans (1989) did recognize that the muscles of the atrial region of cephalochordates might correspond to the vertebrate branchiomeric muscles. He explained that cephalochordates and vertebrates have two patterns of motor innervation: one involves somatic motor neurons located within the basal plate of the spinal cord (somitic muscles); the other is seen in the cranial end of cephalochordates, where somatic motor axons leave the neural tube *via* a dorsal cranial root that proceeds ventrally to innervate the striated pterygial muscle of the atrial floor. Therefore, according to Gans (1989), and contrary to Gans and Northcutt's (1983) new head hypothesis, the pterygial musculature of amphioxus might be homologous with the branchiomeric muscles of vertebrates, which might well have arisen by an invasion of paraxial mesoderm to surround the pharynx laterally and ventrally, instead of by muscularization of hypomeric tissues.

At first glance, the proposed homology between vertebrate and urochordate cardiopharyngeal muscles and cephalochordate muscles might seem counterintuitive: one would expect the urochordate oral siphon and the cephalochordate oral/velar muscles, rather than the urochordate atrial siphon and the cephalochordate atrial muscles, to correspond to, for example, the mandibular (first arch) muscles of vertebrates. In ascidians, water flows into the body through the oral siphon and is then expelled out of the body through the atrial siphon; therefore, it is the oral siphon that most likely corresponds to the mouth of vertebrates (Gans 1989). However, as shown in Figure 3.5, recent studies have shown that the ascidian atrial siphon muscles (ASMs) derive from the cardiopharyngeal field, as do the branchiomeric muscles of vertebrates, but that the ascidian oral siphon muscles (OSMs) do not derive from this field (reviewed by Diogo et al. [2015b]). This fact seems to lend support to the idea that the mandibular arch was not part of the plesiomorphic branchial arch series of chordates, as noted earlier. In cephalochordates and urochordates, the oral/velar region lacks a skeleton, and the branchial bars are positioned a short distance behind the velum in a region that seems to correspond to the second branchial (hyoid) arch of vertebrates (Figures 3.2, 3.3, and 3.5) (e.g., Mallatt and Chen 2003). That is, it makes sense that in early chordate evolution, the oral/velar muscles were not part of the cardiopharyngeal field (as continues to be the case in extant urochordates in the study by Stolfi et al. 2010) (Figure 3.5) and that they only became integrated into this field with the later cooption/homeotic shift of at least some oral structures and their muscles to form the first branchial (mandibular) arch (following Miyashita and Diogo's [2016] hypothesis). Remarkably, in basal vertebrates such as lampreys, *Tbx1/10* is expressed first in the mesodermal core of the branchial arches and pharyngeal muscles and the region of the otic vesicle, which appear to correspond to the atrium of nonvertebrates and, only later in development, becomes expressed in the labial/oral and velar muscles (Sauka-Spengler et al. 2002). If in this case there is a parallel between ontogeny and phylogeny, these data would therefore also support the hypothesis that the inclusion of the velar/oral muscles in the cardiopharyngeal field and in the branchiomeric muscle group was a derived (later) event within chordate evolution.

However, there is at least one alternative scenario: the urochordate OSMs do not correspond to any of the branchiomeric muscle groups present in extant vertebrates; i.e., the urochordate atrial siphon also includes at least some muscles that correspond to/are precursors of the vertebrate's first arch muscles, as suggested by Stolfi et al. (2010). But this suggestion was based on studies showing that in derived vertebrates such as mice and chickens, the cardiopharyngeal field gives rise to mandibular muscles; it is now known that this field also gives rise to muscles of more posterior branchial arches (e.g., of the hyoid arch; e.g., Lescroart et al. 2010, 2015). Remarkably, some oral/velar muscles of adult cephalochordates are innervated by neurons from a region of the brain that is putatively homologous with the region that gives rise to the facial motor neurons innervating the muscles of the second (hyoid) arch in vertebrates (Northcutt 2005). Further studies are needed to investigate whether the cephalochordate oral/velar musculature corresponds to the oral siphon musculature (Figure 3.1; Table 3.1) or instead/also includes part of the atrial musculature of urochordates.

In enteropneust-type hemichordates serially arranged gill openings in the pharynx associated with musculature are found (Cameron 2002). However, this musculature is very different developmentally, anatomically, and histologically from the branchiomeric musculature of chordates (Yasui et al. 2014). In fact, the pharynx of the hemichordate *Saccoglossus* does not express *Tbx1*. According to Gillis et al. (2012), *Tbx1*-expressing pharyngeal mesoderm probably originated along the chordate stem, and the acquisition of cranial paraxial mesoderm within the pharyngeal region is probably a chordate synapomorphy. *Tbx1/10* is expressed in the pharyngeal mesoderm of cephalochordates and the atrial muscles of urochordates. While *Tbx1* expression is found in the branchiomeric muscles of vertebrates. Furthermore, *AmphiPax3/7* is expressed in the anterior and posterior somites of amphioxus and *Pax3* in all somitic muscles of vertebrates (Mahadevan et al. 2004). This distribution of gene expression indicates that the pterygial and oral/velar muscles of basal chordates and the branchiomeric muscles of vertebrates do not derive from the anterior somites and, thus, that the LCA of chordates already had a separation between somitic muscles ("*Pax3*") and branchiomeric muscles ("*Tbx1*"). However, *Tbx1/10* is expressed in the ASMs and in so-called "body wall muscles" of urochordates (Stolfi et al. 2010) and in the pharyngeal mesoderm and the ventral part of some somites of amphioxus (Mahadevan et al. 2004), meaning that this separation was probably not as well defined in early chordates as it is in extant vertebrates.

In fact, although more defined, the separation between branchiomeric and somitic muscles, and between the head and the trunk in general, remains somewhat blurry in living vertebrates. An illustrative example is the cucullaris, one of the best-studied yet most puzzling vertebrate muscles. Its amniote derivatives, the trapezius and sternocleidomastoideus, have played a central role in the studies of the origin and evolution of the vertebrate head and neck. These muscles share characteristics of at least five different muscle types: somitic epibranchial, somitic hypobranchial migratory, somitic limb nonmigratory ("primaxial"), and somitic limb migratory ("abaxial"). Topologically, the cucullaris resembles the epibranchial muscles of lampreys (e.g., Kusakabe et al. 2011), yet its developmental migration is similar to that of somitic hypobranchial migratory muscles (e.g., Matsuoka et al. 2005). Additionally, the trapezius receives contributions from both primaxial and abaxial cells (e.g., Shearman and Burke 2009). However, long-term fate mapping studies have shown that muscles that are generally accepted as branchiomeric, derived not only from posterior (e.g., laryngeal) but also from more anterior (e.g., the hyoid muscle interhyoideus posterior) arches receive a partial contribution from somites (Piekarski and Olsson 2007). These studies not only further complicate the distinction between head/neck and trunk, but they also show that the fact that the trapezius receives some somitic contribution does not contradict its original branchiomeric origin. The balance of available developmental, molecular, and anatomical data strongly supports the idea that the cucullaris and its derivatives are branchial, and thus, branchiomeric, muscles (e.g., Ziermann et al. 2014). The cucullaris anatomically originates from the posterodorsal region of the branchial musculature and is usually innervated by CNXI (accessory nerve) (e.g., Edgeworth 1935; Diogo and Abdala 2010; Ziermann and Diogo 2013, 2014). Also, as will be explained in the following chapters, the levatores arcuum branchialium of osteichthyan fishes (bony fishes), generally considered to be branchial muscles, were clearly derived from the undivided cucullaris of plesiomorphic gnathostomes (Ziermann et al. 2014). Finally, neural crest cells from a caudal branchial arch migrate with trapezius myoblasts and form tendinous and skeletal cells within its zone of attachment (e.g., Noden and Schneider 2006). Even stronger support for the branchiomeric identity of the cucullaris and its derivatives comes from gene expression studies in mammals: *Tbx 1* is expressed in/its lack affects the branchiomeric (e.g., laryngeal, first and second arch) muscles and the trapezius, while *Pax3* is expressed in/its lack affects all the somitic (i.e., limb, diaphragm, tongue, infrahyoid, and trunk) muscles, but not the trapezius (Sambasivan et al. 2011). These data, particularly those obtained from clonal studies in mice by Lescroart et al. (2010, 2015), also emphasize the heterogeneity of the vertebrate neck, which therefore includes branchiomeric (e.g., trapezius and sternocleidomastoideus), hypobranchial (e.g., tongue and infrahyoid), and trunk (somitic epaxial; e.g., deep neck and back) muscles (Figure 3.6).

Despite its profound implications for the new head hypothesis and for evolutionary and developmental biology and human medicine in general, heterogeneity in the vertebrate head and neck is poorly documented in textbooks, academic and medical curricula, and even many specialized research publications. In fact, one of the most crucial implications of recent studies on the cardiopharyngeal field is that they show that the head musculature derives from at least seven developmentally different types of primordia, as shown in Figure 3.6. In addition, Cyclostomata, Selachii, and Holocephali (see Figure 3.1 and the following chapters) possess an eighth group of head muscles, designated epibranchial muscles, which derive from the anterior portion of the somites (Edgeworth 1935) (Table 3.1). Even within the same arch, muscles can follow different genetic programs; for instance, in zebrafish, *Ret* signaling is necessary for the development of only a few specific mandibular and hyoid muscles associated with the movements of the opercula (bony plates supporting the gill covers) (Knight et al. 2011). Likewise, *C-met* is crucial for the development and migration of the mammalian muscles of facial expression, derived from the second (hyoid) arch, but not for the other second-arch muscles (Prunotto et al., 2004).

GENERAL REMARKS

An elongated motile adult stage is likely a representative condition for the adult LCA of Olfactores (Diogo and

Ziermann 2015a) because amphioxus is elongated and motile. Fossil evidence suggests that the hemichordate LCA had an enteropneust-like (wormlike) motile adult stage as well (Caron et al. 2013), which is suggestive of an elongated motile adult as the LCA of chordates (Lowe et al. 2015). The existing scenarios regarding the origin of vertebrates were recently reviewed by Holland et al. (2015). Assuming that the LCA of chordates and the LCA of Olfactores were elongated and motile means that adult sessile ascidians (e.g., *Ciona*) represent a derived condition due to peramorphosis (Diogo and Ziermann 2015a). Furthermore, as argued by Diogo and Ziermann (2015a), it is likely that the amphioxus larva represents a derived feature acquired during the evolutionary history of cephalochordates *via* the addition of early developmental stages. The larval asymmetry was described as secondary or cenogenetic feature without ancestral significance (Willey 1894). One possible explanation is an adaptation to a specialized feeding mode (Presley et al. 1996) in which, during the secondary gill formation, vestigial muscles transiently appeared (Yasui et al. 2014), supporting the addition of early developmental stages in the ancestors of amphioxus. Furthermore, this scenario supports the presence of bilateral muscular gills and segmented branchiomeric muscles in the LCA of extant chordates (Diogo et al. 2015b).

Another piece of evidence supporting the addition of embryonic stages is that the notochord differentiates from anterior to posterior, as in other chordates, but in the embryo with eight paired myocoelomic pouches, the anterior end of the notochord is posterior to the anterior end of the body, while the anterior portion of the archenteron extends beyond the notochord. Later, the notochord extends to the tip of the head, indicating that this is a secondary phenomenon (Willey 1894). As both the larval amphioxus and the adult lamprey are highly specialized due to their feeding mode, it is likely that the LCA of vertebrates probably had muscular features resembling adult amphioxus and other features resembling those of nonadult lampreys. This scenario is also supported by the homology between the endostyle of ascidians and amphioxus and the hypobranchial groove of ammocoetes; the two structures share a similar development, histological structure, and overall position (Willey 1894).

As noted ealier, the circular OSMs in urochordates, which are not part of the cardiopharyngeal field, may correspond to oral/velar muscles of amphioxus and of vertebrates such as cyclostomes and thus to the first arch (mandibular) muscles of gnathostomes, which were incorporated into the branchiomeric muscle series only during vertebrate evolution (Diogo et al. 2015b; Diogo and Ziermann 2015a). The pterygial muscle of amphioxus thus probably corresponds to the atrial siphon and associated muscles of urochordates (Table 3.1), because both these muscles lie in the atrial region and both terminate in the atrial syphon muscles, and therefore, they correspond to the muscles of the second and more posterior branchial arches of vertebrates (Diogo and Ziermann 2015a).

In summary, we infer from our dissections, observations, comparisons, and literature review of developmental processes that the LCA of extant chordates had anterior muscles (muscles of the oral region) that were incorporated into the branchiomeric series only in vertebrates, as mandibular muscles (Table 3.1; Figure 3.1). Other branchiomeric muscles were already present in the LCA of extant chordates. Those muscles correspond to the atrial sphincter and/or pterygial muscle of cephalochordates/urochordates and some transverse and LoMs. However, the genetic/morphogenetic program for differentiation into branchiomeric versus body/somitic muscles was probably not as sharply defined in the LCA of chordates as it was in the LCA of vertebrates. Future developmental, genetic, and comparative studies will be crucial to test this and the other hypotheses and suggestions proposed in the present chapter. Without a detailed knowledge of the muscles of nonchordates, it is difficult to clearly understand the origin and evolution of the vertebrate muscles.

4 General Discussion on the Early Evolution of the Vertebrate Cephalic Muscles

Chapter 4 is mainly based on the general discussion provided in Chapter 3 of the book *Muscles of Vertebrates* (Diogo and Abdala 2010), which was largely based on an extensive review of the available literature. That review included of classical anatomical descriptions such as those provided by Bischoff (e.g., Bischoff 1840), Owen (e.g., Owen 1841), Gegenbaur (e.g., Gegenbaur 1872), Huxley (e.g., Huxley 1876b), Cole (e.g., Cole 1896), Allis (e.g., Allis 1897, 1917, 1919, 1922, 1923, 1931), Alcock (e.g., Alcock 1898), Edgeworth (e.g., Edgeworth 1902, 1911, 1923, 1926a,b,c, 1928, 1935), and Luther (e.g., Luther 1913, 1914, 1938), as well as more recent reviews by authors such as Miyake et al. (1992), Mallatt (1996, 1997), and Anderson (2008), including, importantly, developmental and molecular data obtained in evo-devo studies (e.g., Holland et al. 1993; Kuratani et al. 2002, 2004; Graham 2003; Manzanares and Nieto 2003; Santagati and Rijli 2003; Trainor et al. 2003; Cerny et al. 2004; Kuratani 2004, 2005a,b, 2008a,b; Takio et al. 2004; Helms et al. 2005; Kusakabe and Kuratani 2005; Northcutt 2005; Olsson et al. 2005; Shigetani et al. 2005; Holland et al. 2008; Kuratani and Ota 2008b; Kuratani and Schilling 2008). This chapter is necessary to lay the groundwork for a general discussion of the early evolution of the head and neck muscles of vertebrates. In Chapter 5, we will update this general discussion by presenting data obtained from our own detailed dissections of the cephalic muscles of cyclostomes and of chondrichthyans, as well as from a literature review of the data published by other authors, since the previous book (Diogo and Abdala 2010) was written.

In the discussions presented in this Chapter 4 as well as in the other chapters of this book, we follow the now commonly accepted phylogenetic hypothesis that living hagfishes and lampreys are more closely related to each other than to other living vertebrates (Figure 1.1; see also Chapter 5). Therefore, following this phylogenetic framework, when a certain muscle is absent in non-vertebrate animals and present in both living lampreys and living gnathostomes, then there are two phylogenetically equally parsimonious solutions: the muscle was present in the last common ancestor (LCA) of vertebrates and then secondarily lost in hagfishes or that the muscle was independently acquired in lampreys and in gnathostomes. However, although these solutions are theoretically equally parsimonious in a cladistic sense, if the innervation, topology, blood supply, development, and other features of this muscle in lampreys and gnathostomes is very similar or identical, we would tend to see the secondary loss in hagfishes as more likely than the independent evolution of a very similar muscle in two clades. Before moving on to the discussion of the origin, evolution, ontogeny, functional morphology, and homologies of the muscles of non-osteichthyan vertebrates, we would like to clarify that many of the functional hypotheses mentioned in the following, as well as the figures in this chapter, are based on Mallatt's detailed studies of these vertebrates (e.g., Mallatt 1996, 1997). In our opinion, this is one of the best ways to pay a special, and wholly deserved, tribute to such a remarkable anatomist.

According to Mallatt (1996), the biting, mandibular arch derived jaws of gnathostomes evolved primarily through changes in ventilation. In his view, the jawless ancestors of all living vertebrates were benthonektonic predators that ate slow-moving invertebrates, grasping their prey in a ring of oral cartilage that was squeezed by an oral sphincter muscle (Figures 4.1 and 4.2). Initially, the activity level and ventilation rate of these vertebrates were low. The expiratory phase of their ventilatory cycle resulted from peristaltic contraction of the pharyngeal wall musculature, whereas inspiration resulted from the passive recoil of unjointed internal and external branchial arches. Then, as "pre-gnathostomes" became more active foragers, both expiration and inspiration were strengthened and a capacity for active, forceful inspiration evolved. Correspondingly, many new ventilatory muscles evolved and were attached to the internal arches, which became large, jointed, and highly mobile (Figure 4.3). The most powerful of these ventilatory muscles closed the mouth during forceful expiration to prevent leakage (the adductor mandibulae: see, e.g., Figure 4.3 and the following) and opened the mouth widely during forceful inspiration (the myotomal hypobranchial muscles: see, e.g., Figure 4.4 and the following), and the branchial arch on which these powerful muscles inserted became the largest, forming the mandibular arch jaws (Figures 4.3 and 4.5). Now, for the first time, gnathostomes could capture evasive prey, by sucking it in through forceful "inspiration" and clamping it with biting jaws during "expiration"—the way that extant gnathostome fishes often feed.

Both ammocoetes (lamprey larvae; Figure 4.6) and sharks (Figure 4.3) propel ventilatory water through the pharynx unidirectionally: in through the mouth and out through the external gill openings. Each ventilatory cycle consists of an expiratory then an inspiratory phase. Expiration is effected by the branchial superficial constrictor and the interbranchial muscles, the former being a circular sheet that squeezes water through the pharynx by peristalsis (Figures 4.3 and 4.6),

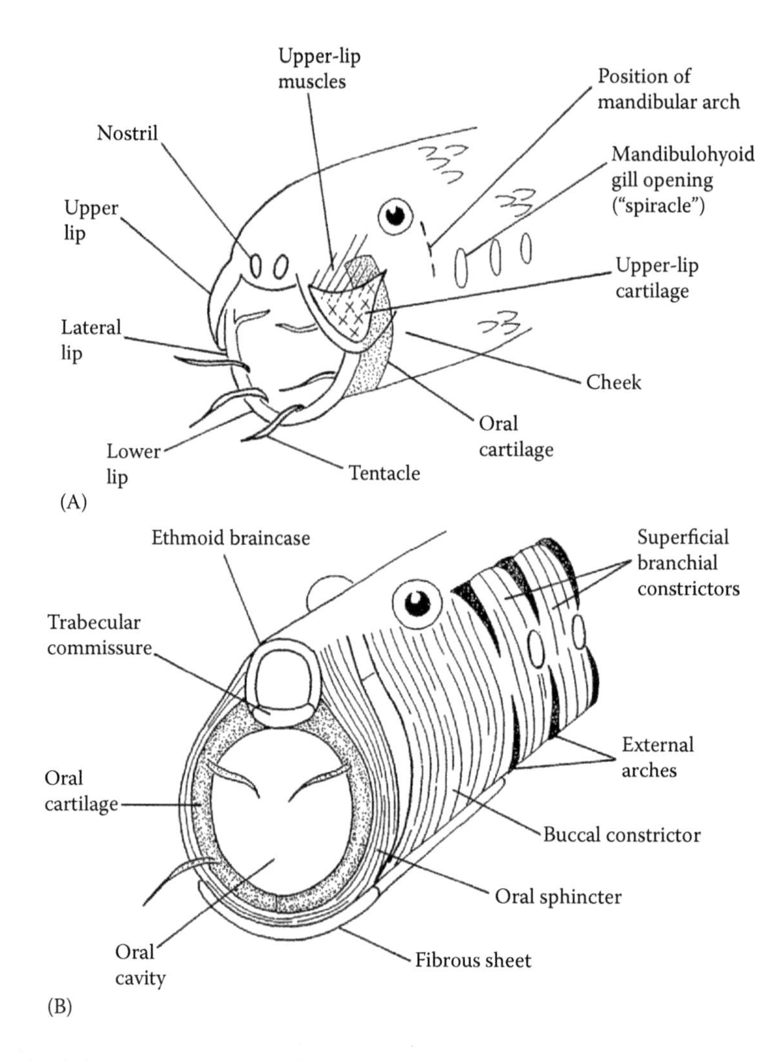

FIGURE 4.1 Lips and mouth of the common ancestor of all living vertebrates, according to Mallatt (1996). (A) External view, but also showing some cartilages and the muscles in the upper lip. (B) View with the skin and the snout removed, emphasizing the muscles around the oral cavity and pharynx. As emphasized in the text, further studies are needed to clarify if the mandibular musculature was already differentiated into labial muscles, such as the oral sphincter, buccal constrictor, and the upper-lip muscles *sensu* Mallatt (1996), as proposed by this author, or if the so-called labial muscles of vertebrate groups such as elasmobranchs, holocephalans, cyclostomes, and osteichthyans are instead the result of an independent differentiation of the mandibular mesoderm in these taxa. (The nomenclature of the structures illustrated basically follows that of Mallat [1996].) (Modified from Mallatt, J., *Zool J Linn Soc*, 117, 329–404, 1996.)

the latter running in the gill septa and acting to decrease the height of the pharynx and compress the gill pouches (Figure 4.7). These expiratory muscles are aided in ammocoetes by a pumping velum and in sharks by muscles to the internal branchial arches. During quiet ventilation, after the expiratory muscles relax, inspiration results from a passive recoil of the pharyngeal skeleton—recoil of the external arches in ammocoetes versus a recoil of the extrabranchial cartilages, internal arches, and fibroelastic membranes in sharks. During forceful ventilation, only in sharks, inspiration is aided by the hypobranchial ventilatory muscles, which actively enlarge the pharynx. According to Mallatt, lamprey and shark ventilation thus share two main features: (a) expiration through the peristaltic action of branchial superficial constrictor and interbranchial muscles and (b) inspiration through passive recoil of the branchial arches. Consequently, the LCA of vertebrates must have possessed the superficial constrictor and interbranchial muscles (Figures 4.1 and 4.2).

As also noted by Mallatt (1996), the velum of lampreys is a pair of cupped, muscular paddles that push water posteriorly into the pharynx during the expiratory phase of each ventilatory cycle. It is a powerful, piston-like pump that can work against back pressure and force ventilatory water through the sand in which ammocoetes live. Projecting posteriorly from each velar paddle is a "medial flap," which is supported by the internal velar bar (Figure 4.6). When the velum starts to contract, its right and left medial flaps come together to form a seal that prevents reflux of water from the pharynx through the mouth. In the embryonic lamprey, the velum develops at the border between the mouth and pharynx from the buccopharyngeal membrane. Its muscles belong to the mandibular branchial segment, being innervated by the mandibular branch (V3) of the trigeminal nerve. The lateral mouth plates of ammocoetes bear a superficial resemblance to the branchial arches behind them and have been called a "premandibular branchial arch" by some authors. However,

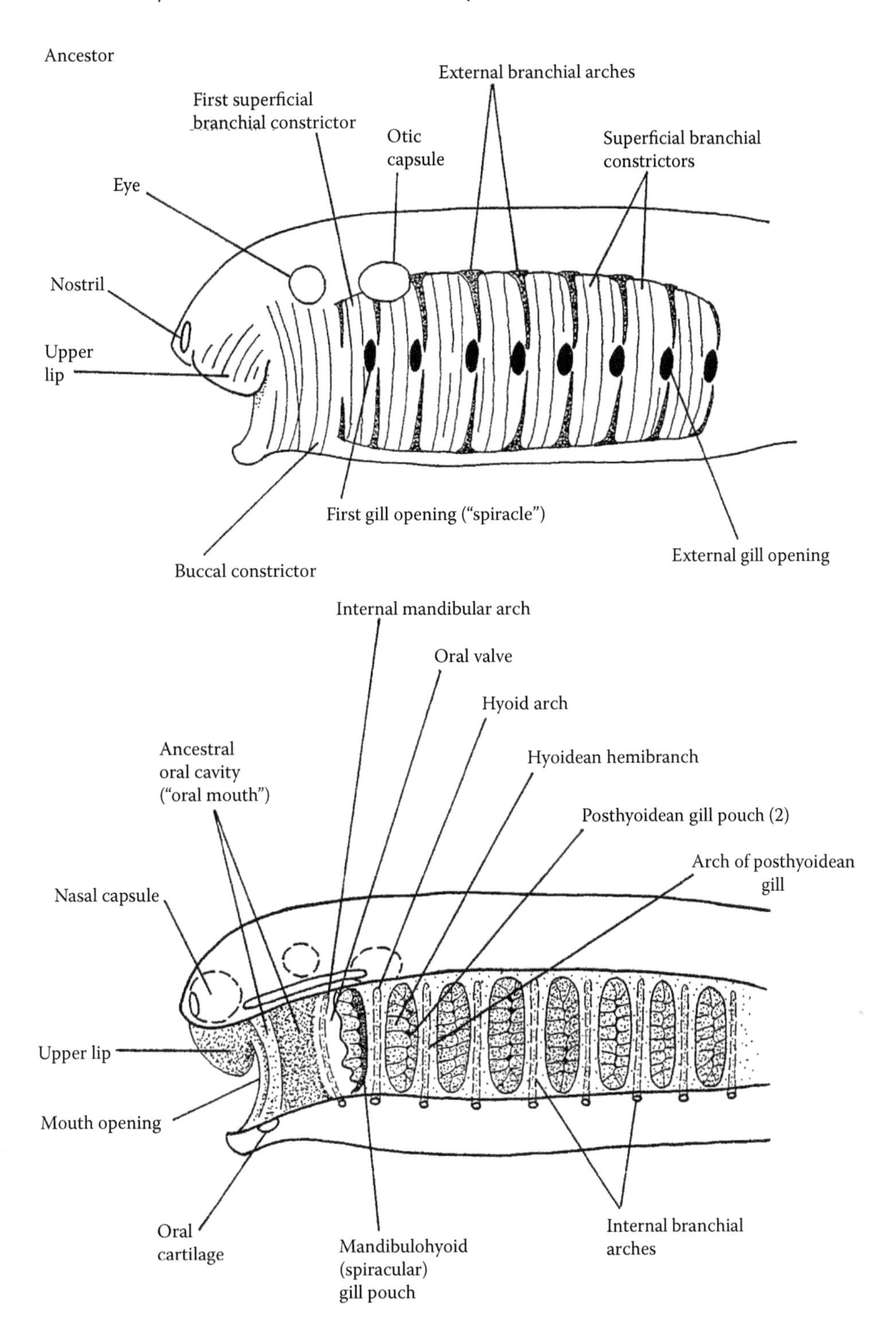

FIGURE 4.2 Head and pharynx of the reconstructed common ancestor of all living vertebrates, according to Mallatt (1996). External and midsagittal views; numbers such as (1) and (2) are used to identify the ancestral/embryonic gill pouches. According to Mallatt (1996), this ancestor may have had more gill pouches than the eight illustrated in this figure. (The nomenclature of the structures illustrated basically follows that of Mallat [1996].) (Modified from Mallatt, J., *Zool J Linn Soc*, 117, 329–404, 1996.)

according to Olsson et al. (2005), "there is no clear fossil evidence that a complete gill arch skeleton ever existed anteriorly to the first gill arch," and most researchers now agree that no such complete "premandibular" arch was present in the LCA of vertebrates or of gnathostomes. Another idea that was often accepted in the past and that has been contradicted by recent findings is that neural crest cells from the first mandibular arch form a dorsal "maxillary" and a ventral "mandibular" condensation, which later give rise to the upper jaw cartilage (palatoquadrate) and the lower jaw cartilage (Meckel's cartilage), respectively. In fact, recent developmental studies using vital-dye labeling in both the Mexican axolotl and the chicken embryo have shown that cells which form the ventral or mandibular condensation give rise to both the upper and lower jaw

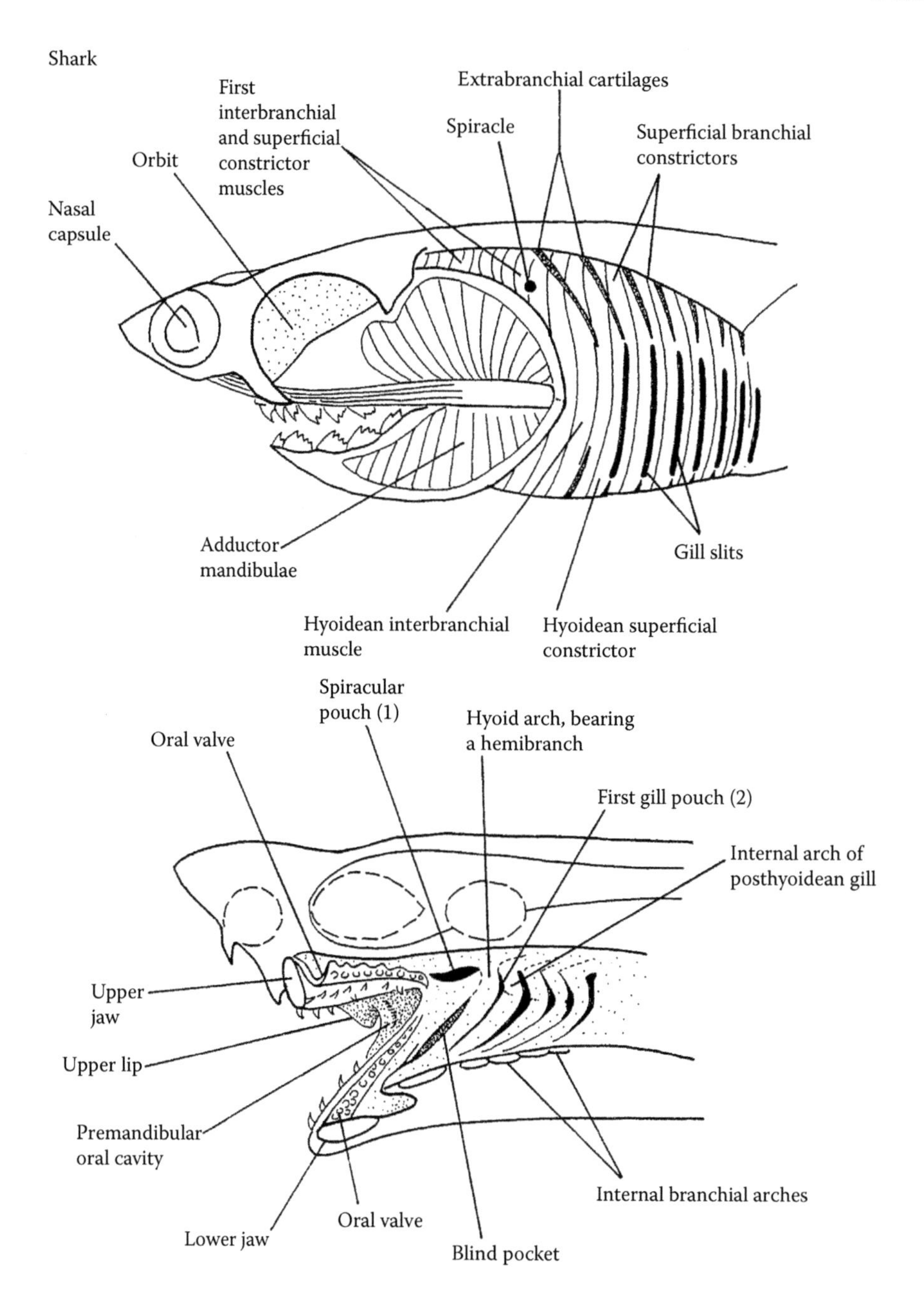

FIGURE 4.3 Head and pharynx of sharks (Elasmobranchii), according to Mallatt (1996); numbers such as "(1)" and "(2)" are used to identify the ancestral/embryonic gill pouches; external and midsagittal views. *Top*: *Heptanchus maculatus*; this species of shark is shown because, according to Mallatt (1996), it has the primitive features of a short snout and a simple type of adductor mandibulae muscle. *Bottom*: *Triakis semifasciatus*. (The nomenclature of the structures illustrated basically follows that of Mallat [1996].) (Modified from Mallatt, J., *Zool J Linn Soc*, 117, 329–404, 1996.)

cartilages (e.g., Cerny et al. 2004; Olsson et al. 2005). The dorsal or maxillary condensation contributes to the trabecular cartilage, but not to the jaw joints as previously assumed. Interestingly, Cerny et al.'s (2004) developmental study provides evidence to support the idea that the jaw cartilages of gnathostomes are homologous to the lower lip and velum of lampreys.

However, the word *mandibular* can be used in different contexts and different development stages, including (a) the *mandibular condensation* is an osteological term that refers to an early stage of development and that thus seems to be inadequate, as it gives rise to both the upper and lower jaw cartilages, contrary to what was thought in the past; (b) the *mandible* is an osteological term that refers to both early and later stages of development and essentially corresponds to the term *lower jaw* (note that the "mandible"/"lower jaw" does not necessarily correspond to "lower jaw cartilages" and/or to the ossifications of these cartilages, because in addition to these cartilages and their ossifications, the mandible/lower jaw of adults may include, and often does, other structures, such as dermal bones); (c) the *mandibular muscular plate sensu* Edgeworth (1935), which, as explained earlier,

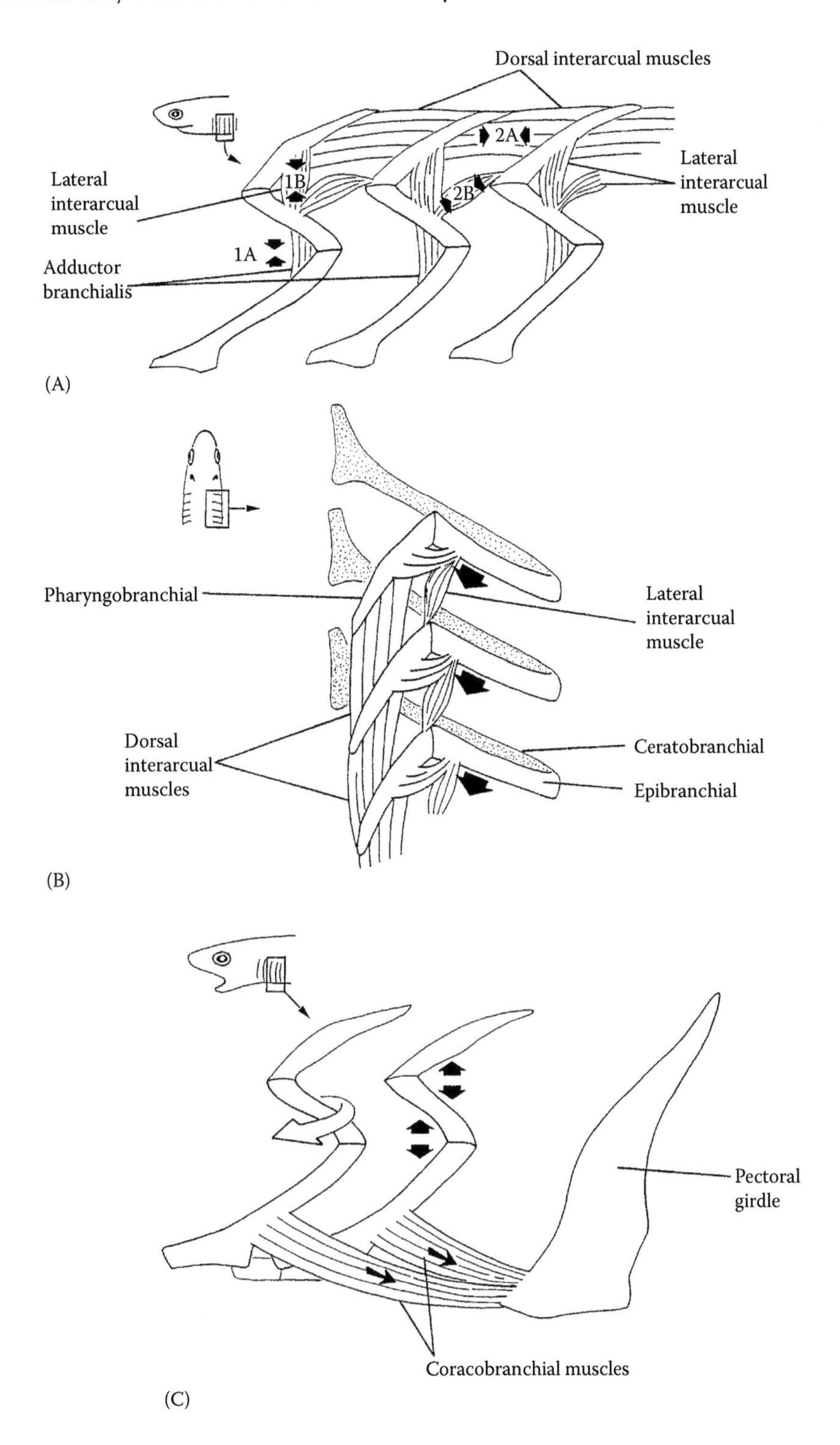

FIGURE 4.4 Probable movements of the internal arches during the ventilatory cycle of sharks (Vertebrata: Elasmobranchii), according to Mallatt (1996). (A) Expiratory movements, lateral view: the segments are flexed and pulled dorsally by adductor branchialis (1A) and lateral interarcual muscles (1B), while successive arches are pulled closer together by the dorsal and lateral interarcuals (2A and 2B) (note: the dorsal interarcuales of sharks are considered to be epibranchial muscles *sensu* Edgeworth 1935). (B) Expiratory movements, dorsal view: the lateral interarcual muscles swing the arches posteromedially (arrows). (C) Forceful inspiration: the coracobranchial muscles swing the arches anterolaterally (large, curved arrow) and abduct the arch segments (dark, diverging arrows). (The nomenclature of the structures illustrated basically follows that of Mallat [1996].) (Modified from Mallatt, J., *Zool J Linn Soc*, 117, 329–404, 1996.)

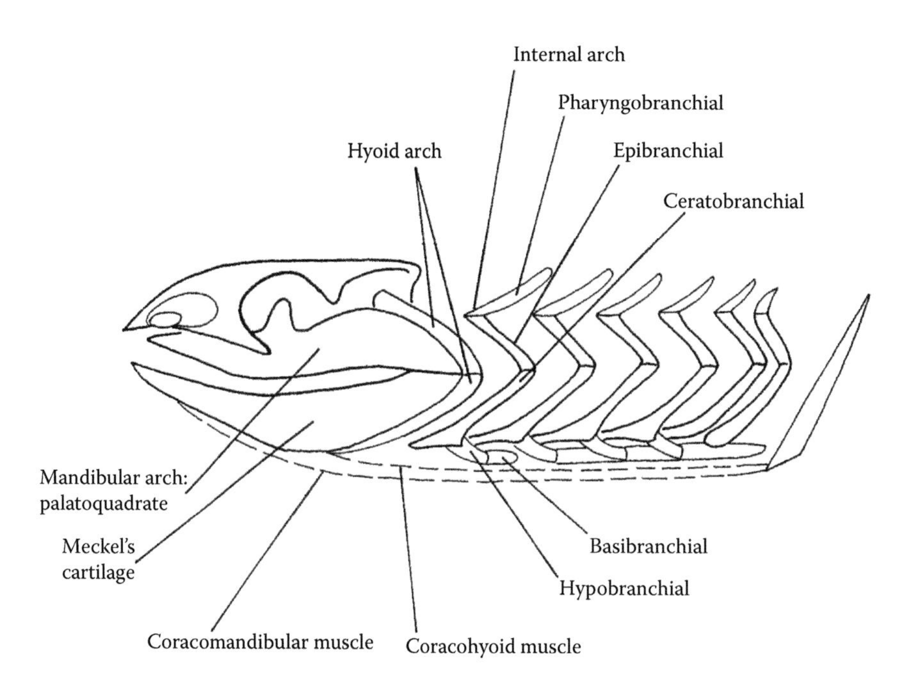

FIGURE 4.5 Head and pharyngeal skeleton of the frilled shark *Chlamydoselachus anguineus* (Vertebrata: Elasmobranchii); the five segments of an "internal arch" (*sensu* Mallatt 1996) are labeled; the extrabranchial cartilages are not shown; the coracomandibular and coracohyoid muscles are shown ventrally as dashed lines (The nomenclature of the structures illustrated basically follows that of Mallat [1996].) (Modified from Mallatt, J., *Zool J Linn Soc*, 117, 329–404, 1996.)

is a myological term and refers to an early stage of development (see Figure 2.1); (d) the *mandibular muscles sensu* Edgeworth (1935; and *sensu* this volume) are myological structures that are found in both early and later stages of development and that derive from the mandibular muscular plate; (e) the *mandibular arch*, which is an osteological term, refers to both early and later stages and corresponds to the "first branchial arch," including both the "palatoquadrate" and the lower jaw (see, e.g., Figure 4.5). Therefore, in this work, we will avoid the use of the term *mandibular condensation*, but we will continue to use the terms *mandible*, *mandibular muscular plate*, *mandibular muscles*, and *mandibular arch* as they are often used in the literature.

As described by Mallatt (1996), in living Chondrichthyes, the robust internal arches are divided into five segments, connected by movable joints: pharyngo-, epi-, cerato-, hypo-, and basibranchial segments (Figure 4.5). The segmentation and jointing allow muscles to attach and pull from many different directions. Unlike the extrabranchial cartilages (external branchial arches *sensu* Mallatt 1996), which are firmly embedded in the pharyngeal wall, the internal arches have an extraordinary range of movement within the gnathostome pharynx. During ventilation in sharks, they are proposed to move as shown in Figure 4.4. During expiration, to decrease pharyngeal volume and expel water, the arch segments are flexed by adductor branchialis and lateral interarcual muscles, and successive arches are pulled closer by dorsal and lateral interarcual muscles (Figure 4.4A). At the same time, the lateral interarcuals swing the arches posteromedially (Figure 4.4B), further decreasing pharyngeal volume. During quiet inspiration, the bent arches passively recoil like springs, thereby helping to enlarge the pharynx and draw in water. On the other hand, active forceful inspiration is effected by the coracobranchial muscles, which rapidly swing the arches anterolaterally and abduct them (Figure 4.4C). At this time, the mouth is widely opened by the coracomandibular and coracohyoid muscles (along with their common base, the coracoarcualis) (see Figures 4.4, 4.5, 4.8, and 4.9).

Mallatt (1996) stated that "the adductor branchialis, lateral interarcual, and coracobranchial muscles develop from 'branchial muscle plates' in the gill septa, indicating they evolved from the interbranchial muscles." In his opinion, the dorsal interarcuals, coracomandibular, and coracohyoid muscles develop from the anterior myotomes and thus evolved from epibranchial and hypobranchial myotomes, which overlie much of the pharyngeal musculature in extant cyclostomes. However, authors such as Johanson (2003) argued that the coracobranchiales of gnathostomes do not correspond to a part of the interbranchiales of lampreys, because, in their view, the interbranchiales are more likely to be homologous to gnathostome muscles involved in branchial arch constriction rather than expansion (see below). Also, authors such as Luther (1938) and Lightoller (1939) argued that the adductor mandibulae of gnathostomes is probably derived from a lateral part of the interbranchialis of the first arch (and not from a medial part of this muscle, as argued by Mallatt 1996) or even from the branchial superficial constrictor of this arch, because in gnathostomes, the adductor mandibulae lies on the lateral rather than the medial surface of its branchial arch.

We agree with Edgeworth (1935), Lightoller (1939), and Lauder (1980a,b) that the labial muscles *sensu* Anderson (2008) are, at least in some cases, likely related to the "adductor mandibulae complex." Mallatt (1996) called these muscles "oral muscles," but he recognized that at least some of them

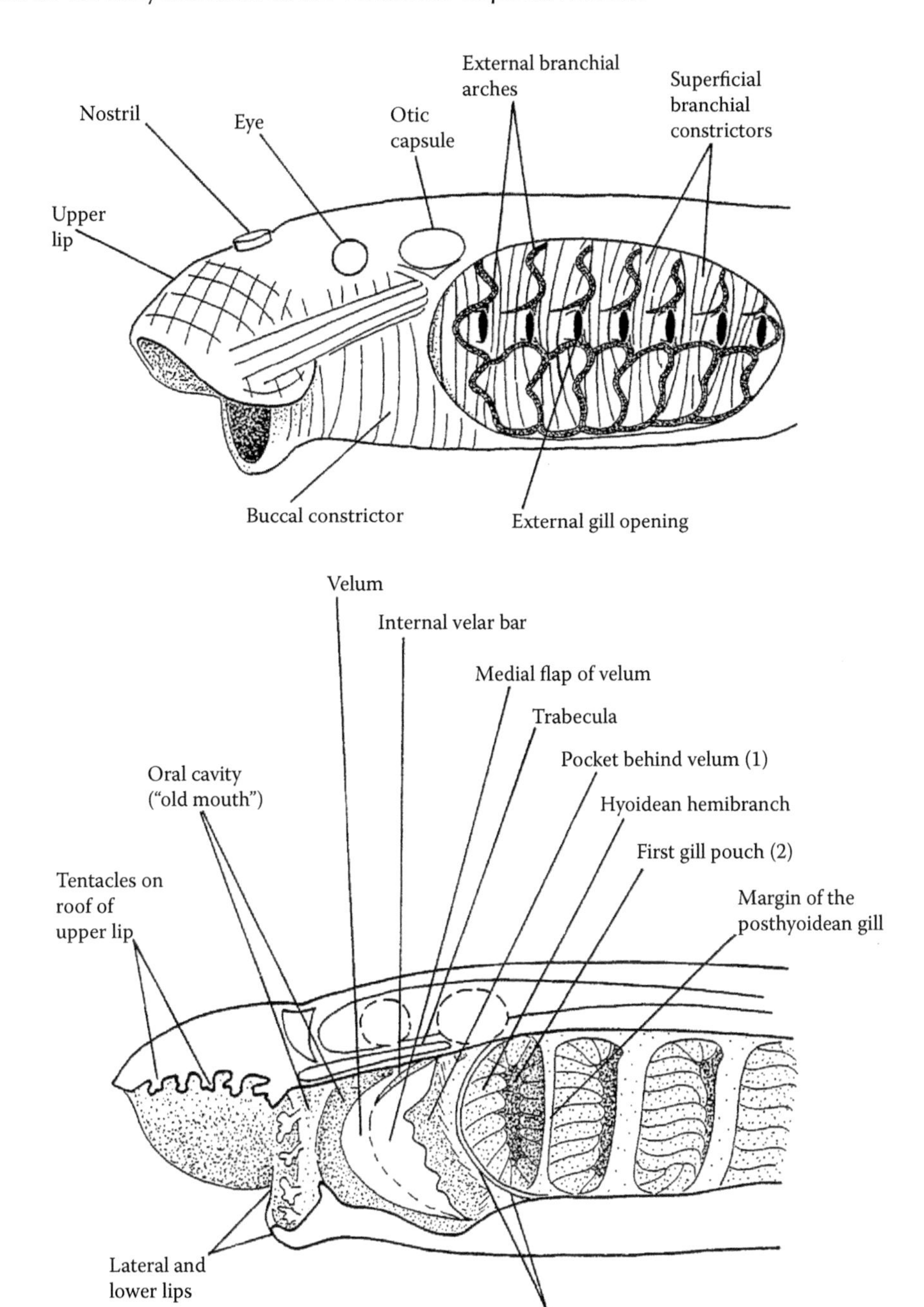

FIGURE 4.6 Head and pharynx of ammocoete lampreys (Cyclostomata), according to Mallatt (1996); numbers such as (1) and (2) are used to identify the ancestral/embryonic gill pouches; external and midsagittal views. *Top*: *Lampetra planeri*. *Bottom*: *Ichthyomyzon fossor*. (The nomenclature of the structures illustrated basically follows that of Mallat [1996].) (Modified from Mallatt, J., *Zool J Linn Soc*, 117, 329–404, 1996.)

(e.g., his "buccal constrictor") develop from the "mandibular branchiomere" in lampreys. The "labial muscles" *sensu* Anderson (2008) also seem to develop from the mandibular plate in elasmobranchs and osteichthyans (e.g., Edgeworth 1935). Therefore, whether these muscles are called labial (e.g., Anderson 2008), "oral" (e.g., Mallatt 1996) or "preorbital/suborbital mandible adductors" (e.g., Edgeworth 1935; Lauder 1980a,b), they seem to develop from the mandibular mesoderm, as does the adductor mandibulae complex. This idea was supported by the developmental work of Kuratani et al. (2004), who concluded that "experiments labeling the mandibular mesoderm of the early lamprey embryo, before the cheek process has differentiated into the upper lip anlage or the premandibular domain, indicate that a part of the mandibular mesoderm secondarily grows anteriorly and laterally and migrates into the upper lip domain." According to Mallatt (1996), the muscles that are derived from this mandibular mesoderm in lampreys, particularly their "labial portion" (*sensu* Anderson 2008), were possibly innervated by both V2 and V3 (divisions of the trigeminal nerve, CNV) in the LCA of vertebrates (within living vertebrates, innervation by V2 and V3 is said to occur in lampreys, holocephalans,

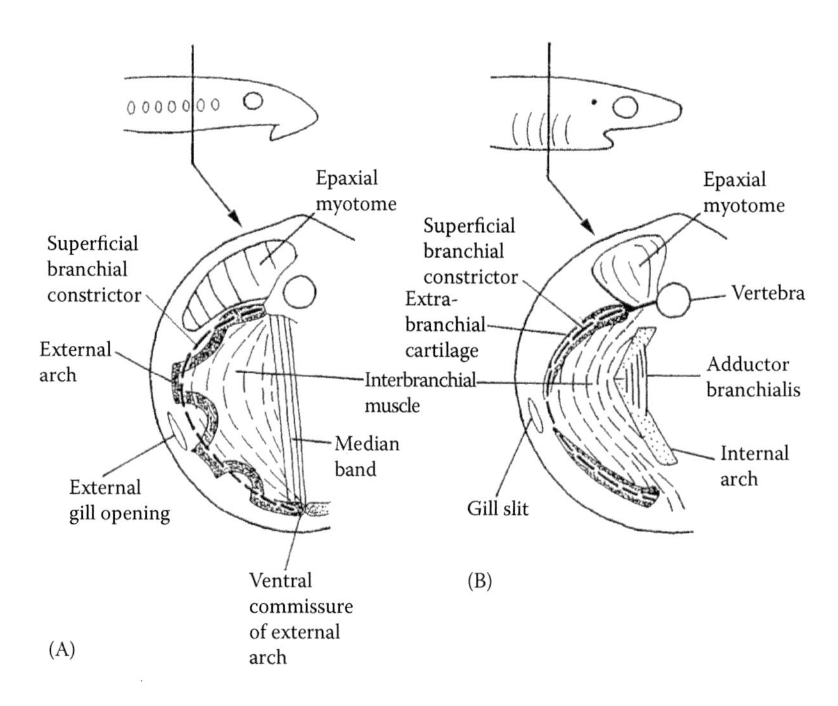

FIGURE 4.7 Basic similarities between the gill muscles of (A) lampreys (Cyclostomata) and (B) sharks (Elasmobranchii), according to Mallatt (1996): in both animals, the superficial branchial constrictors (*dashes*) wrap around the pharynx externally and an interbranchial muscle occupies each gill septum; these two muscles are continuous, separated only by the external branchial arch; a straight band of muscle in the medial part of the lamprey gill (median band) may correspond to the adductor branchialis of sharks (the adductor branchialis of the first arch corresponding to the adductor mandibulae of Figure 4.4). (The nomenclature of the structures illustrated basically follows that of Mallat [1996].) (Modified from Mallatt, J., *Zool J Linn Soc*, 117, 329–404, 1996.)

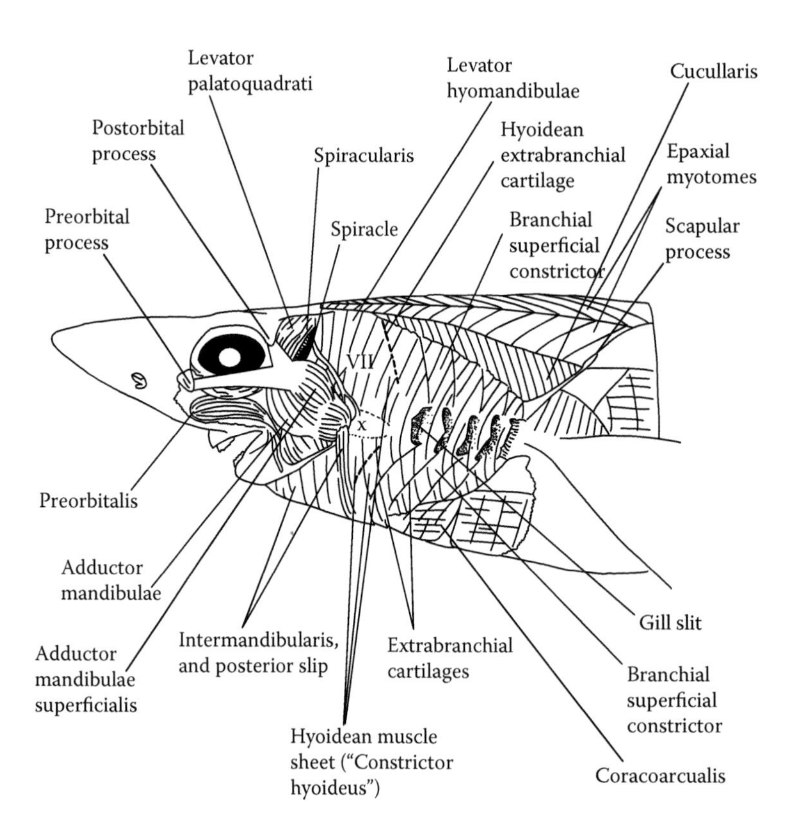

FIGURE 4.8 Pharyngeal muscles of the spiny dogfish, *Squalus acanthias* (Vertebrata: Elasmobranchii), according to Mallatt (1997). VII indicates hyomandibular branch of the facial nerve, while X marks a region behind the jaw from which the expected mandibular musculature is absent and replaced by sheets of dense connective tissue, according to this author. (The nomenclature of the structures illustrated basically follows that of Mallat [1997].) (Modified from Mallatt, J., *Acta Zool*, 78, 279–294, 1997.)

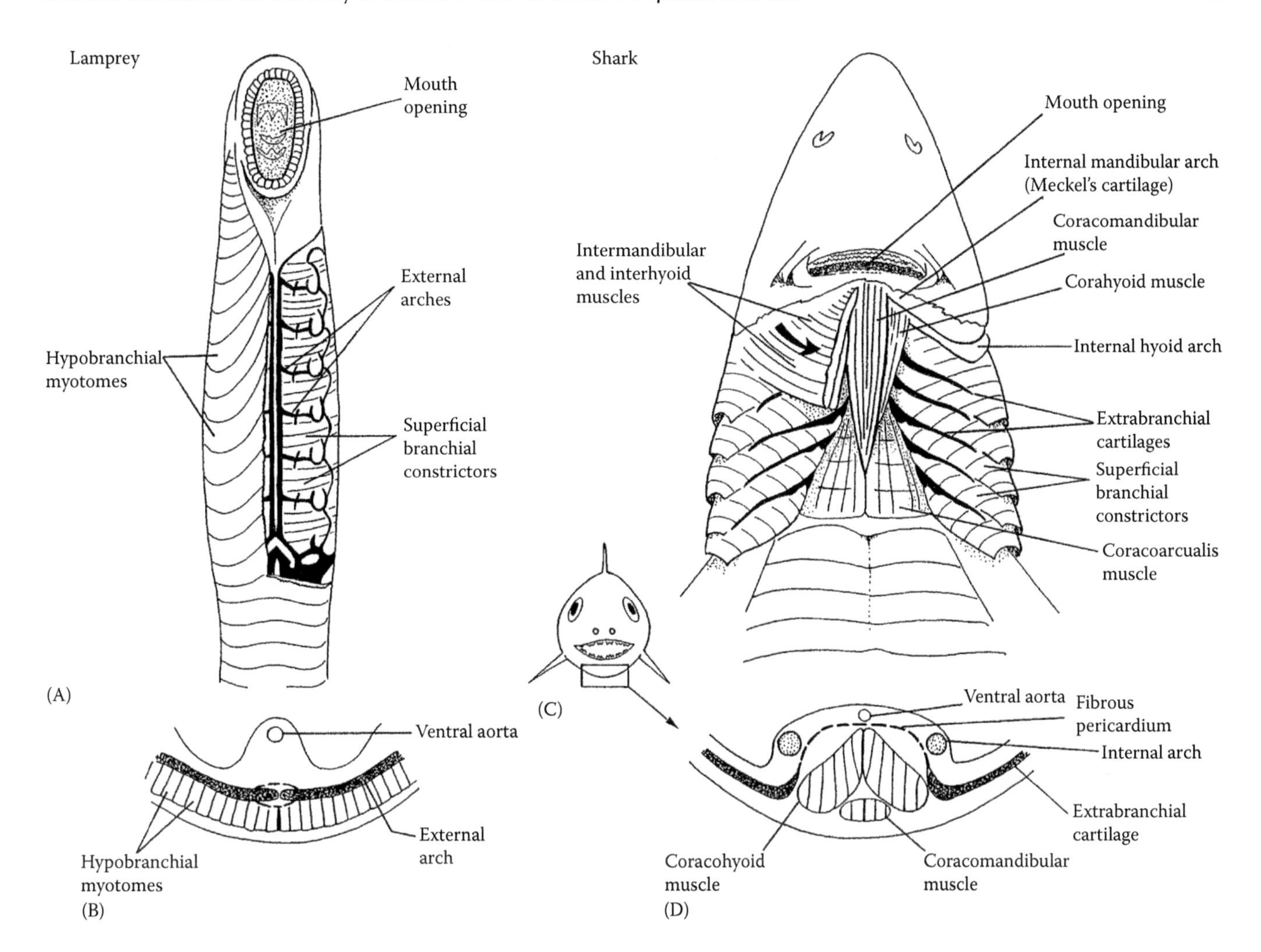

FIGURE 4.9 Spinal group of hypobranchial muscles, according to Mallatt (1996). These ventilatory muscles of gnathostomes evolved from locomotory myotomes. (A) Ventral view of a lamprey, the hypobranchial myotomes have been cut away from the left side of the lamprey, exposing the external branchial arches and the superficial branchial constrictors. (B) Transverse section through the floor of the lamprey pharynx, showing the relationship between the hypobranchial myotomes and the external arches. (C) Ventral view of a shark, the hypobranchial muscles—coracomandibular, coracohyoid, and coracoarcualis—lie much deeper than in lampreys and are concentrated near the ventral midline. (D) Transverse section through the floor of the shark pharynx, showing the right/left separation of the extrabranchial cartilages and the deep location of the coracomandibular and coracohyoid muscles. (The nomenclature of the structures illustrated basically follows that of Mallat [1996].) (Modified from Mallatt, J., *Zool J Linn Soc*, 117, 329–404, 1996.)

and possibly hagfishes: see Figure 4.10), the V2 innervation having been secondarily lost in elasmobranchs and osteichthyans. However, according to Kuratani et al. (2004), the nerve that is often called V2 in cyclostomes such as lampreys possibly does not correspond to the V2 of gnathostomes.

Mallatt (1996) discussed the homologies between the various labial muscles (*sensu* Anderson 2008) present in cephalochordates (Figure 4.11), hagfishes (Figure 4.12), lampreys (Figure 4.13), elasmobranchs (Figure 4.9), and holocephalans (Figure 4.14) and summarized all his hypotheses of homology in a table (see Figure 4.10). A brief description of these muscles, which is mainly based on the work by Mallatt (1996), is given below.

In ammocoetes, the orolabial musculature is complex. In the upper lip, the largest and most important muscle is the buccalis anterior (Figure 4.13A), which originates from the superolateral walls of the oral cavity, the superoanterior surface of the lateral mouth plate, the trabecular commissure just below the nasal capsule, and the nasal capsule itself. It forms most of the mass of the upper lip and inserts onto the entire undersurface of the rostrodorsal plate and onto the lip mucosa. Functionally, the buccalis anterior retracts and constricts the upper lip. Other muscles surround the oral cavity and mouth opening of ammocoetes (Figure 4.13B): buccal constrictor, elevator labialis ventralis, sublabialis, and basalis tentacularis (*sensu* Mallatt [1996]). According to Mallatt, the buccal constrictor encircles the oral cavity from the external hyoid bar posteriorly to the front of the eye. It forms the bulk of the "cheek." Superiorly, it attaches to the trabeculae and the fibrous braincase. The elevator labialis ventralis surrounds the mouth opening; ventrally, it blends with the buccal constrictor (Figure 4.13C). As it ascends, this elevator muscle lies directly anterior to, then directly lateral to, the lateral mouth plate (Figure 4.13B). Superiorly, it extends along the

Ammocoetes	Sharks (*Squalus*)	Chimaera	Vertebrate ancestor	Hagfish	Amphioxus
1. Surface structures					
Lower lip	—	Lower lip	Lower lip		
Lateral lip	Labial fold	Labial fold	Lateral lip		
Upper lip	Lateral fold of upper lip (and region medial to it that likewise contains the anterior labial cartilage)	Nasolabial fold	Upper lip	Skin over cornual cartilage, and cornuo-subnasal muscle	None
Lateral wall of oral (cheek)	Buccal membrane (cheek)	Lateral wall of oral cavity (cheek)	Cheek	Lateral wall or oral cavity	Oral hood
Oral cavity	Slit between lip, cheek, and jaw	Slit between lip, cheek, and jaw	Ancestral oral cavity	Oral cavity	Oral cavity
Mouth opening	Mouth opening	Mouth opening	Mouth opening	Mouth opening	Margin or oral hood
Velum	Jaws, pseudobranch, and first superficial constrictor	Jaws	Mandibular hemibranch, and first superficial constrictor	Velum	Velum (but no superficial constrictor)
Medial flap of velum	Oral valve	Oral valve	Oral valve	? (see text)	Medial part of velum
2. Skeleton					
Rostro-dorsal plate	Anterior labial cartilage	Prelabial cartilage	Upper-lip cartilage	Cornual cartilage	
Lateral mouth plate, with lateral process	Posterior labial cartilages	Maxillary labial cartilage, with lateral process	Oral cartilage	Labial, or coronary, cartilage	Hoop formed by bases of tentacular cartilages
Ventro-lateral plate	Fibrous sheet on Meckel's cartilage	Fibrous mass in lower lip	Fibrous or cartilaginous sheet		
3. Muscles					
Buccalis anterior (V_2)	—	Levator prelabialis, levator anguli oris ant., labialis anterior (V_2)	Upper-lip muscles	Cornuosubnasal muscles	
Elevator labialis ventralis (V_3)	—	Labialis posterior (V_3)	Oral sphincter	Coronarious basitentacularis	Peri-buccal muscles
Buccal constrictor (V_3)	Levator labii superioris (V_3)	Levator anguli oris posterios (V_3)	Buccal constrictor	Levator cartilaginis basalis, protractor cartilaginis basalis, craniobasialis	—

FIGURE 4.10 Proposed homologies in the mouth and lips of ammocoete lampreys and hagfishes (Vertebrata: Cyclostomata), sharks (Elasmobranchii), chimaeroids (Holocephala), and amphioxus (Cephalochordata), according to Mallatt (1996); see also Figures 4.8, 4.11, 4.12, 4.15, and 4.18. (The nomenclature of the structures illustrated basically follows that of Mallat [1996].) (Modified from Mallatt, J., *Zool J Linn Soc*, 117, 329–404, 1996.)

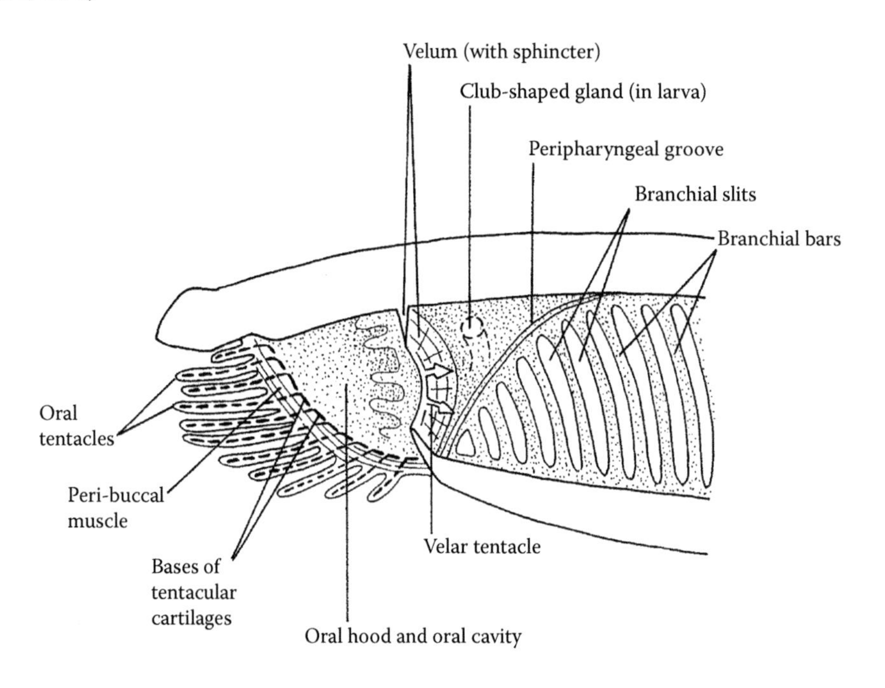

FIGURE 4.11 Cephalochordate amphioxus, according to Mallatt (1996): as in lampreys, a velum defines the anterior boundary of the pharynx; however, the amphioxus velum is not a ventilatory pump; the club-shaped gland is present only in the larval stage and is included in this adult picture for illustrative purposes only. (The nomenclature of the structures illustrated basically follows that of Mallat [1996].) (Modified from Mallatt, J., *Zool J Linn Soc*, 117, 329–404, 1996.)

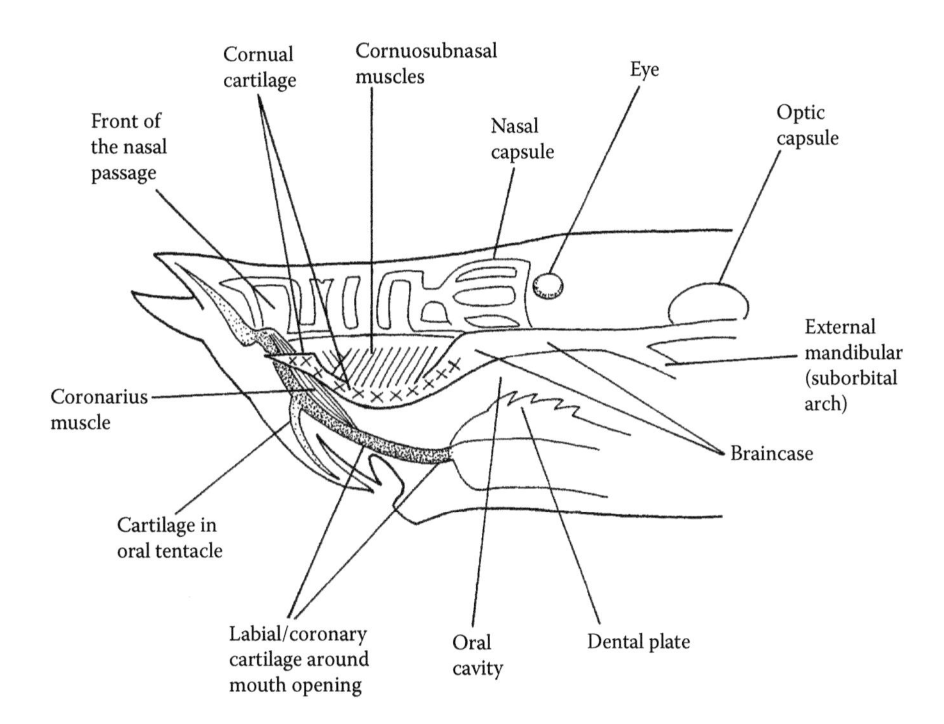

FIGURE 4.12 Anterior part of the hagfish *Myxine glutinosa* (Vertebrata: Cyclostomata). As explained in the text, according to Mallatt (1996), the labeled cartilages and muscles may be homologous to structures in ammocoetes and gnathostomes. (The nomenclature of the structures illustrated basically follows that of Mallat [1996].) (Modified from Mallatt, J., *Zool J Linn Soc*, 117, 329–404, 1996.)

undersurface of the posterior part and subocular process of the rostrodorsal plate and then attaches to the fibrous braincase between the eye and nasal capsule. According to Mallatt, this muscle raises the lower lip and probably acts as an oral sphincter. His description of the sublabialis muscle differs from the literature. Damas (1935), for instance, stated that this muscle originates along a bar of mucocartilage in the ventral longitudinal crest and ascends into the oral tentacles (for this reason, Damas renamed it "muscle retractor papillaris"). Mallatt (1996) confirmed its origin, but stated that "the muscle ascended anterior to the lateral mouth plate and inserted on the rostrodorsal plate lateral to the nasal capsule; it did not send slips to the tentacles; this suggests that the sublabialis does not retract the tentacles, but helps to close the mouth by pulling down on the upper lip and pulling up on the floor of the oral cavity; the true retractor of the tentacles was discovered as a distinct, thin-fibered muscle in the anterior part of the oral cavity, along the bases of the tentacles: this new muscle was named basalis tentacularis" (Figure 4.13B). The maxillary (V2: suboptical) and mandibular (V3) nerves both run through the orolabial region (Figure 4.13B). Mostly, V2 innervates the dorsal muscles and V3 the ventral ones. More specifically, V2 innervates the muscles of the upper lip including buccalis anterior, and V3 innervates the circumoral muscles, i.e., buccal constrictor and sublabialis. As for the elevator labialis ventralis, Mallatt (1996) argued that this muscle is probably innervated by V3, rather than V2. As explained earlier, V2 and V3 in cyclostomes may not correspond to the V2 and V3 of gnathostomes (Kuratani et al. 2004).

In sharks, the oral musculature is simpler than that in ammocoetes (Figure 4.8). Whereas the upper lips of ammocoetes are highly muscular, in sharks, no muscles attach to the anterior labial cartilages. Correspondingly, V3 has no motor axons and is strictly a sensory nerve in sharks and in the teleostome gnathostomes. The only muscle around the mouth opening of sharks is the preorbitalis (often called "levator labii superioris" or "suborbitalis": see Figure 4.8). This muscle originates from the ethmoid region of the braincase between the two eyes just rostral to the palatoquadrate (Figure 4.8). In *Squalus*, it extends caudally behind the posterior labial cartilages and inserts by a thin tendon onto the adductor mandibulae. In other sharks, however, it extends farther ventrally to insert onto the Meckel's cartilage. It is innervated by V3. Functionally, the preorbitalis protrudes the lips, buccal membrane, and labial cartilages to round the mouth opening during suction feeding. It may also help to close the mouth, and in some advanced sharks, it protrudes the palatoquadrate.

According to Mallatt (1996), "in *Chimaera* (Figure 4.14) the lip and mouth muscles seem comparable to those of ammocoetes." Three muscles attach to the prelabial cartilage in the nasolabial fold and can be called "upper-lip" muscles: levator anguli oris anterior, labialis anterior, and levator prelabialis. They extend from the preorbital part of the braincase and the maxillary labial cartilage to the prelabial cartilage, much as the buccalis anterior of ammocoetes extends from the anterior braincase and lateral mouth plate to the rostrodorsal cartilage in the upper lip. Mallatt states that "once the chimaera *Hydrolagus colliei* sucks food from the sediment surface, it initiates a series of chewing movements, rapidly protruding then retracting the cheek and lips." The thin muscle designated "labialis posterior" in holocephalans often curves like a

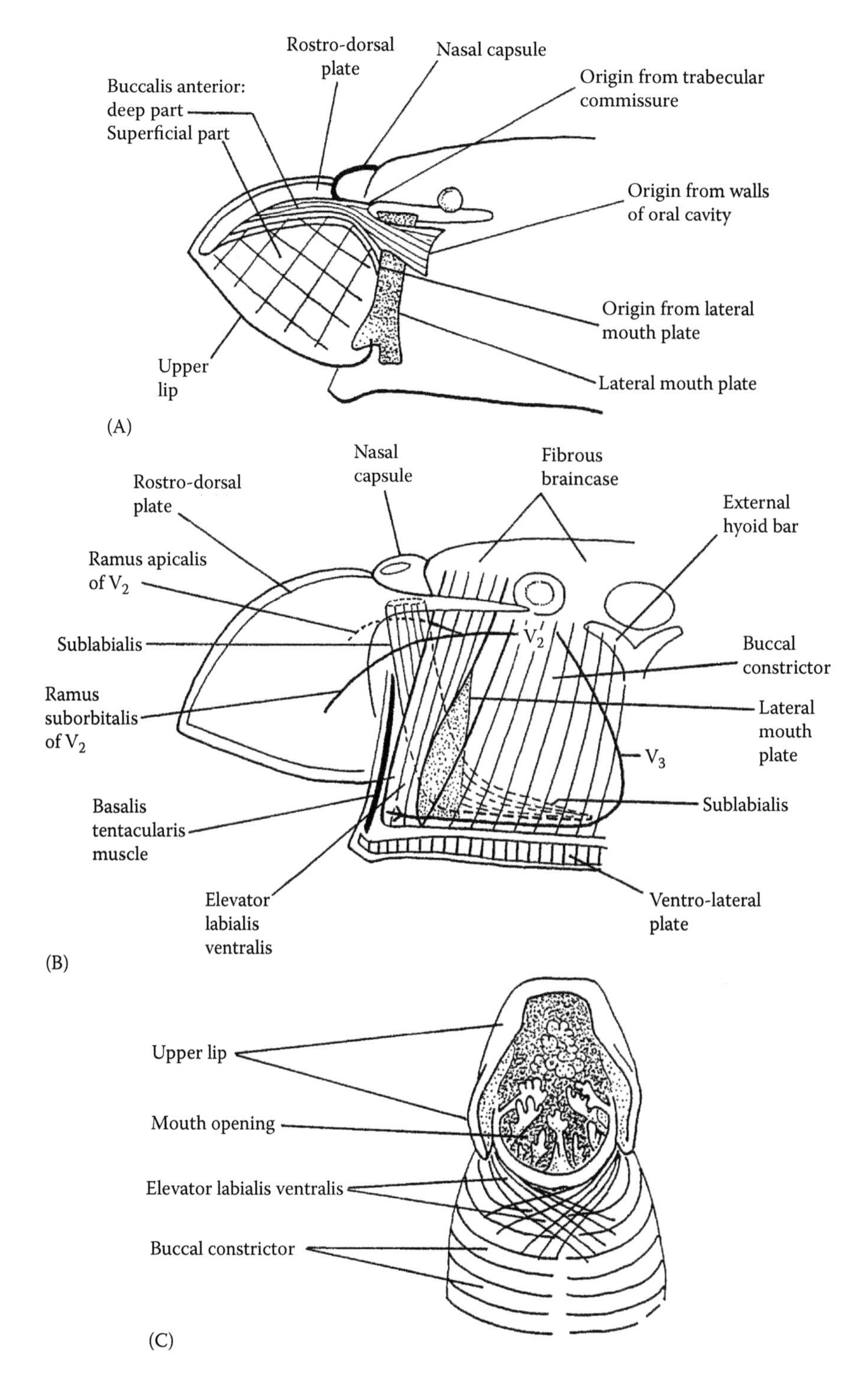

FIGURE 4.13 Muscles and nerves of the lips and mouth of ammocoetes (Vertebrata: Cyclostomata), according to Mallatt (1996). (A) Attachments of buccalis anterior, the main muscle of the upper lip. (B) Muscles around the oral cavity and mouth opening and the paths of the maxillary (V2) and mandibular (V3) nerves (note: as explained in the text, the homology between these nerves and the gnathostome nerves V2 and V3 is questioned by some authors). (C) Ventral view, showing how the elevator labialis ventralis and buccal constrictor overlap on the floor of the oral cavity. (The nomenclature of the structures illustrated basically follows that of Mallat [1996].) (Modified from Mallatt, J., *Zool J Linn Soc*, 117, 329–404, 1996.)

sling around the floor of the mouth just posterior to the lips (a similar muscle called "labialis inferior" is said to lie posterior to this muscle; however, such a labialis inferior was not found in the holocephalan specimens dissected by Mallatt (1996). The labialis posterior extends from the superior anterior margin of the maxillary labial cartilage down across the ventral surface of Meckel's cartilage and deep to the fibrous mass in the lower lip (Figure 4.14). In Mallatt's view, "functionally this muscle could weakly lift the lower lip and mandible; it is innervated by V3; in position and innervation, it resembles the similarly sling-like elevator labialis ventralis of ammocoetes." However, it is now clear that the structures designated labialis posterior and labialis inferior in holocephalans correspond to parts of the muscle intermandibularis of other gnathostomes (Table 5.1). Like the buccal constrictor of ammocoetes and the preorbitalis of sharks, the levator anguli oris posterior (V3) of holocephalans originates from the suborbital braincase, lies directly posterior to the maxillary labial cartilage, and occupies the cheek anterior to the mandibular arch and jaw joint (Figure 4.14).

Based on his observations and comparisons and the proposed homologies shown in Figure 4.10, Mallatt (1996) attempted to reconstruct the lips and mouth of the jawless LCA of living vertebrates. According to Mallatt, in this LCA the "upper lips were well developed, although the condition in chimaeroids and sharks suggests these lips were not as large as in ammocoetes; muscles ran from the braincase into the upper lips (as in ammocoetes and chimaeroids); the mouth opening was circular and smaller than that of the earliest known sharks (based on ammocoetes and chimaeroids); a ring of oral cartilages lay directly behind the continuous lateral and lower lips, and supported both the anterior margin of the mouth and some sensory tentacles." Also, "these oral cartilages were not serially homologous to the branchial arches (not a premandibular arch [see above]). Behind this oral ring, in the cheek, a buccal constrictor circled the premandibular oral cavity. This constrictor graded posteriorly into the branchial superficial constrictors around the pharynx.... The anterior part of the buccal constrictor was thickened into a distinct oral sphincter." According to Mallatt, the reconstructed oral structures of the LCA of vertebrates "seem well suited for 'feeding on semi-sessile soft-bodied invertebrate prey on the ocean floor,' where ingestion involved pushing the head against the prey and forcing it into the oral cavity. In this feeding act, the oral sphincter would have squeezed the ring of oral cartilages to grasp the prey animal, then the buccal

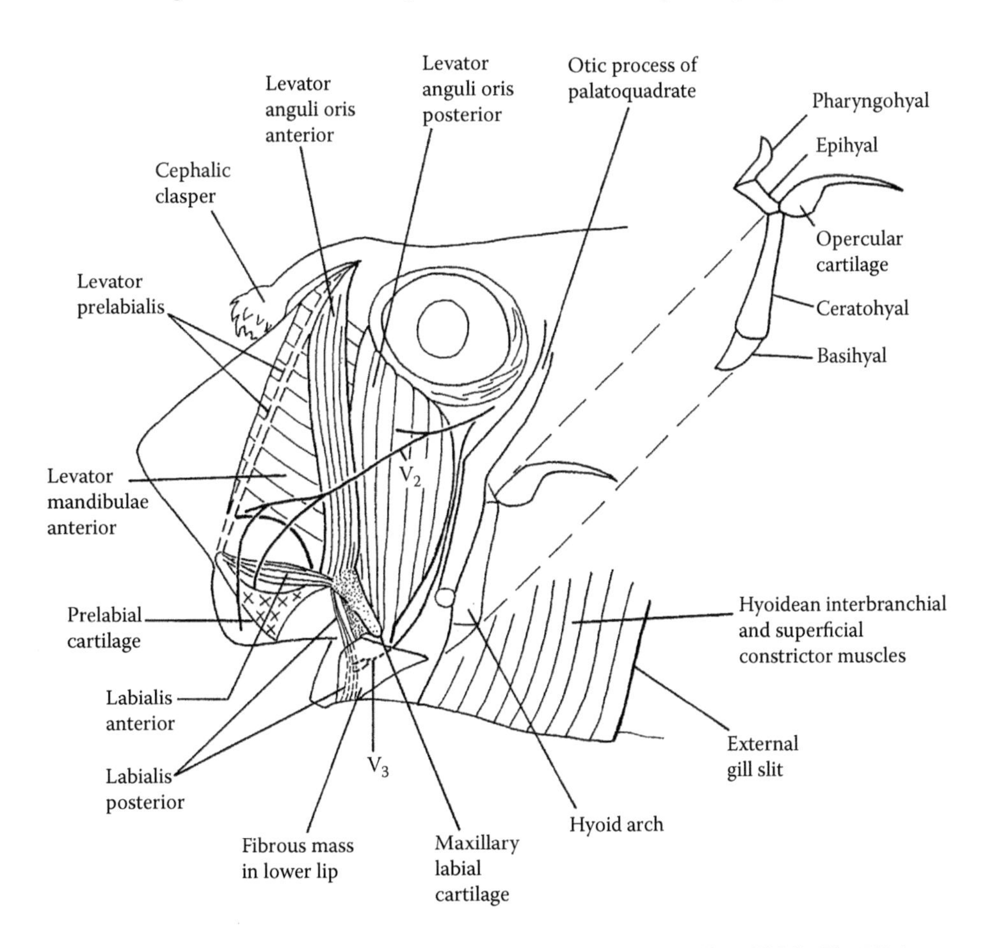

FIGURE 4.14 Orolabial muscles and nerves in *Chimaera monstrosa*, according to Mallatt (1996). The V2 innervated muscles to the upper lip are the levator anguli oris anterior, labialis anterior, and levator prelabialis (the latter being rudimentary in *Chimaera* but large in other chimaeroid genera such as *Callorhinchus*). The inset at the upper right shows the hyoid arch, including its pharyngohyal segment. (The nomenclature of the structures illustrated basically follows that of Mallat [1996].) (Modified from Mallatt, J., *Zool J Linn Soc*, 117, 329–404, 1996.)

and branchial constrictor muscles would have squeezed (swallowed) the prey back to the esophagus through peristalsis. No suction was involved in ingesting and swallowing prey. The upper lips and tentacles would have been tactile and gustatory structures, respectively, for detecting prey on the substrate. The mobile upper lips might have draped over the prey, trapping it against the sediment surface directly in front of the mouth, thereby facilitating ingestion. Covered with tiny denticles (odontodes), this ancestor would have been more mobile and wide-ranging than the more heavily armored osteostracans and heterostracans." The LCA of vertebrates would thus be a nektobenthonic animal that was "actively swimming along the ocean floor in search of food and stopping to rest only occasionally."

Mallatt (1996) argued that the "orolabial region of ammocoetes (Figure 4.13) differs from the reconstructed ancestral vertebrate condition in three ways. First, the upper lip is enlarged, presumably having become so when the ancestors of ammocoetes became burrowers.… Second, the premandibular oral cavity is enlarged, to hold the velum that projects forward into this cavity. The third derived feature in the oral cavity of ammocoetes is the ventral longitudinal crest, which metamorphoses into the lingual apparatus ("tongue") of adult lampreys." He suggested that "the lingual apparatus evolved in the adult ancestors of extant agnathans [cyclostomes] as an appendage that helped pull worms into the mouth and back through the oral cavity—because that is how hagfish use this apparatus and the associated dental plates. The lingual apparatus of adult lampreys pulls and slices food, and its anatomy suggests it is homologous to that of hagfish." Also, "judging from the many resemblances between the orolabial structures of ammocoetes and chimaeroids, the evolution of jaws had little effect on the external anatomy of this premandibular region. That is, in the earliest gnathostomes, the cheeks still reached far forward and the mouth opening was not enlarged. However, the mouth opening had attained its lateral corners, reflecting the fact that it now closed in an up-and-down bite, no longer by sphincter action. Furthermore, the functions of the oro-labial structures had begun to change: Judging from both chimaeroids and sharks, the earliest gnathostomes could protrude and then retract their cheeks during suction feeding, and the upper and lateral lips now served to round the mouth opening during the expansive phase of the feeding strike."

He suggested that "early elasmobranchs began to chase down large, pelagic prey, and the gape of their mouth enlarged to fit this larger prey. To allow this, the lateral corner of the mouth migrated far posteriorly, and the cheek folded into a deep labial pouch as a buccal membrane. Simultaneously, the most anterior part of the upper lip flattened to avoid blocking prey entering the mouth (although the posterior part of this lip remained as the lateral fold). In essence, the premandibular mouth structures became smaller to avoid interfering with the capture of large prey by the jaws. Incidentally, independent enlargements of the gape must have occurred in the ancestor of acanthodians and Osteichthyes, and in the most highly predacious of the arthrodire placoderms." Lastly, regarding the evolution of the mouth and lips of holocephalans, Mallatt (1996) stated that "the relatively small gape and jaws of these animals should be primitive. By this reasoning, chimaeroid feeding must echo a stage after the evolution of jaws but before large, evasive prey were chased down and swallowed whole. Indeed, chimaeroids feed mainly on slow invertebrates such as molluscs, crabs, echinoids, polyps, shrimps, and amphipods. Although their diets contain many hard items, and their jaws and teeth obviously are modified for forceful shearing or durophagy, chimaeroids otherwise retain the benthic feeding mode of ancestral vertebrates."

Mallatt's (1996) ideas are based on a remarkably detailed examination of the configuration of the cartilages, muscles, arteries, nerves, and ligaments of the mouth and lips of living nonosteichthyan vertebrates, and they provide an extremely valuable contribution to understand the evolution and homologies of the mouth and lip structures within basal vertebrates. However, contrary to the diagram shown in Figure 4.10, in the specific case of the myological structures, there may be no direct correspondence between any of the individual labial muscles present in living lampreys (i.e., buccalis anterior, elevator labialis ventralis, and buccal constrictor in ammocoetes, *sensu* Mallatt [1996]), living hagfishes (i.e., coronarious, basitentacularis, levator cartilaginis basalis, protractor cartilaginis basalis, craniobasialis, and cornuosubnasalis muscles *sensu* Mallatt [1996]), living holocephalans (i.e., levator prelabialis, levator anguli oris anterior, levator anguli oris posterior, labialis anterior, and levator posterior *sensu* Mallatt [1996]), living elasmobranchs (i.e., preorbitalis/levator labii superioris *sensu* Mallatt [1996] and Anderson [2008]), and living osteichthyans (levator maxillae superioris 3 and 4 of *Amia* and palatomandibularis major and minor of *Lepisosteus*—*sensu* Diogo [2007] and Anderson [2008]—and possibly "suborbital portion of adductor mandibulae" of acipenseriforms—*sensu* Lauder [1980a,b]—and/or labial muscles of *Latimeria*—*sensu* Millot and Anthony [1958] and Anderson [2008]). That is, in our opinion, one cannot discard the hypothesis that the individual labial muscles present in living lampreys, hagfishes, elasmobranchs, holocephalans, and osteichthyans are the result of an independent differentiation (in the lineages that gave rise to these major vertebrate groups) of the mandibular mesoderm. In fact, as emphasized earlier, the developmental work of Kuratani et al. (2004) indicated not only that at least some of the lamprey labial muscles are the result of a secondary migration of part of the mandibular mesoderm, but also that these muscles are very likely not homologous with the labial muscles of living gnathostomes: "experiments labeling the mandibular mesoderm of the early lamprey embryo, before the cheek process has differentiated into the upper lip anlage or the premandibular domain, indicate that a part of the mandibular mesoderm secondarily grows anteriorly and laterally and migrates into the upper lip domain. No such muscles are known in the gnathostomes, in which all the trigeminal-nerve-innervated muscles are restricted to derivatives of the upper and lower jaws." Also, Konstantinidis et al.'s (2015) study on the development of some head muscles of *Lepisosteus* showed that the muscle palatomandibularis major and minor originates only late in development, being clearly a secondary subdivision of the adductor mandibulae; i.e., it does not seem to be the remaining of a very old, and very different, labial primordium/group of muscles. Further studies, ideally combining dissections and direct comparisons of lampreys, hagfishes, elasmobranchs, holocephalans, osteichthyans, and nonvertebrate taxa, such as cephalochordates and tunicates with molecular techniques such as those employed by Kuratani et al. (2004), will hopefully help clarify the evolution and homologies of the labial muscles within these taxa. We will further discuss these homologies based on our recent dissections and comparisons of various additional vertebrate taxa.

In a more recent study, Miyashita (2012) compared in detail the cephalic musculature of larval and adult lampreys and of adult hagfishes and gnathostomes and stated that none of the direct homologies of the muscles of hagfishes, lampreys, and gnathostomes proposed by Mallatt (1996) are supported by his work. Many developmental and morphological characteristics of the supposedly homologous muscles are incongruent. According to Miyashita (2012), the cyclostome upper lip is developmentally not comparable to the maxillary process of gnathostomes (contra Mallatt 1996), because the latter forms from the more posterior region of the mandibular domain, and therefore, the muscles in the upper lips of holocephalans and elasmobranchs are not homologous with those of hagfishes and lampreys. The implication is that (contra Mallatt 1996) the V2 nerve of gnathostomes may not be a homologue of the anterior branch of the trigeminal nerve in cyclostomes. In fact, the only possible direct homology between muscles of cyclostomes and gnathostomes that Miyashita (2012) mentions is the one between the rectractor lingualis of hagfishes, the constrictor cornualis superficialis and constrictor glossus internus of lampreys, and the intermandibularis of gnathostomes. These muscles are topographically (ventral portion of mandibular domain) and functionally (transverse constriction) similar. In Table 3.2 of his work, he then also suggests that the constrictor pharyngis of hagfishes and the constrictor branchialis dorsalis and ventralis of lamprey larvae (ammocoetes) might be homologous to the superficial constrictors of sharks and that the interbranchialis and constrictor branchialis internus of adult lampreys might be homologues to the interarcuales of sharks (but not to hagfish muscles). In his text, he suggests that the "hypobranchial" muscles of amphioxus and lampreys are not true hypobranchial muscles such as those of gnathostomes. However, in Table 3.2 of his work, he does support the homology of these muscles and suggests that the muscles obliquus and/or rectus of hagfishes might also be homologous to the hypobranchial muscles of lampreys and gnathostomes. Based on his comparisons and the fossil record, he furthermore suggests that a lingual apparatus with protractors and retractors likely represents a plesiomorphic condition of vertebrates, such apparatus having been lost during gnathostome evolution. Miyashita (2012) also stressed that the visceral muscles of adult lampreys differentiate from blastema as those in ammocoete larvae degenerate during metamorphosis (there is disagreement over whether ammocoete muscles are precursors of adult muscles or those present in adults derive from mucocartilage instead) and that it is therefore better to treat these muscles as distinct traits except for several branchial muscles that persist into the adult. He further noted that hagfishes have no extraocular muscles nor motor neurons that would have innervated them (i.e., CNIII, IV, and VI); he suggested that the lack of these muscles might be plesiomorphic, but this is doubtful because, as recognized by him, the ocular muscles of lampreys are very similar to those of gnathostomes (therefore, having such muscles clearly seems to be plesiomorphic for vertebrates).

Miyashita (2012) also emphasized that the gnathostome *Dlx* code is essential in the development of masticatory muscles, as shown by Heude et al. (2010); so, for Miyashita, morphologically and developmentally, the gnathostome jaw muscles represent neomorphs dependent on this code expressed in the cranial neural crests only in that clade. He proposes a new hypothesis on jaw formation, in which the jaw is not originally part of a serial homologous first branchial arch. That is, for him, the original vertebrate branchial arches include only the hyoid and more posterior arches, and because there was no Hox code for the region anterior to those arches, the premandibular and mandibular domains were highly variable in terms of their structures (the mandibular domain mainly simply filled the space between the premandibular and hyoid domains), as seen in hagfishes and lampreys and early gnathostomes. Following this idea, in living gnathostomes, the neural crest cells associated with the mandibular domain were restricted to the region where the jaws are (i.e., limitation of the differentiation and specialization of the mandibular domain). This scenario together with the heterotopy of *Dlx* expression and the gnathostome *Dlx* code (that helps to have ventral and dorsal mandibular structures similar to the ventral and dorsal structures of the other branchial arches) resulted in a derived similarity between the upper and lower jaw and the original branchial arches (i.e., anterior transfer of identity as a pharyngeal arch onto the defaulted mandibular domain). However, interestingly, the duplication of *Dlx* and its downstream genes long preceded the origin of jaws, because these events occurred in the LCA of living vertebrates (those features are also seen in lampreys) and thus cannot explain by themselves the origin of the *Dlx* gnathostomes code. These subjects were recently discussed in a paper by Miyashita and Diogo (2016) and will be discussed further in Chapter 5.

Mallatt (1997) states that the levator arcus palatini ("levator palatoquadrati" in his terminology) of sharks derives from the interbranchialis 1 rather than from the branchial superficial constrictor 1 (as he proposed in 1996). He furthermore states that most fibers of the spiracularis (e.g., Figure 4.8) and intermandibularis of sharks also derive from interbranchialis 1 and that only a small portion of these muscles derives from branchial superficial constrictor 1 (in his 1996 paper, he proposed that these two muscles were mostly derived from the branchial superficial constrictor 1). We tend to agree more with the homologies proposed by Mallatt (1996) than with those proposed by Mallatt (1997), i.e., that the ventral mandibular muscle intermandibularis (e.g., Figure 4.9) of the LCA of gnathostomes probably derived from the ventral region of an ancestral branchial superficial constrictor of the first arch and that the dorsal mandibular muscle levator arcus palatini (= levator palatoquadrati: e.g., Figure 4.8) of this LCA probably derived from the dorsal region of the same constrictor muscle. In fact, the work of several authors (e.g., Marion 1905; Daniel 1928; Luther 1938) has strongly supported the idea that the intermandibularis and the levator arcus palatini *sensu* this book derive from the branchial superficial constrictor of the first arch and that the interhyoideus, adductor operculi, and adductor arcus palatini *sensu* this book derive from the branchial superficial constrictor of the second arch.

Interestingly, according to Holland et al. (1993), the velothyroideus of lampreys might be homologous to the levator arcus palatini + dilatator operculi of teleosts (see following chapters), i.e., to the mandibular constrictor dorsalis *sensu* Edgeworth (1935) and, thus, to the levator arcus palatini of the LCA of gnathostomes, because engrailed immunoreactivity has been detected in both the lamprey's velothyroideus and in the zebrafish's levator arcus palatini. Holland et al. (1993) stated that the mandibular arch mesoderm of lampreys gives rise to one extrinsic eye muscle, three muscles of the velum (velothyroideus, velohyoideus, and velocranialis), and seven muscles of the hood and lip (most, or all, of them probably corresponding to the labial muscles *sensu* Anderson 2008 and to the muscles shown in the table in Figure 4.10: see earlier text). If one accepts the homology between the lamprey's velothyroideus and the levator arcus palatini of the LCA of gnathostomes, then the two other muscles of the lamprey velum (i.e., the velohyoideus and velocranialis *sensu* Holland et al. [1993]) might be homologous to the two other nonlabial mandibular muscles of this LCA (i.e., the adductor mandibulae and the intermandibularis). Mallatt (1996) seems to suggest that this is not the case, because he states that the adductor mandibulae complex *sensu* Anderson (2008) came from a "medial band" of the interbranchialis of the first arch (see earlier text) and thus not from the velohyoideus and/or the velocranialis. It should also be stressed that the detection of engrailed immunoreactivity in certain muscles of a taxon A and in certain muscles of a taxon B does not necessarily imply that the former muscles are directly homologous to the latter (Knight et al. 2008). For instance, as mentioned earlier, in the zebrafish engrailed immunoreactivity is only detected in muscles that are derived from the dorsal portion of the mandibular muscle plate (constrictor dorsalis *sensu* Edgeworth 1935, i.e., the levator arcus palatini + dilatator operculi: see following chapters), while in the mouse, it is detected in mandibular muscles that are very likely derived from the adductor mandibulae portion of that plate (i.e., masseter, temporalis, pterygoideus medialis, and/or pterygoideus lateralis; see Knight et al. 2008 and the following chapters). That is why the detection of immunoreactivity in the lamprey velothyroideus and in the zebrafish levator arcus palatini and dilatator operculi does not necessarily imply that the latter muscles are homologous to, or derived from, the former muscle. Authors such as Kuratani and Ota (2008a) have even suggested that lampreys and probably hagfishes lack "somitomeres," i.e., that cyclostomes probably do not have "mandibular," "hyoid," and "branchial" muscular plates (*sensu* Edgeworth 1935) such as those present in living gnathostomes. However, researchers such as Holland et al. (1993), Mallatt (1996), and Knight et al. (2008) consider at least some of these plates (e.g., the mandibular muscular plate *sensu* Edgeworth 1935) to have been present in extant cyclostomes, thus implying that they were present in the LCA of vertebrates.

Regarding the homologies of the branchial muscles *sensu* Edgeworth, Wiley (1979a,b) suggests that the presence of "ventral branchial musculature in all gill arches" is a synapomorphy of gnathostomes, i.e., that this feature was present in the LCA of gnathostomes but not in the LCA of vertebrates. Johanson (2003) supports this idea. She states that lampreys lack all ventral branchial muscles, including gill arch depressors comparable to the coracobranchiales/pharyngoclaviculares (see Figure 4.4). However, she also states that living lampreys do have branchial muscles, including (a) interbranchiales, (b) external branchial constrictors, (c) internal dorsal and ventral diagonal constrictors, (d) median muscle bands, and (e) isolated muscle fibers associated with the interbranchial septum, as well as hypobranchial muscles (see, e.g., Figures 4.6 and 4.7 and Roberts 1950).

Mallatt (1996) argues that the lateral interarcuales and adductores branchiales (see, e.g., Figures 4.4 and 4.7) of sharks derive from the interbranchiales and their medial bands, respectively (the "adductor branchialis" of the first arch corresponding to the adductor mandibulae, and the adductor branchialis of the second arch being absent: see earlier text). He also suggests that the ventral part of the lamprey interbranchial muscles is homologous to the coracobranchiales/pharyngoclaviculares of gnathostomes, but Johanson (2003) does not agree, because the interbranchiales are mainly related to constrictive movements while the coracobranchiales/pharyngoclaviculares are mainly related to expansive movements (see earlier text). In this case, we tend to agree more with Johanson (2003) than with Mallatt (1996). The coracobranchiales/pharyngoclaviculares of gnathostomes probably do not correspond to the ventral part of the interbranchiales of lampreys, because the cyclostome interbranchiales and gnathostome coracobranchiales/pharyngoclaviculares have quite different overall configurations, attachments, and functions. The coracobranchiales/pharyngoclaviculares of gnathostomes thus may (a) be de novo structures, (b) correspond to muscular structures that have not been described in detail in hagfishes and/or lampreys or that were secondarily lost in both these two groups, or (c) derive from the hypobranchial musculature, as argued by some authors (see, e.g., Johansson's [2003] review). The fact that in some gnathostomes the coracobranchiales/pharyngoclaviculares are innervated by cranial nerve XI (CNXI) (see, e.g., Anderson 2008) would perhaps provide support for the latter hypothesis. However, the results obtained from Edgeworth's (1935) detailed comparative study of several vertebrate embryos and adults strongly support the idea that the gnathostome coracobranchiales/pharyngoclaviculares are branchial muscles derived from the ventral parts of the branchial muscle plates rather than from the hypobranchial muscles. Edgeworth (1935) stated that the coracobranchiales (or their derivatives) are always innervated by the vagus nerve (CNX) and its branchial parts and that only in chondrichthyans is there a "spinal" innervation from the cervical plexus, which he regarded as a derived condition for gnathostomes. However, he normally regarded (often erroneously) the condition found in dipnoans as the plesiomorphic condition for gnathostomes, so it is not sure if such a spinal innervation is really derived for gnathostomes. Confusingly, despite considering that such a spinal innervation was derived for gnathostomes, and despite showing that the coracobranchiales ontogenetically derive from the ventral parts of the branchial muscles plates, even in chondrichthyans, Edgeworth (1935) considered the coracobranchiales as "hypobranchial cerebral" muscles (the hypobranchial muscles *sensu* the present work being the "hypobranchial spinal muscles"). However, in all other parts of his work, he always clearly included the coracobranchiales within the true branchial muscles, as we are doing in the present work. Moreover, in the more recent studies of Didier (1995), which were very detailed and exclusively focused on holocephalans, this author also stated that the innervation of the coracobranchiales by CNX in chondrichthyans is probably a derived condition, supporting the idea that these are branchial muscles. More recently, Anderson (2008) has corroborated the view that the coracobranchiales of elasmobranchs are branchial muscles and that they are very likely homologous and/or ancestral to the pharyngoclaviculares of osteichthyans (see Figures 4.15 through 4.17). The protractor pectoralis (often named "trapezius," see, e.g., review of Kuratani [2008a]) is not present as an independent muscle in living nongnathostome animals, its presence thus constituting a potential synapomorphy of gnathostomes (see, e.g., Figure 4.18 and the following).

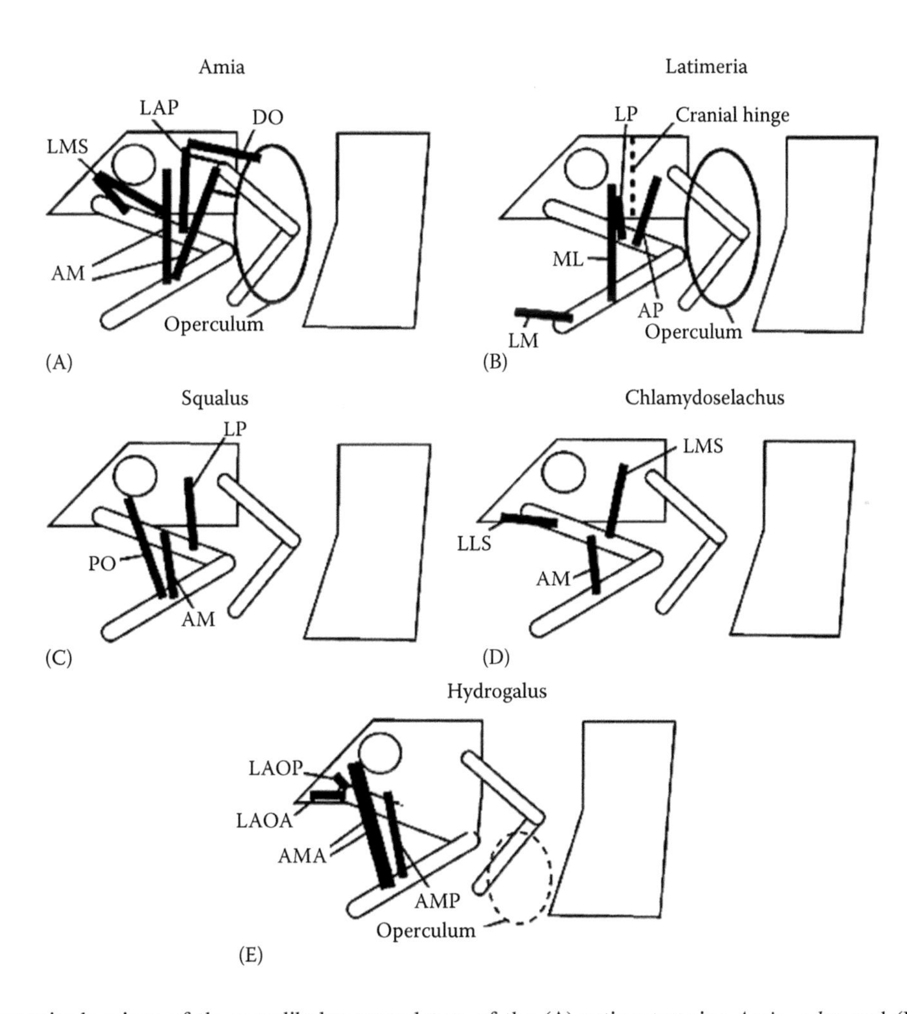

FIGURE 4.15 Schematic drawings of the mandibular musculature of the (A) actinopterygian *Amia calva* and (B) the sarcopterygian *Latimeria chalumnae* (Osteichthyes), the elasmobranchs (C) *Squalus acanthias* and (D) *Chlamydoselachus anguineus*, and (E) the holocephalan *Hydrolagus colliei*, according to Anderson (2008). The thick black lines represent muscles; the intermandibular muscles are not illustrated. For homologies between the illustrated muscles: AM, adductor mandibulae; AMA, adductor mandibulae anterior; AMP, adductor mandibulae posterior; AP, adductor palatoquadrati; DO, dilator operculi; LAOA, labialis anguli oris anterior; LAOA, labialis anguli oris posterior; LAP, levator arcus palatini; LLS, levator labii superioris (preorbitalis); LM, labial muscles; LMS, levator maxillae superioris; LP, levator palatoquadrati; ML, mandibular levator; preorbitalis. (The nomenclature of the structures illustrated basically follows that Anderson [2008].) (Modified from Anderson, P. S. L., *Biol J Linn Soc*, 94, 195–216, 2008.)

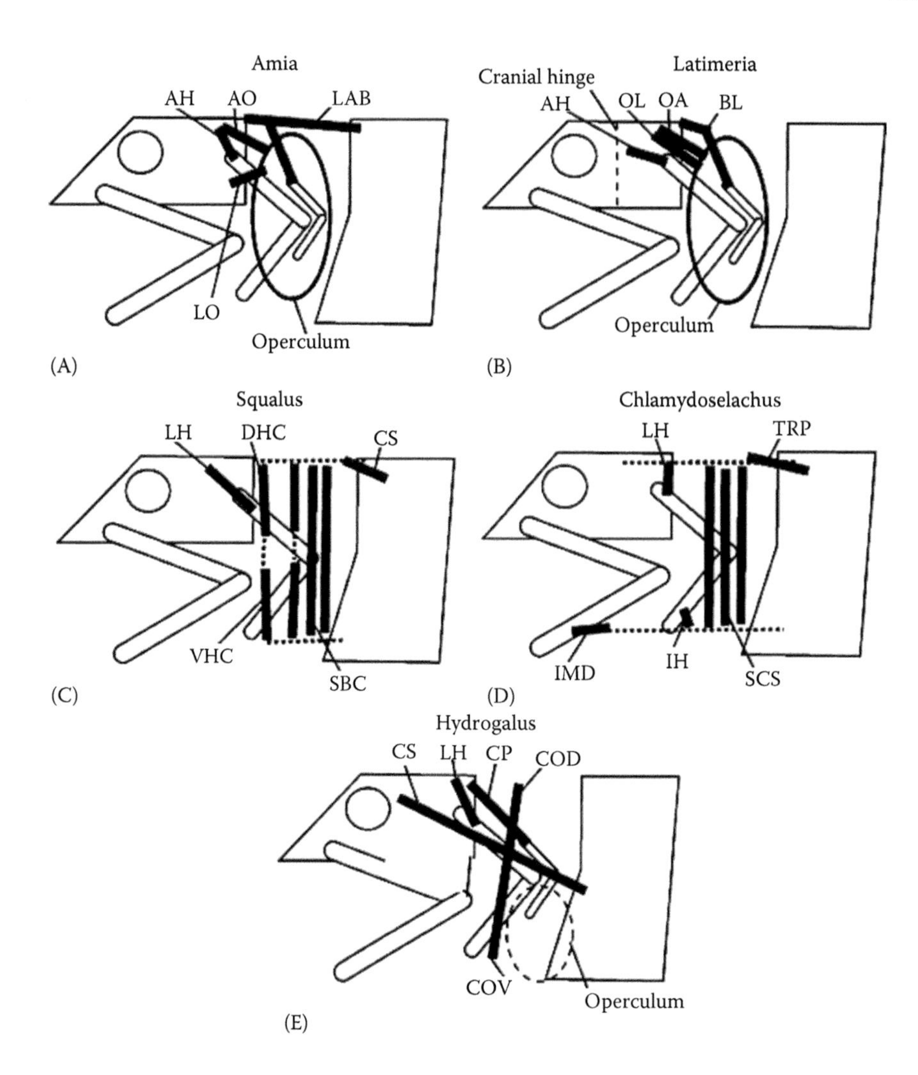

FIGURE 4.16 Schematic drawings of the hyoid and branchial musculature of (A) the actinopterygian *Amia calva* and (B) the sarcopterygian *Latimeria chalumnae* (Osteichthyes), the elasmobranchs (C) *Squalus acanthias* and (D) *Chlamydoselachus anguineus*, and (E) the holocephalan *Hydrolagus colliei*, according to Anderson (2008). The thick black lines represent muscles; the interhyoid type muscles have not been figured here except in *Chlamydoselachus* to show their relation to the superficial constrictor sheet. For homologies between the illustrated muscles: AM, adductor mandibulae; AMA, adductor mandibulae anterior; AH, adductor hyomandibularis; AO, adductor operculi; BL, branchial levators; COD, constrictor operculis dorsalis; COV, constrictor operculis ventralis; CP, cucullaris profundus; CS, cucullaris; DHC, dorsal hyoid constrictors; IH, interhyoideus; IMD, intermandibularis; LAB, levatores arcuum branchialum; LH, levator hyomandibularis; LO, levator operculi; OA, opercular adductor; OL, opercular levators; SBC, superficial constrictors; SCS, superficial branchial constrictors; TRP, trapezius; VHC, ventral hyoid constrictors. (The nomenclature of the structures illustrated basically follows that of Anderson [2008].) (Modified from Anderson, P. S. L., *Biol J Linn Soc*, 94, 195–216, 2008.)

According to Mallatt (1996), in the LCA of vertebrates, the hypobranchial myotomes probably lay superficial to the external branchial arches and pharyngeal musculature (Figure 4.9A and B), because this is the condition in lampreys and hagfishes ("musculus rectus": Marinelli and Strenger [1956]). In hagfishes, these hypobranchial muscles run anteriorly to insert on the floor of the oral cavity, so it seems reasonable to propose that in this LCA, they pulled posteriorly on this floor to widen the oral cavity or open the mouth. In Mallatt's view, these hypobranchial muscles must have shifted, in "pregnathostomes," "to a deeper location, in order to insert on the ventrolateral surfaces of the internal mandibular and hyoid arches (Figure 4.9C). Functionally, this would have allowed faster opening of the mouth through a forceful abduction of the jaw and hyoid segments. To allow this shift in depth, however, the right and left rows of external arches (extrabranchial cartilages) must have separated from one another and moved laterally on the pharyngeal floor (Figure 4.9D)."

According to Edgeworth (1935), among gnathostomes, an epibranchial musculature is present only in the Selachii and Holocephali, both groups having at least some interpharyngobranchiales and a subspinalis. Kesteven (1942) also described epibranchial muscles (namely, the interpharyngobranchiales, the subspinalis being absent) in batoids such as *Dasyatis* (the muscles being reduced to fibrous structures in *Raja*), but more recent works, including those specifically focused on batoids (e.g., Miyake et al. 1992) state that batoids do not have epibranchial muscles. Be that as it may, the fact

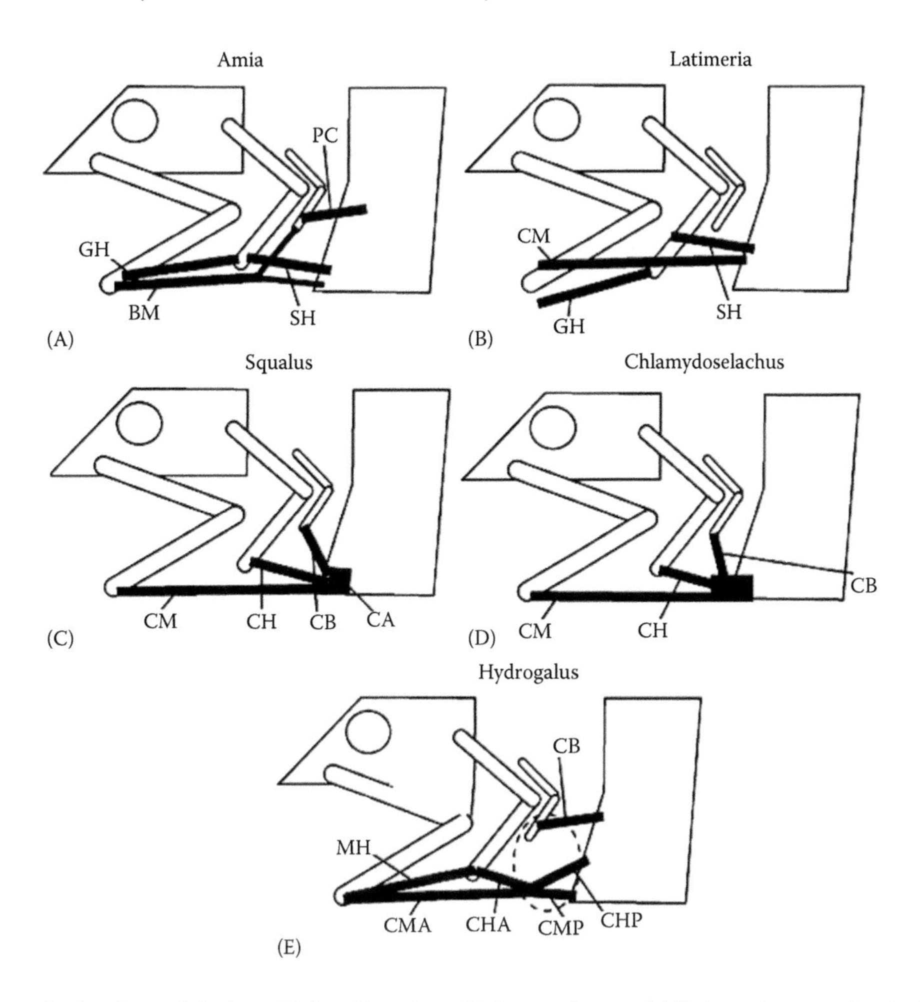

FIGURE 4.17 Schematic drawings of the branchial and hypobranchial musculature of (A) the actinopterygian *Amia calva* and (B) the sarcopterygian *Latimeria chalumnae* (Osteichthyes), the elasmobranchs (C) *Squalus acanthias* and (D) *Chlamydoselachus anguineus*, and (E) the holocephalan *Hydrolagus colliei*, according to Anderson (2008). The thick black lines represent muscles; the geniohyoideus (GH) muscle in *Latimeria* is drawn without an insertion because it is unclear whether it connects to the mandible. For homologies between the illustrated muscles: BM, branchiomandibularis; CA, coracoarcualis; CB, coracobranchials; CH, coracohyoideus; CHA, coracohyoideus anterior; CHP, coracohyoideus posterior; CM, coracomandibularis; CMA, coracomandibularis anterior; CMP, coracomandibularis posterior; GH, geniohyoideus; PC, pharyngo-clavicularis; SH, sternohyoideus. (The nomenclature of the structures illustrated basically follows that of Anderson [2008].) (Modified from Anderson, P. S. L., *Biol J Linn Soc*, 94, 195–216, 2008.)

that holocephalans and sharks have subspinalis and interpharyngobranchiales muscles clearly indicates that the last common ancestor of living chondrichthyans also had these muscles. Mallatt (1996) seems to suggest that this musculature may also be present in living nongnathostome taxa such as lampreys, but Mallatt (1997) states that "dorsal interarcuales" (which are considered to be epibranchial muscles *sensu* Edgeworth [1935]: see Figure 4.4) is "the only group of pharyngeal muscles that is fully unique to chondrichthyans." Kusakabe and Kuratani (2005) state that lampreys have "epibranchial muscles," which "have a peculiar morphology, extending rostrally from the anterior myotomes" (see Figure 4.18), but Kuratani (2008a; pers. comm.) explained that these muscles are likely not homologous with the chondrichthyan epibranchial muscles *sensu* Edgeworth (1935). In view of the information available at the moment, it is thus difficult to discern whether the epibranchial muscles *sensu* Edgeworth (1935) were acquired only in chondrichthyans or were, instead, present in the LCA of gnathostomes and then secondarily lost in osteichthyans.

GENERAL REMARKS

The transition from nonvertebrate to vertebrate animals (the latter of which have numerous cranial ventilatory muscles, unlike the former) and then the transition from nongnathostome to gnathostome vertebrates (the latter of which have many muscles that are not present as independent structures in the former, such as the adductor operculi, interhyoideus, coracomandibularis, and sternohyoideus *sensu* this volume, among others) were crucial events in the evolution and differentiation of the cephalic muscles.

The mandibular musculature of the LCA of gnathostomes was probably divided into a differentiated intermandibularis,

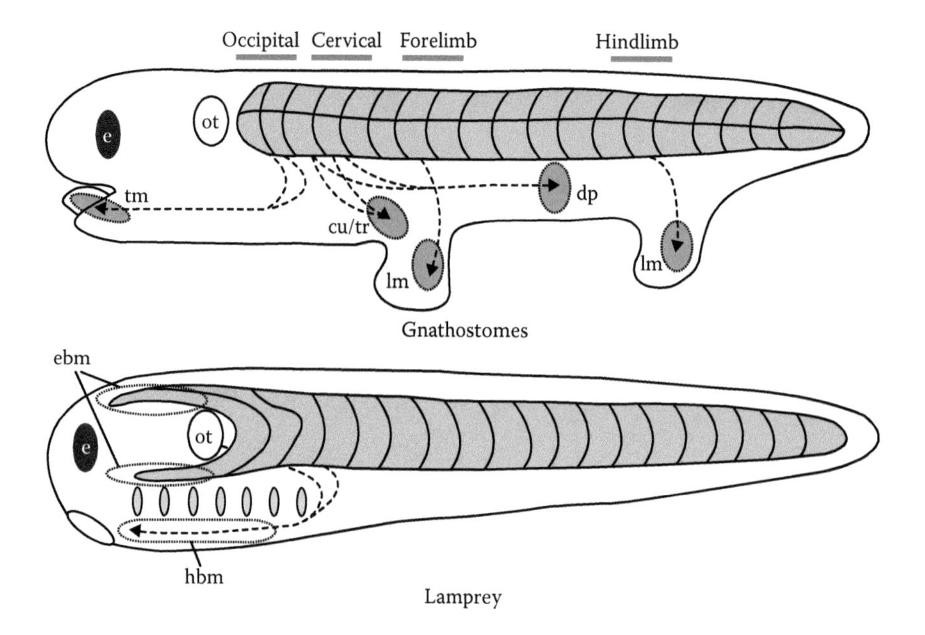

FIGURE 4.18 Schematic representation comparing the skeletal muscle patternings between gnathostomes and the lampreys, according to Kusakabe and Kuratani (2005). In gnathostomes (*top*), some of the hypaxial muscle cells, originating from the four different levels along the anteroposterior axis (indicated by bars at the top), undergo extensive migration (indicated by broken arrows) toward the periphery where they differentiate into the tongue muscle (tm), cucullaris/trapezius muscle (cu/tr), diaphragm (dp), and limb muscles (lm). The cyclostome lamprey (*bottom*) lacks most of these muscles but possesses a hypobranchial musculature (hbm), which resembles the vertebrate tongue muscles, and an "epibranchial" musculature, which has a peculiar morphology, extending (from the anterior myotomes) rostrally to the otic vesicle (ot). (The nomenclature of the structures illustrated basically follows that of Kusakabe and Kuratani [2005].) (Modified from Kusakabe, R., and Kuratani, S., *Dev Dyn*, 234, 824–834, 2005.)

the levator arcus palatini (levator palatoquadrati), the adductor mandibulae complex, and, possibly, a spiracularis and one or more labial muscles (*sensu* Anderson [2008]; see earlier text) (see Table 5.1 and Figure 4.15). In holocephalans, elasmobranchs, and osteichthyans, the intermandibularis is subdivided into an intermandibularis anterior and an intermandibularis posterior (see the following chapter). Such subdivisions were probably acquired more than once in gnathostomes (e.g., Edgeworth 1935; Diogo 2007; Diogo and Abdala 2010), although it is likely that the LCA of gnathostomes already had a divided intermandibularis. The levator arcus palatini became secondarily lost in holocephalans, probably due to the fusion between the first arch and the mandible in these fishes (see Figure 4.15; e.g., Miyake et al. 1992; Mallatt 1996; Anderson 2008). In elasmobranchs, the dorsal part of the mandibular plate gives rise to not only the levator arcus palatini, but also muscles such as the depressor palpebrae superioris, the levator palpebrae nictitantis, the rectractor palpebrae superioris spiracularis, the "spiracularis," and/or the spiracularis *sensu* Miyake et al. (1992) (see Figure 4.8). In gnathostomes, the adductor mandibulae are often subdivided (e.g., adductor mandibulae anterior and posterior in holocephalans; A2, A3′, A3″, A1, A1-OST, Aω, etc., in osteichthyans; see Figure 4.15 and the following chapters). As stated earlier, it is possible that there is no direct correspondence between any of the individual labial muscles (*sensu* Anderson [2008]) present in living holocephalans, living elasmobranchs, and living osteichthyans. That is, the individual labial muscles present in these three major groups of gnathostomes might well be the result of independent differentiation of the mandibular mesoderm.

Regarding the hyoid muscles, the LCA of gnathostomes probably had a ventral muscle, the interhyoideus, and two dorsal muscles, the adductor arcus palatini (often called "levator hyomandibularis" in chondrichthyans, because in these fishes, it usually attaches exclusively on the second arch rather than additionally or exclusively on the first arch, as is the case in many osteichthyans), and the constrictor hyoideus dorsalis (often also named "adductor operculi") in chondrichthyans, because living elasmobranchs, for instance, do not have an operculum: see, e.g., Miyake et al. 1992 (see Table 5.2 and Figure 4.16). Anderson (2008) suggested that holocephalans do not have an interhyoideus. However, the holocephalan muscle that was named "constrictor operculi ventralis" by that author (Figure 4.16E) clearly seems to correspond to the interhyoideus of other living gnathostomes, particularly to that of elasmobranchs and of basal sarcopterygians such as dipnoans. In fact, the interhyoideus and the adductor operculi of fishes such as dipnoans are remarkably similar to the constrictor operculi ventralis and the "constrictor operculi dorsalis" of holocephalans (*sensu* Anderson 2008), respectively: they lie in a superficial position and together form a structure that resembles the continuous, superficial constrictor of the second arch of nongnathostome taxa such as lampreys (see the following chapters). The hyoid musculature may be differentiated into muscles other than the adductor arcus palatini, the adductor opercula, and the interhyoideus in some chondrichthyans (e.g., according to Miyake et al. 1992, batoids may have a "levator rostri," a "depressor hyomandibulae," and/or a "depressor rostri") and in numerous osteichthyans (e.g., levator operculi and facial muscles of mammals: see the following chapters).

The LCA of gnathostomes probably had several branchial muscles, including some "ventral branchial muscles" *sensu* Johanson (2003), "dorsal branchial muscles" *sensu* Miyake et al. (1992), "branchial superficial constrictors" *sensu* Mallatt (1996) (because these muscles are present in elasmobranchs as well as in nongnathostome taxa such as lampreys, it is often inferred that they were present in the LCA of vertebrates and of gnathostomes: e.g., Mallatt 1996; Anderson 2008), coracobranchiales/pharyngoclaviculares (if they are considered branchial rather than hypobranchial muscles; see earlier text), and a cucullaris (see Figures 4.16 and 4.17). The protractor pectoralis of osteichthyan fishes (which gave rise to the amniote trapezius) derives from the cucullaris and is a peculiar branchial muscle in that it often extends directly from the neurocranium to the pectoral girdle, thus protracting this girdle (i.e., it is usually not directly attached to the branchial arches). The primary origin of this muscle is controversial: some authors have stated that it is exclusively derived from the brachial muscle cells, while others have argued that it is derived from hypaxial muscle cells (Figure 4.18), and still others argued that it is derived from branchial and hypaxial muscle cells (see, e.g., the review of Kuratani 2008a). According to Kusakabe and Kuratani (2005), the lack of a pectoral girdle and of protractor pectoralis in living cyclostomes suggests that these basal vertebrates had "not yet established the 'neck region' comparable to that in gnathostomes" (Figure 4.18). (Authors such as Goodrich [1958] argued that the absence of pectoral girdles and fins in extant hagfishes and lampreys may be due to a "degeneration," that is, these structures might have been found in the LCA of vertebrates, but this hypothesis is not commonly accepted today.) The protractor pectoralis will be discussed in further detail in the next chapters, as will be the epibranchial muscles.

The LCA of gnathostomes, as well as the LCA of osteichthyans, probably had two hypobranchial muscles, the coracomandibularis and the sternohyoideus (Table 5.4 and Figure 4.17). As explained by Miyake et al. (1992), although the sternohyoideus ("rectus cervicus" in their terminology) remains undivided in some sharks; in numerous other elasmobranchs, it is subdivided into a coracohyoideus, a coracoarcualis, and occasionally (e.g., batoids) a coracohyomandibularis (see, e.g., Figure 4.9). As explained earlier, based on the information available at the moment, it is difficult to discern whether the epibranchial muscles *sensu* Edgeworth (1935) were acquired only in chondrichthyans or were, instead, present in the LCA of gnathostomes and then secondarily lost in osteichthyans.

The major features characterizing the early evolution of the cephalic muscles in cyclostomes and chondrichthyans are summarized in Figure 3.1.

Before ending this chapter, we will briefly discuss the body musculature of nonosteichthyan vertebrates. Living lampreys and living hagfishes lack paired fins and their body musculature is extremely simple (e.g., in lampreys, the body musculature is not divided into epaxial and hypaxial muscles: see, e.g., Figure 4.18). Living gnathostome fishes also have rather simple and mainly undifferentiated body musculature, but they do have paired fins and median fins, which will be described and discussed in the following chapters. First, in Chapter 5, we will discuss new and crucial findings related to the origin and early evolution of the cephalic muscles of vertebrates based on dissections of representatives of all the major clades of cyclostomes and chondrichthyans, as well as the review of the literature on these taxa, performed since the publication of Diogo and Abdala (2010).

5 Cephalic Muscles of Cyclostomes and Chondrichthyans

In this chapter, which is based on a paper that two of the authors of this book wrote together with Tetsuto Miyashita (Ziermann et al. 2014), we provide an updated, comprehensive overview of the cephalic muscles of nonosteichthyan vertebrates. Some of the muscle features that will be discussed in this chapter are shown in Figure 3.1. The chapter is based on our dissections of cyclostomes (sea lamprey; Atlantic hagfish) and chondrichthyans (spiny dogfish, common smooth-hound, spotted ratfish, and little skate) and a synthesis of primary, including more recent, literature about the evolution and development of the cephalic muscles in vertebrates (for a list of synonyms in the myological terminology of cyclostomes, see Table 5.1). Then, we compare the cephalic muscles of these nonosteichthyan groups with those of osteichthyans, based on our previous dissections and review of the literature (Diogo and Chardon 2000; Diogo et al. 2001a, 2002, 2003, 2004, 2006, 2008a,c, 2009b, 2013, 2015b,c,d,e; Diogo 2004b, 2005b, 2007, 2008a,b; Ziermann and Olsson 2007; Ziermann 2008; Diogo and Abdala 2010; Ziermann et al. 2013; Ziermann and Diogo 2013, 2014). This comparative analysis thus uncovers evolutionary patterns of the cephalic muscles across early vertebrate lineages in order to address long-standing challenges about vertebrate comparative anatomy, such as the evolution of cephalic muscles and the connection between the head and pectoral girdle.

The early evolution of vertebrates—particularly the emergence of the vertebrate head patterning and the origin of the gnathostome jaws—has fascinated comparative anatomists and evolutionary and developmental biologists for centuries (for major reviews, see Gegenbaur 1888; Goodrich 1958; Gans and Northcutt 1983; Hanken and Thorogood 1993; Kuratani et al. 2001, 2013; Cerny et al. 2004; Kuratani 2004, 2005a, 2012; Olsson et al. 2005; Shigetani et al. 2005). This fascination has recently attracted approaches from evo-devo (evolutionary developmental biology) and culminated in numerous studies about the crucial role played by neural crest cells in the development and evolution of the vertebrate skull (e.g., Gans and Northcutt 1983; Kimmel et al. 2001; Santagati and Rijli 2003; Bronner and LeDouarin 2012; and references therein). A resurgence of interest in comparative anatomy has largely kept pace with the progress in evo-devo, with cyclostomes and chondrichthyans highlighted as crucial comparative taxa to understand the origin and early evolution of vertebrate structures (e.g., Janvier 1996, 2007; Kuratani et al. 2001, 2013; Kuratani 2004, 2005a,c, 2008a,b, 2012; Shigetani et al. 2005; Cerny et al. 2010; Medeiros and Crump 2012; Gillis et al. 2013; Green and Bronner 2014).

Despite increasing attention to the skeletal characters of phylogenetically basal vertebrates, few recent works focus on their soft tissues, such as muscles. Consequently, no standard myological nomenclature exists for deep-branching vertebrate taxa. Specific muscles in cyclostomes and chondrichthyans are often referred to under general terms for muscle subgroups (e.g., labial muscles, jaw adductors). The resulting nomenclatural confusion is further exacerbated by the lack of reference to a specific taxon and by the lack of test for phylogenetic congruence within a lineage (e.g., Mallatt 1997; González-Isáis 2003; Clark et al. 2010). For example, there is little information in simply comparing "labial muscles" of "lampreys" and "sharks" (Mallatt 1996), because no specific muscles are identified and because individual muscles may represent conditions unique to individual species rather than conditions general to the lineage.

The conventional nomenclature, primary anatomical descriptions, and comparative analyses of the vertebrate musculature largely stem from pioneering monographs published before 1945 (e.g., Humphry 1872a,b; Edgeworth 1911, 1935; Lubosch 1914; Luther 1914; Huber 1930a,b; Kesteven 1942–1945). However, none of these classics included detailed descriptions of the cyclostomes and, with the exception of Edgeworth (1935), none of them focused on all subgroups of head muscles. Anatomical descriptions of the cyclostome muscles published roughly in parallel did not make comparisons to gnathostomes or between cyclostomes (Fürbringer 1875; Cole 1907; Tretjakoff 1926; Marinelli and Strenger 1954, 1956). With recent evolutionary and developmental insights such as the patterning role of neural crest cells in specifications and configurations of muscles (Olsson et al. 2001; Ericsson et al. 2004; Matsuoka et al. 2005; Rinon et al. 2007), these primary anatomical descriptions now await revision and a synthesis.

Cyclostomes are the main taxa we use for comparison with vertebrate cephalic muscles. Despite having many highly peculiar morphological features, hagfishes and lampreys are crucial to study the evolutionary transitions from nonvertebrate animals to gnathostomes (e.g., Huxley 1876b; Holmgren 1946; Marinelli and Strenger 1954, 1956; De Beer 1959; Fritsch and Northcutt 1993; Horigome et al. 1999; Kuratani et al. 1999; Kuratani 2005c; Clark and Summers 2007; Ota et al. 2007; Clark et al. 2010; Miyashita 2012, and citations within). Comparative data from cyclostomes allow us to take a phylogenetic approach to understand the origin and early evolution of the vertebrate cephalic muscles and predict the muscular configurations in the last common ancestors (LCAs) of living vertebrates and of gnathostomes. The cephalic muscles of cyclostomes have been described in some detail by, for instance, Marinelli and Strenger (1954: *Lampetra fluviatilis*), Marinelli and Strenger (1956: *Myxine glutinosa*), and Miyashita (2012: *Eptatretus stoutii*). Functional analyses of

TABLE 5.1
List of Commonly Used Synonyms for the Muscle Terminology Followed in the Present Chapter

Arch	Name Used in This Chapter	Taxa Studied by Us	Synonyms	Innervation
MA	Subnasonasalis	*M. glutinosa*	Marinelli and Strenger (1956); subnasalis superficialis (Miyashita 2012)	CNV2 anterior branch (Miya12)
MA	Transversus oris	*M. glutinosa*	Marinelli and Strenger (1956); lingual tentacularis (Miyashita 2012)	CNV2 anterior branch (Miya12)
MA	Tentacularis posterior	*M. glutinosa*	Marinelli and Strenger (1956); Miyashita (2012)	CNV2 anterior branch (Miya12)
MA	Nasalis	*M. glutinosa*	Marinelli and Strenger (1956); Miyashita (2012)	CNV2 anterior branch (Miya12)
MA	Palatosubnasalis	*M. glutinosa*	Marinelli and Strenger (1956); subnasalis superficialis and profundus (Miyashita 2012)	CNV2 anterior branch (Miya12)
MA	Tentaculosubnasalis	*M. glutinosa*	Marinelli and Strenger (1956); nasolingualis (Miyashita 2012)	CNV2 anterior branch (Miya12)
MA	Basitentacularis	*M. glutinosa*	Marinelli and Strenger (1956)	CNV externus (MarStr56)
MA	Subnasobasalis	*M. glutinosa*	Marinelli and Strenger (1956); cornual labialis (Miyashita 2012)	CNV2 anterior branch (Miya12)
MA	Coronarius	*M. glutinosa*	Marinelli and Strenger (1956)	
MA	Cornuosubnasalis	*M. glutinosa*	Marinelli and Strenger (1956)	CNV2 anterior branch (Miya12)
MA	Levator cartilagines basalis	*M. glutinosa*	Marinelli and Strenger (1956); cornual lingualis (Miyashita 2012)	CNV (MarStr56)
MA	Craniobasalis	*M. glutinosa*	Marinelli and Strenger (1956); palatolingualis profundus (Miyashita 2012)	CNV2 anterior branch (Miya12)
MA	Palatinalis lateralis	*M. glutinosa*	Marinelli and Strenger (1956)	
MA	Palatocoronarius	*M. glutinosa*	Marinelli and Strenger (1956); Miyashita (2012)	CNV2 anterior branch (Miya12)
MA	Retractor mucosae oris	*M. glutinosa*	Marinelli and Strenger (1956); palatolabialis + ? retractor dentalis lateralis (Miyashita 2012)	CNV2 posterior (Miya12)
MA	Longitudinalis linguae	*M. glutinosa*	Marinelli and Strenger (1956); retractor dentalis major (Miyashita 2012)	CNV2 posterior (Miya12)
MA	Protractor dentium profundus	*M. glutinosa*	Marinelli and Strenger (1956); protractor dentalis medialis (Miyashita 2012)	CNV2 posterior (Miya12)
MA	Protractor dentium superficialis	*M. glutinosa*	Marinelli and Strenger (1956); protractor dentalis lateralis (Miyashita 2012)	CNV2 posterior (Miya12)
MA	Tubulatus	*M. glutinosa*	Marinelli and Strenger (1956); retractor lingualis (Miyashita 2012)	CNV2 posterior (Miya12)
MA	Craniovelaris anterior dorsalis and ventralis	*M. glutinosa*	Marinelli and Strenger (1956); craniovelar anterior dorsalis and ventralis (Miyashita 2012)	CNV2 velar branch (Miya12)
MA	Spinovelaris	*M. glutinosa*	Marinelli and Strenger (1956); Miyashita (2012)	CNV2 velar branch (Miya12)
MA	Craniovelaris posterior	*M. glutinosa*	Marinelli and Strenger (1956); craniovelar posterior (Miyashita 2012)	CNV2 velar branch (Miya12)
MA	Perpendicularis	*M. glutinosa*	Marinelli and Strenger (1956)	CNV2 posterior (MarStr56, Miya12)
HA	Protractor cartilagines basalis anterior	*M. glutinosa*	Marinelli Strenger (1956); palatolingualis superficialis (Miyashita 2012)	CNVII (MarStr56, Miya12)
HA	Protractor cartilagines basalis posterior	*M. glutinosa*	Marinelli and Strenger (1956); palatolingualis profundus (Miyashita 2012)	CNVII (MarStr56, Miya12)
HA	Craniolingualis	*M. glutinosa*	Text: Marinelli and Strenger (1956); Miyashita (2012)	CNVII (MarStr56, Miya12)
HA	Craniohyoideus	*M. glutinosa*	Text: Marinelli and Strenger (1956); craniolingualis (figures: Marinelli and Strenger 1956); otic lingualis (Miyashita 2012)	CNVII (MarStr56, Miya12)
BA	Constrictor branchiarum	*M. glutinosa*	Constrictor branchiarum et cardiae (Marinelli and Strenger 1956); constrictor branchialis + constrictor cardialis + constrictor hepatis (Miyashita 2012)	CNIX (MarStr56)
BA	Constrictor pharyngis	*M. glutinosa*	Marinelli and Strenger (1956); Miyashita (2012)	CNX (MarStr56); CNX + XI (Miya12)

(*Continued*)

TABLE 5.1 (CONTINUED)
List of Commonly Used Synonyms for the Muscle Terminology Followed in the Present Chapter

Arch	Name Used in This Chapter	Taxa Studied by Us	Synonyms	Innervation
BA	Parietalis	*M. glutinosa*	Marinelli and Strenger (1956)	Spinal (MarStr54)
HyB	Obliquus	*M. glutinosa*	Marinelli and Strenger (1956); Miyashita (2012)	spinal (MarStr54; Miya12)
HyB	Decussatus (= ventral portion of obliquus)	*M. glutinosa*	Marinelli and Strenger (1956)	
HyB	Anterior part of rectus	*M. glutinosa*	Miyashita (2012); rectus abdominis (Marinelli and Strenger 1956)	Spinal (MarStr54; Miya12)
MA	Annularis	*P. marinus*	Marinelli and Strenger (1954); Tretjakoff (1926); Fürbringer (1875)	V2 ramus apicalis (MarStr54)
MA	Apicalis lateralis	*P. marinus*	Marinelli and Strenger (1954); Tretjakoff (1926); Fürbringer (1875)	V2 ramus apicalis (MarStr54)
MA	Copuloglossus rectus	*P. marinus*	Marinelli and Strenger (1954); Tretjakoff (1926); Fürbringer (1875)	V ramus mandibularis (MarStr54)
MA	Spinosocopularis	*P. marinus*	Marinelli and Strenger (1954); Tretjakoff (1926); Fürbringer (1875)	V2 ramus basilars (MarStr54)
MA	Tectospinosus anterior and posterior	*P. marinus*	Marinelli and Strenger (1954); tectospinosus posterior (Tretjakoff 1926); spinososemiannularis posterior (Fürbringer 1875)	V2 ramus basilars (MarStr54)
MA	Annuloglossus	*P. marinus*	Marinelli and Strenger (1954); Tretjakoff (1926); Fürbringer (1875)	V ramus mandibularis (MarStr54)
MA	Copuloglossus obliquus	*P. marinus*	Marinelli and Strenger (1954); Tretjakoff (1926); Fürbringer (1875)	V ramus mandibularis (MarStr54)
MA	Basilariglossus	*P. marinus*	Marinelli and Strenger (1954); Tretjakoff (1926); anterior portion of hyoglossus (Fürbringer 1875)	V ramus mandibularis (MarStr54)
MA	Constrictor cornualis superficialis	*P. marinus*	Marinelli and Strenger (1954); mandibularis inferior (Tretjakoff 1926); hyohyoideus anterior (Fürbringer 1875)	V ramus mandibularis (MarStr54)
MA	Tectolateralis	*P. marinus*	Marinelli and Strenger (1954); Tretjakoff (1926); semiannularis (Fürbringer 1875)	V2 ramus apicalis (MarStr54)
MA	Cornuotaenialis	*P. marinus*	Marinelli and Strenger (1954); mandibularibranchialis (Tretjakoff 1926); hyobranchialis (Fürbringer 1875)	V ramus mandibularis (MarStr54)
MA	Constrictor glossae profundus internus and externus	*P. marinus*	Constrictor glossae esternus and internus (Marinelli and Strenger 1954); mandibularis posterior (Tretjakoff 1926); hyohyoideus posterior (Fürbringer 1875)	V ramus mandibularis (MarStr54)
MA	Cornuoglossus	*P. marinus*	Marinelli and Strenger (1954); mandibulariglossus (Tretjakoff 1926); hyoglossus (Fürbringer 1875)	V ramus mandibularis (MarStr54)
MA	Cardioapicalis	*P. marinus*	Marinelli and Strenger (1954); Tretjakoff (1926); longitudinalis linguae (Fürbringer 1875)	V ramus mandibularis (MarStr54)
MA	Tendinoapicalis	*P. marinus*	Marinelli and Strenger (1954); Tretjakoff (1926); tendinoglossus (Fürbringer 1875)	V2 ramus apicalis (MarStr54)
MA	Styloapicalis	*P. marinus*	Marinelli and Strenger (1954); mandibulariapicalis (Tretjakoff 1926); hyomdb. glossus (Fürbringer 1875)	V ramus mandibularis (MarStr54)
MA	Stylotectalis	*P. marinus*	Marinelli and Strenger (1954); tectomandibularis (Tretjakoff 1926); spinososemiannularis anterior (Fürbringer 1875)	V ramus mandibularis (MarStr54)
MA	Levator valvulae velaris	*P. marinus*	Marinelli and Strenger (1954); velomandibularis descendens (Tretjakoff 1926); velo-hyomandibularis externus (Fürbringer 1875)	V ramus mandibularis (MarStr54)
MA	Protractor veli	*P. marinus*	Marinelli and Strenger (1954); velocranialis (Tretjakoff 1926); velopharyngeus (Fürbringer 1875)	V ramus velaris (MarStr54)
MA	Protractor oesophagi	*P. marinus*	Marinelli and Strenger (1954); Tretjakoff (1926)	V ramus velaris (MarStr54)

(Continued)

TABLE 5.1 (CONTINUED)
List of Commonly Used Synonyms for the Muscle Terminology Followed in the Present Chapter

Arch	Name Used in This Chapter	Taxa Studied by Us	Synonyms	Innervation
MA	Depressor veli	*P. marinus*	Marinelli and Strenger (1954); velomandibularis ascendens (Tretjakoff 1926); velohyomandibularis internus (Fürbringer 1875)	V ramus mandibularis (MarStr54)
MA	Basilaris	*P. marinus*	Marinelli and Strenger (1954); Tretjakoff (1926); Fürbringer (1875)	V2 ramus basilars (MarStr54)
MA	Pharyngicus posterior	*P. marinus*	Marinelli and Strenger (1954); pharyngeus posterior (Tretjakoff 1926; Fürbringer 1875)	V ramus velaris (MarStr54)
BA	Constrictores branchiales externi	*P. marinus*	Marinelli and Strenger (1954); Tretjakoff (1926); Johanson (2003)	X (MarStr54)
BA	Constrictor branchiales interni	*P. marinus*	Marinelli and Strenger (1954); internal dorsal and ventral diagonal constrictors (Johanson 2003)	X (MarStr54)
BA	Adductor branchiales dorsales and ventrales	*P. marinus*	Marinelli and Strenger (1954); median muscle bands (Johanson 2003)	X (MarStr54)
BA	Sphincters branchiales externi anteriores and posteriors	*P. marinus*	sphincter branchialis externus, internus (Marinelli and Strenger 1954; Tretjakoff 1926)	X (MarStr54)
BA	Interbranchiales	*P. marinus*	Marinelli and Strenger (1954); Johanson (2003); constrictor int. (Luther 1938)	X (MarStr54)
BA	Compressores bursae circulares et obliqui	*P. marinus*	Marinelli and Strenger (1954); isolated muscle fibers associated with the interbranchial septum (Johanson 2003)	X (MarStr54)
BA	Parietalis and epibranchialis	*P. marinus*	Marinelli and Strenger (1954, 1956)	Spinal (MarStr54)
BA	Probranchialis	*P. marinus*	Marinelli and Strenger (1954); praebranchialis (Tretjakoff 1926; Fürbringer 1875)	Nervus spinooccipitalis radix ventralis 2 (MarStr54)
BA	Supraocularis	*P. marinus*	Marinelli and Strenger (1954); supraocularis (Tretjakoff 1926)	Nervus spinooccipitalis radix ventralis 2 (MarStr54)
HyB	Subocularis	*P. marinus*	Tretjakoff (1926); parietalis p. subocularis (Marinelli and Strenger 1954)	Nervus spinooccipitalis radix ventralis 2 (MarStr54)
BA	Cornealis	*P. marinus*	Marinelli and Strenger (1954); Tretjakoff (1926)	N. spinoccipitalis 2 (own observation); V + occiptal (MarStr54)
HyB	Hypobranchialis	*P. marinus*	Marinelli and Strenger (1954)	Spinal (MarStr54)
MA	Intermandibularis anterior	Gnathostomata	Edgeworth (1935); labialis posterior (Vetter 1878); anterior portion of constrictor superior ventralis 1 (Marion 1905); protractor superior labii inferioris (Kesteven 1933)	CNV3
MA	Intermandibularis posterior	Gnathostomata	Edgeworth (1935); depressor mandibularis = posterior portion of constrictor superior ventralis 1 (Marion 1905); protractor inferior labii inferioris (Kesteven 1933)	CNV3 + CNVII
MA	Intermandibularis	Gnathostomata	Miyake et al. (1992); Anderson (2008); labialis posterior (Didier 1987; Mallatt 1996); 1st ventral constrictor (Kardong and Zalisko 1998); constrtictor superior ventralis 1 (Marion 1905)	CNV3
MA	Adductor mandibulae A2	Gnathostomata	Adductor mandibulae (Vetter 1878; Adams 1919); adductor mandibularis (Marion 1905); pars inscriptionalis, musculus quadrato-mandibularis or cs1b′ (Lightoller 1939); adductor mandibulae medialis + lateralis (for Elasmobranchii: Edgeworth 1935); levator mandibulae anterior + posterior (for Holocephali: Edgeworth 1935; Mallatt 1996); adductor mandibulae anterior and posterior head (Kesteven 1933); adductor mandibulae anterior + posterior (Didier 1987, 1995; Anderson 2008); 1st dorsal constrictor (Kardong and Zalisko 1998)	CNV2,3; V3 (Edgeworth 1935)

(Continued)

TABLE 5.1 (CONTINUED)
List of Commonly Used Synonyms for the Muscle Terminology Followed in the Present Chapter

Arch	Name Used in This Chapter	Taxa Studied by Us	Synonyms	Innervation
MA	Preorbitalis	*L. erinacea, S. acanthias*	Kardong and Zalisko (1998); Kardong (2002); levatores labialis superioris 1-5 (Marion 1905); suborbitalis (Edgeworth 1935: = lev. lab. superioris 1 of Marion 1905); ethmoideo-paraethmoidalis (Edgeworth 1935: = lev. lab. superioris 3 of Marion 1905); levator labii superioris (Soares and Carvalho 2013a,b)	CNV2,3; V2 (Edgeworth 1935; Mallatt 1996)
MA	Prelabialis	*H. colliei*	Levator cartilago prelabialis (Didier 1987); levator anguli oris anterior pars rostralis (Didier 1995); levator cartilaginis prelabialis (for Holocephali: Edgeworth 1935); levator rostri (Kesteven 1933); levator anguli oris pars rostralis (Raikow and Swierczewski 1975); levator prelabialis (Mallatt 1996)	CN2 (Edgeworth 1935; Mallatt 1996)
MA	Levator anguli oris anterior	*H. colliei*	Edgeworth (1935); Didier (1995); Mallatt (1996, 2008); levator anguli oris 1 (Vetter 1878); levator labii superioris (Kesteven 1933); levator anguli oris pars rostralis (Raikow and Swierczewski 1975)	CNV2,3; V2 (Edgeworth 1935; Mallatt 1996)
MA	Preorbitalis	*H. colliei*	Levator anguli oris posterior (Edgeworth 1935; Didier 1995; Mallatt 1996); levator anguli oris 2 (Vetter 1878); levator labii inferioris (Kesteven 1933)	CNV2,3; V2+3 (Edgeworth 1935)
MA	Labialis	*H. colliei*	Edgeworth (1935); labialis anterior (Vetter 1878; Didier 1987, 1995); protractor labii superioris (Kesteven 1933)	CNV2,3; V2 (Edgeworth 1935; Mallatt 1996)
MA	Levator arcus palatini	*L. erinacea*	Levatores maxillae superioris et inferioris (Marion 1905); levator maxillae superioris (Adams 1919) levator palatoquadrati + constrictor dorslais of levator palatoquadrati (Edgeworth 1935); levator palatoquadrati (Kardong 2002; Kardong and Zalisko 1998; Miyake et al. 1992, Anderson 2008)	CN2,3
MA	Spiracularis	*L. erinacea, S. acanthias*	Edgeworth (1935); constrictor superioris dorsalis 1 (Marion 1905; Adams 1919); 1st dorsal constrictor (Kardong and Zalisko 1998)	
HA	Interhyoideus	Gnathostomata	Posterior part of constrictor superioris ventralis II (Marion 1905); posterior part of constrictor hyoideus ventralis (Edgeworth 1935); hyoideus inferior (Vetter 1878); constrictor opcerculi ventralis (Didier 1987, 1995; Anderson 2008); geniohyoideus (Kesteven 1933); hyoidean interbranchial and superficial constrictor muscle (Mallatt 1997)	CNVII
HA	Mandibulohyoideus	*H. colliei*	Anderson (2008); hyoideus inferior (Didier 1987); interhyoideus (Didier 1995)	CNVII (CNVII + IX, Anderson 2008)
HA	Depressor hyomandibulae	*L. erinacea*	Edgeworth (1935); depressor hyomandibularis = anterior part of constrictor superioris ventralis II (Marion 1905);	CNVII
HA	Depressor rostri	*L. erinacea*	Edgeworth (1935); Marion (1905)	CNVII
HA	Levator rostri	*L. erinacea*	Edgeworth (1935); Marion (1905)	CNVII

(Continued)

TABLE 5.1 (CONTINUED)
List of Commonly Used Synonyms for the Muscle Terminology Followed in the Present Chapter

Arch	Name Used in This Chapter	Taxa Studied by Us	Synonyms	Innervation
HA	Constrictor hyoideus dorsalis	Gnathostomata	Edgeworth (1935); posterior part of levator hyomandibularis = posterior part of constrictor superioris dorsalis 2 (Marion 1905); constrictor superficialis (Vetter 1878); levator operculi (Kesteven 1933); constrictor operculi dorsalis (Anderson 2008), constrictor dorsalis (Didier 1987); second dorsal constrictor (Kardong and Zalisko 1998)	CNVII
HA	Adductor arcus palatini	Gnathostomata	Levator hyomandibulae (Edgeworth 1935; Kardong and Zalisko 1998); anterior part of levator hyomandibularis = anterior part of constrictor superioris dorsalis 2 (Marion 1905); hyoideus superior (Vetter 1878; Didier 1987); levator hyoideus (Didier 1995); levator hyomandibulae and epihyoidean (Kardong 2002)	CNVII
BA	Cucullaris	Gnathostomata	Edgeworth (1935); Anderson (2008); trapezius superficialis + cucullaris profundus (Vetter 1878); trapezius (Marion 1905; Allis 1923); trapezius superficialis and profundus (Didier 1987); cucullaris + levator arcuum branchialium posterior (Kesteven 1933); levator scapulae = levator 8 + third to seventh levators (Lightoller 1939)	CNIX+X (Kardong and Zalisko 1998)
BA	Constrictores branchiales superficiales ventrales/dorsales	*L. erinacea, S. acanthias*	Interbranchiales (Vetter 1878); constrictores superiores dorsales/ventrales 3-7 (Marion 1905); constrictores superficiales branchiales I-V (Edgeworth 1935); ventral constrictor hyoideus = constrictor colli (Kardong 2002); hyoidean superficial constrictor (Mallatt 1997); constrictor superior dorsalis and ventralis 3-6 (Kardong and Zalisko 1998)	CNIX+X (Didier 1995)
BA	Subspinalis	*(H. colliei), S. acanthias*	Supbsinalis = medial interarcualis 1 (Marion 1905)	Spinal nerve (Didier 1995)
BA	Interpharyngobranchiales I-III	Chondrichthyes	Medial interarcuales 2-4 (Marion 1905); interpharyngobranchial (Kardong 2002)	CNX, XI, XII (Kardong 2002); XII (our observation); spinal (Edgeworth 1935)
BA	Interarcuales laterales I-IV/V	*L. erinacea, S. acanthias*	Lateral interarcuales 1-4/5 (Marion 1905; Kardong and Zalisko 1998, Kardong 2002); arcuales dorsales (Edgeworth 1935)	CNX (Kardong and Zalisko 1998)
BA	Interbranchiales I-IV/V	*L. erinacea, S. acanthias*	Marion (1905); Edgeworth (1935)	CNX, XI (Kardong 2002)
BA	Adductores branchiales	Gnathostomata	Adductores arcus branchialis (Marion 1905); adductores arcuum branchialium (Vetter 1878; Edgeworth 1935; Didier 1995)	
BA/HyB	Coracobranchiales	Chondrichthyes	Vetter (1878); Kesteven (1933); Didier (1987, 1995); Gudo and Hornberger (2002); coraco-branchiales (Marion 1905; Edgeworth 1935)	Spinal nerves (Edgeworth 1935; Didier 1995)
HyB	Coracomandibularis	Chondrichthyes	Vetter (1878); Kesteven (1933); Didier (1995); coraco-mandibularis (Marion 1905; Kardong and Zalisko 1998; Kardong 2002); genio-coracoideus (Edgeworth 1935); branchiomandibularis	Spinal nerve (Didier 1995); CNX, XI (Kardong 2002)
HyB	Sternohyoideus	Gnathostomata	Coracohyoideus (Didier 1995; Anderson 2008); coraco-hyoideus (Marion 1905, Kardong and Zalisko 1998; Kardong 2002); rectus cervicis (Edgeworth 1935)	Spinal nerve (Didier 1995); CNX, XI (Kardong 2002)

(Continued)

TABLE 5.1 (CONTINUED)
List of Commonly Used Synonyms for the Muscle Terminology Followed in the Present Chapter

Arch	Name Used in This Chapter	Taxa Studied by Us	Synonyms	Innervation
HyB	Coarcoarcualis	*L. erinacea, S. acanthias*	Kardong and Zalisko (1998); Kardong (2002); coraco-arcuales communis (Marion 1905)	CNX, XI (Kardong 2002)
HyB	Coracohyomandibularis	*L. erinacea*	Coraco-hyomandibularis (Marion 1905; Edgeworth 1935)	

Note: Citations without muscles names in front use the same terminology as used in this chapter. A species name in round brackets means that we have not found this muscle in our specimens of that species, but that the muscle was reported to be present in that species by other authors. BA, branchial muscle; CN, cranial nerve; HA, hyoid muscle; HyB, hypobranchial muscles; MA, mandibular arch muscle; MarStr54, MarStr56, Marinelli and Strenger (1954, 1956); Miya, Miyashita 2012.

feeding in hagfishes (Clark and Summers 2007; Clark et al. 2010) provide kinematic data on the muscles of the lingual apparatus. However, these studies are limited in taxonomic sampling, leaving uncertainties about phylogenetic congruence of characters.

We also provide first-hand anatomical data for the cephalic muscles of chondrichthyans. Chondrichthyans have long been used as a model for basal conditions of the vertebrate anatomy (Balfour 1878; Goodrich 1930), because the musculature reconstructed in placoderms (stem gnathostomes) is somewhat similar to that of chondrichthyans (Johanson 2003; Trinajstic et al. 2013) and because various authors have compared their muscles with those of osteichthyans (e.g., Anderson 2008; Diogo and Abdala 2010). Post-Edgeworth (1935) primary literature on chondrichthyan cephalic musculature includes Lazier (1945), Didier (1987), Miyake et al. (1992), Mallatt (1997), González-Isáis and Domínguez (2004), Soares and Carvalho (2013a,b), among other works. We included osteichthyans in the comparative analysis to avoid erroneously treating chondrichthyans as a general gnathostome model. It has become increasingly clear that neither chondrichthyans nor osteichthyans serve as a model for the primitive, general gnathostome pattern. For example, the anatomy of the placoderm *Entelognathus* suggests that the gnathostome LCA had osteichthyan-like marginal jaw bones that are absent in chondrichthyans (Zhu et al. 2013). Therefore, congruence between chondrichthyans and osteichthyans is necessary to establish any synapomorphic conditions for gnathostomes.

As explained in the preceding chapters, the tree topology of Figure 1.1 represents the phylogenetic framework that we use in testing homology and inferring ancestral states. Although morphological, physiological, and paleontological studies have traditionally supported paraphyly of cyclostomes (e.g., Donoghue and Smith 2001; Gess et al. 2006; Near 2009; Miyashita 2012 and citations within), molecular data supporting the cyclostome monophyly have increasingly gained popularity (e.g., Stock and Whitt 1992; Lipscomb et al. 1998; Kuraku et al. 1999; Delarbre et al. 2002; Furlong and Holland 2002; Takezaki et al. 2003; Blair and Hedges 2005; Delsuc et al. 2006; Kuraku and Kuratani 2006; Mallatt and Winchel 2007; Winchel 2009; Heimberg et al. 2010). Developmental studies appear to reinforce the latter hypothesis. In particular, the embryonic development of the hagfish *Eptatretus burgeri* has revealed that developmental features previously considered unique to hagfishes are either experimental artifacts or misinterpretations. Hagfishes and lampreys both exhibit neural crest development comparable to that of gnathostomes, arcualia-like elements derived from sclerotomes, a nasohypophyseal process that is exclusive to cyclostomes and absent in gnathostomes, and other characters typically reduced or less specialized in ontogeny such as lenses, a spleen, and a lateral line system (e.g., Wicht and Northcutt 1995; Ota et al. 2007, 2011; Kuratani and Ota 2008a,b; Oisi et al. 2013b). Therefore, we accept that cyclostomes diverged from gnathostomes early in vertebrate evolution, about 430–520 million years ago (e.g., Kuratani et al. 2002; Kuraku and Kuratani 2006; Sugahara et al. 2013).

Under cyclostome monophyly, five types of phylogenetic distribution of similar muscles are expected: (a) features present both in cyclostomes and gnathostomes likely represent vertebrate synapomorphies (present in the vertebrate LCA); (b) features inferred to be present in the gnathostome LCA (based on comparison among gnathostomes) but absent in cyclostomes may represent gnathostome synapomorphies; (c) features inferred to be present in the cyclostome LCA but absent in gnathostomes may represent cyclostome synapomorphies; (d) features present in two of the three lineages (hagfishes, lampreys, and gnathostomes) may either represent reversals or secondary apomorphies; i.e., they may either have arisen in the vertebrate LCA (subsequently lost in one lineage) or independently evolved in two of the lineages; and (e) features present in only one lineage likely represent specialization within the lineage. Tables 5.2 through 5.5 list the muscles of the adult representatives of the nonosteichthyan extant vertebrate taxa we dissected for this chapter (NB: in these tables and the main text, our descriptions of selachians are based on *Squalus acanthias*, and not on *Mustelus laevis*; any significant differences between the muscles of these two species will be mentioned in the text below). For clarity, we describe each taxon separately in the following.

TABLE 5.2
Adult Mandibular Muscles of Representatives of Various Vertebrate Taxa

LCA of Extant Vertebrata	LCA of Extant Cyclostomata	Myxiniformes: *M. glutinosa* [24 Muscles]	Petromyzontiformes: *P. marinus* [26 Muscles]	LCA of Extant Gnathostomata	LCA of Extant Chondrichthyes	Holocephali: *H. colliei* [7 muscles]	LCA of Extant Elasmobranchii	Squaliformes: *S. acanthias* [5 Muscles]	Rajiformes: *L. erinacea* [6 muscles]	LCA of Extant Osteichthyes
Undifferentiated intermandibularis muscle sheet	Undifferentiated intermandibularis muscle sheet	Tubulatus	Constrictor cornualis superficialis, constrictor glossae profundus internus	Intermandibularis post. Intermandibularis ant.	Intermandibularis post. Intermandibularis ant.	Intermandibularis – (bpio Holocephali)	Intermandibularis post. Intermandibularis ant.	Intermandibularis –	Intermandibularis post. Intermandibularis ant.	Intermandibularis post. Intermandibularis ant.
Mandibular muscles other than the intermandibularis [e.g., at least one labial muscle and, probably, also at least one velar and/or dorsal mandibular muscle, if one accepts the homology between the group including the velothyroideus of lamprey larvae (ammocoetes) (i.e. the velar muscles) and the group including the levator arcus palatini of gnathostomes (i.e., derived from the constrictor dorsalis anlage, which also gives rise to the spiracularis of elasmobranches—and of *Polypterus?*); see text]	–	–	–	Add. mand. A2	Add. mand. A2	Add. mand. A2	Add. mand. A2	Add. mand. A2	Add. mand. A2	Add. mand. A2, Ad. mand. Aω, Add. mand. A3′, Add. mand. A3″
	Labial muscles [at least 1]	Palatocoronarius, retractor mucosae oris	Pharyngicus posterior (levator valvulae velaris)	Labial muscles [probably 1: preorbitalis]	Labial muscles [probably 1: preorbitalis]	Labial muscles [4: levator anguli oris anterior, preorbitalis, levator prelabialis, labialis anterior]	Labial muscles [probably 1: preorbitalis]	Labial muscles [1: preorbitalis]	Labial muscles [1: preorbitalis]	– [labial muscles present? see text]
	Nasal muscles [at least 1]	Transversus oris, tentaculosubnasalis	Tectospinosus anterior (and posterior), (tectolateralis)	–	–	–	–	–	–	–
		Tentacularis posterior	Basilaris	–	–	–	–	–	–	–
		Nasalis	(Basilaris), annularis	–	–	–	–	–	–	–
		Subnasonasalis, palatosubnasalis, cornuosubnasalis	Stylotectalis, tectolateralis	–	–	–	–	–	–	–
		Subnasobasalis	Tectospinosus posterior	–	–	–	–	–	–	–
	Lingual and dental muscles [at least 1]	Levator cartilagines basalis	Basilariglossus	–	–	–	–	–	–	–
		Craniobasalis	(Basilariglossus), cornuoglossus	–	–	–	–	–	–	–
		Longitudinalis linguae	Cardioapicalis	–	–	–	–	–	–	–
		Protractor dentium profundus	Annuloglossus	–	–	–	–	–	–	–
		Protractor dentium superficialis	Copuloglossus rectus	–	–	–	–	–	–	–

(Continued)

TABLE 5.2 (CONTINUED)
Adult Mandibular Muscles of Representatives of Various Vertebrate Taxa

LCA of Extant Vertebrata	LCA of Extant Cyclostomata	Myxiniformes: *M. glutinosa* [24 Muscles]	Petromyzontiformes: *P. marinus* [26 Muscles]	LCA of Extant Gnathostomata	LCA of Extant Chondrichthyes	Holocephali: *H. colliei* [7 muscles]	LCA of Extant Elasmobranchii	Squaliformes: *S. acanthias* [5 Muscles]	Rajiformes: *L. erinacea* [6 muscles]	LCA of Extant Osteichthyes
	Velar muscles [at least 1]	Craniovelaris anterior dorsalis and ventralis	(Protractor veli), levator valvulae velaris	Lev. arcus palatini (+ spiracularis? see text)	Lev. arcus palatini (+ spiracularis? see text)	–	Lev. arcus palatini, spiracularis	Lev. arcus palatini, spiracularis	Lev. arcus palatini, spiracularis	Lev. arcus palatini (+ spiracularis? see text)
		Spinovelaris	Protractor veli							
		Craniovelaris posterior	Depressor veli							
	–	Basitentacularis, coronarius, palatinalis lateralis, perpendicularis	–	–	–	–	–	–	–	–
	–	–	Apicalis lateralis, styloapicalis, spinosocopularis, constrictor glossae profundus externus, copuloglossus obliquus, Tendinoapicalis, protractor oesophagi, vornuotaenalis, pharyngicus anterior	–	–	–	–	–	–	–

Note: The hypothesis of homology in the cranial muscles between *Myxine* and *Petromyzon* is not always unambiguous. Therefore, some muscles of *Petromyzon* appear twice. The hypothesis that is less likely is indicated by round brackets around the muscle. Numbers inside square brackets correspond to the number of muscles present in the respective group of muscles (e.g. *H. colliei* has four labial muscles). Add. mand., adductor mandibulae; ant., anterior; bpio, but present in other; LCA, last common ancestor; post., posterior; lev., levator.
For more details, see text.

TABLE 5.3
Adult Hyoid Muscles of Representatives of Various Vertebrate Taxa

LCA of Extant Vertebrata	LCA of Extant Cyclostomata	Myxiniformes: *M. glutinosa* [4 Muscles]	Petromyzontiformes: *P. marinus* [0 Muscles]	LCA of Extant Gnathostomata	LCA of Extant Chondrichthyes	Holocephali: *H. colliei* (4 Muscles)	LCA of Extant Elasmobranchii	Squaliformes: *S. acanthias* [3 Muscles]	Rajiformes: *L. erinacea* [6 muscles]	LCA of extant Osteichthyes
–	–	–	–	Interhyoideus	Interhyoideus	Interhyoideus, mandibulohyoideus	Interhyoideus	Interhyoideus	Interhyoideus, depressor hyomandibulae, depressor rostri	Interhyoideus
–	–	–	–	Adductor arcus palatini	Adductor arcus palatini	Adductor arcus palatini	Adductor arcus palatini	Adductor arcus palatini	Adductor arcus palatini	Adductor arcus palatini
Constrictores hyoidei [at least 1]	Constrictores hyoidei [at least 1]	Craniolingualis, craniohyoideus	–	Constrictor hyoideus dorsalis	Constrictor hyoideus dorsalis	Constrictor hyoideus dorsalis	Constrictor hyoideus dorsalis	Constrictor hyoideus dorsalis	Constrictor hyoideus dorsalis, levator rostri	Constrictor hyoideus dorsalis [often designated adductor operculi]
–	–	Protractor cartilagines basalis anterior and posterior	–	–	–	–	–	–	–	–

Note: For more details, see text.

TABLE 5.4
Adult Branchial Muscles of Representatives of Various Vertebrate Taxa

LCA of Extant Vertebrata	LCA of Extant Cyclostomata	Myxiniformes: *M. glutinosa* [3 Muscles]	Petromyzontiformes: *P. marinus* [76 Muscles]	LCA of Extant Gnathostomata	LCA of Extant Chondrichthyes	Holocephali: *H. colliei* [16 Muscles]	LCA of Extant Elasmobranchii	Squaliformes: *S. acanthias* [27 Muscles]	Rajiformes: *L. erinacea* [28 Muscles]	LCA of Extant Osteichthyes
–	–	–	–	Cucullaris [could it be derived from epibranchial, and not from true branchial, musculature?] (see caption and text)	Cucullaris	Cucullaris	Cucullaris	Cucullaris	Cucullaris	Protractor pectoralis, levatores arcuum branchialium [5]
Constrictores branchiales [1 or more] [and adductores branchiales and/or interbranchiales? see text)	Constrictores branchiales [1 or more] [and adductores branchiales and/or interbranchiales? see text)	Constrictor branchiarum, constrictor pharyngis	Adductores branchiales dorsales and ventrales [7 + 7], constrictores branchiales externi and interni [7 + 7], Interbranchiales [7], compressores bursae branchiales circulares et obliqui dorsales et ventrales [7 + 7 + 7], sphincters branchiales externi anteriores and posteriores [7 + 7]	Constrictores branchiales [5] [and adductores branchiales and/or interbranchiales? see text]	Adductores branchiales [4 or more], constrictores branchiales [4 or more], interbranchiales [4 or more]	Adductores branchiales [4], constrictores branchiales [4; note: some other holocephalans also have interbranchiales] (protractor pectoralis dorsalis: see text)	Adductores branchiales [5 or more], constrictores branchiales [4 or more], interarcuales laterales [4 or more], interbranchiales [4 or more]	Adductores branchiales [5], constrictores branchiales [4], interarcuales laterales [4], Interbranchiales [4]	Adductores branchiales [6], constrictores branchiales [5], interarcuales laterales [5], interbranchiales [5]	Constrictores branchiales [5]
–	–	–	–	–	–	–	–	–	–	Subarcuales recti [5; often designated recti/ interarcuales ventrales, include rectus communis], transversi ventrales [5; include obliquii ventrales]
–	–	–	–	Coracobranchiales [5 or more]	Coracobranchiales [5 or more]	Coracobranchiales [5]	Coracobranchiales [5 or more]	Coracobranchiales [5]	Coracobranchiales [6]	Coracobranchiales [5]
–	–	–	–	–	–	–	–	–	–	Constrictor laryngis [*Polypterus*, dipnoans]
Undifferentiated epibranchial muscle sheet	Parietalis [anterior part]	Parietalis [anterior part]	Parietalis [anterior part], supraocularis, subocularis, cornealis, probranchialis, epibranchialis	Epibranchial muscles [1 or more]	Interpharyngo-branchiales [1 or more], subspinalis	Interpharyngo-branchiales [1], subspinalis	Interpharyngo-branchiales [3 or more; e.g., 4 in Notidanidae: e.g., Edgeworth 1935], Subspinalis	Interpharyngo-branchiales [3], subspinalis	–	–

Note: For more details, see the footnote of Table 5.2 and text. Based on detailed studies by others (e.g., Edgeworth 1935) and us (e.g., Ziermann and Diogo 2013, 2014) showing that the chondrichthyan cucullaris and the osteichthyan protractor pectoralis develop anatomically from the anlage of the true branchial musculature *sensu* Diogo and Abdala (2010), we consider here these muscles as true branchial muscles, but other authors (e.g., reviews by Sambasivan et al. [2011] and Miyashita [2012]) and in some way, our dissections suggest that they might derived instead/also from the epibranchial musculature (see text).

TABLE 5.5
Adult Hypobranchial Muscles of Representatives of Various Vertebrate Taxa

LCA of Extant Vertebrata	LCA of Extant Cyclostomata	Myxiniformes: *M. glutinosa* [3 Muscles]	Petromyzontiformes: *P. marinus* [1 Muscle]	LCA of Extant Gnathostomata	LCA of Extant Chondrichthyes	Holocephali: *H. colliei* [2 Muscles]	LCA of Extant Elasmobranchii	Squaliformes: *S. acanthias* [3 Muscles]	Rajiformes: *L. erinacea* [4 Muscles]	LCA of Extant Osteichthyes
Undifferentiated hypobranchial muscle sheet	Undifferentiated hypobranchial muscle sheet	Obliquus, decussatus, rectus (anterior part)	Hypobranchialis	Coracomandibularis Sternohyoideus	Coracomandibularis Sternohyoideus [coracoarcualis also present? see text]	Coracomandibularis Sternohyoideus	Coracomanidubularis Sternohyoideus, coracoarcualis	Coracomandiubularis Sternohyoideus, coracoarcualis	Coracomandibularis Sternohyoideus, coracoarcualis, coracohyomandibularis	Coracomanidubularis Sternohyoideus

Note: For more details, see text.

MYXINE GLUTINOSA: ATLANTIC HAGFISH

Concerning the mandibular muscles, the terminal nostril and the subterminal mouth are surrounded by four pairs of tentacles that are moved by muscles (Figure 5.1). The diameter of the nostril can be narrowed by the subnasonasalis (Figure 5.1D and F). The oral cavity has a transverse muscle (transversus oris) that leads to a broadening and flattening of the mouth upon contraction (Figure 5.1C, F, and G), which is involved in the "head depression" stage during feeding (Clark and Summers 2007). The head does not show the typical gnathostome regionalization, but numerous nerves can be found toward the tentacles and on the lateral part of the head (Figure 5.1C and D). A branch of the trigeminal nerve (CNV1—ramus ophthalmicus 1) passes below the skin at the origin of the tentacularis posterior. Below this muscle, another trigeminal branch (CNV2—rami tentaculares 1–4) emerges from below the cornual cartilage and extends anteriorly to split into numerous small branches. Below the nasalis lies yet another trigeminal branch (CNV2—ramus ophthalmicus 2). The nasal capsule provides attachment for three muscles that reach the top of the head. The most dorsal of the three (nasalis) overlaps the nasal capsule, with its origin at the perioptic membrane and the last two nasal arches, and inserts onto the dorsal nasal tube membrane, the second nasal arch, the paranasal tuber, and the paranasal rod (Figure 5.1A and F). The tentacularis posterior is lateroventral to the nasalis (Figure 5.1A) and originates anteriorly to the eye at the perioptic membrane and inserts onto the upper nasohypophyseal process, the tentacular cartilage, and the oral tentacular cartilage. Below the tentacularis posterior is the large palatosubnasalis (Figure 5.1B and F) that connects the rostral end of the subnasal cartilage to the nasal capsule and the parietalis (a trunk/epibranchial muscle). The palatosubnasalis originates from the lateral wall of the nasal capsule, the fascia of the underlying cornuosubnasalis, the commissure of the two palatinal cartilages, and the posterior third of the cornual cartilage. This muscle becomes smaller anteriorly and passes medially to insert onto the subnasal cartilage.

When one removes the nasalis, tentacularis posterior, and palatosubnasalis, numerous underlying mandibular muscles can be seen, which (a) are related to the tentacle apparatus, (b) suspend the ventral head skeleton, or (c) protract or retract the dental plates. The most anterior part of the head houses muscles with diverse fiber directions. Individual muscles are difficult to separate, but collectively extend between the cartilages of the tentacles (cartilagines tentaculi), the subnasal cartilage, and the anterior end of the arcus lingualis. The transversely oriented fiber group that includes some vertical fibers is the transversus oris (Figure 5.1F). Posteriorly, the tentaculosubnasalis (Figure 5.1B, C, and F) consists of dorsoventrally oriented fibers between the subnasal cartilage and the oral tentacle. Likewise, the basitentacularis (Figure 5.1D through F) occurs in this position but extends to the anterior edge of the arcus lingualis. The subnasobasalis (Figure 5.1) is only partially separated from this group of fibers, extending from the subnasal cartilage to the same insertion site with the transversus oris, tentaculosubnasalis, and basitentacularis.

The coronarius (Figure 5.1F and G) consists of a small group of fibers closely attached to the dorsal branch of the coronar cartilage (part of the tentacular cartilage; Figure 5.1B), whereas the cornuosubnasalis has two bundles with different fiber directions between the subnasal cartilage and the cornual cartilage (Figure 5.1C, F, and G). Between the lingual cartilages and the chondrocranium lie two muscles with fibers in a right angle to each other. The anterior one is the subnasobasalis (Figure 5.1D, F, and G). The posterior one (levator cartilagines basalis) originates from the cornual cartilage and the fascia above the craniobasalis and overlaps the subnasobasalis at its insertion (Figure 5.1F and G). The craniobasalis originates from the ventral fascia of the tentacularis posterior, the dorsolateral surface of the palatocoronarius, the perioptic membrane, and the parts of the palatal arch, and inserts to the anterolateral and middle lingual cartilages (Figure 5.1D through G). The dorsal deep muscles of the head originate from the lateral palatine commissure and pila anterior. The muscle fibers (palatinalis lateralis) extend anteroventrally in a tendon, inserting ventrally onto the cornual cartilage (Figure 5.1F and G). The palatocoronarius passes dorsal and lateral to it and inserts onto the coronar cartilage (Figure 5.1F and G). The retractor mucosae oris sits ventrolaterally to it, originating from the anterior border of the arcus lingualis and covering the buccal cavity externally (Figure 5.1F and G).

The next group of mandibular muscles moves the tooth plates and the lingual cartilages (arcus lingualis). It comprises the long longitudinalis linguae (Figure 5.1F and G) that inserts onto the posterior border of the tooth plate (cartilago dentifera) and a muscle with two band-like bundles (protractor dentium profundus) that extends from the posteroventral portion of the lingual apparatus to the anterior border of the tooth plates *via* a tendinous insertion (Figure 5.1E through G). The rectus (the anterior part of which seems to correspond to the hypobranchial musculature of other vertebrates) passes between the two bands. The perpendicularis originates from the cartilago musculi perpendicularis, which is an elongate cartilage at the posterior end of the lingual apparatus. The fibers of the perpendicularis are anterodorsally oriented toward the medial tendon of the longitudinalis linguae. The protractor dentium superficialis inserts *via* a band of connective tissue at the tooth plate (Figure 5.1D through F). The tubulatus is a long tubelike muscle that functions as a constrictor (Figure 5.1F and G). This muscle originates from the lingual arches, surrounds the longitudinalis linguae, and inserts onto the posterior part of the lingual apparatus. Its anterior fibers cover the cartilaginous arch of the metotic fenestra. The velar apparatus extends into the pharynx. The craniovelaris anterior dorsalis and ventralis serve as an attachment for the velar apparatus (Figure 5.1F). Its dorsal portion extends from the perichondrium of the chordal process, nasopharyngeal bar, nasal capsule, and palatal arch, whereas the ventral portion comes from the palatal arch. The anterior continuation of the craniovelaris anterior dorsalis and ventralis gives rise to the palatocoronarius (Figure 5.1F). The craniovelaris posterior

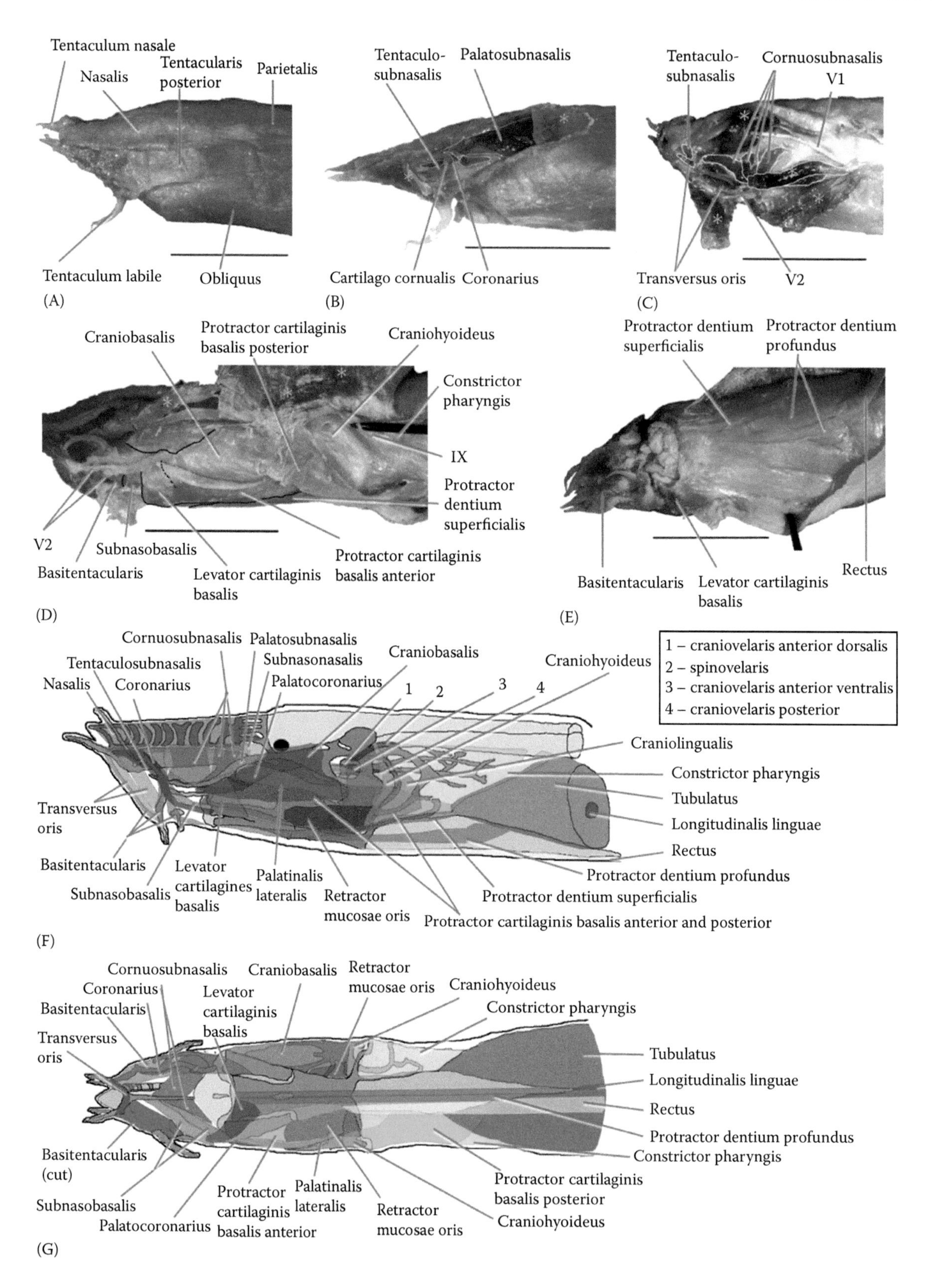

FIGURE 5.1 *M. glutinosa* (Vertebrata, Cyclostomata). (A–D, F) lateral view; (E, G) ventral view. (A–C) Nasal region; (C, D) muscles stained with iodide. (A) Superficial muscles after removing the skin; (B) second layer of muscles, tentacularis posterior reflected (*); (C) deep layer; tentacularis posterior, and palatosubnasalis reflected (*); (D) lateral view of the postnasal (suborbital and branchial) region. Layer with obliquus and decussatus, and tentacularis posterior reflected (*); (E) ventral view after reflecting the obliquus muscle; (F, G) Diagrams of muscles with nandibular (*red, orange, pink*); hyoid (*green*), true branchial (*blue*), epibranchial (*bruin*), and hypobranchial (*green*) muscles. Muscles redrawn after dissection. Chondrocranium of *M. glutinosa* redrawn from Marinelli and Strenger (1956). (F) Tentacularis posterior, parietalis, obliquus, and decussatus not shown; (G) obliquus and decussatus not shown, not all muscles shown on both sides. (A–E) scale bar = 1 cm.

originates from the chordal process and the lateral edge of the nasopharyngeal plate and inserts onto the lateral velar cartilages (Figure 5.1F). The spinovelaris originates from the most anterior region of the notochord and inserts posteromedially onto the velar cartilage (Figure 5.1F).

Concerning the hyoid muscles, the protractor cartilagines basalis anterior (Figure 5.1D, F, and G) is continuous with the levator cartilagines basalis (a mandibular muscle) and attached to the fascia of other muscles (the mandibular muscles tentacularis posterior and craniobasalis and the hyoid muscle craniolingualis) and the perioptic membrane. In its insertion, the muscle is anchored to the fascia of the posterior portion of the craniolingualis and the junction between the lingual cartilages, the middle lingual cartilage, and the first pharyngolingual arch (Figure 5.1D). The protractor cartilagines basalis posterior extends from the pila anterior (cartilage anterior to window/fenestra under otic capsule) to the posterior part of the lingual cartilage (Figure 5.1D, F, and G). The craniolingualis (Figure 5.1F) originates from the lateral edge of the craniobasalis (a mandibular muscle) and the medial surface of the parietalis (a trunk/epibranchial muscle) and inserts onto the fascia of the protractor cartilagines basalis anterior and the protractor dentium superficialis (a mandibular muscle). The craniohyoideus connects the otic capsules and the distal part of the first external arcus lingualis (Figure 5.1D, F, and G).

Regarding the branchial muscles, there is only one pair of external branchial openings in *M. glutinosa*; the constrictor branchiarum originates from the mesentery, ventromedially to the parietalis. The fibers extend ventrally, but not as a continuous sheet muscle, and inserts onto the connective tissue that surrounds the efferent branchial ducts, the surface of the heart, and the fascia of the branchial pouches and the perpendicularis (a mandibular muscle). The most anterior fibers (constrictor pharyngis) of the muscle originate from the lateral tissue of the chorda and the branchial region to the lingual apparatus and insert at the lateral borders of the posterior lingual cartilages (Figure 5.1D, F, and G).

The parietalis is a peculiar muscle because it occupies part of the trunk region and part of the epibranchial region (Figure 5.1A); its anterior portion appears to be corresponding to the epibranchial musculature of lampreys. Thus, the parietalis is treated under this section. The posterior bundles of the parietalis originate from its contralateral pair and from the membrane of the notochord. The anterior bundles of the parietalis (which include the portion that appears to correspond to the epibranchial musculature of lampreys) additionally originate from the spinal membrane, the perioptic membrane, the laterodorsal surface of the facial skeleton, the lateral surface of otic capsule, and the neurocranial membranes. The insertions of the parietalis are onto the neighboring bundles of the same muscle, the membrane over the lateral side of the mandibular muscle protractor dentium profundus (facial to fourth segments), and the dorsal surface of the slime glands. The anterior ventral portion of the parietalis extends anteriorly to approach the tentacle region. The parietalis does not reach the ventral midline, ending at the dorsal surface of the slime glands, and does not show horizontal septa.

With respect to the hypobranchial muscles, the obliquus passes over the lateroventral fascia of the epibranchial muscle parietalis (Figure 5.1A) and extends ventrally to inserts with its contralateral muscle onto a median raphe. Ventrally to, and continuous with, the obliquus is the decussatus. This muscle extends from the ventral border of the parietalis and the slime glands toward the midline where it fuses with its contralateral muscle. The anterior portion of the rectus clearly seems to correspond to the hypobranchial musculature of other vertebrates. The rectus is dorsal to the decussatus (Figure 5.1D, F, and G) and is connected posteriorly to the sphincter cloacae (a somatic muscle) and inserted between the protractor dentium profundus (a mandibular muscle) at the ventral surface of the middle lingual cartilage.

PETROMYZON MARINUS: SEA LAMPREY

Regarding the mandibular muscles, the annularis is a constrictor muscle that inserts onto the cartilago annularis (Figure 5.2A, B, and F). The apicalis lateralis is a small muscle lateral to the insertion of the tendinoapicalis onto the cartilago apicalis. The copuloglossus rectus extends from the cartilago pistoria to the cartilago copularis (Figure 5.2E and F). Its insertion is overlapped by the spinosocopularis (Figure 5.2B, E, and F), which originates at the posterior cartilago spinosa. This cartilage and the anterior cartilago spinosa are connected with the anterior and posterior cartilagines tectoria by the tectospinosus anterior and posterior (Figure 5.2A and E). The annuloglossus (Figure 5.2B, E, and F) originates laterally to the copuloglossus rectus from the cartilago pistoria and inserts *via* a long tendon onto the cartilago annularis. The copuloglossus obliquus sits dorsal to the copuloglossus rectus, originating from the median portion of the cartilago copularis and inserting *via* a long tendon onto the lobus superior (Figure 5.2E). The basilariglossus originates posteriorly to the copuloglossus rectus from the median portion of the cartilago copularis. Its fibers pass lateral to the basilaris muscle. The origin of the basilariglossus (Figure 5.2E) is covered by a vertical band that contains two distinct muscles. The anterior portion is the constrictor cornualis superficialis, which originates from the cartilago styliformis (base of the skull) and inserts with its contralateral muscle onto a median raphe (Figure 5.2F). The posterior portion is the cornuotaenialis, which originates from the planum cornuale (plate that attaches to the base of the skull) and inserts onto the taenia longitudinalis ventralis of the branchial basket (Figure 5.2E and F). The most anterior cartilaginous arch of the branchial basket (extrahyale) is close to the otic capsule. No pharyngeal pouch sits between this element and the following arch. The tectolateralis forms the lateral wall and the floor of the pharynx and spans between the cartilago tectoria and the cartilagines laterales anteriores (Figure 5.2E).

Dorsal to the taenia longitudinalis ventralis, a long muscle spindle is surrounded by the constrictor glossae profundus. This muscle has two parts, the medial (constrictor glossae

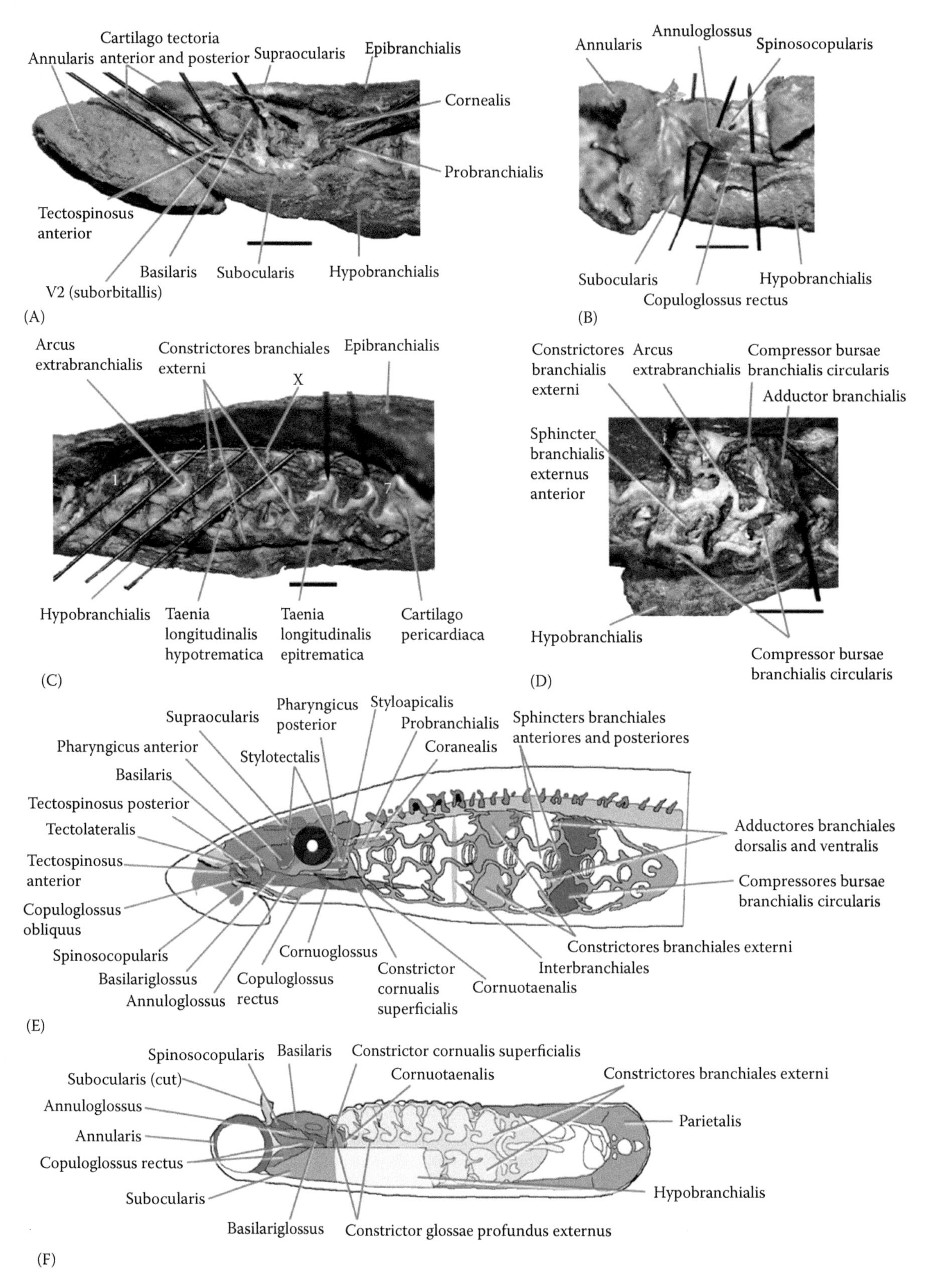

FIGURE 5.2 *P. marinus* (Vertebrata, Cyclostomata). (A, C, D) Left lateral views; (B) ventral view. (A) Lateral superficial muscles; (B) ventral superficial muscles, subocularis on left side reflected (*); (C) superficial branchial muscles, numbers indicate the first and last of segments 1–7 of the constrictores branchiales externi muscle sheet; (D) deep dissection of the branchial muscles. (E, F) Diagrams of muscles with mandibular (*red, orange, pink*); true branchial (*blue*), epibranchial (*bruin*), and hypobranchial (*green*) muscles. Muscles redrawn after dissection. (Chondrocranium of *P. marinus* redrawn from Marinelli, W., and Strenger, A., *Vergleichende Anatomie und Morphologie der Wirbeltiere: I Lieferung* Lampetra fluviatilis *(L)*, pp. 1–80, Deuticke, Vienna, 1954.) (F) Annularis, subocularis, epibranchialis, and hypobranchialis not shown. Not all muscles are shown on both sides. (A–D) Scale bar = 1 cm.

profundus internus) and the lateral (constrictor glossae profundus externus: Figure 5.2F), separated by the cornual plate. The constrictor glossus internus forms the floor of the pharynx from the dorsomedial tip of the cornual plates to the median raphe, where it meets its contralateral muscle. The constrictor glossus externus originates from the cornual plates and inserts with its contralateral pair onto the median raphe, which surrounds several ventral muscles (cornuoglossus, basilariglossus, copuloglossus rectus, and annuloglossus) and the median cartilago pistoria. The cornuoglossus sits in this region and connects between the processus cornualis and the cartilago pistora (Figure 5.2E). Further posteriorly, the cardioapicalis originates from the anterior part of the cartilago apicalis (cartilage at the anterior part of the pericardium) and anchors to a long tendon over the cartilago pistoria. The tendon splits and inserts onto the cartilago copularis and the cartilagines supraapicales. Posterior to this split, the tendinoapicalis extends between the tendon and the cartilago apicalis. The styloapicalis originates from the cartilago styliformis, passes between the splitting tendon of the cardioapicalis, and inserts onto the cartilago supraapicalis lateral to the tendinous insertion of the cardioapicalis (Figure 5.2E). The stylotectalis sits dorsal to the pharynx and inserts onto the cartilago tectoria anterior (Figure 5.2E).

The velar apparatus is framed by the cartilagines styliformes, extending ventrally from the skull. The velar cartilage has velar tentacles and two posterior velar flaps. It is elevated by the levator valvulae velaris, which originates from the dorsolateral border of the cartilago styliformis. The unpaired velar cartilage has several anterior tips and two posterior extensions. The protractor veli from the cartilagines styliformes inserts to each side of these extensions. The depressor veli extends between the cornual plates and the anterior part of the corpus of the velar cartilage. The protractor oesophagi extends from the basis cranii to the wall of the oesophagus. The basilaris (Figure 5.2A and E) is a large lateral muscle with fibers oriented in several different directions. The cartilago pistoria is pinched between the ventral portions of the right and left basilaris. The basilaris inserts onto the cartilagines tectoria anterior and posterior, the cartilago lateralis posterior, the arcus subocularis, and the cartilago pistoria just anterior to its anterior bifurcation. The pharyngicus anterior and posterior surrounds the oesophagus (Figure 5.2E).

We identified no muscle that is innervated by CNVII; therefore, the sea lampreys seem to lack hyoid muscles. Regarding the branchial muscles, the cartilaginous branchial basket is surrounded by an external muscle sheet (constrictores branchiales externi) divided into seven segments by the cartilages of the branchial arches and the branchial openings (Figure 5.2C and D). The horizontal cartilages immediately above and below the branchial openings are the taenia longitudinalis epitrematica and hypotrematica (Figure 5.2D), whereas the midventral element is the taenia longitudinalis ventralis. The fibers of the constrictores branchiales externi extend ventrally to the taenia longitudinalis ventralis (Figure 5.2C, E, and F). The medial portion of this muscle is below the taenia longitudinalis epi- and hypotrematica. The cartilaginous connection between the taenias lies behind the gill openings, each of which is surrounded by the sphincters branchiales externi anteriores and posteriores (Figure 5.2D and E). The superficially visible branchial cartilages extend medially, setting the gill pouches from each other. The compressores bursae circulares et obliqui are at the septum and the walls of the gill pouches (Figure 5.2D). The adductores branchiales dorsales and ventrales insert medially at the extrabranchial cartilages (Figure 5.2D and E). The interbranchiales sit by the septum medianum and the septum interbranchale of the branchial arches (Figure 5.2E). Each bursa branchialis has its muscular lateral wall (compressores bursae circulares) and muscles from the middle of the wall toward the dorsoventral corners of each chamber (compressores bursae branchiales obliqui dorsales/ventrales: Figure 5.2D and E). The dorsales/ventrales originates dorsally/ventrally from the medial portion of the arcus extrabranchialis (dorsally/ventrally to the branchial opening) and extend medially to this arch (Figure 5.2D). The constrictores branchiales interni cover the branchial chamber internally.

As for the epibranchial muscles, the part of the parietalis near the branchial openings is often referred to as epibranchialis (Figure 5.2A and C). The probranchialis is anteriorly continuous with the epibranchialis. This probranchialis sits between the anterior portion of the first branchial opening and the myoseptum of the first hypobranchial myomere (Figure 5.2A and E). The portion of the epibranchial musculature dorsal to the eyes is hereby named supraocularis (Figure 5.2E). This muscle inserts onto the cartilago tectoria posterior (Figure 5.2A). The cornealis is a small muscle just behind the eye and dorsal to the probranchialis (Figure 5.2E). This muscle is continuous with the other epibranchial muscles and attached onto the outer cornea of the eye (Figure 5.2A). The subocularis sits below the eye, meets the anterior portion of the hypobranchialis (a hypobranchial muscle), and inserts *via* a tendon onto the cartilago annualis (Figure 5.2A, B, and F). The position seems to be consistent with the subocularis being a hypobranchial muscle. However, developmental studies of lampreys clearly show that the subocularis represents a ventral extension of the epibranchial musculature, developing from the anlage that gives rise to the epibranchialis, cornealis, supraocularis, and probranchialis. Thus, the contact between the subocularis and hypobranchial muscles develops secondarily (e.g., Sambasivan et al. 2011; Miyashita 2012; Tulenko et al. 2013; see also the following).

Regarding the hypobranchial muscles, as noted earlier, the hypobranchialis is a segmented muscle that is continuous posteriorly with the parietalis (a trunk/epibranchial muscle) and anteriorly with the subocularis (an epibranchial muscle) (Figure 5.2A through D and F).

HYDROLAGUS COLLIEI: SPOTTED RATFISH

The intermandibularis is a mandibular muscle, which originates from the maxillary cartilage, bending ventrally and inserts to the infralabial tissue associated with Meckel's cartilage (NB: in some individuals of *Hydrolagus colliei* and other holocephalan taxa, the intermandibularis may have distinct anterior and posterior parts: e.g., Edgeworth 1935; Kesteven 1942–1945) (Figure 5.3). The adductor mandibulae A2 anterior has three portions originating respectively from the preorbital lamina, preorbital fascia, and suborbital fascia (Figure 5.3C). The adductor mandibulae A2 posterior is a distinct muscle that is deeper with respect to the adductor mandibulae A2 anterior (Figure 5.3A), extending from the suborbital ridge to the lower jaw. The preorbitalis (often named levator anguli oris posterior) extends from the lower edge of the orbit to the maxillary labial cartilage (Figure 5.3A). The levator anguli oris anterior arises from the antorbital crest (Figure 5.3A). The levator prelabialis is reduced to a few fibers in *H. colliei* and is blended with the levator anguli oris anterior (Figure 5.3A). The labialis anterior has a nearly vertical orientation and originates from the maxillary cartilage (Figure 5.3A). The levator anguli oris anterior, levator prelabialis, and labialis anterior insert onto the prelabial cartilage associated with the nasolabial fold. These three muscles are often designated "upper-lip" muscles. Although the intermandibularis and the adductor mandibulae A2 posterior are innervated by the mandibular branch of the trigeminal nerve (CNV3), the other five mandibular muscles receive mixed innervation from the maxillary and mandibular branches (CNV2 + V3). This is in part due to a fiber exchange near the orbit, mainly from fibers of V3 extending toward V2 (Figure 5.3B). At the region of the lower jaw, CNV3 runs in close proximity with the facial nerve (CNVII).

As for the hyoid muscles, the interhyoideus extends from the fibrous operculum (Figure 5.3A) to the ventral midline where it fuses with its contralateral muscle. The mandibulohyoideus connects the ceratohyal and Meckel's cartilage (Figure 5.3D). The constrictor hyoideus dorsalis originates from connective tissue ventrally along the notochord up to the anterior edge of the scapular process and inserts onto the connective tissue of the opercular flap (Figure 5.3A). The adductor arcus palatini originates from the postorbital ridge and attaches onto the hyoid rays/epihyal (Figure 5.3D). The fleshy operculum is connected dorsally to the constrictor hyoideus dorsalis and ventrally to the interhyoideus (Figure 5.3A).

With respect to the branchial muscles, the cucullaris has two heads (superficialis and profundus), which seem to be blended at their origins from the postorbital ridge. The cucullaris superficialis part inserts onto the external part of the

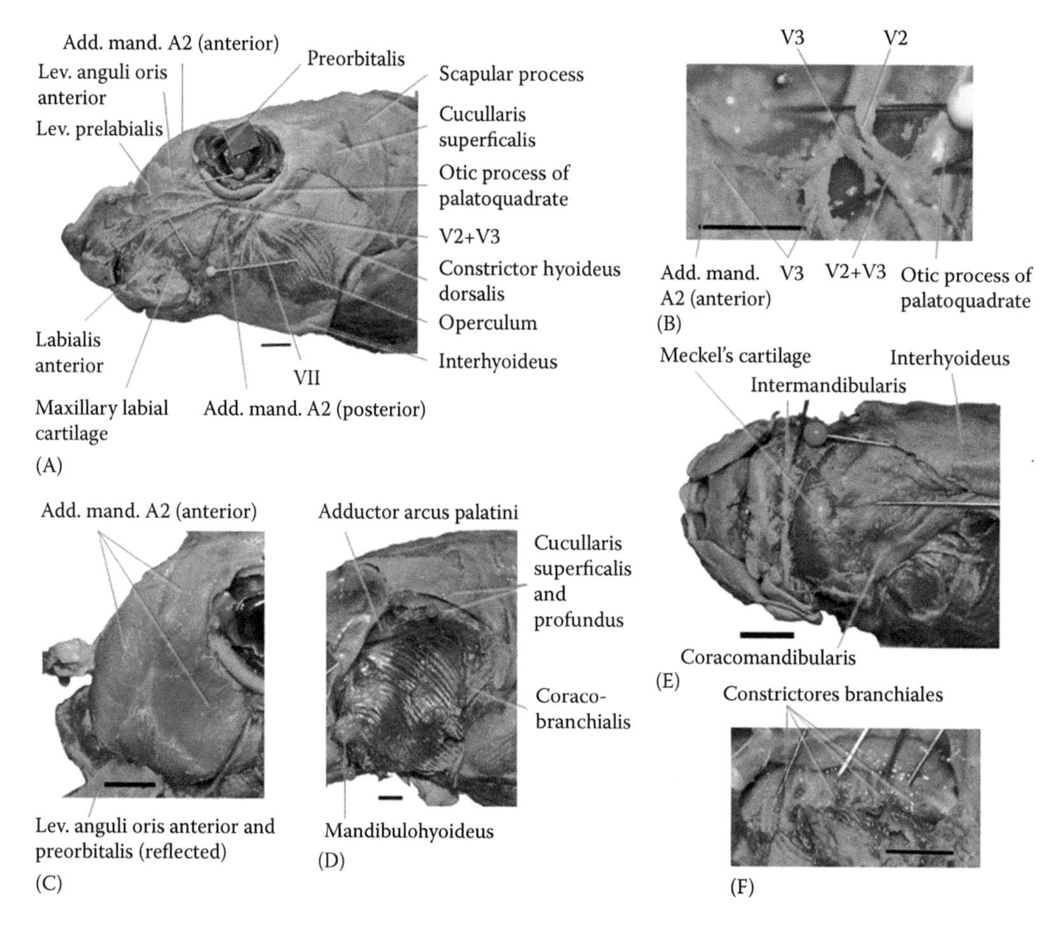

FIGURE 5.3 *H. colliei* (Vertebrata: Holocephali). (A–D, F) Left lateral view; (E) ventral view. Anterior is always to the left. (A) Superficial muscles and nerves after removal of the skin; (B) detail of the ventral part of the orbit to show the fiber exchanges between CNV2 and CNV3; (C) portions of the add. mand. A2 (anterior); (D) Muscles medial to the operculum, the constrictor dorsalis hyoideus, and the interhyoideus were reflected (*); (E) ventral view, on right side the interhyoideus was removed; (F) dorsal parts of the branchial arches. Abbreviations: add. mand., adductor mandibulae; lev., levator. Scale bar = 1 cm.

scapula. In two dissected specimens, we found an attachment of the cucullaris profundus onto the branchial arches (as described by Didier [1987] and Anderson [2008], see the following), but in the other specimens, this bundle inserted instead onto the medial surface of the scapula. The branchial muscles *sensu stricto* are (a) the four constrictores branchiales from epibranchials I–IV to the ceratobranchials of the same arches (Figure 5.3F); (b) the four adductores branchiales from epibranchials I–III and the pharyngobranchial complex of the branchial arches IV and V to ceratobranchials I–IV; and (c) the five coracobranchiales (Figure 5.3D) extending from the coracoid to the ventral region of the ceratobranchials. *H. colliei* and other holocephalan species usually have two epibranchial muscles; i.e., an interpharyngobranchialis and a subspinalis (Edgeworth 1935; see below).

Concerning the hypobranchial muscles, the coracomandibularis is a continuous sheet of muscle originating from the coracoid and fusing with its contralateral muscle in the ventral midline, on the region of the lower jaw (Figure 5.3F). Although the coracomandibularis has been illustrated with anterior and posterior portions in *H. colliei* (Didier 1995; Anderson 2008), this division is artificial (see also Anderson 2008). The sternohyoideus extends from the coracoid and/or the fascia of the coracomandibularis to the basihyal.

SQUALUS ACANTHIAS: SPINY DOGFISH

Regarding the mandibular muscles, the intermandibularis originates from Meckel's cartilage and inserts with its contralateral muscle in a median raphe (Figure 5.4B), being blended to the interhyoideus and constrictor hyoideus dorsalis (hyoid muscles). The adductor mandibulae A2 connects the palatoquadrate to Meckel's cartilage and is divided into superficialis, ventralis, and dorsalis parts (Figure 5.4A). The levator arcus palatini (Figure 5.4C) extends from the posterior edge of the otic capsule to the palatoquadrate in parallel with the spiracularis, which originates from the otic capsule (dorsally to the levator arcus palatini) and inserts onto the palatoquadrate (Figure 5.4C). The preorbitalis is the only labial muscle in *S. acanthias*, originating anterior to the palatoquadrate from the ethmoid region of the chondrocranium and inserting *via* a tendon onto the Meckel's cartilage medial to the insertion of the adductor mandibulae A2 ventralis (Figure 5.4A).

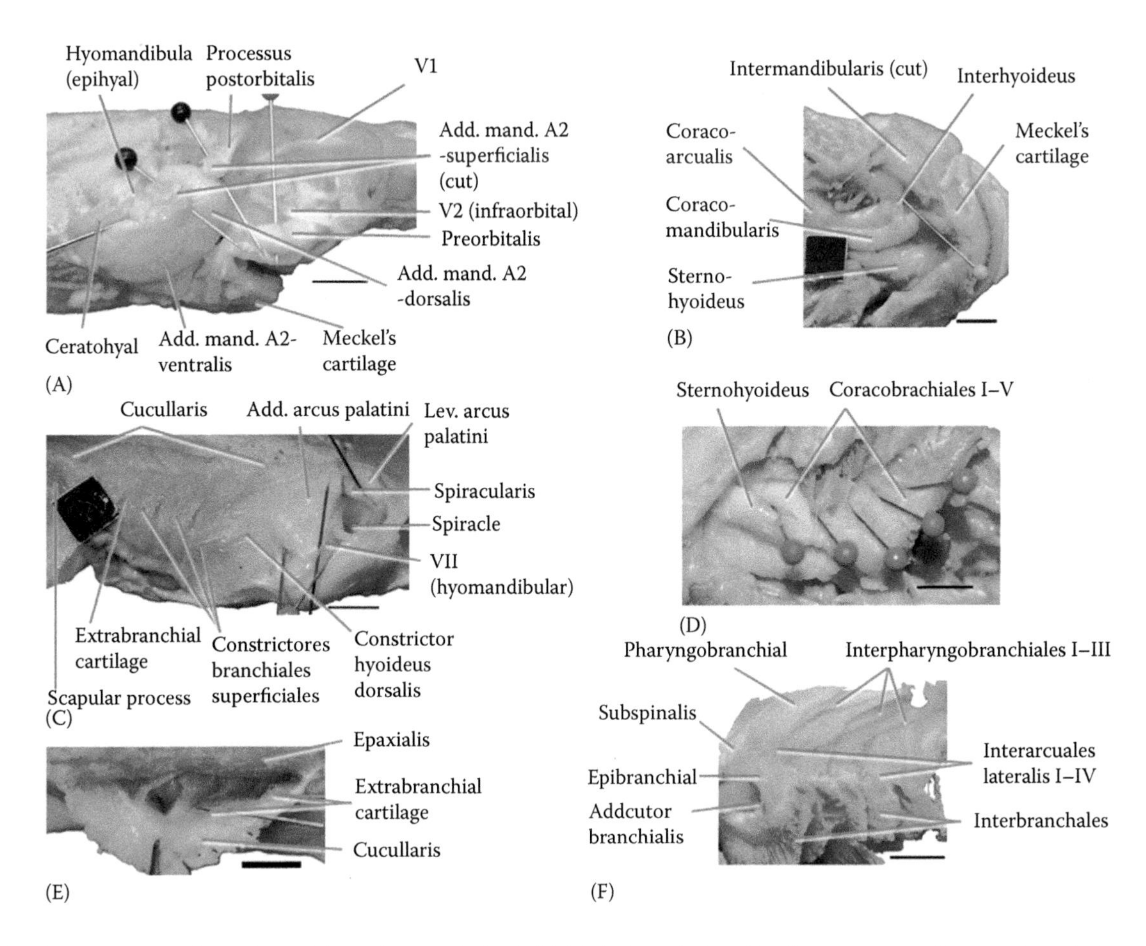

FIGURE 5.4 *S. acanthias* (Vertebrata: Elasmobranchii). (A, C, E) Right lateral view; (B, D) ventral view; (F) anterolateral view. (A–C, E) Anterior to the right; (D, F) anterior to the left. (A) Lateral muscles of the head after removal of the skin. (B) Ventral muscles, the intermandibularis was cut and reflected laterally on both sides, the interhyoideus was reflected on the left side. (C) Superficial postcranial muscles. (D) Deep branchial muscles. (E) Cucullaris reflected. (F) Branchial basket dissected out of the body of the specimen, left lateral view. Abbreviations: add., adductor; lev., levator; mand. = mandibularis. Scale bar = 1 cm.

With respect to the hyoid muscles, the interhyoideus is thin and difficult to separate from the intermandibularis (Figure 5.4B). It extends from the ceratohyal to a median raphe where it fuses with its contralateral muscle. The interhyoideus covers the ventral branchial region posteriorly. The superficial constrictor sheet, i.e., the constrictor hyoideus dorsalis (Figure 5.4C), is also present in this position. This muscle originates at the fascia above the cucullaris and is blended to both the adductor arcus palatini and the interhyoideus without having clear skeletal attachments. The adductor arcus palatini extends from the otic capsule to the hyomandibula (Figure 5.4A and C).

As for the branchial muscles, the cucullaris is a continuous muscle sheet originating from the fascia near the epaxial musculature and inserting at the scapular process of the pectoral girdle, the posterior epibranchial cartilages, and the anterior extrabranchial cartilages (Figure 5.4C and E). After careful dissection of the branchial region in *S. acanthias* (and also in *Mustelus laevis*) we were unable to identify the "levatores arcuum branchialium' described by Kesteven (1942) (see below). There are four constrictores branchiales, each divided into dorsal and ventral parts that laterally overlap the branchial region. The first dorsal part originates anteriorly from the constrictor hyoideus dorsalis (a hyoid muscle), whereas the last dorsal part originates from the pectoral girdle (Figure 5.4C). The intermediate dorsal parts originate from a raphe over one anterior extrabranchial cartilage and insert in a raphe over the next one with the adjacent constrictor, with some fibers also inserting onto the extrabranchial cartilages (Figure 5.4C). Dorsally, some fibers of the dorsal parts of the constrictores branchiales penetrate the cucullaris, whereas ventrally, they reach the gill openings and blend with the ventral parts of the constrictores branchiales. The median fibers of the ventral parts, which have attachments to the dorsal ones, originate additionally from the fascia between the branchial cavities and the coracobranchiales and the hypobranchial muscles (coracoarcualis and coracomandibularis). The four interarcuales laterales (I–IV) extend from the pharyngobranchials to the epibranchial cartilages (Figure 5.4F). The five deep adductores branchiales connect the epibranchial and ceratobranchial cartilages of the same branchial arch (Figure 5.4F). The four interbranchiales sit between the semibranches of the first four gill arches (Figure 5.4F), whereas the five coracobranchiales (I–V) (Figure 5.4) extend from below the sternohyoideus (coracobranchialis I) and the pectoral girdle to the hyal cartilage (coracobranchialis I) and the ventral region of the ceratobranchials. Recently, Kryukova (2017) divided some of the hyoid and branchial muscles of some sharks, including *S. acanthias* described here and the preceding paragraphs into more subsections/muscles, although some of those divisions seem to be somewhat arbitrary and conflict with previous descriptions by other authors.

The subspinalis is an epibranchial muscle *sensu* Edgeworth (1935). It originates from the cranium close to the foramen magnum and inserts onto the first pharyngobranchial (Figure 5.4F). The three interpharyngobranchiales (I–III) are also epibranchial muscles *sensu* Edgeworth (1935). They extend diagonally between two adjacent pharyngobranchial cartilages (Figure 5.4F).

Concerning the hypobranchial muscles, the coracoarcualis originates from the coracoid bar, extends anteriorly, and inserts onto the pericardium (some fibers) and the ceratohyal (most fibers) (Figure 5.4B). The sternohyoideus originates from the aponeurosis with the coracobranchiales (branchial muscles) and the coracoarcualis and inserts onto the hyoid (basihyal) (Figure 5.4B). The coracomandibularis is a paired muscle along the mid-ventral line (Figure 5.4B) from the fascia between the coracoarcuales to the symphysis of the Meckel's cartilage.

LEUCORAJA ERINACEA: LITTLE SKATE

The intermandibularis anterior is a thin muscle connecting the anterior region of the two Meckel's cartilages (Figure 5.5E) (NB: this muscle is absent in some specimens of *Leucoraja erinacea* according to Marion 1905). The intermandibularis posterior extends from Meckel's cartilage to the fascia of the coracomandibularis (a hypobranchial muscle) (Figure 5.5E) and receives innervation not only mainly from CNV3 but also posteriorly from CNVII. The adductor mandibulae A2 connects the upper and lower jaws and has two main portions (Figure 5.5A and B). The small medial portion connects the anterior border of the palatoquadrate and Meckel's cartilage close to its symphysis (Figure 5.5B). The lateral portion splits into superficial and deep bundles. Superficially, the lateral portion is set into anterior and posterior parts by the levator labialis superioris 2. The deep bundle originates from the anterior edge and the processus muscularis of the palatoquadrate. Both bundles insert onto Meckel's cartilage (Figure 5.5A). The labial muscle preorbitalis (which appears to include the "ethmoideo-parethmoidalis" *sensu* Edgeworth [1935]) is a complex muscle that is divided into five bundles (Figure 5.5A and B). Overall, these bundles connect between the chondrocranium and the mandibular cartilage and the adductor mandibulae A2. The levator arcus palatini has a dorsal bundle originating ventrally to the postorbital process and inserting onto the palatoquadrate and a ventral bundle originating from the dorsal portion and inserting onto the membranous wall of the palate and, *via* a tendon, onto the hyomandibular cartilage (Figure 5.5A). The spiracularis extends from the chondrocranium to the dorsal surface of the hyomandibular cartilage.

Moving to the hyoid muscles, the interhyoideus originates partly from the horizontal tendon associated with the hyomandibula. It is continuous posteriorly with the ventral branchial musculature. The depressor hyomandibulae originates from the fascia at contact with the bilateral counterpart (covering dorsally the coracomandibularis, which is a hypobranchial muscle) and from the fascia overlying the sternohyoideus (a hypobranchial muscle). The muscle inserts onto the hyomandibula and the horizontal tendon associated with hyomandibula (Figure 5.5E). The levator rostri originates from the first vertebra and the fascia of the muscles covering this region and inserts *via* a long tendon onto the dense connective tissue between the rostrum and the propterygium (Figure 5.5A,

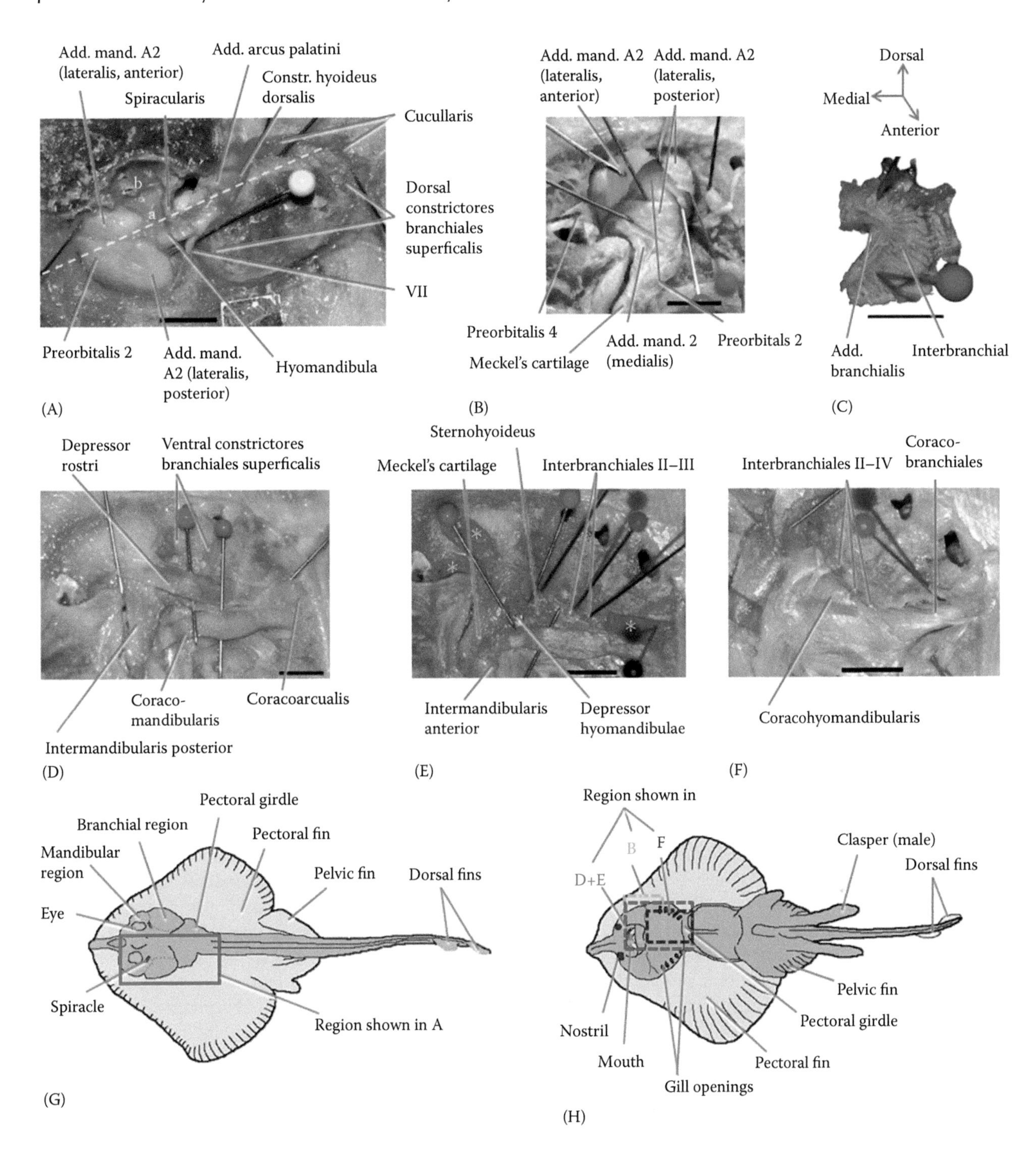

FIGURE 5.5 *L. erinacea* (Vertebrata: Elasmobranchii). (B, D–F) Ventral views; (A) dorsolateral view; (C) anterior view. (A) Left dorsolateral view, the levator rostri (indicated by dashed line) and the eye were removed. (B) Levatores labiales superioris and adductor mandibulae complex; some muscles were partially reflected. (C) A single gill (branchial arch) from the right side of the body; the orientation is given on the top of the picture. (D–F) Ventral muscles from superficial to deep; reflected muscles in (E) and (F) are indicated by an asterisk (*). "Preorbitalis 2, 4" correspond to bundles 2 and 4 of the five bundles that form the preorbitalis muscle in *L. erinacea*. Abbreviations: a, Lmi; b, lms; add., adductor; constr., constrictor; mand., mandibulae. (G) Dorsal view and (H) ventral view of a skate indicating important landmarks and the region shown in the dissections seen earlier indicated by colored rectangles. (A–F) Scale bar = 1 cm.

dashed line). The depressor rostri originates from the fascia covering the coracomandibularis and from the coracoarcualis (both hypobranchial muscles) (Figure 5.5D). Some of its lateral fibers insert onto the fascia covering the adductor mandibulae complex, whereas the remaining fibers extend anteriorly and medially to insert onto the membranous tissue dorsal to the rostral cartilage and ventral to the propterygium. The adductor arcus palatini extends from the chondrocranium to the hyomandibula (Figure 5.5A). The constrictor hyoideus dorsalis originates between the adductor arcus palatini and the anterior edge of the constrictores branchiales (branchial muscles) and inserts onto the horizontal tendon associated with the hyomandibula (Figure 5.5A).

With respect to the branchial muscles, the cucullaris is divided into a medial bundle (from the first vertebra to the suprascapular region) and a lateral bundle (from the region close to the first branchial arch) (Figure 5.5A). In one of the two dissected specimens, the lateral bundle inserts onto some branchial arches and onto the pectoral girdle. In the other specimen an insertion onto the branchial arches cannot be identified, although the ventral fibers over the posterior branchial arches attach to the overlying fascia. The five constrictores branchiales cover the branchial basket dorsally and ventrally (Figure 5.5A and D). The dorsal part connects a tendon located between the neighboring segments to the next (Figure 5.5A), but the fifth dorsal part originates from the shoulder girdle. The tendon between the dorsal segments decreases in size from anterior to posterior. The posterior constrictor muscles extend to reach the horizontal tendon associated with the hyomandibula. The ventral part of the constrictores branchiales has similar sites of origin and insertion as the dorsal part (Figure 5.5D). Here, the fifth ventral part originates from the fascia covering the ventral edge of the coracobranchiales. Each of the five segments of interarcuales laterales originates from the anterior end of the pharyngobranchial and inserts onto the medial end of the epibranchial of the same arch. Each of the six adductores branchiales connects the epibranchial and ceratobranchial elements of the same branchial arch (Figure 5.5C). The five segments of interbranchiales are set between the gill clefts. They originate from the coracomandibularis (a hypobranchial muscle), and some of the ventromedial fibers are very well developed (Figure 5.5F). The six segments of coracobranchiales extend from the pectoral girdle to the basibranchial and the ventral region of the branchial arches (Figure 5.5F).

Lastly, with respect to the hypobranchial muscles, the coracoarcualis originates from the coracoid bar and inserts *via* some medial fibers onto the pericardium and *via* the lateral fibers onto the ceratohyal (Figure 5.5D). The coracomandibularis originates from the fascia between the right and left coracoarcuales and inserts onto the mandibular symphysis along the midline (Figure 5.5D). The sternohyoideus extends from the fascia ventral to the coracohyomandibularis and inserts onto the hypohyal cartilage (Figure 5.5E). The coracohyomandibularis originates from the midline fascia (Figure 5.5F) and inserts *via* a tendon onto the ventral surface of the hyomandibula.

EVOLUTION OF CEPHALIC MUSCLES IN PHYLOGENETICALLY BASAL VERTEBRATES

The number of mandibular muscles is greater in cyclostomes than in extant chondrichthyans and osteichthyans (see Tables 5.2 through 5.5 and Figure 3.1). The intermandibularis derives from the ventral part of the mandibular muscle plate (Edgeworth 1935). Some authors suggested that the intermandibularis of selachians is innervated by CNVII (e.g., Lightoller 1939). However, it is likely that Lightoller (1939) confused at least part of the "intermandibularis" with a facial muscle (the interhyoideus; innervated by CNVII), because the interhyoideus and the intermandibularis occur in a continuous sheet and because Lightoller compared his intermandibularis to some of the facial muscles of mammals. We observed that the intermandibularis in *S. acanthias* is deeply blended with the interhyoideus, which likely led to a misinterpretation that the intermandibularis is innervated by CNVII. Miyake et al. (1992) and Anderson (2008) identified a single intermandibularis in elasmobranchs, but recognized that at least some elasmobranchs (e.g., electric rays) have an "intermandibularis superficialis" and an "intermandibularis profundus" (Edgeworth 1935). In addition, Marion (1905) described anterior and posterior bundles in *S. acanthias*. The intermandibularis posterior in *Leucoraja* and the intermandibularis of *Squalus* are both mainly innervated by CNV3, but innervation by CNVII occurs along and near the posterior margin (Edgeworth 1935). Instead of viewing the intermandibularis posterior (sometimes labeled "depressor mandibulae") as a hyoid muscle (Miyake et al. 1992), we consider the double innervation to be secondary. The innervation by CNVII occurs likely because the muscle sits close to the hyoid muscle interhyoideus, which is also innervated by CNVII. The intermandibularis posterior is clearly a mandibular muscle because of the main innervation by CNV3, because of the insertion to the lower jaw and because of its association with other mandibular muscles such as the preorbitalis. Some electric rays have an additional muscle X that likely derives from the intermandibular anlage, which is designated intermandibularis profundus (e.g., Miyake et al. 1992). Few electric rays also have an intermandibularis superficialis (e.g., *Astrape*, *Torpedo*; Edgeworth 1935).

We identified a single intermandibularis in *H. colliei* as did Anderson (2008). Even though other authors reported an intermandibularis anterior and an intermandibularis posterior in some holocephalans (e.g., Edgeworth 1935; Kesteven 1942–45), these muscles probably represent a single intermandibularis that is partly separated by the premandibular cartilage (Didier 1987, 1995). Alternatively, the intermandibularis of holocephalans may represent the anterior-most remnant of the intermandibularis muscle sheet in elasmobranchs (Edgeworth 1935). A generally accepted hypothesis is that the hypothetical LCA of extant gnathostomes had an undivided intermandibularis muscle sheet (Miyake et al. 1992; Anderson 2008). However, at least some holocephalans, elasmobranchs, and many osteichthyans (including the hypothetical LCA of osteichthyans; Diogo and Abdala 2010) have both an intermandibularis anterior and an intermandibularis posterior. Therefore, it is more parsimonious to support the alternative hypothesis that the LCA of extant gnathostomes had the two muscles. Yet another interpretation is that the distinction of the two muscles is either artificial or insignificant. The intermandibularis anterior and posterior are typically blended with one another so much so that the anterior cannot be readily distinguished from the posterior in many chondrichthyans. The attachment of the muscle to the maxillary (labial) cartilage in holocephalans such as *Hydrolagus* is likely secondary, because the intermandibularis attaches to the mandibles (Meckel's cartilages) in other taxa and because the holocephalan skull has been extensively modified from a typical gnathostome pattern. We propose that the tubulatus of *Myxine* and the constrictor cornualis superficialis and constrictor glossae profundus internus of *Petromyzon* belong to the intermandibularis group, based on the innervation by CNV3, position relative to other muscles and cartilages, and recent studies of the identity of the cranial cartilages of cyclostomes (reviewed by Miyashita [2012] and Oisi et al. [2013a]). Therefore, the LCA of extant vertebrates likely had a muscle sheet that would ultimately give rise to the intermandibularis in gnathostomes (Table 5.2). The proposed similarity partly agrees with the assessment by Miyashita (2012: 295), who listed the lamprey constrictor glossae profundus internus as a potential member of the group that corresponds to the intermandibularis group of gnathostomes. As identified by Miyashita (2012), these muscles are topographically (ventral portion of mandibular arch) and functionally (transverse constriction) similar.

The adductor mandibulae derive from the transversely medial and dorsoventrally intermediate part of the mandibular muscle plate (Edgeworth 1935). A previous interpretation has been that the LCA of gnathostomes has an A2 with a "suborbital" portion, which gave rise to the preorbitals (suborbitalis) in chondrichthyans (Lauder 1980b). In contrast, our observations suggest that the preorbitalis is a labial muscle and distinct from the adductor mandibulae A2 (Table 5.2). Edgeworth (1935) described the adductores mandibulae A2 anterior and posterior in Holocephali as incompletely separated. However, in *H. colliei*, we observed a clear separation between the anterior and posterior parts. Between them, the adductor mandibulae A2 anterior in *H. colliei* has been described as having two portions by Didier (1987, 1995), but our dissections revealed anterior, medial, and posterior portions. The anterior portion corresponds to Didier's (1987) pars nasalis and the medial and posterior portions to Didier's (1987) pars orbitalis. The dorsoventrally flattened and lateromedially expanded skull of *Leucoraja* makes it difficult to identify similarities in the three portions of the adductor mandibulae A2 with those in *Squalus* (Marion 1905; Soares and Carvalho 2013a,b). Based on lateromedial positions with each other, we suggest that (a) the A2 lateralis anterior in *Leucoraja* is similar to the A2 superficialis in *Squalus*, (b) the A2 lateralis posterior to A2 dorsalis, and (c) the A2 medialis to A2 ventralis. An alternative interpretation is that each of these bundles evolved independently in selachians and batoids, which is consistent with the previous assessment that no clear similarity between the bundles of the A2 exist (Marion 1905; Edgeworth 1935). In holocephalans (elasmobranch outgroup), the A2 bundles are different from those of other extant chondrichthyans, as described in the preceding section. The distinct morphology implies that the A2 was not separated into the bundles in the LCA of chondrichthyans and probably in that of gnathostomes (Lauder 1980b). No clear homologue of the gnathostome adductor mandibulae exists in the jawless cyclostomes. In gnathostomes, the proper development of the mandibular muscles is tightly linked to the dorsoventral patterning of the jaw skeleton (Noden 1983a; Rinon et al. 2007; Medeiros and Crump 2012). Many knockdown phenotypes with defects in both jaw and mandibular muscle corroborate this link, where either defects in the jaw or those in the muscle are responsible for the other (e.g., Schilling et al. 1996; Heude et al. 2010; Hinits et al. 2011). Therefore, it is likely that the adductor mandibulae and the jaw evolved in concert in gnathostomes.

The origin and evolution of jaws has been a subject of interest. Recent developmental studies have revealed that dorsoventrally nested expression of *Dlx* genes in cranial neural crest cells is necessary for both the jaw development and masticatory muscle formation. The inactivation of *Dlx5* and *Dlx6* in mice results in reduction or loss of jaw muscles (Heude et al. 2010). Because these genes are not expressed in the myogenic mesoderm, *Dlx5/6*-positive cranial neural crest cells likely pattern the muscles. The muscle defects do not result from the loss of mandibular identity because masticatory muscles are still present in *EdnRA* (–/–) mutants that have similar jaw defects with those in *Dlx5/6*-deficient mice (Heude et al. 2010). Given that the gnathostome *Dlx* code dorsoventrally patterns the jaw skeleton as well (Depew et al. 2001; Medeiros and Crump 2012; Gillis et al. 2013), the gnathostome jaw muscles represent neomorphs that evolved along with the jaws. The *Dlx*-dependent oral musculoskeletal patterning may have preceded the origin of the jaw. Sea lampreys (*Petromyzon marinus*) and Japanese inshore hagfish (*Eptatretus burgeri*) have dorsoventrally polarized expression patterns of *Dlx* cognates in the mandibular region (Cerny et al. 2010; Fujimoto et al. 2013). However, the Japanese river lampreys (*Lethenteron japonicum*) have overlapping, unpolarized expression of *Dlx* cognates (Myojin et al. 2000; Kuraku et al. 2010). The variation among cyclostomes leaves uncertainty about the timing and frequency of the origin of nested *Dlx* expression. It also remains to be tested whether the *Dlx* expression plays a crucial role in oral musculoskeletal patterning in cyclostomes as is the case in gnathostomes. For a recent review and innovative look of the early evolution of jaws in vertebrates, see Miyashita and Diogo (2016).

The levator arcus palatini and spiracularis derive from the dorsal part of the mandibular muscle plate: a specific premyogenic condensation designated "constrictor dorsalis" (Edgeworth 1935). Some batoids also have an accessory muscle associated with the spiracularis (Miyake et al. 1992). Miyake et al. (1992) compared this accessory muscle with the 'spiracularis' in *Polypterus* (Edgeworth 1935). Our previous anatomical and comparative analyses did not identify a "spiracularis" in adults of *Polypterus* or other extant osteichthyan taxa (Diogo and Abdala 2010). Because holocephalans lack a spiracularis, the presence of this muscle in elasmobranchs likely represents a synapomorphy of the clade (Ziermann et al. 2014). However, as will be explained in Chapter 6, a recent study showed a the spiracularis is present in at least some developmental stages of *Polypterus*, so one cannot completely discard the hypothesis—although it is clearly less parsimonious, phylogenetically—that this spiracularis is homologous to that of elasmobranchs. Two hypotheses about the evolutionary origins of the levator arcus palatini, spiracularis, and intermandibularis in chondrichthyans exist in the literature: (a) the intermandibularis and the levator arcus palatini (and later the spiracularis, during chondrichthyan—or gnathostome—evolution) derived, respectively, from the ventral and dorsal portion of the "branchial superficial constrictor 1" in the hypothetical LCA of vertebrates (Mallatt 1996) or (b) the three muscles mainly derived from a "mandibular interbranchial muscle" and partly from branchial superficial constrictor 1 in that hypothetical LCA (Mallatt 1997). However, a hypothetical vertebrate LCA such as that described by Mallatt (1996, 1997) is unlikely to have existed. The reconstructions of such a hypothetical animal have historically depended on the notion of "archetype" in the German romantic school, which later influenced Anglo-Saxon authors such as Owen (Northcutt 2008) and still persist in present times (Jarvik 1980; Mallatt 2008). Recent anatomical studies of cyclostomes and the invertebrate chordates rejected the existence of such vertebrate archetypes (Kuratani 2004; Stolfi et al. 2010; Yasui et al. 2013; Diogo et al. 2015a). There is absolutely no fossil, anatomical, or developmental evidence that the mandibular region was originally a branchial arch with muscles serially homologous with the true, gill-supporting pharyngeal arches (Janvier 1996, 2007; Kuratani 2008b; see also Miyashita and Diogo 2016). Moreover, a comparative analysis of the cephalic musculature of larval and adult lampreys and of adult hagfishes and gnathostomes (Miyashita 2012) did not support the homologies of the muscles of hagfish, lampreys, and gnathostomes proposed by Mallatt (1996, 1997, 2008) (see section about labial muscles in the following, as well as Chapter 4).

The *engrailed* immunoactivity in the velothyroideus of lamprey larvae (ammocoetes) and in the levator arcus palatini of zebrafish has been used to infer the homology between these muscles (Holland et al. 1993). However, we remain cautious about proposing such homologues based only on the immunoactivity. Similar gene expression patterns do not necessarily warrant a homology. For example, engrailed is expressed in mandibular myoblasts that are not part of the dorsal mandibular musculature in some gnathostomes (constrictor dorsalis *sensu* Edgeworth 1935), whereas the expression is in those derived from the adductor mandibulae complex (the masseter, temporalis, pterygoideus medialis, and/or pterygoideus lateralis) in mice (Knight et al. 2008). Furthermore, the velothyroideus is a larval musculature, and it remains uncertain whether the muscle entirely degenerates during metamorphosis or partly becomes incorporated into the adult velar muscles (Miyashita 2012). Therefore, the currently available evidence is insufficient to make an inference about evolutionary conservation of the shared engrailed expression. Instead of uncertain homology, homogenic relationship between the lamprey velar muscles and some of the gnathostome mandibular muscles are a testable hypothesis: some, if not all, of the velar muscles of cyclostomes may derive from the anlage that also (independently) gave rise to the gnathostome levator arcus palatini. The visceral muscles of adult lampreys differentiate from blastema as those in ammocoetes degenerate during metamorphosis, and there is still disagreement over whether the larval muscles are precursors of adult muscles or not (reviewed by Miyashita 2012). Therefore, it is difficult to identify possible correspondence between the larval and adult lamprey muscles. During lamprey metamorphosis, the endostyle and velum are both reduced, and a lingual apparatus develops in the location corresponding to the larval muscle velothyroideus (Miyashita 2012). To explain this, possible interpretations are that the lamprey velothyroideus (a) is reduced at the time of metamorphosis along with the velum or (b) becomes incorporated to adult velar muscles such as the depressor veli. We tentatively favor the latter hypothesis and further suggest that these velar muscles derive from the same anlage with the velar muscles of adult hagfishes and the dorsal mandibular (constrictor dorsalis) muscles of gnathostomes such as the depressor arcus palatini (Table 5.2). This hypothesis of homogenic relationship (evolutionary conservation at the level of anlagen) is consistent with engrailed expression patterns in some of the differentiated mandibular muscles between lampreys and gnathostomes (Holland et al. 1993) and with the development and anatomy of the cyclostome muscles (reviewed by Miyashita [2012] and Oisi et al. [2013b]).

The labial muscles have previously been considered broadly comparable across vertebrates and thus evolutionarily conserved between cyclostomes and gnathostomes (Mallatt 1996, 1997, 2008). However, the modern developmental evidence contradicts this view. In cyclostomes, the "upper lip" forms the cheek process, where the mandibular stream of the ectomesenchyme laterally overlaps the premandibular ectomesenchyme (Horigome et al. 1999; Kuratani et al. 1999; Kuratani 2012; Oisi et al. 2013b). In contrast, the labial muscles in gnathostomes develop in the maxillary process—a secondary anterior extension of the mandibular arch (Kuratani et al. 2002; Noden and Francis-West 2006; Kuratani 2012). This fundamental difference in developmental process contradicts Mallatt's (1996, 1997, 2008) schematic of homology of the upper lip between the cyclostomes and gnathostomes (Kuratani 2012; Miyashita 2012). Following this cyclostome–gnathostome difference in the mandibular patterning, we compare the cyclostome upper lip muscles and the gnathostome labial muscles in detail. The assessment presented here is a tentative one based on our primary anatomical observations, and further tests are necessary to establish the similarity. Among the cyclostome labial muscles, the "buccal constrictor" in larval lampreys appears to develop from the "mandibular branchiomere" (Mallatt 1996) (NB: it is not clear if the buccal constrictor of larval lampreys corresponds, or gives rise, to specific muscles of adult lampreys). A likely gnathostome counterpart of the lamprey mandibular branchiomere is the mandibular muscle plate from which some labial muscles (e.g., the chondrichthyan preorbitalis) derive (Edgeworth 1935). Therefore, the "labial muscles" develop from mandibular mesoderm in both cyclostomes and gnathostomes, as does the adductor mandibulae complex (reviewed by Diogo and Abdala [2010]). Beyond this, however, comparison becomes challenging.

Between the two cyclostome lineages, the innervation by CNV2 is useful in comparing functionally and topographically similar muscles. The ventral branch of CNV2 innervates six muscles in *Myxine* (CNV2 posterior—Table 5.1: retractor mucosae oris, longitudinalis linguae, protractor dentium profundus, protractor dentium superficialis, tubulatus, and perpendicularis). The corresponding muscles, following our analysis and Miyashita's (2012), in *Petromyzon* are all but one innervated by the V2 ramus mandibularis (*sensu* Marinelli and Strenger 1954; V2 ramus velaris: pharyngicus posterior; V2 ramus mandibularis: levator valvulae velaris, cardioapicalis, annuloglossus, copuloglossus rectus, constrictor cornualis superficialis, constrictor glossae profundus internus). Interestingly, Wicht and Northcutt (1995) treated all these trigeminal motoneurons (corresponding to CNV2 posterior and CNV2 velar of the hagfish) as CNV3. Although CNV2 and CNV3 have been used to establish muscle homology in the mandibular arch across vertebrates (Song and Boord 1993; Mallatt 1996), both anatomical and developmental evidence questions the presumed homology of these major branches of CNV between cyclostomes and gnathostomes (Lindström 1949; Wicht and Northcutt 1995; Kuratani et al. 1997; Miyashita 2012; Modrell et al. 2014; Higashiyama and Kuratani 2014). Curiously, the labial muscles are innervated by CNV2 in holocephalans (Song and Boord 1993; Mallatt 1996; but see the following). There are two possible interpretations to this observation: (a) holocephalans conserve the cyclostome-like innervation pattern in the labial muscles, which has been lost in elasmobranchs and osteichthyans, or (b) holocephalans deployed motoneurons to extend with CNV2 and innervate the anteriorly positioned labial muscles, what evolved independently from cyclostomes. We support the latter because motoneurons are supplied from multiple branches (discussed in the following) and because great variations exist in development and neuronal organization of the trigeminal nerve and ganglia between and within cyclostomes and gnathostomes (Modrell et al. 2014). Then, comparison based on the distinction between CNV2 and CNV3 may be uninformative. If this view is correct, some cyclostome mandibular muscles may have functionally and topographically similar counterparts in the mandibular muscles of gnathostomes. One candidate is the tubulatus in *Myxine*, the gnathostome counterpart of which is the intermandibularis.

Within gnathostomes, the preorbitalis is the only labial muscle present in elasmobranchs (e.g., Diogo and Abdala 2010). In *Hydrolagus*, four labial muscles are present: the preorbitalis ("levator anguli oris posterior" *sensu* Mallatt 1996) and three upper-lip muscles (levator anguli oris anterior, levator prelabialis, and labialis anterior) (e.g., Edgeworth 1935; Didier 1987, 1995). We disagree with previous observations that the three upper-lip muscles of *H. colliei* are innervated by CNV2, whereas the preorbitalis is innervated either by (a) V2 only or V3 only (Mallatt 1996) or (b) V2 and V3 (Edgeworth 1935). We identified mixed fibers from CNV2 + 3 innervating all four labial muscles, as well as the adductor mandibulae A2, in *H. colliei*. This pattern of innervation is mainly due to a fiber exchange of those two branches in the suborbital region. Some holocephalans were considered to have a muscle "labialis inferior" lying posteriorly to the intermandibularis ("labialis posterior" *sensu* Mallatt 1996). However, it is now clear that labialis posterior and labialis inferior are part of the intermandibularis group (Diogo and Abdala 2010). Topographically and functionally, these upper-lip muscles are somewhat comparable to the buccalis anterior of ammocoetes (*sensu* Mallatt 1996), which originates from the anterior braincase and lateral mouth plate and inserts onto the anterodorsal upper lip cartilage.

We reject the proposed homology between the holocephalan labial muscles preorbitalis, levator anguli oris anterior, and/or levator prelabialis and the levator arcus palatini in other gnathostomes (Didier 1987). Instead, it is likely that the levator arcus palatini was lost in Holocephali. This loss may be related to the fusion of the palatoquadrate with the neurocranium in this lineage (Miyake et al. 1992; Anderson 2008). Within elasmobranchs, we follow Marion (1905) in identifying the five labial bundles in *Leucoraja* as part of the preorbitalis. However, we also agree with Edgeworth (1935) that some of these five bundles may be associated with the adductor mandibulae A2. As for osteichthyans, some taxa have muscles often designated as labial muscles: the levator maxillae superioris 3 and 4 in *Amia*, the palatomandibularis major and minor in *Lepisosteus*, the suborbital portion of adductor mandibulae in acipenseriforms, and labial muscles in *Latimeria* (Lauder 1980a,b; Anderson 2008; Diogo and Abdala 2010). However, Diogo and Abdala (2010) considered that at least some of these muscles were likely acquired independently in their respective lineages. This does not rule out the possibility that the LCA of extant osteichthyans had labial muscles (Table 5.2).

Concerning the hyoid muscles, they are divided into two groups derived from different anlagen (Edgeworth 1935): dorsomedial (e.g., the adductor arcus palatini and the constrictor hyoideus dorsalis) and ventral (e.g., the interhyoideus; see also, e.g., Marion 1905; Daniel 1928; Luther 1938). The constrictor hyoideus dorsalis is often labeled as "adductor operculi" in osteichthyans (the nonosteichthyan term is preferred because most nonosteichthyan gnathostomes lack an operculum). In batoids, (a) the depressor hyomandibulae develops from the hyoid condensation and (b) the levator rostri and the depressor rostri develop, respectively, from dorsal and ventral portions of the hyoid muscle plate, which are separated from the sheet of myoblasts that gives rise to other hyoid muscles (Edgeworth 1935; Table 5.3).

The mandibulohyoideus in holocephalans is a peculiar muscle. Anderson (2008) posits it as a hypobranchial muscle with no analogue in gnathostomes, which is associated with the holocephalan-specific jaw-opening mechanism intermediate between those of elasmobranchs and osteichthyans. However, the mandibulohyoideus in holocephalans receives innervation by CNVII like other hyoid muscles, and we did not observe the innervation by CNIX as reported by Anderson (2008). Therefore, we treat the mandibulohyoideus as a hyoid muscle. It may have derived from the interhyoideus as suggested by Edgeworth (1935) and Kesteven (1942–1945).

As explained in the preceding chapters, the term *branchial muscles sensu stricto* is one subgroup of "true branchial muscles" (the other subgroup consists of the cucullaris and its derivatives) (Diogo and Abdala 2010). Other "branchial" muscles such as the laryngeal muscles and the epibranchial muscles are neither branchial muscles *sensu stricto* nor true branchial muscles (Table 5.4). As for branchial muscles *sensu stricto*, larval lampreys (ammocoetes) and elasmobranchs may have two shared functions: (a) the expiration is driven by the peristaltic action of superficial constrictores branchiales and interbranchiales and (b) the inspiration is driven by passive recoil of the branchial arches (Mallatt 1996). The similarity implies that the LCA of the living vertebrates had a ventilation system motored by superficial constrictores branchiales and interbranchiales. Our analysis supports the homology of the constrictores branchiales across vertebrates (Table 5.4). However, the interbranchiales have never been identified in hagfishes. *H. colliei* also appears to lack the interbranchiales, but it is present in other holocephalans (Didier 1995, Edgeworth 1935). Osteichthyans do not have interbranchiales or adductores branchiales. Therefore, it is more parsimonious that these muscles appeared independently in lampreys and chondrichthyans (two evolutionary steps) than to assume the presence in the LCA of extant vertebrates with loss in both hagfishes and osteichthyans (three steps). If the constrictors branchiales and the interbranchiales are treated separately, however, one or both may have been present in the LCA of vertebrates and/or of gnathostomes (Table 5.4).

Two conflicting interpretations exist about the branchial muscles *sensu stricto* in holocephalans: (a) the protractor dorsalis pectoralis, retractor lateroventralis pectoralis, and retractor mesio-ventralis pectoralis are branchial muscles innervated by cranial nerves CNIX and CNX (Anderson 2008) or (b) only the protractor dorsalis pectoralis is a branchial muscle, whereas the others represent trunk muscles (Didier 1995). These two conflicting interpretations agree that the protractor dorsalis pectoralis is a branchial muscle, although Edgeworth (1935) and Miyake et al. (1992) did not consider it to be a branchial muscle. As an alternative hypothesis, the position and attachment sites in close association with those of the cucullaris raise an interesting possibility that the protractor dorsalis pectoralis is an epibranchial muscle. This idea presents an alternative to Edgeworth's (1935) assessment that the interpharyngobranchiales and subspinalis are the only epibranchial muscles in holocephalans. All the confusions arise from the unconfirmed innervation of the protractor dorsalis pectoralis by CNIX and/or CNX in holocephalans. The argument in the literature is circular: the protractor dorsalis pectoralis is a branchial muscle and thus should be innervated by CNIX or CNX (Didier 1995; Anderson 2008). We did observe the innervation by CNX, but were unable to confirm whether or not this muscle receives innervation from other nerves. Therefore, we consider the protractor dorsalis pectoralis, retractor lateroventralis pectoralis, and retractor mesioventralis pectoralis to be trunk muscles and that the association between the CNX and the protractor dorsalis pectoralis in holocephalans is probably secondary due to the relatively anterior position of the muscle close to the vagal fibers.

The cucullaris is part of the true branchial muscles, but not branchial muscles *sensu stricto* because it suspends the pectoral girdle (Table 5.4) (Edgeworth 1935; Diogo and Abdala 2010). The cucullaris is perhaps the most puzzling element of vertebrate muscles (Diogo and Abdala 2010). Assuming that the cucullaris is somitic in origin, Kusakabe and Kuratani (2005) suggested that cyclostomes have no muscles that correspond with the cucullaris of gnathostomes. Kuratani (2008a) and Sambasivan et al. (2011) largely follow that assessment, but both note topographical similarity between the cucullaris of gnathostomes and the "infraoptic muscles" of lampreys. Kusakabe et al. (2011) suggest that these "infraoptic" muscles of lampreys are the precursors of the cucullaris, based on the observation that the infraoptic muscles derive from the anterior somites. However, these authors seem to use the term *infraoptic muscles* to refer to different muscles. Three muscles occur ventral and posterior to the eye in lampreys: subocularis, cornealis, and probranchialis. The subocularis is omitted from the infraoptic muscles in the study by Kuratani (2008a), but is included among them along with the cornealis and probranchialis the study by in Kusakabe et al. (2011). In most gnathostomes, the cucullaris and its derivatives typically receive innervation from CNXI (accessory nerve) (Edgeworth 1935). However, these muscles may be partly or entirely innervated by the cervical spinal nerves in some taxa (e.g., skates; Boord and Sperry 1991), as expected for a trunk muscle. In tetrapods, the cucullaris derivatives (e.g., trapezius and sternocleidomastoideus) at develop least partly from somites (Köntges and Lumsden 1996; Matsuoka et al. 2005; Noden and Francis-West 2006; Piekarski and Olsson 2007; Shearman and Burke 2009). However, the connective tissues of the cucullaris and its homologues in gnathostomes have multiple origins: branchial arch neural crest cells, somite, and lateral plate (Shearman and Burke 2009; see the following). As explained in Chapter 3 and, in more detail, in Chapter 8, it is generally accepted that the cucullaris is a true branchial muscle that largely develops from the cardiopharyngeal field, specifically from an anlage that also gives rise to other true branchial muscles and that then extends posteriorly, during development, to assume a more trunk muscle-like position and function. Our new gross anatomical observations of non-osteichthyan vertebrates and comparisons with osteichthyans also support the hypothesis that the cucullaris and its derivatives are true branchial muscles. In adult selachians such as *Squalus*, a single, continuous cucullaris develops from the branchial muscle plate (Edgeworth 1935) and attaches to both the dorsal surface of the branchial arches and the posterior surface of the pectoral girdle. In holocephalans, the cucullaris also develops as in selachians but splits into the superficial and deep bundles (Edgeworth 1935). In batoids, the cucullaris develops only from the dorsal portion of the last branchial muscle plate, often forming a single muscle that then divides during ontogeny into three bundles (Edgeworth 1935): an inner bundle extending to the suprascapula, a middle bundle extending to the scapula, and an external bundle extending to the branchial arches. Osteichthyans have two main homologues of the chondrichthyan cucullaris: (a) the protractor pectoralis (or its derivatives such as trapezius and sternocleidomastoideus, usually innervated by CNXI) develops exclusively or mainly from the last branchial muscle plate and inserts onto the pectoral girdle and (b) the levatores arcuum branchialium develops from the dorsal portion of the branchial muscle plate and extends to the dorsal surface of the branchial arches (absent in amniotes) (reviewed by Diogo and Abdala [2010]). See Chapter 8 for more details on the evolution and developmental origin of the cucullaris.

There is little support for Kesteven's (1942) description of the levatores arcuum branchialium as thin structures over the dorsal surface of the branchial arches in some selachians (*Mustelus* and *Orectolobus*), batoids (*Dasyatis*) and holocephalans (*Callorhynchus* and *Hydrolagus*) and as even thinner and apparently vestigial structures in other selachians and batoids (*Squalus* and *Leucoraja*). Didier (1987, 1995) did not recognize these muscles in holocephalans and suggested that (a) some of the levatores arcuum branchialium *sensu* Kesteven (1942) correspond to a part of the cucullaris profundus and (b) others correspond to a part of the epibranchial muscle subspinalis. However, these proposed synonymies are questionable because Kesteven (1942) distinguishes the subspinalis, the cucullaris, and the levatores arcuum branchialium in chondrichthyans. An intriguing aspect of Kesteven's (1942) description is the innervation of levatores arcuum branchialium in chondrichthyans by "spinal" nerves, whereas those in osteichthyans are mainly innervated by nerves CNX and/or CNIX. This curious discrepancy—coupled with the failure by ourselves and other authors to repeat Kesteven's observation—calls into question the presence of distinct levatores arcuum branchialium in chondrichthyans. The levatores arcuum branchialium sheet described by Lightoller (1939) corresponds to a part of the cucullaris, although it cannot be ruled out that Kesteven's levatores arcuum branchialium is part of the epibranchial musculature (Didier 1995) or even part of the trunk musculature.

It is more parsimonious to assume that the levatores arcuum branchialium were acquired in osteichthyans (one step) than that they were acquired in the LCA of gnathostomes and then lost in chondrichthyans (two steps). The plesiomorphic condition for gnathostomes is thus likely an undivided cucullaris attached to both the pectoral girdle and the dorsal portion of the branchial arches. In osteichthyans, the cucullaris became divided into a protractor pectoralis and the levatores arcuum branchialium. More stem-ward (basal) than either of these groups, placoderms (extinct stem gnathostomes with jaws) had a dermal neck joint capable of vertical rotation between the skull and trunk armor. A movement along the plane of the joint implies levator and depressor muscles (Trinajstic et al. 2013 and citation within). Based on the preserved muscular tissues and osteological correlates of the attachment sites, placoderm arthrodires such as *Compagopiscis* are reconstructed with a cucullaris, a levator arcus palatini, parts of the constrictores branchiales, a levator capitis major and minor, and hypobranchial muscles (Johanson 2003; Sanchez et al. 2013; Trinajstic et al. 2013). The cucullaris appears to have inserted medially to the dermal pectoral girdle or to the anterior dorsolateral plate (Trinajstic et al. 2013). The branchial arches are rarely mineralized or preserved in placoderms. Therefore, it remains unresolved whether or not the cucullaris of these fishes also attached to branchial arches as in chondrichthyans. As for the levator capitis, the reconstructed element would have involved in head elevation and may represent an epibranchial muscle (Trinajstic et al. 2013). Its superficial position is consistent with the epibranchial muscles of cyclostomes, whereas the epibranchial muscles are deep and restricted to the branchial region in chondrichthyans (Edgeworth 1935). Elasmobranchs have separate muscles in the position of the levator capitis of placoderms (e.g., protractor dorsalis pectoralis), which are considered trunk muscles in the present work but may alternatively be epibranchial muscles.

The coracobranchiales are puzzling muscles. They connect the pectoral girdle and the branchial arches on the ventral side of the body and receive the innervation from spinal nerves and/or CNX. Two interpretations exist in the literature: (a) the coracobranchiales/pharyngoclaviculares are branchial muscles derived from the ventral part of the branchial muscle plates (Edgeworth 1935) and (b) the coracobranchiales of gnathostomes correspond to the interbranchiales of cyclostomes (Mallatt 1996). The coracobranchiales (or their derivatives) are always innervated, partly or entirely, by CNX. It is only in chondrichthyans that spinal innervation from the cervical plexus occurs (Edgeworth 1935). Edgeworth (1935) considered the spinal innervation in chondrichthyans as a secondary condition. However, this interpretation rests on two confounding factors: (a) Edgeworth (1935) arbitrarily used dipnoans as a reference taxon and (b) Edgeworth (1935) named the coracobranchiales as "hypobranchial cerebral" muscles, even though he consistently and unambiguously included the coracobranchiales in the branchial muscles. Based on his detailed anatomical study of holocephalans, Didier (1995) contradicted Edgeworth's hypothesis and suggested that the innervation by CNX is a more derived state within chondrichthyans. More recently, Anderson (2008) supported the branchial nature of the coracobranchiales in chondrichthyans and further suggested a homology between the chondrichthyan coracobranchiales and the osteichthyan pharyngoclaviculares. As for Mallatt's (1996) hypothesis about the homology between the chondrichthyan coracobranchiales and the cyclostome interbranchiales, these two groups of muscles have no similarity in attachments and functions (Johanson 2003). Therefore, four hypotheses that are not necessarily mutually exclusive remain to be tested: (a) the coracobranchiales are gnathostome neomorphs (as argued in the present work and shown in Table 5.4); (b) the coracobranchiales have yet unidentified homologues in cyclostomes; (c) the coracobranchiales were lost in cyclostomes; and (d) the coracobranchiales are hypobranchial muscles (Wiley 1979a,b).

The epibranchial muscles develop from the anterior myotomes and migrate into the head (Edgeworth 1935). Based on the spinal innervation, Edgeworth (1935) recognized the interpharyngobranchiales and subspinalis in selachians and holocephalans as the only epibranchial muscles present in extant gnathostomes. Our dissections of selachians supports Marion's (1905) interpretation of the subspinalis as the first interpharyngobranchial muscle. Kesteven (1942) described the epibranchial muscles (the interpharyngobranchiales) in batoids such as *Dasyatis*, although the muscles are reduced to fibrous structures in *Raja*. However, it is likely that the muscles are absent in this group (Miyake et al. 1992). Regardless of whether or not batoids have any epibranchial muscle, the presence in holocephalans, selachians, and cyclostomes suggests that the LCA of extant vertebrates probably had an epibranchial musculature, which was then lost in the node leading to extant osteichthyans.

The hypobranchial muscles are midventral elements that arise in the anterior myotomes, migrate into the head below the pharynx, and retain the spinal innervation (Edgeworth 1935). Both cyclostomes and gnathostomes have a hypobranchial musculature. The coracoarcualis is present in the elasmobranchs (Miyake et al. 1992; Motta et al. 1997, 2008). In *H. colliei*, the sternohyoideus originates from the coracoid and the fascia of the coracomandibularis. Therefore, either (a) the coracoarcualis was present in the LCA of chondrichthyans and secondarily lost in holocephalans or (b) the coracoarcualis evolved in elasmobranchs *via* fusion of the sternohyoideus and coracomandibularis. Although the latter hypothesis is more parsimonious, it remains unresolved whether the absence of the coracoarcualis in *H. colliei* represents a general holocephalan condition. The likely presence of the coracoarcualis in placoderms favors the former hypothesis (Johanson 2003). The coracomandibularis and coracohyoideus of elasmobranchs in Adams (1919) are misidentified hyoid muscles innervated by CNVII (see pages 64–66). The number of individual muscles in the hagfish *M. glutinosa*, the lamprey *P. marinus*, the holocephalan *H. colliei*, the selachian *S. acanthias*, and the batoid *L. erinacea* are, respectively, 24, 26, 7, 5, and 6 for the mandibular muscles; 4, 0, 4, 3, and 6 for the hyoid muscles; 3, 76, 16, 27, and 28 for the branchial muscles; and 3, 1, 2, 3, and 4 for the hypobranchial muscles (Tables 5.2 through 5.5; Figure 3.1). We draw the following conclusions from these counts: (a) a small number of mandibular muscles of the holocephalan, selachian, and batoid taxa has comparable counterparts in the cyclostome taxa, likely with an increase of muscles in cyclostome evolution; (b) *P. marinus* lacks hyoid muscles innervated by CNVII at the adult stage (NB: at the larval stage, the constrictor prebranchialis is innervated by CNVII according to Miyashita (2012), which likely represents a change within the lamprey lineage; (c) a smaller number of the branchial muscles in *M. glutinosa* and a larger number of branchial muscles in *P. marinus* relative to those of the gnathostome taxa suggest loss/reduction in hagfishes and an increase in lampreys; and (d) the number of distinct hypobranchial muscles does not markedly differ between the comparative taxa. The large numbers of mandibular muscles in hagfishes and lampreys relative to those in gnathostomes are a potential synapomorphy of cyclostomes, although independent evolution in each of these lineages cannot be ruled out.

The comparative analysis infers that the LCA of the living vertebrates had a great number of muscles still present in the living taxa (Tables 5.2 through 5.5). As for the adductores branchiales and interbranchiales (absent in hagfishes and osteichthyans), it is more parsimonious that they appeared independently in lampreys and chondrichthyans (two steps) than to assume a common origin and independent loss in hagfishes and osteichthyans (three steps). However, developmental and paleontological insights into the pharyngeal muscles of invertebrate chordates and early vertebrates are needed to test this hypothesis.

The muscles inferred for the LCA of extant Gnathostomata are therefore intermandibularis anterior, intermandibularis posterior, adductor mandibulae A2, levator arcus palatini, and at least some labial muscles; interhyoideus, adductor arcus palatine, and constrictor hyoideus dorsalis; constrictores branchiales, coracobranchiales, at least some epibranchial muscles, and a cucullaris; and coracomandibularis and sternohyoideus. The likely plesiomorphic gnathostome state for the cucullaris is a single bundle of true branchial muscle innervated by CNXI and inserted exclusively onto the pectoral girdle. The cucullaris splits into multiple bundles in batoids and holocephalans toward the adult stage. The general evolutionary trend seen in vertebrates toward a greater separation between the head and pectoral girdle later culminated in the separation of the cucullaris into the protractor pectoralis and levatores arcuum branchialium in osteichthyans and finally in the formation of the tetrapod neck with sternocleidomastoideus and trapezius derived from the protractor pectoralis.

METAMORPHOSIS, LIFE HISTORY, DEVELOPMENT, MUSCLES, AND CHORDATE EARLY EVOLUTION

This section is based on Diogo and Ziermann (2015a). As pioneered by De Beer's (1940) *Embryos and Ancestors*, the origin and early evolution of vertebrates has inspired broader theoretical questions such as the relationship between ontogeny and phylogeny. In the early twentieth century, the LCA of vertebrates was reconstructed as a neotenic chordate (e.g., Garstang 1928; reviewed by Gee [1996]). Specifically, an elongate, motile chordate larva that metamorphosed into a sessile adult was considered to be the plesiomorphic condition for both urochordates and deuterostomes. This scenario suggested swimming capacity and elongated adult body to be derived conditions in cephalochordates and vertebrates due to retardation of somatic development relative to sexual maturity with respect to the ancestor (neoteny). Neoteny, or "juvenilization," is one type of paedomorphosis. It is defined as reduced rate of morphological differentiation (McNamara 1986), in other words, the retention of traits that are ancestrally specific to ontogenetically younger individuals or the deletion of terminal somatic developmental stages. Another type of paedomorphosis is progenesis (precocious sexual maturation), in which sexual development is accelerated relative to somatic growth compared to the ancestors (De Beer 1940; Gould 1977; McNamara 1986, 1990). According to the evolutionary scenario most accepted during De Beer's (1940) time, the occurrence of paedomorphic events during the transitions between the LCA of chordates, the LCA of vertebrates, and the LCA of gnathostomes was considered more likely than the occurrence of opposite peramorphic (often also named gerontomorphic) events, in which ancestral adult characters appear in descendant juveniles due to additions of terminal somatic developmental stages (McNamara 1990). This is because the latter (peramorphic) transitions would include (a) the restriction of the typical adult swimming elongated body adapted to suspension feeding of the LCA of vertebrates to the lamprey larva and (b) the addition of new, derived terminal ontogenetic stages that resulted in the metamorphosis of the structures of the lamprey larva into structures that are mainly related to parasitism, which is the mode of live of many adult lampreys. Therefore, the resemblance between the elongate, motile adult in the LCA of gnathostomes and the motile lamprey larvae (ammocoetes) was usually seen as the result of neotenic events that occurred during the transitions from the LCAs of vertebrates and of gnathostomes, *not* as the result of peramorphic events that occurred during the transitions from the LCAs of vertebrates and of cyclostomes.

However, most authors now accept that urochordates (and not cephalochordates) are the closest living relatives of vertebrates, and moreover, many authors argue that appendicularians (and not ascidians) are the most basal extant urochordates. Because cephalochordates also have a motile adult stage, the adult LCA of urochordates + vertebrates probably did as well. The implication is that sessile adults in extant ascidians represent a derived condition *via* terminal additions (peramorphosis) during the transitions from the LCAs of urochordates to that of ascidians. Either way, as an adult the LCA of vertebrates had an elongated motile body similar to that of the adult members of basal gnathostomes and of amphioxus. Adult amphioxus and lamprey larvae (ammocoetes) were often compared by early anatomists. Similarities include pharynx with pharyngeal slits, a notochord, a dorsal hollow nerve cord, and a series of somites that extend anterior to the otic vesicle. Like amphioxus, the ammocoetes are filter feeders and lack jaws (Holland et al. 2008). However, amphioxus larvae are highly asymmetric, probably due to derived adaptations associated with the feeding mechanism, and, thus, to the addition of early ontogenetic stages (Presley et al. 1996). In contrast, ammocoetes also possess some apparently plesiomorphies like an endostyle, and their early embryonic pattern resembles that of hagfishes, while their oral apparatus, including upper and lower lips, resembles those of some adult fossil gnathostome species more than the adult lamprey does (Kuratani et al. 2002).

As explained earlier, recent molecular and developmental data have contradicted the physiological and/or morphological (including paleontological) data supporting the paraphyly of cyclostomes, for instance, by providing examples of similar ontogenetic skeletal features and their developmental mechanisms shared uniquely by hagfish embryos and lamprey larvae (but not by adults of these two groups), as well as by these embryos and larvae and both the embryos and adults of gnathostomes (e.g., Oisi et al. 2013a,b). By doing so, they further supported the idea that during the transitions leading to extant cyclostomes there was (a) a restriction, *via* developmental acceleration, of the typical features associated with suspension feeding (supposedly seen in adult members of the LCA of vertebrates) to early ontogenetic stages and (b) the addition of derived terminal ontogenetic stages (peramorphosis) in the origin of cyclostomes and then in independently in hagfishes and particularly in lampreys, resulting in the highly peculiar adult features of each of these two groups.

The adult lamprey cannot simply be regarded as a hypermorphic state of hypothetical lamprey larva-like adult LCA of cyclostomes. This is partly because lamprey adults have structures underdeveloped in larvae but shared with hagfish. For example, the lingual apparatus in adult lampreys (also present in adult hagfish) appears after metamorphosis (Yalden 1985). That is why the more likely scenario is that some of the peculiar adult features of hagfishes and lampreys were acquired independently, while others were acquired during the transitions leading to the LCA of cyclostomes. Together with the data mentioned in the previous paragraph, these data thus seem to indicate that metamorphosis might have been acquired independently in cephalochordates (mainly due to early ontogenetic somatic additions), lampreys (mainly due to terminal ontogenetic somatic additions; NB: hagfishes have no true metamorphosis), and some gnathostomes (e.g., actinopterygians and amphibians, in which terminal somatic additions are usually related to transitions to a terrestrial lifestyle). In fact, although in chordates metamorphosis appears always to be regulated by thyroid hormones, it shows very different patterns in different chordate groups; e.g., in ascidians all larval tissues are drastically remodeled into the adult animal; in amphioxus, the highly asymmetric larva transforms into a relatively symmetric adult; and in some flatfish, the symmetric larvae becomes an asymmetric adult (Paris et al. 2008).

Metamorphosis is a process in which a larval form is remodeled into an adult form that differs from the former one in morphology and ecology. Metamorphosis occurs in urochordates (e.g., ascidians), cephalochordates, lampreys, and some ray-finned fishes (Actinopterygii) and lissamphibians (Paris and Laudet 2008; Paris et al. 2008, 2010). It is usually accepted that metamorphosis evolved several times in chordates (e.g., Heyland et al. 2005). However, a common character for all these chordate taxa is that metamorphosis is always tightly linked to changes in thyroid hormone levels (Paris and Laudet 2008). This led some authors to propose that metamorphosis might represent a homologous feature within these taxa (Paris and Laudet 2008; Paris et al. 2008). However, the morphological remodeling process is drastically different in each of the metamorphic chordate taxa. For instance, the remodeling is especially drastic in frogs: a free swimming larva with its own feeding mechanism and without legs reorganizes the cranium and internal organs, develops limbs, and loses its tail (e.g., Rose 2005). In lampreys, structures remodeled during metamorphosis include the eyes, the gut epithelium, the mouth, and the larval kidney (replaced by a juvenile kidney), and the endostyle (a thyroid gland in adults) (Youson 1980, 1997). Hagfishes have a direct development, but their postembryonic development is not yet fully studied (Jørgensen 1998). Thyroid hormones are also involved in the metamorphosis of the cephalochordate amphioxus, in which the bilaterally asymmetric motile larva transforms into a benthic and more bilaterally symmetric juvenile by shifting the left mouth anteroventrally and by duplicating and shifting the pharyngeal slits from the right side to the left side (e.g., Willey 1894; Stokes and Holland 1995; Paris et al. 2008). Within urochordates, ascidians undergo a metamorphosis, whereas members of the Appendicularia keep a larval body plan in adult stages and members of Thaliacea seem to be direct developers as they skip an obvious larval stage (Paris and Laudet 2008).

Moreover, it is clearly more parsimonious to accept that metamorphosis evolved independently than that it was present in the LCA of vertebrates and then secondarily lost. The first possibility requires a minimum of three steps (acquisition in lampreys, actinopterygian fishes, and batrachians [urodeles + anurans]) while the second requires six or seven (loss in hagfishes, chondrichthyans, coelacanths, dipnoans, caecilians, and amniotes, and eventual acquisition in the LCA of vertebrates if not present in the LCA of urochordates + vertebrates). Therefore, it is more parsimonious to assume that the involvement of thyroid hormones in metamorphosis is homoplastic. Furthermore, the degree of changes during metamorphosis differs in and between different clades, and developmental patterns can be profoundly different (e.g., presence of both metamorphosis and direct development in urochordates, urodeles and anurans). Direct development can be interpreted as a suppression of the larval stage. However, in the direct developing frog *Eleutherodactylus coqui*, it was shown that thyroid hormones are still important during embryonic development (e.g., for skin, muscle and gut development), and the comparison of cranial muscle development in *E. coqui* and in metamorphosing frogs supports the idea of a cryptic metamorphosis (not obvious metamorphosis in the egg) (e.g., Elinson 2013; Ziermann and Diogo 2014).

Overall, the evidence suggests that the adult members of basal gnathostome groups share more similarities with lamprey larvae (ammocoetes) and hagfish embryos than with lamprey and hagfish adults due to peramorphic events that occurred in the evolutionary history of cyclostomes and not to neotenic events leading to the LCA of gnathostomes. However, this hypothesis has not been tested in muscle anatomy studies, which have been neglected in analyses of the origin and early evolution of vertebrates and, in particular, on the comparative anatomical analyses of lampreys, hagfishes, and gnathostomes. The data summarized earlier regarding the adult cephalic muscles of nonosteichthyan vertebrates, including hagfish and lampreys, thus give us a more comprehensive understanding of the relationships between development, metamorphosis, heterochrony, and morphological and taxonomical diversity within chordates. In Figures 5.6 through 5.8, we briefly summarize our hypotheses on the homology and evolution of the larval and adult cephalic muscles in cyclostomes and gnathostomes based on that synthesis and on an extensive review of the literature regarding muscle genetics, development, innervation, attachments, and overall topology and relationship to other structures such as cartilages (see also Tables 5.2 through 5.5).

Importantly, the comparison of the muscles of cyclostomes and gnathostomes (Figures 5.6 through 5.8) supports the idea that the lamprey larva is a better model for the cranial muscles of adults of the LCA of gnathostomes and the LCA of vertebrates than is the adult lamprey. In support of this hypothesis, the adult muscles of the LCA of vertebrates inferred on the basis of our observations and, on comparisons between *adult* taxa, are strikingly similar to those of the lamprey larva (Figure 5.8). As explained earlier, we inferred at least one constrictor hyoideus in the vertebrate LCA; ammocoetes have a likely homologue, the constrictor hyoideus ("prebranchialis"), whereas adult lampreys do not seem to have recognizable hyoid muscles (Figure 5.8). We also inferred at least one labial muscle and an undifferentiated intermandibularis muscle sheet (i.e., not separated or divided into several parts) in the vertebrate LCA, and ammocoetes have two muscles within the labial and intermandibularis groups, whereas lamprey adults have many more (Figures 5.6 through 5.8). We postulated at least one velar and/or dorsal mandibular muscle in the vertebrate LCA; ammocoetes have two of these muscles, while adult lampreys do not seem to have muscles that can be clearly recognized as part of this specific group (Figures 5.6 through 5.8; see also below). Furthermore, we inferred at least some constrictores branchiales, and possibly some adductores branchiales as well, in the vertebrate LCA; ammocoetes have adductores and constrictores branchiales, whereas adult lampreys have a much more complex configuration with an impressive number and many different subgroups of true branchial muscles (Figures 5.6 through 5.8). Based on our inference, the LCA of vertebrates had an undifferentiated epibranchial muscle sheet; ammocoetes have three epibranchial muscles, whereas lamprey adults have five. We inferred that a few of the nasal, lingual, and/or dental muscles might have been present in this LCA; ammocoetes have six muscles, while lamprey adults have many more. Lastly, we inferred the presence of one undifferentiated hypobranchial muscle sheet in the vertebrate LCA, and this is the condition found in ammocoetes; this is the only case in which the condition inferred for this LCA is the same as that found in adult lampreys.

As expected, there are more differences between the lamprey larva (ammocoetes) and the adult members of the LCA of gnathostomes than between the lamprey larva and the adults of the vertebrate LCA. However, despite the peculiar gnathostome synapomorphies, many muscles in the lamprey larva are potentially homologous with those inferred for the adult members of the LCA of extant gnathostomes. The major exceptions (i.e., differences between ammocoetes and adults of the gnathostome LCA) are (a) no nasal, lingual, and/or

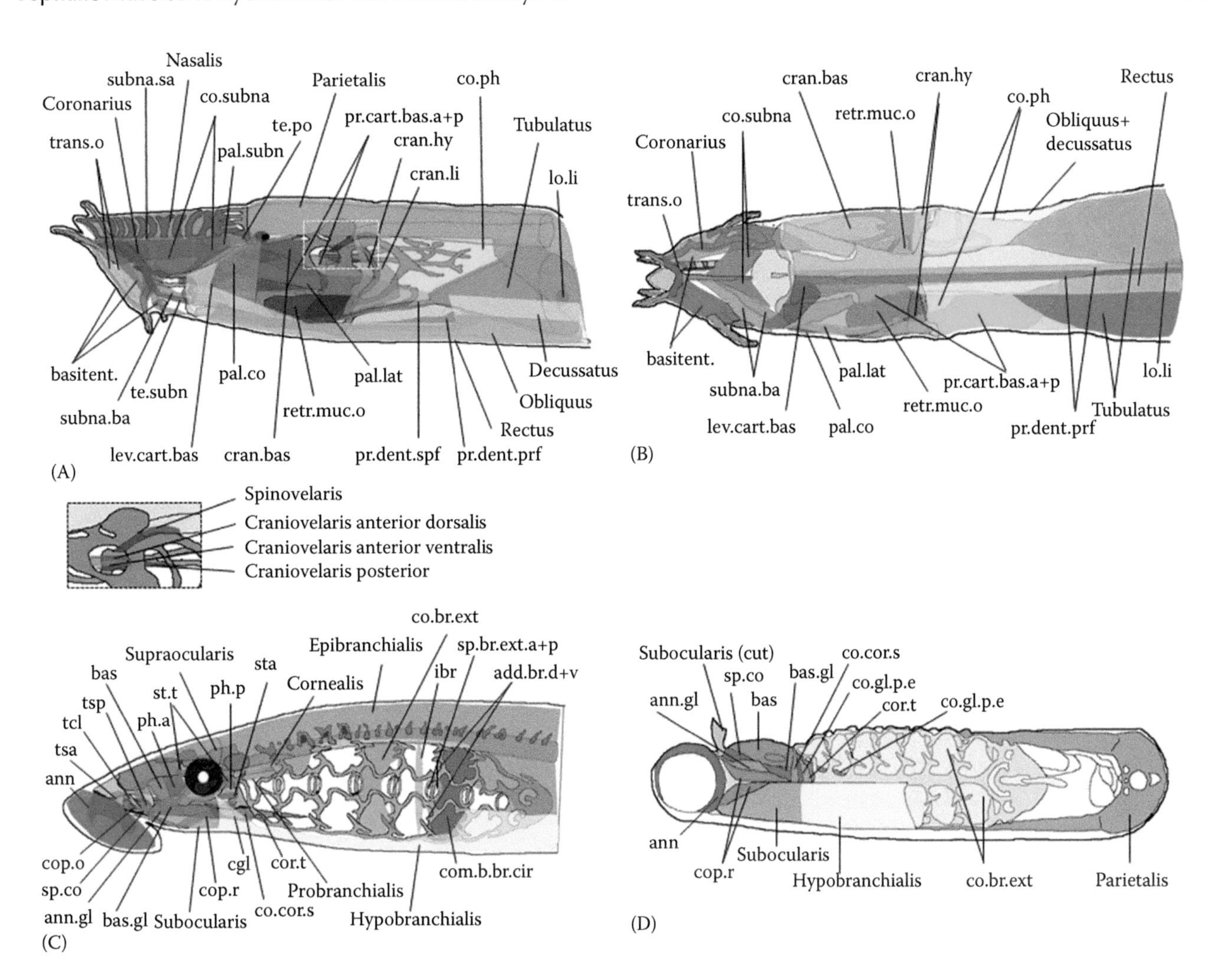

FIGURE 5.6 Our hypothesis, based on the data reviewed here, about the homologies of cephalic muscle groups in Cyclostomata and Chondrichthyes: Mandibular (*red, orange, pink*); hyoid (*green*), true branchial (*blue*), epibranchial (*bruin*), and hypobranchial (*green*) muscles. Left lateral and ventral view of (A, B) *M. glutinosa* and (C, D) *P. marinus.* (Chondrocranium of *M. glutinosa* and *P. marinus* redrawn from Marinelli and Strenger (1954); Marinelli, W., and Strenger, A., *Vergleichende Anatomie und Morphologie der Wirbeltiere: II Lieferung:* Myxine glutinosa *(L)*, pp. 81–172, Deuticke, Vienna, 1956.) (Chondrocranium of *H. colliei* after dissections of J. M. Ziermann.) (Chondrocranium of *S. acanthias* redrawn from Livingstone, I., I.Livingstone@BIODIDAC, 2014.) Specimens not to scale. Not all muscles shown on both sides in ventral view. Not all branchial arch muscles are shown in some views. (A) Parietalis and decussatus cut to enable view to deeper layers. White box—velar muscles in window below the animal. (B) Basitentacularis cut on right side. (C) All branchial muscles (blue) are present in each segment but not shown in each segment. (D) The hyobrachiales extends backward but was cut to show the branchial basket. add.br.d+v, adductores branchiales dorsales + ventrales; add.mand.A2, adductor mandibulae A2; ann, annularis; ann gl, annuloglossus; bas, basilaris; bas.gl, basilariglossus; basitent., basitentacularis; cgl, cornuoglossus; co.br.ext., constrictores branchiales externi; co.cor.s, constrictor cornualis superficialis; co.gl.p.e, constrictor glossae profundus externus; co.ph, constrictor pharyngis; co.subna, cornuosubnasalis; com.b.br.cir, compressores bursae branchiales circulares; cop.r, copuloglossus rectus; cop.o, copuloglossus obliquus; cor.t, cornuotaenalis; cran.bas, craniobasalis; cran.hy, craniohyoideus; cran.li, craniolingualis; ibr, interbranchiales; lev.ang.o.a, levator anguli oris anterior; lev.cart.bas, levator cartilagines basalis; lev.pr, levator prelabialis; lo.li, longitudinalis linguae; pal.co, palatocoronarius; pal.lat, palatinalis lateralis; pal.subn, palatosubnasalis; ph.a, pharyngicus anterior; ph.p, pharyngicus posterior; pr.cart.bas.a+p, protractor cartilagines basalis anterior and posterior; pr.dent.prf, protractor dentium profundus; pr.dent.spf, protractor dentium superficialis; retr.muc.o, retractor mucosae oris; sp.br.ext.a+p, sphincters branchiales anteriores and posteriores; sp.co, spinosocopularis; sta, styloapicalis; st.t, stylotectalis; subna.ba, subnasobasalis; subna.sa, subnasonasalis; tcl, tectolateralis; te.po, tentacularis posterior; te.subn, tentaculosubnasalis; trans.o, transversus oris; tsa, tectospinosus anterior; tsp, tectospinosus posterior.

dental muscles are inferred for the gnathostome LCA and (b) the adult gnathostome LCA likely had an adductor mandibulae A2 (Figures 5.6 through 5.8). However, it is possible that all these muscles derive from the same mandibular anlage: the dorsomedial mandibular *sensu* Edgeworth (1935) (see Diogo and Abdala 2010). The third major difference between adults of the LCA of gnathostomes and ammocoetes is that the former have a cucullaris, which, according to the scenario proposed in the present chapter/book, is part of the branchial musculature. However, according to some authors, the cucullaris may be derived from "infraotic" epibranchial muscles, which are present in the ammocoetes (see preceding section).

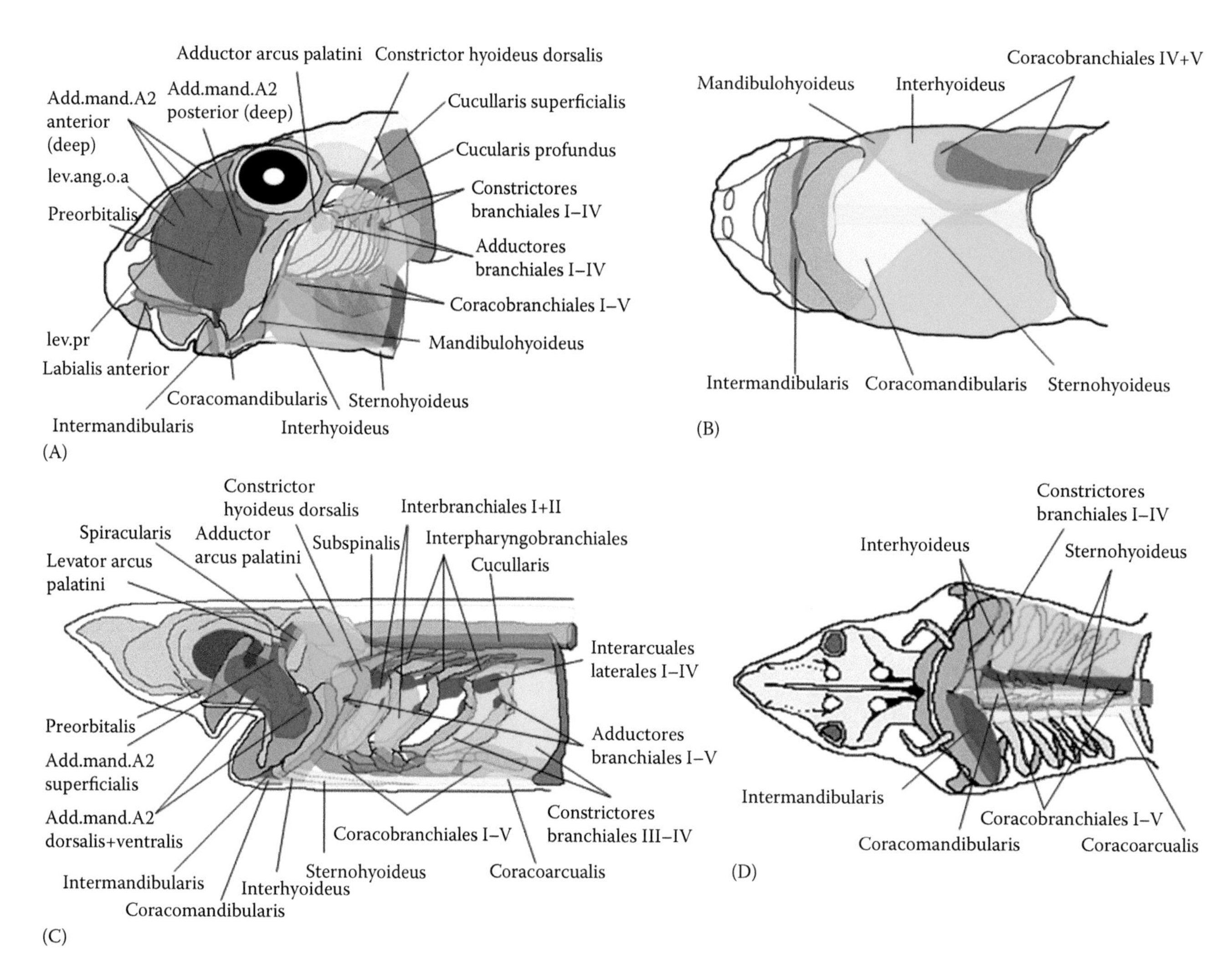

FIGURE 5.7 Our hypothesis, based on the data reviewed here, about the homologies of cephalic muscle groups in Cyclostomata and Chondrichthyes: mandibular (*red, orange, pink*); Hyoid (*green*), true branchial (*blue*), epibranchial (*bruin*), and hypobranchial (*green*) muscles. Left lateral and ventral view of (A, B) *H. colliei* and (C, D) *S. acanthias.* (F) Coracobranchiales I–III not shown. (G) Constrictores branchiales I + II and interbranchiales III + IV not shown. See abbreviations in caption of Figure 5.6.

In addition to the direct comparison between the muscles of larval lamprey and of adults of other vertebrate taxa, support for the hypothesis that lampreys are peramorphic vertebrates also comes from a broader analysis of the number of muscles found in adults of both cyclostome and noncyclostome taxa (Figures 5.6 through 5.8). As noted earlier, the number of mandibular muscles in adults of *M. glutinosa, P. marinus, H. colliei, S. acanthias*, and *L. erinacea* is, respectively, 24, 26, 7, 5, and 6, respectively; for the hyoid muscles is 4, 0, 4, 3, and 6; for the branchial muscles is 3, 76, 16, 27, and 28; and for the hypobranchial muscles is 3, 1, 2, 3, and 4. The total number of adult muscles in *P. marinus* is 103, whereas the total number of cranial muscles in adults of the other taxa ranges between 29 and 44. The mandibular muscle count in adult cyclostomes is therefore much higher than in adult chondrichthyans and osteichthyans, which indicates a significant increase of the number of these muscles during cyclostome evolution. Regarding the muscles of lamprey larvae (ammocoetes), in total, we could identify 44 larval cephalic muscles. Twelve (10 mandibular + 1 hyoid + 1 epibranchial) of larval cephalic muscles seem to be only present in larvae, disappearing during metamorphosis; the other 32 muscles persist until adulthood. Interestingly, the persisting muscles are all hypobranchial, epibranchial, and true branchial. The muscle count per group in ammocoetes is: mandibular, hyoid, true branchial, epibranchial, hypobranchial: 10, 1, 28, 4, 1; these numbers are therefore clearly much closer to the count in the adult gnathostome fishes dissected by us than the total, striking number of adult cephalic muscles found in *P. marinus* (103).

Although in ammocoetes, all five groups of cephalic muscles (mandibular, hyoid, branchial, epibranchial, and hypobranchial) are well separated from each other, most of these groups are in such a close contact to each other that it seems that they became secondarily fused to each other in adults (e.g., the hypobranchial muscle hypobranchialis with

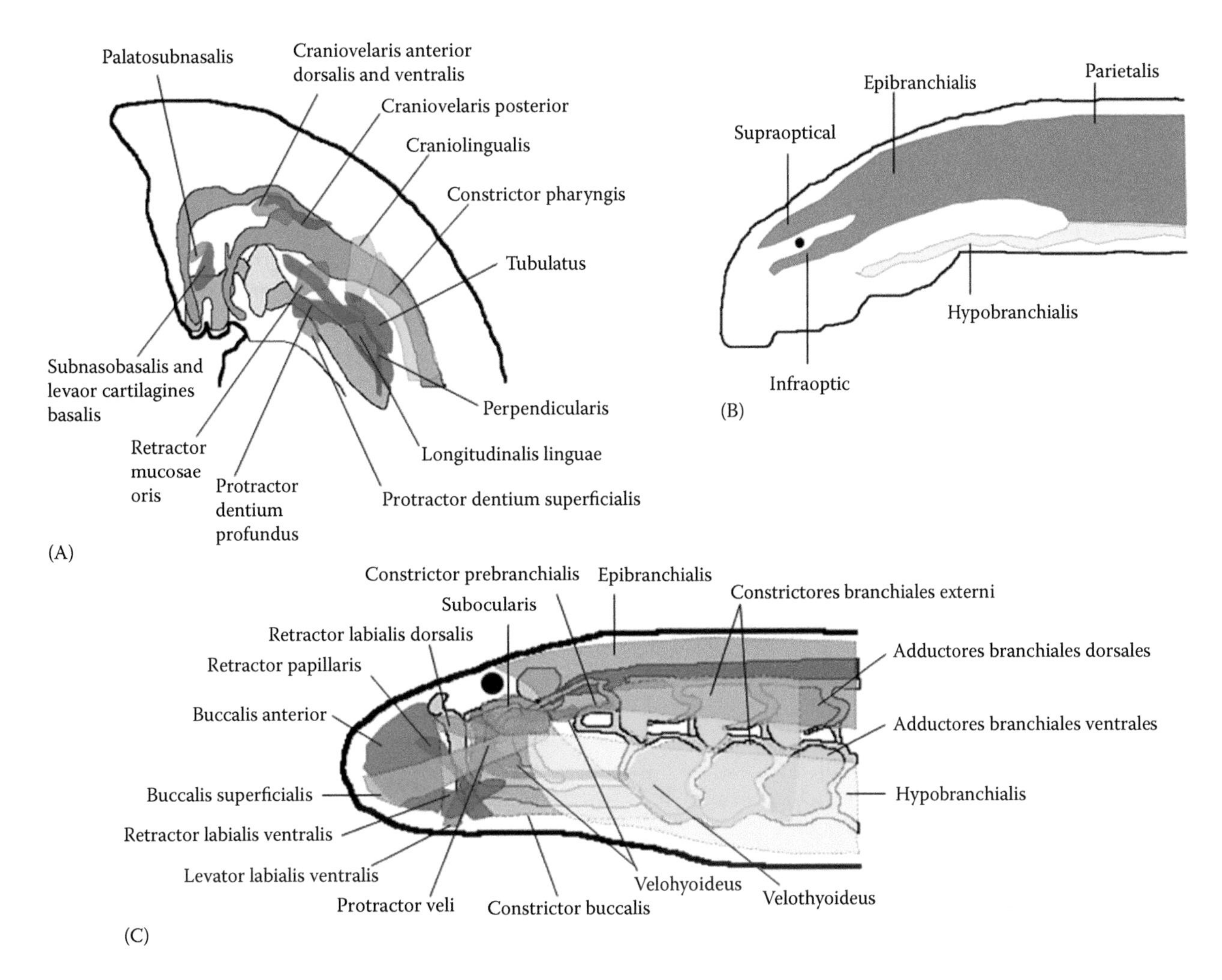

FIGURE 5.8 Our hypothesis, based on the data reviewed here, about the homologies of cephalic muscle groups of embryos/larvae of cyclostomes: mandibular (*red*, *orange*, *pink*); hyoid (*green*), true branchial (*blue*), epibranchial (*bruin*), and hypobranchial (*green*) muscles. Left lateral views. (A) *M. glutinosa* embryo. Somites (removed for clarity) extend to just behind the otic capsule (*M. parietalis* and *M. obliquus* appear not fully differentiated at this stage) (Oisi et al. 2013a,b). (Redrawn from Miyashita, T., *Comparative Analysis of the Anatomy of the Myxinoidea and the Ancestry of Early Vertebrate Lineages*, Thesis, University of Alberta, Edmonton, 2012.) (B) *Petromyzon*: ammocoete larva. (Redrawn from Tulenko, F. J. et al., *PNAS*, 110, 11899–11904, 2013.) (C) Ammocoete larva. Not all branchial muscles shown; constrictores branchiales externi sheet cut to reveal underlying adductores branchiales. Specimens not to scale. (Redrawn from Miyashita, T., *Comparative Analysis of the Anatomy of the Myxinoidea and the Ancestry of Early Vertebrate Lineages*, Thesis, University of Alberta, Edmonton, 2012.)

epibranchial muscle subocularis: Figure 5.6). On the contrary, in adult chondrichthyans, these muscle groups are generally more clearly separated from each other (Figure 5.7). Lamprey larvae (ammocoetes) have lower and upper lips, which resemble the oral apparatus of some adult fossil vertebrate taxa such as heterostracans and osteostracans more than those of the adult lamprey (e.g., Kuratani et al. 2002). These lips are associated with labial muscles that are more similar to adult gnathostome muscles than are the adult lamprey muscles. In addition, the engrailed immunoactivity in the lamprey larval muscle velothyroideus (which is absent in adult lampreys) led to the suggestion that this muscle might be homologous to the adult and embryonic muscles levator arcus palatini and dilatator operculi of teleosts (Holland et al. 1993). These two latter teleost muscles are included in the mandibular constrictor dorsalis group (*sensu* Edgeworth 1935) and, therefore, in the same group as the levator arcus palatini of the adult members of the LCA of gnathostomes. The visceral muscles of adult lampreys differentiate from blastema as those in ammocoetes degenerate during metamorphosis, and it is still not clear whether the larval muscles are precursors of adult muscles or not. Therefore, the identification of possible correspondence, even topological, between the larval and adult lamprey muscles is very difficult because of the dramatic rebuilding processes during metamorphosis (e.g., Marinelli and Strenger 1954). In the case one accepts Holland et al.'s (1993) proposed homology, this would imply that the velum of lampreys and the gnathostome palatoquadrate might be homologous to each other, or at least, homogenic (e.g., Miyashita 2012). During lamprey metamorphosis, the velum and endostyle are both reduced; furthermore, a lingual apparatus develops in the location corresponding to the larval muscle velothyroideus. Therefore, if the homology proposed by Holland et al. (1993) would be accepted, this would provide a further case where the muscles of ammocoetes have a better correspondence to the muscles present in the adult members of the LCA of gnathostomes than have the adult lamprey muscles.

In summary, both the larval chondrocrania (Oisi et al. 2013a,b) and the larval muscles of cyclostomes more closely resemble the adult plesiomorphic vertebrate and gnathostome condition than do the adult cyclostome structures. As noted earlier, in the development of cephalic muscles in the amphioxus *Branchiostoma japonicum* (cephalochordate), the asymmetric larval mouth and unpaired primary pharyngeal slits are associated with asymmetric orobranchial muscles (Yasui et al. 2013). The iterated branchial (pharyngeal) muscles match the dorsal myomeric pattern before metamorphosis. The oral and pharyngeal regions are then dramatically remodeled during metamorphosis. Meanwhile, the orobranchial musculature disappears and the adult muscles in the oral hood, velum, and the pterygial coeloms develop independently, as well as some branchial muscles that were associated with the secondary pharyngeal arch/gill formation which also disappeared. Yasui et al. (2014) suggested that in some ways, the larval, rather than the adult, branchial system of amphioxus better corresponds to the vertebrate system, but then recognized that the amphioxus larva very likely represents a derived feature acquired during the evolutionary history of cephalochordates.

The latter idea is also supported by our comparisons. First of all, the branchial muscles exist mainly (i.e., with the exception of the vestigial muscles associated with the secondary gill formation) when the amphioxus displays a unilateral body pattern that is markedly distinct from the vertebrate bilateral pattern, and their function seems to be specialized for preventing harmful intake in larval life. Because there is no antagonistic system against the larval orobranchial musculature that functions as obturator and levator, the pharynx, once constricted, recovers its dilatation very slowly, and thus the muscle system does not serve as a rhythmic pump or function in active predation. These features support instead the hypothesis that the larval orobranchial musculature is a derived feature for larval adaptation. That is, the highly asymmetric amphioxus larva is probably the result of additions of somatic stages early in cephalochordate ontogeny, and thus, the adult members of this group are very likely a better model for the adult condition of the LCA of vertebrates in this regard. Second, the transient appearance of branchial muscles in association with the secondary gill formation during metamorphosis suggests that the larval unilaterality of extant amphioxus has been derived from a bilateral LCA of chordates that possessed a bilateral muscular branchial system similar to that found in the LCA of vertebrates. Third, the developmental contrast between the labial/velar/pterygial and somatic pattern after the metamorphosis of amphioxus is similar to that found in the head of the ammocoetes. Finally, the postmetamorphic, symmetric amphioxus labial, velar, and pterygial muscles are more similar to the labial, velar, and other branchiomeric muscles of cyclostome larvae/embryos than are the highly asymmetric larval amphioxus cephalic muscles. Notably, the adult pterygial amphioxus muscles develop ventrally in the pharynx and are innervated by peripheral nerves that are similar to the branchiomeric nerves in vertebrates, as explained earlier (Fritzsch and Northcutt 1993; Yasui et al. 2013). In summary, these developmental data suggest that (a) metamorphosis was acquired independently in cephalochordates (mainly due to early ontogenetic somatic additions) and lampreys (mainly due to terminal ontogenetic somatic additions) and (b) in general, adult members of basal gnathostome groups share more similarities with ammocoetes than with lamprey and hagfish adults, due to peramorphic events that occurred in the evolutionary history of cyclostomes and not due to neotenic events leading to the LCA of gnathostomes. This inference conforms to De Beer's (1940) prediction that contrary to what was often believed during the decades in which Haeckel's recapitulation was in vogue, peramorphic events often produce clades with fairly limited taxonomical diversity. While in cases of paedomorphosis, the deletion of terminal ontogenetic stages often leads to more generalized forms that may in turn evolve in completely new ways; in cases of peramorphosis, the addition of terminal stages often results in very specialized forms adapted to very specific environments. The latter case is seen in lampreys, in which most adults have a parasitic life style and have a remarkably specialized cranial musculature comprising a total of 103 muscles. Accordingly, lampreys as well as hagfishes have a remarkably poor taxonomic diversity, with only 38 and 60 extant species (Nelson 2006; Hardisty 2006), in contrast to the roughly 60,000 living gnathostome species.

GENERAL REMARKS

In the light of the recent studies reviewed here, the dichotomy between head and trunk appears less clear than previously thought. Structures often considered in textbooks to be part of the trunk, such as the cucullaris and its derivatives, are developmentally and evolutionarily head structures. In turn, some structures in the adult head are derived from trunk primordia (e.g., the tongue muscles from ventral somites), emphasizing the striking heterogeneity and evolutionary complexity of the vertebrate head. Recent data also undermine the clear distinction previously made between vertebrates and other animals. Urochordates have a cardiopharyngeal field as well as well as neural crest-like cells and very likely have placodes. Another important implication of the data reviewed here is that contrary to previous assumptions and scenarios still often seen in textbooks and even specialized papers, neoteny does *not* seem to have played a major role (in contrast to, e.g., the addition of early/later ontogenetic stages) in the major transitions leading to the origin of chordates, of cephalochordates, of urochordates, of vertebrates, of cyclostomes, and of gnathostomes. In fact, metamorphosis was probably acquired independently in cephalochordates (mainly due to early ontogenetic somatic additions resulting a peculiar asymmetric larva) and lampreys (mainly due to terminal ontogenetic somatic additions, i.e., of peramorphosis, resulting in a highly peculiar parasitic adult). The adults of the LCA of vertebrates were very likely pelagic and motile, as are adult appendicularian urochordates; therefore, the sessile character of adult urochordates (e.g., ascidians) is probably also the result of peramorphosis. The lamprey larvae (ammocoetes) is, therefore, a better model for the cranial muscles of adults of the LCAs of gnathostomes and particularly of vertebrates than is the adult lamprey, due to peramorphic events that have occurred in the evolutionary history of cyclostomes, not to neotenic events leading to the LCA of gnathostomes.

6 Cephalic Muscles of Actinopterygians and Basal Sarcopterygians

The discussions provided in this chapter mainly concern the mandibular, hyoid, branchial, and hypobranchial muscles of actinopterygians and basal sarcopterygians (see cladogram of Figure 1.1, as well as Figures 6.1 and 6.2; as explained in previous chapters, the epibranchial muscles *sensu* Edgeworth [1935] are absent in extant osteichthyans). Figures 6.4 through 6.9 provide an updated version of the diagrams of Miyake et al. (1992), including representatives of some osteichthyan groups that were not considered in Miyake et al.'s paper, such as amphibians and amniotes. The information provided in these figures complements Tables 6.1 through 6.4.

MANDIBULAR MUSCLES

In the sections of this chapter on mandibular, hyoid, branchial, and hypobranchial muscles, we explain Edgeworth's hypotheses regarding the respective plate and discuss the extent to which they are supported by more recent work (Table 6.1; Figures 6.4 and 6.5). According to Edgeworth (1935), in numerous vertebrates, the embryonic mandibular muscle plate gives rise dorsally to the premyogenic condensation constrictor dorsalis, medially to the premyogenic condensation adductor mandibulae, and ventrally to the intermandibularis (no description of a ventral mandibular premyogenic condensation was given by Edgeworth) (Figures 6.4 and 6.5). He suggested that a constrictor dorsalis condensation is not found in extant gnathostome taxa such as holocephalan chondrichthyans, dipnoans, and amphibians. Since he considered that the chondrichthyans, actinopterygians, and tetrapods were derived from an "early dipnoan stock," he concluded that the constrictor dorsalis was plesiomorphically absent in the Gnathostomata and independently acquired in some taxa of this clade.

This is one of the few cases in which one of Edgeworth's conclusions is seriously called into question by current evidence. In fact, very few researchers would now accept that chondrichthyans, actinopterygians, and tetrapods were derived from basal dipnoans (see Figure 1.1 and text). According to the phylogenetic scenario shown in Figures 1.1, 6.1, and 6.2, the constrictor dorsalis was either independently lost within dipnoans and amphibians or lost in the node leading to nonactinistian sarcopterygians and then reacquired in amniotes. Although these scenarios appear to be equally parsimonious, there is reason to favor the first one: the premyogenic condensation of the constrictor dorsalis is very similar in amniotes and in nondipnoan sarcopterygian fishes (e.g., Brock 1938). Therefore, it makes more sense to suggest that such a condensation was lost in dipnoans and in amphibians than to argue that it was lost in nonactinistian sarcopterygians and that a strikingly similar condensation was then independently acquired in amniotes. It is important to stress that it is not only the condensation that is similar in amniotes and nondipnoan bony fishes. The adult muscles derived from it in amniotes such as "lizards"; i.e., the levator pterygoidei and protractor pterygoidei are also very similar to the adult muscle derived from it in sarcopterygians such as *Latimeria*, i.e., the levator arcus palatini (they essentially occupy the same position, running from the neurocranium to the dorsal/dorsolateral margin of the palatoquadrate, and are thus usually related to the elevation of the latter structure: see following chapters). A detailed analysis of the presence/absence of dorsal mandibular muscles in well conserved, plesiomorphic dipnoan and amphibian fossils, as well as in other sarcopterygian fossils, is needed to clarify the actual taxonomic distribution of the dorsal mandibular muscles within the Sarcopterygii.

As stated by, e.g., Edgeworth (1935) and Winterbottom (1974), and supported by molecular developmental studies such as Hatta et al. (1990, 1991), in most extant actinopterygians, the constrictor dorsalis differentiates into a levator arcus palatini and a dilatator operculi (see Table 6.1 and Figures 6.4 and 6.15 through 6.17). The former muscle is usually related to the elevation/abduction of the suspensorium (a structural complex formed by the hyomandibula, quadrate, and pterygoid bones); the latter is mainly associated with the abduction (opening) of the operculum (see, e.g., Stiassny et al. 2000). However, in extant acipenseriforms, the constrictor dorsalis gives rise to a single, peculiar muscle mainly related to the protraction of the hyomandibula, the protractor hyomandibulae (e.g., Danforth 1913; Luther 1913; Sewertzoff 1928; Edgeworth 1935; Kesteven 1942–1945; Miyake et al. 1992; Carroll and Wainwright 2003; in this work, see Table 6.1 and Figure 6.4). The presence of a separate dilatator operculi in adults might thus be seen as a feature acquired in the node leading to all extant actinopterygians and then reverted in a node leading to extant acipenseriforms or, instead, as a feature independently acquired in cladistians and in neopterygians (see Figures 1.1 and 6.1). We agree with authors such as Lauder (1980a,b) and Lauder and Liem (1983) in that there are strong reasons to suggest that a separate dilatator operculi was present in the ancestor of extant actinopterygians. In fact, the dilatator operculi of *Polypterus* and the dilatator operculi of neopterygians have a similar developmental origin (the dorsal part of the mandibular muscle plate), a similar innervation (CNV), a similar function (essentially related to operculum abduction), and a similar overall configuration (e.g., Pollard 1892; Allis 1897, 1922; Edgeworth 1935; Winterbottom 1974; Lauder 1980a,b; Lauder and Liem 1983; Miyake et al. 1992; Diogo and Abdala 2010; Noda et al.

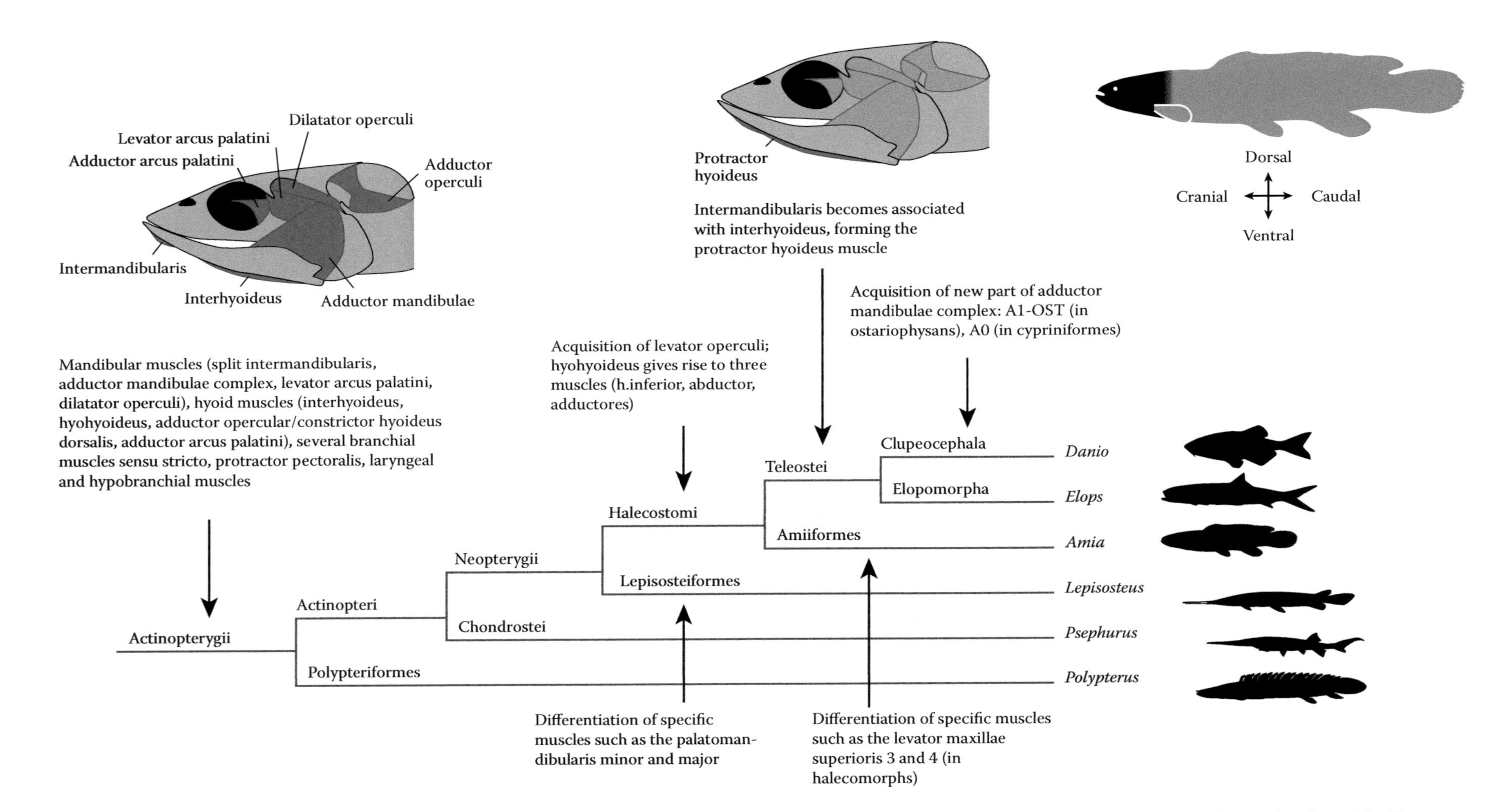

FIGURE 6.1 Some of the major features of the cephalic musculature within the actinopterygians and some of its key subgroups, according to our own data and review of the literature.

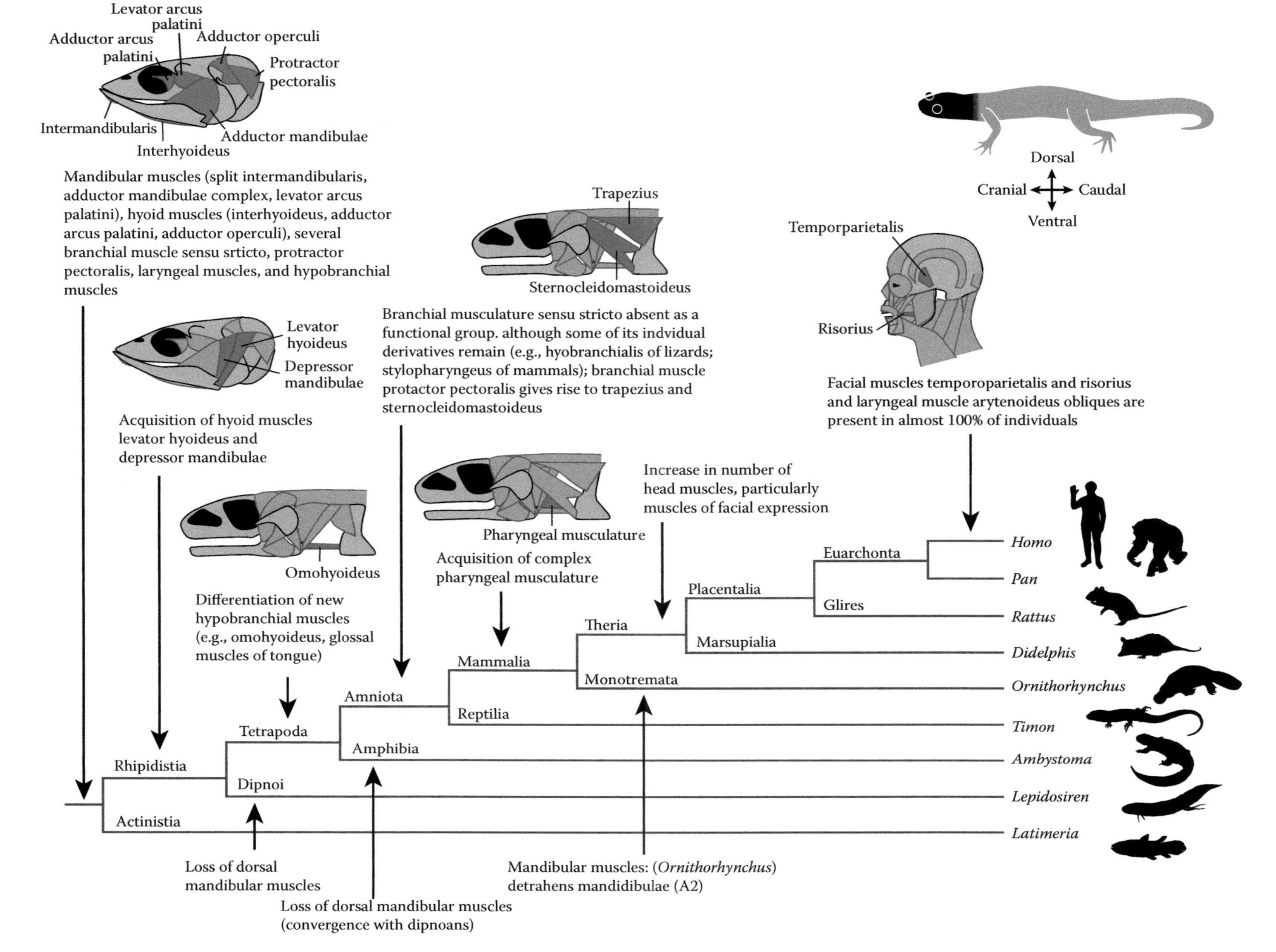

FIGURE 6.2 Some of the major features of the cephalic musculature within the sarcopterygians and some of its key subgroups, according to our own data and review of the literature.

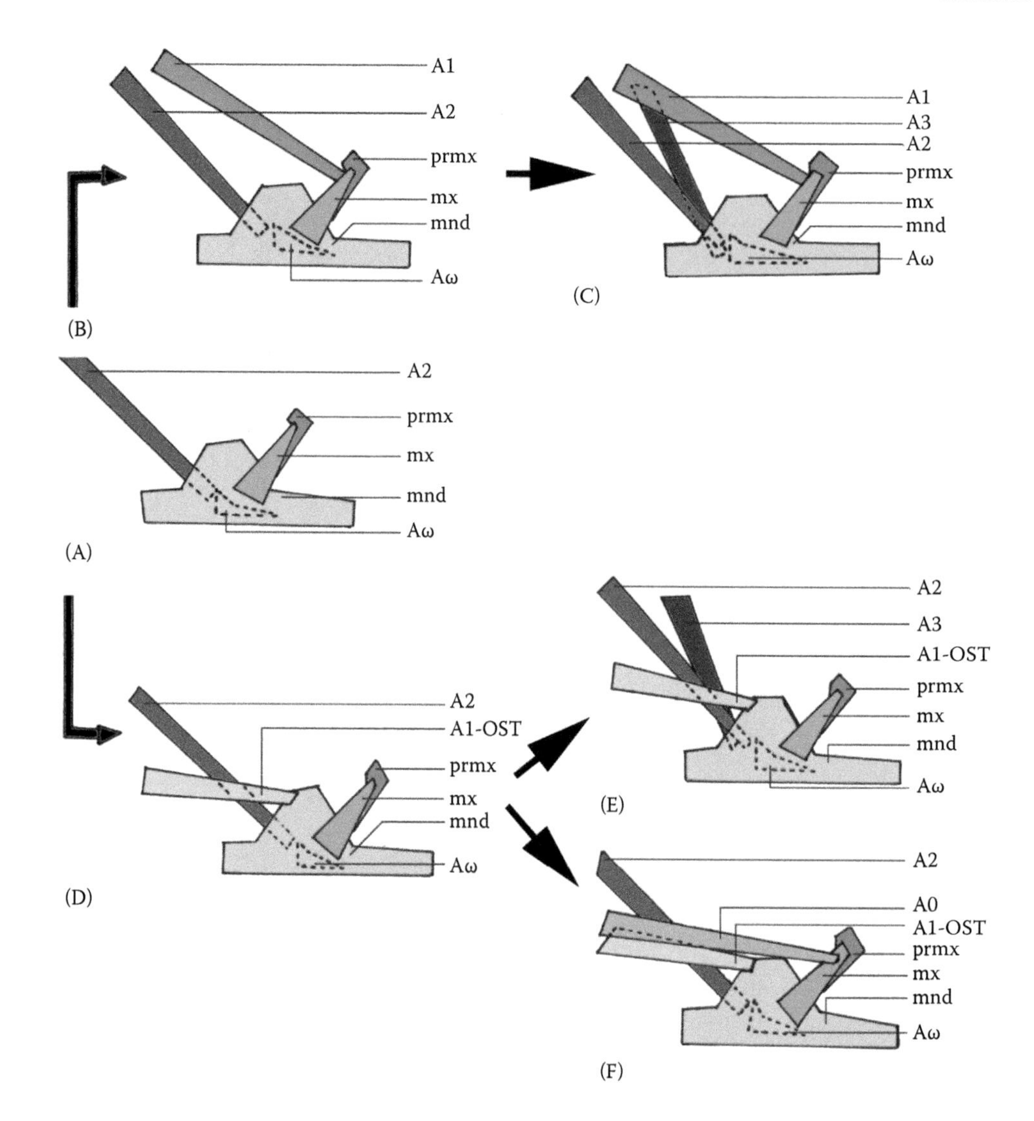

FIGURE 6.3 Diagram illustrating the two patterns of adductor mandibulae differentiation of teleostean fishes, based on Gosline (1989), and following the nomenclature of Diogo and Chardon (2000). (Modified from Diogo, R., and Chardon, M., *J Morphol*, 243, 193–208, 2000.) (A) Basal type in which the cheek muscle is undivided. (B) The neoteleostean pattern in which an *upper* part of the cheek muscle (A1) has become attached to the maxilla. (C) Secondary differentiation in neoteleostean fishes in which a mesial part of the cheek muscle (A3) is present. (D) Initial differentiation in the ostariophysan pattern in which a *lower* part of the cheek muscle (A1-OST) has developed a separate attachment to the back of the mandible. (E) Differentiation in some ostariophysan fishes in which a mesial part of the cheek muscle (A3) is present. (F) Differentiation in some ostariophysan fishes in which an adductor mandibulae section (A0) has developed, *via* the primordial ligament, an attachment to the maxilla. A0, A1, A1-OST, A2, A3, Aω sections of the adductor mandibulae; mnd, mandible; mx, maxilla; prmx, premaxilla.

2017; this work). The absence of a distinct dilatator operculi in adult extant acipenseriforms may reflect the odd lack in these fishes of an opercular bone and/or the fact that they seem to be paedomorphic (e.g., Bemis et al. 1997; Findeis 1997). The latter point might help to explain why, unlike most other living actinopterygians in which the constrictor dorsalis becomes ontogenetically differentiated into two muscles, adult acipenseriforms retain a single, undivided dorsal mandibular muscle, the protractor hyomandibulae.

Miyake et al (1992: 221) stated that a "spiracularis was described in *Polypterus* by Edgeworth (1935), but confirmation of its actual existence in *Polypterus* is needed." We could not find a spiracularis in the adult specimens of this genus that we have dissected for the book by Diogo and Abdala (2010). However, since then a detailed study of the development of the cranial muscles of *Polypterus senegalus* has shown that a muscle spiracularis is present in at least some developmental stages (specifically, after stage L8) of the members of this species (Noda et al. 2017; see also Chapter 7 and Chapter 5). Therefore, Table 6.1 was changed to show the presence of the spiracularis in at least some stages of the ontogeny of *Polypterus*, although Noda et al. (2017) do not clarify whether the muscle is present all the way until the adult stage. Moreover, as adult holocephalans lack a spiracularis, even if it were present in *Polypterus* adults it would still be more parsimonious to conclude that its presence in adult elasmobranchs and *Polypterus* is the result of independent (parallel) acquisition in elasmobranchs and in Polypteriformes.

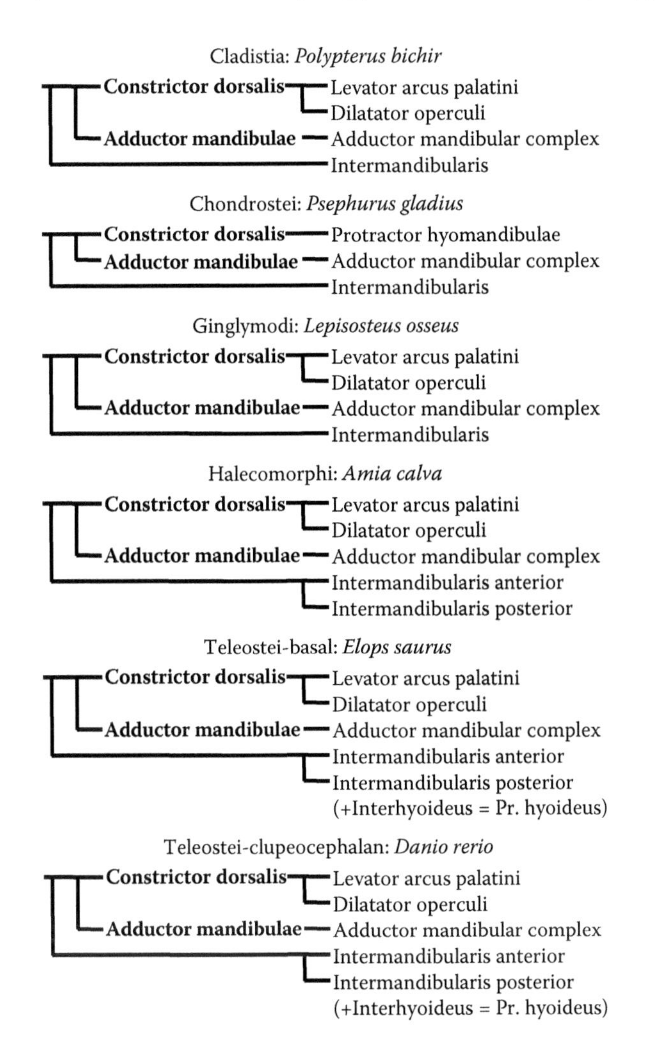

FIGURE 6.4 Developmental lineages of mandibular muscles in actinopterygians; Edgeworth's presumptive premyogenic condensations are in a italix face. (Modified from Miyake, T. et al., *J Morphol*, 212, 213–256, 1992.); the nomenclature of the muscles listed on the right of the figure follows that of the present work, "Pr. Hyoideus" meaning protractor hyoideus. Data compiled from evidence provided by developmental biology, comparative anatomy, experimental embryology and molecular biology, innervation and phylogeny (in studies of the authors and of other researchers). For more details, see text.

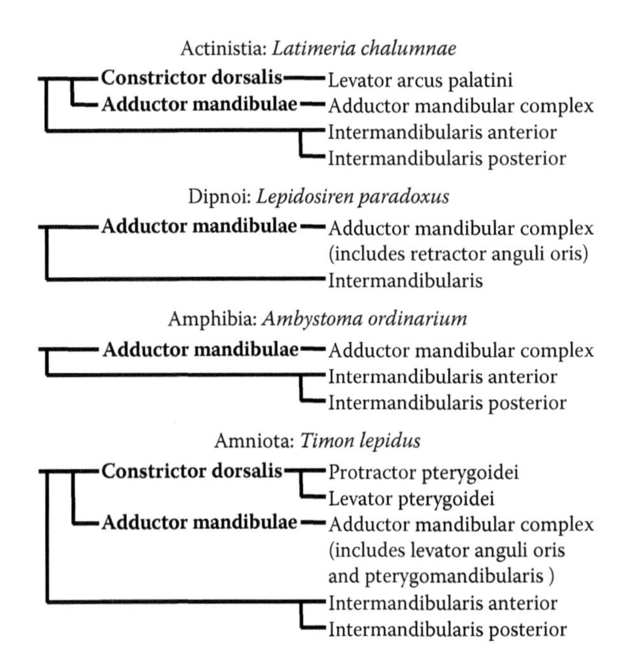

FIGURE 6.5 Developmental lineages of mandibular muscles in sarcopterygians (see caption of Figure 6.4). (Modified from Miyake, T. et al., *J Morphol*, 212, 213–256, 1992.)

The latter scenario requires only two evolutionary steps, while a spiracularis in the adult last common ancestor (LCA) of gnathostomes would require at least four steps: acquisition in gnathostomes and loss in holocephalans, in sarcopterygians, and in Actinopteri (see cladograms of Figures 1.1 and 6.1). However, in muscle evolution secondary loss seems to be much more frequent than the independent acquisition of muscles that are topologically and developmentally as similar as the spiracularis of *Polypterus* and of elasmobranchs. Therefore, we do not discount the idea that this muscle might have been present in the LCA of gnathostomes, and we have updated Table 6.1 accordingly.

In all major osteichthyan groups listed in Table 6.1 and Figures 6.4 and 6.5, the ventral portion of the mandibular muscle plate gives rise to the intermandibularis. In adult extant members of the Actinistia, Chondrostei, Ginglymodi, and Dipnoi, the intermandibularis is mainly undivided (Figure 6.15). In adult specimens of *Amia*, of *Latimeria*, and of numerous amphibian, amniote, and teleostean genera, this structure is divided into an intermandibularis anterior and an intermandibularis posterior (Figure 6.16). It should be noted that the intermandibularis of *Latimeria sensu* Dutel et al. (2015) corresponds to the intermandibularis anterior plus intermandibularis posterior *sensu* the present work (compare their Figure 5 with the figures of *Latimeria* shown in the following chapters). It is difficult to discern whether or not the intermandibularis was divided in adult members of the LCA of osteichthyans. As noted earlier, we tentatively hypothesize that it was divided (Table 6.1), because a divided intermandibularis is found in numerous chondrichthyans (both in elasmobranchs and in some holocephalans), actinopterygians, and sarcopterygians (see also Chapters 4 and 5). However, a detailed analysis of key osteichthyan and nonosteichthyan gnathostome fossils would help to test this hypothesis.

As its name indicates, the intermandibularis is usually a transversal muscle connecting the two mandibles. In most teleosts the intermandibularis posterior forms, together with the interhyoideus (see the following), the protractor hyoideus (Figure 6.18), which is thus derived from both the mandibular and hyoid muscle plates (e.g., Edgeworth 1935; Winterbottom 1974; Schilling and Kimmel 1997; Hernandez et al. 2002, 2005; this chapter) (Figure 6.1). The protractor hyoideus is innervated by both the Vth and the VIIth cranial nerves and functionally is a complex muscle: authors such as Osse (1969) demonstrated that its anterior and posterior portions may contract differently during different phases of respiration. Nonetheless, as stated by Stiassny (2000), as a broad

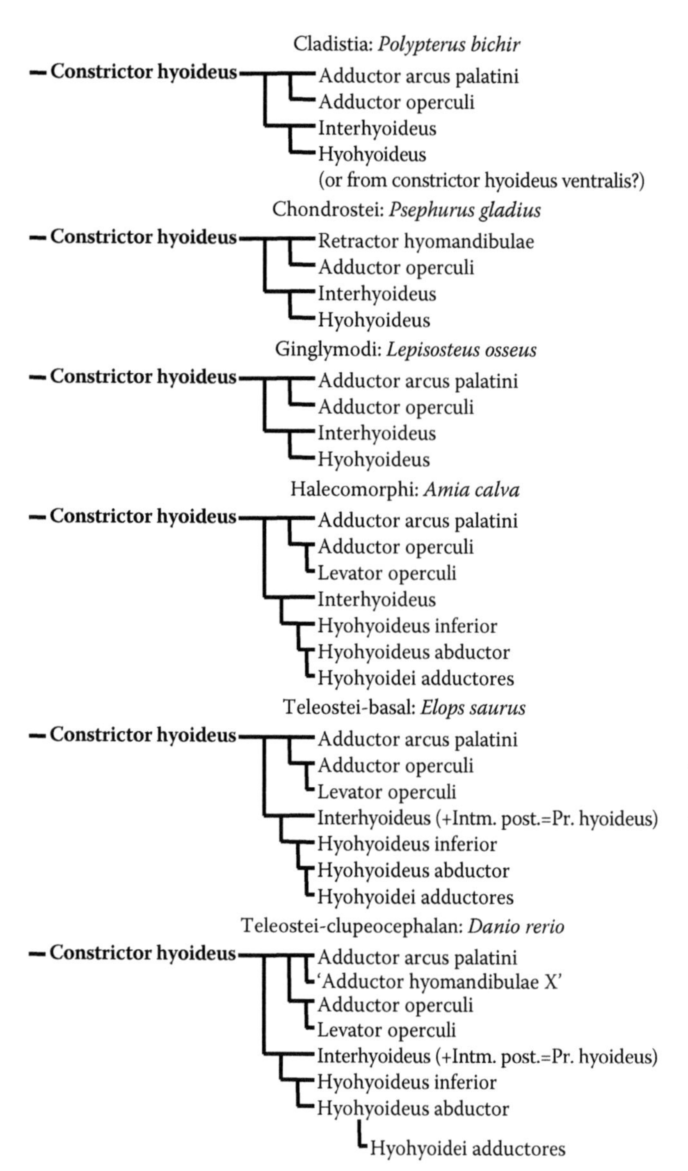

FIGURE 6.6 Developmental lineages of hyoid muscles in actinopterygians, "Pr. hyoideus" meaning protractor hyoideus and "Intm. post." meaning intermandibularis posterior (see caption of Figure 6.4). (Modified from Miyake, T. et al., *J Morphol*, 212, 213–256, 1992.). The question mark concerning *Polypterus* refers to the fact that Noda et al.'s (2017) developmental study suggested that in *Polypterus* the main body of the hyohyoideus muscle derives from the constrictor hyoideus dorsalis and not from the constrictor hyoideus ventralis as has been described for other extant actinopterygians, thus contradicting Edgeworth's (2935) suggestion that in *Polypterus* this structure also develops from the constrictor hyoideus ventralis.

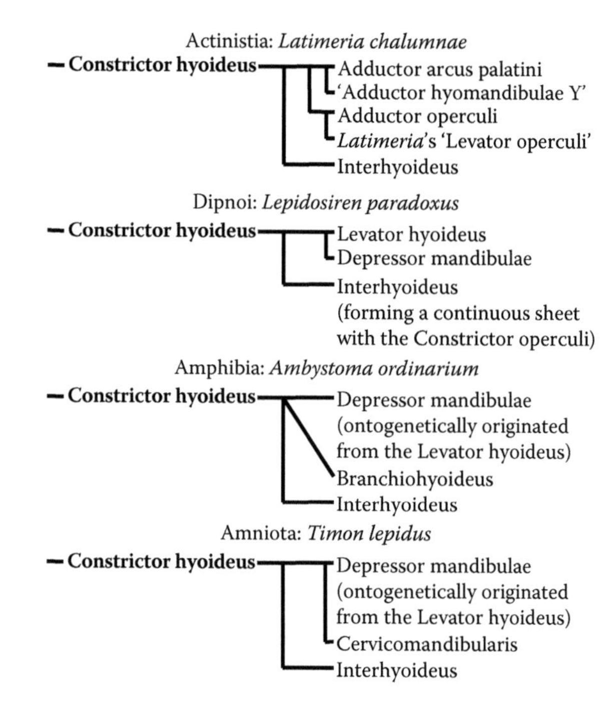

FIGURE 6.7 Developmental lineages of hyoid muscles in sarcopterygians (see caption of Figure 6.4). (Modified from Miyake, T. et al., *J Morphol*, 212, 213–256, 1992.)

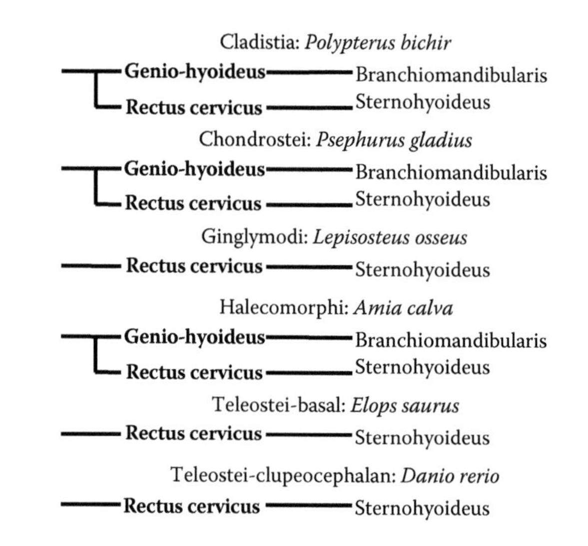

FIGURE 6.8 Developmental lineages of hypobranchial muscles in actinopterygians (see caption of Figure 6.4). (Modified from Miyake, T. et al., *J Morphol*, 212, 213–256, 1992.)

generality, the protractor hyoideus plays a primary role in the elevation (protraction) of the hyoid bars and in the depression of the mandible. According to the phylogenetic scenario shown in Figure 1.1, although a protractor hyoideus is not found in a few teleost taxa such as *Albula* and *Mormyrus* (e.g., Greenwood 1971, 1977; Winterbottom 1974; this work), the ancestors of extant teleosts did probably have a protractor hyoideus. Based on the altered morphology of the protractor hyoideus in morpholino-mediated Hox PG2 (*hoxa2b* and *hoxa2a*) knockdown larvae, Hunter and Prince (2002: 378) suggested that in the zebrafish "the basihyal (cartilage) may be important for the proper ontogenetic organization" of the intermandibularis posterior and the interhyoideus and, thus, for the association of their fibers and the formation of the protractor hyoideus. Further studies are needed to check if this is so and if it is a general feature within the Teleostei.

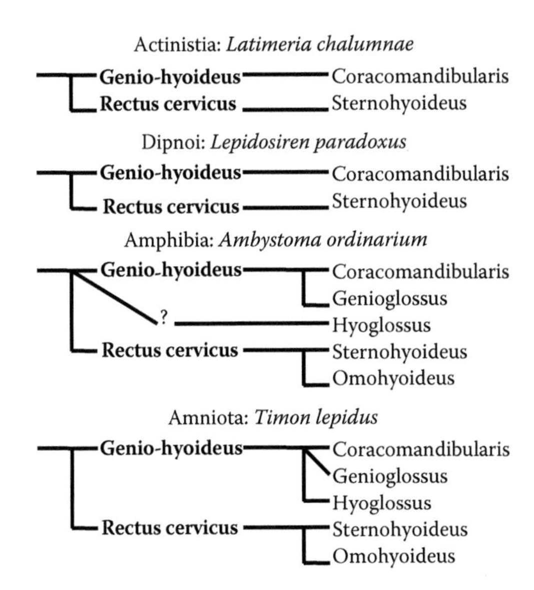

FIGURE 6.9 Developmental lineages of hypobranchial muscles in sarcopterygians (see caption of Figure 6.4). (Modified from Miyake, T. et al., *J Morphol*, 212, 213–256, 1992.).

The adductor mandibulae condensation is found in members of all major osteichthyan groups (Table 6.1; Figures 6.4 and 6.5). The number of structures that originate from this condensation is highly variable within these groups (Figures 6.1 and 6.2). The adductor mandibulae A3′ and A3″ *sensu* this volume (which correspond to the "mesial adductor mandibulae divisions" *sensu* Lauder 1980b) appear to have been present plesiomorphically in osteichthyans (Lauder 1980b) (Table 6.1; Figures 6.15 and 6.16). However, one or both of these bundles may be absent in osteichthyan taxa such as, e.g., dipnoans, acipenseriforms, and various teleosts (Table 6.1). The adductores mandibulae A2 and Aω also appear to have been found in basal osteichthyans (e.g., Lauder 1980a,b) (Table 6.1; Figures 6.19 and 6.20). The Aω may be absent in extant osteichthyans such as chondrosteans, ginglymodians, various teleosts, and most tetrapods (Table 6.1; Figure 6.21). In the adult specimens of the lizard *Timon* that we analyzed, the adductor mandibulae has a large anteroventral portion lodged in the "adductor fossa" (*sensu*, e.g., Lauder 1980b), which is remarkably similar to the Aω of bony fishes. Such an anteroventral portion of the adductor mandibulae was also described in other extant reptiles (see the following chapters). In view of the data available, it is difficult to discern whether or not this anteroventral portion of the adductor mandibulae is homologous to the Aω of bony fishes.

In Lauder's (1980b) Table II, it is suggested that the palatomandibularis minor and major of extant ginglymodians (Figure 6.21) are likely homologous to the levator maxillae superioris 3 and 4 of living halecomorphs (Figure 6.16), since these structures represent an "anterior division" of the adductor mandibulae. However, the overall configuration, position, and attachments of the ginglymodian palatomandibularis minor and major are markedly different from those of the halecomorph levator maxillae superioris 3 and 4. For instance, the *Lepisosteus* palatomandibularis minor and major insert dorsally on the ectopterygoid/entopterygoid and ventrally on the mandible (Figure 6.21), while section 3 of the levator maxillae superioris of *Amia* inserts dorsally on the neurocranium and orbital bones and ventrally mainly on the autopalatine (Figure 6.16). The phylogenetic scenario shown in Figures 1.1 and 6.1 supports a close relationship between the Halecomorphi and the Teleostei, undermining the hypothesis that ginglymodians and halecomorphs are sister-groups. This indicates that the ginglymodian palatomandibularis minor and the halecomorph levator maxillae 3 and 4 may well be nonhomologous. However, if the Halecomorphi and Ginglymodi are sister-groups, as some studies have suggested (see Chapter 1), then these muscles might be homologous.

In addition to the divisions of the adductor mandibulae condensation mentioned earlier, other divisions may be found in adult osteichthyans; e.g., the A1-OST and A0 (*sensu* Diogo and Chardon 2000; e.g., Figures 6.1, 6.3, and 6.17) and the A2-PVM, the pseudotemporalis, retractor anguli oris, the levator anguli oris mandibularis, and the pterygomandibularis (see Figure 6.2 and the following chapters). In the adult dipnoan specimens we dissected, we found a peculiar section of the adductor mandibulae that has some fibers associated with those of the A2 but is well differentiated from it. As this section is somewhat posterior, ventral, and medial to the main body of the A2, we call it adductor mandibulae A2-PVM (the PVM thus meaning posteroventromesial). Authors as Edgeworth (1935), Bemis (1986), Bemis and Lauder (1986), and Miyake et al. (1992) did not mention the presence of such an adductor mandibulae section in extant dipnoans (the A2-PVM should not be confused with the retractor anguli oris of these fishes or to the "retractor anguli oris" of reptiles such as *Sphenodon*, which is usually situated posteroventrolaterally to the A2, being mainly superficial rather than mesial to the A2; both the A2-PVM and the retractor anguli oris can be found in the same taxon; e.g., in *Lepidosiren*). The A2-PVM of dipnoans appears to correspond to the structure that is often named "adductor mandibulae posterior" by researchers working with amphibian and amniote tetrapods (e.g., Brock 1938; Carroll and Holmes 1980; Iordansky 1992; Moro and Abdala 2000; Montero et al. 2002; Abdala and Moro 2003; see the following chapters). However, it is possible that some of the structures that have been, and continue to be, named adductor mandibulae posterior in clades such as turtles are not homologous to the structures that have been designated by the same name in other nonmammalian tetrapod groups and, thus, to the A2-PVM *sensu* the present work (see the following chapters for more details on this subject). All extant nondipnoan bony fishes we dissected lack an A2-PVM. The textual descriptions and illustrations of the "adductor mandibulae complex" of *Latimeria* provided by Millot and Anthony (1958) and Adamicka and Ahnelt (1992) indicate that such an A2-PVM is also absent in that taxon.

Although the adductor mandibulae divisions are usually related to the adduction of the mandible, this is not always the case. For instance, some of them may attach to structures other than the mandible, such as the maxilla (e.g., the adductor mandibulae A0: Figure 6.17) and thus not be directly

TABLE 6.1
Mandibular Muscles of Adults of Representative Actinopterygian Taxa

Probable Plesiomorphic Osteichthyan Condition	Cladistia: *Polypterus bichir* (Bichir)	Chondrostei: *Psephurus gladius* (Chinese Swordfish)	Ginglymodi: *Lepisosteus osseus* (Longnose Gar)	Halecomorphi: *Amia calva* (Bowfin)	Teleostei—Basal: *Elops saurus* (Ladyfish)	Teleostei—Clupeocephalan: *Danio rerio* (Zebrafish)
Intermandibularis anterior [intermandibularis anterior and posterior plesiomorphically present in osteichthyans? See text]	**Intermandibularis**	**Intermandibularis**	**Intermandibularis**	**Intermandibularis anterior**	**Intermandibularis anterior**	**Intermandibularis anterior**
Intermandibularis posterior [see above]	–	–	–	**Intermandibularis posterior**	**Intermandibularis posterior** [forming, together with interhyoideus, the protractor hyoideus]	**Intermandibularis posterior** [see on the left]
–	–	–	–	–	**Protractor hyoideus** [including intermandibularis posterior and interhyoideus; it is thus derived from both the mandibular and hyoid muscle plates]	**Protractor hyoideus** [see on the left]
Ad. mand. A3′	**Ad. mand. A3′** (ad. mand. temporalis *sensu* Lauder [1980a])	–	**Ad. mand. A3′** (preorbitalis superficialis *sensu* Lauder [1980a] and Konstandinidis et al. [2015]; the A3′ and A3″ mainly correspond to the adductor 2 *sensu* Adams [1919])	**Ad. mand. A3′**	–	–
Ad. mand. A3″	**Ad. mand. A3″** (ad. mand. pterygoideus *sensu* Lauder [1980a])	–	**Ad. mand. A3″** (preorbitalis profundus *sensu* Lauder [1980a] and Konstandinidis et al. [2015])	**Ad. mand. A3″**	–	–

(*Continued*)

TABLE 6.1 (CONTINUED)
Mandibular Muscles of Adults of Representative Actinopterygian Taxa

Probable Plesiomorphic Osteichthyan Condition	Cladistia: *Polypterus bichir* (Bichir)	Chondrostei: *Psephurus gladius* (Chinese Swordfish)	Ginglymodi: *Lepisosteus osseus* (Longnose Gar)	Halecomorphi: *Amia calva* (Bowfin)	Teleostei—Basal: *Elops saurus* (Ladyfish)	Teleostei—Clupeocephalan: *Danio rerio* (Zebrafish)
Ad. mand. A2	**Ad. mand. A2** (ad. mand. posterolateral *sensu* Lauder [1980a])	**Ad. mand. A2** (ad. mand. *sensu* Carroll and Wainwright [2003]) [Adams 1919 described both an A2 and an A3′ in *Polyodon*, but only an A2 in *Acipenser*; in the *Psephurus* specimens we dissected, we did not find an separate, distinct A3′ as that found in other osteichthyans]	**Ad. mand. A2** (ad. mand. posterolateral *sensu* Lauder [1980a]; it mainly corresponds to the adductor 1 *sensu* Adams [1919])	**Ad. mand. A2**	**Ad. mand. A2**	**Ad. mand. A2**
Labial muscles plesiomorphically present in osteichthyans? See text	–	–	–	– [levator maxillae superioris 3 and 4 of *Amia* correspond to/derive from the labial muscles of nonosteichthyan vertebrates? See text]	–	–
–	–	–	**Palatomandibularis minor** and **major** (these structures mainly correspond to the adductor 3 *sensu* Adams [1919] and to the palatomandibularis *sensu* and Konstandinidis et al. [2015]) [palatomandibularis minor and major of *Lepisosteus* do not seem to correspond to/derive from the labial muscles of nonosteichthyan vertebrates, because they derive ontogenetically from the adductor mandibulae, as shown by Konstandinidis et al. [2015]]	–	–	–

(Continued)

TABLE 6.1 (CONTINUED)
Mandibular Muscles of Adults of Representative Actinopterygian Taxa

Probable Plesiomorphic Osteichthyan Condition	Cladistia: *Polypterus bichir* (Bichir)	Chondrostei: *Psephurus gladius* (Chinese Swordfish)	Ginglymodi: *Lepisosteus osseus* (Longnose Gar)	Halecomorphi: *Amia calva* (Bowfin)	Teleostei—Basal: *Elops saurus* (Ladyfish)	Teleostei—Clupeocephalan: *Danio rerio* (Zebrafish)
–	–	–	–	**Levator maxillae superioris 3** and **4**	–	–
–	–	–	–	–	–	**Ad. mand. A1-OST**
–	–	–	–	–	–	**Ad. mand. A0**
Ad. mand. Aω	**Ad. mand. Aω** [contrary to what is shown in Adams' [1919] Table I, *Polypterus* does have an Aω, as also recently reported by Noda et al. [2017]]	–	–	**Ad. mand. Aω**	**Ad. mand. Aω**	**Ad. mand. Aω**
Levator arcus palatini	**Levator arcus palatini** (levator maxillae superioris *sensu* Adams [1919])	**Protractor hyomandibulae** [seemingly/apparently originated from the portion of the hyoid muscle plate from which originate the adductor arcus palatini and dilatator operculi of other actinopterygians]	**Levator arcus palatini** (protractor hyomandibularis *sensu* Adams [1919]) [Konstantinidis et al. [2015] seemingly/apparently confused the fact that the levator arcus palatini is a synonym of the "protractor hyomandibularis" in *Lepisosteus*, because in the later stages of development of this taxon that they describe, they refer to a "levator arcus palatine" muscle and to a "protractor hyomandibularis + dilatator opercula" muscle, the former muscle corresponding to the levator arcus palatini *sensu* the present work and the latter to the dilatator operculi *sensu* the present work]	**Levator arcus palatini** (includes the levator arcus palatini + protractor hyomandibularis *sensu* Adams [1919])	**Levator arcus palatini**	**Levator arcus palatini**

(*Continued*)

TABLE 6.1 (CONTINUED)
Mandibular Muscles of Adults of Representative Actinopterygian Taxa

Probable Plesiomorphic Osteichthyan Condition	**Cladistia: *Polypterus bichir* (Bichir)**	**Chondrostei: *Psephurus gladius* (Chinese Swordfish)**	**Ginglymodi: *Lepisosteus osseus* (Longnose Gar)**	**Halecomorphi: *Amia calva* (Bowfin)**	**Teleostei—Basal: *Elops saurus* (Ladyfish)**	**Teleostei—Clupeocephalan: *Danio rerio* (Zebrafish)**
Spiracularis plesiomorphically present in osteichthyans and in actinopterygians? Most parsimonious scenario is that it is not, but it could have been (see text)	**Spiracularis** [present in at least some developmental stages of members of this genus (Edgeworth 1935; Noda et al. 2017), although it is not clear if it is also present in adults: see text]	–	–	–	–	–
–	**Dilatator operculi** (protractor hyomandibularis *sensu* Adams [1919])	– [Dilatator operculi absent as a separate element, but see above]	**Dilatator operculi** (adductor operculi *sensu* Adams [1919]; protractor hyomandibulae + dilatator operculi *sensu* Konstandinidis et al. [2015])	**Dilatator operculi**	**Dilatator operculi**	**Dilatator operculi**

Note: The nomenclature of the muscles shown in bold follows that of the present work, "ad. mand." meaning adductor mandibulae (in order to facilitate comparisons, certain names used by other authors are shown between round brackets; additional comments are given between square brackets); data compiled from evidence provided by developmental biology, comparative anatomy, functional morphology, paleontology, experimental embryology and molecular biology, innervation, and phylogeny (in studies of the author and of other researchers). For more details, see text (see also Figures 6.4 and 6.5).

related to lower jaw adduction (see, e.g., Winterbottom 1974; Stiassny 2000). Certain divisions of the adductor mandibulae may be related to opening the mouth rather than closing it, as is the case of the "abductor mandibulae" of saccopharyngiform teleosts (e.g., Tchernavin 1947a,b, 1953; this work).

HYOID MUSCLES

According to Edgeworth (1935), the constrictor hyoideus condensation usually gives rise to dorsomedial and ventral derivatives throughout the major groups of gnathostomes (Table 6.2; Figures 6.6 and 6.7). As shown in Table 6.2, two dorsomedial hyoid muscles appear to have been present in plesiomorphic osteichthyans: the adductor arcus palatini and the adductor operculi (Figure 6.17). These two muscles, which, as their name indicates, are usually related to the adduction of the suspensorium/palatoquadrate and of the operculum, are found in *Latimeria* and in most living actinopterygians (Table 6.2; Figures 6.6 and 6.7). That the "levator hyomandibularis posterior" *sensu* Dutel et al. (2015) inserts on the medial side of the operculum and also on the hyomandibulae indicates that it probably corresponds to the adductor operculi of *Latimeria, sensu* the present work. Accordingly, the "levator hyomandibularis anterior" *sensu* Dutel et al. (2015) appears to correspond to the "levator operculi" of *Latimeria sensu* the present work, while their "adductor hyomandibularis" seems to correspond to the "adductor hyomandibulae Y" of *Latimeria sensu* the present work. Some actinopterygians may, however, lack an adductor operculi (e.g., saccopharyngiform teleosts: Tchernavin 1947a,b, 1953; this work). In living chondrosteans, the dorsomedial portion of the hyoid muscle plate gives rise to a peculiar retractor hyomandibulae rather than to an adductor arcus palatini as in most other actinopterygians (e.g., Danforth 1913; Luther 1913; Edgeworth 1935; Kesteven 1942–1945; Miyake et al. 1992; Carroll and Wainwright 2003; Diogo and Abdala 2010) (Table 6.2; Figure 6.6).

According to Winterbottom (1974: 239), in addition to the adductor arcus palatini, some osteichthyans have other muscles connecting the neurocranium to the palatoquadrate/suspensorium and promoting the adduction of the latter structures. Examples include the muscles designated "adductor hyomandibulae" by the author. We have found well-differentiated muscles adductor hyomandibulae *sensu* Winterbottom in various teleosts (Table 6.2); an adductor hyomandibulae also appears to be present in *Latimeria* (see Millot and Anthony 1958 and the following chapters). As emphasized by Winterbottom (1974: 239), at least some of these "adductor hyomandibulae" muscles seem to have been acquired independently (i.e., they are nonhomologous), since they may have originated "1) either from the posterior region of the adductor arcus palatini or 2) from the anterior fibers of the adductor operculi." This is the case for the adductor hyomandibulae found in *Latimeria* and in teleosts such as *Danio*, designated in the present book as adductor hyomandibulae Y and adductor hyomandibulae X, respectively. Miyake et al. (1992) suggested that in addition to these taxa, other key osteichthyan genera listed in the tables of this chapter such as *Amia* and *Lepisosteus* also have an adductor hyomandibulae *sensu* Winterbottom (NB: the muscles named adductor hyomandibulae and adductor arcus palatini in Winterbottom's work and in the present study correspond, respectively, to the muscles named "adductor arcus palatini" and adductor hyomandibulae in Miyake et al.'s paper). However, the *Amia* and *Lepisosteus* specimens we dissected lack a separate, well-differentiated adductor hyomandibulae *sensu* Winterbottom. This is also the case in the specimens of these genera that were analyzed by, e.g., Lauder (1980a), who stated that "in *Lepisosteus* the adductor arcus palatini (= Winterbottom's adductor hyomandibulae) and the adductor hyomandibulae (= Winterbottom's 'adductor arcus palatini') form a continuous sheet of parallel-fibered muscle" and that "the adductor arcus palatini (= Winterbottom's adductor hyomandibulae) is absent in *Amia*" (Lauder 1980a: 289). (As explained by Winterbottom (1974), the reason why some authors use the name adductor hyomandibulae to designate his adductor arcus palatini is that in some osteichthyans, the mesial insertion of this muscle is exclusively on the hyomandibula. However, as he pointed out, the name adductor hyomandibulae becomes clearly inappropriate in the numerous cases in which the muscle is expanded anteriorly along the floor of the orbit and attaches to more anterior elements of the suspensorium such as the metapterygoid and/or entopterygoid.)

In addition to the levator operculi, the adductor operculi, and the muscles adductor hyomandibulae, other dorsomedial hyoid muscles may be found in living osteichthyans (e.g., Figures 6.6 and 6.7). For instance, *Amia* and most extant teleosts have a levator operculi (Figures 6.16 and 6.17). The levator operculi of these fishes is usually related to a peculiar mechanism mediating lower jaw depression *via* the so-called four-bar linkage system in which the force of contraction of this muscle is transmitted through the opercular series and the interoperculomandibular ligament to the lower jaw (e.g., Stiassny 2000). A levator operculi may be absent in certain teleosts, such as saccopharyngiforms (e.g., Tchernavin 1947a,b, 1953; Diogo and Abdala 2010). Millot and Anthony (1958) stated that *Latimeria* has a levator operculi. As can be seen in the descriptions and the figures provided by these authors (see, e.g., their plate VII) and as recognized in page 61 of their work, the fibers of this levator operculi are deeply blended with those of the adductor operculi. This has made authors such as Lauder (1980c) to be very skeptical about the presence, in *Latimeria,* of a distinct, well-differentiated muscle levator operculi such as that found in *Amia* and teleosts. However, Adamicka and Ahnelt (1992: 108) reaffirmed, apparently based on their own observations of *Latimeria,* that this taxon "does have a levator operculi muscle differentiated out of the adductor (operculi)." According to the results of Diogo's (2007) cladistic analysis, the levator operculi of *Latimeria* is probably not homologous to the levator operculi of *Amia* and teleosts (see tables in this chapter and Chapter 7). Notably, the function of the levator operculi of *Latimeria* is not similar to that of the levator operculi of *Amia* and teleosts: unlike the latter fishes, *Latimeria* has no

TABLE 6.2
Hyoid Muscles of Adults of Representative Actinopterygian Taxa

Probable Plesiomorphic Osteichthyan Condition	Cladistia: *Polypterus bichir* (Bichir)	Chondrostei: *Psephurus gladius* (Chinese Swordfish)	Ginglymodi: *Lepisosteus osseus* (Longnose Gar)	Halecomorphi: *Amia calva* (Bowfin)	Teleostei—Basal: *Elops saurus* (Ladyfish) (Figure 6.10)	Teleostei—Clupeocephalan: *Danio rerio* (Zebrafish)
Interhyoideus	**Interhyoideus**	**Interhyoideus**	**Interhyoideus**	**Interhyoideus**	**Interhyoideus** [forming, together with intermandibularis posterior, the protractor hyoideus: see Table 6.1]	**Interhyoideus** [see on the left]
–	**Hyohyoideus**	**Hyohyoideus**	**Hyohyoideus**	**Hyohyoideus inferior**	**Hyohyoideus inferior**	**Hyohyoideus inferior**
–	–	–	–	**Hyohyoideus abductor** [often considered to be part of a hyohyoideus superior]	**Hyohyoideus abductor** [see on the left]	**Hyohyoideus abductor** [see on the left]
–	–	–	–	**Hyohyoidei adductores** [often considered as part of a hyohyoideus superior]	**Hyohyoidei adductores** [see on the left]	**Hyohyoidei adductores** [see on the left]
Adductor operculi (or constrictor hyoideus dorsalis?) [it is perhaps more appropriate to designate the muscle of nonholostean actinopterygians as constrictor hyoideus dorsalis, and not as adductor operculi, because (a) it corresponds directly to the constrictor hyoideus dorsalis of nonactinopterygian vertebrates); (b) it does not correspond directly to the adductor operculi of holosteans; i.e., the latter muscle corresponds only to a part of the constrictor hyoideus dorsalis, because this constrictor corresponds to the adductor operculi + levator operculi (and occasionaly to a part or the totality of the adductor hyomandibulae of some teleosts, see below) of holosteans]	**Adductor operculi (or constrictor hyoideus dorsalis?)** (part of adductor hyomandibularis *sensu* Adams [1919]) [see on the left]	**Adductor operculi (or constrictor hyoideus dorsalis?)** (opercularis *sensu* Carroll and Wainwright [2003]) [see on the left]	**Adductor operculi (or constrictor hyoideus dorsalis?)** (levator operculi *sensu* Adams [1919])	**Adductor operculi** [here it is more appropriate to use the name adductor operculi, because the constrictor hyoideus dorsalis is really divided into an adductor operculi and a levator operculi; this is also a way of differentiating the adductor operculi of *Amia* + teleosts from the adductor operculi of *Latimeria*, which was apparently acquired *via* an independent differentiation of the constrictor hyoideus dorsalis into an adductor operculi, an adductor hyomandibulae, and a levator operculi]	**Adductor operculi**	**Adductor operculi**

(Continued)

TABLE 6.2 (CONTINUED)
Hyoid Muscles of Adults of Representative Actinopterygian Taxa

Probable Plesiomorphic Osteichthyan Condition	Cladistia: *Polypterus bichir* (Bichir)	Chondrostei: *Psephurus gladius* (Chinese Swordfish)	Ginglymodi: *Lepisosteus osseus* (Longnose Gar)	Halecomorphi: *Amia calva* (Bowfin)	Teleostei—Basal: *Elops saurus* (Ladyfish) (Figure 6.10)	Teleostei—Clupeocephalan: *Danio rerio* (Zebrafish)
Adductor arcus palatini	**Adductor arcus palatini** (part of adductor hyomandibularis *sensu* Adams [1919]; adductor hyomandibulae *sensu* Noda et al. [2017])	**Retractor hyomandibulae** [apparently originated from the portion of the hyoid muscle plate from which originates the adductor arcus palatini of other actinopterygians]	**Adductor arcus palatini** (adductor arcus palatini + adductor hyomandibulae *sensu* Konstandinidis et al. [2015]) [Adams [1919] failed to describe an adductor arcus palatini (adductor hyomandibularis) in *Lepisosteus*, but in his Table I, he did recognize that this muscle is probably present in this taxon; Konstantinidis et al. [2015] apparently confused the fact that the adductor hyomandibularis is a synonym of the adductor arcus palatini in *Lepisosteus*, because in the later stages of development of this taxon that they describe they refer to an adductor operculi muscle and to a adductor hyomandibularis + adductor arcus palatini muscle, the former muscle corresponding to the adductor operculi *sensu* the present work, and the latter, to the adductor arcus palatini *sensu* the present work;	**Adductor arcus palatini** (adductor hyomandibularis *sensu* Adams [1919])	**Adductor arcus palatini** (adductor hyomandibularis *sensu* Adams [1919])	**Adductor arcus palatini**

(*Continued*)

TABLE 6.2 (CONTINUED)
Hyoid Muscles of Adults of Representative Actinopterygian Taxa

Probable Plesiomorphic Osteichthyan Condition	Cladistia: *Polypterus bichir* (Bichir)	Chondrostei: *Psephurus gladius* (Chinese Swordfish)	Ginglymodi: *Lepisosteus osseus* (Longnose Gar)	Halecomorphi: *Amia calva* (Bowfin)	Teleostei—Basal: *Elops saurus* (Ladyfish) (Figure 6.10)	Teleostei—Clupeocephalan: *Danio rerio* (Zebrafish)
			that is, there is really no distinct adductor hyomandibulae muscle in adult *Lepisosteus*, as there is, for instance, in zebrafish and *Latimeria*; this is supported by their own descriptions, in which they never describe the adductor hyomandibulae and adductor arcus palatini as two separate muscles, describing instead the whole formed by the two muscles as a single "partition"]			
–	–	–	–	–	–	**Adductor hyomandibulae X** [apparently not homologous to the adductor hyomandibulae Y of *Latimeria*]
–	–	– [Adams [1919] stated that chondrosteans and *Lepisosteus* have a separate levator operculi, but the structure he designated under that name corresponds to the adductor operculi *sensu* the present work: compare, e.g., his plate II with Figures 6.11 and 6.12]	– [see on the left]	**Levator operculi** [apparently not homologous to the levator operculi of *Latimeria*]	**Levator operculi** [see on the left]	**Levator operculi** [see on the left]

Note: See footnote of Table 6.1 and Figures 6.6 and 6.7.

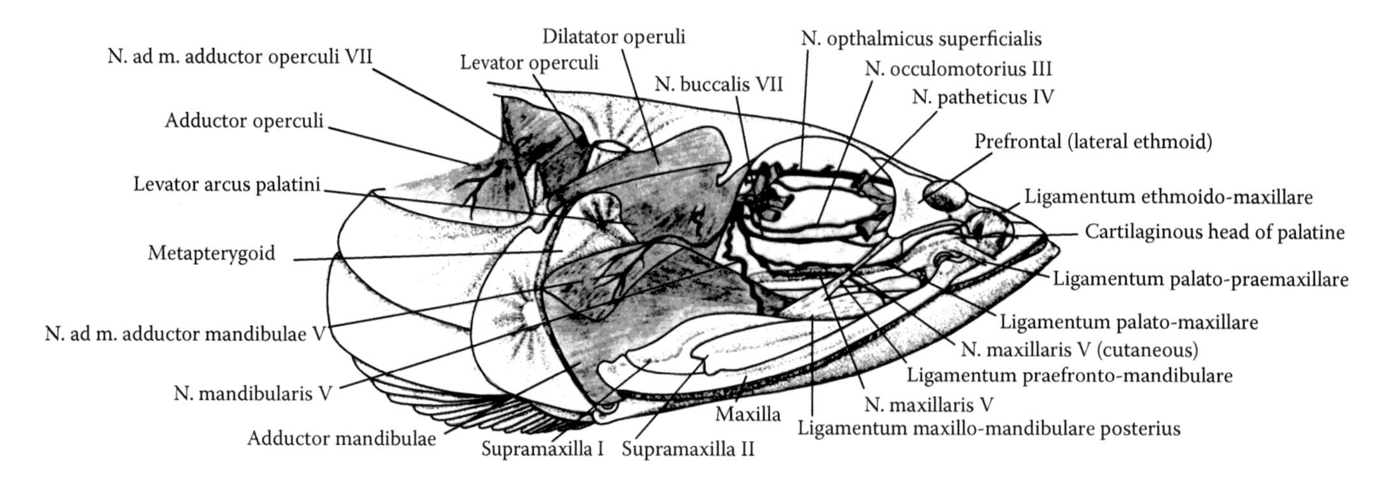

FIGURE 6.10 *Elops saurus* (Teleostei, Elopiformes): lateral view of the cephalic muscles; the dorsal edge of the opercular bone has been folded over to expose the muscle adductor operculi (the nomenclature of the structures illustrated follows that of this author). (Modified from Vrba, E. S., *Zool Afr*, 3, 211–236, 1968.)

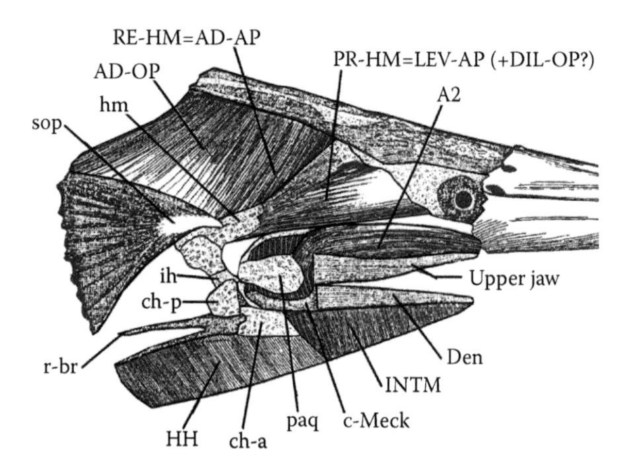

FIGURE 6.11 *Polyodon spathula* (Chondrostei): lateral view of the cephalic muscles (the nomenclature of the structures illustrated follows that used in the present work). (Modified from Danforth, C. H., *J Morphol*, 24, 107–146, 1913.) A2, adductor mandibulae A2; AD-AP, adductor arcus palatini; AD-OP, adductor operculi; ch-a, ch-p, anterior and posterior ceratohyals; c-Meck, Meckelian cartilage; den, dentary bone; DIL-OP, dilatator operculi; hm, hyomandibula; ih, interhyal; INTM, intermandibularis; LEV-AP, levator arcus palatini; paq, palatoquadrate; PR-HM, protractor hyomandibulae; r-br, branchiostegal rays; RE-HM, retractor hyomandibulae; sop, subopercle.

interoperculomandibular ligament and, consequently, has no opercular mechanism mediating mandible depression (e.g., Millot and Anthony 1958; Alexander 1973; Anthony 1980; Lauder 1980c). In order to distinguish the levator operculi of *Latimeria* from the levator operculi of *Amia* and teleosts, the former muscle is called levator operculi of *Latimeria* in the present work (see tables of the following chapters).

A muscle named levator operculi is shown in an illustration of the dipnoan *Neoceratodus* by Kardong (2002; see his Figure 10.39B). However, in the dipnoan specimens we dissected, as well as in those described by, e.g., Bischoff (1840), Owen (1841), Luther (1913–1914), Edgeworth (1935), Kesteven (1942–1945), Bemis (1986), Bemis and Lauder (1986), and Bartsch (1992, 1993, 1994), there is no structure resembling the levator operculi of *Latimeria* or the levator operculi of *Amia* and teleosts. In fact, the levator operculi of Kardong (2002: see his Figure 10.39B) appears to correspond to the constrictor operculi of Bemis and Lauder (1986), which may correspond to the adductor operculi of other bony fishes but forms, in extant adult dipnoans, a continuous sheet of fibers together with other cranial muscles.

Other dorsomedial hyoid muscles found in osteichthyans include the levator hyoideus and the depressor mandibulae (Figure 6.7; see also the following chapters). The levator hyoideus is usually related to the elevation of the posterodorsal portion of the ceratohyal (Figure 6.22), whereas the depressor mandibulae is often related to the opening of the mouth. These muscles are found in at least some developmental stages of extant dipnoans and of numerous extant tetrapods. Contrary to what is suggested in this book, Edgeworth (1935) stated that the depressor mandibulae of adult dipnoans such as *Protopterus* and *Lepidosiren* is not homologous with part or totality of the depressor mandibulae of adult tetrapods. In the dipnoan developmental series he observed, the levator hyoideus and the depressor mandibulae seem to appear at the same time, while in tetrapods, the depressor mandibulae appears as a modification of part, or the totality, of the levator hyoideus. Authors such as Forey (1986) have, however, suggested that the depressor mandibulae/levator hyoideus of adult *Protopterus* and *Lepidosiren* are homologous with the depressor mandibulae/levator hyoideus of adult tetrapods and that this homology provides support for a close relationship between dipnoans and tetrapods. Our observations, comparisons, and phylogenetic results strongly corroborate Forey's hypothesis: the innervation, position, relationships with other structures, and function of the depressor mandibulae/levator hyoideus of adult dipnoans are strikingly similar to those of the depressor mandibulae/levator hyoideus of adult tetrapods, and the results of Diogo's (2007) cladistic analysis strongly suggest that dipnoans are the closest living relatives of tetrapods (Figure 1.1). Even authors who adhere to Edgeworth's (1935) view admit that the depressor mandibulae of adult

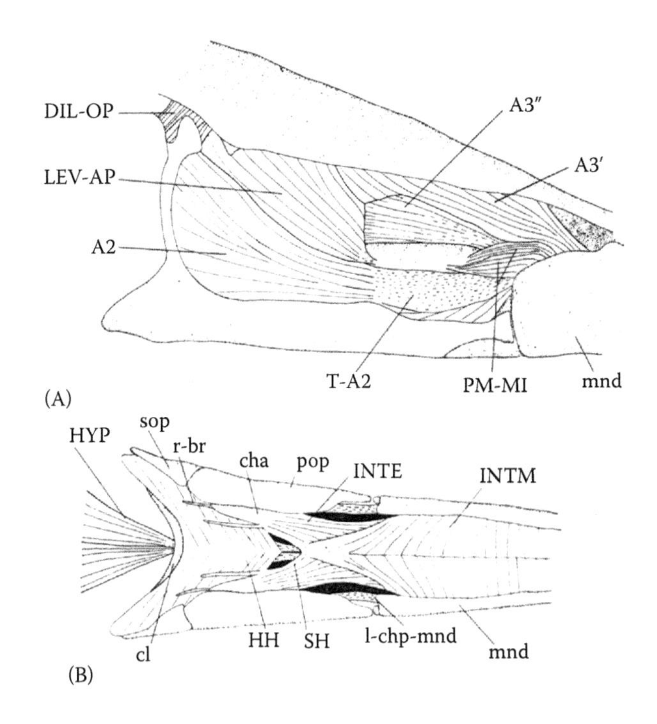

FIGURE 6.12 *Lepisosteus oculatus* (Ginglymodi): lateral (A) and ventral (B) views of the cephalic musculature after removal of the eye (the nomenclature of the structures illustrated basically follows that of Lauder [1980a]). (Modified from Lauder, G. V., *J Morphol*, 163, 283–317, 1980.) A2, A3′, A3″, adductor mandibulae A2, A3′, and A3″; cha, anterior ceratohyal; cl, cleithrum; DIL-OP, dilatator operculi; HH, hyohyoideus; HYP, hypoaxialis; INTE, interhyoideus; INTM, intermandibularis; l-chp-mnd, ligamentum between posterior ceratohyal and mandible; LEV-AP, levator arcus palatini; mnd, mandible; PM-MI, palatomandibularis minor; pop, preopercle; r-br, branchiostegal rays; SH, sternohyoideus; sop, subopercle; T-A2, tendon of adductor mandibulae A2.

dipnoans is "functionally and topographically comparable" to the depressor mandibulae of adult tetrapods such as salamanders (e.g., Bauer 1997: 79).

Therefore, we propose the following hypothesis concerning the evolution of the levator hyoideus and depressor mandibulae. The first evolutionary step was probably the differentiation of the levator hyoideus (possibly from the portion of the hyoid muscle plate that gives rise to the adductor arcus palatini in other osteichthyans, but this is far from clear: see the following chapters). Such a configuration, in which there is no depressor mandibulae and the levator hyoideus consists of a single mass of fibers attaching to the hyoid arch, is found in early development stages of the dipnoan *Neoceratodus* and of various tetrapods (e.g., Edgeworth 1935: see his Figure 313; Bartsch 1994: see his Figure 2B). The second evolutionary step may have been the attachment of some fibers of the levator hyoideus to the mandible and the ultimate differentiation of these fibers into a depressor mandibulae. This configuration, in which the levator hyoideus and the depressor mandibulae inserting, respectively, to the hyoid arch and to the mandible, is found in early developmental stages of various tetrapods and of the dipnoans *Lepidosiren* and *Protopterus*, as well as in juveniles and adults of these two dipnoan genera and of amphibians such as *Siren* (Edgeworth 1935: see his Figure 327; see also the following chapters). These two evolutionary steps seems to have occurred before the phylogenetic split between dipnoans and tetrapods, since at least some members of these groups have depressor mandibulae fibers attaching on the mandible. In other words, the last common ancestors of dipnoans and tetrapods probably had, in at least some stages of their development, both a levator hyoideus and a depressor mandibulae. After that phylogenetic split, the evolution of these muscles appears to have been rather diverse

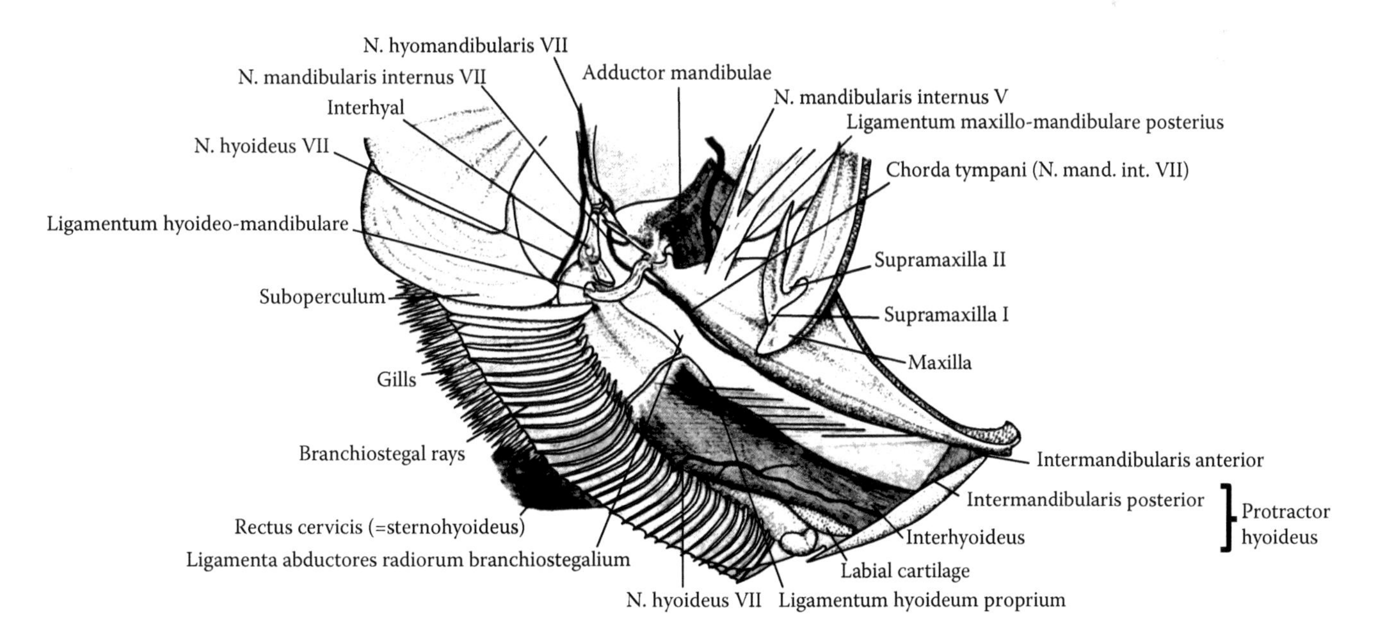

FIGURE 6.13 *Elops saurus* (Teleostei, Elopiformes): ventral head region; right lateral view of fully extended jaw apparatus, the interopercle has been removed (the nomenclature of the structures illustrated follows that of this author). (Modified from Vrba, E. S., *Zool Afr*, 3, 211–236, 1968.)

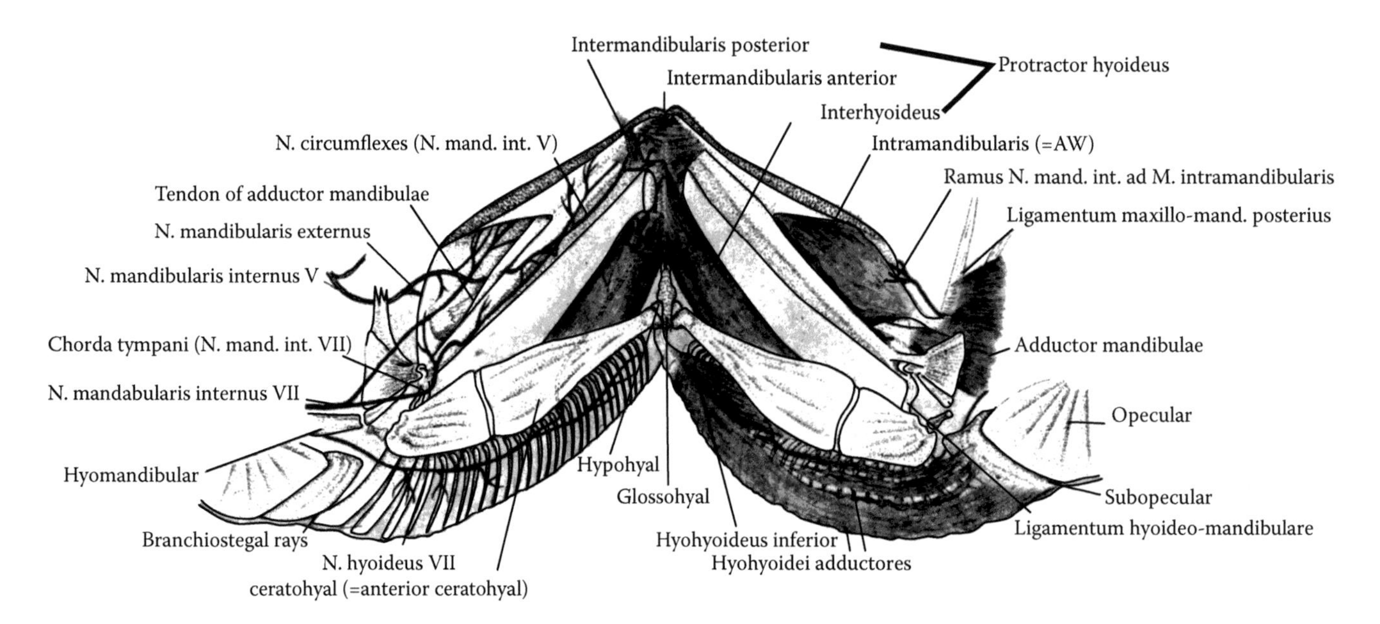

FIGURE 6.14 *Elops saurus* (Teleostei, Elopiformes): dorsal view of the nerves and muscles of the lower jaw; on the left side the adductor mandibulae, intermandibularis, and hyohyoidei have been removed and the posterior gular plate insertion of the intermandibularis posterior and anterior gular plate origin of the interhyoideus have been turned up (the nomenclature of the structures illustrated follows that of this author). (Modified from Vrba, E. S., *Zool Afr*, 3, 211–236, 1968.)

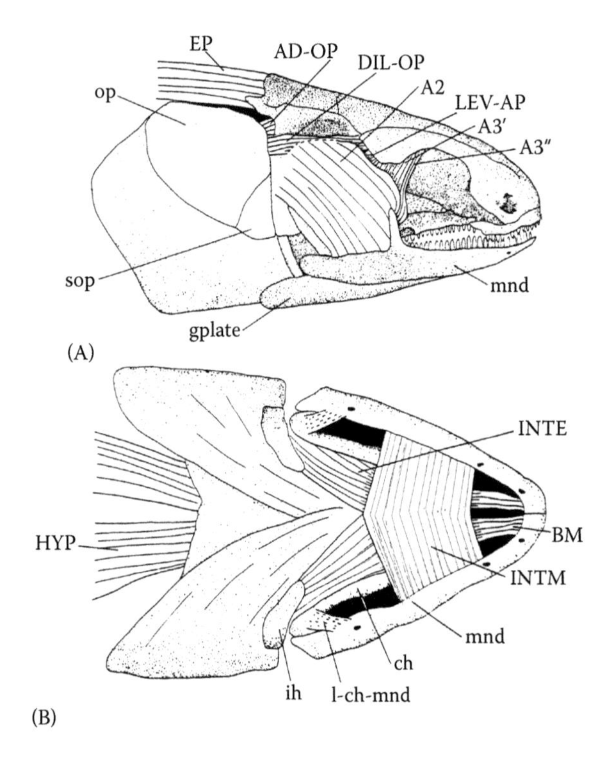

FIGURE 6.15 *Polypterus senegalus* (Cladistia): lateral (A) and ventral (B) views of the head after removal of the eye, suborbital bones, gular plates, and maxilla; in the ventral view, the muscle hyohyoideus is not shown (the nomenclature of the structures illustrated follows that used in the present work). (Modified from Lauder, G. V., *J Morphol*, 163, 283–317, 1980.) A2, A3′, A3″, adductor mandibulae A2, A3′, and A3″; AD-OP, adductor operculi; BM, branchiomandibularis; ch, ceratohyal; DIL-OP, dilatator operculi; EP, epaxialis; gplate, gular plate; HYP, hypoaxialis; ih, interhyal; INTE, interhyoideus; INTM, intermandibularis; l-ch-mand, ligament between ceratohyal and mandible; LEV-AP, levator arcus palatini; mnd, mandible; op, opercle; sop, subopercle.

and complex. For instance, in the dipnoan *Neoceratodus* the levator hyoideus becomes blended with other hyoid muscles during ontogeny and is absent as a separate element in adults; a separate depressor mandibulae is absent in early and late developmental stages of this taxon. In the dipnoans *Lepidosiren* and *Protopterus* and in tetrapods such as *Siren* these two muscles remain as separate elements until the adult stages. As in *Neoceratodus*, the levator hyoideus is also absent as a separate element in numerous adult tetrapods, but for a different reason: as shown by, e.g., Edgeworth (1935), in the course of their development all the fibers of this muscle become attached to the mandible, thus constituting the depressor mandibulae (see the following chapters).

Therefore, the levator hyoideus of adult dipnoans such as *Protopterus* and *Lepidosiren* (Figure 6.22), as well as of adult tetrapods such as *Siren*, appears to correspond to the levator hyoideus of adult amphibians such as *Ambystoma* and to a part of the depressor mandibulae of adult amniotes such as *Timon*. The depressor mandibulae of *Protopterus, Lepidosiren,* and *Siren* appears to correspond to the depressor mandibulae of adult amphibians such as *Ambystoma* and to a part of the depressor mandibulae of adult amniotes such as *Timon* (see the following chapters). The statements of Edgeworth regarding the ontogenetic differences between the dipnoan depressor mandibulae and the depressor mandibulae of tetrapods such as *Ambystoma* have two possible explanations: (a) the mode of appearance of these structures may be somewhat different in dipnoans and in tetrapods (which, in view of the recent discoveries in the field of evolutionary developmental biology, does not completely invalidate the hypothesis that they are homologous: see, e.g., Gould 2002; West-Eberhard 2003; Carroll et al. 2005; Kirschner and Gerhart 2005); (b) the mode of appearance of the depressor mandibulae of dipnoans

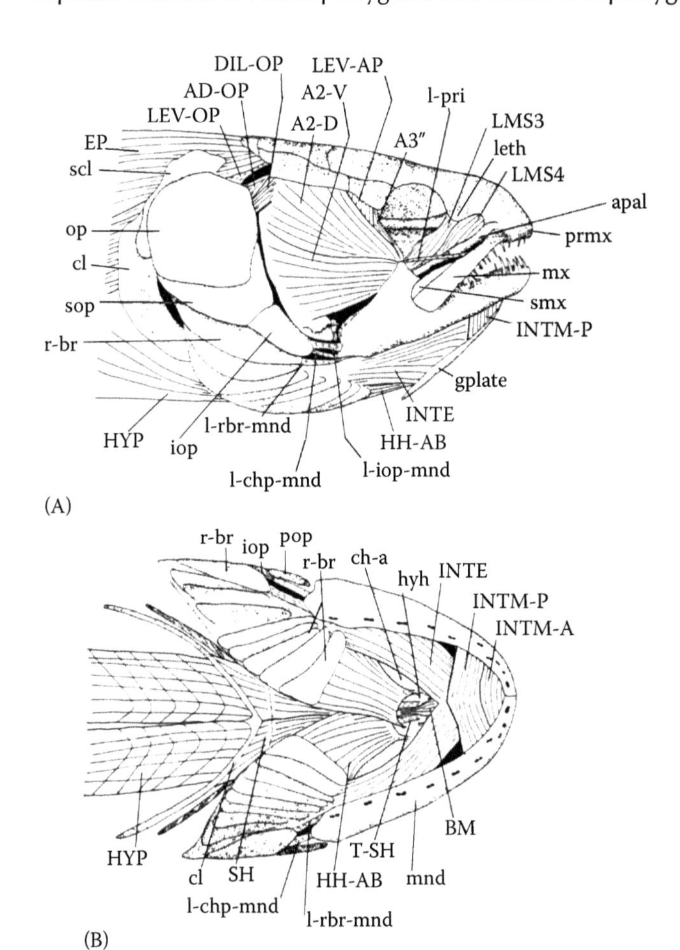

FIGURE 6.16 *Amia calva* (Halecomorphi): lateral (A) and ventral (B) views of the cephalic musculature after removal of the eye, dorsal aspect of the preopercle, and gular plate; in the lateral view, the A3′ is not shown; in the ventral view, the hyohyoideus inferior and the hyohyoidei adductores are not shown (the nomenclature of the structures illustrated follows that used in the present work). (Modified from Lauder, G. V., *J Morphol*, 163, 283–317, 1980.) A2-V, A2-D, A3″, adductor mandibulae A2-V, A2-D and A3′; AD-OP, adductor operculi; AD-SUP, adductor superficialis; apal, autopalatine; BM, branchiomandibularis; ch-a, anterior ceratohyal; cl, cleithrum; DIL-OP, dilatator operculi; EP, epaxialis; gplate, gular plate; HH-AB, hyohyoideus abductor; hyh, hypohyal; HYP, hypoaxialis; INTE, interhyoideus; INTM-A, INTM-P, intermandibularis anterior and posterior; iop, interopercle; l-chp-mand, ligament between posterior ceratohyal and mandible; l-iop-mand, ligament between interopercle and mandible; l-pri, primordial ligament; l-rbr-mand, ligament between branchiostegal rays and mandible; leth, lateral-ethmoid; LEV-AP, levator arcus palatini; LEV-OP, levator operculi; LMS3, LMS4, levator maxillae superioris 3 and 4; branchiomandibularis; mnd, mandible; mx, maxilla; op, opercle; pop, preopercle; prmx, premaxilla; r-br, branchiostegal rays; scl, supracleithrum; SH, sternohyoideus; smx, supramaxilla; sop, subopercle; T-SH, tendon of sternohyoideus.

is similar to that of the depressor mandibulae of urodeles (i.e., it appears ontogenetically after the levator hyoideus, resulting from the differentiation of a part of its fibers), but the youngest dipnoan specimens observed by Edgeworth (1935) were too old to detect such a differentiation (i.e., the differentiation had already occurred, and thus, both the levator hyoideus and the depressor mandibulae were already present in those specimens, giving the impression that these structures originated at the same ontogenetic stage). Further detailed comparative analyses on the development of the hyoid muscles of dipnoans and of other osteichthyans are needed to resolve this question.

Such analyses might also clarify whether the portions of the constrictor hyoideus from which the levator hyoideus/depressor mandibulae and the constrictor operculi of extant dipnoans originate correspond to the portions that give rise to the adductor arcus palatini and the adductor operculi of other bony fishes, as suggested by, e.g., Edgeworth (1935) and Diogo (2007). Edgeworth (1935: 102) stated that the levator hyoideus of dipnoans is derived from the portion of the constrictor hyoideus that gives rise to the retractor hyomandibulae in extant acipenseriforms, which seems to correspond to the adductor arcus palatini of other bony fishes (Table 6.2; see preceding text). In living dipnoans, the hyomandibula is very reduced or even absent and the palatoquadrate is fused to the neurocranium and is thus much less mobile than that of most other bony fishes. Therefore, the portion of the hyoid muscle plate that gives rise to the adductor arcus palatini of other bony fishes may have lost its usual attachments to the hyomandibula and/or palatoquadrate and become attached to the ceratohyal. The dorsal surface of the ceratohyal is more dorsal in dipnoans than in most other bony fishes, occupying a position somewhat similar to that of the hyomandibula of the latter fishes (Figure 6.22) (e.g., Rosen et al. 1981; Forey 1986; Bauer 1997; this chapter). According to our observations, in juvenile and adult specimens of *Lepidosiren*, the levator hyoideus attaches not only to the dorsal surface of the ceratohyal, but also to a part of its dorsomesial margin (Figure 6.11). Thus, as the adductor arcus palatini of other bony fishes usually attaches on the dorsomesial margin of the hyomandibula/palatoquadrate, allowing it to adduct this structure; it seems that the dipnoan levator hyoideus might both elevate and adduct the dorsal portion of the ceratohyal. The dipnoan constrictor operculi may have originated from the same portion of the constrictor hyoideus that gives rise to the adductor operculi in other bony fishes. In support of this hypothesis, some plesiomorphic fossil dipnoans exhibit well-defined scars on the mesial margin of the operculum for the attachment of a muscular structure that appears to correspond to the adductor operculi of other fishes (e.g., Campbell and Barwick 1986). Regardless, in adult extant dipnoans, the constrictor operculi forms a sheet of fibers that is continuous with other muscles such as the interhyoideus. Therefore, even if the constrictor operculi of dipnoans is derived from the portion of the constrictor hyoideus that gives rise to the adductor operculi in other bony fishes, in adult extant dipnoans, there is no separate, distinct adductor operculi such as that found in adult members of, e.g., *Latimeria* and most actinopterygians (see the following chapters).

A single division of the ventral portion of the hyoid muscle plate, which we designate interhyoideus, appears to be plesiomorphic for osteichthyans (Table 6.2). In most extant actinopterygians, part of the interhyoideus becomes ontogenetically

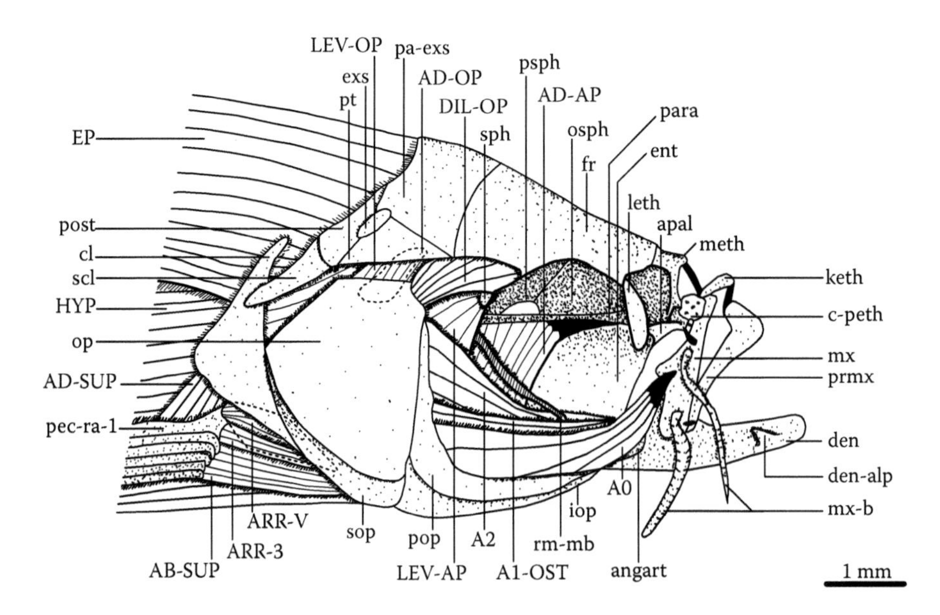

FIGURE 6.17 *Danio rerio* (Teleostei, Cypriniformes): lateral view of the cephalic musculature; all muscles are exposed, the maxillary barbels and the mesial branch of the ramus mandibularis are also illustrated; the nasals, infraorbitals and postcleithra were removed. A0, A1-OST, A2, adductor mandibulae A0, A1-OST, and A2; AB-SUP, abductor superficialis; AD-AP, adductor arcus palatini; AD-OP, adductor operculi; AD-SUP, adductor superficialis; angart, angulo-articular; apal, autopalatine; ARR-3, arrector 3; ARR-V, arrector ventralis; c-peth, preethmoid cartilage; cl, cleithrum; den, dentary bone; den-alp, anterolateral process of dentary bone; DIL-OP, dilatator operculi; ent, entopterygoid; EP, epaxialis; exs, extrascapular; fr, frontal; HYP, hypoaxialis; iop, interopercle; keth, kinethmoid; leth, lateral-ethmoid; LEV-AP, levator arcus palatini; LEV-OP, levator operculi; meth, mesethmoid; mx, maxilla; mx-b, maxillary barbel; op, opercle; osph, orbitosphenoid; pa-exs, parieto-extrascapular; para, parasphenoid; pec-ra-1, pectoral ray 1; pop, preopercle; post, posttemporal; prmx, premaxilla; psph, pterosphenoid; pt, pterotic; rm-mb, mesial branch of ramus mandibularis; scl, supracleithrum; sop, subopercle; sph, sphenotic.

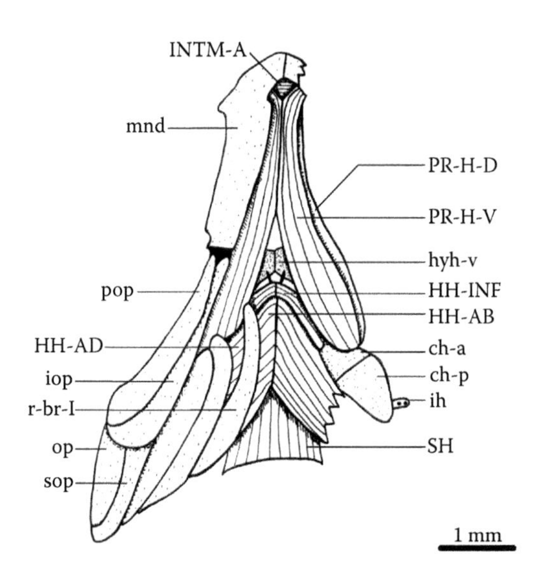

FIGURE 6.18 *Danio rerio* (Teleostei, Cypriniformes): ventral view of the ventral cephalic musculature; on the right side, a portion of the hyohyoidei adductores, as well as of the mandible, was cut, and the opercle, interopercle, subopercle, and preopercle are not represented. ch-a, ch-p, anterior and posterior ceratohyals; HH-AB, hyohyoideus abductor; HH-AD, hyohyoidei adductores; HH-INF, hyohyoideus inferior; hyh-v, ventral hypohyal; ih, interhyal; INTM-A, anterior intermandibularis; iop, interopercle; mnd, mandible; op, opercle; pop, preopercle; PR-H-D, PR-H-V, sections of protractor hyoideus; r-br-I, branchiostegal ray I; SH, sternohyoideus; sop, subopercle.

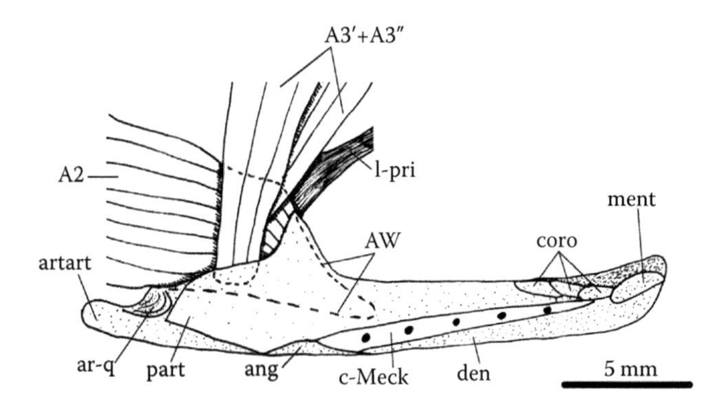

FIGURE 6.19 *Polypterus bichir* (Cladistia): mesial view of adductor mandibulae and mandible; mandibular teeth are not illustrated. A2, A3′, A3″, AW, adductor mandibulae A2, A3′, A3″, and Aω; ang, angular; ar-q, articulatory facet for quadrate; artrart, articulo-retroarticular; c-Meck, Meckelian cartilage; coro, coronoids; den, dentary bone; l-pri, primordial ligament; ment, mentomeckelian bone; part, prearticular.

differentiated into a distinct, separate muscle, the hyohyoideus (Table 6.2; Figure 6.6). However, Noda et al.'s (2017) recent developmental study suggested that in *Polypterus* the main body of the hyohyoideus muscle derives from the constrictor hyoideus dorsalis rather than from the constrictor hyoideus ventralis as has been described in other extant actinopterygians. That is, that study contradicts Edgeworth's

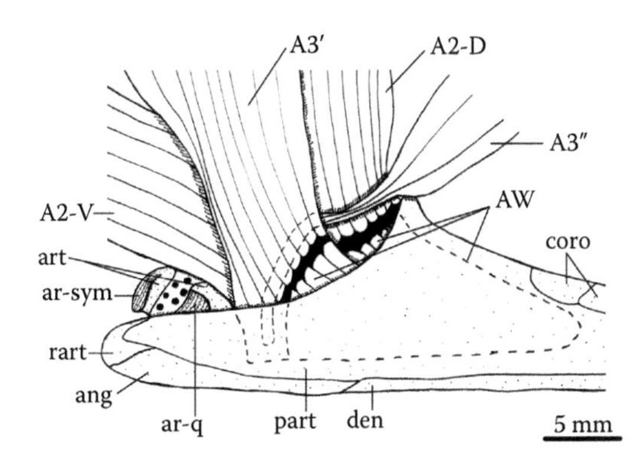

FIGURE 6.20 *Amia calva* (Halecomorphi): mesial view of mandible and of adductor mandibulae sections A2, A3′, A3″, and Aω; the levator maxillae superioris 3 and 4 and the mandibular teeth are not illustrated. A2-D, A2-V, A3′, A3″, AW, adductor mandibulae A2-D, A2-V, A3′, A3″, and Aω; ang, angular; ar-q, ar-sym, articulatory facets for quadrate and for symplectic; art, articular; coro, coronoids; den, dentary bone; part, prearticular; rart, retroarticular.

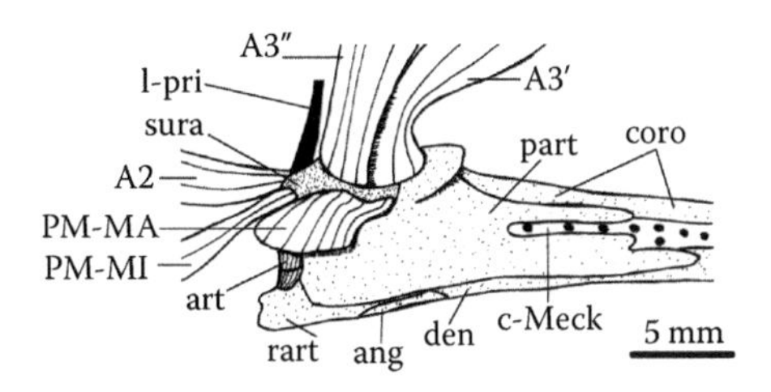

FIGURE 6.21 *Lepisosteus osseus* (Ginglymodi): mesial view of adductor mandibulae and mandible; mandibular teeth are not illustrated. A2, A3′, A3″, adductor mandibulae A2, A3′, A3″; ang, angular; art, articular; c-Meck, Meckelian cartilage; coro, coronoids; den, dentary bone; l-pri, primordial ligament; part, prearticular; PM-MA, PM-MI, palatomandibularis major and minor; rart, retroarticular; sura, surangular.

(1935) suggestion that in *Polypterus* the main part of the hyohyoideus develops from the constrictor hyoideus ventralis, as in other actinopterygians. Noda et al. (2017) even suggest that the hyohyoideus of *Polypterus* might thus not be homologous to that of other actinopterygians, but this seems doubtful to us, particularly because it is well known that many adult structures that are generally considered to be homologous in different taxa might derive from different developmental primordia (see, e.g., recent review by Diogo [2017]).

In *Amia* and most teleosts, the hyohyoideus is subdivided into hyohyoideus inferior, hyohyoideus abductor, and hyohyoidei adductores (Figure 6.18). The hyoideus abductor and hyohyoidei adductores are often considered parts of a hyohyoideus superior (Table 6.2; Figure 6.6). As stated by Stiassny (2000: 122) "there is little commentary in the literature regarding the function of HhI (hyohyoideus inferior) but adduction of the hyoid bar is suggested by its position and presumed line of action." Regarding the hyohyoideus abductor and the hyohyoidei adductores, they are usually related to the expansion and the constriction of the branchiostegal membranes, respectively (Figure 6.18). The interhyoideus, which

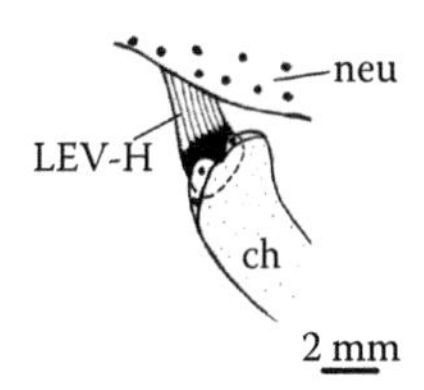

FIGURE 6.22 *Lepidosiren paradoxa* (Dipnoi): lateral view of levator hyoideus; the ventral portion of the ceratohyal was cut. LEV-H, levator hyoideus; neu, neurocranium.

as its name indicates, usually connects the two hyoid bars, has become associated with the intermandibularis posterior in teleosts, forming the peculiar protractor hyoideus (Tables 6.1 and 6.2; see preceding text.).

Miyake et al.'s (1992) Table I suggests that in extant cladistians, chondrosteans, and ginglymodians, the hyohyoideus consists of two well-differentiated divisions: hyohyoideus inferior and hyohyoideus superior. However, in the adult specimens of these three groups that we have dissected, the hyohyoideus is not divided into well-differentiated parts as is the case in *Amia* and teleosts. Instead, it essentially consists of a continuous sheet of fibers. Other authors have recorded similar observations; e.g., Lauder (1980a: 293) wrote that "in *Lepisosteus* the hyohyoideus superioris … also (as in *Polypterus*) is continuous with the fibers of the hyohyoideus inferioris" (i.e. the fibers of the hyohyoideus form a continuous sheet, as is the case in the *Lepisosteus* and *Polypterus* specimens we dissected). Noda et al.'s (2017) recent developmental study of *Polypterus* also supports the idea that there is no clear division between the hyohyoideus inferioris and the hyohyoideus superioris in this taxon. The hyohyoideus of the chondrosteans described by, e.g., Danforth (1913: *Polyodon*) and Carroll and Wainwright (2003: *Scaphirhynchus*) also consists of a continuous sheet of fibers, as is the case in the *Psephurus* and *Acipenser* specimens we analyzed (NB: the interhyoideus of this work corresponds to the "geniohyoideus posterior" and the "constrictor ventralis posterior" *sensu* Danforth [1913] and of Carroll and Wainwright [2003], respectively).

There seems to be no well-differentiated, separate hyohyoideus in extant Sarcopterygii. In some sarcopterygians, the portion of the hyoid muscle plate that gives rise to the interhyoideus and hyohyoideus in actinopterygians might be somewhat differentiated into bundles that may resemble those two structures, but these bundles remain deeply blended throughout ontogeny. For example, this is the case for the interhyoideus anterior and the interhyoideus posterior of salamanders (see the following chapters and, e.g., Luther 1913, 1914; Lubosch 1914; Edgeworth 1935; Kesteven 1942–1945; Jarvik 1963; Larsen and Guthrie 1975; Carroll and Holmes 1980; Bauer 1992; Haas 2001). Similarly, the "géniohyoïdien" and the "hyohyoïdien" described by Millot and Anthony (1958) in *Latimeria* are deeply blended in the adult members of this genus and thus seem to correspond to the interhyoideus *sensu* the present work.

In addition to the muscles mentioned in the preceding paragraphs, osteichthyans may exhibit other hyoid muscles

(Table 6.2 and the following chapters). For example, the levator branchiarium and depressor branchiarium of *Polypterus* are, according to Noda et al. (2017), hyoid muscles derived from the constrictor hyoideus dorsalis (see Figure 7.2). In early developmental stages of this taxon, the levator branchiarium originates from the auditory capsule and runs alongside the dorsal hyoid artery and inserts onto the medial branches of the gill stem. The depressor branchiarium originates medially from the dorsal portion of the muscle hyohyoideus and runs along the ventral hyoid artery, inserting onto the lateral branches of the gill stem. For another example, in at least some developmental stages of urodeles, there is a peculiar muscle branchiohyoideus connecting the hyoid and branchial arches. Ontogenetically, this muscle appears between the interhyoideus and the levator hyoideus/depressor mandibulae, a position that "makes it difficult to determine if it belongs to the dorsal or ventral (hyoid) muscles" (Ericsson and Olsson 2004: 136) (see Figure 6.7). According to Lauder and Shaffer (1985: 308), the function of the branchiohyoideus is to "mediate hyoid retraction and possibly also produce ceratobranchial abduction in the absence of antagonistic activity." Edgeworth (1935) and Lauder and Shaffer (1988) stated that the branchiohyoideus becomes lost after metamorphosis in most urodeles. However, the metamorphosed specimens of *Ambystoma ordinarium* we examined have a distinct muscle branchiohyoideus. As explained by Edgeworth (1935), the structure that is often named "branchiohyoideus" in amniotes appears to be homologous with the branchial muscle subarcualis rectus 1 of amphibians rather than to the hyoid muscle branchiohyoideus of the present work (which Edgeworth designated "branchiohyoideus externus") (see the following chapters). Jarvik (1963) suggested that a branchiohyoideus similar to that of urodeles may have been present in at least some of the sarcopterygian fossil fishes placed in his "porolepiform-stock." However, according to, e.g., Rosen et al. (1981), some of Jarvik's interpretations of fossils, principally those concerning soft structures, should be viewed with much caution, since they were profoundly influenced by certain strong and, rather heterodox, beliefs of this author. For instance, in this specific case regarding the eventual presence of a branchiohyoideus in his "porolepiforms," this has much to do with his profound conviction that the latter fishes were the sister-group of urodeles and, thus, that amphibians are not monophyletic, a view to which almost no authors would adhere nowadays ("these great resemblances ... cannot possibly be due to parallel evolution, and they definitively prove that the urodeles are closely related to and descendants of porolepiforms"; Jarvik 1963: 61). Therefore, a detailed, updated, and, if possible, less biased comparative analysis is needed in order to clarify whether or not a branchiohyoideus is present in sarcopterygian fish fossils.

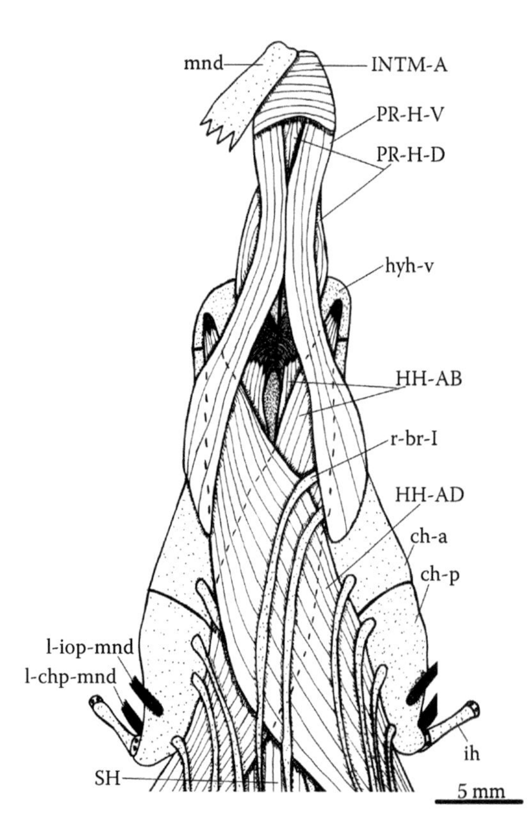

FIGURE 6.23 *Alepocephalus rostratus* (Teleostei, Alepocephaloidea): ventral view of the cephalic musculature; on the right side, the mandible was removed; on the left side, the mandible was cut. ch-a, ch-p, anterior and posterior ceratohyals; HH-AB, hyohyoideus abductor; HH-AD, hyohyoidei adductores; hyh-v, ventral hypohyal; ih, interhyal; INTM-A, anterior intermandibularis; l-chp-mnd, ligament between posterior ceratohyal and mandible; l-iop-mnd, ligament between interopercle and mandible; mnd, mandible; PR-H-D, PR-H-V, sections of protractor hyoideus; r-br-I, branchiostegal ray I; SH, sternohyoideus.

BRANCHIAL MUSCLES

The muscles listed in Table 6.3 correspond to the branchial muscles *sensu lato* of Edgeworth (1935). They can be divided into three groups. The first comprises the "true" branchial muscles, which are subdivided into (a) the branchial muscles *sensu stricto* that are directly associated with the movements of the branchial arches and that in mammals are usually innervated by the glossopharyngeal nerve (CNIX); (b) the protactor pectoralis and its derivatives, which are associated with the pectoral girdle and often innervated by the spinal accessory nerve (CNXI), but which in phylogenetically plesiomorphic gnathostomes such as chondrichthyans are thought to be innervated by the vagus nerve CNX (Edgeworth, 1935; see also Chapters 3 and 4). The second group consists of the pharyngeal muscles, which are only present as independent structures in extant mammals. They are thought to be derived from arches 4–6 and are usually innervated by the vagus nerve (CNX). The mammalian stylopharyngeus is thought to be derived from the third arch and is primarily innervated by the glossopharyngeal nerve and is thus grouped with the true branchial muscles rather than the pharyngeal muscles. The third group is made up of the laryngeal muscles, which are thought to be derived from arches 4–6 and are usually innervated by CNX.

TABLE 6.3
Branchial Muscles of Adults of Representative Extant Actinopterygian Taxa

Probable Plesiomorphic Osteichthyan Condition	Cladistia: *Polypterus bichir* (Bichir)	Chondrostei: *Psephurus gladius* (Chinese swordfish)	Ginglymodi: *Lepisosteus osseus* (Longnose gar)	Halecomorphi: *Amia calva* (Bowfin)	Teleostei basal: *Elops saurus* (Ladyfish) (Figure 6.13)	Teleostei clupeocephalan: *Danio rerio* (Zebrafish)
Branchial muscles *sensu stricto*	**Branchial muscles *sensu stricto*** [according to Miyake et al. [1992] and Noda et al. [2017], *Polypterus* has a dorsal branchial muscle complex (including four muscles according to Noda et al. [2017]), transversi ventrales 2–4, a pharyngoclavicularis, a rectus communis, an interarcualis ventralis 1, and interbranchiales adductores and abductores]	**Branchial muscles *sensu stricto*** [according to Miyake et al. [1992], chondrosteans may have a dorsal branchial muscle complex, transversi ventrales 4 and 5, a pharyngoclavicularis, obliqui ventrales 2 and 3, and possibly an interarcualis ventralis 1]	**Branchial muscles *sensu stricto*** [according to Miyake et al. [1992], ginglymodians may have a dorsal branchial muscle complex, a transversus ventralis 5, a pharyngo-clavicularis, obliqui ventrales 1–4, and a possibly transversi ventrales 4 and 5]	**Branchial muscles *sensu stricto*** [according to Miyake et al. [1992], *Amia* might have a dorsal branchial muscle complex, obliqui ventrales 1–4, a transversus ventralis 5, a rectus ventralis 4, pharyngo-clavicularis internus and externus, and possibly transversi ventrales 3 and 4]	**Branchial muscles *sensu stricto*** [according to Miyake et al. [1992], teleosts may have a dorsal branchial muscle complex, an interbranchialis adductoris, obliqui ventrales 1–4, transversi ventrales 1–4, recti ventrales 2–4, pharyngo-clavicularis internus and externus, a rectus communis, and possibly an interarculalis ventralis 1 and/or a rectus ventralis 1]	**Branchial muscles *sensu stricto*** [see on the left]
Protractor pectoralis	**Protractor pectoralis** (cucullaris *sensu* Edgeworth [1935])	**Protractor pectoralis** (cucullaris *sensu* Edgeworth [1935])	–	**Protractor pectoralis** (cucullaris *sensu* Edgeworth [1935])	–	– [among the zebrafish specimens we dissected, a distinct, independent protractor pectoralis was usually absent in adults, but was found in some old larvae]
Constrictor laryngis present in LCA of osteichthyans? See text and below	Constrictor laryngis and dilatator laryngis of *Polypterus* are homologous to the constrictor laryngis and dilatator laryngis of sarcopterygians? See text and below	–	–	–	–	–
Dilatator laryngis present in LCA of osteichthyans? See text and below	See above	–	–	–	–	–

(*Continued*)

TABLE 6.3 (CONTINUED)
Branchial Muscles of Adults of Representative Extant Actinopterygian Taxa

Probable Plesiomorphic Osteichthyan Condition	Cladistia: *Polypterus bichir* (Bichir)	Chondrostei: *Psephurus gladius* (Chinese swordfish)	Ginglymodi: *Lepisosteus osseus* (Longnose gar)	Halecomorphi: *Amia calva* (Bowfin)	Teleostei basal: *Elops saurus* (Ladyfish) (Figure 6.13)	Teleostei clupeocephalan: *Danio rerio* (Zebrafish)
See on the right	**Dilatator laryngis** [according to Edgeworth [1935], this muscle is not homologous with the dilatator laryngis of *Amia* and *Lepisosteus* nor with the dilatator laryngis of sarcopterygians; this muscle was not described by Noda et al. 2017 in *Polypterus*]	–	–	–	–	–
See on the right	–	–	**Dilatator laryngis** [according to Edgeworth [1935], the dilatator laryngis of *Lepisosteus* and *Amia* is not homologous with the dilatator laryngis of *Polypterus* nor with the dilatator laryngis of sarcopterygians]	**Dilatator laryngis** [see on the left]	–	–
See on the right	**Constrictor laryngis** [according to Edgeworth [1935], this muscle might well be homologous with the constrictor laryngis of sarcopterygians; if this is the case, the presence of this muscle would be plesiomorphic for osteichthyans; this muscle was not described by Noda et al. [2017] in *Polypterus*]	– [apparently absent, according to Edgeworth [1935]]	– [apparently absent, according to Edgeworth [1935]]	–	–	–

Note: See footnote of Table 6.1.

Most adult vertebrates retain some branchial muscles *sensu stricto* (e.g., Bischoff 1840; Owen 1841; Allis 1923; Edgeworth 1935; Millot and Anthony 1958; Wiley 1979a,b; Jollie 1982; Bemis 1986; Miyake et al. 1992; Wilga et al. 2000; Kardong 2002; Carroll and Wainwright 2003; Johanson 2003; this study) (Table 6.3). However, the branchial muscles *sensu stricto* are not present as a group in extant reptiles and extant mammals (see the following chapters). For instance, many adult reptiles have only one branchial muscle *sensu stricto*, the subarcualis rectus I (see preceding text). The two branchial muscles *sensu stricto* found in adult reptiles such as lizards, i.e., the hyobranchialis and "ceratohyoideus," seem to be the result of a subdivision of the subarcualis rectus I (see the following chapters). Adult extant mammals lack all the branchial muscles *sensu stricto* except subarcualis rectus I (which in most adult mammals gives rise to the ceratohyoideus and stylopharyngeus), the subarcualis rectus II (usually present only in adult marsupials), and the subarcualis rectus III (usually present only in adult monotremes) (e.g., Edgeworth 1935).

As will also be explained in the following chapters, the mammalian acromiotrapezius, spinotrapezius, cleido-occipitalis, sternocleidomastoideus, cleidomastoideus, and sternomastoideus correspond to the reptilian trapezius and sternocleidomastoideus and thus to the protractor pectoralis of dipnoans, amphibians, and other osteichthyans. The protractor pectoralis and its derivatives are not branchial muscle *sensu stricto* because they are mainly involved in the movements of the pectoral girdle and not of the branchial arches (e.g., Edgeworth 1935; see the following chapters for more details on the development and homologies of this muscle).

The mammalian laryngeal muscles thyroarytenoideus, vocalis, cricoarytenoideus lateralis, and arytenoideus appear to derive from the laryngeus of nonmammalian tetrapods such as salamanders, which in turn derives from the constrictor laryngis of dipnoans such as *Lepidosiren*; the mammalian cricoarytenoideus posterior corresponds to the dilator laryngis of other tetrapods and of dipnoans (see the following chapters). Some nonsarcopterygian vertebrates such as *Polypterus*, *Lepisosteus*, and *Amia* have a "constrictor laryngis" and/or a "dilatator laryngis," but it is not clear if these muscles do correspond to the constrictor laryngis and dilatator laryngis of sarcopterygians and, thus, if the latter muscles are plesiomorphically present in osteichthyans or not (e.g., Edgeworth 1935). The few descriptions of the laryngeal region of *Latimeria* do not allow us to discern whether or not these laryngeal muscles are present in this taxon (e.g., Millot and Anthony 1958). The laryngeus of tetrapods does not seem to be plesiomorphically found in sarcopterygians because it seems to be absent in sarcopterygian fishes such as dipnoans. However, a detailed study of the laryngeal region of *Latimeria* would be needed to test this hypothesis.

HYPOBRANCHIAL MUSCLES

According to Edgeworth (1935), there are two major lineages of muscles that originated from the hypobranchial muscle plate: his "genio-hyoideus" and his "rectus cervicis" (Figures 6.8 and 6.9). As noted by Miyake et al. (1992), it is not very clear if Edgeworth's genio-hyoideus and rectus cervicis represent separate premyogenic condensations or later states of muscle development (Table 6.4).

The plesiomorphic condition for osteichthyans seems to be similar to that of adult members of Actinistia and Dipnoi, in which there is a coracomandibularis and a sternohyoideus (*sensu* this volume: see the following chapters). According to, e.g., Edgeworth (1935), Kesteven (1942–1945), Wiley (1979a,b), Jollie (1982), Mallatt (1997), Wilga et al. (2000), and Johanson (2003), these muscles were originally mainly related to the opening of the mouth. Extant tetrapods exhibit several hypobranchial muscles that are not found in other living osteichthyans, such as the omohyoideus and the specialized glossal muscles related to tongue movements. The omohyoideus and the genioglossus appear to be derived from the sternohyoideus and the coracomandibularis, respectively (Edgeworth 1935). However, the statements of Edgeworth (1935) regarding the origin of the hyoglossus are somewhat confusing: in page 196 of his study, he states that in amphibians such as salamanders, this muscle originates from the sternohyoideus, but in page 211, he affirms that "the hypobranchial muscles of Amphibia, Sauropsida and Mammalia are essentially similar ..., a genioglossus and a hyoglossus are developed from the genio-hyoideus [= coracomandibularis]." Jarvik (1963: 41) revisited this question and stated that the hyoglossus of salamanders "seems to be an anterior portion of

TABLE 6.4
Hypobranchial Muscles of Adults of Representative Actinopterygian Taxa

Probable Plesiomorphic Osteichthyan Condition	Cladistia: *Polypterus bichir* (Bichir)	Chondrostei: *Psephurus gladius* (Chinese Swordfish)	Ginglymodi: *Lepisosteus osseus* (Longnose Gar)	Halecomorphi: *Amia calva* (Bowfin)	Teleostei—Basal: *Elops saurus* (Ladyfish) (Figure 6.14)	Teleostei—Clupeocephalan: *Danio rerio* (Zebrafish)
Coracomandibularis	Branchiomandibularis [modified/derived from coracomandibularis]	Branchiomandibularis [see on the left]	–	Branchiomandibularis [see on the left]	–	–
Sternohyoideus	Sternohyoideus	Sternohyoideus	Sternohyoideus	Sternohyoideus	Sternohyoideus	Sternohyoideus

Note: See footnore of Table 6.1 and Figures 6.8 and 6.9.

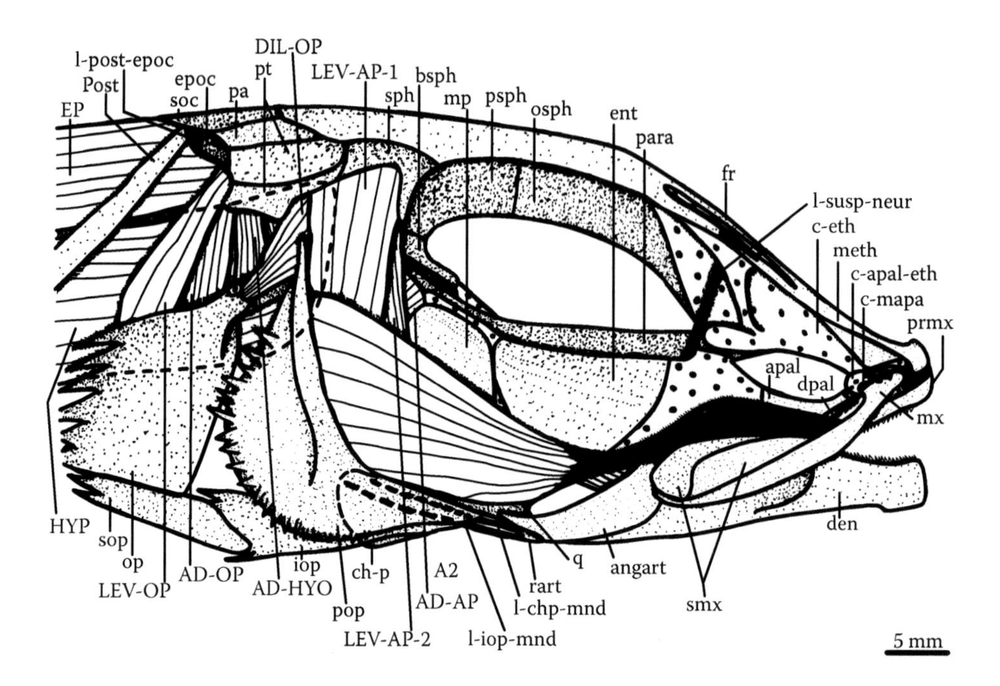

FIGURE 6.24 *Alepocephalus rostratus* (Teleostei, Alepocephaloidea): lateral view of the cephalic musculature; the pectoral girdle muscles are not illustrated; most elements of the pectoral girdle, as well as the nasals and infraorbitals, were removed. A2, adductor mandibulae A2; AD-AP, adductor arcus palatini; AD-HYO, adductor hyomandibulae; AD-OP, adductor operculi; angart, angulo-articular; apal, autopalatine; bsph, basisphenoid; c-apal-eth, cartilage between autopalatine and ethmoid region; c-eth, ethmoid cartilage; c-mapa, small cartilage between maxilla and autopalatine; ch-p, posterior ceratohyal; den, dentary bone; dpal, dermopalatine; DIL-OP, dilatator operculi; ent, entopterygoid; EP, epaxialis; epoc, epioccipital; fr, frontal; HYP, hypoaxialis; iop, interopercle; l-chp-mnd, ligament between posterior ceratohyal and mandible; l-iop-mnd, ligament between interopercle and mandible; l-pri, primordial ligament; l-post-epoc, ligament between posttemporal and epioccipital; l-susp-neur, ligament between suspensorium and neurocranium; leth, lateral-ethmoid; LEV-AP-1, 2, sections of levator arcus palatini; LEV-OP, levator operculi; meth, mesethmoid; mp, metapterygoid; mx, maxilla; op, opercle; osph, orbitosphenoid; pa, parietal; para, parasphenoid; pop, preopercle; post, posttemporal; prmx, premaxilla; psph, pterosphenoid; pt, pterotic; q, quadrate; rtart, retroarticular; smx, supramaxillae; soc, supraoccipital; sop, subopercle; sph, sphenotic.

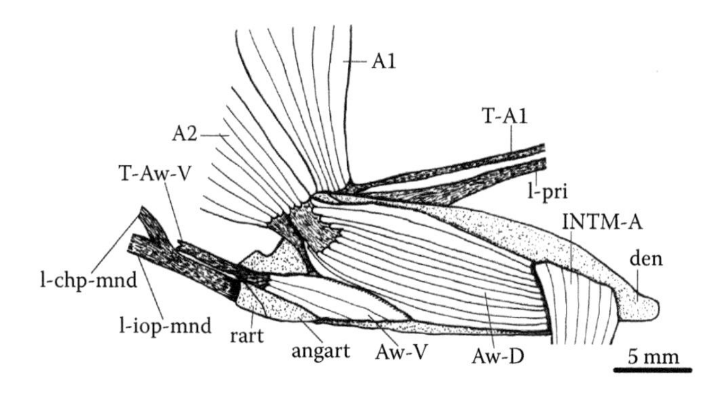

FIGURE 6.25 *Aulopus filamentosus* (Teleostei, Aulopiformes): mesial view of the left mandible and adductor mandibulae; the anterior intermandibularis and the primordial ligament, as well as the ligaments between the mandible, posterior ceratohyal, and interopercle, are also shown; mandibular teeth were removed. A1, A2, Aω-D, Aω-V, sections of adductor mandibulae; angart, angulo-articular; den, dentary bone; l-chp-mnd, ligament between posterior ceratohyal and mandible; l-iop-mnd, ligament between interopercle and mandible; l-pri, primordial ligament; T-A1, tendon of A1; T-Aω-V, tendon of Aω-V; rtart, retroarticular.

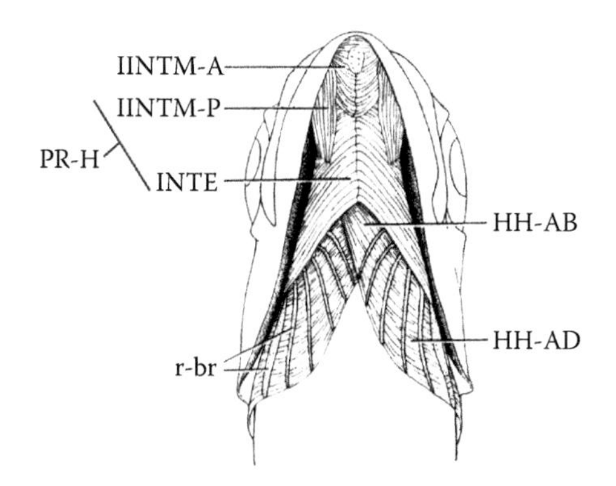

FIGURE 6.26 *Hiodon alosoides* (Teleostei, Hiodontiformes): ventral view of the cephalic musculature (modified from Greenwood 1971). HH-AB, hyohyoideus abductor; HH-AD, hyohyoidei adductores; INTE, interhyoideus; INTM-A, INTM-P, anterior and posterior sections of intermandibularis; PR-H, protractor hyoideus; r-br, branchiostegal rays.

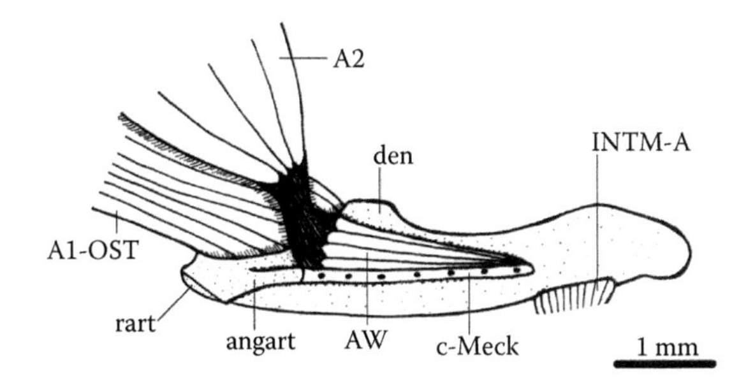

FIGURE 6.27 *Danio rerio* (Teleostei, Cypriniformes): mesial view of the left mandible and adductor mandibulae, the anterior intermandibularis is also shown; the adductor mandibulae A0 was removed. A1-OST, A2, adductor mandibulae A1-OST, A2, and Aω; angart, angulo-articular; c-Meck, Meckelian cartilage; coro, coronoids; den, dentary bone; INTM-A, intermandibularis anterior; rart, retroarticular.

the rectus cervicus" (sternohyoideus *sensu* the present work). Larsen and Guthrie (1975) suggested that the hyoglossus of salamanders may originate from "part of the genioglossus complex" (and, thus, from the coracomandibularis of sarcopterygian fishes), but stated that it was not possible to confirm this hypothesis "without examining early ontogenetic stages." The information provided in more recent publications referring to the cranial musculature of salamanders, e.g., Carroll and Holmes (1980), Lauder and Shaffer (1985, 1988), Reilly and Lauder (1989, 1990, 1991), Bauer (1992, 1997), Iordansky (1992), Kardong and Zalisko (1998), Haas (2001), Kardong (2002), Ericsson and Olsson (2004), and Ericsson et al. (2004), does not allow us to determine whether the hyoglossus of these amphibians is derived from the sternohyoideus or from the coracomandibularis (or from both). The analysis of the salamander specimens presented in the present work does not resolve this question either; detailed comparative analyses of the development of the hyoglossus in salamanders and in other tetrapods are therefore needed.

In extant cladistians, chondrosteans, and halecomorphs, the coracomandibularis is modified into a peculiar muscle branchiomandibularis connecting the branchial arches and the mandible (Figures 6.15 and 6.16), which is absent in living ginglymodians and teleosts (Figure 6.18). Wiley (1979a,b) and Lauder and Liem (1983) argued that the ancestors of extant actinopterygians probably had a branchiomandibularis and that the absence of this muscle in living ginglymodians and teleosts is due to a secondary loss. The phylogenetic results of Diogo (2007) strongly support this hypothesis.

GENERAL REMARKS

Some of the major features concerning the cephalic muscles of actinopterygians and of sarcopterygian fishes are summarized in Figures 6.1 and 6.2. In view of the information summarized earlier and in these figures and Tables 6.1 through 6.4, it appears that plesiomorphically in osteichthyans, the mandibular muscle plate gave rise to the levator arcus palatini (and perhaps to the spiracularis: see preceding text) dorsally, to the adductor mandibulae A2, A3′, A3″, and Aω, medially, and to a divided intermandibularis, ventrally. Within osteichthyan evolution, the posterior part of the intermandibularis become associated with the interhyoideus in teleosts, forming the peculiar muscle protractor hyoideus. The number of divisions of the adductor mandibulae complex has changed during the evolutionary history of osteichthyans. Some divisions are exclusively found in extant members of certain groups; e.g., the A1-OST (ostariophysans), A0 (cypriniforms), palatomandibularis minor and major (ginglymodians), levator maxillae superioris 3 and 4 (halecomorphs), A2-PVM (non-actinistian sarcopterygians), and pterygomandibularis (tetrapods). The number of dorsal mandibular muscles is also variable among adult osteichthyans: some taxa have a single muscle (e.g., *Latimeria*, in which there is only a levator arcus palatini and which seems to represent the plesiomorphic condition for osteichthyans: see preceding text). Some have more than one dorsal mandibular muscle (e.g., the numerous actinopterygians exhibiting a levator arcus palatini and a dilatator operculi and even a spiracularis in at least some developmental stages as is the case of *Polypterus*, or the numerous amniotes exhibiting a levator pterygoidei and a protractor pterygoidei), and still some others have none (e.g., living dipnoans and amphibians).

A single ventral division of the hyoid muscle plate (interhyoideus) and two dorsomedial divisions (adductor arcus palatini and adductor operculi) seem to be plesiomorphic for osteichthyans. In actinopterygians, a portion of the interhyoideus became differentiated into a separate muscle, the hyohyoideus, which then became divided into three separate, well recognizable divisions in halecostomes, the hyohyoideus inferior, the hyohyoideus abductor, and the hyohyoidei adductors (Figures 6.23 through 6.27). Other hyoid muscles may be found in osteichthyans, e.g., the branchiohyoideus of urodeles and the levator operculi of halecostomes. *Latimeria* has a muscle levator operculi that in certain aspects resembles, but does not seem to be homologous to, that of halecostomes. *Latimeria* also has an adductor hyomandibulae that somewhat resembles but does not seem to be homologous with the adductor hyomandibulae found in certain other osteichthyans such as actinopterygians. As explained earlier, there are many uncertainties concerning the homologies and evolution of the levator hyoideus/depressor mandibulae of dipnoans and tetrapods. According to the hypothesis proposed in the present work, both these muscles seem to have been found in at least some ontogenetic stages of the ancestors that latter gave rise to dipnoans and tetrapods (see preceding text). Interestingly, works such as that by Köntges and Lumsden (1996) have shown that in tetrapods such as birds the posterior region of the mandible in which the depressor mandibulae attaches is constituted by neural crest derivatives of the hyoid arch rather than the mandibular arch. This is one of the examples given by these authors to illustrate the highly constrained pattern of cranial skeletomuscular connectivity that they found in these tetrapods, in which each rhombomeric neural crest population remains coherent throughout ontogeny, forming both the connective tissues of specific muscles and their respective attachment sites onto the neuro- and viscerocranium (see the following chapters).

It would thus be interesting to investigate if the depressor mandibulae of dipnoans such as *Protopterus* and *Lepidosiren* also attaches in a region of the mandible constituted by neural crest derivatives of the hyoid arch. If that is the case, and if further investigations support the hypothesis that the mandible of extant nondipnoan bony fishes is exclusively formed by mandibular neural crest derivatives, this would indicate that the presence of a depressor mandibulae in tetrapods and dipnoans might be related to an evolutionary change in which hyoid neural crest derivatives have become incorporated in the formation of the lower jaw.

Regarding the branchial muscles, these can be divided into three groups; i.e., the true branchial muscles (comprising the branchial muscles *sensu stricto* and the protractor pectoralis and its derivatives), the pharyngeal muscles, which are present only as independent structures in extant mammals, and the laryngeal muscles. Most adult vertebrates retain several branchial muscles *sensu stricto*, but these muscles are not present as a group in extant amniotes. As explained earlier, some nonsarcopterygian vertebrates have a constrictor laryngis and/or a dilatator laryngis, but it is not clear if these muscles do correspond to the constrictor laryngis and dilatator laryngis of sarcopterygians and, thus, if the latter muscles are plesiomorphically present in osteichthyans or not.

Concerning the hypobranchial muscles, the plesiomorphic condition for osteichthyans seems to be that found in extant actinistians and dipnoans, in which there is a coracomandibularis and a sternohyoideus. Changes to this plesiomorphic condition occurred during osteichthyan evolution. For instance, in actinopterygians, the coracomandibularis became modified into a peculiar branchiomandibularis, while in sarcopterygians such as tetrapods, the hypobranchial muscle plate became differentiated into muscles that are absent in other extant osteichthyans, such as the omohyoideus and the glossal muscles of the tongue. In mammals, the hypobranchial muscle plate has become divided into an even greater number of muscles; this is also the case of the mandibular and hyoid muscle plates, which, in mammals, usually give rise to more muscles than in other tetrapods (e.g., Edgeworth 1935; Brock 1938; Jarvik 1963, 1980; Gorniak 1985; Pough et al. 1996; Kardong and Zalisko 1998; Kardong 2002; Kisia and Onyango 2005; see the following chapters).

The discussions and hypotheses advanced in this chapter are based on evidence from developmental biology, comparative anatomy, functional morphology, paleontology, experimental embryology and molecular biology, and innervation and phylogeny. According to, e.g., Edgeworth (1935: 222), in order to provide a well-grounded analysis of the homologies and evolution of a certain muscle among different taxa, it is imperative to consider all available lines of evidence, since "no one criterion is sufficient, not even two" (see Chapter 2). We mentioned earlier that even if the development of the depressor mandibulae of dipnoans is not completely similar to the development of the anterior part of the depressor mandibulae of tetrapods such as salamanders, these structures might still be homologous. In this specific case, there are several different lines of evidence supporting their homology, such as (a) innervation (e.g., they are innervated by CNVII); (b) adult anatomy (e.g., they occupy a similar position and have similar relationships to other structures); (c) functional morphology (e.g., depression of the lower jaw); (d) ontogeny (even if their development is not completely similar, as suggested by Edgeworth, most aspects concerning this development are similar; e.g., they originate from the dorsomedial portion of the hyoid muscle plate); and (e) phylogeny (e.g., the phylogenetic results of the present work support the hypothesis that dipnoans are the closest living relatives of tetrapods).

The examples provided in this chapter thus illustrate the risks of discussing the homologies of structures such as muscles on the basis of a single line of evidence, even if it concerns innervation or development. However, they also illustrate that it is possible to establish well-grounded hypotheses of homology when multiple lines of evidence are available and taken into consideration (e.g., developmental biology, comparative anatomy, functional morphology, paleontology, experimental embryology, innervation, and/or phylogeny). In fact, one of the points that we want to stress in this book is that a better understanding of the muscles of a particular taxon allows a much more detailed, integrative analysis of its comparative anatomy, functional morphology, ecomorphology, and evolution.

7 Development of Cephalic Muscles in Chondrichthyans and Bony Fishes

One approach to understanding the evolution of anatomical structures is by comparing development between species. Recently, head muscle development in several gnathostome fishes has been described in some detail. We reviewed some of these studies in Ziermann and Diogo (in press), and Chapter 7 is mainly based on that publication. Developmental patterns of three fish taxa are shown in Tables 7.1 through 7.3: the shark *Scyliorhinus canicula* (Elasmobranchii; Table 7.1, based on Ziermann et al. 2017), the bichir *Polypterus senegalus* (Chondrostei, Actinopterygii; Table 7.2, based on Noda et al. 2017), and the zebrafish *Danio rerio* (Teleostei; Actinopterygii; Table 7.3, based on Diogo et al. 2008c; Schilling and Kimmel 1997). Additional taxa have also been studied recently, such as *Lepisosteus osseus* (Ginglymodi, Actinopterygii: Konstantinidis et al. 2015) and *Neoceratodus forsteri* (Dipnoi, Sarcopterygii: Ericsson et al. 2010; Ziermann 2008). The adult cephalic musculature of the spiny dogfish (*Squalus acanthias*, Selachii), spotted ratfish (*Hydrolagus colliei*, Holocephali), Senegal bichir (*Polypterus senegalus*, Chondrostei), and zebrafish (*Danio rerio*, Cypriniformes) is shown in Figure 7.1.

The development of the cephalic muscles in the zebrafish (*D. rerio*) is often used as a model for the general teleost, or fish, developmental pattern. However, this tendency is probably partly due to ease of study, given that the head of structures zebrafishes are anatomically and evolutionary derived among teleosts (Table 7.3 and Figure 7.1) (Diogo et al. 2008a,c). As can be seen in Table 7.3, many of the zebrafish head muscles are differentiated into several portions. For example, the adductor mandibulae in larval zebrafish gives rise to A2, Aω, and A0/A1-OST in 14-day-old larvae, and the latter muscle separates and can be distinguished by day 35 (Figure 7.2G through I). All adductor mandibulae are clearly separated in the adult zebrafish. At 4 days, two distinct parts of the hyohyoideus can be observed: the anterior hyohyoideus inferior and the posterior hyohyoideus superior (hh.i and hh.s in Figure 7.2G). The latter muscle divides at 9 days into the hyohyoideus abductor and hyohyoideus adductores (hh.ab and hh.ad in Figure 7.2H and I). Other muscles tend to fuse; for example, the fusion of the intermandibularis posterior and interhyoideus in the zebrafish leads to the formation of the protractor hyoideus (ph in Figure 7.2G and H).

By comparing the developmental appearance of cephalic muscles in zebrafish (Table 7.3) with the evolutionary origin of those muscles, Diogo et al. (2008c) found that in general a parallelism exists between both sequences. Most of the mandibular muscles (intermandibularis anterior, intermandibularis posterior, adductor mandibulae, and levator arcus palatini) develop early and appear to have been present in basal osteichthyans. However, the dilatator operculi also develops simultaneously with the other mandibular muscles, i.e., early in ontogeny, but was likely acquired later in actinopterygians. In contrast, the division of the adductor mandibulae seems to parallel the evolutionary appearance of the adductor mandibulae divisions, with A2 and Aω developing first (present in osteichthyans) and A1-OST and AO separating later (appearing in the nodes to ostariophysans and to cypriniforms, respectively).

The hyoid muscles in the zebrafish develop not exactly in the same order as their appearance in evolution. The interhyoideus and hyohyoideus develop simultaneously early in the ontogeny of the zebrafish (Schilling and Kimmel 1997; Diogo et al. 2008c). The former muscle is plesiomorphically present in osteichthyans, while the hyohyoideus is only present in extant actinopterygians. The adductor hyomandibulae X and adductor operculi develop next; the latter is plesiomorphically present in osteichthyans, while the former evolved only in teleosts. The levator operculi develops third and is present in halecomorphs and teleosts. The adductor arcus palatini is the last to develop: this muscle is plesiomorphically present in gnathostomes (Ziermann et al. 2014). That means that the development of adductor operculi and adductor arcus palatini is delayed while the development of hyohyoideus and adductor mandibulae is accelerated during the ontogeny of the zebrafish as compared to the phylogenetic appearance of those muscles.

While there is a general correspondence between the ontogeny and the phylogenetic appearance of muscles in osteichthyans, some further exceptions (heterochrony) should be noted. For example, the levator arcuum branchialium V develops early and is already much larger than other branchial muscles in the 4-day-old zebrafish larva (see lab V in Figure 7.2G), while the adductor mandibulae has not yet differentiated into different sections at that stage. Evolutionarily, by contrast, the hypertrophy of the levator arcuum branchialium V did not occur until the apical node leading to cypriniforms, which is much later than the division of the adductor mandibulae (present in the last common ancestor [LCA] of extant osteichthyans) (Diogo et al. 2008c). This heterochronic change in the muscle's development and in the modification of the ceratobranchial V, which is moved by this muscle, is related to specialized feeding mechanisms of cypriniforms (e.g., Takahasi 1925; Schilling and Kimmel 1997). The ceratobranchial V is accelerated in its ossification compared to the other ceratobranchials, and it bears teeth. The changes during ontogeny (hypertrophy, acceleration) might assure the proper size ratio between the myological and skeletal structures.

Such a parallelism between the order in which the cephalic muscles develop and the order in which those muscles evolved

TABLE 7.1
Head Muscle Development in *S. canicula*

Muscle/Stage	24	26	27	28	29	30	31	32	A*
Extraocular Muscles			o	x	x	x	x	x	x
Mandibular Muscles									
Intermandibularis		o	x	x	x	x	x	x	x
Adductor mandibulae A2			o	x	x	x	x	x	x
Preorbitalis (labial muscle)			o	x	x	x	x	x	x
Levator arcus palatini			o	D	x	x	x	x	x
Spiracularis			o	D	x	x	x	x	x
Hyoid Muscles									
Interhyoideus		o	x	x	x	x	x	x	x
Adductor arcus palatini	CH	CH	CH	x	x	x	x	x	x
Constrictor hyoideus dorsalis	CH	CH	CH	x	x	x	x	x	x
True Branchial Muscles									
Cucullaris				o	x	x	x	x	x
Constrictores branchiales	(3)	(3)	(3)	(4)	(4)	(4)	(4)	(4)	(4)*
Adductores branchiales								o	(5)*
Interarcuales laterales								o	(4)*
Interbranchiales								o	(4)*
Ventral Branchial Muscles									
Coracobranchiales				o	(2)	(3)	(4)	(?)	(5)*
Epibranchial Muscles									
Interpharyngobranchiales								o	(3)*
Subspinalis								o	x
Hypobranchial Muscles									
Coracomandibularis			o	x	x	x	x	x	x
Sternohyoideus				o	x	x	x	x	x
Coracoarcualis			o	x	x	x	x	x	x

Source: Ziermann, J. M. et al., *Frontiers Zool*, 14, 31, 2017.

Note: Stages according to Ballard et al. (1993). Terminology and adult (A*) characteristics follow Ziermann et al. (2014). x, present (independent of status of differentiation); o, absent (or not stained); (number), number of repetitive muscles (usually corresponding to branchial arches and counting from anterior to posterior); D, dorsal constrictor of mandibular arch; CH, constrictor hyoideus; (?), could not be observed because of overlying muscles.

("phylo-ontogenetic" parallelism) was also described for nonfish taxa, such as the tetrapod Mexican axolotl *Ambystoma mexicanum* (Ziermann and Diogo 2013), barring only a few exceptions. Another case of such a phylo-ontogenetic parallelism was also reported in the cephalic (head and neck) and limb muscles of primates (Diogo et al. 2014a). For sharks, it is hard to test for such a parallelism between ontogeny and phylogeny as most of the muscles are not present in nongnathostome extant taxa, and only insufficient detailed reconstructions exists for muscles in fossils. Still, the phylogenetic position of sharks and *P. senegalus*, and the sharks' plesiomorphic muscular structure among gnathostomes (see preceding chapters, as well as Mallatt 1996; Trinajstic et al. 2013), combined with the observation that mandibular muscles develop in both these species after the hyoid muscles (Tables 7.1 and 7.2; see the following), implies that this developmental pattern might be ancestral for the LCA of crown-group gnathostomes.

Comparing the available developmental descriptions in fishes (see preceding citations and Tables 7.1 through 7.3), it becomes clear that there is quite some variability in cranial muscle patterning. The branchial arch muscles first appear: simultaneously with the hyoid muscles in *S. canicula*, after all other cephalic muscles groups have appeared in *P. senegalus* and *D. rerio*, and somewhere in between in *N. forsteri* (Ziermann et al. 2017). Similarly variable is the timely appearance of extraocular eye muscles, which appear simultaneously with mandibular muscles in *S. canicula*, *P. senegalus*, and *D. rerio*, but last in *N. forsteri* (Ziermann et al. 2017). To infer the ancestral sequence of head muscle development in gnathostomes, clearly more fishes have to be analyzed including a detailed description of outgroup species such as hagfishes and lampreys (cyclostomes).

In most extant gnathostomes, the mandibular muscles develop at the same time or even earlier then the hyoid muscles. Simultaneous development was described in the Longnose Gar *Lepisosteus osseus* (Konstantinidis et al. 2015) and is common in amphibians (Platt 1898; Schlosser and Roth 1997a, 1997b; Ericsson and Olsson 2004; Ziermann and Olsson 2007; Ziermann 2008). Earlier development of the mandibular muscles was observed in the zebrafish *D. rerio* (see Table 7.3; Schilling

TABLE 7.2
Head Muscle Development in *P. senegalus*

Muscle/Stage	L1	L2	L3	L4	L5	L6	L7	L8
Extraocular Muscles								
Obliquus superior				o	x			
Obliquus inferior				o	x			
Rectus superior		o	x					
Rectus externus	o	x						
Rectus internus				o	x			
Rectus inferior			o	x				
Retractor lentis				o	x			
Mandibular Muscles								
Intermandibularis		o	x					
Adductor mandibulae A2	o	x						
Anterior subsection A3		o	x					
Medial subsection A3			o	x				
Posterior subsection A3	o	x						
Aω		o	x					
Levator arcus palatini	o	C	C	C	C	C	C	x
Dilatator operculi	o	C	C	C	C	C	C	x
Hyoid Muscles								
Interhyoideus		o	x					
Adductor hyomandibulae		o	x					
Adductor operculi		o	x					
Lb/db	x							
Hyohyoideus		o	x					
Branchial Muscles								
Protractor pectoralis				o	x			
True Branchial Muscles								
Lev. arcuum br. (dbm) 1,2						o	x	
Lev. arcuum br. (dbm) 3,4				o	x			
Pharyngoclavicularis					o	x		
Interarcualis ventralis 1				o	x			
Transversi ventralis 2,3				o	x			
Transversi ventralis 4						o	x	
Rectus communis					o	x		
Hypobranchial Muscles								
Branchiomandibularis				o	x			
Sternohyoideus		o	x					

Source: L1–L8 (L, larval stage) and muscle development modified from Moda, M. et al., *Am J Anat*, 168, 257–276, 2017.

Note: x, present (independent on status of differentiation); o, absent (or not stained); C, constrictor dorsalis; lb/db, levator branchiarum/depressor branchiarum; lev. arcuum br. (dbm), levator arcuum branchialium (termed by the authors dorsal branchial muscle).

and Kimmel 1997) and amniotes (e.g., chick, quail, opossum, mouse; McClearn and Noden 1988; Kaufman and Kaufman 1992; Smith 1994; Noden et al. 1999). However, the recent extensive studies on muscle development in the shark *S. canicula* (Elasmobranchii) and bichir *P. senegalus* (Chondrostei, Actinopterygii) revealed a later development of mandibular muscles than the hyoid muscles in these morphologically plesiomorphic gnathostome and bony fish/actinopterygian taxa, respectively (Tables 7.1 and 7.2; Noda et al. 2017; Ziermann et al. 2017). This has implications for the evolution of the hyoid vs. mandibular regions, so we will now detail the evidence further. In *S. canicula,* the constrictor dorsalis is the only visible muscle anlage of the mandibular arch in stage 28, while the hyoid arch muscle anlage has already differentiated into the adductor arcus palatini and constrictor hyoideus (Figure 7.2A; Table 7.1). In the following stage, the differentiation of mandibular arch muscles (levator arcus palatini, spiracularis, preorbitalis, adductor mandibulae A2) can be observed and the hyoid muscles are clearly distinguishable (Figure 7.2B, Table 7.1). In *P. senegalus*, the levator branchiarum and

TABLE 7.3
Head Muscle Development in *Danio rerio*

Muscle/Stage	53 hpf	58 hpf	62 hpf	68 hpf	72 hpf	85 hpf	4 d	9 d	14 d	35 d
Extraocular Muscles										
Obliquus superior	o	x								
Obliquus inferior		o	x							
Rectus superior	o	x								
Rectus inferior	o	x								
Rectus internus (medial)	x									
Rectus externus (lateral)	o	x								
Mandibular Muscles										
Intermandibularis ant. + post.		o	x				*1			*2
Adductor mandibulae	x								A2, Aω, A0/ A1-OST	A2, Aω, A0, A1-OST
Levator arcus palatini		o	x							
Dilatator operculi		o	x							
Hyoid Muscles										
Interhyoideus (interhyal)	o	x					*1			*2
Hyohyoideus (hypohyal)	o	x					2 parts	3 parts		
Adductor hyomandibulae X			o	x						
Adductor operculi			o	x						
Levator operculi					o	x				
Adductor arcus palatini						?	x			
Branchial Muscles										
Protractor pectoralis				o	x					
True Branchial Muscles										
Lev. arcuum br. (phar. wall)				o	x					
Rectus communis					o	x				
Transversus ventralis		o	x							
Rectus ventralis					o	x				
Hypobranchial Muscles										
Sternohyoideus	x									

Source: Time of appearance and muscle development until 85 hpf (28.5°C) modified from Schilling, T. F., and Kimmel, C. B., *Development*, 124, 2945–2960, 1997; those for later stages modified from Diogo, R. et al. *BMC Dev Biol*, 8, 24–46, 2008.

Note: hpf, hours postfertilization; d, days of larval development. Muscle terminology in round brackets, e.g., (interhyal), corresponds to Schilling and Kimmel (1997) if it differs from Diogo et al. (2008c). x, present (independent of status of differentiation); o, absent (or not stained); ?, it is not certain if this muscle was present earlier; *1, the intermandibularis posterior and interhyoideus fuse and form the protractor hyoideus; *2, the protractor hyoideus has a distinct dorsal and ventral portion; ant., anterior; post., posterior.

depressor branchiarum of the hyoid arch also develop earlier than the mandibular arch muscles and are all developed by larval stage 3 (see L3 in Figure 7.2D) while the mandibular arch muscles are only completely developed by L4 (see L7 in Figure 7.2E), and the separation of levator arcus palatini and dilatator operculi from the common anlage constrictor dorsalis are only visible in L8 (Table 7.2; Noda et al. 2017).

This shark/phylogenetically basal actinopterygian correspondence might indicate that the plesiomorphic developmental pattern in gnathostomes might be that the mandibular muscles developing after the hyoid muscles. If the same pattern was also found in other chondrichthyans, this might support Miyashita's (2015) and Miyashita and Diogo's (2016) idea that the mandibular arch originally was not integrated into or similar to the ancestral pharyngeal arches, and only became secondarily so during the transitions that led to the LCA of gnathostomes (Miyashita 2015) or else to the first vertebrates (Mallatt 1997). In this case, the observation in *S. canicula* and *P. senegalus* might thus be interpreted as an example of parallelism between ontogeny and phylogeny, and the simultaneous or earlier development of mandibular arch muscles would therefore represent a derivation from the ancestral process. Thus, comparing developmental patterns and morphology between species can help us understand the development and the evolution of head muscles in vertebrates, and recent advances in genetics further deepen our evolutionary developmental comprehension of those muscles.

GENERAL REMARKS

Previous studies have described temporal and spatial patterns of muscle development in the zebrafish, Australian lungfish, chick,

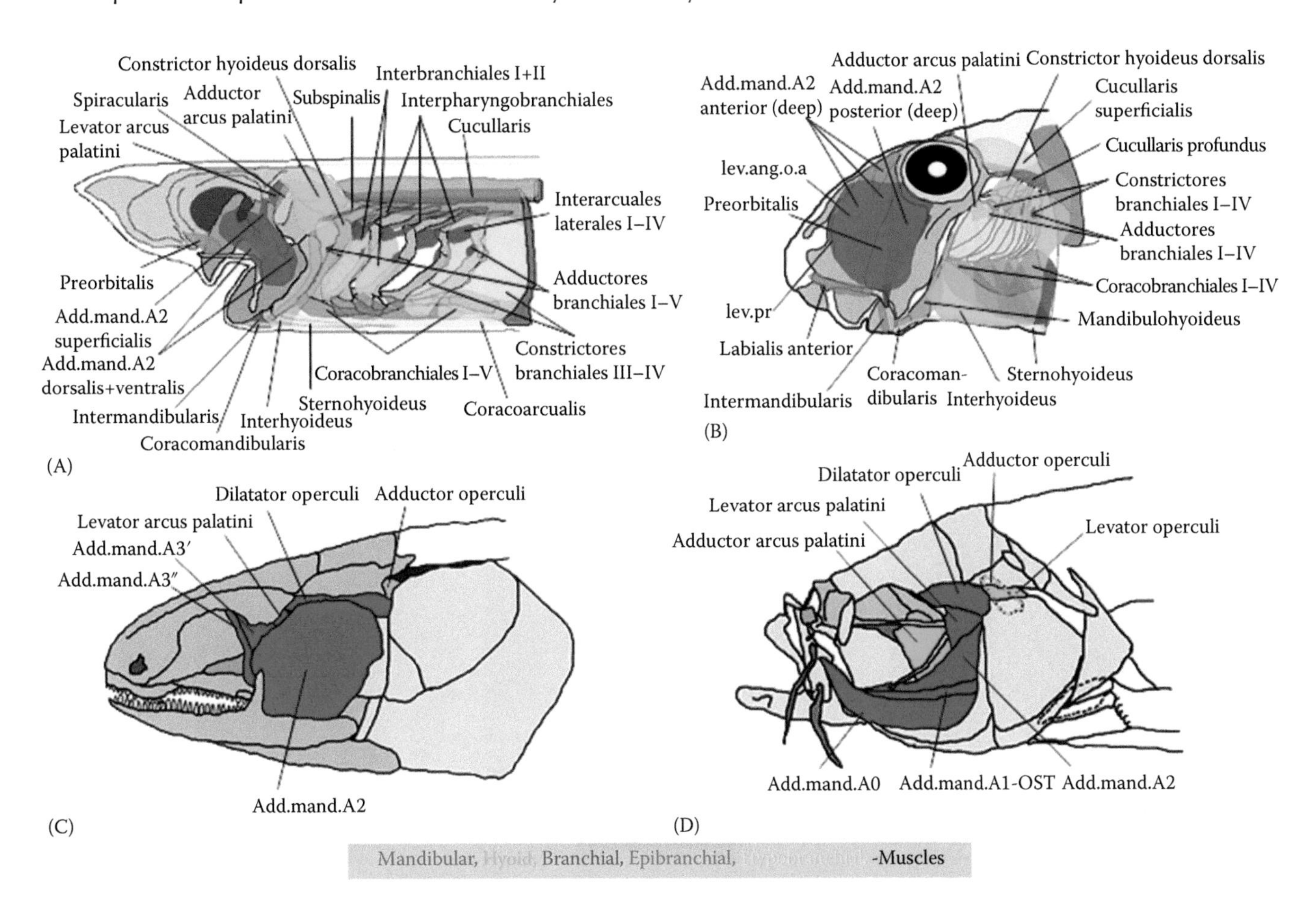

FIGURE 7.1 Schematic lateral views of the cephalic muscles in cartilaginous and bony fishes: mandibular (*red, orange, pink*); hyoid (*green*), true branchial (*blue*), epibranchial (*brown*), and hypobranchial (*yellow*) muscles. (A) Spiny dogfish (*Squalus acanthias*, Selachii) and (B) spotted ratfish (*Hydrolagus colliei*, Holocephali). (Modified from Diogo, R., and Ziermann, J. M., *Dev Dyn*, 244, 1046–1057, 2015.). Not all branchial arch muscles are shown to enable visualization of deeper muscles. Constrictores branchiales I + II and interbranchiales III + IV not shown. (C) Senegal bichir (*Polypterus senegalus*, Chondrostei). (Modified from Lauder, G. V., *J Morphol*, 163, 283–317, 1980.) (D) Zebrafish (*Danio rerio*, Cypriniformes). (Modified from Diogo, R. et al., *BMC Dev Biol*, 8, 24–46, 2008.) (C, D) The branchial and hypobranchial muscles are covered by the operculum and other structures; epibranchial are absent in osteichthyan fishes. add.mand.A0, adductor mandibulae A0; add.mand.A1-OST, adductor mandibulae A1-OST; add.mand. A2, adductor mandibulae A2; add.mand. A3′, adductor mandibulae A3′ (adductor mandibulae temporalis *sensu* Lauder [1980a]); add.mand. A3″, adductor mandibulae A3″ (adductor mandibulae pterygoideus *sensu* Lauder [1980a]); lev.ang.o.a, levator anguli oris anterior; lev.pr, levator prelabialis.

Xenopus, and Mexican Axolotl (Schilling and Kimmel 1997; Noden et al. 1999; Ericsson and Olsson 2004; Ziermann and Olsson 2007; Diogo et al. 2008c; Ziermann 2008; Ziermann and Diogo 2013). These studies describe the development of head muscles as progressing, in general, from anterior to posterior, from lateral/superficial to ventral/medial (outside-in), and from origin to insertion. Furthermore, in several vertebrates, cephalic muscle differentiation has been shown to be closely related to the differentiation of the associated cartilages (Noden 1983a; Schilling and Kimmel, 1997; Hacker and Guthrie 1998). Similar patterns are described for *S. canicula* (Ziermann et al. 2017), but in almost all those patterns, exceptions were found. In the hyoid arch and all branchial arches, the lateral muscles generally develop before the ventral muscles, but the ventral mandibular muscle, intermandibularis, develops before all other lateral mandibular muscles (Table 7.1). Most muscles in *S. canicula* develop from their region of origin to their region of insertion, except for the coracomandibularis, coracoarcualis, and coracobranchiales, which develop from their region of insertion (mandible, ceratohyal, and branchial arches, respectively) to their origin at the pectoral girdle. This insertion-to-origin pattern was previously only described in few other species, for example, in the quail (McClearn and Noden 1988). However, this exception is related to the later development of the cartilages where the muscles originate compared to the cartilages where the muscles insert, which confirms that the muscle development is dependent on the skeletal development (Noden 1983b). As described earlier, the mandibular muscles of *S. canicula* develop after the hyoid muscles, but the remaining head muscles show an anterior to posterior progression in differentiation (Ziermann et al. 2017), as was described for amphibians (e.g., Ziermann and Diogo 2013), birds (e.g., Noden et al. 1999), and mice (Kaufman and Kaufman 1992).

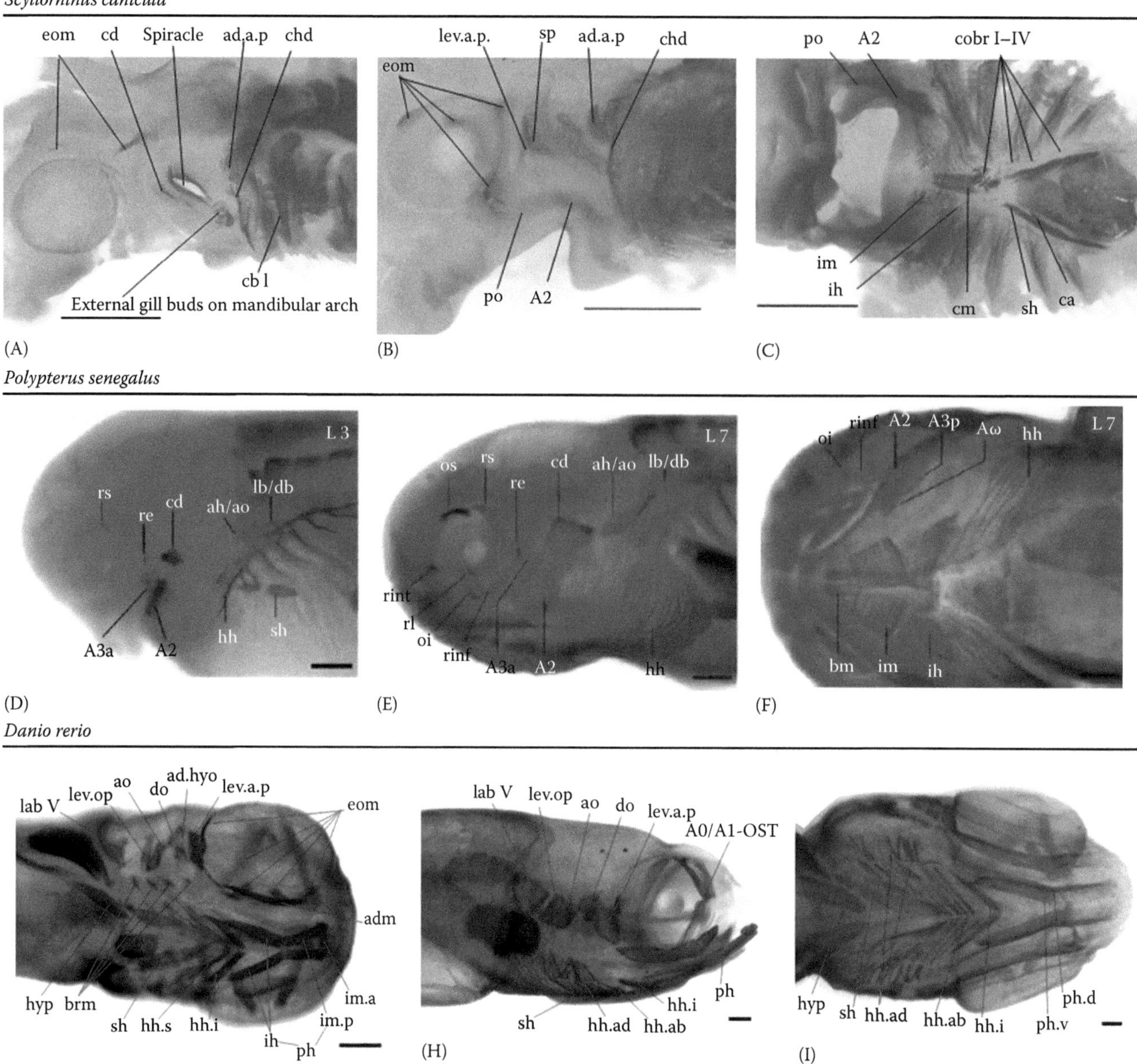

FIGURE 7.2 Cephalic muscle development in representatives of gnathostome fishes. (A–C) *Scyliorhinus canicula*, processed for immunohistochemistry with myosin heavy chain (MyHC; A4–1025, DSHB); scale bar = 1 mm. (A) Lateral view of stage 28 embryo; (B) lateral view of stage 29 embryo; (C) ventral view of stage 31 embryo. (Staging after Ballard et al. 1993.) (D–F) *P. senegalus* processed for immunohistochemistry with antimyosin heavy chain antibody A4.1025; scale bar = 200 μm. (Modified from Noda, M. et al., *J Morphol*, 278, 450–463, 2017.) (D) Lateral view of L3, (E) lateral view of L7, and (F) ventral view of L7. (G–I) *Danio rerio* processed for immunohistochemistry with antimyosin heavy chain antibody A4.1025; scale bar = 1 mm. (Modified from Diogo, R. et al., *BMC Dev Biol*, 8, 24–46, 2008.) (G) Ventrolateral view of 4-day larvae, (H) lateral view of 24-day larvae, and (I) ventral view of 35-day juvenile. A0/A1-OST, adductor mandibulae A0/A1-OST; A2/A3, adductor mandibulae A2/A3; A3a, anterior section of A3; A3p, posterior section of A3; Aω, Aω section of adductor mandibulae complex; ad.a.p, adductor arcus palatini; ad.hyo, adductor hyomandibulae; adm, adductor mandibulae; ah/ao, complex of adductor hyomandibulae and adductor operculi; ao, adductor operculi; bm, branchiomandibularis; brm, branchial muscles; ca, coracoarcualis; cb I, constrictor branchialis I; cd, constrictor dorsalis; chd, constrictor hyoideus dorsalis; cobr I-IV, coracobranciale I-IV; cm, coracomandibularis; do, dilatator operculi; eom, extraocular muscles; hh, hyohyoideus; hh.ab, hyohyoideus abductor; hh.ad, hyohyoideus adductores; hh.i, hyohyoideus inferior; hh.s, hyohyoideus superior; hyp, hypaxialis; im, intermandibularis; im.a, intermandibularis anterior; im.p, intermandibularis posterior; ih, interhyoideus; lab V, levator arcuum branchialium V; lb/db, levator branchiarum/depressor branciarum; lev.a.p, levator arcus palatini; lev.op, levator operculi; os, obliquus superior; oi, obliquus inferior; ph, protractor hyoideus; ph.d, protractor hyoideus dorsal portion; ph.v., protractor hyoideus ventral portion; po, preorbitalis; re, rectus externus; rinf, rectus inferior; rint, rectus internus; rl, retractor lentis; rs, rectus superior; sh, sternohyoideus; sp, spiracularis.

Head muscle development in *P. senegalus* seems to have no anterior to posterior or lateral to ventral gradient: the lateral hyoid muscles develop first, followed by lateral mandibular muscles, then the simultaneous appearance of ventral mandibular and hyoid muscles, followed by lateral and ventral branchial muscles except for the lateral branchial muscles of branchial arches I and II (Table 7.2; Noda et al. 2017). The zebrafish *D. rerio* follows an anterior to posterior pattern, but the ventral hyoid and branchial muscles develop before the lateral muscles (Table 7.3; Schilling and Kimmel 1997).

8 Head and Neck Muscle Evolution from Sarcopterygian Fishes to Tetrapods, with a Special Focus on Mammals

Chapter 6 not only concentrated mainly on the results relevant to actinopterygians (the group that includes extant cladistians, chondrosteans, ginglymods, halecomorphs, and teleosts), but also compared the head and neck muscles of those fishes with the musculature of some sarcopterygians. Chapter 8 focuses mainly on sarcopterygians (the group that includes extant actinistians, dipnoans, and tetrapods: see Figure 1.1) and particularly on how the head and neck muscles evolved during the transitions from sarcopterygian fishes to nonmammalian tetrapods (Figures 8.1 through 8.6) and then to mammals, including modern humans. Tables 8.1 through 8.8 summarize the best-supported hypotheses of homology for the head and neck muscles of the sarcopterygian taxa listed in those tables, and Figure 6.2 summarizes the evolution of these muscles in sarcopterygians. Importantly, Tables 8.1, 8.3, 8.5, and 8.7 also include comments about the development and muscular variations/abnormalities of our own species, *Homo sapiens*, which are presented mainly, but not exclusively, in the right-hand column of those tables.

Mammalia is a diverse clade (Figure 8.7) including the monotremes (Prototheria: e.g., echidnas and platypus; see Figures 8.8 through 8.10), marsupials (Methateria: e.g., opossums; see Figure 8.11), and placentals (Eutheria: e.g., humans, see Figures 8.12 through 8.26). Mammals are a frequent focus of evolutionary and developmental studies because, in addition to the interest in this group per se due to its fascinating diversity and evolutionary history, Mammalia includes the primates and our own species, as well as model organisms, especially mice, which are often used in genetic and medical studies. Figure 8.7 shows just a few among many examples of the striking diversity of the mammalian head and jaws, such as the heads of elephants, with more than 200 bundles of facial muscles, the heads of whales, that can weigh more than 200 t, the minuscule heads of 2 g bumblebee bats (*Craseonycteris thonglongyai*) (Surlykke et al. 1993), and the extremely derived and edentulous jaws of anteaters (suborder Vermilingua).

A weakness of many studies of the adult cephalic muscles of mammals is the lack of comparisons between monotremes, marsupials, and placentals, or between these groups and other tetrapods, which makes it difficult to understand the origin and evolutionary history of the mammalian head muscles. Even Diogo and Abdala's 2010 book *Muscles of Vertebrates*, which provides the most recent review on the subject, did not include marsupials—one of the three major mammalian clades—in its tables summarizing the homology and evolution of mammalian muscles. Some of these omissions, and even factual errors, in the literature are due to a strong historical bias to regard mammals as an example of a "scala naturae" leading from monotremes to marsupials and then to placentals, culminating in humans. The notion of a *scala naturae*, which dates back to thinkers such as Aristotle, represents an evolutionary trend in complexity from "lower" to "higher" taxa, with *Homo sapiens* as the end stage (discussed in Diogo et al. 2015b,c, 2016b). Many authors have, for instance, reported only a few undifferentiated facial muscles in marsupials, more muscles in placentals such as rats, and the "most complex" facial musculature in humans (e.g., Huber 1930a,b, 1931; Lightoller 1940a,b, 1942). Such notions of "scala naturae" are found not only in works from the nineteenth and early twentieth centuries, but even those from the late twentieth century. For instance, Minkoff et al.'s (1979) study of the facial muscles of *Didelphis virginiana* describes 21 muscles of facial expression in this species, including extrinsic ear muscles, i.e., only about 2/3 as many as are found in humans (31; 25 + 6 extrinsic ear muscles: see Table 8.5). In contrast, in our recent dissections, we found exactly the same number of facial muscles in *D. virginiana* as in placentals such as rats, which is very similar to the number found in humans, as we explain in the following (Table 8.5; Figure 8.11).

Here we briefly summarize what is known about the deep evolutionary origin—all the way back to fishes—of the head muscles of marsupials, placentals, and monotremes by providing comparisons with other vertebrates (Tables 8.1 through 8.8). We also provide notes on the development of these muscles, particularly from descriptive and experimental works in mice and other taxa. Specifically, we provide a list and brief description of all the head muscles of representatives of the three major extant mammalian clades, as well as of other tetrapods, and fishes by using an updated, unifying vertebrate myological nomenclature to allow more straightforward comparison between all these taxa.

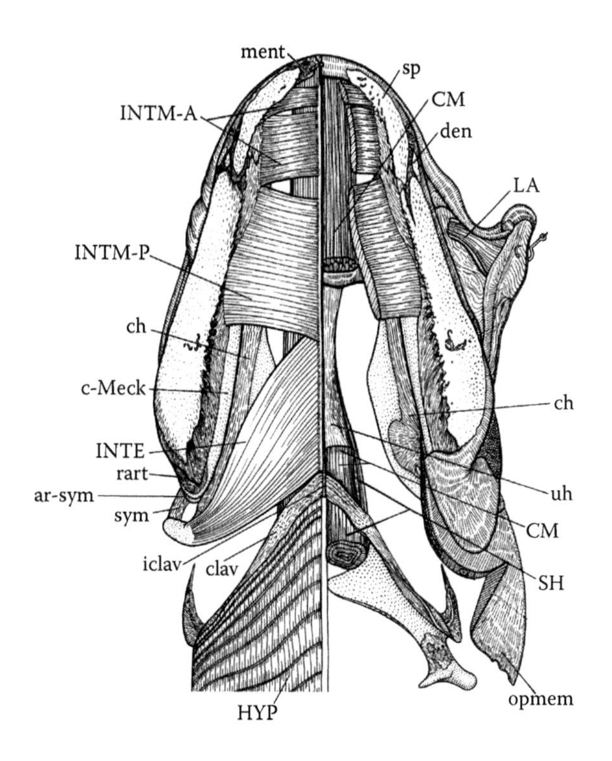

FIGURE 8.1 *Latimeria chalumnae* (Cladistia): ventral view of the cephalic musculature; the most superficial cephalic muscles, after removal of the gular plate are shown on the left side; most muscles were removed or cut in order to shown muscles that are situated more dorsally are shown on the right side (the nomenclature of the structures illustrated follows that used in the present work). ar-sym, articulatory facet for symplectic; c-Meck, Meckelian cartilage; ch, ceratohyal; clav, clavicle; CM, coracomandibularis; dent, dentary bone; HYP, hypoaxialis; iclav, interclavicle; INTE, interhyoideus; INTM-A, INTM-P, intermandibularis anterior and posterior; LA, labialis; ment, mentomeckelian bone; opmem, opercular membrane; rart, retroarticular; SH, sternohyoideus; sp, splenial bone; sym, symplectic, uh, urohyal. (Modified from Millot, J., and Anthony, J., *Anatomie de Latimeria chalumnae—I, squelette, muscles, et formation de soutiens*, CNRS, Paris, 1958.)

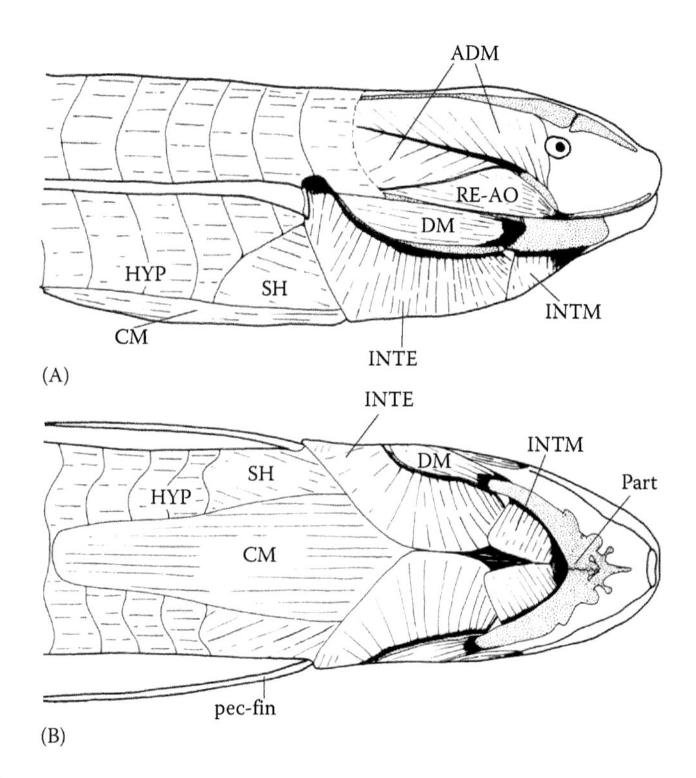

FIGURE 8.2 *Lepidosiren paradoxa* (Dipnoi): (A) lateral view of the cephalic musculature; (B) ventral view of the cephalic musculature (the nomenclature of the structures illustrated follows that used in the present work; anterior is to the right). ADM, adductor mandibulae complex; CM, coracomandibularis; DM, depressor mandibulae; HYP, hypaxialis; INTE, interhyoideus; INTM, intermandibularis; part, prearticular; RE-AO, retractor anguli oris; SH, sternohyoideus. (Modified from Bemis, W. E., and Lauder, C. V., *J Morphol*, 187, 81–108, 1986; Diogo, R., *The Origin of Higher Clades: Osteology, Myology, Phylogeny and Evolution of Bony Fishes and the Rise of Tetrapods*, Science Publisher, Enfield, NH, 2008.)

ORIGIN AND EVOLUTION OF THE MAMMALIAN MANDIBULAR MUSCLES

The plesiomorphic condition for sarcopterygians—the clade which includes sarcopterygian fishes and all tetrapods—is that two ventral mandibular muscles, the intermandibularis anterior and intermandibularis posterior, connect the hemimandibles (Figures 8.1, 8.2, and 8.4) (Tables 8.1 and 8.2). The mylohyoideus and digastricus anterior of mammals (Figures 8.12, 8.14, 8.22, and 8.24 through 8.26) correspond to the intermandibularis posterior of other sarcopterygians (e.g., Bryant 1945; Jarvik 1963, 1980; Saban 1971; Diogo and Abdala 2010). Unlike the condition in monotremes (Figure 8.9), in most marsupials and placentals, including modern humans, the digastricus anterior and the digastricus posterior (a dorsomedial hyoid muscle: see the following) form a compound structure (the "digastricus") that is often related to the depression of the mandible. According to Edgeworth (1935), various tetrapod groups have independently acquired

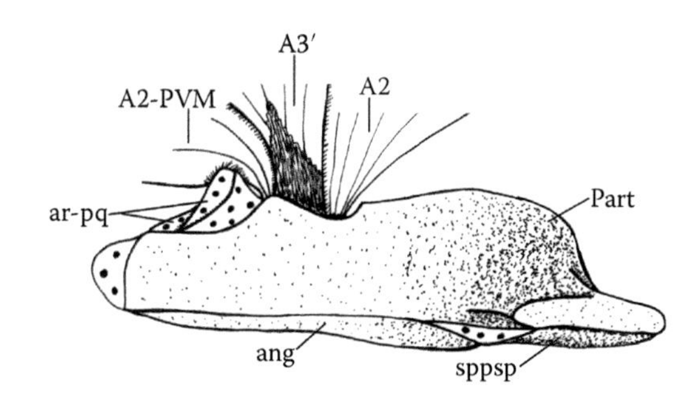

FIGURE 8.3 *Neoceratodus forsteri* (Dipnoi): mesial view of adductor mandibulae and mandible; the mandibular tooth plates are not illustrated (anterior is to the right, dorsal is to the top). A2, A2-PVM, A3′, adductor mandibulae A2, A2-PVM, and A3′; ang, angular; ar-pq, articulatory facet for palatoquadrate; part, prearticular; sppsp, splenio-postsplenial bone.

different mechanisms for depressing the mandible (i.e., to open the mouth) by using muscles other than the hypobranchial ones. For instance, amphibians and reptiles usually have a depressor mandibulae (which is a modified dorsomedial hyoid muscle: see Tables 8.3 and 8.4 and Figures 8.2 and

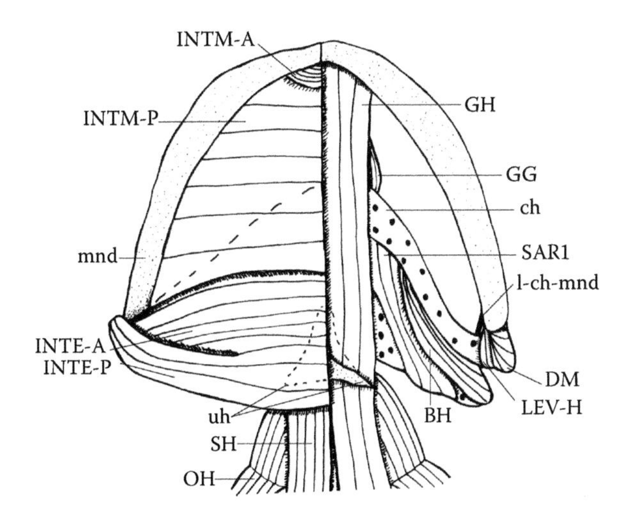

FIGURE 8.4 *Ambystoma ordinarium* (Amphibia, Caudata): ventral view of the cephalic musculature; on the right side, the most ventral muscles were removed (anterior is to the top). BH, branchiohyoideus; ch, ceratohyal; DM, depressor mandibulae; GG, genioglossus; GH, geniohyoideus; INTE-A, INTE-P, anterior and posterior bundles of interhyoideus; INTM-A, INTM-P, intermandibularis anterior and posterior; l-ch-mand, ligament between ceratohyal and mandible; LEV-H, levator hyoideus; mnd, mandible; OH, omohyoideus; SAR1, subarcualis rectus 1; SH, sternohyoideus; uh, urohyal.

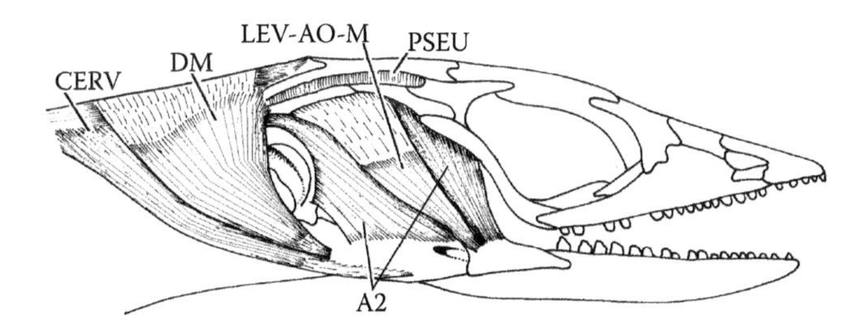

FIGURE 8.5 *Euspondylus acutirostris* (Reptilia, Lepidosauria): lateral view of the cephalic musculature; the adductor mandibulae A2-PVM is not shown (the nomenclature of the structures illustrated follows that used in the present work; anterior is to the right). A2, adductor mandibulae A2; CERV, cervicomandibularis; DM, depressor mandibulae; LEV-AO-M, levator anguli oris mandibularis; PSEU, pseudotemporalis. (Modified from Montero, R. et al, *Russ J Herpetol*, 9, 215–228, 2002; Diogo, R., *The Origin of Higher Clades: Osteology, Myology, Phylogeny and Evolution of Bony Fishes and the Rise of Tetrapods*, Science Publisher, Enfield, NH, 2008.)

8.5), monotremes have a detrahens mandibulae (which is a new division of the adductor mandibulae: Tables 8.1 and 8.2; Figure 8.9), and marsupial and placental mammals usually have the digastricus, i.e., the compound structure formed by the union of the digastricus anterior and digastricus posterior.

The plesiomorphic condition for the sarcopterygian adductor mandibular muscles seems to include an adductor mandibulae A2, an adductor mandibulae Aω, an adductor mandibulae A3′, and possibly an adductor mandibulae A3″ (Diogo and Abdala 2010). The adductor mandibulae Aω was not present as an independent muscle in any of the

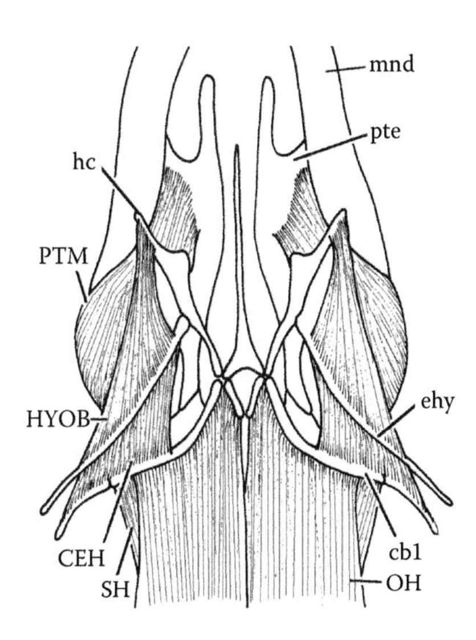

FIGURE 8.6 *Euspondylus acutirostris* (Reptilia, Lepidosauria): ventral view of the deep ventral cephalic musculature; muscles such as the intermandibularis, interhyoideus, geniohyoideus, genioglossus and hyoglossus are not shown (the nomenclature of the structures illustrated follows that used in the present work; anterior is to the top). cb1, ceratobranchial 1; CEH, ceratohyoideus; ehy, epihyal; hc, hyoid cornu; HYOB, hyobranchilais; mnd, mandible; OH, omohyoideus; pte, pterygoid; PTM, pterygomandibularis; SH, sternohyoideus. (Modified from Montero, R. et al, *Russ J Herpetol*, 9, 215–228, 2002; Diogo, R., *The Origin of Higher Clades: Osteology, Myology, Phylogeny and Evolution of Bony Fishes and the Rise of Tetrapods*, Science Publisher, Enfield, NH, 2008.)

mammals we dissected, and to our knowledge, it has also not been found in any extant mammals described in the literature. The adductor mandibulae A2-PVM, retractor anguli oris, and levator anguli oris mandibularis of extant dipnoans and nonmammalian tetrapods correspond to part of the A2 of bony fishes such as *Latimeria* (see preceding chapters and Figures 8.2, 8.3, and 8.5, but see comments on the homology of the A2-PVM in nonmammalian tetrapods, in the following chapters). The masseter, temporalis, pterygoideus lateralis, and detrahens mandibulae of monotremes (Figure 8.9) and the masseter, temporalis, and pterygoideus lateralis of other extant mammals (Figures 8.11, 8.12, 8.16, 8.19, and 8.24 through 8.26) apparently correspond to the A2 of reptiles such as the lizard *Timon* (Tables 8.1 and 8.2). However, although the mammalian temporalis appears to correspond to a part of the A2 of other tetrapods, it may also include a part of other adductor mandibulae structures such as the pseudotemporalis (see, e.g., Barghusen 1968). In two previous papers (Diogo et al. 2008a,b), we stated that the tensor tympani and tensor veli palatini of mammals were probably derived from the adductor mandibulae A2-PVM, as proposed by authors such as Edgeworth (1935) and Saban (1971), but that they might instead have been derived from the pterygomandibularis (see, e.g., Table 1 of Diogo et al. [2008a]). However, work performed since 2008 by us and by other authors indicates that the most likely hypothesis is that the tensor tympani and tensor veli palatini correspond to a part of, or are derived from, the pterygomandibularis, as is accepted by most anatomists

TABLE 8.1
Mandibular Muscles of Adults of Representative Sarcopterygian Taxa

Actinistia: *Latimeria chalumnae* (Coelacanth)	Dipnoi: *Lepidosiren paradoxa* (South American Lungfish)	Amphibia: *Ambystoma ordinarium* (Michoacan Stream Salamander)	Reptilia: *Timon lepidus* (Ocellated Lizard)	Mammalia (Monotremata): *Ornithorhynchus anatinus* (Platypus)	Mammalia (Marsupialia): *Didelphis virginiana* (Virginian Opossum)	Mammalia (Rodentia): *Rattus norvegicus* (Norwegian Rat)	Mammalia (Scandentia): *Tupaia* sp. (Tree Shrew)	Mammalia (Dermoptera): *Cynocephalus volans* (Philippine Colugo)	Mammalia (Primates): *Homo sapiens* (Modern Human)
Intermandibularis posterior	**Intermandibularis**	**Intermandibularis posterior**	**Intermandibularis posterior**	**Mylohyoideus** [as described by, e.g., Lightoller [1942], there is a mylohyoideus profundus, a mylohyoideus superficialis and, superficially to the latter, a digastricus anterior; Saban [1971] states that these three structures come from the same embryological primordium, i.e., that they seem to correspond to the intermandibularis posterior of other vertebrates: this is also supported by, e.g., Jarvik [1963, 1980]	**Mylohyoideus** [as described by Hiiemae and Jenkins [1969], it connects the hemimandibles, being fused to the mylohyoideus of the other side of the body; there is no central raphe and no direct attachment onto the hyoid apparatus, only an indirect attachment *via* fascia, as described by Coues [1872]]	**Mylohyoideus** [the mylohyoideus and digastricus anterior of rats clearly seem to correspond to the posterior, not the anterior, intermandibularis of other sarcopterygians, because the transversus mandibularis of rats corresponds to the intermandibularis anterior of other sarcopterygians; this is also supported by, e.g., Bryant [1945]]	**Mylohyoideus** (posterior part of mylohyoid *sensu* Le Gros Clark [1924])	**Mylohyoideus**	**Mylohyoideus**

(*Continued*)

TABLE 8.1 (CONTINUED)
Mandibular Muscles of Adults of Representative Sarcopterygian Taxa

Actinistia: *Latimeria chalumnae* (Coelacanth)	Dipnoi: *Lepidosiren paradoxa* (South American Lungfish)	Amphibia: *Ambystoma ordinarium* (Michoacan Stream Salamander)	Reptilia: *Timon lepidus* (Ocellated Lizard)	Mammalia (Monotremata): *Ornithorhynchus anatinus* (Platypus)	Mammalia (Marsupialia): *Didelphis virginiana* (Virginian Opossum)	Mammalia (Rodentia): *Rattus norvegicus* (Norwegian Rat)	Mammalia (Scandentia): *Tupaia* sp. (Tree Shrew)	Mammalia (Dermoptera): *Cynocephalus volans* (Philippine Colugo)	Mammalia (Primates): *Homo sapiens* (Modern Human)
–	–	–	–	**Digastricus anterior** [the correspondence between the mammalian digastricus anterior and part of the intermandibularis of other sarcopterygians is strongly supported by, e.g., innervation (the intermandibularis and digastricus anterior are usually innervated by the ramus ventralis of CNV), ontogeny (e.g., the development of the marsupial *Dasyrurus*), and comparative anatomy of adults: see, e.g., Edgeworth 1935]	**Digastricus anterior** [from digastricus posterior to mandible; there is no well-defined intermediate digastric tendon; instead, the anterior and posterior digastric muscles are continuous, and contact the "tendinous arcade" that is in contact with the hyoid apparatus, but none of the two muscles directly attaches to that bone; the anterior digastric is not blended to its counterpart at the midline, although according to Coues [1872] the two muscles might contact each other]	**Digastricus anterior** (anterior belly of digastricus *sensu* Greene [1935])	**Digastricus anterior**	**Digastricus anterior** (part of biventer *sensu* Leche [1886])	**Digastricus anterior** (anterior belly of biventer mandibulae *sensu* Huber [1930a])
Intermandibularis anterior	–	**Intermandibularis anterior** (submentalis *sensu* Iordansky [1992])	**Intermandibularis anterior**	–	–	**Intermandibularis anterior** (transversus mandibularis *sensu* Greene [1935])	**Intermandibularis anterior** (anterior part of mylohyoid *sensu* Le Gros Clark [1924] and Sprague [1944a])	–	–

(Continued)

TABLE 8.1 (CONTINUED)
Mandibular Muscles of Adults of Representative Sarcopterygian Taxa

Actinistia: *Latimeria chalumnae* (Coelacanth)	Dipnoi: *Lepidosiren paradoxa* (South American Lungfish)	Amphibia: *Ambystoma ordinarium* (Michoacan Stream Salamander)	Reptilia: *Timon lepidus* (Ocellated Lizard)	Mammalia (Monotremata): *Ornithorhynchus anatinus* (Platypus)	Mammalia (Marsupialia): *Didelphis virginiana* (Virginian Opossum)	Mammalia (Rodentia): *Rattus norvegicus* (Norwegian Rat)	Mammalia (Scandentia): *Tupaia* sp. (Tree Shrew)	Mammalia (Dermoptera): *Cynocephalus volans* (Philippine Colugo)	Mammalia (Primates): *Homo sapiens* (Modern Human)
Adductor mandibulae A2 (adductor mandibulae "superficiel" *sensu* Millot and Anthony [1958])	**Adductor mandibulae A2** (part of adductor mandidulae posterior *sensu* Bemis and Lauder [1986])	**Adductor mandibulae A2** (adductor mandibulae externus *sensu* Iordansky [1992])	**Adductor mandibulae A2** (adductor mandibulae externus *sensu* Abdala and Moro [2003])	**Masseter** (corresponds to the masseter + zygomatico-mandibularis and, possibly, to the maxillo-mandibularis, *sensu* Saban [1971]) [as shown in, e.g., Saban's 1[971] Figure 569, in the platypus specimens we dissected, the masseter is mainly divided into a deep part with anterior and posterior bundles and a superficial part with anterior and posterior bundles]	**Masseter** (masseter + external adductor *sensu* Hiiemae and Jenkins [1969]) [from squamosal, presphenoid ("alisphenoid"l), maxillary, jugal, frontal, parietal, occipital (from supraoccipital part) bones to mandible, being deeply blended with temporalis and with the pterygoideus medialis; the masseter is subdivided into an inferior/anterior bundle (see, e.g., Figure 6 of Hiiemae and Jenkins [1969]), a superficial bundle, a deep bundle, and a zygomatico-mandibular bundle; the latter bundle is sometimes considered as an independent muscle, but is deeply blended with the other masseter bundles, and partially with the temporalis as well]	**Masseter** [as described by Greene [1935], in the Norwegian rats we dissected, the masseter is mainly divided into a deep part with anterior and posterior bundles and a superficial part with anterior and posterior bundles]	**Masseter** [as described by, e.g., Le Gros Clark [1924] in the *Tupaia* specimens we dissected, the masseter is mainly divided into deep, intermediate, and superficial bundles]	**Masseter** (masseter + zygomatico-mandibularis *sensu* Stafford and Szalay [2000]) [in the colugo specimens we dissected, the masseter is subdivided into a superficial bundle, a deep bundle, and a zygomatico-mandibular bundle; the latter is sometimes considered as an independent muscle, but at least in the case of *Cynocephalus*, it is deeply blended with the other masseter bundles]	**Masseter** [in modern humans the masseter is usually mainly divided into deep and superficial bundles]

(Continued)

TABLE 8.1 (CONTINUED)
Mandibular Muscles of Adults of Representative Sarcopterygian Taxa

Actinistia: *Latimeria chalumnae* (Coelacanth)	Dipnoi: *Lepidosiren paradoxa* (South American Lungfish)	Amphibia: *Ambystoma ordinarium* (Michoacan Stream Salamander)	Reptilia: *Timon lepidus* (Ocellated Lizard)	Mammalia (Monotremata): *Ornithorhynchus anatinus* (Platypus)	Mammalia (Marsupialia): *Didelphis virginiana* (Virginian Opossum)	Mammalia (Rodentia): *Rattus norvegicus* (Norwegian Rat)	Mammalia (Scandentia): *Tupaia* sp. (Tree Shrew)	Mammalia (Dermoptera): *Cynocephalus volans* (Philippine Colugo)	Mammalia (Primates): *Homo sapiens* (Modern Human)
–	–	–	–	**Detrahens mandibulae** [some authors consider that the detrahens mandibulae is homologous to the digastricus anterior of other mammals, but this does not seem to be the case: see Saban [1968: 264]; according to, e.g., Saban [1971], the detrahens mandibulae clearly seems to correspond to part of the adductor mandibulae A2 of nonmammalian tetrapods]	–	–	–	–	–

(*Continued*)

TABLE 8.1 (CONTINUED)
Mandibular Muscles of Adults of Representative Sarcopterygian Taxa

Actinistia: *Latimeria chalumnae* (Coelacanth)	Dipnoi: *Lepidosiren paradoxa* (South American Lungfish)	Amphibia: *Ambystoma ordinarium* (Michoacan Stream Salamander)	Reptilia: *Timon lepidus* (Ocellated Lizard)	Mammalia (Monotremata): *Ornithorhynchus anatinus* (Platypus)	Mammalia (Marsupialia): *Didelphis virginiana* (Virginian Opossum)	Mammalia (Rodentia): *Rattus norvegicus* (Norwegian Rat)	Mammalia (Scandentia): *Tupaia* sp. (Tree Shrew)	Mammalia (Dermoptera): *Cynocephalus volans* (Philippine Colugo)	Mammalia (Primates): *Homo sapiens* (Modern Human)
–	–	–	–	**Temporalis** [corresponds to part of the A2 of nonmammalian tetrapods but may possibly also include part of other adductor mandibulae structures such as the pseudotemporalis: see Barghusen [1968]]	**Temporalis** (posterior adductor + internal adductor *sensu* Hiiemae and Jenkins [1969]) [from squamosal, presphenoid (alisphenoid), parietal, frontal, and occipital (supraoccipital part) bones to mandible; the temporalis is subdivided into a superficial/anterior bundle mainly going to the lateral surface of the coronoid process, a deep/posterior bundle going to medial surface of the coronoid process, and a pars suprazygomatica that is mainly fused with the deep/posterior bundle]	**Temporalis** [Greene [1935] describes the temporalis of rats as an undivided muscle, but as stated by Walker and Homberger [1997], in the specimens we dissected, this muscle is divided into two bundles, one more superficial and anterior and the other more deep and posterior]	**Temporalis** [in the *Tupaia* specimens we dissected, the temporalis is mainly divided into a superficial bundle, a deep bundle, and a pars suprazygomatica *sensu* Saban [1971]]	**Temporalis** [in the *Cynocephalus* specimens we dissected, the temporalis is not clearly divided into superficial and deep bundles, and there is no distinct pars suprazygomatica such as that found in *Tupaia*]	**Temporalis** [the temporalis of modern humans is usually described as an undivided muscle, but various authors, e.g., Gorniak [1985], consider that it is often divided into superficial and deep bundles]

(*Continued*)

TABLE 8.1 (CONTINUED)
Mandibular Muscles of Adults of Representative Sarcopterygian Taxa

Actinistia: *Latimeria chalumnae* (Coelacanth)	Dipnoi: *Lepidosiren paradoxa* (South American Lungfish)	Amphibia: *Ambystoma ordinarium* (Michoacan Stream Salamander)	Reptilia: *Timon lepidus* (Ocellated Lizard)	Mammalia (Monotremata): *Ornithorhynchus anatinus* (Platypus)	Mammalia (Marsupialia): *Didelphis virginiana* (Virginian Opossum)	Mammalia (Rodentia): *Rattus norvegicus* (Norwegian Rat)	Mammalia (Scandentia): *Tupaia* sp. (Tree Shrew)	Mammalia (Dermoptera): *Cynocephalus volans* (Philippine Colugo)	Mammalia (Primates): *Homo sapiens* (Modern Human)
–	–	–	–	**Pterygoideus lateralis** [in some parts of Edgeworth's [1935] work, he seems to suggest that the pterygoideus lateralis and medialis are both included in the "pterygoideus medialis" of monotremes and that the pterygoideus lateralis only becomes separated in other extant mammals; however, in other parts of Edgeworth's (1935) work, he clearly states that the pterygoideus lateralis corresponds to part of the adductor mandibulae externus (= A2) of reptiles; more recent works, e.g., Barghusen 1968 and Jouffroy 1971, support the latter hypothesis; developmental data also indicate that the pterygoideus lateralis and pterygoideus medialis do not develop from the same anlage (e.g., Smith 1994); the platypus specimens we dissected have both a pterygoideus lateralis and a pterygoideus medialis]	**Pterygoideus lateralis** [from presphenoid (alisphenoid) to mandible, being a very thin muscle, much thinner than the pterygoideus medialis, but being still slightly differentiated into an inferior head inserting onto the condylar process and a superior head inserting onto the capsule of the squamodentary joint]	**Pterygoideus lateralis** (pterygoideus externus *sensu* Greene [1935]) [in the Norwegian rats we dissected, the pterygoideus lateralis is constituted by a single bundle]	**Pterygoideus lateralis** (pterygoideus externus *sensu* Le Gros Clark [1924, 1926]) [as described by, e.g., Le Gros Clark [1924], in the *Tupaia* specimens we dissected, the pterygoideus lateralis is constituted by a single bundle]	**Pterygoideus lateralis** [in the *Cynocephalus* specimens we dissected, the pterygoideus lateralis is constituted by a single bundle]	**Pterygoideus lateralis** [in modern humans, the pterygoideus lateralis is usually divided into superior and inferior heads: see, e.g., Birou et al. 1991; Aziz et al. 1998; El Haddioui et al. 2005]

(Continued)

TABLE 8.1 (CONTINUED)
Mandibular Muscles of Adults of Representative Sarcopterygian Taxa

Actinistia: *Latimeria chalumnae* (Coelacanth)	Dipnoi: *Lepidosiren paradoxa* (South American Lungfish)	Amphibia: *Ambystoma ordinarium* (Michoacan Stream Salamander)	Reptilia: *Timon lepidus* (Ocellated Lizard)	Mammalia (Monotremata): *Ornithorhynchus anatinus* (Platypus)	Mammalia (Marsupialia): *Didelphis virginiana* (Virginian Opossum)	Mammalia (Rodentia): *Rattus norvegicus* (Norwegian Rat)	Mammalia (Scandentia): *Tupaia* sp. (Tree Shrew)	Mammalia (Dermoptera): *Cynocephalus volans* (Philippine Colugo)	Mammalia (Primates): *Homo sapiens* (Modern Human)
–	**Adductor mandibulae A2-PVM** (part of adductor mandidulae posterior *sensu* Bemis and Lauder [1986])	**Adductor mandibulae A2-PVM** (adductor mandibulae posterior *sensu* Iordansky [1992]; levator mandibulae posterior *sensu* Edgeworth [1935] and Piatt [1938]) [authors such as Piatt [1938] suggest that the A2-PVM of tetrapods as, e.g., urodeles derives ontogenetically from the A3′ and/or A3″, but the developmental work of Ericsson and Olsson [2004] strongly supports that it derives instead from the A2: see the following chapters for more details on the homologies of the A2-PVM among nonmammalian tetrapods]	**Adductor mandibulae A2-PVM** (adductor mandibulae posterior *sensu* Abdala and Moro [2003] and Holliday and Witmer [2007])	– [the A2-PVM is probably absent as a separated, independent structure in mammals, although it cannot be completely discarded that it was this muscle that gave rise to the mammalian tensor tympani and/or tensor veli palatini: see text and Table 8.2]	–	–	–	–	–
–	**Retractor anguli oris** [seemingly derived from lateral portion of adductor mandibulae]	–	– [but see levator anguli oris mandibularis below]	–	–	–	–	–	–

(*Continued*)

TABLE 8.1 (CONTINUED)
Mandibular Muscles of Adults of Representative Sarcopterygian Taxa

Actinistia: *Latimeria chalumnae* (Coelacanth)	Dipnoi: *Lepidosiren paradoxa* (South American Lungfish)	Amphibia: *Ambystoma ordinarium* (Michoacan Stream Salamander)	Reptilia: *Timon lepidus* (Ocellated Lizard)	Mammalia (Monotremata): *Ornithorhynchus anatinus* (Platypus)	Mammalia (Marsupialia): *Didelphis virginiana* (Virginian Opossum)	Mammalia (Rodentia): *Rattus norvegicus* (Norwegian Rat)	Mammalia (Scandentia): *Tupaia* sp. (Tree Shrew)	Mammalia (Dermoptera): *Cynocephalus volans* (Philippine Colugo)	Mammalia (Primates): *Homo sapiens* (Modern Human)
–	–	–	**Levator anguli oris mandibularis** (levator anguli oris *sensu* Diogo [2007, 2008]) [present, somewhat blended with A2; we use the name "mandibularis" to distinguish this muscle from the levator anguli oris facialis of certain mammals, which is a facial (hyoid), and not a mandibular, muscle; as explained by, e.g., Abdala and Moro [2003] and Wu [2003], some lizards such as *Lanthanotus*, as well as some other lepidosaurs such as *Sphenodon*, have not only a levator anguli oris mandibularis but also a "retractor anguli oris" that occupies the posteroventrolateral region of the adductor mandibulae complex and that is, thus, similar (although it is probably not homologous) to the retractor anguli oris of dipnoans: see the following chapters]	–	–	–	–	–	–

(Continued)

TABLE 8.1 (CONTINUED)
Mandibular Muscles of Adults of Representative Sarcopterygian Taxa

Actinistia: *Latimeria chalumnae* (Coelacanth)	Dipnoi: *Lepidosiren paradoxa* (South American Lungfish)	Amphibia: *Ambystoma ordinarium* (Michoacan Stream Salamander)	Reptilia: *Timon lepidus* (Ocellated Lizard)	Mammalia (Monotremata): *Ornithorhynchus anatinus* (Platypus)	Mammalia (Marsupialia): *Didelphis virginiana* (Virginian Opossum)	Mammalia (Rodentia): *Rattus norvegicus* (Norwegian Rat)	Mammalia (Scandentia): *Tupaia* sp. (Tree Shrew)	Mammalia (Dermoptera): *Cynocephalus volans* (Philippine Colugo)	Mammalia (Primates): *Homo sapiens* (Modern Human)
Adductor mandibulae A3′ (adductor mandidulae "moyen" *sensu* Millot and Anthony [1958])	**Adductor mandibulae A3′** (adductor mandidulae anterior *sensu* Bemis and Lauder [1986])	**Pseudotemporalis** (pseudotemporalis posterior and anterior *sensu* Iordansky [1992]; superficial and deep levator mandibulae anterior *sensu* Edgeworth [1935] and Piatt [1938]; adductor mandibulae A3′ and A3″ *sensu* Diogo [2007; 2008a,b])	**Pseudotemporalis** (pseudotemporalis superficialis and profundus *sensu* Abdala and Moro [2003] and Holliday and Witmer [2007]; adductor mandibulae A3′ and A3″ *sensu* Diogo [2007; 2008a,b])	– [the pseudo-temporalis of nonmammalian tetrapods seems to correspond to a part of the pterygoideus medialis and possibly to a part of the masseter, of extant mammals: see above]	–	–	–	–	–
Adductor mandibulae A3″ (adductor mandidulae "profond" *sensu* Millot and Anthony [1958])	–	– [both the adductor A3′ and A3″ seem to be included in the pseudotemporalis of extant amphibians such as *Ambystoma* and reptiles such as *Timon*: see text; in some species of amphibians and reptiles, the pseudotemporalis is however clearly divided into superficial and deep structures that seemingly correspond directly to the A3′ and A3″ of fishes, respectively]	– [see on the left]	–	–	–	–	–	–

(Continued)

TABLE 8.1 (CONTINUED)
Mandibular Muscles of Adults of Representative Sarcopterygian Taxa

Actinistia: *Latimeria chalumnae* (Coelacanth)	Dipnoi: *Lepidosiren paradoxa* (South American Lungfish)	Amphibia: *Ambystoma ordinarium* (Michoacan Stream Salamander)	Reptilia: *Timon lepidus* (Ocellated Lizard)	Mammalia (Monotremata): *Ornithorhynchus anatinus* (Platypus)	Mammalia (Marsupialia): *Didelphis virginiana* (Virginian Opossum)	Mammalia (Rodentia): *Rattus norvegicus* (Norwegian Rat)	Mammalia (Scandentia): *Tupaia* sp. (Tree Shrew)	Mammalia (Dermoptera): *Cynocephalus volans* (Philippine Colugo)	Mammalia (Primates): *Homo sapiens* (Modern Human)
–	–	–	–	**Pterygoideus medialis** [the pterygoideus medialis seems to correspond to/derive from the pseudotemporalis of amphibians, such as *Ambystoma*, and, thus, to both the pseudotemporalis and pterygomandibularis of reptiles such as *Timon*: see Table 8.2, cells above and text; Adams [1919] stated that monotremes do not have a pterygoideus medialis, and according to Saban [1968], this muscle appears in early embryos of monotremes but then disappears during ontogeny; however, our dissections indicate that, although somewhat mixed, both the pterygoideus lateralis and the pterygoideus medialis are present in *Ornothorhynchus*; Murray [1981] also described a pterygoideus lateralis and a pterygoideus medialis in the Echidna]	**Pterygoideus medialis** [from palatine and presphenoid (alisphenoid)—including pterygoid—bones to mandible]	**Pterygoideus medialis** (pterygoideus internus *sensu* Greene [1935])	**Pterygoideus medialis** (pterygoideus internus *sensu* Le Gros Clark [1924, 1926])	**Pterygoideus medialis**	**Pterygoideus medialis**

(*Continued*)

TABLE 8.1 (CONTINUED)
Mandibular Muscles of Adults of Representative Sarcopterygian Taxa

Actinistia: *Latimeria chalumnae* (Coelacanth)	**Dipnoi: *Lepidosiren paradoxa* (South American Lungfish)**	**Amphibia: *Ambystoma ordinarium* (Michoacan Stream Salamander)**	**Reptilia: *Timon lepidus* (Ocellated Lizard)**	**Mammalia (Monotremata): *Ornithorhynchus anatinus* (Platypus)**	**Mammalia (Marsupialia): *Didelphis virginiana* (Virginian Opossum)**	**Mammalia (Rodentia): *Rattus norvegicus* (Norwegian Rat)**	**Mammalia (Scandentia): *Tupaia* sp. (Tree Shrew)**	**Mammalia (Dermoptera): *Cynocephalus volans* (Philippine Colugo)**	**Mammalia (Primates): *Homo sapiens* (Modern Human)**
–	–	– [at least some caecilian as well as a few urodele amphibians have a "pterygoideus," which seems to correspond to the pterygomandibularis of reptiles: see the following chapters; however, contrary to what was shown in the tables of Chapters 5 and 6 of Diogo and Abdala's [2010] monograph, the pterygomandibularis is usually not present as a distinct muscle in species of the genus *Ambystoma*]	**Pterygomandibularis** (pterygoideus *sensu* Holliday and Witmer [2007]) [seemingly derived from mesial portion of adductor mandibulae; the pterygomandibularis of reptiles such as *Timon* seems to correspond to part of the pterygoideus medialis and, probably, to part/the totality of the tensor tympani and/or tensor veli palatini, of extant mammals: see text]	–	–	–	–	–	–

(*Continued*)

TABLE 8.1 (CONTINUED)
Mandibular Muscles of Adults of Representative Sarcopterygian Taxa

Actinistia: *Latimeria chalumnae* (Coelacanth)	Dipnoi: *Lepidosiren paradoxa* (South American Lungfish)	Amphibia: *Ambystoma ordinarium* (Michoacan Stream Salamander)	Reptilia: *Timon lepidus* (Ocellated Lizard)	Mammalia (Monotremata): *Ornithorhynchus anatinus* (Platypus)	Mammalia (Marsupialia): *Didelphis virginiana* (Virginian Opossum)	Mammalia (Rodentia): *Rattus norvegicus* (Norwegian Rat)	Mammalia (Scandentia): *Tupaia* sp. (Tree Shrew)	Mammalia (Dermoptera): *Cynocephalus volans* (Philippine Colugo)	Mammalia (Primates): *Homo sapiens* (Modern Human)
–	–	–	–	**Tensor tympani** [Maier [2008] confirmed that the tensor tympani is present in *Ornithorynchus*: the chorda tympani passes below the insertion of the muscle (hypotensoric configuration, which, according to this author, probably represents the plesiomorphic configuration for mammals)]	**Tensor tympani** [present, seemingly from temporal bone to malleus: see, e.g., Fig. 225 of Jouffroy and Lessertisseur 1968; Maier [2008] states that in *Didelphis* and other marsupials the chorda tympani passes below the insertion of the muscle (hypotensoric: see on the left)]	**Tensor tympani** [Maier [2008] confirms that the tensor tympani is present in *Rattus norvegicus*: the chorda tympani passes below the insertion of the muscle (hypotensoric: see on the left), but in some other rodents, it passes above the muscle (epitensoric)]	**Tensor tympani** [really present in *Tupaia*? Maier [2008] states that the tensor tympani is not present in *Tupaia* nor in *Ptilocercus*; he examined numerous *Tupaia* specimens and did not find the muscle, and he cites some other authors that also did not find it in this taxon, but he recognizes that the muscle was described in ***Tupaia*** by some authors and, particularly, in the detailed works of Saban (see, e.g., Saban [1968]), although he considers that Saban and other authors were mistaken and that the tensor tympani is autapomorphically absent in the Scadentia; according to Wible [2009], detailed studies on the ear region of tree shrews are thus needed to clarify if this muscle is, or not, present in these mammals]	**Tensor tympani** [Maier [2008] confirms that the tensor tympani is present in *Cynocephalus volans*: the chorda tympani passes below the insertion of the muscle (hypotensoric: see on the left)]	**Tensor tympani**

(Continued)

TABLE 8.1 (CONTINUED)
Mandibular Muscles of Adults of Representative Sarcopterygian Taxa

Actinistia: *Latimeria chalumnae* (Coelacanth)	Dipnoi: *Lepidosiren paradoxa* (South American Lungfish)	Amphibia: *Ambystoma ordinarium* (Michoacan Stream Salamander)	Reptilia: *Timon lepidus* (Ocellated Lizard)	Mammalia (Monotremata): *Ornithorhynchus anatinus* (Platypus)	Mammalia (Marsupialia): *Didelphis virginiana* (Virginian Opossum)	Mammalia (Rodentia): *Rattus norvegicus* (Norwegian Rat)	Mammalia (Scandentia): *Tupaia* sp. (Tree Shrew)	Mammalia (Dermoptera): *Cynocephalus volans* (Philippine Colugo)	Mammalia (Primates): *Homo sapiens* (Modern Human)
–	–	–	–	**Tensor veli palatini** [as described by, e.g., Saban [1971] (and contra the statements of, e.g., Maier et al. [1996]), in the platypus specimens we dissected, the tensor veli palatini is present as an independent muscle: see text]	**Tensor veli palatini** [present, seemingly from sphenoid bone to soft palate: see, e.g., Figure 225 of Jouffroy and Lessertisseur [1968]; in her developmental study of *Monodelphis domestica*, Smith [1994] states that the tensor veli palatini, tensor tympani, and pterygoideus medialis derive from the same anlage, while the pterygoideus lateralis derives from the anlage that gives rise to the temporalis]	**Tensor veli palatini** [in mice, it runs from the pterygoid plate to the sphenoid bone and cartilage of tympanic tube: Grimaldi et al. [2015]]	**Tensor veli palatini**	**Tensor veli palatini**	**Tensor veli palatini**
Adductor mandibulae Aω (intramandibular adductor *sensu* Lauder [1980b])	–	–	**Adductor mandibulae Aω** [in *Timon* and other reptiles the adductor mandibulae has a large and distinct anteroventral division that is lodged in the "adductor fossa" of Lauder [1980b] and that is very similar to the Aω of other osteichthyans: see the following chapters]	–	–	–	–	–	–

(*Continued*)

TABLE 8.1 (CONTINUED)
Mandibular Muscles of Adults of Representative Sarcopterygian Taxa

Actinistia: *Latimeria chalumnae* (Coelacanth)	Dipnoi: *Lepidosiren paradoxa* (South American Lungfish)	Amphibia: *Ambystoma ordinarium* (Michoacan Stream Salamander)	Reptilia: *Timon lepidus* (Ocellated Lizard)	Mammalia (Monotremata): *Ornithorhynchus anatinus* (Platypus)	Mammalia (Marsupialia): *Didelphis virginiana* (Virginian Opossum)	Mammalia (Rodentia): *Rattus norvegicus* (Norwegian Rat)	Mammalia (Scandentia): *Tupaia* sp. (Tree Shrew)	Mammalia (Dermoptera): *Cynocephalus volans* (Philippine Colugo)	Mammalia (Primates): *Homo sapiens* (Modern Human)
Levator arcus palatini [Edgeworth [1935] suggested that the dorsal mandibular musculature was probably acquired independently among gnathostomes, but the presence of this musculature is very likely plesiomorphic for this group and perhaps for vertebrates as a whole: see preceding chapters]	–	– [the only dorsal mandibular muscle present in urodeles such as *Ambystoma* is the levator bulbi; amphibians such as caecilians have a "levator quadrati": see, e.g., Kleinteich and Haas 2007; according to authors such as Edgeworth [1935], the latter muscle is derived from the adductor mandibulae, but authors such as Brocks [1938] argue that it is a dorsal mandibular muscle; see the following chapters]	**Levator pterygoidei** [it is derived from the constrictor dorsalis, so it probably corresponds to part of the levator arcus palatini of *Latimeria*: e.g., Brocks 1938; Holliday and Witmer 2007; this study]	–	–	–	–	–	–
–	–	–	**Protractor pterygoidei** [it is derived from the constrictor dorsalis, so it probably corresponds to part of the levator arcus palatini or of, e.g., *Latimeria*: Brocks 1938; Holliday and Witmer 2007; this study]	–	–	–	–	–	–

(*Continued*)

TABLE 8.1 (CONTINUED)
Mandibular Muscles of Adults of Representative Sarcopterygian Taxa

Actinistia: *Latimeria chalumnae* (Coelacanth)	Dipnoi: *Lepidosiren paradoxa* (South American Lungfish)	Amphibia: *Ambystoma ordinarium* (Michoacan Stream Salamander)	Reptilia: *Timon lepidus* (Ocellated Lizard)	Mammalia (Monotremata): *Ornithorhynchus anatinus* (Platypus)	Mammalia (Marsupialia): *Didelphis virginiana* (Virginian Opossum)	Mammalia (Rodentia): *Rattus norvegicus* (Norwegian Rat)	Mammalia (Scandentia): *Tupaia* sp. (Tree Shrew)	Mammalia (Dermoptera): *Cynocephalus volans* (Philippine Colugo)	Mammalia (Primates): *Homo sapiens* (Modern Human)
–	–	**Levator bulbi** [according to, e.g., Edgeworth [1935], this muscle is derived from the adductor mandibulae; however, our dissections and comparisons support Brocks' [1938] hypothesis, i.e., that the levator bulbi, as well as the levator quadrati of caecilians, are the remains of the constrictor dorsalis group in amphibians; according to Brocks [1938], the constrictor dorsalis group is conserved in many reptiles because of their kinetic skull: see the following chapters]	**Levator bulbi** (the levator bulbi *sensu* Frazzeta [1962], Haas [1973], and Schumacher [1973] seemingly corresponds to the tensor periorbitae *sensu* Holliday and Witmer [2007]; see the following chapters)	–	–	–	–	–	–

Note: The nomenclature of the muscles shown in bold follows that of text; in order to facilitate comparisons, in some cases, names often used by other authors to designate a certain muscle/bundle are given, between round brackets; additional comments are given between square brackets. Data from evidence provided by our own dissections and comparisons and by a review of the literature. See also Table 8.2.

TABLE 8.2

Diagram Illustrating the Authors' Hypotheses Regarding the Homologies of the Mandibular Muscles of Adults of Representative Sarcopterygian Taxa and the Probable Condition for the LCA of Marsupials + Placentals

	Latimeria (7 m.)	*Lepidosiren* (5 m.)	*Ambystoma* (6 m.)	*Timon* (11 m.)	*Ornithorhynchus* (9 m.)	*LCA marsupials + placentals* (9 m.)	*Didelphis* (8 m.)	*Rattus* (9 m.)	*Tupaia* (9 m.)	*Cynocephalus* (8 m.)	*Homo* (8 m.)
VENTRAL	Interm. posterior	Interm.	Interm. posterior	Interm. posterior	Mylohyoideus	Mylohyoideus	Mylohyoideus	Mylohyoideus	Mylohyoideus	Mylohyoideus	Mylohyoideus
	—	—	—	—	Dig. anterior	Dig. anterior	Dig. anterior	Dig. anterior	Dig. anterior	Dig. anterior	Dig. anterior
	Interm. anterior	—	Interm. anterior	Interm. anterior	—	Interm. anterior	—	Interm. anterior	Interm. anterior	—	—
ADDUCTOR MANDIBULAE	Ad. man. A2	Ad. man. A2	Ad. man. A2	Ad. man. A2	Masseter	Masseter	Masseter	Masseter	Masseter	Masseter	Masseter
	—	—	—	—	Detrahens man.	—	—	—	—	—	—
	—	—	—	—	Temporalis	Temporalis	Temporalis	Temporalis	Temporalis	Temporalis	Temporalis
	—	—	—	—	Pterygoideus lat.	Pterygoideus lat.	Pterygoideus lat.	Pterygoideus lat.	Pterygoideus lat.	Pterygoideus lat.	Pterygoideus lat.
	—	Ad. man. A2 -PVM	Ad. man. A2 -PVM	Ad. man. A2 -PVM	—	—	—	—	—	—	—
	—	Retractor ang. oris	—	—	—	—	—	—	—	—	—
	—	—	—	Le. anguli oris mandib.	—	—	—	—	—	—	—
	Ad. mand. A3′	Ad. man. A3′	Pseudotemporalis	Pseudotemporalis	—	—	—	—	—	—	—
	Ad. mand. A3″	—	—	—	—	—	—	—	—	—	—
	—		— [psoa.]	Pterygomandibularis	—	—	—	—	—	—	—
	—		—	—	Pterygoideus med.	Pterygoideus med.	Pterygoideus med.	Pterygoideus med.	Pterygoideus med.	Pterygoideus med.	Pterygoideus med.
	—	—	—	—	Tensor tympani	Tensor tympani	Tensor tympani	Tensor tympani	Tensor tympani	Tensor tympani	Tensor tympani
	—		—	—	Tensor veli palatini	Tensor veli palatini	Tensor veli palatini	Tensor veli palatini	Tensor veli palatini	Tensor veli palatini	Tensor veli palatini
	Ad. mand. A ω	—	—	Ad. man. A ω	—	—	—	—	—	—	—
DORSAL	Le. arcus palatini	—	—	Le. pterygoidei	—	—	—	—	—	—	—
	—	—	—	Protractor pterygoidei	—	—	—	—	—	—	—
	—	—	Le. bulbi	Le. bulbi	—	—	—	—	—	—	—

Note: Data from evidence provided by our own dissections and comparisons and by a review of the literature. The black arrows indicate the hypotheses that are most strongly supported by the evidence available; the gray arrows indicate alternative hypotheses that are supported by some of the data, but overall, they are not as strongly supported by the evidence available as are the hypotheses indicated by black arrows. Ventral and Dorsal means ventral musculature and dorsal constrictor musculature of the mandibular arch sensu Edgeworth (1935). ad., adductor; dig., digastricus; interm., intermandibularis; lat., lateralis; le., levator; m., muscles; man., mandibulae; mandib., mandibularis; psoa., present in some amphibians; psor., present in some other reptiles. See Table 8.1.

TABLE 8.3
Hyoid Muscles of Adults of Representative Sarcopterygian Taxa

Actinistia: *Latimeria chalumnae* (Coelacanth)	Dipnoi: *Lepidosiren paradoxa* (South American Lungfish)	Amphibia: *Ambystoma ordinarium* (Michoacan Stream Salamander)	Reptilia: *Timon lepidus* (Ocellated Lizard)	Mammalia (Monotremata): *Ornithorhynchus anatinus* (Platypus)	Mammalia (Marsupialia): *Didelphis virginiana* (Virginian Opossum)	Mammalia (Rodentia): *Rattus norvegicus* (Norwegian Rat)	Mammalia (Scandentia): *Tupaia* sp. (Tree Shrew)	Mammalia (Dermoptera): *Cynocephalus volans* (Philippine Colugo)	Mammalia (Primates): *Homo sapiens* (Modern Human)
Adductor arcus palatini	– [the portion of the hyoid muscle anlage that gives rise to the levator hyoideus/depressor mandibulae of nonactinistian sarcopterygians probably corresponds to that portion that gives rise to the adductor arcus palatini in other osteichthyans: see text]	–	–	–	–	–	–	–	–
Adductor hyomandibulae Y [seemingly not homologous with the adductor hyomandibulae of actinopterygians such as teleosts: see text]	– [seemingly absent in dipnoans and tetrapods, although it may possibly be included in the levator hyoideus/depressor mandibulae: see text]	–	–	–	–	–	–	–	–
Adductor operculi [its fibers are seemingly deeply blended with those of the adductor arcus palatini: see text]	–	–	–	–	–	–	–	–	–
***Latimeria*'s levator operculi** [seemingly not homologous with the levator operculi of halecomorph and teleostean actinopterygians: see text]	–	–	–	–	–	–	–	–	–

(Continued)

TABLE 8.3 (CONTINUED)
Hyoid Muscles of Adults of Representative Sarcopterygian Taxa

Actinistia: *Latimeria chalumnae* (Coelacanth)	Dipnoi: *Lepidosiren paradoxa* (South American Lungfish)	Amphibia: *Ambystoma ordinarium* (Michoacan Stream Salamander)	Reptilia: *Timon lepidus* (Ocellated Lizard)	Mammalia (Monotremata): *Ornithorhynchus anatinus* (Platypus)	Mammalia (Marsupialia): *Didelphis virginiana* (Virginian Opossum)	Mammalia (Rodentia): *Rattus norvegicus* (Norwegian Rat)	Mammalia (Scandentia): *Tupaia* sp. (Tree Shrew)	Mammalia (Dermoptera): *Cynocephalus volans* (Philippine Colugo)	Mammalia (Primates): *Homo sapiens* (Modern Human)
–	**Depressor mandibulae** [according to Forey [1986], the depressor mandibulae and levator hyoideus of extant dipnoans develop from the same ontogenetic anlage]	**Depressor mandibulae** [depressor mandibulae anterior *sensu* Diogo [2007, 2008b] and Diogo et al. [2008a,b]] [seems to correspond to the "pars noto-gnathica" *sensu* Lightoller [1939], which, contrary to what was suggested by this author, does not seem to directly correspond to the "nucho-maxillaris" of sharks, because such a nucho-maxillaris is not present in any of the bony fishes we dissected; i.e., it was very likely not present in the LCA of osteichthyans]	**Depressor mandibulae** [Haas [1973] described both a depressor mandibulae and a "stylohyoideus" in *Sphenodon* and states that such a muscle is "found elsewhere in reptiles only in Gekkota": see the following chapters]	**Styloideus** (the styloideus *sensu* Huber [1930a] corresponds to the interhyoideus *sensu* Edgeworth [1935] and to the posterior digastric *sensu* Parsons [1898]) [Edgeworth [1935] suggested that the styloideus of monotremes and the stylohyoideus of other mammals derive from the interhyoideus; our observations and comparisons strongly support the interpretations of, e.g., Huber [1930a] and Saban [1968, 1971], i.e., that the monotreme styloideus and stapedius and the therian stylohyoideus, digastricus posterior, jugulohyoideus, stapedius, and possibly mandibulo-auricularis correspond to the depressor mandibulae of reptiles such as *Timon*]	**Stylohyoideus** [from paroccipital process of occipital bone to basihyal, passing deep to—but not being deeply blended with—the digastricus posterior; Reiss [2001] coded the muscle as absent in *Didelphis*, but it is present, as described by Coues [1872] and Minkoff et al. [1979]]	**Stylohyoideus** [see on the left]	**Stylohyoideus** [Sprague [1944a] states that the styloglossus of certain tree shrews is innervated by the hypoglossal nerve, but Le Gros Clark [1926] and Lightoller [1934] claim that this muscle is innervated by the facial nerve, as in other mammals]	– [Saban [1968] states that the stylohyoideus is present in *Cynocephalus*, but Gunnell and Simmons [2005] consider that this muscle is absent in this taxon; our dissections clearly indicate that the stylohyoideus is not present as an independent structure in adult colugos]	**Stylohyoideus** [Gasser's [1967] developmental study indicates that in modern humans. the digastricus posterior, stapedius, and stylohyoideus derive from the same anlage]

(*Continued*)

TABLE 8.3 (CONTINUED)
Hyoid Muscles of Adults of Representative Sarcopterygian Taxa

Actinistia: *Latimeria chalumnae* (Coelacanth)	Dipnoi: *Lepidosiren paradoxa* (South American Lungfish)	Amphibia: *Ambystoma ordinarium* (Michoacan Stream Salamander)	Reptilia: *Timon lepidus* (Ocellated Lizard)	Mammalia (Monotremata): *Ornithorhynchus anatinus* (Platypus)	Mammalia (Marsupialia): *Didelphis virginiana* (Virginian Opossum)	Mammalia (Rodentia): *Rattus norvegicus* (Norwegian Rat)	Mammalia (Scandentia): *Tupaia* sp. (Tree Shrew)	Mammalia (Dermoptera): *Cynocephalus volans* (Philippine Colugo)	Mammalia (Primates): *Homo sapiens* (Modern Human)
–	–	–	–	–	**Digastricus posterior** [paroccipital process of occipital bone to digastricus anterior; as noted above; there is no intermediate digastric tendon]	**Digastricus posterior** [see above]	**Digastricus posterior** [our dissections indicate that the digastricus posterior and digastricus anterior of *Tupaia* are joined by a well-developed tendon, as described by, e.g., Sprague 1944a]	**Digastricus posterior** (part of biventer *sensu* Leche [1886]) [our dissections indicate that the digastricus posterior and digastricus anterior of colugos are joined by a tendinous intersection, as described by Saban [1968]]	**Digastricus posterior** [the digastricus posterior and digastricus anterior of modern humans are usually joined by a well-developed tendon]
–	–	–	–	–	– [seems to be absent in *D. virginiana*, not being described by authors such as Coues [1872], Huber [1931], and Minkoff et al. [1979]; Saban [1968] suggested that it is usually absent in marsupials, and Reiss [2001] coded it as absent in *Didelphis*]	–	**Jugulohyoideus** [see on the left]	**Jugulohyoideus** (mastoideostyloideus *sensu* Saban [1968]) [seemingly corresponds to a part of the stylohyoideus and/or possibly of the digastricus posterior of mammals such as rats: Table 5.3; see, e.g., Huber 1930a, 1931; Saban 1968]	–

(*Continued*)

TABLE 8.3 (CONTINUED)
Hyoid Muscles of Adults of Representative Sarcopterygian Taxa

Actinistia: *Latimeria chalumnae* (Coelacanth)	Dipnoi: *Lepidosiren paradoxa* (South American Lungfish)	Amphibia: *Ambystoma ordinarium* (Michoacan Stream Salamander)	Reptilia: *Timon lepidus* (Ocellated Lizard)	Mammalia (Monotremata): *Ornithorhynchus anatinus* (Platypus)	Mammalia (Marsupialia): *Didelphis virginiana* (Virginian Opossum)	Mammalia (Rodentia): *Rattus norvegicus* (Norwegian Rat)	Mammalia (Scandentia): *Tupaia* sp. (Tree Shrew)	Mammalia (Dermoptera): *Cynocephalus volans* (Philippine Colugo)	Mammalia (Primates): *Homo sapiens* (Modern Human)
		Branchiohyoideus [according to Edgeworth [1935] and Ericsson and Olsson [2004], the branchiohyoideus, interhyoideus, and levator hyoideus appear at about the same time in urodele embryos, thus being difficult to infer if the branchiohyoideus is ontogenetically derived from the ventral or from the dorsomedial hyoid musculature; however, the developmental study of Piatt [1938] indicates that this muscle is in fact part of the dorsomedial hyoid musculature]	– [the branchiohyoideus of reptiles is a branchial muscle that seemingly corresponds to the subarcualis rectus 1 and not to the hyoid muscle branchiohyoideus, of amphibians: see text]						

(*Continued*)

TABLE 8.3 (CONTINUED)
Hyoid Muscles of Adults of Representative Sarcopterygian Taxa

Actinistia: *Latimeria chalumnae* (Coelacanth)	Dipnoi: *Lepidosiren paradoxa* (South American Lungfish)	Amphibia: *Ambystoma ordinarium* (Michoacan Stream Salamander)	Reptilia: *Timon lepidus* (Ocellated Lizard)	Mammalia (Monotremata): *Ornithorhynchus anatinus* (Platypus)	Mammalia (Marsupialia): *Didelphis virginiana* (Virginian Opossum)	Mammalia (Rodentia): *Rattus norvegicus* (Norwegian Rat)	Mammalia (Scandentia): *Tupaia* sp. (Tree Shrew)	Mammalia (Dermoptera): *Cynocephalus volans* (Philippine Colugo)	Mammalia (Primates): *Homo sapiens* (Modern Human)
–	**Levator hyoideus** [see depressor mandibulae above]	**Levator hyoideus** (depressor mandibulae posterior *sensu* Diogo [2007, 2008b] and Diogo et al. [2008a,b]) [seems to correspond to the pars cephalo-gnathica *sensu* Lightoller [1939]]	– [the adult *Timon* specimens dissected do not have an independent levator hyoideus, but as explained by authors such as Edgeworth [1935], some adult reptiles do have this muscle, and it is very likely that the LCA of reptiles did also have this muscle: see following chapters]	**Stapedius** (levator hyoideus *sensu* Edgeworth [1935]) [the mammalian stapedius clearly derives from the levator hyoideus of other tetrapods: e.g., Huber 1930a,b, 1931; Edgeworth 1935; Brocks 1938; Saban 1968, 1971; Kardong 2002; according to Edgeworth [1935] and explained in the following chapters, it is possible, and even likely, that the muscle levator hyoideus that is present in some reptiles is directly homologous to the mammalian stapedius: if this is the case, then it would be probably more appropriate to designate the muscle of those nonmammalian tetrapods as stapedius, as proposed by authors such as Schumacher [1973] or, even better, to designate the muscle of mammals as levator hyoideus, as proposed by authors such as Edgeworth [1935]]	**Stapedius** [from squamosal and/or surrounding regions to stapes; in *Didelphis*, the stapedius is first inserted onto the stylohyal, but then the ventral muscle cells degenerate and new dorsal muscle cells form, resulting in a shift of the insertion from stylohyal (which is moreover not present as a separate cartilage in adults of *D. virginiana*) to stapes]	**Stapedius** [from bulla to stapes]	**Stapedius**	**Stapedius**	**Stapedius**

(Continued)

TABLE 8.3 (CONTINUED)
Hyoid Muscles of Adults of Representative Sarcopterygian Taxa

Actinistia: *Latimeria chalumnae* (Coelacanth)	Dipnoi: *Lepidosiren paradoxa* (South American Lungfish)	Amphibia: *Ambystoma ordinarium* (Michoacan Stream Salamander)	Reptilia: *Timon lepidus* (Ocellated Lizard)	Mammalia (Monotremata): *Ornithorhynchus anatinus* (Platypus)	Mammalia (Marsupialia): *Didelphis virginiana* (Virginian Opossum)	Mammalia (Rodentia): *Rattus norvegicus* (Norwegian Rat)	Mammalia (Scandentia): *Tupaia* sp. (Tree Shrew)	Mammalia (Dermoptera): *Cynocephalus volans* (Philippine Colugo)	Mammalia (Primates): *Homo sapiens* (Modern Human)
–	–	–	**Cervicomandibularis** (cervico-mandibularis posterior *sensu* Edgeworth [1935]) [our dissections and comparisons support the view of, e.g., Huber [1930a] and Edgeworth [1935], i.e., that this muscle corresponds to a part of the depressor mandibulae/levator hyoideus of sarcopterygian fishes such as *Lepidosiren*; see text]	**Platysma cervicale** (the platysma cervicale *sensu* Jouffroy and Saban [1971] corresponds to the pars nuchalis of the platysma *sensu* Saban [1971] and to a part of the platysma *sensu* Lightoller [1942]) [see text]	**Platysma cervicale** (part of platysma *sensu* Minkoff et al. [1979]) [from raphe anchored to ligamentum nuchae to corner of mouth and lower lip, being deeply blended with orbicularis oris and with platysma myoides of the same side of the body, as well as with the platysma cervicale of the other side of the body]	**Platysma cervicale** [the cranial panniculus of Greene [1935] corresponds to the platysma cervicale + zygomaticus major *sensu* this volume; the superficial portion of the cervical panniculus *sensu* Greene [1935] corresponds to the sphincter colli profundus + superficialis *sensu* this volume; the deep cervical panniculus *sensu* Greene [1935] corresponds to the sternofacialis *sensu* this volume, which he describes as an upper limb muscle but, as noted by Jouffroy and Saban [1971], is a facial muscle that is probably derived from the sphincter colli profundus: e.g., Jouffroy and Saban 1971; Ryan 1986, 1989; this is supported by our dissections of canids, in which the pars auricularis of the sphincter colli profundus is very similar to the sternofacialis, running from the ear region to the region near the sternum]	**Platysma cervicale** (part of platysma *sensu* Le Gros Clark [1924] and of notoplatysma *sensu* Lightoller [1934])	**Platysma cervicale** (part of platysma *sensu* Leche [1886])	– [according to, e.g., Gasser [1967], the platysma cervicale (= his nuchal platysma) is present in early developmental stages of modern humans, disappearing in later stages; Aziz [1981] considers that the transversus nuchae found in some humans is a remnant of the platysma cervicale, but Gasser [1967] describes both a platysma cervicale and a transversus nuchae in early human embryos]

(*Continued*)

TABLE 8.3 (CONTINUED)
Hyoid Muscles of Adults of Representative Sarcopterygian Taxa

Actinistia: *Latimeria chalumnae* (Coelacanth)	Dipnoi: *Lepidosiren paradoxa* (South American Lungfish)	Amphibia: *Ambystoma ordinarium* (Michoacan Stream Salamander)	Reptilia: *Timon lepidus* (Ocellated Lizard)	Mammalia (Monotremata): *Ornithorhynchus anatinus* (Platypus)	Mammalia (Marsupialia): *Didelphis virginiana* (Virginian Opossum)	Mammalia (Rodentia): *Rattus norvegicus* (Norwegian Rat)	Mammalia (Scandentia): *Tupaia* sp. (Tree Shrew)	Mammalia (Dermoptera): *Cynocephalus volans* (Philippine Colugo)	Mammalia (Primates): *Homo sapiens* (Modern Human)
				Platysma myoides (seemingly corresponds to pars omoidea *sensu* Saban [1971])	**Platysma myoides** (part of platysma *sensu* Minkoff et al. [1979]) [from skin extending as posteriorly as the acromion of the scapula (no bony attachments) to corner of mouth and lower lip, being deeply blended with orbicularis oris of same side of body and with platysma myoides of other side of the body]	**Platysma myoides** (blended with platysma cervicale)	**Platysma myoides** [blended with platysma cervicale]	**Platysma myoides** (platysma myoides superior + "jugalis propatagii" *sensu* Leche [1886]; dorsal sheet of propatagial complex *sensu* Thewissen and Babcock 1991, 1993) [blended with platysma cervicale; Leche [1886] stated that the propatagial complex of dermopterans has a dorsal muscle formed by the platysma myoides and the jugalis propatagii and a ventral muscle; Thewissen and Babcock [1991, 1993] studied the configuration and innervation of these muscles and concluded that the dorsal one is innervated by the facial nerve and the ventral one by cervical spinal nerves; the dorsal and ventral muscles therefore seem to correspond, respectively, to the platysma myoides and to part of the panniculus carnosus of other mammals]	**Platysma myoides** (platysma *sensu* [Netter 2006]; tracheoplatysma *sensu* Lightoller [1940a])

(Continued)

TABLE 8.3 (CONTINUED)
Hyoid Muscles of Adults of Representative Sarcopterygian Taxa

Actinistia: *Latimeria chalumnae* (Coelacanth)	Dipnoi: *Lepidosiren paradoxa* (South American Lungfish)	Amphibia: *Ambystoma ordinarium* (Michoacan Stream Salamander)	Reptilia: *Timon lepidus* (Ocellated Lizard)	Mammalia (Monotremata): *Ornithorhynchus anatinus* (Platypus)	Mammalia (Marsupialia): *Didelphis virginiana* (Virginian Opossum)	Mammalia (Rodentia): *Rattus norvegicus* (Norwegian Rat)	Mammalia (Scandentia): *Tupaia* sp. (Tree Shrew)	Mammalia (Dermoptera): *Cynocephalus volans* (Philippine Colugo)	Mammalia (Primates): *Homo sapiens* (Modern Human)
–	–	–	–	– [as described by, e.g., Lightoller [1942], in the platypus specimens we dissected, there is a bundle of the platysma that is somewhat similar to the occipitalis of the mammals listed on the right, but this bundle is clearly part of the platysma; i.e., it does not constitute an independent muscle]	**Occipitalis** [from supraoccipital part of occipital and ligamentum nuchae to frontalis muscle anteriorly and to ear and to scutiform cartilage anterolaterally, being deeply blended with auricularis posterior of the same side of the body and with occipitalis of the other side of the body; the occipitalis of *Didelphis* is similar to that of rats and *Tupaia* and *Cynocephalus* in the sense that it has a medial portion (= occipitalis *sensu* Lightoller [1934] and Minkoff et al. [1979]) that extends anteriorly to blend with the frontalis and a lateral portion (= cervico-auriculo-occipitalis *sensu* Lightoller [1934]; cervicoauricularis superficialis + cervicoscutularis *sensu* Minkoff et al. [1979]) that runs anteroventrolaterally to attach on the posterior surface of the ear and to the scutiform cartilage;	**Occipitalis** (part of levator auris longus *sensu* Greene [1935]) [the occipitalis of *Rattus* is similar to that of *Tupaia* and *Cynocephalus*; i.e., it has a medial portion (= occipitalis *sensu* Lightoller [1934]) that extends anteriorly to blend with the frontalis and a lateral portion (= cervico-auriculo-occipitalis *sensu* Lightoller [1934]) that runs anteroventrolaterally to attach on the posterior surface of the ear; these two portions are deeply blended posteriorly, attaching to the dorsal region of the neck, just medially to the posterior attachment of the auricularis posterior]	**Occipitalis** (occipitalis + cervico-auriculo-occipitalis *sensu* Lightoller [1934]: see on the left)	**Occipitalis** [see on the left]	**Occipitalis** [Gasser's [1967] developmental study in modern humans indicates that the occipitalis, auricularis posterior, and transversus nuchae develop from the same anlage]

(Continued)

TABLE 8.3 (CONTINUED)
Hyoid Muscles of Adults of Representative Sarcopterygian Taxa

Actinistia: *Latimeria chalumnae* (Coelacanth)	Dipnoi: *Lepidosiren paradoxa* (South American Lungfish)	Amphibia: *Ambystoma ordinarium* (Michoacan Stream Salamander)	Reptilia: *Timon lepidus* (Ocellated Lizard)	Mammalia (Monotremata): *Ornithorhynchus anatinus* (Platypus)	Mammalia (Marsupialia): *Didelphis virginiana* (Virginian Opossum)	Mammalia (Rodentia): *Rattus norvegicus* (Norwegian Rat)	Mammalia (Scandentia): *Tupaia* sp. (Tree Shrew)	Mammalia (Dermoptera): *Cynocephalus volans* (Philippine Colugo)	Mammalia (Primates): *Homo sapiens* (Modern Human)
					[these two portions are deeply blended posteriorly, attaching to the dorsal region of the neck, just anteriorly to the posterior attachment of the auricularis posterior (which mainly corresponds to the cervicoauricularis medius *sensu* Minkoff et al. [1979]]				
–	–	–	–	–	**Auricularis posterior** [from ligamentum nuchae to ear anterolaterally, being deeply blended with auricularis posterior of the other side of the body; it mainly corresponds to the cervicoauricularis medius (which includes the "transversus nuchae," "interparietoscutularis," "cervicoauricularis posterior profundus," and "interparietoauricularis") *sensu* Minkoff et al. [1979]: see cell above]	**Auricularis posterior** (part of levator auris longus *sensu* Greene [1935]: see above; it does not really correspond to the caudal part of the levator auris longus because that may also/exclusively include the cervico-auriculo-occupitalis)	**Auricularis posterior** [see above]	**Auricularis posterior** [see above]	**Auricularis posterior** [see above]

(Continued)

TABLE 8.3 (CONTINUED)
Hyoid Muscles of Adults of Representative Sarcopterygian Taxa

Actinistia: *Latimeria chalumnae* (Coelacanth)	Dipnoi: *Lepidosiren paradoxa* (South American Lungfish)	Amphibia: *Ambystoma ordinarium* (Michoacan Stream Salamander)	Reptilia: *Timon lepidus* (Ocellated Lizard)	Mammalia (Monotremata): *Ornithorhynchus anatinus* (Platypus)	Mammalia (Marsupialia): *Didelphis virginiana* (Virginian Opossum)	Mammalia (Rodentia): *Rattus norvegicus* (Norwegian Rat)	Mammalia (Scandentia): *Tupaia* sp. (Tree Shrew)	Mammalia (Dermoptera): *Cynocephalus volans* (Philippine Colugo)	Mammalia (Primates): *Homo sapiens* (Modern Human)
–	–	–	–	**Extrinsic muscles of the ear** [our dissections and comparisons indicate that the platypus has at least some extrinsic muscles of the ear, as suggested by Lightoller [1942]; according to Huber [1930a,b, 1931] and Jouffroy and Saban [1971], some of the extrinsic muscles of the ear derive from the platysma while others derive from the sphincter colli profundus]	**Extrinsic muscles of the ear** [examples of extrinsic, facial muscles of the ear present in marsupials are the obliquus auriculae, transversus auriculae, helicis, tragicus, depressor helicis, and/or antitragicus: these muscles, with the exception of the depressor helicis ("auricularis inferior") are seemingly blended with each other in *D. virginiana*, forming the "auricularis externus" muscle complex *sensu* Minkoff et al. [1979]]	**Extrinsic muscles of the ear** [examples of extrinsic, facial muscles of the ear present in therian mammals are the obliquus auriculae, transversus auriculae, helicis, tragicus, and/or antitragicus: see, e.g., Jouffroy and Saban 1971]	**Extrinsic muscles of the ear** [see on the left]	**Extrinsic muscles of the ear** [see on the left]	**Extrinsic muscles of the ear** [see on the left]

(*Continued*)

TABLE 8.3 (CONTINUED)
Hyoid Muscles of Adults of Representative Sarcopterygian Taxa

Actinistia: *Latimeria chalumnae* (Coelacanth)	Dipnoi: *Lepidosiren paradoxa* (South American Lungfish)	Amphibia: *Ambystoma ordinarium* (Michoacan Stream Salamander)	Reptilia: *Timon lepidus* (Ocellated Lizard)	Mammalia (Monotremata): *Ornithorhynchus anatinus* (Platypus)	Mammalia (Marsupialia): *Didelphis virginiana* (Virginian Opossum)	Mammalia (Rodentia): *Rattus norvegicus* (Norwegian Rat)	Mammalia (Scandentia): *Tupaia* sp. (Tree Shrew)	Mammalia (Dermoptera): *Cynocephalus volans* (Philippine Colugo)	Mammalia (Primates): *Homo sapiens* (Modern Human)
–	–	–	–	– [although authors such as Adams et al. [1929] suggest that the therian mandibulo-auricularis is a "preauricular" muscle and thus derives from the sphincter colli profundus, most researchers consider that it is instead a "postauricular" muscle derived from the platysma (e.g., Huber 1930a,b, 1931; Ryan 1986, 1989); however, a few authors, such as Lightoller [1934] and Jouffroy and Saban [1971], have suggested that the mandibulo-auricularis may be ontogenetically and phylogenetically more related to deeper dorso-median muscles such as the stylohyoideus, digastricus posterior, stapedius, and/or jugulohyoideus than to the facial muscles; although we tentatively follow here the most	**Mandibulo-auricularis** [from ear region to mandible]	**Mandibulo-auricularis** [see on the left]	**Mandibulo-auricularis** (auriculo-mandibularis *sensu* Lightoller [1934]) [see on the left]	– [as stated by Lightoller [1934], unlike lemurs and, e.g., *Tupaia*, in primates as, e.g., *Tarsius* and marmosets, the mandibulo-auricularis probably corresponds to a strong sheet connecting the posterior edge of the mandible to the bony external auditory meatus, which might well correspond to the stylo-mandibular ligament of modern humans; such a configuration is found in the colugos dissected; i.e., there is no fleshy muscle mandibulo-auricularis, but instead a strong fascia running from the posterior edge of the mandible to the bony external auditory meatus]	– [seemingly corresponds to the stylo-mandibular ligament, which is usually present in modern humans: see on the left; according to Jouffroy and Saban [1971], it may possibly also correspond to the stylo-auricularis muscle abnormally present in a few modern humans]

(*Continued*)

TABLE 8.3 (CONTINUED)
Hyoid Muscles of Adults of Representative Sarcopterygian Taxa

Actinistia: *Latimeria chalumnae* (Coelacanth)	Dipnoi: *Lepidosiren paradoxa* (South American Lungfish)	Amphibia: *Ambystoma ordinarium* (Michoacan Stream Salamander)	Reptilia: *Timon lepidus* (Ocellated Lizard)	Mammalia (Monotremata): *Ornithorhynchus anatinus* (Platypus)	Mammalia (Marsupialia): *Didelphis virginiana* (Virginian Opossum)	Mammalia (Rodentia): *Rattus norvegicus* (Norwegian Rat)	Mammalia (Scandentia): *Tupaia* sp. (Tree Shrew)	Mammalia (Dermoptera): *Cynocephalus volans* (Philippine Colugo)	Mammalia (Primates): *Homo sapiens* (Modern Human)
				consensual view, we consider that Lightoller's hypothesis should not be completely ruled out, because the mandibulo-auricularis does usually lie deeper to all the other facial muscles and also because its topology, the orientation of its fibers, and its attachments (e.g., on the mandible and/or near the ear region) are similar to those of the deeper dorsomedial hyoid muscles of mammals and to the depressor mandibulae/levator hyoideus of other tetrapods; also, Seiler's [1980] developmental studies of tree shrews and primates seem to suggest that the mandibulo-auricularis does not develop from the anlagen that give rise to most other facial muscles, but instead from a different, deeper anlage]					

(*Continued*)

TABLE 8.3 (CONTINUED)
Hyoid Muscles of Adults of Representative Sarcopterygian Taxa

Actinistia: *Latimeria chalumnae* (Coelacanth)	Dipnoi: *Lepidosiren paradoxa* (South American Lungfish)	Amphibia: *Ambystoma ordinarium* (Michoacan Stream Salamander)	Reptilia: *Timon lepidus* (Ocellated Lizard)	Mammalia (Monotremata): *Ornithorhynchus anatinus* (Platypus)	Mammalia (Marsupialia): *Didelphis virginiana* (Virginian Opossum)	Mammalia (Rodentia): *Rattus norvegicus* (Norwegian Rat)	Mammalia (Scandentia): *Tupaia* sp. (Tree Shrew)	Mammalia (Dermoptera): *Cynocephalus volans* (Philippine Colugo)	Mammalia (Primates): *Homo sapiens* (Modern Human)
–	–	–	–	–	–	–	–	–	**Risorius** [authors such as Huber [1930a,b, 1931] suggest that the risorius derives from the sphincter colli profundus; our dissections and comparisons support the conclusions of Jouffroy and Saban's [1971] review; i.e., that the risorius derives instead from the platysma myoides; the latter hypothesis is also supported by the developmental data of Gasser [1937]: see, e.g., his Figure 10]

(*Continued*)

TABLE 8.3 (CONTINUED)
Hyoid Muscles of Adults of Representative Sarcopterygian Taxa

Actinistia: *Latimeria chalumnae* (Coelacanth)	Dipnoi: *Lepidosiren paradoxa* (South American Lungfish)	Amphibia: *Ambystoma ordinarium* (Michoacan Stream Salamander)	Reptilia: *Timon lepidus* (Ocellated Lizard)	Mammalia (Monotremata): *Ornithorhynchus anatinus* (Platypus)	Mammalia (Marsupialia): *Didelphis virginiana* (Virginian Opossum)	Mammalia (Rodentia): *Rattus norvegicus* (Norwegian Rat)	Mammalia (Scandentia): *Tupaia* sp. (Tree Shrew)	Mammalia (Dermoptera): *Cynocephalus volans* (Philippine Colugo)	Mammalia (Primates): *Homo sapiens* (Modern Human)
Interhyoideus ("géniohyoïdien" + "hyohyoïdien" *sensu* Millot and Anthony [1958], which, as is shown in the illustrations of these authors, are deeply blended: see text]	**Interhyoideus** [see on the left]	**Interhyoideus** (interhyoideus anterior + interhyoideus posterior *sensu* Piatt [1938], Bauer [1992, 1997], and Ericsson and Olsson [2004], which probably correspond respectively, to the pars interhyoidea + pars inscriptionalis of the ventral superficial constrictor of the second arch *sensu* Lightoller [1939]) [as stated by Piekarski and Olsson [2007], developmental works indicate that the interhyoideus posterior of sarcopterygians such as *Ambystoma* might be receive contributions ontogenetically not only from the hyoid region but also possibly from anterior somites: see text]	**Interhyoideus** (constrictor colli *sensu* Herrel et al. [2005])	**Interhyoideus profundus** (sphincter colli profundus *sensu* Lightoller [1942]; hyomandibularis *sensu* Edgeworth [1935]) [Edgeworth [1935] claimed that none of the mammalian facial muscles are derived from the interhyoideus because all of them are de novo structures; however, this view has been abandoned, and it is now commonly accepted that the mammalian muscles correspond to the interhyoideus and possibly dorsomedial muscles such as the cervicomandibularis of reptiles such as the lizard *Timon*: see above and text]	–	–	–	–	–

(Continued)

TABLE 8.3 (CONTINUED)
Hyoid Muscles of Adults of Representative Sarcopterygian Taxa

Actinistia: *Latimeria chalumnae* (Coelacanth)	Dipnoi: *Lepidosiren paradoxa* (South American Lungfish)	Amphibia: *Ambystoma ordinarium* (Michoacan Stream Salamander)	Reptilia: *Timon lepidus* (Ocellated Lizard)	Mammalia (Monotremata): *Ornithorhynchus anatinus* (Platypus)	Mammalia (Marsupialia): *Didelphis virginiana* (Virginian Opossum)	Mammalia (Rodentia): *Rattus norvegicus* (Norwegian Rat)	Mammalia (Scandentia): *Tupaia* sp. (Tree Shrew)	Mammalia (Dermoptera): *Cynocephalus volans* (Philippine Colugo)	Mammalia (Primates): *Homo sapiens* (Modern Human)
–	–	–	–	**Sphincter colli superficialis** (corresponds to the sphincter colli externus of platypus and sphincter colli of echidna—*sensu* Huber [1930a]) [corresponds to part of the interhyoideus of nonmammalian tetrapods: e.g., Huber 1930a; Jouffroy and Saban 1971; Lightoller [1940a] states that Huber's [1930a] sphincter colli superficialis (of, e.g., marsupials and rodents) corresponds to muscle he called "transitus"; i.e., to a part of the sphincter colli profundus that passes superficial to the platysma but that originally was deep to it: Lightoller claims that the rest of the sphincter colli profundus (i.e., everything except the transitus) is absent in all primates and that it is thus this transitus that gives the tracheo-platysma of primates; however, our dissections and	– [as shown in Fig. 7 of Huber [1930a] and Figure 413 of Jouffroy and Saban [1971], the "sphincter colli superficialis" *sensu* Minkoff et al. [1979] corresponds to part of the sphincter colli profundus (the part that extends to the neck, posteroventrally to the platysma myoides) and to part of the platysma (the part that lies more anteriorly, on the ventral region of the head and that, according to Minkoff et al. [1979], was fused with the anterior region of the platysma myoides]	**Sphincter colli superficialis** (transitus *sensu* Lightoller [1942]) [as explained by, e.g., Lightoller [1940a, 1942], although much reduced, in rodents, such as rats, the sphincter colli does have a component that is superficial to the platysma—i.e., a sphincter colli superficialis]	**Sphincter colli superficialis** (seems to correspond to the occipito-cervicalis *sensu* Lightoller [1934] and might correspond to the cervico-mandibularis *sensu* Le Gros Clark [1926], which was originally described as part of the platysma of *Ptilocercus* but seems rather to correspond to the sphincter colli superficialis of *Tupaia* and other mammals)	– [Jouffroy and Saban [1971: 484] state that colugos have a sphincter colli superficialis, but as they explain in page 496, this is because they consider that the ventral sheet of the propatagial muscle complex of colugos probably corresponds to the sphincter colli superficialis of other mammals; Thewissen and Badcock [1991, 1993] have however shown that this ventral sheet is innervated by cervical spinal nerves and not by the facial nerve, as is the sphincter colli superficialis; moreover, as shown in Figure 1 of the latter authors, the position and the orientation of the fibers of that ventral sheet are not similar to those of the sphincter colli superficialis of other mammals (e.g., in a lateral view, it appears deep, not superficial, to the	– [it is commonly accepted that primates such as modern humans and chimpanzees do not have a sphincter colli superficialis, but according to Burrows et al. [2006], this muscle may be found in some chimpanzees and perhaps even in some modern humans; in the modern human cadavers we dissected, the sphincter colli superficialis was not present as an independent muscle]

(Continued)

TABLE 8.3 (CONTINUED)
Hyoid Muscles of Adults of Representative Sarcopterygian Taxa

Actinistia: *Latimeria chalumnae* (Coelacanth)	Dipnoi: *Lepidosiren paradoxa* (South American Lungfish)	Amphibia: *Ambystoma ordinarium* (Michoacan Stream Salamander)	Reptilia: *Timon lepidus* (Ocellated Lizard)	Mammalia (Monotremata): *Ornithorhynchus anatinus* (Platypus)	Mammalia (Marsupialia): *Didelphis virginiana* (Virginian Opossum)	Mammalia (Rodentia): *Rattus norvegicus* (Norwegian Rat)	Mammalia (Scandentia): *Tupaia* sp. (Tree Shrew)	Mammalia (Dermoptera): *Cynocephalus volans* (Philippine Colugo)	Mammalia (Primates): *Homo sapiens* (Modern Human)
				comparisons strongly indicate that the configuration of the platysma of primates such as lemurs is similar to that found in, e.g., colugos and tree shrews; i.e., the latter mammals have both a platysma cervicale and a platysma myoides, although these two latter muscles are blended; in primates such as modern humans the platysma cervicale is usually absent; i.e., the platysma *sensu* Netter [2006] corresponds to the platysma myoides of other mammals: see above; if we accept, as it is nowadays commonly accepted, that the sphincter colli of mammals derives from the interhyoideus of other tetrapods, it makes sense to suppose that plesiomorphically the sphincter colli was a superficial muscle, as is the nonmammalian interhyoideus, and not a deep muscle: see text]				platysma); our dissections indicate that colugos do not have a fleshy, distinct muscle sphincter colli superficialis]	

(Continued)

TABLE 8.3 (CONTINUED)
Hyoid Muscles of Adults of Representative Sarcopterygian Taxa

Actinistia: *Latimeria chalumnae* (Coelacanth)	Dipnoi: *Lepidosiren paradoxa* (South American Lungfish)	Amphibia: *Ambystoma ordinarium* (Michoacan Stream Salamander)	Reptilia: *Timon lepidus* (Ocellated Lizard)	Mammalia (Monotremata): *Ornithorhynchus anatinus* (Platypus)	Mammalia (Marsupialia): *Didelphis virginiana* (Virginian Opossum)	Mammalia (Rodentia): *Rattus norvegicus* (Norwegian Rat)	Mammalia (Scandentia): *Tupaia* sp. (Tree Shrew)	Mammalia (Dermoptera): *Cynocephalus volans* (Philippine Colugo)	Mammalia (Primates): *Homo sapiens* (Modern Human)
–	–	–	–	– [absent as an independent muscle in the platypus, although part of it might have given rise to deep facial muscles such as the orbicularis oris, orbicularis oculi, and mentalis (which clearly seem to correspond to the muscles that are designated under the same names in other mammals) and possibly to the "sphincter bursae buccalis" *sensu* Huber [1930a] (which, contrary to what was stated by the latter author, seems to correspond to the "buccinatorius" of the echidna and to the buccinatorius of other mammals); in the echidna part of the sphincter colli passes deep to other facial muscles, forming the sphincter colli profundus: e.g., Lightoller 1942; Jouffroy and Saban 1971]	**Sphincter colli profundus** (sphincter colli preauricularis + part of sphincter colli superficialis *sensu* Minkoff et al. [1979]) [mainly deep to platysma myoides and platysma cervicale, seemingly having no bony attachments and being deeply blended with zygomaticus minor; it is deeply blended with the sphincter colli profundus of the other side of the body]	**Sphincter colli profundus** (superficial portion of cervical platysma *sensu* Greene [1935]; sphincter colli profundus + "primitive sphincter colli" of Figure 6 of Huber [1930a]; transitus *sensu* Lightoller [1940a]) [deeply blended with the sphincter colli superficialis]	**Sphincter colli profundus**	**Sphincter colli profundus** [absent according to Jouffroy and Saban [1971], but this muscle is clearly present in the specimens dissected by us: dorsally its runs deep to the platysma myoides while ventrally it meets its counterpart in the ventral midline of the head]	–
–	–	–	–	–	–	**Sternofacialis** (deep cervical panniculus *sensu* Greene [1935]: see platysma cervicale above)	–	–	–

(Continued)

TABLE 8.3 (CONTINUED)
Hyoid Muscles of Adults of Representative Sarcopterygian Taxa

Actinistia: *Latimeria chalumnae* (Coelacanth)	Dipnoi: *Lepidosiren paradoxa* (South American Lungfish)	Amphibia: *Ambystoma ordinarium* (Michoacan Stream Salamander)	Reptilia: *Timon lepidus* (Ocellated Lizard)	Mammalia (Monotremata): *Ornithorhynchus anatinus* (Platypus)	Mammalia (Marsupialia): *Didelphis virginiana* (Virginian Opossum)	Mammalia (Rodentia): *Rattus norvegicus* (Norwegian Rat)	Mammalia (Scandentia): *Tupaia* sp. (Tree Shrew)	Mammalia (Dermoptera): *Cynocephalus volans* (Philippine Colugo)	Mammalia (Primates): *Homo sapiens* (Modern Human)
–	–	–	–	**Cervicalis transversus** [its position and the orientation of its fibers are similar to those of the interscutularis of nonmonotreme mammals such as rats; Lightoller [1940a] seems to support the homology of these muscles, because he states that there is a cervicalis transversus in rodents; however, according to Jouffroy and Saban [1971], the interscutularis is derived from the pars intermedia of the sphincter colli profundus, while the cervicalis transversus is derived from the pars cervicalis of the latter muscle; our observations and comparisons indicate that the "cervicalis transversus" of primates such as *Tarsius* is likely homologous to the interscutularis of rats and thus to the interscutularis *sensu* the present work (which is, therefore, also likely homologous to the structure that is often designated cervicalis transversus in monotremes)]	–	– [see on the left]	–	–	–

(*Continued*)

TABLE 8.3 (CONTINUED)
Hyoid Muscles of Adults of Representative Sarcopterygian Taxa

Actinistia: *Latimeria chalumnae* (Coelacanth)	Dipnoi: *Lepidosiren paradoxa* (South American Lungfish)	Amphibia: *Ambystoma ordinarium* (Michoacan Stream Salamander)	Reptilia: *Timon lepidus* (Ocellated Lizard)	Mammalia (Monotremata): *Ornithorhynchus anatinus* (Platypus)	Mammalia (Marsupialia): *Didelphis virginiana* (Virginian Opossum)	Mammalia (Rodentia): *Rattus norvegicus* (Norwegian Rat)	Mammalia (Scandentia): *Tupaia* sp. (Tree Shrew)	Mammalia (Dermoptera): *Cynocephalus volans* (Philippine Colugo)	Mammalia (Primates): *Homo sapiens* (Modern Human)
–	–	–	–	– [see preceding text]	**Interscutularis** [no bony attachments, its fibers cross the top of the head transversely between the ears to attach onto the scutiform cartilage of each side of the head, so it is deeply blended with the interscutularis of the other side of the body; see cervicalis transversus mentioned earlier]	**Interscutularis** [see cervicalis transversus mentioned earlier]	–	–	– [but an interscutularis/ cervicalis transversus might be present in primates such as *Tarsius*: see preceding line]

(*Continued*)

TABLE 8.3 (CONTINUED)
Hyoid Muscles of Adults of Representative Sarcopterygian Taxa

Actinistia: *Latimeria chalumnae* (Coelacanth)	**Dipnoi: *Lepidosiren paradoxa* (South American Lungfish)**	**Amphibia: *Ambystoma ordinarium* (Michoacan Stream Salamander)**	**Reptilia: *Timon lepidus* (Ocellated Lizard)**	**Mammalia (Monotremata): *Ornithorhynchus anatinus* (Platypus)**	**Mammalia (Marsupialia): *Didelphis virginiana* (Virginian Opossum)**	**Mammalia (Rodentia): *Rattus norvegicus* (Norwegian Rat)**	**Mammalia (Scandentia): *Tupaia* sp. (Tree Shrew)**	**Mammalia (Dermoptera): *Cynocephalus volans* (Philippine Colugo)**	**Mammalia (Primates): *Homo sapiens* (Modern Human)**
–	–	–	–	– [according to Seiler [1976], both the zygomarticus major and zygomaticus minor of catarrhine primates derive from the auriculolabialis superior of other mammals; however, our dissections and comparisons support Jouffroy and Saban's [1971] hypothesis; i.e., (a) the zygomaticus major and minor are absent in mammals such as monotremes; (b) in placentals, the zygomaticus is plesiomorphically attached to the zygomatic arch, but in some cases, it extends posteriorly to attach to the ear (that is why it is sometimes named auriculolabialis); (c) in a few mammals, such as some ungulates, pinnipedes, bats, rodents, tree shrews, and primates, the zygomaticus is divided into superficial (= auriculolabialis	**Zygomaticus major** (part of zygomaticus and/or of sphincter colli profundus *sensu* Minkoff et al. [1979]) [from ear to angle of mouth, being deeply blended with zygomaticus minor, platysma cervical, and orbicularis oris; contrary to the descriptions of Minkoff et al. [1979] and statements of authors such as Jouffroy and Saban [1971], there is a zygomaticus major and a zygomaticus minor in *D. virginiana*, similar to those found in many placental mammals, i.e., corresponding to the auriculolabialis inferior and "auricularis superior" of placentals, respectively; as in many placentals, the zygomaticus major runs from the ear to the angle of the mouth, and what happens is that the zygomaticus major is deeply blended	**Zygomaticus major** (lower part of auriculo-labialis of, e.g., Greene 1935 because the "zygomaticus" *sensu* Greene corresponds to the levator labii superioris *sensu* the present work; i.e., to the "maxillo-naso-labialis" *sensu* other authors; therefore in rats and mice the zygomaticus major and minor are deeply blended, forming two bellies of a single auriculolabialis muscle; zygomatico-labialis superficialis and or auriculolabialis inferior *sensu* Jouffroy and Saban [1971]) [our dissections indicate that the zygomaticus major of Norwegian rats are deeply blended with the platysma; does this mean that it is really part of and/or derived from the platysma (see on the left)? Probably not, because as stated by Greene [1935], in other rats and other	**Zygomaticus major** (auriculolabialis inferior or zygomatico-labialis *sensu* Jouffroy and Saban [1971], Le Gros Clark [1926], and Lightoller [1934]) [see on the left]	**Zygomaticus major** [in Huber's [1930a] Figure 27 of the "primate ground plan of superficial facial musculature," he suggests that the auriculabialis inferior (= zygomaticus major) derives from the platysma (see on the left), but at least in the case of colugos, the former muscle is well distinguished from the latter, because these muscles are perpendicular to each other; there is a significant difference between *Cynocephalus*, *Lemur*, and *Tupaia*: in *Cynocephalus*, the auriculolabialis inferior (= zygomaticus major) is superficial to the platysma cervicale; in *Lemur*, these two muscles lie in the same plane; in *Tupaia* the auricularis inferior is deep to the platysma cervicale: e.g., Lightoller 1934; this work]	**Zygomaticus major** [Gasser's [1967] developmental study indicate that the zygomaticus major and zygomaticus minor of modern humans derive from his "infraorbital lamina" and not from his "mandibular lamina"; i.e., they seem to be ontogenetically more related to the facial muscles of the orbit region than with those of the mouth region; interestingly, in earlier stages of human development, these two muscles are more separated from each other than in later stages; i.e., in this respect, the configuration seen in those early stages is more similar to that seen in adult mammals such as colugos, tree shrews, and "lower" primates]

(Continued)

TABLE 8.3 (CONTINUED)
Hyoid Muscles of Adults of Representative Sarcopterygian Taxa

Actinistia: *Latimeria chalumnae* (Coelacanth)	Dipnoi: *Lepidosiren paradoxa* (South American Lungfish)	Amphibia: *Ambystoma ordinarium* (Michoacan Stream Salamander)	Reptilia: *Timon lepidus* (Ocellated Lizard)	Mammalia (Monotremata): *Ornithorhynchus anatinus* (Platypus)	Mammalia (Marsupialia): *Didelphis virginiana* (Virginian Opossum)	Mammalia (Rodentia): *Rattus norvegicus* (Norwegian Rat)	Mammalia (Scandentia): *Tupaia* sp. (Tree Shrew)	Mammalia (Dermoptera): *Cynocephalus volans* (Philippine Colugo)	Mammalia (Primates): *Homo sapiens* (Modern Human)
				inferior and zygomaticus major *sensu* Jouffroy and Saban [1971]) and deep (= auriculolabialis superior and zygomaticus minor *sensu* Jouffroy and Saban [1971]) portions; the former originating ventrally and/or posteriorly to the latter, thus usually lying nearer to the ear and being more associated with the platysma (that is why some authors argue that it might derive from the platysma, although its innervation seems to indicate the contrary); Jouffroy and Saban [1971] explicitly state that these superficial and deep portions very likely correspond to the zygomaticus major and minor of modern humans, respectively; and (d) according to them, in mammals such as tree shrews and primates as lemurs, the zygomatic muscles, and particularly the	with the zygomaticus minor (auriculolabialis superior) and with the superior portion of the platysma cervicale, but that is also the case in many placental mammals; therefore, in the face of the available data, one cannot argue that the zygomaticus major is necessarily less differentiated in marsupials than it is in placentals]	rodents, this auriculolabialis is much more distinct from the platysma, being seemingly a derivative of the sphincter colli profundus; however, it is possible that some of the mammalian structures that are designated "zygomaticus major and minor" and/or "auriculolabialis inferior and superior" in the literature are really part of and/or derive from the platysma, i.e., that they are not really homologous to the zygomaticus major and minor *sensu* this volume: e.g., Boas and Paulli 1908; Huber 1930a; Edgeworth 1935; Jouffroy and Saban 1971; it cannot be completely ruled out, however, that at least in some cases, the "zygomaticus major" and/or auriculolabialis inferior derive from the platysma, while the "zygomaticus minor" and/or auriculolabialis superior derive from, e.g., the orbicularis oculi]			

(Continued)

TABLE 8.3 (CONTINUED)
Hyoid Muscles of Adults of Representative Sarcopterygian Taxa

Actinistia: *Latimeria chalumnae* (Coelacanth)	Dipnoi: *Lepidosiren paradoxa* (South American Lungfish)	Amphibia: *Ambystoma ordinarium* (Michoacan Stream Salamander)	Reptilia: *Timon lepidus* (Ocellated Lizard)	Mammalia (Monotremata): *Ornithorhynchus anatinus* (Platypus)	Mammalia (Marsupialia): *Didelphis virginiana* (Virginian Opossum)	Mammalia (Rodentia): *Rattus norvegicus* (Norwegian Rat)	Mammalia (Scandentia): *Tupaia* sp. (Tree Shrew)	Mammalia (Dermoptera): *Cynocephalus volans* (Philippine Colugo)	Mammalia (Primates): *Homo sapiens* (Modern Human)
				zygomaticus major, often extends posteriorly in order to attach to the ear, but this "trend" is reverted in "higher" primates: for example, in modern humans both the zygomaticus major and zygomaticus minor usually originate relatively far from the ear, although in a few cases at least one of these muscles might originate in the ear region]					
–	–	–	–	– [see zygomaticus major above]	**Zygomaticus minor** (mainly corresponds to zygomaticus *sensu* Minkoff et al. [1979]) [mainly continuous with the sphincter colli profundus and with the orbicularis oris posteriorly and superiorly, it runs anteriorly and inferiorly to attach onto the angle to the mouth, being deeply blended with the orbicularis oris]	**Zygomaticus minor** [upper part of auriculo-labialis of, e.g., Greene [1935] because the zygomaticus *sensu* Greene corresponds to the levator labii superioris *sensu* the present work; i.e., to the maxillo-naso-labialis *sensu* other authors; therefore, in rats and mice, the zygomaticus major and minor are deeply blended, forming two bellies of a singleauriculolabialis muscle]	**Zygomaticus minor** (auriculolabialis superior *sensu* Jouffroy and Saban [1971], Le Gros Clark [1926], and Lightoller [1934])	**Zygomaticus minor**	**Zygomaticus minor**

(Continued)

TABLE 8.3 (CONTINUED)
Hyoid Muscles of Adults of Representative Sarcopterygian Taxa

Actinistia: *Latimeria chalumnae* (Coelacanth)	Dipnoi: *Lepidosiren paradoxa* (South American Lungfish)	Amphibia: *Ambystoma ordinarium* (Michoacan Stream Salamander)	Reptilia: *Timon lepidus* (Ocellated Lizard)	Mammalia (Monotremata): *Ornithorhynchus anatinus* (Platypus)	Mammalia (Marsupialia): *Didelphis virginiana* (Virginian Opossum)	Mammalia (Rodentia): *Rattus norvegicus* (Norwegian Rat)	Mammalia (Scandentia): *Tupaia* sp. (Tree Shrew)	Mammalia (Dermoptera): *Cynocephalus volans* (Philippine Colugo)	Mammalia (Primates): *Homo sapiens* (Modern Human)
–	–	–	–	–	**Frontalis** (frontalis pars epicranialis *sensu* Minkoff et al. [1979]) [mainly continuous with the occipitalis posteriorly and with the frontalis of the other side of the body trough the epicranial aponeurosis, running anteriorly passing between the eyes to fuse anteriorly with the levator labii alaeque nasi (nasolabialis)]	**Orbito-temporo-auricularis** (the frontalis *sensu* Greene [1935] corresponds to the orbito-temporo-auricularis *sensu* Edgeworth [1935])	**Frontalis**	**Frontalis** [our dissections and comparisons indicate that the frontalis and auriculo-orbitalis of *Cynocephalus* and *Tupaia* and the frontalis, temporoparietalis, and auricularis anterior of modern humans, correspond to the orbito-temporo-auricularis of mammals such as rats: see Table 8.4]	**Frontalis**
–	–	–	–	–	**Auriculo-orbitalis** (frontalis pars palpebralis, or frontoscutularis, *sensu* Minkoff et al. [1979]) [from anterior corner of external ear, where its fibers converge onto the scutiform cartilage, to the frontal bone along the supraorbital rim, deep to the orbicularis oculi, anteriorly]	–	**Auriculo-orbitalis** (auriculo-orbitalis or orbito-auricularis *sensu* Lightoller [1934]; it might correspond to *Tupaia*'s attrahens aurem *sensu* Le Gros Clark [1924], and/or to the *Ptilocercus*' scutularis + portio transiens *sensu* Le Gros Clark [1926])	**Auriculo-orbitalis** [see above; this muscle usually runs from the auricular region to the orbital region, being inferior and/or deep to the frontalis: e.g., Figure 8.18]	**Temporoparietalis** [see above; according to Jouffroy and Saban [1971], this muscle is related to, but different from, the auricularis superior; they state that it corresponds to the temporal part of the frontalis, which is also named epicranio-temporal or orbito-temporalis, and which has a longitudinal orientation and covers the temporal aponeurosis, being often fused in primates with the auriculares anterior and superior and also to part of the galea aponeurotica]

(Continued)

TABLE 8.3 (CONTINUED)
Hyoid Muscles of Adults of Representative Sarcopterygian Taxa

Actinistia: *Latimeria chalumnae* (Coelacanth)	Dipnoi: *Lepidosiren paradoxa* (South American Lungfish)	Amphibia: *Ambystoma ordinarium* (Michoacan Stream Salamander)	Reptilia: *Timon lepidus* (Ocellated Lizard)	Mammalia (Monotremata): *Ornithorhynchus anatinus* (Platypus)	Mammalia (Marsupialia): *Didelphis virginiana* (Virginian Opossum)	Mammalia (Rodentia): *Rattus norvegicus* (Norwegian Rat)	Mammalia (Scandentia): *Tupaia* sp. (Tree Shrew)	Mammalia (Dermoptera): *Cynocephalus volans* (Philippine Colugo)	Mammalia (Primates): *Homo sapiens* (Modern Human)
–	–	–	–	–	–	–	– [unlike *Ptilocercus*, *Tupaia* seems to only have an auriculo-orbitalis *sensu* Lightoller [1934]; i.e., it does not have a separate temporoparietalis and a separate auricularis anterior: see preceding text]	–	**Auricularis anterior**
–	–	–	–	–	**Auricularis superior** (auricularis anterior superior, or adductor auris medius, or scutuloauricularis superficialis dorsalis, *sensu* Minkoff et al. [1979]) [lying in region between scutiform cartilage and external ear]	–	**Auricularis superior**	**Auricularis superior** (auricularis anterior superior of Figure 409 of Jouffroy and Saban [1971])	**Auricularis superior**
–	–	–	–	**Orbicularis oculi**	**Orbicularis oculi** [surrounding the eye, its only bony attachment being the maxilla]	**Orbicularis oculi**	**Orbicularis oculi**	**Orbicularis oculi**	**Orbicularis oculi**

(Continued)

TABLE 8.3 (CONTINUED)
Hyoid Muscles of Adults of Representative Sarcopterygian Taxa

Actinistia: *Latimeria chalumnae* (Coelacanth)	Dipnoi: *Lepidosiren paradoxa* (South American Lungfish)	Amphibia: *Ambystoma ordinarium* (Michoacan Stream Salamander)	Reptilia: *Timon lepidus* (Ocellated Lizard)	Mammalia (Monotremata): *Ornithorhynchus anatinus* (Platypus)	Mammalia (Marsupialia): *Didelphis virginiana* (Virginian Opossum)	Mammalia (Rodentia): *Rattus norvegicus* (Norwegian Rat)	Mammalia (Scandentia): *Tupaia* sp. (Tree Shrew)	Mammalia (Dermoptera): *Cynocephalus volans* (Philippine Colugo)	Mammalia (Primates): *Homo sapiens* (Modern Human)
–	–	–	–	–	–	–	**Zygomatico-orbicularis** [Lightoller [1934] states that in *Tupaia javanica*, there is a group of fibers running from the region lying posterodorsally to the eye to the dorsal margin of the zygomatic arch but that it cannot correspond to the zygomatico-orbicularis *sensu* Le Gros Clark [1924] because it lies deep to the orbicularis oculi; however, Le Gros Clark [1924] stated that his zygomatico-orbicularis lies deep to at least some fibers of the orbicularis oculi; the group of fibers described by Lightoller thus seems to correspond to the zygomatico-orbicularis *sensu* Le Gros Clark]	**Zygomatico-orbicularis** [there is a very thin group of fibers attaching to the medial margin of the posterior portion of the orbicularis oculi, anteriorly, and to the dorsal surface of the zygomatic arch, posteriorly; this thin group of fibers lies deep to most fibers of the orbicularis oculi and auriculo-orbitalis, being deeply blended with the temporal fascia covering the temporalis in lateral view; however, our dissections indicate that there are some fleshy fibers, which should thus be considered a zygomatico-orbicularis *sensu* Le Gros Clark [1924, 1926], even if this is a poorly developed muscle; the lemur shown in Figure 4 of Lightoller [1934] also seems to have such a group of fleshy fibers]	–

(*Continued*)

TABLE 8.3 (CONTINUED)
Hyoid Muscles of Adults of Representative Sarcopterygian Taxa

Actinistia: *Latimeria chalumnae* (Coelacanth)	Dipnoi: *Lepidosiren paradoxa* (South American Lungfish)	Amphibia: *Ambystoma ordinarium* (Michoacan Stream Salamander)	Reptilia: *Timon lepidus* (Ocellated Lizard)	Mammalia (Monotremata): *Ornithorhynchus anatinus* (Platypus)	Mammalia (Marsupialia): *Didelphis virginiana* (Virginian Opossum)	Mammalia (Rodentia): *Rattus norvegicus* (Norwegian Rat)	Mammalia (Scandentia): *Tupaia* sp. (Tree Shrew)	Mammalia (Dermoptera): *Cynocephalus volans* (Philippine Colugo)	Mammalia (Primates): *Homo sapiens* (Modern Human)
–	–	–	–	–	– [there is seemingly no separate depressor supercilii, but the medial portion of the orbicularis oculi does have some fibers that run more vertically (superoinferiorly) than the rest of the fibers of the muscle and that can thus correspond to at least part of the orbicularis oculi of other mammals: see, e.g., Figures 2 and 3 of Minkoff et al. [1979]]	–	–	–	**Depressor supercilii** [the depressor supercilii and corrugator supercilii are seemingly derived from the orbicularis oris: see Table 8.4]
–	–	–	–	–	**Corrugator supercilii** (superciliaris *sensu* Minkoff et al. [1979]) [from nasofrontal fascia, deep to the frontalis, and well separated from the corrugator supercilii of the other side of the body, running mainly anteriorly passing deep to the auriculo-orbitalis, to reach the upper eyelid near its medial corner, blending with the orbicularis oculi]	–	**Corrugator supercilii** (superciliaris *sensu* Jouffroy and Saban [1971])	**Corrugator supercilii** [see above]	**Corrugator supercilii**

(Continued)

TABLE 8.3 (CONTINUED)
Hyoid Muscles of Adults of Representative Sarcopterygian Taxa

Actinistia: *Latimeria chalumnae* (Coelacanth)	Dipnoi: *Lepidosiren paradoxa* (South American Lungfish)	Amphibia: *Ambystoma ordinarium* (Michoacan Stream Salamander)	Reptilia: *Timon lepidus* (Ocellated Lizard)	Mammalia (Monotremata): *Ornithorhynchus anatinus* (Platypus)	Mammalia (Marsupialia): *Didelphis virginiana* (Virginian Opossum)	Mammalia (Rodentia): *Rattus norvegicus* (Norwegian Rat)	Mammalia (Scandentia): *Tupaia* sp. (Tree Shrew)	Mammalia (Dermoptera): *Cynocephalus volans* (Philippine Colugo)	Mammalia (Primates): *Homo sapiens* (Modern Human)
–	–	–	–	–	**Retractor anguli oculi lateralis** (retractor anguli oculi *sensu* Minkoff et al. [1979]) [from temporal fascia, posteriorly, to posterolateral portion of orbicularis oculi, anteriorly; according to Minkoff et al. [1979], this muscle is derived from the auriculo-orbitalis]	– [but retractor anguli oculi lateralis is present in many other placental mammals: see, e.g., Jouffroy and Saban [1971]]	–	–	–

(*Continued*)

TABLE 8.3 (CONTINUED)
Hyoid Muscles of Adults of Representative Sarcopterygian Taxa

Actinistia: *Latimeria chalumnae* (Coelacanth)	Dipnoi: *Lepidosiren paradoxa* (South American Lungfish)	Amphibia: *Ambystoma ordinarium* (Michoacan Stream Salamander)	Reptilia: *Timon lepidus* (Ocellated Lizard)	Mammalia (Monotremata): *Ornithorhynchus anatinus* (Platypus)	Mammalia (Marsupialia): *Didelphis virginiana* (Virginian Opossum)	Mammalia (Rodentia): *Rattus norvegicus* (Norwegian Rat)	Mammalia (Scandentia): *Tupaia* sp. (Tree Shrew)	Mammalia (Dermoptera): *Cynocephalus volans* (Philippine Colugo)	Mammalia (Primates): *Homo sapiens* (Modern Human)
–	–	–	–	**Nasolabialis** [Lightoller [1942] and Saban [1971] state that deep to the cranial (anterior) portion of the orbicularis oculi of the echidna lies a small naso-labialis; according to Lightoller [1942], in the platypus, there is a somewhat similar structure, but it is not as differentiated from the other facial musculature as in the echidna; in the platypus we dissected, the naso-labialis does seem to be an independent muscle, being very similar to the naso-labialis of the echidna—see, e.g., Figure 4 of Lightoller [1942], which suggests that the nasolabialis derives from a part of the sphincter colli superficialis (transitus *sensu* Lightoller): Huber [1930a] and Jouffroy and Saban [1971]]	**Levator labii superioris alaeque nasi** (nasolabialis or levator nasolabialis *sensu* Minkoff et al. [1979] and Grant et al. [2013]) [extends anteriorly to the snout from the region just between the eyes, attaching onto the maxilla, nasal bone, and the upper lip and lateral wings of the rhinarium, being deeply blended with the orbicularis oris of the same side of the body and with the levator labii superioris alaeque nasi of the other side of the body]	**Levator labii superioris alaeque nasi** (levator labii superioris *sensu* Parsons 1898 and Greene 1935; pars jugularis of superficial maxillo-naso-labialis *sensu* Lightoller [1940a,b]; nasolabialis *sensu* Diogo [2008b], Diogo et al. [2008a,b, 2009b], and Diogo and Abdala [2010])	**Levator labii superioris alaeque nasi** (levator labii superioris *sensu* Le Gros Clark [1924]; nasolabialis *sensu* Diogo [2008b], Diogo et al. [2008a, 2009b], and Diogo and Abdala [2010]) [not described by Le Gros Clark [1926], but his Figure 49 seems to suggest that it may also be present in *Ptilocercus*]	**Levator labii superioris alaeque nasi** (nasolabialis *sensu* Diogo [2008b], Diogo et al. [2008a, 2009b], and Diogo and Abdala [2010])	**Levator labii superioris alaeque nasi** [contrary to what was suggested by Diogo [2008b], Diogo et al. [2008a, 2009b], and Diogo and Abdala [2010], the levator labii superioris of primates does not seem to correspond to part of the nasolabialis, but instead to the maxillo-naso-labialis of other therian mammals, as proposed by Seiler [1976]; Gasser's [1967] developmental study suggests that at least in modern humans these two muscles appear ontogenetically in the orbital region]

(Continued)

TABLE 8.3 (CONTINUED)
Hyoid Muscles of Adults of Representative Sarcopterygian Taxa

Actinistia: *Latimeria chalumnae* (Coelacanth)	Dipnoi: *Lepidosiren paradoxa* (South American Lungfish)	Amphibia: *Ambystoma ordinarium* (Michoacan Stream Salamander)	Reptilia: *Timon lepidus* (Ocellated Lizard)	Mammalia (Monotremata): *Ornithorhynchus anatinus* (Platypus)	Mammalia (Marsupialia): *Didelphis virginiana* (Virginian Opossum)	Mammalia (Rodentia): *Rattus norvegicus* (Norwegian Rat)	Mammalia (Scandentia): *Tupaia* sp. (Tree Shrew)	Mammalia (Dermoptera): *Cynocephalus volans* (Philippine Colugo)	Mammalia (Primates): *Homo sapiens* (Modern Human)
–	–	–	–	–	–	**Procerus?** (nasolabialis superficialis *sensu* Ryan [1989?]) [present in rats? Not described by Greene [1935] but was seemingly present in the rats we dissected, see also, e.g., Ryan 1989; however, it did not seem to be present in the mice we dissected]	–	–	**Procerus**
–	–	–	–	**Buccinatorius** (the buccinatorius *sensu* Lightoller 1942 corresponds to the sphincter bursae buccalis *sensu* Huber [1930a]) [according to Huber [1930a], there is no buccinatorius in the platypus, and the buccinatorius of echidna may well not be homologous with that of other mammals, because it may be derived from the platysma and not from the sphincter colli profundus; however, as noted in later works as, e.g., Lightoller [1942] and Jouffroy and Saban [1971], the sphincter bursae buccalis of platypus might correspond to the buccinatorius of echidna and/or to the buccinatorius of other mammals]	**Buccinatorius** [from maxilla to mandible and upper lip, being deeply blended with the levator anguli oris facialis]	**Buccinatorius** [not described by Greene [1935] but is clearly present in the rats we dissected, being subdivided into various sections; see also, e.g., Ryan 1989]	**Buccinatorius** [not described by Le Gros Clark [1924], but it is clearly present in the *Tupaia* specimens we dissected; it is also present in *Ptilocercus*, see, e.g., Le Gros Clark 1926]	**Buccinatorius**	**Buccinatorius**

(Continued)

TABLE 8.3 (CONTINUED)
Hyoid Muscles of Adults of Representative Sarcopterygian Taxa

Actinistia: *Latimeria chalumnae* (Coelacanth)	Dipnoi: *Lepidosiren paradoxa* (South American Lungfish)	Amphibia: *Ambystoma ordinarium* (Michoacan Stream Salamander)	Reptilia: *Timon lepidus* (Ocellated Lizard)	Mammalia (Monotremata): *Ornithorhynchus anatinus* (Platypus)	Mammalia (Marsupialia): *Didelphis virginiana* (Virginian Opossum)	Mammalia (Rodentia): *Rattus norvegicus* (Norwegian Rat)	Mammalia (Scandentia): *Tupaia* sp. (Tree Shrew)	Mammalia (Dermoptera): *Cynocephalus volans* (Philippine Colugo)	Mammalia (Primates): *Homo sapiens* (Modern Human)
–	–	–	–	–	**Dilatator nasi** (pars nasalis of maxillonasolabialis *sensu* Minkoff et al. 1979) [from maxilla to skin of nose and upper lip and to alar nasal cartilage, probably acting to dilate the nostrils, as stated by Minkoff et al. 1979; in the specimens we dissected, it has a strong, anterior tendon, as shown in Figure 7E of Huber [1930a]]	**Dilatator nasi** (dilatator naris *sensu* Greene [1935] and Peterka [1936]) [we prefer to use the name dilatator nasi because the name dilatator naris is often used to designate the pars alaris of the nasalis: see, e.g., Jouffroy and Saban 1971; it goes from maxilla to the nasal cartilage]	–	–	–
–	–	–	–	–	**Levator labii superioris** (pars labialis of maxillonasolabialis *sensu* Minkoff et al. [1979]; part or totality of maxillolabialis *sensu* Grant et al. [2013]) [from maxilla to upper lip, being deeply blended posteriorly with the dilatator nasi]	**Levator labii superioris** (maxillo-naso-labialis *sensu* Diogo [2008b], Diogo et al. [2008a, 2009b], and Diogo and Abdala [2010]) [seemingly corresponds to the maxillo-labialis *sensu* Jouffroy and Saban [1971] and Ryan [1989], being clearly present in the rats and mice we dissected, corresponding to the zygomaticus *sensu* Greene [1935], running from zygomatic arch to mouth region: see also, e.g., Ryan 1989]	**Levator labii superioris** (maxillo-naso-labialis *sensu* Diogo [2008b], Diogo et al. [2008a, 2009b], and Diogo and Abdala [2010])	**Levator labii superioris** (maxillo-naso-labialis *sensu* Diogo [2008b], Diogo et al. [2008a, 2009b], and Diogo and Abdala [2010])	**Levator labii superioris** [contrary to what was suggested by Diogo [2008b], Diogo et al. [2008a, 2009b], and Diogo and Abdala [2010], the maxillo-naso-labialis of nonprimate mammals seems to correspond to the levator labii superioris *sensu* the present work, as suggested by Seiler [1976]]

(Continued)

TABLE 8.3 (CONTINUED)
Hyoid Muscles of Adults of Representative Sarcopterygian Taxa

Actinistia: *Latimeria chalumnae* (Coelacanth)	Dipnoi: *Lepidosiren paradoxa* (South American Lungfish)	Amphibia: *Ambystoma ordinarium* (Michoacan Stream Salamander)	Reptilia: *Timon lepidus* (Ocellated Lizard)	Mammalia (Monotremata): *Ornithorhynchus anatinus* (Platypus)	Mammalia (Marsupialia): *Didelphis virginiana* (Virginian Opossum)	Mammalia (Rodentia): *Rattus norvegicus* (Norwegian Rat)	Mammalia (Scandentia): *Tupaia* sp. (Tree Shrew)	Mammalia (Dermoptera): *Cynocephalus volans* (Philippine Colugo)	Mammalia (Primates): *Homo sapiens* (Modern Human)
					Nasalis (nasolabialis profundus *sensu* Grant et al. [2013]) [lies on anterior snout region and is associated with vibrissae]	**Nasalis** (part or totality of nasolabialis profundus *sensu* Jouffroy and Saban [1971] and Ryan [1989]) [from nasal cartilage to the plate of the mystacial pad; not described by Greene [1935], but it is clearly present in the rats we dissected; see also, e.g., Ryan 1989]	**Nasalis** [the nasalis is present, but not highly developed in *Tupaia*, corresponding to the nasolabialis profundus *sensu* Diogo [2008b], Diogo et al. [2008a, 2009b], and Diogo and Abdala [2010] and possibly to part, or the totality, of the dilator naris and/or erector vibrissae *sensu* Le Gros Clark [1924]: see, e.g., Figure 10 of Seiler [1976]]	**Nasalis** [the nasalis is present, but not highly developed, in colugos, corresponding to the nasolabialis profundus *sensu* Diogo [2008b], Diogo et al. [2008a, 2009b], and Diogo and Abdala [2010]]	**Nasalis**

(*Continued*)

TABLE 8.3 (CONTINUED)
Hyoid Muscles of Adults of Representative Sarcopterygian Taxa

Actinistia: *Latimeria chalumnae* (Coelacanth)	**Dipnoi: *Lepidosiren paradoxa* (South American Lungfish)**	**Amphibia: *Ambystoma ordinarium* (Michoacan Stream Salamander)**	**Reptilia: *Timon lepidus* (Ocellated Lizard)**	**Mammalia (Monotremata): *Ornithorhynchus anatinus* (Platypus)**	**Mammalia (Marsupialia): *Didelphis virginiana* (Virginian Opossum)**	**Mammalia (Rodentia): *Rattus norvegicus* (Norwegian Rat)**	**Mammalia (Scandentia): *Tupaia* sp. (Tree Shrew)**	**Mammalia (Dermoptera): *Cynocephalus volans* (Philippine Colugo)**	**Mammalia (Primates): *Homo sapiens* (Modern Human)**
–	–	–	–	–	– [it is possible that the depressor septi nasi is present as a part of the "nasolabialis profundus" *sensu* Grant et al. [2013], but these authors did not describe it as a separate muscle, nor did any other authors, and we did not find it, so it does not seem to be differentiated]	**Depressor septi nasi** [is present in rats and mice, as clearly shown by Hairdarliu et al. [2012], originating from the septum nasi and ventromedial part of the lateral nasal cartilage and directed ventral and passed rostral to the origins of the partes mediae superior et inferior of the nasolabialis profundus (nasalis *sensu* the present work); these fibers are inserted onto the rostral part of the upper lip close to the philtrum bilaterally; contraction will pull the rhinarium and the ventral wall of the nostrils ventrally, uplifting the rostral part of the upper lip, and resulting in nostril dilatation]	– [seemingly absent, but it is difficult to know if this muscle is really absent in mammalian taxa that were not as studied as mice/rats, which are the almost only taxa where such muscles were described in detail]	– [seemingly absent, but it is difficult to know if this muscle is really absent in mammalian taxa that were not as studied as mice/rats, which are almost the only taxa where such muscles were described in detail]	**Depressor septi nasi** [see above]

(*Continued*)

TABLE 8.3 (CONTINUED)
Hyoid Muscles of Adults of Representative Sarcopterygian Taxa

Actinistia: *Latimeria chalumnae* (Coelacanth)	Dipnoi: *Lepidosiren paradoxa* (South American Lungfish)	Amphibia: *Ambystoma ordinarium* (Michoacan Stream Salamander)	Reptilia: *Timon lepidus* (Ocellated Lizard)	Mammalia (Monotremata): *Ornithorhynchus anatinus* (Platypus)	Mammalia (Marsupialia): *Didelphis virginiana* (Virginian Opossum)	Mammalia (Rodentia): *Rattus norvegicus* (Norwegian Rat)	Mammalia (Scandentia): *Tupaia* sp. (Tree Shrew)	Mammalia (Dermoptera): *Cynocephalus volans* (Philippine Colugo)	Mammalia (Primates): *Homo sapiens* (Modern Human)
–	–	–	–	–	**Intrinsic muscles of the snout** [in addition to extrinsic muscles such as the nasalis and levator labii superioris and levator labii superioris alaeque nasalis, Grant et al. [2013] described various intrinsic muscles of the snout/vibrissae in marsupials such as *Monodelphis domestica*, so it is very likely that such muscles are also present in *D. virginiana*]	**Intrinsic muscles of the snout** [examples of these muscles, described by Haidarliu et al. [2012], are the depressor rhinarii and the levator rhinarii, which is a tiny muscle originated from the glabrous skin of the dorsal half of the rhinarium that runs from the border with the nasal tubercles to the skin of the dorsum nasi and can thus pull the rhinarium dorsally; they also describe an intraturbinate muscle running from the lateral wall of the nostrils to the atrioturbinate cartilage; however, it is not clear if the latter muscle is a facial muscle or not]	**Intrinsic muscles of the snout** [it is difficult to know which intrinsic snout muscles are present in mammalian taxa that were not as studied as mice/rats, which are almost the only taxa where such muscles were described in detail]	**Intrinsic muscles of the snout** [it is difficult to know which intrinsic snout muscles are present in mammalian taxa that were not as studied as mice/rats, which are almost the only taxa where such muscles were described in detail]	–

(Continued)

TABLE 8.3 (CONTINUED)
Hyoid Muscles of Adults of Representative Sarcopterygian Taxa

Actinistia: *Latimeria chalumnae* (Coelacanth)	Dipnoi: *Lepidosiren paradoxa* (South American Lungfish)	Amphibia: *Ambystoma ordinarium* (Michoacan Stream Salamander)	Reptilia: *Timon lepidus* (Ocellated Lizard)	Mammalia (Monotremata): *Ornithorhynchus anatinus* (Platypus)	Mammalia (Marsupialia): *Didelphis virginiana* (Virginian Opossum)	Mammalia (Rodentia): *Rattus norvegicus* (Norwegian Rat)	Mammalia (Scandentia): *Tupaia* sp. (Tree Shrew)	Mammalia (Dermoptera): *Cynocephalus volans* (Philippine Colugo)	Mammalia (Primates): *Homo sapiens* (Modern Human)
–	–	–	–	–	**Levator anguli oris facialis** (pars anterior of maxillonasolabialis *sensu* Minkoff et al. [1979]) [from connective tissue of upper lip and from buccinatorius, to region between upper lip and anterior portion of snout, passing deep to the levator labii superioris; it is less differentiated from the surrounding muscles than is the case in placental mammals such as mice/rats]	**Levator anguli oris facialis** (levator anguli oris or caninus *sensu* Lightoller [1934]; bucco-nasolabialis *sensu* Ryan [1986]; buccinatorius *sensu* Greene [1935] and Bryant [1945]; pars profunda of maxillo-naso-labialis *sensu* Lightoller [1940a,b]) [we use the name levator anguli facialis in order to distinguish this muscle from the levator anguli oris mandibularis of some nonmammalian tetrapods, which is a mandibular, and not a hyoid, muscle: see preceding chapters]	**Levator anguli oris facialis** (levator anguli oris *sensu* Le Gros Clark [1926]; incisivus superior + caninus *sensu* Lightoller [1934]) [as noted by Lightoller [1934], this muscle is deeply blended with the orbicularis oris]	**Levator anguli oris facialis**	**Levator anguli oris facialis** [Gasser's [1967] study of human development supports the claim that the levator anguli oris facialis, orbicularis oris, depressor labii inferioris, depressor anguli oris, and mentalis have a common ontogenetic origin, being derived from his "mandibular lamina"; see also, e.g., Sullivan and Osgood 1927 and Jouffroy and Saban 1971]
–	–	–	–	**Orbicularis oris** (plicae anguli oris *sensu* Huber [1930a])	**Orbicularis oris** [surrounding the mouth, being deeply blended with the orbicularis oris of the other side of the body]	**Orbicularis oris**	**Orbicularis oris** [see maxilo-naso-labialis mentioned earlier]	**Orbicularis oris**	**Orbicularis oris**
–	–	–	–	–	–	–	–	–	**Depressor labii inferioris** [see levator anguli oris facialis mentioned earlier]

(Continued)

TABLE 8.3 (CONTINUED)
Hyoid Muscles of Adults of Representative Sarcopterygian Taxa

Actinistia: *Latimeria chalumnae* (Coelacanth)	Dipnoi: *Lepidosiren paradoxa* (South American Lungfish)	Amphibia: *Ambystoma ordinarium* (Michoacan Stream Salamander)	Reptilia: *Timon lepidus* (Ocellated Lizard)	Mammalia (Monotremata): *Ornithorhynchus anatinus* (Platypus)	Mammalia (Marsupialia): *Didelphis virginiana* (Virginian Opossum)	Mammalia (Rodentia): *Rattus norvegicus* (Norwegian Rat)	Mammalia (Scandentia): *Tupaia* sp. (Tree Shrew)	Mammalia (Dermoptera): *Cynocephalus volans* (Philippine Colugo)	Mammalia (Primates): *Homo sapiens* (Modern Human)
–	–	–	–	–	–	–	–	–	**Depressor anguli oris** [see levator anguli oris facialis mentioned earlier]
–	–	–	–	**Mentalis** [present in platypus, but seemingly not in echidna, according to, e.g., Lightoller 1942 and Saban 1971]	– [Minkoff et al. [1979] stated that the mentalis is not differentiated, or is poorly differentiated, in *D. virginiana*, and we did not find a differentiated muscle in our dissections, nor do other authors describe it in this species; however, Jouffroy and Saban [1971] stated that the mentalis is differentiated in at least some other marsupials]	–	**Mentalis** (labiorum profundi inferioris *sensu* Lightoller 1934)	**Mentalis**	**Mentalis** [see levator anguli oris facialis above]

Note: See footnote of Table 8.1 and see Table 8.4.

TABLE 8.4
Diagram Illustrating the Authors' Hypotheses Regarding the Homologies of the Hyoid Muscles

	Latimeria (5 m.)	*Lepidosiren* (3 m.)	*Ambystoma* (4 m.)	*Timon* (3 m.)	*Ornithorhynchus* (12 m.- not ex. ear/int. snout*)	*LCA marsupials + placentals* (27 m.- out ex. ear/int. snout*)	*Didelphis* (25 m.- not ex. ear/int. snout*)	*Rattus* (25 m.- not ex. ear/int. snout*)	*Tupaia* (26 m.- not ex. ear/int. snout*)	*Cynocephalus* (23 m.- not ex. ear/int. snout*)	*Homo* (27 m.- ex. ear*)
DORSO-MEDIAL HYOID MUS.	'Ad.hyo. Y'	---	---	---	---	---	---	---	---	---	---
	Ad. arcus palatini	---	---	---	---	---	---	---	---	---	---
	---	De. man.	De. man.	De. man.	Styloideus	Stylohyoideus	Stylohyoideus	Stylohyoideus	Stylohyoideus	---	Stylohyoideus
	---	---	---	---	---	---	---	---	Jugulohyoideus	Jugulohyoideus	---
	---	---	---	---	---	Digastricus posterior	Digastricus posterior	Digastricus posterior	Digastricus posterior	Digastricus posterior	Digastricus posterior
	Latimeria 's 'le.oper.'	---	---	---	---	---	---	---	---	---	---
	Ad. operculi	---	---	---	---	---	---	---	---	---	---
	---	---	Bra.+Cerat.	---	---	---	---	---	---	---	---
	---	Le. hyoideus	Le. hyoideus	---(Le.hyoideus poar.)	Stapedius	Stapedius	Stapedius	Stapedius	Stapedius	Stapedius	Stapedius
	---	---	---	Cervicomandibularis	Platysma cervicale	Platysma cervicale	Platysma cervicale	Platysma cervicale	Platysma cervicale	Platysma cervicale	---
	---	---	---	---	Platysma myoides	Platysma myoides	Platysma myoides	Platysma myoides	Platysma myoides	Platysma myoides	Platysma myoides
	---	---	---	---	---	Occipitalis	Occipitalis	Occipitalis	Occipitalis	Occipitalis	Occipitalis
	---	---	---	---	---	Auricularis posterior	Auricularis posterior	Auricularis posterior	Aur. posterior	Auricularis posterior	Auricularis posterior
	---	---	---	---	Ex. ear mus. *	Ex. ear mus. *	Ex. ear mus. *	Ex. ear mus. *	Ex. ear mus. *	Ex. ear mus. *	Ex. ear mus. *
	---	---	---	---	---	Mandibulo -auricularis	Mandibulo -auricularis	Mandibulo -auricularis	Mandibulo -auricularis	---	---
	---	---	---	---	---	---	---	---	---	---	Risorius
VENTRAL HYOID MUS.	Interhyoideus	Interhyoideus	Interhyoideus	Interhyoideus	Interhyoideus prof.	---	---	---	---	---	---
	---	---	---	---	Sphincter colli supe.	Sphincter colli supe.	---	Sphincter colli supe.	Sphincter colli supe.	---	---
	---	---	---	---	—(colli prof. in Echidna)	Sphincter colli prof.	Sphincter colli prof.	Sphincter colli prof.	Sphincter colli prof.	Sphincter colli prof.	---
	---	---	---	---	---	---	---	Sternofacialis	---	---	---
	---	---	---	---	Cervicalis tra.	---	---	---	---	---	---
	---	---	---	---	---	Interscutularis	Interscutularis	Interscutularis	---	---	---
	---	---	---	---	---	Zygomaticus major	Zygomaticus major	Zygomaticus major	Zygomaticus major	Zygomaticus major	Zygomaticus major
	---	---	---	---	---	Zygomaticus minor	Zygomaticus minor	Zygomaticus minor	Zygomaticus minor	Zygomaticus minor	Zygomaticus minor
	---	---	---	---	---	Frontalis	Frontalis	Orbito -temporo -aur.	Frontalis	Frontalis	Frontalis
	---	---	---	---	---	Auriculo -orbitalis	Auriculo -orbitalis	---	Auriculo -orbitalis	Auriculo -orbitalis	Temporoparietalis
	---	---	---	---	---	---	---	---	---	---	Auricularis anterior
	---	---	---	---	---	Auricularis superior	Auricularis superior	---	Auricularis superior	Auricularis superior	Auricularis superior
	---	---	---	---	Orbicularis oculi	Orbicularis oculi	Orbicularis oculi	Orbicularis oculi	Orbicularis oculi	Orbicularis oculi	Orbicularis oculi
	---	---	---	---	---	---	---	---	Zygomatico -orbicularis	Zygomatico -orbicularis	---
	---	---	---	---	---	---	---	---	---	---	De. supercilii
	---	---	---	---	---	Corrugator supercilii	Corrugator supercilii	---	Corrugator supercilii	Corrugator supercilii	Corrugator supercilii
	---	---	---	---	---	Re. anguli oculi lateralis	Re. anguli oculi lateralis	--- (poap.)	---	---	---
	---	---	---	---	Naso -labialis	Le. labii sup. al. nasi	Le. labii sup. al. nasi	Le. labii sup. al. nasi	Le. labii sup. al. nasi	Le. labii sup. al. nasi	Le. labii sup. al. nasi
	---	---	---	---	---	---	---	Procerus?	---	---	Procerus
	---	---	---	---	Buccinatorius	Buccinatorius	Buccinatorius	Buccinatorius	Buccinatorius	Buccinatorius	Buccinatorius
	---	---	---	---	---	Dilatator nasi	Dilatator nasi	Dilatator nasi	---	---	---
	---	---	---	---	---	Le. labii sup.	Le. labii sup.	Le. labii sup.	Le. labii sup.	Le. labii sup.	Le. labii sup.
	---	---	---	---	---	Nasalis	Nasalis	Nasalis	Nasalis	Nasalis	Nasalis
	---	---	---	---	---	---	---	Depressor septi nasi	--- (really absent?)	--- (really absent?)	De. septi nasi
	---	---	---	---	---	In. snout mus.*	In. snout mus.*	In. snout mus.*	In. snout mus.*	In. snout mus.*	---
	---	---	---	---	---	Le. anguli oris facialis	Le. anguli oris facialis	Le. anguli oris facialis	Le. anguli oris facialis	Le. anguli oris facialis	Le. anguli oris facialis
	---	---	---	---	Orbicularis oris	Orbicularis oris	Orbicularis oris	Orbicularis oris	Orbicularis oris	Orbicularis oris	Orbicularis oris
	---	---	---	---	---	---	---	---	---	---	Depressor labii inf.
	---	---	---	---	---	---	---	---	---	---	De. anguli oris
	---	---	----	---	Mentalis	Mentalis	--- (poam.)	---	Mentalis	Mentalis	Mentalis

Note: See footnotes of Tables 8.2 and 8.3. ad., adductor; al., alaeque; aur., auricularis; bra. + cerat., branchiohyoideus + ceratomandibularis; de., depressor; ex., extrinsic; hyo., hyomandibulae; inf., inferioris; int., intrinsic; le., levator; m., muscles; man., depressor mandibulae; operc., operculi; poam., present in other adult marsupials; poap., present in some other adult placentals; poar., present in other adult reptiles; prof., profundus; sup., superioris; supe., superficialis; tra., transversus.

TABLE 8.5
Branchial, Pharyngeal, and Laryngeal Muscles of Adults of Representative Sarcopterygian Taxa

Actinistia: *Latimeria chalumnae* (Coelacanth)	Dipnoi: *Lepidosiren paradoxa* (South American Lungfish)	Amphibia: *Ambystoma ordinarium* (Michoacan Stream Salamander)	Reptilia: *Timon lepidus* (Ocellated Lizard)	Mammalia (Monotremata): *Ornithorhynchus anatinus* (Platypus)	Mammalia (Marsupialia): *Didelphis virginiana* (Virginian Opossum)	Mammalia (Rodentia): *Rattus norvegicus* (Norwegian Rat)	Mammalia (Scandentia): *Tupaia* sp. (Tree Shrew)	Mammalia (Dermoptera): *Cynocephalus volans* (Philippine Colugo)	Mammalia (Primates): *Homo sapiens* (Modern Human)
Branchial muscles *sensu stricto* [adult bony fishes and amphibians often have various branchial muscles *sensu stricto*, e.g., the constrictores branchiales, levatores arcuum branchialium, transversi ventrales, and/or subarcuales recti, among others: e.g., Edgeworth 1935; Kesteven 1942–1945]	**Branchial muscles *sensu stricto*** [see on the left]	**Branchial muscles *sensu stricto*** [see on the left]	– [absent as a group; adult lizards such as *Timon* lack all the branchial muscles *sensu stricto*, except the derivatives of the "subarcualis 1," e.g., the hyobranchialis and ceratohyoideus: see text]	– [absent as a group; the only branchial muscles *sensu stricto* that are present as independent structures in adult monotremes such as the platypus are the subarcualis rectus III, the ceratohyoideus, and, seemingly, the stylopharyngeus, the two latter muscles being probably the result of a subdivision of the subarcualis rectus I; a subarcualis rectus II is present in extant mammals such as marsupials: e.g., Edgeworth 1935; see text]	– [absent as a group; the only branchial muscles *sensu stricto* that are present as independent structures in adult marsupials such as the Virginian opossum are the stylopharyngeus and possibly the stylopharyngeus: see text]	– [absent as a group; the only branchial muscles *sensu stricto* that are present as independent structures in adult rodents such as the Norwegian rat are the ceratohyoideus and seemingly the stylopharyngeus: see text]	– [absent as a group; the only branchial muscles *sensu stricto* that are present as independent structures in adult tree shrews as *Tupaia* are the ceratohyoideus and seemingly the stylopharyngeus: see text]	– [absent as a group; the only branchial muscles *sensu stricto* that are present as independent structures in adult colugos are the ceratohyoideus and seemingly the stylopharyngeus: see text]	– [absent as a group; the only branchial muscle *sensu stricto* that is present as an independent structure in adult modern humans is seemingly the stylopharyngeus: see text]

(Continued)

TABLE 8.5 (CONTINUED)
Branchial, Pharyngeal, and Laryngeal Muscles of Adults of Representative Sarcopterygian Taxa

Actinistia: *Latimeria chalumnae* (Coelacanth)	Dipnoi: *Lepidosiren paradoxa* (South American Lungfish)	Amphibia: *Ambystoma ordinarium* (Michoacan Stream Salamander)	Reptilia: *Timon lepidus* (Ocellated Lizard)	Mammalia (Monotremata): *Ornithorhynchus anatinus* (Platypus)	Mammalia (Marsupialia): *Didelphis virginiana* (Virginian Opossum)	Mammalia (Rodentia): *Rattus norvegicus* (Norwegian Rat)	Mammalia (Scandentia): *Tupaia* sp. (Tree Shrew)	Mammalia (Dermoptera): *Cynocephalus volans* (Philippine Colugo)	Mammalia (Primates): *Homo sapiens* (Modern Human)
– [see above]	– [see above]	– [see above]	**Hyobranchialis** (part or totality of subarcualis rectus I or of branchiohyoideus *sensu* Edgeworth [1935] and Herrel et al. [2005]) [see text and following chapters]	**Stylopharyngeus** [the data now available on innervation, development, topology, and comparative anatomy indicate that the mammalian stylopharyngeus is probably not a de novo pharyngeal muscle *sensu* Edgeworth 1935, but instead a derivative of the branchial muscles *sensu stricto*: see text and above]	**Stylopharyngeus** [as noted by Jouffroy and Saban [1971], marsupials often have a broad stylopharyngeus running from the paroccipital process of the occipital bone to the pharynx (pars pharyngea, which is said to be a unique, derived feature of marsupials and to replace functionally the superior constrictor of placentals, because it is a circular constrictor structure) and to the thyroid cartilage (pars thyroidea, which is thus the one that mainly corresponds to the whole stylopharyngeus of placentals because it is a longitundinal constrictor structure)]	**Stylopharyngeus** [see text and above]	**Stylopharyngeus** [see text and above]	**Stylopharyngeus** [see text and above]	**Stylopharyngeus** [see text and above]
– [see above]	– [see above]	– [see above]	**Ceratohyoideus** (probably corresponds to the ceratohyoideus *sensu* Abdala and Moro [2003]) [see text and the following chapters]	**Ceratohyoideus** [see text and above]	**Ceratohyoideus** [usually present in marsupials, from hypohyal to ceratohyal: see, e.g., Jouffroy and Saban [1971]]	**Ceratohyoideus** (the ceratohyoideus *sensu* House [1953] corresponds to the branchiohyoideus *sensu* Sprague [1943] and to the hyoideus latus, keratohyoideus brevis, and intercornualis *sensu* Saban [1968]) [see text and above]	**Ceratohyoideus** (interhyoideus *sensu* Le Gros Clark [1926]) [see text and above]	**Ceratohyoideus** [see text and above]	– [absent as an independent muscle in modern humans, but present in other primates: e.g., Sprague 1944b; Saban 1968; this work]

(Continued)

TABLE 8.5 (CONTINUED)

Branchial, Pharyngeal, and Laryngeal Muscles of Adults of Representative Sarcopterygian Taxa

Actinistia: *Latimeria chalumnae* (Coelacanth)	**Dipnoi: *Lepidosiren paradoxa* (South American Lungfish)**	**Amphibia: *Ambystoma ordinarium* (Michoacan Stream Salamander)**	**Reptilia: *Timon lepidus* (Ocellated Lizard)**	**Mammalia (Monotremata): *Ornithorhynchus anatinus* (Platypus)**	**Mammalia (Marsupialia): *Didelphis virginiana* (Virginian Opossum)**	**Mammalia (Rodentia): *Rattus norvegicus* (Norwegian Rat)**	**Mammalia (Scandentia): *Tupaia* sp. (Tree Shrew)**	**Mammalia (Dermoptera): *Cynocephalus volans* (Philippine Colugo)**	**Mammalia (Primates): *Homo sapiens* (Modern Human)**
–	–	–	–	**Subarcualis rectus III** [see text and above]	–	–	–	–	–
– [really absent? See text]	**Protractor pectoralis** (cucullaris *sensu* Edgeworth [1935])	**Protractor pectoralis** (cucullaris *sensu* Edgeworth [1935])	**Trapezius** (capitodorsoclavicularis *sensu* Tsuihiji [2007])	**Acromiotrapezius** (anterior trapezius *sensu* Saban [1971]; clavoacro-miotrapezius *sensu* Gambaryan et al. [2015]) [the anterior (from parietal and interparietal and neck aponeurosis to clavicle and scapula) and posterior (from thoracic vertebrae 7–13 and ninth rib to scapula, being blended with deltoideus scapularis) trapezius *sensu* Saban 1971, are well separated in platypus and clearly correspond to the acromiotrapezius and spinotrapezius of other mammals; according to Edgeworth [1935], in monotremes, the subarcualis I derives from the branchial arch 1, while the subarcualis III, trapezius, and sternoclei-domastoideus derive from the branchial arch 3; but see text]	**Acromiotrapezius** (anterior trapezius *sensu* Stein [1981]) [from vertebrae, ligamentum nuchae, and occipitalis to spine and acromion of scapula and to lateral half of clavicle]	**Acromiotrapezius** (dorsoscapularis superior, anterior trapezius or trapezius superior *sensu* Greene [1935]) [it is somewhat blended with the spinotrapezius, but is considered as a separate muscle by many authors, see, e.g., Greene 1935; according to Edgeworth [1935], in placental mammals, the subarcualis I usually derives from the branchial arch 1 (in some cases, it may atrophy during development as, e.g., in *Manis* and seemingly in most anthropoids, including modern humans), while the acromiotrapezius, spinotrapezius, cleido-occipitalis, cleidomastoideus, sternomastoideus, and sternoclei-domastoideus derive from the branchial arch 2; also according to	**Trapezius** [in both *Tupaia* and *Ptilocercus*, it is a single, continuous muscle, which seems to correspond to the acromiotrapezius + spinotrapezius of other mammals: e.g., Le Gros Clark 1924, 1926; George 1977; this work]	**Acromiotrapezius** [contrary to what is stated by Macalister [1872] and Gunnell and Simmons [2005] in the colugos we dissected, both the spinotrapezius and the acromiotrapezius are present as independent structures: the former mainly inserts onto the scapular spine, while the latter mainly inserts onto the acromion; this is also the case in the specimens examined by, e.g., Leche 1886: see his Figure 8; as stated by Macalister [1872], in colugos, the trapezius complex (= acromiotrapezius + spinotrapezius *sensu* this volume) does not reach the cranium anteriorly and does not attach on the clavicle posteriorly; i.e., this trapezius complex does not include a "cleido-trapezius"	**Trapezius** [it has three parts; i.e., the acromiotrapezius, claviculotrapezius, and spinotrapezius *sensu* Kardong [2002], which are not differentiated into separate muscles, as is the case in various other mammals; the human "claviculotrapezius" probably corresponds to part of the trapezius of, e.g., *Tupaia*, although it may possibly correspond to the cleido-occipitalis of the latter taxon: e.g., Jouffroy 1971; Table 8.6]

(*Continued*)

TABLE 8.5 (CONTINUED)
Branchial, Pharyngeal, and Laryngeal Muscles of Adults of Representative Sarcopterygian Taxa

Actinistia: *Latimeria chalumnae* (Coelacanth)	Dipnoi: *Lepidosiren paradoxa* (South American Lungfish)	Amphibia: *Ambystoma ordinarium* (Michoacan Stream Salamander)	Reptilia: *Timon lepidus* (Ocellated Lizard)	Mammalia (Monotremata): *Ornithorhynchus anatinus* (Platypus)	Mammalia (Marsupialia): *Didelphis virginiana* (Virginian Opossum)	Mammalia (Rodentia): *Rattus norvegicus* (Norwegian Rat)	Mammalia (Scandentia): *Tupaia* sp. (Tree Shrew)	Mammalia (Dermoptera): *Cynocephalus volans* (Philippine Colugo)	Mammalia (Primates): *Homo sapiens* (Modern Human)
						Edgeworth [1935], in certain adult placentals, e.g., *Sus*, there is a single branchial arch, which gives rise to all the muscles listed earlier; but see text]		*sensu* Kardong [2002] and does not seem to include the cleido-occipitalis *sensu* this volume]	
–	–	– [but see Chapter 6]	– [but see Chapter 7]	**Spinotrapezius** (posterior trapezius *sensu* Saban [1971]: see above) [see above]	**Spinotrapezius** (posterior trapezius *sensu* Stein [1981]) [from vertebrae to spine of scapula, being deeply blended distally to the acromiotrapezius and to the deltoideus scapularis]	**Spinotrapezius** (dorsoscapularis, inferior posterior trapezius or trapezius inferior superior *sensu* Greene [1935]) [see above]	– [see above]	**Spinotrapezius** [see above]	– [see above]
–	–	–	–	**Dorso-cutaneous** [present in monotremes as well as in some other extant mammals; seemingly corresponds to part of the trapezius of tetrapods such as lizards: e.g., Jouffroy 1971; Jouffroy and Lessertisseur 1971; probably corresponds to trapezius auricularis *sensu* Gambaryan et al. [2015], which in platypus runs from ribs 10 and 11 to the auricle, fusing with the platysma cervicale]	– [present in at least some marsupials according to Saban [1968], but we did not find it in our dissections of *D. virginiana*, and it has not been described by authors such as Stein [1981], Coues [1872], and Minkoff et al. [1979], in this species]	–	–	–	–

(Continued)

TABLE 8.5 (CONTINUED)

Branchial, Pharyngeal, and Laryngeal Muscles of Adults of Representative Sarcopterygian Taxa

Actinistia: *Latimeria chalumnae* (Coelacanth)	**Dipnoi: *Lepidosiren paradoxa* (South American Lungfish)**	**Amphibia: *Ambystoma ordinarium* (Michoacan Stream Salamander)**	**Reptilia: *Timon lepidus* (Ocellated Lizard)**	**Mammalia (Monotremata): *Ornithorhynchus anatinus* (Platypus)**	**Mammalia (Marsupialia): *Didelphis virginiana* (Virginian Opossum)**	**Mammalia (Rodentia): *Rattus norvegicus* (Norwegian Rat)**	**Mammalia (Scandentia): *Tupaia* sp. (Tree Shrew)**	**Mammalia (Dermoptera): *Cynocephalus volans* (Philippine Colugo)**	**Mammalia (Primates): *Homo sapiens* (Modern Human)**
–	–	–	–	– [according to Edgeworth [1935], the cleido-occipitalis of mammals as, e.g., *Tatusia* seems to correspond to part of the reptilian trapezius, but the "cleido-occipitalis" of, e.g., the placental carnivores may well correspond to the part of the reptilian sternocleido-mastoideus]	**Cleido-occipitalis** [from occipital region to clavicle, being blended with the cleidomastoideus, but still clearly forming a separate muscle]	**Cleido-occipitalis** (the cleido-occipitalis *sensu* Wood [1870] and Edgeworth [1935] corresponds to the clavotrapezius and cleido-occipitalis cervicalis *sensu* Greene [1935]) [the position and orientation of the fibers of the cleido-occipitalis of, e.g., *Rattus* and *Tupaia* are more similar to those of the monotreme sternocleido-mastoideus than to those of the monotreme trapezius; also, according to Greene [1935] in, e.g., rats the cleido-occipitalis, sternomastoideus, and cleidomastoideus are all innervated by the "spinal accessory and *third* and *fourth* cervical nerves through the subtrapezial plexus," while the spinotrapezius and acromiotrapezius are innervated by the "spinal accessory and *second* and *third* cervical nerves through the subtrapezial plexus"]	**Cleido-occipitalis** [see on the left]	– [see acromiotrapezius above]	– [usually absent as an independent muscle, but may be found in a few modern humans: e.g., Wood 1870]

(Continued)

TABLE 8.5 (CONTINUED)
Branchial, Pharyngeal, and Laryngeal Muscles of Adults of Representative Sarcopterygian Taxa

Actinistia: *Latimeria chalumnae* (Coelacanth)	Dipnoi: *Lepidosiren paradoxa* (South American Lungfish)	Amphibia: *Ambystoma ordinarium* (Michoacan Stream Salamander)	Reptilia: *Timon lepidus* (Ocellated Lizard)	Mammalia (Monotremata): *Ornithorhynchus anatinus* (Platypus)	Mammalia (Marsupialia): *Didelphis virginiana* (Virginian Opossum)	Mammalia (Rodentia): *Rattus norvegicus* (Norwegian Rat)	Mammalia (Scandentia): *Tupaia* sp. (Tree Shrew)	Mammalia (Dermoptera): *Cynocephalus volans* (Philippine Colugo)	Mammalia (Primates): *Homo sapiens* (Modern Human)
–	–	– [but see Chapter 9]	**Sternocleido-mastoideus** (episternocleido-mastoideus *sensu* Herrel et al. [2005]; capiticleidoe-pisternalis *sensu* Tsuihiji [2007]) [see on the left]	**Cleidomastoideus** [as suggested by Howell [1937a] and Saban [1971], in the platypus specimens we dissected, both the sternomastoideus and cleidomastoideus are present as independent structures; the cleidomastoideus runs from the skull to the middle clavicle]	**Cleidomastoideus** [from mastoid process of squamosal to clavicle]	**Cleidomastoideus**	**Cleidomastoideus**	**Cleidomastoideus** [contrary to what is suggested in Leche's [1986] Figure 4, the colugos dissected by us have both a sternomastoideus and a cleidomastoideus, which are well separated; each of these muscles attaches anteriorly on the mastoid process by a thin and long tendon]	**Sternocleido-mastoideus** [including sternal and clavicular heads, which clearly seem to correspond to the sternomastoideus and cleidomastoideus of other mammals, but are not really differentiated into independent muscles]
–	–	–	–	**Sternomastoideus** [from region of zygomatic arch and transverse crest of skull medial one-third of clavicle]	**Sternomastoideus** [from mastoid process of squamosal to sternum]	**Sternomastoideus**	**Sternomastoideus**	**Sternomastoideus**	–

(*Continued*)

TABLE 8.5 (CONTINUED)

Branchial, Pharyngeal, and Laryngeal Muscles of Adults of Representative Sarcopterygian Taxa

Actinistia: *Latimeria chalumnae* (Coelacanth)	**Dipnoi: *Lepidosiren paradoxa* (South American Lungfish)**	**Amphibia: *Ambystoma ordinarium* (Michoacan Stream Salamander)**	**Reptilia: *Timon lepidus* (Ocellated Lizard)**	**Mammalia (Monotremata): *Ornithorhynchus anatinus* (Platypus)**	**Mammalia (Marsupialia): *Didelphis virginiana* (Virginian Opossum)**	**Mammalia (Rodentia): *Rattus norvegicus* (Norwegian Rat)**	**Mammalia (Scandentia): *Tupaia* sp. (Tree Shrew)**	**Mammalia (Dermoptera): *Cynocephalus volans* (Philippine Colugo)**	**Mammalia (Primates): *Homo sapiens* (Modern Human)**
–	–	– [see on the right]	– [plesiomorphically reptiles have no muscular pharynx; reptiles such as crocodylians do possess a secondary palate and means to constrict the pharynx, but this constrictor is a derivative of an hyoid muscle, the interhyoideus: see the following chapters]	**Constrictor pharyngis** [there is only one constrictor of the pharynx in monotremes, but the cricothyroideus and the palatopharyngeus are already differentiated in these mammals; some authors consider that amphibians may have "pharyngeal muscles" lying between the hyoid apparatus and the pharyngeal wall: e.g., Piatt 1938; Smith 1992; however, these pharyngeal muscles seem to be branchial muscles *sensu stricto* as, e.g., the levatores arcuum branchialium and/or the transversi ventrales *sensu* Edgeworth [1935]: see, e.g., Saban 1971: 708; however, the interesting arguments provided by Smith [1992] to support the idea that the mammalian pharyngeal musculature innervated by the vagus nerve (CNX) may be derived from the amphibian pharyngeal musculature innervated by the same nerve cannot be discarded: see text]	**Constrictor pharyngis medius** [from pharyngeal raphe to hyoid apparatus, being blended with inferior constrictor of same side of the body and with middle constrictor of other side of the body]	**Constrictor pharyngis medius** (ceratopharyngeus and/or hyopharyngeus *sensu* House [1953])	**Constrictor pharyngis medius**	**Constrictor pharyngis medius**	**Constrictor pharyngis medius** [including the pars ceratopharyngea and the pars chondropharyngea *sensu Terminologia Anatomica* [1998], which insert onto the hyoid bone and onto the thyroid cartilage, respectively]

(Continued)

TABLE 8.5 (CONTINUED)
Branchial, Pharyngeal, and Laryngeal Muscles of Adults of Representative Sarcopterygian Taxa

Actinistia: *Latimeria chalumnae* (Coelacanth)	Dipnoi: *Lepidosiren paradoxa* (South American Lungfish)	Amphibia: *Ambystoma ordinarium* (Michoacan Stream Salamander)	Reptilia: *Timon lepidus* (Ocellated Lizard)	Mammalia (Monotremata): *Ornithorhynchus anatinus* (Platypus)	Mammalia (Marsupialia): *Didelphis virginiana* (Virginian Opossum)	Mammalia (Rodentia): *Rattus norvegicus* (Norwegian Rat)	Mammalia (Scandentia): *Tupaia* sp. (Tree Shrew)	Mammalia (Dermoptera): *Cynocephalus volans* (Philippine Colugo)	Mammalia (Primates): *Homo sapiens* (Modern Human)
–	–	–	–	–	**Constrictor pharyngis inferior** [from pharyngeal raphe to cricothyroid cartilage, being blended with inferior constrictor of other side of the body]	**Constrictor pharyngis inferior** [as described by Saban [1968], the constrictor pharyngis inferior of therian mammals is often divided into a pars thryropharyngea attaching on the thyroid cartilage, a pars cricopharyngea attaching on the cricoid cartilage, and a pars intermedia lying between these two myological structures; the pars intermedia is often reduced in mammals as, e.g., primates and is often absent in mammals as, e.g., rodents]	**Constrictor pharyngis inferior** [see on the left]	**Constrictor pharyngis inferior** [see on the left]	**Constrictor pharyngis inferior** [see on the left]

(*Continued*)

TABLE 8.5 (CONTINUED)
Branchial, Pharyngeal, and Laryngeal Muscles of Adults of Representative Sarcopterygian Taxa

Actinistia: *Latimeria chalumnae* (Coelacanth)	Dipnoi: *Lepidosiren paradoxa* (South American Lungfish)	Amphibia: *Ambystoma ordinarium* (Michoacan Stream Salamander)	Reptilia: *Timon lepidus* (Ocellated Lizard)	Mammalia (Monotremata): *Ornithorhynchus anatinus* (Platypus)	Mammalia (Marsupialia): *Didelphis virginiana* (Virginian Opossum)	Mammalia (Rodentia): *Rattus norvegicus* (Norwegian Rat)	Mammalia (Scandentia): *Tupaia* sp. (Tree Shrew)	Mammalia (Dermoptera): *Cynocephalus volans* (Philippine Colugo)	Mammalia (Primates): *Homo sapiens* (Modern Human)
–	–	–	–	**Cricothyroideus** [in the platypus specimens we dissected, the cricothyroideus is seemingly not divided into a pars obliqua and a pars recta; in terms of both its ontogeny and phylogeny, the mammalian cricothyroideus is clearly a pharyngeal muscle and not a laryngeal muscle as it is sometimes suggested in the literature: e.g., Edgeworth 1935; Negus 1949; DuBrul 1958; Starck and Schneider 1960; Saban 1968; Wind 1970; Crelin 1987; Harrison 1995; this work]	– [the cricothyroideus is secondarily absent in adult marsupials (present in at least some marsupial embryos), probably as the result of the fusion between the cricoid and thyroid cartilages: see, e.g., Jouffroy and Saban 1971]	**Cricothyroideus** [including a pars obliqua and a pars recta: see on the right]	**Cricothyroideus** [including a pars obliqua and a pars recta: see on the left]	**Cricothyroideus** [including a pars obliqua and a pars recta, which are more separated than in *Rattus* and *Tupaia* but are not as separated as in modern humans]	**Cricothyroideus** [including a pars obliqua and a pars recta: see on the left]

(Continued)

TABLE 8.5 (CONTINUED)
Branchial, Pharyngeal, and Laryngeal Muscles of Adults of Representative Sarcopterygian Taxa

Actinistia: *Latimeria chalumnae* (Coelacanth)	Dipnoi: *Lepidosiren paradoxa* (South American Lungfish)	Amphibia: *Ambystoma ordinarium* (Michoacan Stream Salamander)	Reptilia: *Timon lepidus* (Ocellated Lizard)	Mammalia (Monotremata): *Ornithorhynchus anatinus* (Platypus)	Mammalia (Marsupialia): *Didelphis virginiana* (Virginian Opossum)	Mammalia (Rodentia): *Rattus norvegicus* (Norwegian Rat)	Mammalia (Scandentia): *Tupaia* sp. (Tree Shrew)	Mammalia (Dermoptera): *Cynocephalus volans* (Philippine Colugo)	Mammalia (Primates): *Homo sapiens* (Modern Human)
–	–	–	–	– [according to Edgeworth [1935], the constrictor pharyngis superior is absent in monotremes and was very likely poorly developed in the first placentals, being probably similar to the "glossopharyngeus" of, e.g., rats (see on the right); according to that author the constrictor pharyngis superior, only became a broad muscle as that found in, e.g., modern humans later in evolution]	– [the constrictor pharyngis superior is not present as a well-developed, well-differentiated circular constrictor muscle in marsupials; Maier et al.'s [1996] study of *Monodelphis* has shown that fibers of the palatopharyngeus and of the pterygopharyngeus are somewhat differentiated into parts that seemingly correspond to the superior constrictor of placentals (e.g., the part of the *Didelphis* musculature named "superior constrictor" in Jouffroy and Saban's Figure 300), but there is no true, separated superior constrictor muscle with a median raphe as is the case in placentals; instead, in marsupials the "pars pharyngeal" of the stylopharyngeus and parts of the palatopharyngeus and	**Constrictor pharyngis superior** (glossopharyngeus *sensu* House [1953]) [the constrictor pharyngis superior of rats seemingly includes only a pars glossopharyngea: see on the left]	**Constrictor pharyngis superior** [seemingly includes a pars buccopharyngea, a pars pterygopharyngea (corresponding to the pterygopharyngeus of, e.g., rats and colugos? see on the left), and possibly a pars glossopharyngea: e.g., Sprague 1944a; this work]	**Constrictor pharyngis superior** [our dissections indicate that it includes a pars glossopharyngea and possibly a pars buccopharyngea]	**Constrictor pharyngis superior** [includes a pars buccopharyngea, a pars pterygopharyngea (corresponding to the pterygopharyngeus of, e.g., rats and colugos? see on the left), a pars mylopharyngea, and a pars glossopharyngea]

(Continued)

TABLE 8.5 (CONTINUED)
Branchial, Pharyngeal, and Laryngeal Muscles of Adults of Representative Sarcopterygian Taxa

Actinistia: *Latimeria chalumnae* (Coelacanth)	Dipnoi: *Lepidosiren paradoxa* (South American Lungfish)	Amphibia: *Ambystoma ordinarium* (Michoacan Stream Salamander)	Reptilia: *Timon lepidus* (Ocellated Lizard)	Mammalia (Monotremata): *Ornithorhynchus anatinus* (Platypus)	Mammalia (Marsupialia): *Didelphis virginiana* (Virginian Opossum)	Mammalia (Rodentia): *Rattus norvegicus* (Norwegian Rat)	Mammalia (Scandentia): *Tupaia* sp. (Tree Shrew)	Mammalia (Dermoptera): *Cynocephalus volans* (Philippine Colugo)	Mammalia (Primates): *Homo sapiens* (Modern Human)
					pterygopharyngeus seem to somewhat fullfil the function of the placental superior constrictor; studies such as that of Maier et at. [1996] thus also support the idea that the superior constrictor, palatopharyngeus, and pterygopharyngeus are developmentally closely related; House [1953] and Smith [1992] also suggested that the pterygopharyngeus of, e.g., rats probably correspond to part of the constrictor pharyngis superior of modern humans; recent molecular developmental studies support the idea that the superior constrictor is developmentally closely related to the palatopharyngeus: see, e.g. Grimaldi et al. 2015]				

(Continued)

TABLE 8.5 (CONTINUED)
Branchial, Pharyngeal, and Laryngeal Muscles of Adults of Representative Sarcopterygian Taxa

Actinistia: *Latimeria chalumnae* (Coelacanth)	Dipnoi: *Lepidosiren paradoxa* (South American Lungfish)	Amphibia: *Ambystoma ordinarium* (Michoacan Stream Salamander)	Reptilia: *Timon lepidus* (Ocellated Lizard)	Mammalia (Monotremata): *Ornithorhynchus anatinus* (Platypus)	Mammalia (Marsupialia): *Didelphis virginiana* (Virginian Opossum)	Mammalia (Rodentia): *Rattus norvegicus* (Norwegian Rat)	Mammalia (Scandentia): *Tupaia* sp. (Tree Shrew)	Mammalia (Dermoptera): *Cynocephalus volans* (Philippine Colugo)	Mammalia (Primates): *Homo sapiens* (Modern Human)
–	–	–	–	–	**Palatoglossus** [present in *Didelphis* according to Jouffroy and Saban [1971], running from the soft palate to the tongue, where it blends with the hyoglossus]	– (seems to be absent as an independent muscle in the rats dissected; this is supported by authors such as Barrow and Capecchi [1999])	– [seems to be absent as a separate muscle in the *Tupaia* specimens we dissected; it is also not described by authors such as Le Gros Clark [1924, 1926]]	**Palatoglossus** [seemingly present as an independent muscle in the colugos we dissected, being formed by a group of fibers running from the soft palate and/or the lateral wall of the oropharynx to the posterolateral surface of the tongue]	**Palatoglossus**
–	–	–	–	–	**Pterygopharyngeus** [from pterygoid region of presphenoid and from region of tympanic tube to pharyngeal musculature, being particularly blended with the palatopharyngeus; see superior constrictor above]	**Pterygopharyngeus** [see above]	– [see above]	**Pterygopharyngeus** [see above]	– [see above]

(Continued)

TABLE 8.5 (CONTINUED)
Branchial, Pharyngeal, and Laryngeal Muscles of Adults of Representative Sarcopterygian Taxa

Actinistia: *Latimeria chalumnae* (Coelacanth)	Dipnoi: *Lepidosiren paradoxa* (South American Lungfish)	Amphibia: *Ambystoma ordinarium* (Michoacan Stream Salamander)	Reptilia: *Timon lepidus* (Ocellated Lizard)	Mammalia (Monotremata): *Ornithorhynchus anatinus* (Platypus)	Mammalia (Marsupialia): *Didelphis virginiana* (Virginian Opossum)	Mammalia (Rodentia): *Rattus norvegicus* (Norwegian Rat)	Mammalia (Scandentia): *Tupaia* sp. (Tree Shrew)	Mammalia (Dermoptera): *Cynocephalus volans* (Philippine Colugo)	Mammalia (Primates): *Homo sapiens* (Modern Human)
–	–	–	–	**Palatopharyngeus**	**Palatopharyngeus** [from pterygoid region of presphenoid and soft palate to pharyngeal musculature; see superior constrictor above]	**Palatopharyngeus** [more blended with the salpingopharyngeus than in modern humans]	**Palatopharyngeus** [more blended with the salpingopharyngeus than in modern humans]	**Palatopharyngeus** [more blended with the salpingopharyngeus than in modern humans]	**Palatopharyngeus**
–	–	–	–	–	**Musculus uvuale** (medialis veli palatini *sensu* Jouffroy and Saban [1971]) [lying on inferomedial region of soft palate]	**Musculus uvulae** [the "medialis veli palatini" is clearly present in the mice we dissected and is thus almost surely present in rats also, as it is in most mammals]	–	–	**Musculus uvulae** [according to Edgeworth [1935], this muscle is only found in a few mammals such as primates, corresponding to part of the palatopharyngeus of other mammals]

(Continued)

TABLE 8.5 (CONTINUED)
Branchial, Pharyngeal, and Laryngeal Muscles of Adults of Representative Sarcopterygian Taxa

Actinistia: *Latimeria chalumnae* (Coelacanth)	Dipnoi: *Lepidosiren paradoxa* (South American Lungfish)	Amphibia: *Ambystoma ordinarium* (Michoacan Stream Salamander)	Reptilia: *Timon lepidus* (Ocellated Lizard)	Mammalia (Monotremata): *Ornithorhynchus anatinus* (Platypus)	Mammalia (Marsupialia): *Didelphis virginiana* (Virginian Opossum)	Mammalia (Rodentia): *Rattus norvegicus* (Norwegian Rat)	Mammalia (Scandentia): *Tupaia* sp. (Tree Shrew)	Mammalia (Dermoptera): *Cynocephalus volans* (Philippine Colugo)	Mammalia (Primates): *Homo sapiens* (Modern Human)
–	–	–	–	–	– [the levator veli palatini is not present as a well-developed, well-differentiated muscle in marsupials; Maier et al.'s 1996 study of *Monodelphis* has shown that fibers of the palatopharyngeus/ pterygopharyngeus are somewhat differentiated into what seems to correspond to a very poorly differentiated levator veli palatini, but there is no true, separated muscle as that seen in placentals; studies such as that by Maier et at. [1996] thus also support the idea that the superior constrictor, palatopharyngeus, pterygopharyngeus, and levator veli palatini are developmentally closely related (but see text about the fact that some authors defend a possible origin of the levator veli palatini from the second; i.e., hyoid, arch)]	**Levator veli palatini** [corresponds to a part of the palatopharyngeus of monotremes: e.g., Edgeworth 1935; Saban 1968; this work; in mouse, it extends from palate to petrous part of temporal bone and cartilage of tympanic tube: Grimaldi et al. 2015]	**Levator veli palatini**	**Levator veli palatini**	**Levator veli palatini**

(*Continued*)

TABLE 8.5 (CONTINUED)
Branchial, Pharyngeal, and Laryngeal Muscles of Adults of Representative Sarcopterygian Taxa

Actinistia: *Latimeria chalumnae* (Coelacanth)	Dipnoi: *Lepidosiren paradoxa* (South American Lungfish)	Amphibia: *Ambystoma ordinarium* (Michoacan Stream Salamander)	Reptilia: *Timon lepidus* (Ocellated Lizard)	Mammalia (Monotremata): *Ornithorhynchus anatinus* (Platypus)	Mammalia (Marsupialia): *Didelphis virginiana* (Virginian Opossum)	Mammalia (Rodentia): *Rattus norvegicus* (Norwegian Rat)	Mammalia (Scandentia): *Tupaia* sp. (Tree Shrew)	Mammalia (Dermoptera): *Cynocephalus volans* (Philippine Colugo)	Mammalia (Primates): *Homo sapiens* (Modern Human)
–	–	–	–	–	– [seemingly not differentiated in *D. virginiana* nor in other marsupials: see, e.g., Jouffroy and Saban 1971; Maier et al. 1998]	**Salpingopharyngeus** [corresponds to a part of the palatopharyngeus of monotremes: e.g., Edgeworth 1935; Saban 1968; this work; in mouse, it is present but deeply blended with palatopharyngeus: Grimaldi et al. 2015]	**Salpingopharyngeus**	**Salpingopharyngeus**	**Salpingopharyngeus**

(*Continued*)

TABLE 8.5 (CONTINUED)
Branchial, Pharyngeal, and Laryngeal Muscles of Adults of Representative Sarcopterygian Taxa

Actinistia: *Latimeria chalumnae* (Coelacanth)	Dipnoi: *Lepidosiren paradoxa* (South American Lungfish)	Amphibia: *Ambystoma ordinarium* (Michoacan Stream Salamander)	Reptilia: *Timon lepidus* (Ocellated Lizard)	Mammalia (Monotremata): *Ornithorhynchus anatinus* (Platypus)	Mammalia (Marsupialia): *Didelphis virginiana* (Virginian Opossum)	Mammalia (Rodentia): *Rattus norvegicus* (Norwegian Rat)	Mammalia (Scandentia): *Tupaia* sp. (Tree Shrew)	Mammalia (Dermoptera): *Cynocephalus volans* (Philippine Colugo)	Mammalia (Primates): *Homo sapiens* (Modern Human)
Present? [Some nonsarcopterygian vertebrates as, e.g., *Polypterus* have a "constrictor laryngis" and/or a "dilatator laryngis," but it is not clear if these muscles correspond to the constrictor laryngis and dilatator laryngis of sarcopterygians and thus if the latter muscles are plesiomorphically present in osteichthyans: e.g., Edgeworth 1935; the few descriptions of the laryngeal region of *Latimeria chalumnae* do not allow us to appropriately discern if these muscles are, or are not, present in this taxon: e.g., Millot and Anthony 1958]	**Constrictor laryngis**	**Constrictor laryngis** [see on the right]	**Constrictor laryngis** [see on the right]	– [recent developmental works indicate that in amphibians as, e.g., salamanders and reptiles as, e.g., chickens laryngeal muscles such as the dilatator laryngis are at least partially derived ontogenetically from somites and possibly from branchial mesoderm: e.g., Piekarski and Olsson 2007; the ontogenetic derivation of these muscles is thus similar to that of muscles such as the protractor pectoralis of amphibians and the trapezius/sternocleido-mastoideus of reptiles: see text; according to Piekarski and Olsson [2007], in some cases, the constrictor oesophagus might also be at least partially derived ontogenetically from somites: see text]	–	–	–	–	–

(*Continued*)

TABLE 8.5 (CONTINUED)
Branchial, Pharyngeal, and Laryngeal Muscles of Adults of Representative Sarcopterygian Taxa

Actinistia: *Latimeria chalumnae* (Coelacanth)	Dipnoi: *Lepidosiren paradoxa* (South American Lungfish)	Amphibia: *Ambystoma ordinarium* (Michoacan Stream Salamander)	Reptilia: *Timon lepidus* (Ocellated Lizard)	Mammalia (Monotremata): *Ornithorhynchus anatinus* (Platypus)	Mammalia (Marsupialia): *Didelphis virginiana* (Virginian Opossum)	Mammalia (Rodentia): *Rattus norvegicus* (Norwegian Rat)	Mammalia (Scandentia): *Tupaia* sp. (Tree Shrew)	Mammalia (Dermoptera): *Cynocephalus volans* (Philippine Colugo)	Mammalia (Primates): *Homo sapiens* (Modern Human)
– [the laryngeus of tetrapods does not seem to be plesiomorphically found in sarcopterygians, because it is absent in sarcopterygian fishes as dipnoans; a detailed study of the laryngeal region of *Latimeria* is however needed in order to support, or to contradict, this hypothesis: see preceding text]	– [see on the left]	**Laryngeus** [the laryngeus and constrictor laryngis of amphibians derive ontogenetically from the same anlage: e.g., Edgeworth 1935]	– [see on the right]	**Thyrocricoary-tenoideus** (the thyrocricoary-tenoideus *sensu* Saban [1968] corresponds to the thyroarytenoideus *sensu* Edgeworth [1935]; it has two bundles, which seemingly correspond to the thyroarytenoideus and cricoarytenoideus lateralis of other mammals: that is why we prefer to use the name thyrocricoary-tenoideus for the monotreme muscle) [Smith 1992: 340, states that "the laryngeal muscles of mammals and amphibians are innervated by two homologous branches of cranial nerve X, the superior and inferior (or recurrent) laryngeal nerves; in contrast in reptiles (except in Aves) the innervation of the larynx is *via* a single laryngeal nerve that is a branch of cranial nerve IX"; this supports	**Thyroarytenoideus** [from unpaired cricothyroid cartilage (resulting from fusion of cricoid and thyroid cartilages) to arytenoid cartilage of the respective side]	**Thyroarytenoideus** [mainly divided into superficial and deep bundles]	**Thyroarytenoideus** [mainly divided into superficial and deep bundles]	**Thyroarytenoideus** [divided into a posterior, medial bundle and an anterior, lateral part, which seem to correspond, respectively, to the pars intermedia and pars superioris of Figure 69 of Starck and Schneider [1960]; the latter bundle is in turn subdivided into a medial bundle and a lateral bundle, the latter being fused with the cricoarytenoideus posterior and thus seemingly corresponding to the ceratoarytenoideus lateralis *sensu* Harrison [1995]]	**Thyroarytenoideus** [often includes a pars thyroepiglottica, a pars aryepiglottica, a pars superioris, a pars ventricularis and/or a ceratoarytenoideus lateralis *sensu* Saban [1968] and Harrison [1995]]

(*Continued*)

TABLE 8.5 (CONTINUED)
Branchial, Pharyngeal, and Laryngeal Muscles of Adults of Representative Sarcopterygian Taxa

Actinistia: *Latimeria chalumnae* (Coelacanth)	Dipnoi: *Lepidosiren paradoxa* (South American Lungfish)	Amphibia: *Ambystoma ordinarium* (Michoacan Stream Salamander)	Reptilia: *Timon lepidus* (Ocellated Lizard)	Mammalia (Monotremata): *Ornithorhynchus anatinus* (Platypus)	Mammalia (Marsupialia): *Didelphis virginiana* (Virginian Opossum)	Mammalia (Rodentia): *Rattus norvegicus* (Norwegian Rat)	Mammalia (Scandentia): *Tupaia* sp. (Tree Shrew)	Mammalia (Dermoptera): *Cynocephalus volans* (Philippine Colugo)	Mammalia (Primates): *Homo sapiens* (Modern Human)
				Edgeworth's [1935] view that the "laryngei" of reptiles is not homologous to the laryngeus of amphibians and thus to the thyrocricoarytenoideus + arytenoideus of monotremes: Table 8.6]					
–	–	–	–	–	–	–	–	–	**Vocalis** (thyroarytenoideus inferior *sensu* Saban 1968) [according to, e.g., Edgeworth 1935 the vocalis is only found in a few taxa such as some primates, and corresponds to the medial portion of the thyroarytenoideus of other mammals]

(Continued)

TABLE 8.5 (CONTINUED)
Branchial, Pharyngeal, and Laryngeal Muscles of Adults of Representative Sarcopterygian Taxa

Actinistia: *Latimeria chalumnae* (Coelacanth)	Dipnoi: *Lepidosiren paradoxa* (South American Lungfish)	Amphibia: *Ambystoma ordinarium* (Michoacan Stream Salamander)	Reptilia: *Timon lepidus* (Ocellated Lizard)	Mammalia (Monotremata): *Ornithorhynchus anatinus* (Platypus)	Mammalia (Marsupialia): *Didelphis virginiana* (Virginian Opossum)	Mammalia (Rodentia): *Rattus norvegicus* (Norwegian Rat)	Mammalia (Scandentia): *Tupaia* sp. (Tree Shrew)	Mammalia (Dermoptera): *Cynocephalus volans* (Philippine Colugo)	Mammalia (Primates): *Homo sapiens* (Modern Human)
–	–	–	–	–	–	**Cricoarytenoideus alaris** [described in rats by Inagi et al. [1998], left and right muscles cricoid cartilage to left and right arytenoid cartilages, respectively, extending to single laryngeal alar cartilage, which is not present in humans; they say the muscle appears to function in a protective role in that it is capable of closing the laryngeal opening and depressing the epiglottis, as does the aryepiglottic part of the thyroarytenoideus of humans, although it might well also/instead maintain an open laryngeal vestibule for respiration, as this might be important because the rat laryngeal vestibule is otherwise wide but lax; because of these presumed functions and the fact it is a longitudinal muscle, this muscle seems more likely to be derived from the thyroarytenoideus than from the cricoarytenoideus posterior]	–	–	–

(Continued)

TABLE 8.5 (CONTINUED)
Branchial, Pharyngeal, and Laryngeal Muscles of Adults of Representative Sarcopterygian Taxa

Actinistia: *Latimeria chalumnae* (Coelacanth)	Dipnoi: *Lepidosiren paradoxa* (South American Lungfish)	Amphibia: *Ambystoma ordinarium* (Michoacan Stream Salamander)	Reptilia: *Timon lepidus* (Ocellated Lizard)	Mammalia (Monotremata): *Ornithorhynchus anatinus* (Platypus)	Mammalia (Marsupialia): *Didelphis virginiana* (Virginian Opossum)	Mammalia (Rodentia): *Rattus norvegicus* (Norwegian Rat)	Mammalia (Scandentia): *Tupaia* sp. (Tree Shrew)	Mammalia (Dermoptera): *Cynocephalus volans* (Philippine Colugo)	Mammalia (Primates): *Homo sapiens* (Modern Human)
–	–	–	–	–	**Cricoarytenoideus lateralis** [from unpaired cricothyroid cartilage to arytenoid cartilage of the respective side, being fused with unpaired arytenoideus muscle and with thyroarytenoideus muscle of the same side of the body]	**Cricoarytenoideus lateralis** (cricoarytenoideus ventralis *sensu* Whidden [2000]) [see thyrocricoary-tenoideus above]	**Cricoarytenoideus lateralis**	**Cricoarytenoideus lateralis**	**Cricoarytenoideus lateralis**
–	–	–	–	**Arytenoideus** (interarytenoideus *sensu* Saban [1971] which is divided into crico-pro-arytenoideus and ary-pro-arytenoideus)	**Arytenoideus** [unpaired muscle attached to the left and right arytenoid cartilages and to the unpaired interarytenoid cartilage; as noted by authors such as Symington [1898] and Jouffroy and Saban [1971], the portion going to the interarytenoid cartilage (often also designated procricoid cartilage) is sometimes named "ary-procricoideus"]	**Arytenoideus** (interarytenoideus *sensu* Edgeworth 1935) [Inagi et al. [1998] described in rats a "cricoarytenoideus superioris," the left and right muscles running from the midline tubercle of the cricoid cartilage to the left and right arytenoid cartilages, respectively; but in their discussion, they recognize that the muscle seems to be "analogous" to arytenoideus of humans because it seemingly adducts the arytenoid cartilages and seems to correspond to arytenoideus of mammals in general, which we did find in mice and rats, so probably their left and right cricoarytenoideus superioris correspond to the arytenoideus *sensu* the present work]	**Arytenoideus**	**Arytenoideus**	**Arytenoideus transversus**

(*Continued*)

TABLE 8.5 (CONTINUED)
Branchial, Pharyngeal, and Laryngeal Muscles of Adults of Representative Sarcopterygian Taxa

Actinistia: *Latimeria chalumnae* (Coelacanth)	Dipnoi: *Lepidosiren paradoxa* (South American Lungfish)	Amphibia: *Ambystoma ordinarium* (Michoacan Stream Salamander)	Reptilia: *Timon lepidus* (Ocellated Lizard)	Mammalia (Monotremata): *Ornithorhynchus anatinus* (Platypus)	Mammalia (Marsupialia): *Didelphis virginiana* (Virginian Opossum)	Mammalia (Rodentia): *Rattus norvegicus* (Norwegian Rat)	Mammalia (Scandentia): *Tupaia* sp. (Tree Shrew)	Mammalia (Dermoptera): *Cynocephalus volans* (Philippine Colugo)	Mammalia (Primates): *Homo sapiens* (Modern Human)
–	–	–	–	–	–	–	–	–	**Arytenoideus obliquus** [it is commonly accepted that the arytenoideus transversus and arytenoideus obliquus derive from the arytenoideus of other mammals; however, as the pars aryepiglottica is seemingly derived from the thyroarytenoideus, the possibility that the arytenoideus obliquus also derives from the latter muscle, and not from the arytenoideus, cannot be ruled out: e.g., Saban 1968; Table 8.6]

(Continued)

TABLE 8.5 (CONTINUED)
Branchial, Pharyngeal, and Laryngeal Muscles of Adults of Representative Sarcopterygian Taxa

Actinistia: *Latimeria chalumnae* (Coelacanth)	Dipnoi: *Lepidosiren paradoxa* (South American Lungfish)	Amphibia: *Ambystoma ordinarium* (Michoacan Stream Salamander)	Reptilia: *Timon lepidus* (Ocellated Lizard)	Mammalia (Monotremata): *Ornithorhynchus anatinus* (Platypus)	Mammalia (Marsupialia): *Didelphis virginiana* (Virginian Opossum)	Mammalia (Rodentia): *Rattus norvegicus* (Norwegian Rat)	Mammalia (Scandentia): *Tupaia* sp. (Tree Shrew)	Mammalia (Dermoptera): *Cynocephalus volans* (Philippine Colugo)	Mammalia (Primates): *Homo sapiens* (Modern Human)
Present? [See constrictor laryngis above]	**Dilatator laryngis** [see on the right]	**Dilatator laryngis** [see on the right]	**Dilatator laryngis** [see on the right]	**Cricoarytenoideus posterior** (cerato-crico-arytenoideus *sensu* Saban [1971]) [the cricoarytenoideus posterior of mammals corresponds to the dilatator laryngis of reptiles, amphibians, and dipnoans, which is not homologous to the dilatator laryngis of the actinopterygians *Amia* and *Lepisosteus* nor to the dilatator laryngis of the actinopterygian *Polypterus*, according to Edgeworth [1935]]	**Cricoarytenoideus posterior** (cricoarytenoideus posticus internus + cerato-crico-arytenoideus posticus *sensu* Symington [1898]) [from unpaired cricothyroid cartilage to arytenoid cartilage of the respective side and to unpaired interarytenoid cartilage, its medial margin lying near the midline but not reaching its counterpart medially; as noted by Jouffroy and Saban [1971], in monotremes and marsupials, this muscle is often divided into two bundles, which are often seen as separate muscles; the division of this muscle into bundles was probably present in the last	**Cricoarytenoideus posterior** (cricoarytenoideus dorsalis *sensu* Whidden [2000])	**Cricoarytenoideus posterior**	**Cricoarytenoideus posterior**	**Cricoarytenoideus posterior** [including the ceratocricoideus *sensu* Harrison [1995], which, according to this author, is found in about 63% of modern humans]

(Continued)

TABLE 8.5 (CONTINUED)
Branchial, Pharyngeal, and Laryngeal Muscles of Adults of Representative Sarcopterygian Taxa

Actinistia: *Latimeria chalumnae* (Coelacanth)	Dipnoi: *Lepidosiren paradoxa* (South American Lungfish)	Amphibia: *Ambystoma ordinarium* (Michoacan Stream Salamander)	Reptilia: *Timon lepidus* (Ocellated Lizard)	Mammalia (Monotremata): *Ornithorhynchus anatinus* (Platypus)	Mammalia (Marsupialia): *Didelphis virginiana* (Virginian Opossum)	Mammalia (Rodentia): *Rattus norvegicus* (Norwegian Rat)	Mammalia (Scandentia): *Tupaia* sp. (Tree Shrew)	Mammalia (Dermoptera): *Cynocephalus volans* (Philippine Colugo)	Mammalia (Primates): *Homo sapiens* (Modern Human)
					common ancestor of monotremes and therians, because the two bundles seen in marsupials might also be present in some placentals; as described by Symington [1898], in *D. virginiana*, the two bundles are the "cricoarytenoideus posticus internus," which attaches onto the arytenoid cartilages and the interarytenoid cartilage, and the cerato-crico-arytenoideus posticus, which attaches onto the arytenoid cartilages]				

Note: See footnotes of Tables 8.1 and 8.6.

TABLE 8.6

Diagram Illustrating the Authors' Hypotheses Regarding the Evolution and Homologies of the Branchial, Pharyngeal, and Laryngeal Muscles of Adults of Representative Sarcopterygian Taxa

		Latimeria (4 m.-not st.*)	*Lepidosiren* (3 mus.-not st.*)	*Ambystoma* (4 m.-not str.*)	*Timon* (6 m. total)	*Ornithorhynchus* (14 m. total)	*LCA marsupials + placentals* (19 m. total)	*Didelphis* (17 m. total)	*Rattus* (21 m. total)	*Tupaia* (17 m. total)	*Cynocephalus* (19 m. total)	*Homo* (17 m. total)
'TRUE' BRANCHIAL MUS.	STRICTO*	Fu.m.br.ap.*	Fu.m.br.ap.*	Fu.m.br.ap.	--- (ab. as a group)	--- (ab. as group)	--- (ab. as a group)	--- (ab. as a group)	--- (ab. as a group)	--- (ab. as a group)	--- (abs. as a group)	--- (abs. as a group)
		---	---	---	Hyobranchialis	Stylopharyngeus	Stylopharyngeus	Stylopharyngeus	Stylopharyngeus	Stylopharyngeus	Stylopharyngeus	Stylopharyngeus
		---	---	---	'Ceratohyoideus'	Ceratohyoideus	Ceratohyoideus	Ceratohyoideus	Ceratohyoideus	Ceratohyoideus	Ceratohyoideus	---
		---	---	---	---	Subarcualis rectus III	---	---	---	---	---	---
	OTHER	---	Pro. pectoralis	Pro. pectoralis	Trapezius	Acromiotrapezius	Acromiotrapezius	Acromiotrapezius	Acromiotrapezius	Trapezius	Acromiotrapezius	Trapezius
		---	---	---	---	Spinotrapezius	Spinotrapezius	Spinotrapezius	Spinotrapezius	---	Spinotrapezius	---
		---	---	---	---	Dorsocutaneous	Dorsocutaneous	--- (psom.)	--- (psop.)	---	---	---
		---	---	---	---	---	Cleido -occipitalis	Cleido -occipitalis	Cleido -occipitalis	Cleido -occipitalis	---	---
		---	---	---	Sternocleidomastoideus	Cleidomastoideus	Cleidomastoideus	Cleidomastoideus	Cleidomastoideus	Cleidomastoideus	Cleidomastoideus	Sternocleidomastoideus
		---	---	---	---	Sternomastoideus	Sternomastoideus	Sternomastoideus	Sternomastoideus	Sternomastoideus	Sternomastoideus	---
PHARYNGEAL MUS.		---	---	---	---	Co. pharyngis	Co.pharyngis medius	Co.pharyngis medius	Co.pharyngis medius	Co. pharyngis medius	Co. pharyngis medius	Co. pharyngis medius
		----	---	---	---	---	Co.pharyngis inferior	Co.pharyngis inferior	Co.pharyngis inferior	Co. pharyngis inferior	Co. pharyngis inferior	Co. pharyngis inferior
		----	---	---	---	Cricothyroideus	Cricothyroideus	---(lost in marsupials)	Cricothyroideus	Cricothyroideus	Cricothyroideus	Cricothyroideus
		---	---	---	---	---	---	---	Co. pharyngis superior	Co. pharyngis superior	Co. pharyngis sup.	Co. pharyngis superior
		---	---	---	---	---	Palatoglossus	Palatoglossus	---	---	Palatoglossus	Palatoglossus
		----	---	---	---	---	Pterygopharyngeus	Pterygopharyngeus	Pterygopharyngeus	---	Pterygopharyngeus	---
		----	---	---	---	Palatopharyngeus	Palatopharyngeus	Palatopharyngeus	Palatopharyngeus	Palatopharyngeus	Palatopharyngeus	Palatopharyngeus
		---	---	---	---	---	Musculus uvulae	Musculus uvulae	Musculus uvulae	--- ? (really absent?)	--- ? (really absent?)	Musculus uvulae
		----	---	---	---	---	---	---	Le. veli palatini	Le. veli palatini	Le. veli palatini	Le. veli palatini
		----	---	---	---	---	---	---	Salpingopharyngeus	Salpingopharyngeus	Salpingopharyngeus	Salpingopharyngeus
LARYNGEAL MUS.		---? (see DA)	Co. laryngis	Co. laryngis	Co. laryngis	---	---	---	---	---	---	---
		---	---	Laryngeus	---	Thyrocricoarytenoideus	Thyroarytenoideus	Thyroarytenoideus	Thyroarytenoideus	Thyroarytenoideus	Thyroarytenoideus	Thyroarytenoideus (+vocalis)
		---	---	---	---	---	Cricoarytenoideus lat.	Cricoarytenoideus lat.	Cricoarytenoideus lat.	Cricoarytenoideus lat.	Cricoarytenoideus lat.	Cricoarytenoideus lat.
		---	---	---	---	Arytenoideus	Arytenoideus	Arytenoideus	Arytenoideus	Arytenoideus	Arytenoideus	Arytenoideus transversus
		---	---	---	---	---	---	---	---	---	---	Arytenoideus obliquus
		---	---	---	---	---	---	---	Cricoarytenoideus alaris	---	---	---
		---? (see DA)	Dilator laryngis	Dilator laryngis	Dilator laryngis	Cricoarytenoideus post.	Cricoarytenoideus post.	Cricoarytenoideus post.	Cricoarytenoideus post.	Cricoarytenoideus post.	Cricoarytenoideus post.	Cricoarytenoideus post.

Note: See footnotes of Tables 8.2 and 8.5. ab., absent; ap., apparatus; br., branchial; co., constrictor; fu., functional; le., levator; m., muscles; post., posterior; pro., protractor; psom., present some other marsupials; psop., present some other placentals; see DA, see Diogo and Abdala (2010); st., *sensu stricto*; sup., superior.

TABLE 8.7
Hypobranchial Muscles of Adults of Representative Sarcopterygian Taxa

Actinistia: *Latimeria chalumnae* (Coelacanth)	Dipnoi: *Lepidosiren paradoxa* (South American Lungfish)	Amphibia: *Ambystoma ordinarium* (Michoacan Stream Salamander)	Reptilia: *Timon Lepidus* (Ocellated Lizard)	Mammalia (Monotremata): *Ornithorhynchus anatinus* (Platypus)	Mammalia (Marsupialia): *Didelphis virginiana* (Virginian Opossum)	Mammalia (Rodentia): *Rattus norvegicus* (Norwegian Rat)	Mammalia (Scandentia): *Tupaia* sp. (Tree Shrew)	Mammalia (Dermoptera): *Cynocephalus volans* (Philippine Colugo)	Mammalia (Primates): *Homo sapiens* (Modern Human)
Coracomandibularis	**Coracomandibularis** (geniothoracis *sensu* Bemis and Lauder [1986])	**Geniohyoideus** (coracomandibularis *sensu* Diogo [2007, 2008b])	**Geniohyoideus** (coracomandibularis *sensu* Diogo [2007, 2008b])	**Geniohyoideus** [according to Edgeworth [1935], the geniohyoideus, genioglossus, and hyoglossus of mammals develop ontogenetically at the same time; according to him, the two former muscles develop internally (more medially), while the latter develops externally (more laterally)]	**Geniohyoideus** [from mandible to unpaired basihyal, lying nearby the geniohyoideus of the other side of the body, but not being fused/blended with it]	**Geniohyoideus**	**Geniohyoideus**	**Geniohyoideus**	**Geniohyoideus**
–	–	**Genioglossus** [as explained by, e.g., Edgeworth [1935] and Piatt [1938], the genioglossus of salamanders such as *Ambystoma* and of reptiles such as lizards corresponds to part of the coracomandibularis of sarcopterygian fishes]	**Genioglossus** [see on the left]	**Genioglossus** [as explained by, e.g., Piekarski and Olsson [2007], in mammals such as dogs, the tongue muscles are sometimes innervated by both the hypoglossal (CNXII) and the facial (CNVII) nerves, thus indicating that at least in some cases these muscles may have a dual origin]	**Genioglossus** [from mandible to unpaired basihyal and to tongue, being blended to the hyoglossus of the same side of the body and with the genioglossus of the other side of the body]	**Genioglossus** [not described by, e.g., Greene [1935], but it is present as an independent structure in the rats we dissected; authors such as Bryant [1945] support the claim that this muscle is often present in rodents]	**Genioglossus** (geniohyoglossus *sensu* Le Gros Clark [1924, 1926])	**Genioglossus**	**Genioglossus**

(*Continued*)

TABLE 8.7 (CONTINUED)
Hypobranchial Muscles of Adults of Representative Sarcopterygian Taxa

Actinistia: *Latimeria chalumnae* (Coelacanth)	Dipnoi: *Lepidosiren paradoxa* (South American Lungfish)	Amphibia: *Ambystoma ordinarium* (Michoacan Stream Salamander)	Reptilia: *Timon Lepidus* (Ocellated Lizard)	Mammalia (Monotremata): *Ornithorhynchus anatinus* (Platypus)	Mammalia (Marsupialia): *Didelphis virginiana* (Virginian Opossum)	Mammalia (Rodentia): *Rattus norvegicus* (Norwegian Rat)	Mammalia (Scandentia): *Tupaia* sp. (Tree Shrew)	Mammalia (Dermoptera): *Cynocephalus volans* (Philippine Colugo)	Mammalia (Primates): *Homo sapiens* (Modern Human)
–	–	– [according to, e.g., Piatt [1938] and Saban [1968, 1971], extant amphibians such as salamanders do not have well-developed, well-differentiated intrinsic muscles of the tongue like those found in extant amniotes]	**Intrinsic muscles of the tongue** [according to Saban [1971], Jouffroy and Lessertisseur [1971], Smith [1988, 1992], Meyers and Nishikawa [2000], and Herrel et al. [2005], examples of reptilian intrinsic tongue are the "longitudinalis," the "transversus linguae ventralis," the "transversus linguae dorsalis," the "verticalis linguae," and the "annularis": see on the right and the following chapters]	**Intrinsic muscles of the tongue** [according to Saban [1968], the intrinsic muscles of the tongue of amniotes derive from both the genioglossus and hyoglossus; examples of these muscles are the longitudinalis superior, longitudinalis inferior, transversus linguae, and/or verticalis linguae: e.g., Anderson 1881; Edgeworth 1935; Jarvik 1963; Saban 1968, 1971; Smith 1988, 1992; Sokoloff 2000; Herrel et al. 2005; this work; according to Herrel (Anthony Herrel, pers. comm.), the longitudinalis, transversus linguae, verticalis linguae, and annularis of reptiles are very likely homologous to the longitudinalis, transversus linguae, verticalis linguae, and annularis of mammals, respectively: see on the left and the following chapters]	**Intrinsic muscles of the tongue** [see on the left]	**Intrinsic muscles of the tongue** [see on the left]	**Intrinsic muscles of the tongue** [see on the left]	**Intrinsic muscles of the tongue** [see on the left]	**Intrinsic muscles of the tongue** [see on the left]

(*Continued*)

TABLE 8.7 (CONTINUED)
Hypobranchial Muscles of Adults of Representative Sarcopterygian Taxa

Actinistia: *Latimeria chalumnae* (Coelacanth)	Dipnoi: *Lepidosiren paradoxa* (South American Lungfish)	Amphibia: *Ambystoma ordinarium* (Michoacan Stream Salamander)	Reptilia: *Timon Lepidus* (Ocellated Lizard)	Mammalia (Monotremata): *Ornithorhynchus anatinus* (Platypus)	Mammalia (Marsupialia): *Didelphis virginiana* (Virginian Opossum)	Mammalia (Rodentia): *Rattus norvegicus* (Norwegian Rat)	Mammalia (Scandentia): *Tupaia* sp. (Tree Shrew)	Mammalia (Dermoptera): *Cynocephalus volans* (Philippine Colugo)	Mammalia (Primates): *Homo sapiens* (Modern Human)
–	–	**Hyoglossus** [the statements of Edgeworth [1935] concerning the hyoglossus of salamanders such as *Ambystoma* are confusing: On page 196 of his work, he states that it derives from the sternohyoideus (= his rectus cervicus) but on page 211, he suggests that as in other amphibians, as well as in reptiles and mammals, it derives from the coracomandibularis (= his geniohyoideus); the results of the developmental work of Piatt [1938] support the latter hypothesis]	**Hyoglossus** [according to Edgeworth [1935], the hyoglossus of lizards such as *Timon* corresponds to part of the coracomandibularis of other amphibians and sarcopterygian fishes]	**Hyoglossus**	**Hyoglossus** [from hyoid apparatus—from basihyal and seemingly also from ceratohyal—to tongue]	**Hyoglossus**	**Hyoglossus** (hyoglossus + chondroglossus *sensu* Le Gros Clark [1926]) [according to Saban [1968], in primates and tree shrews, the hyoglossus is divided into a chondroglossus and a ceratoglossus; this is supported by, e.g., Le Gros Clark [1926] and Sprague [1944a], although the latter author erroneously states that the chondroglossus is part of the genioglossus and not of the hyoglossus]	**Hyoglossus** [as described by, e.g., Edgeworth [1935], in colugos, the hyoglossus and thyrohyoideus are seemingly fused]	**Hyoglossus** [see on the left]

(*Continued*)

TABLE 8.7 (CONTINUED)
Hypobranchial Muscles of Adults of Representative Sarcopterygian Taxa

Actinistia: *Latimeria chalumnae* (Coelacanth)	Dipnoi: *Lepidosiren paradoxa* (South American Lungfish)	Amphibia: *Ambystoma ordinarium* (Michoacan Stream Salamander)	Reptilia: *Timon Lepidus* (Ocellated Lizard)	Mammalia (Monotremata): *Ornithorhynchus anatinus* (Platypus)	Mammalia (Marsupialia): *Didelphis virginiana* (Virginian Opossum)	Mammalia (Rodentia): *Rattus norvegicus* (Norwegian Rat)	Mammalia (Scandentia): *Tupaia* sp. (Tree Shrew)	Mammalia (Dermoptera): *Cynocephalus volans* (Philippine Colugo)	Mammalia (Primates): *Homo sapiens* (Modern Human)
–	–	–	–	–	**Styloglossus** [from paraoccipital process of occipital bone to tongue, being deeply blended with hyoglossus; Edgeworth [1935] and Jouffroy and Saban [1971] suggest that marsupials such as *Didelphis* have a "primitive hyoglossus" that is not separated into the hyoglossus and styloglossus muscles found in most placental mammals; however, Smith 1994 refers to a styloglossus in *Monodelphis*, and this muscle was seemingly present in our *D. virginiana* specimens, being however blended with the hyoglossus; Osgood [1921] also described a styloglossus in other marsupials]	**Styloglossus** [our dissections and comparisons support Edgeworth's [1935] suggestion that the styloglossus of therian mammals likely correspond to a part of the hyoglossus of monotremes]	**Styloglossus**	**Styloglossus**	**Styloglossus**

(*Continued*)

TABLE 8.7 (CONTINUED)
Hypobranchial Muscles of Adults of Representative Sarcopterygian Taxa

Actinistia: *Latimeria chalumnae* (Coelacanth)	Dipnoi: *Lepidosiren paradoxa* (South American Lungfish)	Amphibia: *Ambystoma ordinarium* (Michoacan Stream Salamander)	Reptilia: *Timon Lepidus* (Ocellated Lizard)	Mammalia (Monotremata): *Ornithorhynchus anatinus* (Platypus)	Mammalia (Marsupialia): *Didelphis virginiana* (Virginian Opossum)	Mammalia (Rodentia): *Rattus norvegicus* (Norwegian Rat)	Mammalia (Scandentia): *Tupaia* sp. (Tree Shrew)	Mammalia (Dermoptera): *Cynocephalus volans* (Philippine Colugo)	Mammalia (Primates): *Homo sapiens* (Modern Human)
–	–	**Interradialis** [according to Piatt [1938], in at least some adult *Ambystoma*, there is a hypobranchial muscle interradialis, which derives ontogenetically from the genioglossus]	–	–	–	–	–	–	–
Sternohyoideus	**Sternohyoideus** (rectus cervicus *sensu* Bemis and Lauder [1986])	**Sternohyoideus** (rectus cervicus *sensu* Lauder and Shaffer [1988])	**Sternohyoideus** (rectus cervicus *sensu* Kardong 2002; episternohyoideus *sensu* Edgeworth [1935])	**Sternohyoideus** [in the platypus specimens examined by us this muscle is deeply blended posteriorly with the sternothryroideus, as stated by Saban [1971]]	**Sternohyoideus** [from sternum to hyoid apparatus—seemingly only basihyal—being blended with the sternothyroideus of the same side of the body and with the sternohyoideus of the other side of the body; the sternohyoideus has no clear intersections]	**Sternohyoideus**	**Sternohyoideus** [deeply blended with the sternothyroideus]	**Sternohyoideus** [as described by, e.g., Leche [1886] and Saban [1968], in *Cynocephalus*, the sternohyoideus has two bundles: the posterior one extends anteriorly in order to reach the posterior region of the thyroid cartilage and then contacts, *via* a broad but thin tendon, the anterior one that extends anteriorly to attach to the lesser cornu of the hyoid]	**Sternohyoideus**
–	–	**Omohyoideus** [the omohyoideus, sternothyroideus, and thyrohyoideus of tetrapods clearly correspond to a part of the sternohyoideus of sarcopterygian fishes: e.g., Edgeworth 1935; Saban 1968, 1971; Diogo 2007, 2008b; this work]	**Omohyoideus**	**Omohyoideus** [as stated by, e.g., Saban [1971] in the platypus specimens we dissected, the omohyoideus is anteriorly divided into superficial and deep bundles, running from the hyoid cartilage to the scapula]	**Omohyoideus** [from scapula to hyoid apparatus—seemingly only basihyal; it has no defined intermediate tendon, having just a short tendinous intersection]	**Omohyoideus**	**Omohyoideus** [George [1977] states that this muscle has no distinct tendinous intersection, but Le Gros Clark [1924, 1926] and Sprague [1944a] describe such an intersection in tree shrews as in, e.g., *Tupaia* and *Ptilocercus*]	– [not present as an independent structure in the colugos we dissected as well as by authors such as Gunnell and Simmons [2005]]	**Omohyoideus** [it has superior and inferior bellies, which are separated by a distinct tendon]

(Continued)

TABLE 8.7 (CONTINUED)
Hypobranchial Muscles of Adults of Representative Sarcopterygian Taxa

Actinistia: *Latimeria chalumnae* (Coelacanth)	Dipnoi: *Lepidosiren paradoxa* (South American Lungfish)	Amphibia: *Ambystoma ordinarium* (Michoacan Stream Salamander)	Reptilia: *Timon Lepidus* (Ocellated Lizard)	Mammalia (Monotremata): *Ornithorhynchus anatinus* (Platypus)	Mammalia (Marsupialia): *Didelphis virginiana* (Virginian Opossum)	Mammalia (Rodentia): *Rattus norvegicus* (Norwegian Rat)	Mammalia (Scandentia): *Tupaia* sp. (Tree Shrew)	Mammalia (Dermoptera): *Cynocephalus volans* (Philippine Colugo)	Mammalia (Primates): *Homo sapiens* (Modern Human)
–	–	–	– [some authors have described a sternothyroideus in a few reptilian taxa, but this muscle is probably not homologous to the mammalian sternothyroideus: see the following chapters]	**Sternothyroideus** [see above]	**Sternothyroideus** [from sternum to crico-thyroid cartilage, being blended with the sternothyroideus of the other side of the body; the sternothyroideus has no clear intersection]	**Sternothyroideus**	**Sternothyroideus**	**Sternothyroideus**	**Sternothyroideus**
–	–	–	–	–	**Thyrohyoideus** [from crico-thyroid cartilage to hyoid apparatus—basihyal and seemingly thyrohyal—being blended with the sternothyroideus; the anterior margin of the thyrohyoideus is more anterior than the posterior margin of the sternothyroideus]	**Thyrohyoideus** [as explained earlier, the thyrohyoideus of therian mammals clearly corresponds to a part of the sternohyoideus of nonmammalian tetrapods; however, it is not clear if it corresponds to part of the monotreme sternohyoideus or, instead, to part of the monotreme sternothyroideus: e.g., Edgeworth 1935; Saban 1968]	**Thyrohyoideus**	– [not present as an independent structure in the colugos we dissected; it is seemingly fused with the hyoglossus: see above]	**Thyrohyoideus**

Note: See footnote of Table 8.1 and see Table 8.8.

TABLE 8.8
Diagram Illustrating the Authors' Hypotheses Regarding the Evolution and Homologies of the Hypobranchial Muscles of Adults of Representative Sarcopterygian Taxa

	Latimeria (2 m.)	*Lepidosiren* (2 m.)	*Ambystoma* (6 m.- not in. to.*)	*Timon* (5 m.- not in. to.*)	*Ornithorhynchus* (6 m.- not in. to.*)	*LCA marsupials + placentals* (8 m.- not in. to.*)	*Didelphis* (8 m.- not in. to.*)	*Rattus* (8 m.- not in. to.*)	*Tupaia* (8 m.- not in. to.*)	*Cynocephalus* (6 m.- not in. to.*)	*Homo* (8 m.- not in. to.*)
'GENIO-HYOIDEUS'	Coracomandibularis	Coracomandibularis	Geniohyoideus	Geniohyoideus	Geniohyoideus	Geniohyoideus	Geniohyoideus	Geniohyoideus	Geniohyoideus	Geniohyoideus	Geniohyoideus
	---	---	Genioglossus	Genioglossus	Genioglossus	Genioglossus	Genioglossus	Genioglossus	Genioglossus	Genioglossus	Genioglossus
	---	---	---	In. mus. tongue*	In. mus. tongue*	In. mus. tongue*	In. mus. tongue*	In. mus. tongue*	In. mus. tongue*	In. mus. tongue*	In. mus. tongue*
	---	---	Hyoglossus	Hyoglossus	Hyoglossus	Hyoglossus	Hyoglossus	Hyoglossus	Hyoglossus	Hyoglossus	Hyoglossus
	---	---	---	---	---	Styloglossus	Styloglossus	Styloglossus	Styloglossus	Styloglossus	Styloglossus
	---	---	Interradialis	---	---	---	---	---	---	---	---
'RECTUS CERVICIS'	Sternohyoideus	Sternohyoideus	Sternohyoideus	Sternohyoideus	Sternohyoideus	Sternohyoideus	Sternohyoideus	Sternohyoideus	Sternohyoideus	Sternohyoideus	Sternohyoideus
	---	---	Omohyoideus	Omohyoideus	Omohyoideus	Omohyoideus	Omohyoideus	Omohyoideus	Omohyoideus	---	Omohyoideus
	---	---	---	---	Sternothryroideus	Sternothryroideus	Sternothryroideus	Sternothryroideus	Sternothryroideus	Sternothryroideus	Sternothryroideus
	---	---	---	---	---	Thyrohyoideus	Thyrohyoideus	Thyrohyoideus	Thyrohyoideus	---	Thyrohyoideus

Note: See footnotes of Table 8.2 and Table 8.7. Geniohyoideus, Rectus Cervicis mean geniohyoideus and rectus cervicus groups *sensu* Edgeworth (1935). in. to., intrinsic muscles of the tongue; m., muscles.

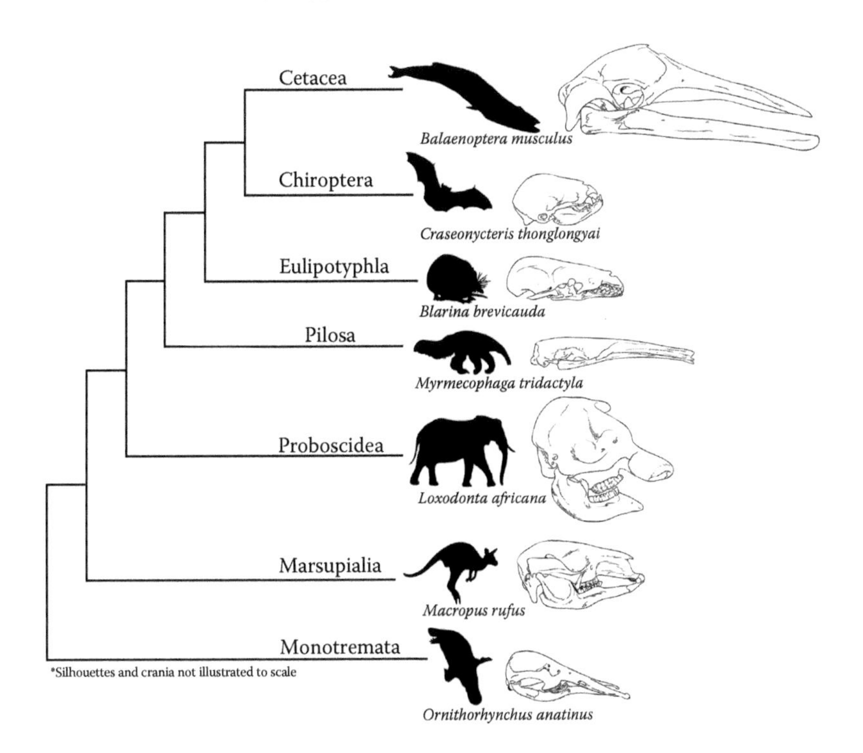

FIGURE 8.7 Examples of mammalian cranio-mandibular diversity using a simplified con*sensu*s phylogeny: blue whale (*Balaenoptera musculus*), bumblebee bat (*Craseonycteris thonglongyai*), Northern short-tailed shrew (*Blarina brevicauda*), giant anteater (*Myrmecophaga tridactyla*), African elephant (*Loxodonta africana*), red kangaroo (*Macropus rufus*), and platypus (*Ornithorhynchus anatinus*). Mammals occupy disparate subterranean, arboreal, and marine habitats, which are related to remarkable size differences among the smallest (i.e., the bumblebee bat and short-tailed shrews) and the largest taxa (i.e., the blue whale). Unique dietary specializations are also reflected in dentognathic adaptations such as the edentulous maxillae and mandibles of blue whales and anteaters that consume exceedingly small prey (e.g., krill and termites, respectively) and the duck-like electroreceptive rostrum of platypus used in foraging.

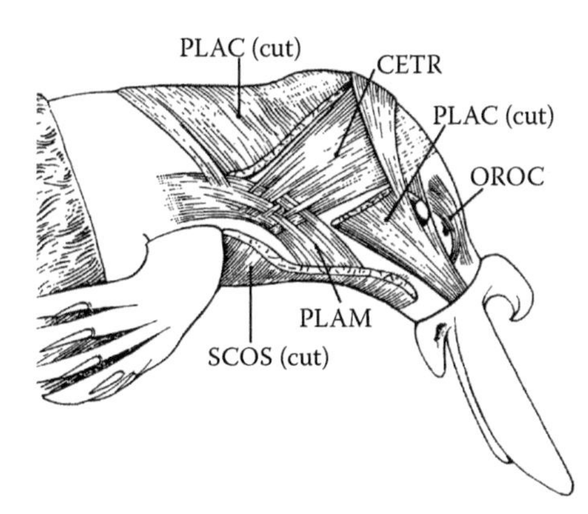

FIGURE 8.8 *Ornithorhynchus anatinus* (Mammalia, Monotremata): lateral view of the deep facial musculature; muscles such as the interhyoideus profundus, buccinatorius, orbicularis oris, and mentalis are not shown (the nomenclature of the structures illustrated basically follows that used in the present work; anterior is to the right). CETR, cervicalis transversus; OROC, orbicularis oculi; PLAC, PLAM, platysma cervicale and platysma myoides; SCOS, sphincter colli superficialis. (Modified from Lightoller, G., *J Anat*, 76, 258–269, 1942; Saban, R., *Traité de Zoologie*, XVI: 3 (Mammifères), pp. 681–732, Masson et Cie, Paris, 1971.)

(e.g., Adams 1919; Brock 1938; Goodrich 1958; Barghusen 1986; Smith 1992, 1994; Witmer 1995a,b; Tables 8.1 and 8.2). Even so, the idea that these two muscles derive instead from the A2-PVM cannot be completely discarded (Table 8.2).

The derivation of these two muscles is related to the fascinating origin of parts of the mammalian inner ear from the jaws of their nonmammalian ancestors. Well-supported theories such as those of Gaupp (1912) and Reichert (1937) demonstrated *via* comparative anatomical and developmental evidence that the middle ear bones (i.e., malleus, incus, gonial, and tympanic ring) of modern mammals evolved from repurposed accessory jawbones (i.e., articular, quadrate, prearticular, and angular, respectively) of their nonmammalian tetrapod ancestors, which are maintained in extant reptilian and avian relatives (Anthwal et al. 2013). Fossils show that the mammalian exaptations of ancestral bony anatomy detectable in *Morganucodon oehleri* were preceded by topological and mechanical alterations to preexisting masticatory and pharyngeal musculature that began within the Eucynodontia (Crompton 1963; Compton and Parker 1978; Lautenschlager et al. 2016). The evolution of the auditory system thus relied not only on osteological reorganization,

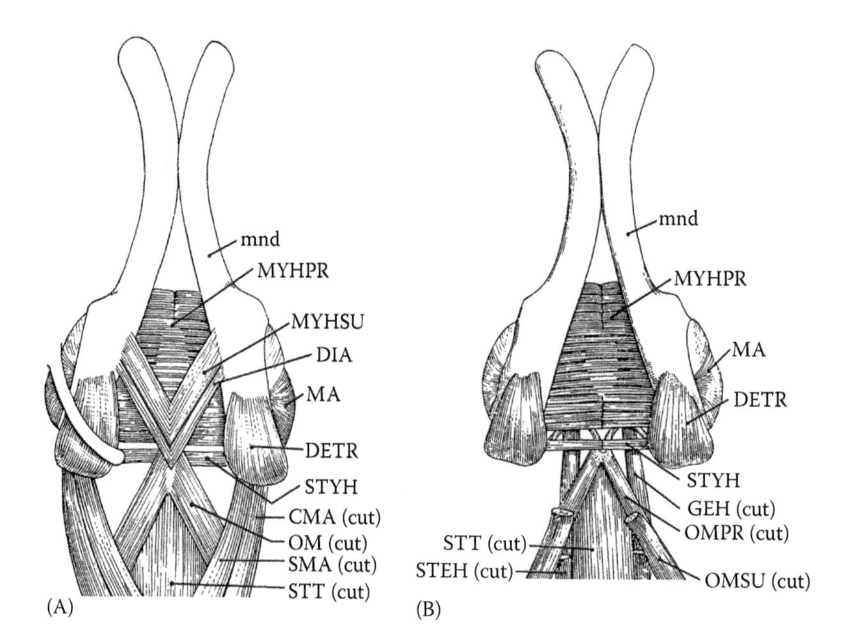

FIGURE 8.9 *Ornithorhynchus anatinus* (Mammalia, Monotremata): (A) ventral view of the head and neck musculature, muscles such as the geniohyoideus and sternohyoideus are not shown; (B) same view, but the digastricus anterior, superficial part of the mylohyoideus, sternomastoideus, and cleidomastoideus were removed and the anterior portion of the sternohyoideus and of the superficial part of the omohyoideus were partially cut (the nomenclature of the structures illustrated basically follows that used in the present work; anterior is to the top). CMA, cleidomastoideus; DETR, detrahens mandibulae; DIA, digastricus anterior; GEH, geniohyoideus; MA, masseter; mnd, mandible; MYHPR, MYHSU, pars profunda and pars superficialis of mylohyoideus; OM, omohyoideus; OMPR, OMSU, pars profunda and pars superficialis of omohyoideus; SMA, sternomastoideus; STEH, sternohyoideus; STT, sternothyroideus; STYH, styloideus. (Modified from Edgeworth, F. H., *The Cranial Muscles of Vertebrates*, Cambridge University Press, Cambridge, UK, 1935; Saban, R., *Traité de Zoologie*, XVI: 3 (Mammifères), pp. 681–732, Masson et Cie, Paris, 1971.)

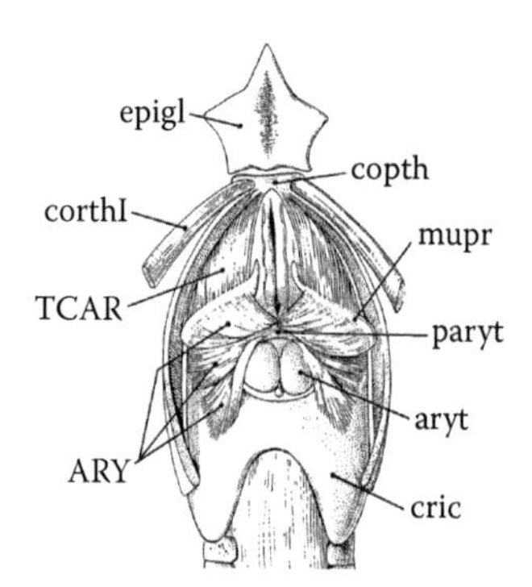

FIGURE 8.10 *Ornithorhynchus anatinus* (Mammalia, Monotremata): dorsal view of the laryngeal musculature; the cricoarytenoideus posterior is not shown (the nomenclature of the structures illustrated basically follows that used in the present work; anterior is to the top). ARY, arytenoideus; aryt, arytenoid cartilage; copth, copula thyroidea; corthI, cornua thyroidea I; cric, cricoid cartilage; epigl, epiglottis; mupr, muscular process; paryt, proarytenoid cartilage; TCAR, thyrocricoarytenoideus. (Modified from Göppert, E., *Handbuch der Vergleichenden, Anatomie der Wirbeltiere*, pp. 797–866, Urban and Schwarzenberg, Berlin, 1937; Saban, R., *Traité de Zoologie*, XVI: 3 (Mammifères), pp. 681–732, Masson et Cie, Paris, 1971.)

but also on myological differentiation of the pterygomandibularis portion of the adductor mandibulae.

The main arguments supporting the differentiation of the tensor tympani and tensor veli palatini from the pterygomandibularis have been clearly summarized in works such as by Barghusen (1986). Developmental data also support this hypothesis. For instance, in Smith's (1994) detailed work on the development of the craniofacial musculature of marsupials, she found that the pterygoideus medialis, the tensor tympani, and the tensor veli palatini of these mammals develop ontogenetically from the same medial anlage, which seems to correspond to the anlage that forms the pterygomandibularis + pseudotemporalis in reptiles (while the pterygoideus lateralis derives instead from the anlage that gives rise to the temporalis). Interestingly, in Figure 3B of Smith's (1994) paper, there appears to be a thin, small muscle connecting the malleus and the incus, which might possibly be a "remnant of the PVM" (Peter Johnson, pers. comm.), but the latter hypothesis clearly needs to be investigated in much more detail. One of the main arguments that authors such as Saban (1971) provided in favor of a derivation of the mammalian tensor veli palatini and tensor tympani from the A2-PVM was that Edgeworth (1935) stated that his development work has clearly shown that the two former muscles were derived ontogenetically from the "levator mandibulae posterior," the name often used in the literature to designate the A2-PVM *sensu* the present work (see Tables 8.1 and 8.2). However, there is much confusion in the literature about the identity and homologies of the components of the "adductor mandibulae complex" of tetrapods and particularly of the structures that are often named "adductor mandibulae posterior" in different nonmammalian tetrapod clades. It is thus possible that in this specific case, Edgeworth (1935) used the name levator

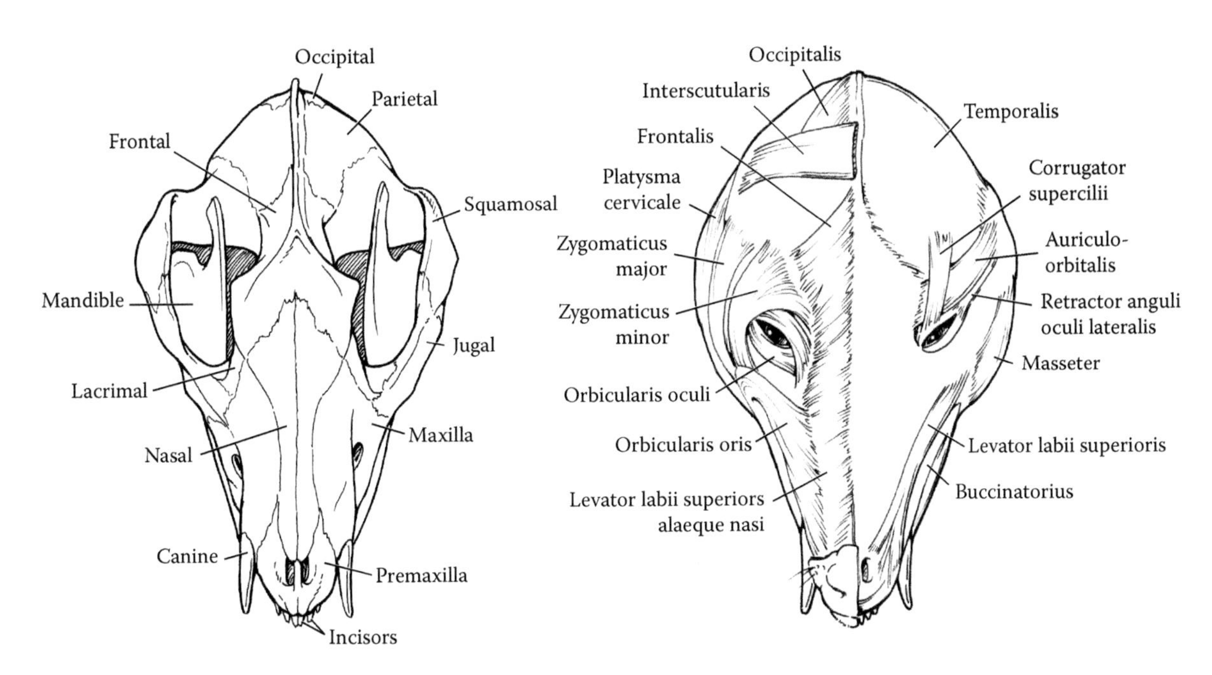

FIGURE 8.11 *Didelphis virginiana* (Mammalia, Marsupialia): dorsofrontal view of head skeleton and muscles (schematic drawing by J. Molnar). (Modified from Diogo, R. et al., *Anat Record*, 299, 1224–1255, 2016.)

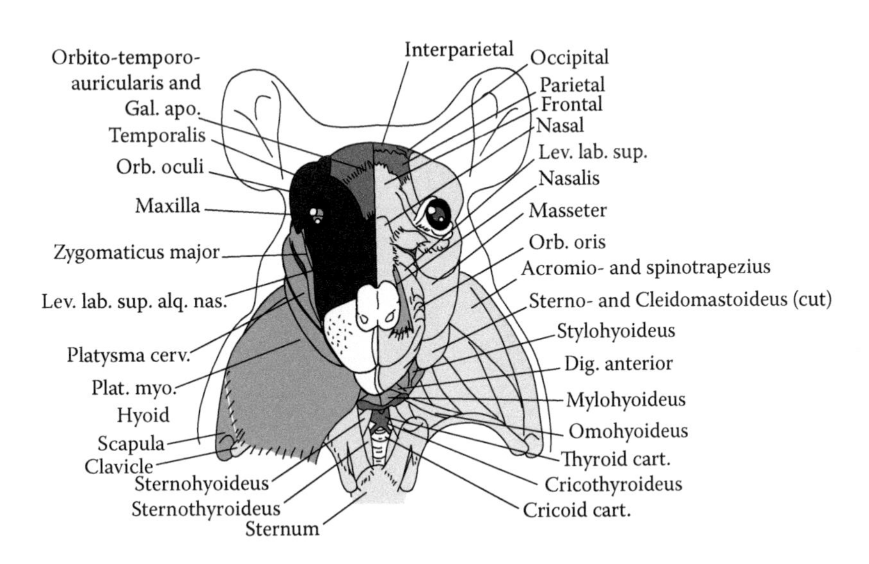

FIGURE 8.12 *Mus musculus* (Mammalia, Rodential): frontal view of head muscles (schematic drawing by V. Powell, the different tone refer to the inclusion of the muscles in different head anatomical network modules, and are not relevant for the present book). (Modified from Powell V. et al., *Nat Sci Rep*, 2018.) cart., cartilage; cerv., cervicalis; dig., digastricus; gal. apo, galea aponeurotica; lev. lab. sup., levator labii superioris; lev. lab. sup. alq. nas., levator labii superioris alaeque nasi; orb., orbicularis; plat. myo., platysma myoides.

mandibulae posterior to designate the pterygomandibularis (and not the A2-PVM) *sensu* the present work, as suggested by authors such as Goodrich (1958). Goodrich (1958) stated that a correct interpretation of Edgeworth's data support the idea that the tensor tympani and tensor veli palatini are derived from the pterygomandibularis because those data show that the mammalian tensor tympani, tensor veli palatini, and pterygoideus medialis derive from the same anlage, as was found in more recent works such as Smith (1994) (see preceding text and Tables 8.1 and 8.2). Rodríguez-Vázquez et al. (2016) recently showed that in early human development, the tensor tympani and tensor veli palatini are (a) tendinously connected to each other, forming a single digastric muscle complex and (b) are closely related to the pterygoideus medialis, as described in developmental works about other mammalian taxa.

The mammalian pterygoideus medialis (Figure 8.14) may not only derive from the pseudotemporalis, but also from part of the pterygomandibularis, of nonmammalian tetrapods (Figure 8.5; Table 8.2). Furthermore, extant mammals lack any dorsal mandibular muscles, *sensu* Edgeworth (1935), which is uncommon among vertebrates (e.g., Saban 1968, 1971;

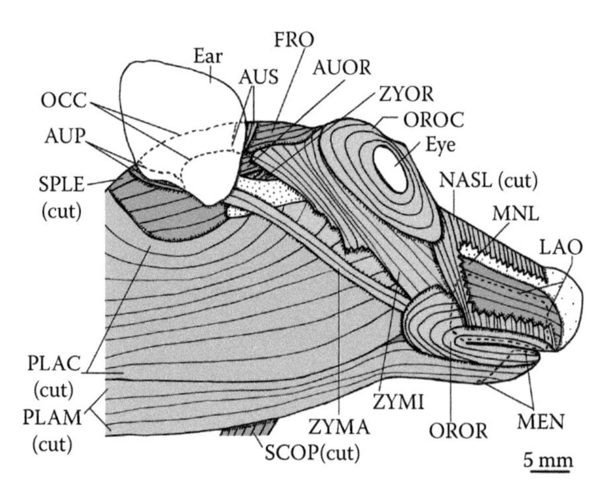

FIGURE 8.13 *Cynocephalus volans* (Mammalia, Dermoptera): lateral view of the facial muscles; the splenius capitis is also shown; anteriorly, the nasolabialis was partially cut in order to show the maxillo-naso-labialis (anterior is to the right; muscles shown in darker gray are deeper than [medial to] those shown in lighter gray). AUOR, auriculo-orbitalis; AUP, auricularis posterior; AUS; auricularis superior; FRO, frontalis; LAO, levator anguli oris facialis; MEN, mentalis; MNL, maxillo-naso-labialis; NASL, nasolabialis; OCC, occipitalis; OROC, orbicularis occuli; OROR, orbicularis oris; PLAC, platysma cervicale; PLAM, platysma myoides; SCOP, sphincter colli profundus; SPLE, splenius capitis; ZYMA, zygomaticus major; ZYMI, zygomaticus minor; ZYOR, zygomatico-orbitalis.

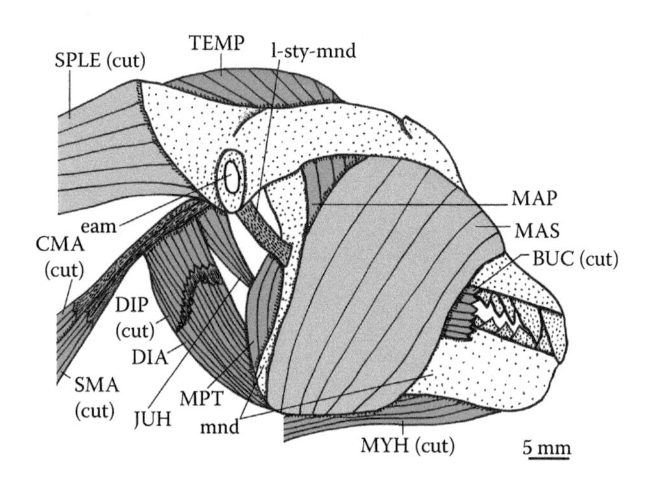

FIGURE 8.14 *Cynocephalus volans* (Mammalia, Dermoptera): postero-ventro-lateral view of the masseter, buccinatorius, pterygoideus medialis, temporalis, digastricus anterior, digastricus posterior, jugulohyoideus, sternomastoideus, cleidomastoideus, mylohyoideus, and splenius capitis (anterior is to the right; muscles shown in darker gray are deeper than [medial to] those shown in lighter gray). BUC, buccinatorius; CMA, cleidomastoideus; DIA, DIP, digastricus anterior and digastricus posterior; eam, external auditory meatus; JUH, jugulohyoideus; l-sty-mnd, stylomandibular ligament; MAP, MAS, pars profunda and pars superficialis of masseter; mnd, mandible; MPT, pterygoideus medialis; MYH, mylohyoideus; SMA, sternomastoideus; SPLE, splenius capitis; TEMP, temporalis.

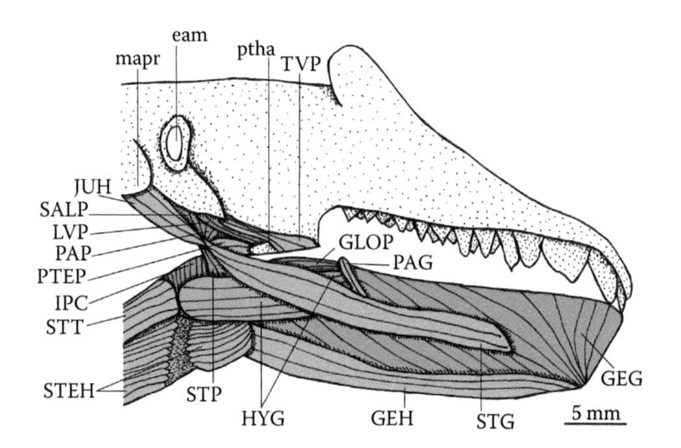

FIGURE 8.15 *Cynocephalus volans* (Mammalia, Dermoptera): ventrolateral view of the genioglossus, geniohyoid, styloglossus, hyoglossus, sternothyroid, sternohyoid, palatoglossus, tensor veli palatini, levator veli palatini, pterygopharyngeus, salpingopharyngeus, stylopharyngeus, jugulohyoideus, and constrictor pharyngis inferior; the mandible, zygomaticus arch, and part of the orbit were removed (anterior is to the right; muscles shown in darker gray are deeper than [medial to] those shown in lighter gray). eam, external auditory meatus; GEG, genioglossus; GEH, geniohyoideus; GLOP, glossopharyngeus; HYG, hyoglossus; IPC, constrictor pharyngis inferior; JUH, jugulohyoideus; LVP, levator veli palatini; mapr, mastoid process; PAG, palatoglossus; PAP, palatopharyngeus; PTEP, pterygopharyngeus; ptha, pterygoid hamulus; SALP, salpingopharyngeus; STEH, sternohyoideus; STG, styloglossus; STP, stylopharyngeus; STT, sternothyroideus; TVP, tensor veli palatini.

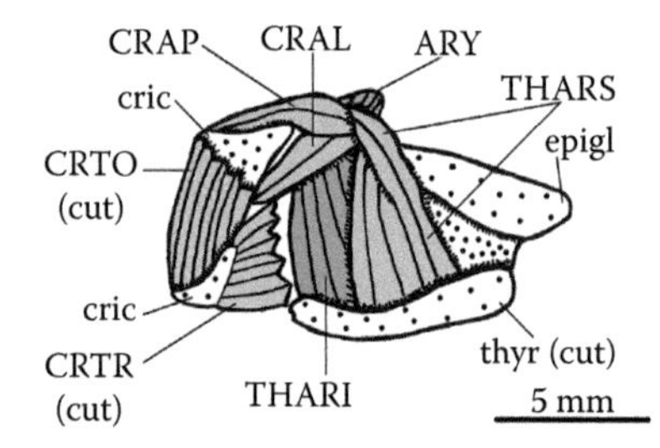

FIGURE 8.16 *Cynocephalus volans* (Mammalia, Dermoptera): lateral view of the laryngeal muscles and of the pharyngeal muscle cricothyroideus; the latter muscle and the lateral surface of the thyroid cartilage were partially cut in order to show the deeper (more medial) muscles (anterior is to the left, dorsal to the top; muscles shown in darker grey are deeper than [medial to] those shown in lighter grey). ARY, arytenoideus; CRAL, cricoarytenoideus lateralis; CRAP, cricoarytenoideus posterior; cric, cricoid cartilage; CRTO, CRTR, pars obliqua and pars recta of cricothyroideus; epigl, epiglottis; THARI, THARS, pars intermedia and pars superioris of thyroarytenoideus; thyr, thyroid cartilage.

Kardong 2002; this work). Interestingly, tetrapod mandibular muscles (derived from the adductor mandibulae plate *sensu* Edgeworth, such as the masseter, temporalis, pterygoideus medialis, and pterygoideus lateralis) of mice display engrailed immunoreactivity (Knight et al. 2008). Teleost fishes (e.g., the zebrafish) also exhibit engrailed immunoreactivity that has only been detected in dorsal mandibular muscles (i.e., in the

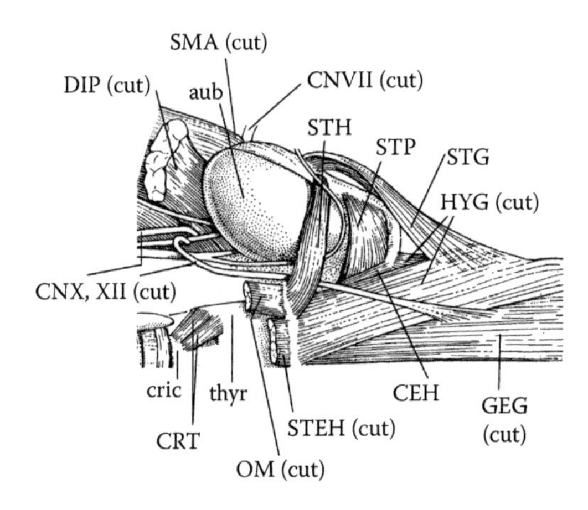

FIGURE 8.17 *Ptilocercus lowii* (Mammalia, Scandentia): ventral view of the musculature of the hyoid region of the right side of the body; muscles such as the geniohyoideus, sternothyroideus, and thyrohyoideus are not shown (the nomenclature of the structures illustrated basically follows that used in the present work; anterior is to the right). aub, auditory bulla; CEH, ceratohyoideus; CNVII,X,XII, cranial nerves VII, X and XII; cric, cricoid cartilage; CRT, cricothyroideus; DIP, digastricus posterior; GEG, genioglossus; HYG, hyoglossus; OM, omohyoideus; SMA, sternomastoideus; STEH, sternohyoideus; STG, styloglossus; STH, stylohyoideus; STP, stylopharyngeus; thyr, thyroid cartilage. (Modified from Le Gros Clark, W. E., *Proc Zool Soc Lond*, 1926, pp. 1178–1309, 1926; Saban, R., *Traité de Zoologie*, XVI: 3 (Mammifères), pp. 229–472, Masson et Cie, Paris, 1968.)

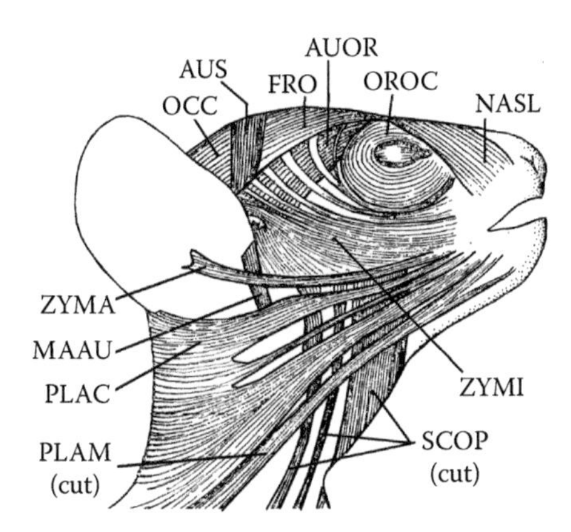

FIGURE 8.18 *Lepilemur* sp. (Mammalia, Primates): lateral view of the facial musculature; muscles such as the buccinatorius, orbicularis oris and mentalis are not shown (the nomenclature of the structures illustrated basically follows that used in the present work; anterior is to the right). AUOR, auriculo-orbitalis; AUS, auricularis superior; FRO, frontalis; MAAU, mandibulo-auricularis; NASL, nasolabialis; OCC, occipitalis; OROC, orbicularis oculi; PLAC, PLAM, platysma cervicale and platysma myoides; SCOP, sphincter colli profundus; ZYMA, ZYMI, zygomaticus major and zygomaticus minor. (Modified from Jouffroy, F. K., and Saban, R., *Traité de Zoologie*, XVI: 3 (Mammifères), pp. 477–611, Masson et Cie, Paris 1971.)

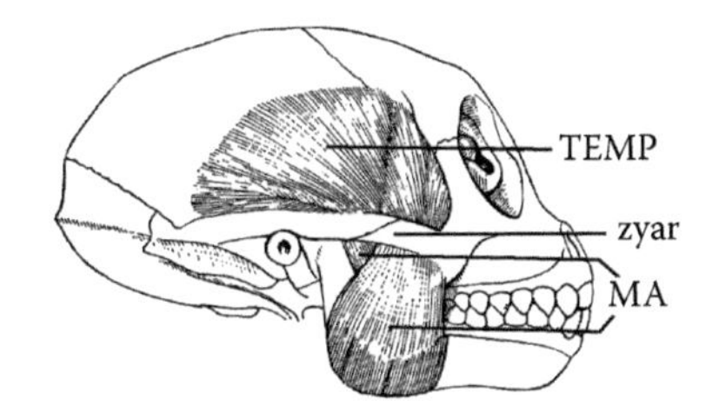

FIGURE 8.19 *Macaca mullata* (Mammalia, Primates): lateral view of the masseter and temporalis (the nomenclature of the structures illustrated basically follows that used in the present work; anterior is to the right). (Modified from Schumacher, G. H., *Funktionelle morphologie der kaumuskulatur*, VEB Gustav Fischer Verlag, Jena, 1961; Saban, R., *Traité de Zoologie*, XVI: 3 (Mammifères), 229–472, Masson et Cie, Paris, 1968.) MA, masseter; TEMP, temporalis; zyar, zygomatic arch.

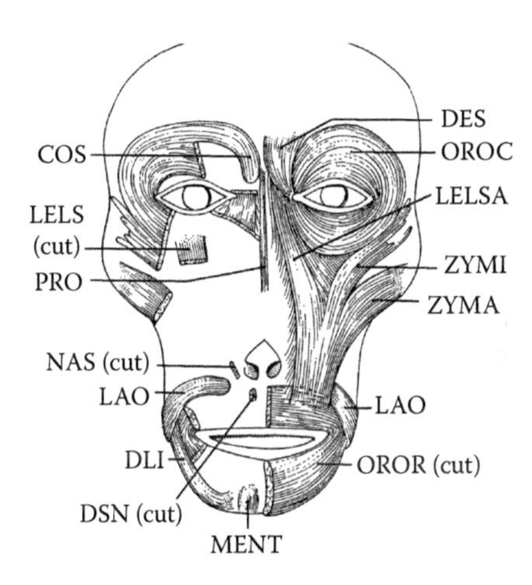

FIGURE 8.20 *Macaca cyclopis* (Mammalia, Primates): anterior view of the facial musculature; muscles such as the buccinatorius, platysma, frontalis, and occipitalis are not shown; on the left side, the depressor supercilii was removed, and the orbicularis oculi, zygomaticus minor, zygomaticus major, levator labii superioris, levator labii superioris alaeque nasi were partially cut (the nomenclature of the structures illustrated basically follows that used in the present work). COS, corrugator supercilii; DES, depressor supercilii; DLI, depressor labii inferioris; DSN, depressor septi nasi; LAO, levator anguli oris facialis; LELS, levator labii superioris; LELSA, levator labii superioris alaeque nasi; MENT, mentalis; NAS, nasalis; OROC, orbicularis oculi; OROR, orbicularis oris; PRO, procerus; ZYMA, ZYMI, zygomaticus major and zygomaticus minor. (Modified from Shibata, S., *OFAJ*, 34, 159–176, 1959; Jouffroy, F. K., and Saban, R., *Traité de Zoologie*, XVI: 3 (Mammifères), pp. 477–611, Masson et Cie, Paris 1971.).

levator arcus palatini and dilatator operculi) *sensu* Edgeworth (Knight et al. 2008). This means that the muscles that arise from cells expressing the same gene in two different vertebrate taxa are not necessarily homologous among those taxa, thus supporting the idea that no single criterion (including the expression of genes such as *engrailed*) is enough to establish myological homologies (for more details on this subject, see Diogo and Abdala (2010) and preceding chapters).

HYOID MUSCLES

Edgeworth (1935) and Huber (1930a,b, 1931) divided the hyoid muscles into two main groups: dorsomedial and ventral (Tables 8.3 and 8.4). The plesiomorphic configuration

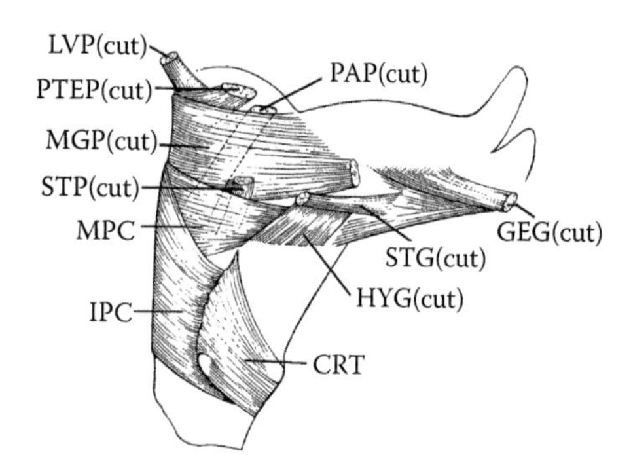

FIGURE 8.21 *Hylobates hoolock* (Mammalia, Primates): lateral view of the pharyngeal musculature (the nomenclature of the structures illustrated basically follows that used in the present work; anterior is to the top, dorsal is to the left: see text). CRT, cricothyroideus; GEG, genioglossus; HYG, hyoglossus; IPC, constrictor pharyngis inferiori; LVP, levator veli palatini; MGP, mylo-glossopharyngeus; MPC, constrictor pharyngis medius; PAP, palatopharyngeus; PTEP, pterygopharyngeus; STG, styloglossus; STP, stylopharyngeus. (Modified from Kanagasuntheram, R., *Ceylon J Sci Sect G*, 5, 11–64, 69–122, 1952–1954; Saban, R., *Traité de Zoologie*, XVI: 3 (Mammifères), pp. 229–472, Masson et Cie, Paris, 1968.)

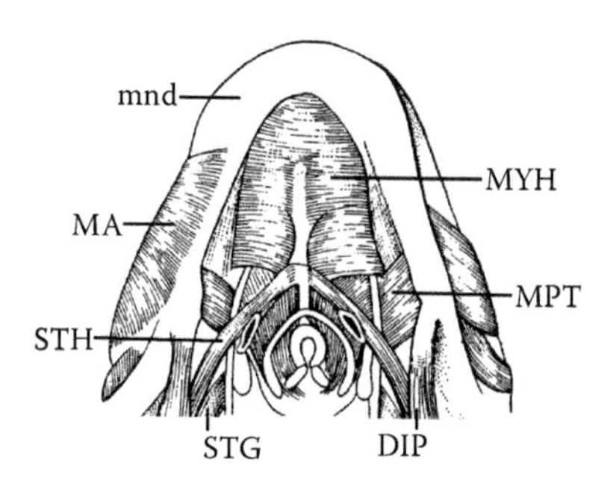

FIGURE 8.22 *Pongo pygmaeus* (Mammalia, Primates): ventral view of the head musculature; on the right side the superficial portion of the masseter was removed (the nomenclature of the structures illustrated basically follows that used in the present work; anterior is to the top). DIP, digastricus posterior; mnd, mandible; MA, masseter; MPT, pterygoideus medialis; MYH, mylohyoideus; STG, styloglossus; STH, stylohyoideus. (Modified from Edgeworth, F. H., *The Cranial Muscles of Vertebrates*, Cambridge University Press, Cambridge, UK, 1935; Saban, R., *Traité de Zoologie*, XVI: 3 (Mammifères), pp. 229–472, Masson et Cie, Paris, 1968.)

for sarcopterygians as a whole is a single ventral hyoid muscle, the interhyoideus, and two dorsomedial hyoid muscles, the adductor arcus palatini, and the adductor operculi (NB: the "adductor hyomandibulae Y" and "levator operculi" of *Latimeria* are not homologues of the adductor hyomandibulae and levator operculi of actinopterygians such as teleosts: see Diogo and Abdala 2010). The depressor mandibulae, levator hyoidei, branchiohyoideus, and cervicomandibularis of extant dipnoans, amphibians, and reptiles seem to develop from the anlage that give rise to the adductor arcus palatini in other osteichthyans (Figures 8.2, 8.4, and 8.5). The adductor operculi is not present as an independent muscle in extant dipnoans, amphibians, and reptiles; at least in dipnoans, it is most likely fused with the ventral hyoid muscle interhyoideus (Figure 8.2). This comparison with nonmammalian taxa is helpful to understand the origin of the mammalian muscles as well as to emphasize that the number of hyoid muscles found in extant mammals, and particularly in therians (placentals + marsupials), is much greater than that found in extant nonmammalian tetrapods (Tables 8.4 and 8.5). Also, in nonmammalian vertebrates, the hyoid muscles are mainly restricted to the region of the second branchial arch and occasionally to the mandibular and/or neck regions (Figures 8.2, 8.4, and 8.5), whereas in extant mammals, these muscles extend more anteriorly, covering much of the anterior region of the head (Figures 8.8, 8.11, 8.12, 8.13, 8.18, 8.20, and 8.24 through 8.26).

Importantly, with the exception of the styloideus, stylohyoideus, digastricus posterior, jugulohyoideus, and stapedius, all the mammalian hyoid muscles listed in Tables 5.3 and 5.4 are usually designated as facial muscles because they attach to freely movable skin and are associated with the display of facial expressions (Figures 8.8, 8.11, 8.12, 8.13, 8.18, 8.20, and 8.24 through 8.27) (e.g., Ruge 1885, 1897, 1910; Boas and Paulli 1908; Lightoller 1928a,b, 1934, 1939, 1940a,b, 1942; Huber 1930a,b, 1931; Edgeworth 1935; Andrew 1963; Gasser 1967; Jouffroy and Saban 1971; Saban 1971; Seiler 1971a,b,c,d,e, 1974a,b, 1975, 1979, 1980; Minkoff 1979; Preuschoft 2000; Schmidt and Cohn 2001; Burrows and Smith 2003; Burrows et al. 2006, 2011; Burrows 2008; Diogo et al. 2009b; Diogo and Wood 2011, 2012a,b; Santana and Diogo 2014; Diogo and Santana 2017). Some researchers have suggested that the mammalian facial muscles derive exclusively from the interhyoideus of nonmammalian tetrapods (e.g., Huber 1930a,b, 1931), but our dissections and comparisons support the ideas of authors such as Lightoller [1942] and Jouffroy and Saban [1971], who claimed that at least some of these muscles (e.g., the extrinsic ear muscles and/or the platysma cervicale, platysma myoides, mandibulo-auricularis) correspond to part of the dorsomedial hyoid musculature (e.g., cervicomandibularis) of other tetrapods (Figure 8.5). The evolution and homologies of the mammalian facial muscles have been, and continue to be, controversial. In light of the overall analysis of the data obtained by our dissections and comparisons and by a review of the literature, some of the hypotheses proposed in Tables 8.3 and 8.4 (black arrows) are well supported by the data that are now available. For instance, data available on topology, functional morphology, development, and innervation strongly suggest that the platysma cervicale, platysma myoides, occipitalis, auricularis posterior, and some of the extrinsic muscles of the ear (e.g., antitragicus, helicis and/or transversus and obliquus auriculae) of mammals (Figures 8.8, 8.11, 8.12, 8.13, 8.18, 8.20, and 8.24 through 8.27) have a common phylogenetic and ontogenetic origin (e.g., Boas and Paulli 1908; Huber 1930a,b, 1931; Gasser 1967; Jouffroy and Saban 1971; Saban 1971; Diogo et al. 2009a; this work). These same lines of evidence also suggest that the interhyoideus profundus, sphincter colli superficialis, sphincter colli profundus, nasolabialis, levator labii superioris, levator labii superioris alaeque nasi, buccinatorius, dilatator nasi, maxillo-nasolabialis, nasalis, depressor septi

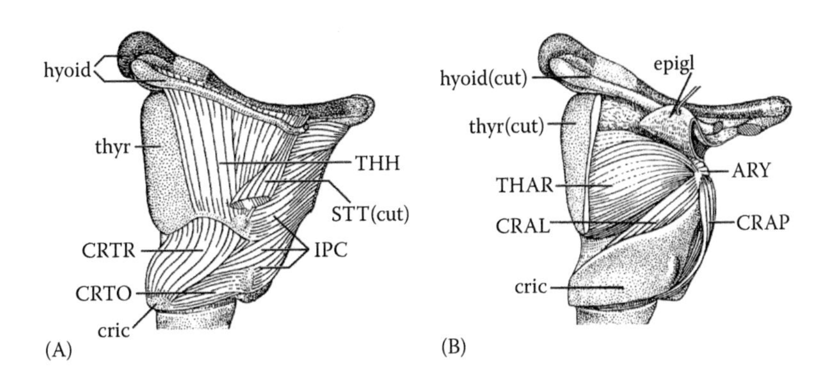

FIGURE 8.23 *Pan troglodytes* (Mammalia, Primates): (A) lateral view of the laryngeal musculature; (B) same view, but the thyrohyoideus, sternothyroideus, constrictor pharyngis inferior, and cricothyroideus were removed, and the lateral portions of the thyroid cartilage and hyoid bone were partially cut (the nomenclature of the structures illustrated basically follows that used in the present work; anterior is to the top, dorsal is to the right: see text). ARY, arytenoideus; CRAL, cricoarytenoideus lateralis; CRAP, cricoarytenoideus posterior; cric, cricoid cartilage; CRTO, CRTR, pars obliqua and pars recta of cricothyroideus; epigl, epiglottis; IPC, constrictor pharyngis inferior; STT, sternothyroideus; THAR, thyroarytenoideus; THH, thyrohyoideus; thyr, thyroid cartilage. (Modified from Starck, D., and Schneider, R., *Primatologia III/2*, pp. 423–587, S. Karger, Basel, 1960; Saban, R., *Traité de Zoologie*, XVI: 3 (Mammifères), 229–472, Masson et Cie, Paris, 1968.)

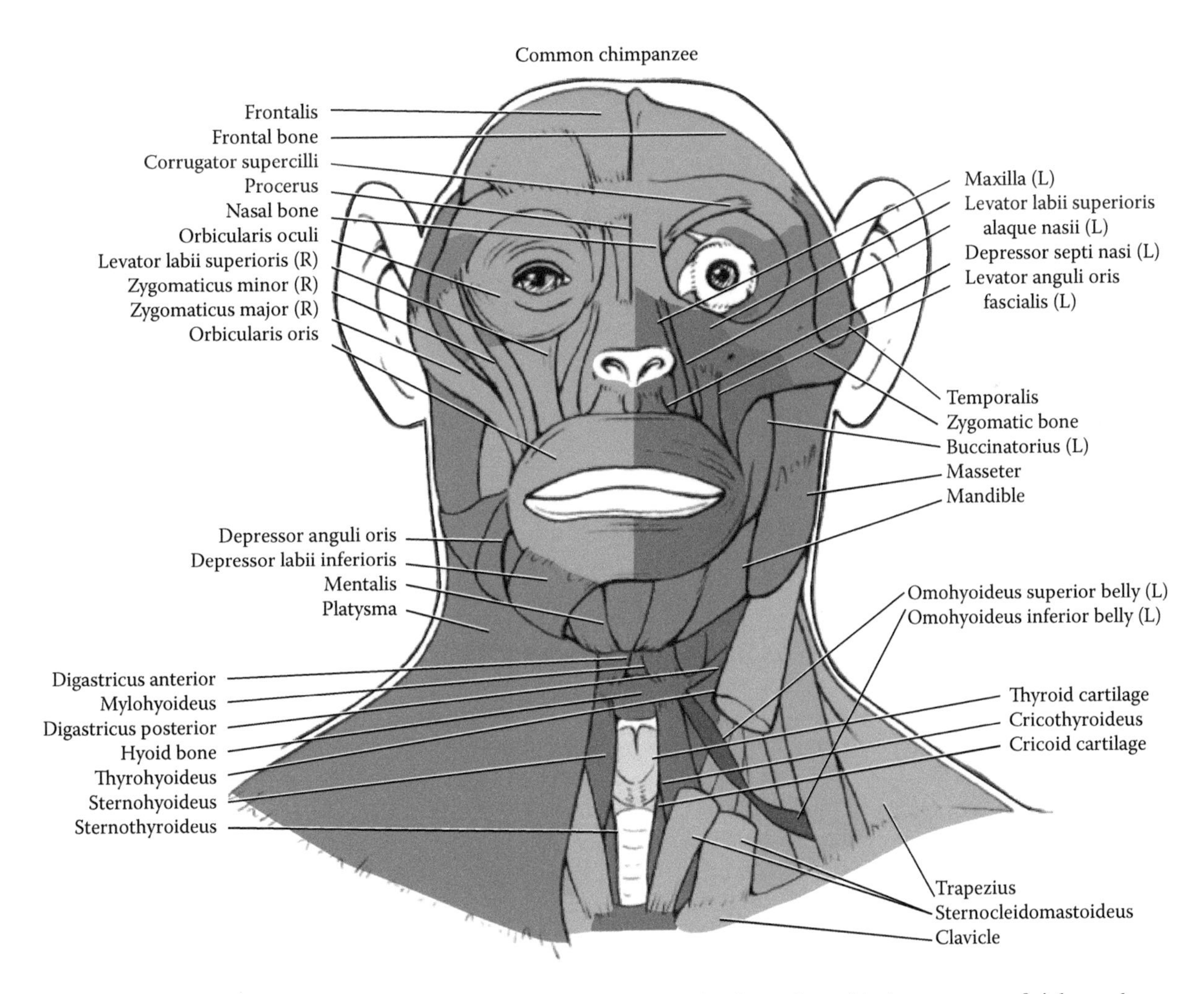

FIGURE 8.24 *Pan troglodytes* (Mammalia, Primates): frontal view of the head muscles, with the more superficial muscles removed on the left side of the head to show the deeper muscles (schematic drawing by V. Powell, the different tone refer to the inclusion of the muscles in different head anatomical network modules, and are not relevant for the present book). (Modified from Powell V. et al., *Nat Sci Rep*, 2018.)

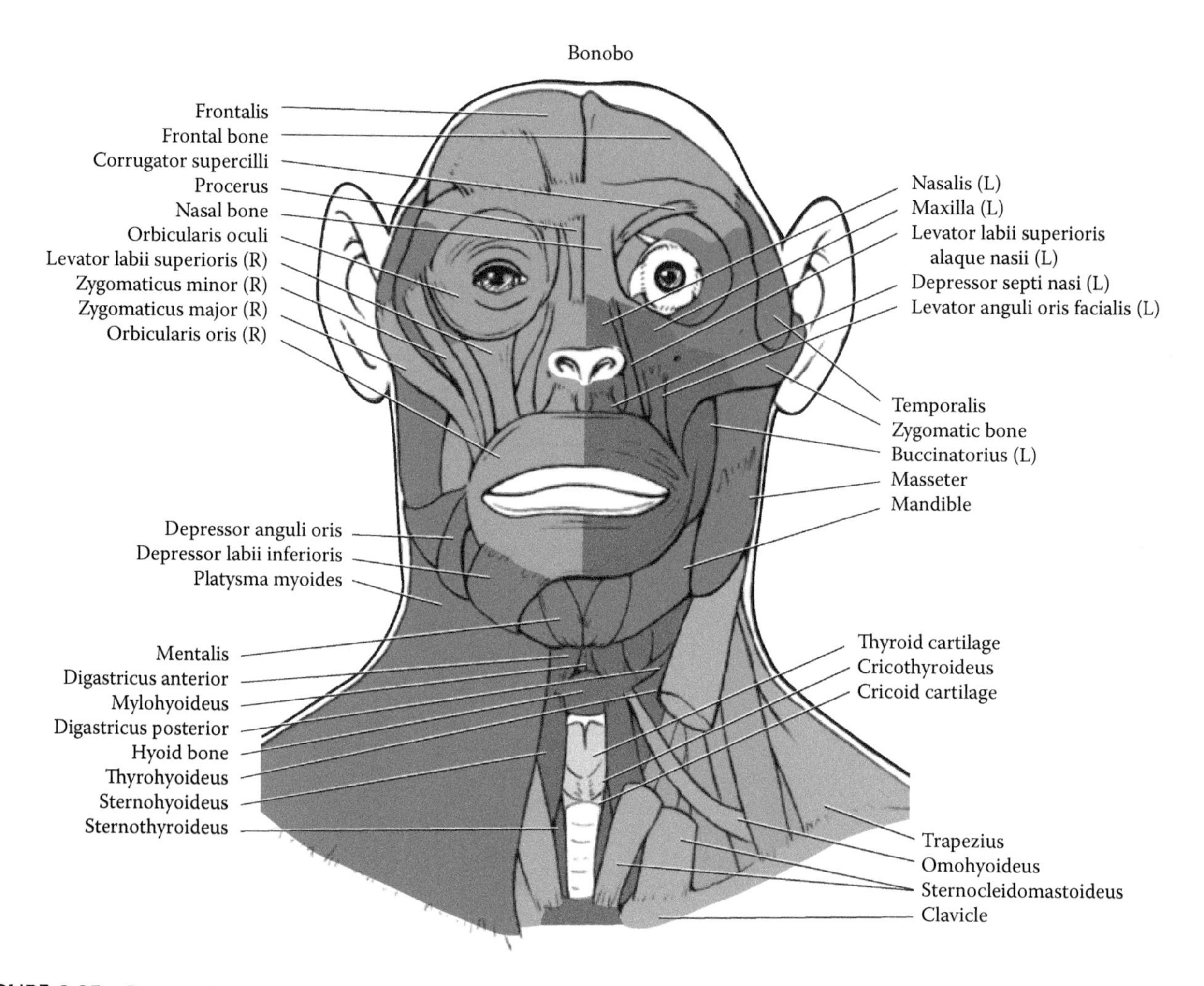

FIGURE 8.25 *Pan paniscus* (Mammalia, Primates): frontal view of the head muscles, with the more superficial muscles removed on the left side of the head to show the deeper muscles (schematic drawing by V. Powell, the different tone refer to the inclusion of the muscles in different head anatomical network modules, and are not relevant for the present book). (Modified from Powell V. et al., *Nature Sci Rep*, 2018.)

nasi, levator anguli oris facialis, orbicularis oris, depressor labii inferioris, depressor anguli oris, and/or mentalis of mammals derive from the interhyoideus of nonmammalian taxa (see also, e.g., Gasser 1967; Jouffroy and Saban 1971; Saban 1971; Seiler 1971a,b,c,d,e, 1974a,b, 1975, 1979, 1980). These hypotheses have also been supported by more developmental studies of mice. For instance, Carvajal et al. (2001) showed that hyoid arch muscle progenitors that migrate out of the hyoid arch from 10.5 days post coitum (dpc) split into dorsal and ventral branches by 11.5 dpc. The dorsal domain divides further at 12.5 dpc and gives rise to the extrinsic facial muscles of the ear and to some auricular muscles (e.g., auricularis posterior; plus the occipitalis according to our interpretation of their figures). The ventral domain elongates rostrally and separates into dorsal and ventral branches that do not divide into different muscle masses until 13.5 dpc and that expand toward the snout and the eye regions where they then form most other facial muscles. However, Carvajal et al.'s (2001) Figure 5 seems to indicate that at least part of the platysma myoides derives from the ventral domain rather than the dorsal domain as hypothesized in our Tables 8.3 and 8.4, although it is not clear from that figure if at least part of the platysma myoides, and/or part or the totality of the platysma cervicale, derives from the dorsal domain.

It is also still not completely clear if, for instance, the therian mandibulo-auricularis (a muscle that is usually deep to all the other mammalian facial muscles) is phylogenetically more closely related to the other facial muscles than to deeper dorsomedial muscles such as the stylohyoideus, digastricus posterior, jugulohyoideus and stapedius (e.g., Lightoller 1934; Jouffroy and Saban 1971; Seiler 1971a,b,c,d,e, 1974a,b, 1975, 1979, 1980; this work) (Tables 8.3 and 8.4). Furthermore, it is commonly accepted that muscles such as the zygomaticus major, zygomaticus minor, orbito-temporo-auricularis, frontalis, auriculo-orbitalis, temporoparietalis, auricularis anterior, and auricularis superior (Figures 8.8, 8.11, 8.12, 8.13, 8.18, 8.20, and 8.24 through 8.27) derive from the sphincter colli profundus and/or superficialis, but Seiler (1971a,b,c,d,e, 1974a,b, 1975, 1979, 1980), based on his comparative and developmental studies, argues that at least some of these

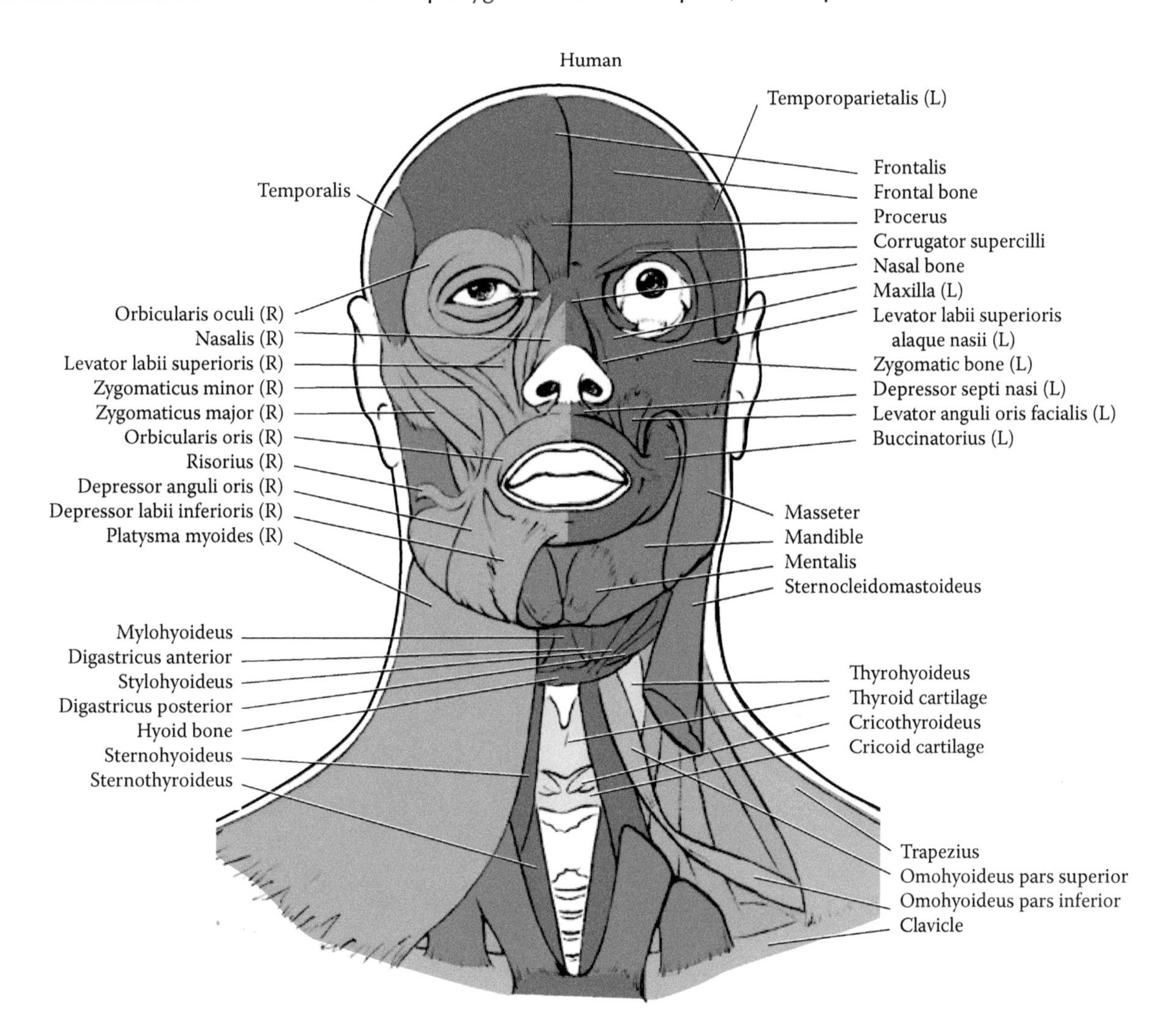

FIGURE 8.26 *Homo sapiens* (Mammalia, Primates): frontal view of the head muscles, with the more superficial muscles removed on the left side of the head to show the deeper muscles (schematic drawing by V. Powell, the different tone refer to the inclusion of the muscles in different head anatomical network modules, and are not relevant for the present book). (Modified from Powell V. et al., *Nat Sci Rep*, 2018.)

muscles may derive from the platysma cervicale and/or myoides (see Tables 8.3 and 8.4).

Seiler did an impressive series of works on the facial muscles of mammals, which are, unfortunately, often unknown to most non-German speaking authors. However, some of Seiler's methods and interpretations are questionable. For example, in his (1980) developmental study of primates and tree shrews, he argues that the facial muscles that are more superficial in early developmental stages are necessarily part of a "platysma anlage" and thus derived phylogenetically from an "ancestral platysma," whereas the majority of the other facial muscles are part of a "sphincter colli profundus" anlage and thus are derived phylogenetically from a "primitive sphincter colli profundus." This contrasts with Gasser's (1967) study of the ontogeny of the facial muscles of modern humans, in which other anlagen are recognized in early developmental stages. Also, it should be stressed that in adult mammals, including monotremes, at least some portions of the platysma (cervicale and/or myoides) lie deep to facial muscles such as the sphincter colli superficialis and even to facial muscles that Seiler categorizes as "sphincter colli profundus derivatives" (e.g., part of the orbicularis oris and/or levator labii superioris) (see, e.g., Figure 8.8). The majority of researchers accept that the sphincter colli of mammals derives from the interhyoideus of other tetrapods, so it is likely that the mammalian sphincter colli was plesiomorphically mainly superficial rather than deep to the other hyoid muscles (the interhyoideus of other tetrapods is usually superficial not only to the other hyoid muscles, but also to most other muscles of the head). Monotremes are plesiomorphic mammals, and both the platypus and the echidna have a well-developed, broad sphincter colli superficialis that is superficial to most of the other facial muscles. The platypus lacks a sphincter colli profundus, although it has an interhyoideus profundus that seems to be derived from the deeper part of the interhyoideus (Figure 8.1); in the echidna, most of the sphincter colli is superficial to the other facial muscles, but part of it passes deep to these muscles, forming a sphincter colli profundus (e.g., Huber 1930a; Lightoller 1942; Jouffroy and Saban 1971; this work). A more detailed comparative analysis of the development and innervation of the

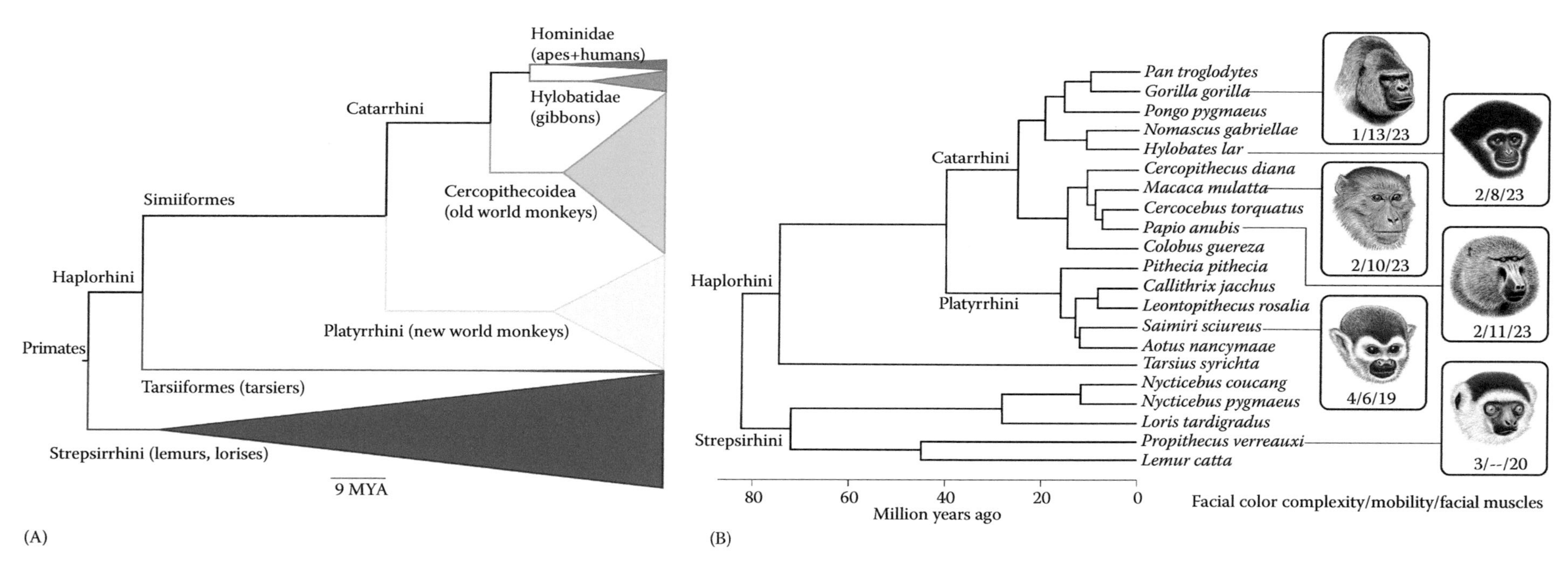

FIGURE 8.27 Phylogenies of (A) the order Primates, showing the major lineages in proportion to their numbers of species and (B) the primate species included in Santana et al.'s (2014) and Diogo and Santana's (2017) works, with examples illustrating major trends in facial color pattern complexity, mobility, and facial muscles; species that are larger and have more plainly colored faces tend to have a larger repertoire of facial expressions (©2012 Stephen D. Nash/IUCN/SSC Primate Specialist Group.; (modified from Diogo, R., and Santana, S. E., *The Science of Facial Expression*, pp. 133–152, Oxford University Press, Oxford, UK, 2017.)

hyoid group of muscles in vertebrates, including key mammalian groups such as monotremes, is needed to clarify these and other controversial issues regarding the origin, homologies, and evolution of the mammalian facial muscles and to test the hypotheses proposed in Tables 8.3 and 8.4. In the following, we will further discuss the evolution of the mammalian facial muscles and focus in particular on the fascinating case study provided by primate and human evolution.

BRANCHIAL, PHARYNGEAL, AND LARYNGEAL MUSCLES

The muscles listed in Tables 8.5 and 8.6 correspond to the branchial muscles *sensu* lato of Edgeworth (1935), which can be divided into three groups: (a) the "true" branchial muscles, which are subdivided into the branchial muscles *sensu stricto* and the protractor pectoralis and its derivatives, e.g., the trapezius and sternocleidomastoideus; (b) the pharyngeal muscles, which are only present as independent structures in extant mammals; and (c) the laryngeal muscles (e.g., Figures 8.10, 8.12, 8.15 through 8.17, 8.21, and 8.23 through 8.26).

Sarcopterygians such as coelacanths, dipnoans, and many amphibians retain various branchial muscles *sensu stricto* (Tables 8.5 and 8.6) (e.g., Bischoff 1840; Owen 1841; Cuvier and Laurillard 1849; Pollard 1892; Gaupp 1896; Allis 1897, 1922; Danforth 1913; Lubosch 1914; Sewertzoff 1928; Edgeworth 1935; Brock 1938; Piatt 1938; Millot and Anthony 1958; Osse 1969; Larsen and Guthrie, 1975; Greenwood 1977; Wiley 1979a,b; Jollie 1982; Bemis et al. 1983, 1997; Lauder and Shaffer 1985, 1988; Bemis 1986; Reilly and Lauder 1989, 1990, 1991; Miyake et al. 1992; Wilga et al. 2000; Kardong 2002; Carroll and Wainwright 2003; Johanson 2003; Kleinteich and Haas 2007; Diogo and Abdala 2010). Most authors agree that the branchial muscles *sensu stricto* are not present as a group in extant reptiles and extant mammals. For instance, many adult reptiles have only one branchial muscle *sensu stricto*, the hyobranchialis (which is often named "branchiohyoideus" or "branchiomandibularis" in the literature: Diogo and Abdala 2010). The two branchial muscles *sensu stricto* seen in adult reptiles such as the "lizard" *Euspondylus*, the hyobranchialis and "ceratohyoideus," seem to be the result of a subdivision of the subarcualis rectus I *sensu* Edgeworth (1935). That is, the "ceratohyoideus" found in these reptiles seems to correspond to/derive from part of the hyobranchialis of other reptiles. This comparison with other vertebrates shows that adult extant mammals lack all the branchial muscles *sensu stricto* except the subarcualis rectus I *sensu* Edgeworth (1935) (present in most adult mammals, being often divided into a ceratohyoideus and a stylopharyngeus: see the following), the subarcualis rectus II (usually present only in adult marsupials), and the subarcualis rectus III (usually present only in adult monotremes) (e.g., Edgeworth 1935; Smith 1992).

Edgeworth (1935) claimed that the pharyngeal muscles of mammals (Figures 8.15 and 8.21) are not derived from branchial muscle plates, but from a separate de novo condensation of myoblasts surrounding the pharyngeal epithelium. He did not consider the mammalian pharyngeal muscles to be homologous with the "pharyngeal muscles" of some amphibians (which probably correspond to branchial muscles *sensu stricto*, such as the levatores arcuum branchialium and/or the transversus ventralis) and some reptiles (which appear to be derived from the hyoid musculature) (e.g., Piatt 1938; Schumacher 1973; Smith 1992; Diogo and Abdala, 2010). Authors such as Jouffroy and Lessertisseur (1971) and Smith (1992) contradicted Edgeworth (1935) by suggesting that the mammalian pharyngeal musculature may derive from the amphibian "pharyngeal musculature" and therefore that this pharyngeal musculature innervated by the vagus nerve (CNX) was lost in reptiles. They state that pharyngeal muscles innervated by this nerve are present in larval amphibians and adult salamanders such as *Ambystoma punctatum* (e.g., "cephalo-dorso-subpharyngeus" *sensu* Piatt [1937]) and *Thorius dubius* ("dorso-pharyngeus"). They argue that these muscles are similar to those of mammals because they lie between the hyoid apparatus and pharyngeal wall, are innervated by CNX, and have two layers—a more or less longitudinal (oblique) layer and a circular muscle layer (Smith 1992). Moreover, they state that monotremes have five branchial arches (mandibular, hyoid, and three more) while no extant reptile has more than four arches (mandibular, hyoid, and two more) and that the laryngeal muscles of mammals and amphibians are innervated by two homologous branches of the vagus nerve, the superior and inferior (or recurrent) laryngeal nerves. In contrast, in reptiles the innervation of the larynx is *via* a single laryngeal nerve that is a branch of the glossopharyngeal nerve (CNIX) instead. Further studies of amphibians, reptiles, and mammals, using state-of-the-art developmental techniques, are needed to test these interesting hypotheses and thus to shed light on the origin of the mammalian pharyngeal musculature.

Our dissections, comparisons, and literature review do support Smith's (1992) and Noden and Francis-West's (2006) claims that at least one of the mammalian muscles included in Edgeworth's pharyngeal group, the stylopharyngeus, is derived from the true branchial musculature of basal tetrapods. That is, this mammalian muscle is not a de novo structure but a derivative of the branchial musculature *sensu stricto*, specifically the subarcualis rectus I (Tables 8.5 and 8.6). The mammalian stylopharyngeus and the reptilian "ubarcualis rectus I" are among the few muscles in either taxon innervated by CNIX: most of the mammalian pharyngeal muscles are innervated, instead, by the vagus nerve (CNX). In fact, in many mammals, including primates such as *Macaca*, the ceratohyoideus and stylopharyngeus are closely related and are innervated by the same ramus of the glossopharyngeal nerve (buccal ramus *sensu* Sprague [1944a, 1994b]; see also Saban 1968). Developmental data from monotremes and marsupials show that early in development, the stylopharyngeus is similar to the nonmammalian subarcualis rectus I in position, function, and connections. According to Smith (1992), although Edgeworth (1935) did not accept that the stylopharyngeus was derived from the branchial musculature *sensu stricto*, he did state that it develops from a muscle primordium that differs from the one that gives rise to the other pharyngeal muscles.

The homology between the mammalian stylopharyngeus and part of the subarcualis rectus I of other tetrapods is further supported by the results of a comparison between adult mammals and adult nonmammalian tetrapods. The stylopharyngeus of mammals usually originates from the styloid process, which is derived from a portion of the second (hyoid) arch; the subarcualis rectus I of nonmammalian taxa is usually associated with this arch (Saban 1971; Smith 1992). Also, as explained earlier, some reptiles (e.g., lizards) have two branchial muscles *sensu stricto*, which apparently are the result of a subdivision of the subarcualis rectus I *sensu* Edgeworth (1935) (Tables 8.5 and 8.6). The more anterior of these muscles, the hyobranchialis (often called branchiohyoideus or branchiomandibularis), usually originates from the hyoid arch (as does the mammalian stylopharyngeus, see preceding texts.) and connects the hyoid cornu to the epihyal, although in various reptilian groups, it extends anterolaterally to attach on the lower jaw (that is why it is usually named branchiomandibularis in these groups). We refer to this muscle as the hyobranchialis because it is not homologous to the hyoid muscle branchiohyoideus of amphibians nor to the hypobranchial muscle branchiomandibularis of actinopterygian fishes such as cladistians, chondrosteans and *Amia* (Diogo and Abdala 2010). The most posteriorly situated muscle in lizards, often named the ceratohyoideus, usually connects the hyoid arch to other (more posterior) branchial arches, as does the mammalian ceratohyoideus (Tables 8.5 and 8.6). In some mammals, such as colugos and tree shrews, the stylopharyngeus does not reach the styloid process; i.e., it may originate from more distal hyoid structures such as the epihyal (as does the reptilian hyobranchialis; e.g., Sprague 1942, 1943, 1944a,b; Saban 1968). This observation, together with the other data available (see preceding text), suggests that the combination of stylopharyngeus and the ceratohyoideus in mammals, and the combination of hyobranchialis and the ceratohyoideus in reptiles, are both the result of the subdivision of the subarcualis rectus I *sensu* Edgeworth (1935; Tables 8.5 and 8.6). However, this does not mean that the stylopharyngeus of mammals is necessarily the homologue of the reptilian hyobranchialis, for one cannot refute the hypothesis that the subdivision of the subarcualis rectus I into two muscles occurred more than once among the amniotes resulting in the hyobranchialis and ceratohyoideus of lizards and in the stylopharyngeus and ceratohyoideus of mammals. However, the mammalian stylopharyngeus is innervated by the glossopharyngeal nerve (CNIX) and not by the vagus nerve (CNX) as noted earlier, so it is not truly a vagus nerve pharyngeal muscle, such as are the other longitudinal pharyngeal constrictors (e.g., palatopharyngeus) and the circular constrictors (middle and inferior, and, in placentals, also superior) of mammals.

The mammalian acromiotrapezius, spinotrapezius, dorsocutaneous, cleido-occipitalis, sternocleidomastoideus, cleidomastoideus, and sternomastoideus correspond to the reptilian trapezius and sternocleidomastoideus and thus to the protractor pectoralis of amphibians and of bony fishes (see recent reviews by Ziermann et al. [2014] and Diogo and Ziermann [2015a]). Most authors agree that the protractor pectoralis is not present as an independent structure in *Latimeria* (e.g., Millot and Anthony 1958). However, Sefton et al. (2016) stated that a muscular structure originating from the pectoral girdle and inserting on the fifth ceratobranchial might be the protractor pectoralis. If this is the case, it would require a change of origin (from pectoral girdle, instead of skull, although they state that it does terminate upon dorsal fascia posterior to the cranium) and of insertion (from pectoral girdle to fifth ceratobranchial). Another recent study, including both computed tomography (CT) scans and histological sections, suggests that this structure is simply part of the levators of the branchial arches and that there is another deeper structure that corresponds to the protractor pectoralis of other fishes running mainly from the skull to the pectoral girdle (Peter Johnston, pers. comm.). A detailed study of the development and/or innervation of these structures is needed to test these hypotheses.

The protractor pectoralis of nonamniote taxa is not a branchial muscle *sensu stricto* because it is mainly involved in the movements of the pectoral girdle and not of the branchial arches. Interestingly, the results of recent developmental and molecular studies indicate that the protractor pectoralis of *Ambystoma* and the trapezius of chickens and mice are at least partially derived from somites (e.g., Köntges and Lumsden 1996; Matsuoka et al. 2005; Noden and Francis-West 2006; Piekarski and Olsson 2007; Shearman and Burke 2009). These studies have also shown that during the ontogeny of mice, some of the cells of the trapezius that are originated from the somites pass the lateral somitic frontier in order to develop within lateral plate-derived connective tissue of the forelimb (e.g., Shearman and Burke 2009). That is, the mammalian trapezius is a rather peculiar muscle that is directly associated with three different types of connective tissue: connective tissue derived from branchial arch neural crest cells, somite-derived connective tissue, and lateral plate-derived (forelimb) connective tissue. Therefore, it has been controversial whether the protractor pectoralis was primarily derived from the paraxial mesoderm, as suggested by Edgeworth (1935), and only later became ontogenetically associated with the cranialmost somites and even with lateral plate-derived connective tissue of the forelimb, or if it was instead primarily derived from somites. In fact, in addition to the hypothesis more strongly defended in the present work, i.e., that the cucullaris and its derivatives (which include the trapezius) are branchial (head) muscles, some recent developmental studies have argued that they are instead epibranchial muscles (e.g., Kuratani and colleagues) or hypaxial migratory muscles (e.g., Burke and colleagues) (see discussion in the following as well as next chapters).

Interestingly, recent works have shown that in addition to branchial muscles (*sensu* Edgeworth 1935, Diogo et al. 2008a, and Diogo and Abdala 2010) such as the protractor pectoralis and the laryngeal muscles constrictor laryngis and dilatator laryngis, even branchial muscles *sensu stricto* such as the levatores arcuum branchialium and hyoid muscles such as the interhyoideus are also partially derived from somites in tetrapods such as amphibians (e.g., Piekarski and Olsson 2007). Thus, that muscles such as the protractor pectoralis have a partial somitic origin does not necessarily mean that they cannot be considered part of the branchial musculature. Matsuoka et al. (2005) not only recognize that the amniote trapezius is partially derived from somites, but also argue that the sum of the data available (i.e., innervation, topology, development, and phylogeny) provides more support for grouping this muscle with the true branchial musculature than for including it in the hypobranchial musculature or in the postcranial axial musculature *sensu* Jouffroy (1971). In fact, it is important to stress that lineage tracing analyses in transgenic mice provide some support for the idea that the trapezius is a branchial muscle: these analyses reveal that neural crest cells from a caudal pharyngeal arch travel with the trapezius myoblasts and form tendinous and skeletal cells within the spine of the scapula (see, e.g., Noden and Schneider 2006). According to Noden and Schneider (2006: 14), "this excursion seemingly recapitulates movements established ancestrally, when parts of the pectoral girdle abutted caudal portions of the skull."

However, Epperlein et al.'s (2012) developmental study of salamanders and review of the literature suggested that neural crest does not serve a general function in vertebrate shoulder muscle attachment sites as predicted by the "muscle scaffold theory," and that it is not necessary to maintain connectivity of the endochondral shoulder girdle to the skull. According to these authors, the contribution of the neural crest to the endochondral shoulder girdle observed in the mouse probably arose de novo in mammals as a developmental basis for their skeletal synapomorphies. On the other hand, the innervation of the trapezius by the accessory nerve (CNXI) and, in many cases, by C3 and C4 spinal cord segments adds weight to the argument that the muscle is derived from both the paraxial mesoderm and somites. According to a recent study, the spinal accessory nerve might be a novel structure specific to living gnathostomes that arose through the repatterning of preexisting spinal motoneurons in the hypothetical ancestor; by de novo upregulation of cranial nerve-specific regulatory genes, the ancestral spinal accessory nerve would have acquired intermediate branchiomeric motoneuron properties (Tada and Kuratani 2015). According to this view, it would not be possible to characterize the accessory nerve based on a simple head/trunk dualism, but rather in a third category of peripheral nerve; it would thus be conceivable that the gnathostome cucullaris muscle would represents a similar intermediate or mixed nature (Tada and Kuratani 2015). However, the innervation of the cucullaris might alternatively provide support for a branchial component because of the position of the accessory nucleus in the ventral horn of the spinal cord, which is in line with the more cranial branchiomotor nuclei (see, e.g., Wilson-Pauwels et al. 2002; Butler and Hodos 2005).

But, according to Ericsson et al.'s (2013) review on the origin and evolution of the neck muscles, the strongest evidence provided so far to support the idea that the trapezius and other derivatives of the cucullaris are following a head muscle developmental program was provided by Tajbakhsh et al. (1997) and Theis et al. (2010). As noted by Ericsson et al., myogenic differentiation in the branchial muscles is regulated by *Pitx2* and *Tbx1*; these two transcription factors act to regulate expression of the myogenic regulatory factors *Myf5* and *MyoD*, while *Myf5* and *MyoD* expressions in the somitic mesoderm are regulated by *Pax3*. Mice lacking *Pax3* function thus show a loss of several trunk and limb muscles, but head muscles are unaffected; *Pax3:Myf5* double mutant mice lack all somitic-derived muscles, including some (but, interestingly, not all: see the following) tongue and infrahyoid muscles, but branchial muscles and the trapezius and sternocleidomastoideus are still present (Tajbakhsh et al. 1997). This suggests that the cucullaris derivatives, trapezius, and sternocleidomastoideus are not developing under the myogenic program functioning in the somites but instead are following a head muscle program. Similar evidence for a head muscle program operating in some neck muscles is observed in *Tbx1* mutant mice, because posterior branchial muscles are absent, as are the trapezius and sternocleidomastoideus (Theis et al. 2010). In contrast, as noted by Ericsson et al. (2013), these mice have been reported to not have any muscle defects in somite-derived muscles of the limb. Moreover, as also noted by these authors, human patients with DiGeorge syndrome have a point mutation in the *Tbx1* gene and show many similar features to those observed in the mouse *Tbx1* mutants and, intriguingly, many of these patients display sloping shoulders due to small shoulder and pectoral muscles, suggesting that *Tbx1* is also important for the development of these somitic-derived muscles in humans. Moreover, the trapezius and sternocleidomastoideus receive cells from an *Isl1*-expressing lineage, providing further evidence that these muscles are following a head muscle program (Theis et al. 2010) because fate mapping of these lateral cells at the posterior extent of the cranial mesoderm expressing *Isl1* in the mouse has revealed that they also contribute to branchial muscles (Theis et al. 2010).

However, the strongest direct, empirical support for the cucullaris and its derivatives being true head (branchial muscles) came from the recent retrospective clonal studies performed in mice by Lescroart et al. (2015), showing that the cardiopharyngeal field includes the trapezius and sternocleidomastoideus muscles (see Chapter 3). Moreover, these clonal studies also supported a close relationship between the trapezius and sternocleidomastoideus and the laryngeal muscles, confirming the idea proposed by Edgeworth (1935) and in our previous works that the trapezius and sternocleidomastoideus are branchial muscles and, thus, true head (i.e., branchiomeric) muscles. The labelled cells in the laryngeal muscles that Lescorart et al. (2015) analyzed (only scored in sectioned embryos) did seem to be ipsilateral to labeling in the trapezius; it is likely that a similar situation for unilateral labeled first arch muscles might be also present but that their analyses just did not allow us to discern which parts of the right ventricle are derived from the left and right sides of the embryo (Robert Kelly, pers. comm.). Interestingly, these clonal studies indicate that the extrinsic ocular muscles (no data were given about their right/left modularity in that study) are closely related to the group formed by the first arch muscles and by the right ventricle, supporting the idea, also proposed by Edgeworth (1935) and followed in our previous works, that the extrinsic eye muscles are developmentally linked to the first arch muscles. Another interesting result of more recent developmental works is that they have confirmed Edgeworth's (1935) hypothesis—based on his embryological observations—that at least part of the esophageal muscles—i.e., the "esophagus striated muscles"—are branchiomeric, i.e., have an origin from the cranial mesoderm (Gopalakrishnan et al. 2015). Also worthy of mention is that the mammalian splenius muscle, which is often seen as a somitic, epaxial muscle innervated by dorsal rami of spinal nerves and appears in these clonal studies as partially somitic and partially cranial,,i.e., related to the trapezius/sternocleidomastoideus. Of the 22 embryos in which the splenius was labeled in the clonal study (Lescroart et al. 2015), 9 also had labeling in somitic neck muscles (40.9%), and 5 showed labeling in the trapezius muscle group (22.7%). Genetic tracing with a Pax3Cre allele in the same study also confirmed that some progenitor cells for the splenius had expressed Pax3 and are therefore somite derived. That is, the splenius seems to be mainly of somitic origin, although it does seem to include a portion derived from the cardiopharyngeal field.

This scenario has become almost universally accepted, but was called into question in a recent study by Czajkowski et al. (2014), who argued that the extrinsic tongue muscles are mainly of cranial origin (the intrinsic tongue muscles being mainly of somitic origin, as expected) (see question marks in Figure 2.2). For instance, they have stated that Mesp1Cre-dependent lineage tracing on the Met mutant background demonstrated that residual muscle and myogenic progenitors in the tongue of Met mutants indeed derive from cranial mesoderm. However, studies such as that by Harel et al. (2009) showed that all tongue muscles are derived from the *Pax3* lineage and not from the *Isl1* lineage and, moreover, that *Mesp1* is not the most appropriate for such analyses because it marks not only cranial muscles but also the occipital somites. Moreover, Kelly's work (Robert Kelly, pers. comm.) has shown that tongue muscles are largely unaffected in *Tbx1* null mutants and mostly absent in *Pax3* null mutants, although they do express *Tbx1* at later stages, showing the dangers of drawing conclusions from gene expression at a single developmental stage. Notably, *Lbx1* expression in the tongue muscle precursors is merely delayed rather than abolished in the *Pax3* mutant Splotch, suggesting that for the occipital somites, which during evolution have been incorporated into the head, the loss of *Pax3* function is compensated and that additional genes may be involved in the mediation of the extrinsic signals (Brand-Saberi 2002). The clonal studies of Lescroart et al. (2015) thus seem to provide further evidence for an exclusive somitic origin for both the extrinsic and intrinsic tongue muscles, as the tongue muscles never appear together with the branchiomeric muscles in that study.

However, unpublished data by Birchmeier (Carmen Birchmeier, pers. comm.) comprising lineage tracing with *Pax3* have shown that some tongue muscles (namely, part of the extrinsic tongue muscles) do not seem to derive from the *Pax3* somitic lineage. Therefore, the idea that part of the tongue muscles are not derived from somites cannot be completely excluded. Recent studies have stressed that the tongue muscles have some hybrid characteristics between branchiomeric and somitic muscles. For instance, unlike somitic muscles, their patterning and attachments seems to be deeply related to neural crest cells (Parada et al. 2012). For example, these cells are not required for myogenic progenitor migration toward their presumptive destinations in the branchial arches and tongue primordium, but as these myogenic progenitors first enter the craniofacial region, they immediately establish intimate contact with these cells. This close association between the two cell types continues throughout the entire course of tongue morphogenesis and suggests that tissue–tissue interactions may play an important role in regulating cell fate determination (Parada et al. 2012). Also, craniofacial myogenesis depends on *Dlx5/6* expression by cranial neural crest cells (CNCCs), because inactivation of *Dlx5* and *Dlx6* results in loss of jaw muscles and compromised tongue development; since *Dlx5/6* are not expressed by the myogenic component, this result indicates an instructive role for *Dlx5/6*-positive CNCCs in muscle formation (Parada et al. 2012). In *Dlx5/6*$^{-/-}$ double mutant mice, the intrinsic muscles of the tongue and sublingual muscles are severely affected; e.g., the genioglossus and geniohyoideus are absent, and intrinsic muscles of the tongue are reduced and disorganized, but the remaining tongue muscles express determination and differentiation markers. Because limb and trunk muscles in *Dlx5/6* mutant mice are not affected, this indicates a specific function of *Dlx* genes in tissue–tissue interactions involving neural crest derivatives; i.e., *Dlx5/6* expression by CNCCs is necessary for interactions between CNCC-derived mesenchyme and mesoderm to occur, which result in myogenic determination, differentiation, and patterning (Parada et al. 2012). According to Parada et al. (2012), the importance of *Dlx* genes in tongue development is twofold: (a) they establish the dorsoventral pattern of the first bracnial arch (mandibular arch) and, indirectly, that of the tongue, and (b) they regulate myogenic determination and differentiation processes, including those affecting the tongue myogenic core. However, this might simply mean that *Dlx5/6* is related to neural crests cells, and because these cells affect the patterning of the tongue muscles, they are affected, so this might have nothing to do with showing that these muscles are not completely somitic.

Importantly, our recent dissections of chondrichthyans and comparisons with other vertebrates (Diogo et al. 2015a; Ziermann et al. 2014) and updated review of the literature clearly support the idea that the cucullaris and its derivatives are mainly branchial muscles. As explained in Chapter 5, in adult sharks such as *Squalus*, there is a single, continuous muscle cucullaris inserting onto both the dorsal surface of the branchial arches and the posterior surface of the pectoral girdle; the latter portion corresponds to the protractor pectoralis of osteichthyans. Edgeworth has shown that this condition is seen from the first embryonic stages until the adult stages of sharks, the cucullaris developing from the dorsal portion of the branchial muscle plates. In adult holocephalans such as *Hydrolagus*, the cucullaris has two bundles, one (superficial) inserting onto the pectoral girdle and the other (deep) inserting onto the dorsal portion of the branchial arches. Edgeworth (1935) has shown that in holocephalans, the cucullaris develops from the dorsal portion of the branchial muscle plates, forming first a single, continuous muscle, and then separating during ontogeny into the deep and superficial bundles. Concerning batoids, the cucullaris develops only from the dorsal portion of the last branchial muscle plate, forming a single muscle that then divides during ontogeny into three bundles, in taxa such as *Leucoraja*: an inner bundle going to the suprascapula, a middle bundle going to the scapula, and an external bundle going to the branchial arches (according to, e.g., Marion [1905]; such bundles were not found by authors such as Kesteven [1942]; NB: as noted by Miyake [1992], some authors argue that the inner and middle bundles of batoids are not really part of the cucullaris because they do not attach onto the branchial arches and they receive innervation from spinal nerves, but as shown by, e.g., Edgeworth [1935] in other gnathostomes, the derivatives of the cucullaris might also receive innervation from spinal nerves and/or not attach to branchial arches). According to Edgeworth, in chondrichthyans, the cucullaris is innervated by cranial nerve CNX; in osteichthyans, there is usually a protractor pectoralis (or its derivatives, e.g., trapezius and sternocleidomastoideus) that derives ontogenetically exclusively from the last branchial muscle plate and that inserts exclusively to the pectoral girdle (in both bony fishes and most tetrapods) and levatores arcuum branchialium developing from the dorsal portion of the branchial muscle plates and going exclusively to the dorsal surface of the branchial arches (in bony fishes and amphibians, the levatores arcuum branchialium being absent in amniotes). Most authors do not describe levatores arcuum branchialium in chondrichthyans, but Kesteven (1942) did describe these muscles as very thin structures in sharks such as *Mustelus* and *Orectolobus*, batoids such as *Dasyatis*, holocephalans such as *Callorhynchus* and *Hydrolagus*, and as even thinner and apparently vestigial structures in sharks such as *Squalus* and batoids such as *Leucoraja*, going to the dorsal surface of five branchial arches (he does not refer to holocephalans). Most authors seem to not agree with Kesteven's descriptions of levatores arcuum branchialium in chondrichthyans. For instance, Didier (1987, 1995) does not describe these muscles in holocephalans, and states that some of the levatores arcuum branchialium *sensu* Kesteven (1942) correspond to part of the protractor pectoralis profundus while others correspond to part of the epibranchial muscle subspinalis. However, Kesteven (1942) described the subspinalis, the protractor pectoralis, and the levatores arcuum branchialium in chondrichthyans, so the synonymies proposed by Didier are questionable. However, to our knowledge, no other author (apart from Kesteven [1942]) has described distinct, fleshy muscles levatores arcuum branchialium in both the elasmobranchs and the holocephalans and that even Kesteven recognized that the structures that designated levatores arcuum branchialium in chondrichthyans are mainly innervated by "spinal" nerves, while those in osteichthyans are mainly innervated by nerves CNX and/or CNIX, calling into question the homology of the muscles of these two major gnathostome taxa proposed by Kesteven. That is, the levatores arcuum branchialium described by Kesteven in chondrichthyans might well be instead part of the epibranchial musculature, as suggested by Didier (1995), or even part of the axial (body) musculature, although one should not discard the hypothesis that they are homologous to the levatores arcuum branchialium of osteichthyans.

This comparison with nonmammalian taxa, including fishes, is crucial to understand the origin of the mammalian head muscles. For instance, because Edgeworth (1935) regarded the dipnoans as the most plesiomorphic group of gnathostomes, he considered the condition found in dipnoans and thus other bony fishes to be the plesiomorphic gnathostome condition. However, the comparison with other fishes and other vertebrates in general clearly indicates that the plesiomorphic gnathostome condition is to have a single branchial muscle cucullaris innervated by cranial nerves and inserting onto both the branchial arches and the pe ctoral girdle, as found in adult sharks and in the early development of sharks, batoids, and holocephalans. That the cucullaris only divides into bundles later during the ontogeny of batoids and holocephalans and in adult batoids (the muscle usually has three bundles while in adult holocephalans the muscle usually has two bundles) clearly indicates that the division of the cucullaris into bundles or muscles is a derived condition independently acquired in batoids and holocephalans. The derived and homoplasic division of the cucullaris in batoids and holocephalans into bundles going exclusively to the pectoral girdle (protractor pectoralis) and exclusively to the branchial arches (levatores arcuum branchialium) seems to be the result of an evolutionary trend seen in gnathostomes. That is, in early gnathostomes, the pectoral girdle basically is, at least functionally and often anatomically, part of the head, while during gnathostome evolution, the pectoral girdle often tended to become functionally and/or anatomically more separated from the head, and thus, it makes sense to not have a continuous muscle inserting onto both the branchial arches and the pectoral girdle. During the evolutionary transitions leading to the origin of osteichthyans, the cucullaris was differentiated into a protractor pectoralis attached onto the pectoral girdle only and levatores arcuum branchialium attached onto the branchial arches only. The protractor pectoralis is a neck muscle in amphibians and became subdivided into the trapezius and sternocleidomastoideus in amniotes. This muscle is therefore intimately related to the origin and evolution of the tetrapod (and thus the mammalian) neck. A recent developmental study including CT datasets, fate-mapping, and vertebrate model organisms provided further evidence confirming that the cucullaris is a branchiomeric muscle (Sefton et al. 2016). Notably, their data suggest that, at least in the axolotl, somite 3 is the posterior limit of mesodermal contribution to cranial structures in both paraxial and lateral mesoderms. This might help explain why some authors have reported a partial contribution from the anterior somites to the cucullaris/protractor pectoralis.

Hypotheses of homology for most of the pharyngeal and laryngeal muscles in sarcopterygians based on the data in this chapter are presented in Tables 8.5 and 8.6. The monotreme pharyngeal muscle constrictor pharyngis corresponds to the constrictor pharyngis medius + constrictor pharyngis inferior and, possibly, the constrictor pharyngis superior and pterygopharyngeus of therian mammals, although it is more likely that the superior constrictor and pterygopharyngeus are derived instead from the palatopharyngeus (Tables 8.5 and 8.6; see Figures 8.15 and 8.21). The pharyngeal muscles salpingopharyngeus + levator veli palatini + musculus uvulae + palatoglossus + palatopharyngeus of therian mammals clearly seem to derive from the primordia that gives rise to the palatopharyngeus in monotremes. In the study by Diogo and Abdala (2010), we stated that the palatoglossus was most likely derived from the hypobranchial muscle hyoglossus and, specifically from the styloglossus, as proposed by Edgeworth (1935). Edgeworth (1935) based this on his own developmental studies, as well as on data provided by other authors, according to which the palatoglossus is usually innervated by the hypoglossal nerve (CNXII), including in humans. Most current atlases of human anatomy, however, refer to an innervation by the vagus nerve (CNX) and, therefore, group this muscle with the true pharyngeal muscles rather than with the tongue hypobranchial muscles; some atlases refer to an innervation by the cranial part of the accessory nerve (CNXI) (see the review by Diogo and Wood [2012a]). In support of Edgeworth's hypothesis of a tongue developmental origin, studies have suggested that in at least some mammals, including nonhuman primates, the palatoglossus is innervated by CNXII (reviewed by Sokoloff and Deacon [1992]). However, in their study of *Macaca fascicularis*, Sokoloff and Deacon [1992] did not find a pattern of innervation truly similar to that found in other tongue muscles such as the styloglossus, which, according to Edgeworth (1935), is the main muscle from which the palatoglossus is derived. Sokoloff and Deacon (1992) pointed out that based on their data and on developmental data on mice, either a palatal or a tongue (or both) origin of the palatoglossus can be considered a plausible hypotheses. In fact, in addition to the well-studied innervation of the palatoglossus in humans indicating that the muscle is innervated by CNX, there are also developmental studies supporting the idea that the muscle derives from the palatopharyngeus/superior constrictor musculature (Schaeffer 1929; Cohen et al. 1993, 1994). Authors such as House (1953), who studied and discussed in detail the pharyngeal region in mammals, suggested that the palatoglossus specifically derives from the glossopharyngeal part of the superior constrictor of the pharynx, i.e., the part that inserts onto the tongue, through an anterior migration of the origin of the muscle from the pharyngeal wall/medial raphe to the soft palate/lateral wall of the oropharynx. More recent developmental studies, including detailed studies of human development (e.g., Cohen et al. 1993) and molecular developmental studies of mice (Grimaldi et al. 2015) strongly support the idea that the palatoglossus derives from the pharyngeal—and not the tongue—musculature and provide stronger support for a close developmental relationship between the palatoglossus and the palatopharyngeus, levator veli palatini, and uvulae than between the palatoglossus and the superior pharyngeal constrictor. Furthermore, our dissections and review of the literature on marsupial myology indicate that the palatoglossus is a well-developed muscle in marsupials and, therefore, that this muscle was already differentiated in the last common ancestor (LCA) of placentals + marsupials, while the superior constrictor only became differentiated in placentals. Based on these data, it is difficult to argue that the evolutionary more recent superior constrictor muscle gave rise to the evolutionary older muscle palatoglossus. Although we believe that more data are needed to settle once and for all the origin of the palatoglossus, the combination of the available literature and our own observations thus strongly supports a pharyngeal origin of the palatoglossus, specifically from the primordia that also give rise to the levator veli palatini, palatopharyngeus, and musculus uvulae (Tables 8.5 and 8.6).

However, some studies, particularly recent ones and/or ones focused on humans, have suggested that the levator veli palatini is at least partially innervated by the facial nerve (CNVII) and might even be primarily a hyoid muscle (i.e., of second branchial arch) (see, e.g., the review of Diogo and Wood [2012a]). Most studies of nonhuman mammals such as *Macaca*, dogs and cats do not refer to an innervation from the facial nerve, but a few studies in mice have suggested an innervation by the vagus (CNX), glossopharyngeal (CNIX), and facial nerve (reviewed by Kishimoto et al. [2016]). Kishimoto et al. (2016), based on their study of human embryos and fetuses, argued that the levator veli palatini primordium appears in the region of the hyoid arch and that the reason most authors do not mention an innervation by the facial nerve is that the pharyngeal plexus (which includes branches of CNX) innervates most of the superior part (near the origin) of the levator veli palatini, while the lesser palatine nerve (related to the facial nerve) innervates only a very small inferior part of the muscle (near its insertion).

Another interesting finding of Grimaldi et al.'s (2015) developmental study of mice is that all the mesenchyme and tendons of the soft palate muscles were derived from the cranial neural crest (CNC), with the exception of the posterior attachment of the palatopharyngeus (which extends posteriorly to the region of the larynx), which was anchored in a mesoderm-derived pharyngeal wall, constituting the posterior border of the CNC-derived mesenchyme domain of the pharynx. The circular pharyngeal constrictor muscles (superior, middle, and inferior) were partially embedded in neural crest-derived mesenchyme, and the pharyngeal wall constituted the posterior border of the neural crest contribution to the craniofacial mesenchymal tissues. That is, the pharyngeal wall may represent an interface between CNC-derived mesenchyme and mesoderm-derived mesenchyme. In birds, the larynx and the area with efferent innervation from the vagus nerve (CNX) also mark the transition between the region of the head where the connective tissues are derived from neural crest cells and the muscles are patterned by these cells; in the region posterior to it, the connective tissues arise from somites (reviewed by Smith [1992]).

With respect to the laryngeal muscles (Figures 8.10, 8.16, and 8.23), the thyroarytenoideus, vocalis, cricoarytenoideus lateralis, and arytenoideus of therian mammals correspond to the thyrocricoarytenoideus and arytenoideus of monotremes and to the laryngeus of nonmammalian tetrapods such as salamanders. The mammalian cricoarytenoideus posterior corresponds to the dilator laryngis of other tetrapods (Tables 8.5 and 8.6). In terms of both its ontogeny and phylogeny, the mammalian cricothyroideus is clearly a pharyngeal rather than a laryngeal muscle as is sometimes suggested in the literature (e.g., *Terminologia Anatomica* [1998]) (Tables 8.5 and 8.6). It should also be noted that according to authors such as Smith (1994), marsupials have no levator veli palatini. Their "functional superior constrictor" is formed by an expansion of the stylopharyngeus, not from the same mass of muscles that give rise to the middle and inferior constrictors, as is the case in placental mammals.

HYPOBRANCHIAL MUSCLES

According to Edgeworth (1935), the hypobranchial muscles (Figures 8.9, 8.15, 8.17, 8.22, and 8.24 through 8.26) are divided into a "geniohyoideus" group and a "rectus cervicis" group (Tables 8.7 and 8.8). However, it is not clear if Edgeworth's groups represent separate premyogenic condensations, or whether they only become apparent at the later stages of muscle development. The plesiomorphic condition for sarcopterygians seems to be that of extant actinistians and dipnoans: there are two hypobranchial muscles that are mainly related to the opening of the mouth, the coracomandibularis and the sternohyoideus (Edgeworth 1935; Kesteven 1942–1945; Wiley 1979a,b; Jollie 1982; Mallatt 1997; Wilga et al. 2000; Johanson 2003; Diogo and Abdala 2010) (Tables 8.7 and 8.8). Amphibians and reptiles have some hypobranchial muscles (e.g., the omohyoideus and the specialized glossal muscles related to tongue movements) that are not present in sarcopterygian fishes (Figure 8.4; Tables 8.7 and 8.8). The geniohyoideus, genioglossus, hyoglossus, and intrinsic muscles of the tongue of nonmammalian tetrapods very likely correspond to the coracomandibularis of sarcopterygian fishes, although it is possible that the "hyoglossus" of, e.g., salamanders is at least partially derived from the sternohyoideus (e.g., Edgeworth 1935; Jarvik 1963; Diogo and Abdala 2010) (Tables 8.7 and 8.8). The styloglossus of therian mammals seems to correspond to part of the hyoglossus of monotremes (Tables 8.7 and 8.8). The mammalian thyrohyoideus and sternothyroideus correspond to part of the sternohyoideus of reptiles such as *Timon* (Tables 8.7 and 8.8; some authors described a "sternothyroideus" in a few reptilian taxa, but that this muscle is probably not homologous to the mammalian sternothyroideus). See also the comments on the mammalian palatoglossus muscle that were provided on the preceding section.

EMBLEMATIC EXAMPLE OF THE REMARKABLE DIVERSITY AND EVOLVABILITY OF THE MAMMALIAN HEAD: THE EVOLUTION OF PRIMATE FACIAL EXPRESSION MUSCLES, WITH NOTES ON THE NOTION OF A SCALA NATURAE

The face of humans and other mammals is a complex morphological structure in which both external and internal parts function in conveying information relevant for social interactions. Externally, facial features bear signals that allow the recognition of conspecifics, individuals within the social group and potential mates. This information is encrypted in traits such as the shape of facial parts and the complexity and hues of its color patterns (Figure 8.27; Waitt et al. 2003; Setchell 2005). Internally, the facial musculature (Figure 8.8) and neural centers control how the external morphology is showcased to other individuals through the production of facial expressions, which are important in communicating behavioral intentions within a social context (e.g., bared teeth communicate the intent to withdraw from an agonistic encounter; Preuschoft and Van Hooff 1997). Therefore, internal and external anatomical features of the face are not only in close physical proximity, but also tightly connected in their function.

Facial coloration patterns evolved in tandem with sociality and sympatry (when two species or populations exist in the same geographic area) in primates (Santana et al. 2012, 2013). In many primate radiations, highly social and sympatric species evolved multicolored faces, while less social species tend to have less colorful faces. Complex facial patterns potentially enable higher interindividual variation within social groups and among species, facilitating recognition at either of these levels. Facial expressions are also linked to sociality; highly gregarious species produce a wider variety of facial movements, which may function in group cohesion by enhancing communication during conflict management and bonding (Dobson 2009a). Facial expressions result from the action of facial muscles that are controlled by neural pathways (facial nucleus of the pons—cranial nerve VII—and the primary motor cortex), and primate species with relatively large facial nuclei tend to have highly dexterous faces (Sherwood et al. 2005). The primate facial musculature is among the most complex across mammals (although not as complex as that of, e.g., elephants; Boas and Paulli 1908, 1925), but it has been unclear if and how it has evolved in response to functional demands associated with ecology and sociality (Burrows 2008; Diogo and Wood 2012a; Diogo et al. 2009a).

The first (mandibular), second (hyoid), and more posterior branchial arches are formed from bilateral swellings on either side of the pharynx. As explained earlier, the muscles of facial expression (e.g., Figure 8.26)—usually designated simply as "facial muscles"—are a subgroup of the hyoid (second arch) muscles and are innervated by the facial nerve (cranial nerve CNVII). This means that all other hyoid muscles (e.g., stapedius and stylohyoideus) are not designated facial muscles, despite being also innervated by the facial nerve. Except for the buccinatorius (and the mandibulo-auricularis present in many nonhuman mammals), the facial muscles are mainly attached to the dermis of the skin and the elastic cartilage of the pinna. They are involved in generating facial expressions during social interactions among conspecifics, as well as in feeding, chemosensation, whisker motility, hearing, vocalization, and human speech. This section, which is a synopsis of one of the sections of Diogo and Santana's (2017) recent review, provides a short summary of the evolution of primate facial muscles as a case study showing the remarkable diversity and evolvability of the mammalian head musculature and its links with external features such as the color of the skin and hair as well as with behavior and ecology.

As explained earlier, the facial muscles are present only in mammals, probably deriving from the ventral hyoid muscle interhyoideus, and likely from at least some dorsomedial hyoid muscles (e.g., cervicomandibularis) of other tetrapods. Monotremes such as the platypus only have 10 distinct facial muscles (not including the extrinsic muscles of the ear) (Figure 8.8). Rodents, such as rats, have up to 24 facial muscles (Figure 8.12). The occipitalis + auricularis posterior, the procerus, and the dilatator nasi + levator labii superioris + levator anguli oris facialis of therian mammals (marsupials + placentals) probably correspond to a part of the platysma cervicale (muscle connecting the back of neck—nuchal region—to the mouth, different from platysma myoides connecting the front of neck and pectoral region to the mouth), of the levator labii superioris alaeque nasi, and of the orbicularis oris of monotremes, respectively. The sternofacialis, interscutularis, zygomaticus major, zygomaticus minor, and orbito-temporo-auricularis of therian mammals probably derive from the sphincter colli profundus, but it is possible that at least some of the former muscles derive from the platysma cervicale and/or platysma myoides. Colugos (Dermoptera or "flying lemurs") and tree shrews (Scandentia), the closest living relatives of primates, have a similar facial musculature (Figure 8.14), but the former lack two muscles that are usually present in the latter, the sphincter colli superficialis and the mandibulo-auricularis. As both these muscles are found in rodents, as well as in tree shrews and at least some primates, they were likely present in the LCA of Primates + Dermoptera + Scandentia. The frontalis, auriculo-orbitalis, and auricularis superior of this LCA very likely derived from the orbito-temporo-auricularis of other mammals, while the zygomatico-orbicularis and corrugator supercilii most likely derived from the orbicularis oculi.

The facial musculature of the LCA of primates was probably very similar to that seen in the extant tree shrew *Tupaia*. Muscles that have been described in the literature as peculiar to primates, e.g., the zygomaticus major and zygomaticus minor, are now commonly accepted as homologues of muscles of other mammals (e.g., of the "auriculolabialis inferior" and "auriculolabialis superior"). The only muscle that is often present as a distinct structure in strepsirrhines (see Figure 8.18)—i.e., the primate group including extant members such as lemurs and lorises (see Figure 8.27)—but not in tree shrews or colugos, is the depressor supercilii, which derives from the orbicularis oris matrix. As the depressor supercilii is present in strepsirrhine and nonstrepsirrhine primates, it is likely that this muscle was present in the LCA of primates. Essentially, the ancestral condition predicted for the LCA of primates is probably similar to that found in some extant strepsirrhines (e.g., *Lepilemur*). Importantly, the number of facial muscles present in living strepsirrhines is higher than that originally reported by authors in the nineteenth and first decades of the twentieth century. For instance, Murie and Mivart (1869) reported only 7 facial muscles in a lemur, grouping all the muscles associated with the nasal region into a single "nasolabial muscle mass." The supposed lack of complexity seen in strepsirrhines was consistent with the anthropocentric, scala naturae, finalistic evolutionary paradigm subscribed to by many anatomists at that time (see preceding text). However, it is now accepted that strepsirrhines often have more than 20 facial muscles, and that although humans have more facial muscles than most primates, the difference is minimal in general. In fact, the total number of facial muscles found in humans is similar to that found in rats, contradicting one of the major myths of human complexity and exceptionalism (Tables 8.3 and 8.4; see Diogo and Wood [2012a, 2013] and Diogo et al. [2015e] for more details on this subject).

In order to give a functional context for these descriptions of the evolution and comparative anatomy of the primate facial muscles, here we provide a brief account of the general function of the facial muscles present in strepsirrhines (Figure 8.16). When, in the next section, we refer to a muscle that is not differentiated in strepsirrhines but is present in anthropoids (monkeys and apes, including humans), we will also briefly describe the general function of that muscle. The platysma myoides most likely draw the oral commissure posteroinferiorly, an action that may be used in social interactions as well as feeding, while the platysma cervicale most likely elevates the skin of the neck. The occipitalis draws the scalp posteriorly toward the nuchal region while the frontalis elevates the skin/brow over the superciliary region. The auriculo-orbitalis may be used to draw the lateral corner of the eyelid posteroinferiorly or the external ear anterosuperiorly. The corrugator supercilii and the depressor supercilii are used to draw the medial edge of the superciliary region inferomedially and inferiorly, respectively. The mandibulo-auricularis may be used to approximate the superior and inferior edges of the external ear, as well as the external ear and the mandible. The muscles clustered around the upper lip, including the zygomaticus major and zygomaticus minor muscles, may be used to draw the upper lip and the posterior region of the mouth posterosuperiorly, functions which may be used in both social interactions and in use of the vomeronasal organ. As their name indicates, the extrinsic muscles of the ear, as well as the auricularis posterior and auricularis superior, are mostly related to movement of the external ear, while the orbicularis oculi and orbicularis oris are primarily associated with movement of the eyes and of the lips, respectively. The buccinatorius mainly pulls the corner of the mouth laterally and presses the cheek against the teeth. The levator labii superioris alaeque nasi, levator labii superioris, and levator anguli oris facialis are most likely used together in drawing the upper lip and the posterior region of the mouth superiorly and medially, which is most likely used in social interactions and in feeding. The mentalis mainly elevates the skin ventral to the lower lip, while the sphincter colli profundus most likely draws the skin of the neck posterosuperiorly.

There are some notable differences between the ancestral condition described earlier for nonanthropoid primates such as *Lepilemur* (Figure 8.18) and the condition found in New World and Old World monkeys (see Figure 8.20). For example, the mandibulo-auricularis is usually not present as an independent, fleshy muscle in most anthropoids, although some of these primates have fleshy vestiges of this muscle as a rare variant. It likely corresponds to the stylo-mandibular ligament seen in hominoids (apes, including humans) such as humans as well as in some monkeys. The sphincter colli profundus is also normally absent in anthropoids, but fleshy vestiges of this muscle have been described in a few macaques as well. Anthropoids often have a depressor anguli oris and a depressor labii inferioris. These muscles are probably derived from the orbicularis oris matrix; some authors suggested that the depressor anguli oris might be the result of a ventral extension of the levator anguli oris. Generally, the depressor anguli oris and depressor labii inferioris function in anthropoids to draw the corner of the mouth posteroinferiorly and to draw the lower lip inferiorly, respectively. These movements are seen in some displays of facial expression and in some feeding contexts.

Among hominoids the platysma cervicale is usually present in hylobatids (lesser apes: gibbons and siamangs) and gorillas, but is often highly reduced or absent in adult orangutans, chimpanzees, and humans (see Figures 8.24 through 8.26). The transversus nuchae, found as a variant in the three latter taxa, is often thought to be a vestigial remain/bundle of the platysma cervicale. Interestingly, the platysma cervicale is present early in the development of humans, but it normally disappears as an independent structure in later stages of development. Unlike the platysma cervicale, the platysma myoides is usually present as a separate structure in adult members of all the major five extant hominoid taxa. The occipitalis is also usually present in these five, but the auricularis posterior is normally not differentiated in orangutans, although it has been described in a few species.

In humans the risorius (Figure 8.26) is usually—but not always—present, pulling the lip corners backward, stretching the lips—a function that is, interestingly, usually associated with the display of fear—being likely derived from the platysma myoides, although it cannot be discarded that it is partly, or even wholly, derived from the zygomaticus major. Among primates a "risorius" is sometimes found in some other hominoids, e.g., chimps, but it does not seem to be present in the normal phenotype (i.e., >50% of the cases) of any of the four major nonhuman hominoid taxa. Moreover, some structures that are often named risorius in these hominoids are probably not homologous to the human risorius, and even to each other, because some apparently derive from the platysma myoides; others, from the depressor anguli oris; and others, from muscles such as the zygomaticus major. All the other facial muscles that are present in macaques are normally present in extant hominoids, but unlike monkeys and other hominoids, humans—and possibly also gorillas—usually also have an auricularis anterior and a temporoparietalis. Both of these muscles are derived from the auriculo-orbitalis, which, in other hominoids such as chimpanzees, has often been given the name "auricularis anterior," although it corresponds to the auricularis anterior plus the temporoparietalis of humans and gorillas. When present, the temporoparietalis stabilizes the epicranial aponeurosis (a tough layer of dense fibrous tissue covering the upper part of the cranium: see, e.g., Figure 8.12), whereas the auricularis anterior draws the external ear superoanteriorly, closer to the orbit.

Each of the three nonprimate taxa listed in Table 8.4 has at least one derived, peculiar muscle that is not differentiated in any other taxa listed in this table. *Ornithorhynchus* has a cervicalis transversus (Figure 8.8), *Rattus* has a sternofacialis and an interscutularis, and *Tupaia* has a zygomatico-orbicularis. This example reinforces the fact that evolution is not directed "toward" a goal, and surely not toward primates and humans; each taxon has its own particular mix of conserved and derived anatomical structures, which is the result of its unique evolutionary history (Diogo and Wood 2013). This is why we encourage the use—in this specific context concerning facial muscles—of the term *correspond* to describe evolutionary relationships among facial muscles, because muscles such as the zygomatico-orbicularis are not *ancestral* to the muscles of primates. The zygomatico-orbicularis simply corresponds to a part of the orbicularis oculi that, in taxa such as *Tupaia*, became sufficiently differentiated to deserve recognition as a separate muscle. Also, strepsirrhines and monkeys have muscles that are usually not differentiated in some hominoid taxa, e.g., the platysma cervicale (usually not differentiated in orangutans, chimps, and humans) and the auricularis posterior (usually not differentiated in orangutans).

Humans, together with gorillas, have the greatest number of facial muscles among primates, and this is consistent with the important role played by facial expression in anthropoids in general, and in humans in particular, for communication. Nevertheless, the evidence presented in this chapter, as well as in recent works by Burrows and colleagues (e.g. Burrows 2008; Burrows et al. 2014), shows that the difference between the number of facial muscles present in humans and in hominoids such as hylobatids, chimpanzees, and orangutans, and between the number of muscles seen in the latter hominoids and in strepsirrhines, is not as marked as previously thought. In fact, as will be shown in the following the display of complex facial expressions in a certain taxon is related not only to the number of facial muscles but also to their subdivisions, arrangements of fibers, topology, biochemistry, and microanatomical mechanical properties, as well as to the peculiar osteological and external features (e.g., color) and specific social group and ecological features of the members of that taxon.

For instance, from bright red to yellow, black, brown, and even blue, the faces of primates exhibit almost every possible hue in the spectrum of mammalian coloration (Figure 8.27; e.g., Santana et al. 2012). In many species, such as mandrills and guenons, facial skin and hair colors are combined to create remarkably complex patterns that are unique to the species. Is there a functional significance to these colors and their patterns? Recently, researchers have harnessed the tools of modern comparative methods and computer simulation to answer this question and investigate the factors underlying the evolution of facial color diversity across primate radiations. Several lines of evidence suggest that facial colors are crucial to the ecology and social communication of primates. Variation in coloration within a species, such as the differences in brightness of red facial patches among male mandrills, appear to be used for the assessment of overall health condition and potential mate quality (Setchell 2005; Setchell et al. 2006). At a broader scale, differences across species in facial color patterns are hypothesized to enable individuals of sympatric and closely related species to identify one another and avoid interbreeding. Phylogenetic comparative studies have demonstrated that social recognition explains trends in the evolution of primate facial color patterns. In the New World primate radiation (Platyrrhini), species that live in small social groups or are solitary (e.g., owl monkeys, *Aotus*) have evolved more complexly patterned faces (Santana et al. 2012). In sharp contrast, diversity trends in Old World groups (Catarrhini) are the opposite, with highly gregarious species having more complexly patterned faces (Santana et al. 2013). These divergent trends may be explained by habitat differences and a higher reliance on facial expressions and displays for intraspecific communication in catarrhines (Dobson 2009a; Mancini et al. 2013), in which facial colors may be further advertised through stereotyped head movements during courtship or appeasement behaviors (Kingdon 1992, 2007).

Across all primates studied to date, the evolution of complexly patterned faces is also tightly linked to high levels of sympatry with closely related species (Santana et al. 2012, 2013). A face that is colorful may present features that are unique and more easily recognizable in the context of multiple sympatric species. Allen et al. (2014) used computational face recognition algorithms to model primate face processing. Their results demonstrated that the evolution of facial color patterns in guenons fits models of selection to become more visually distinctive from other sympatric guenon species. This indicates that facial color patterns function as signals for species recognition in primates and may promote and maintain reproductive isolation among species.

The degree of facial skin and hair pigmentation is also highly variable across primates, and comparative studies suggest that this diversity may illustrate adaptations to habitat. Darker, melanin-based colors in the face and body are characteristic of primate species that inhabit tropical, more densely forested regions (Kamilar and Bradley 2011). It is hypothesized that these darker colors may reduce predation pressure by making individuals more cryptic to visually oriented predators (Zinck et al. 2004; Stevens and Merilaita 2009) and increase resistance against pathogens (Burtt Jr. and Ichida 2004). Darker facial colors may also offer protection against high levels of ultraviolet (UV) radiation and solar glare (Caro 2005) and aid in thermoregulation (Burtt 1986). However, the role of facial pigmentation in these functions remains unclear because primates may use behaviors to regulate their physiology (e.g., arboreal species can move from the upper canopy, which has the highest UV levels, to the middle and lower canopies, which are highly shaded). In catarrhines, ecological trends in facial pigmentation are significant only in African species (Santana et al. 2013), presumably because the African continent presents more distinct habitat gradients than South East Asia. In platyrrhines, darker faces are found in species that live in warmer and more humid areas, such as the Amazon, and darker eye masks are predominant in species that live closer to the Equator. Eye masks likely function in glare reduction in habitats with high UV incidence, and similar trends in this facial feature have also been observed in carnivorans and birds (Burtt 1986; Ortolani 1999). The presence and length of facial hair is highly variable across primate species, but its role in social communication, besides acting as a vehicle to display color, has not been broadly investigated. In platyrrhines, species that live in temperate regions have longer and denser facial hair (Santana et al. 2012), which might aid in thermoregulation (Rensch 1938). Similar trends would be expected in other primate radiations.

To date, the evolutionary connections between external (coloration, facial shape) and internal (musculature) facial traits are poorly known. In recent studies (Santana et al. 2014; Diogo and Santana 2017), we contrasted two major hypotheses that might explain the evolution of primate facial diversity when these traits are integrated. First, if the evolution of facial displays has been primarily driven by social factors, highly gregarious primates would possess both complexly colored and highly expressive faces as two concurrent means for social communication. Alternatively, if external facial features influence the ability of primates to perceive and identify facial expressions (Vick et al. 2007), there would be a trade-off in the evolution of facial mobility and facial color patterning, such that highly expressive faces would have simpler color patterns. We used phylogenetic comparative analyses integrating data on facial mobility, facial musculature, facial color pattern complexity, body size, and orofacial motor nuclei across 21 primate species to test these hypotheses.

The results from our study indicated a significant association between the evolution of facial color patterns and facial mobility in primates. Supporting the second hypothesis, primates evolved plainly colored faces in tandem with an enhanced ability for facial expressions. Thus, while complex facial color patterns may be beneficial for advertising identity (Santana et al. 2013; Allen et al. 2014), a highly "cluttered" face may mask the visibility of facial expressions used to convey behavioral intention. Why a species may rely more on facial color patterns versus facial expressions for communication is still unclear, but it is possible that these different modalities may be differentially selected across primate lineages based on the species' habitat, social systems, or body size. Larger primates (e.g., apes), which have a larger facial nucleus, have more expressive faces than smaller species (Dobson 2009b), which, in turn, seem to use colorful facial patterns and head movements for communication. The evolution of larger bodies, potentially coupled with increased reliance on vision for other ecological tasks (e.g., finding food and avoiding predators), may have allowed a higher reliance on facial expressions, which was not possible at smaller body sizes due to physical constraints on the perception of facial movements. Smaller species are expected to have more difficulty discerning facial expressions because smaller mammalian eyes have lower visual acuity (Moynihan 1967; Veilleux and Kirk 2014).

Although the evolution of facial mobility is linked to facial coloration and body mass, we found that it is not directly related to the number of muscles that produce facial movements. The number of facial muscles per se is a slowly evolving trait that has strong phylogenetic inertia (see preceding sections and Diogo and Wood 2012a). Conversely, the size of the facial nucleus has evolved rapidly in the sample of primates studied. These results indicate that changes in facial mobility are likely to evolve first *via* changes in neurophysiology and body mass, instead of muscle morphology, that is, through motor control of muscles instead of the creation of new divisions of preexisting musculature. That is, it is interesting to note that while the number of facial muscles is rather conservative among primates, the evolution of the facial musculature as a whole and in particular of facial expression in general provides an emblematic case study of the diversity and evolvability of primates and of mammals. Moreover, these patterns of evolution and potential trade-offs give important insight into the simple organismal features, such as body mass, that have a strong relevance for which and how different types of facial cues evolve for social communication.

GENERAL REMARKS

This review, summarized in Tables 8.1 through 8.8, allows us to provide a very detailed list of muscle synapomorphies of mammals, of therians (marsupials + placentals), and of placentals. Based on this comparison, extant mammals share 34 muscle synapomorphies for the head. These numbers illustrate the utility of studying muscles to characterize certain clades, and pave the way for paleontological, developmental, and functional works that investigate the specific evolutionary time of origin/loss and developmental mechanisms that led to the characteristic muscle anatomy of each clade and their functional implications. Specifically, there are 10 synapomorphies of the mandibular muscles of extant mammals: differentiation of mylohyoideus, digastricus anterior, masseter, temporalis, pterygoideus lateralis, pterygoideus medialis, tensor tympani, and tensor veli palatini and loss of adductor mandibulae A2-PVM and of dorsal mandibular muscles (Tables 8.1 and 8.2). There are 12 for the hyoid muscles: differentiation of styloideus, stapedius, platysma cervicale, platysma myoides, extrinsic ear muscles, sphincter colli superficialis, sphincter colli profundus, orbicularis oculi, nasolabialis/levator labii superioris alaeque nasi, buccinatorius, orbicularis oris, and mentalis (Tables 8.3 and 8.4). There are 11 for the branchial muscles: differentiation of acromiotrapezius, spinotrapezius, dorsocutaneous, cleidomastoideus, sternomastoideus, constrictor pharyngis, cricothyroideus, palatopharyngeus, thyrocricoarytenoideus, and arytenoideus and loss of constrictor laryngis (Tables 8.5 and 8.6). There is only one for the hypobranchial muscles: differentiation of sternothyroideus (Tables 8.7 and 8.8). Therefore, in the transitions that led to the LCA of extant mammals, all major groups of head muscles experienced drastic changes with the exception of the much more conserved hypobranchial muscles of somitic origin, which experienced a single synapomorphic change. In addition to describing the drastic changes that occurred in both the head and limbs (see the following chapter) during the transitions leading to extant mammals, our results also show mosaic evolution because some subregions of the head: for example, the hypobranchial muscles changed less than other head muscles.

There were 28 head muscle synapomorphic changes from the LCA of extant mammals to the LCA of extant therians. There were no changes in the mandibular muscles (Tables 8.1 and 8.2), 18 in the hyoid muscles (differentiation of stylohyoideus, digastricus posterior, occipitalis, auricularis posterior, mandibulo-auricularis, interscutularis, zygomaticus major, zygomaticus minor, frontalis, auriculo-orbitalis, auricularis superior, corrugator supercilii, retractor anguli oculi lateralis, dilatator nasi, levator labii superioris, nasalis, and levator anguli oris facialis and loss of remains of original interhyoideus: Tables 8.3 and 8.4), 8 in the branchial muscles (differentiation of cleido-occipitalis, constrictor pharyngis medius, constrictor pharyngis inferior, palatoglossus, pterygopharyngeus, musculus uvulae, thyroarytenoideus, and crico-arytenoideus lateralis: Tables 8.5 and 8.6) and only 2 in the hypobranchial muscles (differentiation of styloglossus and thyrohyoideus: Tables 8.7 and 8.8). These results indicate that the origin of therians was especially marked by evolutionary changes in facial muscles, pharyngeal muscles, and laryngeal muscles. These changes were probably related to specializations in facial and vocal communication through movements of both the larynx and pharynx and new ways of feeding, including mastication and suckling.

There are three or four head muscle synapomorphies of extant placentals. None concerns the mandibular and hypobranchial muscles (Tables 8.1, 8.2, 8.7, and 8.8), none or just one concern the hyoid muscles (possibly the differentiation of depressor septi nasi: Tables 8.3 and 8.4), and three concern the branchial muscles (differentiation of constrictor pharyngis superior, levator veli palatini, and salpingopharyngeus: Tables 8.5 and 8.6). Thus, except for the pharyngeal muscles, the head muscles changed very little from the LCA of extant therians to the LCA of placentals. These changes probably related to further specializations of the movements of the larynx (moved by the pharyngeal muscle salpingopharyngeus) and pharynx for vocal communication and/or feeding mechanisms. Therefore, the few synapomorphic changes from the LCA of extant therians to the LCA of extant placentals are distributed more or less equally among the three major anatomical regions (head three or four and, as will be shown in the followingchapters, forelimb three, hindlimb four). These numbers provide empirical support for a well-defined therian body plan, which can still be easily recognized in most extant placentals and marsupials except very specialized taxa such as bats and whales.

Of course, even among therians that conform to this characteristic therian *Bauplan*, there are minor differences in adult phenotype, especially between taxa from different higher clades such as placentals vs. marsupials. For instance, the larynx in marsupials is clearly derived: the cricoid and thyroid cartilages are fused, leading to the absence of the cricothyroid muscles and an articulation between the two arytenoid cartilages (NB: the articulation between these cartilages and an interarytenoid cartilage seems to be plesiomorphic for mammals). Many of these specific, "minor" differences among adults of different taxa seem to be related to needs of the embryos and/or neonates. For instance, Symington (1898) explained that these differences in larynx morphology might be related to the fact that marsupials remain in the pouch for a long time attached to the teat and thus need to, for instance, have safer ways to drink and breathe simultaneously. This requirement might also explain the expansion of the palatopharyngeus muscle/connective tissue, and perhaps the expansion of the pars pharyngea of the stylopharyngeus, which are also derived characters among marsupials. It is hoped that this review, and our long-term project in general, will contribute toward the multidisciplinary data needed for an integrative synthesis of the anatomical macroevolution of the mammalian head and for future functional and developmental comparative studies.

9 Head and Neck Muscles of Amphibians

The main goal of Chapter 8 was to discuss the homologies and evolution of muscles within Sarcopterygii as a whole and thus to provide a basis for discussions focused on each of the major sarcopterygian groups. In Chapter 9, we specifically analyze the mandibular, hyoid, branchial, and hypobranchial muscles of the three extant amphibian taxa: salamanders (Caudata or Urodela), frogs (Anura), and caecilians (Gymnophiona). In recent decades, new information has become available about the ontogeny of the cephalic muscles in representatives of each of these amphibian groups, which is particularly useful for the analysis of the homologies of these muscles within these groups (e.g., Olsson et al. 2000, 2001; Palavecino 2000; Chanoine and Hardy 2003; Ericsson and Olsson 2004; Ericsson et al. 2004; Kleinteich and Haas 2007; Piekarski and Olsson 2007; Ziermann and Olsson 2007; Schmidt et al. 2013; Ziermann and Diogo 2013). In addition, Carroll (2007) published an excellent, extensive review of the phylogeny and evolution of caecilians, urodeles, and anurans, which provides the phylogenetic background for our discussions about plesiomorphic versus derived traits among the amphibians. As explained by the latter author, extant frogs and salamanders may be more closely related to each other than to caecilians (see Figure 1.1 and the following text).

MANDIBULAR MUSCLES

The adductor mandibulae A2 (Figures 9.1 through 9.5), the adductor mandibulae A2-PVM (Figures 9.1, 9.3 through 9.5), and the pseudotemporalis (Figures 9.1 through 9.4) are included in the so-called adductor mandibulae complex of amphibians, which is often named "levator mandibulae complex" (see, e.g., Carroll 2007) (Table 9.1). In the urodeles we dissected, the adductor mandibulae A2 is a large muscle lying lateral and anterior to the A2-PVM. In the caecilians we dissected, the A2 is covered by the skull bones, which form an adductor chamber: in *Siphonops* and *Chthonerpeton*, this muscle is well-developed (see Figures 9.4 and 9.5). The A2 of the anurans we dissected corresponds to the "levator mandibulae externus," but possibly also to the "levator mandibulae longus," *sensu* authors such as Ziermann and Olsson (2007), the A2-PVM and pseudotemporalis thus possibly corresponding respectively to the "levator mandibulae articularis" and "levator mandibulae internus" *sensu* these authors (Table 9.1 and Figure 9.3). That the "levator mandibulae longus" of anurans corresponds to part of the A2 *sensu* this volume is supported by the fact that in these amphibians, the levator mandibulae longus is often the first element of the "adductor mandibulae complex" to become differentiated during ontogeny (e.g., Ziermann and Olsson 2007). During the development of bony fishes and tetrapods, the first adductor mandibulae muscle to become differentiated is generally the A2 (see preceding chapters). However, the hypothesis proposed by authors such as Carroll and Holmes (1980) and Iordansky (1992), i.e., that the anuran levator mandibulae longus corresponds to part of the pseudotemporalis, and not of the A2, of urodeles, cannot be completely discarded. In anuran larvae and adults the levator mandibulae externus *sensu* Ziermann and Olsson (2007) is often poorly developed; in caecilian larvae, the levator mandibulae externus *sensu* Kleinteich and Haas (2007) is often also a small bundle, which becomes absent as an independent structure in adults, being possibly fused with the levator mandibulae longus *sensu* these authors (e.g., Kleinteich and Haas 2007; Ziermann and Olsson 2007; see Figures 9.3 through 9.5 and Table 9.1).

In the urodeles and anurans we dissected, the pseudotemporalis is often divided into a superficial bundle (which probably corresponds to the "adductor mandibulae A3" *sensu* Diogo and Chardon [2000] and Diogo [2007, 2008a,b], to the "pseudotemporalis posterior" *sensu* Iordansky [1992], and to the "levator mandibulae anterior superficialis" *sensu* Edgeworth [1935] and Piatt [1938]), and a deep bundle (which probably corresponds to the adductor mandibulae A3 *sensu* Diogo and Chardon [2000] and Diogo [2007, 2008a,b], to the "pseudotemporalis anterior" *sensu* Iordansky [1992], and to the "levator mandibulae anterior profundus" *sensu* Edgeworth [1935] and Piatt [1938]) (see Figures 9.1 through 9.3). In the caecilians we dissected, as well as in those caecilians described by authors such as Bemis et al. (1983) and Carroll (2007), the pseudotemporalis is mainly undivided (see Figure 9.4).

The pterygomandibularis is a muscle that is often found in reptiles and that, as its name indicates, usually connects the pterygoid region to the retroarticular process of the mandible (e.g., Oelrich 1956; Abdala and Moro 1996; Herrel et al. 2005; see preceding chapters). Confusingly, Versluys (1904), based on the erroneous (see Chapter 8) supposition that the pterygomandibularis of reptiles corresponds to the pterygoideus medialis and pterygoideus lateralis of mammals, decided to name the pterygomandibularis as "pterygoideus." Since then, both the names pterygoideus and pterygomandibularis have been used to designate the reptilian muscle (see, e.g., Abdala and Moro 1996 and Tables 8.1 and 10.1). This is especially problematic because many reptiles have two other mandibular muscles that are named levator pterygoideus and protractor pterygoideus, but that, unlike the pterygomandibularis, are dorsal mandibular muscles; i.e., they are derived from the constrictor dorsalis anlage *sensu* Edgeworth (1935) (see preceding chapters). In the caecilians we dissected, as well as those described by authors such as Iordansky (1996), Kleinteich and Haas (2007), and Carroll (2007), there is a muscle that is often named pterygoideus and that does seem to correspond to the pterygomandibularis of reptiles, as proposed by Iordansky

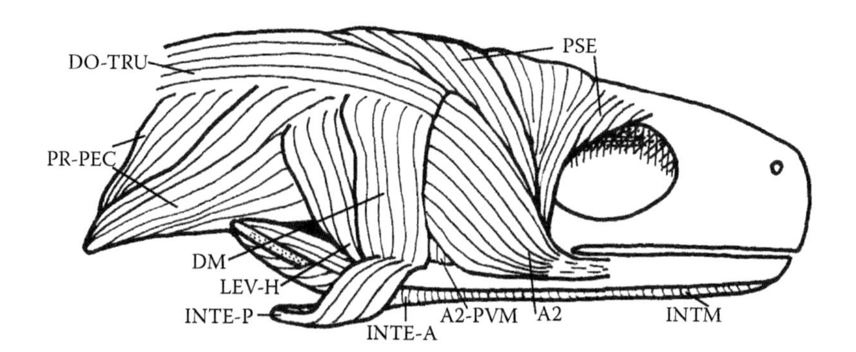

FIGURE 9.1 *Ambystoma tigrinum* (Amphibia, Caudata): lateral view of cephalic musculature (the nomenclature of the myological structures illustrated follows that of the present work; for more details about the osteological structures illustrated, see Larsen and Guthrie [1975]). A2, A2-PVM, adductor mandibulae A2 and A2-PVM; DM, depressor mandibulae; DO-TRU, dorsalis trunci; INTE-A, INTE-P, interhyoideus anterior and posterior; INTM, intermandibularis; LEV-H, levator hyoideus; PR-PEC, protractor pectoralis, PSE, pseudotemporalis. (Modified from Larsen, J. H., and Guthrie, D. J., *J Morphol*, 147, 137–154, 1975.)

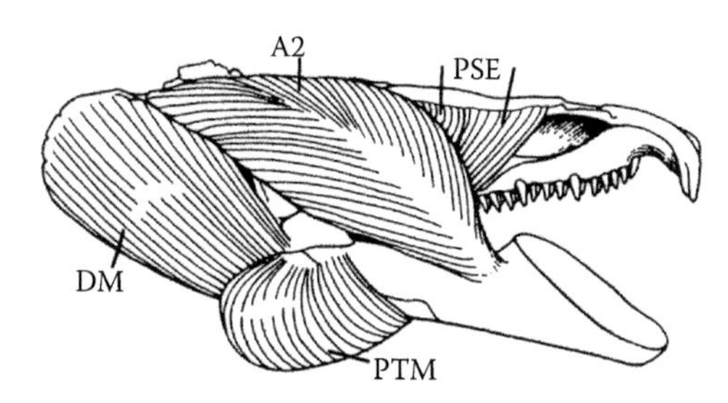

FIGURE 9.2 *Siren lacertina* (Amphibia, Caudata): lateral view of the cephalic musculature (the nomenclature of the myological structures illustrated follows that of the present work; for more details about the osteological structures illustrated, see Carroll and Holmes [1980]). A2, adductor mandibulae A2; DM, depressor mandibulae; PSE, pseudotemporalis, PTM, pterygomandibularis. (Modified from Carroll, R. L., and Holmes, R., *Zool J Linn Soc Lond*, 68, 1–40, 1980.)

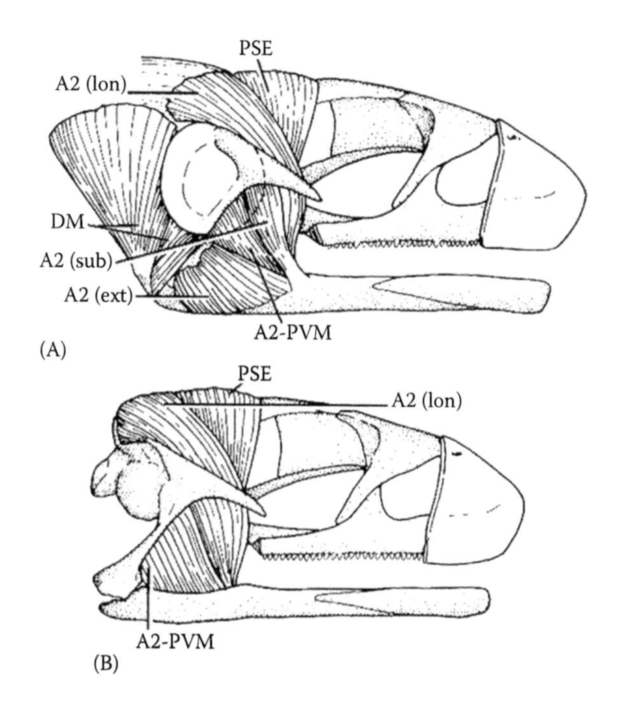

FIGURE 9.3 *Rana temporaria* (Amphibia, Anura): (A) Lateral view of the cephalic musculature; (B) same view, but showing only deeper musculature (the nomenclature of the myological structures illustrated follows that of the present work; for more details about the osteological structures illustrated, see Duellman and Trueb [1986]). A2 (sub), A2 (ext), A2 (lon), "subexternus," "externus" and "longus" portions of adductor mandibulae, which probably correspond to the A2 *sensu* the present work; A2-PVM, adductor mandibulae A2-PVM; DM, depressor mandibulae; PSE, pseudotemporalis. (Modified from Duellman, W. E., and Trueb, L., *The Biology of Amphibians*, McGraw-Hill, New York, 1986.)

(1996) and Kleinteich and Haas (2007) (Figure 9.5). In fact, it is generally agreed that this muscle pterygoideus of caecilians is ontogenetically derived from the pseudotemporalis (Carroll 2007), as is the pterygomandibularis of reptiles (see Chapters 8 and 10). In caecilians, the pterygoideus is functionally related to the opening of the mouth and to the holding of the lower jaw against the quadrate (Carroll 2007). As described by authors such as Carroll and Holmes (1980), urodeles, but not anurans, often have a muscle "pterygoideus" that seems to be homologous to that of caecilians and thus to the pterygomandibularis of reptiles (Figure 9.2), although in taxa such as *Ambystoma*, this muscle is usually deeply mixed with the pseudotemporalis profundus (Table 9.1; see also, e.g., Luther 1914 and Haas 2001).

The intermandibularis is ontogenetically derived from the ventral portion of the mandibular muscle plate (*sensu* Edgeworth [1935]: see preceding chapters) rather than from the central portion of this plate as are the adductor mandibulae A2, adductor mandibulae A2-PVM, pseudotemporalis, and pterygomandibularis. In all extant amphibians, the intermandibularis is a large muscle that connects the two hemimandibles (Figures 9.1 and 9.4 through 9.6). In anurans and urodeles, it is often divided into an intermandibularis anterior (which is often named submentalis) and an intermandibularis posterior (Figure 9.6), and in at least some anuran tadpoles, it is divided into a "mandibulolabialis" *sensu* Carroll (2007). In the caecilians we

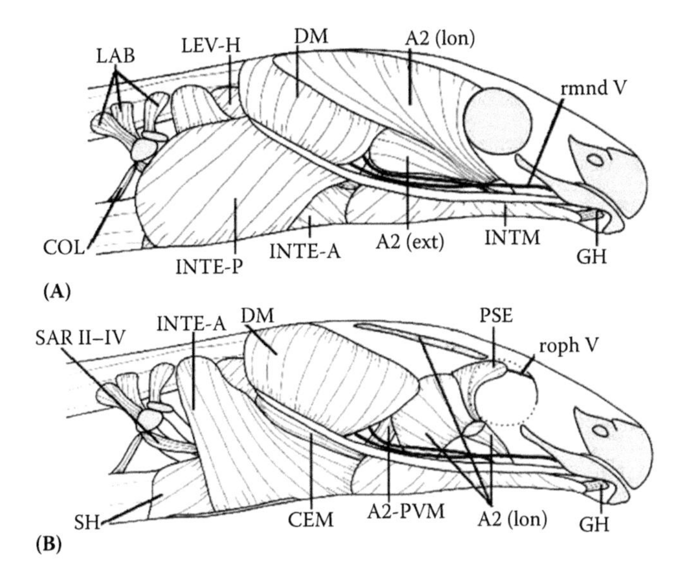

FIGURE 9.4 *Ichthyophis kohtaoensis* (Amphibia, Gimnophiona): (A) Lateral view of the cephalic musculature of a larva; (B) same view, but the eye, the interhyoideus posterior, and the adductor mandibulae "externus" and part of the lateral layer of the adductor mandibulae "longus" *sensu* Kleinteich and Haas (2007) were removed (the nomenclature of the myological structures illustrated follows that of the present work; for more details about the osteological structures illustrated, see Kleinteich and Haas [2007]). A2 (ext), A2 (lon), "externus" and "longus" portions of adductor mandibulae, which probably correspond to the A2 *sensu* the present work; A2-PVM, adductor mandibulae A2-PVM; CEM, ceratomandibularis; COL, constrictor laryngis; DM, depressor mandibulae; GH, geniohyoideus; INTE-A, INTE-P, interhyoideus anterior and interhyoideus posterior; INTM, intermandibularis; LAB, levatores arcuum branchialium; LEV-H, levator hyoideus; PSE, pseudomandibularis; rmnd V, roph V, ramus ophthalmicus and ramus mandibularis V; SAR II-IV, subarcualis rectus II-IV; SH, sternohyoideus. (Modified from Kleinteich, T., and Haas, A., *J Morphol*, 268, 74–88, 2007.)

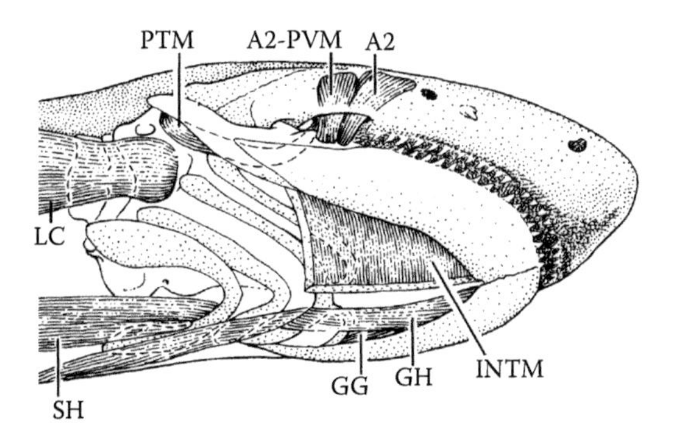

FIGURE 9.5 *Dermophis mexicanus* (Amphibia, Gimnophiona): ventral view of the cephalic musculature, on the right of the animal the A2, A2-PVM, pterygomandibularis and intermandibularis were removed (the nomenclature of the myological structures illustrated follows that of the present work; for more details about the osteological structures illustrated, see Bemis et al. [1983]). A2, A2-PVM, adductor mandibulae A2 and A2-PVM; GG, genioglossus; GH, geniohyoideus; INTM, intermandibularis; LC, longus capitis; PTM, pterygomandibularis; SH, sternohyoideus. (Modified from Bemis, W. E. et al., *Zool J Linn Soc*, 77, 75–96, 1983.)

dissected, as well as in those described by authors such as Kesteven (1942–1945), Bemis et al. (1983), Kleinteich and Haas (2007), and Carroll (2007), the intermandibularis is constituted by a single mass of fibers (see Figures 9.4 and 9.5 and Table 9.1).

Edgeworth (1935) suggested that adult amphibians do not have muscles derived from the "dorsal portion of the mandibular muscle plate," i.e., from the constrictor dorsalis anlage. However, as explained by authors such as Brock (1938) and Iordansky (1996), the levator quadrati and levator bulbi of adult amphibians do seem to derive from this anlage (see Chapter 8). Brock (1938) argued that the levator quadrati of caecilians and the levator bulbi of urodeles and anurans are the remains of the "constrictor dorsalis group" in amphibians. According to this author, the constrictor dorsalis group is usually more fully conserved in extant reptiles because of their kinetic skull (as is the case in the "lizard" *Timon*, many extant reptiles have three dorsal mandibular muscles, the levator pterygoidei, the protractor pterygoidei, and the levator bulbi: see Chapters 8 and 10). Iordansky (1996) stated that the levator quadrati of caecilians corresponds to the levator pterygoideus + protractor pterygoidei of reptiles. Some authors, such as Edgeworth (1935), have suggested that the levator bulbi of urodeles and anurans corresponds to the compressor glandulae orbitalis of caecilians. However, this homology was questioned by authors such as Duellman and Trueb (1986). Carroll (2007: 25) stated that "caecilians have a muscle (the compressor tentaculi) that is homologous with the levator bulbi, but its function has shifted to manipulation of the tentacle (Duellman and Trueb 1986: 385)." However, the caecilian muscle to which Duellman and Trueb (1986) refer on page 385 of their work is an ocular muscle innervated by CNVI (abducens) rather than a mandibular muscle innervated by CNV, as is the levator bulbi of anurans and urodeles (see also Ramaswami 1942).

In fact, among the amphibians we dissected, a distinct levator bulbi was only found in urodeles and anurans, and a distinct levator quadrati was only found in caecilians. As the levator bulbi of urodeles and anurans is indeed similar to that of reptiles (see Chapters 8 and 10), it is very likely that in the last common ancestor (LCA) of amphibians + amniotes, the dorsal mandibular muscles were already divided into (a) an undivided "levator palatini," which was probably similar to, and derived from part of, the levator arcus palatini of sarcopterygian fishes such as *Latimeria* and (b) a levator bulbi, which was probably similar to that of extant urodeles, anurans, and reptiles and derived from part of the levator arcus palatini (Table 9.1). Then, during the evolutionary events leading to the origin of urodeles and anurans, the levator palatini was secondarily lost, probably because these taxa lack a cranial kinesis such as that found in fishes such as coelacanths and in many reptiles (Brock 1938), leaving only the levator bulbi (Table 9.1). As proposed by Iordansky (1996), the levator quadrati of caecilians thus probably corresponds to the levator pterygoidei + protractor pterygoidei of reptiles such as *Timon*,

TABLE 9.1
Mandibular Muscles of Adults of Representative Urodele, Anuran, and Caecilian Taxa

Amphibia (Caudata, or Urodela): *Ambystoma ordinarium* (Michoacan Stream Salamander)	Amphibia (Anura): *Rhinella arenarum* (Argentine Common Toad)	Amphibia (Gymnophiona): *Siphonops paulensis* (Catuchi Caecilian)
Intermandibularis posterior	**Intermandibularis posterior**	**Intermandibularis**
Intermandibularis anterior (submentalis *sensu* Iordansky [1992] and Carroll [2007])	**Intermandibularis anterior** (submentalis *sensu* Iordansky [1992] and Carroll [2007]) [in at least some tadpodes of anurans, the intermandibularis is divided into three portions; i.e., an intermandibularis anterior, an intermandibularis posterior, and a mandibulolabialis: see, e.g., Olsson et al. 2002; Carroll 2007]	– [in this species, the intermandibularis is not divided into intermandibularis anterior and intermandibularis posterior, but see text]
Adductor mandibulae A2 (adductor mandibulae externus *sensu* Carroll and Holmes [1980], Iordansky [1992], and Carroll [2007]; levator mandibulae externus *sensu* Edgeworth [1935] and Piatt [1938, 1939])	**Adductor mandibulae A2** (levator mandibulae externus *sensu* Haas [2001], Ziermann and Olsson [2007], and Johnston [2011]; adductor mandibulae A2 subexternus *sensu* Diogo and Abdala [2010])	**Adductor mandibulae A2** (adductor mandibulae externus *sensu* Carroll [2007], which, according to this author, corresponds to the levator mandibulae anterior *sensu* Bemis et al. [1983]: see his Figure 14G) [according to Kleinteich and Haas [2007, 2011], only the levator mandibulae longus (A3′ or deep pseudotemporalis *sensu* the present work) is present in adults caecilians together with the pterygomandibularis, the larvae also having a levator mandibulae externus (A2 *sensu* the present work)]
– [but the adductor mandibulae lateralis is present in some urodeles according to Johnston [2011]]	**Adductor mandibulae lateralis** [often present in anurans according to, e.g., Iordansky [2010] and Johnston [2011], corresponding to the "levator mandibulae lateralis" of, e.g., Haas [2001], to the "adductor mandibulae A2 externus" of Diogo and Abdala [2010], and to the "levator mandibulae posterior subexternus" *sensu* Duellman and Trueb [1986]]	–
–	– **Levator anguli oris** [Johnston [2011] found a "levator anguli oris" similar to that of reptiles only in a very few anurans, e.g., *Ascaphus*, the muscle being seemingly absent in *Bufo*]	–
Adductor mandibulae A2-PVM (adductor mandibulae posterior *sensu* Carroll and Holmes [1980], Iordansky [1992], and Carroll [2007]; levator mandibulae posterior *sensu* Edgeworth [1935] and Piatt [1938]; levator mandibulae articularis *sensu* Ericsson and Olsson [2004])	**Adductor mandibulae A2-PVM** (adductor mandibulae articularis *sensu* Iordansky [1992]; levator mandibulae articularis *sensu* Ziermann and Olsson [2007]; levator mandibularis posterior *sensu* Sedra and Michel [1957]; adductor mandibulae posterior *sensu* Johnston [2011])	**Adductor mandibulae A2-PVM** (levator mandibulae articularis *sensu* Kleinteich and Haas [2007])

(*Continued*)

TABLE 9.1 (CONTINUED)
Mandibular Muscles of Adults of Representative Urodele, Anuran, and Caecilian Taxa

Amphibia (Caudata, or Urodela): *Ambystoma ordinarium* (Michoacan Stream Salamander)	Amphibia (Anura): *Rhinella arenarum* (Argentine Common Toad)	Amphibia (Gymnophiona): *Siphonops paulensis* (Catuchi Caecilian)
Pseudotemporalis (pseudotemporalis posterior and anterior *sensu* Iordansky [1992]; superficial and deep levator mandibulae anterior *sensu* Edgeworth [1935], Piatt [1938], Piekarsky and Olsson [2007]; adductor mandibulae A3′ and A3″ *sensu* Diogo [2007, 2008b]; adductor mandibulae internus *sensu* Carroll and Holmes [1980] and Lauder and Shaffer [1985]; levator mandibulae longus and levator mandibulae internus *sensu* Haas [2001], Ericsson and Olsson [2004], and Ziermann 2008) [both the adductor A3′ and A3″ seem to be included in the pseudotemporalis of extant nontetrapod mammals such as *Ambystoma mexicanum*; in some species of these amphibians and of reptiles, the pseudotemporalis is however clearly divided into superficial and deep structures that seemingly correspond directly to the A3′ and A3″ of fishes, respectively]	**Pseudotemporalis** (levator mandibulae longus and levator mandibulae internus *sensu* Haas [2001] and Ziermann and Olsson [2007]; adductor mandibulae longus and adductor mandibulae internus *sensu* Johnston [2011])	**Pseudotemporalis** (levator mandibulae internus + part of levator mandibulae longus of Kleinteich and Haas [2007])
– [the muscle "pterygoideus" that is present in some urodeles, *sensu* Haas [2001], is usually not present as a distinct muscle in *Ambystoma*, and seems to correspond to the muscle pterygoideus of caecilians and, thus, to the pterygomandibularis of reptiles, being very likely derived from the deep bundle of the pseudotemporalis, i.e., from the A3″ of fishes; see text and = Chapters 8 and 10; however, some authors use the name pterygoideus as a synonym of the A3″; i.e., the deep pseudotemporalis, of urodeles; authors such as Iordansky [2011] suggested that the pterygomandibularis of caecilians and of reptiles is directly homologous with the A3″ of fishes and the deep pseudotemporalis of anurans and urodeles, but then sometimes Iordansky [2011] instead states in some parts of his paper that the pterygomandibularis *developed from* (i.e., it is *not* directly homologous to) the deep pseudotemporalis/A3″ and was thus homoplasically acquired in caecilians and reptiles; we follow here an idea similar to that of Haas [2001], who noted that Iordansky does not mention the presence of a true pterygoideus muscle in a few urodeles as described by Haas [2001], i.e., seemingly in addition to the deep pseudotemporalis, thus supporting the idea that the pterygomandibularis derives from, but is not directly homologous to, the A3″ of fishes and the deep pseudotemporalis of amphibians; as some urodeles seem to have such a true pterygoideus (pterygomandibularis *sensu* the present work), it is quite possible that this muscle was present in the LCA of tetrapods and then lost in anurans and most urodeles (at least three evolutionary steps), although it is also quite possible that it was independently acquired in caecilians, in some urodeles, and in reptiles (at least three evolutionary steps)]	–	**Pterygomandibularis** [the caecilian muscle pterygoideus *sensu* Iordansky [1996], Carroll [2007], and Kleinteich and Haas [2007] and levator mandibulae posterior *sensu* Bemis et al. [1983] seems to correspond to the pterygomandibularis of reptiles such as the lizard *Timon*: see cell on the left and Chapters 8 and 10]

(*Continued*)

TABLE 9.1 (CONTINUED)
Mandibular Muscles of Adults of Representative Urodele, Anuran, and Caecilian Taxa

Amphibia (Caudata, or Urodela): *Ambystoma ordinarium* (Michoacan Stream Salamander)	Amphibia (Anura): *Rhinella arenarum* (Argentine Common Toad)	Amphibia (Gymnophiona): *Siphonops paulensis* (Catuchi Caecilian)
–	–	**Levator quadrati** (*sensu* Bemis et al. [1983] and Iordansky [1996]) [Iordansky [1996] suggested that the levator quadrati of caecilians corresponds to the levator pterygoidei + protractor pterygoidei of reptiles]
Levator bulbi [it probably corresponds to part of the levator arcus palatini of *Latimeria*, being usually innervated by V2 (the maxillary branch of the trigeminal nerve) in urodeles, supporting the idea that it derives from the dorsal part of the mandibular muscle mass]	**Levator bulbi** [Johnston [2011] states that in some anurans, e.g., *Ascaphus* and *Leiopelma*, there is a small "muscle" derived from the levator bulbi (which in these anurans passes dorsal to the nerve V2), this small muscle being named "depressor membranae nictitantis"; however, authors such as Francis [1934] have shown that in urodeles such as *Salamandra* the levator bulbi has a pars principalis attaching on the posterior end of the lower eyelid (membrana nictitans), a pars sagittalis divided into a caput laterale and a caput mediale and a pars transversalis, the innervation of the whole muscle being by V2; that is, the levator bulbi seems to be often divided into bundles in amphibians, and thus the muscle described by Johnston in anurans very likely corresponds to a bundle of the levator bulbi of urodeles	– [see text]

Note: The nomenclature of the muscles shown in bold follows that of text; in order to facilitate comparisons, in some cases, names often used by other authors to designate a certain muscle/bundle are given in front of that muscle/bundle, between round brackets; additional comments are given between square brackets. Data compiled from evidence provided by our own dissections and comparisons and by a review of the literature (see text and Figures 9.1 through 9.7).

which are derived from the levator palatini of the LCA of amphibians + amniotes (Table 9.1; see Chapter 10). That is, the levator bulbi was probably secondarily lost during the evolutionary events that led to the origin of caecilians. An alternative, less likely hypothesis is that the levator quadrati of caecilians is an undifferentiated muscle corresponding to both the levator palatini and levator bulbi of the LCA of amphibians + amniotes, i.e., to the levator arcus palatini of sarcopterygian fishes such as *Latimeria*.

HYOID MUSCLES

The interhyoideus (Figures 9.1, 9.2, 9.4, and 9.6) is a hyoid muscle that primarily passes from the hyoid bar to the ventral middle line, behind the intermandibularis (Table 9.2) (see preceding chapters). Several urodele and some anuran and caecilian amphibians have an interhyoideus anterior and an interhyoideus posterior, which is often named "interbranchialis" or "sphincter colli" (Figures 9.1 and 9.4; see also, e.g., Edgeworth 1935; Piatt 1938; Duellman and Trueb 1986; Bauer 1992, 1997; Carroll 2007; Kleinteich and Haas 2007). Interestingly, the developmental study of Ericsson and Olsson (2004) suggested that the interhyoideus anterior and interhyoideus posterior of *Ambystoma mexicanum* might derive from distinct, separate anlagen, while the developmental work of Piekarski and Olsson (2007) indicated that in this species, at least part of the interhyoideus posterior receives contribution, ontogenetically, from anterior somites. In the adult caecilian specimens we dissected, the interhyoideus seems to be constituted by a single, continuous mass of fibers. However, authors such as Duellman and Trueb (1986), Carroll (2007), and Kleinteich and Haas (2007) did describe an interhyoideus anterior and an interhyoideus posterior in some caecilians (see Figure 9.4). Bemis et al. (1983: 85) stated that "due to its position directly behind the retroarticular process," the action of the interhyoideus of caecilians "is to close the lower jaw," stressing that "this function of the interhyoideus in caecilians is unique among tetrapods."

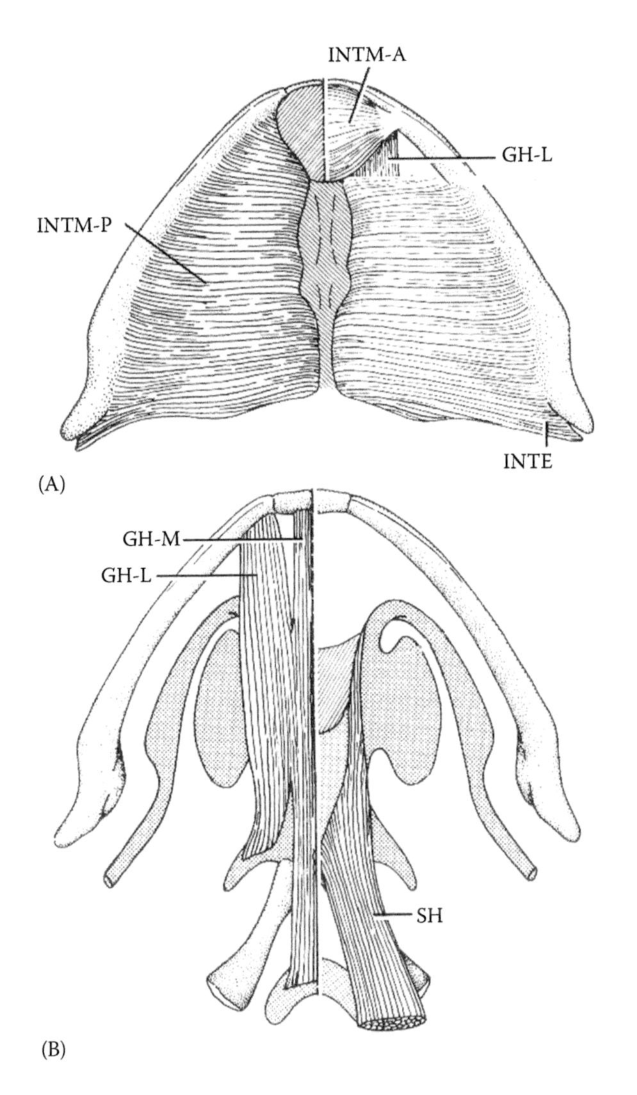

FIGURE 9.6 *Bufo marinus* (Amphibia, Anura): (A) Ventral view of the cephalic musculature; (B) same view, but the superficial muscles were removed in order to show the hypobranchial muscles geniohyoideus (on the left) and sternohyoideus (on the right) (the nomenclature of the myological structures illustrated follows that of the present work; for more details about the osteological structures illustrated, see Duellman and Trueb [1986]). GH-L, GH-M, geniohyoideus lateralis and geniohyoideus medialis; INTE, interhyoideus; INTM-A, INTM-P, intermandibularis anterior and intermandibularis posterior; SH, sternohyoideus. (Modified from Duellman, W. E., and Trueb, L., *The Biology of Amphibians*, McGraw-Hill, New York, 1986.)

In the adult urodele, anuran, and caecilian amphibians we dissected, as well as in those described by other authors, the depressor mandibulae (Figures 9.1 through 9.4) usually originates from the posterolateral surface of the skull and passes to the retroarticular process of the mandible, being mainly a jaw opener (e.g., Ecker 1889; Gaupp 1896; Luther 1914; Duellman and Trueb 1986; Bauer 1992, 1997; Wilkinson and Nussbaum 1997; Carroll 2007). In various urodeles, including *Ambystoma* (Figure 9.1 and Table 9.2), the fibers corresponding to those of the levator hyoideus of dipnoans become attached onto the mandible, forming a muscle that is often named "depressor mandibulae posterior" (see preceding chapters). As this muscle is clearly homologous to the muscle levator hyoideus that inserts on the hyoid bar in numerous dipnoan and tetrapod larvae as well as in adult dipnoans and in some adult tetrapods, including some neotenic urodeles such as *Siren*, in this volume, we designate it as levator hyoideus, as proposed by Edgeworth (1935) (Figure 9.1 and Table 9.2; see also preceding chapters).

The levator hyoideus is usually present as an independent muscle in caecilian larvae (Figure 9.4), but not in caecilian adults (e.g., Edgeworth 1935; Kleinteich and Haas 2007). According to Edgeworth (1935), in at least some caecilian adults, this muscle is fused with the depressor mandibulae. In anuran larvae, the depressor mandibulae and the levator hyoideus are often separated: the depressor mandibulae may be subdivided into a suspensorioangularis, a quadratoangularis, and a hyoangularis, which, as their name indicates, insert onto the angular region of the mandible; the levator hyoideus may be subdivided into an orbitohyoideus and a suspensoriohyoideus, which, as indicated by their names, insert onto the hyoid bar (e.g., Edgeworth 1935; Paterson 1939; Weisz 1945a,b; Sedra 1949; Sedra and Michael 1957; Sokol 1977; Palavecino 2001; Olsson et al. 2001; Kleinteich and Haas 2007; Carroll 2007). However, in postmetamorphic anurans, the orbitohyoideus and suspensoriohyoideus often become inserted on the mandible and fused with the depressor mandibulae, the levator hyoideus thus being often not present as an independent muscle, although it possibly corresponds to the posterior bundle of the depressor mandibulae of some adult anurans (Figure 9.6A) (e.g., Edgeworth 1935; Sedra 1949; Palavecino 2001; Kleinteich and Haas 2007; Carroll 2007; for further considerations about the depressor mandibulae of adult anurans, see Table 9.2 and, e.g., Starret 1968; Hoyos 1999; Manzano et al. 2003).

The branchiohyoideus (Figure 9.7) is a muscle that usually originates on the dorsal margin of the ceratohyal and runs anteroventrally, below the hyoid bar and internal to the interhyoideus, to insert on the mesial surface of the ceratohyal and/or the epihyal (e.g., Edgeworth 1935; see Chapter 8). It is found in urodele larvae and some urodele obligate neotenes such as *Ambystoma ordinarium* (Figure 8.5 and Table 9.2) and is usually absent as an independent structure in fully metamorphosed adult urodeles. However, according to authors such as Piatt (1938: 542), in some metamorphosed urodeles, the branchiohyoideus "may persist as late as the sixth week following the advent of terrestrial life." The branchiohyoideus should not be confused with the ceratomandibularis ("branchiomandibularis" *sensu* Carroll [2007]), which is a muscle that usually runs from the ceratohyal to the mandible (Figure 9.4B). The ceratomandibularis and the branchiohyoideus seem to ontogenetically derive from the same anlage (e.g., Piatt 1938; Bauer 1997). The ceratomandibularis is often absent as an independent structure in fully metamorphosed adult urodeles,

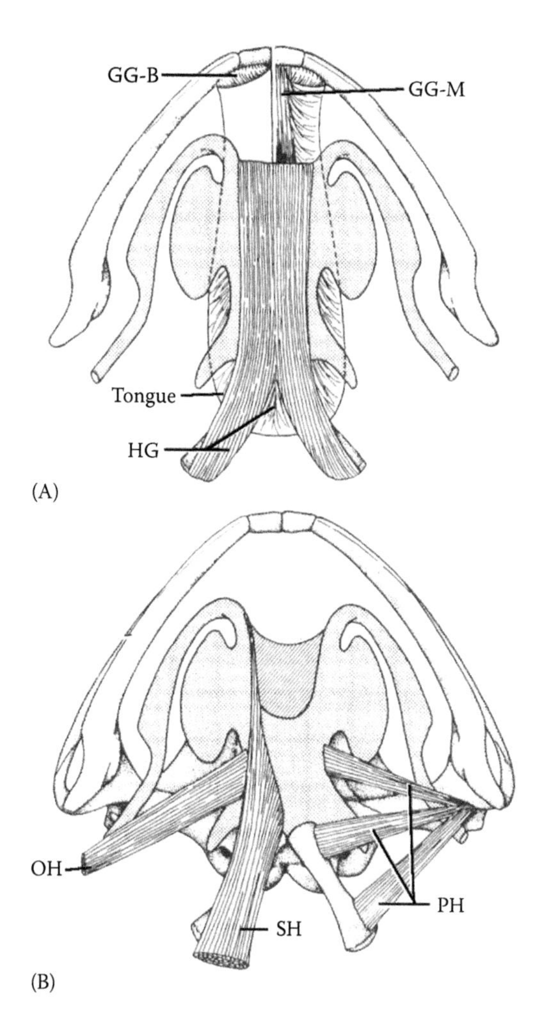

FIGURE 9.7 *Bufo marinus* (Amphibia, Anura): (A) Ventral view showing the hypobranchial tongue muscles genioglossus and hyoglossus; (B) same view showing the hypobranchial muscles sternohyoideus and omohyoideus (on the left) and the branchial muscles petrohyoideus (on the right) (the nomenclature of the myological structures illustrated follows that of the present work; for more details about the osteological structures illustrated, see Duellman and Trueb [1986]). GG-B, GG-M, genioglossus basilis and genioglossus medialis; HG, hyoglossus; OH, omohyoideus; PH, petrohyoideus; SH, sternohyoideus. (Modified from Duellman, W. E., and Trueb, L., *The Biology of Amphibians*, McGraw-Hill, New York, 1986.).

but, as described by Bauer (1997), may persist in some obligate neotenes (e.g., *Ambystoma originarium*: Table 9.2). In his Figure 31B, Carroll (2007) seems to suggest that the branchiohyoideus of urodeles is homologous to the "hyomandibularis" (or "subhyoideus") muscle that is present in larvae and in some adults of caecilians. However, authors such as Edgeworth (1935) and Kleinteich and Haas (2007) stated that the "hyomandibularis" of caecilians corresponds to the ceratomandibularis rather than the branchiohyoideus of urodeles (Figure 9.4). As its name indicates, the caecilian hyomandibularis often attaches anteriorly on the mandible, as does the urodele ceratomandibularis, rather than on the hyoid bar as does the urodele branchiohyoideus. An alternative hypothesis is that the hyomandibularis of caecilians is an undifferentiated muscle corresponding to both the ceratomandibularis and branchiohyoideus of urodeles (Table 9.2). In the adult anurans we dissected, neither the ceratomandibularis nor the branchiohyoideus seem to be present as independent structures. Authors such as Edgeworth (1935) and Jarvik (1963, 1980) stated that these two muscles are absent in anuran adults (Table 9.2).

BRANCHIAL MUSCLES

In Table 9.3, we follow the nomenclature used in the preceding chapters and thus consider the "true" branchial muscles *sensu stricto* as a group. Examples of true branchial muscles *sensu stricto* that are present in urodele larvae and/or adults are the levatores arcuum branchialium, transversi ventrales, and subarcuales recti; examples of these muscles in anuran larvae and/or adults are the levatores arcuum branchialium, subarcuales obliqui, subarcuales recti, and petrohyoideus; examples of these muscles in caecilian larvae and/or adults are the levatores arcuum branchialium, transversi ventrales, subarcuales obliqui, and subarcuales recti (Figures 9.4 and 9.6; for more details on the amphibian true branchial muscles *sensu stricto*, see, e.g., Humphry 1871; Gaupp 1896; Ecker 1889; Edgeworth 1935; Piatt 1938; Kesteven 1942–945; Fox 1959; Bemis et al. 1983; Duellman and Trueb 1986; Haas 1996a,b; Kleinteich and Haas 2007; Carroll 2007; Ziermann and Olsson 2007).

The protractor pectoralis (Figure 9.1) of tetrapods corresponds to the "cucullaris" *sensu* Edgeworth (1935), which, according to the latter author, is a muscle that derives from a caudal levator arcus branchialis of plesiomorphic vertebrates and that became secondarily attached to the pectoral girdle, thus connecting this structure to the skull. Details about the development and evolution of the protractor pectoralis are given in the preceding chapter. According to Edgeworth (1935), the protractor pectoralis is often a large muscle in metamorphosed anurans and in both larvae and adults of urodeles (Figure 6.1); it seems to be absent in the adult caecilians we dissected. Extant caecilians do not have a pectoral girdle (see the review of Carroll 2007), which is the structure to which this muscle usually attaches. However, in a recent study, Sefton et al. (2016) stated that part of the "levator arcus branchiales complex" and/or of the "cephalodorsosubpharyngeus" of caecilians has a posterior division that corresponds to the protractor pectoralis of other tetrapods, originating from the otic capsule and inserts on the fascia separating the "rectus abdominis" from the "interhyoideus." It is possible that this structure does correspond to the protractor pectoralis, its insertion being different because the pectoral appendage is completely absent. However, only detailed developmental and/or innervation—e.g., is it innervated by the accessory nerve, as the protractor pectoralis usually is, or by more anterior nerves, as other branchial muscles usually are—can rigorously test this hypothesis. We provide more details on the development of the protractor pectoralis in urodeles in Chapter 11.

TABLE 9.2
Hyoid Muscles of Adults of Representative Urodele, Anuran, and Caecilian Taxa

Amphibia (Caudata): *Ambystoma ordinarium* (Michoacan Stream Salamander)	Amphibia (Anura): *Rhinella arenarum* (Argentine Common Toad)	Amphibia (Gymnophiona): *Siphonops paulensis* (Catuchi Caecilian)
Depressor mandibulae (depressor mandibulae anterior *sensu* Diogo [2007, 2008b] and Diogo et al. [2008a])	**Depressor mandibulae** [see text]	**Depressor mandibulae** [see text]
Ceratomandibularis (branchiomandibularis *sensu* Edgeworth [1935]; subhyoideus *sensu* Duellman and Trueb [1986]) [the hyomandibularis of caecilians seems to correspond to the ceratomandibularis, or possibly to the ceratomandibularis + branchiohyoideus, of urodeles, which probably correspond to part of the depressor mandibulae/levator hyoideus of dipnoans: see Chapter 8; the ceratomandibularis is not present as a distinct muscle in adult axolotls (see Chapter 11)]	– [see text]	– (subhyoideus or hyomandibularis *sensu* Lawson [1965], Wikinson and Nussbaum [1997], Carroll [2007], and Kleinteich and Haas [2007]) [the hyomandibularis is found not only in larvae, but also in some adults, of caecilians: see, e.g., Edgeworth 1935; Lawson 1965; Kleinteich and Haas 2007]
Branchiohyoideus (ceratohyoideus externus or branchiohyoideus externus *sensu* Edgeworth [1935], Piatt [1938], Lightoller [1939], Bauer [1997], and Ericsson and Olsson [2004]; subhyoideus *sensu* Carroll [2007])	– [see text]	– [see text]
Levator hyoideus (depressor mandibulae posterior *sensu* Diogo [2007, 2008b] and Diogo et al. [2008a]) [unlike the levator hyoideus of some neotenic urodeles such as *Siren*, the levator hyoideus of adults of some species of *Ambystoma*, as well as of various other urodele taxa, inserts on the mandible, and not on the hyoid bar: see text; the ceratomandibularis is not present as a distinct muscle in adult axolotls (see Chapter 11)]	– [see text; the depressor mandibulae posterior is found in various anuran adults, as reported by, e.g., Johnston [2011]; Carroll [2007: 48] states that the "orbitohyoideus" and "suspensoriohyoideus" of anuran tadpoles are homologous with the depressor mandibulae of such anuran adults; by looking to his Figure 3 of the tadpole of *Rana*, it clearly seems that these two muscles correspond to the depressor mandibulae posterior of adults, because they insert onto the ceratohyal, while the hyoangularis, suspensorioangularis, and quadratoangularis seem to correspond to the depressor mandibulae anterior of adults, because they originate near the other two muscles, but insert onto the mandible, as the depressor mandibulae anterior usually does; this idea seems to be directly supported on page 52 of the work by Carroll [2007]]	– [see text]
Interhyoideus [urodeles often have an interhyoideus anterior and an interhyoideus posterior, which is often named interbranchialis or sphincter colli: see, e.g., Piatt 1938; Bauer 1992, 1997; Ericsson and Olsson 2004; Carroll 2007]	**Interhyoideus** [anurans may also have an interhyoideus anterior and an interhyoideus posterior: see, e.g., Carroll 2007]	**Interhyoideus** [caecilians may also have an interhyoideus anterior and an interhyoideus posterior: see, e.g., Nussbaum 1983; Duellman and Trueb 1986; Kleinteich and Haas 2007; Carroll 2007]

Note: See footnote of Table 9.1, text, and Figures 9.1 through 9.7.

As their name indicates, the dilatator laryngis and constrictor laryngis (Figure 9.6) are antagonistic muscles (e.g., Duellman and Trueb 1986). They are often present in larval and adults of urodeles, anurans and caecilians and form, together with the laryngeus, the laryngeal muscles *sensu* this volume (Figure 9.1; Table 9.3). As described by Edgeworth (1935), the laryngeus is present in larvae and adults of urodele and caecilian taxa such as *Ichthyophis*, *Caecilia*, *Hypogeophis*, *Amphiuma*, *Menopoma*, and *Ambystoma* (Table 9.3), but in taxa such as *Salamandra*, *Triton*, and *Siphonops* (Table 6.3), it atrophies during ontogeny, being absent as an independent muscle in adults. As also described by this author, the laryngeus does not seem to be present as a separate muscle in anuran adults (Table 9.3); according to him, this muscle is also absent in anuran larvae.

TABLE 9.3
Branchial Muscles of Adults of Urodele, Anuran, and Caecilian Taxa

Amphibia (Caudata): *Ambystoma ordinarium* (Michoacan Stream Salamander)	Amphibia (Anura): *Rhinella arenarum* (Argentine Common Toad)	Amphibia (Gymnophiona): *Siphonops paulensis* (Catuchi Caecilian)
True branchial muscles *sensu stricto* [see text]	**True branchial muscles *sensu stricto*** [see text; note that the petrohyoidei *sensu* Johnston [2011] correspond to the levatores arcuum branchialium]	**True branchial muscles *sensu stricto*** [see text]
Protractor pectoralis (cucullaris or cucullaris major *sensu* Edgeworth [1935] and Carroll [2007]; trapezius *sensu* Jollie [1962]; seemingly corresponds to part of levator branchiarum 4 *sensu* Piatt [1939]) [in his Figure 14.16, Jollie [1962] shows a *Cryptobranchus* specimen with a "sternomastoideus" and a "trapezius" divided into at least two parts, i.e., a "pars spinotrapezius" and a "pars clavotrapezius"; however, both his illustration and his captions suggest that at least part of the trapezius is deeply blended with the sternomastoideus, so it is not clear if in this taxon the protractor pectoralis is really differentiated into a separate, distinct trapezius and a separate, distinct sternocleidomastoideus *sensu* the present work, as is the case in many extant amniotes; according to Howell [1933a,b, 1935, 1936a,b,c,d, 1937a,b,c,d], the sternocleidomastoideus is present only as a separate muscle in reptiles and mammals; see Chapter 8]	**Protractor pectoralis** (cucullaris *sensu* Edgeworth 1935)	– [but see text, including comments about the work by Sefton et al. [2016]]
–	**Interscapularis** [according to Howell [1935], unlike other tetrapods, anurans have a peculiar muscle interscapularis, which clearly seems to be a branchial muscle and which, according to him, is probably derived from the protractor pectoralis; this muscle is present in, e.g., *Rana*, being innervated by the vagus nerve, connecting the two scapulas, see, e.g., Duellman and Trueb 1986]	–
Constrictor laryngis [Duellman and Trueb (1986) state that the constrictor laryngis is present in adults of all neotenic salamanders; it is unclear if this muscle is or not present in axolotl adults, see Chapter 11]	**Constrictor laryngis** [confirmed to be often present in adult anurans by Johnston [2011], who reports that in some frogs, e.g., *Leiopelma*, the constrictor laryngis is divided into three muscle structures, i.e., the "constrictor laryngis," the "constrictor laryngis ventralis," and the "constrictor laryngis dorsalis"; Edgeworth [1935] stated that anurans have no laryngeus, but could one of these three structures of anurans and/or the two structures present in at least some caecilians—constrictor laryngis dorsalis and constrictor laryngis ventralis—correspond to the laryngeus of urodeles, which is also closely related topologically and ontogenetically with the constrictor laryngis anlage?]	**Constrictor laryngis** [as in anurans, in at least some caecilians the constrictor laryngis is divided into a dorsal head and a ventral head]
Laryngeus [the laryngeus and constrictor laryngis of amphibians are ontogenetically derived from the same anlage; unlike other urodeles, in which the laryngeus is divided into laryngeus dorsalis and laryngeus ventralis *sensu* Piatt [1938], in *Ambystoma*, the laryngeus is often constituted by a single mass of fibers, which corresponds to the laryngeus ventralis, according to Edgeworth [1935]; it is clear if this muscle is or not present in axolotl adults, see Chapter 11 (note that it does not correspond to the constrictor oesephagi *sensu* Edgeworth [1935], as both muscles are shown in his Figure 370)]	– [see text; but see cell above]	– [but present in some caecilian adults: see text]
Dilatator laryngis [seem to be absent as a distinct muscle in axolotl adults, see Chapter 11]	**Dilatator laryngis**	**Dilatator laryngis**

Note: See footnote of Table 9.1, text, and Figures 9.1 through 9.7.

HYPOBRANCHIAL MUSCLES

According to Edgeworth (1935), the hypobranchial muscles are divided into a "geniohyoideus" group and a "rectus cervicis" group (Table 9.4; Chapter 2). Extant sarcopterygian fishes have two hypobranchial muscles, the coracomandibularis and the sternohyoideus, which are included in the geniohyoideus and rectus cervicis groups *sensu* Edgeworth, respectively (e.g., Edgeworth 1935; Kesteven 1942–1945; Wiley 1979a,b; Wilga et al. 2000; this work) (see preceding chapters). These muscles connect the pectoral girdle to the mandible and the hyoid bar and are mainly related to the opening of the mouth. Amphibians and amniotes have multiple hypobranchial muscles (Table 9.4). The amphibian muscles geniohyoideus (Figures 9.4 through 9.6), genioglossus (Figures 9.5 and 9.7), and hyoglossus (Figure 9.7) probably derive from the coracomandibularis of sarcopterygian fishes, while the sternohyoideus (Figures 9.4 through 9.6) and omohyoideus (Figure 9.7) derive from the sternohyoideus of these fishes (see Chapter 8). It is, however, possible that the "hyoglossus" of some urodeles is at least partially derived from the sternohyoideus (e.g., Edgeworth 1935; Jarvik 1963; Chapter 8).

TABLE 9.4
Hypobranchial Muscles of Adults of Urodele, Anuran, and Caecilian Taxa

Amphibia (Caudata): *Ambystoma ordinarium* (Michoacan Stream Salamander)	Amphibia (Anura): *Rhinella arenarum* (Argentine Common Toad)	Amphibia (Gymnophiona): *Siphonops paulensis* (Catuchi Caecilian)
Geniohyoideus (*sensu* Lauder and Shaffer [1988])	**Geniohyoideus** (geniobranchialis *sensu* Haas [1996a,b]; includes the geniohyoideus lateralis and geniohyoideus mesialis *sensu* Carroll [2007])	**Geniohyoideus**
Genioglossus [according to Edgeworth [1935] and Piatt [1938], the genioglossus of salamanders such as *Ambystoma* is derived from the coracomandibularis; it may be absent in urodele taxa such as *Siren* and *Spelerpes*; this muscle is not present in axolotl adults, see Chapter 11]	**Genioglossus** [confirmed to be often present in adult anurans by Johnston [2011]; see hyoglossus, about *Xenopus*]	**Genioglossus**
– [as explained in Chapter 8, extant amphibians do seemingly not have well-developed, independent "intrinsic muscles of the tongue" like those found in extant amniotes; however, authors such as Schwenk [2001] do not agree with the definition of "extrinsic" and "intrinsic" muscles of the tongue, arguing that the latter should not be considered independent muscles]	– [see left column]	– [see left column]
Hyoglossus [the statements of Edgeworth [1935] concerning the hyoglossus of urodeles such as *Ambystoma* are confusing: on page 196 of his work, he states that it derives from the sternohyoideus but on page 211, he suggests that as in other amphibians and in reptiles and mammals, it derives from the coracomandibularis; the results of the developmental work of Piatt [1938] support the latter hypothesis; this muscle is not present in axolotl adults, see Chapter 11]	**Hyloglossus** [confirmed to be often present in adult anurans by Johnston [2011]; Horton [1982] stated that, although *Xenopus* has no tongue, specimens of this genus—of *Xenopus muelleri* and *X. capensis*—had traces of the hyoglossus and genioglossus, fused to the ventral surface of the floor of the mouth; i.e., the two muscles were vestigial but present]	– [as described by Edgeworth 1935, the hyoglossus does not seem to be present as a separate muscle in *Siphonops*]
Interradialis [see text; this muscle is not present in axolotl adults, see Chapter 11]	–	–
Sternohyoideus (rectus cervicis *sensu* Lauder and Shaffer [1988], Kardong [2002], and Carroll [2007])	**Sternohyoideus** (rectus cervicis *sensu* Carroll [2007])	**Sternohyoideus** (rectus cervicis *sensu* Kleinteich and Haas [2007] and Carroll [2007])
Omohyoideus (pectoriscapularis *sensu* Edgeworth [1935], Walker [1954], and Kardong [2002]) [according to Edgeworth [1935], the omohyoideus is derived from the sternohyoideus; the abdomino-hyoideus *sensu* Piatt [1938] probably corresponds to the omohyoideus *sensu* this volume; this muscle is deeply blended with the sternohyoideus in axolotl adults, see Chapter 11]	**Omohyoideus** [confirmed to be often present in adult anurans by Johnston [2011]]	–

Note: See footnote of Table 9.1, text, and Figures 9.1 through 9.7.

Authors such as Ecker (1889) stated that the geniohyoideus of adult anurans is often divided into a geniohyoideus lateralis and a geniohyoideus medialis, which run from different regions of the hyoid bar to different regions of the mandible. Our observations of adult anurans match this description of the spatial arrangement of the geniohyoideus (Figure 9.6). However, because in some amphibians this muscle does not originate from the hyoid bar but instead on more posterior branchial arches, authors such as Haas (1996a,b) proposed the name "geniobranchialis" to replace geniohyoideus. However, most authors prefer to continue using the name geniohyoideus because the muscle is clearly homologous to the geniohyoideus of other tetrapods (see Chapter 8; Table 9.4).

The "tongue" muscles genioglossus and hyoglossus are often present in urodeles and anurans, although the first muscle may be absent in urodeles such as *Siren* and *Pseudotriton* (see Figures 9.5 and 9.7 and, e.g., Ecker 1998; Edgeworth 1935; Duellman and Trueb 1986). According to Duellman and Trueb (1986), in caecilians, the hyoglossus is not present as an independent structure. We were also unable to find this muscle in the adult caecilians we dissected. In fact, the tongue in *Siphonops* and *Chthonerpeton* is very much like that described in *Dermophis* by Bemis et al. (1983), being large but only slightly free at its margins and being associated with a single glossal muscle, the genioglossus (Figure 9.5; see also Carroll 2007). As described by Piatt (1938), in at least some adult *Ambystoma*, there is a hypobranchial muscle interradialis, which probably derives from the genioglossus (Table 9.4).

The sternohyoideus and omohyoideus usually connect the pectoral girdle to the hyobranchial apparatus (Figures 9.4 through 9.7). Carroll (2007: 40) stated that "the omohyoideus (of anurans) cannot be compared to any muscle in salamanders." However, urodeles do have a muscle omohyoideus (Figure 9.2), which clearly seems to be homologous to that of anurans (e.g., Albrecht 1876; Edgeworth 1935; Diogo and Abdala 2010). Caecilians do not have a pectoral girdle, and in the specimens we dissected, as well as in those described by other authors, the omohyoideus is not present as an independent structure (Table 9.4). However, these amphibians do have a muscle sternohyoideus, which usually arises directly from the fascia of the muscle rectus abdominis (Figures 9.4 and 9.5 and Table 9.4; e.g., Edgeworth 1935; Kleinteich and Haas 2007; Diogo and Abdala 2010).

GENERAL REMARKS

Regarding the mandibular muscles, the LCA of amphibians probably had an adductor mandibulae A2, an adductor mandibulae A2-PVM, a pseudotemporalis, a pterygomandibularis, an intermandibularis anterior, an intermandibularis posterior, a levator palatini, and a levator bulbi. Caecilians usually have a single intermandibularis, either because one of the intermandibularis muscles was secondarily lost or, more likely, because their intermandibularis is an undifferentiated muscle that does not become subdivided into anterior and posterior portions during ontogeny. As explained above, the levator palatini was probably secondarily lost in urodeles and anurans, and the levator bulbi was probably secondarily lost in caecilians, although the possibility cannot be discounted that the levator quadrati of caecilians is an undifferentiated muscle corresponding to both the levator palatini and levator bulbi of the LCA of amphibians + amniotes, i.e., to the levator arcus palatini of sarcopterygian fishes such as *Latimeria*. The secondary loss of a levator palatini in anurans and urodeles might constitute a potential synapomorphy of the clade including these two amphibian groups, if it is accepted that these two groups are more closely related to each other than to caecilians (see the overview of Carroll [2007]). Iordansky (1996) and Kleinteich and Haas (2007) suggested that the caecilian muscle that they call pterygoideus probably corresponds to the pterygomandibularis of reptiles, and thus that the pterygomandibularis was probably present in the LCA of tetrapods and in the LCA of amphibians. Some urodeles have a muscle pterygoideus that seems to be homologous to the pterygoideus of caecilians. According to this interpretation, it would be more parsimonious to assume that the pterygomandibularis was present in the LCA of tetrapods and then secondarily lost (i.e., probably not differentiated from the pseudotemporalis) in anurans than to assume that it was independently acquired in caecilians, in urodeles, and in amniotes. For more details on this subject, see Table 9.1. Future detailed studies, ideally including information about the development of these muscles in multiple representatives of amphibians and amniotes and data on plesiomorphic, fossil members of these two tetrapod groups, are clearly needed to clarify whether or not the pterygoideus of caecilians, the pterygoideus of urodeles, and the pterygomandibularis of amniotes are homologous structures.

Concerning the hyoid muscles, the LCA of urodeles, anurans, and caecilians probably had an interhyoideus, a depressor mandibulae, a levator hyoideus, and possibly a ceratomandibularis, if one accepts the idea, defended by authors such as Edgeworth (1935), that the hyomandibularis of caecilians is homologous to the ceratomandibularis of urodeles (i.e., that this muscle was present in the LCA of these two groups) and that anurans and urodeles are sister-groups. That is, under this scenario, one would consider the ceratomandibularis to have been present in the LCA of extant amphibians and secondarily lost in anurans. However, as explained earlier, one cannot discard the hypothesis that the hyomandibularis corresponds to both the ceratomandibularis and the branchiohyoideus of urodeles (i.e., that the LCA of extant amphibians had a hyomandibularis similar to those of caecilians and that only in urodeles, this hyomandibularis became differentiated into ceratomandibularis and branchiohyoideus). The levator hyoideus is present as an independent structure in the closest living relatives of tetrapods, i.e., the dipnoans, as well as in at least some adult members of the three main extant tetrapod groups, i.e., the amphibians, reptiles, and mammals (e.g., Edgeworth 1935; this work). Therefore, it is likely that this muscle was present as an independent structure in the LCA of extant tetrapods and in the LCA of extant amphibians. The interhyoideus of the LCA of caecilians, urodeles, and anurans possibly included an interhyoideus anterior and an interhyoideus posterior, because both latter structures are present in at least some members of these three amphibian groups (see preceding text).

With respect to the branchial muscles, the following muscles were probably present in the LCA of extant amphibians: a protractor pectoralis, a constrictor laryngis, a dilatator laryngis, a laryngeus, and various true branchial muscles *sensu stricto* (see preceding text). As explained in Chapter 8, the laryngeus was probably present in the LCA of extant tetrapods (Table 9.3). As this muscle is present in some urodeles and caecilians, it was very likely also present in the LCA of extant amphibians, its absence in anurans being probably due to a secondary loss. The apparent absence, in extant caecilians, of a muscle protractor pectoralis such as that found in other amphibian and nonamphibian sarcopterygians is very likely related to the fact that these amphibians lack a pectoral girdle, which is the usual site of insertion of this muscle (but see comments about Sefton et al. [2016] and the preceding text).

Regarding the hypobranchial muscles, the LCA of urodeles, anurans, and caecilians probably had a geniohyoideus, a genioglossus, a hyoglossus, a sternohyoideus, and an omohyoideus. Among extant sarcopterygians a hypobranchial muscle interradialis has only been described in some urodeles; therefore, this muscle was very likely absent in the LCA of extant amphibians (see preceding text).

Before ending this chapter, we would like to stress that the suggested homologies summarized in Tables 9.1 through 9.4 are simply scientific hypotheses that should be tested by data obtained in the future. Detailed comparative ontogenetic studies of key caecilian taxa as well as a range of other amphibian and nonamphibian tetrapods are clearly needed to address some controversial questions, such as the following: Is the hyomandibularis of caecilians homologous to the ceratomandibularis of urodeles? Is the pterygoideus of caecilians homologous to the pterygoideus of urodeles and to the pterygomandibularis of amniotes? Or does the caecilian hyomandibularis correspond to both ceratomandibularis + branchiohyoideus of urodeles? Is the intermandibularis of caecilian adults an undifferentiated muscle corresponding to the intermandibularis anterior + intermandibularis posterior of other amphibians, which does not become subdivided into these two structures during the ontogeny of caecilians? Might the levator quadrati of adult caecilians be also an undifferentiated muscle corresponding not only to the levator pterygoidei + protractor pterygoidei, but also to the levator bulbi, of reptiles such as the lizard *Timon*? And is the interhyoideus posterior partially derived from somites as suggested by the study of Piekarski and Olsson (2007)? Hopefully, the present work will stimulate, and pave the way for, future studies on the comparative anatomy, development, functional morphology, and evolution of the amphibian muscles.

10 Head and Neck Muscles of Reptiles

In Chapter 10, we focus on the comparative anatomy, evolution, and homologies of the head and neck muscles of the major extant clades of reptiles: Testudines (turtles), Lepidosauria (including *Sphenodon*, "lizards," mosasaurs, snakes, and amphisbaenians *sensu* Conrad [2008]: see below), Crocodylia (crocodylians), and Aves (birds). Many anatomical works have provided information about the head and neck musculature of reptiles (e.g., Fürbringer 1874, 1876; Albrecht 1876; Versluys 1904; Edgeworth 1911, 1935; Phisalix 1914; Adams 1919; Camp 1923; Lakjer 1926; Brock 1938; Engels 1938; Lightoller 1939; Walker 1954; Oelrich 1956; Schumacher 1961, 1973; Frazzetta 1962; Jollie 1962; Jarvik 1963, 1980; Iordansky 1964, 2000, 2004, 2008; Barghusen 1968, 1986; Gaunt and Gans 1969; Haas 1973; Vanden Berge 1975; Rieppel 1980, 1981, 1984, 1990; Ghetie et al. 1981; Busbey 1989; Elzanowski 1987; Smith 1988; Witmer 1995a,b, 1997; Abdala and Moro 1996, 2003; Moro and Abdala 1998, 2000; Herrel et al. 1999, 2005; Wyneken 2001; Montero et al. 2002; Sedlmayr 2002; Holliday and Witmer 2007; Tsuihiji 2007; Conrad 2008; Tsukahara et al. 2009; Abdala et al. in press). However, as is the case with other vertebrate groups, most of these works focused on a specific reptilian taxon and/or a specific head and neck region, and none has provided detailed information about the homologies of the whole head and neck musculature of turtles, lepidosaurs, crocodylians, and birds. In the present work, we tried to take into account as much bibliographical information as possible, from classic anatomical descriptions (e.g., Fürbringer 1874, 1876; Albrecht 1876; 1904; Edgeworth 1911, 1935; Phisalix 1914; Adams 1919; Lakjer 1926; Engels 1938; Lightoller 1939) to more recent reviews (e.g., Haas 1973; Schumacher 1973; Vanden Berge 1975; Ghetie et al. 1981; Müller and Weber 1998; Noden et al. 1999; Iordansky 2000, 2008; Abdala and Moro 2003; Holliday and Witmer 2007; Conrad 2008; Holliday 2009; Diogo and Abdala 2010), including, importantly, the developmental and molecular data obtained in the numerous evo-devo studies that have been undertaken in recent decades with both reptilian and nonreptilian tetrapods such as chickens, quails, salamanders, frogs, and mice (e.g., Noden 1983a, 1984, 1986; McClearn and Noden 1988; Davis et al. 1991; Couly et al. 1992; Gardner and Barald 1992; Köntges and Lumsden 1996; Huang et al. 1999; Marcucio and Noden 1999; Olsson et al. 2000, 2001, 2005; Ellies et al. 2002; Mootoosamy and Dietrich 2002; Borue and Noden 2004; Ericsson and Olsson 2004; Ericsson et al. 2004; Le Douarin et al. 2004; Prunotto et al. 2004; Tokita 2004; O'Gorman 2005; Yamane 2005; Noden and Francis-West 2006; Noden and Schneider 2006; Piekarski and Olsson 2007; Ziermann and Olsson 2007; Knight et al. 2008; Kundrát et al. 2009; Shearman and Burke 2009; Tzahor 2009; Tokita and Schneider 2009; Tokita et al. 2012; Lescroart et al. 2010, 2015). The results of our observations and comparisons are summarized in the tables in this chapter, which present the best-supported hypotheses of homology for the muscles discussed in this chapter. As explained in Chapter 1, we have updated Chapter 10 to reflect new data suggesting that lepidosaurs rather than turtles are the sister-group of all other extant reptiles (Figure 1.1).

As stressed in previous chapters, one of the major problems researchers face when they compare the muscles of a certain tetrapod taxon with those of other taxa is the use of different names by different authors to designate the same muscle in the members of different clades and even of the same clade. In this respect, some of the names that are often used to designate the avian muscles in the literature (e.g., *Nomina Anatomica Avium*: Baumel et al. [1979]) are especially problematic, because they are different from the names used to designate homologous muscles in other reptiles and/or similar to the names that are used to designate nonhomologous muscles in mammals (e.g., "masseter," "mylohyoideus," "stylohyoideus," "ceratoglossus," "thyrohyoideus": see the tables in the chapter and the text that follows). In order to reconcile the nomenclatures used for turtles, lepidosaurs, crocodylians, birds, and mammals, as well as for other extant sarcopterygian groups such as caecilians, anurans, urodeles, dipnoans, and coelacanths, we thus use a unifying nomenclature for the head and neck muscles of the Vertebrata as a whole (see Chapters 1 and 2). The muscle names that we propose are shown in bold letters in these tables, which also provide a list of more than 100 synonyms that have been used by other authors to designate these reptilian muscles in the literature.

MANDIBULAR MUSCLES

The intermandibularis is a ventral mandibular muscle (e.g., Edgeworth 1935; Diogo 2007, 2008b; Diogo et al. 2008a,b) that usually connects the two hemimandibles (e.g., Figures 10.1 and 10.2). In numerous lepidosaurs, including *Timon*, the anlage that gives rise to the intermandibularis becomes differentiated into two separated muscles, the intermandibularis anterior and intermandibularis posterior (Table 10.1; Figure 10.2). In turtles, including *Trachemys scripta* (Table 10.1), the intermandibularis is often not clearly divided into an intermandibularis anterior and an intermandibularis posterior, although some authors stated that both these muscles might be present in a few turtles (see, e.g., Schumacher 1973; Wyneken 2001). As explained by Schumacher (1973), in crocodylians the intermandibularis might be slightly differentiated into bundles, but these reptiles usually do not have distinct, well-differentiated muscles intermandibularis anterior and intermandibularis posterior such as those found in lepidosaurs such as *Timon* (Table 10.1; Figures 10.1 and 10.3,

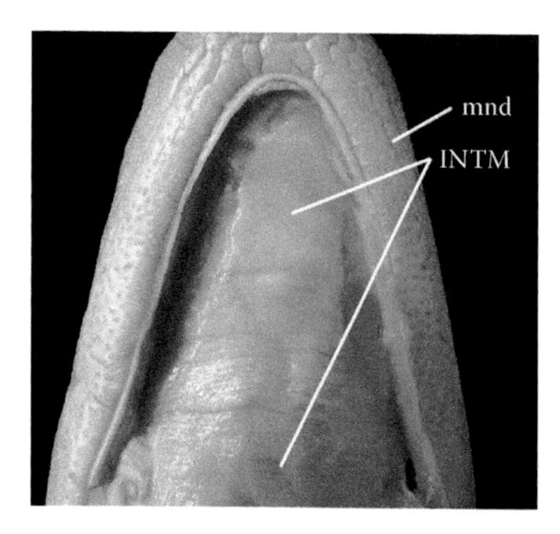

FIGURE 10.1 *Caiman latirostris* (Reptilia, Crocodylia): ventral view of the adult cephalic musculature showing the intermandibularis muscle (anterior is to the top). INTM, intermandibularis; mnd, mandible.

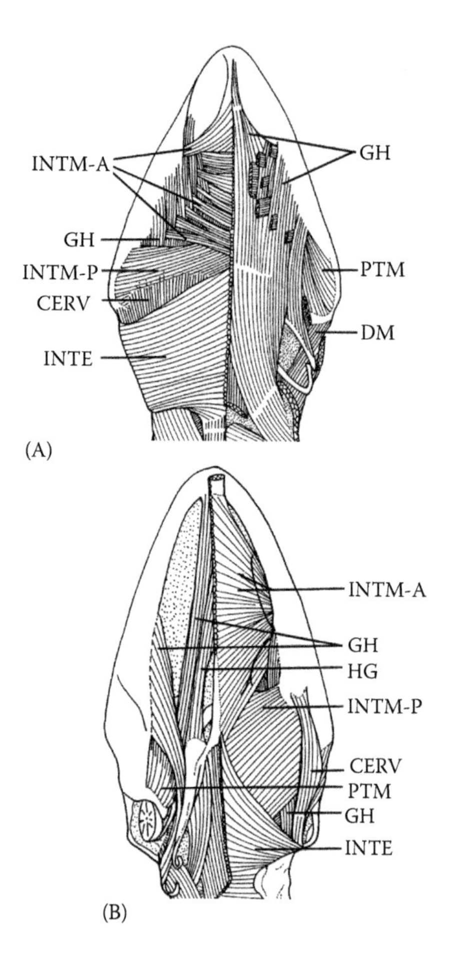

FIGURE 10.2 *Coleonyx variegatus* and *Uroplatus fimbriatus* (Reptilia, Lepidosauria): ventral view of the adult cephalic musculature of *C. variegatus* (A) and *U. fimbriatus* (B). (The nomenclature of the myological structures illustrated follows that of the present work; anterior is to the top.) CERV, cervicomandibularis; DM, depressor mandibulae; GH, geniohyoideus; HG, hyoglossus; INTE, interhyoideus; INTM-A, INTM-P, anterior and posterior bundles of intermandibularis; PTM, pterygomandibularis. (Modified from Camp, C. L., *Bull Amer Mus Nat Hist*, 48, 289–481, 1923; Haas, G., Muscles of the jaws and associated structures in the Rhynchocephalia and Squamata. In *Biology of the Reptilia*, Vol. 14 (eds. Gans, C., and Parsons, T. S.), pp. 285–490, Academic Press, New York, 1973.)

compare with Figure 10.2). Most birds, including *Gallus*, also have a mainly undivided intermandibularis (Table 10.1; Figure 10.4). Some authors described a "caudal mylohyoideus" (often also named "pars intermandibularis of the constrictor colli") in birds such as chickens, but this structure is part of the hyoid muscle interhyoideus rather than the mandibular musculature (e.g., Baumel et al. 1979; McClearn and Noden 1988; this work; see Figure 10.4, Table 10.2, and text that follows). As explained in Chapter 8, the last common ancestor (LCA) of sarcopterygians, the LCA of tetrapods, and the LCA of amniotes likely had both an intermandibularis anterior and an intermandibularis posterior, because these two muscles are present in at least some members of most of the major extant sarcopterygian clades; i.e., of coelacanths, amphibians, reptiles, and mammals. According to this hypothesis, the fact that turtles, crocodylians, and birds often have a single intermandibularis might therefore be the result of a secondary loss of one of the two muscles that were plesiomorphically present in the LCA of amniotes. Or, more likely, the single intermandibularis of these reptiles might be an undifferentiated muscle that corresponds to the intermandibularis anterior + intermandibularis posterior of lepidosaurs such as *Timon*, but that does not become subdivided into two separate muscles during ontogeny. However, as also noted in previous chapters, one cannot completely discard the hypothesis that at least some of the structures that are designated "intermandibularis anterior" and "intermandibularis posterior" in sarcopterygians such as coelacanths, salamanders, anurans, lepidosaurs, and various mammals are, instead, the result of an independent (convergent/parallel) division of a mainly undivided muscle intermandibularis into two separate muscles.

The "adductor mandibulae complex" of reptiles includes mandibular muscles such as the adductor mandibulae A2-PVM, the adductor mandibulae A2, the levator anguli oris mandibularis, the "retractor anguli oris," the pseudotemporalis, the pterygomandibularis ventralis, the pterygomandibularis dorsalis, and the adductor mandibulae Aω. The A2 is well developed in extant reptiles, being usually situated laterally and rostrally to the maxillary and mandibular nerves, respectively (e.g., Holliday and Witmer 2007; this work; Figures 10.4 and 10.5). This muscle is often subdivided into three sections: "profundus," "medialis," and "superficialis" (Table 10.1; Figure 10.6). Holliday and Witmer (2007) stated that in birds, the A2 ("adductor mandibulae externus") is often divided into only two main sections: profundus (which includes the "adductor mandibulae externus coronoideus" and "adductor mandibulae externus zygomaticus" *sensu* some authors) and superficialis (which includes a muscle slip that is often designated "adductor externus pars profundus" or "articularis externus") (see also Tokita 2004). According to these authors, the medialis section of the A2 is not sufficiently distinct to be reliably identified in birds. However, the latter section is clearly recognizable in the *Gallus* specimens we dissected and also in the specimens described by other authors (see, e.g., Lakjer 1926). In the turtles that we dissected, the profundus section is the more developed portion of the A2 (Figure 10.6).

TABLE 10.1
Mandibular Muscles of Adults of Representative Reptilian Taxa

Reptilia, Testudines: *Trachemys scripta* (Red-Eared Slider Turtle)	Reptilia, Lepidosauria: *Timon lepidus* (Ocellated Lizard)	Reptilia, Crocodylia: *Caiman latirostris* (Brown-Snouted Caiman)	Reptilia, Aves: *Gallus domesticus* (Domestic Chicken)
Intermandibularis [see text]	**Intermandibularis posterior**	**Intermandibularis** [see text])	**Intermandibularis** (mylohyoideus or mylohyoideus anterior *sensu* Engels [1938]) [see text]
– [see text]	**Intermandibularis anterior**	– [see text]	– [see text]
Adductor mandibulae A2-PVM (adductor mandibulae posterior *sensu* Schumacher [1973], Rieppel [1990], and Holliday and Witmer [2007]) [Adams [1919] stated that the "adductor mandibulae complex" of turtles is mainly undivided, but authors such as Edgeworth [1935], Schumacher [1973], Rieppel [1990], and Holliday and Witmer [2007] have shown that in turtles this complex does comprise various separate muscles, including the adductor mandibulae A2-PVM]	**Adductor mandibulae A2-PVM** (adductor mandibulae posterior *sensu* Haas [1973], Abdala and Moro [2003], Wu [2003], and Holliday and Witmer [2007]; part of the adductor mandibulae externus *sensu* Iordansky [2004], which also includes the A2)	**Adductor mandibulae A2-PVM** (adductor mandibulae posterior *sensu* Busbey [1989] and Holliday and Witmer [2007])	**Adductor mandibulae A2-PVM** (adductor caudalis *sensu* Baumel et al. [1979] and McClearn and Noden [1988]; adductor mandibulae posterior *sensu* Elzanowski [1987]; seemingly corresponds to the adductor mandibulae posterior, adductor mandibulae ossis quadrati, or adductor mandibulae caudalis *sensu* Holliday and Witmer [2007]; adductor mandibulae caudalis *sensu* Vanden Berge [1975]) [the A2-PVM is not always considered to be a separate muscle in galliforms (the avian order that includes *Gallus domesticus*), but this muscle was described in the specimens described by authors such as Vanden Berge [1975] and is present in the specimens we dissected; see also, e.g., McClearn and Noden [1988]]
Adductor mandibulae A2 (adductor mandibulae externus *sensu* Schumacher [1973], Rieppel [1990], and Holliday and Witmer [2007]) [in turtles the A2 is often divided into three sections: "profundus," "medialis," and "superficialis": e.g., Schumacher [1973], Rieppel [1990], and Holliday and Witmer [2007]; Schumacher [1973] stated that trionychid turtles have both an A2 ("musculus adductor mandibulae externus") and a "musculus zygomaticomandibularis," the latter structure being a "separate division" of the A2 according to this author]	**Adductor mandibulae A2** (adductor mandibulae externus *sensu* Abdala and Moro [1996, 2003]; part of the adductor mandibulae externus *sensu* Haas [1973] and Wu [2003], which also includes the levator anguli oris mandibularis and the "retractor anguli oris"; part of the adductor mandibulae externus *sensu* Iordansky [2004], which also includes the A2-PVM) [as explained by authors such as Wu [2003] and Holliday and Witmer [2007], in lepidosaurs the A2 ("adductor mandibulae externus") is often divided into three sections: profundus, medialis, and superficialis']	**Adductor mandibulae A2** [according to authors such as Busbey [1989] and Holliday and Witmer [2007], in crocodylians the A2 ("adductor mandibulae externus") is often divided into three sections: profundus, medialis, and superficialis']	**Adductor mandibulae A2** (adductor externus *sensu* Baumel et al. [1979], Elzanowski [1987], McClearn and Noden [1988], and Köntges and Lumsden [1996]; masseter and quadratomandibularis complex of, e.g., Ghetie et al. [1981]) [Adams [1919] stated that birds such as chickens have a small muscle entotympanicus that is a "remnant" of the "adductor mandibulae complex": see, e.g., his plate VIII; however, this small muscle is not reported in more recent works: see, e.g., Vanden Berge [1975], Noden [1983a,b], McClearn and Noden [1988], and Noden and Francis-West [2006]; Vanden Berge [1975] confirms that chickens have superficial, ventral, and medial heads of the A2, as do crocodylians]

(*Continued*)

TABLE 10.1 (CONTINUED)
Mandibular Muscles of Adults of Representative Reptilian Taxa

Reptilia, Testudines: *Trachemys scripta* (Red-Eared Slider Turtle)	Reptilia, Lepidosauria: *Timon lepidus* (Ocellated Lizard)	Reptilia, Crocodylia: *Caiman latirostris* (Brown-Snouted Caiman)	Reptilia, Aves: *Gallus domesticus* (Domestic Chicken)
– [not present in turtles, according to, e.g., Adams [1919] and Schumacher [1973]; Rieppel's [1990] study corroborated the idea that, unlike lepidosaurs, turtles such as *Chelydra* do not form a levator anguli oris mandibularis at any developmental stage]	**Levator anguli oris mandibularis** (levator anguli oris *sensu* Diogo [2007, 2008b]) [present, somewhat blended with A2; we use the name "mandibularis" to distinguish this muscle from the levator anguli oris facialis of certain mammals, which is a facial (hyoid), not a mandibular, muscle; according to authors such as Wu [2003] and Holliday and Witmer [2007], the levator angulis oris mandibularis of reptiles is derived from the A2 (adductor mandibulae externus), as proposed in previous chapters; see text]	– [as explained by, e.g., Haas [1973], the levator anguli oris mandibularis is not present in adult crocodylians and adult birds]	– [see on the left]
Pseudotemporalis [not present in turtles according to Adams [1919], but it is present in the specimens we dissected and described by authors such as Schumacher [1973], Rieppel [1990], and Holliday and Witmer [2007]]	**Pseudotemporalis** [both the adductor A3′ and A3″ seem to be included in the pseudotemporalis of extant amphibians such as *Ambystoma* and reptiles such as *Timon*: see text; in some species of these two groups, the pseudotemporalis is, however, clearly divided into superficial and deep structures that seemingly correspond directly to the A3′ and A3″ of fish, respectively: see, e.g., cell on the right]	**Pseudotemporalis** [according to Holliday and Witmer [2007], crocodylians and birds have a "pseudotemporalis profundus" (which corresponds to the "adductor mandibulae intermedius" *sensu* Iordansky [1964, 2000], and to the "quadratomandibularis" or "adductor mandibulae caudalis" of avian literature, and may correspond to the A3″ of at least some bony fish) and a "pseudotemporalis superficialis" (which may thus correspond to the A3′ of at least some bony fish, and which, in some birds, includes a small belly that is sometimes named "musculus caput absconditum" in avian literature)]	**Pseudotemporalis** (probably includes the quadratomandibularis *sensu* Elzanowski [1987]: see text and on the left) [Vanden Berge [1975] confirms that chickens have a profundus or A3″ head and a superficialis or A3′ head, which are usually considered as separate muscles]
Pterygomandibularis ventralis (pterygoideus ventralis *sensu* Schumacher [1973]; pterygoideus typicus *sensu* Wu [2003]) [authors such as Adams [1919], Haas [1973], and Rieppel [1990] suggested that the pterygomandibularis ("pterygoideus" in their nomenclature) of turtles is mainly undivided; however, it is now commonly accepted that turtles do have a pterygomandibularis ventralis and a pterygomandibularis dorsalis such as those of other reptiles: see, e.g., Schumacher [1973], Witmer [1995b], and Holliday and Witmer [2007]]	**Pterygomandibularis** (pterygoideus anterior *sensu* Adams [1919]; pterygoideus *sensu* Lightoller [1939], Goodrich [1958], Iordansky [2004], and Holliday and Witmer [2007]) [according to Adams [1919], in lepidosaurs such as *Sphenodon* and *Iguana* there is a single pterygomandibularis, but in other lepidosaurs, such as *Varanus*, there is a pterygomandibularis ventralis and a pterygomandibularis dorsalis; contrary to the statements of Adams [1919], it is now commonly accepted that *Sphenodon*, as well as some other lepidosaurian taxa (e.g., Haas 1973), have both a pterygomandibularis dorsalis and a pterygomandibularis ventralis: see, e.g., Haas [1973], Witmer [1995b], Wu [2003], Holliday and Witmer [2007]]	**Pterygomandibularis ventralis** (pterygoideus posterior *sensu* Adams [1919], Busbey [1989], and Iordansky [2000]; pterygoideus typicus or pterygoideus superficialis *sensu* Haas [1973] and Wu [2003])	**Pterygomandibularis ventralis** (pterygoideus posterior *sensu* Adams [1919] and Iordansky [2000]; probably corresponds to the pars ventromedialis of the pterygoideus *sensu* Baumel et al. [1979] and McClearn and Noden [1988]; part of pterygoideus *sensu* Vanden Berge [1975] and Köntges and Lumsden [1996]) [in various birds, the pterygomandibularis ventralis *sensu* this volume includes a small section that is sometimes designated "muscle retractor palatini" in the literature: see, e.g., Holliday and Witmer [2007]]

(*Continued*)

TABLE 10.1 (CONTINUED)
Mandibular Muscles of Adults of Representative Reptilian Taxa

Reptilia, Testudines: *Trachemys scripta* (Red-Eared Slider Turtle)	Reptilia, Lepidosauria: *Timon lepidus* (Ocellated Lizard)	Reptilia, Crocodylia: *Caiman latirostris* (Brown-Snouted Caiman)	Reptilia, Aves: *Gallus domesticus* (Domestic Chicken)
Pterygomandibularis dorsalis (pterygoideus dorsalis *sensu* Schumacher [1973]; pterygoideus atypicus *sensu* Wu [2003]) [based on his own interpretation of Edgeworth's [1935] developmental study, Wu [2003] suggested that the pterygomandibularis dorsalis ("anterior" or "atypicus" in his terminology) differentiates earlier than the pterygomandibularis ventralis during the ontogeny of reptiles and, thus, that the pterygomandibularis ventralis may be a more "primitive" subdivision of the adductor mandibular complex]	– [but present in lepidosaurs such as *Sphenodon* and seemingly also in some squamates: see pterygomandibularis above and text]	**Pterygomandibularis dorsalis** (pterygoideus anterior *sensu* Adams [1919], Schumacher [1973], Busbey [1989], and Iordansky [2000]; pterygoideus atypicus or pterygoideus profundus *sensu* Haas [1973] and Wu [2003])	**Pterygomandibularis dorsalis** (pterygoideus anterior *sensu* Adams [1919] and Iordansky [2000]; probably corresponds to the pars dorsolateralis of the pterygoideus *sensu* Baumel et al. [1979] and McClearn and Noden [1988]; part of pterygoideus *sensu* Vanden Berge [1975] and Köntges and Lumsden [1996])
Adductor mandibulae Aω (intramandibularis *sensu* Schumacher [1973], Rieppel [1990], and Iordansky [2008]: see on the right) [see on the right]	**Adductor mandibulae Aω** (intramandibular *sensu* Iordansky [2008]) [in *Timon*, the adductor mandibulae have a large and distinct anteroventral portion that is lodged in the "adductor fossa" *sensu* Lauder [1980a,b] and that seems very similar to the Aω of other osteichthyans; Iordansky [2008] reviewed this subject and stated that crocodylians, some turtles, some lepidosaurs, and charadriiform and procellariiform birds do have "intramandibular" muscles; he argued that there are two main types of these muscles in reptiles (the "lacertiloidan" and "crocodiloidan" types), that some turtles have rudiments of both muscle types, and that probably these two types of muscles "originated from different parts of primitive fish jaw adductors; they are not homologues to each other"]	**Adductor mandibulae Aω** (intramandibular *sensu* Schumacher [1973], Busbey [1989], Holliday and Witmer [2007], and Iordansky [2008]: see on the left)	– [absent as a distinct structure in *Gallus*, but present in other birds, such as Palaeognathae, Sphenisciformes, Pelecaniformes, and Procellariiformes, according to Holliday and Witmer [2007] and Iordansky [2008]: see on the left]
– [see text]	**Levator bulbi** (retractor pterygoidei *sensu* Moro and Abdala [2000] and Abdala and Moro [2003]; part or totality of depressor palpebrae *sensu* Edgeworth [1935]; tensor periorbitae or periorbitalis *sensu* Holliday and Witmer [2007])	**Levator bulbi** (part or totality of depressor palpebrae *sensu* Edgeworth [1935]; tensor periorbitae or periorbitalis *sensu* Holliday and Witmer [2007])	**Levator bulbi** (part or totality of depressor palpebrae *sensu* Edgeworth [1935] and Elzanowski [1987]; palpebral depressor *sensu* Noden et al. [1999] and Noden and Francis-West [2006]; and tensor periorbitae or periorbitalis *sensu* Holliday and Witmer [2007])
– [see text]	**Levator pterygoidei** (part or totality of pterygo-parietalis *sensu* Adams [1919])	– [but see text]	– [but see text]

(*Continued*)

TABLE 10.1 (CONTINUED)
Mandibular Muscles of Adults of Representative Reptilian Taxa

Reptilia, Testudines: *Trachemys scripta* (Red-Eared Slider Turtle)	Reptilia, Lepidosauria: *Timon lepidus* (Ocellated Lizard)	Reptilia, Crocodylia: *Caiman latirostris* (Brown-Snouted Caiman)	Reptilia, Aves: *Gallus domesticus* (Domestic Chicken)
– [see text]	**Protractor pterygoidei** (part or totality of pterygo-sphenoidalis *sensu* Adams [1919])	– [really absent in adults of *Caiman latirostris*? See text]	**Protractor pterygoidei** (part of protractor pteryoidei et quadrati *sensu* Elzanowski [1987]; part of protractor pterygoquadrati *sensu* Köntges and Lumsden [1996] and seemingly of quadrate protractor *sensu* Noden et al. [1999] and Noden and Francis-West [2006]; part of protractor quadratopterygoidei or cranio-pterygoquadratus *sensu* Vanden Berge [1975]) [see text]
–	–	–	**"Protractor quadratus"** (part of protractor pteryoidei et quadrati *sensu* Elzanowski [1987]; part of protractor pterygoquadrati *sensu* Köntges and Lumsden [1996] and seemingly of quadrate protractor *sensu* Noden et al. [1999] and Noden and Francis-West [2006]; part of protractor quadratopterygoidei or cranio-pterygoquadratus *sensu* Vanden Berge [1975])

Note: The nomenclature of the muscles shown in bold follows that of the text; in order to facilitate comparisons, in some cases names often used by other authors to designate a certain muscle/bundle are given in front of that muscle/bundle, between round brackets; additional comments are given between square brackets. Data compiled from evidence provided by our own dissections and comparisons and by an extensive review of the literature (see text).

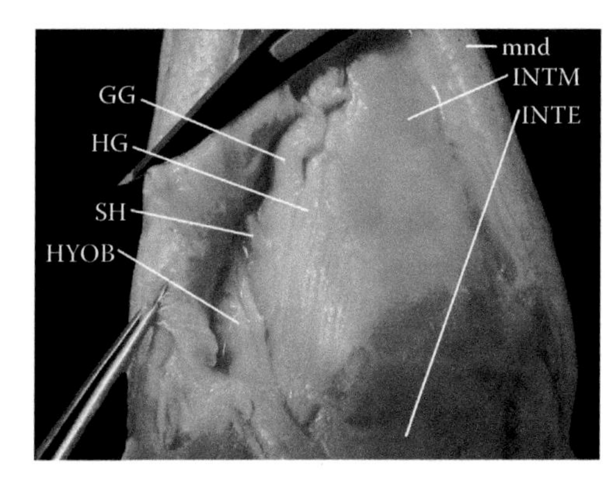

FIGURE 10.3 *Caiman latirostris* (Reptilia, Crocodylia): ventral view of the adult cephalic musculature; on the right side of the body (left side of the picture), superficial structures such as the intermandibularis and the interhyoideus are reflected to show the deeper musculature (anterior is to the top). GG, genioglossus; HG, hyoglossus; HYOB, hyobranchialis; INTE, interhyoideus; INTM, intermandibularis; mnd, mandible; SH, sternohyoideus.

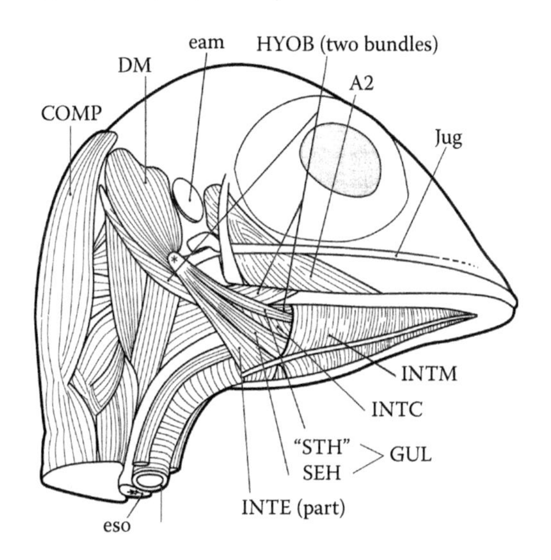

FIGURE 10.4 *Coturnix coturnix* (Reptilia, Aves): ventrolateral view of the superficial head musculature of a 15-day embryo; see configuration of this musculature in an adult member of this species in Figure 10.5. (The nomenclature of the myological structures illustrated follows that of the present work, while that of the skeletal structures follows McClearn and Noden [1988]; anterior is to the right). A2, adductor mandibulae A2; COMP, complexus (not a mandibular, hyoid, branchial, or hypobranchial muscle); DM, depressor mandibulae; eam, external auditory meatus; eso, esophagus; HYOB, hyobranchialis; INTC, interceratobranchialis; jug, jugal bar; SEH, serpihyoideus (bundle of gularis); "STH," "stylohyoideus" (bundle of gularis); tra, trachea. (Modified from McClearn, D., and Noden, D. M., *Amer J Anat*, 183, 277–293, 1988.)

There is much confusion and controversy in the literature regarding the homology of the structures that are often named "adductor mandibulae posterior" in nonmammalian tetrapod groups such as caecilians, urodeles, anurans, turtles, lepidosaurs, crocodylians, and birds. As explained by Rieppel (1990), the adductor mandibulae posterior seems to derive ontogenetically from the "adductor mandibulae internus" (which corresponds to the A3′/A3″ of bony fishes) in turtles and from the adductor mandibulae externus (which corresponds to the A2 of bony fishes) in other reptilian groups such as lepidosaurs. Therefore, this might indicate that the adductor mandibulae posterior of turtles is not homologous to the A2-PVM *sensu* the present work, because evolutionarily the latter muscle is supposedly derived from the A2 (see preceding chapters). However, according to Rieppel (1990), the posterior (but not the anterior) head of the adductor mandibulae posterior of adult turtles does seem to be the topological counterpart of the adductor mandibulae posterior of adult lepidosaurs. So, for Rieppel (1990: 52), this might provide "another example for the well-known phenomenon that structures judged to be homologous by their topological relationship in the adult may develop (ontogenetically) along different pathways; the problem then is to decide whether the relationship of homology should be based on the static adult topography or rather on the dynamic developmental patterns." But it should also be stressed that ontogenetic studies have their own problems and that it is not always easy to determine the specific anlage from which a certain muscle develops. For instance, some developmental studies indicated that the adductor mandibulae posterior of nontestudine reptiles such as birds also develops ontogenetically from the adductor mandibulae internus (A3′/A3″) (e.g., McClearn and Noden 1988). That is, these studies contradict the statements of Rieppel (1990), who suggested that among extant reptiles, this happens only in turtles. To add to the confusion, the studies of Piatt (e.g., Piatt 1938) also suggested that the adductor mandibulae posterior of the urodele *Ambystoma* develops from the adductor mandibulae internus (A3′/A3″), but the work of Ericsson and Olsson (2004), using more recent techniques, indicated that it develops from the adductor mandibulae externus (A2), thus supporting the idea that the urodele adductor mandibulae posterior is homologous to the A2-PVM *sensu* Diogo et al. (2008a,b) and *sensu* the present work. In fact, a related problem concerns the homology of the "intramandibularis" (Aω *sensu* the present work: see Table 10.1) in reptiles, because the latter muscle seems to derive from the adductor mandibulae externus (A2) in lepidosaurs and from the adductor mandibulae internus (A3′/A3″) in turtles, as explained by Rieppel (1990). To further complicate this subject, the adductor mandibulae posterior and the intramandibularis of turtles are partially blended to each other, as are the adductor mandibulae posterior and the intramandibularis of lepidosaurs. This similarity between the connection of these two muscles in turtles and in lepidosaurs could be used as an argument to favor the hypothesis that, despite the apparent (real or erroneously suggested by the developmental studies that have addressed this subject so far) differences regarding the developmental pathways leading to the formation of both the adductor mandibulae posterior and the intramandibularis in turtles and in other reptilian groups such as lepidosaurs, these structures are homologous among extant reptiles. Holliday and Witmer (2007) supported the idea that at least part of the adductor mandibulae posterior of turtles is homologous to the adductor mandibulae posterior of lepidosaurs, crocodylians, and birds and, thus, to the A2-PVM *sensu* Diogo et al. (2008a,b) and *sensu* this book. This idea is also

TABLE 10.2
Hyoid Muscles of Adults of Representative Reptilian Taxa

Reptilia, Testudines: *Trachemys scripta* (Red-Eared Slider Turtle)	Reptilia, Lepidosauria: *Timon lepidus* (Ocellated Lizard)	Reptilia, Crocodylia: *Caiman latirostrisi* (Brown-Snouted Caiman)	Reptilia, Aves: *Gallus domesticus* (Domestic Chicken)
Depressor mandibulae [see text]	**Depressor mandibulae** (pars notognathica *sensu* Lightoller [1939], which, contrary to what was suggested by this author, does not seem to correspond directly to the nucho-maxillaris of sharks, because such a "nucho-maxillaris" is not present in any of the bony fishes we dissected; i.e., it was very likely not present in the LCA of osteichthyans, in the LCA of sarcopterygians, nor in the LCA of tetrapods: see Chapter 8) [according to Edgeworth [1935], the anlage that gives rise to the depressor mandibulae and cervicomandibularis in lizards such as *Timon* gives rise to three muscles in snakes, which he called "occipito-quadrato-mandibularis," "cervico-mandibularis," and "neuro-costo-mandibularis"]	**Depressor mandibulae**	**Depressor mandibulae** [according to Adams [1919], *Gallus* have a small muscle parietomandibularis that is probably derived from the depressor mandibulae: see, e.g., his plate VIII; however, this muscle is not mentioned in more recent descriptions of this taxon: see, e.g., Köntges and Lumsden [1996]; Noden and Francis-West [2006]]
– [but see text]	**Cervicomandibularis** (cervicomandibularis posterior *sensu* Edgeworth [1935] seems to correspond to the pars cephalognathica *sensu* Lightoller [1939]) [see text]	– [but see text]	– [but see text]
– [as explained by authors such as Edgeworth [1935] and Schumacher [1973], the levator hyoideus is not present as a separate muscle in adult turtles; according to Edgeworth [1935], in turtles such as *Chrysemys* the levator hyoidei is not formed even as a temporary structure during ontogeny: see text]	– [see text]	**Levator hyoideus** [according to authors such as Edgeworth [1935], Lightoller [1939], and Schumacher [1973], the levator hyoideus is usually present as a separate muscle in adult crocodylians]	**Levator hyoideus** (muscle of the columella *sensu* Baumel et al. [1979] and McClearn and Noden [1988]; stapedius sensu Köntges and Lumsden [1996]) [the levator hyoideus (often named "laxator tympani," "occipito-tympanicus," "stapedius," "extracolumellaris," "columellae," "muscle of the columella," or "extra-stapedial muscle" in the literature) is often present in adult birds, including galliforms (which include *Gallus*): see, e.g., Edgeworth [1935] and McClearn and Noden [1988]]
Interhyoideus (constrictor colli *sensu* Schumacher [1973] and Wyneken [2001])	**Interhyoideus** [Lightoller [1939] states that in numerous lepidosaurs, including *Sphenodon*, the interhyoideus is mainly constituted by a single mass, that is, the structure that he designates as "pars interhyoideia" and that corresponds to the interhyoideus anterior of amphibians such as *Ambystoma* is not differentiated in these reptiles]	**Interhyoideus** (constrictor colli *sensu* Schumacher [1973]; constrictor colli + constrictor profundus *sensu* Li and Clarke [2015]) [see text]	**Interhyoideus** (constrictor colli *sensu* Edgeworth [1935]; part or totality of cutaneous colli *sensu* Ghetie et al. [1981] includes the pars intermandibularis as well as the other parts of the constrictor colli *sensu* Baumel et al. [1979], the caudal mylohyoideus and the constrictor colli *sensu* McClearn and Noden [1988], and the mylohyoideus posterior and constrictor colli *sensu* Vanden Berge [1975] and Köntges and Lumsden [1996])
–	–	–	**Gularis** (serpihyoideus plus stylohyoideus *sensu* Vanden Berge [1975], Baumel et al. [1979], McClearn and Noden [1988], Köntges and Lumsden [1996], and Huang et al. [1999]: see interhyoideus above) [see text]
–	–	–	**Interceratobranchialis** (interkeratoideus *sensu* Edgeworth [1935]; ceratohyoideus *sensu* Engels [1938]) (see text)

Note: See footnote of Table 10.1 and text. LCA, last common ancestor.

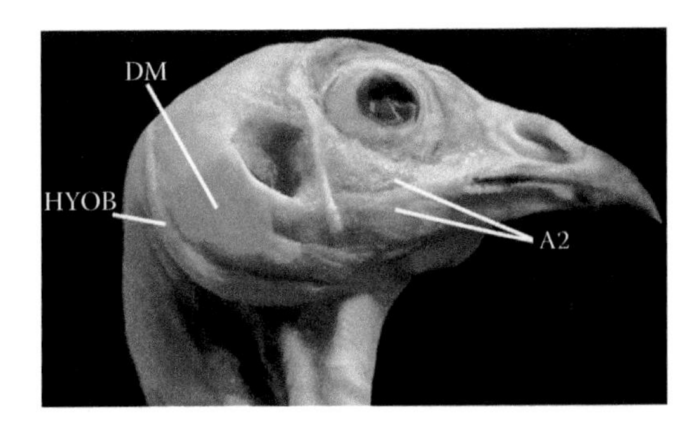

FIGURE 10.5 *Coturnix coturnix* (Reptilia, Aves): lateral view of the adult cephalic musculature showing the superficial section of the adductor mandibulae A2 (anterior is to the right). A2, adductor mandibulae A2; DM, depressor mandibulae; HYOB, hyobranchialis.

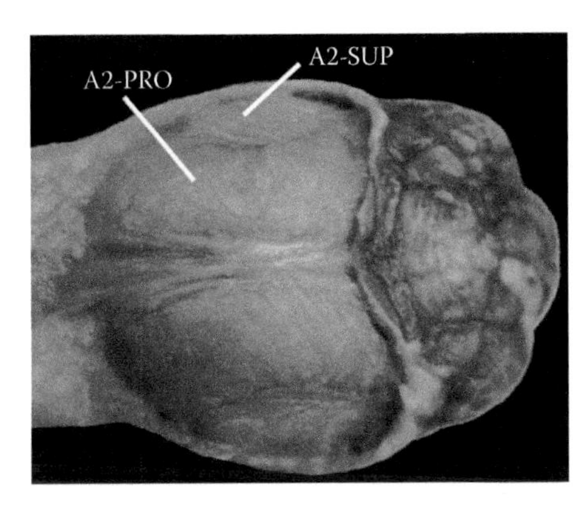

FIGURE 10.6 *Geochelone chilensis* (Reptilia, Testudines): dorsal view of the adult cephalic musculature showing the superficial and deep sections of the adductor mandibulae A2 (anterior is to the right). A2-PRO, A2-SUP, pars profunda and pars superficialis of the adductor mandibulae A2.

strongly supported by our dissections of members of these four major extant reptilian clades (see Table 10.1).

Unlike birds, turtles, and crocodylians, some "lizards" and some other lepidosaurs such as amphisbaenians and *Sphenodon* have a levator anguli oris mandibularis (Figure 8.6) and also a retractor anguli oris (e.g., Haas 1973; Rieppel 1980; Moro and Abdala 2000; Abdala and Moro 2003; Wu 2003; this work; see Table 10.1). The latter muscle occupies the posteroventrolateral region of the adductor mandibulae complex and, thus, is topologically similar to the retractor anguli oris of dipnoans (see preceding chapters). According to Wu (2003), the levator anguli oris mandibularis of lizards such as *Timon* corresponds to the retractor anguli oris + levator anguli oris mandibularis of taxa such as *Sphenodon*. Although phylogenetically it is more parsimonious to assume two independent acquisitions (i.e., during the evolution of dipnoans and the evolution of reptiles) than one acquisition (i.e., in the LCA of dipnoans + tetrapods) plus numerous secondary losses (i.e., in amphibians, in mammals, and in various reptilian groups), future studies are needed to investigate if the retractor anguli oris and/or possibly the levator anguli oris mandibularis of reptiles such as *Sphenodon* are, or are not, homologous to the retractor anguli oris of dipnoans.

The pseudotemporalis (Figure 8.7) is divided into superficialis and profundus layers in at least some members of the four major extant reptilian clades. In turtles, the pseudotemporalis is visible only after resection of the temporal roof. Elzanowski (1987) described a "musculus quadratomandibularis" in birds, but, based on its topology, we suggest that this structure is very likely homologous to/derived from the profundus layer of the pseudotemporalis of other extant reptiles. Both a pterygomandibularis ventralis and a pterygomandibularis dorsalis are found in at least some members of the four major extant reptilian clades, including the phylogenetically basal lepidosaur *Sphenodon* (Table 8.1; Figures 8.5, 8.6, and 10.7). However, most other lepidosaurs have a single pterygomandibularis (Table 10.1; Figures 10.2 and 10.8). As explained in Chapter 9, early authors such as Versluys (1904), based on the erroneous (see Diogo et al. 2008a) supposition that the pterygomandibularis of reptiles corresponds to the pterygoideus medialis and pterygoideus lateralis of mammals, decided to name the reptilian pterygomandibularis as "pterygoideus." Since then, both the names pterygoideus and pterygomandibularis have been used to designate the reptilian muscle (see Table 10.1). This is particularly problematic because many reptiles have two other mandibular muscles that are named levator pterygoideus and protractor pterygoideus but that, unlike the pterygomandibularis, are dorsal mandibular muscles, i.e., are derived from the constrictor dorsalis anlage *sensu* Edgeworth (1935) (see Table 10.1 and the following text).

Because turtles, birds, crocodylians, and lepidosaurs such as *Sphenodon* and some squamates (e.g., Haas 1973) have a pterygomandibularis ventralis and a pterygomandibularis dorsalis, the LCA of extant reptiles very likely had both these muscles (Witmer 1995b; Wu 2003; Holliday and Witmer 2007; Diogo and Abdala 2010). This plesiomorphic condition implies not only more muscular divisions, but also more complex relationships between them. Thus, in turtles, these two muscles are further subdivided into smaller bundles, and some of these bundles lie in a more rostral position than do the pterygomandibularis derivatives in other reptiles. Elzanowski (1987) also described an especially complex configuration of the pterygomandibularis derivatives in birds, with different grades of pennation. The descriptions of this author were corroborated by our dissections. Interestingly, the relationships among the pterygomandibularis derivatives, the depressor mandibulae, and the posterior process of the mandible are quite distinct among different reptilian groups. So, for instance, in lizards the posterior process of the mandible is mainly covered laterally by the pterygomandibularis (Figures 8.7 and 10.8), and the depressor mandibulae lies in a more dorsal location (Figures 8.6, 10.8, and 10.9). In turtles, the depressor mandibular usually lies in a more ventral position and often covers the posterior mandibular process. In crocodylians, the pterygomandibularis ventralis and the depressor mandibulae meet at the midline of the posterior portion of the mandible. In birds (Figures 10.4 and 10.5), like in turtles, the depressor mandibulae covers almost all the posterior mandibular process, and the pterygomandibularis ventralis

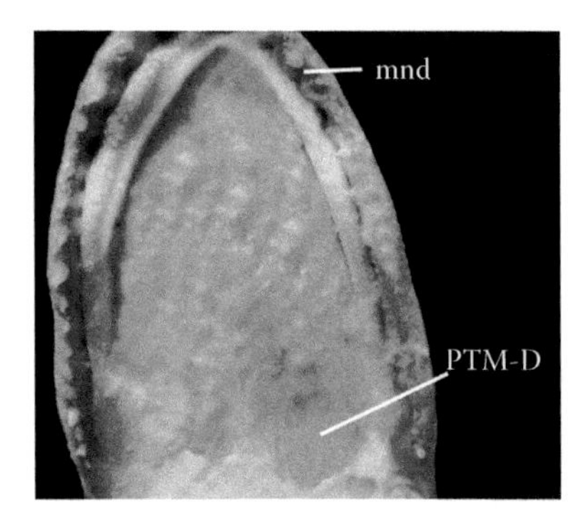

FIGURE 10.7 *Caiman latirotris* (Reptilia, Crocodylia): ventral view of the adult cephalic musculature with a portion of the thick aponeurosis cover reflected to show the pterygomandibularis dorsalis muscle (anterior is to the top). Mnd, mandible; PTM-D, pterygomandibularis dorsalis.

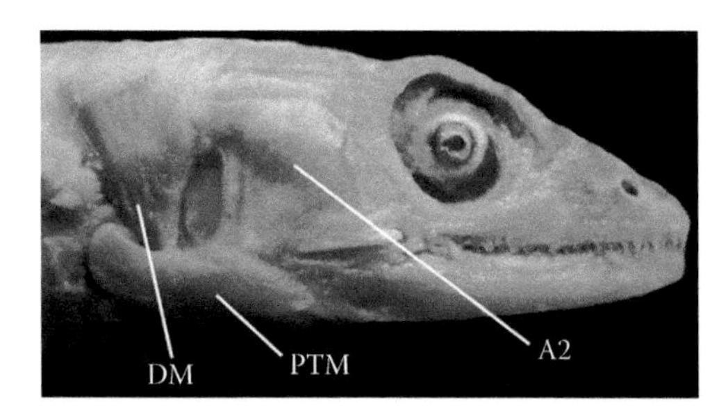

FIGURE 10.8 *Polychrus acutirostris* (Reptilia, Lepidosauria): lateral view of the adult cephalic musculature showing the posterior process of the jaw covered by the pterygomandibularis and the depressor mandibulae (anterior is to the right). A2, adductor mandibulae A2; DM, depressor mandibulae; PTM, pterygomandibularis.

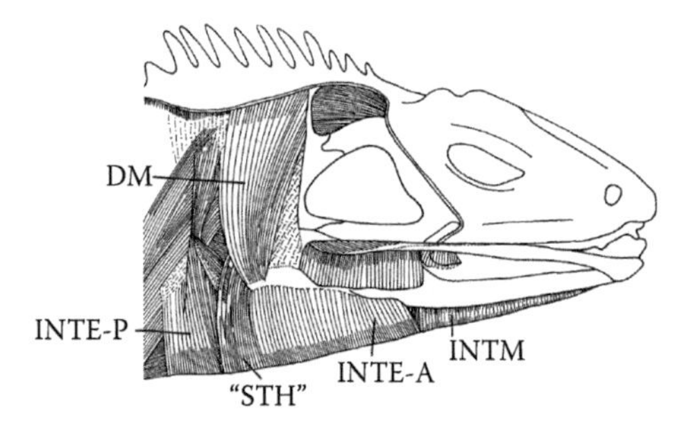

FIGURE 10.9 *Sphenodon punctatus* (Reptilia, Lepidosauria): lateral view of the adult cephalic musculature. (The nomenclature of the myological structures illustrated follows that of the present work; anterior is to the right.) DM, depressor mandibulae; INTE-A, INTE-P, anterior and posterior bundles of interhyoideus; INTM, intermandibularis; "STH," "stylohyoideus." (Modified from Haas, G., Muscles of the jaws and associated structures in the Rhynchocephalia and Squamata. In *Biology of the Reptilia*, Vol. 14 (eds. Gans, C., and Parsons, T. S.), pp. 285–490, Academic Press, New York, 1973.)

and pterygomandibularis dorsalis lie in a more rostral position. Authors such as Haas (1973), Wu (2003), and Holliday and Witmer (2007) suggested that the pterygomandibularis of squamates such as *Timon* corresponds to the pterygomandibularis ventralis of other reptiles. However, more detailed ontogenetic and comparative studies are needed to clarify if the pterygomandibularis dorsalis is completely absent in these squamates or, instead, if the pterygomandibularis of these taxa corresponds to both the pterygomandibularis ventralis + pterygomandibularis dorsalis of other reptiles.

The dorsal mandibular muscles (i.e., muscles derived from the constrictor dorsalis anlage *sensu* Edgeworth [1935]; see preceding chapters) present in extant reptiles are the levator bulbi, protractor pterygoidei, levator pterygoidei, "protractor quadrati," and "protractor quadratus" (Table 10.1). According to researchers such as Edgeworth (1935), Brock (1938), and Schumacher (1973), the constrictor dorsalis is formed in early ontogenetic stages of turtles, but then atrophies, and none of these dorsal mandibular muscles are thus present in adult turtles. Interestingly, in his developmental study of the turtle *Chelydra serpentina*, Rieppel (1990) stated that no constrictor dorsalis rudiment could be identified unequivocally during any developmental stage of this species. The adult turtle specimens we dissected and that were described by other authors (e.g., Schumacher 1973) lack dorsal mandibular muscles (Table 10.1). Edgeworth (1935) suggested that in *Sphenodon*, the constrictor dorsalis persists until the adult stage as a single mass, which he called "spheno-pterygo-quadratus." However, as explained in more recent works, this taxon often has a protractor pterygoidei, a levator pterygoidei, and a levator bulbi, as do most other nonophidian lepidosaurs (see, e.g., Wu 2003). In ophidians (snakes), the dorsal mandibular musculature is well developed and is often divided into a levator pterygoidei, a protractor pterygoidei plus a protractor quadrati (according to Haas [1973], these two muscles derive from the protractor pterygoidei of other reptiles), and a retractor pterygoidei plus a retractor vomeris (according to Haas [1973], these two muscles derive from the levator bulbi of other reptiles). Authors such as Abdala and Moro (2003) stated that *Sphenodon* also has a "retractor pterygoidei," but the latter muscle corresponds to the levator bulbi *sensu* the present work, as explained by Moro and Abdala (2000).

According to Brock (1938), the constrictor dorsalis is formed in early ontogenetic stages of crocodylians, but then atrophies, and the only dorsal mandibular muscle that is present in crocodylian adults is the levator bulbi. In the adult specimens of *Caiman latirostris* that we dissected, we were unable to find a separate protractor pterygoidei or a separate levator pterygoidei such as those seen in lepidosaur taxa such as *Timon* (Table 10.1). However, in their Figures 4 and 5, Holliday and Witmer (2007) show not only a levator bulbi ("periorbitalis" or "tensor periorbitae" in their terminology) but also a protractor pterygoidei in crocodylians such as *Alligator*, although in a later paper Holliday (2009: 1256) stated that, unlike birds, "crocodyliforms lost m. protractor pterygoideus." More studies are clearly needed to clarify the taxonomic distribution and homologies of the dorsal mandibular muscles in crocodylians. In their Figures 4 and 6, Holliday and Witmer (2007) also show three dorsal mandibular muscles in birds such as

Struthio: protractor pterygoidei, levator bulbi (periorbitalis or tensor periorbitae), and protractor quadratus. These three muscles are present in other birds, including *Gallus* (e.g., Noden 1983a; Elzanowski 1987; McClearn and Noden 1988; Holliday and Witmer 2007; Knight et al. 2008; this work; Table 10. 1).

As previously mentioned, adult turtles do not have dorsal mandibular muscles. Authors such as Brock (1938) argued that the plesiomorphic condition for tetrapods, and for reptiles, is to have one or more dorsal mandibular muscles. This idea is supported by the studies of Edgeworth (1935) and Schumacher (1973), which indicated that the constrictor dorsalis anlage is formed in early ontogenetic stages of turtles, but then atrophies in later stages (see previous text). Moreover, extant amphibians also have dorsal mandibular muscles: our dissections of numerous amphibian specimens confirmed that a levator bulbi is present in urodeles and anurans and that a levator quadrati is present in caecilians (Chapter 8). According to Brock (1938), the "constrictor dorsalis group" is usually more fully conserved in extant reptiles such as lepidosaurs because of their kinetic skull and tends to be secondarily reduced in taxa with a less kinetic skull such as turtles, crocodylians, birds, and amphibians. Moreover, according to the currently preferred idea that lepidosaurs are the sister-group of other extant reptiles (Figure 1.1), it is also more parsimonious, phylogenetically, to consider the loss of dorsal mandibular muscles in turtles to be a secondary evolutionary loss. Iordansky (1996) stated that the levator quadrati of caecilians corresponds to the levator pterygoideus + protractor pterygoidei of reptiles. As the levator bulbi of urodeles and anurans is indeed similar to that of reptiles, it is very likely that in the LCA of tetrapods, the dorsal mandibular muscles were already divided into (a) a levator bulbi and (b) an undivided "levator palatini," which was probably similar to, and derived from part of, the levator arcus palatini of sarcopterygian fishes such as *Latimeria* (see Chapters 8 and 9). Then, during the evolutionary events leading to the origin of urodeles and anurans, the levator palatini was secondarily lost, probably because these taxa lack a cranial kinesis such as that found in fishes such as coelacanths and in reptiles such as lepidosaurs (Brock 1938), leaving only the levator bulbi. As proposed by Iordansky (1996), the levator quadrati of caecilians thus probably corresponds to the structure that gave rise to reptilian muscles such as the levator pterygoidei, the protractor pterygoidei, the protractor quadrati (of snakes), and the protractor quadratus (of birds) (see Table 10.1). The levator bulbi was probably secondarily lost during the evolutionary events that led to the origin of caecilians. An alternative, a less likely hypothesis is that the levator quadrati of caecilians is an undifferentiated muscle corresponding to both the levator palatini and levator bulbi of the LCA of amphibians + amniotes. Either way, the protractor quadratus of birds and the protractor quadrati of snakes are not homologues to the levator quadrati of caecilians (Table 10.1; see also Chapter 9). As the protractor quadrati of snakes and the protractor quadratus of birds are derived from the protractor pterygoidei, these structures might hypothetically be homologous to each other. However, it is phylogenetically more parsimonious to accept that these structures were acquired independently in each of these groups (two evolutionary steps) than to assume that a protractor quadratus/protractor quadrati was present in the LCA of lepidosaurs, birds, and crocodylians and then was secondarily lost in crocodylians and in numerous lepidosaurian clades (see Conrad [2008] for a recent review of the phylogenetic position of snakes among lepidosaurs). In a paper, Holliday (2009) argues that, although the levator pterygoidei is absent in extant crocodylians and birds, there is evidence supporting the idea that this muscle was present in archosaurs such as dinosaurs, thus indicating that both the levator pterygoidei and the levator pterygoidei were present in the LCA of lepidosaurs, crocodylians, and birds. A major question that remains open, however, is whether in the LCA of all extant reptiles (i.e., also including turtles) the levator palatini *sensu* Diogo and Abdala (2010) was still mainly undivided (as was very likely that of the LCA of tetrapods: see preceding text) or was, instead, already divided into different muscles, such as the levator pterygoidei and the protractor pterygoidei. Obviously, some specialized mandibular muscles in reptiles are not discussed here or listed in Table 10.1, e.g., ethmomandibularis and pseudomasseter muscles that are specific to parrots (see, e.g., Tokita 2004).

HYOID MUSCLES

According to Edgeworth (1935), the hyoid musculature comprises dorso-medial and ventral muscles. The LCA of tetrapods probably had two dorso-medial hyoid muscles, the depressor mandibulae and levator hyoideus, and a ventral hyoid muscle, the interhyoideus (see preceding chapters). In addition to these three muscles (Figures 10.3, 10.4, and 10.8 through 10.10), extant reptiles might have other muscles such as the cervicomandibularis (Figures 8.6 and 10.2), which

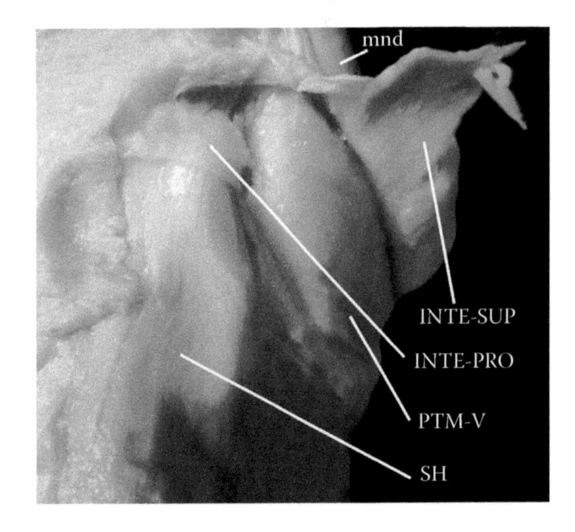

FIGURE 10.10 *Caiman latirostris* (Reptilia, Crocodylia): ventral view of the adult ventral cephalic musculature; on the left side of the body (right side of the figure), superficial structures such as the pars superficialis of the interhyoideus were reflected to show the deeper musculature (anterior is to the top). INTE-PRO, INTE-SUP, pars profunda and pars superficialis of interhyoideus; mnd, mandible; PTM-V, pterygomandibularis ventralis; SH, sternohyoideus.

is a dorso-medial hyoid muscle that probably derived from the depressor mandibulae, and the gularis (Figure 10.4) and interceratobranchialis (Figure 10.4), which are likely ventral hyoid muscles derived from the interhyoideus (Table 10.2). As explained by authors such as Schumacher (1973) and Smith (1992), plesiomorphically in reptiles there is no muscular pharynx, but crocodylians do possess a secondary palate and a means to constrict the pharynx by using a deep bundle (Figure 10.10) of the interhyoideus ("constrictor colli profundus" or "constrictor pharyngis" in Schumacher's [1973] terminology), which, interestingly, is topologically similar to the muscle interhyoideus profundus of monotreme mammals (which is also often designated in the literature as constrictor colli profundus) (see Chapter 8). The remaining, more superficial part of the interhyoideus of crocodylians is blended with the mandibular muscle intermandibularis, as can be seen in *C. latirostris* (Figure 10.3). The avian interceratobranchialis is a small muscle that usually connects the medial and ventromedial surfaces of the ceratobranchial cartilages to the ventral midline raphe (Figure 10.4) and that, as stated in the preceding text, is probably derived from the interhyoideus (see the section "Hypobranchial Muscles" for more details on this subject).

As noted by Edgeworth (1935), numerous birds, including *Gallus*, have a muscle gularis (Figure 10.4), which is often described as part of the "tongue musculature" but is innervated by cranial nerve (CN) VII and very likely derived from the anterior part of the interhyoideus. The gularis *sensu* Edgeworth (1935) and *sensu* the present work corresponds to the "serpihyoideus" plus stylohyoideus *sensu*, e.g., Huang et al. (1999) and McClearn and Noden (1988). Interestingly, Versluys (1904) and Haas (1973) also described a stylohyoideus in the lepidosaur *Sphenodon* (Figure 10.9), which is somewhat similar to the stylohyoideus of therian mammals, running from the dorsal end of the hyoid arch to the basal part of the "cornu branchiale I" and/or the interhyoideus. As the stylohyoideus of birds (Figure 10.4), this muscle stylohyoideus also seems to be derived from the interhyoideus (Figure 10.9). However, Haas (1973) stated that this muscle is found only in *Sphenodon* and in the squamate clade Gekkota, thus suggesting that the muscle is not homologous to the avian stylohyoideus.

Köntges and Lumsden (1996: 3241) stated that the "reptilian stylohyoideus" "shifted its attachment point from the lower jaw to the styloid process of the otic capsule" in order to form the mammalian stylohyoideus and that "the 2nd arch derived retroarticular process and attached depressor mandibulae were lost" in mammals. However, phylogenetically basal mammals such as monotremes do not have a separate stylohyoideus, and the therian stylohyoideus derives most likely from the depressor mandibulae (Chapter 8) rather than from the interhyoideus as does the avian stylohyoideus (Figure 10.4) and the stylohyoideus of *Sphenodon* and the Gekkota (Figure 10.9). Moreover, extant reptiles such as crocodylians, turtles, and the vast majority of lepidosaurs do not have a stylohyoideus such as that found in birds nor a stylohyoideus such as that found in *Sphenodon* and Gekkota (Table 10.1). Therefore, it seems unlikely that a muscle stylohyoideus was present in the LCA of reptiles (as this would oblige to assume a great number of secondary losses, e.g., in turtles and crocodylians—or in archosaurs as a whole, with a secondary reversion in birds—as well as in numerous squamate groups) and, even more unlikely, in the LCA of the amniotes as a whole (as this would oblige to assume even more secondary losses, e.g., in groups such as the monotremes). However, further studies on a wide range of taxa from all the major extant reptilian groups, especially birds and lepidosaurs, are needed to clarify the taxonomic distribution of the stylohyoideus muscles among reptiles.

The dorso-medial hyoid muscle depressor mandibulae is mainly a jaw opener that is present in birds, turtles, crocodylians, and lepidosaurs and that usually runs from the posterolateral surface of the skull to the retroarticular process of the mandible (Figures 10.2 through 10.4, 10.8, and 10.9; Table 10.2). In some turtles, e.g., *Dermochelys* and *Testudo*, the depressor mandibulae has a small bundle that is attached to the auditory tube and that is sometimes designated "musculus dilator tubae" in the literature (e.g., Edgeworth 1935; Schumacher 1973; Wyneken 2001). This supports the hypothesis, defended in Chapter 8, that mammalian hyoid muscles that are associated with the ear/auditory region, such as the stylohyoideus, the stapedius, and/or the mandibulo-auricularis, are probably derived from the anlage that gives rise to the depressor mandibulae/levator hyoideus of sarcopterygian fishes such as *Lepidosiren* (see Table 8.4 and also the following text).

Anatomists such as Saban (1968, 1971) suggested that the cervicomandibularis (Figure 8.6) derives from the interhyoideus. However, our dissections, comparisons, and review of the literature support the view defended by, e.g., Huber (1930a,b, 1931) and Edgeworth (1935), i.e., that the cervicomandibularis corresponds instead to part of the depressor mandibulae/levator hyoideus of sarcopterygian fishes such as *Lepidosiren*. Table 2 of Diogo et al. (2008a,b) suggests that the cervicomandibularis probably corresponds specifically to part of the levator hyoideus ("depressor mandibulae posterior" in their terminology) of amphibians, but the cervicomandibularis seems to derive from the depressor mandibulae (Figure 8.6; Table 8.4). Tsuihiji (2007) corroborated the latter idea and stated that in lepidosaurs such as *Sphenodon*, the depressor mandibulae and the cervicomandibularis are completely blended to each other, as previously described by Haas (1973), who suggested that the condition present in this genus represents the plesiomorphic condition for the Lepidosauria and for the Reptilia as a whole. According to Tsuihiji (2007), the avian depressor mandibulae is usually undivided. However, three bundles of this muscle have been reported in some birds (e.g., tinamiforms: Elzanowski [1987]), and we found two branches of this muscle in birds such as *Gallus*, *Nothura*, and *Pitangus*. Edgeworth (1935) and Schumacher (1973) stated that the cervicomandibularis is usually not present as a separate muscle in turtles. However, Schumacher's (1973) illustrations do show a "cervicomandibularis" in some turtles (see, e.g., his Figure 39). Wyneken (2001) reported that in sea turtles, the depressor mandibulae is often divided into different bundles, but did also not describe a separate muscle cervicomandibularis in these reptiles. Researchers such as Edgeworth (1935), Schumacher (1973), Noden (1983a), and McClearn and Noden (1988) also stated that in crocodylians and birds, the depressor mandibulae is often divided into bundles, but that these reptiles do not have a separate muscle cervicomandibularis as that found in lepidosaurs such as *Timon* (see Figure 8.6).

The more posterior bundles of at least some of the depressor mandibulae muscles described previously in turtles, crocodylians, and birds might well correspond to the cervicomandibularis of taxa such as *Timon*. These bundles clearly derive from the same anlage that gives rise to the cervicomandibularis (i.e., they derive from the depressor mandibulae anlage, namely, from its posterior portion, as does the cervicomandibularis: see previous text), and their overall configuration and function is very similar to that of the latter muscle. For instance, the topology of the posterior bundle of the depressor mandibulae that we found in our dissections of the turtle *Trachemys* is very similar to that of the cervicomandibularis of "lizards" such as *Timon*. The vast majority of anatomists have described this structure in turtles, crocodylians, and birds as a posterior bundle of the depressor mandibulae (as we do in the present work) rather than as a distinct muscle cervicomandibularis because in many if not most of the taxa of these three groups, this structure does not seem to be as differentiated from the main body of the depressor mandibulae as is the cervicomandibularis of most lizards (e.g., Edgeworth 1935; Schumacher 1973; Wyneken 2001; Tsuihiji 2007; Abdala and Diogo 2010). Therefore, future studies including a wider range of reptilian taxa should clarify whether in the LCA of extant reptiles, the posterior fibers of the depressor mandibulae were still deeply blended with the main body of this muscle, as suggested by Haas (1973), or whether these fibers were already somewhat differentiated from the anterior portion of the muscle, forming a posterior bundle of the depressor mandibulae such as that found in many turtles, crocodylians, and birds (which then became—apomorphically—further differentiated in order to form the cervicomandibularis of squamates such as *Timon*: Figure 8.6; Table 10.2).

As noted by authors such as Edgeworth (1935), the levator hyoideus is present as a separate muscle in early ontogenetic stages of most reptilian taxa (but not all, e.g., not in turtles such as *Chrysemys*), but then, in later stages, fuses with the depressor mandibulae in taxa such as *Sphenodon* and many squamates (Table 10.2). However, in squamate taxa such as the Gekkota, as well as in crocodylians and birds, the muscle usually persists as a separate structure until the adult stage and is often designated "laxator tympani" (e.g., Brock 1938), "occipito-tympanicus" (e.g., Edgeworth 1935), "stapedius" (e.g., Lightoller 1939; Schumacher 1973), "extracolumellaris" (e.g., Lee 2000; Conrad 2008), "columellae" (e.g., Witmer 1995b), "muscle of the columella" (e.g., Baumel et al. 1979; McClearn and Noden 1988), or "extra-stapedial muscle" (e.g., Goodrich 1958). The developmental work of McClearn and Noden (1988) on quails strongly supports the idea that the muscle of the columella of reptiles corresponds to the levator hyoideus of sarcopterygian fishes such as *Lepidosiren*, because this reptilian muscle differentiates from the same anlage that gives rise to the depressor mandibulae, as does the dipnoan levator hyoideus (see Diogo 2007, 2008b; Diogo et al. 2008a). In fact, according to Edgeworth (1935), it is very likely that the muscle levator hyoideus present in adult nonmammalian tetrapods such as these reptiles is directly homologous to the mammalian stapedius (see Chapter 8). If this is the case, then it would be probably more appropriate to designate the muscle of these nonmammalian tetrapods the stapedius, as proposed by Schumacher (1973), or, even better, to designate the muscle of mammals levator hyoideus, as proposed by Edgeworth (1935).

BRANCHIAL MUSCLES

As explained in the preceding chapters, in this volume we divide the branchial muscles into three main groups: the "true" branchial muscles (Table 10.3), the pharyngeal muscles, and the laryngeal muscles. The second group is present only in mammals (as stated previously, crocodylians possess a secondary palate and a means to constrict the pharynx, but they do that by using a deep bundle of the interhyoideus, which is a hyoid rather than a branchial muscle). Following this nomenclature, two groups can be recognized in reptiles: the true branchial muscles and the laryngeal muscles. The true branchial muscles are subdivided into (a) the true branchial muscles *sensu stricto* (e.g., hyobranchialis and "ceratohyoideus"), which are directly associated with the movements of the branchial arches and are often innervated by the glossopharyngeal nerve (CNIX), and (b) the derivatives of the protractor pectoralis ("cucullaris") of bony fishes and amphibians (e.g., trapezius, sternocleidomastoideus), which are instead mainly associated with the movement of postcranial structures such as the pectoral girdle and are primarily innervated by the spinal accessory nerve (CNXI). The laryngeal muscles (e.g., constrictor laryngis, dilatator laryngis) are usually innervated by the vagus nerve (CNX).

According to Edgeworth (1935), the only element of the true branchial musculature *sensu stricto* that is present in adult reptiles is the "subarcualis rectus I," which corresponds to the hyobranchialis of most reptiles and which is differentiated into an additional muscle in *Sphenodon* and other squamates, the ceratohyoideus (Figure 8.6 and Table 10.3, e.g., Haas 1973; Montero et al. 2002; Abdala and Moro 2003; Abdala and Diogo 2010). As mentioned in Chapter 8, we prefer to use the name hyobranchialis instead of "branchiohyoideus" or "branchiomandibularis" because this reptilian branchial muscle is clearly not homologous to the hyoid muscle branchiohyoideus that is often found in urodelan amphibians or to the hypobranchial muscle branchiomandibularis that is often found in cladistians, chondrosteans, and *Amia* (see preceding chapters). Schumacher (1973) described two true branchial muscles *sensu stricto* in turtles, the "branchiomandibularis pars visceralis" and branchiohyoideus, and stated that the branchiohyoideus is often designated in the literature as ceratohyoideus. However, this structure attaches anteriorly

TABLE 10.3
Branchial Muscles of Adults of Representative Reptilian Taxa

Reptilia, Testudines: *Trachemys scripta* (Red-Eared Slider Turtle)	Reptilia, Lepidosauria: *Timon lepidus* (Ocellated Lizard)	Reptilia, Crocodylia: *Caiman latirostris* (Brown-Snouted Caiman)	Reptilia, Aves: *Gallus domesticus* (Domestic Chicken)
Hyobranchialis (corresponds to the branchiohyoideus plus branchiomandibularis visceralis *sensu* Schumacher [1973]) [see text]	**Hyobranchialis** (corresponds to a part of the subarcualis rectus 1 and to the branchiohyoideus *sensu* Edgeworth [1935] and Herrel et al. [2005]) [according to Edgeworth [1935], snakes do not have branchial muscles *sensu stricto*; i.e., unlike lepidosaurs such as *Timon*, they do not have a hyobranchialis nor a "ceratohyoideus"]	**Hyobranchialis** (corresponds to the branchiomandibularis visceralis and probably also to the branchiomandibularis spinalis, *sensu* Schumacher [1973]; branchiohyoideus *sensu* Diogo and Abdala [2010]) [see text]	**Hyobranchialis** (corresponds to the branchiomandibularis visceralis and probably also to the branchiomandibularis spinalis, *sensu* Schumacher [1973]; branchiomandibularis *sensu* Engels [1938], McClearn and Noden [1988], and Köntges and Lumsden [1996]; geniohyoideus *sensu* Gethie et al. [1981]; mandibularis epibranchialis *sensu* Vanden Berge [1975], who states that it is innervated by the glossopharyngeal nerve = CNIX)
– [seems to be absent as a separate muscle in turtles: see hyobranchialis above]	**"Ceratohyoideus"** [Haas [1973] stated that the lepidosaurian "ceratohyoideus" is a hyoid muscle (innervated by CNVII), but this muscle clearly seems to be part of the branchial musculature: see text and Chapter 8]	– [seems to be absent as a separate muscle in crocodylians: see hyobranchialis above]	– [seems to be absent as a separate muscle in birds: see hyobranchialis above]
Trapezius (often designated cucullaris or testo-capitis) [see text]	**Trapezius** (capitidorsoclavicularis or dorsoscapularis *sensu* Fürbringer [1876, 1900], Edgeworth [1935], Holmes [1977], and Tsuihiji [2007]) [authors such as Edgeworth [1935] and Tsuihiji [2007] stated that in *Sphenodon* and some other lepidosaurs. there is no differentiation between the trapezius and the sternocleidomastoideus, but that most other lepidosaurs have these two muscles; Figure 14.13 of Jollie [1962] shows an *Iguana* specimen with the trapezius divided into, at least, a "pars clavotrapezius" and a "pars spinotrapezius" (see Figure 10.11); the names used by Jollie [1962] thus suggest that these are simply bundles of the trapezius, and not really separate muscles such as the acromiotrapezius and the spinotrapezius of some mammals: this idea is also supported by authors such as Kardong [2002]; as noted by Sefton et al. [2016], it is often stated that the trapezius and sternocleidomastodeus are absent in snakes, but they might potentially be present but shifted to the body wall due to absence of the pectoral girdle]	**Trapezius** (dorsoscapularis *sensu* Fürbringer and Tsuihiji [2007]) [as described by authors such as Meers [2003] and Tsuihiji [2007], in birds and crocodylians, including *Caiman*, the trapezius is not directly originated from the skull; according to Tsuihiji [2007], this feature might constitute a synapomorphy of birds + crocodylians)	**Trapezius** (cervical part of cranio-cervicalis *sensu* Edgeworth [1935]; part or totality of cucullaris *sensu* Elzanowski [1987]; cucullaris cervicus *sensu* Dilkes [1999] and Tsuihiji [2007]) [Edgeworth [1935] suggested that in birds, the muscles that perform the functions that are usually undertaken by the trapezius and the sternocleidomastoideus are not homologous to these two latter muscles, because they "are developed from occipital myotomes"; however, although recent studies have confirmed that muscles such as the trapezius are partially originated from somites in birds such as chickens as well as in other taxa such as mice, the avian sternocleidomastoideus and trapezius clearly seem to correspond to sternocleidomastoideus and trapezius of other amniotes: see text and, e.g., Dilkes [1999], Noden and Francis-West [2007], Tsuihiji [2007], and Shearman and Burke 2009; Vanden Berge [1975] confirmed that the "cucullaris cervicalis" or "cervical part of the cranio-cervicalis" is mainly innervated by CNXI and by some spinal nerves, as is the trapezius of other amniotes, although it may possibly also receive innervation from CNVII]

(*Continued*)

TABLE 10.3 (CONTINUED)
Branchial Muscles of Adults of Representative Reptilian Taxa

Reptilia, Testudines: *Trachemys scripta* (Red-Eared Slider Turtle)	Reptilia, Lepidosauria: *Timon lepidus* (Ocellated Lizard)	Reptilia, Crocodylia: *Caiman latirostris* (Brown-Snouted Caiman)	Reptilia, Aves: *Gallus domesticus* (Domestic Chicken)
Sternocleidomastoideus (often designated sternomastoideus, capiti-plastralis or plastrosquamosus) [see text]	**Sternocleidomastoideus** (capiticleidoepisternalis or episternocleidomastoideus *sensu* Fürbringer [1876], Edgeworth [1935], Holmes [1977], Herrel et al. [2005], and Tsuihiji [2007]) [see trapezius above and text]	**Sternocleidomastoideus** (capitisternalis *sensu* Fürbringer [1876, 1900] and capitiepisternalis *sensu* Tsuihiji [2007]) [Tsuihiji [2007] states that in Crocodylia, this muscle is divided into an anterior part, atlantimastoideus (iliocostalis capitis *sensu* Seidel [1978]), running from the atlas to the mastoid process, and a posterior part, sternoatlanticus, running from the atlas to the episternum; these divisions do not seem to correspond with divisions of the muscle found in humans and other mammals, they seem to be characteristic of Crocodylia]	**Sternocleidomastoideus** (cranial part of cranio-cervicalis *sensu* Edgeworth [1935]; cucullaris capitis or dermotemporalis *sensu* Dilkes [1999] and Tsuihiji [2007]) [Tsuihiji [2007] states that in Aves this muscle may be divided into as many as three slips depending on the species, e.g., pars interscapularis, pars propatagialis, and pars clavicularis; these three divisions do not seem to correspond to divisions of the muscle found in humans and other mammals, they seem to be characteristic of Aves; Vanden Berge [1975] confirmed that the "cucullaris cranialis" or "cranial part of the cranio-cervicalis" is mainly innervated by CNXI and by some spinal nerves, as is the sternocleidomastoideus of other amniotes, although it may possibly also receive innervation from CNVII]
Constrictor laryngis (probably corresponds to the cricohyoideus *sensu* Gaunt and Gans [1969]: see, e.g., Schumacher [1973])	**Constrictor laryngis** [see text and, e.g., Diogo et al. [2008a]]	**Constrictor laryngis** [see text and, e.g., Edgeworth [1935] and Schumacher [1973]]	**Constrictor laryngis** [see text and, e.g., Edgeworth [1935])
Dilatator laryngis (probably corresponds to the cricoarytenoideus *sensu* Gaunt and Gans [1969])	**Dilatator laryngis** [see text and, e.g., Diogo et al. [2008a]]	**Dilatator laryngis** [see text and, e.g., Edgeworth [1935] and Schumacher [1973]	**Dilatator laryngis** [see text and, e.g., Edgeworth [1935]]

Note: See footnote of Table 10.1 and text.

on the cornu hyale, where the hyobranchialis rather than the ceratohyoideus typically inserts anteriorly in lepidosaurs (see Figure 8.6). In fact, the testudine branchiohyoideus and branchiomandibularis pars visceralis *sensu* Schumacher (1973) clearly seem to correspond to the branchiomandibularis pars visceralis and "branchiomandibularis pars spinalis" of crocodylians and to the two bundles of the muscle branchiomandibularis of birds, and, thus, of the muscle hyobranchialis *sensu* the present work (Figure 10.4; Table 10.3). That is, the turtles described by Schumacher (1973) and by other authors (e.g., Lakjer 1926; Edgeworth 1935; Wyneken 2001) and those we dissected do not seem to have a separate muscle ceratohyoideus such as that found in lepidosaurs (Figure 8.6; Table 10.3).

Schumacher (1973) used the names branchiomandibularis pars visceralis and branchiomandibularis pars spinalis to describe the two bundles of the hyobranchialis in crocodylians because he thought that these two structures have different origins and innervations. According to the author, the branchiomandibularis pars visceralis is derived from the branchial musculature and innervated by the glossopharyngeal nerve (CNIX), while the branchiomandibularis pars spinalis is derived from the hypobranchial musculature and innervated by the hypoglossal nerve (CNX II]). However, these two structures were thought to be bundles of the same muscle by Edgeworth (1935), and, topologically, they are similar to each other and are often deeply blended posteriorly. Moreover, developmental studies have reported that at least in birds (Figure 10.4), the two bundles of the branchiomandibularis (which clearly seem to correspond to the branchiomandibularis pars spinalis and branchiomandibularis pars visceralis *sensu* Schumacher [1973]: see previous text) are innervated by the same nerve (CNIX) and differentiated from the same anlage (which derives from the third branchial arch; in quails, these two structures become only partially separated by day 8: see, e.g., McClearn and Noden [1988 and Noden and Francis-West [200). These studies thus strongly support the idea that the branchiomandibularis pars spinalis and branchiomandibularis pars visceralis *sensu* Schumacher (1973) are part of the same muscle, as is now often accepted in the literature, and, thus, that each of these two structures corresponds to part of the hyobranchialis *sensu* the present work. Further studies, ideally including new techniques in conjunction with motor neuron axon labeling data, are needed to determine whether or not the hyobranchialis is at least partially innervated by the hypoglossal nerve in some reptilian taxa such as crocodylians, as suggested by Schumacher (1973).

As mentioned by Edgeworth (1935) and Straus and Howell (1936), the topology (e.g., extending from the postcranial region of the body to the squamosal bone) and innervation (e.g., innervated by the "ramus accessorius of the vagus nerve" and by the "branches of the cervical nerve III") of the testudine muscle "plastrosquamosus" are remarkably similar to those of the trapezius + sternocleidomastoideus of other reptiles. Therefore, in the previous version of this book (Diogo and Abdala 2010), we suggested that this testudine muscle probably corresponded to an undivided "protractor pectoralis," i.e., to the trapezius + sternocleidomastoideus of taxa such as *Timon* (Table 7.3). This idea had been proposed by Holmes (1977) and Dilkes (1999), who stated that all major groups of living reptiles have a sternocleidomastoideus, except snakes, which also lack the trapezius, and turtles, which do have a trapezius. Holmes (1977) suggested that plesiomorphically in reptiles, the sternocleidomastoideus was not present as a separate muscle. However, this seems unlikely, because this muscle is present in lepidosaurs, crocodylians, birds, and mammals, so it was very likely present in the LCA of amniotes and in the LCA of reptiles. Fürbringer (1874) described a "testoscapuloprocoracoideus" and a "capitiplastris" in turtles and thought that these muscles are homologous to the trapezius and sternocleidomastoideus of other reptiles, respectively (if this were the case, then the "capitiplastralis" *sensu* Edgeworth [1935]—i.e., the plastrosquamosus *sensu* Schumacher [1973]—would not correspond to an undivided protractor pectoralis [trapezius + sternocleidomastoideus] but to the sternocleidomastoideus *sensu* the present work: see preceding text). We also found these two muscles in the turtles we dissected. However, it is now commonly accepted that the muscles "testoscapularis" and "testocoracoideus" of turtles are postcranial (pectoral) muscles, not head and neck muscles *sensu* the present work, although it is not completely clear whether or not the testoscapuloprocoracoideus described by Fürbringer (1874) corresponds to one or both of these muscles (e.g., Wyneken 2001). As stated by Diogo and Abdala (2010), further studies were needed to clarify the identity/homology of the testoscapuloprocoracoideus *sensu* Fürbringer (1874). Fortunately, such a study was published by Lyson et al. (2013). The authors dissected several extant turtles and, based on comparisons with other tetrapods, stated that turtles usually have (a) a trapezius muscle (often designated by other authors as "splenius capitis," cucullaris, or "testo-capitis") innervated by the vagus/spinal accessory complex and extending from the parietal and/or squamosal and the anteroventral surface of the nuchal bone and (b) a sternocleidomastoideus muscle (often designated "sternomastoideus," "capiti-plastralis," "plasto-squamosus," "rectus capitis," or "plastrosquamosus") innervated by the spinal accessory nerve (CNXI) and extending from the squamosal and/or parietal to the epiplastra (clavicles) and entoplastron (interclavicle).

As their name indicates, the dilatator laryngis and constrictor laryngis are antagonistic laryngeal muscles that are present in members of all the four major extant reptilian clades (Figure 10.11; Table 10.3). According to Edgeworth (1935), in birds such as *Gallus*, the constrictor laryngis has a pars ventralis and a pars dorsalis early in ontogeny, but in later stages the pars ventralis atrophies and only the pars dorsalis persist until the adult stage. Also according to this author, some birds have "laryngei" muscles (which in his view are not homologous to the laryngei muscles of amphibians: see Chapters 8 and 9), but these muscles are not present as separate structures in adults of the genus *Gallus*. We did not find such muscles in the chickens we dissected. Such muscles also seem to be absent in adults of crocodylians such as *Caiman*, of lepidosaurs such as *Timon*, and of turtles such as *Trachemys* (Figure 10.11; Table 10.3; e.g., Edgeworth 1935;

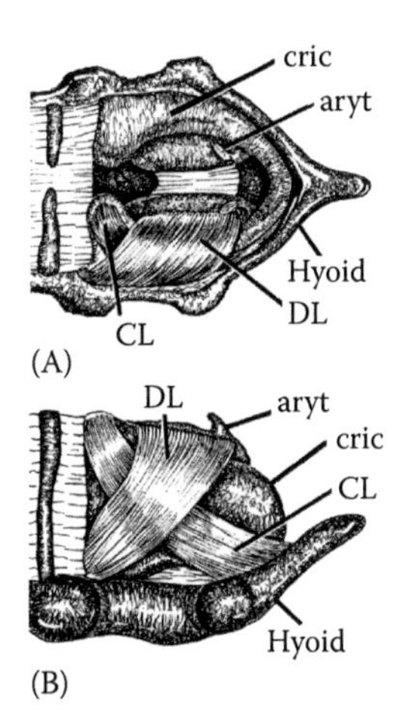

FIGURE 10.11 *Chelydra serpentina* (Reptilia, Testudines): (A) Dorsal view of the adult laryngeal musculature; the muscles of the left side of the body were removed. (B) Lateral view showing the adult laryngeal muscles on the sagitally dissected hyoid bone. (The nomenclature of the myological structures illustrated follows that used in the present work, while that of the skeletal structures follows Gaunt and Gans [1969]; anterior is to the right). aryt, arytenoid cartilage; CL, constrictor laryngis; cric, cricoid cartilage; DL, dilatator laryngis. (Modified from Gaunt A. S., Gans C., *J Morphol*, 128, 195–228, 1969.)

Gaunt and Gans 1969; Schumacher 1973; Diogo and Abdala 2010).

HYPOBRANCHIAL MUSCLES

As explained in the previous chapters, according to Edgeworth (1935), there are two major lineages of muscles originated from the hypobranchial muscle plate: the "geniohyoideu," which in tetrapods gives rise to structures such as the tongue muscles (e.g., "geniohyoideus," genioglossus, hyoglossus, and intrinsic muscles of the tongue), and the "rectus cervicis," which gives rise to structures such as the infrahyoid muscles (e.g., sternohyoideus, omohyoideus) (Table 10.4; as noted by Miyake et al. [1992], it is not clear if Edgeworth's "geniohyoideus" and "rectus cervicis" represent separate premyogenic condensations or later states of muscle development). It should be stressed that the musculature that is often designated tongue musculature of birds includes not only hypobranchial muscles, as is often the case in other amniotes, but also mandibular (the intermandibularis), hyoid (the gularis, which includes the stylohyoideus and serpihyoideus *sensu* Huang et al. [1999], and also the interceratobranchialis; although some authors have argued that the latter muscle might be a mandibular muscle or a branchial muscle: see below), and branchial (the hyobranchialis; i.e., the branchiomandibularis *sensu* Huang et al. [1999]) muscles (see Figure 10.4 and Table 10.4). The developmental work of Huang et al. (1999) showed that somites 2–6 participate in the formation of the hypobranchial muscles of chickens, but not of the other, nonhypobranchial "tongue muscles" previously listed, with the exception of the interceratobranchialis (Figure 10.4), which included a few somite myogenic cells. As explained by these authors, since myoblasts from more cranially located somites and those from more caudally located somites are not different in the extent of their contribution to the individual avian tongue muscles, it seems that myoblasts from different somites have the same ability to invade the "tongue primordia" and contribute to the tongue muscles. This means that myoblasts from different somites intermingle during their migration to the "tongue anlagen." Huang et al.'s (1999) results showed that cells from multiple somites stream to a focal point and then migrate as a single population. In this context, the mechanisms of myogenic cell migration of the tongue differ from those in the limb bud. In the limb bud, myoblasts initially appear to emigrate from the somites strictly laterally along projections of the original segmental borders into the limb bud mesoderm. The obvious mixing of myoblasts from different somites takes place much later at the central part of the forearm and hand. According to Huang et al. (1999), the reason for this difference between tongue and limb might be that the distances between the source and destination of muscle precursor cells are longer for the tongue and that the cells that migrate to the tongue are restricted to a narrow band that is often named "hypoglossal cord" in the literature.

One interesting result of Huang et al.'s (1999) developmental work is that it showed that the avian muscle interceratobranchialis (Figure 10.4; Table 10.2) included a few somitic myogenic cells. Edgeworth (1935) stated that this avian muscle ("interkeratoideus" in his terminology) is a mandibular muscle innervated by CNV and, very likely, derived from the intermandibularis. However, this muscle has also been reported to have a CNXII innervation (e.g., Vanden Berge 1975), and some authors thus suggest that it is a branchial muscle derived from the third and/or fourth branchial arches (see, e.g., Table 1 of Marcucio and Noden 1999). But more recent developmental studies indicated that the interceratobranchialis is derived ontogenetically from the same axial level as the hyoid muscles depressor mandibulae and gularis (serpihyoideus + stylohyoideus), and, because of this, this muscle is now commonly thought to be a hyoid muscle (see, e.g., Noden and Francis-West 2006). A careful analysis of the topology of the interceratobranchialis in adult birds such as chickens also indicates that this structure is derived from the hyoid musculature and probably from the interhyoideus, because it lies just dorsal to, and is associated with, the anteroventral portion of the interhyoideus ("constrictor colli, pars intermandibularis" *sensu* Baumel et al. [1979]) and its derivatives (e.g., gularis, i.e., serpihyoideus + stylohyoideus *sensu* Baumel et al. [1979]) (see the section "Hyoid Muscles," Table 10.2, and Figure 10.4). The data obtained in Huang et al.'s (1999) work, showing that the interceratobranchialis receives some (although very few) contributions from somitic myogenic cells, also support the idea that this muscle is more likely part of the hyoid musculature than of the mandibular musculature. In other tetrapod taxa such as amphibians, some hyoid muscles (e.g., the interhyoideus posterior) might also receive a contribution of somitic myogenic cells (some branchial muscles might also include such cells, and hypobranchial muscles almost always do), while mandibular muscles almost never do (see, e.g., the recent work of Piekarski and Olsson

TABLE 10.4
Hypobranchial Muscles of Adults of Representative Reptilian Taxa

Reptilia, Testudines: *Trachemys scripta* (Red-Eared Slider Turtle)	Reptilia, Lepidosauria: *Timon lepidus* (Ocellated Lizard)	Reptilia, Crocodylia: *Caiman latirostris* (Brown-Snouted Caiman)	Reptilia, Aves: *Gallus domesticus* (Domestic Chicken)
Geniohyoideus	**Geniohyoideus** (geniohyoideus and/or mandibulohyoideus *sensu* Edgeworth [1935] and Herrel et al. [2005])	**Geniohyoideus** [see text]	– [see text]
Genioglossus	**Genioglossus** [see text]	**Genioglossus** [see text]	– [see text]
Hypoglossoglossus [as explained by Schumacher [1973], this muscle is often present in turtles and is probably derived from the genioglossus; it often originates from the dorsal side and lateral margin of the hypoglossum and inserts onto the lingual process and the anterior part of the tongue]	–	–	–
Entoglossoglossus [as explained by Schumacher [1973], this muscle is also often present in turtles and is also probably derived from the genioglossus; it corresponds to the "protrusor linguae" or to the "hyoentoglossus" of some authors and often extends posteriorly from the apex of the lingual process to the lateral margins of the hypoglossum]	–	–	–
– [see on the right]	**Intrinsic muscles of the tongue** [according to Saban [1971], Smith [1988, 1992], and Herrel et al. [2005], examples of reptilian intrinsic tongue are the "longitudinalis" (which seemingly derives from the hyoglossus; NB: some reptilian taxa have both a dorsal longitudinal layer and a ventral longitudinal layer, which are similar to the longitudinalis dorsalis and longitudinalis ventralis of mammals such as humans, respectively), the "transversus linguae ventralis," the "transversus linguae dorsalis," the "verticalis linguae," and the "annularis" (the "verticalis linguae" corresponds to the "accelarator" *sensu* Herrel et al. [2005]); Edgeworth [1935] suggested that the "accelerator" corresponds to part or the totality of the "annulus of the lingual process of the basihyobranchiale" and, thus, seemingly to the "annularis"/"ring muscle" *sensu* Smith [1988, 1992]; however, the works of authors such as Saban [1968, 1971] seem to indicate that the "verticalis linguae" and "annularis" may be present at the same time in the same taxon; at least part of the "verticalis linguae" derives from the genioglossus according to Saban [1971]]	– [according to Edgeworth [1935], crocodylians do not have a longitudinalis linguae nor a transversus linguae, but Schumacher [1973] stated that crocodylians do often have a transversalis linguae; see on the left]	**Intrinsic muscles of the tongue** [according to authors such as Noden and Francis-West [2006], birds lack intrinsic tongue muscles: see on the left; however, authors such as Vanden Berge [1975] clearly show that chickens do have intrinsic muscles of the tongue, namely, a "superficial intrinsic muscle" and a "deep intrinsic muscle"]

(*Continued*)

TABLE 10.4 (CONTINUED)
Hypobranchial Muscles of Adults of Representative Reptilian Taxa

Reptilia, Testudines: *Trachemys scripta* (Red-Eared Slider Turtle)	Reptilia, Lepidosauria: *Timon lepidus* (Ocellated Lizard)	Reptilia, Crocodylia: *Caiman latirostris* (Brown-Snouted Caiman)	Reptilia, Aves: *Gallus domesticus* (Domestic Chicken)
Hyoglossus	**Hyoglossus**	**Hyoglossus**	**Hyoglossus** (paratoglossoceratobasi-branchialis *sensu* Vanden Berge [1975], which is innervated by CNXII) [corresponds to the "ceratoglossus" *sensu* Engels [1938], Köntges and Lumsden [1996], and Huang et al. [1999], which is thus not homologous to the ceratoglossus of some mammals because the latter structure corresponds to a bundle of the hyoglossus, not to the hyoglossus as a whole, as does the "ceratoglossus" of birds: see Chapter 8]
–	–	–	**Hypoglossus** (paratoglossobasibranchialis lateralis or hypoglossus obliquus or lateralis plus paratoglossobasibranchialis medialis or hypoglossus rostralis or medialis *sensu* Vanden Berge [1975], innervated by CNXII) [as described by authors such as Edgeworth [1935] and Huang et al. [1999], birds often have a muscle hypoglossus, which derives, together with the hyoglossus, from the posterior part of the "geniohyoideus anlage," the anterior part of this anlage thus giving rise to the geniohyoideus and the genioglossus (as explained in the text, both these two latter muscles are often absent in adults of the genus *Gallus*); Vanden Berge [1975] reports a "paratoglossobasibranchialis lateralis" or "hypoglossus obliquus or lateralis" head and a "paratoglossobasibranchialis medialis" or "hypoglossus rostralis or medialis" head in chicken]

(*Continued*)

TABLE 10.4 (CONTINUED)
Hypobranchial Muscles of Adults of Representative Reptilian Taxa

Reptilia, Testudines: *Trachemys scripta* (Red-Eared Slider Turtle)	Reptilia, Lepidosauria: *Timon lepidus* (Ocellated Lizard)	Reptilia, Crocodylia: *Caiman latirostris* (Brown-Snouted Caiman)	Reptilia, Aves: *Gallus domesticus* (Domestic Chicken)
Sternohyoideus (rectus cervicis *sensu* Edgeworth [1935]; coracohyoideus *sensu* Schumacher [1973] and Wyneken 2001)	**Sternohyoideus** (episternohyoideus *sensu* Edgeworth [1935] and Holmes [1977]; rectus cervicis *sensu* Kardong [2002]) [see text]	**Sternohyoideus** (episternohyoideus *sensu* Edgeworth [1935] and Holmes [1977]; probably corresponds to the episternobranchiotendineus and/or episternobranchialis *sensu* Schumacher [1973], as suggested by Li and Clarke [2015]: see, e.g., Sedlmayr [2002]) [according to authors such as Sedlmayr [2002], in some crocodylians the rectus cervicis gives rise to the sternohyoideus, to the omohyoideus, and to a structure that is often named musculus sternomandibularis in the literature; according to Edgeworth [1935], this "sternomandibularis" is mainly derived from the rectus cervicis, and, namely, from the sternohyoideus, but its anterior portion is formed by a part or the whole geniohyoideus]	**Tracheolateralis** [according to authors such as Sedlmayr [2002], in birds the rectus cervicis (often named "tracheo-laryngo-hyoideus" in the literature) gives rise to the sternohyoideus (which is said to mainly correspond to the "tracheolateralis" or "sternotracheolaryngeus" *sensu* King [1975] and thus to the "tracheolateralis" *sensu* Huang et al. [1999] and the "trachealis laterali" *sensu* Tsukahara et al. [2009), to the omohyoideus (which is often designated "cleidohyoideus" in the literature but might include both the "cleidohyoideus" and the "cleidotrachealis" *sensu* Huang et al. [1999), and to a structure that is often named "musculus cricohyoideus" (often also designated "musculus thyrohyoideus" in the literature), which in turn gives rise to a "sternotrachealis" and a "tracheohyoideus," the latter structure corresponding to the "sternotracheolaryngeus medialis" *sensu* King [1975]; however, authors such as Engels [1938] suggested that the "cricohyoideus" *sensu* Edgeworth [1938] corresponds to both the structures that are often named "thyrohyoideus" and "tracheohyoideus" in the literature; Tsukahara et al. [2009] used names such as "tracheobronchialis ventralis," "syringeus ventromesialis," "syringeus ventrolateralis," and "syringeus dorsolateralis" to describe some of the "syringeal" muscles of birds such as *Corvus*, which are controlled by the hypoglossal nerve (CNXII) and seem to originate from the rectus cervicis *sensu* Edgeworth and, thus, to derive from/correspond to part of the sternohyoideus/omohyoideus of other reptiles, but King [1975] states that chickens have no true syringeal muscles; studies on the development and innervation of the tongue and syringeal muscles of various representatives of each of the major avian clades are clearly needed to clarify the exact taxonomic distribution and homologies of the derivatives of the rectus cervicis *sensu* Edgeworth [1935] among these clades; King [1975] provided a

(Continued)

TABLE 10.4 (CONTINUED)
Hypobranchial Muscles of Adults of Representative Reptilian Taxa

Reptilia, Testudines: *Trachemys scripta* (Red-Eared Slider Turtle)	Reptilia, Lepidosauria: *Timon lepidus* (Ocellated Lizard)	Reptilia, Crocodylia: *Caiman latirostris* (Brown-Snouted Caiman)	Reptilia, Aves: *Gallus domesticus* (Domestic Chicken)
			summary of the laryngeal and tracheal muscles that are present in chickens—he stated that these birds have no true syringeal muscles—and described six muscles, which are hypobranchial muscles derived from the rectus cervicis anlage, with perhaps the exception of the basibranchialis laryngeus (part of the rostral extrinsic musculature of the larynx/trachea), which is a more anterior muscle that can be thus derived from the geniohyoideus anlage or even be a branchial muscle; all the other five muscles are more posterior and are associated with a dorsal mass of muscle; these five muscles are the tracheolateralis ("sternotracheolaryngeus lateralis," often designated "sternohyoideus"), the tracheohyoideus ("sternotracheolaryngeus medialis"), the tracheolaryngeus dorsalis, the tracheolaryngeus ventralis, and the sternolaryngeus]
–	–	–	**Tracheohyoideus** [see tracheolaterais above]
–	–	–	**Sternolaryngeus** [see tracheolaterais above]
–	–	–	**Tracheolaryngeus dorsalis** [see tracheolaterais above]
–	–	–	**Tracheolaryngeus ventralis** [see tracheolaterais above]
–	–	–	**Basibrachialis laryngeus** [not sure if it is a derivative of the sternohyoideus anlage, or instead of the geniohyoideus anlage, or even if it might be a branchial muscle: see tracheolaterais above]
– [the descriptions of Edgeworth [1935], Schumacher [1973], and Wyneken [2001] indicate that in turtles the omohyoideus is absent or, more likely, is not differentiated from the sternohyoideus; in the turtles we dissected, we were unable to find a separate omohyoideus such as that found in other extant reptiles]	**Omohyoideus** [present in most lepidosaurs, including *Iguana* and *Sphenodon*: see, e.g., Haas [1973] and Holmes [1977]]	**Omohyoideus** [as explained by authors such as Edgeworth [1935 and Holmes [1977], the omohyoideus is often present as a separate muscle in crocodylians; it probably corresponds to the coracohyoideus *sensu* Schumacher [1973]: see, e.g., Sedlmayr [2002]]	**Omohyoideus** [according to Sedlmayr [2002], the omohyoideus corresponds to the structure that is often named "cleidohyoideus" in the avian literature, although it might include both the "cleidohyoideus" and the "cleidotrachealis" *sensu* Huang et al. [1999], and, thus, the "cleidolaryngeal" *sensu* Noden et al. [1999]: see text]

Note: See footnote of Table 10.1 and text.

[2007] for more details on this subject; see also Chapter 8). The most convincing argument supporting this idea was, however, provided in works such as Köntges and Lumsden (1996), which have conclusively shown that (a) the connective tissues/fasciae associated with the interceratobranchialis are derived from hyoid crest cells and (b) this muscle attaches exclusively to hyoid crest-derived skeletal domains. Further studies may clarify whether this muscle is innervated by CNV (as proposed by, e.g., Edgeworth [1935]), CNXII (as proposed by, e.g., Vanden Berge [1975]), and/or by CNVII (like almost all hyoid muscles).

Schumacher (1973) suggested that crocodylians do not have a separate, independent muscle geniohyoideus. However, Edgeworth (1935) clearly described this muscle in crocodylian taxa, and his descriptions were corroborated by the study of Sedlmayr (2002) (who stated that the geniohyoideus of crocodylians includes, or is deeply associated with, the branchiomandibularis pars spinalis *sensu* Schumacher [1973]; i.e., to part of the hyobranchialis *sensu* the present work: see previous text) (Table 10.4). Regarding birds, as explained by Edgeworth (1935) in earlier ontogenetic stages, *Gallus* has a muscle that corresponds to the geniohyoideus + genioglossus of other reptiles but then this muscle disappears later in ontogeny, so both the geniohyoideus and genioglossus are absent in adults of this genus and of most other avian taxa (see Figure 10.4 and Table 10.4; particularly in the older literature, the name "geniohyoideus" has been erroneously used for avian structures such as the hyobranchialis: see, e.g., Müller and Weber [1998]). Developmental studies confirmed that the "genioglossus" (which probably corresponds to the geniohyoideus–genioglossus *sensu* Edgeworth [1935] and Engels [1938] and *sensu* the present work) is absent in adult chickens but persists through at least 18 days of development in passeriform birds such as quails (see, e.g., Köntges and Lumsden 1996; Huang et al. 1999; Marcucio and Noden 1999). Müller and Weber (1998) reported both a geniohyoideus and a genioglossus in some adult paleognathous birds, and, although they stated that these structures are often deeply blended, they argued that both were probably present in the LCA of extant birds and then were secondarily lost, being absent in most adult neognathous birds. Interestingly, authors such as Engels (1938) have described a "vestigial geniohyoideus–genioglossus muscle" in a few adult neognathous passeriform birds, e.g., *Corvus* and *Toxostoma*. Some reptiles have glossal muscles other than the genioglossus (Figure 10.3), the intrinsic muscles of the tongue, and the hyoglossus (Figure 10.3). For instance, turtles often have a hypoglossoglossus and an entoglossoglossus, and birds often have a hypoglossus (Table 10.4; e.g., Edgeworth 1935; Engels 1938; Schumacher 1973; Müller and Weber 1998; Huang et al. 1999; Abdala and Diogo 2010).

According to Diogo et al. (2008a), the LCA of amniotes very likely had two derivatives of the rectus cervicis *sensu* Edgeworth (1935): the sternohyoideus and the omohyoideus (see, e.g., Figure 8.7). Authors such as Jollie (1962) and Abdala and Moro (2003) described and illustrated a "sternothyroideus" in some lepidosaurian taxa, including *Iguana*. However, it is commonly accepted in the literature that most, if not all, turtles, crocodylians, and birds do not have a separate, well-defined muscle sternothyroideus such as that found in mammals (see, e.g., Saban 1968, 1971; Schumacher 1973; Wyneken 2001; Sedlmayr 2002). In our dissections, we were unable to find such a muscle in the member of these three groups. This thus supports the idea that the sternothyroideus was not present in the LCA of reptiles and, therefore, in the LCA of amniotes, i.e., that the structure that is designated sternothyroideus in some lepidosaurs is not homologous to the mammalian sternothyroideus. Further studies are needed to clarify the exact taxonomic distribution and homologies of the former structure among reptiles.

The peculiar syringeal muscles of birds are usually innervated by the hypoglossal nerve (CNXII) and are part of the hypobranchial musculature, namely, the rectus cervicis group *sensu* Edgeworth (1935) (see, e.g., Tsukahara et al. 2009). According to authors such as Sedlmayr (2002), in these reptiles the rectus cervicis (which is often called "tracheo-laryngo-hyoideus" in the literature) gives rise to the sternohyoideus (which might correspond to, or include the, "tracheolaryngealis" *sensu* Noden et al. [1999]), to the omohyoideus (which is often designated "cleidohyoideus" in the literature but might include both the cleidohyoideus and the "cleidotrachealis" *sensu* Huang et al. [1999]), and to a structure that is often named "musculus cricohyoideus" (often also designated "musculus thyrohyoideus" in the literature), which in turn gives rise to a "sternotrachealis" and a "tracheohyoideus" (the latter muscle probably corresponds to the "tracheolateralis" *sensu* Huang et al. [1999] and/or to the "trachealis laterali" *sensu* Tsukahara et al. [2009]). However, authors such as Engels (1938) suggested that the "cricohyoideus" *sensu* Edgeworth (1935) corresponds to both the structures that are often named thyrohyoideus and tracheohyoideus in the avian literature. In their recent work, Tsukahara et al. (2009) used names such as "tracheobronchialis ventralis," "syringeus ventromesialis," "syringeus ventrolateralis," and "syringeus dorsolateralis" to describe some of the components of the complex syringeal musculature of birds such as *Corvus*. Detailed studies of the development and innervation of the "tongue" and syringeal muscles of numerous representatives of each of the major avian clades are clearly needed to clarify the exact taxonomic distribution and homologies of the derivatives of the rectus cervicis (*sensu* Edgeworth 1935) within these clades.

GENERAL REMARKS

The number of cranial muscles in reptiles varies within narrow limits. Much of the anatomical variation between the major extant clades seems to be linked to general skull shape. For instance, some of the cephalic muscles tend to be more vertically oriented as the skull becomes more globe-like (e.g., Figure 10.4). A list of the muscles that were probably present in the LCA of turtles, lepidosaurs, crocodylians, and birds is given in Tables 10.1 through 10.4.

Regarding the mandibular muscles, this LCA likely had an adductor mandibulae A2, an adductor mandibulae A2-PVM, a pseudotemporalis, a pterygomandibularis ventralis, a pterygomandibularis dorsalis, a levator bulbi, and, possibly, an adductor mandibulae Aω, an intermandibularis anterior and an intermandibularis posterior, a protractor pterygoidei, and/or other dorsal mandibular muscles. As explained earlier, one cannot completely discard the hypothesis that the levator anguli oris mandibularis and/or the retractor anguli oris of reptiles such as lepidosaurs corresponds to/derive from the retractor anguli oris of sarcopterygian fishes such as dipnoans. If this were the case, this would mean that the LCA of tetrapods and the LCA of reptiles had at least one of these structures. However, phylogenetic parsimony indicates that this hypothesis is rather unlikely because these structures are absent in extant amphibians, turtles, crocodylians, birds, and mammals (see preceding text). Regarding the hyoid muscles, the LCA of extant reptiles probably had an interhyoideus and a depressor mandibulae. As previously noted, the presence of a stylohyoideus in lepidosaurs such *Sphenodon* and Gekkota, a gularis and a interceratobranchialis in birds, and a separate, well-differentiated muscle cervicomandibularis in numerous lepidosaurs probably represent apomorphic features of each clade rather than plesiomorphic features among reptiles. With respect to the branchial muscles, this LCA likely had a hyobranchialis, a trapezius, a sternocleidomastoideus, a constrictor laryngis, and a dilatator laryngis. Lastly, the hypobranchial musculature of this LCA probably included a sternohyoideus, an omohyoideus, a geniohyoideus, a genioglossus, a hyoglossus, and at least some intrinsic muscles of the tongue.

Detailed comparative studies on the development, innervation, and adult configuration of the head and neck muscles of an even wider range of reptilian and numerous other tetrapod and nontetrapod sarcopterygian taxa are clearly needed to address some controversial questions, such as the following: Do the levator anguli oris mandibularis and/or retractor anguli oris of reptiles such as lepidosaurs correspond to/derive from the retractor anguli oris of dipnoans? Is the intermandibularis of most adult birds, crocodylians, and turtles an undifferentiated muscle that corresponds to the intermandibularis anterior + intermandibularis posterior of lepidosaurs such as *Timon* and that does not become subdivided during ontogeny? What is the exact innervation of the interceratobranchialis among the different clades of extant birds? Was the levator palatini of the LCA of reptiles still mainly undivided as was very likely that of the LCA of tetrapods and of the LCA of amphibians, or was it instead already divided into different muscles, such as the levator pterygoidei and protractor pterygoidei? Is the pterygomandibularis dorsalis really completely absent in squamates such as *Timon* or does the pterygomandibularis of these reptiles correspond to both the pterygomandibularis ventralis + pterygomandibularis dorsalis of other reptiles? What is the exact taxonomic distribution of the structures that are often named stylohyoideus and sternothyroideus in reptiles? Were the posterior fibers of the depressor mandibulae of the LCA of extant reptiles still deeply blended to the main body of this muscle, or were these fibers already somewhat differentiated from the anterior portion of the muscle, forming a posterior bundle of the depressor mandibulae such as that found in various turtles, crocodylians, and birds?

11 Development of Cephalic Muscles in Tetrapods

In this chapter, we use urodeles as the main example to illustrate the development of cephalic muscles in tetrapods. This chapter is not intended to summarize all the numerous works on muscle development in tetrapods, which are discussed throughout the book. Instead, this chapter is mainly based on the paper of Ziermann and Diogo (2013) on the development of the axolotl (Urodela: *Ambystoma mexicanum*), because our overall analyses and comparisons indicate that urodeles are the best model to understand the early evolution and thus basal configuration of tetrapod cephalic muscles (see, e.g., Diogo et al. 2015d, 2015e). In addition to being considered the most useful model organism to discuss the early evolutionary history of tetrapods (see, e.g., Schmidt et al. 2013), the axolotl is also a crucial model for regenerative and evolutionary developmental biology (e.g., Diogo and Tanaka 2012; Diogo et al. 2014b,c).

Among amphibians, several types of metamorphosis are known that involve greater or lesser change between the larval and adult stages (Lynn 1961). The morphology of the jaw apparatus in salamanders is relatively stable during ontogeny until adulthood (Iordansky 1992). Therefore, it is easier to compare the muscles in larvae and adult salamanders than in anurans, in which massive changes in the cranium occur during development (Haas 1996a). The family Ambystomatidae includes several salamander species that are neotenic, which means that they fail to undergo a full metamorphosis. The head morphology of some members of this family was described over the past century (Luther 1914; Edgeworth 1935; Piatt 1938), and the development of larval cranial muscles was also investigated but to a lesser extent (e.g., Edgeworth 1935; Ericsson and Olsson 2004; Piekarski and Olsson 2007; Ziermann 2008). The Mexican axolotl (*A. mexicanum*) is a neotenic salamander species of which the adults display many larval characters (e.g., external gills, tail fin, no eyelids, branchial arch levators). As explained in the preceding text, this salamander has become an important model organism in developmental, regenerative, comparative, and evolutionary works, but paradoxically there is not a single paper explicitly dedicated to the study of the development of the mandibular, hyoid, branchial, and hypobranchial muscles of this species. Most studies focus only on a few muscles that are of interest for providing discussions about specific subjects such as functional morphology (Lauder and Reilly 1988), the evolution of novelties (Schmidt et al. 2013), the role of the neural crest in cranial development (Ericsson et al. 2004), or the origin of cranial muscles derived from somites (Piekarski and Olsson 2007).

In this chapter, we describe the development and adult configuration of the mandibular, hyoid, branchial, and hypobranchial muscles of *A. mexicanum*, from early ontogenetic stages to the adult stage. Staging follows Bordzilovskaya et al. (1989) and Nye et al. (2003). The materials and methods of this study were provided in detail in Ziermann and Diogo (2013). The nomenclature of adult bones mainly follows Carroll (2007). Larval cartilages were called as in Haas (1996a,b, 2001, 2003), while the developmental stages of muscle development (Table 11.1) follow Ziermann and Olsson (2007) and Ziermann (2008). A list of muscle synonyms used by other authors is given in Table 11.2.

DEVELOPMENT OF MANDIBULAR MUSCLES

During stage 34, an elongated cluster is visible lateral to the brain reaching the ventral portion of the head. The ventral part is the anlage of the intermandibularis posterior muscle, while the dorsal part is the adductor mandibulae anlage. By stage 36 both anlagen can be clearly divided (Figures 11.1A and 11.2A). At the beginning of stage 38, all mandibular arch muscles, except the intermandibularis anterior, are visible as muscle cell bands (Table 11.1). In stage 40, the fiber development is clearly visible, and by the end of stage 42, the mandibular arch muscles are fully developed.

The intermandibularis anterior is the only muscle that is always a bit delayed in its development compared to most other cranial muscles (Table 11.1; Figure 11.2D). It is visible by the end of stage 38, stretching anteriorly between Meckel's cartilages and with a clear gap to the intermandibularis posterior. The intermandibularis anterior connects with the contralateral muscle in a median raphe. In the adult, this muscle band stretches between the most anterior medial sides of the dentary (Figure 11.3E). The intermandibularis posterior develops from its origin at Meckel's cartilage to its insertion where it meets the contralateral muscle in median raphe (Figure 11.2). During the early development, the fibers are posteromedially oriented. By stage 40 (Figure 11.2E), the fibers have turned to their final lateromedial orientation and connect with the contralateral muscle. In the adults, this muscle covers the anterior ventral half of the head, running from the dentary toward the medial raphe and forming a continuous sheet with the interhyoideus (Figure 11.3C).

In stage 36 (Figure 11.1A), a small anlage of lateral mandibular muscles is visible just posterior to the eye anlage. This stretches during the following stages, and by stage 39 (Figure 11.1D) it can be distinguished between the anterior developing pseudotemporalis and the posterior superficial developing adductor mandibulae (A2) (Figure 11.1D). The deep adductor mandibulae A2-PVM (posteroventromedial) is not visible until stage 43, in our antibody-stained sample (Figure 11.4C).

TABLE 11.1
Updated and Completed Diagram Showing the Development of Cranial Muscles in *Ambystoma mexicanum*

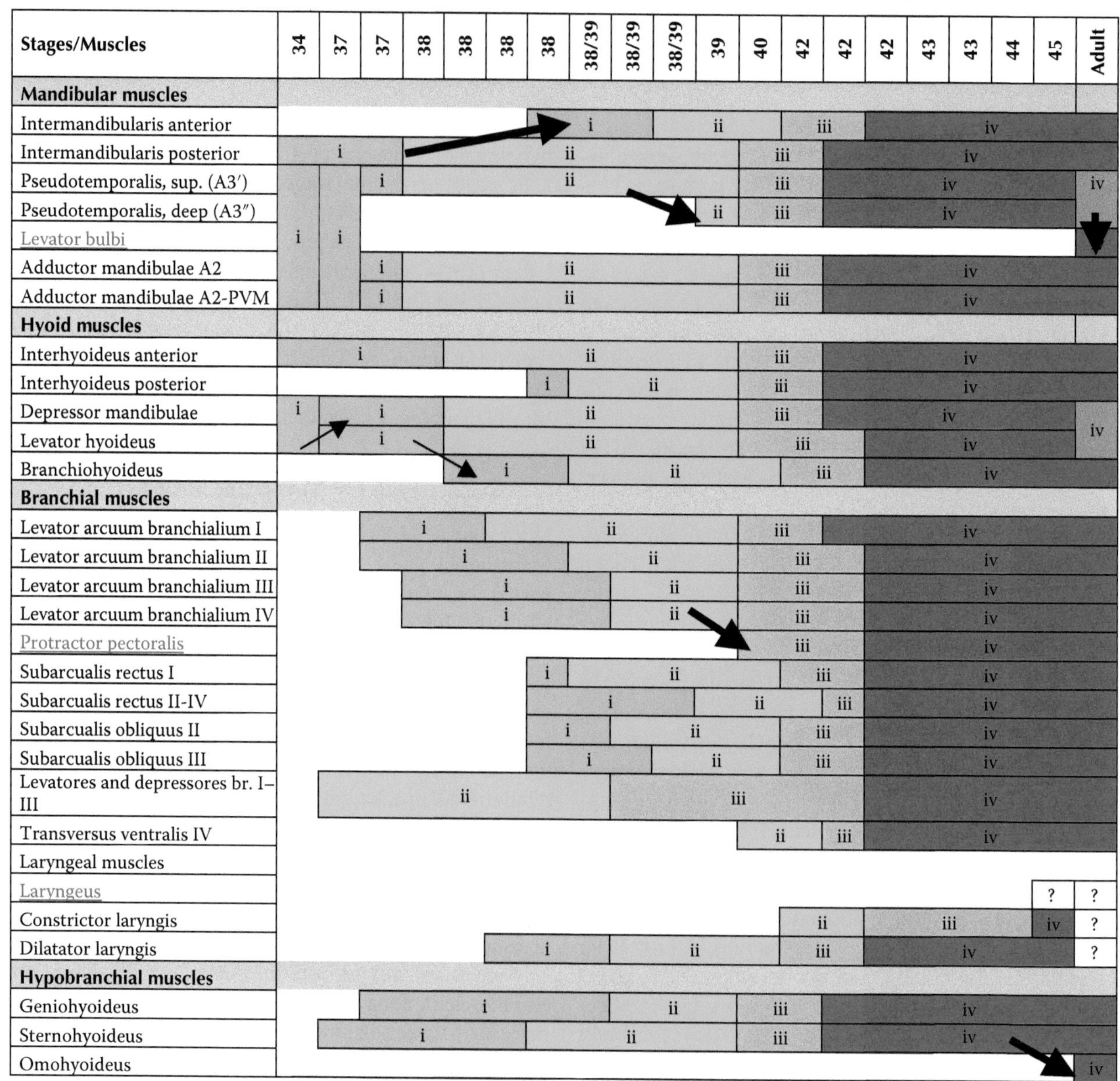

Stages/Muscles	34	37	37	38	38	38	38	38/39	38/39	38/39	39	40	42	42	42	43	43	44	45	Adult
Mandibular muscles																				
Intermandibularis anterior								i			ii		iii				iv			
Intermandibularis posterior		i					ii					iii					iv			
Pseudotemporalis, sup. (A3′)			i				ii					iii				iv				iv
Pseudotemporalis, deep (A3″)											ii	iii				iv				
Levator bulbi	i	i																		
Adductor mandibulae A2			i				ii					iii					iv			
Adductor mandibulae A2-PVM			i				ii					iii					iv			
Hyoid muscles																				
Interhyoideus anterior		i						ii				iii					iv			
Interhyoideus posterior							i		ii			iii					iv			
Depressor mandibulae	i		i					ii				iii				iv				iv
Levator hyoideus			i					ii					iii				iv			
Branchiohyoideus						i				ii			iii				iv			
Branchial muscles																				
Levator arcuum branchialium I				i				ii				iii					iv			
Levator arcuum branchialium II					i				ii				iii				iv			
Levator arcuum branchialium III						i				ii			iii				iv			
Levator arcuum branchialium IV						i				ii			iii				iv			
Protractor pectoralis													iii				iv			
Subarcualis rectus I							i			ii			iii				iv			
Subarcualis rectus II-IV								i				ii		iii			iv			
Subarcualis obliquus II							i			ii			iii				iv			
Subarcualis obliquus III								i			ii		iii				iv			
Levatores and depressores br. I–III					ii						iii						iv			
Transversus ventralis IV												ii		iii			iv			
Laryngeal muscles																				
Laryngeus																			?	?
Constrictor laryngis													ii			iii			iv	?
Dilatator laryngis							i			ii			iii				iv			?
Hypobranchial muscles																				
Geniohyoideus						i				ii		iii					iv			
Sternohyoideus				i					ii			iii					iv			
Omohyoideus																				iv

Note: Staging following Bordzilovskaya et al. (1989) and Nye et al. (2003). I, anlage (myoblasts); II, myocytes visible; III, fiber development starts; IV, functional muscle (modified from Ziermann 2008). Thicker arrows indicate development from another muscle. The small arrows showing the observed contribution of the levator hyoideus to the branchiohyoideus and depressor mandibulae. The magenta cells refer to the fact that in adults some muscles are integrated into each other. This is the case with the levator hyoideus and the deep pseudotemporalis (A3″) which are not present as distinct muscle structures because they are integrated in the adult depressor mandibulae and the adult pseudotemporalis, respectively. In stages 34 and 37, the long blue boxes (merged cells) indicate a common anlage. When anlage or muscle can be seen individually, the cells are separated. Red font highlights muscles that were not included in the detailed developmental table Ziermann (2008) but that were found in the present work in the adult axolotl. "sup." means superficial, "br." means branchiarum, and "?" refers to muscles that we could not identify in the adult axolotls dissected for the present project but that are often present in salamanders and in *Ambystoma mexicanum* as reported by Ziermann (2008); the only exception is the laryngeus (see text).

However, in histological sections, this muscle can be distinguished from the adductor mandibulae A2 and the pseudotemporalis from stage 38 onward (Table 11.1). The A2-PVM is blended with both the A2 and the pseudotemporalis in the adult, its posterior fibers being particularly blended with the fibers of the A2 (Figure 11.1F). Between A2 and A2-PVM runs the mandibular branch of the trigeminal nerve (CNV3) (Figure 11.4B). The anterior fibers of the A2-PVM in both the larvae and the adult look like a part of the pseudotemporalis. In larvae the A2-PVM originates from the palatoquadrate

TABLE 11.2
Synonyms Used for the Adult Mandibular, Hyoid, Branchial, and Hypobranchial Arch Muscles

A. mexicanum	Synonyms Used in Urodela
	Mandibular Muscles
Intermandibularis anterior	Submentalis (Iordansky 1992)
Intermandibularis posterior	–
Adductor mandibulae A2	Levator mandibulae externus (Edgeworth 1935; Piatt 1938; Larsen and Guthrie 1975; Carroll and Holmes 1980; Ericsson and Olsson 2004; Piekarski and Olsson 2007; Ziermann 2008); adductor mandibulae externus (Carroll and Holmes 1980; Lauder and Shaffer 1985; Iordansky 1992)
Adductor mandibulae A2-PVM	Levator mandibulae posterior (Edgeworth 1935; Piatt 1938); adductor mandibulae posterior (Piatt 1938; Carroll and Holmes 1980; Iordansky 1992); levator mandibulae articularis (Ziermann 2008)
Pseudotemporalis, superficial (A3′)	Superficial levator mandibulae anterior (Edgeworth 1935; Piekarski and Olsson 2007); levator mandibulae anterior superficialis (Piatt 1938, 1939); part of adductor mandibulae internus (Carroll and Holmes 1980; Lauder and Shaffer 1985); levator mandibulae longus (Haas 2001; Ericsson and Olsson 2004; Ziermann 2008); adductor mandibulae A3′ (Diogo 2007, 2008a)
Pseudotemporalis, deep (A3″)[a]	Deep levator mandibulae anterior (Edgeworth 1935; Piatt 1938; Piekarski and Olsson 2007); part of adductor mandibulae internus (Carroll and Holmes 1980; Lauder and Shaffer 1985); levator mandibulae internus or pseudopterygoideus (Haas 2001; Ericsson and Olsson 2004; Ziermann 2008); adductor mandibulae A3″ (Diogo 2007, 2008a)
Levator bulbi	–
	Hyoid Muscles
Interhyoideus anterior + posterior	Interhyoideus anterior + posterior (Piatt 1938; Bauer 1992, 1997; Ericsson and Olsson 2004)
Depressor mandibulae	Depressor mandibulae anterior (Diogo 2007, 2008a; Diogo et al. 2008a,b)
Branchiohyoideus	Branchiohyoideus externus (Piekarski and Olsson 2007)
Levator hyoideus[a]	Depressor mandibulae posterior (Diogo 2007, 2008a; Diogo et al. 2008a,b)
Ceratomandibularis[a]	Branchiomandibularis (Edgeworth 1935); ceratohyoideus externus (Piatt 1938); subhyoideus (Duellman and Trueb 1986)
	Branchial Muscles
True branchial muscles s.st.	–
Protractor pectoralis	Cucullaris or cucullaris major (Edgeworth 1935; Carroll 2007; Piekarski and Olsson 2007)
Constrictor laryngis[a]	–
Dilatator laryngis[a]	–
Laryngeus[a]	Laryngeus ventralis (Edgeworth 1935)
	Hypobranchial Muscles
Geniohyoideus	–
Sternohyoideus	Rectus cervicus (Carroll 2007); rectus cervicis (Ziermann 2008)
Omohyoideus	Pectoriscapularis (Edgeworth 1935); abdomino-hyoideus (Piatt 1938)

[a] The deep pseudotemporalis and levator hyoideus are present as distinct muscles in some ontogenetic stages of *A. mexicanum* but are completely integrated in the pseudotemporalis and depressor mandibulae of the adult, respectively (see text); the ceratomandibularis is not present as a distinct muscle in all observed developmental stages of *A. mexicanum* (see text); the constrictor laryngis and dilatator laryngis are clearly usually present in *A. mexicanum*, but the laryngeus was not identified in the present work either because it is absent in this species or because it is a very small and deep laryngeal muscle (see text). s.st., *sensu stricto*.

(commissura quadrato-cranialis anterior) and inserts onto Meckel's cartilage. The adductor mandibulae A2 originates in larvae from the palatoquadrate and the orbital cartilage and in the adult from the squamosal (Figure 11.4A). During development, the A2 changes its orientation from vertical (from dorsal to ventral posterior to the eye) toward diagonal (from caudo-dorsal to antero-ventral). Its insertion is dorsal at Meckel's cartilage (larvae) or dentary (adult) (Figure 11.4B and C).

In larvae a small deeper part of the anlage that gives rise to the pseudotemporalis superficialis (corresponding to the A3′ of fishes) gives rise to the pseudotemporalis profundus (corresponding to the A3″ of fishes); this anlage is visible from stage 39 onward (Table 11.1). During larval development, the pseudotemporalis profundus will become completely integrated in the main body of the pseudotemporalis, forming a mainly undifferentiated muscle in the adult. The origin of the pseudotemporalis superficialis moves dorsal during development (Figure 11.1D through F): in larvae it originates dorsolaterally from the palatoquadrate (orbital cartilage), and in adults it originates from the midline of the parietal and caudal from the first vertebrae (Figures 11.3A and 11.4A). In larvae,

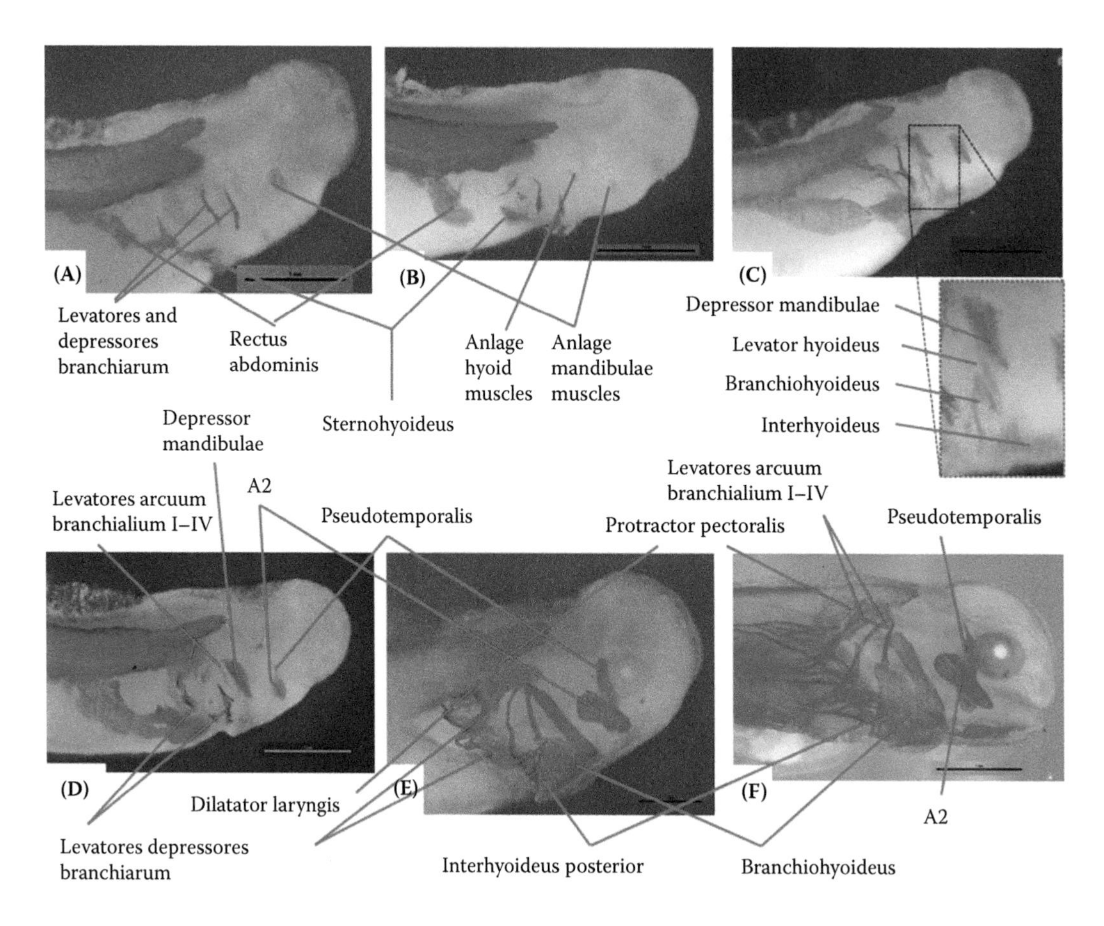

FIGURE 11.1 Lateral views of different stages during development of *Ambystoma mexicanum* (Amphibia: Caudata). Muscles stained with 12/101 antibody. The muscle differentiation can be observed from anterior (mandibular arch; A) toward posterior (branchial arches; D). The lateral anlage of the mandibular and hyoid arch splits up in several lateral muscles, while the branchial arch muscles develop more or less one by one. For details, see text. Stages: A536, B536, C537, D539, E540, and F543. Scale bar = 1 mm.

the pseudotemporalis profundus originates ventrolaterally from the palatoquadrate, being blended with medial fibers of the pseudotemporalis; both portions insert onto Meckel's cartilage. In the adult, the undivided pseudotemporalis inserts caudo-dorsally onto the dentary.

The levator bulbi is a muscle that is related to the eye and that we could identify only in the adults (Figure 11.4A). Superior to this muscle runs the maxillary branch of the trigeminal nerve, i.e., CNV2 (Figure 11.4A; see the following). Medial to CNV2 lies a branch that runs to the levator bulbi, probably innervating it. However, we were not able to confirm whether this branch is part of CNV2, as it seems to be, or of the abducens nerve (CNVI). The levator bulbi is usually thought to be derived from a dorsal mandibular anlage, and an innervation by CNV2 would support this idea, although the orientation of its fibers in the adult suggests that it might perhaps derive also/exclusively from the anlage of the pseudotemporalis (see the following). During development, the pseudotemporalis lies close to the posterior edge of the eye, and the fiber orientation of the deeper layer of the dorsal part of the pseudotemporalis in the adult appears similar to that of the levator bulbi.

DEVELOPMENT OF HYOID MUSCLES

In histological sections of stage 34 embryos, the anlage of the hyoid arch muscles is visible lateral to the buccal cavity (Table 11.1). The ventral part of this anlage stretches caudally, later during development, forming the anlage of the interhyoideus posterior at stage 38 (Figure 11.2D through F). The dorsomedial part becomes the anlage of the branchiohyoideus, levator hyoideus, and depressor mandibulae (Figure 11.1C), this anlage being distinct at stage 38 (Table 11.1; Figure 11.1D through F). Muscle cells are visible in all hyoid muscles at stage 38, in the interhyoideus posterior at stages 38 and 39; during stage 40, the fiber development increases, and by stage 42, all hyoid arch muscles are visible. The interhyoideus originates from the ceratohyal and meets the contralateral muscle in a median raphe (Figure 11.3C). It develops from its origin to its insertion and the anterior part before the posterior part (Figure 11.2). An anterior thin part (interhyoideus anterior) and a posterior thicker part (interhyoideus posterior) can be distinguished even if the border between both parts is not always clear (Figure 11.2E and F).

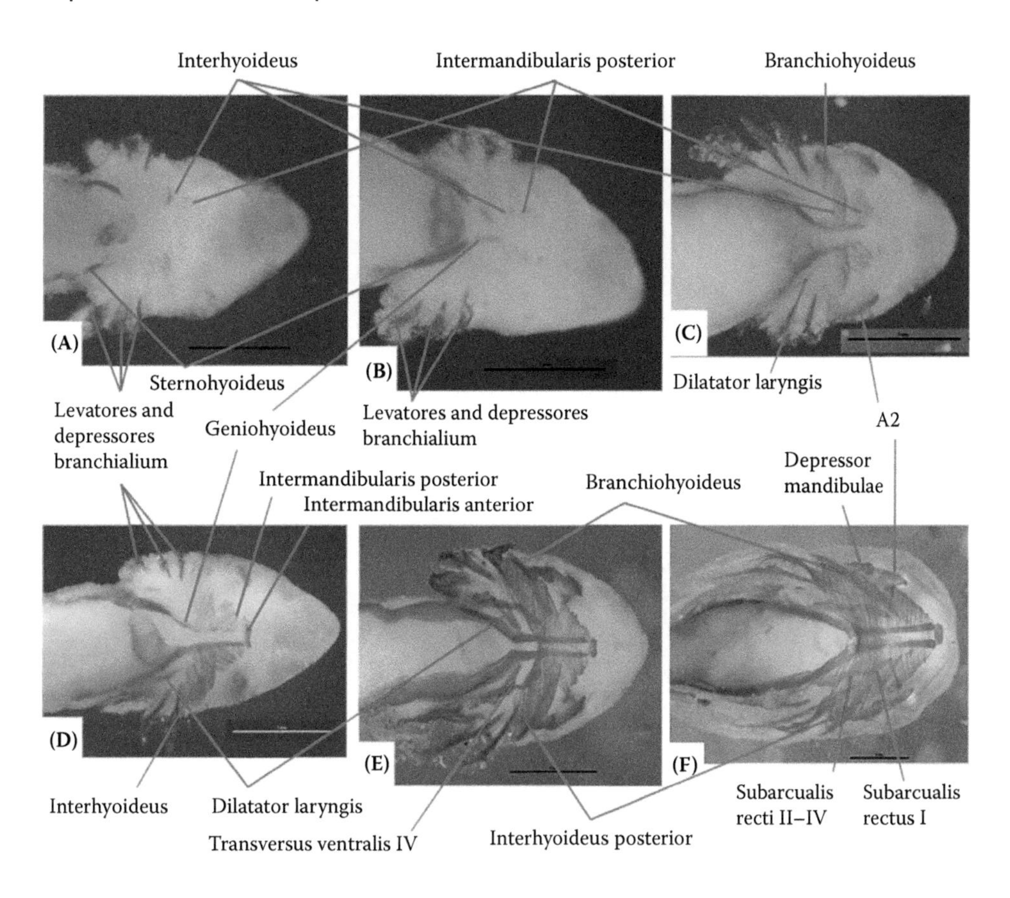

FIGURE 11.2 Ventral views of different stages during the development of *Ambystoma mexicanum* (Amphibia: Caudata). Same specimens and stages as shown in Figure 11.1. Many muscles develop from their region of origin toward their future insertion site, which can be best observed in ventral muscles such as the geniohyoideus. The ventral muscles intermandibularis posterior and interhyoideus (including interhyoideus posterior) are clearly separated early in development (A–D), but can hardly be distinguished anymore in advanced larvae (E–F). For details, see text. Scale bar = 1 mm.

In larvae, the fibers of the interhyoideus posterior stretch lateral to the depressor mandibulae, ventral to the opercular fold, and posteroventrolaterally to the ceratobranchial I. In adults the interhyoideus posterior is mostly restricted to an attachment onto the ceratohyal and only a few fibers still originate laterally at the ceratobranchial and in the opercular fold. The interhyoideus posterior is the thickest portion of the ventral muscle sheet covering the adult head.

In larvae the dorsomedial anlage of the hyoid muscles stretches, giving rise to three muscles (Figure 11.1C). The branchiohyoideus seems to develop from the caudal ventralmost part of this dorsomedial anlage (Figure 11.1C), inserting only onto the ceratohyal (Figure 11.2C). Interestingly, during the larval development, both origin and insertion are changed. The origin shifts toward the lateral part of the ceratobranchial I (Figure 11.1E and F) while the insertion will be ventral from the whole length of the ceratohyal, ventral to the origin of the interhyoideus (Figure 11.2E and F). The anlage of the branchiohyoideus is visible at stage 37 (Table 11.1). Shortly thereafter, muscle cells are visible, followed by fiber development, and at the end of stage 42, the muscle is fully developed.

The depressor mandibulae and the levator hyoideus originate from the otic capsule. From the developmental investigation, it seems that the ventral fibers of the levator hyoideus continue into the depressor mandibulae, which attaches to Meckel's cartilage in larvae and to the dentary in the adult (Figure 11.1C). The depressor mandibulae extends anteriorly and ends caudal to the anlage of Meckel's cartilage (stages 38 and 39). In later stages (43 onward), there seems to be a turn in the fiber orientation of the depressor mandibulae (Figures 11.1F and 11.4C): the dorsal fibers have a more vertical orientation, while the ventral fibers are more diagonal. This might indicate that the dorsomedial fibers still represent the levator hyoideus and the dorsolateral and ventral fibers represent the part of the depressor mandibulae that derives directly from the original depressor mandibulae anlage. By definition, an origin from the otic capsule and an insertion onto the ceratohyal corresponds to the configuration of the levator hyoideus. The dorsomedial fibers of the levator hyoideus do reach the

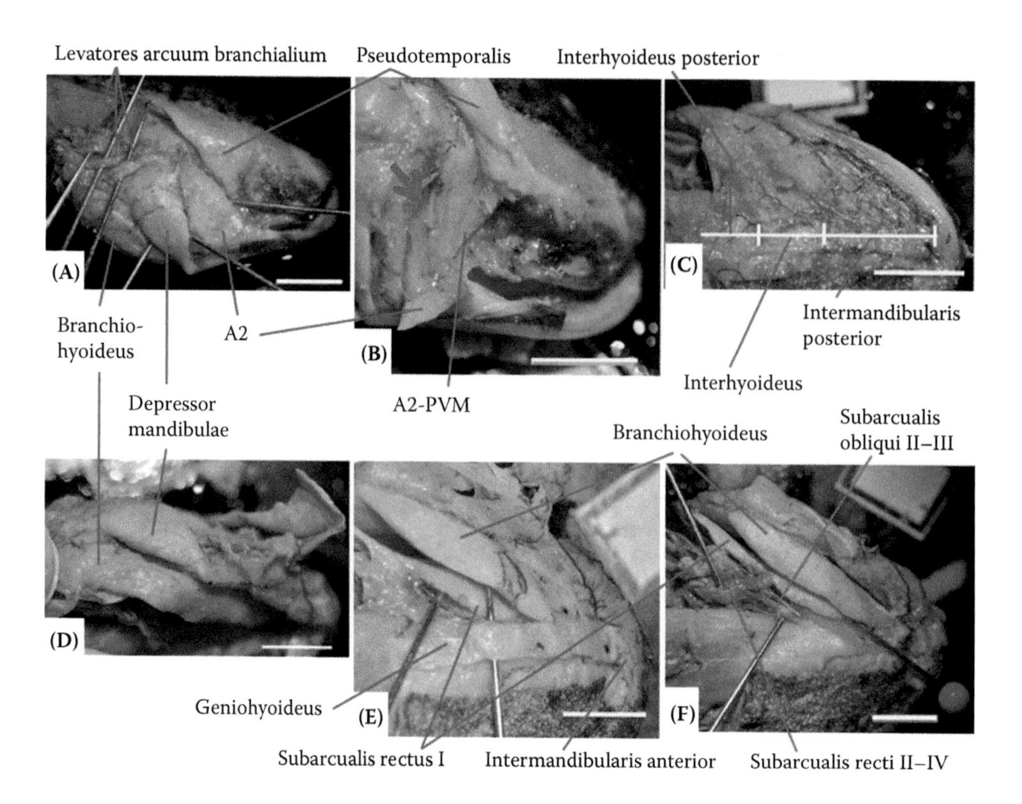

FIGURE 11.3 Dissection of cranial muscles of an adult *Ambyostoma mexicanum* (Amphibia: Caudata). Anterior is on the right in all pictures. (A, B): Right lateral view. (A) superficial muscles, (B) to show the A2-PVM, the pseudotemporalis and A2 are flipped dorsally and ventrally, respectively. The red arrow is pointing toward the mandibular branch of the trigeminus nerve (n. V3). (C, E, F): Ventral views. (C) All superficial ventral muscles are fused into one thin sheet and can only be divided by their origins at the mandible (intermandibularis posterior) and the ceratohyal (interhyoideus). (E and F): Deeper ventral muscles. (D): Ventrolateral view to show the branchiohyoideus originating laterally and curving posterior to the jaw articulation to insert along the ceratohyal. Scale bar = 1 cm.

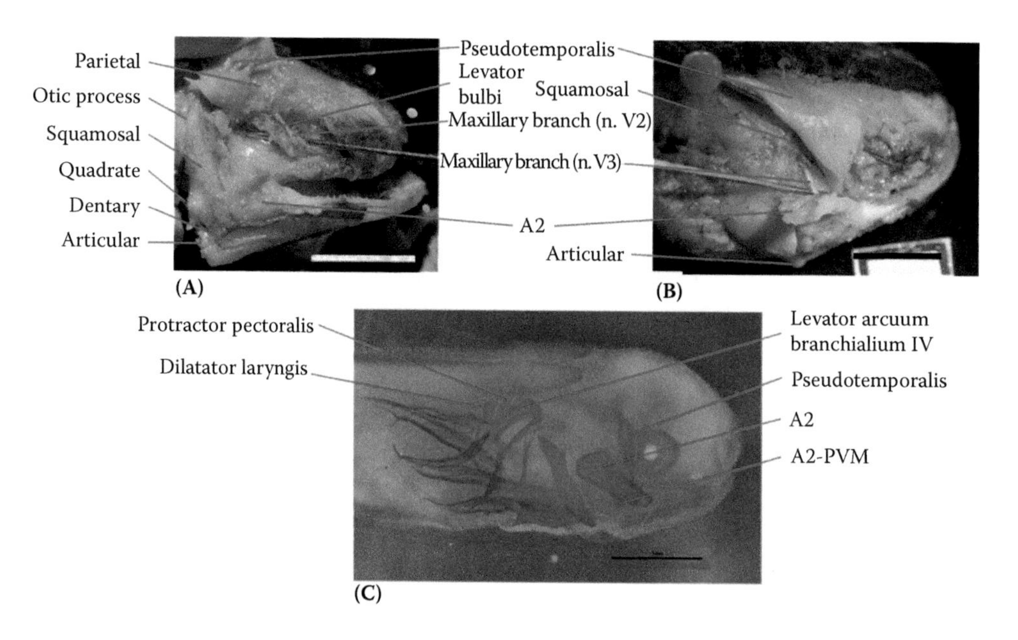

FIGURE 11.4 Lateral views of adult (A, B) and advanced larvae (C) of *Ambystoma mexicanum* (Amphibia: Caudata). (A, B): Branches of the trigeminus nerve between the A2 and the A2-PVM (n. V3) and dorsal to the levator bulbi (n. V2, see text). (C): Muscles stained with 12/101, stage 44. The protractor pectoralis separates from the posterior fibers of the levator arcuum branchialium IV.

hyoid arch in the early stages of development (Figure 11.1C), but during development, this muscle shifts its insertion completely to the mandible and is hardly, or not at all, distinguished from the depressor mandibulae in later stages (Figure 11.1E and F). In stage 44, some medial fibers of the levator hyoideus/depressor mandibulae complex still reach in a steep angle of the ceratohyal, which makes this part recognizable as the levator hyoideus. The lateral fibers attach to Meckel's cartilage and are probably a mix of the lateral fibers of the levator hyoideus and the fibers of the depressor mandibulae (see preceding text).

DEVELOPMENT OF BRANCHIAL MUSCLES

The branchial arch muscles can be divided into the true branchial muscles *sensu stricto*, the protractor pectoralis, and the laryngeal muscles (see Chapter 2). The anlagen of the levatores arcuum branchialium (LAB) I and II can be identified in stage 37 (Table 11.1). LAB III and IV are visible by stage 38. Muscle cells are first visible in LAB I (stage 38), followed by LAB II–IV (stages 38 and 39) (Figure 11.1D). The fiber development in LAB I–IV starts by stage 40, and by stage 42, LAB I–IV are fully developed (Figure 11.1F), originating from the fascia cephalodorsalis covering the otic capsule and lateral to the epaxial muscles, each inserting onto its respective ceratobranchial. LAB I–IV develop slightly before the true ventral branchial muscles *sensu stricto* (Figure 11.1D; cf. with Figure 11.2D and E). These include the anlagen of subarcualis rectus I, subarcualis recti II–IV, and subarcuali obliqui II and III, which are visible at stage 38 (Table 11.1). Shortly thereafter, there are muscle cells visible in subarcualis rectus I, followed by subarcualis obliquus II and then subarcualis obliquus III, while at stage 39 subarcualis recti II–IV also have visible muscle cells. Subarcualis rectus I attaches from stage 42 onward onto the ceratohyal. Subarcualis recti II–IV only reach toward ceratobranchial III in stage 40. Fiber development of subarcualis rectus I starts in stage 40, followed by development of subarcualis obliqui II and III and then of subarcualis recti II–IV. The anlagen of the levatores et depressores branchiarum are visible at stage 36 (Table 11.1; Figure 11.1A), the first muscle cells being observed at stage 39 and fiber development at stage 42. Fully developed are these muscles at stage 44. Transversus ventralis IV is visible from stage 40 onward (Figure 11.2E), its fiber development starting at stage 42. Therefore, at the end of stage 42, all true ventral branchial muscles *sensu stricto* are fully developed, although with the 12/101 antibody staining, we were not able to identify the subarcualis obliquus muscles that we could identify in the histological sections (see preceding text and Table 11.1).

Subarcualis rectus I extends from the ceratobranchial I to the ceratohyal in larvae and in adults. Subarcualis recti II–IV originate from ceratobranchial IV and inserts medially onto ceratobranchial I and ventrally onto ceratobranchialia II and III. Subarcualis obliqui II and III originate from ceratobranchialia II and III, respectively. Toward their insertion, their fibers fuse, inserting together onto the basibranchial. Levatores et depressores branchiarum I–III originate from comissurae terminales I, II, and III and extend to the external gills (Figure 11.4C). Transversus ventralis IV originates ventrally from the proximal end of ceratobranchial IV and extends medially to connect with its contralateral muscle in a medial raphe. In the adults, transversus ventralis IV is a thin muscle sheet originating along the whole ceratobranchial IV. In stage 40, the posterior fibers of levator arcuum branchialium IV become more separated from the anterior fibers of this muscle and start to differentiate into a protractor pectoralis muscle (Figure 11.1E and F; see the following). In the adult axolotls, the protractor pectoralis is broad and very thin, originating caudal to the head from the fascia cephalodorsalis and attaching onto the scapula.

Regarding the laryngeal muscles, the dilatator laryngis starts as a band of mesodermal cells lateroventrally to the esophagus. Muscle cells become obvious first in the dilatator laryngis (stages 38 and 39, Figure 11.2D) and then in the constrictor laryngis (stage 39) (Table 11.1). During stage 39, the dilatator laryngis ends in connective tissue situated ventrolaterally to the esophagus (Figure 11.2D), while the constrictor laryngis becomes distinct only at stage 42. The constrictor esophagi, which is not a laryngeal nor a head muscle by definition, lies near the laryngeal muscles and is visible at stage 38 as a ring of mesodermal cells around the esophagus. The fiber development in the constrictor esophagi and dilatator laryngis is clearly visible at stage 42 and shortly thereafter starts also in the constrictor laryngis. The dilatator laryngis is fully developed by stage 42; the constrictor esophagi by stage 43; and the constrictor laryngis by stage 45 (Table 11.1). The dilatator laryngis originates from the fascia cephalodorsalis lateral to the epaxial muscles in close proximity to the protractor pectoralis and levator arcuum branchialium IV (Figure 11.1E). It turns ventral and attaches laterally onto the larynx (Figure 11.2E). The constrictor laryngis originates from a raphe ventral to the larynx and dorsal to transversus ventralis IV and inserts ventrally to the dilatator laryngis onto the larynx. The constrictor laryngis could be observed only in the larval histological sections, while the dilatator laryngis could be observed in both the sections and 12/101 antibody staining; the laryngeus, which is present in at least some salamanders and some specimens of the genus *Ambystoma*, could not be identified in the sections nor in the antibody staining (see Table 11.1). The dilatators laryngis, constrictor laryngys, and laryngeus were not observed in the adult specimens.

DEVELOPMENT OF HYPOBRANCHIAL MUSCLES

The hypobranchial muscles are usually thought to be mainly derived from the body musculature, followed by a migration to the ventral region of the head, and are divided into an anterior "geniohyoideus' group and a posterior "rectus cervicis" group (see Chapter 2). In stage 37, first the sternohyoideus (from the rectus cervicis group) and then the geniohyoideus (from the geniohyoideus group) are visible (Table 11.1, Figure 11.2). In stage 38 the anlage of the geniohyoideus is ventral to the ceratobranchial I anlage, is a band of muscle cells visible dorsally to the interhyoideus anterior at stages 38 and 39, and reaches the caudal parts of the intermandibularis posterior by stage 40. At stage 38 the muscle cells of the sternohyoideus are visible, and by stage 40 the fiber development of both these muscles starts, which are fully developed at stage 42. The geniohyoideus develops from its origin from the basibranchial to its insertion onto Meckel's cartilage and/or dentary, close to the origin of the intermandibularis anterior (Figure 11.2). In adults the muscle is wider and closes the gap existing between both contralateral muscles, medially. The sternohyoideus is blended posteriorly to the rectus abdominis and develops lateroventrally (Figures 11.1 and 11.2) toward its insertion onto the hypobranchial, dorsal onto the basibranchial, and medial onto the rostral end of ceratobranchial I. The omohyoideus in the adult is deeply blended with the sternohyoideus, from which it derives ontogenetically, extending posterolaterally to attach onto the pectoral girdle (Table 11.1). Other hypobranchial muscles such as the tongue muscles hyoglossus and genioglossus are absent in both the larvae and the adult.

DEVELOPMENT OF CEPHALIC MUSCLES IN THE AXOLOTL IN A BROADER COMPARATIVE TEXT

As noted earlier, *A. mexicanum* is thought to be the best model organism to discuss the early evolutionary history of tetrapods. The axolotl larval muscles identified in the present work are in general similar to the descriptions of other salamanders (e.g., Edgeworth 1935) and to the partial myological descriptions of *A. mexicanum* (e.g., Ericsson and Olsson 2004; Ericsson et al. 2004; Schmidt et al. 2013). However, our descriptions are more detailed and include information collected from a wide range of methods, which is important because there are, for instance, differences in the observation of timing of muscle appearance using histological sections and antibody staining with the 12/101 antibody (i.e., this antibody stains against myosin, a protein that appears after the anlage of the muscle is visible for a while using other techniques).

There are only a few major topological changes during the development from larvae to the adult *A. mexicanum*. This is expected because the aquatic adults and the larvae have a similar feeding style as the larvae (Lauder and Shaffer 1985; Lauder and Reilly 1988). *Ambystoma* species do not go through a complete metamorphosis; i.e., they are neotenic, which explains the occurrence of larval characters in adults, such as the presence of branchial arch levators and external gills, and the absent tongue and tongue muscles. Most muscles simply gain muscle mass and change slightly their orientation due to the growth of the bony elements, which leads to an increase in unidirectional pressure forces in adult axolotls (Lauder and Reilly 1988). For example, the adductor mandibulae muscles, which have in feeding larvae a diagonal orientation between the palatoquadrate and the posterior end of Meckel's cartilage, have their insertions shifted backward due to the caudal elongation of the lower jaw, leading to a more vertical orientation of those muscles in the adult (Figure 11.3). In the following, we provide a brief discussion of each of the major muscle groups, focusing particularly on those that have more or less clear homologies with the muscles of other vertebrates and/or that are of special importance for the understanding of the early evolutionary history of tetrapods.

Regarding the mandibular muscles, in some adult amphibians and reptiles the pseudotemporalis is divided into superficial and deep structures that clearly seem to correspond directly to the adductor mandibulae A3′ and adductor mandibulae A3″ of fishes, respectively (Tables 11.1 and 11.2; see, e.g., Diogo and Abdala 2010; Iordansky 2010). Those two pseudotemporalis portions have been described in several *Ambystoma* species (e.g., Larsen and Guthrie 1975; Carroll and Holmes 1980; Iordansky 1992; Haas 2001). They are often recognized due to a different fiber orientation and in some cases cannot be separated from each other. Interestingly, the deep bundle seems to derive first from the superficial bundle and then becomes again integrated into it, forming a single, undivided pseudotemporalis muscle in the adult (Table 11.1). The relatively short jaw present in axolotl larvae might cause this pattern; with the steep angle of the pseudomandibularis profundus, a stronger force to open the mouth can be achieved. This is not necessary in adults, where all the adductor mandibulae muscles have a similar vertical fiber orientation. In summary, both the adductor mandibulae A3′ and A3″ seem to be included in, and usually inseparable from, the pseudotemporalis of adult members of extant tetrapod nonmammalian species such as *A. mexicanum*. According to Haas (2001), the deep and superficial pseudotemporalis bundles in axolotl larvae (his "levator mandibulae longus" and "levator mandibulae internus or pseudopterygoideus," respectively; Table 11.2) correspond to the "levator mandibulae internus" of anurans and to the levator mandibulae internus plus "pterygoideus" of caecilians.

The muscle pterygoideus that is present in some urodeles, *sensu* Haas (2001), is usually not present as a separate muscle in *Ambystoma* and seems to be homologous to the pterygoideus of caecilians and, thus, to the pterygomandibularis of reptiles, being probably derived from the pseudotemporalis profundus, i.e., from the A3″ of fishes. In the preceding chapters, we already discussed the problematic use of the name "pterygoideus" to designate this muscle. Iordansky (2010) suggested that the pterygomandibularis of caecilians and of reptiles is directly homologous to the A3″ of fishes and the pseudotemporalis profundus of anurans and urodeles. However, he then confusingly states that the pterygomandibularis developed from the deep pseudotemporalis/A3″ and thus was acquired homoplastically in caecilians and reptiles. Haas (2001) noted that Iordansky (1996) did not mention the presence of a true pterygoideus muscle in a few urodeles, i.e., in addition to the pseudotemporalis profundus. This also supports the observations and comparisons made in the present work, according to which the pterygomandibularis derives from, but is not directly homologous to the A3″ of fish and thus the pseudotemporalis profundus. As some urodeles seem to have such a true pterygoideus (pterygomandibularis *sensu* the present work), it is quite possible that this muscle was present in the last common ancestor (LCA) of tetrapods and then was lost in anurans and most urodeles, although it is also possible that it was independently acquired in caecilians, in some urodeles, and in reptiles (at least three evolutionary steps in both hypotheses).

The A2-PVM of salamanders was suggested to be ontogenetically derived from the A3′ and/or A3″ of fishes (Piatt 1938) or from the fish A2 (Ericsson and Olsson 2004; Diogo and Abdala 2010). The homologies of the ventral mandibular muscles intermandibularis anterior and posterior are relatively well established, as noted in the preceding chapters. However, the homology of the levator bulbi has been a subject of much controversy, as also noted earlier. It was suggested that this muscle is derived from the adductor mandibulae anlage (Edgeworth 1935) or from the levator arcus palatini of fishes, which is in turn derived from the dorsal mandibular anlage (Brock 1938; Iordansky 1996; Diogo and Abdala 2010; see preceding chapters). Diogo and Abdala (2010) proposed that in the LCA of amphibians + amniotes the dorsal mandibular muscles were already divided into (a) a levator bulbi and (b) an undivided "levator palatini" (see preceding chapters).

Our developmental study supports the hypothesis that the levator bulbi is derived from the dorsal mandibular anlage, but it also indicates that it might be partially derived from the pseudotemporalis because the position and orientation of the fibers of the levator bulbi are somewhat similar to those of the dorsal portion of the pseudotemporalis profundus of adult axolotls. An origin from the pseudotemporalis would support the idea defended by Edgeworth (1935). However, our study of the nerves of adult axolotls contradicts this origin. The levator bulbi found in urodeles and anurans, and the levator quadrati found in caecilians, is usually innervated by the CNV (Ramaswami 1942; Fox 1959; Kleinteich and Haas 2007). Coghill (1902) reported that in *Ambystoma tigrinum*, the levator bulbi is innervated by the ophthalmic branch of the trigeminal nerve (CNV1) as well as by the abducens nerve (CNVI), which normally innervates the eye muscles. That is, the levator bulbi of urodeles and anurans might potentially be a complex structure derived from both the anlage of the eye muscles and the anlage of the mandibular muscles, as indicated by our morphological analyses. Gaupp (1899) and Francis (1934) stated that the levator bulbi of salamander and frogs is innervated by the maxillary (CNV2), not the ophthalmic (CNV1), branch of the trigeminal nerve. In the adults we dissected, CNV1 lies superficial (as shown by Piatt [1939]) but parallel to CNV2 (Figure 11.4), and just medial to the main body of CNV2 is a branch that is far from the mandibular branch (CNV3) and that goes to the levator bulbi. However, we could not confirm that this branch is really part of CNV2 rather than part or all of the abducens nerve (CNVI) because the latter nerve usually runs close to CNV2 and often fuses with it (see, e.g., Ramaswami 1942). It is, however, clear that in the axolotl adults we dissected, the levator bulbi is not innervated by CNV1 or CNV3. Our observations thus give some support to the hypothesis that this muscle derives from the dorsal mandibular anlage, because all the other mandibular muscles of axolotls are exclusively innervated by CNV3. However, the levator arcus palatini of fishes, including coelacanths, is derived from the dorsal mandibular anlage and is usually said to be innervated by CNV2 and/or by CNV3 (e.g., Schilling and Kimmel 1997; Diogo and Abdala 2010). Notably, the levator bulbi lies deep (so more difficult to penetrate) and develops late (long duration of breeding of the animals), unfortunately giving us too few samples of the older stages to resolve its origin and innervation. Further detailed studies are clearly needed to clarify the evolutionary and developmental origin and innervation of the levator bulbi.

Concerning the hyoid muscles, the ventral muscle sheet in adult axolotl resembles closely the organization for adult *Necturus*, which is also a neotenic salamander (Bauer 1992). The interhyoideus posterior develops from a separate anlage (Ericsson and Olsson 2004) and might, besides being derived from the hyoid region, possibly receive contributions from anterior somites (Piekarski and Olsson 2007). During metamorphosis in nonneotenic salamanders, the first ceratobranchial remodels and at least the dorsal part of it degenerates (Bauer 1992). As a result, the origin is shifted toward other structures such as the mandible in *Salamandra* (Francis 1934). In the adult axolotl, the interhyoideus posterior still shows the larval characterization with fibers toward ceratobranchial I and into the opercular fold, which are clearly neotenic characters. The interhyoideus gave rise to numerous mammalian muscles of facial expression, as explained in the preceding chapters.

As also explained in the preceding text, the homologies of the derivatives of the dorsomedial hyoid anlage have been controversial, but are crucial to understanding tetrapod origin and evolution because this anlage gives rise to muscles important for the functional morphology of both the hyoid and masticatory apparatuses, of the inner ear, and of facial expressions in mammals. The depressor mandibulae is a single bundle in adult *A. mexicanum*. In other *Ambystoma* and in other salamander species, the muscle has anterior and posterior portions (e.g., Diogo 2008a). Diogo (2008a) suggested that the depressor mandibulae anterior largely corresponds to the depressor mandibulae of dipnoans because several salamander species have fibers of the depressor mandibulae posterior that correspond to the levator hyoideus of dipnoans, which also attaches to the mandible. Forey (1986) described that both the levator hyoideus and the depressor mandibulae of extant dipnoans develop from the same ontogenetic anlage. Our developmental study shows that in *A. mexicanum* the larval levator hyoideus becomes completely incorporated in the depressor mandibulae during development, changing its insertion toward the posterior end of the mandible. In *Lissotriton helveticus* and *Ichthyosaura alpestris*, both nonneotenic taxa, some fibers of the larval depressor mandibulae change their orientation to become part of the larval branchiohyoideus (Ziermann 2008). In caecilians the levator hyoideus is an independent muscle in larvae (e.g., Edgeworth 1935; Kleinteich and Haas 2007, 2011), and in some (perhaps in all) adult caecilians, it is fused with the depressor mandibulae (Edgeworth 1935). Both muscles are usually separated in anuran larvae, with the depressor mandibulae often subdivided into suspensorio-, quadrato-, and hyo-angularis, and the levator hyoideus possibly subdivided into suspensorio- and orbito-hyoideus (e.g., Edgeworth 1935; Olsson et al. 2001; Kleinteich and Haas 2007; Carroll 2007; see preceding chapters).

Possible homologies between the branchiohyoideus and depressores branchiales of salamanders and the muscles of anurans have been discussed by several researchers (e.g., Haas 1996a; Cannatella 1999). One hypothesis, mainly based on muscle origins and insertions, is that the salamander branchiohyoideus is homologous to the constrictor branchialis I of anurans (Ziermann 2008). Schlosser and Roth (1995) showed that the innervation of the four constrictores branchiales of anurans is through the main branch of the n. glossopharyngeal nerve (CNIX) and three branchial branches of the vagus nerve (CNX). However, the branchiohyoideus is innervated by the facial nerve (CNVII) and, therefore, is not likely to be homologous with the constrictor branchialis I of anurans (e.g., Diogo and Abdala 2010). It has also been proposed that the branchiohyoideus might be a caudal portion of the depressor mandibulae (Lightoller 1939).

The new data obtained in the present work indicate that the branchiohyoideus develops from the ventro-medial fibers of the depressor mandibulae in *A. mexicanum*, supporting the observations of *I. alpestris* and *L. helveticus* by Ziermann (2008). The branchiohyoideus and the other hyoid muscles appear at about the same ontogenetic stages in salamanders (Edgeworth 1935; Ericsson and Olsson 2004), which makes it difficult to determine whether the branchiohyoideus is ontogenetically a derivative from the ventral or the dorsomedial hyoid musculature. The results from Piatt's (1938) study of *Ambystoma punctatum* indicate a dorsomedial origin, as does our study of *A. mexicanum*. Interestingly, our study shows that both the origin and the insertion of the branchiohyoideus change during larval development, and the adult muscle extends from ceratobranchial I to the ceratohyal. Those changes are one of the main reasons for confusion about the homology of this muscle and for the erroneous suggestion that it might be a branchial muscle.

Therefore, we subscribe to the hypothesis that the "hyomandibularis" (or "subhyoideus" *sensu* Carroll [2007]) of caecilians corresponds to both the ceratomandibularis ("branchiomandibularis") and the branchiohyoideus of urodeles (for more details about this hypothesis, see preceding chapters). The ceratomandibularis and branchiohyoideus both seem to derive ontogenetically from the same anlage (e.g., Piatt 1938; Bauer 1997). In fully metamorphosed urodeles, the branchiohyoideus is not present as an independent muscle, nor is the ceratomandibularis, which may be fused with the depressor mandibulae in some species (Edgeworth 1935; Bauer 1997). However, the ceratomandibularis may be present in adult obligate neotene salamanders (e.g., *Ambystoma ordinarium*: Bauer 1997). The ceratomandibularis or the branchiohyoideus might be absent in adult anurans as an independent structure (Edgeworth 1935; Jarvik 1963, 1980; Chapter 9).

With respect to the branchial muscles, the dilatator laryngis is present in the axolotl larvae we studied, in full-grown larvae of *Ambystoma maculatum* (Piatt 1939), and in axolotl adults (e.g., Piekarski and Olsson 2007). The latter authors showed that the dilatator laryngis is at least partially derived from anterior somites. Of course, this result does not exclude the possibility that the dilatator laryngis is a head muscle and, specifically, a branchial muscle (because these authors also showed that the hyoid muscle interhyoideus and the branchial muscles levatores arcuum branchialium receive contributions from somites). However, the authors did not provide detailed information about any somitic contribution to the other branchial muscles (Table 11.1). Levator arcuum branchialium IV is anterior to the protractor pectoralis and the dilatator laryngis. The figures from Piekarski and Olsson (2007) clearly show that the dilator laryngis crosses the protractor pectoralis dorsally, while the protractor pectoralis extends posteroventrally to attach onto the pectoral girdle. Although the protractor pectoralis and levator arcuum branchialium IV were shown in close proximity in Piekarski and Olsson's (2007) figures, these authors did not propose a common ontogenetic origin of these two muscles.

The protractor pectoralis ("cucullaris" *sensu* Edgeworth [1935]) derives at least partially from somites (Piekarski and Olsson 2007) and from a caudal levator arcus branchialis of plesiomorphic vertebrates and secondarily inserts to the pectoral girdle (Edgeworth 1935). The evolution of the protractor pectoralis has been the subject of recent studies, mainly due to its implications for the origin of and evolution of the neck in vertebrates (e.g., Diogo and Abdala 2010; Ericsson et al. 2012; preceding chapters). Importantly, our observations clearly show that the protractor pectoralis in *A. mexicanum* derives ontogenetically from the anlage of levator arcuum branchialium IV (e.g., Figures 11.1 and 11.4 and Table 11.1). The protractor pectoralis is present in adult anurans and urodeles but absent in adult caecilians (Edgeworth 1935). Its absence is likely related to the loss of the pectoral girdle during evolution toward extant caecilians (e.g., Carroll 2007). The protractor pectoralis gave rise to the trapezius and sternocleidomastoideus of mammals, in which the only muscle that derives from the anlage of the true branchial muscles *sensu stricto* is the stylopharyngeus (Chapter 8).

The constrictor laryngis and dilatator laryngis (Table 11.1) are antagonistic muscles in amphibians (e.g., Duellman and Trueb 1986), and form, together with the laryngeus, the laryngeal muscles *sensu* the present book, as explained in the preceding chapters. The laryngeus is absent in anurans, but present in caecilian and most urodeles (Edgeworth 1935). The laryngeus dorsalis is absent in *Ambystoma*, so only the laryngeus ventralis can be found in species of this genus (Edgeworth 1935). The dilatator laryngis and constrictor laryngis are present in axolotl larvae (Table 11.1; Piatt 1939), but we could not identify these muscles in the adult, although Duellman and Trueb (1986) state that the constrictor laryngis is present in adults of all neotenic salamanders. Neither could we identify the laryngeus in any specimen, but it is not clear whether this is because the muscle is absent or because it is a small and very deep muscle that is usually very difficult to detect (see, e.g., Diogo and Abdala 2010). Notably, we did not find any cartilages in the laryngeal region, an observation in line with Tucker (1993) but contradicted by Sasaki (2006), who states that lateral cartilages may be observed in certain amphibians such as axolotls in which they form bars for the attachment of the dilatator laryngis. Piatt (1939) shows all three muscles in full-grown larvae of *A. maculatum*.

Lastly, concerning the hypobranchial muscles, Edgeworth (1935) divides them into a geniohyoideus group and a rectus cervicis group as explained in Chapter 2. The coracomandibularis of extant sarcopterygian fishes is included in the former group and the sternohyoideus in the latter group *sensu* Edgeworth (1935). Our study shows that the tongue and tongue muscles (genioglossus, hyoglossus, and interradialis in amphibians) are absent in *A. mexicanum*. A hypobranchial muscle interradialis, probably derived from the genioglossus, is present in at least some *Ambystoma* (Piatt 1938). The "hyoglossus" of some urodeles, however, might at least partially derive from the sternohyoideus (e.g., Edgeworth 1935) and is absent as an independent structure in caecilians following Duellman and Trueb (1986). The omohyoideus in urodeles seems to be homologous to that of anurans as noted in Chapter 9 and is also present in the adult axolotl according to our new observations, although it is partially blended with the sternohyoideus (Table 11.1).

GENERAL REMARKS

The confusion about the homology and evolution of cranial muscles in amphibians comes partly from the disappearance or fusion of many larval muscles during development, especially during metamorphosis. Even in a neotenic species such as *A. mexicanum,* lacking a full metamorphosis and showing many larval characters (Figure 11.3), some muscles become completely indistinct during ontogeny: for instance, the pseudotemporalis profundus and the levator hyoideus become completely integrated in the pseudotemporalis superficialis and in the depressor mandibulae, respectively (Table 11.1). Furthermore, muscles that are often considered distinct, such as the intermandibularis posterior and the interhyoideus, can be distinguished only by their attachments in adults (Figure 11.3C). This developmental trend contradicts the commonly accepted view that during ontogeny the tendency is almost always toward the differentiation of muscles rather than the reverse. Recent studies in modern humans have provided similar data and also contradicted the latter view (Diogo and Wood 2012a,b).

We found three main morphogenetic trends in the ontogeny of the axolotl cranial muscles. First, the order of cranial muscle development mainly proceeds from anterior to posterior (Table 11.1). The muscle anlagen can be observed in histologically stained embryos from stage 34 onward; the first anlage to be observed in the 12/101-stained specimens is the mandibular one (Figure 11.1A, stage 36), followed immediately by the hyoid anlage (Figure 11.1B, stage 36), and then by the branchial muscles (Figure 11.1C, stage 37) including the laryngeal muscles (Figure 11.1D, stage 39). Second, larval cranial muscle development follows an outside-in pattern, with more lateral muscles generally developing before more medial muscles (Figure 11.1). Third, several muscles develop from their origin toward their insertion (e.g., the development of the geniohyoideus; Figure 11.2).

These detailed data about the muscle development make it possible to compare the order in which the muscles appeared during axolotl ontogeny and during phylogeny. According to Diogo et al. (2008c), in zebrafish head muscles, there is a general parallel between ontogeny and phylogeny with a few differences regarding, e.g., the early ontogenetic appearance of muscles that evolved late in phylogeny but that play a particularly important role in the feeding mechanisms of the zebrafish. A similar general parallel was observed in the axolotl studied in the present work.

Regarding the mandibular muscles, from stage 34 two anlagen were distinguished: the ventral one of the intermandibularis and a dorsomedial one including the A2 (A2/A2-PVM) and the A3 (pseudotemporalis superficialis/profundus, mainly corresponding to A3′/A3″). In stage 37, those dorsomedial muscles are already differentiated and three main events follow: (a) differentiation of the intermandibularis anterior from the intermandibularis muscle mass (late stage 38); (b) differentiation of the pseudotemporalis profundus from the main A3 muscle mass (stage 39); and (c) differentiation of the levator bulbi (Table 11.1). These events follow the phylogenetic order: a division into intermandibularis anterior and posterior is present in early gnathostomes, while a division between A3′ and A3″ (which mainly give rise to the pseudotemporalis superficialis and to the pseudotemporalis profundus, respectively) is present only in osteichthyans, and the levator bulbi is present only in tetrapods, as noted in previous chapters.

Concerning the hyoid muscles, beginning at stage 34 in which the anlagen of the interhyoideus and levator hyoideus/depressor mandibulae are already differentiated, there are two main events: (a) the differentiation of the branchiohyoideus (early stage 38) and (b) the differentiation of the interhyoideus posterior (later stage 38; Table 11.1). This order of events does not follow the order of phylogenetic events, because the interhyoideus is divided into bundles in phylogenetically basal gnathostomes, while the branchiohyoideus is present only in amphibians (see preceding chapters and Figure 6.2). The true branchial muscles *sensu stricto* are present in phylogenetically basal vertebrates such as lampreys, while the protractor pectoralis appears only in gnathostomes, and the laryngeal muscles apparently only in osteichthyans, although few data are available regarding the latter muscles in fishes (Diogo and Abdala 2010). We do not have detailed data about the timing of differentiation of the protractor pectoralis in *A. mexicanum*, but it is clear that this muscle, as well as the laryngeal muscles dilatator laryngis and constrictor laryngis, also appears later in development than the true branchial muscles *sensu stricto* (Table 11.1). Regarding the hypobranchial muscles, the sternohyoideus appears at early stage 37 and the geniohyoideus at late stage 37 (Table 11.1). Both the sternohyoideus and coracomandibularis were present in the LCA of gnathostomes, and the geniohyoideus mainly derives from the coracomandibularis, although the geniohyoideus itself appeared only in tetrapods (see preceding chapters).

Comparing the ontogeny and phylogeny of the cranial muscles in their entirety, the main sequence of ontogenetic differentiations and respective times of phylogenetic origin is as follows: (a) sternohyoideus (early 37; LCA of gnathostomes); (b) geniohyoideus (late 37; tetrapods, but mainly corresponds to coracomandibularis which evolved at LCA of gnathostomes); (c) branchiohyoideus (early 38; amphibians); (d) dilatator laryngis (mid-38; early osteichthyans); (e and f) interhyoideus posterior and intermandibularis anterior (late 38; early gnathostomes); (g) pseudotemporalis profundus (39; mainly corresponds to A3″ which evolved in LCA of osteichthyans); (h) constrictor laryngis (early osteichthyans); and (i) levator bulbi (tetrapods). Among these nine ontogenetic events, only the differentiation of branchiohyoideus (c) and of dilatator laryngis (d) directly contradict the order of phylogenetic events. These numbers are similar to those obtained by Diogo et al. (2008c) in the zebrafish and point out that in at least the case of the zebrafish and of the axolotl, the ontogeny of cranial muscle ontogeny parallels, in general, the order in which these muscles evolved in phylogeny, with just a few, but probably functionally and ontogenetically important (see Diogo et al. 2008c), exceptions (see Chapter 7).

We hope that this detailed description of the development of the cephalic muscles of the model organism *A. mexicanum* will not only provide an illustrative example of the ontogeny of these muscles in tetrapods, but also pave the way for future comparative evolutionary developmental works. This will enable addressing of crucial questions regarding the origin, homology, and evolution of the muscles of the Amphibia, the evolutionary history of the tetrapods as a whole, and, importantly, the origin of the ground plan of reptiles and mammals.

12 Pectoral and Pelvic Girdle and Fin Muscles of Chondrichthyans and Pectoral-Pelvic Nonserial Homology

Chapter 12 is mainly based on Diogo and Ziermann (2015b). In addition to providing an updated description of the muscles of the paired appendages of chondrichthyans (see Figures 1.1 and 12.1 and 12.2), it discusses the idea that the structures of the fore- and hindlimbs are serial homologues, which was first proposed by authors such as Vicq-d'Azyr (1774), Oken (1843), and Owen (1849, 1866). This idea is still accepted in most textbooks and scientific papers. However, an examination of the original works of these authors reveals that their fore- and hindlimb comparisons were almost exclusively based on bones, and almost no attention was paid to soft tissues such as muscles, nerves, and blood vessels. Moreover, in most cases, the term *limb serial homology* as used by these authors clearly referred to what is currently viewed as evolutionary parallelism or convergence, i.e., homoplasy, not to serial homology. A true, ancestral morphological or developmental serial homology of these appendages would imply that they were the result of an ancestral duplication and were therefore originally similar and then diverged anatomically/functionally (Figure 12.3).

The historical context of the development of the forelimb–hindlimb serial homology hypothesis was discussed in some detail by Diogo et al. (2013) and Diogo and Molnar (2014). Briefly, the striking similarity of several individual structures of the fore- and hindlimbs did not greatly trouble earlier nonevolutionary comparative anatomists, because they could argue that the design of animals followed an "archetype," created by a supranatural or a vital power. For instance, within the *Naturphilosophie* paradigm followed by authors such as Oken (e.g., year), the term *serial homology* did not necessarily imply an anatomical duplication of a primitive "limb archetype." Instead, the romantic definition of an archetype, which was partially followed by Owen (e.g., 1849), often referred to all the possible phenotypes that a structure or organ could assume, thus somewhat resembling the current definition of *reaction norm* (see, e.g., Russel 1916; Richards 2002). In *On the Nature of Limbs* (Owen 1849), the examples of strong similarity between the fore- and hindlimbs refer mainly to tetrapods with highly derived limbs, such as horses and plesiosaurs. When referring to these examples, Owen, who was influenced by the German *Naturphilosophie* (Richards 2002), uses the word *parallelism* more often than *serial homology*. Moreover, when Owen discusses phylogenetically more plesiomorphic taxa, e.g., chondrichthyans, he states that those taxa confuse the notion of archetype and serial homology. That is, he was always referring to derived, peculiar cases of homoplasy, not to the concept of serial homology that is commonly accepted nowadays (see, e.g., Roth 1994; Wagner 1994; Willmer 2003; Diogo et al. 2013).

For early evolutionary comparative anatomists, the similarity between the fore- and hindlimbs ("fore–hindlimb enigma" *sensu* Diogo et al. [2013]) was an especially important subject of discussion, and they approached it in two main ways (see, e.g., Humphry 1872a,b; Quain et al. 1894; McMurrich 1905; Bardeen 1906; and references therein). One approach was the romantic idea that the vertebrate body is perfectly segmented (e.g., Goodrich 1930) and thus that the hindlimb is nothing more than a second forelimb, which was in turn often viewed as a derivative of a posterior branchial arch (e.g., Gegenbaur 1878). The other way was to argue that all the hard and soft tissue structures that are present in the fore- and hindlimbs of basal tetrapods were already present in some form in plesiomorphic gnathostome fishes, as the result of a unique, ancestral duplication of the paired appendages. An illustrative example is Humphry's (1872b) study, in which multiple vertebrate animals were compared and in which it was argued that not only dipnoans but also fishes such as sharks have, for instance, a "latissimus dorsi" muscle or a "pectoralis" muscle similar to those of extant tetrapods. Since the publication of these classical studies, a large body of evidence has been published on the subject and has contradicted both the view that the limbs are the result of supranatural or vital phenomena and the romantic idea of perfect segmentation of the vertebrate body (Diogo et al. 2013).

However, authors have continued to cite these and other classical studies to "show" that the structures of the pelvic and pectoral appendages, including the ones of the tetrapod fore- and hindlimbs, are serial homologues. Instead of questioning this idea, researchers preferred to focus on more specific details about the origin and evolution of the paired appendages within—and thus accepting *a priori*—the serial homology paradigm. More recent developmental studies have provided some evidence that apparently contradicted the hypotheses of Goodrich's and Gegenbaur's hypotheses about the origin of the paired appendages, suggesting that pectoral and pelvic fins evolved from continuous stripes of competency for appendage formation located ventrally and laterally along the embryonic flank (see, e.g., Don et al. 2013; but see Gillis et al. 2009, and the following text). A continuation of this theory proposed that the paired appendages evolved with a shift in the zone of competency to the lateral plate mesoderm in conjunction with the establishment of the lateral somitic frontier,

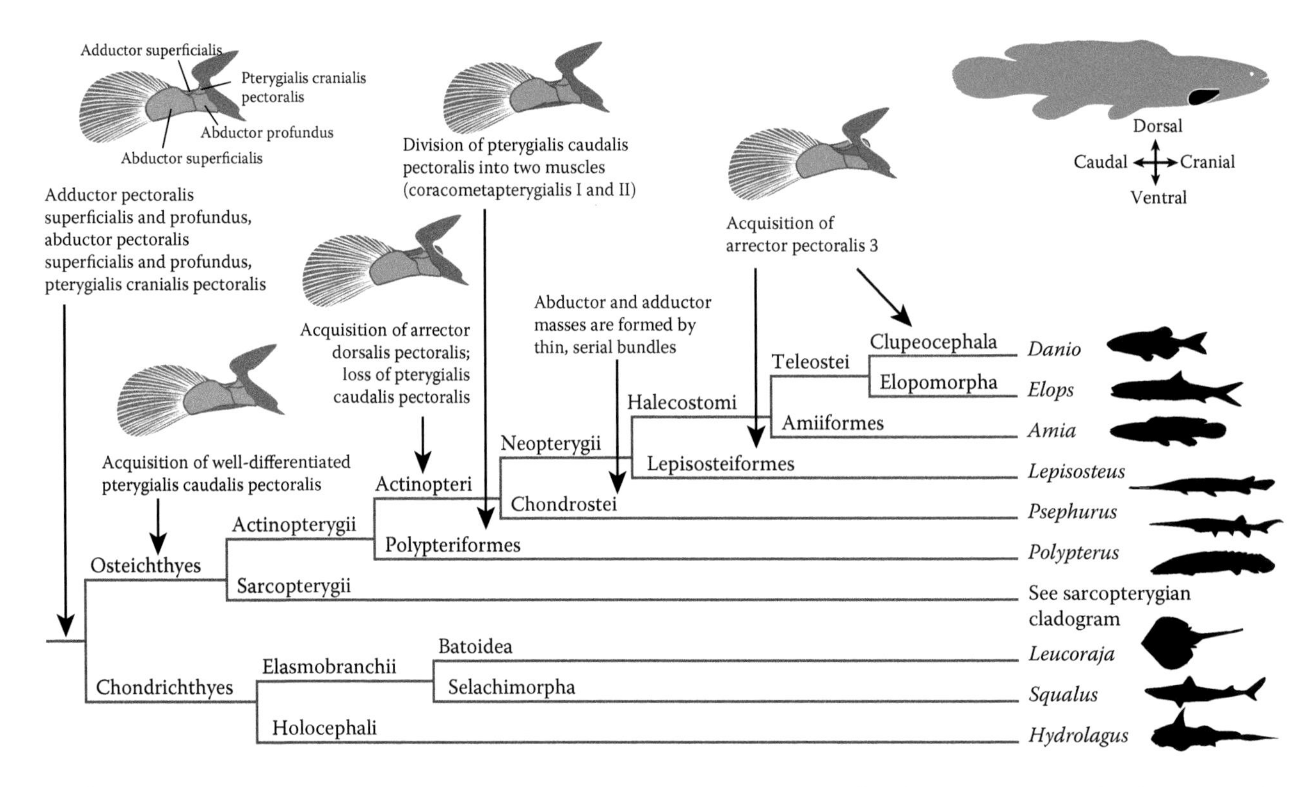

FIGURE 12.1 Some of the major features of the musculature of the pectoral girdle and fin within nonsarcopterygian fishes, according to our own data and review of the literature.

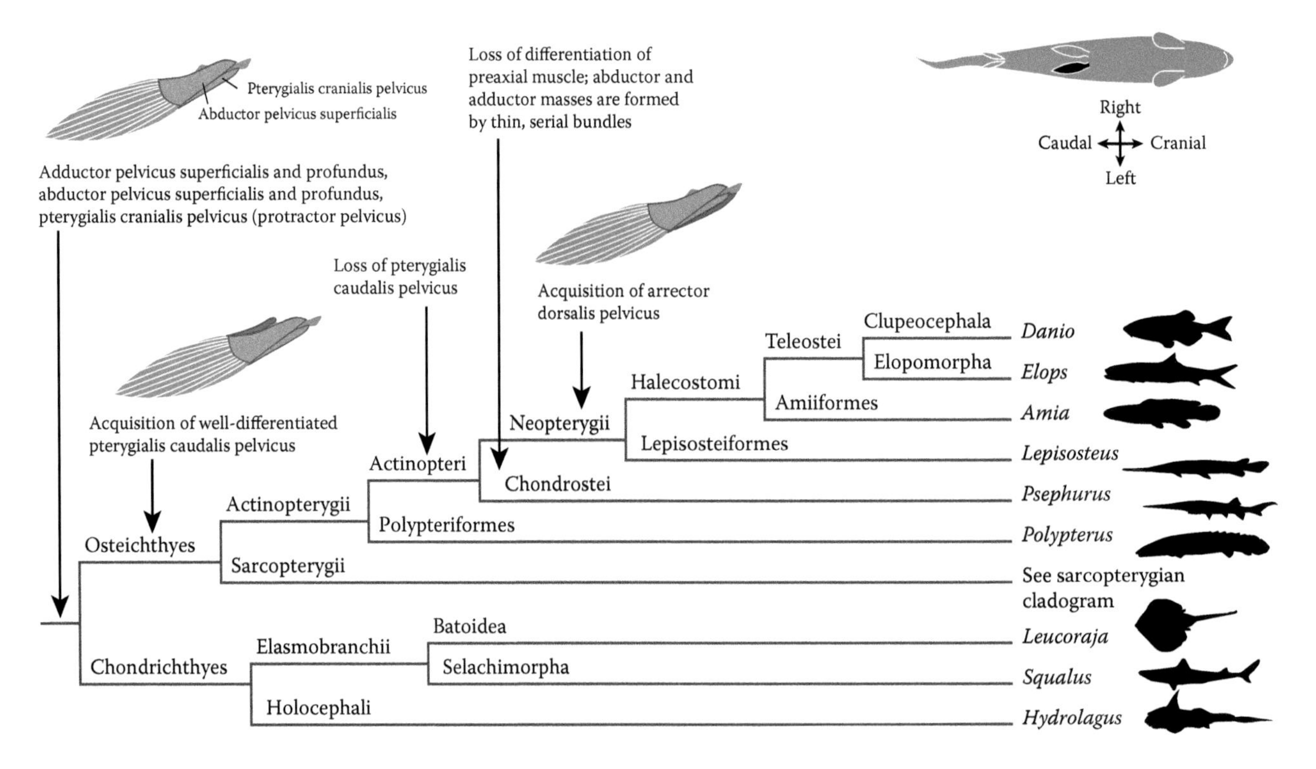

FIGURE 12.2 Some of the major features of the musculature of the pelvic girdle and fin within nonsarcopterygian fishes, according to our own data and review of the literature.

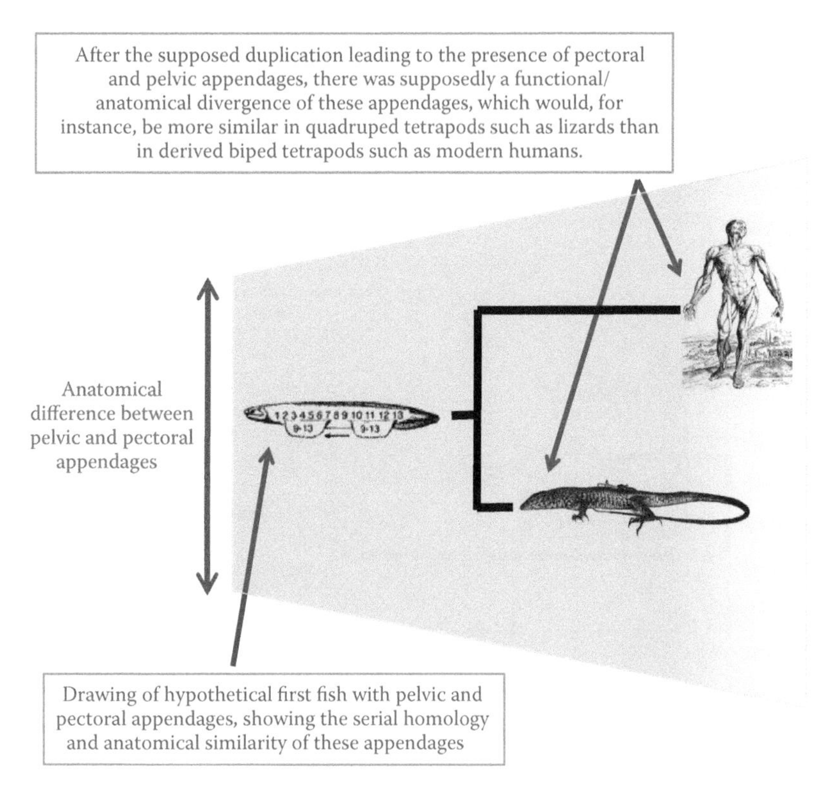

FIGURE 12.3 Simplified diagram illustrating the "serial homology followed by functional/anatomical divergence" hypothesis often shown in textbooks and followed in more technical papers, particularly within the fields of developmental biology and evo-devo. The picture of the hypothetical fish (modified from the diagram of Shubin, N. et al., *Nature*, 388, 639–648, 1997) shows the origin and evolution of paired appendages. According to Shubin et al.'s (1997) diagram, establishment of serially homologous appendages was proposed to result from gene co-option during the evolution of Paleozoic vertebrates. That is, *Hox* genes were initially involved in specifying regional identities along the primary body axis, particularly in caudal segments, and then during the origin of jawed fish there was a co-option of similar nested patterns of *HoxD* expression in the development of both sets of paired appendages (numbers shown within the fish body). According to this diagram, the co-option may have happened in both appendages simultaneously, or *Hox* expression could have been initially present in a pelvic appendage and co-opted in the development of an existing pectoral outgrowth.

which allowed for the formation of fin/limb buds with internal supporting skeletons (Don et al. 2013).

The idea that the pectoral and pelvic appendages are serial homologues is generally associated with the notion that these appendages were originally similar to each other and that there was a subsequent functional/anatomical divergence between them (Diogo et al. 2013). For instance, Don et al. (2013) explain that the ancestral *Tbx4/5* cluster of vertebrates probably underwent a duplication event and that now *Tbx4* is related to the hindlimb and *Tbx5* with the forelimb of tetrapods; they then state that "pectoral fins evolved first and then duplicated to form pelvic fins" (see their Figure 3). This illustrates the confusion, in evo-devo studies, between (a) a duplication of the *Tbx4/5* cluster and subsequent co-option for the genetic pathways associated with the ontogeny of the different paired appendages and (b) the morphological duplication of the appendages themselves and of their individual structures (e.g., of their muscles and bones). That similar genes operate to facilitate an outgrowth that gives rise to different limbs in the same animal does not necessarily mean that these limbs are serial homologues; similar genes and gene cascades/networks may have been simply recruited homoplasically as organizers of limb development (see, e.g., Willmer 2003). In fact, the detailed analysis of recent developmental, genetic, paleontological, and functional studies also raises serious questions about the serial homology hypothesis. An extensive review of these lines of evidence was provided by Diogo et al. (2013), who strongly contradicted that hypothesis based on a review of the literature and on the results of the long-term study of the muscles of osteichthyans compiled in the previous version of the present book (see Figure 12.4; Diogo and Abdala 2010). This long-term project was recently expanded to include both developmental and regenerative studies of tetrapod taxa such as salamanders and frogs, which also provided new insights into the forelimb–hindlimb enigma and further evidence contradicting the forelimb–hindlimb serial homology idea (Diogo et al. 2014b,c; Diogo and Tanaka 2014; Diogo and Ziermann 2015b; see Chapter 20).

In Chapter 12, we summarize the results of our dissections of the muscles of the paired appendages of representatives of the three major chondrichthyan extant clades (Holocephali and the Elasmobranchii clades Selachii, or sharks, and Batoidea, or rays and skates: see Figure 1.1, as well as Figures 12.1 and 12.2). The materials and methods of this study were provided in detail in Diogo and Ziermann (2015b). Numerous anatomical and, more recently, developmental studies of chondrichthyans have been performed because these fishes are crucial for understanding the origin and early evolution of vertebrate

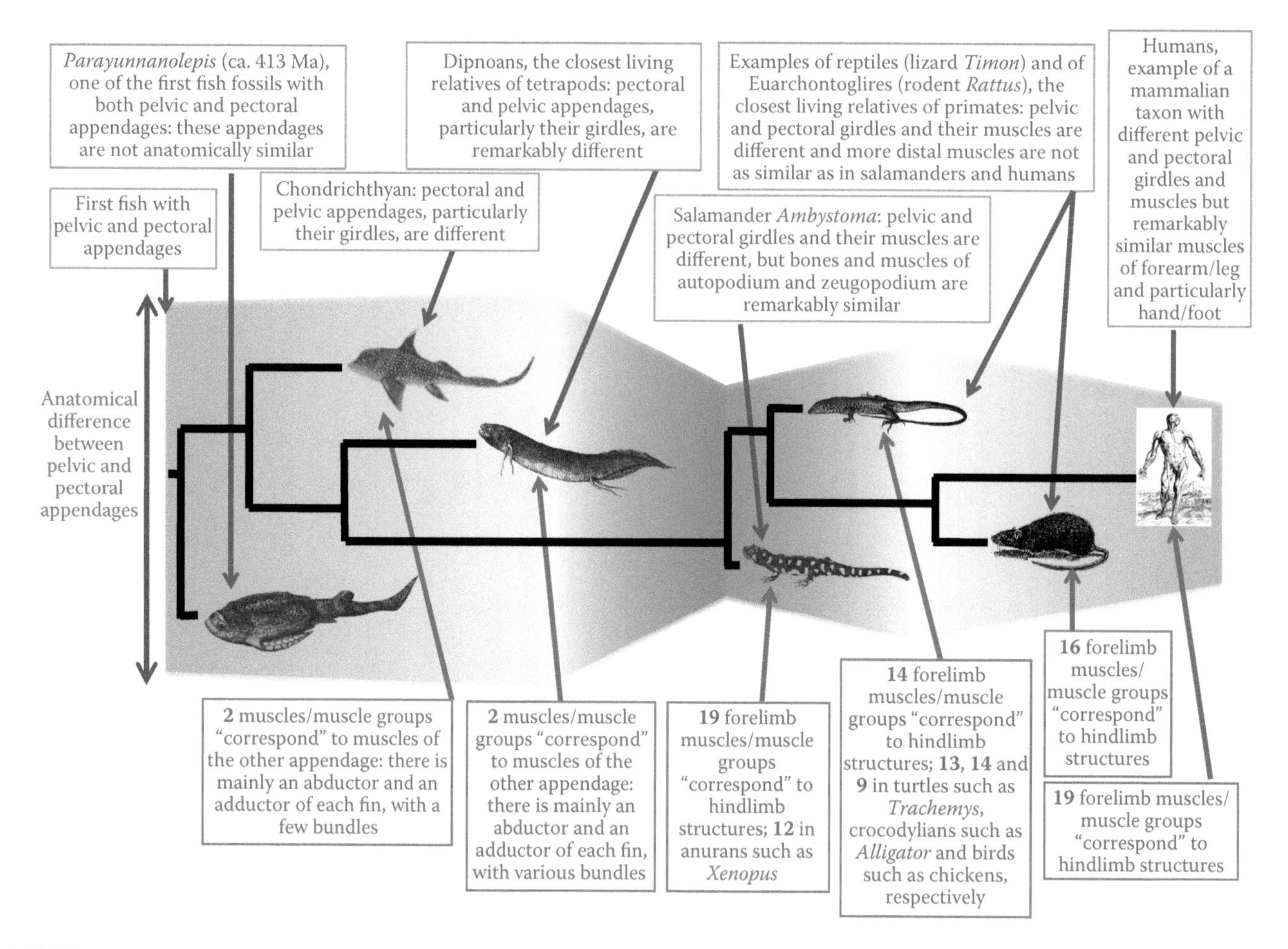

FIGURE 12.4 The evolutionary history of the pelvic and pectoral appendages was much more complex than what the "serial homology followed by functional/anatomical divergence" hypothesis shown in Figure 12.3 suggests. This is because it was the result of a complex interplay between ontogenetic, functional, topological, and phylogenetic constraints leading to cases of anatomical divergence followed by cases of anatomical convergence ("similarity bottlenecks"). This is exemplified in this simplified diagram of the evolutionary muscle transitions leading to modern humans (number of muscles from Diogo, R. et al., *Biol Rev*, 88, 196–214, 2013; Diogo, R., and Molnar, J., *Anat Rec Hoboken*, 297, 1047–1075, 2014).

and gnathostome structures in the context of evo-devo and evolutionary biology. However, those studies have mainly focused on hard tissues and/or on the cephalic region, and the few studies that have focused on appendicular muscles have often used their own nomenclature and did not really provide comparisons with older, classical anatomical and evolutionary works and/or those focused on other gnathostomes, thus lacking a broader historical, comparative, and evolutionary framework. Therefore, in this chapter, we combine the results of our recent dissections of chondrichthyans and an extensive review of the literature, including both recent anatomical and developmental papers and older, classical works, in order to provide such a broad framework, which is necessary to pave the way for further developmental and evolutionary discussions. In addition to providing new, updated textual and visual descriptions of the muscles of the pelvic and pectoral girdles and fins of chondrichthyans, the data obtained from our dissections will allow us to discuss which muscles were likely present in the last common ancestor (LCA) of extant gnathostomes, chondrichthyans, and osteichthyans and thus to specifically test the hypothesis (proposed by authors such as Humphry [1872b]: see preceding text) that at least some forelimb and/or hindlimb muscles of tetrapods were already present in early gnathostomes.

MUSCLES OF PAIRED APPENDAGES OF *SQUALUS ACANTHIAS*

The adductor muscle of the pectoral fin (Figure 12.5; "extensor or levator or pterygii dorsalis" *sensu*, e.g., Walker 1954; Jarvik 1965) arises from the pectoral girdle (mainly scapular region) to the dorsal surfaces of the radial cartilages. The name *adductor* follows the nomenclature of Diogo and Abdala (2010), as it corresponds to the adductor of osteichthyan fishes. Some of its fibers also contact the propterygium, metapterygium, and mesopterygium, because the muscle has a superficial bundle (adductor superficialis; Figure 12.5) but also a deep bundle (adductor profundus) that attaches onto a region of the pectoral fin that is more proximal than the insertion of the superficial bundle. The abductor of the pectoral fin (Figure 12.5; "flexor or depressor or pterygii ventralis" *sensu*, e.g., Walker 1954; Jarvik 1965) spreads mainly

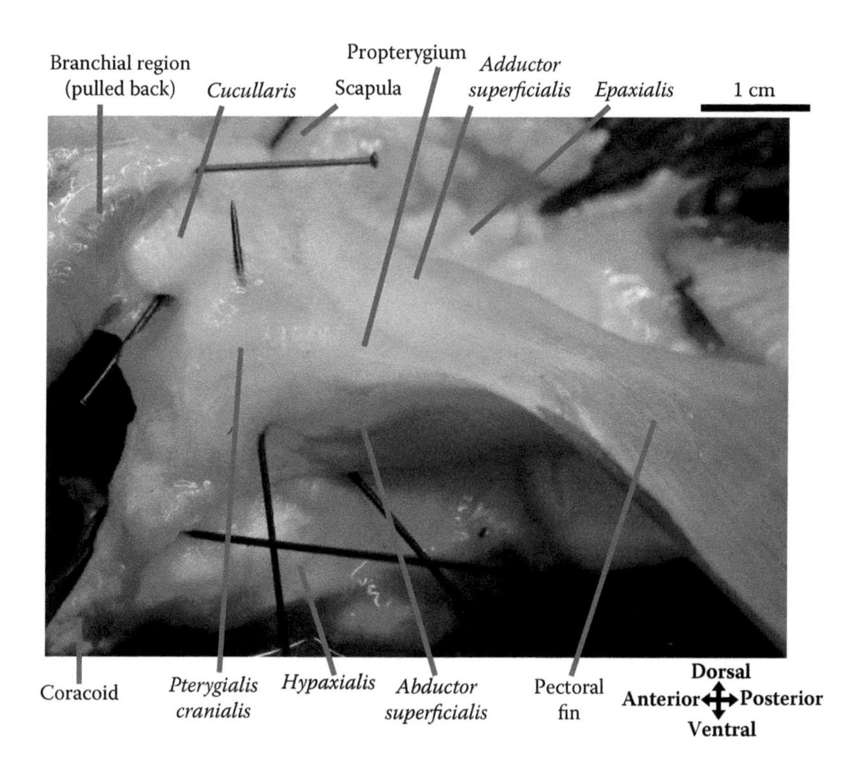

FIGURE 12.5 *Squalus acanthias* (Vertebrata, Elasmobranchii): lateral view of the left pectoral girdle and fin muscles.

from the coracoid bar to the ventral surfaces of the radial cartilages, also having a superficial bundle (abductor superficialis; Figure 12.5) and a deep bundle (abductor profundus) that extends from the propterygium, metapterygium, and mesopterygium to an insertion point that is proximal to the insertion of the superficial bundle. The pterygialis cranialis (Figure 12.5; *sensu*, e.g., Marinelli and Strenger 1959), which is essentially a protractor of the pectoral fin, although it was designated "adductor" by Jarvik (1965), is clearly distinct from the abductor of the pectoral fin, from which it seems to derive, running from the pectoral girdle (both scapular region and coracoid bar) to the propterygium (the ontogeny of this protractor muscle is shown by Jarvik [1965], who also designated it as a "modified anterior radial muscle"). The epaxialis and hypaxialis (Figure 12.5) are not differentiated into individual muscles (e.g., pectoralis, latissimus dorsi), contrary to what was suggested by Humphry (1872a,b; see preceding text). For the context of the present work, it is worthy to also describe the muscle cucullaris (Figure 12.5; "trapezius" *sensu* Marion [1905]), which is a head (branchial) muscle (usually innervated by cranial nerve CNX in chondrichthyans: e.g., Edgeworth 1935; see preceding chapters) but is partially inserted onto the pectoral girdle. In the *Squalus acanthias* specimens we dissected, the cucullaris is a continuous muscle originating from the fascia of the epibranchial and/or epaxial musculature and inserting onto both the pectoral girdle and the dorsal surface of all branchial arches. The levatores arcuum branchialium were not found in our dissections, and the cucullaris seems to correspond to the protractor pectoralis plus the levatores arcuum branchialium of osteichthyan fishes, and that is why we use the name *cucullaris* to designate this muscle (see following text).

Regarding the muscles of the pelvic appendage, the adductor of the pelvic fin (Figure 12.6; "extensor or levator" *sensu*, e.g., Walker [1954]; Jarvik [1965]) arises mainly from posterior trunk myomeres, particularly its superficial bundle (adductor superficialis). Its deep bundle (adductor profundus) originates mainly from the iliac process and metapterygium and inserts less distally than the superficial bundle, both bundles inserting onto the dorsal surfaces of the radial cartilages; therefore, the adductor of the pelvic fin is anatomically somewhat similar to the adductor of the pectoral fin. However, the abductor of the pelvic fin (Figure 12.6; "flexor or depressor" *sensu*, e.g., Walker [1954]; Jarvik [1965]) is anatomically different from the abductor of the pectoral fin in that it consists of proximal and distal bundles (abductor proximalis, abductor distalis; Figure 12.6). Its proximal bundle runs from the puboischiac bar to the metapterygium, while its distal bundle runs from the metapterygium to the ventral surfaces of the radial cartilages. The protractor of the pelvic fin (Figure 12.6) is anatomically similar to the pterygialis cranialis of the pectoral appendage but somewhat less differentiated; i.e., it is more deeply blended to the abductor of the pelvic fin from which it seems to derive (that is probably why Jarvik [1965] did not label this muscle in the pelvic fin shown in his Figure 9E, even though the muscle is somewhat visible in that figure). The protractor of the pelvic fin runs from the pelvic girdle to the propterygium of the pelvic fin.

MUSCLES OF PAIRED APPENDAGES OF *LEUCORAJA ERINACEA*

The abductor of the pectoral fin (Figure 12.7) extends from the pectoral girdle to the ventral surfaces of the radial cartilages,

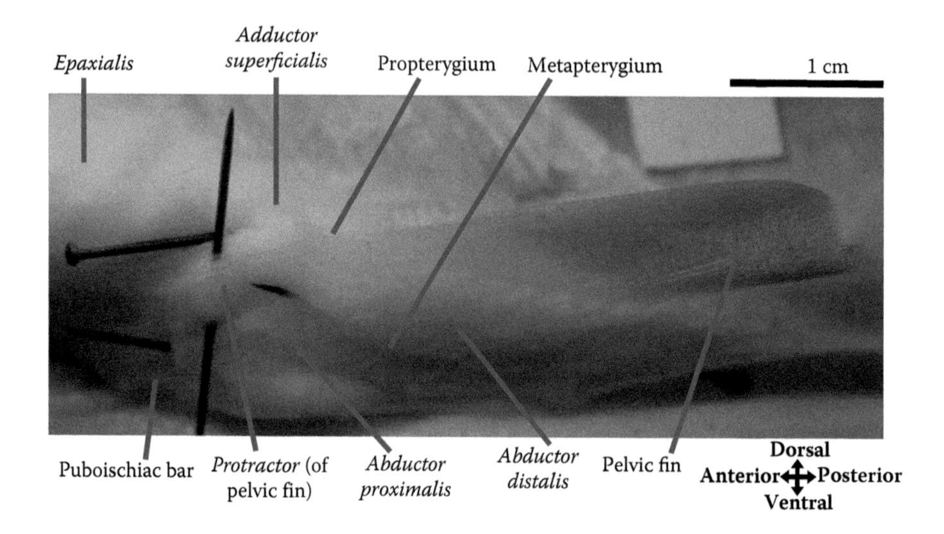

FIGURE 12.6 *Squalus acanthias* (Vertebrata, Elasmobranchii): lateral view of the left pelvic girdle and fin muscles.

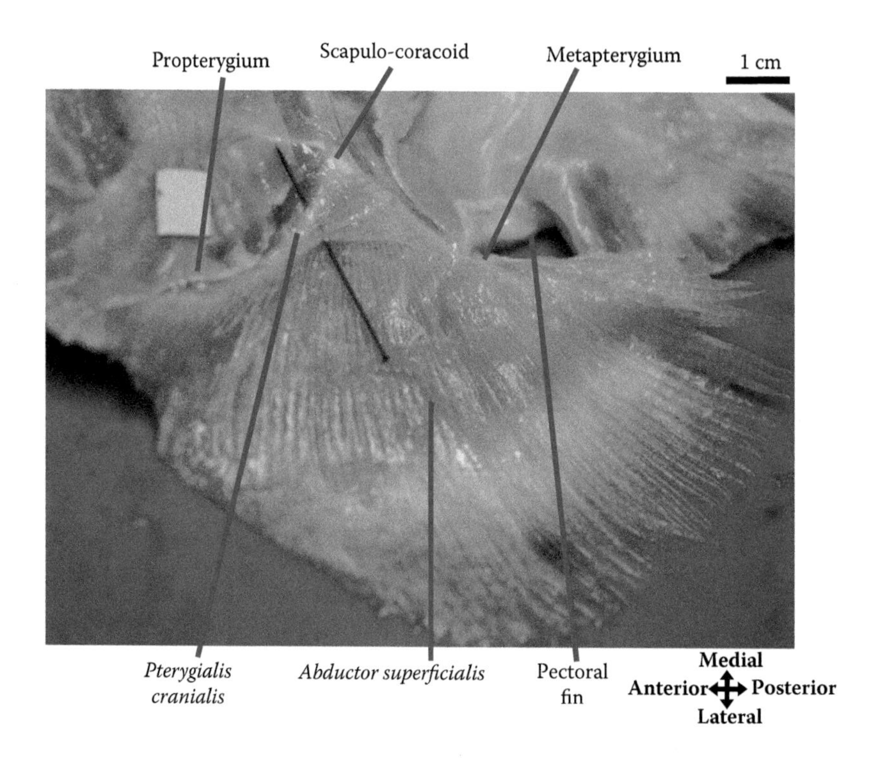

FIGURE 12.7 *Leucoraja erinacea* (Vertebrata, Elasmobranchii): ventrolateral view of the pectoral girdle and fin muscles.

having a superficial bundle and a deep bundle that originates from the propterygium and metapterygium and inserts at a point proximal to the insertion of the superficial bundle. There is a distinct muscle that corresponds topologically, and seems to be homologous, to the pterygialis cranialis of sharks, running (mainly anteriorly, not posteriorly as in sharks, due to the peculiar position of the propterygium of *Leucoraja*) from the pectoral girdle to the propterygium (Figure 12.7). The adductor of the pectoral fin (Figure 12.8) runs from the pectoral girdle to the dorsal surfaces of the radial cartilages, some fibers also contacting the propterygium and metapterygium, because the muscle has superficial and deep bundles similar those found in *S. acanthias*. There is also a muscle levator 4 of the pectoral fin (Figure 12.8), which is somewhat similar to the levator 3 of the pectoral fin of *Hydrolagus* because it originates from the fascia of, and is superficial to, the epaxial/hypaxial musculature and mainly inserts onto the metapterygium, but is much broader and more vertical than the relatively thin and oblique (antero-ventrally directed) levator 3 of *Hydrolagus* (see the following text). On the adductor (dorsal) side of the pectoral fin of *Leucoraja*, there is a mesial muscular band running from the pectoral girdle to the metapterygium that is similar to that described in sharks by Jarvik (1965) but that is more differentiated in *Leucoraja* than in the

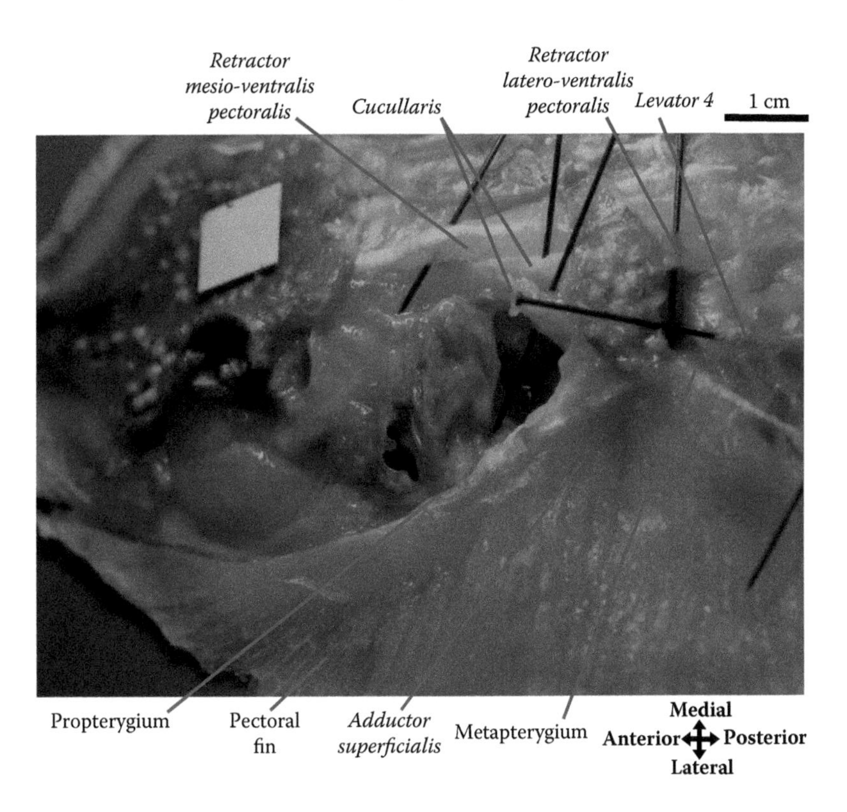

FIGURE 12.8 *Leucoraja erinacea* (Vertebrata, Elasmobranchii): dorsolateral view of the pectoral girdle and fin muscles.

Squalus specimens we dissected. In *Leucoraja* there is also a similar muscular band on the adductor side of the pectoral fin extending from the pectoral girdle to the propterygium, but both this structure and the mesial muscular band attaching onto the metapterygium are much less differentiated than is the pterygialis cranialis of the abductor side of the pectoral fin; i.e., these muscular bands seem to be simply bundles of the abductor superficialis of this fin. Apart from levator 4, in *Leucoraja* the only structures related to the epaxial/hypaxial musculature and to the pectoral appendage that seem to constitute separate muscles are the retractor mesio-ventralis pectoralis (from skull to epaxial musculature, being loosely connected to medial side of pectoral girdle) and the retractor latero-ventralis pectoralis (from epaxial musculature to lateral surface of the pectoral girdle) (Figure 12.8). The head muscle cucullaris (Figure 12.8) is divided into a medial bundle that runs from the first vertebra to the suprascapula and a lateral bundle (not two lateral bundles as described by Marion 1905) that inserts onto at least some of the most posterior branchial arches and also onto the pectoral girdle.

Concerning the muscles of the pelvic appendage, the abductor of the pelvic fin (Figure 12.9; "flexor or depressor" *sensu*, e.g., Macesic and Kajiura 2010) has two distinct bundles, i.e., a proximal bundle extending from the pelvic girdle and hypaxial musculature to the metapterygium and a distal bundle extending from the metapterygium to the ventral surfaces of the radial cartilages; i.e., this muscle is anatomically somewhat similar to the abductor of the pelvic fin of *Squalus*. The adductor of the pelvic fin (Figure 12.9; extensor or levator *sensu*, e.g., Macesic and Kajiura 2010) is also anatomically similar to that of *Squalus*. It arises mainly from posterior trunk myomeres, particularly its superficial bundle, its deep bundle arising largely from the metapterygium and inserting less distally than the superficial bundle, both bundles going to the dorsal surfaces of the radial cartilages. In addition to these two muscles of the pelvic fin, *Leucoraja* also has four ventral and four dorsal muscles attaching onto the propterygium, which are probably derived structures related to the peculiar type of locomotion of benthic batoids such as *Leucoraja erinacea*, as suggested by Macesic and Kajiura (2010). At least some of these additional muscles, especially some of the ventral ones, seem to correspond to parts of the pelvic fin protractor of sharks. The four additional ventral muscles are the ventral protractor of the proximal propterygium (mainly from body musculature to the proximal portion of propterygium), the depressor of the proximal propterygium (from pelvic girdle to the proximal portion of propterygium), the depressor of the distal propterygium (from pelvic girdle to the distal portion of propterygium), and the retractor of the propterygium (from pelvic girdle to propterygium) (Figure 12.9). The four additional dorsal muscles are the dorsal protractor of the proximal propterygium (mainly from body musculature to the proximal portion of propterygium), the flexor of the proximal propterygium (from pelvic girdle to the proximal portion of propterygium), the flexor of the distal propterygium (from pelvic girdle to the distal portion of propterygium), and the dorsal retractor of the propterygium (from pelvic girdle to propterygium).

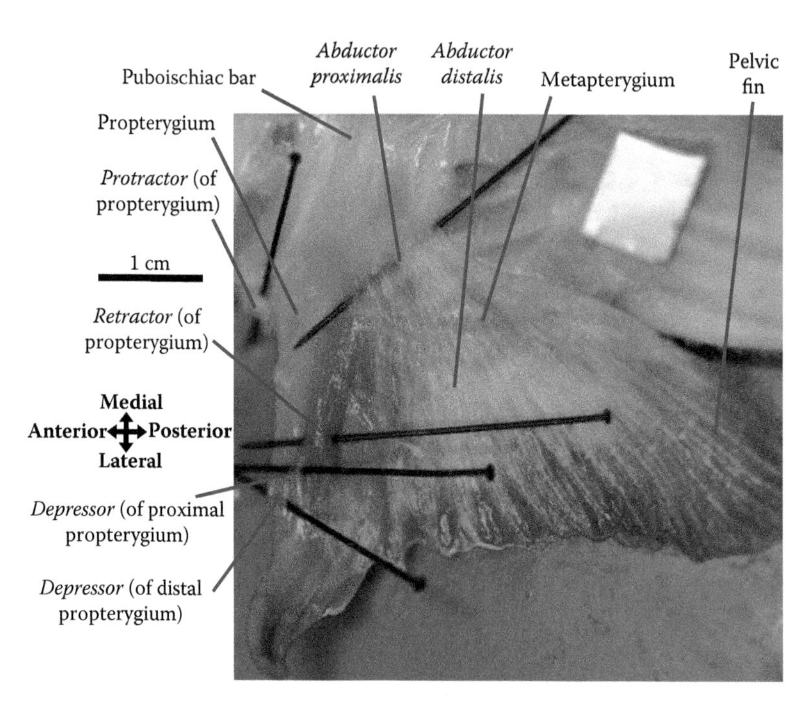

FIGURE 12.9 *Leucoraja erinacea* (Vertebrata, Elasmobranchii): ventrolateral view of the pelvic girdle and fin muscles.

MUSCLES OF PAIRED APPENDAGES OF *HYDROLAGUS COLLIEI*

The adductor of the pectoral fin (Figure 12.10) extends from the pectoral girdle (mainly scapular region) to the dorsal surfaces of the radial cartilages, some fibers also contacting the propterygium and metapterygium, because the muscle has superficial and deep bundles similar to those of sharks. Levator 3 (Figure 12.10) was designated by Didier (1987, 1995) as "adductor superficialis" but is very likely not homologous to the adductor superficialis of osteichthyans because it extends primarily from the fascia of the epaxial musculature to the dorsal surface of the propterygium. That is, it elevates and adducts the pectoral fin but also rotates it laterally (in the sense of human terminology, i.e., it displaces the lateral portion of the pectoral fin dorsally). The origin of levator 2 (Figure 12.10) is ventromedial to that of levator 3; i.e., the muscle arises from the fascia of the hypaxial musculature

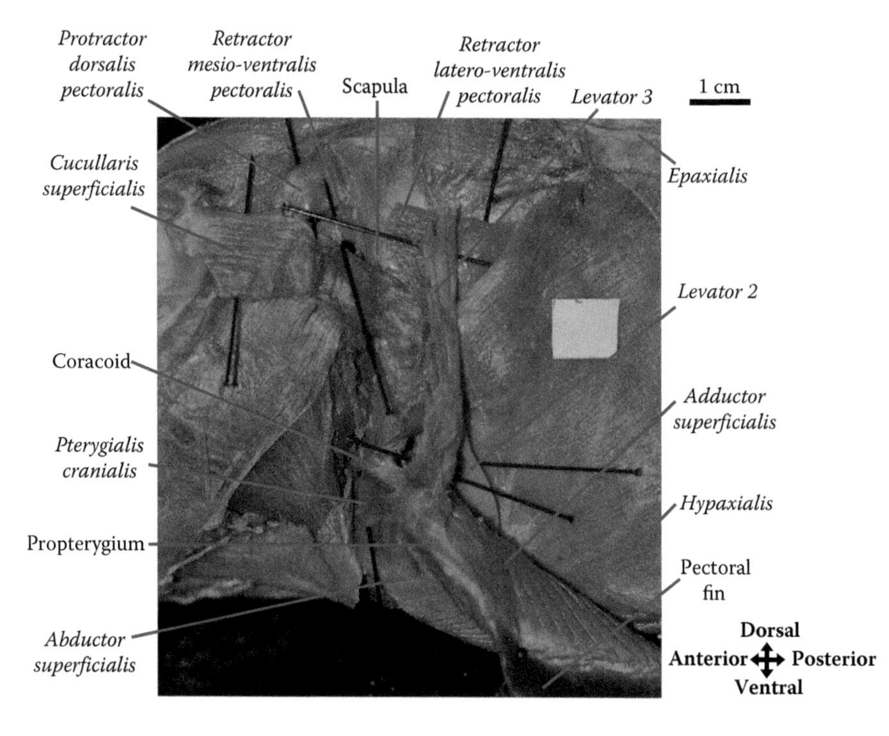

FIGURE 12.10 *Hydrolagus colliei* (Vertebrata, Holocephali): lateral view of the left pectoral girdle and fin muscles.

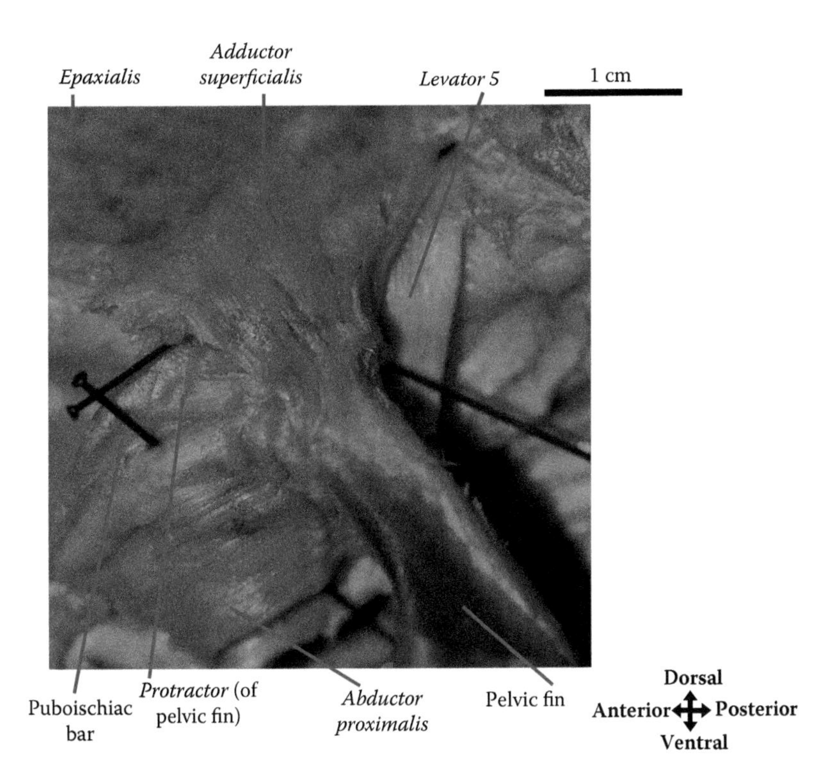

FIGURE 12.11 *Hydrolagus colliei* (Vertebrata, Holocephali): lateral view of the left pelvic girdle and fin muscles.

to insert onto the dorsal surface of the metapterygium and the more medial radial cartilages. So besides elevating and adducting the pectoral fin, it also rotates it "medially" (in the sense that it pulls the medial portion of the pectoral fin dorsally). The abductor of the pectoral fin (Figure 12.10) extends mainly from the coracoid bar to the ventral surfaces of the radial cartilages, also having a superficial bundle and a deep bundle that are similar to those of sharks. The pterygialis cranialis is anatomically similar to that of sharks but is more blended to the abductor of the pectoral fin than in the case of sharks. It extends from the pectoral girdle (both scapular region and coracoid bar) to the propterygium of the pectoral fin. Apart from levators 2 and 3, the epaxial and hypaxial musculature is divided into multiple muscles. The retractor latero-ventralis pectoralis (Figure 12.10) is constituted by a single bundle (as described by Anderson [2008], and contrary to Didier [1987], who described two bundles) extending from the epaxial musculature to the lateral surface of the pectoral girdle. The retractor mesio-ventralis pectoralis (Figure 12.10) is deep to the cucullaris superficialis and extends from the epaxial musculature to the medial side of the pectoral girdle and mainly to the back of the chondrocranium. The protractor dorsalis pectoralis (Figure 12.10) arises from the postorbital ridge of the chondrocranium and inserts onto the scapular process of the pectoral girdle and has a single bundle as described by Anderson (2008) (not two as described by Didier [1987]). The "retractor dorsalis pectoralis" and "latero-ventralis" *sensu* Didier (1987) are simply part of the body musculature going to the pectoral girdle and chondrocranium rather than separate, distinct muscles; this fact is also suggested by Didier (1995), who listed the retractor dorsalis pectoralis as part of the epaxial musculature. The head muscles cucullaris superficialis and cucullaris profundus (Figure 12.10; "trapezius superficialis and profundus" *sensu* Didier [1987]) extend from the postorbital ridge and surrounding structures to the pectoral girdle and to at least some branchial arches, respectively.

With respect to the pelvic appendage, the adductor of the pelvic fin (Figure 12.11) arises from the fascia of the body musculature, particularly its superficial bundle, its deep bundle originating mainly from the basipterygium and going less distal than the superficial bundle, both bundles attaching onto the dorsal surfaces of the radial cartilages. There is a levator 5 of the pelvic fin (Figure 12.11) running medially to the adductor of the fin and connecting the fascia of the body musculature to the basipterygium and medial radial cartilages, so its function seems to be somewhat analogue to that of the levator 2 of the pectoral fin. The abductor of the pelvic fin (Figure 12.11) has a proximal bundle extending from the puboischiac bar to the basipterygium and a distal bundle extending from the basipterygium to the ventral surfaces of the radial cartilages. The protractor of the pelvic fin (Figure 12.11) arises from the pelvic girdle and inserts into the propterygium, being similar to that of sharks.

PLESIOMORPHIC CONFIGURATION FOR CHONDRICHTHYANS AND EVOLUTION OF THE CUCULLARIS

The comparison between the muscles attached onto the pelvic and pectoral appendages of the representatives of the three major chondrichthyan extant clades (Holocephali, Selachii, and Batoidea: see the introduction) we dissected and described in the literature allows us to provide the first detailed discussion about which is the plesiomorphic configuration of the pelvic and pectoral muscles in chondrichthyans (see Figures 12.1 and 12.2) and also about the similarity between these muscles in these fishes. In general, in chondrichthyans the adductor of the pelvic fin is somewhat similar anatomically and functionally to the adductor of the pectoral fin, because both muscles predominantly extend the fins and have similar superficial and deep bundles, the deep bundle (adductor profundus) running mainly from the principal cartilages of the fins (propterygium, metapterygium, and/or mesopterygium) to a region of the radial cartilages of the fins that is more proximal than the insertion of the superficial bundle (adductor superficialis). The pterygialis cranialis of the pectoral appendage and the protractor of the pelvic fin are also similar anatomically and functionally, because both muscles predominantly protract the fin and run from the girdles to the propterygium and apparently are differentiated from the abductor superficialis. The abductor of the pectoral fin of chondrichthyans has superficial and deep bundles that are similar to those of the adductor of the pelvic and pectoral fins, but the abductor of the pelvic fin has instead proximal and distal bundles, the proximal mainly inserting onto the metapterygium and the distal mainly originating from the latter bone. This stresses that there are significant anatomical differences between not only the hard tissues (e.g., Coates and Cohn 1998; Diogo et al. 2013) but also the soft tissues of the pelvic and pectoral appendages even in phylogenetically plesiomorphic gnathostomes such as chondrichthyans.

We are not arguing that fishes such as sharks are ideal models of the first fish with both pelvic and pectoral appendages. As can be seen in Figure 12.4, the oldest fossil fish discovered so far with both these appendages (e.g., *Parayunnapolepis*: see Zhu et al. 2012) are remarkably different morphologically from sharks. Sharks display numerous derived, peculiar anatomical features (autapomorphies). However, as Diogo et al. (2015b) argued, the opposite extreme view that sharks and other chondrichthyans cannot tell us much about the plesiomorphic condition of, e.g., the soft tissues of the first gnathostomes is not correct either. In fact, based on strict cladistic phylogenetic studies, with respect to, e.g., their head muscles, sharks are indeed one of the best living models for the LCA of gnathostomes (Diogo and Abdala 2010; Diogo et al. 2015b). This statement also seems to apply to the appendicular muscles, as will be explained later.

In fact, Coates and Cohn (1998) argued that the close similarity between the hard tissues of the pelvic and pectoral appendages seen in nonchondrichthyan extant fishes—i.e., in osteichthyans—is a derived, specialized feature among gnathostomes. Regarding the soft tissues, the differences between the pelvic and pectoral appendages of chondrichthyans and of other gnathostomes is also illustrated by the partial attachment of the head muscle cucullaris to the pectoral girdle; no similar muscle is found on the pelvic region. The understanding of the evolution of the cucullaris in gnathostomes is of crucial importance to comprehend the origin and evolution of the paired fins of these vertebrates. The developmental origin of the cucullaris and its osteichthyan derivative protractor pectoralis and amniote derivatives trapezius and sternocleidomastoid has been subject of much debate (see preceding chapters). As noted in the preceding chapters, our dissections of chondrichthyans and comparisons with other vertebrates, our recent developmental studies of tetrapods (e.g., Diogo and Ziermann 2014; Ziermann and Diogo 2013, 2014), and our updated review of the literature allow us to solve the controversy regarding a body versus head origin of the cucullaris. In adult sharks such as *Squalus*, there is a single, continuous muscle cucullaris inserting onto both the dorsal surface of the branchial arches and the posterior surface of the pectoral girdle. Edgeworth (1935) showed that this condition is seen from the first embryonic stages until the adult stages of sharks, the cucullaris developing from the dorsal portion of the branchial muscle plates. In adult holocephalans such as *Hydrolagus*, the cucullaris has two bundles, one (superficial) inserting onto the pectoral girdle and the other (deep) inserting onto the dorsal portion of the branchial arches. This is further supported by the development in holocephalans where the cucullaris develops from the dorsal portion of the branchial muscle plates, forming first a single, continuous muscle and then separating during ontogeny into the deep and superficial bundles (Edgeworth 1935). This can be also observed in batoids with the slight difference that the cucullaris develops only from the dorsal portion of the last branchial muscle plate before dividing into bundles that attach onto the pectoral girdle and branchial arches, in taxa such as *Leucoraja* (Edgeworth 1935; see also, e.g., Miyake et al. 1992). According to Edgeworth, in chondrichthyans the cucullaris is mainly innervated by CNX; in osteichthyans, there are usually (a) a protractor pectoralis (or its amniote derivatives, e.g., trapezius and sternocleidomastoideus) that derives ontogenetically exclusively from the last branchial muscle plate and inserts exclusively onto the pectoral girdle (in bony fish and most tetrapods) and (b) levatores arcuum branchialium developing from the dorsal portion of the branchial muscle plates and attaching exclusively onto the dorsal surface of the branchial arches (in bony fishes and amphibians, these muscles being absent in amniotes). Our recent developmental studies strongly support the idea that the protractor pectoralis and its amniote derivatives trapezius and sternocleidomastoideus develop anatomically exclusively from the last branchial plate (e.g., Ziermann and Diogo 2013, 2014; see Chapters 7 and 11).

We did not find the levatores arcuum branchialium muscles in any of the chondrichthyan specimens we dissected, and almost all authors agree that these are not present as distinct muscles in chondrichthyans. However, Kesteven (1942) did describe these muscles as very thin structures in sharks such as *Mustelus* and *Orectolobus* and batoids such as *Dasyatis* and holocephalans such as *Callorhynchus* and *Hydrolagus* and as even thinner and apparently vestigial structures in sharks such as *Squalus* and batoids such as *Leucoraja*, going to the dorsal surface of five branchial arches. Didier (1987, 1995) does not describe these muscles in any of the numerous holocephalan taxa dissected by her and states that some of the levatores arcuum branchialium *sensu* Kesteven (1942) correspond to part of the cucullaris profundus, while others correspond to part of the epibranchial muscle subspinalis of holocephalans. But Kesteven (1942) described the subspinalis, the cucullaris, and the levatores arcuum branchialium in chondrichthyans, so the synonymies proposed by Didier are questionable. However, to our knowledge, no other author (apart from Kesteven 1942) has described distinct, fleshy muscles levatores arcuum branchialium in chondrichthyans, and even Kesteven recognized that the structures that he designated levatores arcuum branchialium in these fishes are mainly innervated by "spinal" nerves, while those in osteichthyans are mainly innervated by CNX and/or CNIX, calling into question the homology of the muscles of these two major gnathostome taxa proposed by Kesteven. In fact, our dissections and direct comparisons with Kesteven's (1942) descriptions point out that the levatores arcuum branchialium muscles described by Kesteven are very likely just part of the cucullaris *sensu* the present work or, alternatively, part of the epibranchial musculature, as suggested by Didier (1995), but clearly do not seem to correspond to the true levatores arcuum branchialium of osteichthyans.

As explained in the preceding text, the epaxial and hypaxial muscles are not divided into distinct muscles attached to the paired appendages in sharks (e.g., pectoralis, latissimus dorsi) as suggested by Humphry (1872a,b). In fact, even sharks with peculiar locomotor behaviors (e.g., *Hemiscyllium ocellatum*, which has a walking/crawling locomotion) have essentially the same type of muscle configuration as *Squalus*, with only a few minor changes (Goto et al. 1999). However, batoids and holocephalans do have some muscles that are apparently derived from the epaxial and/or hypaxial musculature, including the protractor dorsalis pectoralis (e.g., in *Hydrolagus*), the retractor latero-ventralis pectoralis and retractor mesio-ventralis pectoralis (in, e.g., *Leucoraja* and *Hydrolagus*), the levators 2 and 3 of the pectoral fin (e.g., in *Hydrolagus*), the levator 4 of the pectoral fin (e.g., in *Leucoraja*), and the levator 5 of the pelvic fin (e.g., in *Hydrolagus*). Authors such as Anderson (2008) suggested that the protractor dorsalis pectoralis, retractor latero-ventralis pectoralis, and retractor mesio-ventralis pectoralis of holocephalans are branchial muscles innervated by CNIX and CNX. However, anatomically these muscles do not seem to be branchial muscles. Moreover, Didier (1995) stated that only the protractor dorsalis pectoralis might be a branchial muscle innervated by nerves CNIX and/or CNX and that the other muscles are clearly body muscles. Therefore, the protractor dorsalis pectoralis might be a branchial or epibranchial muscle because it lies in the epibranchial region, just deep to the cucullaris and connecting, as does the cucullaris, the cephalic region to the pectoral girdle. However, Edgeworth (1935) studied the development of the head muscles of holocephalans and stated that the only epibranchial muscles of these fishes are the interpharyngobranchiales and subspinalis, so the protractor dorsalis pectoralis does not seem to be an epibranchial muscle. According to Edgeworth (1935), among gnathostomes an epibranchial musculature is present only in the Selachii and Holocephali. Kesteven (1942) also described epibranchial muscles (namely, the interpharyngobranchiales, the subspinalis being absent) in batoids such as *Dasyatis* (the muscles being reduced to fibrous structures in *Leucoraja*), but more recent works, including those focused specifically on batoids (e.g., Miyake et al. 1992), have stated that batoids do not have epibranchial muscles. A characteristic feature of the epibranchial muscles is that they are innervated by spinal nerves. Therefore, if the protractor dorsalis pectoralis is innervated by the nerves CNIX and/or CNX, as described by Didier (1995) and Anderson (2008), this muscle is probably not part of the epibranchial musculature. However, neither Edgeworth (1935) nor Miyake et al. (1992) list this muscle as a branchial muscle in holocephalans. Moreover, our review of the literature reveals that most authors refer to an innervation of the muscle by the nerves CNIX and/or CNX primarily because they consider it to be a branchial muscle *a priori* rather than because they conducted a detailed study of its innervation. Therefore, based on all the data currently available, the most likely hypothesis is that the protractor dorsalis pectoralis, retractor latero-ventralis pectoralis, and retractor mesio-ventralis pectoralis are derived from the body muscles. Regardless, only the retractor latero-ventralis pectoralis and retractor mesio-ventralis pectoralis seem to be consistently found in batoids and holocephalans (our study, see also, e.g., Shann 1920, 1924, and the following text), so the protractor dorsalis pectoralis, as well as levators 2, 3, 4, and 5, does not seem to have been present in the LCA of extant chondrichthyans (i.e., of elasmobranchs and holocephalans).

In summary, the plesiomorphic condition for chondrichthyans (as well as extant gnathostomes as a whole, see following chapters) is that the pectoral appendage is associated with the head muscle cucullaris (inserted onto both the pectoral girdle and at least some branchial arches), the trunk muscles retractor latero-ventralis pectoralis and retractor mesio-ventralis pectoralis, and the abductor and adductor muscles of the pectoral fin (both divided into superficial and deep bundles), as well as the pterygialis cranialis (Figure 12.1). Regarding the pelvic appendage, the plesiomorphic chondrichthyan (and gnathostome: see the following chapters) condition is to have abductor the and adductor muscles of the pelvic fin (the latter divided into superficial and deep bundles and the former divided, very likely, into proximal and distal bundles) as well as a protractor muscle of this fin, i.e., a pterygialis cranialis pelvicus (Figure 12.2).

FORELIMB–HINDLIMB SERIAL HOMOLOGY DOGMA

As explained in the preceding text, plesiomorphically chondrichthyans have three distinct muscles related to the movements of the pectoral girdle: the head muscle cucullaris and the trunk muscles retractor latero-ventralis pectoralis and retractor mesio-ventralis pectoralis. Based on the ideas of authors such as Humphry (1872a,b) that at least some of the specific muscles of the pectoral and/or pelvic appendages found in chondrichthyans are homologous to the muscles of the forelimb and/or hindlimb of tetrapods (see preceding text),

Shann (1920, 1924) stated that the retractor latero-ventralis pectoralis and the retractor mesio-ventralis pectoralis of holocephalans are homologous to muscles with the same names in sharks and respectively to the "serratus" and the "obliquus/rectus abdominis" of dipnoans and tetrapods. However, we did not find a retractor latero-ventralis pectoralis nor a retractor mesio-ventralis pectoralis in sharks, and these muscles are also not described by most authors that have studied the muscles of these fishes in some detail (e.g., Marion 1905; Kesteven 1942; Walker 1954; Jarvik 1965; Anderson 2008). In fact, a comparison of Shann (1920, 1924) with Humphry's (1872a,b) descriptions and the results of our own dissections of all major extant gnathostomes groups reveals that, as recognized by Shann himself, the only living fish taxon in which these muscles are truly differentiated is the Holocephali. That is, Shann (1920, 1924) recognized that in the other living fishes in which he listed these muscles, including sharks, he was referring to parts of the epaxial/hypaxial musculature that do not form distinct, separate muscles. As explained in the preceding text, we did find the retractor latero-ventralis pectoralis and the retractor mesio-ventralis pectoralis in batoids. However, we did not find these muscles in most osteichthyans we dissected in the past. Therefore, the plesiomorphic condition for the first gnathostomes that had both pelvic and pectoral appendages appears to be that there were almost no distinct muscles differentiated from the body musculature that attached to the paired girdles and that were specifically related to the movements of these girdles. However, many muscles associated with the movements of these girdles were then independently acquired during gnathostome evolution.

Therefore, few muscles are plesiomorphically associated with either pectoral or pelvic appendages in gnathostomes and osteichthyans (Figures 12.1 and 12.2). This result contradicts the notion of authors such as Humphry (1872b) that some muscles present in the fore- and hindlimbs of basal tetrapods were already present in some form in plesiomorphic gnathostome fishes, as the result of a unique, ancestral duplication of the paired appendages (see preceding text). Rather, it strongly supports the scenario shown in Figure 12.4, i.e., that all the numerous, and in many cases strikingly similar, muscles of the tetrapod forelimb and hindlimb were undoubtedly acquired *independently* (*homoplasically*) at some stage during the evolutionary transitions that occurred from early gnathostomes to tetrapods. Consequently, under a historical (phylogenetic) definition of homology, these fore- and hindlimb structures cannot be considered serial homologues. One might argue that they are homologues under the morphological or developmental definitions of homology (see Wagner 1994), but this view is also contradicted by the integration of the results of the present work on chondrichthyans and of our previous works on osteichthyans (Figure 12.4). As explained in the preceding text, ancestral morphological serial homology of these appendages would imply that these appendages were originally similar and then diverged anatomically/functionally, as shown in Figure 12.3. This is not the case (Figure 12.4): in the more plesiomorphic fishes found so far with pectoral and pelvic appendages, the hard tissues of these appendages are anatomically very different (e.g., Coates and Cohn 1998). Both the hard and soft tissues of the proximal region of the paired appendages (particularly the girdles) remained, in general, markedly different throughout all gnathostome clades (Coates and Cohn 1998; Diogo et al. 2013). For instance, among all tetrapod clades studied by Diogo and Molnar (2014), including anatomically plesiomorphic taxa such as urodeles (caudates), not one pelvic–thigh muscle has a clear "topological equivalent" in the pectoral region and arm. This lack of equivalence seems to be the result of a strong phylogenetic constraint; i.e., the pelvic and pectoral girdles of fishes were anatomically markedly different from the start, and the girdles of tetrapods are consequently also quite different anatomically. In contrast, the more derived distal regions of the tetrapod fore- and hindlimbs, especially the autopodium (hand/foot), include at least some new bones and a developmental plan that is quite different from those of the fish paired appendages (Diogo et al. 2013). That is, the evolution of the autopodium represents a major "evolutionary novelty," meaning that fewer phylogenetic constraints were involved, and the developmental constraints/factors resulting from further co-option of some similar genes in the ontogeny of the fore- and hindlimbs thus led to a more marked similarity between the distal regions of these limbs in basal tetrapods (similarity bottleneck: Figure 12.4). However, the co-option of similar genes during the "fins–limbs transitions" is not sufficient to explain the strong similarity between the forearm–hand and leg–foot muscles of, and thus the other similarity bottlenecks leading to, phylogenetically and anatomically derived tetrapods such as our bipedal species, *Homo sapiens* (Figure 12.4), and teleosts (see following chapters). As explained by Diogo et al. (2013), Diogo and Molnar (2014), and Miyashita and Diogo (2016), those derived similarity bottlenecks were instead the result of a complex process also including topological and functional constraints.

GENERAL REMARKS

Plesiomorphically, the paired appendages of gnathostomes probably had only abductor and adductor muscles, plus a pterygialis cranialis (Figures 12.1 and 12.2). Therefore, all the similar muscles of the zeugopodia and autopodia of the tetrapod forelimb and hindlimb were necessarily the result of homoplasy. These data contradict the forelimb–hindlimb serial homology hypothesis because they contradict the idea that some muscles that are present in the more proximal regions of the tetrapod fore- and hindlimbs (e.g., latissimus dorsi, serratus anterior) were already present in some form in plesiomorphic gnathostome fishes, as the result of a unique, ancestral duplication of the paired appendages. In fact, the present chapter emphasizes that the soft tissues associated with the pectoral and pelvic girdles were very different from the beginning of gnathostome evolution, further contradicting the serial homology hypothesis. One major difference between the soft tissues associated with the pelvic and pectoral girdles is that in nontetrapod gnathostomes, including chondrichthyans, the movements of the pectoral girdle are associated with the cucullaris muscle, while no similar muscle is associated with the pelvic girdle. This difference is probably related to the deep anatomical and functional relationship between the origin of the pectoral girdle and the evolution of the head. In gnathostomes such as chondrichthyans, the cucullaris is a continuous muscle inserting onto both the branchial arches and the pectoral girdle. Thus, this chapter provides new insights not only into the forelimb–hindlimb enigma, but also into the evolution of the vertebrate neck, by identifying three major transitions concerning the evolution of the muscles lying between the head and the pectoral girdle: (a) Plesiomorphically in gnathostomes, there is a single, continuous muscle (cucullaris) running from the cephalic region to the branchial arches and pectoral girdle. (b) In osteichthyan (bony) fishes, the pectoral girdle is still anatomically and functionally deeply associated with the head, but the cucullaris differentiated into the muscle protractor pectoralis (attaching only onto the pectoral girdle) and the muscles levatores arcuum branchialium (attaching only onto the branchial arches), thus allowing more mobility between the head and pectoral girdle. (c) In tetrapods the pectoral girdle is anatomically and functionally more independent from the head movements than in bony fishes, and in amniotes the protractor pectoralis gave rise to the neck muscles trapezius and sternocleidomastoideus. According to Coates and Cohn (1998), there are numerous evolutionary and functional reasons for the deep spatial relation between the pectoral girdle and the skull in early gnathostomes: the girdle forms the rear wall of the internal gill chamber, a shield for the pericardial cavity, and a secure insertion for the pectoral fins. Recent developmental studies of chondrichthyans are also pointing out not only that the pectoral girdle of early gnathostomes was anatomically and functionally deeply associated with the head, but also that the developmental mechanisms leading to the formation of this girdle are strikingly similar to those leading to the formation of the branchial arches (e.g., Gillis et al. 2009).

The evolutionary history of the retractor mesio-ventralis pectoralis provides another insight into the separation between the head and trunk in derived vertebrates. As explained in the preceding text, this muscle is present in at least some chondrichthyans and has been also described as, and confirmed to be in developmental studies, a distinct muscle in various osteichthyan fishes (e.g., Shann 1924), including the zebrafish (e.g., Windner et al. 2011). In a study in the zebrafish, Windner et al. (2011) provided interesting information about the migratory (abaxial) origin of this muscle from the hypaxial musculature. In the zebrafish, this muscle lies posterior to the pectoral fin and attaches to the cleithrum, the first-formed element of the teleost pectoral girdle. The founder cells of this muscle, which Windner et al. (2011) designated "posterior hypaxial muscle" and is often designated "ventral-most hypaxial muscle," "inferior obliquus muscle," or "anterior hypaxial muscle" in teleosts, arise from somites 5 and 6 and express the same marker genes of cell migration (*lbx1* and *met*) as pectoral fin muscle precursors. As noted by these authors, the muscle probably has a transient importance during early larval suction feeding, in coaction with the hypobranchial sternohyoideus muscle, which becomes one of the neck muscles of tetrapods. As noted in the preceding text, according to Shann (1925) the retractor mesio-ventralis pectoralis of fishes gave rise to the rectus and/or oblique abdominal muscles of tetrapods, a hypothesis supported by our comparative studies between fishes and tetrapods. In amphibians these abdominal muscles also attach onto the pectoral girdle and are also continuous with the sternohyoideus and other infrahyoid (neck) muscles, in a condition similar to that found in the zebrafish. However, in a further example of the evolutionary trend toward the separation of the head from the trunk during vertebrate evolution, these abdominal muscles became increasingly separated from both the pectoral girdle and the infrahyoid muscles during amniote evolution. This derived condition is for instance found in our own species *H. sapiens*, in which the only infrahyoid muscle that is attached to the pectoral girdle is the thin omohyoideus. This derived condition therefore allows these abdominal muscles to contract without interfering with the movements of the head and the infrahyoid muscles to contract without interfering with the main postcranial axial or limb movements. Hopefully, these data and discussions will contribute to deconstruction of the forelimb–hindlimb serial homology dogma and also promote future and more integrative anatomical, functional, and developmental studies on the puzzling and fascinating evolutionary history of the paired appendages and the vertebrate neck under a new paradigm that is more complex, but also more realistic.

13 Pectoral and Pelvic Muscles of Actinopterygian Fishes

The most comprehensive comparative analyses of the muscles of the pectoral and pelvic appendages of bony fishes based on a direct observation of members of a wide range of taxa (e.g., Teleostei, Halecomorphi, Ginglymodi, Chondrostei, and Cladistia) rather than on previous literature were performed many decades ago (reviewed in Diogo and Abdala [2010]). Therefore, despite the quality of some of those works, their authors did not have access to information currently available concerning, e.g., the pectoral and pectoral fin musculature of *Latimeria chalumnae* and the development and patterning of the muscles of the pectoral and pelvic appendages. Also, some of the hypotheses proposed in those works regarding the homologies and evolution of osteichthyan pectoral muscles were based on phylogenetic hypotheses that have been contradicted by numerous studies. For instance, Romer (1944) argued that the cladistian *Polypterus* is more closely related to tetrapods than are extant dipnoans, a view to which very few authors subscribe today, as explained in the preceding chapters. Diogo and Abdala (2010) provided a detailed review of the muscles of the pectoral appendage of actinopterygians. Since then, we and other authors (e.g., Wilhelm et al. 2015) have undertaken new studies of these muscles, using dissections but also other techniques, such as the study of antibody staining of muscles, scans, and three-dimensional imaging. These analyses and new methods have revealed that some taxa have more muscles than described in previous works. Therefore, in Chapter 13, we provide an updated review of the homologies and evolution of the pectoral and pectoral fin muscles of actinopterygians.

MUSCLES OF THE PECTORAL AND PELVIC APPENDAGES OF ACTINOPTERYGIANS

The plesiomorphic condition for living actinopterygians and for living bony fishes seems to be that there were six muscles related to the movements of the pectoral fins: the abductor superficialis and abductor profundus, the adductor superficialis and adductor profundus, a preaxial muscle (pterygialis cranialis, apparently derived from the ventral/abductor muscle mass), and a postaxial muscle (pterygialis caudalis, apparently derived from the dorsal/adductor muscle mass) (Table 13.1; see also Chapters 12 and 16). This condition is found for instance in the cladistian *Polypterus*, the only difference being that in this genus, the pterygialis caudalis is further divided into two muscles, the coracometapterygialis I and II (see Figure 12.1 and Table 13.1). The preaxial and postaxial muscles of *Polypterus* were thought to be bundles of the abductor and adductor masses by Diogo and Abdala (2010), but the study of Wilhelm et al. (2015) showed that they are differentiated muscles. The pterygialis caudalis is not present in any other actinopterygians we dissected, but it is present in dipnoans, so it was very likely phylogenetically acquired in bony fishes and then was lost in the Actinopteri (Figure 12.1). However, a few teleosts not dissected by us may have muscles attaching to the last pectoral rays, which thus correspond topologically to the pterygialis caudalis of dipnoans and of *Polypterus*, but this similarity is due to homoplasy (i.e., their presence in those few teleosts is due either to reversion or to convergence/parallelism).

It is interesting to note that in the node in which the pterygialis caudalis was phylogenetically lost—i.e., in the node leading to the Actinopteri—the arrector dorsalis was gained. This arrector dorsalis usually originates on the mesial surface of the pectoral girdle, laterally to the adductor of the fin and to the mesocoracoid arch (when this structure is present), and inserts onto the proximal head of the first and of the second pectoral fin rays (e.g., Figures 13.3, 13.4, 13.5B, 13.11, and 13.13). That these two events happened in the node leading to the Actinopteri would indicate that the two muscles may be homologous, as, moreover, both of them seem to derive ontogenetically mainly from the dorsal (adductor) mass, unlike the pterygialis cranialis/arrector ventralis, which derives from the ventral (abductor) mass (e.g., Jessen 1972; Winterbottom 1974; Thorsen and Hale 2005; Diogo and Abdala 2010; Diogo et al. 2016b). However, the arrector dorsalis of the Actinopteri does not seem to be homologous to the pterygialis caudalis of other bony fishes because the former muscle usually inserts onto the first ray, while the latter muscle usually inserts onto the last ray/rays. In the family Acipenseridae, we observe a variation of the muscle configuration of the pectoral fin. The arrector dorsalis is present in *Acipenser brevirostrum* that we dissected, and it was also described in *A. gueldenstaedtii* by Grom et al. (2016). On the other hand, the *A. sturio* described by Diogo and Abdala (2010) and the *A. fulvescens* dissected by us lack the arrector dorsalis. Instead, *A. fulvescens* has a strong thick ligament that fixes the first ray with a bony plate on the dorsal side. Also, in *A. fulvescens*, the abductor is formed by a single but thick layer of the abductor superficialis. It is unlikely that the observed variation in the muscle composition within the *Acipenseridae* is related to phylogenetic relationships within this family, because sturgeons as closely related as *A. brevirostrum* and *A. fulvescens* (Bemis et al. 2002; Birstein et al. 2002; Krieger et al. 2008) display significantly different muscle configurations, while phylogenetically distant species such as *A. fulvescens* and *A. sturio* (Bemis et al. 2002; Birstein et al. 2002; Krieger et al. 2008) share similar muscle patterns and lack the arrector dorsalis.

TABLE 13.1

Pectoral and Pectoral Fin Muscles of Adults of Representative Actinopterygian Taxa

Probable Plesiomorphic Osteichthyan Condition	Cladistia: *Polypterus bichir* (Bichir)	Chondrostei: *Acipenser brevirostrum* (Shortnose Sturgeon)	Ginglymodi: *Lepisosteus osseus* (Longnose Gar)	Halecomorphi: *Amia calva* (Bowfin)	Teleostei—Basal: *Elops saurus* (Ladyfish)	Teleostei—Clupeocephalan: *Danio rerio* (Zebrafish)
Abductor (of pectoral fin, with superficial and deep muscles)	**Abductor** (of pectoral fin, with superficial and deep muscles) [Wilhelm et al. [2015] suggest that the abductor superficialis of *Polypterus* is different from that of other fishes, but our comparisons and literature review do not support this view]	**Abductor** (of pectoral fin, with superficial and deep muscles, in other species of the Acipenseridae family; only the superficial layer may be present in, e.g., *Acipenser fulvescens*)	**Abductor** (of pectoral fin, with superficial and deep muscles)	**Abductor** (of pectoral fin, with superficial and deep muscles)	**Abductor** (of pectoral fin, with superficial and deep muscles)	**Abductor** (of pectoral fin, with superficial and deep muscles) (abductores superficiales 1 and 2 *sensu* Diogo et al. [2001a,b] and Diogo [2004a,b])
Adductor (of pectoral fin, with superficial and deep muscles)	**Adductor** (of pectoral fin, with superficial and deep muscles)	**Adductor** (of pectoral fin, with superficial and deep muscles)	**Adductor** (of pectoral fin, with superficial and deep muscles)	**Adductor** (of pectoral fin, with superficial and deep muscles)	**Adductor** (of pectoral fin, with superficial and deep muscles)	**Adductor** (of pectoral fin, with superficial and deep muscles) (adductores superficiales 1 and 2 *sensu* Diogo et al. [2001a,b] and Diogo [2004a,b])
Pterygialis cranialis (preaxial muscle)	**Pterygialis cranialis** (preaxial muscle, or "dilator anterior," or "zonopropterygialis" *sensu* Wilhelm et al. [2015]) [probably corresponds to arrector ventralis of teleosts, as this muscle is also a preaxial muscle and was seemingly also derived from abductor musculature]	**Arrector ventralis** [this structure was not described as a separate muscle for *Psephurus* nor for *Polyodon* by Diogo and Abdala [2010]; our more recent dissections of *Acipenser* and of *Polyodon* indicate that it is present as a separate muscle in these two taxa]	**Arrector ventralis** [this structure was not described as a separate muscle for *Lepisosteus* by Diogo and Abdala [2010]; our more recent dissections of *Lepisosteus oculatus* indicate that it is present as a separate muscle in at least some members of this genus]	**Arrector ventralis** [this structure was not described as a separate muscle for *Amia* by Diogo and Abdala [2010]; our more recent dissections indicate that it is present as a separate muscle in this taxon]	**Arrector ventralis**	**Arrector ventralis** (arrector dorsalis *sensu* Diogo et al. [2001a,b] and Diogo [2004a,b]) [according to authors such as Jessen [1972], Winterbottom [1974], and Thorsen and Hale [2005], the arrector ventralis is ontogenetically derived from the abductor of the pectoral fin]
–	–	–	**Arrector 3** [this structure was not described as a separate muscle for *Lepisosteus* by Diogo and Abdala [2010]; our more recent dissections of *L. oculatus* indicate that it is present as a separate muscle in at least some members of this genus, probably not homologous to arrector 3 of clupeocephalans; see text]	–	–	**Arrector 3** (large external bundle of superficial abductor *sensu* Brousseau [1978a,b]; arrector ventralis *sensu* Diogo et al. [2001a,b] and Diogo [2004a,b]) [this small muscle has frequently been overlooked or considered as a bundle of the arrector ventralis or of the abductor of the pectoral fin; it is, thus, more likely that arrector 3 derives from the abductor, and not from the adductor, of the pectoral fin]

(*Continued*)

TABLE 13.1 (CONTINUED)
Pectoral and Pectoral Fin Muscles of Adults of Representative Actinopterygian Taxa

Probable Plesiomorphic Osteichthyan Condition	Cladistia: *Polypterus bichir* (Bichir)	Chondrostei: *Acipenser brevirostrum* (Shortnose Sturgeon)	Ginglymodi: *Lepisosteus osseus* (Longnose Gar)	Halecomorphi: *Amia calva* (Bowfin)	Teleostei—Basal: *Elops saurus* (Ladyfish)	Teleostei—Clupeocephalan: *Danio rerio* (Zebrafish)
–	–	**Arrector dorsalis** [this structure was not described as a separate muscle for *Psephurus*, nor for *Polyodon*, by Diogo and Abdala [2010]; our more recent dissections indicate that it is present as a separate muscle in some species of the *Acipenseridae* family [e.g., *A. brevirostrum*] and in *Polyodon*)	**Arrector dorsalis**	**Arrector dorsalis**	**Arrector dorsalis**	**Arrector dorsalis** (adductor profundus *sensu* Diogo et al. [2001a,b] and Diogo [2004a,b) [according to authors such as Jessen [1972], Winterbottom [1974], and Thorsen and Hale [2005], the arrector dorsalis is ontogenetically derived from the adductor of the pectoral fin]
Pterygialis caudalis (postaxial muscle) [pterygialis caudalis probably only became consistently present as separate muscle in the LCA of extant osteichthyans]	**Pterygialis caudalis** (postaxial muscle, or "dilator posterior," or "coracometapterygialis I and II" *sensu* Wilhelm et al. [2015]) [does not seem to have given rise/correspond to the arrector dorsalis of phylogenetically more derived actinopterygians, because although the arrector dorsalis does seem to also have derived from the adductor musculature, it is a preaxial muscle, going mainly to ray 1, and not a postaxial muscle as the pterygialis caudalis is]	–	–	–	–	–

Note: The nomenclature of the muscles shown in bold follows that of the text; in order to facilitate comparisons; in some cases, names often used by other authors to designate a certain muscle/bundle are given in front of that muscle/bundle, between round brackets. Data compiled from evidence provided by our own dissections and comparisons and by an overview of the literature (see text and Figures 13.1 through 13.13). LCA, last common ancestor.

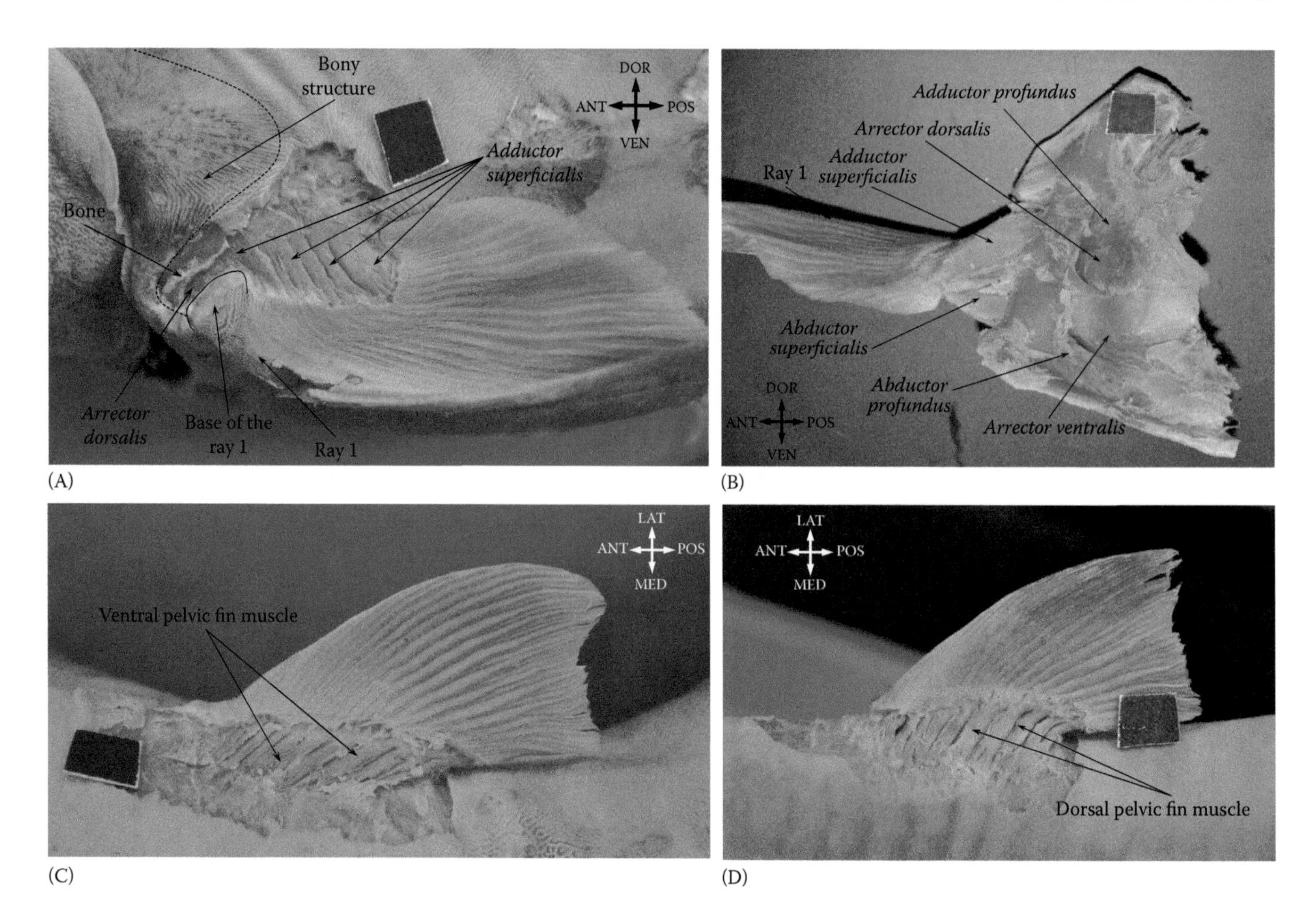

FIGURE 13.1 *Acipenser brevirostrum* (Chondrostei): lateral (A) and medial (B) views showing muscles of pectoral appendage; ventral (C) and dorsal (D) views showing muscles of pelvic appendage. Scale bar = 1 cm.

As their names indicate, in nontetrapod vertebrates the adductor and the abductor are mainly related to the adduction and with the abduction of the pectoral fin, respectively (e.g., Bischoff 1840; Owen 1841; Pollard 1892; Romer 1924; Millot and Anthony 1958; Greenwood and Thomson 1960; Jessen 1972; Winterbottom 1974; Brousseau 1978a,b; Lauder and Liem 1983; Adriaens et al. 1993; Kardong and Zalisko 1998; Diogo et al. 2001a,b; Kardong 2002; Westneat et al. 2004; Thorsen and Westneat 2005; Thorsen and Hale 2005). The arrector ventralis appears to be derived from the abductor mass as noted in the preceding text and usually originates lateral to the abductor and inserts on the first pectoral ray (e.g., Figure 13.5A). Another small muscle is arrector 3, which usually connects the pectoral girdle to the ventrolateral surface of the first pectoral ray (e.g., Figures 13.3, 13.7, and 13.8). This small muscle has been frequently neglected in the literature, as it was overlooked or considered to be a bundle of the arrector ventralis or of the abductor of the fin (e.g., Brousseau 1978a,b). However, as explained by Diogo (2007), arrector 3 is found in numerous otocephalans and apparently in at least some euteleosts, thus constituting a potential synapomorphy of the Clupeocephala (see Figure 12.1; Table 13.1). Interestingly, a small arrector 3 was also present in the *Lepisosteus* specimens we dissected recently (Table 13.1; N.S., pers. obs.). As an arrector 3 seems to be plesiomorphically absent in teleosts (Winterbottom 1974; Diogo and Abdala 2010), as it is also absent in *Amia* according to our most recent dissections (Figure 13.4), it is phylogenetically more parsimonious that it was independently acquired in the evolutionary history of the Ginglymodi and of the Clupeocephala (see Figure 12.1). As explained by Thorsen and Hale (2005: 149), the arrectors of the pectoral fin "initiate the movement of the fin at the leading edge" while the adductor and the abductor "power the upstroke and downstroke." Besides the abductor, the adductor, and the arrector muscles, some derived teleosts (e.g., certain neoteleosts) may exhibit other pectoral muscles, such as the coracoradialis, adductor radialis, interradialis pectoralis, and adductor medialis (e.g., Figure 13.13; see, e.g., Winterbottom [1974] for more details).

As mentioned in Chapters 12 and 19, the muscles of the pelvic appendage in teleosts (Figures 13.1 through 13.4 and 13.14) are very similar to those of the pectoral appendage (Tables 13.1 and 13.2). Therefore, we will not describe them here; readers can simply refer to Figures 13.1 through 13.4 and 13.14 and Table 13.2 for details. In fact, in adult teleosts such as zebrafishes, the six muscles inserting onto the pelvic fin have a clear topological correspondence, with six of the seven muscles inserting onto the pectoral fin, with the sole exception

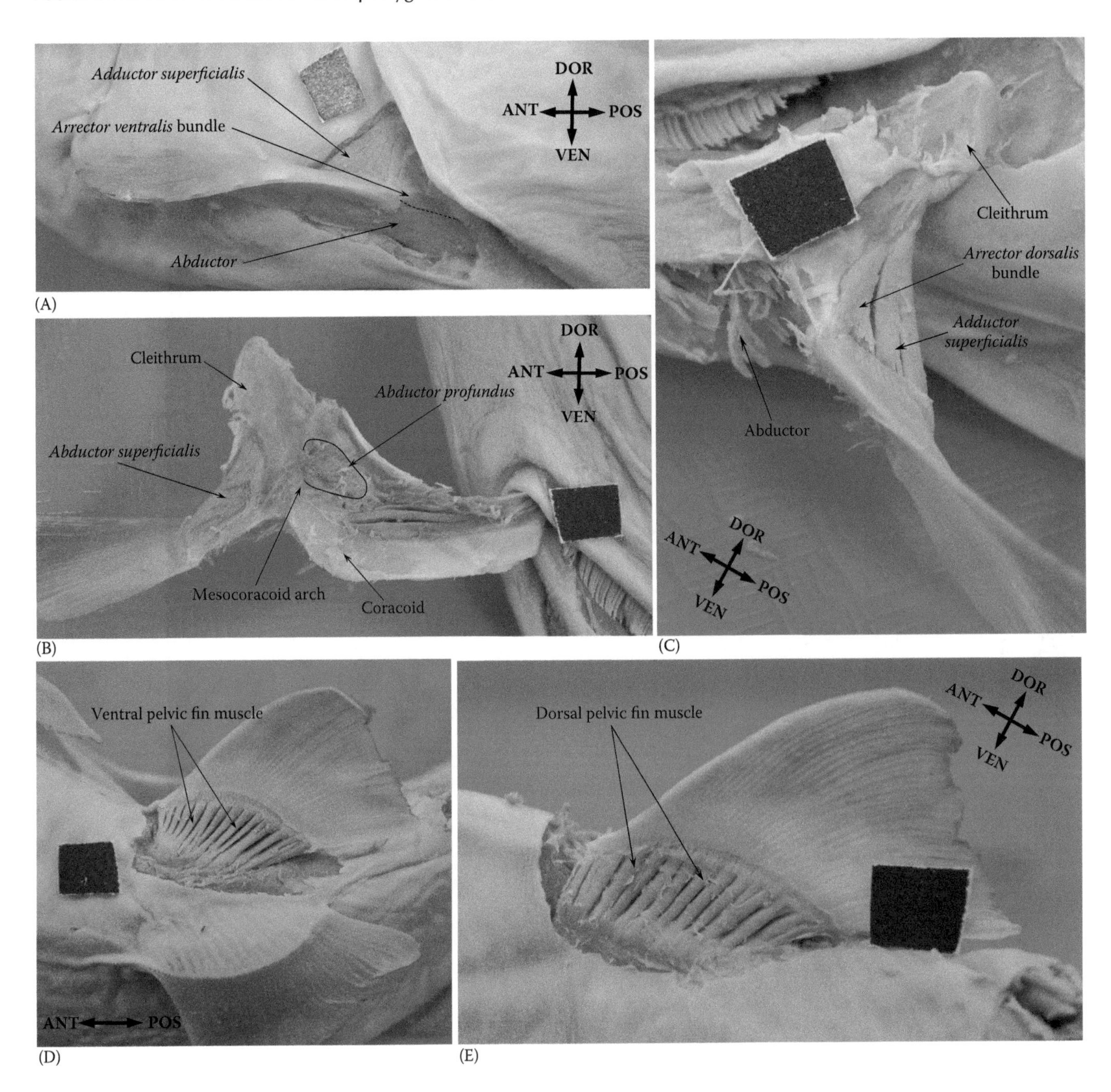

FIGURE 13.2 *Polyodon spathula* (Chondrostei): superficial (A) and deep (C) lateral, as well as medial (B), views showing muscles of pectoral appendage; ventral (D) and dorsal (E) views showing muscles of pelvic appendage. Scale bar = 1 cm.

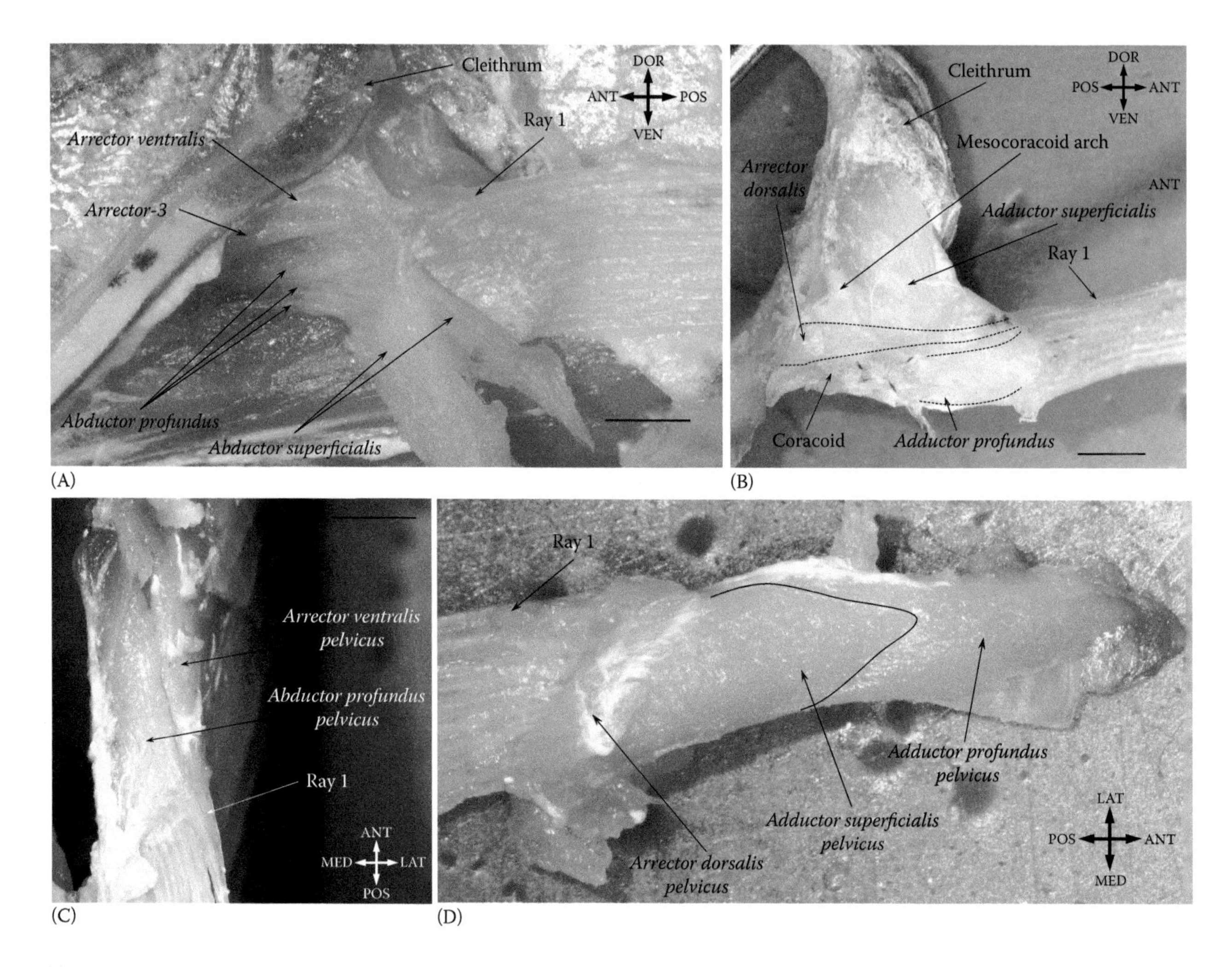

FIGURE 13.3 *Lepisosteus osseus* (Ginglymodi): lateral (A) and medial (B) views with the left pectoral fin and girdle reflected to show muscles of pectoral appendage; ventral (C) and dorsal (D) views with the left pelvic fin reflected to show muscles of pelvic appendage. Scale bar = 1 cm.

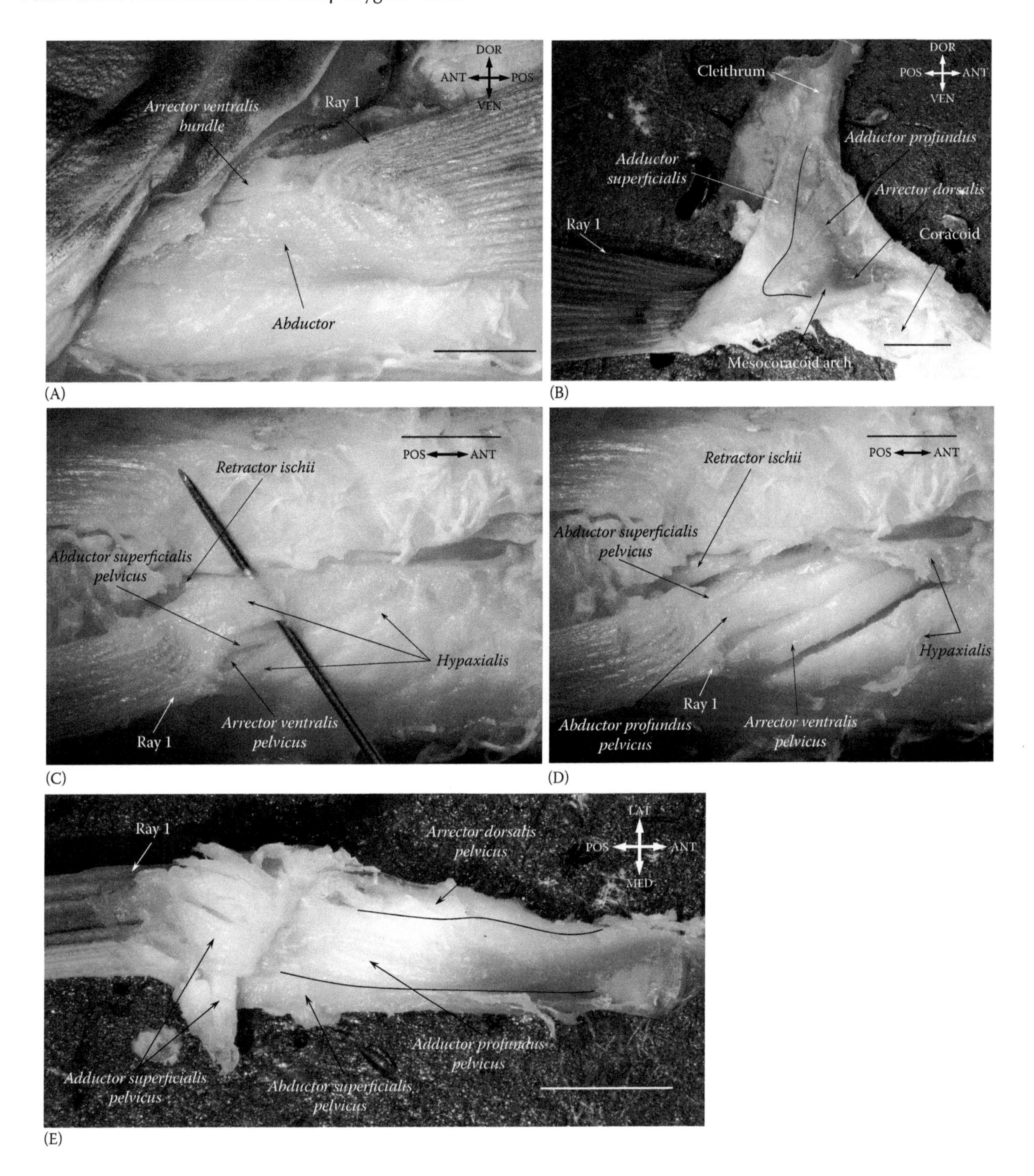

FIGURE 13.4 *Amia calva* (Halecomorphi): lateral (A) and medial (B) views with the left pectoral fin and girdle reflected to show muscles of pectoral appendage; ventral superficial (C) and deep (D), as well as dorsal (E), views with the left pelvic fin reflected to show muscles of pelvic appendage. Scale bar = 1 cm.

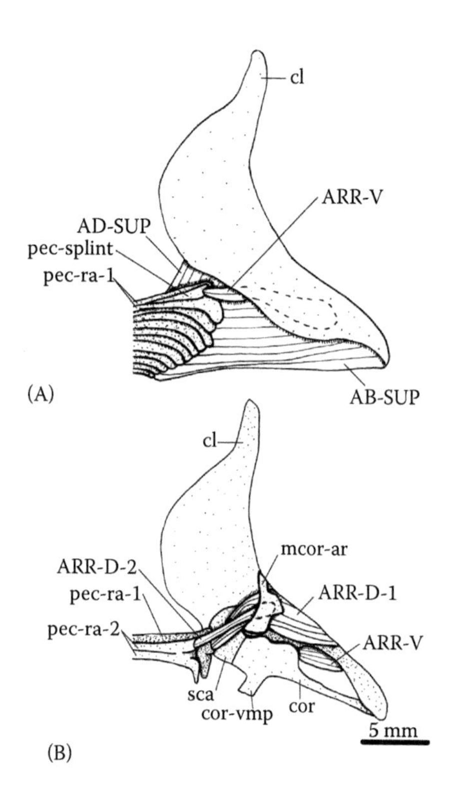

FIGURE 13.5 *Elops saurus* (Teleostei, Elopiformes): lateral (A) and mesial (B) view of the pectoral musculature; in the mesial view the adductor superficialis and abductor superficialis are not shown. AB-SUP, abductor superficialis; AD-SUP, adductor superficialis; ARR-D-1 and ARR-D-2, arrector dorsalis 1 and 2; ARR-V, arrector ventralis; cl, cleithrum; cor, coracoid; cor-vmp, ventromesial process of coracoid; mcor-ar, mesocoracoid arch; pec-ra-1, 2, pectoral rays 1 and 2; pec-slint, pectoral splint; sca, scapula.

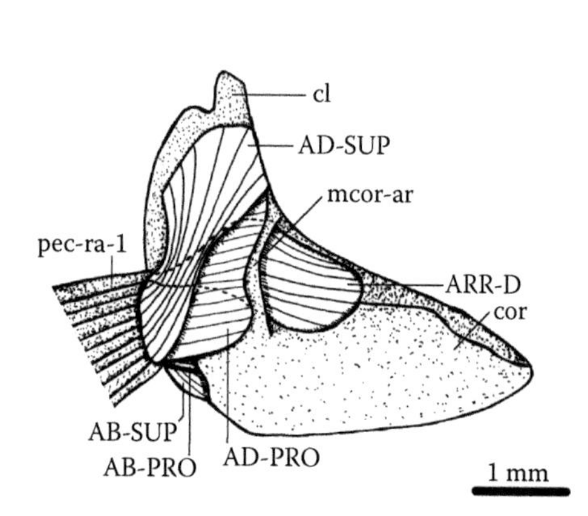

FIGURE 13.6 *Denticeps clupeoides* (Teleostei, Clupeiformes): mesial view of the pectoral girdle musculature; the lateral muscles abductor superficialis and abductor profundus are also shown. AB-PRO, abductor profundus; AB-SUP, abductor superficialis; AD-PRO, adductor profundus; AD-SUP, adductor superficialis; ARR-D, arrector dorsalis; cl, cleithrum; cor, coracoid; mcor-ar, mesocoracoid arch; pec-ra-1, pectoral ray 1.

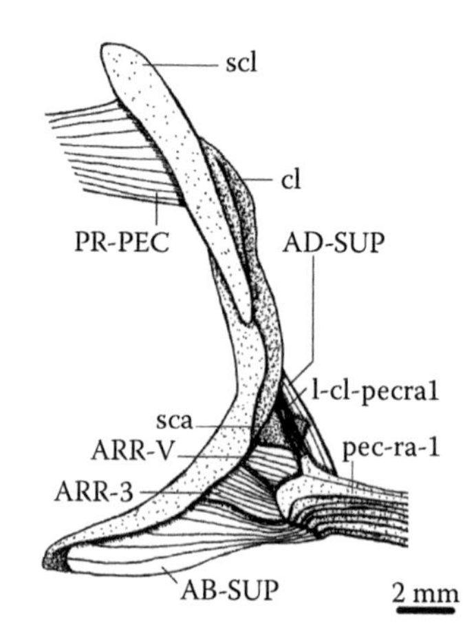

FIGURE 13.7 *Chanos chanos* (Teleostei, Gonorynchiformes): lateral view of the pectoral girdle musculature. AB-SUP, abductor superficialis; AD-SUP, adductor superficialis; ARR-3, arrector 3; ARR-V, arrector ventralis; cl, cleithrum; l-cl-pecral, ligament between cleithrum and pectoral ray 1; pec-ra-1, pectoral ray 1; PR-PEC, protractor pectoralis; sca, scapula; scl, supracleithrum.

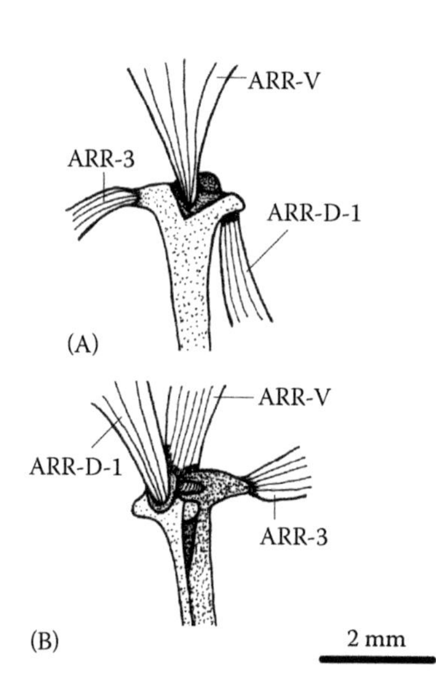

FIGURE 13.8 *Chanos chanos* (Teleostei, Gonorynchiformes): lateral (A) and mesial (B) views of the anterior portion of the first pectoral ray and the insertions of the section 1 of the arrector dorsalis, of arrector 3, and of the arrector ventralis. ARR-3, arrector 3; ARR-D-1, section 1 of arrector dorsalis; ARR-V, arrector ventralis.

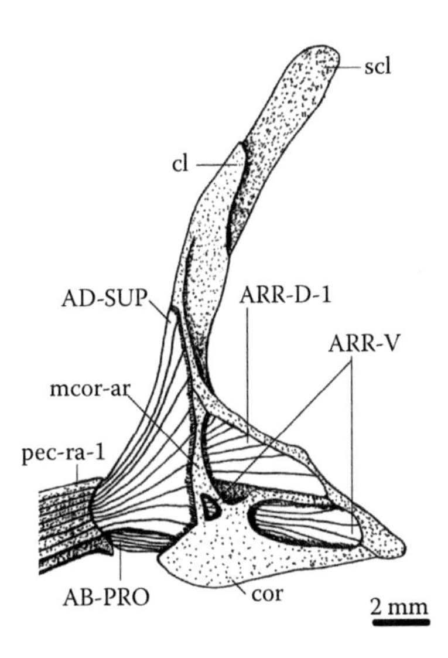

FIGURE 13.9 *Chanos chanos* (Teleostei, Gonorynchiformes): mesial view of the pectoral girdle musculature; the lateral muscle abductor profundus is also shown. AB-PRO, abductor profundus; AD-SUP, adductor superficialis; ARR-D-1, arrector dorsalis 1; ARR-V, arrector ventralis; cl, cleithrum; cor, coracoid; mcor-ar, mesocoracoid arch; pec-ra-1, pectoral ray 1; scl, supracleithrum.

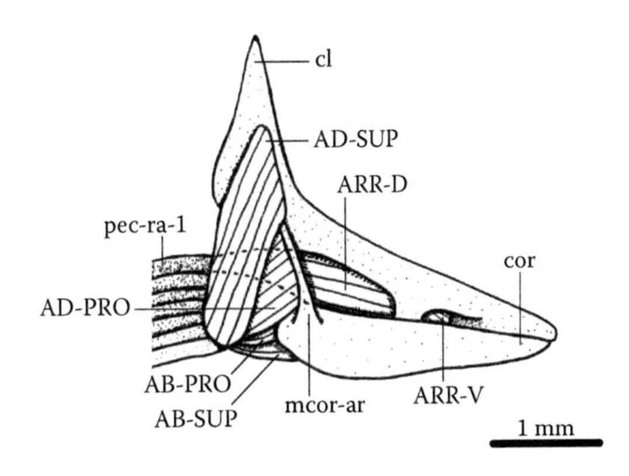

FIGURE 13.11 *Danio rerio* (Teleostei, Cypriniformes): mesial view of the pectoral musculature; despite being lateral structures, the abductor superficialis and abductor profundus are also shown. AB-PRO, abductor profundus; AB-SUP, abductor superficialis; AD-PRO, adductor profundus; AD-SUP, adductor superficialis; ARR-D, arrector dorsalis; ARR-V, arrector ventralis; cl, cleithrum; cor, coracoid; mcor-ar, mesocoracoid arch; pec-ra-1, pectoral ray 1.

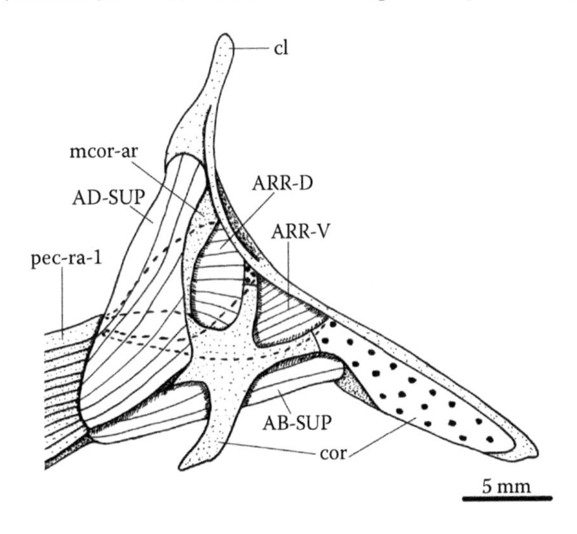

FIGURE 13.12 *Alepocephalus rostratus* (Teleostei, Alepocephaloidea): mesial view of the pectoral girdle musculature; the lateral muscle abductor superficialis is also shown. AB-SUP, abductor superficialis; AD-SUP, adductor superficialis; ARR-D, arrector dorsalis; ARR-V, arrector dorsalis; cl, cleithrum; cor, coracoid; mcor-ar, mesocoracoid arch; pec-ra-1, pectoral ray 1.

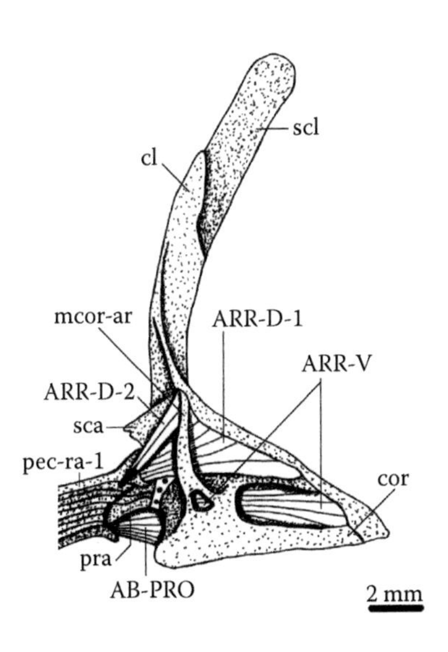

FIGURE 13.10 *Chanos chanos* (Teleostei, Gonorynchiformes): mesial view of the pectoral girdle musculature; the lateral muscle abductor profundus is also shown, and the adductor superficialis was removed. AB-PRO, abductor superficialis, ARR-D-1 and ARR-D-2, arrector dorsalis 1 and 2; ARR-V, arrector ventralis; cl, cleithrum; cor, coracoid; mcor-ar, mesocoracoid arch; pec-ra-1, pectoral ray 1; pra, proximal radials; sca, scapula; scl, supracleithrum.

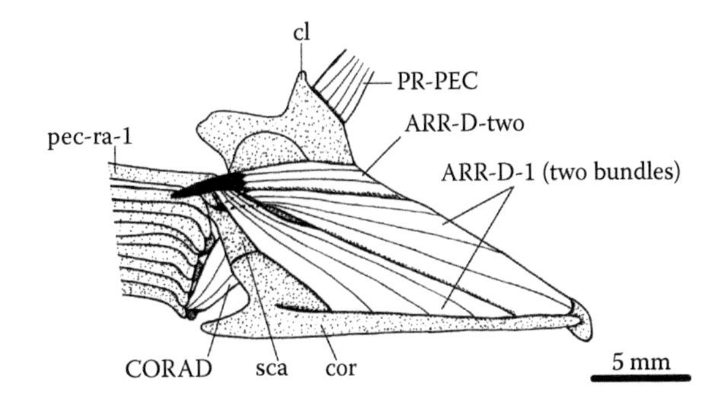

FIGURE 13.13 *Aulopus filamentosus* (Teleostei, Aulopiformes): mesial view of the pectoral girdle musculature. ARR-D-1 and ARR-D-2, arrector dorsalis 1 and 2; cl, cleithrum; cor, coracoid; CORAD, coracoradialis; PR-PEC, protractor pectoralis; pec-ra-1, pectoral ray 1; sca, scapula.

TABLE 13.2
Pelvic and Pelvic Fin Muscles of Adults of Representative Actinopterygian Taxa

Probable Plesiomorphic Osteichthyan Condition	Cladistia: *Polypterus bichir* (Bichir)	Chondrostei: *Acipenser brevirostrum* (Shortnose Sturgeon)	Ginglymodi: *Lepisosteus osseus* (Longnose Gar)	Halecomorphi: *Amia calva* (Bowfin)	Teleostei—Basal: *Elops saurus* (Ladyfish)	Teleostei—Clupeocephalan: *Danio rerio* (Zebrafish)
Abductor (of pelvic fin, with superficial and deep muscles)	**Abductor** (of pelvic fin, with superficial and deep muscles)	**Abductor** (ventral pelvic fin muscle) [in *Acipenser*, as well as in *Polyodon*, the abductor and adductor masses are mainly formed by thin, serial bundles (Figures 13.1 and 13.2) that resemble the serial muscle units seen in the anal and dorsal fins of most fishes: see Chapter 14]	**Abductor** (of pelvic fin, with superficial and deep muscles)	**Abductor** (of pelvic fin, with superficial and deep muscles)	**Abductor** (of pelvic fin, with superficial and deep muscles)	**Abductor** (of pelvic fin, with superficial and deep muscles)
Adductor (of pelvic fin, with superficial and deep muscles)	**Adductor** (of pelvic fin, with superficial and deep muscles)	**Adductor** (dorsal, pelvic fin muscle) [see cell above]	**Adductor** (of pelvic fin, with superficial and deep muscles)	**Adductor** (of pelvic fin, with superficial and deep muscles)	**Adductor** (of pelvic fin, with superficial and deep muscles)	**Adductor** (of pelvic fin, with superficial and deep muscles)
–	–	**Inner pelvic muscle** [in addition to the peculiar, "serial" configuration of the superficial abductor and adductor masses (see cells above), in the chondrostean fishes dissected there is also a peculiar inner, also "serial," pelvic muscle]	–	–	–	–
Pterygialis cranialis (preaxial muscle)	**Pterygialis cranialis** (preaxial muscle, or "dilator anterior")	–	**Arrector ventralis**	**Arrector ventralis**	**Arrector ventralis**	**Arrector ventralis**
–	–	–	**Arrector dorsalis**	**Arrector dorsalis**	**Arrector dorsalis**	**Arrector dorsalis**
Pterygialis caudalis (postaxial muscle) [pterygialis caudalis probably only became consistently present as separate muscle in LCA of extant osteichthyans]	**Pterygialis caudalis** (postaxial muscle, or "dilator anterior" or "coracometapterygialis" *sensu* Wilhelm et al. [2015]) [does not seem to have given rise/correspond to the arrector dorsalis of phylogenetically more derived actinopterygians, because although the arrector dorsalis does seem to also have derived from the adductor musculature, it is a preaxial muscle, going to ray 1, and not a postaxial muscle as the pterygialis caudalis is]	–	–	–	–	–

Note: The nomenclature of the muscles shown in bold follows that of the text; in order to facilitate comparisons; in some cases, names often used by other authors to designate a certain muscle/bundle are given in front of that muscle/bundle, between round brackets. Data compiled from evidence provided by our own dissections and comparisons and by an overview of the literature (see text and Figures 13.1 through 13.4 and 13.14).

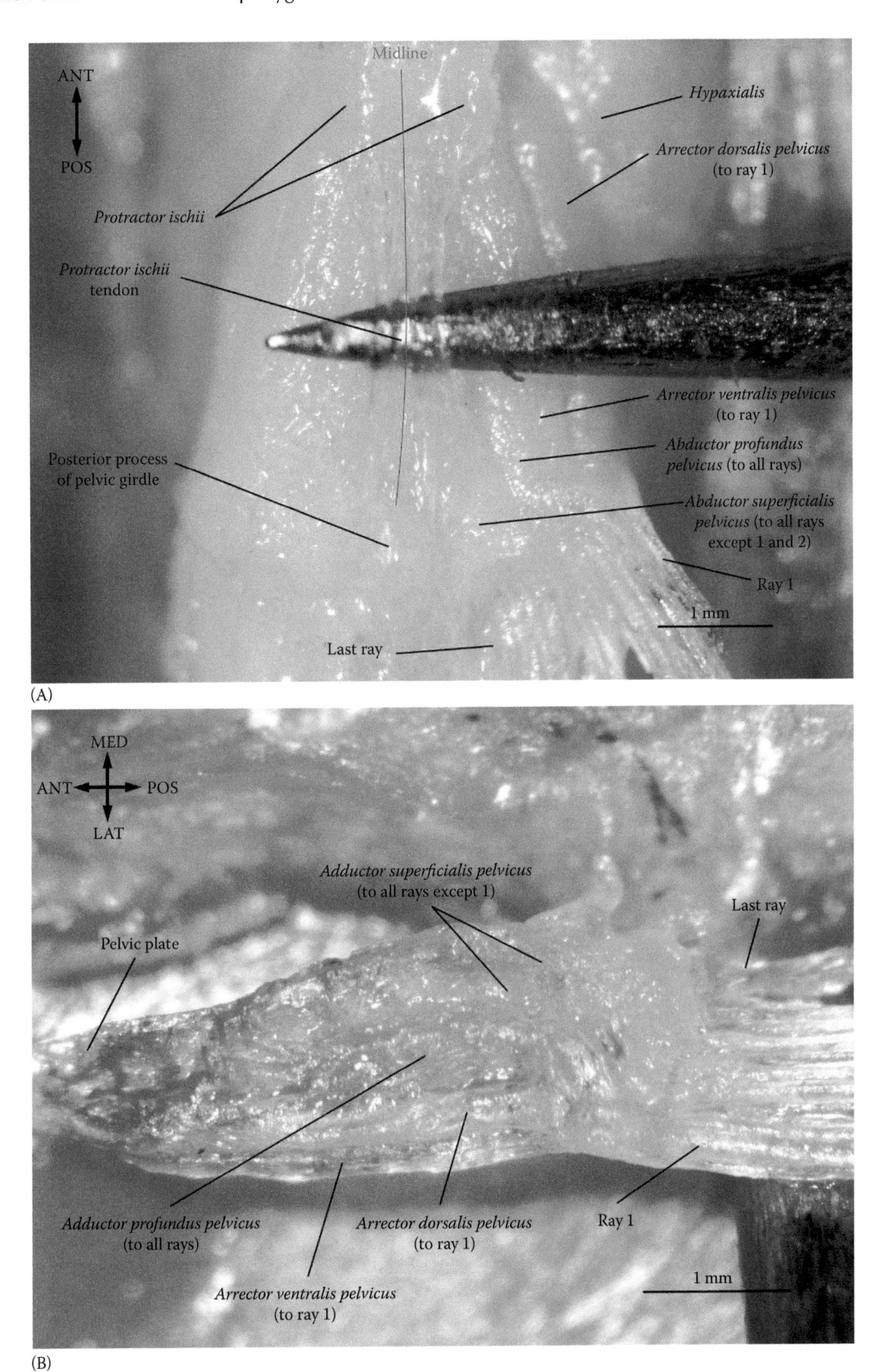

FIGURE 13.14 *Danio rerio* (Teleostei, Cypriniformes): ventral (A) and dorsal (B) views showing muscles of pelvic appendage.

of arrector 3 (compare Figures 13.11 and 13.14). However, as will be explained in Chapters 14 and 15, this pectoral–pelvic similarity is mainly superficial, in the sense that it is found only in adults, and, moreover, it does not apply to the muscles that insert onto the girdles (i.e., protractor pectoralis vs. the protractor ischii). Lastly, our more recent dissections of chondrostean fishes have indicated that the main abductor and adductor masses of the paired fins of these fishes strongly resemble the serial units of muscles seen in the anal and dorsal median fins of most fishes; i.e., they do not form continuous muscle sheets as seen in the pectoral and pelvic abductor/abductor masses of other fishes (Figures 13.1 and 13.2; Tables 13.1 and 13.2; N.S., pers. obs.). We will further discuss the similarities between the pelvic and pectoral fin muscles, as well as between these muscles and those of the median fins, in Chapters 14 and 15.

GENERAL REMARKS

From a similar plesiomorphic overall configuration (see Chapter 12), the evolution of the musculature of the paired fins has been rather different in the actinopterygian and sarcopterygian clades (see Figures 12.1 and 12.2, and more details in Chapter 16). For instance, within Actinopterygii and Sarcopterygii, arrector muscles were acquired at different evolutionary stages: an arrector dorsalis is present only in extant Actinopteri, and an arrector 3 is present only in extant clupeocephalans and in Ginglymodi. These two arrector muscles, the adductor superficialis and profundus and the abductor superficialis and profundus, as well as the arrector ventralis, which seems to directly correspond to the pterygialis cranialis of other fishes, are present in one of the model organisms that is most studied among actinopterygians as well as among osteichthyan fishes in general, the zebrafish *Danio rerio*. However, as a result of the different evolutionary routes followed within Actinopterygii and Sarcopterygii, none of the individual muscles found for example in derived actinopterygians such as teleosts are found in derived sarcopterygians such as tetrapods (see Chapter 16). Thus, concerning the pectoral girdle and pectoral fin/forelimb musculature, a model actinopterygian organism such as the teleostean zebrafish should be taken to represent osteichthyan (bony) fishes as a whole only with great caution—such caution is unfortunately lacking in some recent developmental and molecular studies. Also, much caution should be taken in comparing the results obtained in developmental and molecular studies of model organisms such as the zebrafish with those of model tetrapods such as salamanders, chickens, or rats. However, the musculature of the paired fins of zebrafish does seem to be a good model for that of actinopterygians as a whole (see Chapter 14). Accurate knowledge of the osteichthyan pectoral musculature is thus important not only to increase our general understanding of the comparative anatomy, functional morphology, and evolution of this group, but also to provide a solid basis for the comparisons and extrapolations made in developmental and molecular studies.

14 Muscles of Median Fins and Origin of Pectoral vs. Pelvic and Paired vs. Median Fins

Diogo and Abdala (2010) did not describe the muscles of the median fins of fishes, largely because there were almost no available, detailed comparative data about the configuration and attachments of those muscles in a wide range of fish taxa. Therefore, since then, we have paid special attention to these muscles and (a) dissected them in several taxa representing all the major extant clades of fishes and (b) undertook an extensive literature review of the scarce data published by other authors about these muscles. As a result, we are publishing here, in Chapter 14, the first detailed overview of these muscles and their homologies and evolution. This information is briefly summarized in the cladogram shown in Figure 14.1, as well as in Tables 14.1 through 14.7 and Figures 14.2 through 14.9. Moreover, our analyses and comparisons allow us to provide, at the end of this chapter, a brief discussion on current crucial topics in developmental and evolutionary biology, such as the origin of pelvic vs. pectoral appendages and of paired vs. median appendages.

DORSAL AND ANAL FINS

The muscles of these fins are described together because they share a similar pattern. The appendicular musculature of the first dorsal fin located on the back of sharks is usually described as a "lateral" dorsal fin muscle (Maia and Wilga 2013a,b) or "radial" muscle (Lingham-Soliar 2005)—present on the right and left sides of the fin (Figure 14.2A, Table 14.1). Although there is no clear separation between the superficial and deep layers, the deeper posterior muscle fibers of the first dorsal fins of *Mustelus laevis* we dissected are more inclined than the superficial fibers in the same region (Figure 14.2A, empty arrowhead). Another difference between these deep and superficial layers observed in *M. laevis* is the clear segmentation of the latter. In contrast to the *Squalus acanthias* specimens we dissected, which have a uniform dorsal fin muscle bilaterally, the superficial layer of this muscle in *M. laevis* is clearly segmented into muscular units that are not separated by any kind of membrane or ligament, in contrast to chondrosteans (see following text). The dorsal fin muscle is attached to the ceratotrichia of the fin and attaches to the dorsal side of the fin ligament that separates the fin from the epaxialis. The protractor dorsalis and retractor dorsalis are usually not listed as part of the dorsal fin appendicular musculature by other authors (Lingham-Soliar 2005; Maia and Wilga 2013a,b). However, the protractor dorsalis is directly attached to the spike of the dorsal fin and partially to the anterior portion of the dorsal fin ligament. Therefore, we believe that it should also be considered an appendicular muscle of the dorsal fin, as it is often the case in other fishes such as teleosts (see following text). In contrast to the protractor dorsalis, the retractor dorsalis has an indirect connection with the first dorsal fin *via* the dorsal fin ligament. The retractor dorsalis inserts onto the ventral side of the ligament and is also, in a way, a protractor dorsalis of the second dorsal fin, which has almost the same anatomy as the first dorsal fin except that it lacks a retractor dorsalis (Figure 14.2B). Alternatively, the musculature of both dorsal fins of these sharks might be identical and consist of the main dorsal fin muscle plus a protractor, merely with different anterior insertion sites. This difference might be explained by the topological position of the fins and immediate proximity of the first dorsal fin to the head. Further developmental studies are required to attribute a precise identity and origin to the muscle connecting the two dorsal fins and to answer the question of whether it belongs (a) to the first dorsal fin and is thus a retractor dorsalis of this fin, (b) to the second dorsal fin and is thus a protractor dorsalis of this fin, or (c) to both dorsal fins.

Polypterus is, phylogenetically, the most basal extant actinopterygian (see Figure 1.1). The anatomy of the dorsal fin of the *Polypterus senegalus* specimens we dissected is strikingly different from those of Chondrichthyes. Instead of ceratotrichia, *Polypterus* has bony fin rays like most actinopterygians have. There is no clear external distinction between the dorsal and caudal fins. Dorsal rays go along the back of the fish and join the caudal fin. Such morphology can thus be related to the absence of the retractor dorsalis, a muscle present in most other actinopterygians (see, e.g., Figure 14.8 and Siomava and Diogo in press). Moreover, unlike in sharks, there is no protractor dorsalis muscle in *Polypterus* (Table 14.2). Instead, erectors and depressors of the most anterior ray stretch anteriorly along the back (Figure 14.3A), in a configuration that somewhat resembles that of the protractor dorsalis in other fishes. Similar erectores dorsales and depressores dorsales extend to each ray of the dorsal fin in *Polypterus*; i.e., each half ray (left and right) has its serial unit—an erector dorsalis, a depressor dorsalis, and sometimes also an inclinator dorsalis—but the attachment sites of these muscle units gradually change posteriorly. Erectors of anterior rays insert onto the posterior base of the previous ray. Depressors do not extend that long; they run parallel with the erectors and then attach onto the erectors, as well as onto the median membrane. Posterior rays are less connected to each other *via* muscles. Erectors and depressors of those rays extend deep

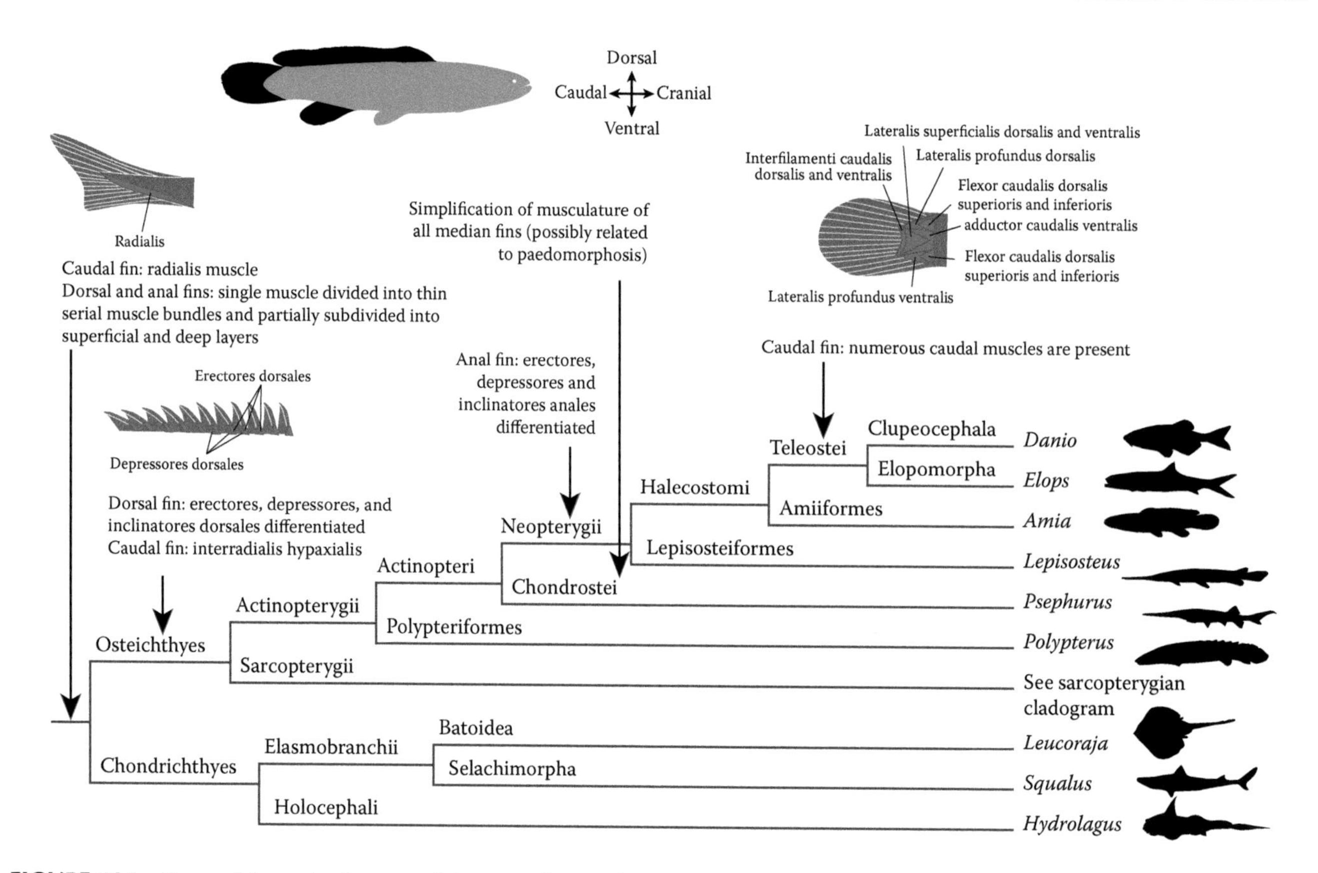

FIGURE 14.1 Some of the major features of the musculature of median fins within nonsarcopterygian fishes, according to our own data and review of the literature.

TABLE 14.1
List of Muscles of Median Fins and Their Attachment Sites of the Chondrichthyan *Mustelus laevis*

Muscle Name	Distal Attachment	Proximal Attachment
	First Dorsal Fin	
Dorsal fin muscle	Ceratotrichia	Dorsal side of the first dorsal fin ligament, radials
Protractor dorsalis	Spike of the first dorsal fin, anterodorsal part of the first dorsal fin ligament	Posterior skull
Retractor dorsalis (protractor dorsalis of the second dorsal fin)	Posteroventral side of the first dorsal fin ligament	Spine of the second dorsal fin, anterodorsal part of the second dorsal fin ligament
	Second Dorsal Fin	
Dorsal fin muscle	Ceratotrichia	Dorsal side of the first dorsal fin ligament, radials
Protractor dorsalis (retractor dorsalis of the first dorsal fin)	Posteroventral side of the first dorsal fin ligament	Spine of the second dorsal fin, anterodorsal part of the second dorsal fin ligament
	Caudal Fin	
Radialis	Vertebrae, ventral ligament of hypaxialis	Lateral base of ceratotrichia

Note: No anal fins were present in the specimens of the species that we dissected.

and mainly attach to neural spines. Additionally, small parts of these muscles fuse with the fibers of the inclinators, which are present only in the region of the posterior rays and are themselves fused with the epaxial trunk musculature (Figure 14.3C). All rays above the notochord have the same serial unit configuration and are therefore considered here to be dorsal rays, not caudal rays as these might appear at first sight. All true caudal rays are thus below the notochord and do not have erectors and depressors. Unlike in the zebrafish and other Actinopterygii (see following text), the dorsal and anal fins are very different morphologically in *Polypterus*. The musculature of the anal fin is subdivided into deep (analis profundus) and superficial (analis superficialis) layers, which form a single muscle sheet that is not differentiated into clear half-ray

TABLE 14.2
List of Muscles of Median Fins and Their Attachment Sites of the Cladistian *Polypterus senegalus*

Muscle Name	Distal Attachment	Proximal Attachment
	Dorsal Fin	
Erector dorsalis of ray 1	Epaxialis, median membrane	Anteromedial base of ray 1
Erectores dorsales of anterior rays	Posterior base of the previous ray	Anteromedial base of rays
Erectores dorsales of posterior rays	Neural spines (posterior rays), epaxialis	Anteromedial base of rays
Depressores dorsales of anterior rays	Erector dorsalis, medial ligament	Posterior base of rays
Depressores dorsales of posterior rays	Neural spines (posterior rays), epaxialis	Posterior base of rays
Inclinator fibers	Lateral base of rays	Epaxialis
	Anal Fin	
Analis superficialis	Lateral base of each ray and spike(s)	Hypaxialis
Analis profundus	Anterolateral base of each ray and spike(s)	Hemal spines
Protractor analis	Anterior base of ray 1/spike(s)	Hypaxialis
	Caudal Fin	
Lateralis superficialis caudalis	Posterior dorsal rays fused with inclinator fibers, caudal rays	Myomeres, caudal vertebrae
Interradialis hypaxialis (flexor ventralis *sensu* Lauder 1989)	Between neighboring caudal rays	

TABLE 14.3
List of Muscles of Median Fins and Their Attachment Sites of the Chondrostean *Acipenser brevirostrum*

Muscle Name	Distal Attachment	Proximal Attachment
	Dorsal Fin	
Dorsal fin muscle		
1. Superficial layer	Epaxialis	Ceratotrichia
2. Deep layer	Dorsal side of the fin ligament, radials	Ceratotrichia (*via* tendons)
	Anal Fin	
Anal fin muscle		
3. Superficial layer	Epaxialis	Ceratotrichia
4. Deep layer	Dorsal side of the fin ligament, radials	Ceratotrichia (*via* tendons)
	Caudal Fin	
Radialis	Ventral ligament of hypaxialis, caudal cartilage	Lateral base of ceratotrichia

TABLE 14.4
List of Muscles of Median Fins and Their Attachment Sites, in the Chondrostean *Polyodon spathula*

Muscle Name	Distal Attachment	Proximal Attachment
	Dorsal Fin	
Dorsal fin muscle		
Superficial layer	Epaxialis	Ceratotrichia
Deep layer	Dorsal side of the fin ligament, radials	Ceratotrichia (*via* tendons)
	Anal Fin	
Anal fin muscle		
Superficial layer	Epaxialis	Ceratotrichia
Deep layer	Dorsal side of the fin ligament, radials	Ceratotrichia (*via* tendons)
	Caudal Fin	
Radialis	Ventral ligament of hypaxialis, caudal cartilage	Lateral base of ceratotrichia

muscle units. Anterior rays of the anal fin gradually decrease in size, making it difficult to distinguish anterior spikes and the true first anal ray. The protractor analis attaches to the first anterior rays/spikes and is fused with the hypaxialis anteriorly. No retractor analis has been observed in the dissected specimens; it is possibly missing due to the very short distance between the anal fin and the ventral part of the caudal fin.

The composition of the dorsal and anal fin musculature in the *Acipenser brevirostrum* (Figure 14.4A and B, Table 14.3) and *Polyodon spathula* (Figure 14.5A and B, Table 14.4) specimens we dissected seems to be paedomorphic in the sense that it is composed of a single dorsal/anal fin muscle on each side of each fin. Each muscle, however, shows clear signs of segmentation: it is composed of serial, thin muscle bundles. These bundles are very similar to the serial bundles that were described for the main adductor and abductor masses of the pectoral and pelvic fins of these chondrosteans in Chapter 13 (see Figures 13.1 and 13.2). Moreover, as in the dorsal fin muscle in sharks, the deep and superficial muscle fibers of each muscle of each side of the dorsal and anal fins are differentiated in these chondrostean fishes (Figure 14.4A and B). The most superficial fibers mainly attach to trunk muscles, while the deep fibers bend at almost 90° when reaching the trunk muscles and go deep to the dorsal side of the fin ligament and radials. Another criterion for distinguishing deep muscle fibers of the anal and dorsal fins is their attachment to ceratotrichia *via* long tendons, which are absent in the superficial layer. In both chondrostean taxa analyzed by us, the dorsal and anal fins lack protractors and retractors.

The musculature of the dorsal and anal fins of *Lepisosteus oculatus* (Figure 14.6A and B, Table 14.5) is rather similar to that of teleosts such as the zebrafish *Danio rerio*, which will be described in the following text (Figure 14.8A and B). The muscles of both fins are segmented into muscle units that

TABLE 14.5
List of Muscles of the Median Fins and Their Attachment Sites, in *Ginglymodian Lepisosteus oculatus*

Muscle Name	Distal Attachment	Proximal Attachment
	Dorsal Fin	
Inclinatores dorsales	Lateral base of each ray, retractor dorsalis (posterior rays)	Epaxialis
Erectores dorsales	Anterior base of each ray	Proximal pterygiophores
Depressores dorsales	Posterolateral base of each ray	Proximal pterygiophores
Protractor dorsalis	Anterior base of ray 1	Posterior skull
Retractor dorsalis	Epaxialis	Anterior region of caudal fin
	Anal Fin	
Inclinatores anales	Lateral base of each ray (no spike)	Hypaxialis
Erectores anales	Anterolateral base of each ray and spike	Hemal spines
Depressores anales	Posterolateral base of each ray	Hemal spines
Protractor analis	Anterior base of ray 1	Hypaxialis
Retractor analis	Posterior base of the last ray	Hypaxial musculature
	Caudal Fin	
Lateralis superficialis caudalis	All caudal rays	Myomeres
Hypochordal longitudinalis	Dorsal rays 1–4	Hemal spines along notochord
Interradialis hypaxialis (flexor ventralis *sensu* Lauder [1989])	Between neighboring caudal rays	

TABLE 14.6
List of Muscles of Median Fins and Their Attachment Sites in the Halecomorph *Amia calva*

Muscle Name	Distal Attachment	Proximal Attachment
	Dorsal Fin	
Inclinatores dorsales	Anterolateral and posterolateral base of rays	Epaxialis
Erectores dorsales	Anterolateral base of rays and spike	Dorsal side of the fin ligament
Depressores dorsales	Posterior base of each ray and spike	Dorsal side of the fin ligament
Protractor dorsalis	Anterior base of the spike	Posterior skull
	Anal Fin	
Inclinatores anales	Lateral base of rays	Hypaxialis
Erectores anales	Anterolateral base of each ray	Proximal pterygiophores
Depressores anales	Posterior base of each ray	Proximal pterygiophores
Protractor analis	Anterior base of ray 1	Hypaxialis
	Caudal Fin	
Lateralis superficialis caudalis	All caudal rays	Myomeres
Hypochordal longitudinalis	Last epaxial ray and first hypaxial ray	Hemal spines along notochord
Interradialis hypaxialis	Between neighboring hypaxial caudal rays	
Interradialis dorsalis (Lauder [1989])	Between neighboring hypaxial caudal rays 1–5	

support fin rays. The half ray of each side has three types of muscles: erector, depressor, and inclinator. Inclinators of the dorsal and anal fins strongly attach to the epaxialis and attach, less strongly, to the skin. In addition to ray muscles, both fins have protractors and retractors. Like in teleosts such as *Danio*, the protractor analis and protractor dorsalis attach to the first ray of the anal and dorsal fins, respectively. The retractor analis starts from the posterior base of the last anal ray (Figure 14.6B), while the retractor dorsalis extends under the dorsal rays and fuses with the epaxialis at the middle of the dorsal fin, approximately between rays 4 and 5.

As is the case in *Polypterus*, the dorsal fin of *Amia calva* extends all the way to the caudal fin, but the most posterior dorsal rays of *Amia* do not belong to the caudal fin like they do in *Polypterus*. As in other Neopterygii dissected for this study (see Figures 1.1 and 14.2), the dorsal fin is segmented in muscular units (Figure 14.7A, Table 14.6). Each unit consists of an erector, a depressor, and the inclinator dorsalis, which, however, belongs to different rays, as will be explained in the following text (Figure 14.7B). Unlike in other fishes, inclinatores dorsales do not attach to the lateral base of dorsal rays but instead to their anterolateral and posterolateral surfaces. Two heads of each inclinator dorsalis fuse and envelop the erector dorsalis from the lateral side. Posteriorly, these erector and inclinator dorsales muscles are separated from the depressor dorsalis of the same rays by a ligament. This ligament comes together with the dorsal fin ligament, which seems to topologically correspond to the dorsal fin ligament in sharks and Chondrostei. The depressor dorsalis is attached to the posterior base of the ray and extends deep together with the wrapped erector dorsalis of the next ray, except in the last ray, where the depressor dorsalis is separated from the erector dorsalis by the ligament. Thus, muscle units of the dorsal rays consist of the respective erector and inclinator dorsales and the depressor dorsalis of the previous ray. The muscle unit of the first dorsal ray includes the depressor dorsalis of the dorsal spike, and the muscle unit of the spike consists of the erector and inclinator only. The protractor dorsalis inserts onto the spike, and the retractor dorsalis is absent. The muscles of the anal fin in *Amia* are very similar to those of its dorsal fin, with only one minor difference: the protractor analis inserts onto the first anal ray.

TABLE 14.7
List of Muscles of Median Fins and Their Attachment Sites in the Teleost *Danio rerio*

Muscle Name	Distal Attachment	Proximal Attachment
	Dorsal Fin	
Inclinatores dorsales	Lateral base of each ray	Skin and epaxial musculature
Erectores dorsales	Anterolateral base of each ray (lepidotrichia *sensu* Bird and Mabee [2003])	Proximal pterygiophores (radials *sensu* Bird and Mabee [2003])
Depressores dorsales	Posterolateral base of each ray	Proximal pterygiophores
Protractor dorsalis (supracarinalis anterior *sensu* Winterbottom [1974])	Anterior base of ray 1	Posterior skull
Retractor dorsalis (supracarinalis posterior *sensu* Winterbottom [1974])	Posterior base of last ray	Anterior region of caudal fin
	Anal Fin	
Inclinatores anales	Lateral base of each ray (no spike)	Hypaxialis
Erectores anales	Anterolateral base of each ray and spike	Hemal spines
Depressores anales	Posterolateral base of each ray (not to spike)	Hemal spines
Protractor analis (infracarinalis medius *sensu* Winterbottom [1974])	Anterior base of first ray	Posterior pelvis
Retractor analis (infracarinalis posterior *sensu* Winterbottom [1974])	Posterior base of last ray	Hypaxial musculature
	Caudal Fin	
Lateralis profundus dorsalis (dorsal slip of lateralis superficialis *sensu* Greene and Greene [1914])	Rays 6–8	Myomeres, proximal caudal fin bones
Lateralis superficialis dorsalis	Ray 8	Midline
Lateralis superficialis ventralis	Ray 22	Myomeres, proximal caudal fin bones
Lateralis profundus ventralis (ventral slip of lateralis superficialis *sensu* Greene and Greene [1914])	Rays 21–24	Midline, proximal caudal fin bones
Interradialis caudalis	Ray 5	Rays 6 and 7
Interfilamenti caudalis dorsalis (interradialis caudalis *sensu* Winterbottom [1974])	Rays 7–12	Base of the dorsomedial fin rays
Interfilamenti caudalis ventralis (interradialis caudalis *sensu* Winterbottom [1974])	Rays 16–22	Base of the ventromedial fin rays and dorsomedial ray 9
Flexor caudalis dorsalis superioris (flexor dorsalis *sensu* Winterbottom [1974])	Rays 6 and 7	Vertebral column, proximal caudal fin bones
Flexor caudalis dorsalis inferioris (flexor dorsalis *sensu* Winterbottom [1974])	Rays 7 and 8	Vertebral column, proximal caudal fin bones
Adductor caudalis ventralis (hypochordal longitudinalis *sensu* Winterbottom [1974])	Rays 8–10	Vertebral column, proximal caudal fin bones
Flexor caudalis ventralis superior (flexor ventralis *sensu* Winterbottom [1974])	Rays 15–23	Vertebral column, proximal caudal fin bones
Flexor caudalis ventralis inferior (flexor ventralis *sensu* Winterbottom [1974])	Ray 24	Vertebral column, proximal caudal fin bones

As in *Amia* and *Lepisosteus*, the dorsal and anal fins of *D. rerio* consist of a continuous series of units decreasing in size posteriorly (Figure 14.8A and B). Each muscular unit is composed of three muscles (inclinator, erector, and depressor) that attach to the base of a single half ray, on each side of the body (Table 14.7). Both the dorsal and anal fins have protractors and retractors. The protractor dorsalis of the dorsal fin originates mainly from the posterior region of the skull and inserts onto the anterior base of the first dorsal ray. The retractor dorsalis extends from the anterior region of the dorsal fin to the posterior base of the last dorsal ray (note: the number of dorsal rays is variable among zebrafishes). The protractor analis is continuous with the retractor ischii anteriorly and inserts onto the anterior base of the first anal ray (Figure 14.8B). The retractor analis attaches anteriorly to the posterior base of the last anal ray and becomes continuous with the hypaxial musculature. Within the series of similar units, the superficial layer of the dorsal and anal fins is composed of inclinator muscles (inclinatores dorsales and anales, respectively). Unlike in the anal fin of the zebrafish and the dorsal and anal fins of other fishes described in the preceding text, the inclinatores dorsales are strongly attached to the skin and are often

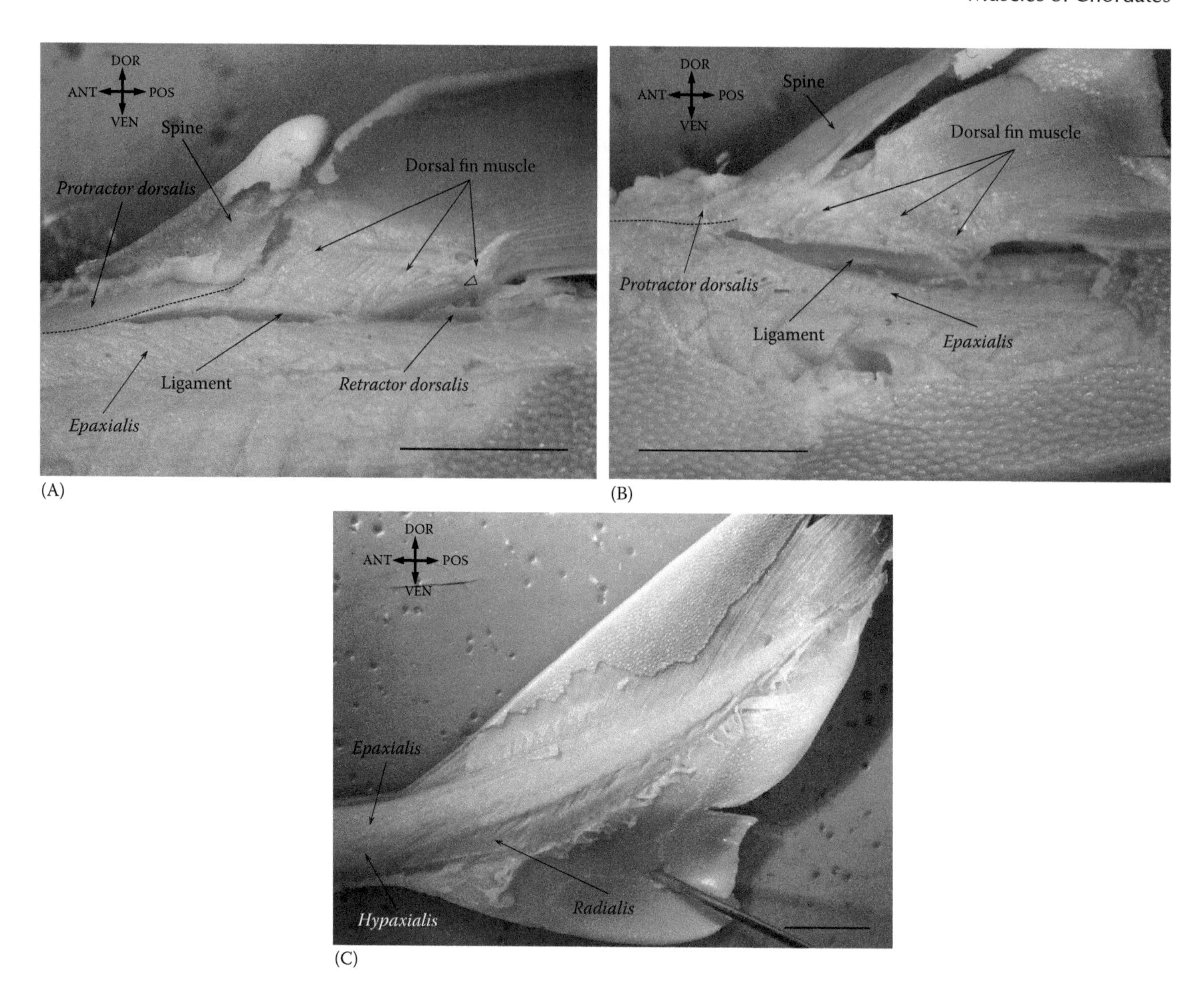

FIGURE 14.2 *Mustelus laevis* (Chondrichthyes): musculature of the first dorsal (A), second dorsal (B), and caudal (C) fins. Scale bar = 5 mm.

easily broken after skin removal. The deep layer of the serial muscles in the anal fin includes the erectores anales and the depressores anales. As the dorsal/anal fin ligament is absent in the zebrafish fins, erectors and depressors extend from the anterior (erectores) and posterior (depressores) regions of the base of each ray to the radials.

CAUDAL FINS

Unlike the pectoral and the other median fins (anal and dorsal), which are locally connected to the body, the caudal fin is a posterior continuation of the trunk. Therefore, although the caudal fin musculature is often considered to be part of the appendicular musculature (namely, of the musculature of the median fins), it is mainly a continuation of/derived from the epaxial/hypaxial axial body musculature (e.g., Figure 14.2C [Flammang 2009]). In sharks (Figure 14.2C), long tendons originate from the posterior myomeres and extend posterodorsally along the notochord. They attach to connective tissue overlying the notochord and tail cartilage. The only intrinsic caudal muscle present in sharks is the radialis muscle. It is a thin strip of muscle fibers that lies ventral to the hypaxialis, extending from the ventral lobe of the caudal fin to the ligament that separates it from the hypaxialis. Deep fibers of the radialis muscle also attach to the hemal arches of the caudal vertebrae.

In the caudal fin of *Polypterus*, lateral myotomes become thin and attach to all caudal and posterior dorsal rays (Figure 14.3C). Hence, the lateralis superficialis caudalis is a continuation of the axial musculature, which keeps the myotomal composition. Deep to the lateralis superficialis caudalis lies the interradialis hypaxialis (flexor ventralis *sensu* Lauder [1989]). Its fibers connect neighboring hypaxial (below the notochord) rays. As it has been noted in the preceding text, the epaxial portion of the caudal rays in this taxon comes from the dorsal fin and their anatomy follows the pattern of the dorsal fin rays (Figure 14.3D).

The caudal musculature of the chondrosteans we dissected is basically similar to that we found in sharks (Figures 14.4C

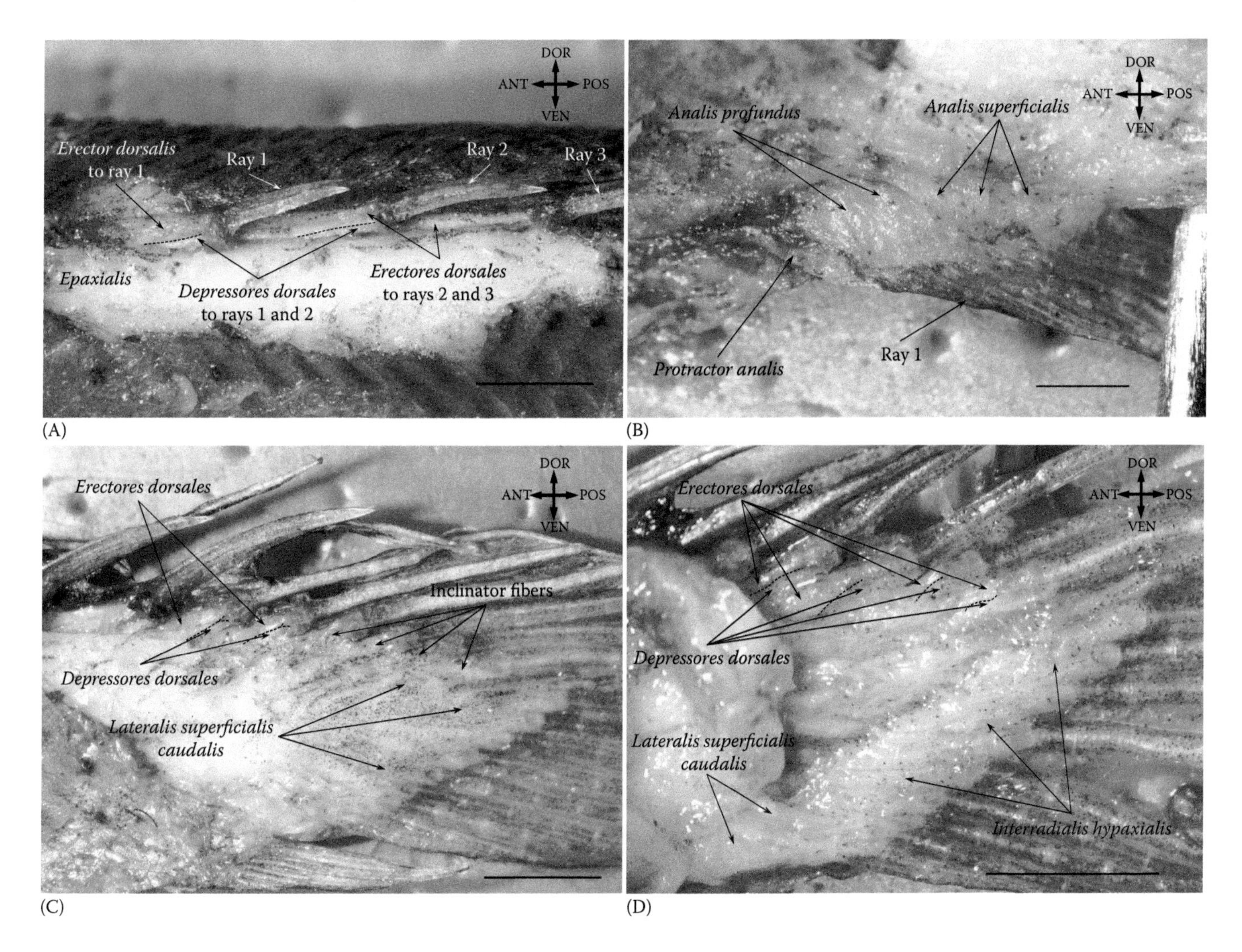

FIGURE 14.3 *Polypterus senegalus* (Cladistia): most anterior rays of the dorsal fin and their muscle units (A); musculature of the anal fin (B) and superficial (C) and deep (D) layers of the caudal fin. Scale bars in (A), (C), and (D) = 2.5 mm. Scale bar in (B) = 1 mm.

and 14.5C). The radialis muscle of *Acipenser* (not described by Lauder [1989]) and *Polyodon* does not reach the notochord but attaches instead onto the ventral ligament of the hypaxialis and of the tail cartilage.

Both muscles, the lateralis superficialis caudalis and interradialis hypaxialis, are also present in the caudal fin of *Lepisosteus* (Figure 14.6C and D) and *Amia* (Figure 14.7D). In *Lepisosteus* and *Amia*, another intrinsic caudal muscle is also present: the hypochordal longitudinalis (Figures 14.6D and 14.7D). This muscle has been reported by Lauder (1989) but has not been mentioned in some more recent studies (e.g., Flammang and Lauder 2009). In *Lepisosteus*, the hypochordal longitudinalis stretches from the first four dorsal rays to the dorsoventral midline of the tail and attaches to hemal spines along the notochord. In *Amia*, the hypochordal longitudinalis has basically the same position but attaches to the last epaxial and first hypaxial rays (Figure 14.7D). Additionally, *Amia* has a pronounced interradialis caudalis dorsalis (interradialis *sensu* Lauder [1989]) between the hypaxial caudal rays 1–5. Some studies describe a "supracarinalis" in the caudal fin of *Amia* (Lauder 1989; Flammang and Lauder 2009), but we did not find a clearly separate muscle in this taxon: it can only be implicitly allocated by the direction of muscle fibers, not by its separation/differentiation from surrounding muscles.

In *D. rerio*, the most used model organism among teleosts, the lateralis superficialis caudalis seems to be replaced by/give rise to muscles of the superficial caudal musculature layer (Figure 14.9A). These muscles are often fused with the trunk musculature and, therefore, do not have clearly defined borders. The lateralis profundus dorsalis and lateralis profundus ventralis originate from the myoseptum of the caudal musculature, proximal caudal fin bones, and the membrane connecting these bones. Posteriorly, these two muscles insert onto caudal fin rays 6–8 and 21–24, respectively. Similar to its configuration in *Amia*, the interradialis caudalis is present only on the dorsal side of the caudal fin and connects the bases of rays 5–7 (the first three long rays in the zebrafish shown in Figure 14.9A). The lateralis superficialis dorsalis and lateralis superficialis ventralis form a complex with the trunk musculature. They originate from the dorsoventral midline of the caudal fin and lie medially. Posteriorly, they insert onto dorsal ray 8 and ventral ray 22, in the zebrafishes we dissected.

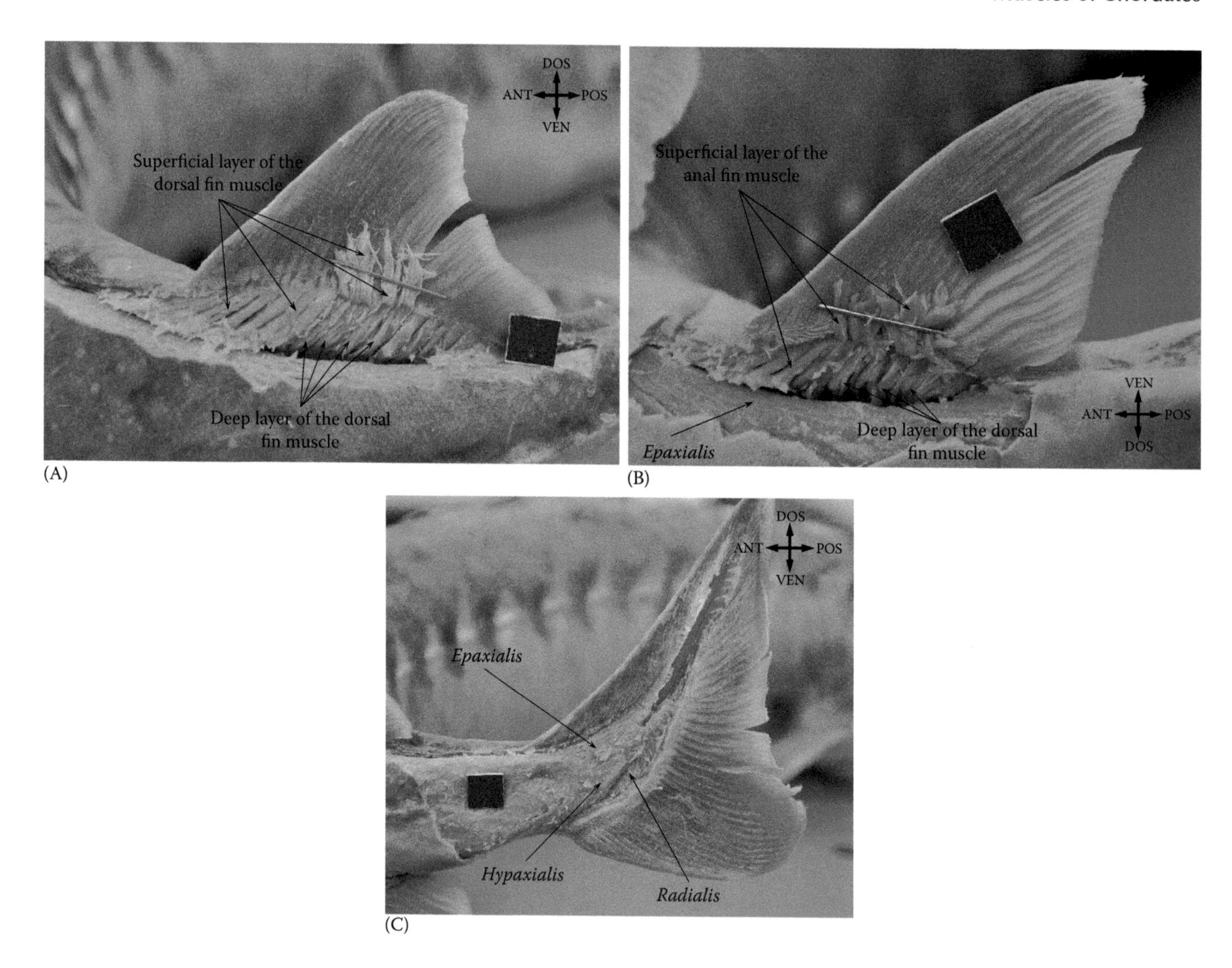

FIGURE 14.4 *Acipenser brevirostrum* (Chondrostei): musculature of the dorsal (A), anal (B), and caudal (C) fins. Scale bar = 1 cm.

At the bases of the fin rays, there are interfilamenti caudalis dorsalis and interfilamenti caudalis ventralis that fan out asymmetrically and insert onto caudal rays 8–12 and 16–21 in our specimens. They mainly connect adjacent caudal rays but can also extend over more than one ray. The deep layer of caudal muscles includes some flexors and the muscle adductor caudalis ventralis (Figure 14.9B). These muscles originate from the vertebral column, proximal caudal fin bones, and the membrane between them. The adductor caudalis ventralis extends mainly from the dorsoventral midline to caudal fin rays 8–10. The flexor caudalis dorsalis superioris and flexor caudalis dorsalis inferioris mainly originate from the second last caudal vertebra and lie dorsomedially to the adductor caudalis ventralis, inserting onto caudal fin rays 6–8 in our specimens. On the ventral side, the flexor caudalis ventralis superioris and flexor caudalis ventralis inferioris originate from the last two vertebrae and the caudal bones and insert onto all the long ventral caudal rays. In the following sections, we compare the zebrafish median fin muscles with those of other teleosts and discuss whether using the zebrafish as *the* key teleost and actinopterygian model organism is appropriate or not.

EVOLUTION OF MUSCLES OF MEDIAN FINS

The scarcity of available myological information on the anatomy and development of median fin muscles in fishes makes it difficult to discuss their evolution. Nevertheless, our recent dissections of specimens from all major extant nondipnoan fish clades—several shark specimens (Chondrichthyes), *Polypterus* (Cladistia), *Acipenser* and *Polyodon* (Chondrostei), *Lepisosteus* (Ginglymodi), *Amia* (Halecomorphi), and the model teleost *D. rerio* (see Figure 1.1)—and extensive literature review allow us to provide a brief, preliminary discussion on this topic (see Figure 14.1). It should be noted that in addition to the three types of median fins that we cover in this book—anal, dorsal, and caudal—some fishes have other median fins, e.g., adipose fins, which usually lack muscles but may be connected to musculature: see, e.g., Stewart and Hale (2013).

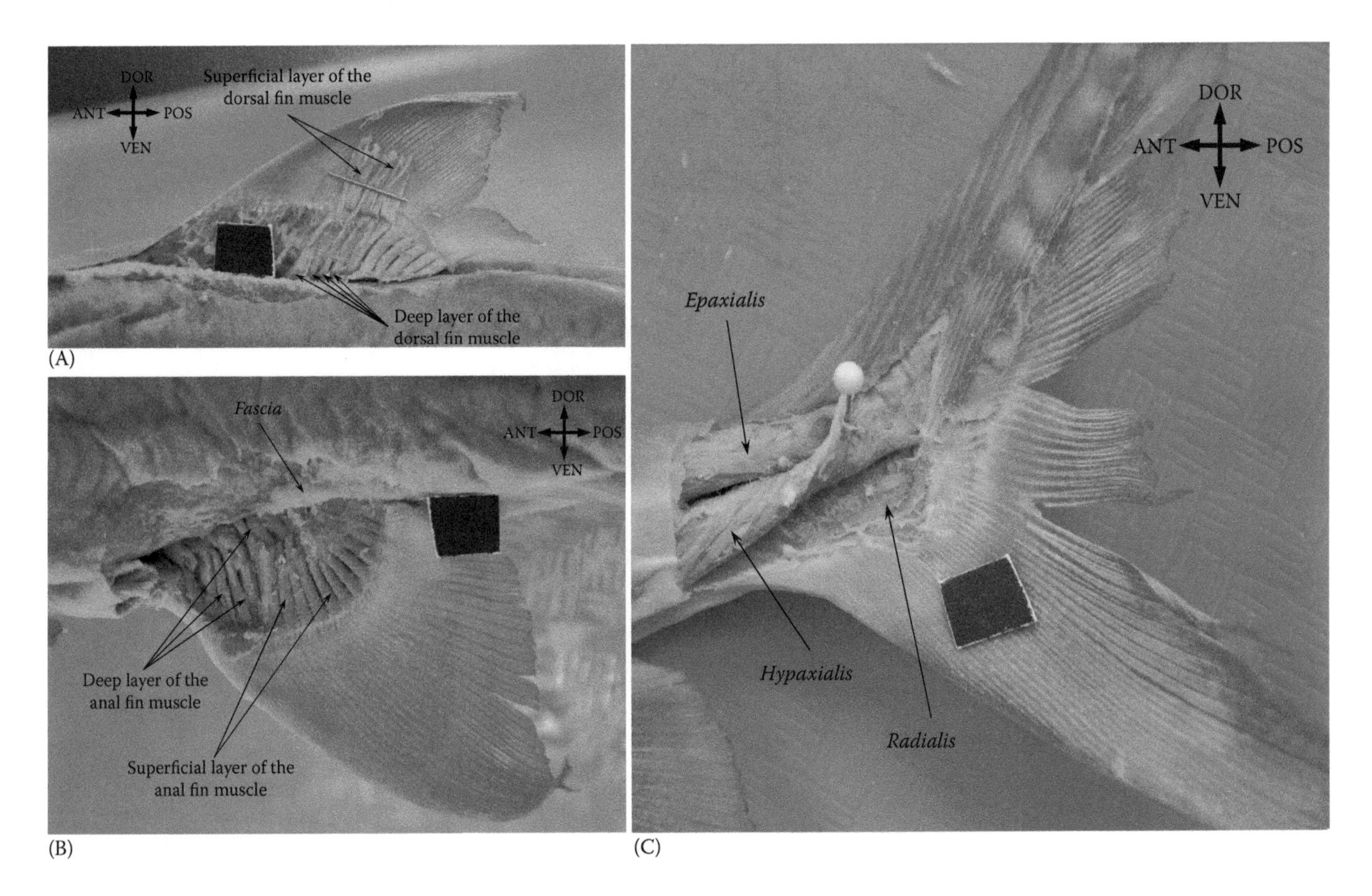

FIGURE 14.5 *Polyodon spathula* (Chondrostei): musculature of the dorsal (A), anal (B), and caudal (C) fins. Scale bar = 1 cm.

The body musculature of the phylogenetically most basal extant chordate—amphioxus—consists of myomeres that extend into the tip of the tail, which is surrounded by the caudal fin fold (Kusakabe and Kuratani 2005; Mansfield et al. 2015). In amphioxus, there is no true caudal fin musculature *per se*, but the trunk musculature that lies in the region of the tail plays an important role in locomotion. In the heterocercal caudal fins of sharks, the axial trunk musculature remains the main component of the tail and of the caudal fin, as noted in the preceding text. The only true intrinsic caudal muscle seen in sharks is the radialis, located ventrally to the hypaxial musculature (Figure 14.2C). *Acipenser* and *Polyodon* are chondrosteans, which are phylogenetically basal actinopterygian fishes and thus phylogenetically more derived than chondrichthyans. Nevertheless, they retain a similar caudal fin morphology, probably because they often display paedomorphic features within their musculature, as they also seem to have a paedomorphic configuration within their pelvic, pectoral, dorsal, and anal musculature as noted in the preceding text. The shark *M. laevis* lacks an anal fin, but the dorsal fins of sharks and of chondrosteans are very similar: the dorsal fin muscle (and the anal fin muscle in the case of chondrosteans) is subdivided into thin serial muscle bundles and can be partially subdivided into superficial and deep layers (Figures 14.2, 14.4, and 14.5). These two layers are also present in the anal fin of *Polypterus*, in which they form two distinct muscles—analis profundus and analis superficialis (Figure 14.3). However, the dorsal fin of *Polypterus* is remarkably different from that of the dorsal fins of sharks and of chondrosteans. In *Polypterus*, each dorsal half ray (left and right) has at least two different muscles (erector dorsalis and depressor dorsalis) and the posterior rays have, in addition, inclinator muscle fibers. Concerning the caudal fin, in addition to the radialis seen in sharks *Polypterus* also has an interradialis hypaxialis, which connects adjacent rays and likely allowed more mobility to each caudal fin ray, and which seems to be derived from the radialis. As an interradialis hypaxialis is also present in nonchondrostean actinopterygian fishes (see, e.g., Figure 14.6), it is likely that the nondifferentiation of this muscle in the adult chondrosteans we dissected is related to paedomorphic events, which often affect the musculature of these fishes as noted in the preceding text (see Figure 14.1).

Examined specimens of the Neopterygii fish clade exhibited only little variations concerning their dorsal and anal fin muscle morphology. *Lepisosteus*, *Amia*, and *Danio* have three types of muscles inserting onto the rays of the dorsal fin, as is the case in *Polypterus* as well as onto the anal fin: inclinator, erector, and depressor (see Figure 14.1). In *Lepisosteus* and *Danio*, the dorsal and anal fins are far from the caudal fin and—probably related to this distance—they have both protractors and retractors in the dorsal and in the anal fins. In *Amia*, where dorsal and anal fins are in close proximity to the caudal fin, retractors are absent and the protractor is present only in the dorsal fin (protractor dorsalis).

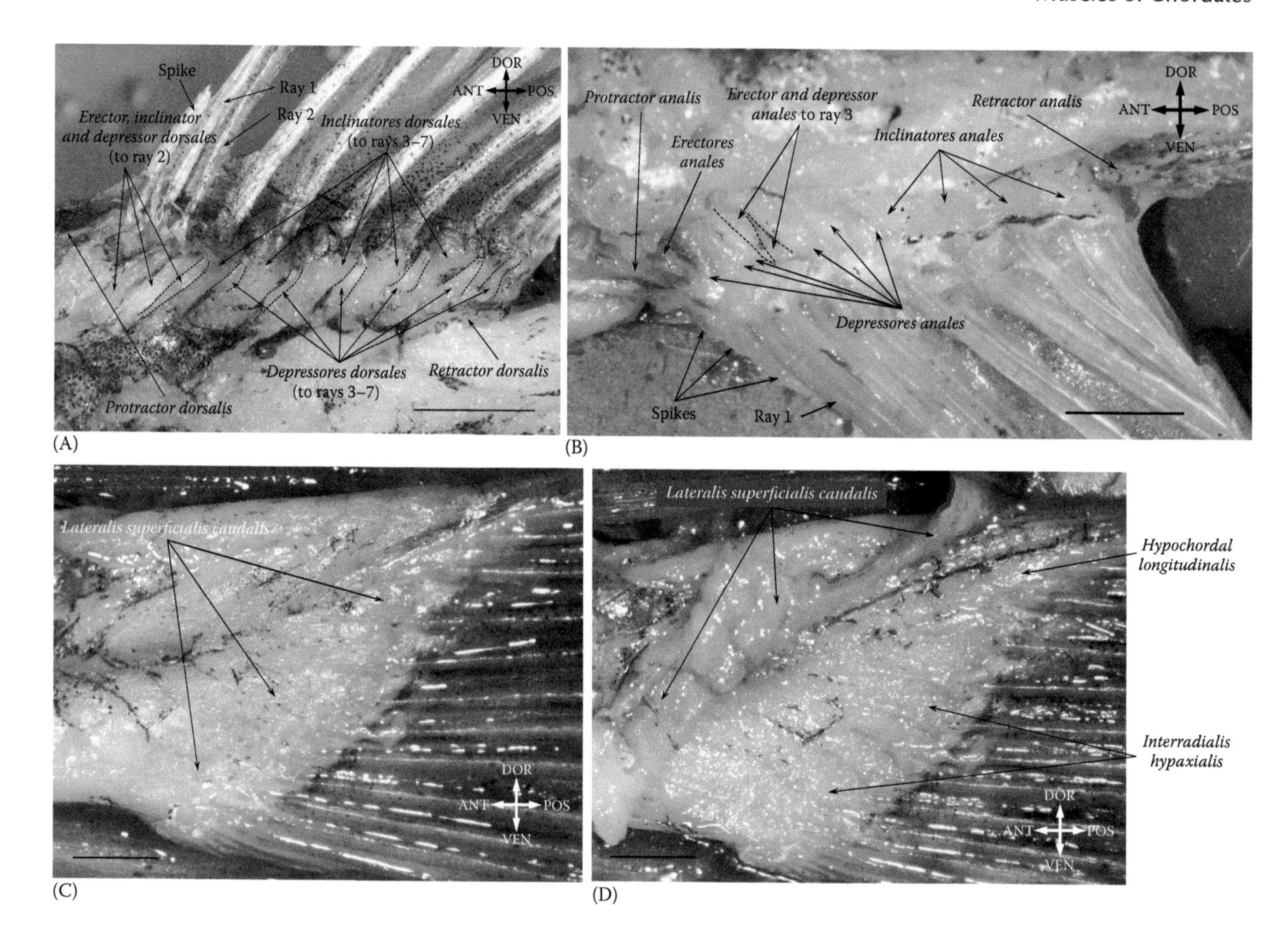

FIGURE 14.6 *Lepisosteus oculatus* (Ginglymodi): musculature of the dorsal (A) and anal (B) fins; musculature of the caudal fin is shown in superficial (C) and deep (D) layers. Scale bar = 2 mm.

The hypochordal longitudinalis in the caudal fin of *Lepisosteus* seems to topologically correspond to the radialis of sharks and of Chondrostei as it lies just below the notochord (see Figure 14.1). Thus, both the hypochordal longitudinalis and the interradialis hypaxialis muscles seen in Neopterygii fishes such as *Lepisosteus* appear to be derived from the primordium that gives rise to the radialis in sharks. *Amia* has, in addition, two more muscles of the caudal fin: the interradialis dorsalis and the supracarinalis (Figure 14.7). The differentiation of such muscles in the dorsal parts of the caudal fin results into the deep dorsoventral muscle asymmetry of a caudal fin that is apparently symmetrical (homocercal) in superficial view. This asymmetry is even more marked in the zebrafish, in which the caudal fin has many more independent muscles (Figures 14.1 and 14.7).

SIMILARITIES AND DIFFERENCES BETWEEN THE MUSCULATURE OF PAIRED FINS

This section as well as the following sections are mainly based on Siomava and Diogo (in press). The pectoral fin musculature of the adult zebrafish is in general very similar to that of the pelvic fin. Both fins have three types of muscles, which are organized in layers: superficial and deep layers of abductors and adductors for all/most rays and arrectors reaching only to the first ray of each fin. This similarity had been noted in teleosts by previous authors (e.g., Winterbottom 1974) and was recently discussed by (Diogo et al. 2013) and Ziermann et al. (2017). In those papers we and our colleagues have reviewed comparative, paleontological, and developmental data and suggested that the pectoral–pelvic musculature similarity seen in teleosts is a derived, homoplastic feature acquired independently in the evolutionary transition leading to these fishes and that somewhat parallels the homoplastic similarity seen in other derived gnathostome groups, such as tetrapods. For instance, in cartilaginous fishes, such as sharks, and in the living sister-group of all ray-finned fishes (actinopterygians), *Polypterus*, there is a single well-defined preaxial muscle in both the pelvic and pectoral appendages (Diogo et al. 2016b). As this single preaxial muscle is mainly a part of the ventral (abductor) musculature, it might correspond to the arrector ventralis of the pelvic and pectoral fins of teleosts and therefore of the zebrafish. However, even if this is the case, this means that at least the arrector dorsalis of each fin arose independently in each of these fins during the evolutionary history of other actinopterygians. That is, derived actinopterygians

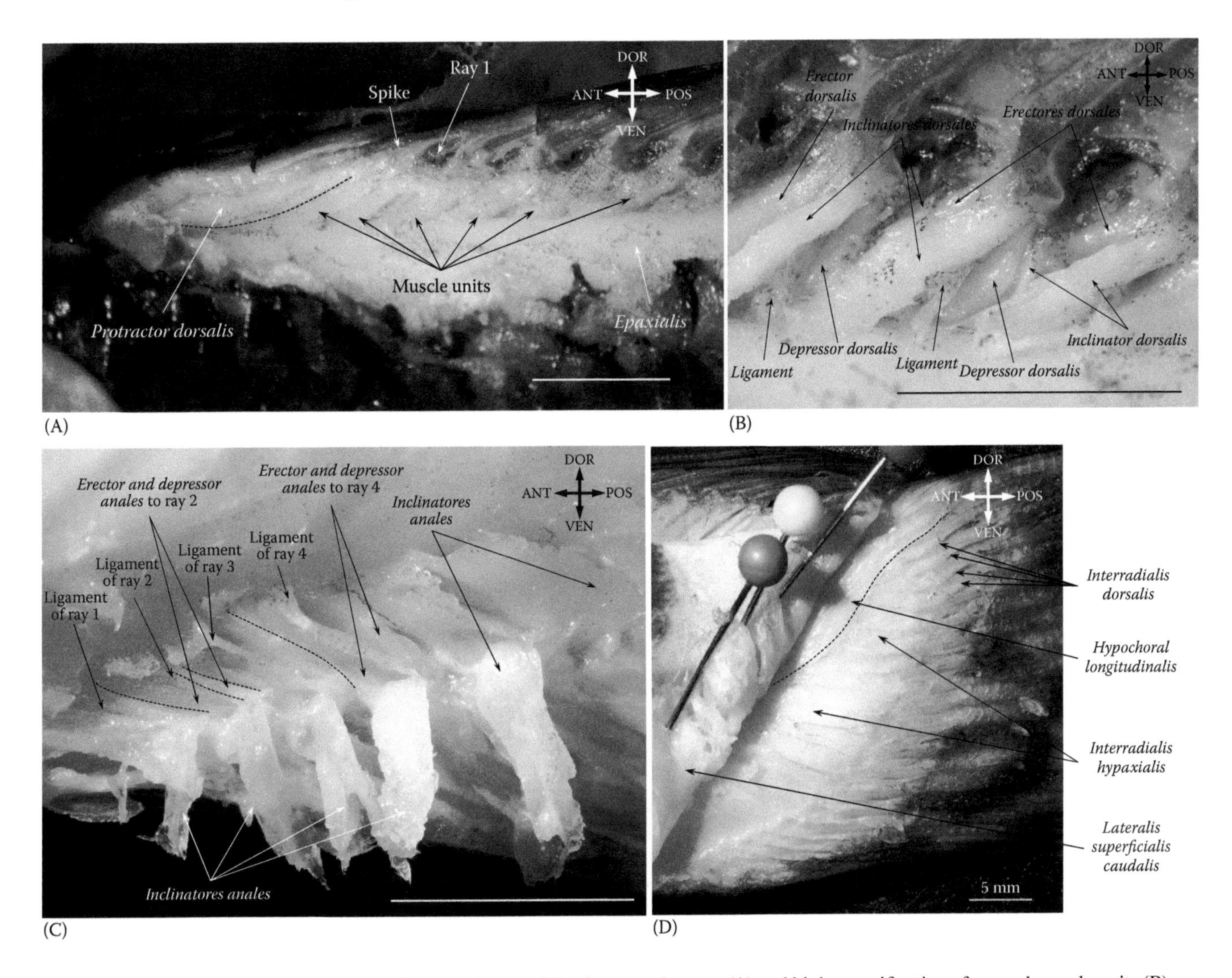

FIGURE 14.7 *Amia calva* (Halecomorphi): musculature of the first anterior rays (A) and high magnification of several muscle units (B) are shown for the dorsal fin; musculature of the anterior anal rays (C) and of the entire caudal fin (D). Scale bar = 5 mm.

such as teleosts display an increased number of muscles that are topologically similar between the pectoral and pelvic fins.

The main difference between the pectoral and pelvic musculature concerns the protractors and retractors of the girdles. Namely, in the zebrafish as well as in most teleosts, there is not a well-defined retractor of the pectoral girdle that would topologically correspond to the retractor ischii of the pelvic girdle (Winterbottom 1974; Diogo and Abdala 2010). Moreover, the protractor ischii of the pelvic fin is an appendicular/trunk muscle, while the protractor of the pectoral girdle—the protractor pectoralis—is a head (branchiomeric) muscle derived from the cardiopharyngeal field (Diogo et al. 2015b; Lescroart et al. 2015). The immediate proximity between the pectoral girdle and the skull might explain why the protractor pectoralis (reduced in size or absent in adult zebrafishes) and many other muscles connect the pectoral girdle to the cranium in teleosts. Following this argument, pelvic fins might lack such a possibility because of their remote posterior position, leading to the pelvic girdle being fixed to the body *via* the hypaxial muscle mass attached to the pleural ribs. In other cyprinid fishes, this connection may be additionally strengthened by a ligament from the seventh pleural rib to the dorsal surface of the pelvic plate (Saxena and Chandy 1966). On the other hand, our comparative data indicate that this difference is not merely a question of the topological position. Instead, the different developmental and evolutionary origin of the protractor pectoralis (head muscle) and protractor ischii (appendicular/trunk muscle) seem to reflect a profound, major difference between protractors of each girdle. A sharp distinction between these two muscles can also be seen in anatomically generalized teleosts such as *Elops* and *Oncorhynchus*. In these fishes, the protractor ischii is short and fused with the hypaxialis before reaching the pectoral girdle (Greene and Greene 1914; Winterbottom 1974). According to Winterbottom (1974), in basal teleosts this muscle is likely undifferentiated/more fused with the hypaxialis; i.e., the paired fins were probably not connected *via* a differentiated protractor ischii in early teleosts. However, more information is needed in order to determine the exact ancestral condition, and we plan to address this subject in future works. These differences, together with the very different position and anatomy of the pelvic and pectoral

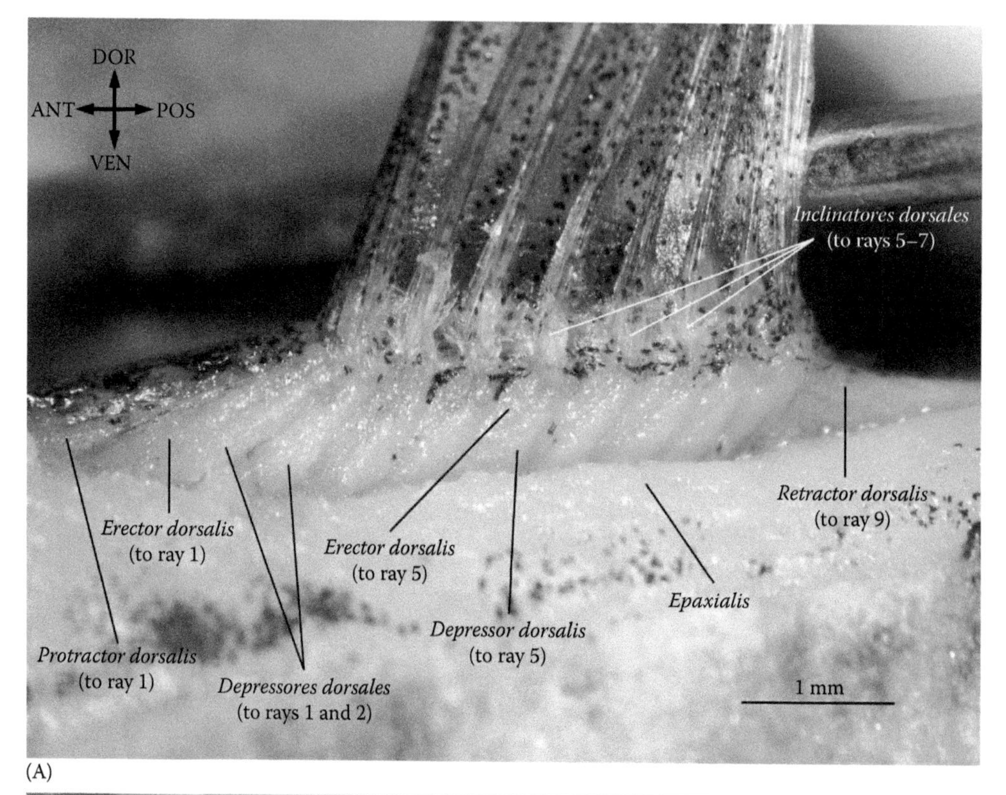

(A)

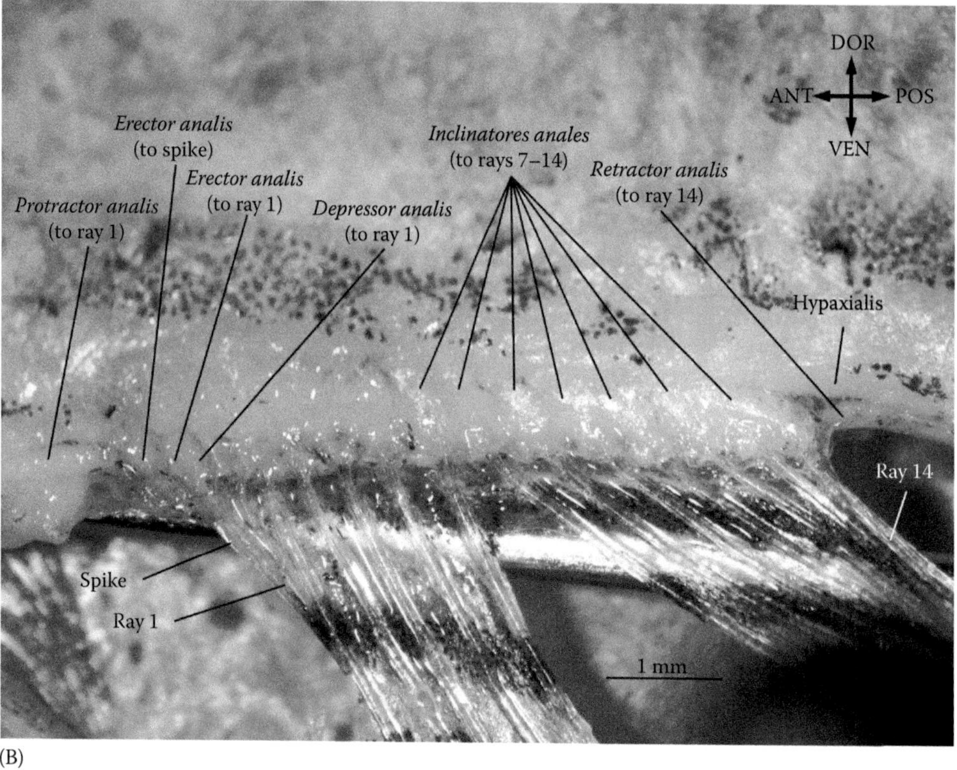

(B)

FIGURE 14.8 *Danio rerio* (Teleostei): musculature of the dorsal (A) and anal (B) fins. Scale bar = 1 mm.

girdles, support the increasingly popular idea that the pectoral appendage is at least partly derived from the head region and thus it is very different originally/ancestrally from the pelvic appendage (see recent works by Diogo and Ziermann [2015b] and Miyashita and Diogo [2016]).

Another evident difference in the muscular composition of the paired fins in the adult zebrafish is the presence of two arrector muscles on the ventral side of the pectoral fin (the arrector ventralis and arrector 3) and only one arrector muscle on the ventral side of the pelvic fin

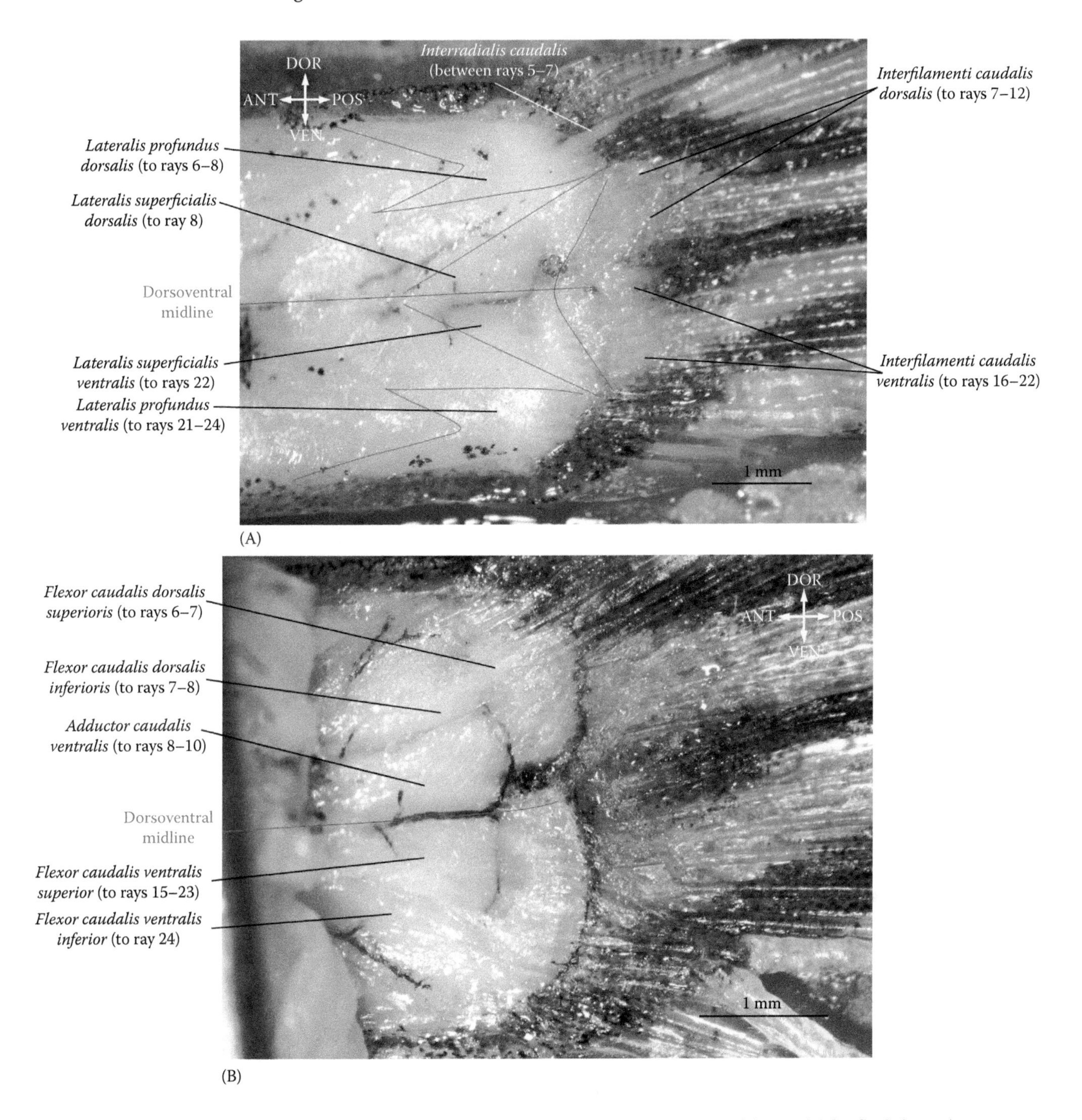

FIGURE 14.9 *Danio rerio* (Teleostei): musculature of the superficial (A) and deep (B) layers of the caudal fin. Scale bar = 1 mm.

(arrector ventralis pelvicus). The arrector 3 of the pectoral appendage has been frequently overlooked or treated as a bundle of the arrector ventralis or the fin abductor (e.g., Brousseau 1976a,b). However, arrector 3 is found in various teleosts, constituting a potential synapomorphy of the teleost clade Clupeocephala, which was acquired independently in *Lepisosteus* as noted in the preceding text (Figure 12.1). This muscular reinforcement of the first pectoral ray may possibly be related to the fact that in many fishes, particularly teleosts, the first pectoral and pelvic fin rays are functionally uncoupled from other rays, being, for instance, related to sound production or defensive tasks (e.g., Hadjiaghai and Ladich 2015).

The derived presence of three distinct arrectors in the pectoral fins of fishes such as zebrafish may also be related to the directive function of these fins in swimming. In many fishes, the pectoral fins act as the primary propulsors during rhythmic swimming (Webb 1973; Blake 1979; Drucker and Jensen 1996a,b; Walker and Westneat 1997; Hale 2006), braking (Drucker and Lauder 2003; Higham 2005), and maneuvering (Drucker and Lauder 2003). Being usually the longest and the thickest, the first ray of the pectoral fin experiences high resistance in water. In contrast, the pelvic fins of cyprinoid fishes seem to be less important for the equilibrium and power of locomotion (Harris 1938). Harris (1938) proposed that these fins are mainly used to neutralize the lift force after stopping and to produce slow elevation and depression forces, which result in less resistance during movements. Therefore, in contrast to the leading pectoral fins, pelvic fins do not usually require additional reinforcement in usual conditions.

SIMILARITIES AND DIFFERENCES BETWEEN THE MUSCULATURE OF THE MEDIAN FINS

Like the musculature of the paired fins, the anal and dorsal fins of most actinopterygian fishes have three major types of muscles: erectors, depressors, and inclinators (see Figure 14.1). In addition, the anal and dorsal fins of most actinopterygians have longitudinal protractors and retractors. They are somewhat similar to the protractor and retractor ischii of the pelvic appendage, but they insert onto rays rather than onto a girdle (as the median fins are not associated with girdles). The general architecture of the dorsal and anal fins is very similar: with the exception of the protractors and retractors, both fins consist of serially repeated units of rays and corresponding muscles as explained in the preceding text. Such a composition of serial units suggests that either *Hox* genes or another class of unidentified genes with a similar function contribute to their positioning (see, e.g., Mabee et al. 2002) and development. Interestingly, while dissecting and reviewing information about the dorsal fin in the zebrafish, we noticed some variation in the number of units. For example, Schneider and Sulner (2006) described 10 muscular units and 10 dorsal rays in their specimens, while the dorsal fins of our specimens were composed of 9 units. The variation of units in the anal fin is unknown because detailed studies describing and comparing this particular fin type among zebrafishes are not available. All zebrafishes we dissected had 14 units. While units of the dorsal fin significantly decreased in size posteriorly, in the anal fin they remained of a similar size. Another slight difference between the anal and dorsal fins is that the inclinatores dorsales attach strongly to the skin and only small portions of their fibers are blended with the epaxial trunk muscles. The inclinatores dorsales were largely destroyed during removal of the skin, while the inclinatores anales remained intact because they are mainly attached to the hypaxial trunk muscles. However, these differences are relatively minor, and there is little doubt that the anal and dorsal fins are close copies of each other, both anatomically (this study) and developmentally (e.g., Freitas et al. 2006). In contrast, the muscle pattern of the caudal fin is very different from those of the other fins, supporting the idea that it is developmentally and evolutionarily unique (Quint et al. 2002; Agathon et al. 2003). Thus, the caudal fin is in a sense an "axial" postcranial fin (i.e., it is associated with the posterior elements of the axial skeleton), not a true appendicular structure like the other fins.

CAN THE MUSCLES OF THE MEDIAN FINS CORRESPOND TO THOSE OF THE PAIRED FINS?

Studies have pointed out profound similarities in developmental and genetic mechanisms involved in the formation of the paired and median fins (Freitas et al. 2006), with the exception of the median caudal fin, which as noted just earlier seems to be developmentally very different from other fins. The sharing of similar mechanisms between the paired fins and the median dorsal and anal fins may support the fin-fold theory describing the origin of the paired fins from a median fin fold (Thacher 1877; Mivart 1879; Balfour 1881). This theory was accepted by most authors for decades, but it has been called into question recently (e.g., Gillis et al. 2009; Gillis and Hall 2016). Some of these authors have returned to an old—and for many decades discredited—idea that the paired appendages, or at least the pectoral ones, may be a derivative of the posterior pharyngeal region (Gegenbaur 1859). Indeed, there are numerous evolutionary and functional reasons for the deep spatial relationship between the skull and pectoral girdle in early gnathostomes: the girdle forms the rear wall of the internal branchial chamber—a shield for the pericardial cavity and a secure insertion for the pectoral fins (Coates and Cohn 1998; Matsuoka et al. 2005). Moreover, the pectoral appendage is closely associated with the head developmentally. The two structures use very similar developmental mechanisms, including a *Shh* dependence in both this appendage and the pharyngeal arches in chondrichthyans, a commonality also seen in the pelvic appendage (Gillis et al. 2009).

There is a possible hybrid hypothesis that takes into account the fin-fold theory, Gegenbaur's idea, and developmental, genetic, and comparative data: the pectoral appendage is at least partially derived from the head, while the pelvic appendage is derived from the median fins. A similar idea has been recently advanced in the developmental work by Nagashima et al. (2016), who argued that the pectoral appendage develops partially from the head region using a head "program," while the pelvic appendage fully uses a trunk program (H. Nagashima, pers. comm.).

How do the new anatomical data obtained in the present work relate to this crucial debate in the fields of comparative, developmental, and evolutionary biology? As noted in the preceding text, the pelvic and pectoral appendages of teleosts share some striking similarities, but (a) at least some of them seem to be derived (homoplastic), i.e., they were not present ancestrally, and (b) there are some major differences between the musculature of these appendages, e.g., branchiomeric muscles are related only to the pectoral appendage. These observations fit the hybrid scenario, asserting that at least part of the pectoral appendage (e.g., its girdle, as proposed by Nagashima et al. [2016]) might be developmentally/evolutionarily related to the head. But do the muscles of the pelvic fin correspond, in some way, to those of the dorsal and anal fins and support the idea that the pelvic fins are derived from the median fins, as proposed by authors such as Freitas et al. (2006) and by the fin-fold theory?

At first, it may be difficult to see how pelvic muscles can be compared with the muscles of the dorsal and anal fins, which are composed of serially repetitive units. Nevertheless, we have already noted some similarities between the protractors and retractors of the anal and dorsal fins and the protractor and retractor ischii of the pelvic appendage. An additional potential similarity between the dorsal/anal fins and the pelvic fin is the excessive development of the most anterior (preaxial) erector in the dorsal fin/anal fin. Such an overdeveloped erector might thus be compared, both topologically and functionally, to the arrector ventralis of the pelvic (or pectoral) fin. Therefore, it is possible that during fish evolution, the erectors of the more posterior units (from the second to the last) of the anal and dorsal fins became fused with other muscles, e.g., with the depressors, to form a single muscle mass. An example of such fusion between the muscle units associated with most rays of the dorsal/anal fins can be seen in adult *Elops* and *Argyropelecus* (Winterbottom 1974). In these fishes, the erectors and depressors of the posterior rays fuse to form a single compound muscle. Alternatively, the erectors of the posterior rays might have become reduced and then disappeared, as they have in the seahorse *Hippocampus zosterae*. In this seahorse, the first ray retains three types of muscles (the erector, depressor, and inclinator muscles), while posterior rays have only two types—the depressors and inclinators (Consi et al. 2001). Following this line of reasoning, muscles of the posterior units of the dorsal and anal fins might have become undifferentiated during development/evolution and given rise to the abductor/adductor muscle masses of the pelvic and pectoral fins. In the early development of the pelvic and pectoral fin musculature of fishes such as sharks, the abductor and adductor muscle masses are formed from a series of individual muscles that clearly reflect a segmented pattern, which later in development becomes imperceptible (Jarvik 1965). Moreover, the adductor and abductor masses of both the pelvic and pectoral appendages of adult chondrostean fishes, which in some respects of muscle development are often considered to be paedomorphic, strongly resemble the clearly segmented, serial thin muscle bundles that can be seen in early developmental stages of fishes such as sharks (Figures 13.1 and 13.2). In fact, in adult chondrosteans, there is a strong similarity between the configuration of the main abductor and adductor masses of the pectoral, pelvic, anal, and dorsal fins (compare Figures 13.1 and 13.2 with Figures 14.4 and 14.5). Therefore, it is possible that this was the plesiomorphic condition for adult basal gnathostomes with pectoral, pelvic, dorsal, and anal fins, whereas in extant fishes the individual muscles inserting exclusively onto the first ray remain separate and form the adult arrectors due to the functional uncoupling of this ray, particularly in the paired fins (Hadjiaghai and Ladich 2006, 2015). We plan to test these hypotheses in further comparative studies by undertaking studies of muscle development of both the paired and median fins from the earliest to the adult stages in other fishes, as we recently did for the zebrafish (see Chapter 15).

IS THE ZEBRAFISH AN APPROPRIATE MODEL FOR THE APPENDICULAR MUSCULATURE OF TELEOSTS?

As a basis for future studies and for discussions of whether the zebrafish is a good model for teleosts in comparative, developmental, functional, evolutionary, and evo-devo studies, we contrast here our results with descriptions of the fin muscles in other teleosts. As described in Chapter 13, the pectoral fin muscles of the zebrafish are similar to those of other teleosts, which usually have six muscles: two arrectors (dorsalis and ventralis), two abductors (superficialis and profundus), and two adductors (superficialis and profundus) (Winterbottom 1974). The seventh pectoral fin muscle in the zebrafish—arrector 3—is often present in other teleosts, as noted in the preceding text. In this sense, the overall configuration seen in the zebrafish is very similar to that of numerous teleosts (Table 13.1). Moreover, the phylogenetically basal actinopterygian *Polypterus* has six pectoral fin muscles, of which five clearly correspond to muscles present in the zebrafish (Table 13.1; Molnar et al. 2017b). The postaxial muscle present in the *Polypterus* pectoral fin has no correspondence in the zebrafish, and the arrector dorsalis and arrector 3 of the zebrafish pectoral fin have no correspondence in *Polypterus*. These comparisons indicate that concerning the pectoral fin muscles the adult zebrafish is a very good model for teleosts in general and for actinopterygians (ray-finned) fishes as a whole.

The configuration of the six pelvic muscles in the zebrafish is also very similar to that found in other teleosts, including generalized forms such as *Elops* (Winterbottom 1974). Interestingly, in certain Indian hill stream fishes of the cyprinid family, the abductor superficialis can be complemented by an extra separate bundle, attached to the first ray (Saxena and Chandy 1966). The reinforcement of the first pelvic fin ray with this extra muscle might occur due to the high resistance in swift current. It resembles the situation with the reinforcement of the first ray in the pectoral fins with arrector 3 and provides a further case of derived (homoplastic) similarity between these two types of fins. As recently noted by Molnar et al. (2017b) and Diogo et al. (2016b), of the six pelvic fin muscles present in the adult zebrafish, five might have been ancestrally present in actinopterygians, as they are found in the phylogenetically basal *Polypterus* (Table 13.2). Similar to the muscle configuration in the pectoral fin, the postaxial pelvic fin muscle present in *Polypterus* has no correspondence in the zebrafish, and the arrector dorsalis pelvicus of the zebrafish has no correspondence to the *Polypterus* muscles (Molnar et al. 2017b). Therefore, regarding the pelvic fin musculature, the zebrafish is also a good model for both teleosts and actinopterygians as a whole.

The dorsal and anal fins are very similar anatomically to each other, and their musculature is, in general, conservative within teleosts, with only minor differences between some taxa. For instance, in *Elops* the erector muscles of the first two dorsal fin rays are absent and the first erector is thus attached to the third ray (Winterbottom 1974). The same pattern can be seen in the anal fin of *Elops*: the first two rays are not associated with erectors. These observations support the idea that these two fins are deeply related/integrated developmentally and evolutionarily. In the seahorse *H. zosterae*, erectors are severely reduced and there is only a single pair of erector muscles attached to the anteriormost fin ray (Consi et al. 2001). In this case, the erection of the entire fin is achieved by pulling the first ray. All subsequent rays are pulled up *via* the interray ligament, which mechanically couples all fin rays into one unit (Consi et al. 2001). Another type of muscular modification is a fusion of an erector and depressor into one muscle, which is usually attached medially to the ray and maintains both functions (Winterbottom 1974). Such compound muscles are connected to the last two rays of the dorsal fin in *Elops* and to the last ray of the anal fin in *Argyropelecus* (Winterbottom 1974). In contrast to these species, both the king salmon (*Oncorhynchus tshawytscha*) and zebrafish (*D. rerio*) have more evenly distributed muscles. In the dorsal and anal fins of the salmon and zebrafish, the number of muscle series corresponds to the number of rays (Figure 14.8; Greene and Greene 1914). In this sense, the zebrafish is likely to be a very good model for teleost anal and dorsal fins. First, it has the same five types of muscles that are normally present in the dorsal and anal fins (inclinators, erectors, depressors, protractor, and retractor). Second, it might reflect an ancestral condition of having serial units for each ray.

The caudal fin of *Danio* is very similar to that of other teleosts, including generalized taxa such as the elopiform *Elops* (Greene and Greene 1914; Winterbottom 1974; Schneider and Sulner 2006). The superficial layer of the caudal fin consists of several muscles tightly consolidated with the hypaxial and epaxial trunk muscles, which makes it difficult to separate them and to define their sites of origin. Some fishes, e.g., the bluegill sunfish (*Lepomis macrochirus*), are known for the advanced articulation of their caudal fin. Thus, *L. macrochirus* can move each caudal fin ray independently (Flammang 2014), but despite the high number of discrete caudal muscles, the overall structure of muscles in the tail fin resembles the deep layer in other teleosts, particularly zebrafish. Furthermore, zebrafishes and other teleosts exhibit dorsoventral asymmetry of the deep musculature (e.g., interradialis caudalis and adductor caudalis ventralis). This muscle asymmetry resembles the skeleton asymmetry seen in the caudal fin of other *Danio* and teleost species and contrasts with the symmetrical superficial muscle layer and the even ray distribution within *Danio* and most teleosts (Sanger and McCune 2002). Altogether, these observations show that the adult zebrafish is a good model for the caudal fin musculature of teleosts. The highly conservative pattern of caudal muscles and their clear anatomical difference from those of other fins align with the idea that the caudal fin is a developmentally and evolutionarily distinct unit (Quint et al. 2002; Agathon et al. 2003) that is regulated *via* different mechanisms than the head and trunk (Flammang 2014). However, as shown in Tables 14.1 through 14.7, the zebrafish is not a very good model for the muscles of the median fin in Actinopterygii as a whole because phylogenetically basal actinopterygians such as chondrosteans or cladistians display a very different number and configuration of muscles (compare Tables 14.1 through 14.3 to Table 14.7).

GENERAL REMARKS

Our comparative data show that *D. rerio* is a very good model for the appendicular muscles of teleosts. It is also an appropriate model for the paired fin muscles of actinopterygians as a whole, but less so for the muscles of the median fins. Therefore, the zebrafish should be increasingly used for comparative, developmental, and macroevolutionary studies of appendicular muscles. Moreover, the presence in the zebrafish of the few paired fin muscles and the many median fin muscles that do not seem to have been ancestrally present in ray-finned fishes makes *D. rerio* a good model for comparative and macroevolutionary studies focused on locomotion and muscular segregation.

15 Development of Muscles of Paired and Median Fins in Fishes

In Chapter 15 we use the zebrafish *Danio rerio* as a case study to illustrate the development of the muscles of all the five types of fins (pectoral, pelvic, caudal, anal, and dorsal) covered in this book. One reason is that *D. rerio* is the only fish for which the development of the muscles of all these types of fins was studied in detail in the same project, namely, by ourselves and our colleagues (see the following text), which allows a better comparison between the ontogeny of all these fins. Another reason is that, as noted in the preceding chapters, *D. rerio* is one of the most popular model organisms in various fields of biological research, particularly developmental biology. A significant percentage of evolutionary and developmental studies use this fish for evo-devo comparisons with different vertebrate taxa and for general discussions on the evolution of the appendages and even on paired fin–limb transitions that occurred during the origin of the tetrapod lineage (Zhang et al. 2010; Yano et al. 2012; Leite-Castro et al. 2016; Nakamura et al. 2016; Saxena and Cooper 2016). However, most of such studies are based on gene expressions and anatomical comparisons of the skeleton, usually not including details about soft tissues such as muscles (Nakamura et al. 2016; Saxena and Cooper 2016). Accordingly, despite the common use of the zebrafish as a model organism for developmental works and discussions on both paired and median appendages, almost nothing is known about the development of the fin musculature in these fishes. Patterson et al. (2008) studied the growth of the pectoral fin and trunk musculature and looked at different fiber types that constitute the abductor and adductor muscles, but the differentiation of these muscles and development of other pectoral muscles were not studied by them in detail. Cole et al. (2011) provided a general discussion on the development and evolution of the musculature of the pelvic appendage, but their study was mainly focused on developmental mechanisms and migration of muscle precursors and not on specific muscles. Thorsen and Hale (2005) did refer to specific muscles in their report on the development of the pectoral fin musculature of zebrafishes, but they omitted some muscles such as the arrector 3. Surprisingly, the development of the musculature of the median fins in the zebrafish has never been studied. In order to tackle this scarcity of information on the development of the zebrafish appendicular musculature, we thus studied in detail and briefly describe in the following text, the ontogeny of each muscle in each fin from the time it first becomes visible until it displays a configuration basically similar to that of the adult stage (summarized in Table 15.1 and Figure 15.1). This work was performed and published with our colleagues Fedor Shkil and Elena Voronezhskaya (Siomava et al. in press), and the following sections are mainly based on that paper, which should be consulted for more details on the specific methodology used in that work.

DEVELOPMENT OF THE PAIRED AND MEDIAN MUSCLES OF THE ZEBRAFISH

The caudal fin is the first fin to develop in the zebrafish (Table 15.1). It appears as a continuation of the zebrafish postcranial axial skeleton and is surrounded by the caudal fin fold with the mesenchyme condensation ventrally, where the first caudal muscles and bones will later develop (Figure 15.2A). At stage 2.95 mm notochord length (NL), it is already associated with muscles. By 3.30 mm NL it includes, two muscle masses—dorsal and ventral caudal muscles—are continuous with the trunk muscles (epaxialis and hypaxialis, respectively). They expand posteriorly to the tip of the tail (Figure 15.2A). Even though there is no clear border between these early caudal muscles and the trunk muscles, they can be distinguished from myotomes by the absence of segmentation.

At 4.4 mm NL, three new ventral muscles can be seen (Figure 15.2B and C). Myofibrils of the adductor caudalis ventralis and flexor caudalis ventralis, which at this time included both the flexor caudalis ventralis superior and inferior, start bifurcating from the ventral caudal muscle (Figure 15.2B). The flexor caudalis ventralis extends ventrally toward the caudal fin fold. The adductor caudalis ventralis mainly follows the direction of the ventral caudal muscle but has shorter fibers that end halfway to the tip of the tail. Several short muscle fibers of the lateralis profundus ventralis begin separating from the hypaxialis at this stage. The adductor caudalis ventralis becomes more distinguishable but still keeps the direction of the ventral caudal muscle. By 5.0 mm standard length (SL; tip of snout to posterior end of last vertebra or to posterior end of midlateral portion of hypural plate), when the notochord starts bending dorsally, both the adductor caudalis ventralis and flexor caudalis ventralis substantially increase in size (Figure 15.3A). The flexor caudalis ventralis is attached to the ventral rays. The adductor caudalis ventralis becomes more separated from the ventral caudal muscle and changes the direction towards the dorsal fin rays. At this stage, fibers of the lateralis profundus ventralis are relatively short and they do not insert onto the caudal rays.

At stage 5.2 mm SL, the flexor caudalis dorsalis inferioris can be seen for the first time, arising deeply from the dorsal side of the ventral caudal muscle (Figure 15.3B). This flexor runs medial to the adductor caudalis ventralis, which is well developed by this stage, but does not insert onto the fin rays

TABLE 15.1
Diagram of Development of Appendicular Muscles in the Zebrafish

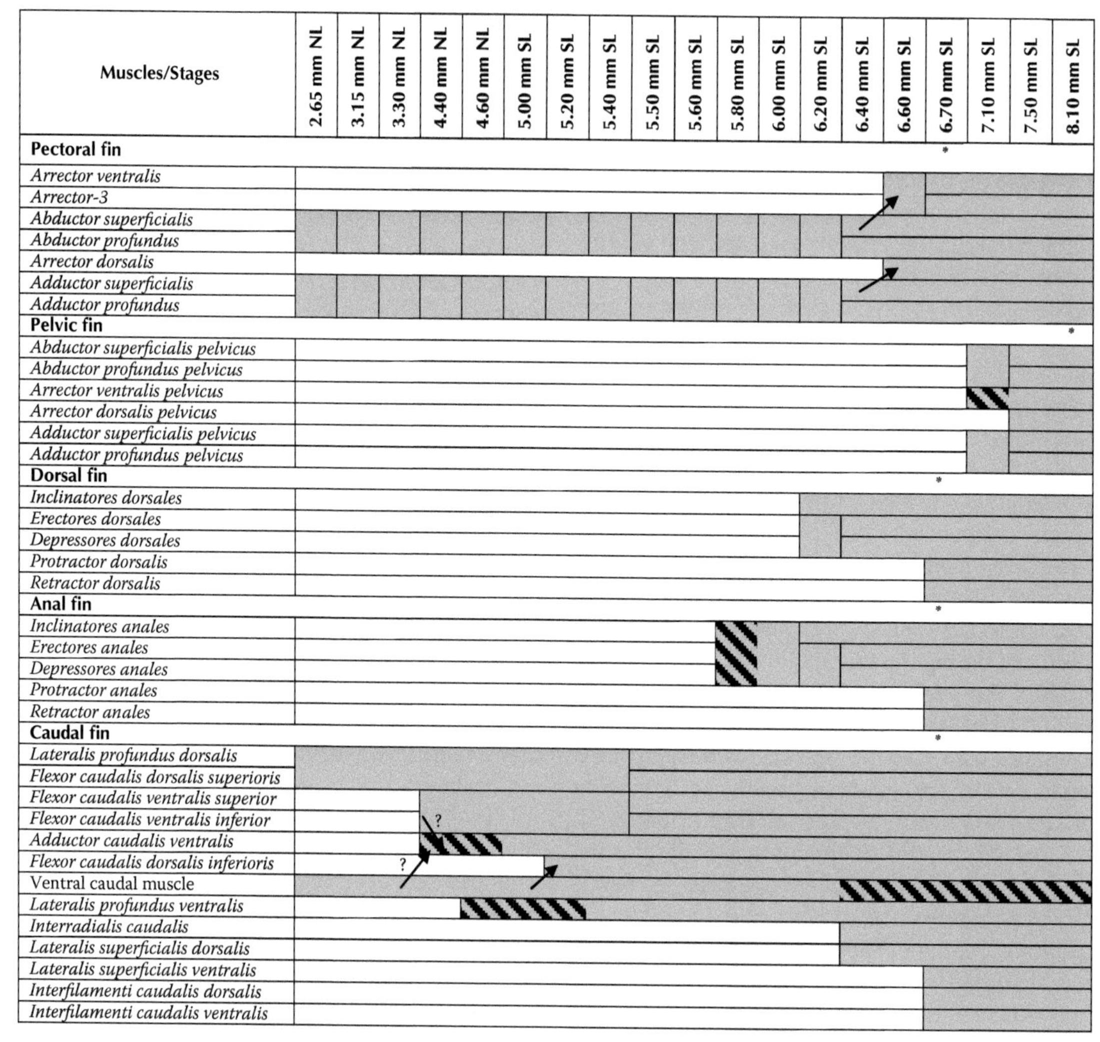

Note: Arrows indicate development from another muscle. Stars mark the stage when the adult muscle configuration is achieved. "?" refers to the question of whether the ventral caudal muscle and/or the flexor caudalis ventralis inferioris contribute or not fibers to the adductor caudalis ventralis. Shaded cells show stages when muscles have no attachment to fin rays.

as it does in later stages. At stage 5.5 mm SL, new muscles appear *via* rearrangements of the previous ones. Thus, the flexor caudalis ventralis splits into the large flexor caudalis ventralis superior and small flexor caudalis ventralis inferior, which inserts onto one ventral ray (Figure 15.3C). The dorsal caudal muscle breaks up into the flexor caudalis dorsalis superioris and lateralis profundus dorsalis overlying the former. The lateralis profundus ventralis stretches closer to the fin rays. During the growth of the tail, the ventral caudal muscle splits into superficial and deep layers (at 5.6 mm SL), which shift toward the midline. Lastly, superficial fibers become reduced, while deep fibers increase in number, and now instead of inserting onto fin rays they insert onto proximal caudal bones and vertebrae (Figure 15.4A and B). At 6.4 mm SL, long and very thin fibers of the lateralis superficialis dorsalis are visible (Figure 15.4A), and the interradialis caudalis already connects the bases of the all three long dorsal rays (Figure 15.4B). The last muscles to develop are the lateralis superficialis ventralis, interfilamenti dorsalis, and interfilamenti ventralis. The three muscles can be seen at 6.7 mm SL (Figure 15.4C). At this stage, basically all muscles are present and have a configuration that resembles the adult condition (see Figure 14.9 and Table 14.7). Interestingly, in addition to these muscles, in young specimens the space between the hypural bones was filled with muscle fibers (Figure 15.5) that then disappear before the adult stage, when this space becomes smaller. We observed these fibers between hypurals 1–2, 2–3, and 3–4 from 5.0 to 7.1 mm SL, forming very thin muscles that we designate here as interhypurales.

Concerning the pectoral fins, they are already formed by 2.65 mm NL. Our results and previous studies have shown

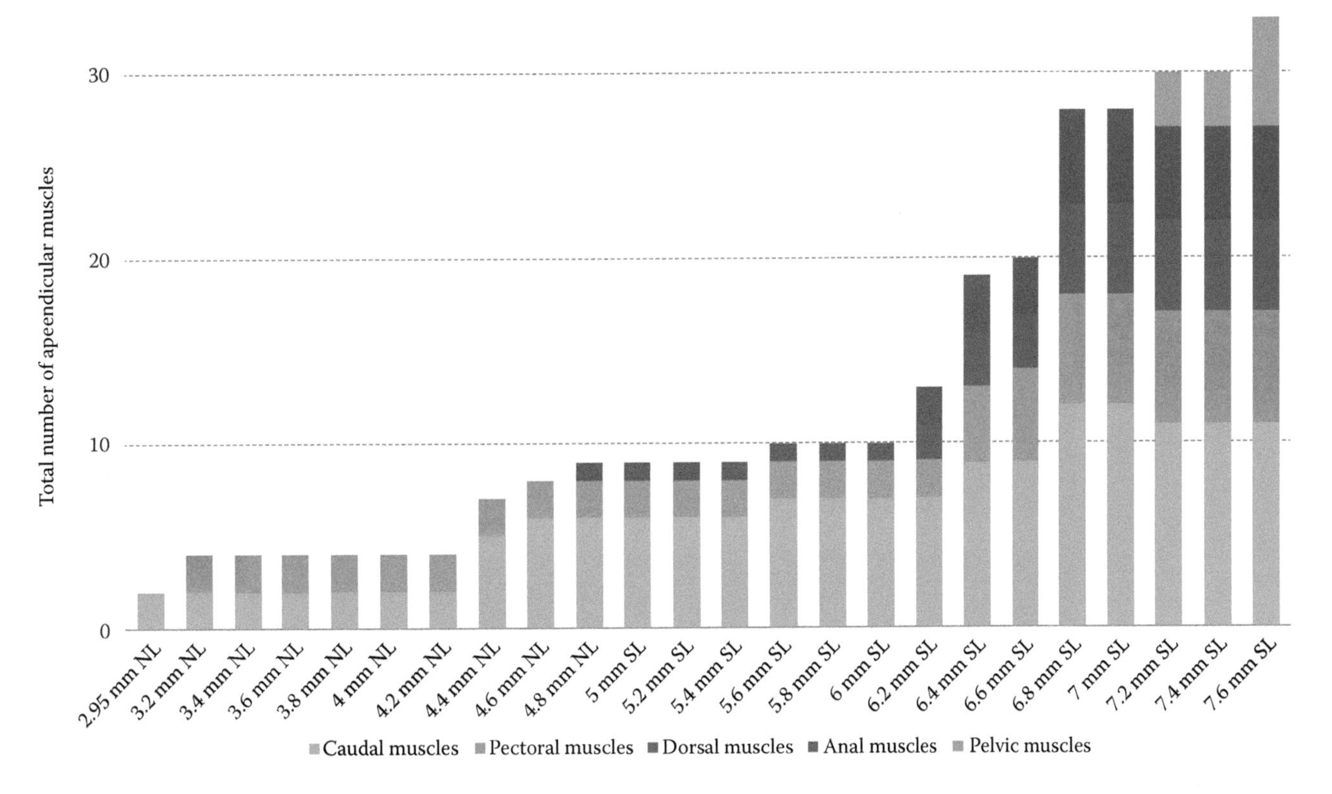

FIGURE 15.1 *Danio rerio* (Teleostei): total number of appendicular muscles during different developmental stages. NL, notochord length; SL, standard length.

that pectoral fin musculature start developing during early embryogenesis (Figure 15.6) and both abductor and adductor muscle masses differentiate as early as 2.8 mm NL (~46 hours post fertilization). Before (2.65-2.9 mm NL) and after (3.15 mm NL) hatching of larvae, we observe continuous fibers of the abductor and adductor masses (Figure 15.6). By 3.3 mm NL fibers extend to the edge of the endoskeletal disc and attach to actinotrichia (Figure 15.7). Along with the growth of the fin, the abductor and adductor extensively increase in size until 6.4 mm SL, when they split into deep and superficial layers (adductor profundus and superficialis; abductor profundus and superficialis). At 6.6 mm SL, a small bundle attached to the first pectoral ray starts separating from the abductor superficialis (Figure 15.8). This bundle later gives rise to both the arrector ventralis and arrector 3 (Figure 15.9A). On the medial side, only one arrector (arrector dorsalis) develops, apparently from the adductor mass (Figure 15.9D). At 6.7 mm SL (Figure 15.9), all seven pectoral fin muscles are present and display the adult configuration (see Figure 13.11 and Table 13.1).

We describe the dorsal and anal fin muscle development together here, because of the striking similarity of both their adult anatomy and ontogenetic development. At 5.8 mm SL, muscle fibers appear in the region of several middle rays of the anal fin (Figure 15.10A). By 6.0 mm SL, these fibers elongate proximally and distally towards the body and fin rays respectively and myofibrils appear in more anterior and posterior serial units of the fin (Figure 15.10B). The dorsal fin musculature develops slightly later than the anal fin musculature, and by 6.0 mm SL, no muscle fibers can be seen (Figure 15.10B). By this stage, any structural rearrangements mainly occur in the same proximo-distal axis, but at 6.2 mm SL, muscle differentiation into deep and superficial layers is visible (Figure 15.11). The deep layer soon gives rise to the erectors and depressors of each ray (Figure 15.12) that are covered by the overlying superficial inclinators (Figure 15.12C). The development of muscles corresponding to different rays is asynchronous, resulting in muscle units in the dorsal and anal fins (i.e., including an inclinator, depressor, and erector going to both the left and right sides of each half ray on each side of the body, each ray therefore receiving six muscles in total), which are developed to a different extent at the same stage. Thus, muscle units of the rays that are more central antero-posteriorly can have all six muscles differentiated into superficial-deep layers while the outermost rays may have undifferentiated developing muscle fibers or even single myofibrils (Figure 15.12). By 6.7 mm SL, development of each muscle unit is accomplished and the depressors partially overlie the erectors, which are subdivided into two small heads attached to the dorsal fin rays (Figure 15.13A). In addition to these units of six muscles going to each ray, there are also two longitudinal muscles that develop by 6.7 mm SL within each fin to move the first and last rays, which are named the protractor and retractor anales and the protractor and retractor dorsales (Figure 15.13). Therefore, by 6.7 mm SL, all dorsal and anal fin muscles are already present and have a configuration that resembles that seen in adults (see Figure 14.8 and Table 14.7).

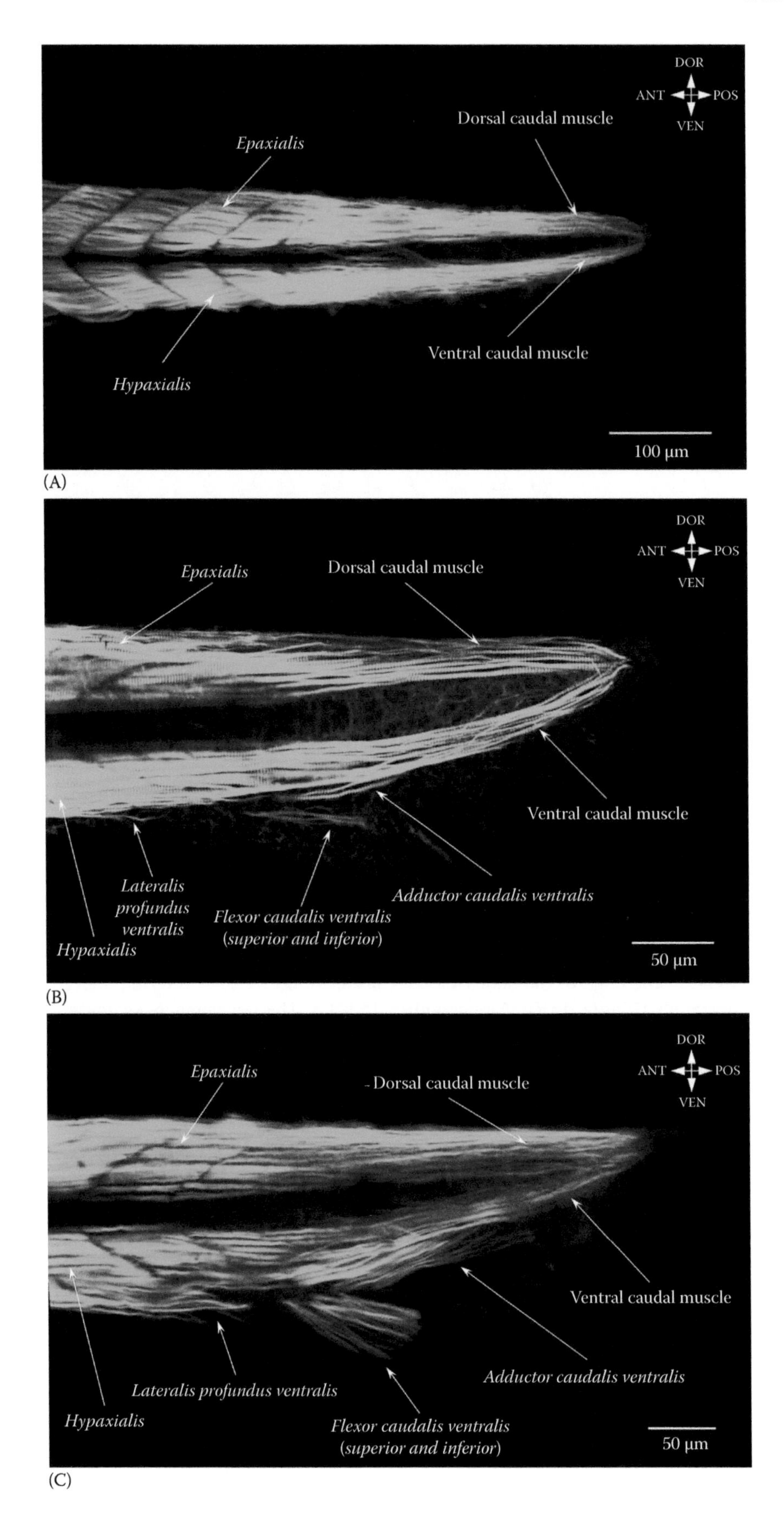

FIGURE 15.2 *Danio rerio* (Teleostei): early development of the caudal fin musculature. At 3.3 mm NL, two muscles are present in the caudal fin (A). Ventral caudal muscles develop before the dorsal muscles. At 4.4 mm NL, the first fibers of ventral caudal muscles can be seen (B). At 4.6 mm NL, ventral caudal muscles grow toward the caudal fin rays (C).

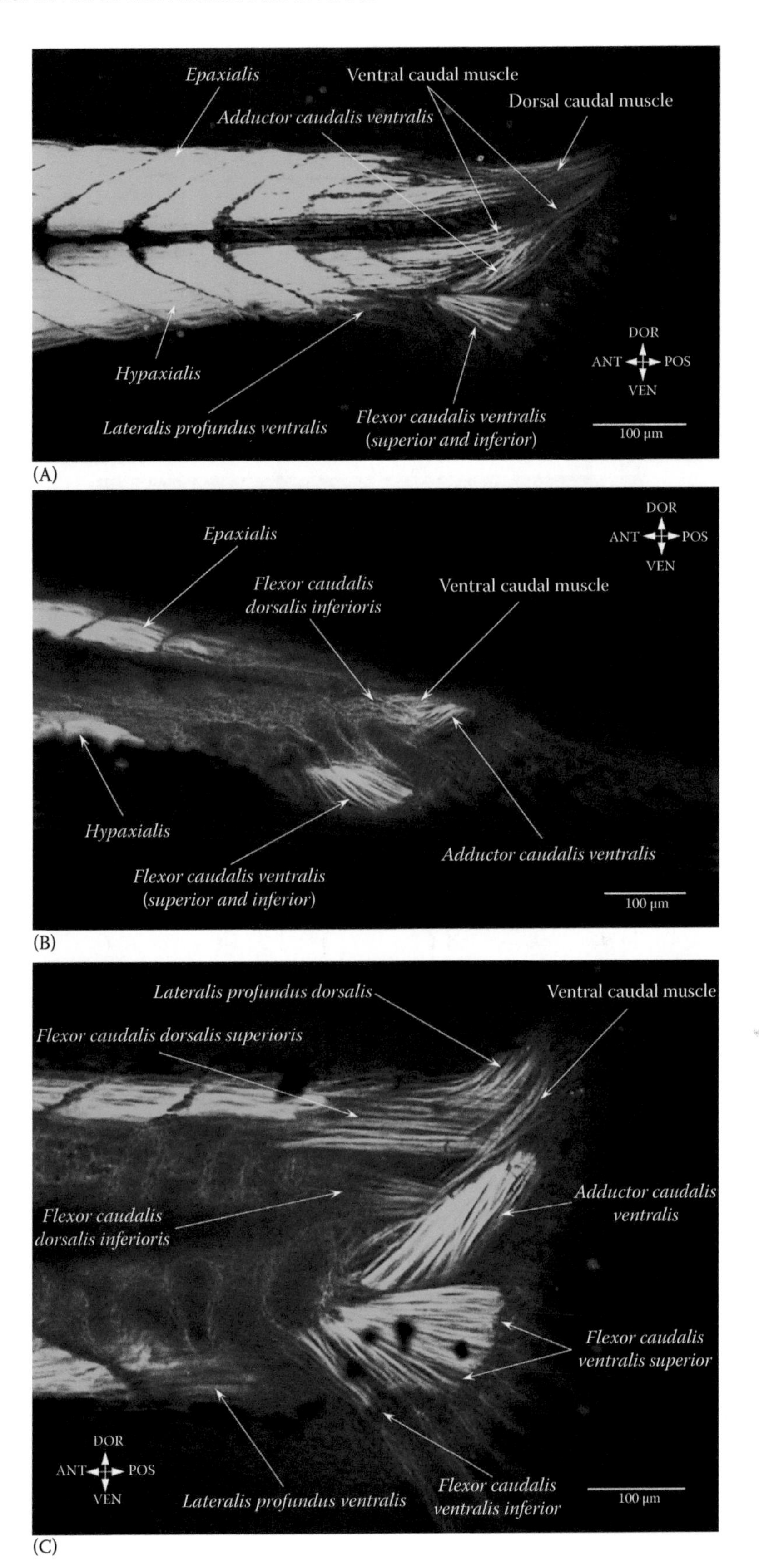

FIGURE 15.3 *Danio rerio* (Teleostei): by 5.0 mm SL, ventral caudal muscles reach the caudal fin rays (A) and development of the deep dorsal fin muscles starts (B and C). The flexor caudalis dorsalis inferioris can be seen at 5.2 mm SL (B), and the flexor caudalis dorsalis superioris appears at 5.5 mm SL (C). The ventral caudal muscle is still attached to the dorsal caudal fin rays, and the flexor caudalis ventralis splits into superior and inferior portions (C).

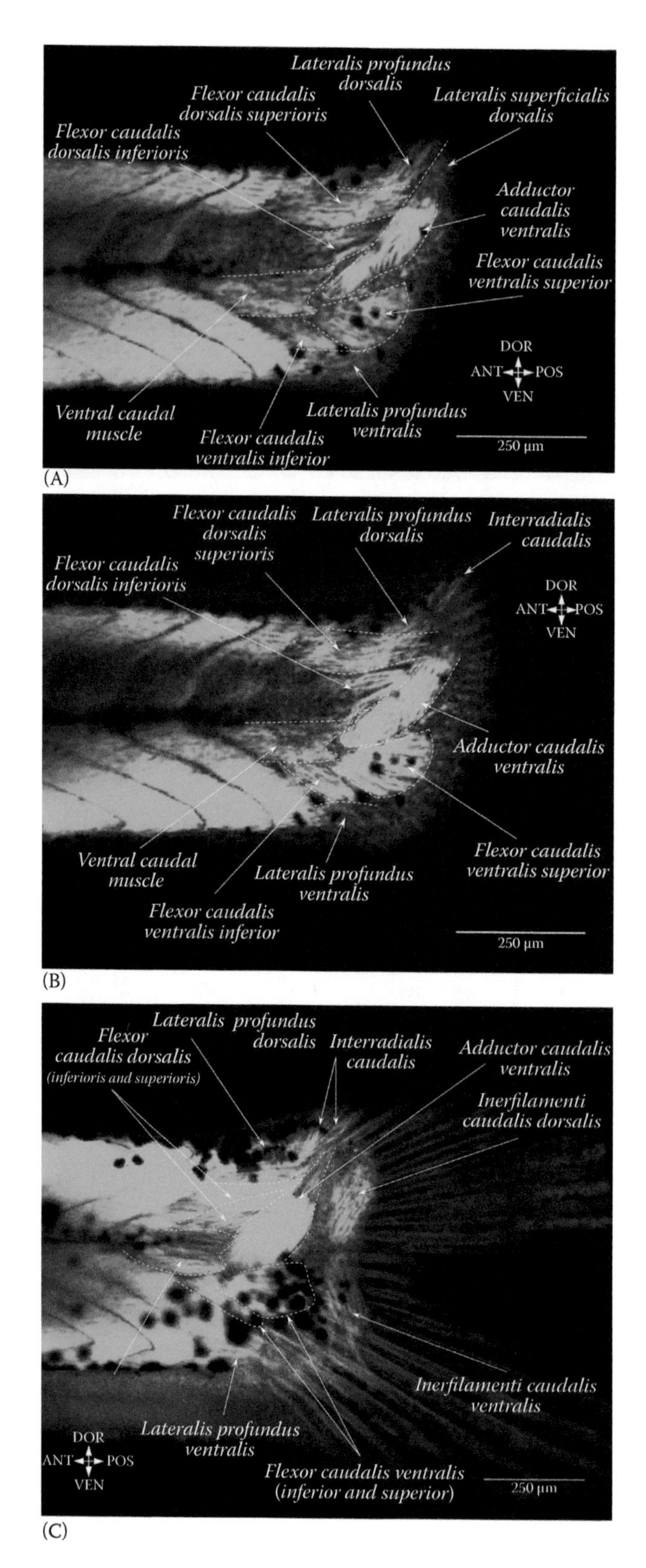

FIGURE 15.4 *Danio rerio* (Teleostei): at 6.4 mm SL, caudal muscles lateralis superficialis dorsalis (A) and interradialis caudalis (B) are formed. The ventral caudal muscle is shifted backward and attaches to the proximal caudal bones and vertebrae. At 6.7 mm SL, interfilamenti caudalis dorsalis and ventralis are formed and thus all caudal muscles are present (C).

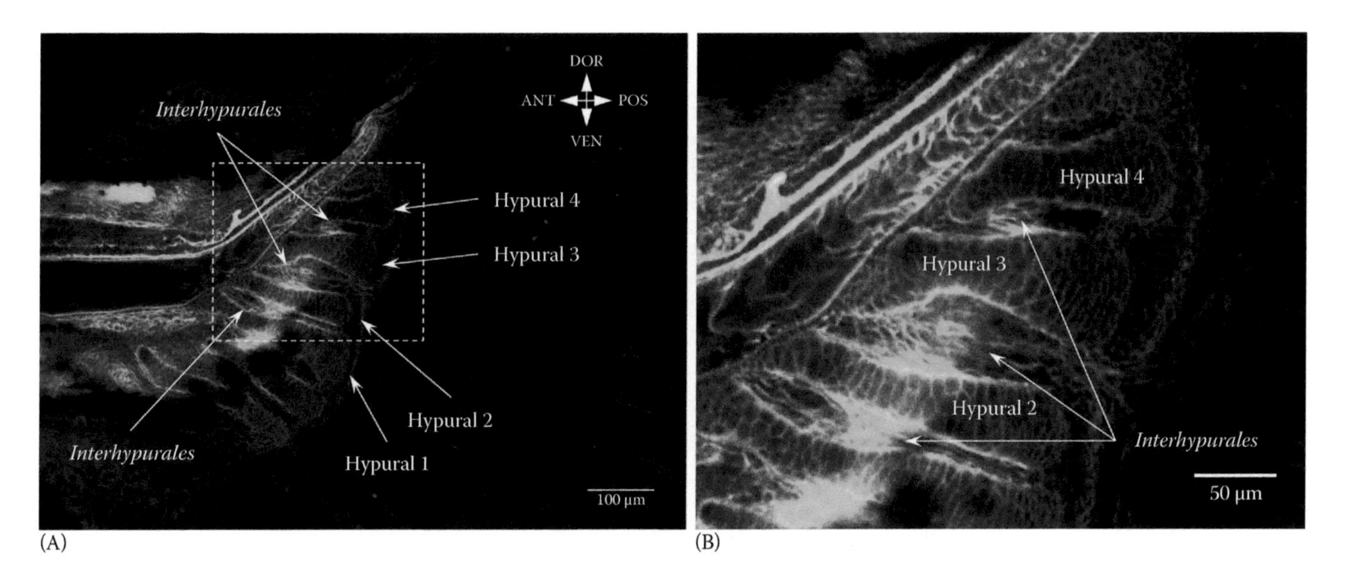

FIGURE 15.5 *Danio rerio* (Teleostei): deep interhypural fibers at 6.0 mm SL. Interhypurales were present between the hypural bones of the caudal fin from 5.00 to 7.1 mm SL. Panel (B) shows a high magnification of the rectangle outlined in the main image (A).

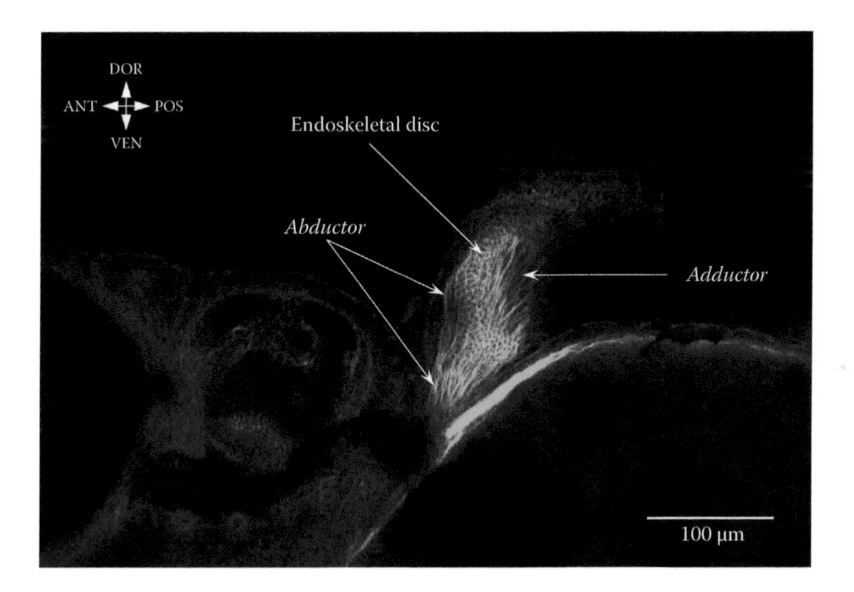

FIGURE 15.6 *Danio rerio* (Teleostei): undifferentiated abductor and adductor muscle masses of the zebrafish pectoral fin at 3.15 mm NL.

The pelvic fins are the last to develop in the zebrafish (Table 15.1). Their buds become visible after 6.7 mm SL. At 7.1 mm SL each pelvic fin already has three differentiated muscle masses (Figure 15.14): the undifferentiated abductor and adductor consist of long thin muscle fibers that stretch proximo-distally along the fin for approximately one-third of its length (Figure 15.14B), and in addition there is also an arrector ventralis pelvicus (Figure 15.14A). Notably, fibers of the arrector dorsalis pelvicus cannot be seen until 7.1 mm SL. The growth of the muscles proceeds quickly, and at 7.5 mm SL, both arrectors are well developed and attach to the base of the first ray (Figure 15.15A). The abductor and adductor muscle masses differentiate into the deep and superficial layers (abductor superficialis and profundus pelvicus; adductor superficialis and profundus pelvicus), which are still difficult to recognize at this stage (Figure 15.15B). By 8.1 mm SL (Figure 15.16), all pelvic muscles are clearly present and have an adult configuration (see Figure 13.14 and Table 13.2).

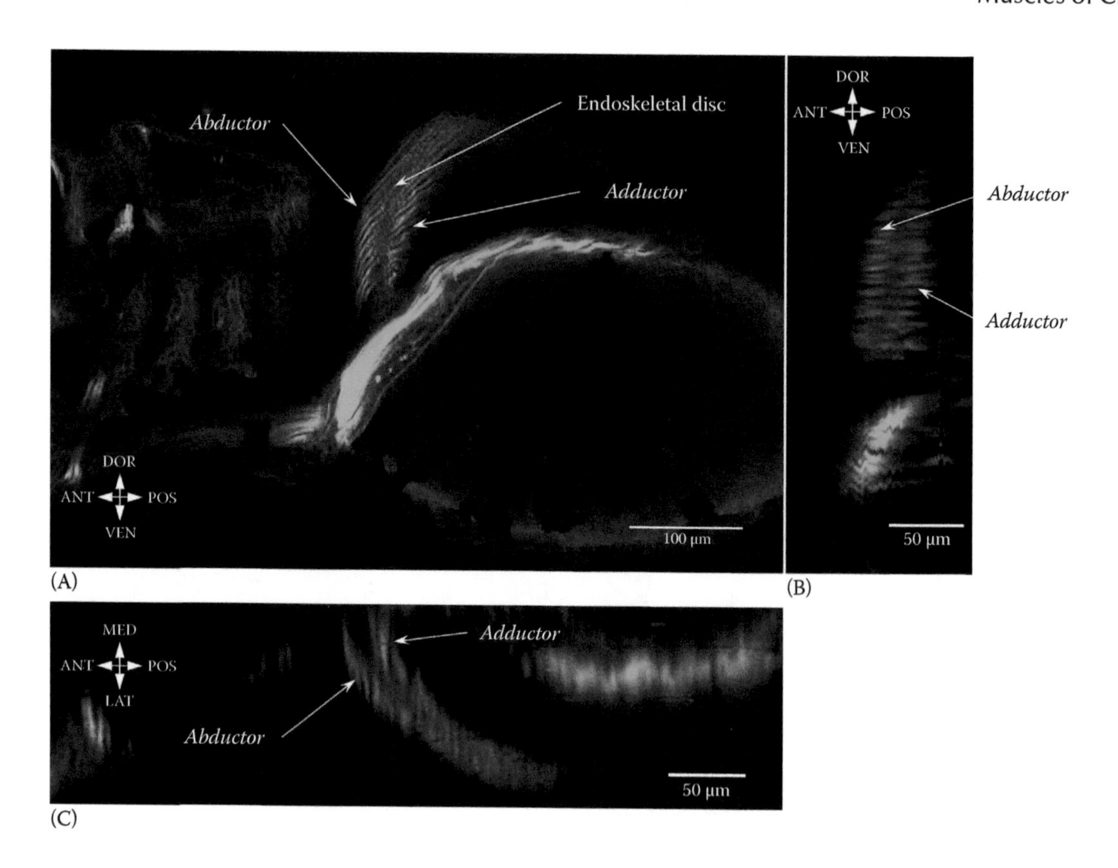

FIGURE 15.7 *Danio rerio* (Teleostei): abductor and adductor masses form two muscle layers, each, in the zebrafish pectoral fin. Lateral view (A) and dorsoventral (B) and anteroposterior (C) cross sections showing that abductor and adductor muscles extend to the edge of the endoskeletal disk and form two muscle layers at 3.3 mm NL.

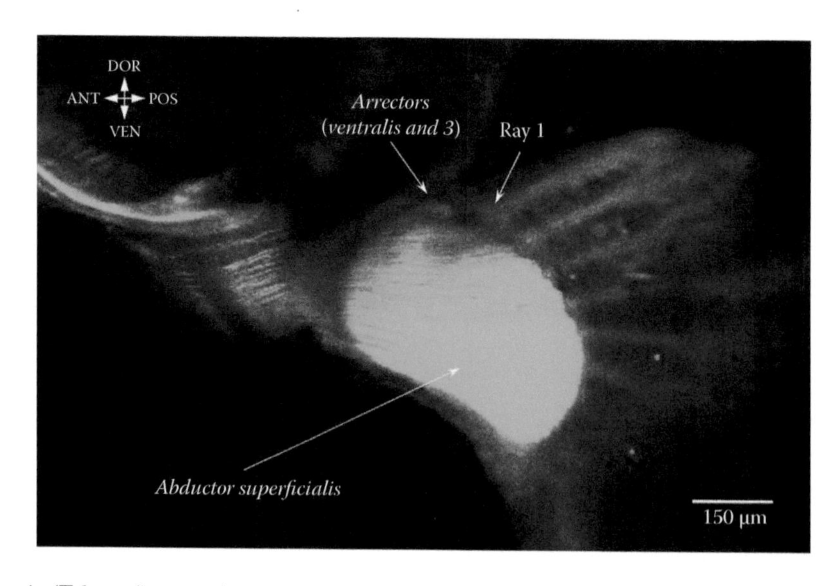

FIGURE 15.8 *Danio rerio* (Teleostei): ventral arrector complex of the pectoral fin in the zebrafish at 6.6 mm SL. The ventral arrector complex will later give rise to the arrector ventralis and arrector 3.

DEVELOPMENTAL AND EVOLUTIONARY UNIQUENESS OF THE CAUDAL FIN

As previously noted, unlike other fins, which are functionally and developmentally distinct structures locally connected to the body, the caudal fin is mainly a posterior continuation of the trunk and of the vertebral column in particular. It has been suggested that such a peculiar position and association with the posterior elements of the postcranial axial skeleton make this fin developmentally and evolutionary distinct from other

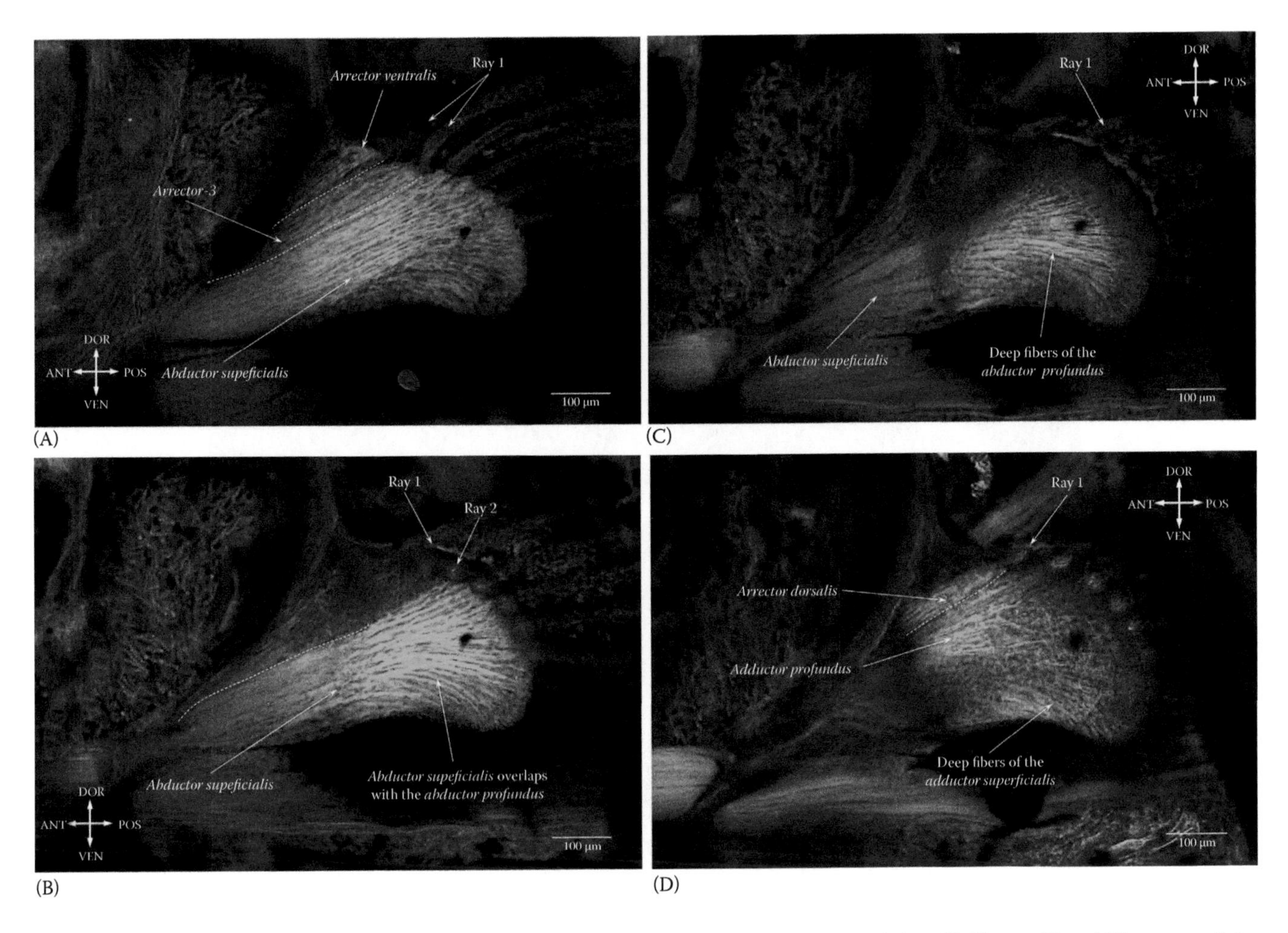

FIGURE 15.9 *Danio rerio* (Teleostei): all the muscles of the ventral/abductor (A and B) and dorsal/adductor (C and D) masses of the zebrafish pectoral fin are developed by 6.7 mm SL.

fins (Quint et al. 2002; Agathon et al. 2003). Our observations and comparisons provide additional evidence supporting this idea.

From very early development, the caudal fin is supported by musculature, while other fins appear as relatively simple homogeneous structures—fin folds or fin buds—that grow and acquire muscles much later in development (exceptionally, pectoral fins develop muscles during embryogenesis: see above) (see Table 15.1). In contrast to such gradual development, the caudal fin at 2.95 mm NL already has two muscles (the dorsal and ventral caudal muscles) that differ from the trunk muscles by their orientation and composition (i.e., absence of myomeric pattern) (Figure 15.2A). Interestingly, even though the caudal fin is the first fin to appear in the zebrafish, it does not reach the adult configuration before other fins do, probably because of its complexity as it includes more muscles and skeletal elements than any other fin (Table 15.1). That is, the caudal fin develops gradually along with fish growth and the adult muscle configuration of the caudal fin is attained at a similar developmental stage as in the pectoral, dorsal, and anal fins (i.e., by about 6.7 mm SL) (Figure 15.4C).

Another distinguishing feature of the caudal fin muscles is the proximal shift during development, that is, away from the fin rays. While muscles of all other fins mainly grow toward the rays and insert onto their bases, the position of the ventral caudal muscle changes from more dorsal to more medial and later the connection between this muscle and the caudal fin rays is lost (Figures 15.3C and 15.4A). In adult zebrafishes, the ventral caudal muscle becomes a deep trunk muscle, attached to the caudal vertebrae and proximal caudal bones. This particular muscle rearrangement along with the intensive growth of the ventral caudal muscles and notochord bending (i.e., the adductor caudalis ventralis and flexor caudalis ventralis superior and inferior) results into the peculiar marked dorsoventral asymmetry of the caudal fin, which initially was mainly symmetrical (until 4.4 mm NL: Figure 15.2A). The presence of the temporary interhypural muscles on the dorsal side only, development of the lateralis superficialis dorsalis before the lateralis superficialis ventralis, and appearance of the interradialis dorsalis at 6.4 mm SL enhance the difference between the dorsal and ventral sides of this fin. Thus, what appears (externally) to be a dorsoventral symmetrical caudal fin, with roughly equal dorsal and ventral lobes and somewhat evenly distributed fin rays, is a fin with muscles that display a marked dorsoventral asymmetry (Schneider and Sulner 2006; see Chapter 14 and Figure 14.9).

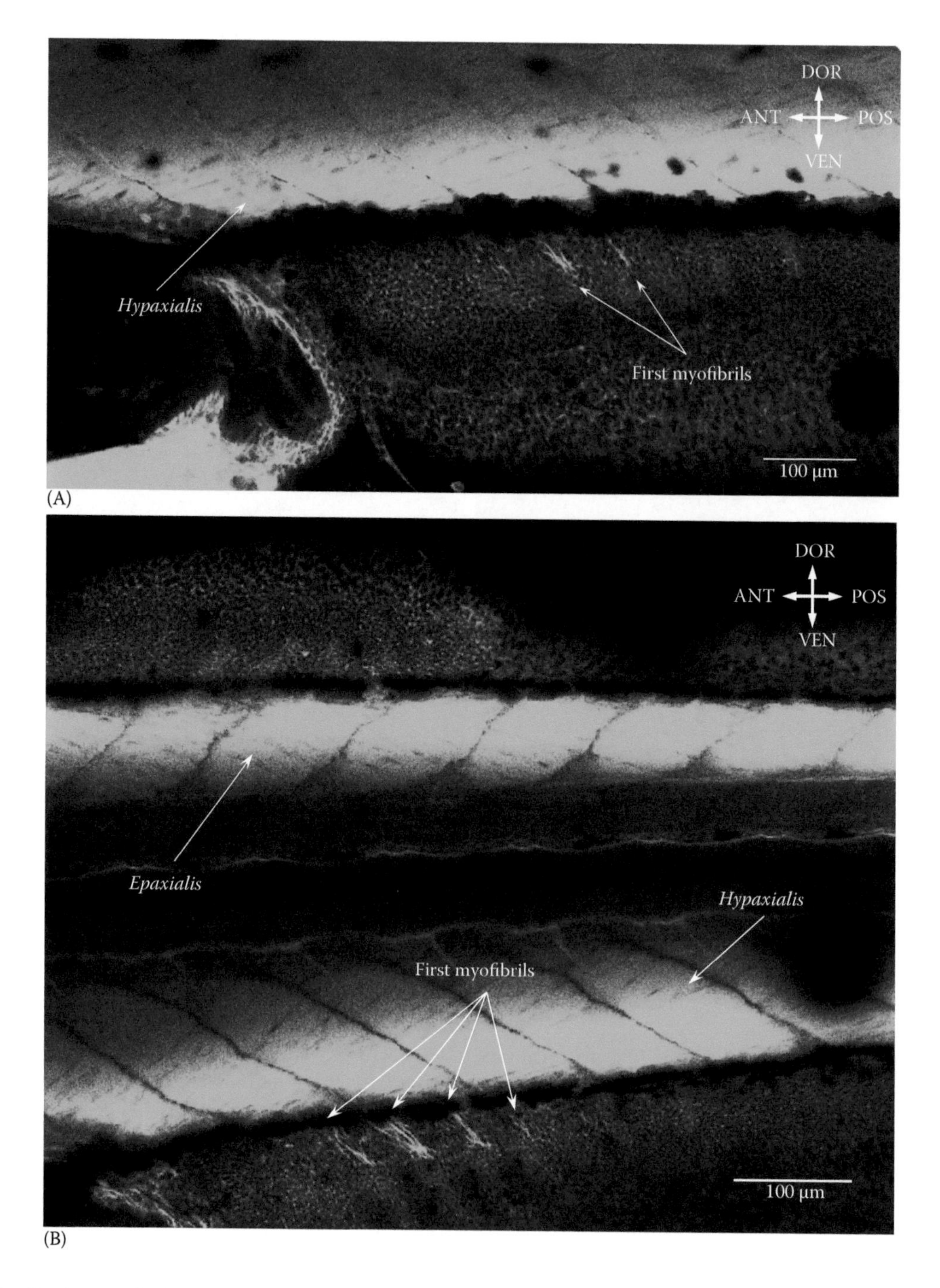

FIGURE 15.10 *Danio rerio* (Teleostei): early development of the dorsal and anal fin musculature. The first muscle fibers are seen in the anal fin at 5.8 mm SL (A). Number of fibers and serial units with the muscle fibers increases by 6.0 mm SL (B).

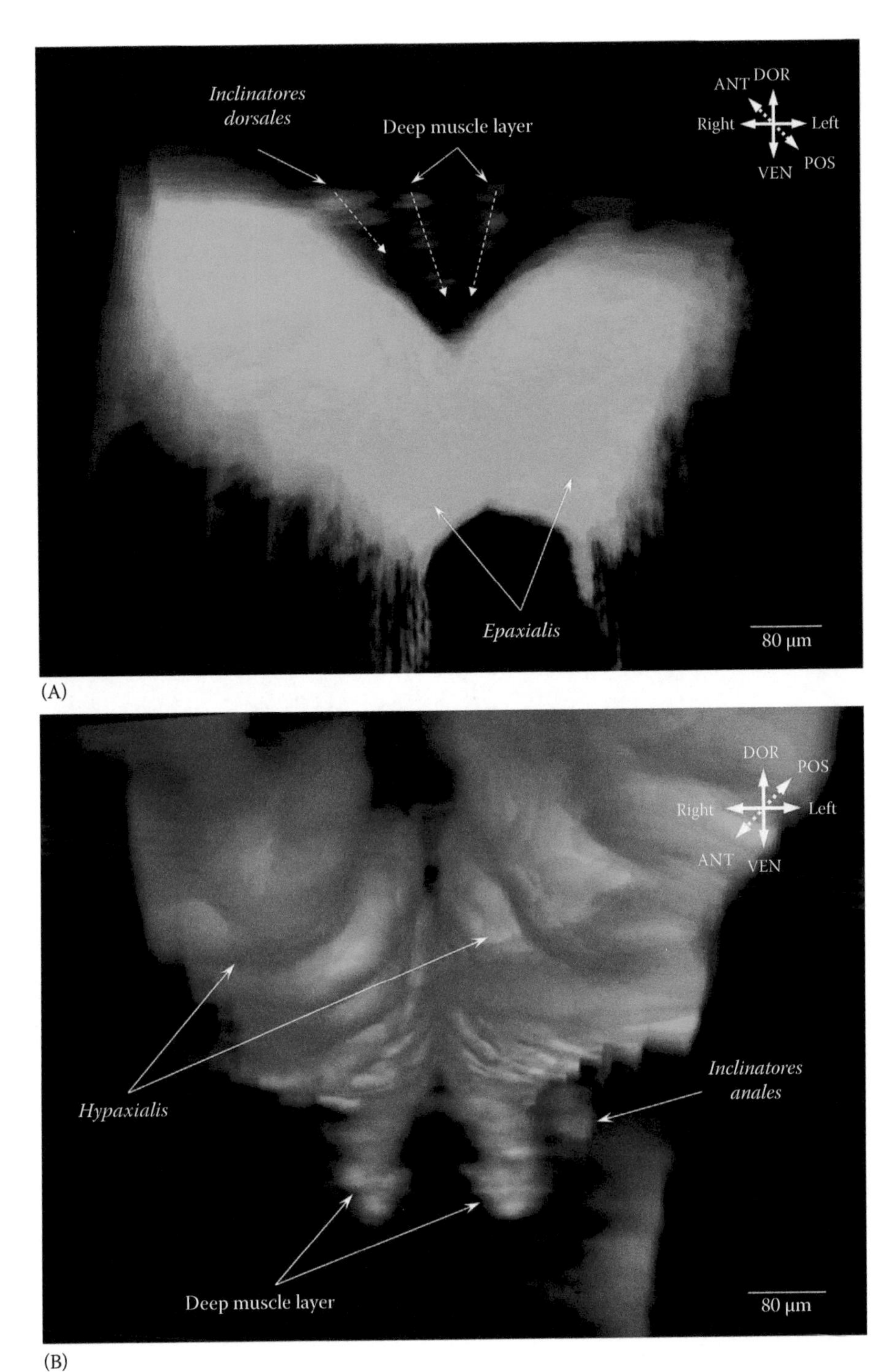

FIGURE 15.11 *Danio rerio* (Teleostei): two muscle layers in the dorsal and anal fins at 6.2 mm SL. The superficial muscle layer of the dorsal (A) and anal (B) fins corresponds/gives rise to the inclinators present in later stages. The deep muscle layer will later split into erectors and depressors.

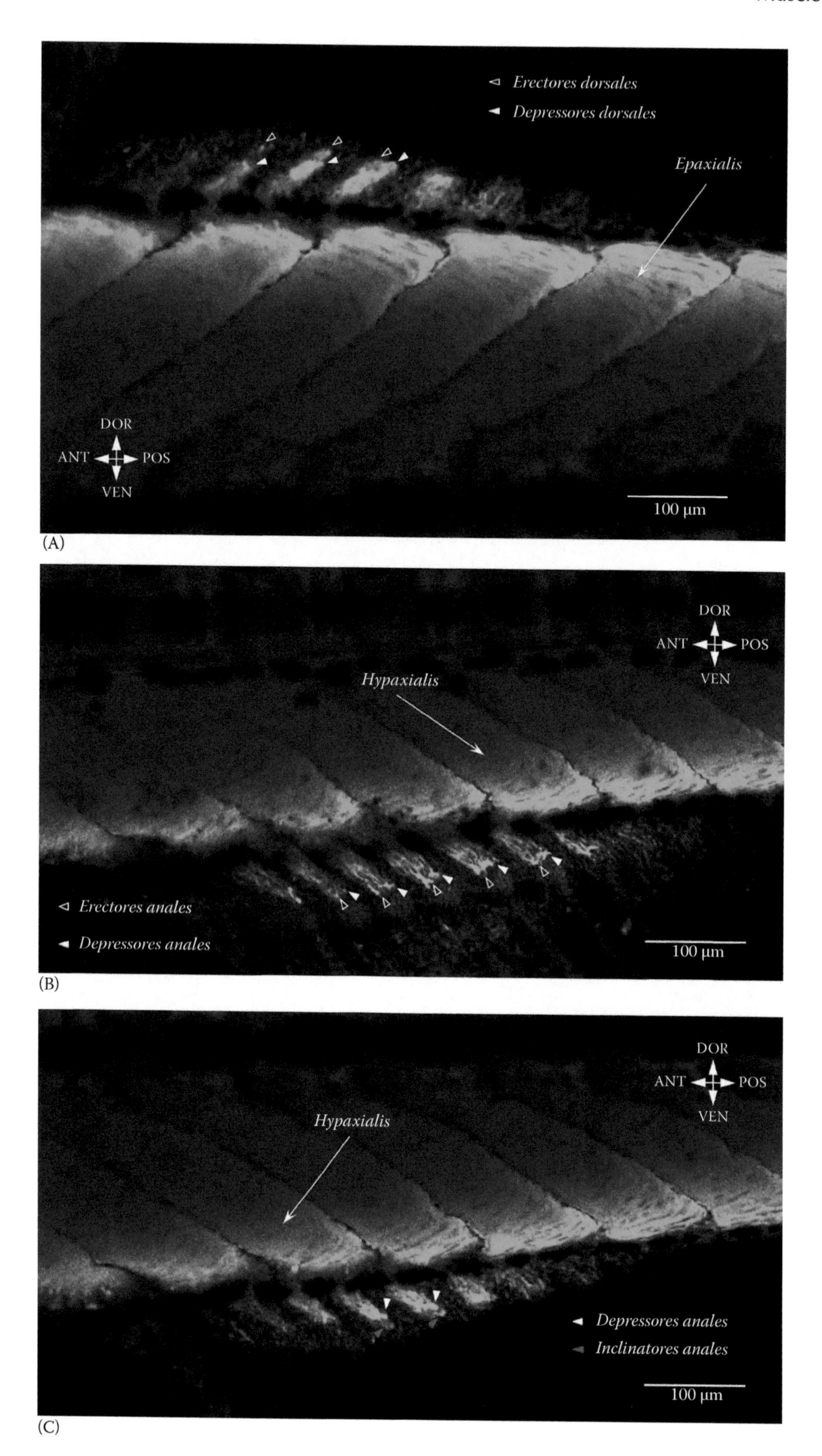

FIGURE 15.12 *Danio rerio* (Teleostei): muscular composition of serial units in the dorsal and anal fins of the zebrafish at 6.4 mm SL. Erectors and depressors of the dorsal (A) and anal (B) fins form the deep muscle layer, covered with superficial inclinators (C) at 6.4 mm SL.

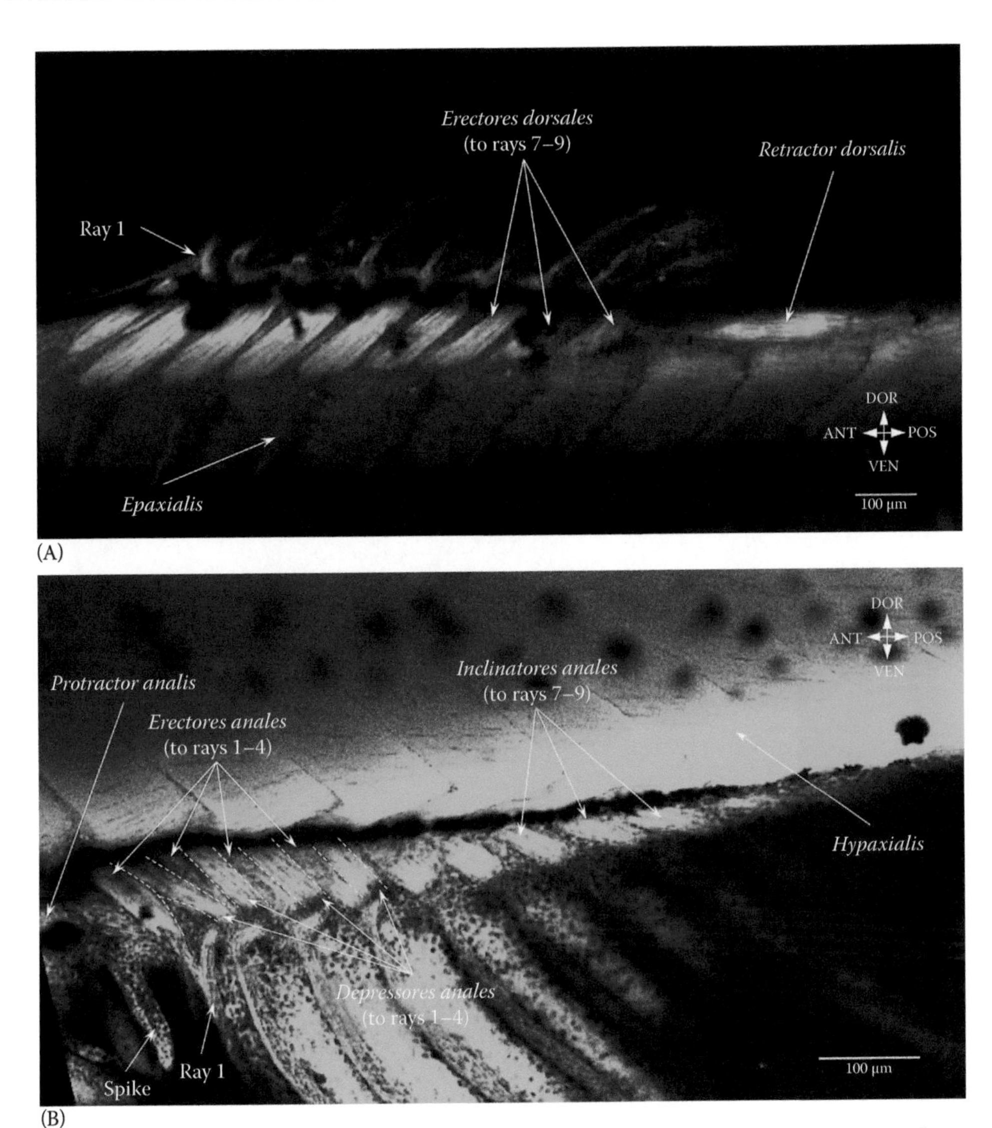

FIGURE 15.13 *Danio rerio* (Teleostei): origin of the protractors and retractors in the dorsal (A) and anal (B) fins at 6.7 mm SL led to a configuration where all adult muscles of these fins are present.

This dorsoventral asymmetry of the caudal fin is also seen in its bone architecture (Sanger and McCune 2002) as well as in the development of its rays (Parichy et al. 2009). In fact, it is interesting to note that even dorsal and ventral caudal muscles that are relatively symmetric in adults, and are hence commonly described under similar names, display very different developmental patterns. For instance, the lateralis profundus dorsalis, which basically corresponds to the dorsal caudal muscle, is well developed at very early stages (at 3.3 mm SL) and extends to the tip of the tail (Figure 15.2A). In contrast, the lateralis profundus ventralis develops later and derives from the hypaxial, segmented musculature, not from the ventral caudal muscle (Figure 15.2B and C). A similar discrepancy can also be seen in the development of the dorsal and ventral flexors. On the ventral side, a single flexor caudalis ventralis splits into the flexor caudalis ventralis superior and flexor caudalis ventralis inferior. In contrast, the flexor caudalis dorsalis superioris and flexor caudalis dorsalis inferioris appear at different stages from different sources: from the dorsal and ventral caudal muscles, respectively (Figure 15.3B and C).

GENERAL REMARKS

As shown in Figure 15.1, the development of the zebrafish appendicular muscles does not proceed at a constant rate during embryonic development: there are clear cases of developmental acceleration and steady states between them. At 3.2 mm NL, zebrafish embryos acquire first four appendicular muscle masses: two caudal ones (the dorsal and ventral caudal muscles) and two pectoral ones (the abductor and adductor muscles). From this stage starts the first steady condition that lasts for 1 mm NL change, until 4.2 mm NL, in which these four muscles grow along with the overall size of the tail and pectoral fins. During this period, no new muscles arise. At 4.2 mm NL, the first small developmental burst starts, and by 4.8 mm NL,

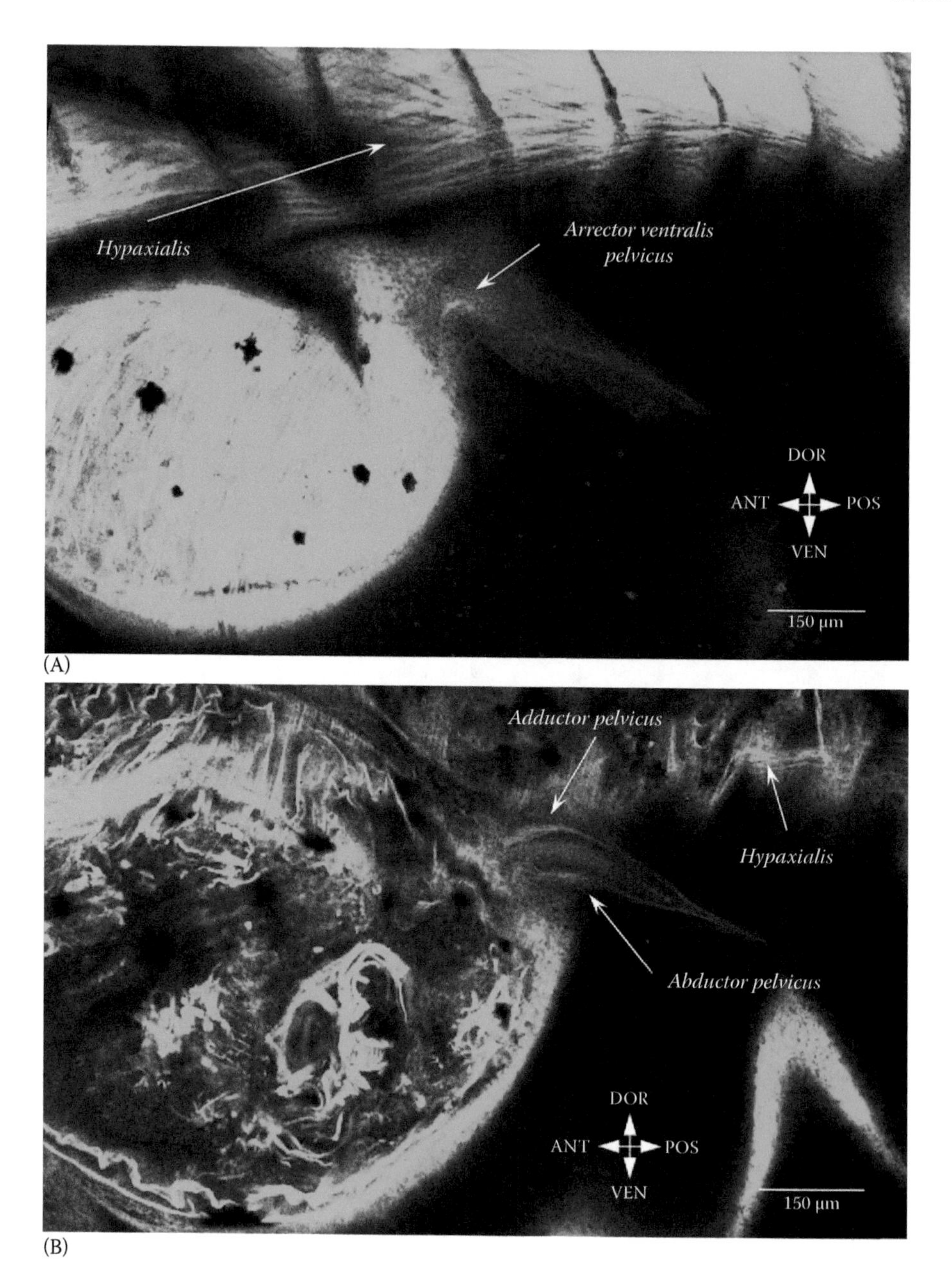

FIGURE 15.14 *Danio rerio* (Teleostei): early development of the pelvic fin musculature in the zebrafish. At 7.1 mm SL, fibers of the initial three muscles are present: arrector ventralis pelvicus (shown in A), abductor, and adductor pelvicus (shown in B). Fibers of the arrector ventralis pelvicus do not reach the fin rays. Tissue condensation can be seen distal to the arrector fibers (A).

five new muscles appear. The first muscle fibers of the anal fin appear at the end of this acceleration time (at 4.8 mm NL), and during the next steady period (4.8 mm NL–6.0 mm SL), the caudal, pectoral, and anal fins keep a constant number of muscles. The second developmental burst is seen by the muscle development in the dorsal fin at 6.2 mm SL and leads to the adult muscle configuration in all five fin types. During the next 1.4 mm SL change, the number of muscles increases from 13 to 32. The pelvic musculature development is continuous without passing through a steady state and thus falls within the last third of the second developmental burst.

We found a discrepancy between the order of fin development within the zebrafish and the order of origin of the fins in the phylogenetic history of vertebrates. Over evolutionary history, the median fins appear before the pectoral and pelvic ones. Contrary to this, in zebrafish ontogeny the caudal and pectoral fins, including their muscles, start developing during embryogenesis much earlier than all other fins. Interestingly, in contrast to the order of fin development in general, the order of individual muscle development of the appendicular musculature in the zebrafish coincides with the order in which muscles appeared in the evolutionary history that lead to teleosts. As noted in the preceding chapters, this parallel between phylogeny and ontogeny was also reported in the head muscles of zebrafish (Diogo et al. 2008c) as well in studies of muscles of other vertebrate

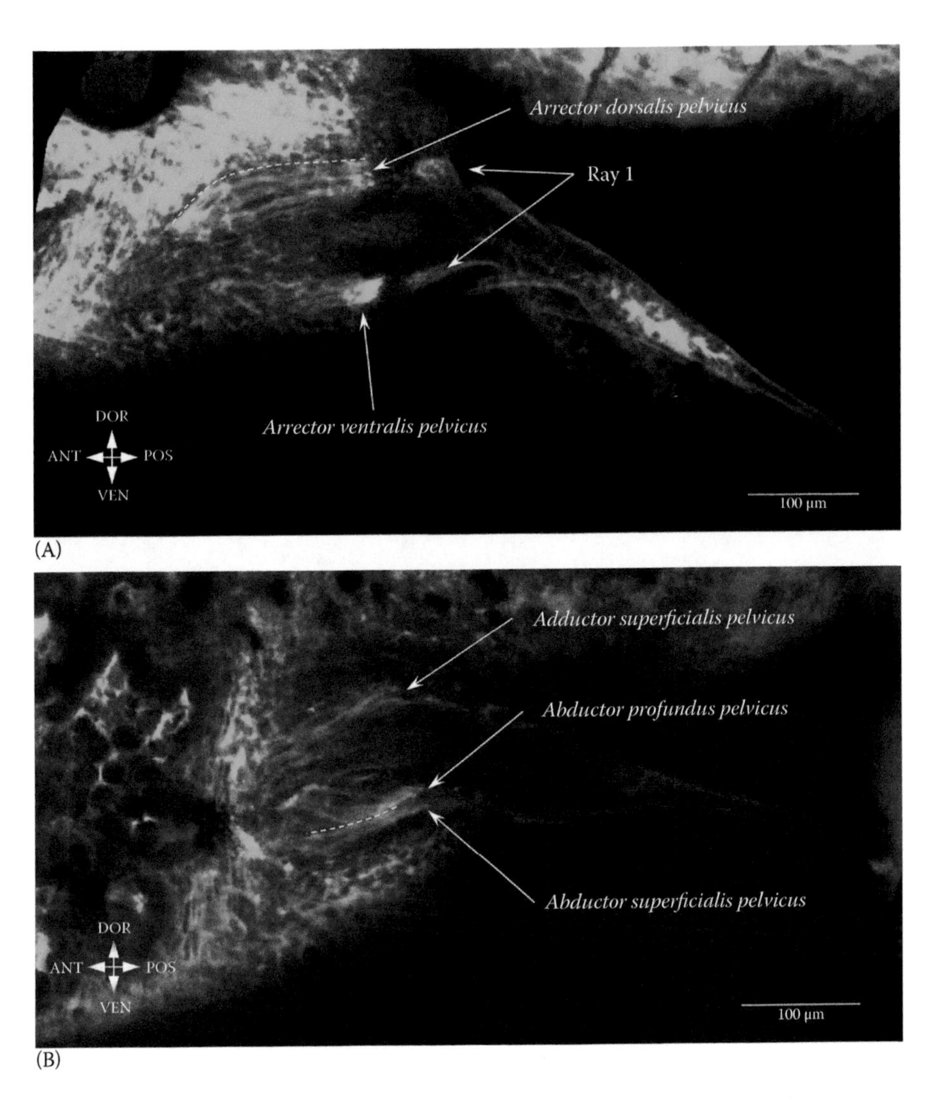

FIGURE 15.15 *Danio rerio* (Teleostei): two arrectors of the pelvic fin are formed by 7.5 mm SL. Both the arrector dorsalis pelvicus and arrector ventralis pelvicus are well developed (A). The abductor and adductor muscles start segregating into the deep and superficial layers (B).

taxa, being thus designed as the "phylo-devo parallelism" by Diogo et al. (2015c) (see Chapter 7). For instance, the ontogenetic development of the caudal musculature in the zebrafish reflects all major evolutionary transitions seen in the caudal fin from nonvertebrate chordates such as amphioxus (see Figure 1.1) to teleosts such as the zebrafish. The dorsal and ventral caudal muscles of the zebrafish are in a certain sense a continuation of the trunk musculature, as noted in the preceding text. Even though fibers of these two muscles are not segmented into myomeres (Figure 15.2A), they are inextricably linked to the trunk muscles and form the tail of the animal, which lacks a proper caudal fin at this stage but has a caudal fin fold similar to that of amphioxus (Mansfield et al. 2015). As explained in Chapter 14 and shown in Figure 14.1, the next evolutionary step is the development of the ventral intrinsic caudal muscle (radialis) seen in fishes such as sharks (Flammang 2009). Similar to the phylogenetic sequence, the first intrinsic muscles of the caudal fin in the zebrafish develop on the ventral side at 4.4 mm (Figure 15.2B and C).

Such an ontogenetic–phylogenetic parallelism is also seen in the development of the pectoral and pelvic muscles of the zebrafish. Thus, the pectoral fin musculature at early stages comprises two muscle masses attached to the endoskeletal disk—the abductor and adductor. Such simple muscle composition has been hypothesized to represent a plesiomorphic state for the pectoral appendages (Diogo and Abdala 2010; Diogo et al. 2016b). Further differentiation of muscles from the abductor mass going to the first ray in the zebrafish (the arrector ventralis) was acquired in the last common ancestor of extant gnathostomes (Diogo et al. 2016b), while the acquisition of arrector 3 is a late evolutionary event that occurred during teleost evolutionary history and was apparently independently acquired in the Ginglymodi (see Figure 12.1). Accordingly, the separation of arrector 3 from the arrector ventralis complex is the last ontogenetic event in the developing pectoral fins of the zebrafish (Table 15.1).

Concerning the pelvic appendage, the first three muscles to form are the adductor and abductor masses and the

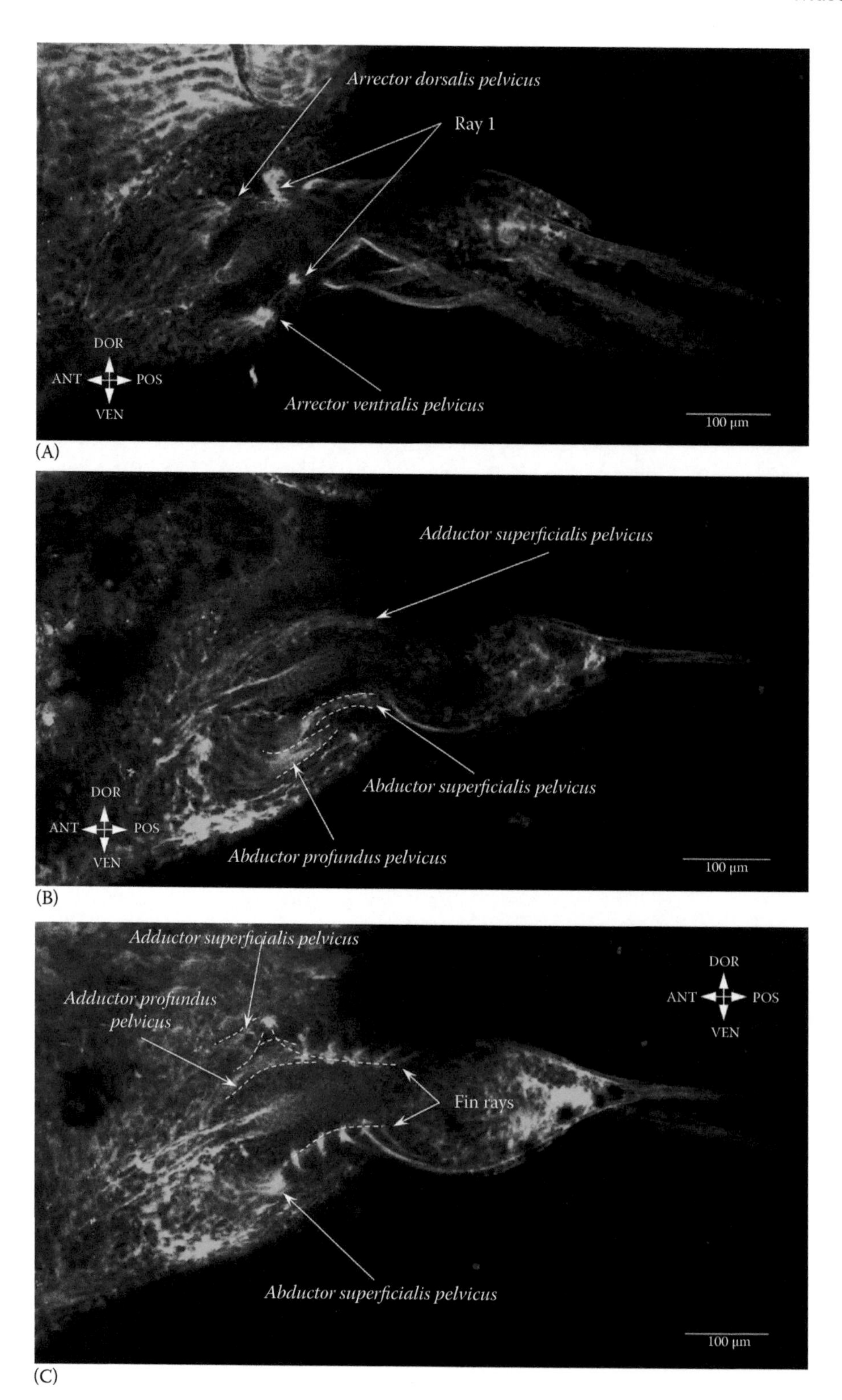

FIGURE 15.16 *Danio rerio* (Teleostei): all muscles of the pelvic fin musculature are present at 8.1 mm SL, i.e., arrectors (shown in A), abductors (shown in B), and adductors (shown in C).

arrector ventralis pelvicus. These three muscles were present in the last common ancestor of extant gnathostomes (Diogo et al. 2016b). The arrector dorsalis pelvicus was apparently acquired during the evolutionary history of actinopterygians, because it is absent in chondrichthyans, sarcopterygians, and the phylogenetically basal actinopterygians such as *Polypterus* and chondrosteans (see Figure 12.2). Accordingly, the arrector dorsalis also develops ontogenetically later than the other muscles of the pelvic fins of the zebrafish (Table 15.1).

16 Pectoral and Pelvic Appendicular Muscle Evolution from Sarcopterygian Fishes to Tetrapods

Most studies on the origin of limbs focus on fossil skeletal structures (e.g., Coates et al. 2002; Shubin et al. 2004, 2006; Ahlberg 2011; Pierce et al. 2012), mainly because fossils usually do not preserve soft tissues, and because it is difficult to compare fish fins and tetrapod limbs as they are morphologically very different (e.g., in orientation of axes and number/configuration of muscles). Classic comparative anatomy works provided in-depth descriptions of the major rotation of the paired appendages that occurred during the early stages of the fin–limb transition: the preaxial (radial/tibial) side, directed anterodorsally in extant fishes such as *Polypterus*, *Latimeria*, and living dipnoans, became directed anteroventrally (e.g., Humphry 1872a; Braus 1941; Romer 1924; Gregory and Raven 1941; Romer 1942, 1944). However, these descriptions are not always taken into account in recent works, leading to errors and terminological problems (see the following text). Although numerous appendicular muscles have been described in the coelacanth *Latimeria* (Millot and Anthony 1958; Miyake et al. 2016), these descriptions are often excluded from recent discussions about the fin–limb transition because dipnoans are phylogenetically closer to tetrapods than are coelacanths (e.g., Brinkmann et al. 2004). Therefore, most authors agree that a transition occurred after the dipnoan–tetrapod divergence, from a very simple fin configuration with only two major muscle masses (adductor/abductor) to the highly complex tetrapod limbs that can have more than 50 muscles (reviewed in Diogo and Abdala [2010]). Accordingly, extant phylogenetic bracketing (Witmer 1995a), one of the most powerful tools for soft tissue reconstruction, has never been used to study this fin–limb transition (Bishop 2014), despite the fact that the relationships of extant sarcopterygians have long been well established (Meyer and Dolven 1992).

Original data and comparisons obtained from extant taxa are thus crucial to pave the way for the use of this method in muscle reconstructions of key tetrapod and nontetrapod sarcopterygian extinct taxa. For a paper done by two of us (R. D. and J. M.) together with Peter Johnson and Borja Esteve-Altava (Diogo et al. 2016b), we obtained new musculoskeletal data from dissections, magnetic resonance imaging scans, three-dimensional reconstructions, and histological sections of coelacanths (*Latimeria*) and dipnoans (*Neoceratodus*) (see the following text and Tables 16.1 through 16.4, which show all muscle–bone attachments in these taxa) and combined them with data gathered during our long-term study on chordate muscles that led to the writing of this book. This chapter is thus mainly based on that paper. Regarding the muscular anatomy of lobe-finned fishes, the major contributions of the data presented in this chapter are (a) description of new muscles; (b) reappraisal of evolutionary origin (e.g., from ventral/abductor vs. dorsal/adductor masses) and identity of previously described muscles; and (c) first comprehensive comparisons of pelvic and pectoral appendages among these and other fishes and in tetrapods, leading to proposal of new names, evolutionary origins, and one-to-one homology hypotheses for all muscles of these taxa.

Specifically, in Tables 16.5 and 16.6 and Figures 16.1 through 16.3, we present one-to-one homology hypotheses for the muscles of the paired appendages across five major extant gnathostome clades: chondrichthyans (shark, *Squalus*), as the extant sister-group of osteichthyans (bony fishes); actinopterygians (bichir, *Polypterus*), as the extant sister-group of sarcopterygians; coelacanths (*Latimeria*), as the extant sister-group of dipnoans plus tetrapods; dipnoans (*Neoceratodus*); and tetrapods (*Ambystoma*), as salamanders are, anatomically, the most plesiomorphic extant tetrapods (i.e., the most similar to the last common ancestor (LCA) of extant tetrapods (Millot and Anthony 1958; Bardeen 1906; Diogo and Ziermann 2015b). These evolutionary hypotheses are summarized in Figures 16.4 and 16.5. As explained in the following text, our hypotheses of homology between fin muscles in *Latimeria* and *Neoceratodus* are very straightforward because three out of the four fins studied have very similar muscle configurations. Here we summarize the major points supporting key homology hypotheses. The same points can be applied to support similar homology hypotheses between other muscles. These hypotheses combine developmental, anatomical, and paleontological evidence and multiple cross comparisons with other muscles from the same and from other paired appendages in different taxa, and embryonic primordia, following strict standards of homology such as (a) positional equivalence, determined by bony attachments; (b) special quality, determined by, e.g., the orientation of fibers and innervation of muscles; (c) transition, determined by paleontological and/or developmental evidence of intermediate conditions; and (d) congruence, determined by applying the previous criteria to adjacent muscles and muscles of both the dorsal and ventral sides of each appendage and of the two paired appendages (i.e., pectoral vs. pelvic). For example, regarding the use of paleontological data, the homology hypotheses shown in Table 16.5 and Figure 16.2 are consistent with microanatomical evidence that the humerus of the early tetrapodomorph

TABLE 16.1
Origins and Insertions of Pectoral Muscles of *Neoceratodus*

Muscle	Origin	Insertion
Retractor lateralis ventralis pectoralis	Cranial rib	Medial face of cleithrum
Adductor superficialis (including dorsal superficial segmented muscular layer)	Cleithrum and scapulocoracoid dorsal to articular process	*Via* aponeurosis onto distal radials and bases of lepidotrichia; divided by tendinous sheets that insert on joints between axial elements
Adductor profundus	Scapulocoracoid dorsal to articular process	Dorsal face of the first element
Abductor superficialis (including ventral superficial segmented muscular layer)	Lateral face of clavicle, cleithrum, scapulocoracoid ventral to articular process	Distal radials and bases of lepidotrichia; divided by tendinous sheets that insert on joints between axial elements
Abductor profundus	Scapulocoracoid adjacent to and ventral to articular process	Postaxial border of the first element

TABLE 16.2
Origins and Insertions of Pelvic Muscles of *Neoceratodus*

Muscle	Origin	Insertion
Abductor dorsolateralis ("superficial ventrolateral abductor")	Body wall muscles dorsal to pelvis	Distal, lateral edge of the first element
Adductor superficialis ("mesial abductor" + superficial dorsal segmented layer that corresponds to "dorsal lepidotrichial flexors + radial flexors")	Midline raphe connecting with adductor superficialis on the contralateral side	Distal radials and bases of lepidotrichia; divided by tendinous sheets that insert on joints between axial elements
Pterygialis caudalis (postaxial muscle, or "superficial ventrolateral + ventromesial adductor")	Midline raphe connecting with pterygialis caudalis on the contralateral side	Distal, medial edge of the first element
Adductor profundus ("dorsomesial adductor–levator")	Dorsal face of pubic ramus	Joint between first and second elements *via* tendinous sheet
Pronator 1 (dorsolateral abductor–levator)	Caudolateral face of pubic ramus	Proximal, lateral edge of the first element and joint between first and second elements with adductor profundus
Pronators 2–9 (dorsal "radial–axial" muscles)	All axial elements	Distal ends of radials
Abductor superficialis ("superficial ventromesial abductor" + superficial ventral segmented layer that corresponds to "ventral lepidotrichial flexors + radial flexors")	Anterolateral process of pelvis and adjacent (lateral) body wall	Distal radials and bases of lepidotrichia; divided by tendinous sheets that insert on joints between axial elements
Pterygialis cranialis (preaxial muscle, or part of "superficial ventromesial abductor")	Caudolateral face of pubic ramus	Distal end of first preaxial radial
Abductor profundus ("deep ventral abductor–depressor")	Ventrolateral face of pelvis caudal to anterolateral process	Distal, ventral edge of first element
Supinator 1 ("deep ventral adductor–depressor")	Medial face of pubic ramus	Distal, medial edge of first element
Supinators 2–9 (ventral "radial–axial" muscles)	All axial elements	Distal ends of radials

Source: Names in parentheses from Young, G. C. et al., Pelvic girdles of lungfish (Dipnoi). In *Pathways in Geology: Essays in Honour of Edwin Sherbon Hills* (ed. LeMaitre, R. W.), pp. 59–75, Blackwell Scientific, Melbourne, 1989.

fish *Eusthenopteron* had osteological correlates of a muscle corresponding to pronator 1 in *Latimeria* (Sanchez et al. 2013). Regarding the use of ontogenetic data, an illustrative example concerns the pterygialis cranialis of the pelvic fin of *Latimeria* and *Neoceratodus*, which is similar to the pelvic muscle ischioflexorius of salamanders because developmental evidence supports the idea that both the pterygialis cranialis of fishes and the ischioflexorius of tetrapods are derived from the ventral embryonic muscle mass (Diogo and Tanaka 2014; Diogo and Ziermann 2015b). In fact, distally the ischioflexorius includes the ancestral leg muscle flexor cruris et tarsi tibialis, which is a preaxial muscle (like the pterygialis cranialis) that corresponds topologically to the preaxial muscle flexor antebrachii et carpi radialis of the salamander forelimb (Diogo and Tanaka 2014) (Figures 16.2 and 16.3; Tables 16.5 and 16.6). Moreover, in tetrapods such as salamanders, the flexor antebrachii et carpi radialis differentiates from a different primordium than do the more ulnar/postaxial muscles flexor antebrachii et carpi ulnaris and flexor digitorum communis, suggesting that the former preaxial forearm

TABLE 16.3
Origins and Insertions of Pectoral Muscles of *Latimeria*

Muscle	Origin	Insertion
Adductor superficialis ("levator superficialis")	Posteromedial border of cleithrum between anocleithrum and endoskeleton	*Via* a broad tendon that attaches onto bases of lepidotrichia; bundles insert onto preaxial radials with pronators 2 and 3
Adductor profundus ("levator profundus")	Medial face of cleithrum and endoskeleton in region of articular process	Deep face of adductor superficialis
Pronator 1	Medial face of endoskeleton adjacent to articular process	First preaxial radial and adjacent joint between first and second elements; a bundle continues with pronator 2
Pronator 2	Postaxial border of first element	Second preaxial radial and adjacent joint between second and third elements; a bundle continues with pronators 3 and 4
Pronator 2a	Preaxial border of first element	With pronator 2
Pronator 3	Postaxial border of second element	Bases of first 8–10 preaxial lepidotrichia
Pronator 3a	Preaxial border of second element	With pronator 3
Pronator 4 + 4a	Pre- and postaxial borders of third element	Bases of preaxial lepidotrichia distal to pronator 3 and small cartilages distal to the fourth element
Abductor superficialis ("abaisseur" superficialis)	Bedial face of cleithrum, extracleithrum and clavicle ventral to the articular process	*Via* a broad aponeurosis onto bases of lepidotrichia; bundles insert onto preaxial radials with supinators 2 and 3
Abductor profundus ("abaisseur" profundus)	Medial face of endoskeleton ventral to articular process	Deep face of abductor superficialis
Supinator 1	Medial face of endoskeleton immediately adjacent and ventral to articular process	First preaxial radial and dorsolateral aspect of the joint between first and second elements
Supinator 2	Postaxial border of first element	Second preaxial radial and adjacent joint between second and third elements; partially fused with supinator 3
Supinator 2a	Preaxial border of first element	With supinator 2
Supinator 3	Postaxial border of second element	Bases of first 8–10 preaxial lepidotrichia; partially fused with supinators 2 and 4
Supinator 3a	Preaxial border of second element	With supinator 3
Supinator 4	Postaxial border of third element	Bases of preaxial lepidotrichia distal to supinator 3 and small cartilages distal to fourth element; partially fused with supinator 3
Supinator 4a	Preaxial border of third element	With supinator 4
Pterygialis caudalis (postaxial muscle, or "supinator 5 and/or pronator 5")	Postaxial borders of first–third elements together with pronators and supinators 2–4	Postaxial border between aponeuroses of adductor and abductor superficialis
Pterygialis cranialis (preaxial muscle)	From abductor superficialis	Preaxial radials and bases of lepidotrichia

Source: Names in parentheses from Millot, J., and Anthony, J., *Anatomie de Latimeria chalumnae—I, squelette, muscles, et formation de soutiens*, CNRS, Paris, 1958.

muscle, as well as the corresponding preaxial leg muscle flexor cruris et tarsi tibialis, derives from the pterygialis cranialis muscles of the pectoral and pelvic appendages, respectively. Accordingly, the superficial and postaxial muscles of the ventral zeugopodium, such as the forearm muscle flexor antebrachii et carpi ulnaris and the leg muscle flexor cruris et tarsi fibularis, derive from the fish abductor superficialis, as do the flexor digitorum communis and other ventral superficial muscles.

A similar line of reasoning leads to the hypothesis of homology between the pelvic fin muscle pterygialis caudalis and the tetrapod muscles tenuissimus and extensor cruris et tarsi fibularis. The latter muscle is the mirror image (dorsal instead of ventral and fibular instead of tibial) of the flexor cruris et tarsi tibialis in other salamanders which, because of its evolutionary and developmental history, is probably included in the tenuissimus of *Ambystoma* (Diogo and Tanaka 2014). Therefore, because both muscles are probably derived from a single ancestral muscle, lie on the postaxial side of the limb, and develop from the dorsal muscle mass, they are probably derived from the pelvic postaxial muscle pterygialis caudalis. The same argument supports homology between the pectoral fin muscle pterygialis caudalis and a part of triceps plus the extensor antebrachii et carpi ulnaris of tetrapods. The latter muscle is the mirror image of the flexor antebrachii et carpi radialis (Diogo and Tanaka 2014), corresponding topologically to the extensor cruris et tarsi fibularis of the tetrapod hindlimb (Diogo and Molnar 2014).

The homology between the *Neoceratodus* retractor lateralis ventralis pectoralis of fishes and the serratus anterior and levator scapulae of salamanders rests on the fact that these are the only muscles in the two taxa that connect the axial skeleton to the pectoral girdle; i.e., they are primaxial muscles (Figure 16.2; Table 16.5). Homology between the retractor lateralis ventralis pectoralis of fishes and the serratus anterior complex of tetrapods has been proposed by previous authors

TABLE 16.4
Origins and Insertions of Pelvic Muscles of *Latimeria*

Muscle	Origin	Insertion
Levator lateralis	Fascia of body wall muscles	Preaxial edge of first element
Adductor superficialis ("levator superficialis")	Dorsal face of lateral process of pelvis	Bases of lepidotrichia *via* aponeurosis, distal to pronator insertion
Pterygialis caudalis (postaxial muscle, or "pelvic adductor")	Distal extremity of longitudinal shaft of pelvis, passes along postaxial border	Bases of postaxial lepidotrichia
Adductor profundus ("levator profundus")	Proximal two-thirds of dorsolateral face of pelvis	Bases of lepidotrichia *via* aponeurosis, distal to supinator insertion
Pronator 1	Dorsal face of pelvis anterior to articular process	Preaxial cartilages and bases of lepidotrichia; partially fused with pronator 2
Pronator 2	Postaxial border proximal to lepidotrichia	Preaxial cartilages and bases of lepidotrichia distal to pronator 1; partially fused with pronators 1 and 3
Pronator 3	Postaxial border proximal to lepidotrichia and distal to pronator 2	Preaxial cartilages and bases of lepidotrichia distal to pronator 2; partially fused with pronators 2 and 4
Pronator 4	Postaxial border proximal to lepidotrichia and distal to pronator 3	Preaxial cartilages and bases of lepidotrichia distal to pronator 3; partially fused with pronator 3
Abductor superficialis ("abaisseur" superficialis)	Ventral face of the pelvis in two bundles	Covers the ventral face of the fin, gives way to a broad tendon at the level of the third–fourth element joint, inserts onto bases of lepidotrichia
Pterygialis cranialis (preaxial muscle, or "pelvic abductor")	Ventral face of lateral process of pelvis, passes along preaxial border	Bases of preaxial lepidotrichia
Abductor profundus ("abaisseur" profundus)	Medial side of the longitudinal shaft (medial component) of pelvis	Lies deep to abductor superficialis, breaks into poorly defined tendons at a more proximal level than that muscle and these insert into the aponeurosis of that muscle
Supinator 1	Medial border of pelvis at the level of the articular process	Preaxial radials and bases of lepidotrichia; partially fused with supinator 2
Supinator 2	Postaxial border proximal to lepidotrichia	Preaxial radials and bases of lepidotrichia; partially fused with supinators 1 and 3
Supinator 3	Postaxial border proximal to lepidotrichia and distal to supinator 2	Preaxial radials and bases of lepidotrichia distal to supinator 2; partially fused with supinators 2 and 4
Supinator 4	Postaxial border proximal to lepidotrichia and distal to supinator 3	Preaxial radials and bases of lepidotrichia distal to supinator 3; partially fused with supinator 3

Source: Names in parentheses from Millot, J., and Anthony, J., *Anatomie de Latimeria chalumnae—I, squelette, muscles, et formation de soutiens*, CNRS, Paris, 1958.

(e.g., Shann 1920, 1924). The caudofemoralis in *Ambystoma*, levator lateralis in *Latimeria*, and abductor dorsolateralis in *Neoceratodus* are included in the abaxial/primaxial group of muscles because all originate from the axial skeleton and/or axial muscles (Table 16.5). However, we do not conclude that these muscles are directly homologous because the levator lateralis in *Latimeria* seems to be part of the dorsal musculature, while the tetrapod caudofemoralis is part of the ventral musculature (Diogo and Molnar 2014; Diogo and Tanaka 2014).

We thus propose that the abductor and adductor superficialis are homologous with the superficial muscles that extend all the way from the body wall or girdles to the autopodia in tetrapods. This idea was presented in a more theoretical way by Gadow (1882). The author suggested that muscles running all the way from the axial skeleton/musculature and/or the girdles to the distal region of the fins became proximo-distally partitioned in the region of major joints during the fin–limb transitions. The homology hypotheses in the present work combine Gadow's evolutionary scenario with developmental and comparative data that were not available in his time. For instance, developmental data for *Ambystoma* show that the superficial layer of the ventral muscles of the pectoral girdle, arm, and forearm comprise the pectoralis, flexor digitorum communis, flexor antebrachii et carpi ulnaris, coracobrachialis, humeroantebrachialis, and flexor antebrachii et carpi radialis (Diogo and Tanaka 2014). Therefore, we propose that the fish abductor superficialis gave rise to and is homologous with all of these developmentally ventral superficial muscles. The only exceptions are the humeroantebrachialis and flexor antebrachii et carpi radialis in *Ambystoma* which, as explained in the preceding text, most likely correspond to the pterygialis cranialis, derived from the superficial ventral (abductor) musculature, in fishes (Table 16.5; Figure 16.2).

Also, on the basis of topology and developmental history (in salamanders), we propose that the second most superficial ventral muscle of the pectoral fin (abductor profundus) is homologous with the second most superficial ventral pectoral

TABLE 16.5
Our Interpretations of the Homologies between the Muscles (Including Synonyms) of the Pectoral Appendage of the Shark *Squalus* (Chondrichthyes), the Bichir *Polypterus* (Actinopterygii: Cladistia), and Sarcopterygii: the Coelacanth *Latimeria* (Coelacanthimorpha), Lungfish *Neoceratodus* (Dipnomorpha), and Salamander *Ambystoma* (Tetrapoda)

Muscle Groups	*Squalus* (5 Muscles)	*Polypterus* (6 Muscles)	*Latimeria* (20 Muscles)	*Neoceratodus* (5 Muscles)	*Ambystoma* (28 Muscles; 48 with Hand Muscles)
Primaxial musculature	– (retractor lateralis ventralis pectoralis poorly differentiated)	– (retractor lateralis ventralis pectoralis poorly differentiated)	– (retractor lateralis ventralis pectoralis apparently undifferentiated)	Retractor lateralis ventralis pectoralis (“muscle connecting cranial rib to girdle”)	Serratus anterior
					Levator scapulae
Adductor superficialis (superficial dorsomesial musculature)	Adductor superficialis	Adductor superficialis	Adductor superficialis (“levator superficialis”)	Adductor superficialis (including dorsal superficial segmented muscular layer)	Deltoideus scapularis
					Latissimus dorsi (or it is instead homologous with “levators 2/3” and/or “retractor dorsalis pectoralis” of e.g., some chondrichthyans (as it is single appendicular muscle with both abaxial and primaxial developmental features)?
					Part of triceps (i.e., triceps scapularis and triceps humeralis lateralis; and perhaps triceps medialis?)
					Extensor digitorum
					Extensor carpi radialis + supinator
		Pterygialis caudalis (postaxial muscle, or “dilatator posterior” or “coracometapterygialis I–II”)	Pterygialis caudalis (postaxial muscle, or “supinator 5 and/or pronator 5”)		Extensor antebrachii et carpi ulnaris
					Part of triceps (i.e., triceps coracoideus)
Adductor profundus (deep dorsomesial musculature)	Adductor profundus	Adductor profundus	Adductor profundus (“levator profundus”)	Adductor profundus	Procoracohumeralis
			Pronator 1		Subcorascapularis
			Pronator 2		Abductor et extensor digit 1
			Pronator 2a		Extensores breves digitorum 2–4
			Pronator 3		
			Pronator 3a		
			Pronator 4 + 4a		
Abductor superficialis (superficial ventrolateral musculature)	Abductor superficialis	Abductor superficialis	Abductor superficialis (“abaisseur” superficialis)	Abductor superficialis (including ventral superficial segmented muscular layer)	Pectoralis
					Flexor digitorum communis
					Flexor antebrachii et carpi ulnaris
					Coracobrachialis
	Pterygialis cranialis (preaxial muscle)	Pterygialis cranialis (preaxial muscle, or “dilatator anterior” or “zonopropterygialis”)	Pterygialis cranialis (preaxial muscle)		Flexor antebrachii et carpi radialis
					Humeroantebrachialis
					+ Some/all intrinsic hand muscles?

(*Continued*)

TABLE 16.5 (CONTINUED)
Our Interpretations of the Homologies between the Muscles (Including Synonyms) of the Pectoral Appendage of the Shark *Squalus* (Chondrichthyes), the Bichir *Polypterus* (Actinopterygii: Cladistia), and Sarcopterygii: the Coelacanth *Latimeria* (Coelacanthimorpha), Lungfish *Neoceratodus* (Dipnomorpha), and Salamander *Ambystoma* (Tetrapoda)

Muscle Groups	*Squalus* (5 Muscles)	*Polypterus* (6 Muscles)	*Latimeria* (20 Muscles)	*Neoceratodus* (5 Muscles)	*Ambystoma* (28 Muscles; 48 with Hand Muscles)	
Abductor profundus (deep ventrolateral musculature)	Abductor profundus	Abductor profundus	Abductor profundus ("abaisseur" profundus)	Abductor profundus	Supracoracoideus	
			Supinator 1		Coracoradialis	
			Supinator 2		Flexor accessorius medialis	+ Some/all intrinsic hand muscles?
					Palmaris profundus 1	
					Pronator quadratus	
			Supinator 2a		–	
			Supinator 3		Flexor accessorius lateralis	
			Supinator 3a		–	
			Supinator 4		Contrahentium caput longum	
			Supinator 4a		–	

TABLE 16.6

Our Interpretations of the Homologies between (Including Synonyms of) the Muscles of the Pelvic Appendage of the Shark *Squalus* (Chondrichthyes), the Bichir *Polypterus* (Actinopterygii: Cladistia), and Sarcopterygii: the Coelacanth *Latimeria* (Coelacanthimorpha), the Lungfish *Neoceratodus* (Dipnomorpha), and the Salamander *Ambystoma* (Tetrapoda)

Muscle Groups	*Squalus* (5 Muscles)	*Polypterus* (6 Muscles)	*Latimeria* (15 Muscles)	*Neoceratodus* (25 Muscles)	*Ambystoma* (27 Muscles; 59 with Foot Muscles)	
Abaxial, and partially primaxial? (because these muscles originate proximally from axial skeleton and/or musculature)	–	–	Levator lateralis	Abductor dorsolateralis ("superficial ventrolateral abductor")	Caudofemoralis (included here because of origin from axial skeleton/muscles, but direct homology with *Latimeria*'s lateral levator and/or *Neoceratodus*' dorsolateral abductor is not assumed, as e.g., the muscle of *Latimeria* is apparently part of dorsal musculature, while caudofemoralis is part of ventral musculature)	
Adductor superficialis (superficial dorsomesial musculature)	Adductor superficialis	Adductor superficialis	Adductor superficialis ("levator superficialis")	Adductor superficialis ("mesial adductor" + superficial dorsal segmented layer that corresponds to "dorsal lepidotrichial flexors + radial flexors")	Extensor iliotibialis ("iliotibialis")	
					Extensor cruris tibialis	
					Extensor tarsi tibialis	
					Extensor digitorum longus	
		Pterygialis caudalis (postaxial muscle: present in our micro-CT scans and dissections of *Polypterus*)	Pterygialis caudalis (postaxial muscle, or 'pelvic adductor')	Pterygialis caudalis (postaxial muscle, or "superficial ventrolateral + ventromesial adductor")	Extensor cruris et tarsi fibularis	
					Tenuissimus ("iliofibularis")	
Adductor profundus (deep dorsomesial musculature)	Adductor profundus	Adductor profundus	Adductor profundus ("levator profundus")	Adductor profundus ("dorsomesial adductor-levator")	Puboischiofemoralis internus	
			Pronator 1	Pronator 1	Iliofemoralis	
			Pronator 2	Pronators 2–9 (dorsal "radial–axial" muscles)	Abductor et extensor digit 1	
			Pronator 3		Extensores breves digitorum 2–5	
			Pronator 4			
Abductor superficialis (superficial ventrolateral musculature)	Abductor superficialis	Abductor superficialis	Abductor superficialis ("abaisseur" superficialis)	Abductor superficialis ("superficial ventromesial abductor" + superficial ventral segmented layer that corresponds to "ventral lepidotrichial flexors + radial flexors")	Gracilis ("puboischiotibialis")	
					Flexor digitorum communis	+ Some/all intrinsic foot muscles?
					+ Pubotibialis? (or pubotibialis derived from pterygialis cranialis or, more likely, from abductor profundus?)	
	Pterygialis cranialis (preaxial muscle, or "pelvic protractor")	Pterygialis cranialis (preaxial muscle, or "dilatator anterior")	Pterygialis cranialis (preaxial muscle, or "pelvic abductor")	Pterygialis cranialis (preaxial muscle, or part of "superficial ventromesial abductor")	Ischioflexorius (which likely includes flexor cruris et tarsi tibialis) and perhaps femorofibularis (+ pubotibialis? see above)	

(Continued)

TABLE 16.6 (CONTINUED)
Our Interpretations of the Homologies between (Including Synonyms of) the Muscles of the Pelvic Appendage of the Shark *Squalus* (Chondrichthyes), the Bichir *Polypterus* (Actinopterygii: Cladistia), and Sarcopterygii: the Coelacanth *Latimeria* (Coelacanthimorpha), the Lungfish *Neoceratodus* (Dipnomorpha), and the Salamander *Ambystoma* (Tetrapoda)

<table>
<tr><th>Muscle Groups</th><th>Squalus (5 Muscles)</th><th>Polypterus (6 Muscles)</th><th>Latimeria (15 Muscles)</th><th>Neoceratodus (25 Muscles)</th><th colspan="2">Ambystoma (27 Muscles; 59 with Foot Muscles)</th></tr>
<tr><td rowspan="7">Abductor profundus (deep ventrolateral musculature)</td><td rowspan="7">Abductor profundus</td><td rowspan="7">Abductor profundus</td><td>Abductor profundus (“abaisseur” profundus)</td><td>Abductor profundus (“deep ventral abductor–depressor”)</td><td colspan="2">Puboischiofemoralis externus + adductor femoris (“pubofemoralis”) (+ pubotibialis? see above)</td></tr>
<tr><td>Supinator 1</td><td>Supinator 1 (“deep ventral adductor–depressor”)</td><td colspan="2">Ischiotrochantericus (“ischiofemoralis”)</td></tr>
<tr><td rowspan="3">Supinator 2</td><td rowspan="5">Supinators 2–9 (ventral “radial–axial” muscles)</td><td>Flexor accessorius medialis</td><td rowspan="5">+ Some/all intrinsic foot muscles?</td></tr>
<tr><td>Tibialis posterior (“pronator profundus”)</td></tr>
<tr><td>Interosseous cruris</td></tr>
<tr><td>Supinator 3</td><td>Flexor accessorius lateralis</td></tr>
<tr><td>Supinator 4</td><td>Contrahentium caput longum</td></tr>
</table>

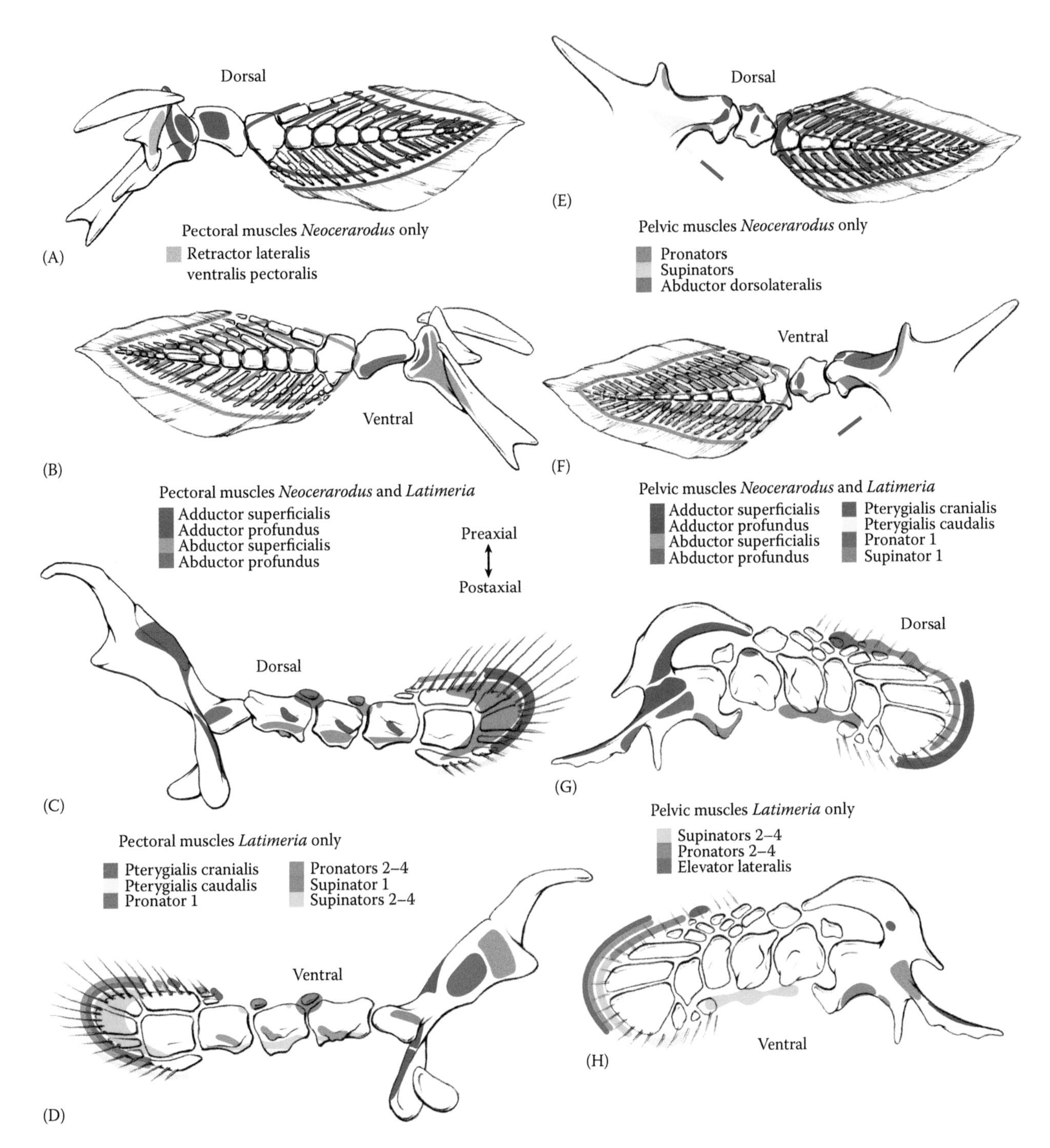

FIGURE 16.1 Right pectoral (A–D) and pelvic (E–H) appendages of *Neoceratodus* (A, B, E, F) and *Latimeria* (C, D, G, H) in dorsal (A, C, E, G) and ventral (B, D, F, H) views. Note that the use of similar colors in the pectoral and pelvic muscles does not indicate ancestral serial homology between the structures of these paired appendages, but instead the result of derived similarity (see text). (Modified from Diogo, R. et al., *Sci Rep*, 6, 1–9, 2016.)

muscle of salamanders, developmentally (supracoracoideus: Table 16.5; Figures 16.2 and 16.3). In fact, in its attachments, fiber orientation, and overall configuration, the abductor profundus of the dipnoan pectoral fin is strikingly similar to the supracoracoideus of salamanders (Figure 16.2). Both are short, parallel-fibered triangular muscles running from the ventral aspect of the pectoral girdle near the shoulder joint to the ventral surface of the proximal humerus (Figure 16.2). Likewise, supinator 1, which is the most proximal of the deeper muscles of the pectoral fin in *Latimeria* and connects the girdle to both the first and second fin elements, is probably homologous with the coracoradialis, which is the most proximal of the developmentally deeper muscles of salamander and originates from the girdle and runs along the humerus to insert onto the radius (Table 16.5; Figures 16.2 and 16.3). The same reasoning supports homology between the remaining pronators of

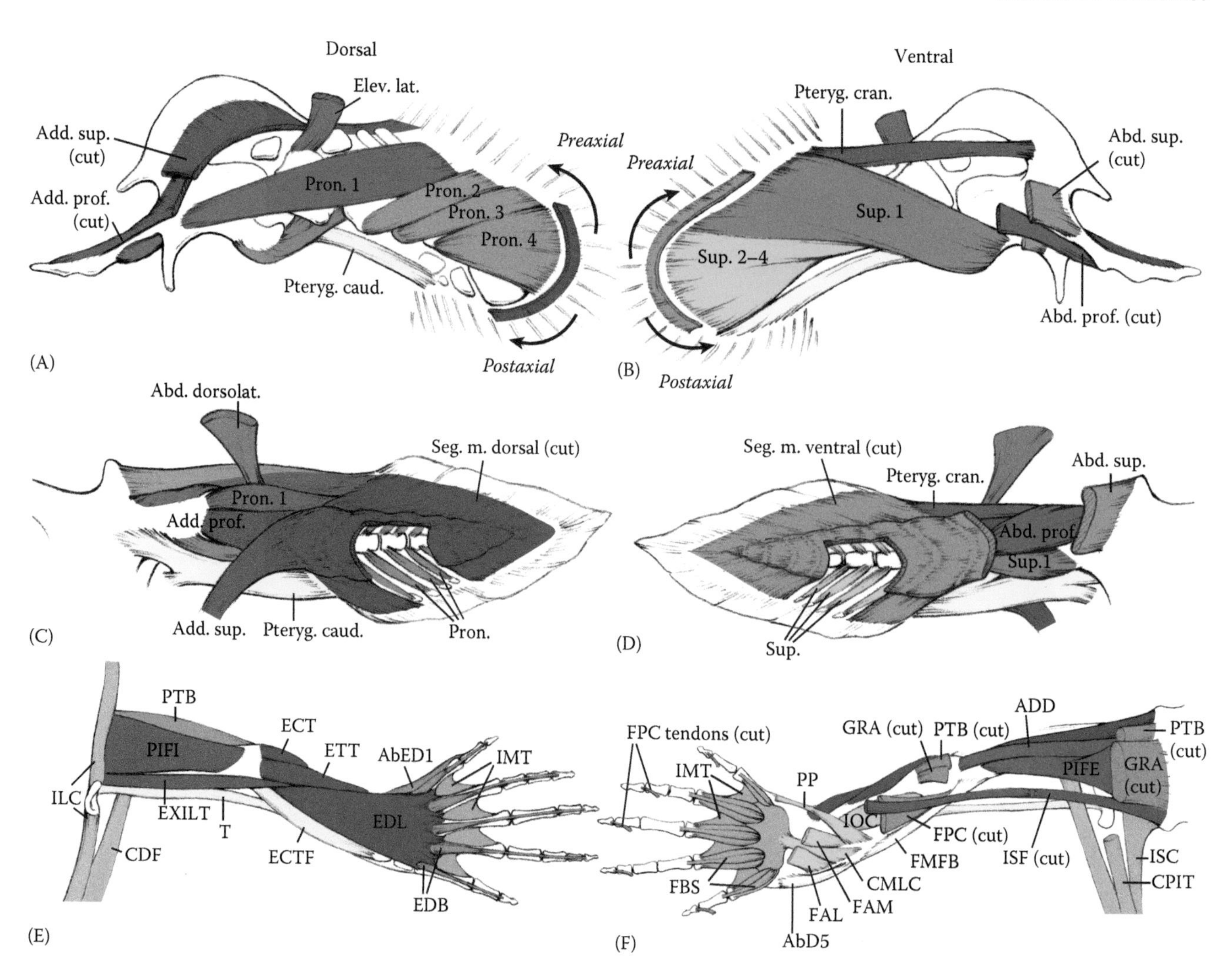

FIGURE 16.2 Hypotheses of muscle homology within the pelvic appendage. *Latimeria* (a and b), *Neoceratodus* (C and D), and *Ambystoma* (E and F). Dorsal views (A, C, E) and ventral views (B, D, F). Colors indicate homologous muscles. Abbreviations: abd. dorsolat., abductor dorsolateralis; abd. prof., abductor profundus; abd. sup., abductor superficialis; AbD5, abductor digiti minimi; AbED1, abductor et extensor digiti I; ADD, adductor femoris; add. prof., adductor profundus; add. sup., adductor superficialis; CCL, contrahentium caput longum; CDF, caudofemoralis; CPIT, caudalipuboischiotibialis; ECT, extensor cruris tibialis; ECTF, extensor cruris et tarsi fibularis; EDB, extensores digitorum breves; EDL, extensor digitorum longus; elev. lat., elevator lateralis; ETT, extensor tarsi tibialis; EXILT, extensor iliotibialis; FAL, flexor accessorius lateralis; FAM, flexor accessorius medialis; FBS, flexores breves superficiales; FDC, flexor digitorum communis; FMFB, femorofibularis; GRA, gracilis; ILC, iliocaudalis; IMT, intermetatarsales; IOC, interosseus cruris; ISC, ischiocaudalis; PIFE, puboischiofemoralis externus; PIFI, puboischiofemoralis internus; PIT, puboischiotibialis; PP, pronator profundus; PTB, pubotibialis; pteryg. caud., pterygialis caudalis; pteryg. cran., pterygialis cranialis; seg. m., segmented muscle; T, tenuissimus. (Modified from Diogo, R. et al., *Sci Rep*, 6, 1–9, 2016.)

the pectoral fin in *Latimeria* and the more distal deep ventral muscles of the salamander (Table 16.5). The same topological reasoning was applied to reach the homology hypotheses proposed for the adductor superficialis, adductor profundus, and pronators of the pectoral fin and also for the abductor and adductor superficialis, abductor and adductor profundus, and supinators and pronators of the pelvic fin because, as explained in the main text, the four paired appendages of each taxon essentially include eight copies of the same model. Among the relatively straightforward homology hypotheses between the fish and tetrapod muscles, the most speculative concern the autopodial muscles. As shown in Tables 16.5 and 16.6, none of the intrinsic autopodial muscles of tetrapods seem to be present as a separate, distinct structure in extant sarcopterygian fishes. However, future work on intermediate fossils might reveal that some supinators and/or pronators in fishes directly correspond to tetrapod autopodial muscles.

MUSCLE ANATOMY AND REDUCTION OF THE PECTORAL FIN OF *NEOCERATODUS*

To reconstruct the configuration of the paired fins of the LCA of extant sarcopterygians, we must consider whether the very simplified muscle anatomy of the pectoral fin of *Neoceratodus*

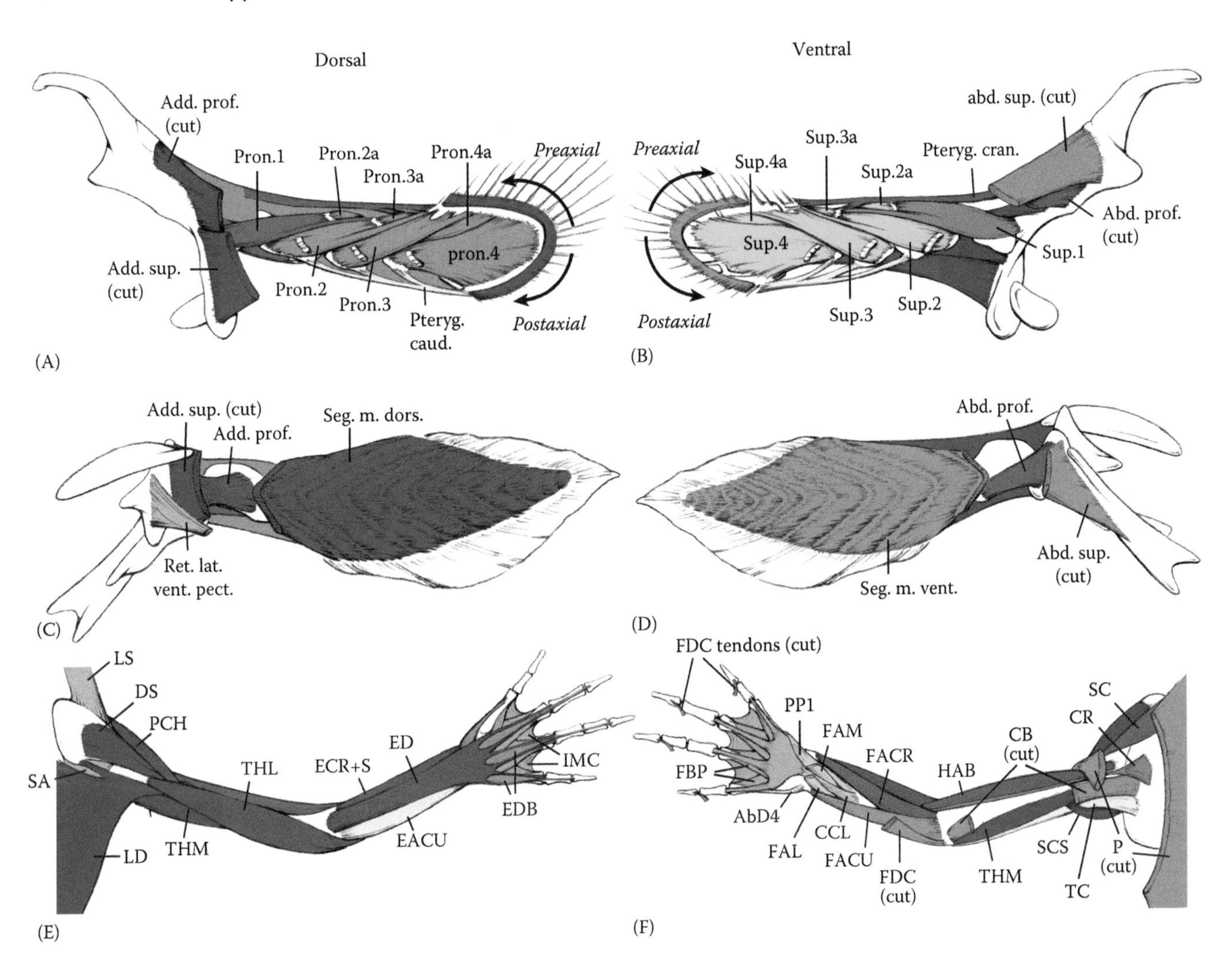

FIGURE 16.3 Hypotheses of muscle homology within the pectoral appendage. *Latimeria* (A and B), *Neoceratodus* (C and D), and *Ambystoma* (E and F). Dorsal views (A, C, E) and ventral views (B, D, F). Colors indicate homologous muscles. Abbreviations: AbD4, abductor digiti minimi; AbED1, abductor et extensor digit 1; CB, coracobrachialis; CCL, contrahentium caput longum; CD, contrahentes digitorum; DS, deltoideus scapularis; EACU, extensor antebrachii et carpi ulnaris; ECR+S, extensor carpi radialis + supinator; ED, extensor digitorum; EDB, extensores digitorum breves; FACR, flexor antebrachii et carpi radialis; FACU, flexor antebrachii et carpi ulnaris; FAL, flexor accessorius lateralis; FAM, flexor accessorius medialis; FBP, flexores breves profundi; FDC, flexor digitorum communis; HAB, humeroantebrachialis; IMC, intermetacarpales; LD, latissimus dorsi; LS, levator scapulae; P, pectoralis; PCH, procoracohumeralis; PP1, palmaris profundus 1; SA, serratus anterior; SC, supracoracoideus; TC, triceps coracoideus; Thindlimb, triceps humeralis lateralis; THM, triceps humeralis medialis; TSM, triceps scapularis medialis; CR, coracoradialis; SCS, subcoracoscapularis; ret. lat. vent. pect., retractor lateralis ventralis pectoralis; add. sup., adductor superficialis; ret. lat. vent. pect., retractor lateralis ventralis pectoralis; add. prof., adductor profundus; seg. m., segmented muscle; abd. sup., abductor superficialis; abd. prof., abductor profundus; pteryg. cran., pterygialis cranialis; pteryg. caud., pterygialis caudalis. (Modified from Diogo, R. et al., *Sci Rep*, 6, 1–9, 2016.)

is most likely representative of the LCA of dipnoans + tetrapods or a derived characteristic of dipnoans. Several lines of evidence indicate that the nondifferentiation of the pterygialis caudalis and pterygialis cranialis in the pectoral fin of *Neoceratodus*, as well as of other muscles inferred to have been acquired during the transitions from the LCA of bony fishes to the LCA of sarcopterygians, such as the pronators and supinators, results from secondary simplification of this fin. First, pronators and supinators are present in the pectoral fin of the phylogenetically most basal extant sarcopterygian, *Latimeria*, and they appear to be homologous with muscles/muscle groups of the deep forelimb musculature of tetrapods (Figures 16.1 through 16.3 and 16.6; Tables 16.5 and 16.7). Second, the pelvic fin of *Neoceratodus* contains many more muscles than the pectoral fin, including muscles that appear to be homologous with pronators and supinators of coelacanths and tetrapods (deep, segmentally arranged muscles with a fiber direction diagonal to the main axis of the fin), and the pterygialis cranialis and pterygialis caudalis of coelacanths (strap-like muscles with parallel fibers running the length of the preaxial and postaxial edges of the fin). Third, it is generally accepted that the fins of the two other extant

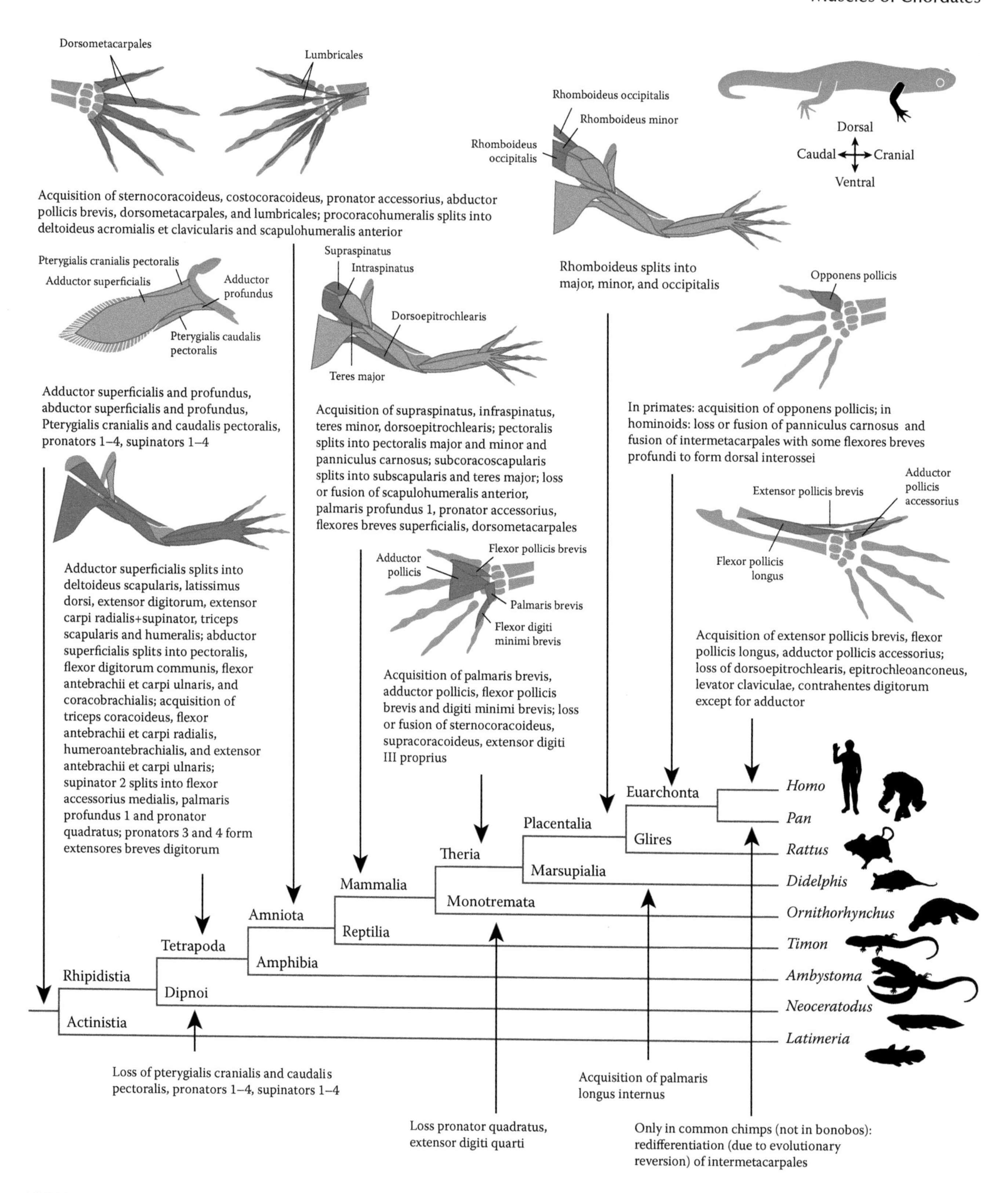

FIGURE 16.4 Some of the major features of the musculature of the pectoral girdle and fin within nonsarcopterygian fishes, according to our own data and review of the literature.

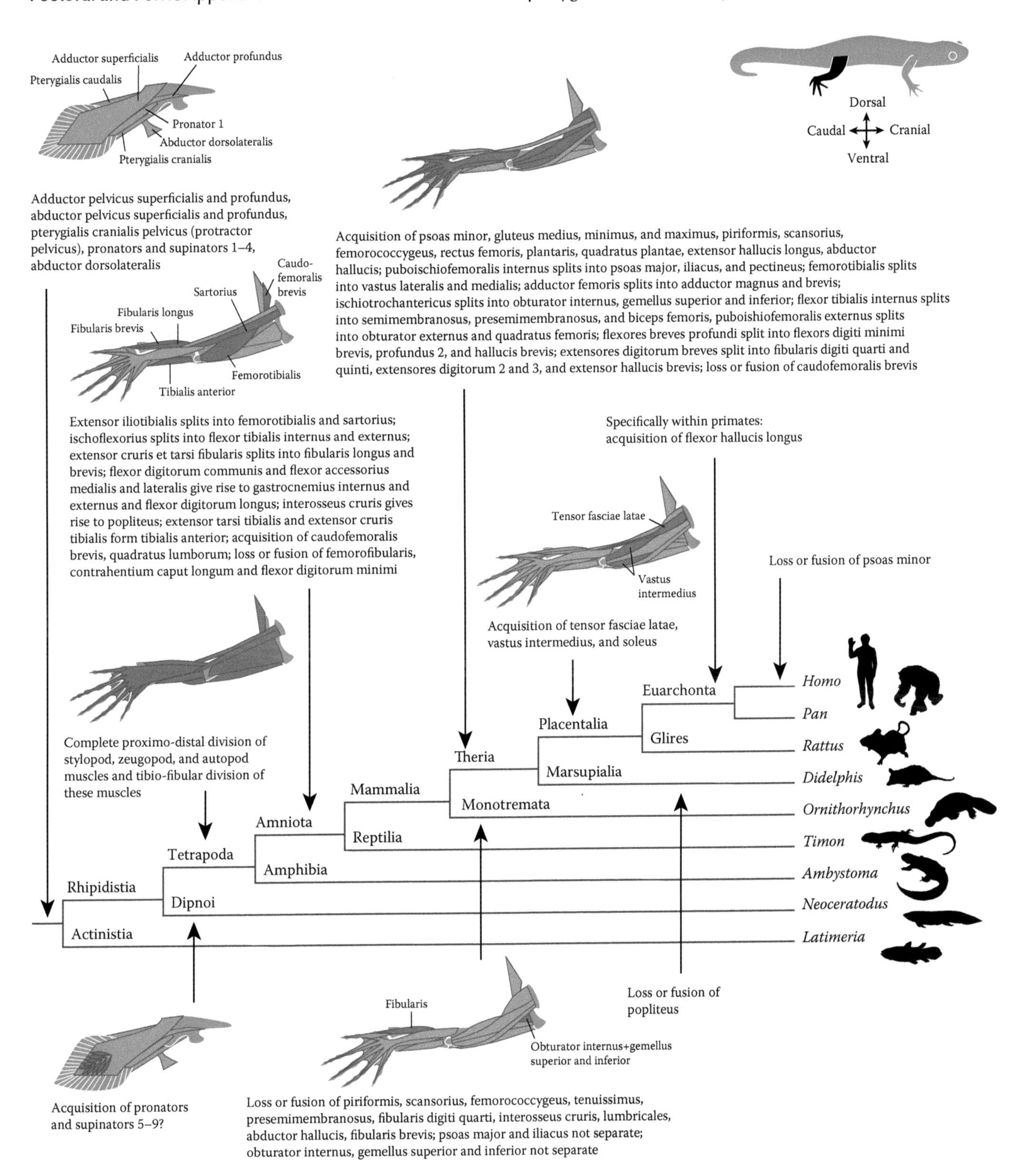

FIGURE 16.5 Some of the major features of the musculature of the pelvic girdle and fin within nonsarcopterygian fishes, according to our own data and review of the literature.

TABLE 16.7
Our Interpretations of Topological Correspondences between Forelimb and Hindlimb Muscles in Salamanders and in Humans

Muscle Groups	*Ambystoma* Forelimb	*Ambystoma* Hindlimb	*Homo* Forelimb	*Homo* Hindlimb	Notes on Human Fore–Hindlimb Comparison
Primaxial	–	– (or quadratus lumborum present?)	–	Quadratus lumborum	–
	Serratus anterior + levator scapulae	–	Serr. ant. + lev. scapulae + rhomboideus major and minor + subclavius	–	Triceps = quadriceps and deltoideus = gluteus maximus supported by Coues (1872); latissimus dorsi = gluteus maximus by Humphry (1872b) and Quain et al. (1894_; deltoideus = sartorius by Humphry (1872b)
Abaxial + partially primaxial?	–	Caudofemoralis	–	–	
	– (latissimus dorsi?)	–	– (latissimus dorsi?)	–	
Adductor superficialis	Deltoideus scapularis + latissimus dorsi (?) + part of triceps brachii (e.g., scapularis)	Extensor iliotibialis	Part of deltoideus + lat. dor. (?) + part of triceps (e.g., long head)	Gluteus maximus + quadriceps femoris + sartorius	
	Extensor digitorum	**Extensor digitorum longus**	Extensor digitorum	Ex. dig. and hallucis longus + fib. tertius	Supported by Diogo et al. (2013)
	Supinator	**Extensor cruris tibialis**	Supinator	Part of tibialis anterior	Supported by Diogo et al. (2013)
	Extensor carpi radialis	**Extensor tarsi tibialis**	Ex. ca. ra. br. and lon. + brachioradialis	Part of tibialis anterior	Supported by Diogo et al. (2013)
Pterygialis caudalis (postaxial)	**Extensor antebrachii et carpi ulnaris**	**Extensor cruris et tarsi fibularis**	Extensor carpi ulnaris + anconeus	Fibularis longus and brevis	Supported by Diogo et al. (2013)
	Part of triceps brachii (i.e., t. coracoideus)	Tenuissimus ("iliofibularis")	Part of triceps brachii	Part of biceps femoris	Ontogeny: tenuissimus migrates to ventral side to fuse with biceps fem.
Adductor profundus	**Procoracohumeralis**	**Puboischiofemoralis internus**	Part of deltoideus + teres minor	Iliopsoas + part of pectineus	Supported by Humphry (1872b)
Pronator 1	**Subcorascapularis**	**Iliofemoralis**	Subscapularis + teres major	Glu. med. and min. + pirif.. +ten. fas. lat.	T.ma. = part of iliofem. by Coues (1872); sub. = glut.medandmin. + pir. + t.f.l. by Quain et al. (1894)
Other pronators	**Abductor et extensor digit 1**	**Abductor et extensor digit 1**	Abd. pol. longus + ex. pol. bre. andlong.	Ex. hallucis brevis	Supported by Diogo et al. (2013)
	Extensores breves digitorum 2–4	**Extensores breves digitorum 2–5**	Ex.indicis + ex. digiti minimi	Extensor digitorum brevis	Supported by Diogo et al. (2013)
Abductor superficialis	Pectoralis + coracobrachialis	Gracilis ("puboischiotibialis") + pubotibialis (?)	Pect. major and minor + coracobr. + part of biceps brachii	Gracilis + ad. long. + part of pectineus	pect. = pectin. and coracob. = adductors and part of bi. bra. = graci. by Coues (1872_; pect. = graci. +add. lon. + part of pect. by Humphry (1872b) and Quain et al. (1894)
	Flexor digitorum communis + flexor antebrachii et carpi ulnaris	Flexor digitorum communis	**Fl. dig. profundus + fl. pollicis longus + palmaris longus** + fl. carpi ulnaris + part of fl. dig. superficialis	**Fl. dig. longus + fl. hallucis longus + plantaris** + soleus + part of quadr. plantae + gastrocnemius	Supported by Diogo et al. (2013)

(Continued)

TABLE 16.7 (CONTINUED)
Our Interpretations of Topological Correspondences between Forelimb and Hindlimb Muscles in Salamanders and in Humans

Muscle Groups	*Ambystoma* Forelimb	*Ambystoma* Hindlimb	*Homo* Forelimb	*Homo* Hindlimb	Notes on Human Fore–Hindlimb Comparison
Pterygialis cranialis (preaxial)	Flexor antebrachii et carpi radialis + humeroantebrachialis	Ischioflexorius (which likely includes flexor cruris et tarsi tibialis) + femorofibularis (?)	Fl. car. radi. + pronat. ter. + brachialis (+ part of bic. brachii, if partially derives from humeroantebrachialis?)	Semimembranosus + semitendinosus + part of biceps femoris + part of adductor magnus	bic. br. = bic. fe. and brachialis = semiten. + semimem. by Coues (1872) and roughly by Quain et al. (1894) and Humphry (1872b)
Abductor profundus	Supracoracoideus	Puboischiofemoralis externus + adductor femoris ("pubofemoralis")	Infraspinatus + supraspinatus	Obtur. externus + quadratus fem. + ad. brevis + part of ad. magnus	Ontogeny: mostly ventral muscles that seem to move dorsally, supporting correspondence
Supinator 1	**Coracoradialis**	**Ischiotrochant.** ("ischiofemoralis")	Part of biceps brachii	Obtur. internus + gemellus inf. andsup.	Supported by Diogo et al. (2013)
Other supinators	**Flexor accessorius medialis**	**Flexor accessorius medialis**	– (part of fl. dig. profundus)	– (part of fl. dig. longus and of qua. plant.)	Supported by Diogo et al. (2013)
	Palmaris profundus 1	**Tibialis posterior ("prona. profundus")**	–	Tibialis posterior	Supported by Diogo et al. 2013
	Pronator quadratus	**Interosseus cruris**	Pronator quadratus	Popliteus	Supported by Diogo et al. (2013)
	Flexor accessorius lateralis	**Flexor accessorius lateralis**	– (part of fl. dig. profundus)	– (Part of fl. dig. longus)	Supported by Diogo et al. (2013)
	Contrahentium caput longum	**Contrahentium caput longum**	– (part of fl. dig. profundus)	– (Part of fl. dig. longus)	Supported by Diogo et al. (2013)
Intrinsic autopod muscles in tetrapods	**Flexores breves superficiales**	**Flexores breves superficiales**	Part of **fl. dig. sup.** + palmaris brevis	**Flexor digitorum brevis**	Supported by Diogo et al. (2013)
	Contrahentes digitorum	**Contrahentes pedis**	**Add. pollicis** + **ad. pol. accessorius**	**Add. pollicis** + **ad. hal. accessorius**	Supported by Diogo et al. (2013)
	–	–	**Lumbricales**	**Lumbricales**	Supported by Diogo et al. (2013)
	Flexores breves profundi	**Flexores breves profundi**	**Fl. pol. br.** and **dig,. min.** + oppo. pol. and di. mi. + **int. palm.** + part of **int. dor.**	**Fl. hal. br.** and **dig,. min.** + **Int. plant.** + part of **int. dorsales**	Supported by Diogo et al. (2013)
	Fls. digitorum minimi + interphalangeus 3	**Fls. digito. min. + interphalangei 3 and 4**	–	–	Supported by Diogo et al. (2013)
	Abductor digiti minimi	**Abductor digiti minimi**	**Abductor digiti minimi**	**Abductor digiti minimi**	Supported by Diogo et al. (2013)
	–	–	**Abductor pollicis brevis**	**Abductor hallucis**	Supported by Diogo et al. (2013)
	Intermetacarpales	**Intermetacarpales**	Part of **interossei dorsales**	Part of **interossei dorsales**	Supported by Diogo et al. (2013)

Note: Names of muscles shown in bold highlight cases in which there is a one-to-one muscle correspondence in forelimb–hindlimb of salamanders and/or in forelimb–hindlimb of humans. For the full name of the muscles that have abbreviated names in these tables, see Tables 17.2, 17.3, 19.1, 19.2, 19.3, and 19.4.

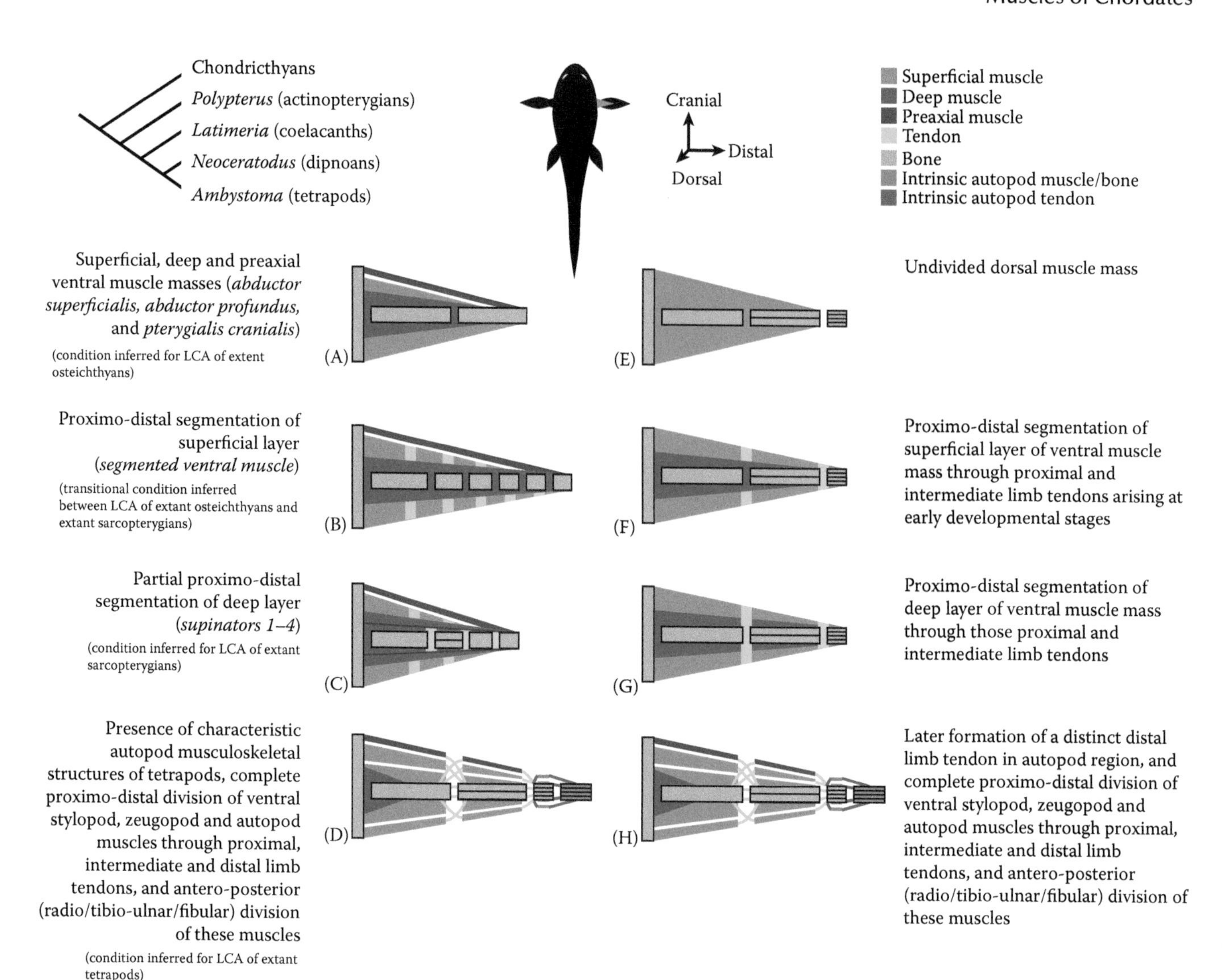

FIGURE 16.6 Evolutionary and developmental transitions leading to the modern adult tetrapod limb. (Modified from Diogo, R. et al., *Sci Rep*, 6, 1–9, 2016.) Evolutionary transitions in adult morphology exemplified by ventral musculature, based on the present paper ((A) LCA of extant osteichthyans; (B) stem sarcopterygians; (C) LCA of extant sarcopterygians; (D) LCA of extant tetrapods: see cladogram on top left), and developmental transitions from early stages to adult morphology in tetrapods (E–H), exemplified by ontogeny of ventral musculature in chicken hindlimb (based on Kardon, G., *Development*, 125, 4019–4032, 1998). All images show dorsal views with dorsal muscles (therefore including the dorsal, postaxial muscle pterygialis caudalis) removed.

dipnoan species (*Protopterus*, *Lepidosiren*) were secondarily simplified.

The pectoral fin muscles of *Lepidosiren* are very similar to the pectoral fin muscles of *Neoceratodus*, including a primaxial retractor lateralis ventralis pectoralis that corresponds to the tetrapod "serratus" (*sensu* Humphry [1872a]), adductor superficialis ("latissimus dorsi"), abductor superficialis ("pectoralis"), abductor profundus, and adductor profundus ("coracobrachialis"). The pelvic fin of *Lepidosiren* displays an even more extreme case of secondary simplification, having only two muscles, one adductor and one abductor. Such a simplified configuration is also found in the pelvic fin of *Protopterus*: the "protractor + anterior circumradials" and the "retractor + posterior circumradials" described by King and Hale (2014) are bundles of fibers of the continuous adductor/abductor muscles that are just slightly separated superficially by connective tissue attaching onto the skin. Fourth, the muscle configuration of the pelvic fins of *Protopterus* and *Lepidosiren* just described is strikingly similar to that of tetrapod limbs at early developmental stages (Figure 16.6) and in some adult tetrapods with marked secondary limb reduction, which have only adductor and abductor limb muscle masses (Abdala et al. 2015). Finally, recent developmental studies of the *Neoceratodus* pectoral fin showed that at early stages there is a radius and an ulna, as is the case in basal adult sarcopterygians (e.g., *Sauripterus*) and likely in adults of the extinct dipnoan genus *Pentlandia*; at later developmental stages, these two cartilages fuse into a single element (Johanson et al. 2007; Jude et al. 2014), mirroring the evolutionary trend toward secondarily simplification of the fins in dipnoans.

PREVIOUS ANATOMICAL STUDIES OF *LATIMERIA* AND *NEOCERATODUS*

Very few studies have described the muscles of the paired appendages of *Latimeria* and *Neoceratodus* in detail. Previous studies on *Neoceratodus* (Humphry 1872a; Braus 1941) reported only two muscle masses on the pectoral appendage: an adductor mass subdivided into superficial and deep muscles and an abductor mass also subdivided into superficial and deep muscles. Diogo and Abdala (2010) also described a muscle "connecting the cranial rib to pectoral girdle" which corresponds to the retractor lateralis ventralis pectoralis *sensu* the present work (Table 16.5). Because a similarly simple configuration is found in the fins of other extant dipnoans (*Protopterus* and *Lepidosiren*) which are the closest extant relatives of tetrapods, most authors agreed that this configuration was shared by the LCA of extant dipnoans and tetrapods as well (Diogo and Abdala 2010). The results of the present study of the pectoral appendage of *Neoceratodus* agree with those of Diogo and Abdala (2010).

Concerning the muscles of the pelvic appendage of *Neoceratodus*, the most detailed previous descriptions are those of Young et al. (1989) and Boisvert et al. (2013). The data presented here mainly agree with those of Young et al. (1989). However, the authors did not include comparisons with other fishes, so the nomenclature they used was mainly descriptive. For instance, Young et al. (1989) describe "radial flexors" and "lepidotrichial flexors" as separate muscles, but these structures clearly correspond topologically to part of the segmented muscles abductor superficialis and adductor superficialis found in the pectoral fin of *Neoceratodus* and in both the pectoral and pelvic fins of *Protopterus* (King and Hale 2014) and *Lepidosiren* (Humphry 1872a) (Tables 16.5 and 16.6; Figures 16.2 and 16.3; see main text). Also, we disagree with some of the muscle groups assigned by Young et al. (1989). For instance, their "deep ventral adductor-depressor" appears to be a ventral muscle and thus part of the abductor rather than the adductor musculature. It originates mainly from the ventral side of the girdle and inserts exclusively on the ventral side of the fin, as shown in Figure 16.1 and in their Figure 12.

Boisvert et al. (2013) compare the pelvic appendicular muscles of *Neoceratodus* with those of other taxa (*Ambystoma* and *Latimeria*, the same taxa used for the present work). However, their discussion is mainly focused on skeletal rather than muscle homologies, and they do not discuss the broader evolutionary and phylogenetic implications of their muscle comparisons. We agree with some but not all of the authors' muscle homology hypotheses. For instance, they suggest that the ventral pelvic muscle "superficial ventrolateral adductor" (*sensu* Young et al. [1989]; lateral part of pterygialis caudalis *sensu* the present work) in *Neoceratodus* is homologous with the muscle "puboischiotibialis" (gracilis) in *Ambystoma*, which is a dorsal muscle. Similarly, they suggest that the dorsal pronators 1–3 in *Latimeria* are homologous with the ventral muscle "deep ventral adductor depressor" (supinator 1) in *Neoceratodus*.

The only previous detailed account of the pectoral and pelvic muscles of *Latimeria* is Millot and Anthony's (1958) monograph. The monograph is detailed and beautifully illustrated, but it is not very accessible because it is in French, uses uncommon terminology, and is not easy to find. Like Young et al. (1989), the authors did not make detailed comparisons with other fishes and tetrapods or discuss the origin and early evolution of limbs. The results of the present work agree with those of Millot and Anthony (1958) with two exceptions. First, the authors described a "pronator 5" and a "supinator 5" in the pectoral fin, but these muscles appear to be very different from the other pronators and supinators. While the latter are short muscles that run diagonally between the pre- and postaxial edges and span one to two mesomeres, their pronator 5 and supinator 5 are very large, long muscles that run along the postaxial edge of the fin and seem to correspond to the postaxial muscles of fishes such as *Polypterus* ("pterygialis caudalis": Figure 16.3). Second, we found several additional muscles in the pectoral fin: small, preaxial muscles that span two mesomeres (pronators 2a, 3a, and 4a and supinators 2a, 3a, and 4a; Figure 16.3).

Miyake et al. (2016) recently redescribed the pectoral muscles of *Latimeria*. The authors disagreed with Millot and Anthony (1958) on several points, including the orientation of the fin axis, the number and placement of pronators and supinators, and the attachments of muscles onto the dorsal and ventral processes of the axial elements of the fin. According to Miyake et al. (2016), the "ventral ridges" of Millot and Anthony (1958) ("*crochets du bord inferior*") are oriented more laterally (presumably in neutral position). We acknowledge that the fin may be habitually held in this position during life, but we agree with the anatomical axes of Millot and Anthony (1958). Miyake et al. (2016) found nine pronators and nine supinators; as stated in the preceding text, we found seven of each, while Millot and Anthony (1958) found five of each. Miyake et al. (2016) also stated that, contrary to Millot and Anthony (1958), the pronators are located on the lateral side and the supinators on the mesial side. However, this distinction seems to relate to the position of the fin rather than to the identity of the muscles. Both Figure 3a of Miyake et al. (2016) and plate LXIX of Millot and Anthony (1958) seem to show the pronators on the same side of the fin as the abductors. Finally, Miyake et al. (2016) stated that a portion of the first pronator and supinator muscles is attached to the ridges on the humerus, but they do not mention or figure an attachment of the superficial adductors and abductors onto these ridges. Like Millot and Anthony (1958), we found that the superficial adductors and abductors were attached to these ridges *via* the tendinous intersections.

EVOLUTION AND HOMOLOGY OF APPENDICULAR MUSCLES IN SARCOPTERYGIANS

Our analysis reveals that the pectoral and pelvic appendages of *Latimeria* and *Ambystoma* and the pelvic appendage of *Neoceratodus* share a very similar, complex configuration of homologous (between pectoral appendages) and topologically corresponding (between pectoral and pelvic appendages) muscles (Tables 16.5 through 16.7; Figures 16.1 through 16.3). Among several striking similarities, the two limbs (pectoral and pelvic appendages of *Ambystoma*) and three fins (pectoral and pelvic appendages of *Latimeria* and pelvic appendages of *Neoceratodus*) share dorsal and ventral superficial muscle masses that extend from the girdles to the distal regions of the fins/limbs, a series of similar dorsal and ventral deep muscles (supinators and pronators and their derivatives), and pre- and postaxial muscles that often span more than one joint. Based on this evidence and on the strong evidence that the very simplified pectoral fin muscle anatomy of *Neoceratodus* is a derived characteristic of dipnoans (see preceding text; see also Ahlberg [1989]), we propose that the characteristic muscle configuration of the tetrapod limbs arose through a series of stepwise changes from the LCA of extant osteichthyans to the LCA of tetrapods. The LCA of extant gnathostomes most likely had five muscles in each paired fin: ventrally, the abductor superficialis, abductor profundus, and a preaxial muscle pterygialis cranialis; dorsally, the adductor superficialis and adductor profundus (Diogo and Ziermann 2015b) (Tables 16.5 and 16.6). The LCA of extant bony fishes probably had the same five muscles plus a postaxial muscle (pterygialis caudalis) in both the pectoral and pelvic appendages, because the plesiomorphic extant osteichthyan *Polypterus* (see Wilhelm et al. 2015; our observations) and *Latimeria* share the presence of this muscle in each appendage (Tables 16.5 and 16.6, Figures 16.1 through 16.3). This postaxial muscle ("coracometapterygialis I and II" in *Polypterus* pectoral fin *sensu* Wilhelm et al. [2015]), present in both its pectoral and pelvic appendages according to our observations) and the preaxial muscle pterygialis cranialis ("zonopropterygialis" in *Polypterus* pectoral fin *sensu* Wilhelm et al. [2015]), present in both its pectoral and pelvic appendages according to our observations) are thought to be derived from the dorsal (adductor/"levator") and ventral (abductor/"depressor") fin musculature, respectively (Diogo and Ziermann 2015b). However, in some fishes, ventral muscles may be differentiated postaxially, and dorsal muscles preaxially (Thorsen and Hale 2005).

One implication of our synthesis is that the LCA of extant sarcopterygians probably already had the basic tetrapod limb phenotype in both pectoral and pelvic appendages, with the exception of the characteristic tetrapod autopodium (hand/foot) (Tables 16.5 through 16.7; Figure 16.6C). Specifically, this LCA probably had at least two layers of adductor and abductor muscles that were partially segmented proximo-distally at the level of each joint. That is, the dramatic changes between the LCA of extant bony fishes and the LCA of extant sarcopterygians affected in a markedly similar way the ventral and dorsal sides of both the pectoral and pelvic appendages. In particular, the deep musculature (adductor profundus dorsally; abductor profundus ventrally) gave rise to a series of smaller muscles (pronators dorsally; supinators ventrally) (Figures 16.1 through 16.3 and 16.6; Tables 16.5 through 16.7). An illustrative example of the pronounced overall pectoral–pelvic similarity of sarcopterygians is the almost identical configuration of the *Latimeria* pectoral and pelvic fins, which is in turn strikingly similar to that of the *Neoceratodus* pelvic fin (Tables 16.5 and 16.6; Figures 16.1 through 16.3). Because of this marked pectoral–pelvic similarity, most hypotheses of homology shown in Tables 16.5 and 16.6 are straightforward.

As shown in Figure 16.6, the inferred order of phylogenetic events leading to the origin of tetrapod limbs is very similar to that of the ontogeny of the limbs of extant tetrapods. Moreover, the rotation of the paired appendages (internal rotation *sensu* human anatomy) that occurred over the fin–limb transition, turning the ventrolateral abductor ("depressor") fin musculature toward the body to become the limb "flexor musculature" in tetrapods, is also paralleled by a similar rotation during the ontogeny of tetrapods such as salamanders (Chen 1935). The chief exception to this developmental–phylogenetic similarity is that the preaxial and postaxial muscles pterygialis cranialis and caudalis were differentiated evolutionarily long before the appearance of clear tendinous intersections segmenting proximo-distally the main abductor/adductor fin musculature. In contrast, such intersections appear at very early stages of tetrapod limb development, before any observable antero-posterior (i.e., radio-ulnar or tibio-fibular) division of the musculature is evident. However, this difference makes sense from a biomechanical perspective: segmented or divided muscles that cross only one joint are effective only when the fin skeleton is elongated and segmented proximo-distally into numerous bones connected by numerous and/or more mobile joints, as is the case in lobe-finned fishes but not in most other fishes (Figure 16.6B). Accordingly, early morphogenesis of limb skeletal cartilages and joints in tetrapods is associated with early morphogenesis of proximal and intermediate tendons lying in the region of the major limb joints: the elbow/knee and wrist/ankle joints, respectively (Figure 16.6F). In fact, this is probably a chief developmental constraint in extant tetrapods, as such limb tendons likely can only develop ontogenetically in the neighborhood of joints (Kardon 1998). Contrary to the usual condition in the ontogeny of extant tetrapods, in sarcopterygian fishes such as *Latimeria*, the intersections of the superficial layer mainly lie *at the level* of the major fin bones rather than *between* these bones, i.e., in the region of the joints connecting them (Figures 16.1 and 16.6B and C).

Some aspects of our evolutionary hypothesis (Figure 16.6A through D) are similar to those proposed more than 120 years ago by Gadow (1882). He suggested that muscles running all the way from the axial skeleton/musculature and/or girdles to the distal region of the fins became proximo-distally partitioned in the region of major joints—particularly those related to the overall internal rotation of the fins—during the fin–limb transition. This view, which is supported by the present work, contradicts the statements of more recent works, particularly paleontological ones. For example, in Bishop's (2014) detailed reconstruction of the shoulder/arm/forearm muscles of a stem tetrapod, it was assumed that ancestrally these muscles did not cross more than one joint. However, a few paleontologists did propose a proximo-distal partition of muscles that originally crossed more than one joint, during the fin–limb transition, as suggested by Gadow (e.g., Ahlberg 1989).

Most authors agree that the bones of the tetrapod stylopod and zeugopod bones are homologous with the proximal bones of sarcopterygian fins, but whether the tetrapod autopodia are neomorphic structures or include structures homologous to specific fin structures remains controversial (e.g., Johanson et al. 2007; Coates et al. 2008; Shubin et al. 2009; Schneider et al. 2011). Some evidence from soft tissue development favors the neomorphic hypothesis. For example, during tetrapod development, the distal tendon primordium that gives rise to most tendons of the intrinsic hand/foot muscles appears later than the primordia of the proximal and intermediate tendons associated with girdle, stylopod (arm/thigh), and zeugopod muscles (Figure 16.6H). Additionally, there are significant differences between the morphogenesis of the proximal/intermediate tendons vs. the distal tendon (Kardon 1998) (see also, e.g., more recent works from Schweitzer's group, reviewed in Huang et al. [2015]). While the segregation of the primordia of the former tendons depends on interactions with muscle, the distal tendons (a) develop by a two-step process in which their primordium segregates into various tendon blastemas—each associated with a digit—that in turn subdivide into individual tendons; (b) develop mainly in spatial isolation from, and likely independently of interactions with, the muscles to which they will attach; and (c) express the transcription factors *six-1* and *six-2* and the *eph*-related receptor tyrosine kinase *cek-8*, while proximal/intermediate tendons do not (reviewed by Kardon [1998]). These developmental data, combined with our comparative anatomical data, support the idea that the overall musculotendinous configuration of the hand/foot constitutes a tetrapod evolutionary novelty (Kardon 1998), probably acquired later in evolution than were most of the girdle/stylopod/zeugopod muscles (Figure 16.6D).

A recent compilation of comparative anatomical, paleontological, and developmental data strongly suggests that the pectoral and pelvic appendages were markedly different from each other anatomically in the earliest fishes that had both and that their most proximal regions (i.e., pelvic vs. pectoral girdles) have remained anatomically, developmentally, and genetically quite different (Coates and Cohn 1998; Diogo et al. 2013; Sears et al. 2015a,b). In contrast, the co-option of various similar genes in the development of the more distal, and phylogenetically more recent, stylopodial/zeugopodial and particularly autopodial regions of the pectoral and pelvic appendages of tetrapods led to a marked *derived* anatomical and developmental similarity between these structures in both appendages (i.e., a "similarity bottleneck" *sensu* Diogo et al. [2013]; Diogo and Molnar [2014], and *sensu* the present book: see Chapter 12). These more distal limb regions, principally the autopodia, display developmental patterns that are quite different from those of the fins of plesiomorphic gnathostomes (Diogo and Ziermann 2015b) and of more proximal limb regions in tetrapods. This information agrees with the notes in the previous paragraph regarding the distal vs. proximal/intermediate tendons (Figure 16.6B) and with data on the development and genetic networks of tetrapod limbs (recently reviewed in Sears et al. [2015a,b]). However, it remained an open question whether such a co-option and/or other (e.g., functional/topological) factors leading to the pectoral–pelvic similarity bottlenecks might have occurred even before the rise of tetrapods. Our results suggest that there was a second, much earlier major similarity bottleneck between the muscles of the pectoral and pelvic appendages: during the transition from the LCA of extant bony fishes to the LCA of extant sarcopterygians. The latter LCA probably already displayed strong muscular similarities not only between the dorsal and ventral sides of each fin, but also between the two pectoral and two pelvic appendages, thus essentially having eight copies of the same highly complex configuration. This condition is exemplified by *Latimeria*, in which 14 pectoral fin muscles have clear, straightforward one-to-one topological correspondences with pelvic fin muscles and the dorsal muscles of each fin have clear one-to-one correspondences with ventral muscles on the same fin (Figures 16.2 and 16.3; Tables 16.5 through 16.7). In contrast, the similarity between six of the muscles of each paired appendage of plesiomorphic actinopterygians and osteichthyans, such as *Polypterus*, is because these fins display a very simple, basic condition that was acquired much earlier in gnathostome evolution: the presence of poorly differentiated deep and superficial abductor/adductor masses (Diogo and Ziermann 2015b; Figure 16.6; Tables 16.5 and 16.6).

Our study therefore allows us, for the first time, to propose a detailed list of topological correspondences between all pectoral vs. pelvic appendicular muscles, including girdle/stylopodial muscles, based on the same empirical comparative, evolutionary, and developmental data used for the homology hypotheses (Table 16.7). Such schematics have previously been attempted, mostly in the nineteenth/early twentieth centuries, but they were strongly biased by the old Romantic "archetypal," idealistic view of evolution (reviewed in Diogo et al. [2013]). As seen in Table 16.7, the topological correspondences inferred here between the girdle/stylopodial muscles of the pectoral and pelvic appendages are, in both salamanders and humans, mainly between groups of muscles, without clear one-to-one equivalences, while those between the zeugopodial/autopodial muscles are mainly one to one. Therefore, the data provided in this chapter reinforce the idea that muscles associated with the pectoral and pelvic girdles have remained more different from each other since the appearance of these appendages in basal gnathostome fishes in comparison to the more distal muscles, which were affected by similarity bottlenecks during the transitions leading to sarcopterygians and then to tetrapods.

GENERAL REMARKS

In summary, the fin–limb transition was a long, stepwise process, and the characteristic tetrapod musculoskeletal limb configuration was very likely present in the Silurian LCA of extant sarcopterygians, more than 400 million years ago. In addition to the fact that proximal bones and numerous muscles of the paired appendages of *Latimeria* and *Neoceratodus* have clear homologues in tetrapods, the absolute numbers of muscles in each appendage suggest that the muscle configuration of extant sarcopterygian fishes is more similar to that of tetrapods than to that of any other extant fishes. Chondrichthyans such as sharks have 5 muscles in each paired appendage (total [*T*] = 10) and plesiomorphic osteichthyans such as *Polypterus* have 6 pectoral and 6 pelvic muscles (*T* = 12), while *Latimeria* has 20 and 15 (*T* = 35), *Neoceratodus* 5 and 25 (*T* = 30), and anatomically plesiomorphic tetrapods such as *Ambystoma* 48 and 59 (*T* = 107) (Tables 16.5 and 16.6; see cladogram of Figure 16.6). If we exclude intrinsic hand/foot muscles, which do not seem to be directly homologous to any specific fish muscles, *Ambystoma* has 28 and 27 (*T* = 55), only 20 more than the number found in *Latimeria*, so the difference between *Polypterus* and *Latimeria* (35 − 12 = 23) is, surprisingly, larger than that that between *Latimeria* and *Ambystoma*. Moreover, the data provided here point out that the major transitions that led to the characteristic phenotype of tetrapod limbs (one leading to sarcopterygians and the other to tetrapods) corresponded to the two major similarity bottlenecks that led to the striking derived myological similarity between the pectoral and pelvic appendages. Finally, by providing one-to-one homology hypotheses for each muscle of the paired appendages of all these taxa, we lay the foundation for the use of extant phylogenetic bracketing in musculoskeletal reconstructions in paleontological studies on the origin/early evolution of limbs.

17 Forelimb Muscles of Tetrapods, Including Mammals

The few comparative forelimb myological analyses directly based on dissections of members of diverse taxa (e.g., sarcopterygian fish, amphibians, reptiles, monotremes, and therian mammals, including humans) were published decades ago by Humphry (1872a,b), Brooks (1886a,b, 1887, 1889), Ribbing (1907), Romer (1922, 1924, 1927, 1942, and 1944), Howell (1933a,b, 1935, 1836a,b,c,d, 1937a,b,c,d), Haines (1939, 1946, 1950, 1951, 1952, 1955, 1958), and Straus (1942), among others. Thus, these authors did not have access to information that is now available concerning the development of the pectoral and forelimb muscles of taxa such as marsupials, chickens, mice, and humans or the molecular and other evidence that has accrued about the phylogenetic relationships of some groups (e.g., Cheng 1955; Cihak 1972; Shellswell and Wolpert 1977; Kardon 2002; Durland et al. 2008; Shearman and Burke 2009; Diogo and Ziermann 2014; Diogo and Tanaka 2014). Moreover, although the authors mentioned earlier did compare a wide range of taxa, the results of their comparisons were usually published in papers that were mainly focused on a localized group of muscles (e.g., forearm extensors: Haines 1939; forearm flexors: Straus 1942; Haines 1950; muscles of pectoral girdle and arm: Romer 1924, 1944; Howell 1935, 1936b, 1937a; muscles of forearm and hand: Howell 1936c,d), or on a specific subgroup of tetrapods (e.g., amphibians: Howell 1935; reptiles: Howell 1936; monotremes: Howell 1937a,b,d).

Chapter 17 focuses on tetrapods, particularly how the pectoral and forelimb muscles have evolved during the transitions from nonmammalian tetrapods to monotreme, marsupial, and placental mammals, including primates and modern humans (see Figures 1.1 and 16.2). It includes new detailed data and figures of the forelimb muscles of opossums and other marsupials that are based on our own recent dissections (Diogo et al. 2016a) and works performed by others (e.g., Warburton and Marchal 2017), as well as more details on primate and human evolution (mainly based on the works by Diogo and Wood [2011, 2012a,b] and Diogo et al. [2012]) including recent data on the least studied apes, the bonobos (Diogo et al. 2017a) (see Figure 17.1). This information is synthesized in Tables 17.1 through 17.3. Comments about the development and muscular variations/abnormalities of our own species, *Homo sapiens*, are presented mainly, but not exclusively, in the right column of these tables.

PECTORAL MUSCLES DERIVED FROM THE POSTCRANIAL AXIAL MUSCULATURE

The plesiomorphic condition for the pectoral and forelimb muscles of sarcopterygians and of tetrapods was discussed in Chapter 16 (see Figure 16.2 and Table 16.5) (see also Tables 17.1 and 17.2). As explained in that chapter, unlike the configuration found in actinopterygians and in nonosteichthyan gnathostomes such as living chondrichthyans, in sarcopterygian fish, the muscles of the pectoral (and pelvic) appendage extend far into the fin, giving it a characteristic "lobed" or "fleshy" appearance (see also, e.g., Bischoff 1840; Owen 1841; Romer 1924; Howell 1933a,b; Millot and Anthony 1958; Jessen 1972; Kardong 2002; Diogo 2007; Diogo and Abdala 2007). The majority of the pectoral and forelimb muscles of tetrapods derive from the adductor and abductor musculature of basal fish. However, a few of these muscles derive instead from the postcranial axial (epaxial and hypaxial) musculature, which is highly specialized in tetrapods (e.g., Jouffroy 1971). As noted in the preceding chapters, the appendicular musculature of the pectoral girdle, arm, forearm, and hand that derives from the main adductor and abductor musculature (see Tables 17.2 and 17.3) essentially corresponds to the "abaxial musculature" *sensu* authors such as Shearman and Burke (2009), while the axial pectoral girdle musculature (see Table 17.2) that is derived from the postcranial axial musculature, as well as most of the remaining epaxial and hypaxial muscles of the body (with the exception of, e.g., the muscles of the pectoral girdle and of the hind limb), corresponds to the "primaxial musculature" *sensu* these authors. As they explained, the muscles of the vertebrate body are classically described as epaxial or hypaxial according to the innervation from either the dorsal or ventral rami of the spinal nerves, respectively, while the terms *abaxial musculature* and *primaxial musculature* reflect embryonic criteria that are used to distinguish domains relative to embryonic patterning. The "primaxial" domain comprises somitic cells that develop within somite-derived connective tissue, and the "abaxial" domain includes muscle and bone that originates from somites but then mixes with, and develops within, lateral plate-derived connective tissue. Interestingly, recent developmental and molecular studies have shown that most of the cells contributing to the latissimus dorsi (which clearly seems to have derived, in evolution, from the adductor musculature

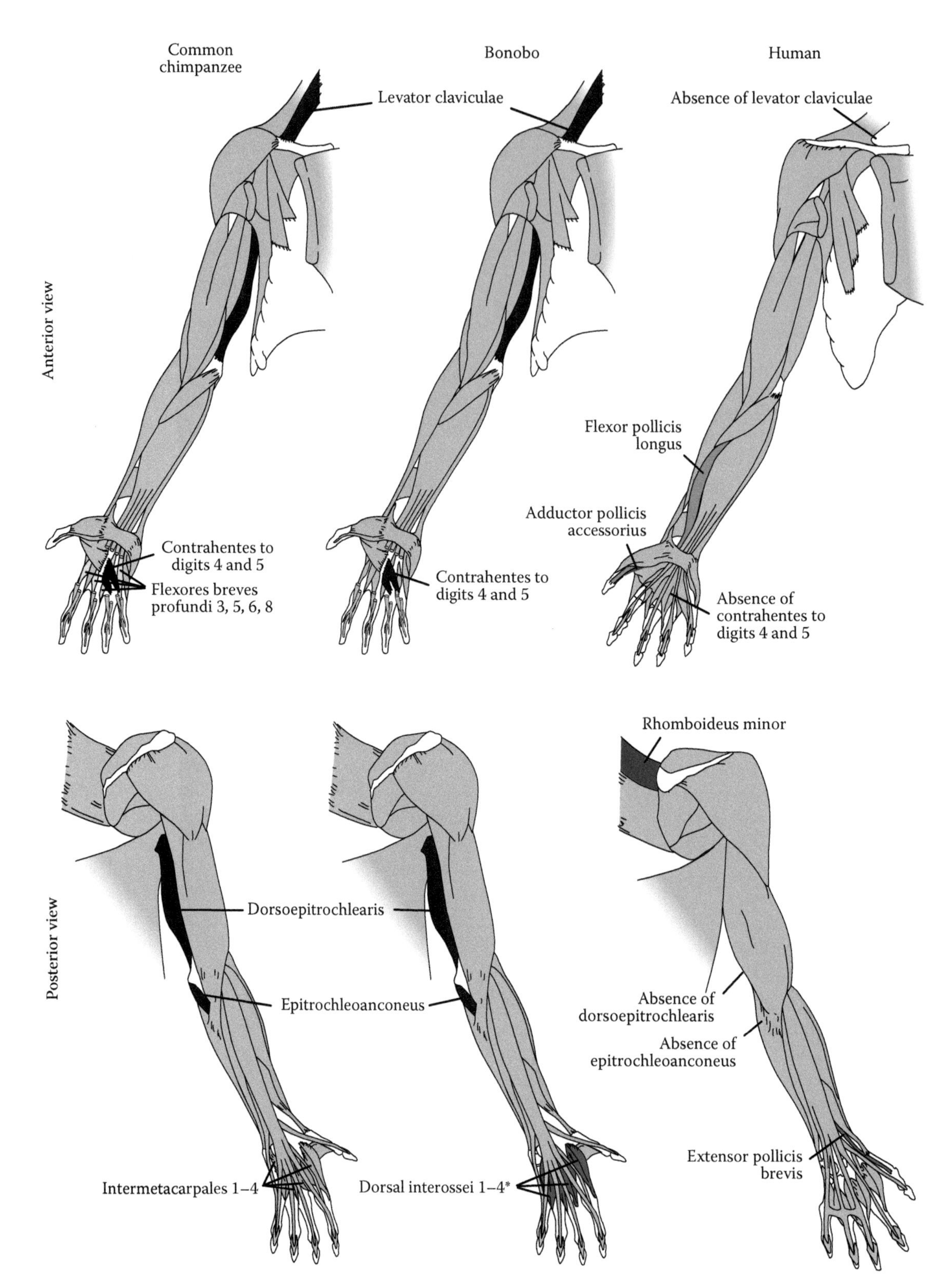

FIGURE 17.1 Schematic drawing (by J. Molnar) showing the major differences between the forelimb muscles of common chimps, bonobos, and modern humans. The only consistent difference between bonobos (center) and common chimps (left) concerning the presence/absence of muscles (shown in colors in the common chimps and bonobos figures) is that in the former, the intermetacarpales 1–4 are usually fused with the flexores breves profundi 3, 5, 6, and 8 to form the dorsal interossei muscles 1–4 (* in bonobo) figure, as is the case in humans. In contrast, there are many differences between bonobos and humans (*right*) concerning the presence/absence of muscles (shown in colors and/or with labels in the human figure; muscles present in chimps and not in humans are shown in black, in chimps).

TABLE 17.1
Pectoral and Forelimb Muscles of Adults of Representative Tetrapod Taxa

Amphibia (Caudata): *Ambystoma mexicanum* (Axolotl)	Reptilia (Lepidosauria): *Timon lepidus* (Ocellated Lizard)	Mammalia (Monotremata): *Ornithorhynchus anatinus* (Platypus)	Mammalia (Marsupialia): *Didelphis virginiana* (Virginian Opossum)	Mammalia (Rodentia): *Rattus norvegicus* (Norwegian Rat)	Mammalia (Scandentia): *Tupaia* sp. (Tree shrew)	Mammalia (Dermoptera): *Cynocephalus volans* (Philippine colugo)	Mammalia (Primates): *Homo sapiens* (Human)
Serratus anterior (part of serrati *sensu* Howell [1937a,b])	**Serratus anterior** (serratus ventralis *sensu* Kardong and s [1998] and Kardong [2002])	**Serratus anterior** (serrati scapulae *sensu* Howell [1937a]; serratus ventralis thoracis or serratus anterior superficialis *sensu* Gambaryan et al. [2015]) [from three first ribs to scapula, being thus much less developed than in therians]	**Serratus anterior** (part of serratus magnus *sensu* Coues [1872]; part of serratus ventralis *sensu* Jenkins and Weijs [1979]; serratus magnus *sensu* Stein [1981]) [from ribs 1-7, 1-8, or 1-9 to scapula, being deeply blended with levator scapulae, forming a single, continuous structure with it]	**Serratus anterior** (serratus magnus *sensu* Greene [1935] and cervical part of serratus ventralis *sensu* Walker and Homberger [1998]) [from ribs 1-7 to scapula; blended with levator scapulae]	**Serratus anterior** (serratus anticus *sensu* Le Gros Clark [1926]; ventral portion of serratus anticus and of serratus magnus *sensu* Le Gros Clark [1924] and George [1977])	**Serratus anterior** (serratus anticus major *sensu* Leche [1886])	**Serratus anterior**
– [but see text and Chapter 18]	– [but see text and Chapter 18]	**Rhomboideus** [from occipital ("rhomboideus capitis bundle" *sensu* Gambaryan et al. [2015]) and thoracic vertebrae ("rhomboideus thoracis bundle" *sensu* Gambaryan et al. [2015]) to scapula; as noted by Howell [1937a], there is a single rhomboideus muscle in monotremes; our dissections indicate that this muscle is poorly differentiated into dorsal and ventral portions, the former being somewhat posterior to the latter; Jouffroy and Lessertisseur [1971] use the terms *rhomboideus cervicus* and *capitis* to designate these bundles, but their configuration are more similar to those of the rhomboideus major and minor/occipitalis of placentals, respectively]	**Rhomboideus** [from occipital crest of occipital bone, nuchal ligament, cervical vertebrae, and first two to four thoracic vertebrae, to medial border of scapula; Jouffroy [1971] states that most marsupials have a single, undivided rhomboideus, as seen in *Didelphis*; this might suggest that a division into more than one rhomboid muscle is present in at least some marsupials, but the fact that most marsupials have a single rhomboid, as generally stated by most authors, together with the fact that monotremes have also a single rhomboid muscle, indicates that having a single muscle was likely the plesiomorphic condition for both mammals as a whole and for the LCA of marsupials and placentals; therefore the presence of a rhomboideus occipitalis in anurans is probably due to convergence]	**Rhomboideus major** (rhomboideus thoracis *sensu* Walker and Homberger [1998]; rhomboideus posticus *sensu* Peterka [1936]) [from C4-C7 to scapula; posterior and somewhat dorsal to rhomboideus minor]	**Rhomboideus major** (part of rhomboideus cervicus *sensu* Le Gros Clark [1924, 1926] and George [1977])	**Rhomboideus** [as described by Macalister [1872], the rhomboideus is undivided; it seems to correspond to the rhomboideus minor + major of humans, because none of its fibers extends anteriorly to the anterior margin of the medial side of the scapula: see Figure 8 of Leche [1886]]	**Rhomboideus major**

(Continued)

TABLE 17.1 (CONTINUED)
Pectoral and Forelimb Muscles of Adults of Representative Tetrapod Taxa

Amphibia (Caudata): ***Ambystoma mexicanum*** **(Axolotl)**	**Reptilia (Lepidosauria):** ***Timon lepidus*** **(Ocellated Lizard)**	**Mammalia (Monotremata):** ***Ornithorhynchus anatinus*** **(Platypus)**	**Mammalia (Marsupialia):** ***Didelphis virginiana*** **(Virginian Opossum)**	**Mammalia (Rodentia):** ***Rattus norvegicus*** **(Norwegian Rat)**	**Mammalia (Scandentia):** ***Tupaia*** **sp. (Tree shrew)**	**Mammalia (Dermoptera):** ***Cynocephalus volans*** **(Philippine colugo)**	**Mammalia (Primates):** ***Homo sapiens*** **(Human)**
–	–	– [see preceding text]	–	**Rhomboideus minor** (rhomboideus cervicus *sensu* Walker and Homberger [1998]; rhomboideus anticus *sensu* Peterka [1936]) [from C1-C3 to scapula]	**Rhomboideus major** (part of rhomboideus cervicus *sensu* Le Gros Clark [1924, [1926] and George [1977])	– [see preceding text]	**Rhomboideus minor**
–	–	– [see preceding text]	–	**Rhomboideus occipitalis** (occipitoscapularis or levator scapulae dorsalis *sensu* Greene [1935]) [from occipital to scapula],	**Rhomboideus occipitalis** (rhomboideus capitis *sensu* Le Gros Clark [1924, 1926] and George [1977])	–	– [found in various primates and occasionally in humans, being the rhombo-atlantoid *sensu* Wood [1867a,b, 1870]; see also Aziz 1981)
Levator scapulae	**Levator scapulae** [according to authors such as Holmes [1977] and Dilkes [1999], in lepidosaurs, including *Sphenodon*, the levator scapulae is usually divided into superficial and deep heads; the "levator scapulae ventralis" of *Iguana* shown in Figure 56A of Jouffroy [1971] corresponds to the ventral head of the levator scapulae *sensu* most authors (e.g., Russell and Bauer 2008); i.e., in *Iguana*, various other lizards, and *Sphenodon*, the levator claviculae is present, but is usually seen as a head of the levator scapulae (that often runs from cervical vertebrae to the suprascapular cartilage in the region of the acromion process and to the clavicular bar) and not	**Levator scapulae** (levator scapulae dorsalis *sensu* Howell [1937a,b] and Jouffroy and Lessertisseur [1971]; serratus ventralis cervicis or serratus anterior profundus *sensu* Gambaryan et al. [2015]) [from C2-7 to scapula, being blended with levator claviculae and with serratus anterior; as described by Jouffroy and Lessertisseur [1971], the levator scapulae and the levator claviculae are somewhat mixed in platypus, thus supporting the idea that these muscles derive from the same structure, which seems to be an anterior (cranial) part of the serratus anterior; Jouffroy and Lessertisseur [1971] explain that the innervation of the levator claviculae is similar to	**Levator scapulae** (part of serratus magnus *sensu* Coues [1872]; levator anguli scapulae *sensu* Stein [1981]) [from C3-C7 or C4-C7 vertebrae to scapula, being deeply blended with the serratus anterior, forming a single, continuous structure with it]	**Levator scapulae** (levator anguli scapulae *sensu* Greene [1935]) [from C4-C7 to scapula]	**Levator scapulae** (levator anguli scapulae *sensu* Le Gros Clark [1926]; dorsal part of serratus anticus *sensu* Le Gros Clark [1924] and of serratus magnus *sensu* George [1977])	**Levator scapulae** [in *Tupaia* and *Cynocephalus* this muscle is deeply mixed with the serratus anterior]	**Levator scapulae**

(Continued)

TABLE 17.1 (CONTINUED)
Pectoral and Forelimb Muscles of Adults of Representative Tetrapod Taxa

Amphibia (Caudata): *Ambystoma mexicanum* (Axolotl)	Reptilia (Lepidosauria): *Timon lepidus* (Ocellated Lizard)	Mammalia (Monotremata): *Ornithorhynchus anatinus* (Platypus)	Mammalia (Marsupialia): *Didelphis virginiana* (Virginian Opossum)	Mammalia (Rodentia): *Rattus norvegicus* (Norwegian Rat)	Mammalia (Scandentia): *Tupaia* sp. (Tree shrew)	Mammalia (Dermoptera): *Cynocephalus volans* (Philippine colugo)	Mammalia (Primates): *Homo sapiens* (Human)
	as a distinct muscle; but according to Gregory and Camp [1918], these correspond, at least in *Sphenodon*, to the levator claviculae and levator scapulae, respectively, of placentals; our dissections of tetrapods such as anurans have supported the idea that these animals have a distinct levator claviculae muscle; it thus seems that this muscle was present in the LCA of amphibians and amniotes]	that of the rhomboideus while that of the levator scapulae is similar to that of the serratus anterior, but that there is still more evidence supporting that the levator claviculae derives from the levator scapulae and not from the rhomboideus]					
–	**Levator claviculae** [see levator scapulae]	**Levator claviculae** (atlantoscapularis inferior, omotrachelien, omoatlantic, atlanto-acromialis, cervico-humeralis, trachelo-acromial, acromio-atlantal, occipito-acromial, and levator scapulae ventralis *sensu* Howell [1937a] and Jouffroy and Lessertisseur [1971]; levator scapulae *sensu* Gambaryan et al. [2015]) [from atlas to scapula and distal one-fifth of clavicle; our dissections of *Ornithorhynchus* support the descriptions of Jouffroy and Lessertisseur [1971]:	**Levator claviculae** (levator scapulae ventralis *sensu* Jouffroy [1971]; atlanto-acromialis *sensu* Coues [1872] and Stein [1981]) [from transverse process of atlas to acromion and lateral third of scapular spine]	**Levator claviculae** (levator scapulae ventralis *sensu* Greene [1935]; omotransversalis *sensu* Walker and Homberger [1998]) [from C1 to scapula]	**Levator claviculae** (levator scapulae *sensu* Le Gros Clark [1924]; levator scapulae anticus *sensu* Le Gros Clark [1926]; atlantoscapularis ventralis *sensu* George [1977])	**Levator claviculae** (omocervicalis *sensu* Gunnell and Simmons [2005])	– [found in various primates and occasionally in humans, see Wood [1970]]

(*Continued*)

TABLE 17.1 (CONTINUED)
Pectoral and Forelimb Muscles of Adults of Representative Tetrapod Taxa

Amphibia (Caudata): *Ambystoma mexicanum* (Axolotl)	Reptilia (Lepidosauria): *Timon lepidus* (Ocellated Lizard)	Mammalia (Monotremata): *Ornithorhynchus anatinus* (Platypus)	Mammalia (Marsupialia): *Didelphis virginiana* (Virginian Opossum)	Mammalia (Rodentia): *Rattus norvegicus* (Norwegian Rat)	Mammalia (Scandentia): *Tupaia* sp. (Tree shrew)	Mammalia (Dermoptera): *Cynocephalus volans* (Philippine colugo)	Mammalia (Primates): *Homo sapiens* (Human)
		the levator scapulae and the levator claviculae are somewhat mixed in this taxon; this seems to indicate that the levator claviculae derives from the levator scapulae (and thus from the rhomboid-levator scapulae complex, as suggested by Cheng's [1955] developmental study of *Didelphis*), as defended by most anatomists: see, e.g., Howell 1937a,b, Jouffroy and Lessertisseur 1971, and Jouffroy 1971]					
–	–	– [absent in extant monotremes, but found in several marsupial and placental mammals, including many primates: see Jouffroy 1971 and Warburton 2003]	**Atlantoscapularis posticus** (atlanto-scapularis *sensu* Coues [1872] and Stein [1981]) [from transverse process of atlas to medial third of scapular spine]	– [see on the left]	**Atlantoscapularis posticus** (part of rhomboideus capitis *sensu* Le Gros Clark [1924]; levator scapulae posticus *sensu* Le Gros Clark [1926]; atlantoscapularis dorsalis *sensu* George [1977]) [as suggested by Jouffroy [1971], the configuration found in gorillas, in which the levator claviculae and the atlantoscapularis posticus are deeply blended—e.g., Figure 36 of Jouffroy [1961]—as well as the origin from both of these structures from the atlas, seems to suggest that the latter is derived from the former; however, the configuration found in other primates,	– [see on the left]	– [the atlantoscapularis posticus is usually not present as a distinct muscle in humans and other hominoids, but a comparison with other primates seems to indicate that at least some fibers of this muscle might be included in the levator scapulae of hominoids; see on the left]

(*Continued*)

TABLE 17.1 (CONTINUED)
Pectoral and Forelimb Muscles of Adults of Representative Tetrapod Taxa

Amphibia (Caudata): *Ambystoma mexicanum* (Axolotl)	Reptilia (Lepidosauria): *Timon lepidus* (Ocellated Lizard)	Mammalia (Monotremata): *Ornithorhynchus anatinus* (Platypus)	Mammalia (Marsupialia): *Didelphis virginiana* (Virginian Opossum)	Mammalia (Rodentia): *Rattus norvegicus* (Norwegian Rat)	Mammalia (Scandentia): *Tupaia* sp. (Tree shrew)	Mammalia (Dermoptera): *Cynocephalus volans* (Philippine colugo)	Mammalia (Primates): *Homo sapiens* (Human)
					in which the atlantoscapularis posticus, when present as a distinct muscle, is blended with, and has an insertion similar to that of, the levator scapulae, indicates that the atlantoscapularis is probably derived from the levator scapulae; however, as the levator claviculae is probably derived from the levator scapulae anyway, this is basically a similar scenario]		
–	**Sternocoracoideus**	**Sternocoracoideus** (sterno-epicoracoideus *sensu* Lander [1918]; *sensu* Howell [1937a,b] and Jouffroy and Lessertisseur [1971]) [from sternum and first rib to procoracoid, being blended with the costocoracoideus muscle]	**Subclavius** [from first rib to clavicle, acromion, and cleidoacromial ligament; blended with cleidoacromialis muscle]	**Subclavius** [from C4-C7 to scapula; from rib 1 to middle of clavicle; there is some confusion regarding the homologies of the therian subclavius; Jouffroy and Lessertisseur [1971] suggest that it possibly corresponds to both the sternocoracoideus and costocoracoideus of nontherian tetrapods such as *Ornithorhynchus*; as the therian subclavius may originate on the sternum, the ribs, or both and may insert on the clavicle, the scapula, or both; one may accept that it corresponds to both the sternocoracoideus and costocoracoideus;	**Subclavius**	**Subclavius**	**Subclavius**

(Continued)

TABLE 17.1 (CONTINUED)
Pectoral and Forelimb Muscles of Adults of Representative Tetrapod Taxa

Amphibia (Caudata): *Ambystoma mexicanum* (Axolotl)	Reptilia (Lepidosauria): *Timon lepidus* (Ocellated Lizard)	Mammalia (Monotremata): *Ornithorhynchus anatinus* (Platypus)	Mammalia (Marsupialia): *Didelphis virginiana* (Virginian Opossum)	Mammalia (Rodentia): *Rattus norvegicus* (Norwegian Rat)	Mammalia (Scandentia): *Tupaia* sp. (Tree shrew)	Mammalia (Dermoptera): *Cynocephalus volans* (Philippine colugo)	Mammalia (Primates): *Homo sapiens* (Human)
				however, humans have a subclavius and a costocoracoid ligament, and the configuration of the latter is somewhat similar to that of the muscle costocoracoideus of nontherian tetrapods, what could indicate that the subclavius corresponds exclusively to the sternocoracoideus of the latter tetrapods; Cheng's 1955 developmental study supports the idea that the "costoscapularis" (which apparently corresponds to the costocoracoideus *sensu* this volume) and the subclavius derive from the same anlage]			
–	**Costocoracoideus**	**Costocoracoideus** [from first rib to coracoid]	– [seems to be absent in the *D. virginiana* specimens we dissected, and was not described by authors such as Coues [1872] and Stein [1981]; however, Warburton [2003] states that it is present in various other marsupials, corresponding to the sternoscapularis of, e.g., Howell [1937b] and Kyou-Jouffroy and Lessertisseur [1971a] and to sternochondro-scapularis of, e.g., Wood [1870] and Kotoglu [2005]]	– [seems to be absent in the rat specimens we dissected: see preceding text; however, Wood [1870: 108] states that the Norwegian rat has a "sternoscapular" muscle: does this muscle correspond to the costoscapularis (which, as explained earlier, apparently corresponds to the costocoracoideus *sensu* this volume)?]	–	–	– (corresponds to the costocoracoid ligament of most humans? See preceding text) [might be present as an anomaly in humans, if it is homologous with a part or the totality of the "sternoclavicularis," "sternocostalis," "scapuloclavicularis," and/or "sternoscapularis" of Wood [1870], although it might well not correspond to a part or the totality of any of these muscles; the observation of the human specimen shown in Huntington's [1904] plate 8, which has both a sternoclavicularis and a sternoscapularis,

(Continued)

TABLE 17.1 (CONTINUED)
Pectoral and Forelimb Muscles of Adults of Representative Tetrapod Taxa

Amphibia (Caudata): *Ambystoma mexicanum* (Axolotl)	Reptilia (Lepidosauria): *Timon lepidus* (Ocellated Lizard)	Mammalia (Monotremata): *Ornithorhynchus anatinus* (Platypus)	Mammalia (Marsupialia): *Didelphis virginiana* (Virginian Opossum)	Mammalia (Rodentia): *Rattus norvegicus* (Norwegian Rat)	Mammalia (Scandentia): *Tupaia* sp. (Tree shrew)	Mammalia (Dermoptera): *Cynocephalus volans* (Philippine colugo)	Mammalia (Primates): *Homo sapiens* (Human)
							clearly seems to indicate that the sternoclavicularis, often extending from the sternum and/or first rib to the clavicle, corresponds to the subclavius and thus to the sternocoracoideus of monotremes, and that the sternoscapularis, often extending from the first and/or second ribs to the scapula, corresponds to the costocoracoideus of monotremes]
Pectoralis	**Pectoralis**	**Pectoralis major** (anterior portion of pectoralis *sensu* Jouffroy and Lessertisseur [1971]; pectoralis *sensu* Gambaryan et al. [2015]) [from interclavicle and first five ribs to humerus; the pectoralis major of monotremes is not subdivided into clavicular, sternocostal, and abdominal heads: see text]	**Pectoralis major** (part of pectoralis *sensu* Coues [1872] and Jenkins and Weijs [1979]; ectopectoralis *sensu* Lander [1918]; pectoralis superficialis *sensu* Langworthy [1932]) [the clavicular head originates from the sternum and sometimes from a small portion of the clavicle, being superficial to, and inserting more distally onto the humerus than, the sternocostal head, which mainly originates from the sternum and ribs; the abdominal part originates from the sternum; these three heads are shown as part of the pectoralis superficialis of Langworthy [1932] (see his Figures 2 and 3) and insert onto the humerus; the pectoralis major contacts the contralateral muscle in the midline]	**Pectoralis major** (ectopectoralis *sensu* Lander [1918]) [from clavicle, sternum, and ribs to humerus, being blended with pectoralis major of the other side of the body; our dissections indicate that it is divided into three heads, and not two as stated by Greene [1935]; these three heads clearly correspond to the clavicular, sternocostal, and abdominal heads of the pectoralis major of humans; contrary to the descriptions of Greene [1935], in the rats we dissected, the pectoralis major did originate from the clavicle; the figures of Greene [1935] and Walker and Homberger [1997] suggest that the pectoralis major contacts its counterpart at the midline]	**Pectoralis major** (ectopectoralis *sensu* Lander [1918]) [the sternocostal and abdominal heads are present as independent structures, but as noted by Jouffroy [1971], the clavicular head is apparently fused with the deltoideus clavicularis]	**Pectoralis major** (ectopectoralis *sensu* Lander [1918]) [this author describes a single head of the pectoralis major in colugos, while Leche [1886] describes two; our dissections indicate that it is divided into clavicular, sternocostal, and abdominal heads]	**Pectoralis major** [divided into clavicular, sternocostal, and abdominal heads; about 3–5% of humans have a muscle sternalis, which should not be confused with the sternocostalis or "supracostalis" muscles, because these two latter muscles are deep (dorsal) to the pectoralis, while the true sternalis *sensu* Jouffroy [1971] is completely superficial (ventral) to the pectoralis; Jouffroy [1971] states that some of the structures that are often named supracostalis in the literature may derive from the rectus thoracis, others from the scaleni, and still others from the external oblique of the thoracic region, but not from the pectoralis, thus contradicting Lander's [1918] hypothesis; Parsons [1898],

(*Continued*)

TABLE 17.1 (CONTINUED)
Pectoral and Forelimb Muscles of Adults of Representative Tetrapod Taxa

Amphibia (Caudata): ***Ambystoma mexicanum*** **(Axolotl)**	**Reptilia (Lepidosauria):** ***Timon lepidus*** **(Ocellated Lizard)**	**Mammalia (Monotremata):** ***Ornithorhynchus anatinus*** **(Platypus)**	**Mammalia (Marsupialia):** ***Didelphis virginiana*** **(Virginian Opossum)**	**Mammalia (Rodentia):** ***Rattus norvegicus*** **(Norwegian Rat)**	**Mammalia (Scandentia):** ***Tupaia*** **sp. (Tree shrew)**	**Mammalia (Dermoptera):** ***Cynocephalus volans*** **(Philippine colugo)**	**Mammalia (Primates):** ***Homo sapiens*** **(Human)**
							Huntington [1904], and Jouffroy [1971] suggest that the true sternalis probably derives from the pectoralis, and specifically from the pectoralis major, because it is superficial (ventral) to the latter structure, and it is usually innervated by nerves of the brachial plexus, and namely by "pectoral nerves," according to Huntington [1904]]
–	–	**Pectoralis minor** (posterior portion of pectoralis *sensu* Jouffroy and Lessertisseur [1971]; pectoralis abdominalis *sensu* Gambaryan et al. [2015]) [midline of abdomen to humerus, being blended with pectoralis major; our dissections indicate that, as suggested by Jouffroy and Lessertisseur [1971], the pectoralis of *Ornithorhynchus* is differentiated into an anterior, ventral portion—i.e., the pectoralis major—and a posterior, dorsal portion—i.e., the pectoralis minor]	**Pectoralis minor** (part of pectoralis *sensu* Coues [1872] and Jenkins and Weijs [1979]; pectoralis abdominalis + pectoralis profundus *sensu* Langworthy [1932]) [Lander [1918] stated that the entopectoralis is absent in *Didelphis*, but the configuration seen in *D. virginiana* is very similar to that seen in rats having three heads; i.e., the pectoralis abdominalis *sensu* Langworthy [1932], which is deep to the other two heads (of the pectoralis profundus *sensu* Langworthy [1932]) and thus cannot correspond to the abdominal head of the pectoralis major seen in, e.g., humans, because this head is never deep to the pectoralis minor; this supports the idea of Jouffroy [1971, that the pectoralis abdominalis *sensu* different authors in different taxa probably corresponds to different structures; i.e.,	**Pectoralis minor** (entopectoralis *sensu* Lander [1918]) [from ribs and sternum (xiphoid process) to humerus and to coracoid process of scapula; as described by Greene [1935], it is divided into three sections corresponding to the cephalic part of the entopectoralis (= "pectoralis minor"), the caudal part of the entopectoralis (= pectoralis abdominalis) and the xiphihumeralis (= "pectoralis tertius") *sensu* Lander 1918]	**Pectoralis minor** (entopectoralis *sensu* Lander [1918]) [divided into two heads corresponding to the pectoralis minor and abdomino-humeralis *sensu* Le Gros Clark [1924, 1926], to the pectoralis minor and the pectoralis abdominalis *sensu* George [1977], and to the pectoralis minor and the pars abdominalis of the pectoralis *sensu* Kladetzky and Kobold [1966]]	**Pectoralis minor** (entopectoralis *sensu* Lander [1918]) [as described by this author and by Leche [1886], it includes a single section]	**Pectoralis minor** [includes a single section]

(Continued)

TABLE 17.1 (CONTINUED)
Pectoral and Forelimb Muscles of Adults of Representative Tetrapod Taxa

Amphibia (Caudata): *Ambystoma mexicanum* (Axolotl)	Reptilia (Lepidosauria): *Timon lepidus* (Ocellated Lizard)	Mammalia (Monotremata): *Ornithorhynchus anatinus* (Platypus)	Mammalia (Marsupialia): *Didelphis virginiana* (Virginian Opossum)	Mammalia (Rodentia): *Rattus norvegicus* (Norwegian Rat)	Mammalia (Scandentia): *Tupaia* sp. (Tree shrew)	Mammalia (Dermoptera): *Cynocephalus volans* (Philippine colugo)	Mammalia (Primates): *Homo sapiens* (Human)
			when the structure is superficial to the pectoralis minor and adjacent to the main body of the pectoralis major, it corresponds to the abdominal head seen in, e.g., humans, while when the structure is deep to the pectoralis minor and adjacent to the main body of the pectoralis minor, it corresponds to a part of the pectoralis minor instead, as seen in *Didelphis* and rats, for instance; the three heads extend from the xiphoid process and abdominal muscles to the humerus only (not to the coracoid process as stated by Langworthy [1932]; the noninsertion onto the scapula was confirmed by authors such as Coues [1872] and Jenkins and Weijs [1979]); NB: the "fourth pectoral layer" *sensu* Langworthy [1932] corresponds to the muscle sternalis, not to part of the pectoralis major or pectoralis minor; the pectoralis minor contacts the contralateral muscle in the midline]				

(*Continued*)

TABLE 17.1 (CONTINUED)
Pectoral and Forelimb Muscles of Adults of Representative Tetrapod Taxa

Amphibia (Caudata): *Ambystoma mexicanum* (Axolotl)	Reptilia (Lepidosauria): *Timon lepidus* (Ocellated Lizard)	Mammalia (Monotremata): *Ornithorhynchus anatinus* (Platypus)	Mammalia (Marsupialia): *Didelphis virginiana* (Virginian Opossum)	Mammalia (Rodentia): *Rattus norvegicus* (Norwegian Rat)	Mammalia (Scandentia): *Tupaia* sp. (Tree shrew)	Mammalia (Dermoptera): *Cynocephalus volans* (Philippine colugo)	Mammalia (Primates): *Homo sapiens* (Human)
– [as noted by authors such as Jollie [1962], the panniculus carnosus is not present as a distinct muscle in amphibians and reptiles]	– [see on the left]	**Panniculus carnosus (part?)** (cutaneous trunci *sensu* Gambaryan et al. [2015]) [mainly from soft tissues to humerus, ulna, and distal phalanx, i.e., to the claws, of digits 1–5 according to Gambaryan et al. [2015]; some authors consider that the panniculus carnosus is the muscle that gave rise to many of the facial muscles of humans, as well as to muscles such as the palmaris brevis; however, Huber (1930a), Howell (1936c,d), Cheng (1955), and Jouffroy (1971) contradict this idea and argue that the panniculus carnosus is a muscle that is at least partially derived from the pectoralis (being often at least partially innervated by "thoracic nerves") and that is, in some mammals such as rats, cranially extended to reach the head region]	**Panniculus carnosus (part?)** [from soft tissues such as the abdominal muscles and fascia, inserting only—as bony attachment—onto the humerus; the panniculus carnosus contacts the panniculus carnosus of the other side of the body, in the midline, and is blended with the pectoralis minor and the latissimus dorsi of the same side of the body]	**Panniculus carnosus (part?)** (cutaneous maximus, but not platysma nor superficial and deep portions of cervical panniculus—which are hyoid muscles—*sensu* Greene 1935) [blended with pectoralis major; attached to humerus] [Greene [1935] states that his "cervical panniculus," which includes the "tracheoplatysma" and the sternofacialis, is innervated by facial + "superficial cervical" nerves; Jouffroy and Saban [1971] state that the sternofacialis corresponds to the "cervical portion of the sphincter colli profundus"; as noted earlier, this is supported by our dissections of canids, in which the pars auricularis of the sphincter colli profundus is very similar to the sternofacialis, running from the ear region to the region near the sternum; Prunotto et al.'s [2004] study of Met mutants shows that the panniculus carnosus (which includes the structure that they designate as "cutaneous maximus") is derived from "migratory hypaxial muscles" (i.e., from "true appendicular muscles," in this case), and not from nonmigratory (body) hypaxial muscles)]	**Panniculus carnosus (part?)** [the pectoralis abdominalis corresponds to the "abdomino-humeralis" *sensu* Le Gros Clark [1924, 1926], so only the "dorso-humeralis" *sensu* the latter author is considered to be part of the panniculus carnosus *sensu* Jouffroy [1971]; the panniculus carnosus of *Tupaia* is deeply blended with the pectoralis abdominalis, thus supporting the idea that at least part of the panniculus carnosus does derive from the pectoralis: see on the left]	**Panniculus carnosus (part?)** [dermopterans have a panniculus carnosus, which makes part of their "flying membrane," which includes the "ventral sheet of the propatagial complex," the "humero-patagialis" and the "coraco-patagialis" (these three structures might be considered part of the "panniculus carnosus pectoralis"; i.e., they are apparently derived from the pectoralis), as well as other structures such as the "plagiopatagialis dorsalis" and the "plagiopatagialis ventralis"; see Jouffroy and Saban [1971]]	– [but present in some primates and present as an anomaly in a few humans]

(Continued)

TABLE 17.1 (CONTINUED)
Pectoral and Forelimb Muscles of Adults of Representative Tetrapod Taxa

Amphibia (Caudata): *Ambystoma mexicanum* (Axolotl)	Reptilia (Lepidosauria): *Timon lepidus* (Ocellated Lizard)	Mammalia (Monotremata): *Ornithorhynchus anatinus* (Platypus)	Mammalia (Marsupialia): *Didelphis virginiana* (Virginian Opossum)	Mammalia (Rodentia): *Rattus norvegicus* (Norwegian Rat)	Mammalia (Scandentia): *Tupaia* sp. (Tree shrew)	Mammalia (Dermoptera): *Cynocephalus volans* (Philippine colugo)	Mammalia (Primates): *Homo sapiens* (Human)
–	–	**Infraspinatus** (teres minor *sensu* Diogo and Abdala [2010]) [from scapula to greater tubercle of humerus; see text]	**Infraspinatus** [from scapula to humerus, not fused with teres minor]	**Infraspinatus** [from scapula to humerus, being blended with teres major and teres minor]	**Infraspinatus**	**Infraspinatus**	**Infraspinatus** [see cells on the left]
–	–	**Supraspinatus** (coracohumeralis profundus *sensu* Howell [1937a,b]; suprascapularis *sensu* Jouffroy and Lessertisseur [1971]) [from acromion of scapula to greater tubercle of humerus; the infraspinatus and supraspinatus are derived from the supracoracoideus; as in a few reptiles such as chameleons the supracoracoideus occupies a more dorsal position than it usually does in nonmammalian tetrapods, some authors consider that it is transformed into an infraspinatus and a supraspinatus as in mammals (these two muscles are usually dorsal to the pectoral girdle in mammals), but this is not accepted by, e.g., Jouffroy and Lessertisseur [1971]; see infraspinatus mentioned earlier]	**Supraspinatus** [from scapula to humerus]	**Supraspinatus** [from scapula to humerus]	**Supraspinatus**	**Supraspinatus**	**Supraspinatus**

(Continued)

TABLE 17.1 (CONTINUED)
Pectoral and Forelimb Muscles of Adults of Representative Tetrapod Taxa

Amphibia (Caudata): *Ambystoma mexicanum* (Axolotl)	Reptilia (Lepidosauria): *Timon lepidus* (Ocellated Lizard)	Mammalia (Monotremata): *Ornithorhynchus anatinus* (Platypus)	Mammalia (Marsupialia): *Didelphis virginiana* (Virginian Opossum)	Mammalia (Rodentia): *Rattus norvegicus* (Norwegian Rat)	Mammalia (Scandentia): *Tupaia* sp. (Tree shrew)	Mammalia (Dermoptera): *Cynocephalus volans* (Philippine colugo)	Mammalia (Primates): *Homo sapiens* (Human)
–	–	–	**Cleidoacromialis** (probably corresponds to scapulo-clavicularis *sensu* Wood [1870]) [not derived from subclavius, being apparently an appendicular muscle derived from the supracoracoid group, i.e., the group that gives rise to the supraspinatus and infraspinatus, according to, e.g., Warburton [2003]; this was supported by Cheng's [1955] developmental study of *Didelphis*, which strongly supported an origin of the cleidoacromialis from the supraspinatus, and is usually accepted by most authors since then: see, e.g., Jouffroy [1971]; in *D. virginiana*, the cleidoacromialis extends from the distal end of the clavicle not only to the acromion and lateral third of the scapular spine but also to the superficial aponeurosis of the supraspinatus, as described by Jenkins and Weijs [1979], although the muscle is partially blended with the subclavius; this muscle seems to be often absent in placentals, and Wood [1870] stated that rats have a scapulo-clavicularis, which might correspond to the cleidoacromialis *sensu* Warburton [2003];	–	–	–	–

(Continued)

TABLE 17.1 (CONTINUED)
Pectoral and Forelimb Muscles of Adults of Representative Tetrapod Taxa

Amphibia (Caudata): *Ambystoma mexicanum* (Axolotl)	Reptilia (Lepidosauria): *Timon lepidus* (Ocellated Lizard)	Mammalia (Monotremata): *Ornithorhynchus anatinus* (Platypus)	Mammalia (Marsupialia): *Didelphis virginiana* (Virginian Opossum)	Mammalia (Rodentia): *Rattus norvegicus* (Norwegian Rat)	Mammalia (Scandentia): *Tupaia* sp. (Tree shrew)	Mammalia (Dermoptera): *Cynocephalus volans* (Philippine colugo)	Mammalia (Primates): *Homo sapiens* (Human)
			Wood [1870] also stated that a few humans may have a scapulo-clavicularis as a variation; if Wood [1870] is right about the presence of the cleidoacromialis in some placentals, this muscle was probably present in the LCA of placentals + marsupials]				
Deltoideus scapularis [the deltoideus scapularis *sensu* Jouffroy [1971] corresponds to the dorsalis scapulae *sensu* Howell [1937a,b], Romer [1944], Walthall and Ashley-Ross [2006], Diogo and Abdala [2007], and Diogo [2007]]	**Deltoideus scapularis** (dorsalis scapulae *sensu* Diogo [2007] and Diogo and Abdala [2007])	**Deltoideus scapularis** (scapulodeltoideus *sensu* Jollie [1962]; spinodeltoideus *sensu* Gambaryan et al. [2015]) [from scapula to proximal humerus]	**Deltoideus scapularis** (part of deltoid *sensu* Coues [1872]) [Stein [1981] describes only a clavodeltoid and a spinodeltoid in *D. virginiana*; the clavodeltoid is only partially differentiated into a pars acromialis (from acromion) and a pars clavicularis (from clavicle), which are blended with each other and thus constitute the deltoideus acromialis et scapularis *sensu* the present work; the spinodeltoid corresponds to the deltoideus scapularis *sensu* the present work; therefore, the deltoideus scapularis runs from the scapula to the humerus]	**Deltoideus scapularis** (spinodeltoid *sensu* Greene [1935] and Walker and Homberger [1998]) [from scapula to humerus, being blended with teres minor; Greene [1935], describes only an "acromiodeltoid" and a spinodeltoid in *Rattus*; in the rats we dissected, the acromiodeltoid is differentiated into a pars acromialis and a pars clavicularis, which are blended with each other and thus constitute the deltoideus acromialis et scapularis *sensu* the present work; the spinodeltoid corresponds to the deltoideus scapularis *sensu* the present work]	**Deltoideus scapularis** (teres minor *sensu* Le Gros Clark [1924, 1926])	**Deltoideus scapularis**	**Deltoideus, part** [scapular head of deltoideus; as noted by, e.g., Parsons [1898] and Jouffroy [1971]; humans, various other primates, and a few other mammals have a single deltoideus, which corresponds to the deltoideus scapularis, deltoideus clavicularis, and deltoideus acromialis of many mammals]
Procoracohumeralis [according to Romer [1944], the procoracohumeralis longus of urodeles corresponds to the deltoideus clavicularis + humeroradialis of reptiles (see Chapter 18) and to the deltoideus clavicularis et acromialis of mammals;	**Deltoideus acromialis et clavicularis** (procoracohumeralis *sensu* Diogo [2007] and Diogo and Abdala [2007]) [in this case using the name deltoideus clavicularis,	**Deltoideus acromialis et clavicularis** (epicoracohumeralis *sensu* Lander [1918]; deltoideus clavicularis + coracohumeralis superficialis *sensu* Howell [1937a];	**Deltoideus acromialis et clavicularis** (part of deltoid *sensu* Coues [1872]) [extends from the acromion of the scapula and the clavicle to the humerus and is blended with the pectoralis major: see preceding text]	**Deltoideus acromialis et clavicularis** (acromiodeltoideus us *sensu* Greene [1935]; cleidobrachialis *sensu* Walker and Homberger [1998]) [from scapula and clavicle to humerus]	**Deltoideus acromialis et clavicularis** [fused with clavicular head of pectoralis major]	**Deltoideus acromialis et clavicularis**	**Deltoideus, part** [clavicular and acromial heads of deltoideus]

(*Continued*)

TABLE 17.1 (CONTINUED)
Pectoral and Forelimb Muscles of Adults of Representative Tetrapod Taxa

Amphibia (Caudata): *Ambystoma mexicanum* (Axolotl)	Reptilia (Lepidosauria): *Timon lepidus* (Ocellated Lizard)	Mammalia (Monotremata): *Ornithorhynchus anatinus* (Platypus)	Mammalia (Marsupialia): *Didelphis virginiana* (Virginian Opossum)	Mammalia (Rodentia): *Rattus norvegicus* (Norwegian Rat)	Mammalia (Scandentia): *Tupaia* sp. (Tree shrew)	Mammalia (Dermoptera): *Cynocephalus volans* (Philippine colugo)	Mammalia (Primates): *Homo sapiens* (Human)
according to him, the procoracohumeralis brevis of urodeles corresponds to the scapulo-humeralis anterior of reptiles and to the teres minor of mammals: see scapulo-humeralis anterior in the following and Chapter 18]	which is used by most authors working with amniotes, is justified because this muscle does not directly correspond to the procoracohumeralis of *Ambystoma*: it corresponds only to a part of it, the other part corresponds to the scapulo-humeralis anterior of *Timon*: see on the left and Chapter 18]	clavodeltoideus + acromiodeltoideus *sensu* Gambaryan et al. [2015]) [from scapula, clavicle, and interclavicle to distal portion and shaft of humerus, being blended with deltoideus scapularis; our dissections clearly suggest that as proposed by Jouffroy and Lessertisseur [1971], the deltoideus acromialis et clavicularis of *Ornithorhynchus* corresponds to the deltoideus clavicularis + coracohumeralis superficialis *sensu* Howell [1937a]; the latter author stated that the coracohumeralis superficialis corresponds to part of the pectoralis of other mammals, but in the *Ornithorhynchus* we dissected, it clearly corresponds to the deltoideus acromialis et clavicularis of other mammals, which is often mixed with, but not part of, the pectoralis; Gambaryan et al. [2015] describe a "clavodeltoideus" and an "acromiodeltoideus" in the platypus, but they state that the muscles are inseparable, forming a single, continuous muscle structure]					

(*Continued*)

TABLE 17.1 (CONTINUED)
Pectoral and Forelimb Muscles of Adults of Representative Tetrapod Taxa

Amphibia (Caudata): *Ambystoma mexicanum* (Axolotl)	Reptilia (Lepidosauria): *Timon lepidus* (Ocellated Lizard)	Mammalia (Monotremata): *Ornithorhynchus anatinus* (Platypus)	Mammalia (Marsupialia): *Didelphis virginiana* (Virginian Opossum)	Mammalia (Rodentia): *Rattus norvegicus* (Norwegian Rat)	Mammalia (Scandentia): *Tupaia* sp. (Tree shrew)	Mammalia (Dermoptera): *Cynocephalus volans* (Philippine colugo)	Mammalia (Primates): *Homo sapiens* (Human)
– [see cell on the right]	**Scapulo-humeralis anterior** (proscapulo-humeralis brevis *sensu* Romer [1924, 1944]) [the scapulo-humeralis anterior of *Timon* apparently corresponds to a part of the procoracohumeralis of *Ambystoma*: see preceding text]	**Teres minor** (proscapulo-humeralis *sensu* Howell [1937a,b] and Jouffroy and Lessertisseur [1971]; scapulo-humeralis anterior *sensu* Diogo and Abdala [2010]) [from scapula to lesser tubercle of humerus; see text]	**Teres minor** [Coues [1872] stated he could not find a teres minor, but as stated by Stein [1981] and Jenkins and Weijs [1979], there is a teres minor, from scapula to humerus, which is not blended with the infraspinatus]	**Teres minor** [from scapula to humerus]	– [as noted by Kladetsky and Kobold [1966] and George [1977], in *Tupaia* the teres minor might be present as an independent muscle, but in the vast majority of the cases, it is completely fused to the infraspinatus and/or possibly with the deltoideus scapularis]	**Teres minor**	**Teres minor**
Subcoracoscapularis [see Chapter 18]	**Subcoracoscapularis**	**Subscapularis** [from scapula to proximal humerus; although some nonmammalian tetrapods do have both a subcoracoscapularis and a muscle teres major that is probably homologous to the mammalian teres major, their muscle subcoracoscapularis is often divided into a "subscapularis" head and a "subcoracoideus" head; therefore, we prefer to keep the term *subscapularis* for mammals, instead of using the term *subcoracoscapularis*, because the mammalian subscapularis might well correspond only to the subscapularis head, and not to both this head and the subcoracoideus head, of the nonmammalian subcoracoscapularis of nonmammalian tetrapods: see Chapter 18]	**Subscapularis** [from scapula to humerus]	**Subscapularis** [from scapula to humerus]	**Subscapularis**	**Subscapularis** [the configuration of the teres major and of the subscapularis in colugos clearly indicates that these muscles are derived from the same anlage, because they are deeply blended at their origins on the scapula]	**Subscapularis**

(Continued)

TABLE 17.1 (CONTINUED)
Pectoral and Forelimb Muscles of Adults of Representative Tetrapod Taxa

Amphibia (Caudata): ***Ambystoma mexicanum*** **(Axolotl)**	**Reptilia (Lepidosauria):** ***Timon lepidus*** **(Ocellated Lizard)**	**Mammalia (Monotremata):** ***Ornithorhynchus anatinus*** **(Platypus)**	**Mammalia (Marsupialia):** ***Didelphis virginiana*** **(Virginian Opossum)**	**Mammalia (Rodentia):** ***Rattus norvegicus*** **(Norwegian Rat)**	**Mammalia (Scandentia):** ***Tupaia*** **sp. (Tree shrew)**	**Mammalia (Dermoptera):** ***Cynocephalus volans*** **(Philippine colugo)**	**Mammalia (Primates):** ***Homo sapiens*** **(Human)**
– [but see Chapter 18]	– [but see Chapter 18]	**Teres major** [from scapula to lesser tubercle of humerus; see text]	**Teres major** [from scapula to humerus, not blended with latissimus dorsi]	**Teres major** [from scapula to humerus, being blended with latissimus dorsi]	**Teres major** [one of the proximal heads (medial) of the dorsoepitrochlearis is associated with the teres major while the other (lateral) is associated with the latissimus dorsi]	**Teres major**	**Teres major**
Latissimus dorsi	**Latissimus dorsi**	**Latissimus dorsi** [according to Gambaryan et al. [2015], it has a "latissimus spinalis" bundle from neural spines of thoracic vertebrae 3–11 to humerus (including its medial epicondyle, which according to these authors, is a derived condition seen only in monotremes among amniotes) and a "latissimus costalis" (which according to them might be the precursor of the therian dorsoepitrochlearis or be simply lost) from ribs 8–15 to shaft of the humerus]	**Latissimus dorsi** [from T4 or T5 to L3 or L4 vertebrae, to humerus, being blended with the dorsoepitrochlearis muscle]	**Latissimus dorsi** [from thoracic vertebrae to humerus]	**Latissimus dorsi** (latissimus dorsi + spinohumeralis *sensu* Le Gros Clark [1924])	**Latissimus dorsi**	**Latissimus dorsi**

(Continued)

TABLE 17.1 (CONTINUED)
Pectoral and Forelimb Muscles of Adults of Representative Tetrapod Taxa

Amphibia (Caudata): *Ambystoma mexicanum* (Axolotl)	Reptilia (Lepidosauria): *Timon lepidus* (Ocellated Lizard)	Mammalia (Monotremata): *Ornithorhynchus anatinus* (Platypus)	Mammalia (Marsupialia): *Didelphis virginiana* (Virginian Opossum)	Mammalia (Rodentia): *Rattus norvegicus* (Norwegian Rat)	Mammalia (Scandentia): *Tupaia* sp. (Tree shrew)	Mammalia (Dermoptera): *Cynocephalus volans* (Philippine colugo)	Mammalia (Primates): *Homo sapiens* (Human)
– [but see text and Table 18.1]	– [but see text and Table 18.1]	**Dorsoepitrochlearis** (dermo-flexor antebrachii or dorso-antebrachialis *sensu* Jouffroy and Lessertisseur [1971]; tensor fasciae antebrachii *sensu* Gambaryan et al. [2015]) [from distal portion of latissimus dorsi to lateral crest of ulna; see text; further lines of evidence that support the idea that the dorsoepitrochlearis derives from the triceps brachii are (a) plesiomorphically in therian mammals the dorsoepitrochlearis often inserts onto the olecranon process of the ulna, as usually does the triceps brachii, and (b) in some mammals, including primates such as new world monkeys and nonhuman hominoids, the dorsoepitrochlearis is often partially originated from the scapula, as the triceps brachii often is; Gambaryan et al. [2015] stated that the dorsoepitrochlearis is completely continuous with the distal portion of a bundle of the latissimus dorsi in *Zaglossus* and *Tachyglossus* and that this condition is probably plesiomorphic for mammals and supports the idea that the former muscle derives from the latter;	**Dorsoepitrochlearis** (omo-anconeus *sensu* Coues [1872]; tensor fasciae antebrachii *sensu* Stein [1981]) [as described by Jenkins and Weijs [1979], Coues [1872], and Stein [1981], it extends from the latissimus dorsi—being thus fused to it—to the olecranon process of the ulna and adjacent fascia]	**Dorsoepitrochlearis** (epitrochleoanconeus, extensor antebrachii longus, and extensor parvus antebrachii *sensu* Greene [1935]) [from latissimus dorsi to medial epicondyle of humerus and olecranon process of ulna; according to Greene [1935], the dorsoepitrochlearis is innervated by the ulnar nerve, but this may be a confusion of Greene, because at least in mammals such as primates, this muscle is usually innervated by the radial nerve and because Greene confuses	**Dorsoepitrochlearis**	**Dorsoepitrochlearis** (part of dorsoepitrochlearis *sensu* Macalister 1872, which also included the dorso-brachialis *sensu* Leche 1886)	– [probably corresponds to the tensor fasciae antebrachii of humans]

(Continued)

TABLE 17.1 (CONTINUED)
Pectoral and Forelimb Muscles of Adults of Representative Tetrapod Taxa

Amphibia (Caudata): *Ambystoma mexicanum* (Axolotl)	Reptilia (Lepidosauria): *Timon lepidus* (Ocellated Lizard)	Mammalia (Monotremata): *Ornithorhynchus anatinus* (Platypus)	Mammalia (Marsupialia): *Didelphis virginiana* (Virginian Opossum)	Mammalia (Rodentia): *Rattus norvegicus* (Norwegian Rat)	Mammalia (Scandentia): *Tupaia* sp. (Tree shrew)	Mammalia (Dermoptera): *Cynocephalus volans* (Philippine colugo)	Mammalia (Primates): *Homo sapiens* (Human)
		however, as they recognized, the condition found in platypus is similar to that found in therians; i.e., the dorsoepitrochlearis splits from the latissimus dorsi, not being completely continuous to it; therefore, it is more parsimonious to accept that the condition found in *Zaglossus* and *Tachyglossus* is derived (one step, in the node leading to the family including these two genera, i.e., the Tachyglossidae) than it is to accept that the conditions seen in platypus and therians are homoplastic (two steps); that is, although the condition seen in the Tachyglossidae is interesting, it does not provide, in a phylogenetic sense, new information to support for the evolutionary origin of the dorsoepitrochlearis from the latissimus dorsi]		(see e.g., his list of synonyms) the dorsoepitrochlearis with the epitrochleoanconeus, which is different muscle that is usually innervated by the ulnar nerve: see epitrochleoanconeus in the following]			
Humeroantebrachialis (part of brachialis *sensu* Howell [1937b] and of biceps *sensu* Romer [1944])	**Brachialis** (brachialis inferior *sensu* Diogo [2007] and Diogo and Abdala [2007]) [see the following]	**Brachialis** (brachialis anticus *sensu* Shrivastava [1962a,b]) ["ulnar head" from proximal humerus to ulna and "radial head" from proximal humerus to radius]	**Brachialis** (brachialis anticus *sensu* Coues [1872]) [from humerus to ulna]	**Brachialis** (brachialis anticus *sensu* Greene [1935]) [from humerus to ulna]	**Brachialis** (brachialis anticus *sensu* Le Gros Clark [1924, 1926])	**Brachialis** (brachialis anticus *sensu* Macalister [1872]; brachialis internus *sensu* Leche [1886])	**Brachialis** (brachialis anticus and flexor brachii brevis *sensu* Parsons [1898])
Coracoradialis	–	–	–	–	–	–	–
	Biceps brachii (humeroantebrachialis *sensu* Diogo [2007] and Diogo and Abdala [2007];	**Biceps brachii** [both short (from coracoid to radius) and long (from procoracoid and interclavicle to radius)	**Biceps brachii** [Coues [1872] questioned if the two heads of the biceps of *D. virginiana* really correspond to the long	**Biceps brachii** [from scapula to radius; both short and long heads are	**Biceps brachii** [both short and long heads are present]	**Biceps brachii** [both short and long heads are present]	**Biceps brachii** [both short and long heads are present]

(*Continued*)

TABLE 17.1 (CONTINUED)
Pectoral and Forelimb Muscles of Adults of Representative Tetrapod Taxa

Amphibia (Caudata): *Ambystoma mexicanum* (Axolotl)	Reptilia (Lepidosauria): *Timon lepidus* (Ocellated Lizard)	Mammalia (Monotremata): *Ornithorhynchus anatinus* (Platypus)	Mammalia (Marsupialia): *Didelphis virginiana* (Virginian Opossum)	Mammalia (Rodentia): *Rattus norvegicus* (Norwegian Rat)	Mammalia (Scandentia): *Tupaia* sp. (Tree shrew)	Mammalia (Dermoptera): *Cynocephalus volans* (Philippine colugo)	Mammalia (Primates): *Homo sapiens* (Human)
–	short and long heads of biceps brachii *sensu* Jouffroy [1971])	heads are present (no bicipital aponeurosis was described by Gambaryan et al. 2015), corresponding to the coracoid and procoracoid heads *sensu* Jouffroy and Lessertisseur [1971]]	and short heads of other mammals, as they mainly originate from the coracoid process and surrounding regions and not really from the capsular ligament of the shoulder joint or the supraglenoid region; i.e., for him, both heads can well be simply part of the short head of other mammals; as a whole the biceps extends from scapula to both the ulna and radius (without a well-defined bicipital aponeurosis), so another peculiarity of the muscle in this species is that it also attaches onto the ulna]	present, as is the bicipital aponeurosis]			
Coracobrachialis [corresponds to the coracobrachialis longus/ superficialis *sensu* Walthall and Ashley-Ross [2006], Diogo [2007], and Diogo and Abdala [2007]; the coracobrachialis medius/ proprius and coracobrachialis profundus/ brevis seem to be absent in urodeles as, e.g., *Taricha*, but are present in various other amphibians according to, e.g., Howell [1935] and Romer [1944]; see Chapter 18]	**Coracobrachialis** [corresponds to the coracobrachialis longus *sensu* Diogo [2007] and Diogo and Abdala [2007]; as in many other reptiles, the coracobrachialis is divided into two bundles that seem to correspond to the coracobrachialis longus/superficialis and coracobrachialis profundus/brevis *sensu* Parsons [1898] and George [1977]: see Chapter 18]	**Coracobrachialis** (coracobrachialis + subcoracoideus *sensu* Gambaryan et al. [2015]) [Parsons [1898] states that monotremes have a coracobrachialis longus/ superficialis, a coracobrachialis medius/ proprius, and a coracobrachialis brevis/ profundus; however, our dissections support the descriptions of Howell [1937a], Jollie [1962], and Jouffroy and Lessertisseur [1971]: i.e., the brevis/ profundus section does not seem to be present as an independent muscle in *Ornithorhynchus*; however, Gambaryan et al. [2015] did describe, in the platypus, a subcoracoideus	**Coracobrachialis** [has a single head, which could in theory correspond to the coracobrachialis medius/ proprius of other mammals, but, in reality, is very short, going from coracoid process to proximal humerus only, just distal to lesser tubercle, and moreover being mainly deep, and not fused, with the biceps brachii; in Figure 93 of Jouffroy [1971], this head is designated as coracobrachialis brevis/profundus]	**Coracobrachialis** [from scapula to humerus, being blended with biceps brachii; has a single section, which apparently corresponds to the coracobrachialis medius/ proprius of other mammals]	**Coracobrachialis** [has two sections, which apparently correspond to the coracobrachialis medius/ proprius and coracobrachialis brevis/ profundus of other mammals: see, e.g., George [1977]]	**Coracobrachialis** [has two sections, which apparently correspond to the coracobrachialis medius/ proprius and coracobrachialis brevis/ profundus of other mammals]	**Coracobrachialis** [has a single section, which apparently correspond to the coracobrachialis medius/proprius of other mammals: see, e.g., Parsons [1898] and Jouffroy [1971]]

(Continued)

TABLE 17.1 (CONTINUED)
Pectoral and Forelimb Muscles of Adults of Representative Tetrapod Taxa

Amphibia (Caudata): *Ambystoma mexicanum* (Axolotl)	Reptilia (Lepidosauria): *Timon lepidus* (Ocellated Lizard)	Mammalia (Monotremata): *Ornithorhynchus anatinus* (Platypus)	Mammalia (Marsupialia): *Didelphis virginiana* (Virginian Opossum)	Mammalia (Rodentia): *Rattus norvegicus* (Norwegian Rat)	Mammalia (Scandentia): *Tupaia* sp. (Tree shrew)	Mammalia (Dermoptera): *Cynocephalus volans* (Philippine colugo)	Mammalia (Primates): *Homo sapiens* (Human)
		(coracobrachialis profundus/brevis *sensu* the present work, which as they stated is deeply blended at its origin, but somewhat separate at its insertion, with the coracobrachialis medius/proprius) from procoracoid to proximal humerus, a coracobrachialis brevis (coracobrachialis medius/proprius *sensu* the present work) from coracoid to shaft and proximal region of humerus, and a coracobrachialis longus (coracobrachialis superficialis/longus *sensu* the present work) from coracoid to shaft and distal region of humerus]					
Pronator quadratus (interosseous antebrachii *sensu* Eisler [1935] and Ribbing [1907, 1938]) [the true pronator quadratus was not reported by Walthall and Ashley-Ross [2006], Diogo [2007], and Diogo and Abdala [2007], and this led Diogo and Abdala [2010] to erroneously consider that the "pronator profundus" *sensu* those reports thus corresponded to the pronator quadratus of other tetrapods; however, both the pronator profundus (often also named as palmaris profundus I by, e.g., Eisler [1895],	**Pronator quadratus** (pronator profundus *sensu* Moro and Abdala [2004], Abdala and Moro [2006], Diogo [2007], and Diogo and Abdala [2007])	–	**Pronator quadratus** [from distal ulna to distal radius]	**Pronator quadratus** [from ulna to radius]	**Pronator quadratus**	– [Leche [1886] describes a few fibers running from the ulna to the radius in the colugos dissected by him; he stated that they could correspond to a vestigial pronator quadratus; in the colugo specimens we dissected, there is no well-defined, distinct pronator quadratus as that found in most other mammals]	**Pronator quadratus**

(*Continued*)

TABLE 17.1 (CONTINUED)
Pectoral and Forelimb Muscles of Adults of Representative Tetrapod Taxa

Amphibia (Caudata): *Ambystoma mexicanum* (Axolotl)	Reptilia (Lepidosauria): *Timon lepidus* (Ocellated Lizard)	Mammalia (Monotremata): *Ornithorhynchus anatinus* (Platypus)	Mammalia (Marsupialia): *Didelphis virginiana* (Virginian Opossum)	Mammalia (Rodentia): *Rattus norvegicus* (Norwegian Rat)	Mammalia (Scandentia): *Tupaia* sp. (Tree shrew)	Mammalia (Dermoptera): *Cynocephalus volans* (Philippine colugo)	Mammalia (Primates): *Homo sapiens* (Human)
McMurrich [1903a], Blount [1935], and Grim and Carlson [1974a,b], a name that we adopt here) and the pronator quadratus (often named as "interosseous antebrachii," by, e.g., Eisler [1895] and Ribbing [1907]) are often present in urodeles, as reported by these authors and by Straus [1942], including in *Ambystoma*, as reported by, e.g., Grim and Carlson 1974a,b; therefore, based on our review of the literature and of our new observations of wildtype and GFP transgenic *A. mexicanum*, here we correct the nomenclature of Diogo and Abdala [2010] and recognize the presence of the two muscles in *Ambystoma*, i.e., of the pronator quadratus and of the palmaris superficialis I (we prefer to use the latter name instead of pronator profundus as suggested by Ribbing [1907], to avoid any confusion with the pronator accessorius of reptiles)]							
Palmaris profundus I (pronator profundus *sensu* Ribbing [1907]) [see notes in the preceding cells; "ronator profundus, or palmaris profundus I, is present as a distinct muscle in at least some amphibians such as *A. mexicanum* according to Eisler [1895], McMurrich [1903a], Ribbing [1907], Bliunt [1935], and Grim and Carlson [1974a,b];	**Palmaris profundus I** (pronator profundus *sensu* Russell and Bauer [2008], who however failed to describe a pronator quadratus, which we saw in *Timon*—but they did describe the pronator accessorius)	– [not present, see preceding cell]	– [not present, see preceding cell]	– [not present, see preceding cell]	– [not present, see preceding cell]	– [not present, see preceding cell]	– [not present, see preceding cell]

(*Continued*)

TABLE 17.1 (CONTINUED)
Pectoral and Forelimb Muscles of Adults of Representative Tetrapod Taxa

Amphibia (Caudata): *Ambystoma mexicanum* (Axolotl)	Reptilia (Lepidosauria): *Timon lepidus* (Ocellated Lizard)	Mammalia (Monotremata): *Ornithorhynchus anatinus* (Platypus)	Mammalia (Marsupialia): *Didelphis virginiana* (Virginian Opossum)	Mammalia (Rodentia): *Rattus norvegicus* (Norwegian Rat)	Mammalia (Scandentia): *Tupaia* sp. (Tree shrew)	Mammalia (Dermoptera): *Cynocephalus volans* (Philippine colugo)	Mammalia (Primates): *Homo sapiens* (Human)
Ribbing [1907] states that the "abductor pollicis" *sensu* Gaupp [1896] corresponds to the pronator profundus, i.e., to the palmaris profundus I *sensu* the present study, and not to the abductor pollicis brevis *sensu* the present work]							
–	**Pronator accessorius** [see preceding cell]	– [as noted by, e.g., Howell [1936c], Straus [1942], and Lewis [1989], the pronator accessorius is apparently absent as an independent muscle in mammals]	–	–	–	–	–
Contrahentium caput longum (ulnocarpalis *sensu* Bunnell [1942], Straus [1942], and Grim and Carlson [1974a,b]) [Ribbing [1907] states that the contrahentium caput longum and the contrahentes digitorum may originally form a primitive, continuous unit extending from the ulna to the digits; Howell [1936c], however, considers that the contrahentium caput longum is simply a contrahentes digitorum that has migrated proximally; Straus [1942] states that none of these theories are satisfactory and that the contrahentium caput longum derives from his "flexor palmaris profundus" layer, which also included muscles such as the flexor accessorius lateralis and medialis of the present volume (see the following);	– [see on the left; according to Diogo and Abdala [2007], a few lizards may have a small muscle that somewhat resembles the contrahentium caput longum of *Ambystoma*]	– [see on the left]	–	–	–	–	– [some authors, such as McMurrich [1903a,b], consider that the contrahentium caput longum might appear as an anomalous muscle in some humans, which is often designated the "ulnocarpeus," but this scenario is considered "exceedingly doubtful" by Straus [1942], because the contrahentium caput longum is only consistently found as a separate, distinct muscle in amphibians: see on the left]

(*Continued*)

TABLE 17.1 (CONTINUED)
Pectoral and Forelimb Muscles of Adults of Representative Tetrapod Taxa

Amphibia (Caudata): *Ambystoma mexicanum* (Axolotl)	Reptilia (Lepidosauria): *Timon lepidus* (Ocellated Lizard)	Mammalia (Monotremata): *Ornithorhynchus anatinus* (Platypus)	Mammalia (Marsupialia): *Didelphis virginiana* (Virginian Opossum)	Mammalia (Rodentia): *Rattus norvegicus* (Norwegian Rat)	Mammalia (Scandentia): *Tupaia* sp. (Tree shrew)	Mammalia (Dermoptera): *Cynocephalus volans* (Philippine colugo)	Mammalia (Primates): *Homo sapiens* (Human)
according to him, the contrahentium caput longum is possibly not separated from his flexor palmaris profundus in extant amniotes, thus forming part of the flexor digitorum longus/profundus *sensu* this volume: see text and Table 17.3]							
Flexor accessorius lateralis (caput dorsale of flexor palmaris profundus *sensu* Straus 1942 and palmaris profundus III *sensu* Eisler [1895], McMurrich [1903a], Blount [1935], and Grim and Carlson [1974a,b]) [as stated by Diogo and Abdala [2007], this muscle is absent as an independent structure in extant amniotes; these authors hypothesized that it could correspond to part of the pronator accessorius of reptiles, but it is very likely that the flexor accessorius medialis and lateralis correspond instead to part of the flexor digitorum longus of reptiles and monotremes and of the flexor digitorum profundus of therian mammals: see, e.g., Straus 1942, Lewis 1989, and Chapter 18]	– [absent as an independent muscle in extant amniotes, but see on the left and Chapter 18]	–	–	–	–	–	–

(*Continued*)

TABLE 17.1 (CONTINUED)
Pectoral and Forelimb Muscles of Adults of Representative Tetrapod Taxa

Amphibia (Caudata): *Ambystoma mexicanum* (Axolotl)	Reptilia (Lepidosauria): *Timon lepidus* (Ocellated Lizard)	Mammalia (Monotremata): *Ornithorhynchus anatinus* (Platypus)	Mammalia (Marsupialia): *Didelphis virginiana* (Virginian Opossum)	Mammalia (Rodentia): *Rattus norvegicus* (Norwegian Rat)	Mammalia (Scandentia): *Tupaia* sp. (Tree shrew)	Mammalia (Dermoptera): *Cynocephalus volans* (Philippine colugo)	Mammalia (Primates): *Homo sapiens* (Human)
Flexor accessorius medialis (caput volare of flexor palmaris profundus *sensu* Straus [1942]; palmaris profundus II and palmaris profundus I *sensu* Eisler [1895], McMurrich [1903a], Blount [1935], and Grim and Carlson [1974a,b]) [see flexor accessorius laterali mentioned earlier and Chapter 18]	–	–	–	–	–	–	–
Flexor digitorum communis	**Flexor digitorum longus** (flexor digitorum communis *sensu* Diogo [2007] and Diogo and Abdala [2007]) [we prefer to use here the name flexor digitorum longus, because this muscle does not seem to directly correspond to the flexor digitorum communis of *Ambystoma*; i.e., it probably also includes part, or the totality, of the flexor accessorius lateralis and medialis and the contrahentium caput longum of the latter taxon: see preceding text]	**Flexor digitorum longus + part of flexores breves superficiales** [from humerus and ulna to distal phalanges of digits 1–5, the tendons to digits 1 (metacarpophalangeal joint), 2 (proximal phalanx), 3 (proximal phalanx), and 4 (proximal phalanx) being the only place of origin of the flexores breves superficiales of digits 1–4, respectively, supporting the idea that in the LCA of therians and monotremes, the flexores breves superficiales became completely integrated onto the flexor digitorum longus, and that then in the LCA of placentals and marsupials, the complex formed by these structures differentiated into a flexor digitorum superficialis and a flexor digitorum profundus; in the *Ornithorhynchus* specimens we dissected,	**Flexor digitorum profundus** [fused with flexor digitorum superficialis, extending from the humerus, radius, and ulna to distal phalanges of digits 1–5]	**Flexor digitorum profundus** [from humerus, ulna and radius to distal phalanges of digits 1–5, being blended with flexor digitorum superficialis; as stated by, e.g., Lewis [1989], much of the confusion regarding the homologies of the flexor digitorum profundus and superficialis of mammals is that these names derive from the human anatomy; for instance, in the rats, colugos, and tree shrews we dissected, the flexor digitorum superficialis is less developed than in humans, being deeply mixed with and/or having a significant part deep to the flexor digitorum profundus: see also, e.g., Leche 1886; Le Gros Clark 1924, 1926; Greene 1935; Haines 1950, 1955]	**Flexor digitorum profundus** [see on the left]	**Flexor digitorum profundus** [see on the left]	**Flexor digitorum profundus** [see on the left]

(Continued)

TABLE 17.1 (CONTINUED)
Pectoral and Forelimb Muscles of Adults of Representative Tetrapod Taxa

Amphibia (Caudata): *Ambystoma mexicanum* (Axolotl)	**Reptilia (Lepidosauria): *Timon lepidus* (Ocellated Lizard)**	**Mammalia (Monotremata): *Ornithorhynchus anatinus* (Platypus)**	**Mammalia (Marsupialia): *Didelphis virginiana* (Virginian Opossum)**	**Mammalia (Rodentia): *Rattus norvegicus* (Norwegian Rat)**	**Mammalia (Scandentia): *Tupaia* sp. (Tree shrew)**	**Mammalia (Dermoptera): *Cynocephalus volans* (Philippine colugo)**	**Mammalia (Primates): *Homo sapiens* (Human)**
		the flexor digitorum longus seems to be fused to tendons of the flexores breves superficiales; this is supported by Jouffroy and Lessertisseur [1971], who stated that the compound formed by these structures corresponds to the flexor digitorum profundus, flexor digitorum superficialis, and likely the palmaris longus of most other mammals]					
–	–	–	–	–	–	–	**Flexor pollicis longus** [in most mammals, including rats, tree shrews, and colugos, the flexor digitorum profundus often attaches to digit 1; however, the differentiation of this muscle in a well-developed, distinct flexor pollicis longus is very rare; according to McMurrich [1903a,b], this occurs in humans, *Hylobates*, and a few other mammals such as dogs and hyenas]
–	–	–	**Flexor digitorum superficialis** [fused with flexor digitorum profundus, extending from the humerus to middle phalanges of digits 2–5, although in a few specimens of this species, the muscle apparently only attaches to digits 2–4, as a variant]	**Flexor digitorum superficialis** [from humerus to middle phalanges of digits 2–4, usually, in both mice and rats (as a variant, it can also go to digit 5); see flexor digitorum profundus mentioned earlier]	**Flexor digitorum superficialis**	**Flexor digitorum superficialis**	**Flexor digitorum superficialis** (flexor sublimis *sensu* Windle [1889])

(*Continued*)

TABLE 17.1 (CONTINUED)
Pectoral and Forelimb Muscles of Adults of Representative Tetrapod Taxa

Amphibia (Caudata): *Ambystoma mexicanum* (Axolotl)	Reptilia (Lepidosauria): *Timon lepidus* (Ocellated Lizard)	Mammalia (Monotremata): *Ornithorhynchus anatinus* (Platypus)	Mammalia (Marsupialia): *Didelphis virginiana* (Virginian Opossum)	Mammalia (Rodentia): *Rattus norvegicus* (Norwegian Rat)	Mammalia (Scandentia): *Tupaia* sp. (Tree shrew)	Mammalia (Dermoptera): *Cynocephalus volans* (Philippine colugo)	Mammalia (Primates): *Homo sapiens* (Human)
–	– [there is a palmaris longus in other lizards, as well as in other reptiles as, e.g., turtles (see Chapter 18); authors, such as Howell [1936c], suggest that this muscle is probably not homologous with the palmaris longus of mammals; however, authors, such as Haines 1950, suggested that the palmaris longus of at least some reptiles and the palmaris longus of at least some mammals are probably homologous, i.e., that the LCA of mammals and reptiles probably had a palmaris longus: see text and Chapter 18]	–	**Palmaris longus** (corresponds to palmaris longus externus *sensu* Jouffroy [1971]) [fused with palmaris longus internus and also with the flexor digitorum superficialis, as stated by Coues [1872] and Stein [1981], extending mainly from the humerus to the palmar aponeurosis; Straus [1942] states that the palmaris longus muscles of placentals + marsupials may be derived from the flexor carpi radialis (this would be an exception), from the flexor digitorum superficialis (this is the most frequent case), and/or from the flexor carpi ulnaris (this is somewhat frequent) and that in certain mammals, such as some marsupials and some Carnivora, there are two palmaris longus, probably derived from the flexor carpi ulnaris and/or flexor digitorum superficialis;	**Palmaris longus** (corresponds to palmaris longus externus *sensu* Jouffroy [1971]) [from humerus to soft tissue (palmar aponeurosis), being blended with flexor digitorum superficialis; for Straus [1942] and McMurrich [1903a]: see cell on the left and text]	**Palmaris longus** (corresponds to palmaris longus externus *sensu* Jouffroy [1971]) [see cell on the left and text]	**Palmaris longus** (corresponds to palmaris longus externus *sensu* Jouffroy [1971]) [see cell on the left and text]	**Palmaris longus** (corresponds to palmaris longus externus *sensu* Jouffroy [1971]) [see cell on the left and text]

(*Continued*)

TABLE 17.1 (CONTINUED)
Pectoral and Forelimb Muscles of Adults of Representative Tetrapod Taxa

Amphibia (Caudata): *Ambystoma mexicanum* (Axolotl)	Reptilia (Lepidosauria): *Timon lepidus* (Ocellated Lizard)	Mammalia (Monotremata): *Ornithorhynchus anatinus* (Platypus)	Mammalia (Marsupialia): *Didelphis virginiana* (Virginian Opossum)	Mammalia (Rodentia): *Rattus norvegicus* (Norwegian Rat)	Mammalia (Scandentia): *Tupaia* sp. (Tree shrew)	Mammalia (Dermoptera): *Cynocephalus volans* (Philippine colugo)	Mammalia (Primates): *Homo sapiens* (Human)
			Windle and Parsons 1897 refer to a palmaris longus externus and a palmaris longus internus, both apparently present in *D. virginiana*, and they state that the externus is the more commonly found in mammals; see also text; McMurrich [1903a] states that the palmaris longus of mammals corresponds to a part of the flexor digitorum longus of reptiles; Jouffroy [1971] reviews the literature and states that although some authors refer to three different palmaris longus, i.e., radial derived from flexor carpi radialis, intermedius from flexor digitorum superficialis, and ulnaris from flexor carpi ulnaris, in reality, the intermedius and radialis nerve seem to coexist, so he prefers to refer simply to externus, derived from the flexor digitorum superficialis and often innervated by median nerve, and the "internus" derived from the flexor carpi ulnaris and often innervated by ulnar nerve, which are both present in *D. virginiana*]				

(*Continued*)

TABLE 17.1 (CONTINUED)
Pectoral and Forelimb Muscles of Adults of Representative Tetrapod Taxa

Amphibia (Caudata): *Ambystoma mexicanum* (Axolotl)	Reptilia (Lepidosauria): *Timon lepidus* (Ocellated Lizard)	Mammalia (Monotremata): *Ornithorhynchus anatinus* (Platypus)	Mammalia (Marsupialia): *Didelphis virginiana* (Virginian Opossum)	Mammalia (Rodentia): *Rattus norvegicus* (Norwegian Rat)	Mammalia (Scandentia): *Tupaia* sp. (Tree shrew)	Mammalia (Dermoptera): *Cynocephalus volans* (Philippine colugo)	Mammalia (Primates): *Homo sapiens* (Human)
–	–	–	**Palmaris longus internus** (corresponds to palmaris longus internus *sensu* Jouffroy [1971]) [fused with palmaris longus externus and with the flexor carpi ulnaris, as stated by Coues [1872] and Stein [1981], extending mainly from the humerus to the palmar aponeurosis; see preceding cell]	–	–	–	–
Flexor antebrachii et carpi ulnaris (flexor carpi ulnaris plus epitrochleoanconeus *sensu* Diogo and Abdala [2010])	**Flexor carpi ulnaris** (flexor antebrachii et carpi ulnaris *sensu* Diogo [2007] and Diogo and Abdala [2007])	**Flexor carpi ulnaris** [from ulna and humerus to pisiform]	**Flexor carpi ulnaris** [from humerus and ulna to pisiform bone and surrounding aponeurosis; this seems to be the common situation for *D. virginiana*, as also described by Coues [1872] and Stein [1981], although authors such as Straus 1942 refer to an attachment to both the pisiform and metacarpal V in *Didelphis*]	**Flexor carpi ulnaris** [from humerus and ulna to pisiform]	**Flexor carpi ulnaris**	**Flexor carpi ulnaris**	**Flexor carpi ulnaris**
–	**Epitrochleoanconeus** (flexor antebrachii ulnaris *sensu* Jouffroy and Lessertisseur, [1971])	**Epitrochleoanconeus** (epitrochleoanconeus medialis *sensu* Gambaryan et al. [2015]) [from medial epicondyle of humerus to olecranon of ulna; Straus [1942] states that the epitrochleo-anconeus is much more developed in monotremes than in other mammals, i.e., that in monotremes it is more similar to that of nonmammalian tetrapods than to that of therian mammals]	**Epitrochleoanconeus** (anconeus internus *sensu* Stein [1981]) [from medial epicondyle of humerus to olecranon process of ulna; Straus [1942] suggested that the epitrochleoanconeus of opossums did not correspond to the anconeus internus, but the anconeus internus *sensu* authors such as Stein [1981] clearly corresponds with the epitrochleoanconeus of other mammals]	**Epitrochleoanconeus** [from humerus to ulna; Greene's [1935] work suggests that this muscle is not present in *Rattus*, but this may well be due to Greene's confusion between the dorsoepitrochlearis and the epitrochleoanconeus; the latter muscle seems to be present in the *Rattus norvegicus* specimens we dissected]	**Epitrochleoanconeus** (condylo-olecranonis *sensu* Le Gros Clark [1924, 1926])	**Epitrochleoanconeus**	– [as noted by authors such as Straus [1942], this muscle may be occasionally found in humans; when it is not present as an independent muscle, it likely corresponds to the fibrous arcade spanning the interval between the epicondylar and olecranon heads of the flexor carpi ulnaris]

(*Continued*)

TABLE 17.1 (CONTINUED)
Pectoral and Forelimb Muscles of Adults of Representative Tetrapod Taxa

Amphibia (Caudata): *Ambystoma mexicanum* (Axolotl)	Reptilia (Lepidosauria): *Timon lepidus* (Ocellated Lizard)	Mammalia (Monotremata): *Ornithorhynchus anatinus* (Platypus)	Mammalia (Marsupialia): *Didelphis virginiana* (Virginian Opossum)	Mammalia (Rodentia): *Rattus norvegicus* (Norwegian Rat)	Mammalia (Scandentia): *Tupaia* sp. (Tree shrew)	Mammalia (Dermoptera): *Cynocephalus volans* (Philippine colugo)	Mammalia (Primates): *Homo sapiens* (Human)
Flexor antebrachii et carpi radialis [as explained by authors such as Straus [1942], in most urodeles, the flexor antebrachii et carpi radialis is usually not differentiated into a pronator teres (flexor antebrachii radialis) and a flexor carpi radialis: see Chapter 18]	**Flexor carpi radialis** (flexor antebrachii et carpi radialis *sensu* Diogo [2007] and Diogo and Abdala [2007])	**Flexor carpi radialis** [from humerus to proximal phalanx of prepollex, being blended with the pronator teres]	**Flexor carpi radialis** [from humerus to metacarpal II; this seems to be the common situation for *D. virginiana*, as also described by Coues [1872] and Stein [1981], although authors such as Straus [1942] refer to an attachment to both metacarpals I and II in *Didelphis*]	**Flexor carpi radialis** [from humerus to metacarpal III]	**Flexor carpi radialis**	**Flexor carpi radialis**	**Flexor carpi radialis** [as noted by Lewis [1989], occasionally in humans. the flexor carpi radialis may be divided into a flexor carpi radialis longus and a flexor carpi radialis brevis]
– [see flexor antebrachii et carpi radialis mentioned earlier]	**Pronator teres** [corresponds to part of the flexor antebrachii et carpi radialis of *Ambystoma*: see flexor antebrachii et carpi radialis mentioned earlier]	**Pronator teres** [from humerus to shaft of radius; as explained by Straus [1942], Haines [1950], and Lewis [1989], the pronator teres of mammals corresponds to the flexor antebrachii radialis/pronator teres of taxa such as *Timon*; i.e., it is not partially derived from the pronator quadratus as suggested by authors such as Howell [1936c]]	**Pronator teres** [from humerus to middle third or distal two thirds of radius; Straus [1942] says that only some marsupials, e.g., phalanger, have humeral and ulnar heads separated by the median nerve, as in humans]	**Pronator teres** [from humerus to radius]	**Pronator teres** (pronator radii teres *sensu* Le Gros Clark [1924, 1926])	**Pronator teres** [as in some bats, in colugos, this muscle does not pronate the forearm, but mainly flexes it, together with the supinator and the brachioradialis: see, e.g., Leche 1886]	**Pronator teres**
Flexores breves superficiales	**Flexores breves superficiales** (part of flexores digitorum breves superficiales *sensu* McMurrich [1903a,b]; flexores breves sublimes *sensu* Holmes [1977]; flexores digiti brevis superficialis *sensu* Abdala and Moro [2006]; flexores digitores breves *sensu* Russell and Bauer 2008)	– [the flexores breves superficiales are absent as a group in mammals—they are present in monotremes, but exclusively originate from the tendons of the flexor digitorum longus, thus no longer being separate structures originating from skeletal elements; see flexor digitorum longus mentioned earlier—but some of them apparently correspond to mammalian structures such as the flexor brevis digitorum manus and/or the palmaris brevis,	–	–	–	–	–

(Continued)

TABLE 17.1 (CONTINUED)
Pectoral and Forelimb Muscles of Adults of Representative Tetrapod Taxa

Amphibia (Caudata): *Ambystoma mexicanum* (Axolotl)	Reptilia (Lepidosauria): *Timon lepidus* (Ocellated Lizard)	Mammalia (Monotremata): *Ornithorhynchus anatinus* (Platypus)	Mammalia (Marsupialia): *Didelphis virginiana* (Virginian Opossum)	Mammalia (Rodentia): *Rattus norvegicus* (Norwegian Rat)	Mammalia (Scandentia): *Tupaia* sp. (Tree shrew)	Mammalia (Dermoptera): *Cynocephalus volans* (Philippine colugo)	Mammalia (Primates): *Homo sapiens* (Human)
		as well as to a part of the flexor digitorum superficialis of most mammals: see flexor digitorum superficialis mentioned earlier]					
–	– [see Chapter 18]	–	– [as in the specimens dissected by Coues [1872] and Stein [1981], there was no palmaris brevis in the specimens we dissected; Jouffroy and Saban [1971] however explain that the muscle is present in some other marsupials]	**Palmaris brevis** [from humerus to ulna; it is difficult to discern if this small muscle is present in the specimens we dissected, but authors such as Peterka [1936], state that rats do have this muscle, and Jouffroy [1971] states that rodents usually have this muscle]	**Palmaris brevis** (muscle of hypothenar pad *sensu* Haines [1955]) [some authors describe a distinct muscle "palmaris superficialis" in tree shrews such as *Tupaia*, but in the most recent descriptions, such as that by George [1977], the palmaris superficialis is considered to be a "simple fibrous plate"]	**Palmaris brevis** [not described by Leche [1886], but it was present and well developed in the colugos we dissected]	**Palmaris brevis** [as explained by, e.g., Howell 1936c and Lewis 1989, the palmaris brevis seems to be absent as an independent muscle in nonmammalian tetrapods such as lizards and salamanders; it likely corresponds to part of their flexores breves superficiales]
–	–	–	**Flexor brevis digitorum manus** [from hamate to middle phalanx of digit 5; numerous authors have studied and discussed the hand muscles of marsupials, including *D. virginiana*, e.g., Coues 1872, Young 1880, Brooks 1886a, McMurrich 1903b, Brandell 1935, Campbell 1939, Jouffroy 1971, Stein 1981, and Lewis 1989; among these studies, Brooks [1886a] seems to be the most complete and accurate, describing four lumbricales, an abductor pollicis brevis (ab2), flexor pollicis brevis (= superficial head of human anatomy: f2r), adductor pollicis (a1), contrahentes to digits 2 (a2), 4 (a4) and 5 (a5, plus a somewhat distinct	– [unlike the "opponens digiti V" of colugos (see on the right), the "opponens digiti quinti" (*sensu* Greene [1935]) of rats does seem to correspond to the opponens digiti minimi of the present work, because it is deep, and not superficial to the flexor digiti minimi brevis]	**Flexor brevis digitorum manus** (primitive flexor brevis manus *sensu* Howell [1936c] and Straus [1942]; flexor brevis manus *sensu* Le Gros Clark [1924, 1926], and George [1977])	**Flexor brevis digitorum manus** (opponens digiti V *sensu* Leche [1886]) [the opponens digiti V *sensu* this author is very similar, and seems to correspond to the flexor brevis digitorum manus of, e.g., *Tupaia*, being superficial, and not deep, to the flexor digiti minimi brevis]	–

(*Continued*)

TABLE 17.1 (CONTINUED)
Pectoral and Forelimb Muscles of Adults of Representative Tetrapod Taxa

Amphibia (Caudata): *Ambystoma mexicanum* (Axolotl)	Reptilia (Lepidosauria): *Timon lepidus* (Ocellated Lizard)	Mammalia (Monotremata): *Ornithorhynchus anatinus* (Platypus)	Mammalia (Marsupialia): *Didelphis virginiana* (Virginian Opossum)	Mammalia (Rodentia): *Rattus norvegicus* (Norwegian Rat)	Mammalia (Scandentia): *Tupaia* sp. (Tree shrew)	Mammalia (Dermoptera): *Cynocephalus volans* (Philippine colugo)	Mammalia (Primates): *Homo sapiens* (Human)
			bundle, a5a1), an abductor digiti minimi (abd5), a flexor digiti minimi with two heads (a5a, mainly corresponding to the flexor digiti minimi of humans, and a deeper f5u, which probably derives from the same anlage that gives rise to the opponens digiti minimi of humans, because it is deeper to the main body of the flexor digiti minimi brevis), as well as a flexor brevis digitorum manus to the middle phalanx of digit 5, and the normal 8 flexores breves profundi plus an adductor pollicis accessorius (which makes a "third" short flexor of digit 1, but, as stated by him, seems to derive instead from the adductor pollicis), and the normal four intermetacarpales]				
–	**Lumbricales** (palmar heads of flexor digitorum longus *sensu* Russell and Bauer [2008])	**Lumbricales** [according to Jouffroy and Lessertisseur [1971], the platypus usually has three lumbricales (from tendons of flexor digitorum longus), to proximal phalanx of digits 3–5 (and then fusing with extensor expansions/structures at the level of the proximal interphalangeal joints, as in many therians), the lumbrical to digit 5 being fused with the flexor brevis profundi 10; the echidna often has 4 or 5 lumbricales, to digits 2–5 or to digits 1–5]	**Lumbricales** [from tendons of flexor digitorum profundus, directly to proximal phalanges of digits 2-5—NB: indirectly, they also attach onto the middle and distal phalanges of these digits, because they attach on the extensor expansions of these digits]	**Lumbricales** [as described by Greene [1935], *Rattus* usually has four lumbricales, from tendons of flexor digitorum profundus to proximal phalanges of digits 2–5 (plus to soft tissues, i.e., to extensor expansions, as in humans)]	**Lumbricales** [as described by Le Gros Clark [1924] and George [1977], *Tupaia* usually has four lumbricales, to digits 2–5]	**Lumbricales** [as described by Leche [1886], *Cynocephalus* usually has seven lumbricales, to the radial sides of digits 2–5 and to the ulnar sides of digits 2–4]	**Lumbricales** [as described by Netter [2006], humans usually have four lumbricales, to digits 2–5]

(*Continued*)

TABLE 17.1 (CONTINUED)
Pectoral and Forelimb Muscles of Adults of Representative Tetrapod Taxa

Amphibia (Caudata): *Ambystoma mexicanum* (Axolotl)	Reptilia (Lepidosauria): *Timon lepidus* (Ocellated Lizard)	Mammalia (Monotremata): *Ornithorhynchus anatinus* (Platypus)	Mammalia (Marsupialia): *Didelphis virginiana* (Virginian Opossum)	Mammalia (Rodentia): *Rattus norvegicus* (Norwegian Rat)	Mammalia (Scandentia): *Tupaia* sp. (Tree shrew)	Mammalia (Dermoptera): *Cynocephalus volans* (Philippine colugo)	Mammalia (Primates): *Homo sapiens* (Human)
Contrahentes digitorum [see Chapter 18]	**Contrahentes digitorum** (include the flexor digitorum V transversus I and flexor digitorum V transversus II *sensu* Abdala and Moro [2006], Diogo [2007], and Diogo and Abdala [2007]: see Chapter 18)	– [according to Jouffroy and Lessertisseur [1971], the contrahens to digit 1 (= adductor pollicis) is apparently present, but there are no other contrahentes digitorum in monotremes, with the exception of an eventual contrahentes going to digit 5; according to Howell [1937d], even the adductor pollicis is absent in monotremes; the latter idea was supported by the study of the monotreme forelimb muscles by Gambaryan et al. [2015], as no adductor pollicis nor any other contrahentes were described in that study]	**Contrahentes digitorum** [there are three contrahentes digitorum, other than the adductor pollicis: contrahens of digit 2 extends from metacarpal III to proximal phalanx of digit 2; contrahens of digit 4 extends from metacarpal III to proximal phalanx of digit 4; contrahens of digit 5 extends from metacarpal III and capitate to proximal phalanx of digit 5, having a small, somewhat separate head; see description of flexor brevis digitorum manus mentioned earlier]	**Contrahentes digitorum** [contrahens to digit 2 from metacarpal III and contrahens raphe to proximal phalanx of digit 2; contrahens to digit 5 from metacarpal III and contrahens raphe to proximal phalanx of digit 5; the contrahentes digitorum, other than the adductor pollicis, are really present in *Rattus norvegicus*? It was not possible to appropriately discern if these muscles are present, or not, in the rats we dissected; Greene [1935] does not describe contrahentes other than the adductor pollicis in this taxon, but authors such as Peterka [1936] seem to suggest that they may be present in at least some rats; this is supported by Cihak [1972], who states that rats have three well-developed contrahentes, i.e., an adductor pollicis to digit 1 and two other contrahentes,	**Contrahentes digitorum** (part of the contrahentes manus *sensu* George [1977], which also include the adductor pollicis) [Le Gros Clark [1924] does not describe the contrahentes digitorum to digits 2 and 5 in *Tupaia minor*, but they are very likely present: they were found in all the *Tupaia* specimens analyzed by us and by, e.g., Haines 1955 and George 1977]	**Contrahentes digitorum** (adductor indicis and adductor digiti V *sensu* Leche [1886]) [as described by Leche [1886], colugos usually have contrahentes to digits 1 (adductor pollicis), 2, and 5]	– [humans usually only have a contrahens, which goes to digit 1 and thus corresponds to the adductor pollicis *sensu* this volume: see adductor pollicis in the following and text]

(Continued)

TABLE 17.1 (CONTINUED)
Pectoral and Forelimb Muscles of Adults of Representative Tetrapod Taxa

Amphibia (Caudata): *Ambystoma mexicanum* (Axolotl)	**Reptilia (Lepidosauria): *Timon lepidus* (Ocellated Lizard)**	**Mammalia (Monotremata): *Ornithorhynchus anatinus* (Platypus)**	**Mammalia (Marsupialia): *Didelphis virginiana* (Virginian Opossum)**	**Mammalia (Rodentia): *Rattus norvegicus* (Norwegian Rat)**	**Mammalia (Scandentia): *Tupaia* sp. (Tree shrew)**	**Mammalia (Dermoptera): *Cynocephalus volans* (Philippine colugo)**	**Mammalia (Primates): *Homo sapiens* (Human)**
				which thus probably correspond to the contrahentes to digits 2 and 5, as suggested by Peterka [1936]; this idea is also supported by McMurrich [1903a,b], who states that mice have contrahentes to digits 1, 2, and 5; our dissections confirm that the contrahentes to digits 2 and 5 are present in mice]			
–	– [see cell above]	– [see preceding cell]	**Adductor pollicis** [contrahens of digit 1, which extends from metacarpal III and capitate to proximal phalanx of digit 1; the adductor pollicis of mammals clearly corresponds to a part of the contrahentes digitorum of other tetrapods, but most anatomists working with mammals describe it as a muscle that is somewhat independent from the other contrahentes, thus deserving a distinct name]	**Adductor pollicis** [from metacarpals II and III capitate to proximal phalanx of digit 1]	**Adductor pollicis** (part of the contrahentes manus *sensu* George [1977], which also include the contrahentes digitorum described earlier)	**Adductor pollicis** (adductor pollicis or part of the contrahentes *sensu* Leche [1886], which also include the contrahentes digitorum described earlier)	**Adductor pollicis**
–	–	–	**Adductor pollicis accessorius** (volaris primus of Henle) [from trapezoid and metacarpal I to proximal phalanx of digit 1; see flexor brevis digitorum manus mentioned earlier]	–	–	–	**Adductor pollicis accessorius** (volaris primus of Henle) [see text]

(*Continued*)

TABLE 17.1 (CONTINUED)
Pectoral and Forelimb Muscles of Adults of Representative Tetrapod Taxa

Amphibia (Caudata): *Ambystoma mexicanum* (Axolotl)	Reptilia (Lepidosauria): *Timon lepidus* (Ocellated Lizard)	Mammalia (Monotremata): *Ornithorhynchus anatinus* (Platypus)	Mammalia (Marsupialia): *Didelphis virginiana* (Virginian Opossum)	Mammalia (Rodentia): *Rattus norvegicus* (Norwegian Rat)	Mammalia (Scandentia): *Tupaia* sp. (Tree shrew)	Mammalia (Dermoptera): *Cynocephalus volans* (Philippine colugo)	Mammalia (Primates): *Homo sapiens* (Human)
Flexores breves profundi [see Chapter 18]	**Flexores breves profundi** (flexores digitorum breves profundi *sensu* McMurrich [1903a,b]; flexores digiti brevis profundus of, e.g., Abdala and Moro 2006, include lumbricales and/or interossei ventrales *sensu* Russell and Bauer [2008]) [see Chapter 18]	**Flexores breves profundi** (interossei *sensu* Jouffroy and Lessertisseur [1971]) [the platypus often has 8–10 flexores breves profundi, inserting onto digits 1–5; 10 muscles were described in the platypus study of Gambaryan et al. [2015], with two muscles for each digit, one to the medial side and one to the radial side of the proximal phalanx of each digit; the flexores 1, 2, 3, and 4 originate from the prepollex, flexores 5, 6, 7, and 8 from soft tissues, and 9 and 10 from the pisiform; see also text]	**Flexores breves profundi** (8 muscles of the total 10, because the numbers 1 and 10 correspond to the flexor pollicis brevis and to the flexor digiti minimi brevis, respectively, so they are described in other cells; thus, the 8 described here are number 2, from trapezium to proximal phalanx of digit 1; numbers 3 and 4 from metacarpal II to proximal phalanx of digit 2; numbers 5 and 6 from metacarpal III to proximal phalanx of digit 3; numbers 7 and 8 from metacarpal IV to proximal phalanx of digit 4; and number 9 from metacarpal V to radial side of proximal phalanx of digit 5]	**Flexores breves profundi** (palmar interossei *sensu* Greene [1935]) [exactly as in *Didelphis*; i.e., 8 muscles of the total 10, because the number 1 correspond to the flexor pollicis brevis and the number 10 corresponds to the flexor digiti minimi brevis + opponens digiti minimi, so these 3 latter muscles are described in other cells; thus, the 8 described here are see cell on the left; we use the name interossei for *Cynocephalus* and *Tupaia* because in these taxa, unlike rats, the flexores breves are deeply mixed with the intermetacarpales, forming the "interossei externi" *sensu* Leche [1886], and the dorsal interossei *sensu* Le Gros Clark [1924]; as noted by, e.g., Haines [1955], in *Tupaia*, the 3 interossei attaching on the ulnar side of digit 2 and radial sides of digits 4 and 5 are somewhat ventral to, but not really separated from, the four interossei attaching on the radial sides of digits 2 and 3 and ulnar sides of digits 3 and 4;	**Interossei** [see text; see also on the left] [in addition to the flexor pollicis brevis and the flexor digiti minimi brevis (which correspond to the flexores breves profundi 1 and 10 *sensu* this volume), *Tupaia* apparently has three (to digits 2, 4, and 5) or possibly four (to digits 1, 2, 4, and 5) flexores breves profundi; that is, it apparently has a total of five, or possibly six, flexores breves profundi; see cells on the left and text]	**Interossei** (interossei externi and interni *sensu* Leche [1886]) [in addition to the flexor pollicis brevis and the flexor digiti minimi brevis (which correspond to the flexores breves profundi 1 and 10 *sensu* this volume), colugos apparently have three (to digits 2, 4, and 5) or possibly four (to digits 1, 2, 4, and 5) flexores breves profundi, that is, they apparently have a total of five, or possibly six, flexores breves profundi; see cells on the left and text]	**Interossei palmares** [see text; see also cells on the left]

(*Continued*)

TABLE 17.1 (CONTINUED)
Pectoral and Forelimb Muscles of Adults of Representative Tetrapod Taxa

Amphibia (Caudata): *Ambystoma mexicanum* (Axolotl)	Reptilia (Lepidosauria): *Timon lepidus* (Ocellated Lizard)	Mammalia (Monotremata): *Ornithorhynchus anatinus* (Platypus)	Mammalia (Marsupialia): *Didelphis virginiana* (Virginian Opossum)	Mammalia (Rodentia): *Rattus norvegicus* (Norwegian Rat)	Mammalia (Scandentia): *Tupaia* sp. (Tree shrew)	Mammalia (Dermoptera): *Cynocephalus volans* (Philippine colugo)	Mammalia (Primates): *Homo sapiens* (Human)
				that is why we prefer to refer here to interossei [=flexores breves profundi + intermetacarpales) and to not subdivide the latter muscles into interossei palmares and interossei dorsales, as we do in humans: see cells on the right]			
Flexores digitorum minimum	–	–	–	–	–	–	–
Interphalangeus digiti III	–	–	–	–	–	–	–
–	–	–	–	–	–	–	**Interossei dorsales** [see interossei palmares]
–	–	– [our dissections support that, as described by Howell [1937d] and Jouffroy and Lessertisseur [1971], there is are no independent, distinct flexor pollicis brevis nor digiti minimi brevis in the platypus; that is, the flexor brevis profundus 1 and flexor brevis profundus 10, which give rise, respectively, to the flexor pollicis brevis and the flexor digiti minimi brevis of other mammals, are not really independent and distinct from the remaining flexores breves profundi]	**Flexor pollicis brevis** [from trapezium to radial side of proximal phalanx of thumb, corresponding to the superficial head of the flexor pollicis brevis of human anatomy; see flexor brevis digitorum manus mentioned earlier, as well as cell on the right]	**Flexor pollicis brevis** [from trapezium to radial side of proximal phalanx of thumb, corresponding to the superficial head of the flexor pollicis brevis of human anatomy]	**Flexor pollicis brevis** (radial head of flexor brevis profundus 1 *sensu* Haines [1955])	**Flexor pollicis brevis**	**Flexor pollicis brevis** [the results of Cihak's [1972] developmental studies of human fetuses suggest that both the superficial and deep heads of the flexor pollicis brevis of humans are derived from the flexores brevis profundus 1]

(Continued)

TABLE 17.1 (CONTINUED)
Pectoral and Forelimb Muscles of Adults of Representative Tetrapod Taxa

Amphibia (Caudata): *Ambystoma mexicanum* (Axolotl)	Reptilia (Lepidosauria): *Timon lepidus* (Ocellated Lizard)	Mammalia (Monotremata): *Ornithorhynchus anatinus* (Platypus)	Mammalia (Marsupialia): *Didelphis virginiana* (Virginian Opossum)	Mammalia (Rodentia): *Rattus norvegicus* (Norwegian Rat)	Mammalia (Scandentia): *Tupaia* sp. (Tree shrew)	Mammalia (Dermoptera): *Cynocephalus volans* (Philippine colugo)	Mammalia (Primates): *Homo sapiens* (Human)
–	–	–	– [Figure 9.8A of Lewis [1989] suggests that at least some marsupials have an opponens pollicis, but the consensus is that at least *D. virginiana* does not have an opponens pollicis; see flexor brevis digitorum manus mentioned earlier]	– [apparently absent as an independent muscle in *Rattus*, *Tupaia*, and *Cynocephalus*, but is found in other nonprimate therian mammals: see, e.g., Jouffroy 1971]	– [see on the left]	– [see on the left]	**Opponens pollicis** [authors, such as Howell [1936c], stated that the opponens pollicis and opponens digiti minimi probably derive from part of the flexores breves superficiales; however, it is now commonly accepted that these two muscles derive, respectively, from the flexor pollicis brevis and the flexor digiti minimi brevis, as stated by Lewis [1989], and supported by Cihak's [1972] ontogenetic studies of human fetuses]
–	–	– [see preceding text]	**Flexor digiti minimi brevis** [with two heads, from pisiform to proximal phalanx of digit 5: one more superficial and one more deep, corresponding topologically to the flexor digiti minimi brevis and to the opponens digiti minimi of humans, respectively, but the latter is not homologous to the opponens digiti minimi of humans, because this muscle was acquired only later in evolution, among placental mammals, so the condition of marsupials is a rough parallelism, as the deeper head attaches onto the proximal phalanx of digit 5, and not onto the metacarpal V as is the case of a true opponens digiti minimi as seen, e.g., in humans; see flexor brevis digitorum manus mentioned earlier]	**Flexor digiti minimi brevis** (flexor digiti quinti brevis *sensu* Greene [1935]) [from hamate to ulnar side of proximal phalanx of thumb, corresponding to a part of the flexor brevis profundus 10; see preceding text]	**Flexor digiti minimi brevis** (flexor digiti quinti manus *sensu* George [1977]; ulnar head of flexor brevis profundus 5 *sensu* Haines [1944])	**Flexor digiti minimi brevis** (flexor brevis digiti V *sensu* Leche [1886])	**Flexor digiti minimi brevis**

(*Continued*)

TABLE 17.1 (CONTINUED)
Pectoral and Forelimb Muscles of Adults of Representative Tetrapod Taxa

Amphibia (Caudata): *Ambystoma mexicanum* (Axolotl)	Reptilia (Lepidosauria): *Timon lepidus* (Ocellated Lizard)	Mammalia (Monotremata): *Ornithorhynchus anatinus* (Platypus)	Mammalia (Marsupialia): *Didelphis virginiana* (Virginian Opossum)	Mammalia (Rodentia): *Rattus norvegicus* (Norwegian Rat)	Mammalia (Scandentia): *Tupaia* sp. (Tree shrew)	Mammalia (Dermoptera): *Cynocephalus volans* (Philippine colugo)	Mammalia (Primates): *Homo sapiens* (Human)
–	–	–	– [see cell above]	**Opponens digiti minimi** (opponens digiti quinti *sensu* Greene [1935]) [from triquetrum and pisiform to ulnar side of distal portion of metacarpal V]	–	–	**Opponens digiti minimi** [see preceding text]
– [see Chapter 18]	**Abductor pollicis brevis** (abductor brevis pollici *sensu* Abdala and Moro [2006])	**Abductor pollicis brevis** [contrary to most authors, Gambaryan et al. [2015] do not describe this muscle in the platypus, but their figures of this taxon and the other monotreme genera clearly seem to indicate that it is present, probably corresponding to, e.g., the structure they designate as "flexor brevis superficialis 1" in their Figure 16.B; that is, this muscle was almost surely included in their "flexores breves superficiales" and/ or "flexores breves profundi" series; see descriptions of these muscles mentioned earlier]	**Abductor pollicis brevis** [from trapezium to radial side of proximal phalanx of thumb]	**Abductor pollicis brevis** (abductor pollicis *sensu* Greene [1935]) [from falciform to radial side of proximal phalanx of thumb]	**Abductor pollicis brevis**	**Abductor pollicis brevis**	**Abductor pollicis brevis**
Abductor digiti minimi (extensor lateralis digiti IV *sensu* Walthall and Ashley-Ross [2006], Diogo [2007], Diogo and Abdala [2007], and Diogo et al. [2009a]) [see Chapter 18]	**Abductor digiti minimi** (abductor digitorum V *sensu* Abdala and Moro [2006], Diogo [2007], and Diogo and Abdala [2007]) [see Chapter 18]	**Abductor digiti minimi** [contrary to most authors, Gambaryan et al. [2015] do not describe this muscle in the platypus, but their figures of this taxon and the other monotreme genera clearly seem to indicate that it is present, probably corresponding to, e.g., the structure they designate as "interosseous V lateralis" in their Figure 16.A1; that is,	**Abductor digiti minimi** [from pisiform to ulnar side of proximal phalanx of digit 5]	**Abductor digiti minimi** (abductor digiti quinti *sensu* Greene [1935]; abductor digiti V *sensu* Rocha-Barbosa et al. [2007]) [from pisiform to ulnar side of proximal phalanx of digit 5]	**Abductor digiti minimi** (abductor digiti quinti manus *sensu* George [1977]; abductor minimi digiti *sensu* Le Gros Clark [1924, 1926])	**Abductor digiti minimi**	**Abductor digiti minimi**

(*Continued*)

TABLE 17.1 (CONTINUED)
Pectoral and Forelimb Muscles of Adults of Representative Tetrapod Taxa

Amphibia (Caudata): *Ambystoma mexicanum* (Axolotl)	Reptilia (Lepidosauria): *Timon lepidus* (Ocellated Lizard)	Mammalia (Monotremata): *Ornithorhynchus anatinus* (Platypus)	Mammalia (Marsupialia): *Didelphis virginiana* (Virginian Opossum)	Mammalia (Rodentia): *Rattus norvegicus* (Norwegian Rat)	Mammalia (Scandentia): *Tupaia* sp. (Tree shrew)	Mammalia (Dermoptera): *Cynocephalus volans* (Philippine colugo)	Mammalia (Primates): *Homo sapiens* (Human)
		this muscle was almost surely included in their flexores breves superficiales and/or flexores breves profundi series; see descriptions of these muscles mentioned earlier]					
Intermetacarpales [see Chapter 17]	**Intermetacarpales** (intermetacarpales I and II *sensu* Abdala and Moro [2006], Diogo [2007], and Diogo and Abdala [2007]) [as described by authors such as Abdala and Moro [2006], lizards often have four including four "intermetacarpales I" heads connecting the metacarpals of the five digits and four "intermetacarpales 2" heads, also connecting the metacarpals of these five digits]	**Intermetacarpales** [contrary to other authors, which stated that the platypus usually has no intermetacarpales—an idea followed by Diogo and Abdala [2010]—Gambaryan et al. [2015] described intermetacarpales in platypus (but not in the other two extant monotreme genera); they refer to four muscles, but lying in two intermetacarpal spaces (i.e., two muscles between metacarpals II and III and two other muscles lying between metacarpals III and IV) only, so they seem to correspond to two bipennate intermetacarpales *sensu* the present work, in total; intermetacarpalis 1 thus originate on ulnar side of metacarpal II and radial side of metacarpal III and inserts onto the ulnar side of proximal phalanx of digit 2 and radial side of proximal phalanx of digit 3,	**Intermetacarpales** [as described by Young [1880], the intermetacarpalis 1 runs from metacarpal I to radial side of proximal phalanx of digit 2; intermetacarpalis 2 runs from metacarpals II and III to ulnar side of proximal phalanx of digit 2 as well as to radial side of proximal phalanx of digit 3; intermetacarpalis 3 runs from metacarpals III and IV to ulnar side of proximal phalanx of digit 3 as well as to radial side of proximal phalanx of digit 4; intermetacarpalis 4 runs from metacarpals IV and V to ulnar side of proximal phalanx of digit 4 as well as to radial side of proximal phalanx of digit 5]	**Intermetacarpales** [our dissections of *Rattus norvegicus* indicate that this taxon has four intermetacarpales, the first from metacarpals I and II to radial side of digit 2, the second from metacarpals II and III to radial side of digit 3, the third from metacarpals III and IV to ulnar side of digit 3, and the fourth from metacarpals IV and V to ulnar side of digit 4; see also text]	– [see text]	– [see text]	– [see text]

(*Continued*)

TABLE 17.1 (CONTINUED)
Pectoral and Forelimb Muscles of Adults of Representative Tetrapod Taxa

Amphibia (Caudata): *Ambystoma mexicanum* (Axolotl)	Reptilia (Lepidosauria): *Timon lepidus* (Ocellated Lizard)	Mammalia (Monotremata): *Ornithorhynchus anatinus* (Platypus)	Mammalia (Marsupialia): *Didelphis virginiana* (Virginian Opossum)	Mammalia (Rodentia): *Rattus norvegicus* (Norwegian Rat)	Mammalia (Scandentia): *Tupaia* sp. (Tree shrew)	Mammalia (Dermoptera): *Cynocephalus volans* (Philippine colugo)	Mammalia (Primates): *Homo sapiens* (Human)
		while intermetacarpalis 2 originates on ulnar side of metacarpal III and radial side of metacarpal IV and inserts onto the ulnar side of proximal phalanx of digit 3 and radial side of proximal phalanx of digit 4; see text]					
Extensor carpi radialis [see Chapter 18]	**Extensor antebrachii et carpi radialis** (extensor carpi radialis *sensu* Abdala and Moro [2006]) [see Chapter 18]	**Extensor carpi radialis** [from humerus to metacarpals II–IV]	**Extensor carpi radialis longus** [from humerus to metacarpal II]	**Extensor carpi radialis longus** [humerus to metacarpal II; this muscle as well as the extensor carpi radialis brevis are derived from the extensor carpi radialis]	**Extensor carpi radialis longus**	**Extensor carpi radialis longus**	**Extensor carpi radialis longus**
– [see Chapter 18]	– [see Chapter 18]	–	**Extensor carpi radialis brevis** [from humerus to metacarpal III; as a variant in rats it can attach also to metacarpal II: see Straus 1941a,b]	**Extensor carpi radialis brevis** [from humerus to metacarpal II; see extensor carpi radialis mentioned earlier]	**Extensor carpi radialis brevis**	**Extensor carpi radialis brevis**	**Extensor carpi radialis brevis**
– [see Chapter 18]	– [see Chapter 18]	**Brachioradialis** [from humerus to radiale, intermedium, and centrale and distal radius, being blended with the abductor pollicis longus]	**Brachioradialis** (supinator longus *sensu* Coues [1872]) [from humerus to scaphoid bone, lunate, and triquetrum, as described by Haines [1939] and Straus [1941a,b] and contra Coues [1872] and Stein [1981]]	– [not present as an independent muscle in rats nor in mice]	**Brachioradialis** [apparently absent as an independent structure in tree shrews such as *Ptilocercus* and even in some specimens of *Tupaia javanica*: see Le Gros Clark 1926 and George 1977]	**Brachioradialis** (supinator longus *sensu* Macalister [1872] and Leche [1886]) [Gunnell and Simmons [2005] stated that *Cynocephalus* lacks a brachioradialis, but this is clearly an error: this muscle is present and welldeveloped in the specimens dissected by Macalister [1872], by Leche [1886], and by us]	**Brachioradialis**

(*Continued*)

TABLE 17.1 (CONTINUED)
Pectoral and Forelimb Muscles of Adults of Representative Tetrapod Taxa

Amphibia (Caudata): *Ambystoma mexicanum* (Axolotl)	Reptilia (Lepidosauria): *Timon lepidus* (Ocellated Lizard)	Mammalia (Monotremata): *Ornithorhynchus anatinus* (Platypus)	Mammalia (Marsupialia): *Didelphis virginiana* (Virginian Opossum)	Mammalia (Rodentia): *Rattus norvegicus* (Norwegian Rat)	Mammalia (Scandentia): *Tupaia* sp. (Tree shrew)	Mammalia (Dermoptera): *Cynocephalus volans* (Philippine colugo)	Mammalia (Primates): *Homo sapiens* (Human)
Supinator	– [supinator not present as a distinct muscle in the *Timon* specimens we dissected, and this is confirmed by Russell and Howell [2008], who stated that in *Lacerta* and most lizards, the extensor antebrachii et carpi radialis have three heads that seem to correspond to the supinator, brachioradialis (or supinator longus), and extensor carpi radialis of mammals, but these heads are deeply fused to each other forming a single, complex muscle]	**Supinator** [from humerus to radius; as noted by Jouffroy and Lessertisseur [1971] and Lewis [1989], the supinator of mammals corresponds to a part of the extensor antebrachii et carpi radialis of reptiles such as *Timon*, and not of the abductor pollicis longus ("supinator manus"), as suggested by Howell [1936c]]	**Supinator** (supinator brevis *sensu* Coues [1872]) [from humerus to proximal radius, being a well-defined, separated muscle, as described by, e.g., Stein [1981] and contra Coues [1872]]	**Supinator** [from humerus and ulna to radius]	**Supinator**	**Supinator** (supinator brevis *sensu* Leche [1886])	**Supinator** (supinator brevis *sensu* Parsons [1898])
Extensor antebrachii et carpi ulnaris	**Extensor antebrachii et carpi ulnaris**	**Extensor carpi ulnaris** [from humerus to metacarpal V, being blended with the abductor digiti minimi and with the extensor digiti minimi]	**Extensor carpi ulnaris** [humerus and ulna to metacarpal V]	**Extensor carpi ulnaris** [from humerus and ulna to metacarpal V]	**Extensor carpi ulnaris**	**Extensor carpi ulnaris**	**Extensor carpi ulnaris**
– [see Chapter 18]	– [see Chapter 18]	**Anconeus** (epitrochleoanconeus lateralis *sensu* Gambaryan et al. [2015]) [from lateral epicondyle of humerus to olecranon process of ulna; see text]	**Anconeus** (anconeus externus *sensu* Stein [1981]) [from lateral epicondyle humerus to olecranon process of ulna]	**Anconeus** [from humerus to ulna, being blended with triceps brachii]	**Anconeus**	**Anconeus** (anconeus quartus *sensu* Leche [1886])	**Anconeus**
Extensor digitorum [see Chapter 18]	**Extensor digitorum** [see Chapter 18]	**Extensor digitorum** (extensor digitorum communis *sensu* Jouffroy and Lessertisseur [1971], Stein [1981], and Warburton [2003]) [from humerus to distal phalanges of digits 1–4 (one head) and 3–5 (the other head),	**Extensor digitorum** (extensor digitorum communis *sensu* Coues [1872]) [humerus to distal phalanges of digits 2–5; as a variant, the tendon to digit 5 might be absent]	**Extensor digitorum** (extensor digitorum communis *sensu* Greene 1935) [according to Greene [1935], in *Rattus*, the extensor digitorum usually extends from the humerus to the distal phalanges of digits 2–5]	**Extensor digitorum** (extensor digitorum communis *sensu* George [1977]; extensor communis digitorum *sensu* Le Gros Clark [1924, 1926]) [as described by Le Gros Clark [1924] and George [1977], in *Tupaia*,	**Extensor digitorum** (extensor digitorum communis *sensu* Leche [1886]) [as described by Leche [1886], in colugos, the extensor digitorum usually sends tendons to digits 2–5]	**Extensor digitorum** [as described by authors such as Lewis [1989], in humans, this muscle usually sends tendons to digits 2–5, although it can occasionally also send a tendon to digit 1]

(Continued)

TABLE 17.1 (CONTINUED)
Pectoral and Forelimb Muscles of Adults of Representative Tetrapod Taxa

Amphibia (Caudata): *Ambystoma mexicanum* (Axolotl)	Reptilia (Lepidosauria): *Timon lepidus* (Ocellated Lizard)	Mammalia (Monotremata): *Ornithorhynchus anatinus* (Platypus)	Mammalia (Marsupialia): *Didelphis virginiana* (Virginian Opossum)	Mammalia (Rodentia): *Rattus norvegicus* (Norwegian Rat)	Mammalia (Scandentia): *Tupaia* sp. (Tree shrew)	Mammalia (Dermoptera): *Cynocephalus volans* (Philippine colugo)	Mammalia (Primates): *Homo sapiens* (Human)
		the tendons being deeply blended with those of the panniculus carnosus; contrary to authors such as Ribbing [1907], authors such as Straus [1941a,b] consider that plesiomorphically in tetrapods, the extensor digitorum did not reach the phalanges of the digits and, thus, that the extensor digitorum of mammals—which does insert on the phalanges—includes part of the extensores digitorum breves; according to Jouffroy and Lessertisseur [1971], in the platypus, the extensor digitorum usually sends tendons to the five digits]			the extensor digitorum usually sends tendons to digits 2–5; see Kaneff [1979, 1980a,b] for variations of this muscle in this genus]		
Extensores digitorum breves [see Chapter 18]	**Extensores digitorum breves** [see Chapter 18]	– [it is commonly accepted that the extensores digitorum breves are absent as a group in extant mammals; they gave rise to independent, distinct muscles as, e.g., the extensor pollicis longus, extensor indicis, extensor digiti III proprius, extensor digiti quarti, and extensor digiti minimi: see text and the following]	–	–	–	–	–

(*Continued*)

TABLE 17.1 (CONTINUED)
Pectoral and Forelimb Muscles of Adults of Representative Tetrapod Taxa

Amphibia (Caudata): *Ambystoma mexicanum* (Axolotl)	Reptilia (Lepidosauria): *Timon lepidus* (Ocellated Lizard)	Mammalia (Monotremata): *Ornithorhynchus anatinus* (Platypus)	Mammalia (Marsupialia): *Didelphis virginiana* (Virginian Opossum)	Mammalia (Rodentia): *Rattus norvegicus* (Norwegian Rat)	Mammalia (Scandentia): *Tupaia* sp. (Tree shrew)	Mammalia (Dermoptera): *Cynocephalus volans* (Philippine colugo)	Mammalia (Primates): *Homo sapiens* (Human)
–	–	**Extensor digiti minimi** (extensor lateralis *sensu* Jouffroy and Lessertisseur [1971] and Gambaryan et al. [2015]) [from extensor digitorum to distal phalanges of digits 4 and 5, being blended also with the extensor carpi ulnaris; this contrasts with the description of Jouffroy and Lessertisseur [1971], according to which in platypus the extensor digiti minimi has a single tendon, inserting onto digit 5, but in echidna, it has two tendons inserting onto digits 4 and 5; some developmental studies, e.g., Cihak [1972], indicate that this muscle derives from the extensor digitorum, while others, e.g., Gräfenberg [1906], indicate that it corresponds to a part of the extensores digitorum breves; the analysis of other lines of evidence strongly supports the latter hypothesis: (a) both the extensor digitorum and extensor digiti minimi usually have tendons to digits 4 and 5 in mammals: as the extensor digitorum of other tetrapods usually has a single tendon to each digit, it is unlikely that it corresponds to both the former muscles;	**Extensor digiti minimi** (extensor digitorum ulnaris *sensu* Haines [1939]; extensor digitorum lateralis *sensu* Stein [1981]) [humerus to distal phalanges of digits 4 and 5; as a variant, it can also attach onto distal phalanx of digit 3: see, e.g., Haines 1939]	**Extensor digiti minimi** (extensor digiti quinti proprius *sensu* Greene [1935] and Howell [1936c]; extensor digiti quinti *sensu* Peterka [1936]) [in rats, from humerus to distal phalanx of digit 5, being blended with extensor digiti quarti; in mice, from humerus to distal phalanges of digits 4 and 5, as shown in wildtype mouse of Figure 4 of Duboc and Logan [2011]; see cell on the left and text]	**Extensor digiti minimi** (extensor digitorum lateralis *sensu* Le Gros Clark [1924, 1926]; extensor digitorum ulnaris *sensu* George [1977]) [as described by Le Gros Clark [1924] and George [1977], in *Tupaia*, this muscle usually inserts onto digits 4 and 5; see cells on the left and text, and see Kaneff [1979, 1980a,b] for variations of this muscle in *Tupaia*]	**Extensor digiti minimi** (extensor digitorum secundus *sensu* Leche [1886]) [as described by Leche [1886], in colugos, this muscle usually inserts onto digits 3, 4, and 5; see cells on the left and text]	**Extensor digiti minimi** [as described by authors such as Netter [2006], in humans, this muscle usually inserts onto digit 5; see on the left and text]

(Continued)

TABLE 17.1 (CONTINUED)
Pectoral and Forelimb Muscles of Adults of Representative Tetrapod Taxa

Amphibia (Caudata): *Ambystoma mexicanum* (Axolotl)	Reptilia (Lepidosauria): *Timon lepidus* (Ocellated Lizard)	Mammalia (Monotremata): *Ornithorhynchus anatinus* (Platypus)	Mammalia (Marsupialia): *Didelphis virginiana* (Virginian Opossum)	Mammalia (Rodentia): *Rattus norvegicus* (Norwegian Rat)	Mammalia (Scandentia): *Tupaia* sp. (Tree shrew)	Mammalia (Dermoptera): *Cynocephalus volans* (Philippine colugo)	Mammalia (Primates): *Homo sapiens* (Human)
		(b) there are known homologies for the mammalian extensores digitorum breves 1 (ext pollicis longus), 2 (extensor indicis proprius), and 3 (extensor digiti III proprius), but not for those of digits 4 and 5: the extensor digiti minimi goes to digits 4 and 5 in many mammals; (c) as noted by Lewis [1989], the tendons of the mammalian extensor digiti minimi often occupy the deep plane on the dorsum of the hand; i.e., they are usually deep to the tendons of the extensor digitorum; see text]					
–	–	–	–	**Extensor digiti quarti** (extensor digiti quarti proprius *sensu* Howell [1936c]) [as its name indicates, in *Rattus*, this muscle usually extends from humerus to distal phalanx of digit 4; as noted earlier, in mice, there is a single muscle extensor digiti minimi, with a distally bifurcated tendon, from humerus to distal phalanges of digits 4 and 5, as shown in wildtype mouse of Figure 4 of Duboc and Logan [2011]; see extensores digitorum breves mentioned earlier and text]	–	–	–

(*Continued*)

TABLE 17.1 (CONTINUED)
Pectoral and Forelimb Muscles of Adults of Representative Tetrapod Taxa

Amphibia (Caudata): *Ambystoma mexicanum* (Axolotl)	Reptilia (Lepidosauria): *Timon lepidus* (Ocellated Lizard)	Mammalia (Monotremata): *Ornithorhynchus anatinus* (Platypus)	Mammalia (Marsupialia): *Didelphis virginiana* (Virginian Opossum)	Mammalia (Rodentia): *Rattus norvegicus* (Norwegian Rat)	Mammalia (Scandentia): *Tupaia* sp. (Tree shrew)	Mammalia (Dermoptera): *Cynocephalus volans* (Philippine colugo)	Mammalia (Primates): *Homo sapiens* (Human)
–	–	**Extensor digiti III proprius** (part of extensor profundus *sensu* Jouffroy and Lessertisseur [1961] and of extensor pollicis et indicis *sensu* Gambaryan et al. [2015]) [from ulna to distal phalanx of digit 3, being blended with the extensor indicis and extensor digitorum; as its name indicates, in platypus and echidna, this muscle usually inserts onto digit 3; see extensores digitorum breves mentioned earlier and text]	–	–	–	–	–
–	–	**Extensor indicis** (part of extensor pollicis et indicis *sensu* Gambaryan et al. [2015]) [from ulna to distal phalanx of digit 2, being blended with the extensor digitorum and extensor pollicis longus; in platypus, this muscle usually inserts onto digit 2; see extensores digitorum breves mentioned earlier and text]	**Extensor indicis** (extensor digitorum profundus *sensu* Straus [1941a,b]) [ulna to distal phalanges of digits 1, 2, and 3, as described by Haines [1939] and Straus [1941a,b]; as a variant connection to digit 3 might be absent: see, e.g., Coues 1872 and Stein 1981]	**Extensor indicis** (extensor indicis proprius *sensu* Greene [1935]) [as described by Greene [1935], in *Rattus*, this muscle usually runs from the ulna to distal phalanges of digits 2 and 3; in mice, this muscle also goes to digits 2 and 3, as described by Duboc and Logan [2011]; see extensores digitorum breves mentioned earlier and text]	**Extensor indicis** [as described by Leche [1886], Le Gros Clark [1924], George [1977], and Kaneff [1979, 1980a,b]; in *Tupaia*, this muscle usually inserts onto digits 2 and 3, although sometimes it also attaches to digits 1 and/or 4; see extensores digitorum breves mentioned earlier and text]	**Extensor indicis** [as described by Leche [1886], in colugos, this muscle usually inserts onto digits 1, 2, and 3; see extensores digitorum breves mentioned earlier and text]	**Extensor indicis** [as described by authors such as Netter [2006], in humans, this muscle usually inserts onto digit 2; see extensores digitorum breves mentioned earlier and text]

(*Continued*)

TABLE 17.1 (CONTINUED)
Pectoral and Forelimb Muscles of Adults of Representative Tetrapod Taxa

Amphibia (Caudata): *Ambystoma mexicanum* (Axolotl)	Reptilia (Lepidosauria): *Timon lepidus* (Ocellated Lizard)	Mammalia (Monotremata): *Ornithorhynchus anatinus* (Platypus)	Mammalia (Marsupialia): *Didelphis virginiana* (Virginian Opossum)	Mammalia (Rodentia): *Rattus norvegicus* (Norwegian Rat)	Mammalia (Scandentia): *Tupaia* sp. (Tree shrew)	Mammalia (Dermoptera): *Cynocephalus volans* (Philippine colugo)	Mammalia (Primates): *Homo sapiens* (Human)
–	–	**Extensor pollicis longus** (part of extensor pollicis et indicis *sensu* Gambaryan et al. [2015]) [from ulna to distal phalanx of digit 1; in platypus, this muscle usually inserts onto digit 1; see extensores digitorum breves mentioned earlier and text]	– [probably became integrated in extensor indicis complex]	**Extensor pollicis longus** [as described by Greene [1935], in *Rattus*, this muscle usually extends from the ulna to the distal phalanx of digit 1, being blended with the abductor pollicis longus; see extensores digitorum breves a mentioned earlier and text; however, according to our dissections this muscle is not present as a distinct muscle in mice, being perhaps completely fused with the abductor pollicis longus: this is also seen in wildtype mouse of Figure 4 of Duboc and Logan [2011]]	**Extensor pollicis longus** (part of extensor digitorum radialis *sensu* George [1977] and of extensor profundus digitorum *sensu* Le Gros Clark [1924, 1926], which also include the extensor indicis) [as described by George [1977] and Kaneff [1979, 1980a,b], in *Tupaia*, this muscle usually inserts onto digit 1; see extensores digitorum mentioned earlier and text]	**Extensor pollicis longus** (extensores pollicis longus et brevis *sensu* Leche [1886]) [as described by Leche [1886], in colugos, this muscle usually inserts onto digit 1; see extensores digitorum breves mentioned earlier and text]	**Extensor pollicis longus** [as described by authors such as Netter [2006], in humans, this muscle usually inserts onto digit 1; see extensores digitorum breves mentioned earlier and text]
– [Straus [1941a,b] states that urodeles such as *Salamandra* and *Cryptobranchus* do have dorsometacarpales and shows a *Necturus* specimen with "dorsometacarpales" in his Figure 1 and explains that they correspond to the "extensores digitorum breves" *sensu* other authors; Haines [1939] states that anurans such as *Rana* and urodeles such as *Salamandra* clearly have extensores digitorum breves and dorsometacarpales, but see just the following; in our more recent dissections of *A. mexicanum*,	**Dorsometacarpales** [see Chapter 18]	– [not present as a group in extant mammals, but part or the totality of some of them may be fused to the dorsal interossei, intermetacarpales and/or flexores breves profundi of mammals: see, e.g., Cihak 1972 and Lewis 1989]	–	–	–	–	–

(Continued)

TABLE 17.1 (CONTINUED)
Pectoral and Forelimb Muscles of Adults of Representative Tetrapod Taxa

Amphibia (Caudata): *Ambystoma mexicanum* (Axolotl)	Reptilia (Lepidosauria): *Timon lepidus* (Ocellated Lizard)	Mammalia (Monotremata): *Ornithorhynchus anatinus* (Platypus)	Mammalia (Marsupialia): *Didelphis virginiana* (Virginian Opossum)	Mammalia (Rodentia): *Rattus norvegicus* (Norwegian Rat)	Mammalia (Scandentia): *Tupaia* sp. (Tree shrew)	Mammalia (Dermoptera): *Cynocephalus volans* (Philippine colugo)	Mammalia (Primates): *Homo sapiens* (Human)
there are deep and superficial layers of the extensores digitorum breves; however, as described by authors such as Francis [1934] and Grimm and Carlson [1974a,b], these bundles originate from different bones but are then fused distally, and then each whole muscle formed by the two bundles sends a single tendon to the respective digit, so there are no distinct, separate muscles dorsometacarpales as is the case in, e.g., lizards, as also stated by Howell [1936c]]							
Abductor et extensor digiti I [see Chapter 18]	**Abductor pollicis longus** [see Chapter 18; as explained by Russell and Bauer [2008], in *Iguana*, there is a short extensor to digit 1, as we have seen in *Timon*, so the abductor pollicis longus of lizards does not seem to include the short extensor to digit 1, as it does not in mammals; probably such a fusion is a derived feature of some amphibians only]	**Abductor pollicis longus** [from radius and ulna to ulnare; Jouffroy and Lessertisseur [1971] seem to suggest that the plesiomorphic condition for mammals is that in which the abductor pollicis longus has a single tendon to digit 1, as is usually the case in monotremes, although the muscle of platypus sometimes has two tendons to this digit]	**Abductor pollicis longus** (extensor ossis metacarpi pollicis *sensu* Coues [1872]) [from ulna and radius to metacarpal I]	**Abductor pollicis longus** (extensor pollicis brevis *sensu* Greene [1935]) [from radius to metacarpal I in both mice and rats (not proximal phalanx of digit 1 as stated by Greene [1935]); in rats, there is apparently no sign of a differentiation of the abductor pollicis longus into two bundles corresponding to the abductor pollicis longus and extensor pollicis brevis of humans; in rats, the abductor pollicis longus is often deeply mixed with the extensor pollicis longus, although for Greene [1935],	**Abductor pollicis longus** (abductor pollicis *sensu* Le Gros Clark [1924, 1926]) [the *Tupaia* specimens we dissected lack an extensor pollicis brevis; George [1977] stated that an extensor pollicis brevis was reported in *Tupaia picta*, but this muscle may not be homologous to that of humans, because the latter is said to be plesiomorphically absent in primates: see, e.g., Lewis 1989; Gibbs 1999; Gibbs et al. 2000, 2002]	**Abductor pollicis longus** (extensor pollicis brevis plus abductor pollicis longus of Leche [1886]; extensor pollicis brevis of Chapman [1902])	**Abductor pollicis longus**

(Continued)

TABLE 17.1 (CONTINUED)
Pectoral and Forelimb Muscles of Adults of Representative Tetrapod Taxa

Amphibia (Caudata): *Ambystoma mexicanum* (Axolotl)	Reptilia (Lepidosauria): *Timon lepidus* (Ocellated Lizard)	Mammalia (Monotremata): *Ornithorhynchus anatinus* (Platypus)	Mammalia (Marsupialia): *Didelphis virginiana* (Virginian Opossum)	Mammalia (Rodentia): *Rattus norvegicus* (Norwegian Rat)	Mammalia (Scandentia): *Tupaia* sp. (Tree shrew)	Mammalia (Dermoptera): *Cynocephalus volans* (Philippine colugo)	Mammalia (Primates): *Homo sapiens* (Human)
				it is clear that the tendon of the extensor pollicis longus goes to the distal phalanx of the thumb and that the tendon of the abductor pollicis longus goes only to the proximal phalanx of this digit, although our dissections indicate that in both rats and mice, the abductor pollicis longus usually goes to metacarpal I instead]			
–	–	–	–	–	– [see above]	–	**Extensor pollicis brevis** [derived from a part of the abductor pollicis longus: see, e.g., Jouffroy 1971; Kaneff 1979, 1980a,b; Lewis 1989; Diogo and Wood 2011, 2012a]

Note: The nomenclature of the muscles shown in bold follows that of the text; in order to facilitate comparisons, in some cases, names often used by other authors to designate a certain muscle/bundle are given in front of that muscle/bundle, between round brackets; additional comments are given between square brackets. Data compiled from evidence provided by our own dissections and comparisons and by an overview of the literature (see text, Tables 17.2 through 17.3 and Figures 17.1 through 17.7).

TABLE 17.2

Diagram Illustrating the Authors' Hypotheses Regarding the Evolution and Homologies of the Pectoral and Arm Muscles of Adults of Representative Tetrapod Taxa

	Ambystoma (12 m.)	*Timon* (16 m.)	*Ornithorhynchus* (23 m.)	*LCA mar. + plac.* (23 m.)	*Didelphis* (23 m.)	*Rattus* (23 m.)	*Tupaia* (23 m.)	*Cynocephalus* (21 m.)	*Homo* (18 m.)
Axial (primaxial): pectoral girdle	Serratus anterior	Serratus anterior	Serratus anterior	Serratus anterior	Serratus anterior	Serratus anterior	Serratus anterior	Serratus anterior	Serratus anterior
	---	---	Rhomboideus	Rhomboideus	Rhomboideus	Rhomboideus major	Rhomboideus major	Rhomboideus	Rhomboideus major
	---	---	---	---	---	Rhomboideus minor	Rhomboideus minor	---	Rhomboideus minor
	---	---	---	---	---	Rhomboideus occipitalis	Rhomboideus occipitalis	---	---
	Levator scapulae	Levator scapulae	Levator scapulae	Levator scapulae	Levator scapulae	Levator scapulae	Levator scapulae	Levator scapulae	Levator scapulae
	---	Levator claviculae	Levator claviculae	Levator claviculae	Levator claviculae	Levator claviculae	Levator claviculae	Levator claviculae	---
	---	---	---	Atlantoscapularis posticus	Atlantoscapularis posticus	---	Atlantoscapularis posticus	---	---
	---	Sternocoracoideus	Sternocoracoideus	Subclavius	Subclavius	Subclavius	Subclavius	Subclavius	Subclavius
	---	Costocoracoideus	Costocoracoideus	---	---	---	---	---	---
Appendicular (abaxial): pectoral girdle and arm — From ventral mass	Pectoralis	Pectoralis	Pectoralis major	Pectoralis major	Pectoralis major	Pectoralis major	Pectoralis major	Pectoralis major	Pectoralis major
	---	---	Pectoralis minor	Pectoralis minor	Pectoralis minor	Pectoralis minor	Pectoralis minor	Pectoralis minor	Pectoralis minor
	---	---	Panniculus carnosus (part)	Panniculus carnosus (part)	Panniculus carnosus (part)	Panniculus carnosus (part)	Panniculus carnosus (part)	Panniculus carnosus (part)	---
	Supracoracoideus	Supracoracoideus	Supracoracoideus	---	---	Infraspinatus	---	---	---
			Infraspinatus	Infraspinatus	Infraspinatus	Infraspinatus	Infraspinatus	Infraspinatus	Infraspinatus
	---	---	Supraspinatus	Supraspinatus	Supraspinatus	Supraspinatus	Supraspinatus	Supraspinatus	Supraspinatus
	---	---	---	Cleidoacromialis	Cleidoacromialis	--- (but apparently psop.)	---	---	---
	Humeroantebrachialis	Brachialis	Brachialis	Brachialis	Brachialis	Brachialis	Brachialis	Brachialis	Brachialis
	Coracoradialis	---	---	---	---	---	---	---	---
	---	Biceps brachii	Biceps brachii	Biceps brachii	Biceps brachii	Biceps brachii	Biceps brachii	Biceps brachii	Biceps brachii
	Coracobrachialis	Coracobrachialis	Coracobrachialis	Coracobrachialis	Coracobrachialis	Coracobrachialis	Coracobrachialis	Coracobrachialis	Coracobrachialis
Appendicular (abaxial): pectoral girdle and arm — From dorsal mass	Deltoideus scapularis	Deltoideus scapularis	Deltoideus scapularis	Deltoideus scapularis	Deltoideus scapularis	Deltoideus scapularis	Deltoideus scapularis	Deltoideus scapularis	Deltoideus
	Procoracohumeralis	Deltoideus acro. et clav.	Deltoideus acro. et clav.	Deltoideus acro. et clav.	Deltoideus acro. et clav.	Deltoideus acro. et clav.	Deltoideus acro. et clav.	Deltoideus acro. et clav.	---
	---	Scapulo-humeralis anterior	Teres minor	Teres minor	Teres minor	Teres minor	---	Teres minor	Teres minor
	Subcoracoscapularis	Subcoracoscapularis	Subscapularis	Subscapularis	Subscapularis	Subscapularis	Subscapularis	Subscapularis	Subscapularis
	---	---	Teres major	Teres major	Teres major	Teres major	Teres major	Teres major	Teres major
	Latissimus dorsi	Latissimus dorsi	Latissimus dorsi	Latissimus dorsi	Latissimus dorsi	Latissimus dorsi	Latissimus dorsi	Latissimus dorsi	Latissimus dorsi
	Triceps brachii	Triceps brachii	Triceps brachii	Triceps brachii	Triceps brachii	Triceps brachii	Triceps brachii	Triceps brachii	Triceps brachii
	---	---	Dorsoepitrochlearis	Dorsoepitrochlearis	Dorsoepitrochlearis	Dorsoepitrochlearis	Dorsoepitrochlearis	Dorsoepitrochlearis	---

Note: See caption of tables of preceding chapters and Table 17.1). acro., acromialis; clav., clavicularis; m., muscles; mar. + plac., marsupials + placentals; psop., present in some other placentals.

TABLE 17.3

Hypotheses Regarding the Evolution and Homologies of the Forearm/Hand Muscles of Adults of Tetrapod Taxa

	Ambystoma (37 m.)	*Timon* (51 m.)	*Ornithorhynchus* (31 m.)	*LCA mar. + plac.* (48 m.)	*Didelphis* (45 m.)	*Rattus* (44 m.)	*Tupaia* (40 m.)	*Cynocephalus* (42 m.)	*Homo* (40 m.)
Abaxial: ventral forearm	Palmaris profundus I	Palmaris profundus I	---	---	---	---	---	---	---
	Pronator quadratus	Pronator quadratus	---	Pronator quadratus	Pronator quadratus	Pronator quadratus	Pronator quadratus	---	Pronator quadratus
	---	Pronator acessorius	---	---	---	---	---	---	---
	Contrahentium caput longum	---	---	---	---	---	---	---	---
	Fle. accessorius lateralis	---	---	---	---	---	---	---	---
	Fle. accessorius medialis	---	---	---	---	---	---	---	---
	Fle. digitorum communis	Fl. digitorum longus	Fle. digitorum longus+ t. fbs	Fle. digitorum profundus	Fl. digitorum profundus	Fle. digitorum profundus	Fle. digitorum profundus	Fle. digitorum profundus	Fle. digitorum profundus
	---	e---	---	---	---	---	---	---	Fle. pollicis longus
	---	---	---	Fle. digitorum superficialis	Fl. digitorum superficialis	Fle. digitorum superficialis	Fle. digitorum superficialis	Fle. digitorum superficialis	Fle. digitorum superficialis
	---	---	---	Palmaris longus	Palmaris longus	Palmaris longus	Palmaris longus	Palmaris longus	Palmaris longus
	---	---	---	--- (really absent?)	Palmaris longus internus	---	---	---	---
	Fle. antebrachii et carpi ulnaris	Fle. carpi ulnaris	Fle. carpi ulnaris	Fle. carpi ulnaris	Fl. carpi ulnaris	Fle. carpi ulnaris	Fle. carpi ulnaris	Fle. carpi ulnaris	Fle. carpi ulnaris
	---	Epitrochleoanconeus	Epitrochleoanconeus	Epitrochleoanconeus	Epitrochleoanconeus	Epitrochleoanconeus	Epitrochleoanconeus	Epitrochleoanconeus	---
	Fle. antebrachii et carpi radialis	Fle. carpi radialis	Fle. carpi radialis	Fle. carpi radialis	Fle. carpi radialis	Fle. carpi radialis	Fle. carpi radialis	Fle. carpi radialis	Fle. carpi radialis
	---	Pronator teres	Pronator teres	Pronator teres	Pronator teres	Pronator teres	Pronator teres	Pronator teres	Pronator teres
Abaxial: hand	Fles. breves superficiales	Fles. breves superficiales	--- (absent as a group)	---	---	---	---	---	---
	---	---	---	Palmaris brevis?	--- (present other marsupials)	Palmaris brevis	Palmaris brevis	Palmaris brevis	Palmaris brevis
	---	---	---	Fle. brevis digitorum manus	Fle. brevis digitorum manus	---	Fle. brevis digitorum manus	Fle. brevis digitorum manus	---
	---	Lumbricales	Lumbricales	Lumbricales	Lumbricales	Lumbricales	Lumbricales	Lumbricales	Lumbricales
	Contrahentes digitorum	Contrahentes digitorum	---	Contrahentes digitorum	Contrahentes digitorum	Contrahentes digitorum	Contrahentes digitorum	Contrahentes digitorum	---
	---	---	---	Adductor pollicis	Adductor pollicis	Adductor pollicis	Adductor pollicis	Adductor pollicis	Adductor pollicis
	---	---	---	Adductor pollicis accessorius (H.)	Adductor pollicis accessorius (H.)	---	---	---	Add pollicis accessorius (H.)
	Fles. breves profundi	Fles. breves profundi	Fles. breves profundi	Fles. breves profundi	Fles. breves profundi	Fles. breves profundi	Fl. brevis profundis 2	Fle. brevis profundis 2	Fle. brevis profundis 2
	---	---	---	---	---	---	Interossei (fbp + int)	Interossei (fbp + int)	Interossei pal. 1-3
	---	---	---	---	---	---	---	---	Interossei dor. 1-4
	---	---	---	Fle. pollicis brevis	Fle. pollicis brevis	Fle. pollicis brevis	Fle. pollicis brevis	Fle. pollicis brevis	Fle. pollicis brevis
	---	---	---	---	---	---	---	---	Opponens pollicis
	---	---	---	Fl.e. digiti minimi brevis	Fle. digiti minimi brevis (2 heads)	Fle. digiti minimi brevis	Fle. digiti minimi brevis	Fle. digiti minimi brevis	Fle. digiti minimi brevis
	---	---	---	---	---	Opponens digiti minimi	---	---	Opponens digiti minimi
	Fles. digitorum minimi	---	---	---	---	---	---	---	---
	Interphalangeus digiti III	---	---	---	---	---	---	---	---
	---	Abductor pollicis brevis	Abductor pollicis brevis	Abductor pollicis brevis	Abductor pollicis brevis	Abductor pollicis brevis	Abductor pollicis brevis	Abductor pollicis brevis	Abductor pollicis brevis
	Abductor digiti minimi	Abductor digiti minimi	Abductor digiti minimi	Abductor digiti minimi	Abductor digiti minimi	Abductor digiti minimi	Abductor digiti minimi	Abductor digiti minimi	Abductor digiti minimi
	Intermetacarpales	Intermetacarpales	Intermetacarpales	Intermetacarpales	Intermetacarpales	Intermetacarpales	---	---	--
Abaxial: dorsal forearm	Ex. carpi radialis	Ex. antebrachii et carpi radialis	Ex.r carpi radialis	Ex. carpi radialis longus	Ex. carpi radialis longus	Ex. carpi radialis longus	Ex. carpi radialis longus	Ex carpi radialis longus	Ex. carpi radialis longus
	---	--- (but ex.c.r.b psor)	---	Ex. carpi radialis brevis	Ex. carpi radialis brevis	Ex. carpi radialis brevis	Ex. carpi radialis brevis	Ex. carpi radialis brevis	Ex. carpi radialis brevis
	---	--- (but brachioradialis psor)	Brachioradialis	Brachioradialis	Brachioradialis	---	Brachioradialis	Brachioradialis	Brachioradialis
	Supinator	--- (but supinator psor)	Supinator	Supinator	Supinator	Supinator	Supinator	Supinator	Supinator
	Ex. antebrachii et carpi ulnaris	Ex. antebrachii et carpi ulnaris	Ex. carpi ulnaris	Ex. carpi ulnaris	Ex. carpi ulnaris	Ex. carpi ulnaris	Ex. carpi ulnaris	Ex. carpi ulnaris	Ex. carpi ulnaris
	---	---	Anconeus	Anconeus	Anconeus	Anconeus	Anconeus	Anconeus	Anconeus
	Ex. digitorum	Ex. digitorum	Ex. digitorum	Ex. digitorum	Ex. digitorum	Ex. digitorum	Ex. digitorum	Ex. digitorum	Ex. digitorum
	---	Dorsometacarpales	---	---	---	---	---	---	---
	Exs. digitorum breves	Ex. digitorum breves	---	---	---	---	---	---	---
	---	---	Ex. digit minimi	Ex. digiti minimi	Ex. digiti minimi	Ex. digiti minimi	Ex. digiti minimi	Ex. digiti minimi	Ex. digiti minimi
	---	---	---	Ex. digiti quarti	---	Ex. digiti quarti	---	---	---
	---	---	Ex. digiti III proprius	Ex. digiti III proprius	---	---	---	---	---
	---	---	Ex. indicis	Ex. indicis	Ex. indicis	Ex. indicis	Ex. indicis	Ex. indicis	Ex. indicis
	---(edbI included in muscle below	---	Ex. pollicis longus	Ex. pollicis longus	---	Ex. pollicis longus (ab. mice)	Ex. pollicis longus	Ex. pollicis longus	Ex. pollicis longus
	Abductor et ex. digit I	Abductor pollicis longus	Abductor pollicis longus	Abductor pollicis longus	Abductor pollicis longus	Abductor pollicis longus	Abductor pollicis longus	Abductor pollicis longus	Abd. pollicis longus
	---	---	---	---	---	---	---	---	Ex. pollicis brevis

Note: See caption of tables of preceding chapters and Table 17.1). ab., absent; dmc., dorsometacarpales; or., dorsales; ex., extensor; exs., extensores; fbp., flexores breves profundi; fle., flexor; fles., flexores;.c.r.b., extensor carpi radialis brevis; H., "volaris primus of Henle"; int., intermetacarpales; m., muscles; mar. + plac., marsupials + placentals; t.fbs., tendons of flexores breves superficiales.

of the pectoral fin of sarcopterygian fish, and should therefore, at least in theory, be considered an appendicular muscle and thus an abaxial muscle *sensu* Shearman and Burke [2009]: see Tables 16.5 and 17.2 and the following) as well as to the trapezius (which, in the present work, is considered to be a branchial muscle of the head and neck musculature: see preceding chapters and the following) of mice are primaxial, and that only a small part of these cells are abaxial (e.g., Shearman and Burke 2009). According to Shearman and Burke (2009: 609–610), this might indicate that "primitively these muscles were entirely primaxial and associated with the axial musculoskeletal system; during the expansion of the lateral plate and the evolution of the appendicular system, the lateral plate may have recruited putative latissimus and spinotrapezius cells to the limb, thereby altering the insertion points and functions of these muscles."

However, examining Durland et al.'s (2008) paper by the same team, and comparing with our own work and the work of others included in our literature review, indicate that the latissimus dorsi is mainly an abaxial muscle, thus confirming the opinion of most authors that it is an appendicular, and not an axial, limb muscle (reviewed by Diogo and Abdala [2010]). The latissimus dorsi, like other muscles often considered to be hypaxial migratory muscles (e.g., intercostals, panniculus carnosus) always originate from/near the midline, which is unlabeled with *Prx1/Cre/Z/AP* (an abaxial marker). The insertion of these muscles onto the lateral portion of the ribs (intercostals) or onto the humerus (panniculus carnosus) is labeled with this abaxial marker, as expected. In contrast, true axial limb muscles, such as the serratus anterior, rhomboids, and levator scapulae, are almost, or completely, unlabeled. The same pattern occurs in the hindlimb: the axial muscle quadratus lumborum is almost completely unlabeled, while appendicular muscles such as the gluteus maximums are unlabeled at their origin from/near the midline but labeled at their insertion onto the femur. There is a remarkable match between the unlabeling/labeling with *Prx1/Cre/Z/AP* and the grouping of axial *versus* appendicular muscles of the fore- and hindlimbs, in which an almost complete, or complete, unlabeling is seen only in axial limb muscles. In the light of this information, the results of the study by Durland et al. (2008) indicate that the latissimus dorsi is indeed an appendicular forelimb muscle, in agreement with other developmental and comparative anatomical data and innervation patterns. Because the trapezius is also unlabeled at its origin from the midline, but labeled at its insertion onto the scapula/clavicle, the studies of Burke and colleagues (e.g., Shearman and Burke 2009) indicate that this muscle is, like the latissimus dorsi, mainly an abaxial muscle, i.e., a migratory "hypaxial" muscle derived from somites, using the older terms. However, Winder et al. (2011) studied the development of the hypaxial muscles in teleosts using anatomical studies and expression molecular markers, and in none of their images or text do they refer to the protractor pectoralis (a teleost derivative of the cucullaris). This includes their Figure 1, which shows the hypaxial migratory (i.e., abaxial) muscles, thus supporting the idea that the protractor pectoralis is not derived from the hypaxial musculature. This would support our own developmental studies of teleosts: as explained in the preceding chapters and shown in Figure 7.2 of the present book, in the teleost zebrafish, the protractor pectoralis is anatomically derived from the branchial region of the head (rather than from the hypaxial musculature) and then migrates posteriorly to attach onto the pectoral girdle. Did the developmental mechanisms associated with the formation of the cucullaris derivatives in mammals change, resulting in derived genetic similarity between these muscles and the primaxial and particularly abaxial muscles of mammals, as suggested by Durland et al.'s (2008) labeling of the mouse trapezius with *Prx1/Cre/Z/AP*? Either way, the cucullaris and its derivatives are truly branchiomeric muscles that are derived from the cardiopharyngeal field, as shown by the recent clonal studies in mice by Lescroart et al. (2015) and by numerous other recent studies (see Chapter 3 as well as other preceding chapters).

Among the muscles of nonmammalian taxa listed in Tables 17.1 through 17.3, four thus seem to derive from this axial musculature: serratus anterior, levator scapulae, sternocoracoideus, and costocoracoideus. These four muscles connect the axial skeleton to the pectoral girdle and are therefore associated with the movements of the pectoral girdle. Most textbooks state that the rhomboideus (Figures 17.2 and 17.3), a muscle derived from the postcranial axial musculature that also connects the axial skeleton to the pectoral girdle, is only consistently found in mammals (e.g., Kardong 2002). In reptiles such as *Timon* and amphibians such as *Ambystoma*, the rhomboideus does not seem to be present as an independent muscle (Tables 17.1 through 17.3). However, Dilkes (1999) stated that a "rhomboideus" is found in numerous reptiles and that the unresolved question is whether the plesiomorphic reptilian condition for amniotes is to have one rhomboideus muscle or both a "rhomboideus superficialis" and a "rhomboideus profundus." According to some authors, a "rhomboideus" is also found in some anurans (e.g., Howell 1937a,b). We found a rhomboideus in certain nonmammalian tetrapods we dissected, such as anurans, crocodylians, and birds, as shown in Tables 18.1 and 18.2; this subject will be discussed in more detail in Chapter 18.

Other examples of mammalian pectoral muscles that are derived from the postcranial axial musculature and connect the pectoral girdle to the axial skeleton are the levator claviculae, atlantoscapularis anticus, atlantoscapularis posticus, and subclavius (see Figure 17.3). The former three muscles very likely correspond to parts of the levator scapulae of nonmammalian tetrapods such as *Timon*; the latter muscle corresponds to the sternocoracoideus and/or costocoracoideus of those tetrapods (Tables 17.1 through 17.3). The hypaxial musculature *sensu* the present volume corresponds to the "nonmigratory hypaxial musculature" *sensu* Yamane (2005), while the appendicular musculature *sensu* the present book is included in the "migratory hypaxial musculature" *sensu* the latter author. That is, according to Yamame the appendicular muscles are "hypaxial migratory muscles" that migrated to the limbs. Also according to him, the hypobranchial muscles *sensu* this volume, as well as the tongue muscles, the diaphragm,

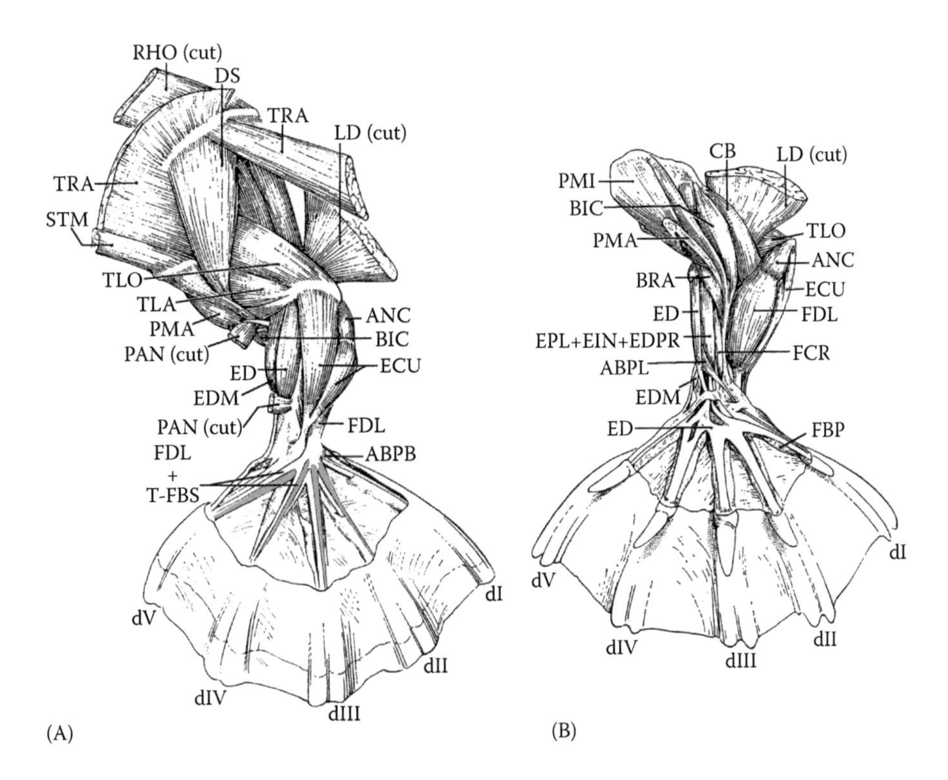

FIGURE 17.2 *Ornithorhynchus anatinus* (Mammalia, Monotremata). (A) Showing dorsal musculature of the pectoral girdle, arm, forearm, and ventral (palmar) musculature of hand; (B) showing ventral musculature of the pectoral girdle, arm, forearm, and dorsal musculature of hand (anterior is toward the top and the left of the figure) (the nomenclature of the structures illustrated follows that of the present work). (Modified from Cuvier, G., and Laurillard, L., *Anatomie comparée* 3, Paris, 1849; Jouffroy, F. K., and Lessertisseur, J., *Traité de Zoologie, XVI: 3 (Mammifères)*, 679–836, Masson et Cie, Paris, 1971.) ABPB, abductor pollicis brevis; ABPL, abductor pollicis longus; ANC, anconeus; BIC, biceps brachii; BRA, brachialis; CB, coracobrachialis; dI, dII, dIII, dIV, dV, digits I, II, III, IV and V; DS, deltoideus scapularis; ECU, extensor carpi ulnaris; ED, extensor digitorum; EDM, extensor digiti minimi; EDPR, extensor digiti III proprius; EIN, extensor indicis; EPL, extensor pollicis longus; FBP, flexores breves profundi; FCR, flexor carpi radialis; FDL, flexor digitorum longus; LD, latissimus dorsi; PAN, panniculus carnosus (cutaneous muscle); PMA, pectoralis major; PMI, pectoralis minor; RHO, rhomboideus; STM, sterno-mastoideus (branchial muscle); T-FBS, tendons of flexores breves superficiales; TLA, TLO, lateralis and longus sections of triceps brachii; TRA, trapezius (branchial muscle).

and possibly the protractor pectoralis, are also hypaxial migratory muscles (but as noted in the preceding paragraphs and chapters, the protractor pectoralis and at least a substantial part of the tongue musculature are clearly derived from the branchiomeric musculature and thus from the cardiopharyngeal field rather than from somites/hypaxial muscles).

APPENDICULAR MUSCLES OF THE PECTORAL GIRDLE AND ARM

As explained earlier, all the tetrapod muscles discussed in this and the next section are derived from the abductor and adductor musculature of basal fishes. With a few exceptions, all these muscles directly insert on the forelimb (arm, forearm, and/or hand) (Tables 17.1 and 17.2). One of the exceptions is the pectoralis minor, which, in many mammals, inserts on both the humerus and the pectoral girdle, but in others, including modern humans, inserts exclusively onto the pectoral girdle (usually onto the coracoid process). The homologies of the pectoralis minor and pectoralis major of mammals (Figure 17.3) have been the subject of much controversy in the past. Some authors suggested that the mammalian pectoralis major corresponds to the pectoralis of other tetrapods (Figure 17.4), the pectoralis minor being derived from axial musculature (e.g., from the rectus abdominis) and being plesiomorphically attached to the pectoral girdle, not to the humerus (e.g., Lander 1918). Other authors suggested that it is the pectoralis minor that corresponds to the pectoralis of other tetrapods, the pectoralis major being derived from other appendicular muscles (e.g., from the "supracoracoideus" and/or "axillary" groups *sensu* Jouffroy [1971]; e.g., Howell 1937a,b). However, most authors now accept that both the pectoralis major and pectoralis minor derive from the pectoralis of nonmammalian tetrapods (e.g., Kardong 2002; Warburton 2003; see also Chapter 18). In fact, the data now available on the innervation and development of the pectoralis major and pectoralis minor clearly support the latter view (e.g., in the vast majority of mammals, both these muscles are innervated by pectoral nerves, and they derive from the same anlage during the development of, e.g., the marsupial *Didelphis*: Romer 1944; Cheng 1955; Jouffroy 1971; Warburton 2003). This view is also supported by our own dissections (Tables 17.1 through 17.3). For instance, in monotremes such as *Ornithorhynchus*,

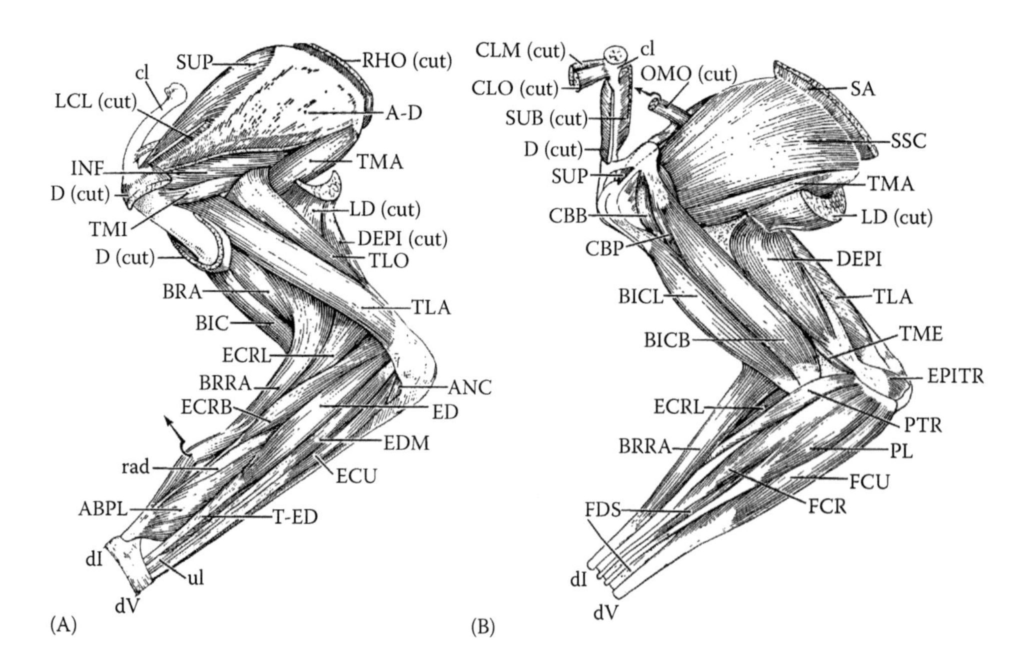

FIGURE 17.3 *Macaca mulata* (Mammalia, Primates). (A) Dorsal view of the musculature of the pectoral girdle, arm, and forearm; (B) ventral view of the musculature of the pectoral girdle, arm and forearm (anterior is toward the top and the left of the figure) (the nomenclature of the structures illustrated follows that of the present work). (Modified from Howell, A. B., and Straus, W. L., *The Anatomy of the Rhesus Monkey*, 89–175. Williams and Wilkins Co., Baltimore, 1933; Jouffroy, F. K., *Traité de Zoologie, XVI: 3 (Mammifères)*, 1–475, Masson et Cie, Paris, 1971.) ABPL, abductor pollicis longus; A-D, aponeurosis of deltoideus; ANC, anconeus; BIC, biceps brachii; BICL, BICB, longus and brevis (short) sections of biceps brachii; BRA, brachialis; BRRA, brachioradialis; CBB, CBP, brevis and proprius sections of the coracobrachialis; cl, clavicle; CLM, cleido-mastoideus (branchial muscle); CLO, cleido-occipitalis (branchial muscle); D, deltoideus; dI, dV, digits I and V; DEPI, dorsoepitrochlearis; ECRB, extensor carpi radialis brevis; ECRL, extensor carpi radialis longus; ECU, extensor carpi ulnaris; ED, extensor digitorum; EDM, extensor digiti minimi; EPITR, epitrochleoanconeus; FCR, flexor carpi radialis; FCU, flexor carpi ulnaris; FDS, flexor digitorum superficialis; INF, infraspinatus; LCL, levator claviculae; LD, latissimus dorsi; OMO, omohyoideus (hypobranchial muscle); PL, palmaris longus; PTR, pronator teres; rad, radius; RHO, rhomboideus; SA, serratus anterior; SSC, subscapularis; SUB, subclavius; SUP, supraspinatus; T-ED, tendon of extensor digitorum; TLA, TLO, TME, lateralis, longus and medialis sections of triceps brachii; TMA, teres major; TMI, teres minor; ul, ulna.

the pectoralis is similar to that of nonmammalian tetrapods such as *Ambystoma* (Figure 17.4). However, it is differentiated into an anterior, superficial component (pectoralis major) that is undivided and inserts onto the humerus and a posterior, deeper component (pectoralis minor) that is also undivided and that attaches onto the humerus, not to the pectoral girdle. In therian mammals such as the Norwegian rat, the pectoralis major attaches onto the humerus and is divided into three sections, which appear to correspond to the clavicular, sternocostal, and abdominal components of the pectoralis major of modern humans (Table 17.1). The pectoralis minor is also divided into three components corresponding to the "cephalic" (attaching on the pectoral girdle and humerus, and equivalent to the "pectoralis minor" of some authors), "caudal" (attaching on the humerus, and equivalent to the "pectoralis abdominis" of some authors), and "tertius" (attaching on the pectoral girdle and equivalent to the "xiphiohumeralis" of some authors) components of the "entopectoralis" *sensu* Lander (1918; Table 17.1). In the tree shrew *Tupaia*, the pectoralis major attaches to the humerus and is divided into two sections that seem to correspond to the sternocostal and abdominal sections of the modern human pectoralis minor (the clavicular component being apparently fused with the deltoideus clavicularis: Table 17.1). The pectoralis minor in *Tupaia* is divided into a "cephalic" and a "caudal" component *sensu* Lander (1918). The former attaches to the humerus and shoulder capsule; the latter attaches exclusively onto the humerus. Contrary to the descriptions of Lander (1918), in the *Tupaia* specimens we dissected, neither of these two sections directly attaches onto the coracoid process (see also Le Gros Clark 1924; George 1977). The major subdivisions and distal attachments of the pectoralis major and pectoralis minor of colugos and modern humans are essentially similar: the pectoralis major attaches to the humerus and is subdivided into clavicular, sternocostal, and abdominal components; the undivided pectoralis minor attaches on the coracoid process of the scapula (Table 17.1). The data obtained from our dissections, comparisons, and review of the literature therefore indicates that the plesiomorphic condition for extant mammals is to have a pectoralis minor inserting on the humerus rather than on the pectoral girdle as suggested by Lander (1918).

It is now accepted that the mammalian supraspinatus and infraspinatus, which usually connect the dorsal region of the pectoral girdle to the proximal region of the arm (Figure 17.3), derive from the supracoracoideus, a muscle that lies ventral rather than dorsal to the pectoral girdle in most other extant tetrapods (Figures 17.3 and 17.4; Tables 17.1 through 17.3; e.g., Kardong 2002). It is also accepted that the coracobrachialis,

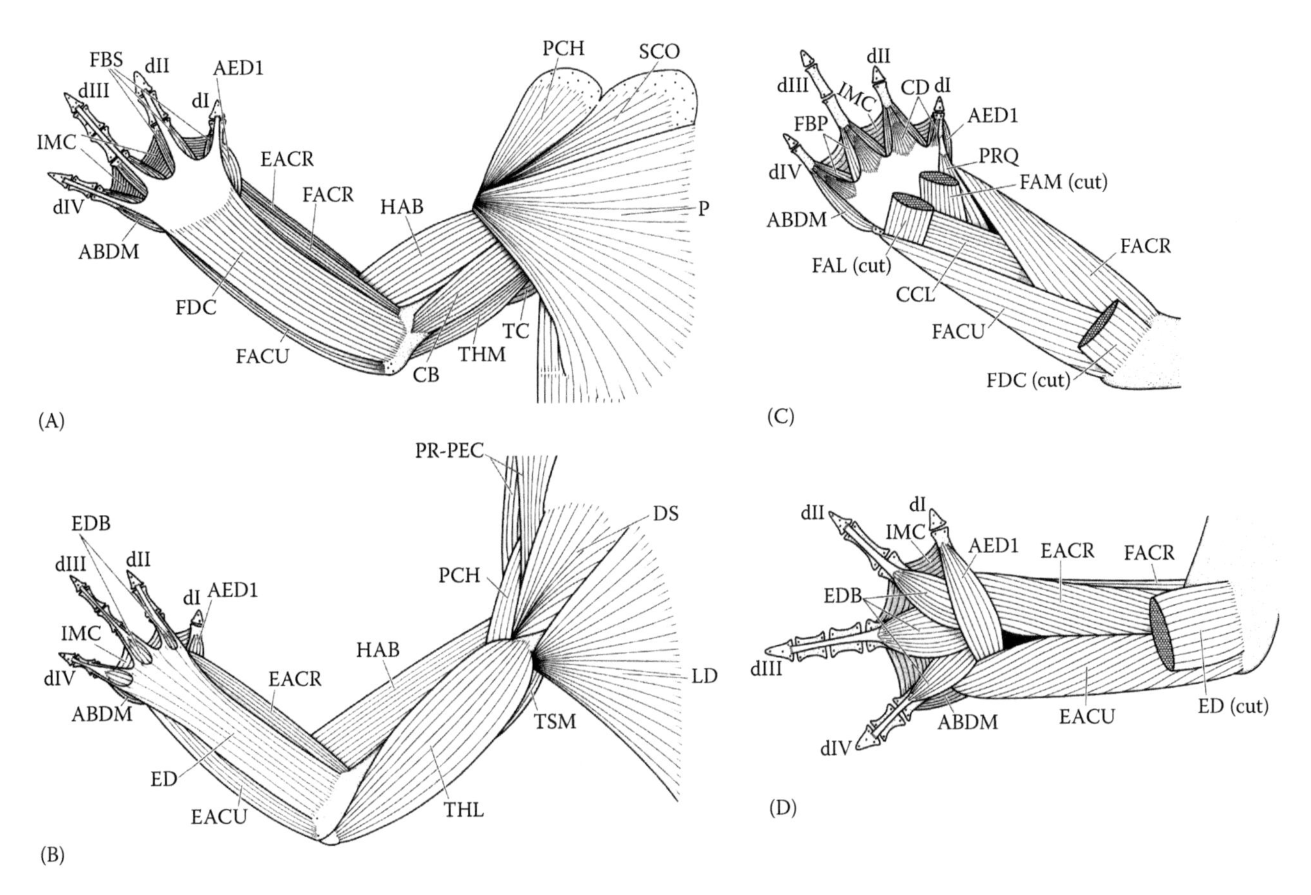

FIGURE 17.4 *Taricha torosa* (Amphibia, Caudata). (A) Ventral view of the superficial musculature of the pectoral girdle and forelimb; (B) dorsal view of the superficial musculature of the pectoral girdle and forelimb; (C) ventral view of the deep musculature of the forearm; (D) dorsal view of the deep musculature of the forearm (anterior is towards the top of the figure) (the nomenclature of the structures illustrated follows that of the present work). (Modified from Walthall, J. C., and Ashley-Ross, M. A., *Taricha torosa. Anat Rec A*, 288, 46–57, 2006.) ABDM, abductor digiti minimi; AC, anconaeus coracoideus; AED1, abductor et extensor digiti I; CB, coracobrachialis; CCL, contrahentium caput longum; CD, contrahentes digitorum; dI, dII, dIII, dIV, digits I, II, III and IV; DS, deltoideus scapularis; EACR, extensor antebrachii et carpi radialis; EACU, extensor antebrachii et carpi ulnaris; ED, extensor digitorum; EDB, extensores digitorum breves; FACR, flexor antebrachii et carpi radialis; FACU, flexor antebrachii et carpi ulnaris; FAL, flexor accessorius lateralis; FAM, flexor accessorius medialis; FBP, flexores breves profundi; FBS, flexores breves superficiales; FDC, flexor digitorum communis; HAB, humeroantebrachialis; IMC, intermetacarpales; LD, latissimus dorsi; P, pectoralis; PCH, procoracohumeralis; PR-PEC, protractor pectoralis; PRQ, pronator quadratus; SCO, supracoracoideus; TC, Thindlimb, THM, TSM, coracoideus, humeralis lateralis, humeralis medialis and scapularis medialis sections of triceps brachii.

brachialis, and biceps brachii of mammals correspond to/derive from the coracobrachialis, humeroantebrachialis, and coracoradialis (which is often found in both anuran and urodele amphibians: see also Chapter 18) of nonmammalian tetrapods such as amphibians and that the deltoideus scapularis, deltoideus clavicularis, deltoideus acromialis, teres minor, and scapulo-humeralis anterior of the former derive from the deltoideus scapularis and procoracohumeralis of the latter (Figures 17.2 through 17.4; Tables 17.1 through 17.3; see also Chapter 18). The deltoideus scapularis, clavicularis, and acromialis are fused into a single muscle in mammals such as modern humans (Tables 17.1 through 17.3). In Diogo and Abdala (2010), we stated that the teres minor appears to correspond to a part of the deltoideus scapularis of nonmammalian tetrapods, being possibly directly homologous to the reptilian scapulo-humeralis posterior. This statement was partially based on Jouffroy's (1971) statement that the supposed homology between the mammalian teres minor and the scapulo-humeralis anterior proposed by, e.g., Romer (1944) and by Cheng's (1955) detailed developmental study of marsupials had two main problems: (a) both the scapulo-humeralis anterior and teres minor are present in monotremes; (b) in reptiles, such as "lizards," the nerve innervating the scapulo-humeralis anterior is related to the radial nerve rather than the axillary nerve, which usually innervates the teres minor in mammals (and the deltoideus scapularis in mammals and in reptiles). However, this second point is not a strong one, because in placentals such as humans, both the radial and axillary nerves are branches of the posterior cord of the brachial plexus, and moreover, the scapulo-humeralis anterior is very likely derived from the procoracohumeralis, which gave rise to the deltoideus acromialis et clavicularis, i.e., to a muscle that is also innervated by the axillary nerve. Regarding the first point, Gambaryan et al. (2015) recently reviewed this subject and stated that the structure that is often designated "teres minor" in monotremes (in addition to the true scapulo-humeralis anterior) corresponds to the infraspinatus, because it is innervated by the supracoracoid nerve. They therefore concluded that monotremes have three muscles derived from the ancestral supracoracoideus—i.e., the infraspinatus, supraspinatus, and the remaining of the original supracoracoideus—the latter being lost in eutherians, which have only a supraspinatus and an infraspinatus. That is, the teres minor, "scapulo-humeralis anterior," and "infraspinatus" *sensu* Diogo and Abdala (2010) thus correspond to the infraspinatus, teres minor, and supracoracoideus *sensu* the present work, respectively; this updated scenario combines in a better way the available data on the development and innervation of the shoulder muscles, as pointed out by Gambaryan et al. (2015).

There is still controversy regarding the origin of the mammalian dorsoepitrochlearis (Figure 17.3). This is one of the examples where different lines of evidence apparently support different hypotheses of homology, thus stressing the importance of taking into account all the data available (see preceding text). Some authors, mainly influenced by Cheng's (1955) study of the ontogeny of the muscles of the marsupial *Didelphis*, argue that the mammalian dorsoepitrochlearis corresponds to part of the latissimus dorsi of other tetrapods, because both these muscles seem to originate from the same developmental anlage (Jouffroy 1971; Jouffroy and Lessertisseur 1971; Warburton 2003). Regarding innervation, the situation found in the phylogenetically plesiomorphic monotremes is ambiguous; the dorsoepitrochlearis is apparently innervated by branches of the nerves innervating both the triceps brachii (radial nerve) and the latissimus dorsi (subscapular nerves) (Jouffroy and Lessertisseur 1971). In the vast majority of other mammals, including marsupials; however, the dorsoepitrochlearis is innervated solely by the radial nerve (Jouffroy 1971). This has led some authors to argue that the mammalian dorsoepitrochlearis corresponds to part of the triceps brachii rather than the latissimus dorsi of other tetrapods (e.g., Howell 1937a,b; Gibbs 1999). Our dissections and comparisons indicate that when all the lines of evidence are taken into account, there is more support for an origin of the dorsoepitrochlearis from the triceps brachii. In most mammals, the dorsoepitrochlearis is deeply mixed with the latissimus dorsi proximally and the triceps brachii distally (Figure 17.3B). However, regarding its overall configuration and the direction of its fibers, the dorsoepitrochlearis is in general more similar to the triceps brachii than to the latissimus dorsi. As shown in, e.g., Figure 17.3B, the dorsoepitrochlearis usually runs parallel to the triceps brachii, being almost, or completely, perpendicular, to the proximal fibers of the latissimus dorsi with which it is associated. Its overall configuration, the direction of its fibers, and its relationship with other soft as well as hard tissues all suggest that the dorsoepitrochlearis is a medial, superficial component of the triceps brachii that instead of proximally attaching onto the pectoral girdle attaches onto the latissimus dorsi. In phylogenetically plesiomorphic therian mammals, the dorsoepitrochlearis often inserts onto the olecranon process of the ulna (as usually does the triceps brachii). Moreover, in mammalian taxa including primates such as new world monkeys and apes, the dorsoepitrochlearis is often partially originated from the scapula (the "triceps coracoideus" of nonmammalian tetrapods usually has a scapular origin: see the following).

The hypothesis that the dorsoepitrochlearis derives from the triceps brachii is also supported by phylogenetic parsimony. As shown in Table 17.1 and explained in Chapter 18, in some extinct mammals, the triceps brachii is thought to have had four distinct components, or heads, not three as in most extant mammals. According to several authors, the long, medial, and lateral heads of the triceps brachii of mammals correspond, respectively, to the "scapularis," "humeralis medialis," and "humeralis lateralis" components of the triceps brachii of other tetrapods (e.g., Howell 1937a,b). In other words, the component that is absent in extant mammals, the "triceps coracoideus," is a medial, superficial head of the triceps brachii that runs from the elbow to the pectoral girdle in most nonmammalian tetrapods (see, e.g., the descriptions of Walthall and Ashley-Ross [2006] and Figure 17.4C). Thus, it is phylogenetically more parsimonious to assume that during the evolutionary transition leading to mammals, the "triceps coracoideus" was simply modified into a dorsoepitrochlearis, than to assume that it was completely lost, and that in the course of the same evolutionary transition, a new muscle (that in many ways is very similar to the triceps coracoideus) was acquired through the differentiation of the latissimus dorsi. The former hypothesis does not force us to assume the loss of a certain structure, nor the emergence of a new one, whereas the latter hypothesis forces us to assume both. That in many mammals the dorsoepitrochlearis is proximally attached to muscles other than the latissimus dorsi (e.g., teres major in tree shrews: Table 17.1) also indicates that the association between these muscles was secondarily acquired. That is, originally the structure that became the mammalian dorsoepitrochlearis was very likely not attached to the latissimus dorsi. In summary, the available evidence supports the latter hypothesis, but the possibility that the dorsoepitrochlearis originated from the latissimus dorsi cannot be completely rejected (Table 17.2). We plan to address this question in more detail in a future study.

Some authors have suggested that the teres major (Figure 17.3) corresponds to part of the latissimus dorsi of nonmammalian tetrapods such as *Ambystoma* (e.g., Romer 1924, 1944; Howell 1935, 1937a,b). This correspondence was probably suggested because in many mammals, the distal portion of the latissimus dorsi blends with the distal portion of the teres major and/or that the latter is attached to the proximal portion of the dorsoepitrochlearis (Table 17.1). Some textbooks continue to follow this view (e.g., Kardong 2002). However, most researchers now agree that the mammalian teres major corresponds to part of the subcoracoscapularis (Tables 17.1 through 17.3). Several lines of evidence, including development, innervation, and topology, strongly support this hypothesis (for instance, the subscapularis and teres major develop from the same anlage in mammals such as *Didelphis*; these two muscles are innervated by similar subscapular nerves in most mammals and are intimately related or even fused in many mammals: e.g., Cheng 1955; Jouffroy 1971; Jouffroy and Lessertisseur 1971; Warburton 2003; this work).

The evolutionary/homology hypotheses in Diogo and Abdala (2010) and the present book are supported by our recent detailed study of the adult *Didelphis* muscles (see Diogo et al. 2016a) as well as by comparative anatomical data and developmental data (e.g., Cheng 1955). In fact, the only difference in the most likely evolutionary origin of the muscles between Table 17.2 and the table on page 469 of the work by Cheng (1955) (mainly based on developmental studies of marsupials and salamanders, and comparisons with the literature) is that in the latter, the dorsoepitrochlearis of mammals derives—developmentally, in *Didelphis*—from the latissimus dorsi (as shown in our pointed arrows, as a likely hypothesis), while in the latter, it most likely derives from the triceps brachii.

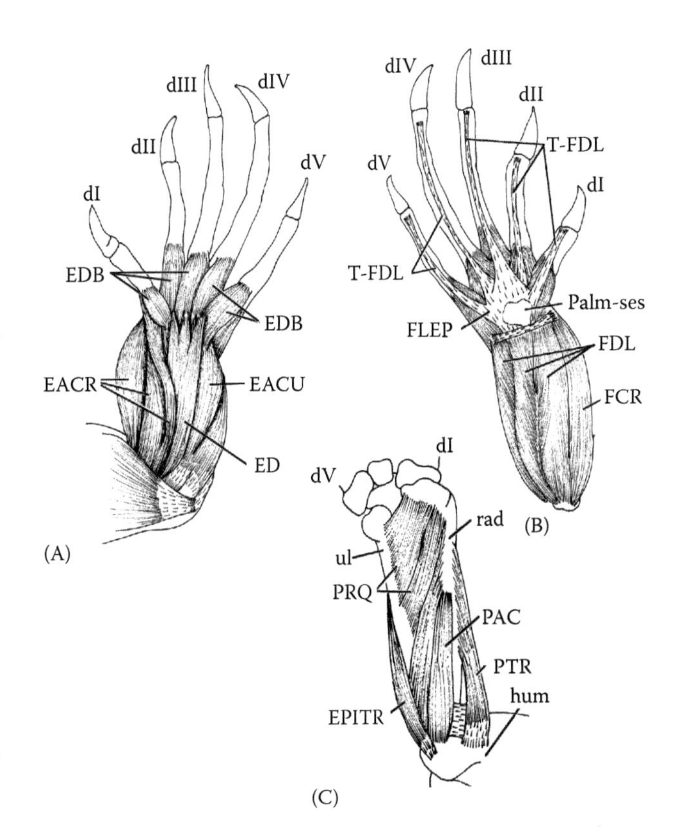

FIGURE 17.5 *Phymaturus* sp. (Reptilia, Lepidosauria). (A) Dorsal view of the superficial musculature of the forearm; (B) ventral view of the superficial musculature of the forearm; (C) ventral view of the deep musculature of the forearm (anterior is toward the top of the figure) (the nomenclature of the structures illustrated follows that of the present work). (Modified from Abdala, V., and Moro, S., *Acta Zool*, 87, 1–12, 2006.) dI, dII, dIII, dIV, dV, digits I, II, III, IV and V; EACR, extensor antebrachii et carpi radialis; EACU, extensor antebrachii et carpi ulnaris; ED, extensor digitorum; EDB, extensores digitorum breves; EPITR, epitrochleoanconeus; FCR, flexor carpi radialis; FDL, flexor digitorum longus; forelimbEP, flexor plate; hum, humerus; PAC, pronator accessorius; palm-ses, palmar sesamoid; PRQ, pronator quadratus; PTR, pronator teres; rad, radius; T-FDL, tendons of flexor digitorum longus; ul, ulna.

APPENDICULAR MUSCLES OF THE FOREARM AND HAND

The muscles of the forearm and hand of tetrapods may be divided into three main groups: the ventral muscles of the forearm (usually flexors of the hand/digits and/or pronators of the forearm), the muscles of the hand (mainly ventral muscles and are often associated with the flexors of the hand/digits), and the dorsal muscles of the forearm (usually extensors of the hand/digits and/or supinators of the forearm) (Tables 17.1 and 17.3).

As explained in Table 17.1 and shown in Table 17.3, the data obtained from our dissections, comparisons and review of the literature allow well-supported hypotheses of homology to be established for most of the ventral forearm muscles. The mammalian pronator quadratus corresponds to the pronator quadratus of nonmammalian taxa such as *Ambystoma* (Figure 17.4C), and the pronator accessorius found in lizards and some other reptiles (Figure 17.5C; see Chapter 18) is the result of the differentiation of part of this muscle (Tables 17.1 through 17.3). The flexor digitorum longus of lizards such as *Timon* (Figure 17.5C) corresponds to the flexor digitorum communis, flexor accessorius lateralis, flexor accessorius medialis, and likely contrahentium caput longum of urodeles such as *Ambystoma* (Figure 17.4). Diogo and Abdala (2010) stated that the contrahentium caput longum of amphibians may also correspond to part of the flexor digitorum longus of reptiles and monotremes, but that it was more likely, in view of the evidence available at that time, that the former muscle is completely absent in amniotes. However, the study of the forelimb musculature of monotremes by Gambaryan et al. (2015) revealed that one of the heads of the flexor digitorum longus of monotremes is very similar to the contrahentium caput longum of amphibians, lying between structures that clearly seem to correspond to the flexor accessorius lateralis and flexor accessorius medialis of amphibians. By integrating this new information with all the other data available, we now consider it more likely that the contrahentium caput longum became integrated into the flexor digitorum longus of amniotes, together with the flexor accessorius lateralis and flexor accessorius medialis (Table 17.3). In monotremes, the flexor digitorum longus blends with the tendons of the flexores breves superficiales, which are ventral muscles of the hand (Figure 17.2; Tables 17.1 through 17.3). In most other extant mammals, the compound structure formed by the former muscle, and the latter tendons are subdivided into a flexor digitorum profundus, a palmaris longus, and a flexor digitorum superficialis. The flexor digitorum superficialis is not exclusively comprised of the tendons of the flexores breves superficiales, for it also incorporates part of the fleshy belly of the flexor digitorum longus of, e.g., monotremes (Figure 17.3; Tables 17.1 through 17.3). In a few mammals, including modern humans, there is a flexor pollicis longus, which is derived from the part of the flexor digitorum profundus that attaches to the first digit, or pollex (Lewis 1989; Diogo and Wood 2011, 2012a,b; Diogo et al. 2012) (Tables 17.1 through 17.3; Figure 17.1). The flexor carpi radialis, pronator teres ("flexor antebrachii radialis"), flexor carpi ulnaris, and epitrochleoanconeus ("flexor antebrachii ulnaris") of mammals correspond to the same muscles in other tetrapods. Urodeles such as *Ambystoma* have an undivided flexor antebrachii et carpi radialis, but a distinct pronator teres is found in other amphibians, such as various anurans (Tables 17.1 through 17.3 and Figures 17.2 and 17.4 through 17.6; see Chapter 18).

Empirical evidence suggests that at least some of the "palmaris longus" muscles of tetrapods are not homologous. For instance, some mammalian taxa have two "palmaris longus" muscles, one derived from the flexor digitorum superficialis and innervated by the median nerve, the other derived from the flexor carpi ulnaris and innervated by the ulnar nerve (e.g., Straus 1942; Jouffroy 1971). According to Straus (1942), most of the palmaris longus muscles of mammals are derived from the flexor digitorum superficialis; some are derived from the flexor carpi ulnaris, and a few, from the flexor carpi radialis

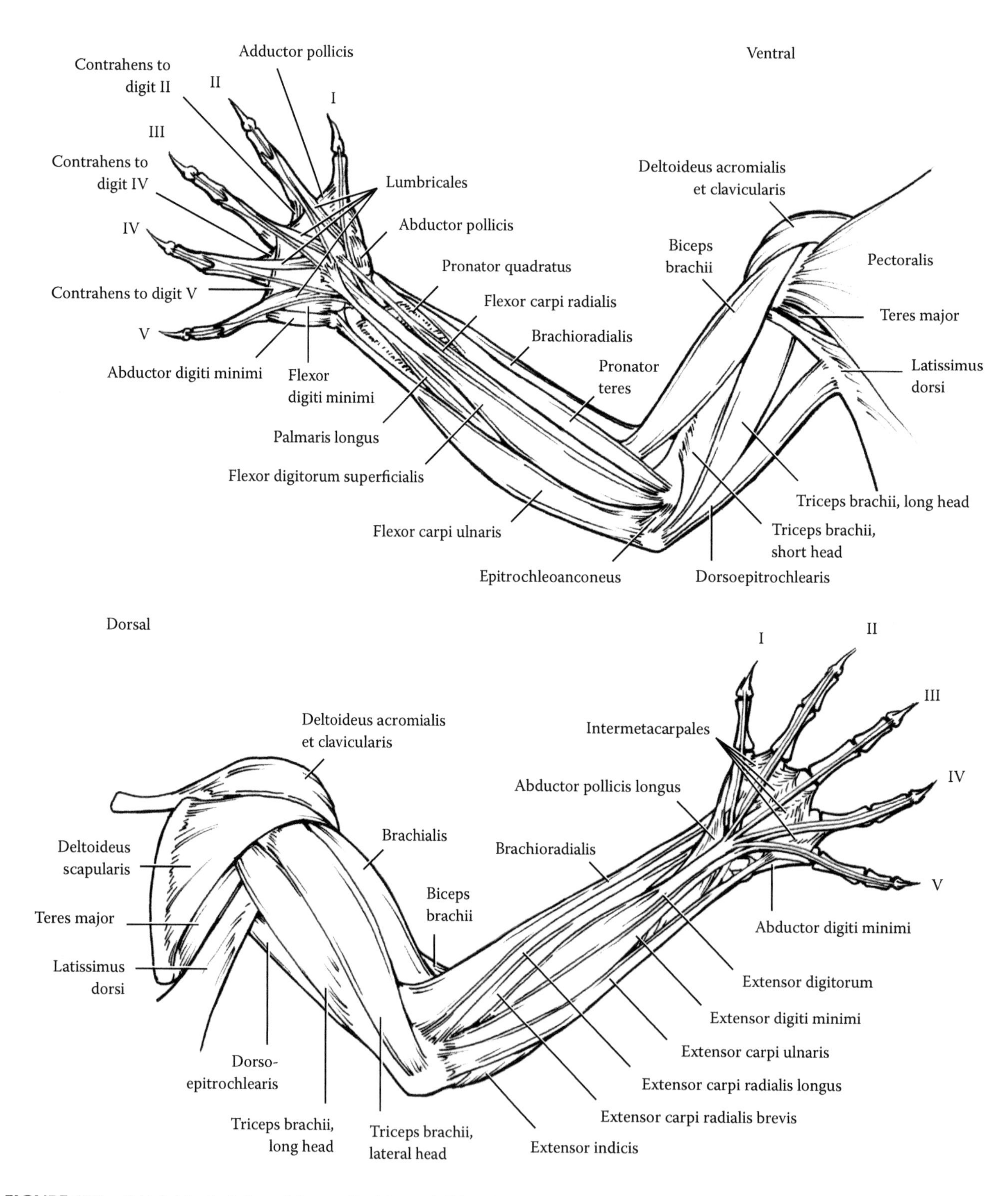

FIGURE 17.6 *Didelphis virginiana* (Mammalia, Marsupialia): schematic drawing (by J. Molnar) of forelimb muscles.

(in his view, this is the case in primates). Jouffroy (1971) states that when the mammalian palmaris longus is derived from either the flexor digitorum superficialis or the flexor carpi radialis, it is always innervated by the median nerve. Only when it is derived from the flexor carpi ulnaris is it at least partially innervated by the ulnar nerve. Our dissections suggest that most of the palmaris longus muscles of the therian mammals listed in Tables 17.1 through 17.3 are very likely homologous and are derived from neither the flexor carpi ulnaris nor the flexor carpi radialis, although the latter hypothesis cannot be completely discarded in the case of modern humans. For instance, the palmaris longus of rats blends with the proximal portion of the flexor carpi ulnaris, but it is innervated by the median nerve so it is very likely not derived from the latter muscle (e.g., Greene 1935; Diogo and Abdala 2010). The palmaris longus of tree shrews and colugos are also innervated by the median nerve, and their configuration suggests that they derive from the flexor digitorum superficialis (e.g., Le Gros Clark [1924, 1926]; George [1977]; Diogo and Abdala [2010]). It is therefore not parsimonious to assume that the common ancestor of colugos, tree shrews, and primates had a palmaris longus muscle derived from the flexor brevis superficialis and that during the course of the transition to primates that muscle was lost, and a new palmaris longus, morphologically similar to the former, but derived from the flexor carpi radialis, was acquired.

The homologies of the hand muscles have been the subject of numerous discussions and remain controversial. Examples of mammalian hand muscles include the palmaris brevis and flexor brevis digitorum manus, which are ventral (palmar, superficial) to the other muscles and are often poorly developed or absent; the abductor pollicis brevis and abductor digiti minimi, which usually lie on the ventrolateral (radial) and ventromesial (ulnar) surface of the hand and abduct the most lateral and most medial digits, respectively; the lumbricales, which are deeper and are usually associated with the tendons of the flexor digitorum profundus, being often related to the extension and/or flexion of different parts of the digits; the contrahentes digitorum, which are deep to the lumbricales and often adduct the digits; the flexores breves profundi, which usually are deep to the contrahentes digitorum and which are bicipital, inserting, respectively, on the radial (lateral) and ulnar (medial) sides of the digits and being mainly associated with the abduction, adduction, flexion, and/or extension of different parts of the digits; and the intermetacarpales, which are the deepest (most dorsal) muscles of the hand and are usually related to the abduction of the digits. The dorsometacarpales are not present as independent muscles in extant therian mammals (Tables 17.1 through 17.3 and Figures 17.1 through 17.3 and 17.7). As shown in Tables 17.1 and 17.3 and explained earlier, in extant mammals, the flexores breves superficiales are not present as a group, but some of the mammalian muscles do include/correspond to part of these muscles. This is the case with the palmaris brevis and flexor brevis digitorum manus. This is also the case with the tendons that are fused with the monotreme flexor digitorum longus and incorporated into the therian flexor digitorum superficialis (Figure 17.2A; see preceding text and, e.g., Howell 1936c; Straus 1942; Jouffroy 1971; Lewis 1989; Diogo and Wood 2011, 2012a,b). As convincingly argued by Lewis (1989), the abductor pollicis brevis and abductor digiti minimi of mammals (Figure 17.7) do not correspond to part of the flexores breves superficiales of other tetrapods. In fact, our dissections confirm the observations in the literature that the flexores breves superficiales, abductor pollicis brevis, and abductor digiti minimi, may coexist in various nonmammalian tetrapods (Tables 17.1 through 17.3). The lumbricales, contrahentes digitorum, flexores breves profundi, and intermetacarpales are also found in tetrapods other than mammals (Tables 17.1 through 17.3; see also Chapter 18).

The first contrahens digitorum (to digit 1) is highly developed in many mammals and is sometimes divided into transverse and oblique heads; many researchers designate this muscle as adductor pollicis (Figure 17.3). In various mammals, including modern humans, the other contrahentes digitorum are aponeurotic (Figures 17.1 and 17.7) or absent as independent structures. Early in their ontogeny, modern humans have four contrahentes digitorum. That of digit 1 gives rise to the well-developed adductor pollicis, with

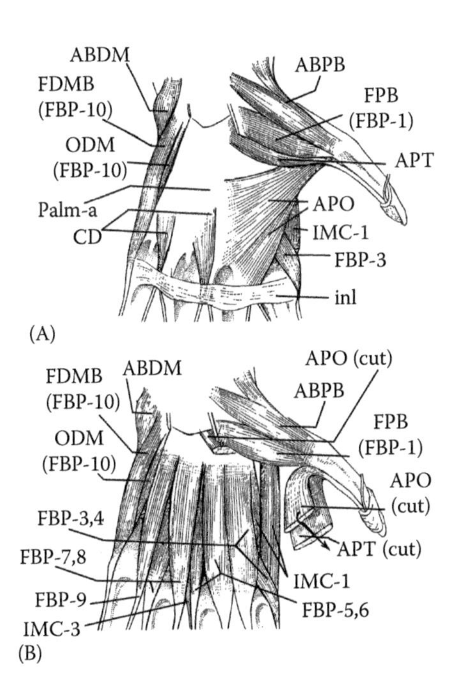

FIGURE 17.7 *Pan troglodytes* (Mammalia, Primates). (A) Ventral (palmar) view, contrahentes layer; (B) Ventral (palmar) view; the contrahentes to digits IV and V were removed and that to digit 1 was cut (the proximal region of the hand is towards the top of the figure) (the nomenclature of the structures illustrated follows that of the present work). (Modified from Forster, A., *Arch Anat Physiol Anat Abt*, 1916, 101–378, 1917; Jouffroy, F. K., *Traité de Zoologie, XVI: 3 (Mammifères)*, 1–475, Masson et Cie, Paris, 1971.) ABDM, abductor digiti minimi; ABPB, abductor pollicis brevis; APO, APT, obliquus and transversus sections of adductor pollicis; CD, vestigial, aponeurotic contrahentes digitorum to digits IV and V; FBP-1,3,4,5,6,7,8,9,10, flexores breves profundi 1, 3, 4, 5, 6, 7, 8, 9, and 10; FDMB, flexor digiti minimi brevis; FPB, flexor pollicis brevis; inl, intercapitular ligaments; IMC-1,3, intermetacarpales 1 and 3; ODM, opponens digiti minimi; palm-a, palmar aponeurosis.

perhaps some contribution from that of digit 2; those of digits 4 and 5, as well as part of that of digit 2, apparently become incorporated into the dorsal interossei (e.g., Cihak 1972). Therefore, the main difference between adult mammals such as common chimpanzees (Figures 17.1 and 17.7) and modern humans may be that in the former, the contrahentes digitorum to digits 4 and 5 do not completely become ontogenetically incorporated into the interossei muscles, persisting as independent, although highly reduced and aponeurotic structures in later developmental stages (Diogo and Wood 2012b). According to Lewis (1989), the plesiomorphic condition for mammals, and probably for primates, is to have 10 flexores breves profundi inserting on the lateral (radial) and medial (ulnar) sides of the five digits. It should be noted that each of the 10 flexores breves profundi *sensu* Lewis corresponds to one of the two heads of each of the five bicipital flexores breves profundi *sensu* authors such as Haines (1950, 1955). That is, the flexores breves profundi 1 and 2 *sensu* Lewis correspond, respectively, to the radial and ulnar head of the flexor brevis profundus 1 (to digit 1) *sensu* Haines, the flexores breves profundi 3 and 4 of Lewis correspond to the radial and ulnar head of the flexor brevis profundus 2 (to digit 2) *sensu* Haines, and so on. According to Lewis (1989) each of the palmar interossei of primates such as chimpanzees directly corresponds to one of the flexores breves profundi of nonmammalian tetrapods. In his view, mammals such as chimpanzees therefore have 9 flexores breves profundi (Figure 9.6): the flexor pollicis brevis + opponens pollicis inserting on digit 1 and metacarpal I (= "flexor brevis profundus 1"), a first palmar interosseous inserting on the lateral side of digit 2 (= "flexor brevis profundus 3"), a second palmar interosseous inserting on the medial side of digit 2 (= "flexor brevis profundus 4"), a third palmar interosseous inserting on the lateral side of digit 3 (= "flexor brevis profundus 5"), a fourth palmar interosseous inserting on the medial side of digit 3 (= "flexor brevis profundus 6"), a fifth palmar interosseous inserting on the lateral side of digit 4 (= "flexor brevis profundus 7"), a sixth palmar interosseous inserting on the medial side of digit 4 (= "flexor brevis profundus 8"), a seventh palmar interosseous inserting on the lateral side of digit 5 (= "flexor brevis profundus 9"), and the flexor digiti minimi brevis + opponens digiti minimi inserting on the medial side of digit 5 and of metacarpal V (= "flexor brevis profundus 10"). A palmar interosseous inserting on the medial side of digit 1 (= "flexor brevis profundus 2") may be found in a few Pan specimens; these specimens would thus exhibit all the 10 flexores breves profundi (Lewis 1989). According to this scenario, the three palmar interossei inserting respectively on the medial side of digit 2, the lateral side of digit 4, and the lateral side of digit 5 in modern humans correspond to the flexores breves profundi 4, 7, and 9 (i.e., to the second, fifth, and seventh of the seven palmar interossei found in most chimpanzees). Flexores breves profundi 3, 5, 6, and 8 of modern humans (which correspond to the first, third, fourth, and sixth of the seven palmar interossei found in most chimpanzees), fuse with the intermetacarpales (and ultimately with some contrahentes digitorum and/or some dorsometacarpales) to form the four dorsal interossei (Lewis 1989) (Tables 17.1 through 17.3). However, Cihak (1972), based on his studies of human embryos, suggested that while the flexores breves profundi 5, 6, and 8 do fuse with intermetacarpales 2, 3, and 4 to form the interossei dorsales 2, 3, and 4 of adult humans, the first dorsal interosseus is actually the fusion of flexores breves profundi 2 and 3, with the intermetacarpalis never being well developed in human ontogeny. In our recent anazlyzes of human embryos that had the muscles stained with antibodies—a work done by the team of our colleague Alain Chedotal—we did find the intermetacarpalis, 1 but it is in fact a small muscle that is clearly degenerating by 9.5 gestational weeks. This observation supports Cihak's idea that the first dorsal interosseous of humans is the result of a fusion of flexores breves profundi 2 and 3, and thus that both the deep and the superficial heads of the flexor pollicis brevis derive—at least in humans—from the flexor brevis profundus 1, together with the opponens pollicis. Be that as it may, what is clear is that the so-called dorsal interossei of mammals are not necessarily equivalent. For example, the four dorsal interossei of chimpanzees are not equivalent to the four dorsal interossei of modern humans, because they receive little or no contribution from the flexores breves profundi (Figures. 17.6 and 17.7). Instead, they essentially correspond to the intermetacarpales of nonmammalian tetrapods, although they may also include part of the contrahentes digitorum and/or of the dorsometacarpales, as explained earlier. Adult modern humans often have a small muscle inserting on the ulnar side of digit 1, which is often designated "interosseous volaris primus of Henle" (Tables 17.1 through 17.3). According to some authors, this small muscle probably corresponds to the flexor brevis profundus 2 of other tetrapods (see, e.g., Abramowitz 1955; Lewis 1989; Susman et al. 1999). However, our dissections, comparisons, and review of the literature indicate that it is more likely derived from the adductor pollicis, and we accordingly designated this muscle as "adductor pollicis accessorius" (Bello-Hellegouarch et al. 2013; see Tables 17.1 through 17.3 and Figure 17.1). In the studies of Abramowitz (1955), Lewis (1989), Susman et al. (1999), and Henkel-Kopleck and Schmidt (2000), the interosseous volaris primus of Henle was found in 100%, in 92%, in 86%, and in 69% of the adult human bodies examined, respectively. That is, this muscle does seem to be present in the majority of adult modern humans; that is why it is listed in Tables 17.1 and 17.3 (see also Bello-Hellegouarch et al. 2013, for a more recent discussion on this subject). Some authors consider that the intermetacarpales, contrahentes digitorum, and dorsometacarpales to be absent in monotremes (e.g., Howell 1937c; Jouffroy and Lessertisseur 1971). However, recently, Gambaryan et al. (2015) described intermetacarpales in the platypus (they did not find them in other extant monotremes) (see Tables 17.1 and 17.3). In mammals such as modern humans at least some of these muscles fuse with the flexores breves profundi (e.g., to form the dorsal interossei: see preceding text); the possibility that in monotremes at least parts of them are incorporated into the flexores breves profundi cannot be ruled out (Tables 17.1 through 17.3).

A detailed ontogenetic study of the hand muscles in monotremes, such as that undertaken by Cihak (1972) in modern humans, is needed to answer this question.

The detailed analysis of the data obtained from our dissections, combined with the information provided in the literature, has allowed us to develop robust hypotheses of homology for most of the dorsal muscles of the forearm. The extensor carpi radialis longus, extensor carpi radialis brevis, brachioradialis, and supinator of mammals (Figure 17.3) clearly correspond to the extensor antebrachii et carpi radialis of tetrapods such as *Ambystoma* (Figure 17.4; Tables 17.1 through 17.3; NB: some reptiles do have a muscle that seems to be homologous to the mammalian brachioradialis: see Chapter 18). In fact, according to some authors, in some urodeles, e.g., *Necturus*, part of the extensor antebrachii et carpi radialis is differentiated into at least one of the four former muscles (e.g., Haines 1939; Jouffroy 1971; Jouffroy and Lessertisseur 1971; Lewis 1989; Meers 2003) (see Table 17.1 and Chapter 18). In mammals such as modern humans, part of the abductor pollicis longus forms the extensor pollicis brevis (e.g., Jouffroy 1971; Kaneff 1980a; Lewis 1989; Diogo and Abdala 2010) (Tables 17.1 through 17.3). Most authors agree that unlike the condition in other tetrapods, in extant mammals, the extensores digitorum breves are not present as a group. Mammals often lack some of these muscles, and the ones that remain are usually considered to be functionally independent from each other (e.g., modern humans usually lack short extensors to digits 3 and 4; Tables 17.1 through 17.3, see also, e.g., Howell 1936c; Haines 1939; Jouffroy 1971; Jouffroy and Lessertisseur 1971; Lewis 1989). However, the outcome of our dissections, comparisons, and review of the literature suggest that the first mammals may well have had four, or even five, muscles corresponding to the extensores digitorum breves of other tetrapods. In fact, the phylogenetically most plesiomorphic extant mammals, the monotremes, have four muscles that appear to correspond to the extensores digitorum breves of other tetrapods (the extensor pollicis longus, extensor indicis, extensor digiti III proprius, and extensor digiti minimi: Tables 17.1 through 17.3). Moreover, a fifth short extensor, the extensor digiti quarti, is found in some therian mammals such as rats. The extensor digiti minimi of mammals such as tree shrews (which usually inserts on digits 4 and 5) seems to correspond to the extensores digitorum breves of digits 4 and 5 of other tetrapods and, thus, to the extensor digiti minimi and extensor digiti quarti of rats Tables 17.1-17.3. The extensor indicis of mammals such as rats and tree shrews (which usually inserts on digits 2 and 3) seems to correspond to the extensores digitorum breves of digits 2 and 3 of other tetrapods and thus to the extensor digiti III proprius and extensor indicis of monotremes (Tables 17.1 through 17.3). A detailed comparative investigation of the development and innervation of the short extensors in tetrapods is however needed to clarify the exact homologies between the muscles of amphibians, reptiles, monotremes, marsupials, and placentals.

Some authors have stated that the anconeus, a small muscle situated on the dorsal region of the elbow (Figures 17.2 and 17.3), is derived from the triceps brachii (e.g., Howell 1936c,d, 1937b). However, it is now commonly accepted that this small muscle corresponds to part of the extensor antebrachii et carpi ulnaris of taxa such as *Ambystoma* and *Timon* (e.g., Haines 1939; Jouffroy 1971; Jouffroy and Lessertisseur 1971; Lewis 1989; Tables 17.1 through 17.3 and Figures 17.4 and 17.5). In fact, in some nonmammalian tetrapods, the latter muscle is also divided into an extensor carpi radialis and an anconeus (the latter muscle is also referred to as "extensor antebrachii ulnaris"; Haines 1939; Jouffroy 1971; Jouffroy and Lessertisseur 1971; see Chapter 18). According to Lewis (1989: 133), the intimate relationship between the anconeus and the triceps brachii is "clearly a secondary feature" because in reptiles such as *Sphenodon*, the anconeus and the triceps brachii are clearly distinct. Lewis exposes the flaw in the argument that the mammalian anconeus is derived from the triceps brachii because it is innervated by the radial nerve, by pointing out that "the nerve that supplies the anconeus of mammals is merely the attenuated remnant of a branch (sometimes called nervus extensorius caudalis) which enters the forearm to join the posterior interosseous nerve in more primitive tetrapods and participate in the nerve supply of the forearm extensor musculature." Shellswell and Wolpert's (1977) study of the development of the chicken forelimb muscles strongly supports the proposal that at least in this tetrapod taxon, the anconeus is derived from the extensor antebrachii et carpi ulnaris rather than the triceps brachii (see also Chapter 18).

MARSUPIALS AND THE EVOLUTION OF MAMMALIAN FORELIMB MUSCULATURE

Marsupial opossums (order Didelphimorpha) are frequently the focus of evolutionary developmental studies because this order is the sister-group of a clade including all other extant marsupials and is therefore a good model to study the origin and early evolution of marsupials as a whole (e.g., Horovitz and Sánchez-Villagra 2003). Moreover, opossums are a very useful model to investigate the diversity of mammalian development because marsupials share very peculiar developmental features that are not seen in placentals, many of which are related to remarkable heterochronic and heterotopic changes in marsupials (Smith 1994, 2001, 2006; Sánchez-Villagra et al. 2002; Vaglia and Smith 2003; Sears 2004; Keyte and Smith 2010; Kelly and Sears 2011; Moustakas et al. 2011; Goswami et al. 2012; Hübler et al. 2013; Chew et al. 2014; Wakamatsu et al. 2014). Interestingly, despite these marked developmental differences, the skeletal structures of adult marsupials such as opossums and placentals such as mice are, in general, quite similar (e.g., Smith 2006; Goswami et al. 2012). However, with the exception of very few studies (e.g., Cheng 1955, Smith 1994, Keyte and Smith 2010), publications on the developmental biology of opossums do not refer to muscles. This lack of muscular developmental studies limits our understanding of mammalian evolution because the scarce data available indicate that muscle development might also considerably differ between marsupials and placentals (e.g., Keyte and Smith 2010). For instance, in most tetrapods, the skeletal and muscular systems develop over the same period. However, in marsupials, many skeletal elements are delayed in development; therefore, some muscles appear far in advance of the skeletal elements that will form their points of attachment (Smith 2006). However, we do not know whether the adult muscular phenotype of marsupials such as opossums is very different from that found in most placentals or is instead fairly similar (as is the skeletal system). The latter case would be a further example of different developmental patterns leading to a similar adult configuration.

In fact, a major weakness of most of the studies on the adult muscles of the Virginia opossum (*Didelphis virginiana*) and other marsupials is that they do not provide comparisons with placentals or other tetrapods. For instance, numerous authors have studied and discussed the hand muscles of marsupials, including *D. virginiana* (e.g., Young 1880; Brooks 1886a; McMurrich 1903a,b; Brandell 1935; Campbell 1939; Jouffroy 1971; Coues 1872; Stein 1981; Lewis 1989). However, their lack of comparative context and common nomenclature rendered them confusing or simply "useless," as stated by Lewis (1989). Moreover, most studies on the muscles of adult opossums mainly refer to a specific region of the head or limbs (e.g. Young 1880; Huber 1930a,b, 1931; Langworthy 1932; Haines 1939; Lightoller 1940a,b, 1942; Straus 1941a,b, 1942; Shrivastava 1962a,b; Brandell 1965; Hiiemae and Jenkins 1969; Minkoff et al. 1979; Grant et al. 2013). Even those studies that focus on both the head and limbs (e.g., Coues 1870, 1872) omit some muscles.

As explained in the preceding chapters, in our paper (Diogo et al. 2015d), we discussed related notions, such as that of *Bauplan*, or "body plan," which was originally related to a preevolutionary notion of "archetype" in the sense that it refers to a "plan." We also argued that the (justified) refutation of old notions such as scala naturae does not mean that we should abandon terms such as *phylogenetically basal* and particularly *anatomically plesiomorphic* to refer to groups such as the urodeles among the Tetrapoda, or lemurs among the Primates. Here we investigate whether the term *anatomically plesiomorphic* might also apply to the limbs of marsupials such as the opossum, among the Mammalia, by providing, for the first time, a rigorous comparative framework between the adult myology of marsupials, placentals, monotremes, and other tetrapods, for both the head and limbs. Specifically, we provide a list and brief description of all the *D. virginiana* head and limb muscles—based on Diogo et al.'s (2016a) paper—using an updated, unifying vertebrate myological nomenclature to allow more straightforward comparison between marsupials and other taxa. We combine the new anatomical data obtained from our dissections of *D. virginiana* with the myological information obtained in our previous works, and a detailed literature review of works performed by others on the myology of marsupials and other mammals. The present work is therefore the culmination of this 20-year project because we include marsupials and the detailed data on monotreme hindlimb musculature of Gambaryan et al. (2002) in the evolutionary/homology tables (Tables 17.1 through 17.3) elaborated in our previous works. Due to the importance of marsupials to understand mammalian evolution, we will therefore provide here a detailed description of the forelimb muscles of the opossum *D. virginiana*, in the following paragraphs; this textual description is complemented with Figure 17.6 and with Tables 17.1 through 17.3.

Concerning the proximal forelimb muscles, six muscles connect the axial skeleton to the pectoral girdle in the adult specimens of this species. The serratus anterior (part of serratus magnus of Coues 1872; part of serratus ventralis of Jenkins and Weijs 1979; serratus magnus of Stein 1981) and levator scapulae (part of serratus magnus of Coues 1872; levator anguli scapulae of Stein 1981) are fused to form a single, continuous structure that inserts on the scapula. The former originates from ribs 1-7, 1-8, or 1-9 and the latter from vertebrae C3-C7 or C4-C7. The levator claviculae (levator scapulae ventralis of Jouffroy 1971; atlanto-acromialis of Stein 1981 and Coues 1872) and atlantoscapularis posticus (atlanto-scapularis of Stein 1981 and Coues 1872) connect the transverse process of the atlas to the lateral and medial third of the scapular spine, respectively. The subclavius extends from the first rib to the clavicle, acromion, and cleidoacromial ligament and is blended with the cleidoacromialis. Regarding the pectoral musculature and its derivatives, the pectoralis major (ectopectoralis of Lander 1918; pectoralis superficialis of Langworthy 1932; part of pectoralis of Coues 1872 and Jenkins and Weijs 1979) has a clavicular head originating from the sternum and sometimes from a small portion of the clavicle, a sternocostal head mainly originating from the sternum and ribs, and an abdominal head originating from the sternum. All insert on the humerus, the clavicular head being superficial to and inserting more distally than the sternocostal head. These three heads are shown as part of the "pectoralis superficialis" of Langworthy (1932: see his figs. 2 and 3). The pectoralis major contacts the pectoralis major of the opposite side of the body at the midline. The pectoralis minor (entopectoralis of Lander 1918; pectoralis abdominalis + pectoralis profundus of Langworthy 1932; part of pectoralis of Coues 1872 and Jenkins and Weijs 1979) has three heads extending from the xiphoid process and abdominal muscles to the humerus only. Langworthy (1932) stated that the pectoralis minor inserts onto the coracoid process of the scapula, but other authors confirmed that it does not (e.g. Coues 1872; Jenkins and Weijs 1979). Lander (1918) stated that the entopectoralis is absent in *Didelphis*. On the contrary, the configuration seen in *D. virginiana* is very similar to that seen in rats: in both species, the entopectoralis has three heads. Because the "pectoralis abdominalis" of Langworthy (1932) is deep to the other two heads (of the "pectoralis profundus" of Langworthy 1932), it cannot correspond to the abdominal head of the pectoralis major seen in, e.g., humans, which is never deep to the pectoralis minor. This supports the idea of Jouffroy (1971) that the pectoralis abdominalis of different authors in different taxa probably corresponds to different structures. That is, when the structure is superficial to the pectoralis minor and adjacent to the main body of the pectoralis major, it corresponds to the abdominal head seen in, e.g., humans, while when the structure is deep to the pectoralis minor and adjacent to the main body of the pectoralis minor, it corresponds to a part of the pectoralis minor instead, as seen in, e.g., *Didelphis* and rats. The "fourth pectoral layer" of Langworthy (1932) corresponds to the muscle sternalis, not to a part of the pectoralis major or pectoralis minor. The pectoralis minor contacts the contralateral muscle at the midline. The panniculus carnosus extends from soft tissues such as the abdominal muscles and fascia, its only bony attachment being onto the humerus. It contacts its counterpart at the midline and is blended with the pectoralis minor and the latissimus dorsi of the same side of the body.

The posterior shoulder muscles infraspinatus, supraspinatus, teres minor, subscapularis, and teres major connect the scapula to the proximal humerus. Coues (1872) stated that he could not find a teres minor, but according to Stein (1981) and Jenkins and Weijs (1979), this muscle is present. The cleido-acromialis (probably corresponds to scapulo-clavicularis of Wood [1870]) extends from the distal end of the clavicle to the acromion and lateral third of the scapular spine and to the superficial aponeurosis of the supraspinatus, as described by Jenkins and Weijs (1979). It is partially blended with the subclavius. The deltoideus scapularis (part of deltoid of Coues 1872) runs from the scapula to the humerus. Stein (1981) describes only a "clavodeltoid" and a "spinodeltoid" in *D. virginiana*. The clavodeltoid is only partially differentiated into a pars acromialis (from acromion) and a pars clavicularis (from clavicle), which are blended with each other and thus constitute the deltoideus acromialis et clavicularis of the present work. The spinodeltoid corresponds to the deltoideus scapularis of the present work. The deltoideus acromialis et clavicularis (part of deltoid of Coues 1872) extends from the acromion of the scapula and the clavicle to the humerus and is blended with the pectoralis major. The latissimus dorsi extends from vertebrae T4 or T5 to L3 or L4 to the humerus and is blended with the dorsoepitrochlearis muscle.

Regarding the arm muscles, as described by Coues (1872), Jenkins and Weijs (1979), and Stein (1981) and as found in humans, the triceps brachii has scapular, medial, and lateral heads. It connects the scapula and humerus to the olecranon process of the ulna and is blended with the dorsoepitrochlearis. As described by these authors, the dorsoepitrochlearis (omo-anconeus of Coues 1872; tensor fasciae antebrachii of Stein 1981) extends from the latissimus dorsi to the olecranon process of the ulna and adjacent fascia. The brachialis (brachialis anticus of Coues 1872) connects the humerus to the ulna. Coues (1872) suggested that the two heads of biceps brachii of *D. virginiana* may both correspond to the short head of other mammals because they mainly originate from the coracoid process and surrounding regions rather than from the capsular ligament of the shoulder joint or the supraglenoid region. However, we agree with Jouffroy (1971) that these heads correspond to both the long and short heads of other mammals. Another peculiarity of the biceps is that it runs from the scapula to both the ulna and radius (without a well-defined bicipital aponeurosis). Lastly, the coracobrachialis has a single head, which might, in theory, correspond to the coracobrachialis medius/proprius of other mammals. However, it is very short, extending from the coracoid process to the proximal humerus, just distal to the lesser tubercle. Moreover, it is mainly deep to, rather than fused with, the biceps brachii. Jouffroy (1971) agreed with this interpretation, designating this head "coracobrachialis brevis/profundus" (see his fig. 93).

With respect to the flexor (ventral) forearm musculature, the pronator quadratus connects the distal ulna to the distal radius. The flexor digitorum profundus is fused with the flexor digitorum superficialis, extending from the humerus, radius, and ulna to the distal phalanges of digits 1 to 5. The flexor digitorum superficialis connects the humerus to the middle phalanges of digits 2–5, although in a few specimens of this species, the muscle seems to go only to digits 2, 3, and 4, as a variant (e.g., Coues 1872; Stein 1981). The palmaris longus (palmaris longus externus of Jouffroy 1971) and palmaris longus internus (Jouffroy 1971) are fused with each other and with the flexor digitorum longus and flexor carpi ulnaris, respectively. Both extend mainly from the humerus to the palmar aponeurosis. The flexor carpi ulnaris connects the humerus and ulna to the pisiform bone and surrounding aponeurosis. This seems to be the common condition for *D. virginiana*, as it is also described by Coues (1872) and Stein (1981), although authors such as Straus (1942) refer to an attachment to both the pisiform and metacarpal 5 in *Didelphis*. The epitrochleo-anconeus (anconeus internus of Stein 1981) extends from the medial epicondyle of the humerus to the olecranon process of the ulna. Straus (1942) suggested that the epitrochleoanconeus of opossums does not correspond to the "anconeus internus" of other mammals, but the "anconeus internus" of authors such as Stein (1981) clearly does correspond to the epitrochleoanconeus of other mammalian taxa. The flexor carpi radialis connects the humerus to metacarpal 2. This seems to be the common condition for *D. virginiana*, as it was also reported by Coues (1872) and Stein (1981), but Straus (1942) reported an attachment to both metacarpals 1 and 2 in *Didelphis*. The pronator teres extends from the humerus to the middle third or distal two thirds of the radius. A palmaris brevis was not present in the specimens dissected by us, Coues (1872), or Stein (1981), but Jouffroy and Saban (1971) explain that this muscle is present in some other marsupials.

Among the numerous studies that referred to the hand muscles of *D. virginiana* (see preceding text), Brooks (1886a,b) is the most complete and accurate. Brooks describes four lumbricales, an abductor pollicis brevis (ab2), flexor pollicis brevis (f2r), adductor pollicis (a1), contrahentes to digits 2 (a2), 4 (a4), and 5 (a5, plus a somewhat distinct bundle, a5a1), an abductor digiti minimi (abd5), a flexor digiti minimi with two heads (a5a and f5u), a flexor brevis digitorum manus to the middle phalanx of digit 5, the usual eight flexores breves profundi plus an adductor pollicis accessorius, and the usual four intermetacarpales. Of the two heads of the flexor digiti minimi brevis, the first (a5a) mainly corresponds to the flexor digiti minimi of humans, while the deeper f5u probably derives from the same anlage that gives rise to the opponens digiti minimi of humans because it is deeper to the main body of the flexor digiti minimi brevis. The adductor pollicis accessorius constitutes a "third" short flexor of digit 1 but seems to derive instead from the adductor pollicis (Brooks 1886a,b). Hopefully, by providing one-to-one comparisons with other mammals and tetrapods, the present section will help to resolve once and for all the controversy and misunderstandings concerning these muscles. The flexor brevis digitorum manus connects the hamate to the middle phalanx of digit 5. The four lumbricales extend from tendons of the flexor digitorum profundus to the ventral side of the proximal phalanges of digits 2 to 5. They also indirectly attach onto the dorsal side of the middle and distal phalanges of these digits *via* the extensor expansions. There are three contrahentes digitorum other than the adductor pollicis: the contrahens of digit 2 extends from metacarpal III to the proximal phalanx of digit 2; contrahens of digit 4 extends from metacarpal III to the proximal phalanx of digit 4; and the contrahens of digit 5 extends from metacarpal 3 and capitate to the proximal phalanx of digit 5, having a small, somewhat separate head. The adductor pollicis (contrahens of digit 1) connects metacarpal III and the capitate to the proximal phalanx of digit 1. The adductor pollicis accessorius (volaris primus of Henle) connects the trapezoid and metacarpal I to the proximal phalanx of digit 1. All 10 flexores breves profundi are present, but number 1 corresponds to the flexor pollicis brevis and number 10 corresponds to the flexor digiti minimi brevis, both of which are described in the following. The other eight flexores breves profundi are number 2 from trapezium to ulnar side of proximal phalanx of digit 1; numbers 3, 4, 5, 6, 7, and 8 from metacarpals III, IV, and V to the radial and ulnar sides of the respective digit and number 9 from metacarpal V to the radial side of proximal phalanx of digit 5. The flexor pollicis brevis connects the trapezium to the radial side of proximal phalanx of thumb, corresponding to the "superficial head of the flexor pollicis brevis" of human anatomy. Although Figure 9.8A of Lewis (1989) suggests that at least some marsupials have an opponens pollicis, the consensus is that this muscle is not present in *D. virginiana*, as noted earlier. The flexor digiti minimi brevis runs from the pisiform to the proximal phalanx of digit 5 and has two heads, one superficial and one deep, which correspond topologically to the flexor digiti minimi brevis and to the opponens digiti minimi of humans, respectively. However, the latter head does not seem to be directly homologous to the opponens digiti minimi of humans because this muscle was acquired only later in evolution, among placental mammals. The condition seen marsupials reflects only a rough evolutionary parallelism because the deeper head attaches onto the proximal phalanx of digit 5 rather than onto the metacarpal V as does the opponens digiti minimi of, e.g., humans (see the following). The abductor pollicis brevis connects the trapezium to the radial side of the proximal phalanx of digit 1. The abductor digiti minimi runs from the pisiform to the ulnar side of the proximal phalanx of digit 5. Lastly, there are four intermetacarpales: as described by Young (1880), number 1 extends from metacarpal I to radial side of proximal phalanx of digit 2; number 2 extends from metacarpals II and III to ulnar side of proximal phalanx of digit 2 and radial side of proximal phalanx of digit 3; number 3 extends from metacarpals III and IV to ulnar side of proximal phalanx of digit 3 and radial side of proximal phalanx of digit 4; and number 4 connects metacarpals IV and V to the ulnar side of proximal phalanx of digit 4 and radial side of proximal phalanx of digit 5. There are 10 muscles in the extensor layer of the forearm; all originate from the humerus if not otherwise noted. The extensor carpi radialis longus and brevis insert onto metacarpals II and III, respectively (the latter may attach to both, as a variant: see Straus 1941a,b). The brachioradialis (supinator longus of Coues 1872) inserts onto the scaphoid, lunate, and triquetrum, as described by Haines (1939) and Straus (1941) and contra Coues (1872) and Stein (1981). The supinator (supinator brevis of Coues 1872) inserts onto the proximal radius. It is a well-defined, separated muscle, as described by Stein (1981) and contra Coues (1872). The extensor carpi ulnaris extends from the humerus and ulna to metacarpal V. The anconeus (anconeus externus of Stein 1981) connects the lateral epicondyle of the humerus to the olecranon process of the ulna. The extensor digitorum (extensor digitorum communis of Coues 1872) inserts onto the distal phalanges of digits 2, 3, 4, and 5 (as a variant, the tendon to digit 5 might be absent: e.g., Coues 1872; Stein 1981). The extensor digiti minimi (extensor digitorum lateralis of Stein 1981; extensor digitorum ulnaris of Haines, 1939) inserts onto the distal phalanges of digits 4 and 5 (and 3, as a variant: e.g., Haines, 1939). The extensor indicis (extensor digitorum profundus of Straus 1941a,b) extends from the ulna to the distal phalanges of digits 1, 2, and 3, as described by Straus (1941a,b) and Haines (1939) (or 1 and 2 only, as a variant: e.g., Coues 1872; Stein 1981). The abductor pollicis longus (extensor ossis metacarpi pollicis of Coues 1872) connects the ulna and radius to metacarpal I.

By combining these anatomical data on *D. virginiana* with developmental studies (e.g., Cheng 1955), we were able to test the homology/evolutionary hypotheses proposed by Cheng (1955) and Diogo and Abdala (2010), which agree on most points. As noted earlier, a major exception is that Diogo and Abdala (2010) considered it more likely that the dorsoepitrochlearis derives from the triceps brachii than from the latissimus dorsi, as suggested by Cheng (1955) (see Tables 17.1 through 17.3 and preceding text). In their recent study on monotreme forelimb musculature, Gambaryan et al. (2015) stated that the dorsoepitrochlearis is continuous with the distal portion of a bundle of the latissimus dorsi in *Zaglossus* and *Tachyglossus* and that this condition is probably plesiomorphic for mammals. They thus proposed that the dorsoepitrochlearis derives from the latissimus dorsi, as did Cheng (1955). However, they recognized that the platypus (*Ornithorhynchus*) condition is similar to that found in therians; i.e., the dorsoepitrochlearis is not completely continuous with the latissimus dorsi. Therefore, it is more parsimonious to accept that the *Zaglossus* and *Tachyglossus* condition is derived (one step in branch leading to family Tachyglossidae, including these two genera) than it is to accept that the conditions seen in platypus and therians are homoplastic (two steps). That is, the Tachyglossidae, while interesting, do not provide new phylogenetic information to challenge the hypothesis that the dorsoepitrochlearis derived from the triceps brachii. Moreover, the dorsoepitrochlearis is usually innervated by the radial nerve (which usually innervates the triceps brachii) rather than by the subscapular nerves (which usually innervate the latissimus dorsi; for other pieces of evidence, see Diogo and Abdala 2010 and preceding text).

Regarding the mammalian teres minor (Tables 17.1 through 17.3), as noted earlier, the results of our reanalysis contradict Diogo and Abdala (2010) and support Cheng (1955). In addition, the results of our dissections and comparisons agree with studies such as those by Jouffroy (1971) and Warburton (2003) that the cleidoacromialis is not derived from the subclavius, but instead is an appendicular muscle derived from the supracoracoid group (i.e., the group that gives rise to the supraspinatus and infraspinatus). This idea was also supported by Cheng's (1955) developmental study of *Didelphis*, which strongly suggested that the cleidoacromialis originated from the supraspinatus. The cleidoacromialis seems to be often absent in placentals, but Wood (1870) stated that at least some rats, and humans as variants, have a "scapulo-clavicularis" that might correspond to the cleidoacromialis. The presence of the cleidoacromialis as a distinct muscle in some adult placentals would support the idea that this muscle was present in the last common ancestor (LCA) of placentals + marsupials (Tables 17.1 through 17.3). Jouffroy (1971) stated that most (so, supposedly not all) marsupials have a single, undivided rhomboideus, as seen in *Didelphis*. However, most authors agree that marsupials have a single rhomboid, as do monotremes, indicating that this was probably the plesiomorphic condition for both mammals as a whole and for the LCA of marsupials and placentals (Tables 17.1 through 17.3). The presence of a "rhomboideus occipitalis" in anurans is therefore probably homoplastic; see Diogo and Ziermann (2014).

Our dissections of *Didelphis* revealed a single deltoideus acromialis et clavicularis (rather than two separate deltoid muscles). After dissecting more mice and rats and reviewing our notes on *Tupaia*, *Cynocephalus*, and other mammals, we conclude that this was also the most common condition in mammals and, therefore, very likely the condition in the LCA of placentals and marsupials (Tables 17.1 through 17.3). Also, we conclude that the atlantoscapularis posticus of placentals corresponds to the "atlanto-scapularis" of marsupials and was thus probably present in the LCA of placentals and marsupials (Tables 17.1 through 17.3). As shown by Cheng (1955), the atlantoscapularis posticus seems to be developmentally derived from the levator scapulae, not from the levator claviculae. In fact, the levator claviculae itself is derived from the levator scapulae. Our dissections of lizards, particularly *Timon* and *Lacerta*, lead us to conclude that there is a levator claviculae in *Timon* and many lepidosaurs, meaning that it is probably an ancient muscle (Tables 17.1 through 17.3). Consequently, in the present book, we made a slight change in nomenclature: in the work by Diogo and Abdala (2010), we stated that the levator claviculae gave rise to the atlantoscapularis posticus and anticus of, e.g., *Tupaia*, but here for taxa such as *Tupaia*, we simply use the names levator claviculae, levator scapulae, and atlantoscapularis posticus (Tables 17.1 through 17.3). Because the levator claviculae does not seem to have changed much from the LCA of placentals + marsupials to *Tupaia*, we decided not to use the term *atlantoscapularis anticus* for the levator claviculae of *Tupaia*, as we did in the work by Diogo and Abdala (2010).

Straus (1942) stated that the palmaris longus muscles of placentals + marsupials may be derived from the flexor carpi radialis (as an exception), from the flexor digitorum superficialis (most frequently), and/or from the flexor carpi ulnaris (somewhat frequently). In certain mammals, such as some marsupials and some Carnivora, there are two palmaris longus muscles, probably derived from the flexor carpi ulnaris and/or flexor digitorum superficialis. Windle and Parsons (1897) refer to a "palmaris longus externus" and a "palmaris longus internus," stating that both are present in *D. virginiana* and that the palmaris longus externus is more commonly found in mammals. McMurrich (1903a) states that the palmaris longus of mammals corresponds to part of the flexor digitorum longus of reptiles. Jouffroy (1971) reviewed the literature and stated that although some authors refer to three different palmaris longus muscles ("radial" derived from flexor carpi radialis, "intermedius" from flexor digitorum superficialis, and "ulnaris" from flexor carpi ulnaris), the intermedius and radialis never seem to coexist. Therefore, Jouffroy preferred to refer simply to the palmaris longus externus, derived from the flexor digitorum superficialis and often innervated by median nerve, and the palmaris longus internus, derived from the flexor carpi ulnaris and often innervated by the ulnar nerve. Our dissections and comparisons support the idea that the palmaris longus internus and externus are both present in *D. virginiana*, and Jouffroy's idea that the presence of a palmaris longus internus (in addition to the palmaris longus) in some marsupials and some placentals is probably due to homoplasy (Tables 17.1 through 17.3). In contrast, the palmaris longus ("externus") very likely was present in the LCA of marsupials and placentals because it is also present in other tetrapods (Tables 17.1 through 17.3).

The presence/absence and homologies of some hand muscles in various tetrapods are controversial. As explained earlier, Diogo and Abdala (2010) stated that the contrahentium caput longum of amphibians may correspond to a part of the flexor digitorum longus of reptiles and monotremes. However, based on the evidence available at the time, they considered it more likely that the former muscle is completely absent in amniotes. On the contrary, the detailed study of the forelimb musculature of monotremes by Gambaryan et al. (2015) revealed that one of the heads of the flexor digitorum longus of monotremes is extremely similar to the contrahentium caput longum of amphibians, lying between structures that clearly seem to correspond to the flexor accessorius lateralis and flexor accessorius medialis of amphibians. As also explained earlier, while some authors consider the intermetacarpales, contrahentes digitorum, and dorsometacarpales to be absent in monotremes (e.g., Howell 1937d; Jouffroy and Lessertisseur 1971)—an idea followed by Diogo and Abdala (2010)—Gambaryan et al. (2015) recently described intermetacarpales in the platypus (they did not find them in other extant monotremes, however) (see Tables 17.1 through 17.3). Similarly, following Jouffroy (1971), Diogo and Abdala (2010) stated that monotremes appear to have an adductor pollicis, but in Gambaryan et al.'s (2015) study, no such muscle was described, supporting Howell's (1937d) suggestion that this muscle is probably absent in monotremes.

The data summarized earlier and in Tables 17.1 through 17.3 thus allow us to provide a very detailed list of muscle synapomorphies shared by extant marsupials, extant therians, and extant mammals as a whole. Based on this comparison, extant mammals share 11 muscle synapomorphies for the forelimb. These data illustrate the utility of studying muscles to characterize certain clades and pave the way for paleontological, developmental, and functional works that investigate the specific evolutionary time of origin/loss and developmental mechanisms that led to the characteristic muscle anatomy of each clade and their functional implications. Regarding the proximal (girdle/arm) forelimb muscles, during these transitions, there were eight synapomorphic changes in the transitions that led to the LCA of extant mammals: differentiation of pectoralis major, pectoralis minor, panniculus carnosus, infraspinatus, supraspinatus, subscapularis, teres major, and dorsoepitrochlearis (Tables 17.1 through 17.3). In contrast, there were only three synapomorphic changes within the distal (forearm/hand) forelimb muscles: loss of palmaris profundus 1, of flexores breves superficiales as a group, and of dorsometacarpales (Tables 17.1 through 17.3). There were 11 synapomorphic changes from the LCA of extant mammals to the LCA of extant therians in the forelimb. Four relate to the proximal (girdle/arm) forelimb muscles (differentiation of atlantoscapularis posticus and cleidoacromialis; loss of costocoracoideus and of remaining of original supracoracoideus) and seven to the distal (forearm/hand) forelimb muscles (differentiation of flexor digitorum superficialis, palmaris longus, palmaris brevis, flexor brevis digitorum manus, extensor carpi radialis longus, and extensor carpi radialis brevis) (Tables 17.1 through 17.3). Therefore, unlike the transitions to mammals, those leading to therians were more dramatic in the distal forelimb, consistent with the idea that the former were mainly related to posture and locomotion. Lastly, there are three forelimb synapomorphies of extant placentals, all concerning the proximal (girdle/arm) forelimb (differentiation of rhomboideus major, rhomboideus minor, and rhomboideus occipitalis) (Tables 17.1 through 17.3). These numbers provide empirical support for a well-defined therian body plan, which can still be easily recognized in most extant placentals and marsupials except very specialized taxa such as bats and whales.

GENERAL REMARKS

In order to discuss morphological complexity and make myological comparisons such as those listed in Tables 17.1 through 17.3, we must be able to measure complexity. Diogo et al. (2016a) used the concept of gross morphological complexity. Specifically, we use the morphological definition of complexity of Eble (2004, 2005), which corresponds to "pure complexity" as defined by McShea and Brandon (2010). These authors argued that pure complexity—i.e., the number of part types (e.g., different muscles) within a body—is a much more appropriate, and practical, measure of complexity than is "colloquial complexity," which refers not only to part types but also to their functionality, sophistication, and integration, among other characteristics. According to them, "colloquial complexity" is so vague, subjective, and/or difficult to measure that it has become the source of many problems in biology, including the paucity of empirical treatments of complexity in the biological literature. Therefore, here we refer to "pure complexity" by taking into account the number of muscles found in *Didelphis*, rats/mice, and humans with the number most likely found in the LCA of these three taxa (i.e., of marsupials + placentals), and the number of evolutionary changes that probably occurred between the LCA of marsupials + placentals and each of those three taxa. This method is particularly valuable because (a) the number of muscles is an objective measure (e.g., most researchers/anatomical atlases describe the same number of muscles in the human body) and (b) it combines macroevolutionary and developmental definitions of complexity, including the notion that each muscle is the result of parcellation, i.e., of innovation through differentiation leading to a morphogenetic semi-independence of the muscle (e.g., Vermeij 1973; Bonner 1988; Wagner and Altenberg 1996).

Regarding the forelimb muscles, there are 23 proximal (pectoral girdle/arm) muscles in the LCA of extant therians, 23 in *Didelphis* and rats, and 18 in humans (Tables 17.1 through 17.3). There were no changes from LCA to *Didelphis*, four to rats (rhomboids minor differentiated; rhomboid occipitalis differentiated; atlantoscapularis posticus lost; cleidoacromialis lost), and seven to humans (rhomboid minor differentiated; levator claviculae lost; atlantoscapularis posticus lost; panniculus carnosus lost; cleidoacromialis lost; two deltoids fused into one; dorsoepitrochlearis lost) (Tables 17.1 through 17.3). The number of distal (forearm/hand) muscles was 48 in the LCA of extant therians, 45 in *Didelphis*, 44 in *Rattus*, and 41 in *Homo*. *Didelphis* has 4 changes since the LCA of extant therians (gain of palmaris longus internus; loss of palmaris brevis; fusion of extensor digiti minimi and extensor digiti quarti into extensor digiti minimi; fusion of extensor digit 3 proprius, extensor indicis and extensor pollicis longus into extensor indicis), rats have 6 (loss of flexor brevis digitorum manus; loss of one contrahens; loss of adductor pollicis accessorius; gain of opponens digiti minimi; loss of brachioradialis; undifferentiated extensor digiti 3 proprius), and humans have 12 (gain of flexor pollicis longus; loss of epitrochleoanconeus; loss of flexor brevis digitorum manus; loss of contrahens to digit 2; loss of contrahens to digit 4; loss of contrahens to digit 5; intermetacarpales integrated into interossei; gain of opponens pollicis; gain of opponens digiti minimi; undifferentiated extensor digiti quarti; undifferentiated extensor digiti 3 proprius; gain of extensor pollicis brevis) (Tables 17.1 through 17.3).

In the review provided by Diogo et al. (2016a), including also the head, neck, and hindlimb muscles, the total numbers of head and limb muscles are 206 for the LCA of extant therians, 193 for *Didelphis*, 194 for rats, and 180 for humans. Therefore, those results show no support for a scala naturae or increase in pure morphological complexity of McShea and Brandon (2010) between the LCA of therians and humans. In fact, if a trend exists, it is toward fewer muscles and less pure complexity. Specifically, serial muscles such as the contrahentes, intermetacarpales, and flexores breves profundi were lost or fused, leading toward anisomerism (see Diogo et al. 2015e). Humans have 26 fewer muscles than the LCA of extant eutherians, 14 fewer muscles than placentals such as rats, and 13 fewer muscles than marsupials such as *Didelphis*. These results contradict previous reports, which suggested that humans have many more muscles than opossums (e.g. Coues 1872; Huber 1930a,b, 1931; Minkoff et al. 1979). Similarly, Grant et al. (2013) showed that contrary to previous assumptions, all major muscles present in the snout region of placentals such as mice are present in opossums. From the LCA of extant eutherians there were 14 changes to *Didelphis*, 37 to rats, and 63 to humans. Although even more muscles were lost during the evolution of humans than during that of opossums or rats, some muscles also became differentiated, because there are 63 changes toward humans but humans actually have only 26 fewer muscles than the LCA of extant eutherians. These numbers are important, because they show that being anatomically plesiomorphic and being morphologically less complex are two very different things. This distinction is often neglected or downplayed by evolutionary biologists, partly because of the continuing profound effect of scala naturae in our field (Diogo et al. 2015d). Anatomical changes do not always produce greater complexity, particularly in phylogenetically derived animals such as mammals. *Didelphis* provides an illustrative example, because it is anatomically more plesiomorphic (only 14 changes from LCA of extant therians) than are rats (37 changes) and humans (63 changes), but in terms of pure morphology, its musculature is more complex than that of humans. Thus, *Didelphis* is a very good model for the LCA of extant therians. Similarly, our previous study showed that monotremes provide a very good model for the head and forelimb musculature of basal mammals (Diogo et al. 2015d). Another recent study independently reached the same conclusion: "among living tetrapods, one cannot find a better model for reconstruction of the locomotorium of the mammal-like reptiles than monotremes" (Gambaryan et al. 2015: 54).

As also pointed out by Diogo et al. (2016a), of the 194 muscles of the head, neck, and limbs of *Didelphis*, 172 (89%) are present, with exactly the same one-to-one identity/homology, in rats. Therefore, the adult anatomy of placentals and marsupials is very similar despite their very different early developmental patterns, which include not only marked heterochronies but also differences in gene upregulation and expression between the two taxa (e.g., Hübler et al. 2013). As explained in the introduction, this pattern is also seen in the skeletal system. For instance, marsupials have a transient shoulder arch that results from the fusion of several elements of the shoulder girdle complex, including the scapula, coracoid, and manubrium. This structure provides support and important muscle attachment sites for the crawl to the teat (reviewed by, e.g., Hübler et al. [2013]). The early morphology of the marsupial shoulder arch is thus similar to that seen in embryonic and adult monotremes. Immediately before or shortly after parturition, the coracoid often detaches from the sternum, leading to the adult marsupial shoulder girdle morphology. This morphology is shared with placentals, despite the fact that placentals never form a shoulder arch during their ontogeny (Hübler et al. 2013). A similar pattern is seen in the nervous system. Spinal nerve outgrowth into the opossum forelimb bud is expanded, including cervical nerves 3–8 and thoracic nerves 1 and 2. This outgrowth is probably related to the expansion of the opossum forelimb early in development (Keyte and Smith 2010). The contribution from the third cervical nerve is then reduced later in ontogeny, resulting in an adult phenotype like that of adult placentals such as mice in which there is no connection between C3 and the brachial plexus (Keyte and Smith 2010). This result led Keyte and Smith (2010: 4290) to point out that "although the anterior expansion of the forelimb bud seems to have impacted early spinal nerve growth into the limb, this is corrected as development proceeds so that in the adult a typical mammalian pattern is observed."

Smith (2001:129) also briefly referred to this topic, using marsupials and placentals to suggest "that development, even at its earliest stages, is highly plastic. The observation of significant early plasticity, even in animals in which the adults are quite similar, is important for understanding the ways in which development and evolution interact." This is because the changes observed in marsupials "occur immediately before, during, and after the phylotypic stage and include shifts in some of the major patterning events in the body. These observations thus confirm recent studies that suggest that the degree of conservation at a phylotypic stage in vertebrates has been overestimated" (Smith 2001:129). Such examples reinforce the notion of a *Bauplan* (body plan), although this interpretation is something of a paradox because the "phylotypic stage" is precisely the stage at which earlier authors might have thought that a *Bauplan* would be most obvious. Smith's (2001) case study shows that many evolutionary changes are driven by the needs of the embryos/neonates rather than by a "goal" toward a specific adult plan or "archetype." That is, within a clade such as Theria, even drastic changes in early development do not produce, except in very rare cases (e.g., whales and bats), major changes in the adult body plan. This new evo-devo interpretation of *Bauplan* is not adult driven, as suggested by its name, related to a preevolutionary notion of "archetype" and "plan." It is instead development driven in that it emphasizes the importance of developmental constraints. These constraints are so prevalent and interdependent (e.g., through well-defined genomic networks) that even drastic developmental changes, such as those seen in early marsupial ontogeny, do not break the whole "spider web" created by the constraints and lead to completely new phenotypes (see, e.g., Wagner 2014). Therefore, the ancient term *Bauplan* with its modern evo-devo interpretation does reinforce both the importance of developmental constraints and the existence of a common, easily identifiable phenotype shared by some clades, but not by others. Such developmental constraints may explain why the adult musculature of marsupials is similar to that of other mammals even though their development is very different. In this sense, it is hoped that our long-term project will contribute toward the multidisciplinary data needed for an integrative synthesis of the anatomical macroevolution of vertebrates and for future functional and developmental comparative studies.

18 Forelimb Muscles of Limbed Amphibians and Reptiles

Chapter 17 focused on the tetrapods as a whole, particularly how the pectoral and pectoral fin/forelimb muscles have evolved during the transitions from nonmammalian tetrapods to monotremes, to therian mammals, and to placentals, including modern humans. Therefore, most taxa included in Tables 17.1 through 17.3 were mammals. The two other taxa, *Ambystoma* and *Timon*, were included in those tables to show the general hypotheses of homology between the muscles of the mammalian taxa and the muscles of amphibians and lepidosaurs. In Chapter 18, we will turn our focus to the major groups of limbed amphibians and reptiles: urodeles, anurans, lepidosaurs, crocodylians, birds, and turtles. The tables included in Chapter 18 will include *Ambystoma* and *Timon* as representative members of urodelan amphibians and of lepidosaurian reptiles as well as *Bufo*, *Caiman*, *Gallus*, and *Trachemys* as representative members of anuran amphibians and of crocodylians, birds, and turtles, respectively (adult extant caecilian amphibians lack a pectoral girdle and lack limbs, and thus, their pectoral and forelimb musculature is extremely reduced: these amphibians will therefore not be discussed in the present chapter; see, e.g., Carroll 2007). Many anatomical works have provided information about the pectoral and forelimb muscles of amphibians and reptiles (e.g., Mivart 1869; Humphry 1872a,b; Fürbringer 1876; Ecker 1889; Gaupp 1896; McMurrich 1903a,b; Ribbing 1907, 1938; Romer 1922, 1924, 1944; Howell 1935, 1936b,c; Haines 1939, 1950; Straus 1942; Sullivan 1962, 1967; Grim 1971; Hudson et al. 1972; Walker 1973; Holmes 1977; Ghetie et al. 1981; Duellman and Trueb 1986; Russell 1988; Manzano 1996; Burton 1998; Dilkes 1999; Wyneken 2001; Meers 2003; Walthall and Ashley-Ross 2006; Maxwell and Larsson 2007; Russell and Bauer 2008). However, most of these works have focused on a specific taxon and/or a specific pectoral or forelimb region, and none of them has provided detailed information about the homologies of all the pectoral and forelimb muscles of urodeles, anurans, lepidosaurs, crocodylians, birds, and turtles.

In the following paragraphs, we briefly summarize the results of our observations and comparisons, presented in Tables 18.1 through 18.3, paying special attention to issues that remain controversial among morphologists. Chapter 18 is considerably updated from Diogo and Abdala (2010) in that it includes data from dissections we performed since that book was written (e.g., of urodeles, frogs, and lepidosaurs such as chameleons: e.g., Diogo and Tanaka 2012; Diogo et al. 2016c; Molnar et al. 2017a) and data recently published by other authors. Additional illustrations of the salamander forelimb muscles are provided in Chapter 16, and additional illustrations of the forelimb muscles of chameleons are provided in Figure 18.14.

In this chapter we follow the interpretation, supported in studies of fossils and of *Hox* genes, that the three digits that are usually present in adult birds are digits 1, 2, and 3 rather than digits 2, 3, and 4, as is often suggested by the authors of embryological studies (see, e.g., Burke and Feduccia 1997; Galis et al. 2003, 2005; Vargas and Fallon 2005a,b; Vargas et al. 2008; Kundrát 2009). For clarity, we also give in brackets the number of the digit according to most embryologists. For instance, the most radial digit of adult chickens would be described as "digit 1 (i.e., digit 2 according to most embryologists)." We consider this to be a clear, simple, and neutral way of referring to the avian digits.

PECTORAL MUSCLES DERIVED FROM THE POSTCRANIAL AXIAL MUSCULATURE

Amphibian and reptilian taxa have numerous pectoral muscles derived from the postcranial axial musculature, such as the serratus anterior, rhomboideus, levator scapulae, opercularis, sternocoracoideus, and costocoracoideus (Tables 18.1 and 18.2). These muscles mainly connect the axial skeleton to the pectoral girdle and are thus associated with the movements of this girdle (see Chapter 17). Authors such as Kardong (2002) suggested that reptiles do not have a rhomboideus, but Howell (1935, 1936b, 1937a,b), Sullivan (1962, 1967), Hudson et al. (1972), Duellman and Trueb (1986), and Dilkes (1999), among others, argued that crocodylians, birds, and at least some anurans do have a "rhomboideus." For instance, Howell (1935, 1937b) and Duellman and Trueb (1986) described a "rhomboideus anterior" and a "rhomboideus posterior" in anurans such as *Rana*. Our dissections confirmed the presence of these two muscles in frogs such as *Bufo* and *Rana* (Tables 18.1 through 18.3). Howell (1936b), Dilkes (1999), and Meers (2003) also described a rhomboideus in crocodylians, and Howell (1937b), Sullivan (1962, 1967), Hudson et al. (1972), and Dilkes (1999) reported a "rhomboideus superficialis" and a "rhomboideus profundus" in birds. These descriptions are also confirmed by our dissections (Tables 18.1 through 18.3). To our knowledge, a rhomboideus has never been described in lepidosaurs or turtles, and we were unable to find this structure in the turtles and lepidosaurs we dissected (Tables 18.1 through 18.3). In one of the most detailed reviews of mammalian pectoral and forelimb muscles, Jouffroy (1971) stated that the mammalian rhomboideus is homologous to the "basiscapularis" of "lower tetrapods," thus supporting the idea that at least some extant nonmammalian tetrapods do have a muscle that is homologous

TABLE 18.1
Pectoral and Forelimb Muscles of Adults of Representative Amphibians and Reptilian Taxa

Amphibia (Caudata): *Ambystoma mexicanum* (Axolotl)	Amphibia (Anura): *Rana pipiens* (Northern Leopard Frog)	Reptilia (Testudines): *Trachemys scripta* (Red-Eared Slider Turtle)	Reptilia (Lepidosauria): *Timon lepidus* (Ocellated Lizard)	Reptilia (Crocodylia): *Caiman latirostris* (Brown-Snouted Caiman)	Reptilia (Aves): *Gallus domesticus* (Chicken)
Serratus anterior (part of serrati *sensu* Howell 1937a,b)	**Serratus anterior** (part of serrati *sensu* Howell [1935, 1937a,b] and Duellman and Trueb [1986]) [Duellman and Trueb [1986] state that *Rana* has a "serratus superior" innervated by spinal nerve 3 and extending from presacral IV to the dorsomedial region of the suprascapula; a "serratus medius" innervated by spinal nerve 3 and extending from presacral III to the central, medial region of the suprascapula; and a "serratus inferior" innervated by spinal nerve 4 and extending from presacrals III and IV to the medial region of the suprascapula]	**Serratus anterior** (the "testocoracoideus" *sensu* authors such as Gaunt and Gans [1969] probably corresponds to a part or the totality of the serratus anterior *sensu* this volume: see, e.g., Wyneken [2001]; but Nagashima et al. [2009] contradict this, saying that it corresponds to the "serratus" *sensu* Fürbringer [1874] and to the carapaco-scapularis *sensu* Ogushi [1913], which makes sense because the muscle does run from the carapace to the scapula and is innervated by the long thoracic nerve as shown by Nagashima et al. [2009]; as also shown by these authors, the serratus anterior of turtles does not change its attachments, but becomes folded with the dorsal encapsulation of the scapula by the ribs and carapace, so instead of lying mainly medial to the scapula, as is usually the case in other amniotes, the muscle now mainly lies lateral to the scapula)	**Serratus anterior** (serratus ventralis *sensu* Kardong and Zalisko [1998] and Kardong [2002]) [according to authors such as Holmes [1977], Dilkes [1999], and Tsuihiji [2007], lepidosaurs, crocodylians and birds have a "serratus superficialis," and a "serratus profundus"; according to Holmes [1977], the serratus profundus corresponds to the serratus anterior]	**Serratus anterior** (serratus ventralis *sensu* Meers [2003])	**Serratus anterior** (serratus superficialis, which in carinate birds, but not in ratite birds, includes three parts: pars anterior, posterior, and metapatagialis *sensu* Fürbringer [1902] and Nagashima et al. [2009]) [Vanden Berge [1975] states that chickens have just a pars anterior and a pars posterior and a pars metapatagialis]

(*Continued*)

TABLE 18.1 (CONTINUED)
Pectoral and Forelimb Muscles of Adults of Representative Amphibians and Reptilian Taxa

Amphibia (Caudata): *Ambystoma mexicanum* (Axolotl)	Amphibia (Anura): *Rana pipiens* (Northern Leopard Frog)	Reptilia (Testudines): *Trachemys scripta* (Red-Eared Slider Turtle)	Reptilia (Lepidosauria): *Timon lepidus* (Ocellated Lizard)	Reptilia (Crocodylia): *Caiman latirostris* (Brown-Snouted Caiman)	Reptilia (Aves): *Gallus domesticus* (Chicken)
–	**Rhomboideus** (rhomboideus posterior *sensu* Duellman and Trueb [1986]) [according to Howell [1935, 1937a,b], there is a rhomboideus in anurans such as *Rana*, but not in urodeles; see text; Duellman and Trueb [1986] describe two rhomboid muscles in *Rana*, one (rhomboideus anterior) extending from the skull to the pectoral girdle and the other (rhomboideus posterior) running from the presacral IV to the dorsomedial surface of the suprascapula; both these muscles are innervated by spinal nerve 3 so they might well correspond to the rhomboideus occipitalis and rhomboideus of mammals, although it is not clear if the LCA already had these muscles or if this is instead the result of a convergence/parallelism; Duellman and Trueb [1986] also state that the complex formed by the serrati plus the rhomboideus posterior is probably homologous to the "thoraciscapularis" of urodeles, thus supporting the idea that the rhomboideus is not present as a separate muscle in urodeles]	– [seem to be absent as a distinct muscle in turtles, but see levator scapulae in the following]	– [see text]	**Rhomboideus** [rhomboideus is present in crocodylians according to Howell [1936a], Dilkes [1999], Meers [2003], and Lima et al. [2016]; see text]	**Rhomboideus** [rhomboideus is present in birds according to Howell [1937b], Sullivan [1962, 1967], and Dilkes [1999], see on the left; Sullivan [1962, 1967] and Vanden Berge [1975] describe a "rhomboideus superficialis" and a "rhomboideus profundus" in chickens, the superficialis sometimes also being divided into "clavicularis and scapularis"; see text; Nagashima et al. [2009] support the idea that in both ratite and carinate birds, there is a rhomboid complex extending from the vertebrae to the scapula]

(Continued)

TABLE 18.1 (CONTINUED)
Pectoral and Forelimb Muscles of Adults of Representative Amphibians and Reptilian Taxa

Amphibia (Caudata): ***Ambystoma mexicanum*** **(Axolotl)**	**Amphibia (Anura):** ***Rana pipiens*** **(Northern Leopard Frog)**	**Reptilia (Testudines):** ***Trachemys scripta*** **(Red-Eared Slider Turtle)**	**Reptilia (Lepidosauria):** ***Timon lepidus*** **(Ocellated Lizard)**	**Reptilia (Crocodylia):** ***Caiman latirostris*** **(Brown-Snouted Caiman)**	**Reptilia (Aves):** ***Gallus domesticus*** **(Chicken)**
–	**Rhomboideus occipitalis** (rhomboideus anterior *sensu* Duellman and Trueb [1986]) [it is not clear if the LCA of tetrapods already had a separated rhomboideus occipitalis or if the presence of this muscle in anurans is the result of homoplasy: see rhomboideus mentioned earlier and text]	–	–	–	–
Levator scapulae (part of thoracico-scapularis *sensu* Jouffroy [1971]; levator scapulae superioris *sensu* Hetherington and Tugaoen [1990]; opercularis *sensu* Walthall and Ashley-Ross [2006] and Piekarski and Olsson [2007]) [see text]	**Levator scapulae superioris** (levator scapulae *sensu* Howell [1935, 1937b]) [see text]	**Levator scapulae** [Walker [1973] describes two sheets of muscle just next to each other; one is the testoscapuloprocoracoideus extending from the scapular prong and adjacent parts of the scapular prong to the acromion (which was designated by Fürbringer as procoracoid, hence the name testoscapulo-procoracoideus, but if it attaches to the acromion, it may well correspond to the levator claviculae part of the levator scapulae complex); the other is the colloscapularis extending from the scapular prong to cervical vertebrae, which might correspond to the levator scapulae part of the levator scapulae complex or to a rhomboid part; Walker [1973] described however these as separate muscles, and	**Levator scapulae** [according to authors such as Holmes [1977] and Dilkes [1999], in lepidosaurs, including *Sphenodon*, the levator scapulae is usually divided into superficial and deep heads; the "levator scapulae ventralis" of *Iguana* shown in Figure 56A of Jouffroy [1971] corresponds to the ventral head of the levator scapulae *sensu* most authors (e.g., Russell and Bauer [2008]); i.e., in *Iguana* and various other lizards and in *Sphenodon*, the levator claviculae is present, but is usually seen as a head of the levator scapulae (that often extends from cervical vertebrae to the suprascapular cartilage in the region of the acromion process and to the clavicular bar) and not as a distinct muscle; but according	**Levator scapulae** [according to authors such as Holmes [1977], Dilkes [1999], Meers [2003], Tsuihiji [2007], and Lima et al. [2016], all the Crocodylia have a mainly undivided levator scapulae; according to Holmes [1977], the "levator scapulae profundus" portion of the "collothoraci-scapularis profundus" *sensu* Fürbringer [1876] appears to be a part of the serratus musculature and not of the levator scapulae *sensu* this volume]	– [according to authors such as Dilkes [1999], in birds, the levator scapulae is not present as an independent muscle; Nagashima et al. [2009] support this idea because although they talk about a levator scapulae–rhomboid complex in carinate birds—but not in ratite birds, where they only talk about a rhomboid complex—the so-called levator-scapulae rhomboid shown in their Figre S1 seems to be a single, mainly undivided muscle that is innervated by the suprascapular nerve and that topologically corresponds to the rhomboid complex of mammals); Fürbringer [1902] also did not describe a distinct levator scapulae in birds, referring instead to a rhomboid–serratus profundus

(Continued)

TABLE 18.1 (CONTINUED)
Pectoral and Forelimb Muscles of Adults of Representative Amphibians and Reptilian Taxa

Amphibia (Caudata): *Ambystoma mexicanum* (Axolotl)	Amphibia (Anura): *Rana pipiens* (Northern Leopard Frog)	Reptilia (Testudines): *Trachemys scripta* (Red-Eared Slider Turtle)	Reptilia (Lepidosauria): *Timon lepidus* (Ocellated Lizard)	Reptilia (Crocodylia): *Caiman latirostris* (Brown-Snouted Caiman)	Reptilia (Aves): *Gallus domesticus* (Chicken)
		taking into account the content of the cell just on the right, we also consider these as distinct muscles; Nagashima et al. [2009] refer to a levator scapulae-rhomboid complex in turtles, which is innervated by the suprascapular nerve and corresponds to the levator scapulae *sensu* the present work and according to them corresponds to the testoscapulo-procoracoideus *sensu* Fürbringer [1874] and Walker [1973] and to the collo-claviculo-plastralis plus carapaco-plastralis *sensu* Ogusgi [1913]; as shown in Nagashima et al.'s Figure S1, this complex mainly extends ventrally from the anterior vertebrae to the scapula but mainly to the plastron (derived from clavicle and interclavicle); this supports the idea that these are axial muscles that are first connected medially to the axial skeleton and only attach later to more lateral elements; i.e. unlike the appendicular girdle muscles (see latissimus dorsi below) in these axial muscles, the proximal origin is more evolutionarily stable than the distal insertion, which, in the case of this complex, changed from being only onto the scapula to be onto both the scapula and the clavicle+interclavicle]	to Gregory and Camp [1918], these correspond, at least in *Sphenodon*, to the levator claviculae and levator scapulae, respectively, of placentals; as our further dissections of tetrapods such as anurans have supported the idea that these animals have a distinct levator claviculae muscle, it thus seems that this muscle was present in the LCA of amphibians and amniotes]		complex; Vanden Berge [1975] confirms that the levator scapulae is not present as a distinct muscle, probably corresponding to the anterior head of the serratus anterior]

(*Continued*)

TABLE 18.1 (CONTINUED)
Pectoral and Forelimb Muscles of Adults of Representative Amphibians and Reptilian Taxa

Amphibia (Caudata): *Ambystoma mexicanum* (Axolotl)	Amphibia (Anura): *Rana pipiens* (Northern Leopard Frog)	Reptilia (Testudines): *Trachemys scripta* (Red-Eared Slider Turtle)	Reptilia (Lepidosauria): *Timon lepidus* (Ocellated Lizard)	Reptilia (Crocodylia): *Caiman latirostris* (Brown-Snouted Caiman)	Reptilia (Aves): *Gallus domesticus* (Chicken)
– [absent as an independent muscle, see text]	**Opercularis** [apparently only found as an independent muscle in anurans; see text]	–	–	–	–
–	**Colummelaris** [as stated and shown by Duellman and Trueb (1986) (e.g., their Figure 4.11), anurans often have both a muscle opercularis and a muscle colummelaris, both involved in the auditory system, the opercularis being clearly derived from the levator scapulae and the colummelaris being likely derived from the levator scapulae: see levator scapulae mentioned earlier]	–	–	–	–
–	**Levator claviculae** (levator scapulae inferioris *sensu* Duellman and Trueb [1986]) [it is not clear if the LCA of tetrapods already had a separated levator claviculae or if the presence of this muscle in anurans is the result of homoplasy; however, it is clear that anurans often have a "levator scapulae inferioris" innervated by spinal nerve 2 and running from the prootic and exoccipital bones to the posteroventral corner of the suprascapula, which thus has at least some features similar to those of the "levator scapulae ventralis"; i.e., of the levator claviculae, of other tetrapods]	**Levator claviculae** [see on the left]	**Levator claviculae** [see on the left]	–	–

(*Continued*)

TABLE 18.1 (CONTINUED)
Pectoral and Forelimb Muscles of Adults of Representative Amphibians and Reptilian Taxa

Amphibia (Caudata): *Ambystoma mexicanum* (Axolotl)	**Amphibia (Anura): *Rana pipiens* (Northern Leopard Frog)**	**Reptilia (Testudines): *Trachemys scripta* (Red-Eared Slider Turtle)**	**Reptilia (Lepidosauria): *Timon lepidus* (Ocellated Lizard)**	**Reptilia (Crocodylia): *Caiman latirostris* (Brown-Snouted Caiman)**	**Reptilia (Aves): *Gallus domesticus* (Chicken)**
– [Mivart [1869] suggested that the subclavius could be part of the procoracohumeralis of amphibians, but according to Romer [1924], the latter muscle gives rise to mammalian muscles such as the teres minor and the deltoideus clavicularis instead; according to Howell [1937a,b], (a) the sternocoracoideus and costocoracoideus are not present as distinct muscles in urodeles and anurans; (b) reptiles such as *Iguana* have a costocoracoideus and a sternocoracoideus superior and inferior; (c) mammals such as *Ornithorhynchus* have a costocoracoideus and a sternocoracoideus; and (d) placental mammals have a "costoscapularis" and a subclavius, which thus seems to correspond to the sternocoracoideus of other tetrapods]	– [see on the left; according to Duellman and Trueb [1986], among anurans, only discoglossids and *Leiopelma* have a "sternocoracoideus" (also named "sternoepicoracoideus") innervated by spinal nerve 3 and extending from the sternal horn to the epicoracoid cartilage, the muscle being apparently derived from the rectus abdominis; therefore, this muscle does not seem to be homologous to the sternocoracoideus *sensu* the present work; i.e., the LCA of tetrapods apparently did not have a sternocoracoideus, the sternocoracoideus of discoglossids and *Leiopelma* being apparently a derived, peculiar structure of these taxa]	– [see on the left]	**Sternocoracoideus** (sternocoracoid superior and inferior *sensu* Howell [1937a,b]) [Holmes [1977] and Dilkes [1999] state that in lepidosaurs, including *Iguana* and *Sphenodon*, the sternocoracoideus is present and is divided into superficial and deep heads]	– [according to authors such as Fürbringer [1876], Walker [1973], Holmes [1977], and Dilkes [1999], the sternocoracoideus is not present as an independent muscle in turtles and in crocodylians]	**Sternocoracoideus** [according to authors such as Vanden Berge [1975] and Dilkes [1999], the sternocoracoideus is present in birds]

(*Continued*)

TABLE 18.1 (CONTINUED)
Pectoral and Forelimb Muscles of Adults of Representative Amphibians and Reptilian Taxa

Amphibia (Caudata): *Ambystoma mexicanum* (Axolotl)	Amphibia (Anura): *Rana pipiens* (Northern Leopard Frog)	Reptilia (Testudines): *Trachemys scripta* (Red-Eared Slider Turtle)	Reptilia (Lepidosauria): *Timon lepidus* (Ocellated Lizard)	Reptilia (Crocodylia): *Caiman latirostris* (Brown-Snouted Caiman)	Reptilia (Aves): *Gallus domesticus* (Chicken)
– [see preceding text]	– [see preceding text]	– [according to Walker [1973], the costocoracoideus is not present as a distinct muscle in turtles]	**Costocoracoideus** (costoscapularis *sensu* Howell [1936b, 1937b] and Holmes [1977]; costosternocoracoideus *sensu* Holmes [1977] and Dilkes [1999]) [authors such as Howell [1937b], Homes [1977], and Dilkes [1999] state that in lepidosaurs, including *Sphenodon* and *Iguana*, the costocoracoideus is present and is subdivided into pars superficialis and pars profundus"]	**Costocoracoideus** (costosternocoracoideus *sensu* Holmes [1977] and Dilkes [1999]) [according to Meers [2003] and Lima et al. [2016] in crocodylians the costocoracoideus is divided into pars superficialis and pars profundus]	– [according to authors such as Vanden Berge [1975] and Dilkes [1999], the costocoracoideus is not present as an independent muscle in birds]
Pectoralis	**Pectoralis** (corresponds only to the sternal and abdominal heads of the pectoralis *sensu* Duellman and Trueb [1986], the epicoracoid head of the pectoralis *sensu* these authors corresponding to the muscle supracoracoideus *sensu* the present work) [the sternal and abdominal heads *sensu* Duellman and Trueb [1986] are very similar, respectively, to the main body of the pectoralis and to the abdominal head of the pectoralis and apparently also to a part of the panniculus carnosus of mammals]	**Pectoralis** (pectoralis major *sensu* Wyneken [2001]) [in turtles, the pectoralis major has a peculiar configuration; i.e., its insertion is onto the humerus, as in most other amniotes, but its origin is not from the superficial—ventral—side of bones such as the sternum and clavicle but instead from the deep surface of the axial skeleton and plastron, which is derived from the clavicle and interclavicle (Nagashima et al. 2009); i.e., in this specific case, the attachments did not change so much]	**Pectoralis** [according to authors such as Fürbringer [1876], Holmes [1977], Walker [1983], and Dilkes [1999], the pectoralis is present in all the major extant reptilian clades; according to authors such as Romer [1944] and Kardong [2002], the plesiomorphic condition for reptiles is that in which this muscle was mainly undivided]	**Pectoralis** [see text]	**Pectoralis** (includes pectoralis pars thoracica *sensu* Vanden Berge [1975] and Maxwell and Larsson [2007]) [see text; it should however be noted that Vanden Berge [1975] clearly describes a pars thoracica, a pars propatagialis, and pars abdominalis, the latter head being strikingly similar to the panniculus carnosus of mammals, being innervated by both the intercostal nerves and the pectoral nerves]

(*Continued*)

TABLE 18.1 (CONTINUED)
Pectoral and Forelimb Muscles of Adults of Representative Amphibians and Reptilian Taxa

Amphibia (Caudata): *Ambystoma mexicanum* (Axolotl)	Amphibia (Anura): *Rana pipiens* (Northern Leopard Frog)	Reptilia (Testudines): *Trachemys scripta* (Red-Eared Slider Turtle)	Reptilia (Lepidosauria): *Timon lepidus* (Ocellated Lizard)	Reptilia (Crocodylia): *Caiman latirostris* (Brown-Snouted Caiman)	Reptilia (Aves): *Gallus domesticus* (Chicken)
Supracoracoideus (coracohumeralis *sensu* Howell [1935, 1937b])	**Supracoracoideus** (coracohumeralis *sensu* Howell [1935, 1937b]; epicoracoid head of the pectoralis *sensu* Duellman and Trueb [1986]: see pectoralis mentioned earlier) [as in urodeles, the supracoracoideus lies mainly superficial (ventral) to the coracoradialis, as shown by Duellman and Trueb [1986]]	**Supracoracoideus** [according to Walker [1973], in turtles, the supracoracoideus often consists of "anterior" and "posterior" parts]	**Supracoracoideus** (coracohumeralis *sensu* Howell [1936b])	**Supracoracoideus** [according to Meers [2003], in crocodylians, the supracoracoideus is often divided into three heads: longus, intermedius, and brevis; according to Lima et al. [2016], in *Caiman crocodilus crocodiles*, the intermedius head is however not present as a distinct structure]	**Supracoracoideus** (pectoralis secundus *sensu* Jollie [1962])
Deltoideus scapularis [the deltoideus scapularis *sensu* Jouffroy [1971] corresponds to the dorsalis scapulae *sensu* Howell [1937b], Romer [1944], Walthall and Ashley-Ross [2006], Diogo [2007], and Diogo and Abdala [2007]]	**Deltoideus scapularis** (dorsalis scapulae *sensu* Howell [1935, 1937b] and Duellman and Trueb [1986])	**Deltoideus scapularis** (dorsal, or scapular, part of deltoideus *sensu* Walker [1973] and Wyneken [2001])	**Deltoideus scapularis** (dorsalis scapulae *sensu* Howell [1936b], Diogo [2007], and Diogo and Abdala [2007]) [according to Holmes [1977] and Dilkes [1999], the deltoideus scapularis is present in turtles, crocodylians, birds, and lepidosaurs]	**Deltoideus scapularis** (dorsalis scapulae *sensu* Fürbringer [1876])	**Deltoideus scapularis** (part of acromialis *sensu* Ribbing [1938]; part of deltoideus *sensu* Jollie [1962]) [according to Dilkes [1999], the deltoideus scapularis probably corresponds to the muscle that is often designated, in birds, as "deltoideus major," and not to both the deltoideus major and "deltoideus minor," as suggested by Romer [1944]; as explained by Sullivan [1962, 1967], the avian muscles that he designates as "deltoideus" and "coracobrachialis anterior" correspond to the muscles that are often called deltoideus major and deltoideus minor, respectively, by other authors]

(*Continued*)

TABLE 18.1 (CONTINUED)
Pectoral and Forelimb Muscles of Adults of Representative Amphibians and Reptilian Taxa

Amphibia (Caudata): *Ambystoma mexicanum* (Axolotl)	Amphibia (Anura): *Rana pipiens* (Northern Leopard Frog)	Reptilia (Testudines): *Trachemys scripta* (Red-Eared Slider Turtle)	Reptilia (Lepidosauria): *Timon lepidus* (Ocellated Lizard)	Reptilia (Crocodylia): *Caiman latirostris* (Brown-Snouted Caiman)	Reptilia (Aves): *Gallus domesticus* (Chicken)
–	–	– [see on the right]	– [according to Romer [1944], Jollie [1962], Jouffroy [1971], and Holmes [1977], the scapulo-humeralis posterior is present in *Sphenodon*, crocodylians, and birds and absent in turtles and all lizards except *Agama*; according to Dilkes [1999], this muscle is also present in squamates; see text]	**Scapulo-humeralis posterior** (scapulo-humeralis caudalis *sensu* Meers [2003]) [see on the left and text]	**Scapulo-humeralis posterior** (scapulo-humeralis caudalis *sensu* Vanden Berge [1975], Dilkes [1999], and Maxwell and Larsson [2007]) [see on the left and text]
Procoracohumeralis [see text]	**Procoracohumeralis** [corresponds to the deltoideus *sensu* Ecker [1864] and Duellman and Trueb [1986], which is divided into "clavicular," "episternal," and "scapular" heads according to these authors]	**Deltoideus clavicularis** (ventral, or clavicular, part of deltoideus *sensu* Walker [1973] and Wyneken [2001]) [Dilkes [1999] states that in turtles, the deltoideus clavicularis is partially fused with the deltoideus scapularis; these two structures are described as "part of the deltoideus" by Walker [1973] and Wyneken [2001], and Walker [1973] states that in some turtles, such as trionychids, the deltoideus is undivided, i.e., that the "dorsal, scapular, or head" is not differentiated in these turtles]	**Deltoideus clavicularis** (procoracohumeralis *sensu* Diogo [2007] and Diogo and Abdala [2007]) [in the case of reptiles, using the name deltoideus clavicularis, which is used by most authors working with amniotes, is justified because this muscle does not directly correspond to the procoracohumeralis of amphibians such as *Ambystoma*: it corresponds only to a part of it, the other part corresponding to the scapulo-humeralis anterior of *Timon*: see on the left; according to Holmes [1977] and Dilkes [1999], the deltoideus clavicularis is present in turtles, crocodylians, lepidosaurs, and birds]	**Deltoideus clavicularis** (scapularis inferior *sensu* **Fürbringer** [1876])	**Deltoideus clavicularis** (part of acromialis *sensu* Ribbing [1938]) [according to Dilkes [1999], in birds, the deltoideus clavicularis is sometimes divided into a "pars cranialis" and a "pars caudalis"; as stated by Dilkes [1999], the deltoideus minor of birds probably corresponds to a part or the totality of the deltoideus clavicularis of other tetrapods and not to a part of the deltoideus scapularis, as suggested by Romer [1944]: see deltoideus scapularis mentioned earlier]

(Continued)

TABLE 18.1 (CONTINUED)
Pectoral and Forelimb Muscles of Adults of Representative Amphibians and Reptilian Taxa

Amphibia (Caudata): *Ambystoma mexicanum* (Axolotl)	Amphibia (Anura): *Rana pipiens* (Northern Leopard Frog)	Reptilia (Testudines): *Trachemys scripta* (Red-Eared Slider Turtle)	Reptilia (Lepidosauria): *Timon lepidus* (Ocellated Lizard)	Reptilia (Crocodylia): *Caiman latirostris* (Brown-Snouted Caiman)	Reptilia (Aves): *Gallus domesticus* (Chicken)
–	–	– [see on the left]	– [the humeroradialis does not seem to be present as a distinct muscle in *Timon*, but contrary to the statements of Meers [2003], Romer [1944] and Jollie [1962] argued that in addition to birds and crocodylians, the humeroradialis is also present in *Sphenodon*]	**Humeroradialis** [according to Meers [2003], the humeroradialis (Figure 18.13) is mainly a flexor of the antebrachium that is only present in living archosaurs, i.e., in birds and crocodylians and that was probably derived from the dorsal musculature, being perhaps developmentally related to the deltoid muscles (e.g., it is innervated by the axillary nerve: see brachioradialis); authors such as Romer [1944], Jollie [1962], and Sullivan [1962, 1967] do support the idea that the humeroradialis is related to the deltoid group (see Figure 18.13) and specifically to the deltoideus clavicularis, thus corresponding to part of the procoracohumeralis longus of amphibians]	**Humeroradialis** (deltoides propatagialis *sensu* Romer [1944]; tensor patagii *sensu* Jollie [1962]; tensor propatagii *sensu* Sullivan [1962, 1967]; tensor propatagialis *sensu* Vanden Berge [1975] and Meers [2003]) [see on the left; Vanden Berge [1975] reported longus and a brevis heads of this muscle in chickens]
–	–	– [according to Holmes [1977] and Dilkes [1999], the scapulo-humeralis anterior is not present as a separate muscle in turtles]	**Scapulo-humeralis anterior** [according to Romer [1944], Jollie [1972], Holmes [1977], and Dilkes [1999], the scapulo-humeralis anterior is present in *Sphenodon*, lizards and birds]	– [as explained by authors such as Fürbringer [1876], Romer [1944], Holmes [1977], Dilkes [1999], Meers [2003], and Lima et al. [2016], in crocodylians, the scapulo-humeralis anterior is not present as an independent muscle; i.e., it is not differentiated from the deltoideus clavicularis]	**Scapulo-humeralis anterior** (scapulohumeralis cranialis *sensu* Vanden Berge [1975], Dilkes [1999], and Maxwell and Larsson [2007])

(*Continued*)

TABLE 18.1 (CONTINUED)
Pectoral and Forelimb Muscles of Adults of Representative Amphibians and Reptilian Taxa

Amphibia (Caudata): *Ambystoma mexicanum* (Axolotl)	Amphibia (Anura): *Rana pipiens* (Northern Leopard Frog)	Reptilia (Testudines): *Trachemys scripta* (Red-Eared Slider Turtle)	Reptilia (Lepidosauria): *Timon lepidus* (Ocellated Lizard)	Reptilia (Crocodylia): *Caiman latirostris* (Brown-Snouted Caiman)	Reptilia (Aves): *Gallus domesticus* (Chicken)
Subcoracoscapularis	**Subcoracoscapularis** (subscapularis *sensu* Ecker[1864, 1889)	**Subcoracoscapularis** (subscapularis *sensu* Walker [1973], Holmes [1977], Dilkes [1999], and Wyneken [2001]) [according to Walker [1973], the subcoracoscapularis is usually undivided in turtles, but may be divided into a shorter, "medial head" and a longer "lateral head" in taxa such as sea turtles, *Testudo*, and *Hydromedusa*]	**Subcoracoscapularis** (subscapularis plus subcoracoideus *sensu* Holmes [1977] and Dilkes [1999]) [according to Holmes [1977] and Dilkes [1999], the subcoracoscapularis is mainly undivided in turtles and crocodylians, corresponding to the muscle that is often designated, in these two groups, as "subscapularis"; in *Sphenodon*, squamates, and birds, the subcoracoscapularis is divided into a subscapularis and a "subcoracoideus," each of these two structures being in turn often subdivided into two heads in various birds]	**Subcoracoscapularis** (subscapularis *sensu* Dilkes and Meers [2003] and Lima et al. [2016]) [see on the left]	**Subcoracoscapularis** (subscapularis plus subcoracoideus *sensu* Vanden Berge [1975], Sullivan [1962, 1967], Dilkes [1999], and Maxwell and Larsson [2007]) [see on the left]
– [see text]	–	**Teres major** [see text]	– [see text]	**Teres major** [see text; Lima et al. [2016] further support the homology between the teres major of mammals and of reptiles such as crocodylians, as proposed in our previous works]	– [see text]

(Continued)

TABLE 18.1 (CONTINUED)
Pectoral and Forelimb Muscles of Adults of Representative Amphibians and Reptilian Taxa

Amphibia (Caudata): *Ambystoma mexicanum* (Axolotl)	Amphibia (Anura): *Rana pipiens* (Northern Leopard Frog)	Reptilia (Testudines): *Trachemys scripta* (Red-Eared Slider Turtle)	Reptilia (Lepidosauria): *Timon lepidus* (Ocellated Lizard)	Reptilia (Crocodylia): *Caiman latirostris* (Brown-Snouted Caiman)	Reptilia (Aves): *Gallus domesticus* (Chicken)
Latissimus dorsi	Latissimus dorsi	**Latissimus dorsi** [in turtles, the latissimus dorsi has a peculiar configuration; i.e., its insertion is onto the humerus, as in most other amniotes, but its origin is not from the posterior ribs/vertebrae/lumbar region, but instead anteriorly from the nuchal bone of the neck region, the muscle thus mainly passing ventral to the carapace and ribs and not dorsal to the ribs as is the case in most amniotes (Nagashima et al. 2009); Nasashima et al. [2009] state that the appendicular girdle muscles such as the latissimus dorsi and pectoralis major changed their origins and not their insertions because these muscles are formed through the in–out mechanism; i.e., they develop in the limb bud and only then extend medially to attach to the axial skeleton]	**Latissimus dorsi** [the latissimus dorsi is present in all living reptiles, being mainly undivided in crocodylians, lepidosaurs, and turtles, as described by authors such as Holmes [1977], Dilkes [1999], and Meers [2003]]	**Latissimus dorsi** [see on the left]	**Latissimus dorsi** [according to authors such as Vanden Berge [1975] and Dilkes [1999], in birds the latissimus dorsi is often divided into a pars cranialis and a pars caudalis; these two bundles are clearly shown in Figure S1 of Nagashima et al. [2009]]

(*Continued*)

TABLE 18.1 (CONTINUED)
Pectoral and Forelimb Muscles of Adults of Representative Amphibians and Reptilian Taxa

Amphibia (Caudata): *Ambystoma mexicanum* (Axolotl)	Amphibia (Anura): *Rana pipiens* (Northern Leopard Frog)	Reptilia (Testudines): *Trachemys scripta* (Red-Eared Slider Turtle)	Reptilia (Lepidosauria): *Timon lepidus* (Ocellated Lizard)	Reptilia (Crocodylia): *Caiman latirostris* (Brown-Snouted Caiman)	Reptilia (Aves): *Gallus domesticus* (Chicken)
Triceps brachii [the triceps brachii of urodeles usually includes coracoideus ("coracotriceps"), scapularis medialis ("dorsitriceps"), humeralis lateralis ("humerotriceps lateralis") and humeralis medialis (humerotriceps lateralis) sections, which correspond, respectively, to the "anconeus coracoideus," "anconeus scapularis medialis," "anconeus humeralis lateralis," and "anconeus humeralis medialis" *sensu* Walthall and Ashley-Ross [2006], Diogo [2007], and Diogo and Abdala [2007]; Howell [1935, 1937b] seems to suggest that the coracotriceps of urodeles such as *Necturus* might correspond to triceps coracoideus of reptiles such as *Iguana* and thus to the dorso-epitrochlearis of mammals]	**Triceps brachii** [according to Howell [1935, 1937b], in anurans such as *Rana* the coracotriceps is not present as a distinct structure, but the dorsitriceps (or anconeus scapularis) is present, and the humerotriceps is divided into three divisions comprising "laterale," "mediale," and "profundum," the profundus division being merely a separable part of the mediale division; Duellman and Trueb [1986] confirm that there is no triceps coracoideus in anurans such as *Rana* and only describe a triceps scapularis, triceps humeralis medialis, and triceps humeralis lateralis in this genus]	**Triceps brachii** [according to Walker [1973], Holmes [1977], Dilkes [1999], and Wyneken [2001], in turtles, the triceps brachii usually has a scapular head and a humeral head (which are designated "long lateral head and short lateral head" by Holmes [1977]), and in some taxa, such as *Dermochelys*, only one head (the humeral head according to Wyneken [2001]) is present]	**Triceps brachii** [according to Holmes 1977 and Dilkes 1999, there are usually four heads of the triceps brachii (scapular, coracoid, lateral humeral, and medial humeral) in lepidosaurs, including *Sphenodon*; Holmes [1977] argues that having four heads is the plesiomorphic condition for reptiles, and Chapter 17 supports the idea that this is also the plesiomorphic condition for amniotes and for living tetrapods as a whole, because extant amphibians often have four heads of the triceps, and mammals usually have three heads of the triceps plus a dorsoepitrochlearis, which clearly seems to derive from/correspond to the coracoid head of the triceps of other tetrapods]	**Triceps brachii** [according to Dilkes [1999], there are usually five heads of the triceps brachii (scapular, coracoid, lateral humeral, medial humeral, and an extrahumeral head known as the posticum) in crocodylians; Holmes [1977] suggests that crocodylians usually only have four heads, but Meers [2003] and Lima et al. [2016] do describe five heads (which Meers designated "triceps longus lateralis," "triceps longus caudalis," "triceps brevis cranialis," "triceps brevis intermedius," and "triceps brevis caudalis"), as did Dilkes 1999]	**Triceps brachii** [according to Vanden Berge [1975] and Dilkes [1999], there are usually two or three heads of the triceps brachii ("scapulotriceps," "humerotriceps," and occasionally a greatly reduced coracotriceps) in Aves; authors such as Grim [1971] and Haninec et al. [2009] state that Aves such as chickens have a "dorsoepitrochlearis," which is usually named "metapatagial latissimus dorsi," and which would correspond to the triceps coracoideus *sensu* this volume and thus to the coracotriceps *sensu* Dilkes [1999]; Sullivan [1962, 1967] only describes a humerotriceps and a scapulotriceps in chickens; however, in addition to these two heads, Vanden Berge [1975] describes a "dorsocutaneous head" and a "metapatagialis head" of the latissimus dorsi, but they are both innervated by cutaneous branches and originated from the axial skeleton so they cannot correspond to the triceps coracoideus of other tetrapods; but Vanden Berge [1975] reports an expansor secundariorum in chicken (see following cell) and states that it is probably derived in many birds, the muscle

(*Continued*)

TABLE 18.1 (CONTINUED)
Pectoral and Forelimb Muscles of Adults of Representative Amphibians and Reptilian Taxa

Amphibia (Caudata): *Ambystoma mexicanum* (Axolotl)	Amphibia (Anura): *Rana pipiens* (Northern Leopard Frog)	Reptilia (Testudines): *Trachemys scripta* (Red-Eared Slider Turtle)	Reptilia (Lepidosauria): *Timon lepidus* (Ocellated Lizard)	Reptilia (Crocodylia): *Caiman latirostris* (Brown-Snouted Caiman)	Reptilia (Aves): *Gallus domesticus* (Chicken)
					having some resemblance to the coracoradialis of urodeles, having two tendons of origin, one from the scapula, coracoid and sternum and surrounding muscles and giving rise to a long tendon and the other from the distal portion of the humerus, the muscle then attaching onto the elbow region and extending proximally into the metapatagium, but being innervated by the anconeal branch of the radial nerve; so it might be a coracoid head of the triceps corresponding to the dorsoepitrochlearis of mammals and does not seem to be coracoradialis because biceps has already both the humeral and coracoid heads]
–	–	–	–	–	**Expansor secundariorum** [see preceding cell]
Humeroantebrachialis (part of biceps *sensu* Romer [1944] and of brachialis *sensu* Howell [1937b])	– [as noted by Howell [1937b], there is no distinct humeroantebrachialis in anurans, the only ventral arm present being the coracobrachialis and the only ventral long muscle passing the elbow joint to insert onto the forearm being the coracoradialis; according to this author, the humeroantebrachialis of urodeles might however correspond to part of the "caput superior of the flexor carpi radialis" of anurans]	**Brachialis** (brachialis inferior *sensu* Romer [1944], Walker [1973], Holmes [1977], Dilkes [1999], and Wineken [2001])	**Brachialis** (brachialis inferior *sensu* Romer [1944], Holmes [1977], and Dilkes [1999])	**Brachialis** (brachialis inferior *sensu* Romer [1944], Holmes [1977], and Dilkes [1999])	**Brachialis** (brachialis inferior *sensu* Holmes 1977 and Dilkes 1999)

(Continued)

TABLE 18.1 (CONTINUED)
Pectoral and Forelimb Muscles of Adults of Representative Amphibians and Reptilian Taxa

Amphibia (Caudata): *Ambystoma mexicanum* (Axolotl)	Amphibia (Anura): *Rana pipiens* (Northern Leopard Frog)	Reptilia (Testudines): *Trachemys scripta* (Red-Eared Slider Turtle)	Reptilia (Lepidosauria): *Timon lepidus* (Ocellated Lizard)	Reptilia (Crocodylia): *Caiman latirostris* (Brown-Snouted Caiman)	Reptilia (Aves): *Gallus domesticus* (Chicken)
Coracoradialis	**Coracoradialis** [the coracoradialis is often designated "biceps" in the literature about anurans, being designated by "sternoradialis" by authors such as Duellman and Trueb [1986]; the innervation of the coracoradialis and the coracobrachialis in anurans is similar, i.e., by the spinal nerve 3, so this contradicts the idea that the innervation of the coracoradialis is very different from that of the other arm muscles and because of that, this muscle cannot have contributed to the formation of the biceps brachii]	–	–	–	–

(Continued)

TABLE 18.1 (CONTINUED)
Pectoral and Forelimb Muscles of Adults of Representative Amphibians and Reptilian Taxa

Amphibia (Caudata): *Ambystoma mexicanum* (Axolotl)	Amphibia (Anura): *Rana pipiens* (Northern Leopard Frog)	Reptilia (Testudines): *Trachemys scripta* (Red-Eared Slider Turtle)	Reptilia (Lepidosauria): *Timon lepidus* (Ocellated Lizard)	Reptilia (Crocodylia): *Caiman latirostris* (Brown-Snouted Caiman)	Reptilia (Aves): *Gallus domesticus* (Chicken)
–	–	**Biceps brachii** [clearly seems to directly correspond to coracoradialis of amphibians, with the exception of the very distal part of the biceps superficialis, which might integrate the humero-antebrachialis and/or coracobrachialis as in lizards, see on the right; Walker [1973] states that turtles often have a superficial head and a deep head of the biceps brachii, which usually originate from the coracoid; he states that in testudines and sea turtles, the biceps brachii is mainly undivided or only partially divided, but Wyneken [2001] states that in most sea turtles, the biceps is clearly divided into superficial and deep heads]	**Biceps brachii** (humeroantebrachialis *sensu* Diogo [2007] and Diogo and Abdala [2007]; short and long heads of biceps brachii *sensu* Jouffroy [1971]) [Holmes [1977] and Dilkes [1999] state that apart from some birds, in which there is usually an origin from the humerus and the coracoid, the biceps brachii of reptiles normally originates from the coracoid only; as described by these authors, the biceps brachii has more than one belly in nonavian reptiles such as some lepidosaurs and some turtles: see also cells on the right; we now have almost no doubt that the coracoradialis is the main contributor to the biceps brachii of amniotes, because the configuration seen in axolotls is exactly the same as that seen in the lizard *Timon*; i.e., when one removes the more superficial ventral muscles of the pectoral region (e.g., pectoralis), one sees in salamanders the procoracohumeralis and in lizards the scapulohumeralis anterior that phylogenetically derives from the procoracohumeralis; just near to/associated with these structures, one sees a fleshy belly of the coracoradialis in	**Biceps brachii** [Meers [2003] states that occasionally in some crocodylians, a poorly developed "short head" of the biceps may originate from the shoulder joint capsule]	**Biceps brachii** [see on the left; Vanden Berge [1975] stated that the biceps has a humeral head (probably derived from the humeroantebrachialis) and a coracoid head (almost surely derived from coracoradialis) in chicken]

(Continued)

TABLE 18.1 (CONTINUED)
Pectoral and Forelimb Muscles of Adults of Representative Amphibians and Reptilian Taxa

Amphibia (Caudata): *Ambystoma mexicanum* (Axolotl)	Amphibia (Anura): *Rana pipiens* (Northern Leopard Frog)	Reptilia (Testudines): *Trachemys scripta* (Red-Eared Slider Turtle)	Reptilia (Lepidosauria): *Timon lepidus* (Ocellated Lizard)	Reptilia (Crocodylia): *Caiman latirostris* (Brown-Snouted Caiman)	Reptilia (Aves): *Gallus domesticus* (Chicken)
			salamanders that then gives rise to a long tendon running through all the length of the arm between the humeroantebrachialis and the coracobrachialis to attach onto the forearm; exactly as in lizards, one sees a fleshy belly of the biceps brachii that then gives rise to a long tendon running through all the length of the arm between the brachialis and the coracobrachialis to attach onto the forearm, the only difference being that in lizards, the distal tendon of the biceps brachii is associated with fleshy fibers (see, e.g., Figure 1.26B of Russell and Bauer [2008]), as is the case in many regenerated salamanders (see Diogo and Tanaka 2010). Either those fleshy fibers are derived from the anlage of the coracoradialis or they originate from one of the surrounding muscles, i.e., the humeroantebrachialis or the coracobrachialis, and that is why we keep the name humeroantebrachialis, although this muscle is clearly homologous to at least a part of the brachialis of mammals, i.e., we keep these names because the humeroantebrachialis not only gave mainly a rise to the brachialis but may also have contributed to part of the biceps brachii of amniotes]		

(*Continued*)

TABLE 18.1 (CONTINUED)

Pectoral and Forelimb Muscles of Adults of Representative Amphibians and Reptilian Taxa

Amphibia (Caudata): *Ambystoma mexicanum* (Axolotl)	Amphibia (Anura): *Rana pipiens* (Northern Leopard Frog)	Reptilia (Testudines): *Trachemys scripta* (Red-Eared Slider Turtle)	Reptilia (Lepidosauria): *Timon lepidus* (Ocellated Lizard)	Reptilia (Crocodylia): *Caiman latirostris* (Brown-Snouted Caiman)	Reptilia (Aves): *Gallus domesticus* (Chicken)
Coracobrachialis (corresponds to the coracobrachialis longus/superficialis *sensu* Walthall and Ashley-Ross [2006], Diogo [2007], and Diogo and Abdala [2007]) [the coracobrachialis medius/proprius and coracobrachialis profundus/brevis seem to be absent in urodeles such as *Taricha*, but are present in various other urodeles according to authors such as Howell [1935], Romer [1944], and Jollie [1962]]	**Coracobrachialis** [Howell [1937b] refers to three heads of the coracobrachialis in anurans, but in anurans such as *Rana*, there are mainly two heads, i.e., a coracobrachialis longus and a coracobrachialis brevis (e.g., Duellman and Trueb 1986), as is usually the case in urodeles]	**Coracobrachialis** (coracobrachialis magnus plus coracobrachialis brevis *sensu* Walker [1973] and Wyneken [2001], which correspond, respectively, to the coracobrachialis longus plus coracobrachialis brevis *sensu* Holmes [1977] and Dilkes [1999])	**Coracobrachialis** (coracobrachialis superficialis/longus plus coracobrachialis profundus/brevis *sensu* Howell [1936b], Romer [1944], Jollie [1962], Holmes [1977], and Dilkes [1999]) [according to Jollie [1962], Holmes [1977], and Dilkes [1999], turtles, *Sphenodon*, lizards, and birds have a "coracobrachialis brevis" and a "coracobrachialis longus" (see, e.g., Fig 18.12)—in birds, these two structures are usually designated as "coracobrachialis cranialis" (or anterior) and "coracobrachialis caudalis" (or posterior), respectively]	**Coracobrachialis** (coracobrachialis brevis *sensu* Holmes [1977] and Dilkes [1999]) [according to Romer [1944], Holmes [1977], and Dilkes [1999] crocodylians have only a coracobrachialis brevis; Meers [2003] describes a "coracobrachialis brevis ventralis" and a "coracobrachialis brevis dorsalis" in crocodylians, but Lima et al. [2016] only describe a single muscle, i.e. the "coracobrachialis brevis ventralis," in *Caiman crocodilus crocodilus*]	**Coracobrachialis** (coracobrachialis longus plus coracobrachialis brevis, or coracobrachialis cranialis plus coracobrachialis caudalis, or coracobrachialis anterior plus coracobrachialis posterior, *sensu* Jollie [1972], Holmes [1977], and Dilkes [1999]) [as explained by Sullivan [1962, 1967], the avian muscles that he designates as coracobrachialis anterior and coracobrachialis correspond to the muscles that are often called deltoideus minor and coracobrachialis anterior by other authors, respectively; also according to that author, the avian muscle that he designates as "coracobrachialis posterior" has no separate homologue in other, nonavian reptiles; see on the left and text; the coracobrachialis "caudalis" of chicken *sensu* Vanden Berge [1975] is different from the brevis in the sense that it originates from the coracoid and the sternum and that it is innervated by a branch from the root of the pectoral nerve trunk, but the insertion of the muscle is similar to that of the coracobrachialis brevis of mammals, i.e., on the very proximal portion of the humerus; the coracobrachialis "cranialis" of chicken *sensu*

(*Continued*)

TABLE 18.1 (CONTINUED)
Pectoral and Forelimb Muscles of Adults of Representative Amphibians and Reptilian Taxa

Amphibia (Caudata): *Ambystoma mexicanum* (Axolotl)	Amphibia (Anura): *Rana pipiens* (Northern Leopard Frog)	Reptilia (Testudines): *Trachemys scripta* (Red-Eared Slider Turtle)	Reptilia (Lepidosauria): *Timon lepidus* (Ocellated Lizard)	Reptilia (Crocodylia): *Caiman latirostris* (Brown-Snouted Caiman)	Reptilia (Aves): *Gallus domesticus* (Chicken)
					Vanden Berge [1975] extends from the coracoid process to the deltoid tuberosity of the humerus and is innervated by the branch that innervates the biceps; i.e., one cannot exclude the hypothesis that the cranialis corresponds to the brevis of other tetrapods and that the caudalis is a new muscle, particularly because crocodylians only seem to have a coracobrachialis brevis, but our comparisons indicate that birds have the two heads that are present in, e.g., lizards and indicate that probably the brevis head just extended more medially to attach onto the sternum and perhaps gained an innervation from branches of the pectoralis muscle because of that extension]

(Continued)

TABLE 18.1 (CONTINUED)
Pectoral and Forelimb Muscles of Adults of Representative Amphibians and Reptilian Taxa

Amphibia (Caudata): *Ambystoma mexicanum* (Axolotl)	Amphibia (Anura): *Rana pipiens* (Northern Leopard Frog)	Reptilia (Testudines): *Trachemys scripta* (Red-Eared Slider Turtle)	Reptilia (Lepidosauria): *Timon lepidus* (Ocellated Lizard)	Reptilia (Crocodylia): *Caiman latirostris* (Brown-Snouted Caiman)	Reptilia (Aves): *Gallus domesticus* (Chicken)
Pronator quadratus	**Pronator quadratus** (part or totality of pronator profundus *sensu* Ribbing [1907]; abductor pollicis *sensu* Gaupp [1896] and Duellman and Trueb [1986]) [Ribbing [1907] states that the "abductor pollicis" *sensu* Gaupp [1896] corresponds to the pronator quadratus *sensu* this volume and not to the abductor pollicis brevis *sensu* this volume]	**Pronator quadratus** (probably corresponds to a part of the pronator profundus *sensu* Walker [1973], the other part corresponding to the palmaris profundus 1 *sensu* the present work, because the "pronator profundus" of *Trachemys scripta sensu* Abdala et al. [2008a] has two layers, one proximal that apparently corresponds to the pronator quadratus extending from the radius to the ulna and one distal that apparently corresponds to the palmaris profundus I because it extends from the distal 2/3 of the ulna to the extreme of the radius and the distal aspect of the carpals 1, 2, and 3; these two muscles should not be confused with the pronator accessorius of Abdala et al. [2008], which is much more proximal as usual, originating from the humerus; a further line of evidence supporting the presence of a palmaris profundus I is that Walker [1973] stated that his pronator profundus corresponds to the "pronator quadratus plus pronator profundus" of some authors, showing that the pronator profundus *sensu* Walker [1973] and Abdala et al. [2008] does correspond to	**Pronator quadratus** (pronator profundus *sensu* Moro and Abdala [2004], Abdala and Moro [2006], Diogo [2007], and Diogo and Abdala [2007])	**Pronator quadratus** (pronator profundus *sensu* Straus [1942])	**Pronator quadratus** (ulnimetacarpalis *sensu* Vanden Berge [1975]) [see text, and pronator accessorius in the following]

(*Continued*)

TABLE 18.1 (CONTINUED)
Pectoral and Forelimb Muscles of Adults of Representative Amphibians and Reptilian Taxa

Amphibia (Caudata): *Ambystoma mexicanum* (Axolotl)	Amphibia (Anura): *Rana pipiens* (Northern Leopard Frog)	Reptilia (Testudines): *Trachemys scripta* (Red-Eared Slider Turtle)	Reptilia (Lepidosauria): *Timon lepidus* (Ocellated Lizard)	Reptilia (Crocodylia): *Caiman latirostris* (Brown-Snouted Caiman)	Reptilia (Aves): *Gallus domesticus* (Chicken)
		both the pronator profundus plus palmaris profundus I *sensu* the present work) [according to Walker [1973] and Holmes [1977], the pronator quadratus is absent in some turtles]			
Palmaris profundus I	– [apparently absent; see flexor accessorius in the following]	**Palmaris profundus I**	**Palmaris profundus I** [the palmaris profundus I is clearly present in the *Timon* specimens we dissected after the publication of Diogo and Abdala's 2010 book; it clearly corresponds to the pronator profundus *sensu* Russell and Bauer 2008, who however failed to describe a pronator quadratus, which we also found in our *Timon* specimens (but Russell and Bauer did describe the pronator accessorius)]	–	**Palmaris profundus I** (pronator profundus *sensu* Vanden Berge [1975])
–	–	**Pronator accessorius** [according to Abdala et al. [2008], turtles often have a pronator accessorius]	**Pronator accessorius** [as explained by authors such as Straus [1942] and in the preceding chapters, the pronator accessorius is a peculiar reptilian muscle that very likely corresponds to part of the pronator quadratus of tetrapods such as amphibians]	– [Straus [1942] states that the only major group of living reptiles where the pronator accessorius is absent is the Crocodylia; Meers [2003] and Lima et al. [2016] corroborated that this muscle is absent in crocodylians, and this muscle is apparently absent in the *Caiman* specimens that we dissected]	– [really absent? In the chickens we dissected, the pronator accessorius did not seem to be present as a distinct muscle, and this muscle was also not described in the chickens and the other birds analyzed by authors such as Meyers [1996], Shellswell and Wolpert [1977], and Maxwell and Larsson [2007]; however, according to Straus [1942], the only major group of living reptiles where the pronator accessorius is absent is the Crocodylia: see on the left and pronator quadratus mentioned earlier]

(*Continued*)

TABLE 18.1 (CONTINUED)
Pectoral and Forelimb Muscles of Adults of Representative Amphibians and Reptilian Taxa

Amphibia (Caudata): *Ambystoma mexicanum* (Axolotl)	Amphibia (Anura): *Rana pipiens* (Northern Leopard Frog)	Reptilia (Testudines): *Trachemys scripta* (Red-Eared Slider Turtle)	Reptilia (Lepidosauria): *Timon lepidus* (Ocellated Lizard)	Reptilia (Crocodylia): *Caiman latirostris* (Brown-Snouted Caiman)	Reptilia (Aves): *Gallus domesticus* (Chicken)
Contrahentium caput longum (ulnocarpalis *sensu* McMurrich [1903a,b], Bunnell [1942], and Straus [1942]) [McMurrich [1903a] and Ribbing [1907] support the idea that the flexor accessorius lateralis and medialis of urodeles correspond to a part of the flexor digitorum longus of reptiles, and although they state that the contrahentium caput longum of urodeles may also correspond to a part of the flexor digitorum longus of reptiles, they consider that it is more likely, based on topology and innervation, that the former muscle is completely absent in reptiles; but see preceding chapters]	**Contrahentium caput longum** (ulnocarpalis plus intercarpalis *sensu* Gaupp [1896])	–	– [see on the left; according to Diogo and Abdala [2007], a few lizards may have a small muscle that somewhat resembles the contrahentium caput longum of *Ambystoma*]	–	–

(Continued)

TABLE 18.1 (CONTINUED)
Pectoral and Forelimb Muscles of Adults of Representative Amphibians and Reptilian Taxa

Amphibia (Caudata): *Ambystoma mexicanum* (Axolotl)	Amphibia (Anura): *Rana pipiens* (Northern Leopard Frog)	Reptilia (Testudines): *Trachemys scripta* (Red-Eared Slider Turtle)	Reptilia (Lepidosauria): *Timon lepidus* (Ocellated Lizard)	Reptilia (Crocodylia): *Caiman latirostris* (Brown-Snouted Caiman)	Reptilia (Aves): *Gallus domesticus* (Chicken)
Flexor accessorius lateralis (palmaris profundus III *sensu* McMurrich [1903a]; caput dorsale of flexor palmaris profundus *sensu* Straus [1942]) [the flexor accessorius medialis and lateralis very likely correspond to a part of the flexor digitorum longus of reptiles and monotremes and of the flexor digitorum profundus of therian mammals: see, e.g., Straus [1942] and Lewis [1989]; see also preceding chapters]	**Flexor accessorius** (*sensu* Ribbing [1907]; palmaris profundus *sensu* Straus [1942] and Manzano et al. [2008]) [as explained by authors such as Ribbing [1907], anurans have a single flexor accessorius, which corresponds to the flexor accessorius lateralis plus flexor accessorius medialis of urodeles; however, because the "palmaris profundus" of anurans is deep (dorsal) to the flexor digitorum communis and superficial (ventral) to at least a part of the pronator quadratus, it could correspond instead to the palmaris profundus I of urodeles: see, e.g., Figure 13.38D of Duellman and Trueb [1986]; however, the study of Ribbing [1907] and the fact that normally the palmaris profundus 1 of urodeles connects the ulna and radius, which are fused in anurans, indicate that probably the anuran muscle is a flexor accessorius and not a palmaris profundus I]	–	– [absent as an independent muscle in extant amniotes, but see on the left]	–	– [absent in chicken as shown by, e.g., Vanden Berge [1975[]
Flexor accessorius medialis (caput volare of flexor palmaris profundus *sensu* Straus 1942) [see flexor accessorius lateralis mentioned earlier]	– [see flexor accessorius mentioned earlier]	–	–	–	–

(*Continued*)

TABLE 18.1 (CONTINUED)
Pectoral and Forelimb Muscles of Adults of Representative Amphibians and Reptilian Taxa

Amphibia (Caudata): *Ambystoma mexicanum* (Axolotl)	Amphibia (Anura): *Rana pipiens* (Northern Leopard Frog)	Reptilia (Testudines): *Trachemys scripta* (Red-Eared Slider Turtle)	Reptilia (Lepidosauria): *Timon lepidus* (Ocellated Lizard)	Reptilia (Crocodylia): *Caiman latirostris* (Brown-Snouted Caiman)	Reptilia (Aves): *Gallus domesticus* (Chicken)
Flexor digitorum communis (palmaris superficialis *sensu* McMurrich [1903a]; flexor primordialis communis *sensu* Ribbing [1907])	**Flexor digitorum communis** (palmaris longus *sensu* Gaupp [1896] and Duellman and Trueb [1986]; flexor primordialis communis *sensu* Ribbing [1907]; flexor digitorum communis longus *sensu* Manzano [1996] and Manzano et al. [2008])	**Flexor digitorum longus** (flexor primordialis communis plus flexor accessorius communis *sensu* Ribbing [1907]) [according to authors such as Ribbing [1907] and Abdala et al. [2008] in turtles the flexor digitorum longus (not including the "palmaris longus": see palmaris longus in the following) is often divided into a superficial bundle and a deep bundle; normally attaches to digits 1–5 in turtles]	**Flexor digitorum longus** (palmaris communis *sensu* Holmes [1977] and Dilkes [1999]; flexor digitorum communis *sensu* Diogo [2007] and Diogo and Abdala [2007]; it probably also includes the pronator radii teres *sensu* McMurrich [1903a] and Holmes [1977], at least in taxa such as *Sphenodon* and lizards: see flexor carpi radialis) [see text; the muscle attaches to digits 1–5 in the *Timon* specimens we dissected]	**Flexor digitorum longus** (palmaris communis *sensu* Holmes [1977] and Dilkes [1999]) [according to Meers [2003], in crocodylians, the flexor digitorum longus has humeral, ulnar, and carpal heads (the humeral head clearly corresponds to the superficial head *sensu* Holmes [1977], while the two other heads seem to correspond to the deep humeral and deep ulnar heads *sensu* Holmes [1977], see text) and inserts onto the penultimate phalanx of digits 1, 2,and 3 (and not to digits 2, 3, and 4 as stated by Dilkes [1999]; Lima et al. [2016] also refer to the distal phalanges of digits 1, 2, and 3); contrary to Meers [2003], Holmes [1977], and Dilkes [1999] state that in lepidosaurs, turtles, and crocodylians, the flexor pollicis longus usually inserts onto the distal phalanges of the digits, and in the crocodylian shown in Figure 16 of Meers [2003], this muscle does seem to insert onto the distal phalanges]	**Flexor digitorum longus, to digit 2 of paleontology according to Vanden Berge (1975)** (flexor accessorius communis *sensu* Ribbing [1938]; palmaris communis *sensu* Holmes [1977] and Dilkes [1999]) [according to authors such as Sullivan [1962], Shellswell and Wolpert [1977], Meyers [1996], and Dilkes [1999], Aves usually have a superficial head and a deep head of the flexor digitorum longus, which, in neognath birds, are usually inserted onto the two phalanges of digit 2 (i.e., digit 3 according to embryology) and onto the distal phalanx of the same digit, respectively; also according to Dilkes 1999, the kiwi *Apteryx* lacks the superficial head and has a mostly tendinous deep head that inserts on the terminal phalanx of digit 2; see text]

(*Continued*)

TABLE 18.1 (CONTINUED)
Pectoral and Forelimb Muscles of Adults of Representative Amphibians and Reptilian Taxa

Amphibia (Caudata): ***Ambystoma mexicanum*** **(Axolotl)**	**Amphibia (Anura):** ***Rana pipiens*** **(Northern Leopard Frog)**	**Reptilia (Testudines):** ***Trachemys scripta*** **(Red-Eared Slider Turtle)**	**Reptilia (Lepidosauria):** ***Timon lepidus*** **(Ocellated Lizard)**	**Reptilia (Crocodylia):** ***Caiman latirostris*** **(Brown-Snouted Caiman)**	**Reptilia (Aves):** ***Gallus domesticus*** **(Chicken)**
–	– [see flexor digitorum communis]	**Palmaris longus** [as described by authors such as Walker [1973] and Abdala et al. [2008], turtles often have a broad muscle palmaris longus, which is possibly homologous to the palmaris longus of mammals: see on the left; normally attaches to digits 1–5 in turtles]	– [see text]	– [Meers 2003 does not describe a palmaris longus in crocodylians; Haines [1950] stated that crocodylians do not have a palmaris longus; see text]	–
Flexor antebrachii et carpi ulnaris (flexor carpi ulnaris plus epitrochleoanconeus *sensu* Diogo and Abdala [2010])	**Flexor carpi ulnaris**	**Flexor carpi ulnaris** (part of the flexor carpi ulnaris *sensu* Walker [1973] and Abdala et al. [2008]: see epitrochleoanconeus in the following)	**Flexor carpi ulnaris** (part of flexor antebrachii et carpi ulnaris *sensu* Diogo [2007] and Diogo and Abdala [2007]) [according to Dilkes [1999], the flexor carpi ulnaris is present in all major extant groups of reptiles]	**Flexor carpi ulnaris**	**Flexor carpi ulnaris**
–	**Epitrochleoanconeus** (epitrochleo-cubitalis *sensu* Gaupp [1896], Duellman and Trueb [1986], and Manzano et al. [2008]; flexor antebrachii ulnaris *sensu* McMurrich [1903a] and Ribbing [1907])	**Epitrochleoanconeus** (flexor antebrachii ulnaris *sensu* Ribbing [1907, 1938]; part of the flexor carpi ulnaris *sensu* Walker [1973] and Abdala et al.[2008]) [see text]	**Epitrochleoanconeus** [Holmes [1977] stated that the epitrochleoanconeus is differentiated in lepidosaurs, including *Sphenodon*, and argues that the epitrochleoanconeus was probably plesiomorphically present as a distinct muscle in reptiles]	**Epitrochleoanconeus** (really present in *Caiman*? See text)	**Epitrochleoanconeus** (flexor antebrachii ulnaris *sensu* Ribbing [1938]; entepicondylo-ulnaris *sensu* Maxwell and Larsson [2007]) [Ribbing [1938] described a flexor carpi ulnaris and a flexor antebrachii ulnaris in birds, the latter muscle being very likely homologous to the epitrochleoanconeus *sensu* this volume]

(Continued)

TABLE 18.1 (CONTINUED)
Pectoral and Forelimb Muscles of Adults of Representative Amphibians and Reptilian Taxa

Amphibia (Caudata): *Ambystoma mexicanum* (Axolotl)	Amphibia (Anura): *Rana pipiens* (Northern Leopard Frog)	Reptilia (Testudines): *Trachemys scripta* (Red-Eared Slider Turtle)	Reptilia (Lepidosauria): *Timon lepidus* (Ocellated Lizard)	Reptilia (Crocodylia): *Caiman latirostris* (Brown-Snouted Caiman)	Reptilia (Aves): *Gallus domesticus* (Chicken)
Flexor antebrachii et carpi radialis (flexor carpi radialis *sensu* McMurrich [1903]) [see text]	**Flexor carpi radialis**	**Flexor carpi radialis** [as described by Walker [1973], Holmes [1977], and Abdala et al. [2008], in turtles, the pronator teres (which Holmes designates as a "head of the flexor carpi radialis") and the flexor carpi radialis are usually present as distinct muscles]	**Flexor carpi radialis** (part of flexor carpi radialis *sensu* Holmes [1977] and Dikes [1999]) [see text]	**Flexor carpi radialis** (part of flexor carpi radialis *sensu* Holmes [1977] and Dilkes [1999] and of pronator teres *sensu* Meers [2003]) [see text]	**Flexor carpi radialis** (part of flexor carpi radialis *sensu* Holmes [1977] and Dilkes [1999]; seems to correspond to the pronator superficialis *sensu* Sullivan [1962], Shellswell and Wolpert [1977], Meyers [1996], and Dilkes [1999]) [see text]
–	**Pronator teres** (flexor antebrachii radialis *sensu* Ribbing [1907], who stated that it corresponds to the flexor antebrachii medialis *sensu* Gaupp [1896])	**Pronator teres** (pronator radii teres *sensu* Holmes [1977])	**Pronator teres** (flexor antebrachii radialis *sensu* Ribbing [1907]; part of flexor carpi radialis *sensu* Holmes [1977] and Dikes [1999])	**Pronator teres** (part of flexor carpi radialis *sensu* Holmes]1977] and Dilkes [1999] and of pronator teres *sensu* Meers [2003])	**Pronator teres** (part of flexor carpi radialis *sensu* Holmes [1977] and Dilkes [1999]; corresponds to the pronator superficialis *sensu* Sullivan [1962], Vanden Berge [1975], Shellswell and Wolpert [1977], Meyers [1996], and Dilkes [1999])

(Continued)

TABLE 18.1 (CONTINUED)
Pectoral and Forelimb Muscles of Adults of Representative Amphibians and Reptilian Taxa

Amphibia (Caudata): *Ambystoma mexicanum* (Axolotl)	Amphibia (Anura): *Rana pipiens* (Northern Leopard Frog)	Reptilia (Testudines): *Trachemys scripta* (Red-Eared Slider Turtle)	Reptilia (Lepidosauria): *Timon lepidus* (Ocellated Lizard)	Reptilia (Crocodylia): *Caiman latirostris* (Brown-Snouted Caiman)	Reptilia (Aves): *Gallus domesticus* (Chicken)
Flexores breves superficiales (flexores digitorum breves superficiales *sensu* McMurrich [1903a,b]) [four muscles attaching to digits 1–4, respectively, in the *A. mexicanum* specimens dissected by us]	**Flexores breves superficiales** (part of flexores digitorum breves superficiales *sensu* McMurrich [1903a,b]; flexores breves sublimes *sensu* Holmes [1977]; flexores digiti brevis superficialis *sensu* Abdala and Moro [2006]) [they are often subdivided into bundles as shown by Ecker 1864; four muscles attaching to digits 2–5, respectively, in the *Rana* specimens observed by us and as described by, e.g., Ecker 1864]	**Flexores breves superficiales** (flexor brevis superficialis *sensu* Walker [1973]; flexores digiti brevis superficiales *sensu* Abdala et al. [2008]) [according to Walker [1973], in turtles, including *Trachemys*, the flexores breves superficiales often include eight slips, there being two slips to each of the three middle digits and one to the first and fifth digits; however, as explained by Abdala et al. [2008], turtles such as *Trachemys* only have five flexores breves superficiales *sensu* this volume, one to each digit; according to Walker [1973], the insertion of the flexores breves superficiales is variable in turtles; i.e., it may be onto the proximal phalanges (as is the case in *Trachemys*; this is corroborated by Abdala et al. [2008]), onto the sheaths of the long flexor tendon, or onto the "penultimate" phalanges]	**Flexores breves superficiales** (part of flexores digitorum breves superficiales *sensu* McMurrich [1903a,b]; flexores breves sublimes *sensu* Holmes [1977]; flexores digiti brevis superficialis *sensu* Abdala and Moro [2006]; flexores digitores breves *sensu* Russell and Bauer [2008]) [as described by authors such as McMurrich [1903a,b], lepidosaurs such as lizards often have five flexores breves superficiales, inserting onto digits 1–5; four muscles attaching to digits 1–5, respectively, in the *Timon* specimens observed by us]	**Flexores breves superficiales** (part of flexores breves sublimes *sensu* Holmes [1977]; flexores digitorum breves superficiales and probably flexor digiti quinti and/or transversus palmaris *sensu* Meers [2003] and Lima et al. [2016]) [see text; five muscles attaching to digits 1–5, respectively, in the *Caiman* specimens dissected by us]	**Flexores breves superficiales** [present in chicken, corresponding to the "flexor digitorum superficialis" *sensu* Vanden Berge [1975], which is somewhat similar to the flexor digitorum superficialis of humans because it originates from the medial epicondyle of the humerus, and mirrors what happens in the leg, where the flexores breves superficiales also migrate proximally; only one muscle present in chicken, going to digit 3 (*sensu* paleontologists; i.e., digit 4 *sensu* embryology)]

(Continued)

TABLE 18.1 (CONTINUED)
Pectoral and Forelimb Muscles of Adults of Representative Amphibians and Reptilian Taxa

Amphibia (Caudata): *Ambystoma mexicanum* (Axolotl)	Amphibia (Anura): *Rana pipiens* (Northern Leopard Frog)	Reptilia (Testudines): *Trachemys scripta* (Red-Eared Slider Turtle)	Reptilia (Lepidosauria): *Timon lepidus* (Ocellated Lizard)	Reptilia (Crocodylia): *Caiman latirostris* (Brown-Snouted Caiman)	Reptilia (Aves): *Gallus domesticus* (Chicken)
– [really absent in *Ambystoma ordinarium*? In the *Ambystoma ordinarium* specimens we dissected, the lumbricales seem to be absent as distinct, separate muscles, and these muscles were not described in other urodeles such as *Taricha* by authors such Walthall and Ashley-Ross [2006]; however, McMurrich [1903a] shows an *Ambystoma tigrinum* specimen where the lumbricales are present as distinct muscles, and these muscles are present in other amphibians, such as anurans: see on the right and text]	**Lumbricales** (mainly correspond to flexor digiti longus sublimi 2, 3, 4, and 5 *sensu* Ecker [1864]) [our dissections of anurans such as *Phyllomedusa bicolor* show that these anurans have both "lumbricales breves" and "lumbricales longi": see also, e.g., Gaupp [1896] and Manzano [1996]; however, it is possible that part of the "lumbricales" correspond to bundles of the flexores breves superficiales *sensu* the present work; four muscles attaching to digits 2–5 in the *Rana* specimens we dissected]	**Lumbricales** (lumbricalis *sensu* Abdala et al. [2008]) [according to Walker [1973], in *Trachemys*, there are two lumbricales to the penultimate phalanx of each of the three middle digits and one to the penultimate phalanx of the first and fifth digits; however, as explained by Abdala et al. [2008], turtles such as *Trachemys* have only five lumbricales *sensu* this volume, each attaching to the "ungual" phalanx of each digit]	**Lumbricales** [according to authors, such as McMurrich [1903a,b], lizards often have five lumbricales inserting onto digits 1–5, but some lizards have fewer lumbricales; note that Russell and Bauer [2008] designate the lumbricales *sensu* this volume as "the palmar head of the flexor digitorum longus," and the flexores breves profundi and contrahentes *sensu* this volume as the lumbricales; three muscles attaching to digits 2, 3, and 4, respectively in the *Timon* specimens dissected by us; see notes in the following cell]	**Lumbricales** [according to Meers 2003, crocodylians have 5 lumbricales, the first and second attaching to digit 2, the third and fourth to digit 3, and the fifth to digit 5]	– [absent in *Gallus domesticus* according to Vanden Berge [1975]]

(Continued)

TABLE 18.1 (CONTINUED)
Pectoral and Forelimb Muscles of Adults of Representative Amphibians and Reptilian Taxa

Amphibia (Caudata): *Ambystoma mexicanum* (Axolotl)	Amphibia (Anura): *Rana pipiens* (Northern Leopard Frog)	Reptilia (Testudines): *Trachemys scripta* (Red-Eared Slider Turtle)	Reptilia (Lepidosauria): *Timon lepidus* (Ocellated Lizard)	Reptilia (Crocodylia): *Caiman latirostris* (Brown-Snouted Caiman)	Reptilia (Aves): *Gallus domesticus* (Chicken)
Contrahentes digitorum (part of flexores digitorum breves medii *sensu* McMurrich [1903a,b]) [as described by authors such as McMurrich [1903a,b], urodeles such as *Ambystoma* often have four contrahentes, each inserting onto each of the four digits]	**Contrahentes digitorum** (part of flexores digitorum breves medii *sensu* McMurrich [1903a,b]; probably includes the adductor pollicis *sensu* Manzano et al. [2008]) [according to Ribbing [1907], anurans such as *Discoglossus* have four contrahentes digitorum *sensu* this volume, which probably include the “flexor teres indicis,” the “caput volare des [des = of the] m. flexor teres digiti V” and the “adductor proprius digiti V” *sensu* Gaupp 1896; apparently four muscles attaching to digits 2–5, respectively, in the *Rana* specimens we dissected]	**Contrahentes digitorum** (includes adductor digiti minimi *sensu* Walker [1973]) [Walker [1973] describes a single contrahens (“adductor digiti minimi”) in turtles such as *Trachemys*, which attaches to digit 5; according to him, some turtles have contrahentes to digits 4 and 5, and some other turtles lack contrahentes; Abdala et al. [2008] stated that turtles such as *Trachemys* have “contrahentes” to the proximal phalanx of each digit, but they stated that these contrahentes are the deepest ventral (palmar) muscles of the hand, so these contrahentes probably do not correspond to the contrahentes *sensu* this volume, which are deep (dorsal) to the flexores breves superficiales, but superficial (ventral) to the flexores breves profundi and to the intermetacarpales; these contrahentes cannot correspond to dorsometacarpales because Abdala et al. [2008] describe these muscles; following the current evidence, it is here assumed that they are layers of the intermetacarpales and/or flexores breves profundi?	**Contrahentes digitorum** (part of flexores digitorum breves medii *sensu* McMurrich [1903a,b] and Lewis [1989]; include the flexor digitorum V transversus I and II of, e.g., Abdala and Moro 2006; part of lumbricales *sensu* Russell and Bauer [2008]) [as explained by Lewis [1989], the “flexores digitorum intermedii” *sensu* authors such as Holmes [1977] and Meers [2003] or “flexores digitorum breves medii” *sensu* authors such as McMurrich [1903a,b] clearly seem to correspond to the contrahentes digitorum *sensu* this volume, because, as indicated by the names used by these authors, these muscles are ventral to the flexores breves superficiales and dorsal to the flexores breves profundi; this idea is also supported by authors such as Howell [1936a,b], who explicitly designates the “flexores digitorum breves intermedii/medii” of reptiles as contrahentes; fice muscles attaching to digits 1–5, respectively, in the *Timon* specimens we dissected]	**Contrahentes digitorum1** (part of flexores digitorum breves medii *sensu* authors such as McMurrich [1903a,b]; flexores digitorum intermedii *sensu* Holmes [1977] and Meers [2003]) [according to Meers [2003], crocodylians usually have a “flexor digitorum intermedius digiti IV et V” (that is, a contrahens *sensu* this volume, see on the left), which commonly inserts onto the distal end of the proximal phalanx of digit 4 and, sometimes, also onto the distal metacarpal of digit 5; Meers [2003] describes an additional muscle in *Alligator mississippiensis*, the “flexor digitorum intermedius digiti V,” which was absent in all the other crocodylian species examined by him and which, according to him, possibly derives from the flexores breves profundi and not from the contrahentes layer]	**Contrahentes digitorum** [Holmes [1977] seems to suggest that the contrahentes digitorum are present in all major extant groups of reptiles; the adductor indicis *sensu* Sullivan [1962] and Shellswell and Wolpert [1977] (which attaches to digit 1—i.e., digit 2 according to embryology—and corresponds to the “adductor alulae” *sensu* Meyers [1996]) is possibly a part of the contrahentes digitorum *sensu* this volume; according to Vanden Berge [1975], there is only a muscle going to digit 1, because in the present work, we include the “abductor digiti majoris” *sensu* Vanden Berge [1975] in the flexores breves profundi: see following cell]

(Continued)

TABLE 18.1 (CONTINUED)
Pectoral and Forelimb Muscles of Adults of Representative Amphibians and Reptilian Taxa

Amphibia (Caudata): *Ambystoma mexicanum* (Axolotl)	Amphibia (Anura): *Rana pipiens* (Northern Leopard Frog)	Reptilia (Testudines): *Trachemys scripta* (Red-Eared Slider Turtle)	Reptilia (Lepidosauria): *Timon lepidus* (Ocellated Lizard)	Reptilia (Crocodylia): *Caiman latirostris* (Brown-Snouted Caiman)	Reptilia (Aves): *Gallus domesticus* (Chicken)
Flexores breves profundi (flexores digitorum breves profundi *sensu* McMurrich [1903a,b]) [as described by authors such as McMurrich [1903a,b], urodeles such as *Ambystoma* usually have eight flexores breves profundi *sensu* this volume, inserting onto the ulnar and radial sides of each of the four digits (the muscles that insert onto the ulnar and radial side of each digit are often considered as "heads" of a single, "bicipital" muscle, so authors such as McMurrich often refer to four bicipital muscles, which thus correspond to the eight flexores breves profundi *sensu* this volume: see preceding chapters]	**Flexores breves profundi** [according to Ribbing [1907], there are eight flexores breves profundi *sensu* this volume (often described as "four bicipital muscles": see on the left) in anurans such as *Rana*, which include the "opponens indicis," "flexor ossis metacarpi III," "flexor ossis metacarpi IV," and "opponens digiti V," and possibly the "abductor secundus digiti V" *sensu* Gaupp [1896]; we confirmed the presence of these eight muscles in *Rana*]	**Flexores breves profundi** [the flexores breves profundi *sensu* the present work possibly correspond to a part or the totality of the "interossei volares" *sensu* Walker [1973] and/or of the "flexores digiti brevis profundus" *sensu* Abdala et al. [2008a]); however, Walker [1973] states that the "interossei volaris" insert onto the proximal phalanges, and it suggests they are not differentiated in *Trachemys*, although he admits that little is known about these muscles in turtles, while Abdala et al. [2008a] state that the "flexores digiti brevis profundi" insert onto the metacarpals 1–5 in this genus]	**Flexores breves profundi** (flexores digitorum breves profundi *sensu* McMurrich [1903a,b]; flexores digiti brevis profundus of, e.g., Abdala and Moro, [2006]; include lumbricales and/or interossei ventrales *sensu* Russell and Bauer [2008]) [as described by authors such as McMurrich [1903a,b, lizards usually have 10 flexores breves profundi *sensu* this volume (often described as "5 bicipital muscles": see on the left)]	**Flexores breves profundi** (flexores digitorum profundus, and possibly flexor digitorum intermedius digiti V, *sensu* Meers [2003]: see contrahentes digitorum) [Meers [2003] described 5 "flexores breves profundi" (or 6, if the muscle that he called "flexor digitorum intermedius digiti V" is also part of the deep flexor layer: see contrahentes digitorum mentioned earlier), so crocodylians clearly seem to have the full series of deep flexors; i.e., they have 10 flexores breves profundi *sensu* this volume, each digit receiving two of these muscles (i.e., each of the 5 "muscles" described by Meers [2003] corresponds to 2 of the flexores breves profundi *sensu* this volume: see on the left)]	**Flexores breves profundi** (flexores digitorum breves profundi *sensu* Ribbing [1938] and Holmes [1977]) [according to authors such as Ribbing [1938] and Holmes [1977], birds do have flexores breves profundi; the flexor indicis and/or flexor digiti quarti *sensu* Sullivan [1962] and Shellswell and Wolpert]1977] and thus the flexor alulae and flexor digiti minoris *sensu* Vanden Berge [1975], which go, respectively, to digits 1 and 3 and correspond to a part of the flexores breves profundi *sensu* this volume; note that we also include the abductor digiti majoris *sensu* Vanden Berge [1975] in the flexores breves profundi: see abductor pollicis brevis in the following]

(Continued)

TABLE 18.1 (CONTINUED)
Pectoral and Forelimb Muscles of Adults of Representative Amphibians and Reptilian Taxa

Amphibia (Caudata): *Ambystoma mexicanum* (Axolotl)	Amphibia (Anura): *Rana pipiens* (Northern Leopard Frog)	Reptilia (Testudines): *Trachemys scripta* (Red-Eared Slider Turtle)	Reptilia (Lepidosauria): *Timon lepidus* (Ocellated Lizard)	Reptilia (Crocodylia): *Caiman latirostris* (Brown-Snouted Caiman)	Reptilia (Aves): *Gallus domesticus* (Chicken)
Flexores digitorum minimi [two muscles in axolotls]	**Flexores digitorum minimi** [the anuran flexores digitorum minimi (*sensu*, e.g., Ribbing 1907) are flexors of the digits and are often, but not always, superficial (ventral) to the intermetacarpales according to Burton [1998] and correspond to the "flexor teres digitorum III, IV, and V" *sensu* Gaupp [1896] and to the "flexores teretes I, II, III and IV" *sensu* Burton [1998]]	–	–	–	–
Interphalangeus digiti III [one muscles to digit 3 in axolotls]	**Interphalangeus digiti IV** (interphalangeus digiti IV *sensu* Gaupp 1896) [we confirmed the presence of this muscle in *Rana*, mainly from the metacarpophalangeal joint to the proximal phalangeal joint of digit 4 of the 2–5 nomenclature]	–	–	–	–
–	**Interphalangeus digiti V** (interphalangeus digiti V *sensu* Gaupp [1896]) [we confirmed the presence of this muscle in *Rana*, mainly from the metacarpophalangeal joint to the proximal phalangeal joint of digit 5 of the 2–5 nomenclature]	–	–	–	–

(*Continued*)

TABLE 18.1 (CONTINUED)
Pectoral and Forelimb Muscles of Adults of Representative Amphibians and Reptilian Taxa

Amphibia (Caudata): *Ambystoma mexicanum* (Axolotl)	**Amphibia (Anura): *Rana pipiens* (Northern Leopard Frog)**	**Reptilia (Testudines): *Trachemys scripta* (Red-Eared Slider Turtle)**	**Reptilia (Lepidosauria): *Timon lepidus* (Ocellated Lizard)**	**Reptilia (Crocodylia): *Caiman latirostris* (Brown-Snouted Caiman)**	**Reptilia (Aves): *Gallus domesticus* (Chicken)**
– absent in urodeles	– [as explained earlier, the abductor pollicis *sensu* Gaupp [1896] and Manzano et al. [2008] probably corresponds to the pronator quadratus) *sensu* the present work (see pronator quadratus); our dissections of *Rana* and other anurans, after Diogo and Abdala's [2010] book, strongly indicate that anurans such as *Rana* do not have a distinct abductor pollicis brevis]	**Abductor pollicis brevis**	**Abductor pollicis brevis** (abductor brevis pollicis *sensu* Abdala and Moro [2006])	**Abductor pollicis brevis** (abductor metacarpi I *sensu* Meers [2003])	**Abductor pollicis brevis** [the "abductor alulae" *sensu* Vanden Berge [1975] and Meyers [1996] corresponds to the "abductor indicis" *sensu* Sullivan [1962] and Shellswell and Wolpert [1977] and goes to digit 1 (i.e., digit 2 according to embryology), thus corresponding to the abductor pollicis brevis *sensu* this volume; the abductor digiti majoris described by Vanden Berge [1975] seems to be either a contrahens to digit 2 or, more likely, a flexor brevis profundi to digit 2; that is, it does not seem to be part of the abductor pollicis brevis because it is deep to the tendons of the flexor digitorum longus as described by Vanden Berge [1975]]
Abductor digiti minimi (extensor lateralis digiti IV *sensu* Walthall and Ashley-Ross 2006 and preceding chapters) [see text]	**Abductor digiti minimi** (abductor primus digiti V *sensu* Gaupp [1896])	**Abductor digiti minimi, to digit 5** (abductor digitorum V *sensu* Abdala et al. [2008])	**Abductor digiti minimi** (abductor digitorum V *sensu* Abdala and Moro [2006])	**Abductor digiti minimi** (abductor metacarpi V *sensu* Meers [2003])	– [see text and extensores breves in the following]

(*Continued*)

TABLE 18.1 (CONTINUED)
Pectoral and Forelimb Muscles of Adults of Representative Amphibians and Reptilian Taxa

Amphibia (Caudata): *Ambystoma mexicanum* (Axolotl)	Amphibia (Anura): *Rana pipiens* (Northern Leopard Frog)	Reptilia (Testudines): *Trachemys scripta* (Red-Eared Slider Turtle)	Reptilia (Lepidosauria): *Timon lepidus* (Ocellated Lizard)	Reptilia (Crocodylia): *Caiman latirostris* (Brown-Snouted Caiman)	Reptilia (Aves): *Gallus domesticus* (Chicken)
Intermetacarpales [as described by authors such as McMurrich [1903a,b], urodeles such as *Ambystoma* usually have three intermetacarpales connecting the metacarpals of the four digits]	**Intermetacarpales** (transversi metacarpi *sensu* Gaupp [1896]; transversi metacarporum *sensu* Burton [1998]; transversus metacarpi superficialis plus transversus metacarpi profundus *sensu* Manzano [1996]) [as described by authors such as Ribbing [1907] and Burton [1998], anurans usually have three intermetacarpales connecting the metacarpals of the four digits]	**Intermetacarpales** [Abdala et al. [2008a] state that turtles such as *Trachemys* have four intermetacarpales, which connect metacarpals I and II, II and III, III and IV, and of digits IV and V; Walker [1973] did not describe intermetacarpales in turtles; see also on the left]	**Intermetacarpales** (intermetacarpales I and II *sensu* Abdala and Moro [2006], Diogo [2007], and Diogo and Abdala [2007]) [as described by authors such as Abdala and Moro [2006], lizards often have four "intermetacarpales I" heads connecting the metacarpals of the five digits and four "intermetacarpales 2" heads, also connecting the metacarpals of these five digits]	**Intermetacarpales** (part or totality of interossei dorsale *sensu* Meers [2003[) [Meers [2003] described eight "interossei," the "ventralis" and "dorsalis" names being simply related to their function of adduction and abduction, so it seems that there are only four intermetacarpales, which are however maybe divided into different bundles so we will consider four intermetacarpales as a whole here]	**Intermetacarpales** [the intermetacarpales *sensu* this volume probably correspond to a part of the interossei dorsales and/or "interossei ventrales" *sensu* authors such as Sullivan [1962], Shellswell and Wolpert [1977], and Meyers [1996], although they might also/instead include the abductor medius *sensu* Sullivan [1962] and Shellswell and Wolpert [1977], which corresponds to the abductor digiti majoris *sensu* Meyers [1996], but that muscle seems to be instead a part of the contrahens and/or flexor breves profundi layer because it is just deep to the flexor digitorum longus and lies on the ventral surface of the metacarpal region, see abductor pollicis brevis; the intermetacarpales clearly seem to correspond to the interosseous ventralis *sensu* Vanden Berge [1975], which is only one muscle attaching to digit 2; see also lumbricales mentioned earlier]

(Continued)

TABLE 18.1 (CONTINUED)
Pectoral and Forelimb Muscles of Adults of Representative Amphibians and Reptilian Taxa

Amphibia (Caudata): *Ambystoma mexicanum* (Axolotl)	Amphibia (Anura): *Rana pipiens* (Northern Leopard Frog)	Reptilia (Testudines): *Trachemys scripta* (Red-Eared Slider Turtle)	Reptilia (Lepidosauria): *Timon lepidus* (Ocellated Lizard)	Reptilia (Crocodylia): *Caiman latirostris* (Brown-Snouted Caiman)	Reptilia (Aves): *Gallus domesticus* (Chicken)
Extensor carpi radialis [Haines [1939] describes three heads of the extensor antebrachii et carpi radialis in urodeles such as *Salamandra*: superficialis, intermedius, and profundus; these three heads possibly correspond to the three heads that are often designated by the same names in reptiles such as lizards, but see text]	**Extensor carpi radialis** (mainly corresponds to "extensor radialis superficialis" *sensu* Haines [1939] and to "extensor carpi radialis, caput inferius" *sensu* Gaupp [1896] and Duellman and Trueb [1975]) [according to Haines [1939], the extensor antebrachii et carpi radialis *sensu* this volume is divided into five divisions in anurans such as *Rana*, which he designates as "extensor radialis profundus" ("flexor antebrachii lateralis profundus" *sensu* Gaupp [1896]), "extensor radialis intermedius," or "brachioradialis" ("flexor antebrachii lateralis superficialis, caput inferius" *sensu* Gaupp [1896]), extensor radialis superficialis ("extensor carpi radialis, caput inferius" *sensu* Gaupp [1896]), and two "small accessory slips" ("extensor carpi radialis caput superius" and "flexor antebrachii lateralis superficialis caput superius" *sensu* Gaupp [1896]); according to Haines [1939], the extensor radialis superficialis, extensor radialis intermedius, and extensor radialis profundus mainly correspond to the extensor	**Extensor carpi radialis longus** (extensor radialis superficialis *sensu* Haines [1939] and Walker [1973]) [the extensor carpi radialis longus and the extensor carpi radialis brevis *sensu* the present volume correspond to the muscles that are designated "extensor carpi radialis superficialis" and "extensor carpi radialis intermedius" *sensu* authors such as Walker [1973], which are clearly part of the extensor carpi radialis of taxa such as anurans, and are present in addition to the brachioradialis (tractor radii *sensu* Walker [1973]) and supinator extensor radialis profundus *sensu* Walker [1973]); that is, the structure that is often designated as extensor radialis intermedius in turtles does not correspond to the structure designated by the same name in, e.g., anurans, which corresponds instead to the brachioradialis *sensu* the present work; the extensor carpi radialis longus and extensor carpi radialis brevis of turtles are thus very similar to the extensor carpi radialis longus and extensor carpi radialis brevis of mammals such as humans,	**Extensor antebrachii et carpi radialis** (including extensor radialis longus + extensor radialis brevis *sensu* Straus [1941a,b]; extensor carpi radialis *sensu* Abdala and Moro [2006]) [see on the left, and text and brachioradialis in the following]	**Extensor carpi radialis longus** (extensor radialis superficialis, extensor radialis intermedius and extensor radialis profundus *sensu* Haines [1939]) [includes the extensor carpi radialis longus, the "extensor carpi radialis brevis, and the abductor radialis *sensu* Meers [2003], although part of the extensor carpi radialis brevis *sensu* Meers [2003] (his pars ulnaris) might correspond, to the abductor pollicis longus *sensu* the present work: see abductor pollicis longus in the following; the extensor carpi radialis longus, extensor carpi radialis brevis (or at least its pars radialis: see preceding text), and the abductor radialis *sensu* Meers [2003] might well correspond to the "pars superficialis, pars intermedia and pars profunda of the extensor carpi radialis" *sensu* Holmes [1977] and, thus, to the extensor carpi radialis longus, extensor carpi radialis brevis, and supinator *sensu* the present work, because the former two crocodylian structures insert onto hand bones, as usually do the mammalian extensor carpi radialis longus and extensor carpi radialis brevis,	**Extensor carpi radialis** ("extensor metacarpi radialis" *sensu* Meyers [1996]) [the extensor metacarpi radialis is often divided into two heads in birds, which probably correspond either to the extensor carpi radialis and brachioradialis or to the extensor carpi radialis longus and extensor carpi radialis brevis of crocodylians]

(Continued)

TABLE 18.1 (CONTINUED)
Pectoral and Forelimb Muscles of Adults of Representative Amphibians and Reptilian Taxa

Amphibia (Caudata): *Ambystoma mexicanum* (Axolotl)	Amphibia (Anura): *Rana pipiens* (Northern Leopard Frog)	Reptilia (Testudines): *Trachemys scripta* (Red-Eared Slider Turtle)	Reptilia (Lepidosauria): *Timon lepidus* (Ocellated Lizard)	Reptilia (Crocodylia): *Caiman latirostris* (Brown-Snouted Caiman)	Reptilia (Aves): *Gallus domesticus* (Chicken)
	carpi radialis, brachioradialis, and supinator *sensu* the present work, and because these three structures are clearly present as distinct muscles in anurans, we consider them as different muscles in the present work; see also on the right and on the left]	and because these are present as distinct muscles in marsupials and placentals, in crocodylians and in turtles, and possibly as distinct heads of the extensor carpi radialis of birds (see birds), it is possible that the LCA of reptiles and mammals already had these two structures differentiated as muscles; hence, we use similar muscle names for reptiles and for mammals]		while the latter head does not reach the hand bones, inserting distally exclusively onto the forearm bones, as usually does the mammalian supinator; this hypothesis is supported by the fact that some authors have designated the "extensor carpi radialis profundus" *sensu* Holmes [1977] as supinator or supinator brevis: see, e.g., Walker 1973; see also brachioradialis in the following]	
–	–	**Extensor carpi radialis brevis** (extensor radialis intermedius *sensu* Haines [1939] and Walker [1973]) [see extensor carpi radialis longus]	–	**Extensor carpi radialis brevis** (extensor radialis intermedius *sensu* Walker [1973] and Haines [1939]) [see extensor carpi radialis longus]	– [not present as a distinct muscle, but might be distinct bundle of the extensor carpi radialis]
– [not present as an independent muscle but instead as a bundle of the extensor carpi radialis in *Ambystoma*; see text]	**Brachioradialis** (mainly corresponds to flexor antebrachii lateralis superficialis *sensu* Gaupp [1896] and Duellman and Trueb [1986] and to extensor radialis intermedius or brachioradialis *sensu* Haines [1939]) [see extensor carpi radialis mentioned earlier]	**Brachioradialis** (tractor radii *sensu* Haines [1939], Walker [1973], Wyneken [2001], and Abdala et al. [2008]) [see extensor carpi radialis longus]	– [our dissections indicate that *Timon* does not have a distinct, separate brachioradialis muscle such as that found in mammals, but Holmes [1977] suggests that the supinator longus (which clearly seems to correspond to the mammalian brachioradialis: see on the right) is usually present in extant reptiles and was probably already differentiated in the LCA of all extant reptiles; Haines [1939] states that the supinator longus/tractor radii is present as a distinct muscle in *Sphenodon*; see text and Figure 18.12]	**Brachioradialis** (seems to correspond to the supinator *sensu* Meers [2003] and Lima et al. [2016]) [see extensor carpi radialis longus]	– [none of the muscles described in chickens and other birds by authors such as Sullivan [1962], Shellswell and Wolpert [1977], Vanden Berge [1975], and Meyers [1996] seem to correspond to the brachioradialis *sensu* this volume, unless the humeroradialis/"tensor propatagii" of birds does correspond to the brachioradialis *sensu* this volume: see text]

(*Continued*)

TABLE 18.1 (CONTINUED)
Pectoral and Forelimb Muscles of Adults of Representative Amphibians and Reptilian Taxa

Amphibia (Caudata): *Ambystoma mexicanum* (Axolotl)	Amphibia (Anura): *Rana pipiens* (Northern Leopard Frog)	Reptilia (Testudines): *Trachemys scripta* (Red-Eared Slider Turtle)	Reptilia (Lepidosauria): *Timon lepidus* (Ocellated Lizard)	Reptilia (Crocodylia): *Caiman latirostris* (Brown-Snouted Caiman)	Reptilia (Aves): *Gallus domesticus* (Chicken)
Supinator	**Supinator** (mainly corresponds to the extensor radialis profundus *sensu* Haines [1939] and to the flexor antebrachii lateralis profundus *sensu* Gaupp [1896] and Duellman and Trueb [1986]) [see extensor carpi radialis mentioned earlier]	**Supinator** (extensor radialis profundus *sensu* Walker [1973] and Haines [1939]) [see extensor carpi radialis longus]	– [the supinator is not present as a distinct muscle in the *Timon* specimens we dissected, which supports Russell and Howell [2008] who stated that in *Lacerta* and most lizards the extensor antebrachii et carpi radialis clearly have three heads that seem to correspond to the supinator, brachioradialis (or supinator longus), and extensor carpi radialis of mammals; but they also noted that these heads are deeply fused to each other forming s single, complex muscle]	**Supinator** (extensor radialis profundus *sensu* Walker [1973] and Haines [1939]) [see extensor carpi radialis longus]	**Supinator** [present in Aves according to Meywes [1996] and Remes [2008]; its presence was confirmed in chicken by Vanden Berger [1975]]
Extensor antebrachii et carpi ulnaris (extensor ulnaris *sensu* Straus [1941a,b]) [see text]	**Extensor carpi ulnaris** [Ribbing [1907] states that, unlike urodeles, in anurans the "extensor carpi ulnaris" and "extensor antebrachii carpi ulnaris" (anconeus *sensu* this volume) are often present as distinct muscles; see on the left]	**Extensor antebrachii et carpi ulnaris** (extensor ulnaris *sensu* Haines [1939]; extensor carpi ulnaris *sensu* Walker [1973] and Abdala et al. [2008]) [as described by Ribbing [1907], Walker [1973], and Abdala et al. [2008], in *Trachemys*, the extensor antebrachii et carpi ulnaris is mainly undivided, but in some turtles such as *Testudo*, *Pelomedusa*, *Chelodina* and *Emys*, it is divided into an extensor carpi ulnaris connecting the humerus and ulna and an extensor carpi ulnaris accessorius connecting the ulna and carpus, which probably correspond to the anconeus and extensor carpi radialis *sensu* Holmes [1977], respectively; see on the left]	**Extensor antebrachii et carpi ulnaris** (extensor carpi ulnaris plus extensor antebrachii ulnaris *sensu* Haines [1939]; extensor ulnaris *sensu* Straus [1941a,b]; extensor carpi ulnaris + anconeus quartus *sensu* Holmes [1977]) [see on the left and on the right]	**Extensor antebrachii et carpi ulnaris** (flexor ulnaris *sensu* Meers [2003] and Lima et al. [2016]; extensor carpi ulnaris + anconeus quartus *sensu* Holmes [1977]) [the extensor antebrachii et carpi ulnaris *sensu* this volume clearly corresponds to the "flexor ulnaris" *sensu* Meers [2003]; as described by authors such as Holmes [1977] and Dilkes [1999] and corroborated by our dissections, in crocodylians this muscle clearly seems to extend the antebrachium, and not to flex it, as proposed by Meers [2003]]	**Extensor carpi ulnaris** (apparently corresponds to the extensor metacarpi ulnaris *sensu* Shellswell and Wolpert [1977] and Meyers [1996]) [It is questionable if the extensor carpi ulnaris and the anconeus of birds are really homologous with the extensor carpi ulnaris and the anconeus of mammals. See text]

(Continued)

TABLE 18.1 (CONTINUED)
Pectoral and Forelimb Muscles of Adults of Representative Amphibians and Reptilian Taxa

Amphibia (Caudata): *Ambystoma mexicanum* (Axolotl)	Amphibia (Anura): *Rana pipiens* (Northern Leopard Frog)	Reptilia (Testudines): *Trachemys scripta* (Red-Eared Slider Turtle)	Reptilia (Lepidosauria): *Timon lepidus* (Ocellated Lizard)	Reptilia (Crocodylia): *Caiman latirostris* (Brown-Snouted Caiman)	Reptilia (Aves): *Gallus domesticus* (Chicken)
– [see extensor antebrachii et carpi ulnaris]	**Anconeus** (epicondylo-cubitalis *sensu* Gaupp [1896]; extensor antebrachii ulnaris *sensu* Ribbing [1907]) [see extensor carpi ulnaris]	– [see extensor antebrachii et carpi ulnaris]	– [see extensor antebrachii et carpi ulnaris]	– [see extensor antebrachii et carpi ulnaris]	**Anconeus** (corresponds to the ectepicondylo-ulnaris *sensu* Vanden Berge [1975], Meyers [1996], and Maxwell and Larsson [2007]; anconeus quartus *sensu* Holmes [1977]) [see extensor carpi ulnaris]
Extensor digitorum (extensor digitorum communis *sensu* Ribbing [1907] and Walthall and Ashley-Ross [2006], Diogo [2007], and Diogo and Abdala [2007]; humerodorsalis *sensu* Haines [1939] [the muscle attaches to digits 1–4 in the axolotls we dissected]	**Extensor digitorum** (extensor digitorum communis longus *sensu* Gaupp [1896] and Manzano et al. [2008]; extensor digitorum communis *sensu* Ribbing [1907]; humerodorsalis *sensu* Haines [1939]) [the muscle attaches to digits 2–5 in the *Rana* specimens we dissected]	**Extensor digitorum** (humerodorsalis *sensu* Haines [1939]; extensor digitorum communis *sensu* Walker [1973]; extensor digitorum longus *sensu* Abdala et al. [2008]) [as described by Walker [1973] and Abdala et al. [2008], in turtles such as *Trachemys*, the extensor digitorum has eight tendons attaching onto the ulnar and radial sides of the distal end of each metacarpal, except for digit 1 and 5, which lack tendons to their radial and ulnar sides, respectively]	**Extensor digitorum** (humerodorsalis *sensu* Haines [1939]; extensor digitorum longus *sensu* Abdala and Moro [2006]; extensor digitorum communis *sensu* Diogo [2007] and Diogo and Abdala [2007]) [as explained by Howell [1936a,b], Haines [1939], Straus [1941a,b], Holmes [1977], and Dilkes [1999], lepidosaurs, turtles, crocodylians, and birds have an extensor digitorum, which usually inserts onto the metacarpals but, in some taxa such as birds, extends distally to insert onto the phalanges]	**Extensor digitorum** (extensor digitorum longus *sensu* Holmes [1977] and Dilkes [1999]; extensor carpi ulnaris longus *sensu* Meers [2003]) [as described by Holmes [1977] and Dilkes [1999], crocodylians have an extensorum digitorum ("longus"), which originates on the distal portion of the humerus and usually inserts variably onto the metacarpals II, III, and/or IV, and clearly seems to correspond to the extensor carpi ulnaris longus *sensu* Meers [2003]]	**Extensor digitorum** (extensor digitorum longus *sensu* Ribbing [1938], Holmes [1977], and Dilkes [1999]; extensor longus communis *sensu* Vanden Berge [1975] and Dilkes [1999]) [probably includes the "extensor digitorum communis" *sensu* Sullivan [1962], Shellswell and Wolpert [1977], and Meyers [1996], which attaches to digits 1 and 2 (i.e., 2 and 3 according to embryology), but usually does not extend distally to the proximal phalanges of these digits; in chickens, the "extensor metacarpi longus digiti majoris" *sensu* Meyers [1996] ("extensor medius

(Continued)

TABLE 18.1 (CONTINUED)
Pectoral and Forelimb Muscles of Adults of Representative Amphibians and Reptilian Taxa

Amphibia (Caudata): *Ambystoma mexicanum* (Axolotl)	Amphibia (Anura): *Rana pipiens* (Northern Leopard Frog)	Reptilia (Testudines): *Trachemys scripta* (Red-Eared Slider Turtle)	Reptilia (Lepidosauria): *Timon lepidus* (Ocellated Lizard)	Reptilia (Crocodylia): *Caiman latirostris* (Brown-Snouted Caiman)	Reptilia (Aves): *Gallus domesticus* (Chicken)
					longus" *sensu* Sullivan [1962] and Shellswell and Wolpert [1977]) often goes from the proximal portion of the radius and/or ulna to the distal phalanx of digit 2 (i.e., digit 3 according to embryology); according to Sullivan [1962] and Shellswell and Wolpert [1977], their "extensor indicis longus" corresponds to part of the long extensors of the hand, i.e., of the extensor digitorum *sensu* this volume, because these authors also describe an "extensor medius brevis" connecting the metacarpal region to digit 2 (i.e., digit 3 according to embryology), which would thus correspond to part of the extensores digitorum breves *sensu* this volume; this description is confirmed by Vanden Berge [1975]]

(Continued)

TABLE 18.1 (CONTINUED)
Pectoral and Forelimb Muscles of Adults of Representative Amphibians and Reptilian Taxa

Amphibia (Caudata): *Ambystoma mexicanum* (Axolotl)	Amphibia (Anura): *Rana pipiens* (Northern Leopard Frog)	Reptilia (Testudines): *Trachemys scripta* (Red-Eared Slider Turtle)	Reptilia (Lepidosauria): *Timon lepidus* (Ocellated Lizard)	Reptilia (Crocodylia): *Caiman latirostris* (Brown-Snouted Caiman)	Reptilia (Aves): *Gallus domesticus* (Chicken)
Extensores digitorum breves [three muscles to digits 2–4, because the one to digit 1 is included in the abductor et extensor pollicis I]	**Extensores digitorum breves** [probably correspond to the extensores breves superficiales, and, possibly, to the extensores breves medii, *sensu* Gaupp [1896], the "profundi" apparently corresponding to the dorsometacarpales; according to authors such as Haines [1939], anurans such as *Rana* usually have two muscle structures inserting onto each of the four digits]	**Extensores digitorum breves** (extensores digiti brevis *sensu* Abdala et al. [2008]) [according to Walker [1973] and Abdala et al. [2008], turtles such as *Trachemys* have five extensores digitorum breves, each going to each digit; Walker [1973] stated that the insertion of these muscles is onto the "penultimate phalanges" of the digits, while Abdala et al. [2008] stated it is onto the "first phalanx" of the digits]	**Extensores digitorum breves** (extensor digitorum brevis communis *sensu* Holmes [1977]; extensores digiti brevis *sensu* Abdala and Moro [2006]; extensores digitores breves superficiales *sensu* Russell and Bauer [2008]) [according to Holmes [1977], in *Sphenodon* and lizards, the extensores digitorum breves insert onto the distal phalanges of the digits; Haines [1939] argues that the plesiomorphic condition for reptiles, found, for instance, in turtles and in lepidosaurs such as *Sphenodon* and numerous lizards, is to have five extensores digitorum breves to digits 1-5; see text]	**Extensores digitorum breves** (extensor digitorum brevis communis *sensu* Holmes [1977]) [usually five muscles to digits 1–5; see text]	**Extensores digitorum breves** (extensores breves digitorum superficiales *sensu* Ribbing [1938]; extensor digitorum brevis communis *sensu* Holmes [1977]; includes unimetacarpalis dorsalis to digit 3, extensor brevis digiti majoris to digit 2 and extensor brevis alulae to digit 1 *sensu* Vanden Berge [1975] and may include or not extensor longus digiti majoris *sensu* this author as explained in the preceding cell) [in chickens, the extensores digitorum breves include the "extensor indicis brevis" *sensu* Sullivan [1962] and Shellswell and Wolpert [1977] (which corresponds to the "extensor brevis alulae" *sensu* Meyers [1996] and attaches to digit 1, i.e., to digit 2 according to embryology) and the extensor medius brevis *sensu* Sullivan [1962] and Shellswell and Wolpert [1977] (which attaches to digit 2, i.e., to digit 3 according to embryology); as explained earlier, the "ulnimetacarpalis dorsalis" *sensu* Sullivan [1962], Meyers [1996], and Shellswell and Wolpert [1977] probably corresponds to a reduced short extensor ("extensor digiti brevis") of digit 3 (i.e., of digit 4 according to embryology);

(*Continued*)

TABLE 18.1 (CONTINUED)
Pectoral and Forelimb Muscles of Adults of Representative Amphibians and Reptilian Taxa

Amphibia (Caudata): *Ambystoma mexicanum* (Axolotl)	Amphibia (Anura): *Rana pipiens* (Northern Leopard Frog)	Reptilia (Testudines): *Trachemys scripta* (Red-Eared Slider Turtle)	Reptilia (Lepidosauria): *Timon lepidus* (Ocellated Lizard)	Reptilia (Crocodylia): *Caiman latirostris* (Brown-Snouted Caiman)	Reptilia (Aves): *Gallus domesticus* (Chicken)
					it cannot correspond to a reduced abductor digiti minimi *sensu* this volume because it is innervated by the radial nerve as stated by Vanden Berger [1975]: see, e.g., Figure 1 of Shellswell and Wolpert [1977] and abductor digiti minimi mentioned earlier]
– [Straus [1941a,b] states that urodeles such as *Salamandra* and *Cryptobranchus* do have dorsometacarpales and shows in his Figure 1 a *Necturus* specimen with dorsometacarpales and explains that they correspond to the "extensores digitorum breves" *sensu* other authors; Haines [1939] states that anurans such as *Rana* and urodeles such as *Salamandra* clearly have extensores digitorum breves and dorsometacarpales, but see dissection notes in the following; our dissections of axolotls indicate that there are deep and superficial layers of the extensores digitorum breves; however as described by authors such as Francis [1934] and Grim and Carlson [1974a,b], these bundles originate from different bones but are then fused distally, and each whole muscle formed by the two bundles sends a single	**Dorsometacarpales** [correspond to the extensores breves profundi *sensu* Gaupp [1896]; according to Haines [1939], the dorsometacarpales are highly developed in anurans such as *Rana*: see text; our dissections of *Rana* clearly show that the dorsometacarpales are present as a separate group because the dorsometacarpales of reptiles also correspond to the extensores breves profundi of Ribbing [1938], so in *Rana* there are four muscles to digits 2–5 of 2–5 nomenclature]	**Dorsometacarpales** (extensores digitorum breves profundi *sensu* Ribbing [1938]; dorsal interossei *sensu* Walker [1973], which might also include the intermetacarpales *sensu* the present work) [according to Abdala et al. [2008], turtles such as *Trachemys* have five dorsometacarpales, each covering the dorsal surface of each of the five digits and sending a tendon that attaches from the second phalanx to the ungual phalanx of each digit]	**Dorsometacarpales** (extensores digitores breves profundi *sensu* Russell and Bauer [2008] [according to Holmes [1977], the dorsometacarpales are usually found in the major extant reptilian groups] [our dissections indicate that lizards usually have 5 dorsometacarpales inserting onto digits 1–5]	**Dorsometacarpales** (extensores digitorum breves profundi *sensu* Meers [2003]) [usually five muscles inserting onto digits 1–5; see extensores digitorum breves mentioned earlier]	**Dorsometacarpales** (extensores breves digitorum profundi *sensu* Ribbing [1938]) [in birds, the dorsometacarpales *sensu* this volume correspond very likely to part, or the totality, of the interossei dorsalis *sensu* authors such as Sullivan [1962], Shellswell and Wolpert [1977], and Meyers [1996]; see on the left and intermetacarpales mentioned earlier; chicken apparently have one muscle going to digit 2 (*sensu* paleontologists), clearly corresponding to the interosseous dorsalis *sensu* Vanden Berge [1975]]

(Continued)

TABLE 18.1 (CONTINUED)
Pectoral and Forelimb Muscles of Adults of Representative Amphibians and Reptilian Taxa

Amphibia (Caudata): *Ambystoma mexicanum* (Axolotl)	Amphibia (Anura): *Rana pipiens* (Northern Leopard Frog)	Reptilia (Testudines): *Trachemys scripta* (Red-Eared Slider Turtle)	Reptilia (Lepidosauria): *Timon lepidus* (Ocellated Lizard)	Reptilia (Crocodylia): *Caiman latirostris* (Brown-Snouted Caiman)	Reptilia (Aves): *Gallus domesticus* (Chicken)
tendon to the respective digit, so there are no distinct, separate muscles dorsometacarpales as is the case in, e.g., lizards, as also stated by Howell [1936b]; our more recent dissections and comparisons also indicate that, as stated by Howell [1936b], the dorsometacarpales are clearly dorsal/extensor muscles innervated by the radial nerve and not ventral/plantar muscles, i.e., intrinsic muscles of the hand innervated by the ulnar/median nerves; this is the opinion of most authors, e.g., McMurrich [1903, 1904], Jouffroy [1971], and Lewis [1989], who stated that some of the "extensor manus" muscles sometimes present as anomalies in some primates correspond to vestigial dorsometacarpals; the dorsometacarpales of, e.g., reptiles clearly correspond to the extensores digitorum breves of urodeles such as axolotls, the difference being that in these urodeles, these form simply a deep layer of the extensores breves digitorum that is fused and shares tendons with the superficial layer of these muscles, unlike what happens in reptiles such as lizards, where the bundles are completely separated to form the extensores breves digitorum (the superficial layer) and the dorsometacarpales *sensu* the present work]					

(*Continued*)

TABLE 18.1 (CONTINUED)
Pectoral and Forelimb Muscles of Adults of Representative Amphibians and Reptilian Taxa

Amphibia (Caudata): *Ambystoma mexicanum* (Axolotl)	Amphibia (Anura): *Rana pipiens* (Northern Leopard Frog)	Reptilia (Testudines): *Trachemys scripta* (Red-Eared Slider Turtle)	Reptilia (Lepidosauria): *Timon lepidus* (Ocellated Lizard)	Reptilia (Crocodylia): *Caiman latirostris* (Brown-Snouted Caiman)	Reptilia (Aves): *Gallus domesticus* (Chicken)
Abductor et extensor digiti I (supinator manus *sensu* Brooks [1889], Ribbing [1907], Howell [1936a,b], Haines [1939], and Straus [1941a,b]) [see extensores digitorum breves mentioned earlier]	**Abductor pollicis longus** (abductor indicis longus *sensu* Gaupp [1896] and Manzano et al. [2008]; supinator manus *sensu* Haines [1939]) [we confirmed that the abductor pollicis longus (which attaches to metacarpal II) is clearly separated from the short extensor to the most radial digit; see text]	**Abductor pollicis longus** (supinator manus *sensu* Haines [1939] and Walker [1973]) [according to Haines [1939], Walker [1973], and Abdala et al. [2008], turtles usually have an abductor pollicis longus extending from the ulna to the metacarpal I]	**Abductor pollicis longus** (supinator manus *sensu* Haines [1939] and Holmes [1977]; abductor longus pollicis *sensu* Abdala and Moro [2006]; abductor et extensor digiti I *sensu* Diogo [2007] and Diogo and Abdala [2007]) [according to Holmes [1977], all major groups of living reptiles have an abductor pollicis longus, which usually originates from the distal end of the ulna and often inserts onto the radial side of the carpus and the metacarpal I (in crocodylians, turtles, and *Sphenodon*) and to the distal end of the radius (in lizards); as described by Russell and Bauer [2008] in *Iguana*, there is a short extensor to digit 1 in the *Timon* specimens that we dissected, so the abductor pollicis longus of lizards does not seem to include the short extensor to digit 1, as is also the case in mammals]	**Abductor pollicis longus** (supinator manus *sensu* Holmes [1977]) [see on the left, and also extensores digitorum breves mentioned earlier]	**Abductor pollicis longus** (abductor digit 1 *sensu* Ribbing [1938]; supinator manus *sensu* Holmes [1977]) [this muscle attaches to digit 1 (i.e., digit 2 according to embryology) and probably corresponds to the extensor indicis longus *sensu* Sullivan [1962] and Shellswell and Wolpert [1977] and thus to the extensor longus alulae *sensu* Vanden Berge [1975] and Meyers [1996], which, as noted by the latter author, has often been designated "extensor pollicis longus," "extensor longus pollicis," or "extensor ossis metacarpi pollicis" by other authors]

Note: The nomenclature of the muscles shown in bold follows that of the text; in order to facilitate comparisons, in some cases ,names often used by other authors to designate a certain muscle/bundle are given in front of that muscle/bundle, between round brackets; additional comments are given between square brackets. Data compiled from evidence provided by our own dissections and comparisons and by an overview of the literature (see text, Tables 18.2 and 18.3, and Figures 18.1 through 18.14).

TABLE 18.2
Diagram Illustrating the Authors' Hypotheses Regarding the Homologies of the Pectoral and Arm Muscles of Adults of Representative Amphibian and Reptilian Taxa

		Ambystoma 13 muscles	*Rana*, 17 muscles	*Trachemys*, 15 muscles	*Timon*, 17 muscles	*Caiman*, 18 muscles	*Gallus*, 18 muscles
Axial: pectoral girdle		Serratus anterior	Serratus anterior	Serratus anterior	Serratus anterior	Serratus anterior	Serratus anterior
		---	Rhomboideus	---	---	Rhomboideus	Rhomboideus
		---	Rhomboideus occipitalis	---	---	---	---
		Levator scapulae	Levator scapulae sup.	Levator scapulae	Levator scapulae	Levator scapulae	---
		---	Opercularis	---	---	---	---
		---	Colummelaris	---	---	---	---
		---	Levator claviculae	Levator claviculae	Levator claviculae	---	---
		---	---	---	Sternocoracoideus	---	Sternocoracoideus
		---	---	---	Costocoracoideus	Costocoracoideus	---
Appendicular: pectoral girdle and arm	From ventral mass	Pectoralis	Pectoralis	Pectoralis	Pectoralis	Pectoralis	Pectoralis
		Supracoracoideus	Supracoracoideus	Supracoracoideus	Supracoracoideus	Supracoracoideus	Supracoracoideus
		Humeroantebrachialis	---	Brachialis	Brachialis	Brachialis	Brachialis
		Coracoradialis	Coracoradialis	---	---	---	---
		---	---	Biceps brachii	Biceps brachii	Biceps brachii	Biceps brachii
		Coracobrachialis	Coracobrachialis	Coracobrachialis	Coracobrachialis	Coracobrachialis	Coracobrachialis
		---	---	---	---	Scapulo-humeralis posterior	Scapulo-humeralis posterior
	From dorsal mass	Deltoideus scapularis	Deltoideus scapularis	Deltoideus scapularis	Deltoideus scapularis	Deltoideus scapularis	Deltoideus scapularis
		Procoracohumeralis	Procoracohumeralis	Deltoideus clavicularis	Deltoideus clavicularis	Deltoideus clavicularis	Deltoideus clavicularis
		---	---	---	---	Humeroradialis	Humeroradialis
		---	---	---	Scapulo-humeralis anterior	---	Scapulo-humeralis anterior
		Subcoracoscapularis	Subcoracoscapularis	Subcoracoscapularis	Subcoracoscapularis	Subcoracoscapularis	Subcoracoscapularis
		---	---	Teres major	---	Teres major	---
		Latissimus dorsi	Latissimus dorsi	Latissimus dorsi	Latissimus dorsi	Latissimus dorsi	Latissimus dorsi
		Triceps brachii	Triceps brachii	Triceps brachii	Triceps brachii	Triceps brachii	Triceps brachii
		---	---	---	---	---	Expansor secundariorum
		Coracobrachialis	Coracobrachialis	Coracobrachialis	Coracobrachialis	Coracobrachialis	Coracobrachialis

Note: The nomenclature of the muscles follows that used in the text. Data compiled from evidence provided by our own dissections and comparisons and from a review of the literature. The black arrows indicate the hypotheses that are most strongly supported by the evidence available; the gray arrows indicate alternative hypotheses that are supported by some data, but that in overall are not as strongly supported by the evidence as are the hypotheses indicated by the black arrows (see text, Tables 18.1 and 18.3, and Figures 18.1 through 18.13). sup, superioris.g.

TABLE 18.3

Diagram Illustrating the Authors' Hypotheses Regarding the Homologies of the Forearm and Hand Muscles of Adults of Representative Amphibian and Reptilian Taxa

	Ambystoma, 8 ventral forearm, 8 dorsal forearm and 23 hand muscles	*Rana*, 8 ventral forearm, 15 dorsal forearm and 29 hand muscles	*Trachemys*, 9 ventral forearm muscles, 17 dorsal forearm muscles and 22 hand muscles	*Timon*, 8 ventral forearm, 14 dorsal forearm and 29 hand	*Caiman*, 6 ventral forearm muscles, 17 dorsal forearm muscles and 25 hand muscles	*Gallus*, 7 ventral forearm muscles, 10 dorsal forearm muscles and 7 hand muscles
Ventral forearm	Palmaris profundus I	---	Palmaris profundus I	Palmaris profundus I	---	Palmaris profundus I
	Pronator quadratus	Pronator quadratus	Pronator quadratus	Pronator quadratus	Pronator quadratus	Pronator quadratus
	---	---	Pronator accessorius	Pronator accessorius	---	---
	Contrahentium caput longum	Contrahentium caput longum	---	---	---	---
	Flexor accessorius lateralis	Flexor accessorius	---	---	---	---
	Flexor accessorius medialis	---	---	---	---	---
	Flexor digitorum communis	Flexor digitorum communis	Flexor digitorum longus (d.1-5)	Flexor digitorum longus	Flexor digitorum longus	Flexor digitorum longus (d.2)
	---	---	'Palmaris longus' (d.1 -5)	---	---	---
	Flexor an. et carpi ulnaris	Flexor carpi ulnaris	Flexor carpi ulnaris	Flexor carpi ulnaris	Flexor carpi ulnaris	Flexor carpi ulnaris
	---	Epitrochleoanconeus	Epitrochleoanconeus	Epitrochleoanconeus	Epitrochleoanconeus	Epitrochleoanconeus
	Flexor an. et carpi radialis	Flexor carpi radialis	Flexor carpi radialis	Flexor carpi radialis	Flexor carpi radialis	Flexor carpi radialis
	---	Pronator teres	Pronator teres	Pronator teres	Pronator teres	Pronator teres
Hand	Flexores breves superficiales (4m)	Flexores breves superficiales (4m)	Flexores breves superficiales (5m, to d.1-5)	Flexores breves superficiales (5m)	Flexores breves superficiales (5m)	Flexores breves superficiales (1m)
	---	Lumbricales (4m)	Lumbricales (5m)	Lumbricales (3m)	Lumbricales (3m)	---
	Contrahentes digitorum (4m)	Contrahentes digitorum (4m)	Contrahentes digitorum (1m)	Contrahentes digitorum (5m)	Contrahentes digitorum (1m)	Contrahentes digitorum (1m)
	Flexores breves profundi (8m)	Flexores breves profundi (8m)	Flexores breves profundi (5m)	Flexores breves profundi (10m)	Flexores breves profundi (10m)	Flexores breves profundi (3m)
	Flexores digitorum minimi (2m)	Flexores digitorum minimi (3m)	---	---	---	---
	Interphalangeus digiti III (1m)	Interphalangeus digiti IV (1m)	---	---	---	---
	---	Interphalangeus digiti V (1m)	---	---	---	---
	---	---	Abductor pollicis brevis	Abductor pollicis brevis	Abductor pollicis brevis	Abductor pollicis brevis
	Abductor digiti minimi	Abductor digiti minimi	Abductor digiti minimi	Abductor digiti minimi	Abductor digiti minimi	---
	Intermetacarpales (3m)	Intermetacarpales (3m)	Intermetacarpales (4m)	Intermetacarpales (4m)	Intermetacarpales (4m)	Intermetacarpales (1m)
Dorsal forearm	Extensor carpi radialis	Extensor carpi radialis	Extensor carpi radialis longus	Extensor an. et carpi radialis	Extensor carpi radialis longus	Extensor carpi radialis
	---	Brachioradialis	Brachioradialis	---	Brachioradialis	---
	---	---	Extensor carpi radialis brevis	---	Extensor carpi radialis brevis	---
	Supinator ('extensor ant radialis')	Supinator	Supinator	---	Supinator	Supinator
	Extensor an. et carpi ulnaris	Extensor carpi ulnaris	Extensor an. et carpi ulnaris	Extensor an. et carpi ulnaris	Extensor an. et carpi ulnaris	Extensor carpi ulnaris
	---	Anconeus	---	---	---	Anconeus
	Extensor digitorum	Extensor digitorum	Extensor digitorum	Extensor digitorum	Extensor digitorum	Extensor digitorum
	---	Dorsometacarpales (4m)	Dorsometacarpales (5m)	Dorsometacarpales (5m)	Dorsometacarpales (5m)	Dorsometacarpales (1m)
	Extensores digitorum breves (3m)	Extensores digitorum breves (4m)	Extensores digitorum breves (5m)	Extensores digitorum breves (5m)	Extensores digitorum breves (5m)	Extensores digitorum breves (3m)
	Abductor et extensor digiti 1	Abductor pollicis longus	Abductor pollicis longus	Abductor pollicis longus	Abductor pollicis longus	Abductor pollicis longus

Note: See caption of Tables 18.1 and 18.2 and Figures 18.1 through 18.14. an., antebrachia; d, digit; m, muscle(s).

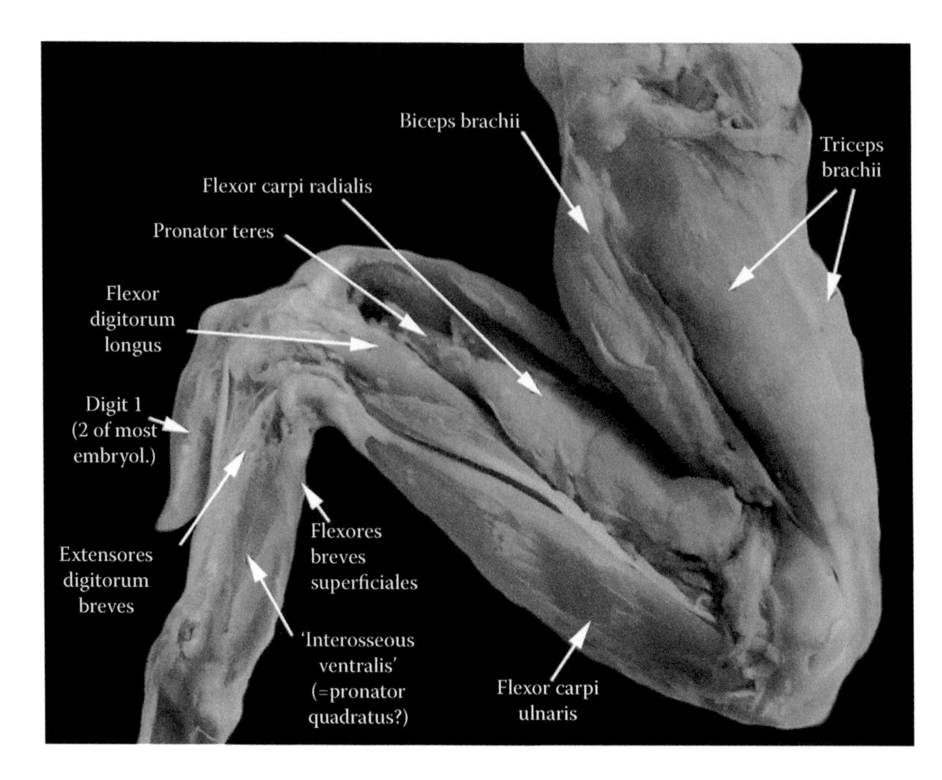

FIGURE 18.1 *Gallus domesticus* (Reptilia, Aves): ventral view of the superficial musculature of the wing. embryol., embryologists.

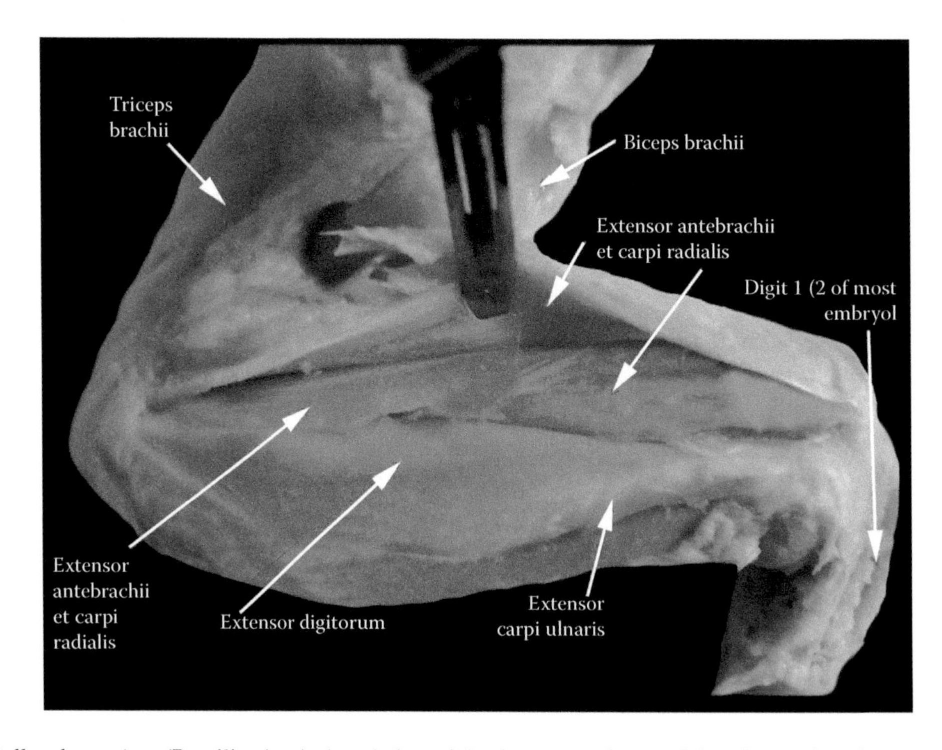

FIGURE 18.2 *Gallus domesticus* (Reptilia, Aves): dorsal view of the deep musculature of the wing. embryol., embryologists.

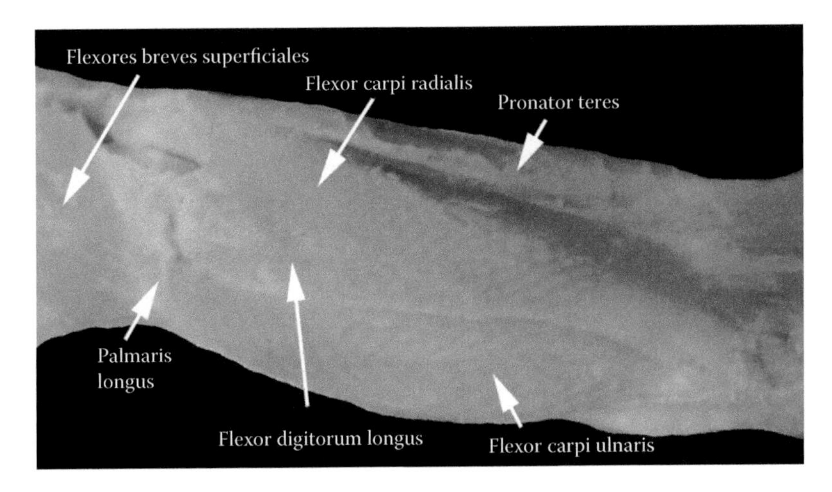

FIGURE 18.3 *Tupinambis meriane* (Reptilia, Lepidosauria): ventral view of the superficial musculature of the forelimb, showing the continuous layer between the flexores digitorum breves and the palmaris longus.

to the rhomboideus of mammals. Our dissections and comparisons do also corroborate that the overall configuration and the proximal and distal attachments of the rhomboideus of anurans, crocodylians, and birds are similar to those of the rhomboideus of mammals and, thus, that these structures are probably homologous (see Tables 18.1 through 18.3). Further studies, ideally including a detailed analysis of the innervation and development of the rhomboideus of numerous amphibians and reptiles, could determine whether or not the rhomboideus of these taxa is homologous to the mammalian rhomboideus (see also Chapter 17 and the preceding text).

The levator scapulae is not present as a distinct muscle in birds (Tables 18.1 through 18.3). As noted by Hetherington and Tugaoen (1990), in urodeles such as *Ambystoma*, the muscle that is often named "opercularis" clearly corresponds to part of the levator scapulae *sensu* this volume, which, in anurans such as *Rana*, is completely differentiated into multiple distinct muscles, including the levator scapulae superioris and the opercularis *sensu* this volume (Tables 18.1 through 18.3). Therefore, the name opercularis should only be used for anurans (according to Carroll [2007], it is possible that the last common ancestor (LCA) of all caecilians had a levator scapulae extending from the margin of an "operculum"-like structure to the suprascapula, but this muscle is absent in extant caecilians). Piatt (1938), based on his developmental study of *Ambystoma*, suggested that the levator scapulae of this taxon derives from somites 2–4, together with the hypobranchial muscles. The ontogenetic work of Piekaski and Olsson (2007) indicated, in turn, that in *Ambystoma*, the levator scapulae derives mainly from somite 3, being innervated by the first spinal nerve and by the nerve hypoglossus (CNXII), what is somewhat unexpected because the latter nerve is usually associated with the hypobranchial muscles (see preceding chapters). However, Piekaski and Olsson (2007) did show that the development and innervation of the levator scapulae are different from the innervation and development of the branchial muscle protractor pectoralis ("cucullaris"), thus contradicting that the levator scapulae of urodeles derives from the protractor pectoralis, as was often suggested in the older literature (for more details about this subject, see Piekaski and Olsson 2007). Regarding the sternocoracoideus and costocoracoideus, the former muscle is present in various lepidosaurs and birds, while the latter is found in various lepidosaurs and crocodylians, often being subdivided into a "pars superficialis" and a "pars profunda" (Tables 18.1 through 18.3). The evolution and homologies of these muscles and of the mammalian subclavius are discussed in Chapter 17.

APPENDICULAR MUSCLES OF THE PECTORAL GIRDLE AND ARM

The pectoralis muscle of amphibians and reptiles is an intrinsic muscle of the forelimb (i.e., an appendicular muscle: see, e.g., Romer 1944, Russell and Bauer 2008, and Chapter 17) and is differentiated in mammals, into the pectoralis major, the pectoralis minor, and into at least a part of the panniculus carnosus (Tables 18.1 and 18.2). The pectoralis muscle in amphibians and reptiles is usually divided into superficial and deep heads (e.g., Russell and Bauer 2008). Our dissections show three heads of the pectoralis in anurans such as Bufo. Manzano (1996) also described three heads of this muscle in pseudid frogs, which she designated "epicoracoideus," "esternalis," and "abdominalis." Interestingly, in Ambystoma, as well as in other urodeles such as Taricha (Walthall and Ashley-Ross 2006), the pectoralis is mainly undivided. According to authors such as Romer (1944) and Kardong (2002), in reptiles, the pectoralis is plesiomorphically undivided, as is often the case in lepidosaurs. In crocodylians, the pectoralis is however usually subdivided into two or three heads: "cranial" and "caudal" or "cranial," "caudal," and "deep," sensu Meers (2003). In birds, the pectoralis is often divided into a "pectoralis superficialis" and a "pectoralis profundus" (e.g., Dilkes 1999), although authors such as Hudson et al. (1972) refer to a "pars thoracica," a "pars propatagialis," and a "pars abdominalis." The avian pectoralis profundus does not seem to correspond to the entopectoralis of mammals. It seems to correspond instead to part of the mammalian "ectopectoralis," i.e., of the pectoralis major sensu this volume, which is also often divided, in the mammalian literature, into "profundus" (abdominal head sensu this volume) and "superficialis" (sternocostal and/or clavicular head sensu this volume). The three divisions of the mammalian entopectoralis, i.e., the pectoralis abdominalis, pectoralis minor, and "pectoralis tertius" sensu this volume, thus seem to be absent as distinct structures in birds and in most, if not all, nonmammalian tetrapods (Tables 18.1 through 18.3; see also Chapter 17). It is now accepted that the mammalian supraspinatus and infraspinatus, which usually connect the dorsal region of the pectoral girdle to the proximal region of the arm, derive from the supracoracoideus (Tables 18.1 through 18.3), a muscle that lies ventral rather than dorsal to the pectoral girdle in most nonmammalian tetrapods (e.g., Kardong 2002; Chapter 17). In a few nonmammalian taxa, such as chameleons, the supracoracoideus does also occupy a somewhat more dorsal position than in most reptiles (Figure 18.14), thus leading some authors to propose that these reptiles do have an "infraspinatus" and a "supraspinatus" (see, e.g., Jouffroy 1971). However, this idea is not accepted by authors such as Romer (1922, 1924, 1944) and Jouffroy (1971), who argue that the somewhat more dorsal position of the supracoracoideus of chameleons was acquired inside the reptilian clade, while the change from a ventral supracoracoideus to a dorsal infraspinatus and a dorsal supraspinatus was acquired during the evolutionary transitions that led to extant mammals. This idea has been supported by our recent anatomical studies of chameleons (Molnar et al. 2017a).

The deltoideus scapularis is consistently present in amphibians and reptiles. In turtles, the deltoid musculature has been described as one of the most variable of the shoulder muscles (Walker 1973). The scapulo-humeralis posterior is present as a distinct muscle only in birds, crocodylians, and possibly in a few lepidosaurs such as Agama (Tables 18.1 through 18.3; see also Chapter 17). As explained by Romer (1944), the "longus" head of the amphibian procoracohumeralis corresponds to the deltoideus clavicularis plus humeroradialis of reptiles such as Sphenodon, birds, and crocodylians (see Figure 18.13), and the "brevis" head of the amphibian procoracohumeralis corresponds to the scapulo-humeralis anterior of reptiles such as lepidosaurs and birds (Tables 18.1 through 18.3; see also Chapter 17). According to Jouffroy (1971), the reptilian scapulo-humeralis posterior might be homologous to the mammalian teres minor, because according to him both these muscles derive from the deltoideus scapularis (see Table 18.2). However, authors such as Holmes [1977] argued that the scapulo-humeralis anterior and scapulo-humeralis posterior were acquired during the evolution of reptiles, i.e., that these muscles were not differentiated in the LCA of extant reptiles and, thus, that the mammalian teres minor cannot be directly homologous to the scapulo-humeralis posterior of some reptilian taxa. This subject was discussed in detail in Chapter 17.

The subcoracoscapularis is consistently present in amphibians and reptiles. This muscle was not described in *Taricha torosa* by Walthall and Ashley-Ross (2006), but our recent dissections showed that it is present in urodeles such as *Ambystoma ordinarium* (Tables 18.1 through 18.3); authors such as Romer (1944) and Kardong (2002) had stated that the subcoracoscapularis is found in many urodeles. As explained in Chapter 17, the mammalian teres major is probably derived from the subcoracoscapularis. According to Dilkes (1999), there is a "teres major" in turtles, crocodiles, and many "lizards," but not in lepidosaurs such as *Sphenodon* and *Iguana* and in birds. Jouffroy (1971) and Meers (2003) confirm that crocodylians have a teres major. Romer (1944) also states that there is a teres major in crocodylians and that this muscle is absent in *Sphenodon* and Aves, but contrary to Dilkes (1999), he argues that the teres major is also absent in the whole Lacertilia. In our dissections, we were unable to find a distinct teres major in lizards such as *Timon*. Walker (1973) and Wyneken (2001) state that turtles often have a teres major, although this structure is often blended with the latissimus dorsi. Howell (1937b) defends that only mammals have a "true" teres major, thus suggesting that the teres major of reptiles such as crocodylians and turtles is not homologous to the mammalian teres major. In view of our dissections, comparisons, and review of the literature, we however see no reasons to completely discard, a priori, the hypothesis that the teres major of reptiles such as crocodylians and turtles might be homologous to the teres major of mammals. If future studies do reveal that a teres major is present in at least some lepidosaurs, as stated by Dilkes (1999), it would be phylogenetically more parsimonious to assume that the LCA of amniotes had a teres major and that this muscle was secondarily lost in the node leading to birds (two steps), than to assume that a teres major was independently acquired in lepidosaurs, turtles, and crocodylians and that, in addition, a remarkably similar muscle, the mammalian teres major, was independently acquired in mammals (four steps).

The latissimus dorsi is a dorsal muscle of the pectoral girdle, and the triceps brachii is mainly an extensor of the forearm; both these muscles are consistently present in urodeles, anurans, turtles, lepidosaurs, crocodylians, and birds (Tables 18.1 through 18.3). As explained in Chapter 17, the mammalian dorsoepitrochlearis was very likely derived from the coracoid head of the triceps brachii of nonmammalian tetrapods. The details about the specific subdivisions of the triceps brachii in urodeles, anurans, turtles, lepidosaurs, crocodylians, and birds are given in Table 18.1. The humeroantebrachialis of urodeles such as *Ambystoma* very likely corresponds to the brachialis and might have given rise to a part (e.g., possibly the long head) of the biceps brachii of amniotes; in many anuran amphibians, as well as in urodeles (Diogo and Tanaka 2012), there is also a coracoradialis, which probably corresponds to a part (possibly the short head) of the biceps brachii of amniotes, although it is possible that this short head derives instead/also from the coracobrachialis (see Diogo and Tanaka 2012 and Diogo et al. 2014a,b and Chapter 20). Cheng's (1955) developmental study of *Didelphis* strongly suggests that at least in mammals the biceps is developmentally more closely related to the coracobrachialis than it is to the brachialis (see Tables 18.1 through 18.3). Romer (1944) suggested that the "coracobrachialis externus" plus "coracobrachialis internus" of birds correspond to the coracobrachialis brevis of crocodylians and, thus, that the coracobrachialis longus is absent in birds. However, as explained by authors such as Jollie (1962), Holmes (1977), and Dilkes (1999) and corroborated by our dissections, birds do seem to have both a coracobrachialis brevis and a coracobrachialis longus *sensu* this volume (see Tables 18.1 through 18.3).

APPENDICULAR MUSCLES OF THE FOREARM AND HAND

As explained in Chapter 17, the muscles of the forearm and hand of tetrapods may be divided into three main groups: the ventral muscles of the forearm (usually flexors of the hand/digits and/or pronators of the forearm), the muscles of the hand, and the dorsal muscles of the forearm (usually extensors of the hand/digits and/or supinators of the forearm) (Tables 18.1 and 18.3). In both amphibian and reptiles, the extensor (dorsal) and flexor (ventral) layers of the forearm have the same basic structure, being both superficially composed by three muscular complexes: the "ulnar extensors/flexors," the "radial extensors/flexors," and the "extensor/flexor digitorum communis/longus." These muscular complexes usually arise from the humerus and insert onto the distal portion of the radius, the distal portion of the ulna, and the hand (carpal, metacarpal, and/or phalangeal) bones, respectively. These six muscular complexes are present in all major extant clades of limbed amphibians and reptiles. We prefer to use the name muscular complexes, because all these six complexes include more than one muscle in at least one of these clades (see Tables 18.1 through 18.3 and the following). Unlike most other nonmammalian tetrapods, crocodylians have a more distal insertion of the radial extensors/flexors and the ulnar extensors/flexors. For instance, the pars superficialis and pars intermedia (*sensu* Holmes [1977]) of the extensor antebrachii et carpi radialis of crocodylians insert onto the radial bone (i.e., a carpal bone) rather than onto the radius (see Table 18.1). In birds, this tendency is still more acute, e.g., part of the extensor antebrachii et carpi radialis extends distally to insert onto the proximal end of metacarpus I (e.g., Hudson et al. 1972; Diogo and Abdala 2010). In mammals, the insertion of the muscles of the forearm onto hand bones is common (e.g., Jouffroy 1971; see Chapter 17). Interestingly, a similar trend is also found in some anurans (e.g., *Phyllomedusa*: Manzano 1996; see the following).

Concerning the "flexor digitorum communis/longus" muscular complex, the homologies of the flexor accessorius lateralis and flexor accessorius medialis and of the flexor digitorum communis of urodeles and the muscles of amniotes was discussed in Chapter 17. According to Holmes (1977) and Dilkes (1999), in reptiles such as lizards, *Sphenodon*, and crocodylians, the flexor digitorum longus usually has "superficial," "deep ulnar," and "deep humeral" heads, the latter being very likely fused with the structure that Holmes (1977) designated as "pronator radii teres" (see Tables 18.1 through 18.3 and the following). Therefore, the superficial head might derive from/correspond to the amphibian flexor digitorum communis, and the two reptilian deep heads derive from/correspond to the amphibian flexor accessorius lateralis and flexor accessorius medialis. Also according to Holmes (1977) and Dilkes (1999), the plesiomorphic condition for reptiles is to have a flexor digitorum longus inserted onto the distal phalanges of digits 1–5, although in some taxa such as birds and crocodylians, the muscle inserts only onto some of these digits. For instance, the flexor digitorum longus of birds usually inserts exclusively onto digit 2 (i.e., digit 3 according to most embryologists) (see Tables 18.1 through 18.3 and the following). Authors such as Ribbing (1938) described a "flexor digitorum sublimis" in birds, but this structure clearly seems to correspond to the superficial head of the flexor digitorum longus *sensu* Holmes (1977) and Dilkes (1999) rather than the flexor digitorum superficialis (often called "sublimis") of marsupial and placental mammals (see Chapter 17). The flexor accessorius of anurans such as *Bufo* corresponds to the flexor accessorius medialis plus flexor accessorius lateralis of urodeles such as *Ambystoma* (e.g., Ribbing 1907; see Tables 18.1 through 18.3).

One muscle that also makes part of the flexor digitorum communis/longus complex is the palmaris longus, which is variable among tetrapods and is often the most superficial ventral muscle of the forearm. As explained in Chapter 17, some of the structures that are designated "palmaris longus" in different tetrapod groups are probably not homologous to each other. For instance, Gaupp (1896), described a palmaris longus in anurans. However, Howell (1935, 1936c,d) and Straus (1942) stated that a "true palmaris longus" is only seen as a variant in some reptiles such as *Iguana* (see Figure 18.12) and is only consistently present in mammals. In fact, the flexor digitorum communis of amphibians is often designated, in the older literature, as "palmaris communis" and/or as "flexor digitorum longus." Therefore, it would not be surprising if Gaupp (1896) would have simply combined these names and thus used the name palmaris longus to designate the flexor digitorum communis *sensu* this volume (Tables 18.1 through 18.3). Regarding reptiles, there is no palmaris longus in *Timon*, but there is a palmaris longus in some other lizards, as well as in other clades such as turtles, according to authors such as Howell (1936c,d), Haines (1939, 1950), Walker (1973), Russell and Bauer (2008), and Abdala et al. (2008), as well as to our dissections (see Tables 18.1 through 18.3 and Figure 18.12). Howell (1936b) argued that the palmaris longus found in some reptiles is probably derived from a part of the flexor carpi radialis, although he also stated that some reptiles may have a palmaris longus derived from the flexor carpi ulnaris, thus supporting the idea that at least some of these palmaris longus muscles are not homologous to each other. Straus (1942) also states that the structures that are designated palmaris longus in therian mammals may be derived from the flexor carpi radialis, from the flexor digitorum superficialis and/or from the flexor carpi ulnaris (see Chapter 17). Russell and Bauer (2008) considered the palmaris longus of lizards to be an additional "humeral" head of the flexor digitorum longus. According to our dissection, the palmaris longus is a muscle occasionally present in lizards such as *Tupinambis* (Figure 18.3), *Teyus*, *Ameiva*, and varanids, but tends to be absent in iguanids (but see Figure 18.12 and preceding text). In the lepidosaurian specimens we dissected, the palmaris longus often inserts superficially to the distal insertion of the common tendon of the flexor digitorum longus and to the flexor plate, being the only ventral forearm muscle that has some connection with the most superficial muscles of the hand. As can be seen in Figure 18.3, this palmaris longus forms a continuum with the layer of the flexores breves superficiales. Therefore, we agree with the statements of authors such as Howell (1935, 1936c,d) and Straus (1942) that anurans do not have a palmaris longus because the only forearm muscle that connects the forearm to the most superficial layer of the hand muscles in anurans is the flexor digitorum communis (Figure 18.4). In the lizards that we have dissected, when the palmaris longus is present, it tends to have a more ulnar topology (Figure 18.3). According to Walker (1973) and Abdala et al. (2008), as well as to our dissections, a palmaris longus with the same overall configuration as that found in lepidosaurs is present in turtles such as *Trachemys* and *Chelonoidis* (Tables 18.1 through 18.3). Haines (1950) and Lewis (1989) stated that the palmaris longus might have been part of the muscular equipment of the LCA of amniotes, thus suggesting that at least some of the structures that are designated palmaris longus in reptiles

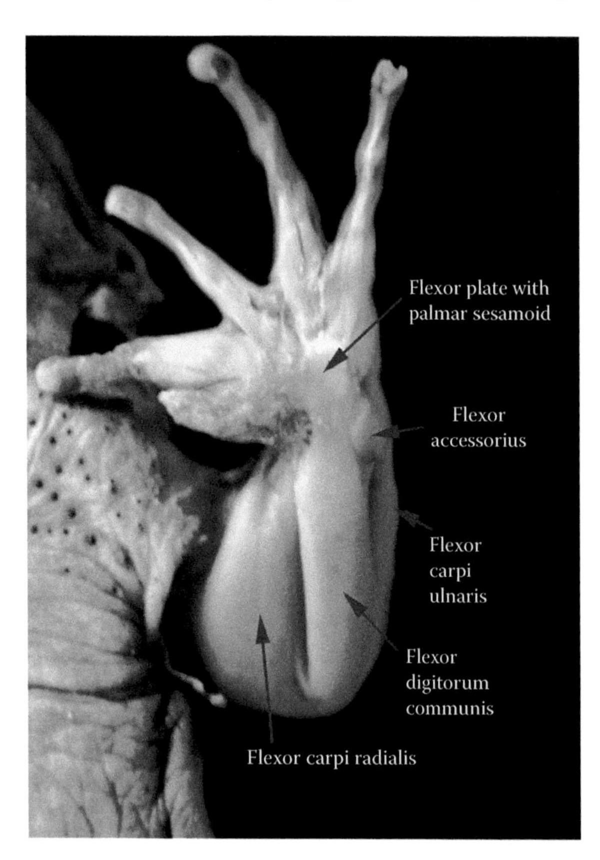

FIGURE 18.4 *Telmatobius laticeps* (Amphibia, Anura): ventral view of the superficial musculature of the forelimb and hand showing the flexor plate with the embedded sesamoid.

such as lepidosaurs and turtles is homologous to at least some of the structures that are designated palmaris longus in mammals (see Chapter 17 and Tables 18.1 through 18.3).

Regarding the ulnar ventral (flexor) muscular complex of the forearm, in amphibians, reptiles, and mammals, this usually includes a flexor carpi ulnaris and an epitrochleoanconeus (Tables 10.1 and 10.3). The latter muscle, which is often also designated "flexor antebrachii ulnaris," usually runs from the medial epicondyle of the humerus to the proximal portion of the ulna, being very thin proximally and being very easily not detected or confused with the flexor carpi ulnaris in dissections of the forearm. According to Walthall and Ashley-Ross (2006), there is a "flexor antebrachii et carpi ulnaris" in the urodele *Taricha*. However, McMurrich (1903a,b), Ribbing (1907), and Straus (1942) argued that these two muscles may be differentiated in at least some urodeles. The epitrochleoanconeus is commonly present in reptiles. Authors such as Walker (1973) and Abdala et al. (2008) did not recognize a distinct epitrochleoanconeus in the turtle *Trachemys*, but Holmes (1977) stated that he did find this muscle in a specimen of this genus, dissected by him. Straus (1942) and Meers (2003) suggested that the epitrochleoanconeus is absent in crocodylians, but we did find it in the specimens of *Caiman latirostris* we dissected (Tables 18.1 through 18.3).

Regarding the radial ventral (flexor) muscular complex, Macalister (1869) stated that in most amphibians, including urodeles, the flexor antebrachii et carpi radialis is usually differentiated into a "flexor antebrachii radialis" (pronator teres *sensu* this volume) and a flexor carpi radialis. However, according to authors such as McMurrich (1903a,b) and Straus (1942), these structures are usually not present as separate, distinct muscles in urodeles, and this seems to be the case in the *Ambystoma* specimens we dissected (Tables 18.1 through 18.3). Ribbing (1907) also supported this idea, stating that the flexor carpi radialis and the pronator teres are usually present as distinct muscles in anurans, but not in urodeles (Tables 18.1 through 18.3). According to Walthall and Ashley-Ross (2006), the flexor antebrachii et carpi radialis of urodeles such as *Taricha* not only flexes, but also helps to pronate, the hand; i.e., it does the function of the flexor carpi radialis and of the pronator teres of other tetrapods. There is some confusion regarding the identity of the flexor carpi radialis and of the pronator teres in reptiles. The "two heads of the flexor carpi radialis" *sensu* authors such as Holmes (1977) and Dilkes (1999) and of the "pronator teres" *sensu* Meers (2003), which are usually present in lepidosaurs, including *Sphenodon* and *Iguana* and other taxa such as *Timon*, as well as in some crocodylians, in turtles, and in birds (according to Dilkes 1999 these "two heads" are often designated "pronator superficialis and pronator profundus" in birds), clearly seem to correspond to the flexor carpi radialis and pronator teres of mammals. However, the structure that McMurrich (1903a) and Holmes (1977) described as pronator radii teres in taxa such as *Sphenodon* and lizards clearly seems to derive from the flexor digitorum longus, as recognized by these two latter authors. That is, this pronator radii teres probably does not correspond to the pronator teres *sensu* this volume, which derives from the flexor antebrachii et carpi radialis (Tables 18.1 through 18.3; Chapter 17). As described by Straus (1942) and Meers (2003), and corroborated by our dissections of *Caiman latirostris*, in some crocodylians, the pronator teres and the flexor carpi radialis are deeply blended, while in other crocodylians, these two muscles are well separated, corresponding to the two heads of the flexor carpi radialis *sensu* Holmes (1977) and Dilkes (1999).

The remaining ventral muscles of the forearm are the palmaris profundus 1, the pronator quadrates, and pronator accessorius (Tables 18.1 through 18.3; see Chapter 17). The pronator quadratus is present in urodeles, anurans, turtles, lepidosaurs, crocodylians, birds, and mammals, while the palmaris profundus 1 may be absent in some of these taxa (e.g., anurans and crocodylians: see Tables 18.1 through 18.3). A fully differentiated pronator accessorius is only found in reptiles (Tables 18.1 through 18.3; Figure 18.14). Birds have a ventral forearm muscle that is often designated "ulnimetacarpalis ventralis" (e.g., Sullivan 1962; Shellswell and Wolpert 1977; Meyers 1996) that usually connects the distal portion of the ulna to the metacarpal region, which probably corresponds to the pronator quadratus and/or possibly (less likely) the surrounding muscles *sensu* this volume. This idea is supported by authors such as Straus (1942) and Holmes (1977), who state that the pronator quadratus is present as a distinct muscle in all major extant groups of reptiles.

Regarding the dorsal muscles of the forearm, the homologies of the extensor antebrachii et carpi radialis, and its derivatives in nonmammalian tetrapods and the mammalian extensor carpi radialis longus, extensor carpi radialis brevis, brachioradialis, and supinator have caused much confusion in the literature (Tables 18.1 through 18.3). Authors such as Howell (1936c) and Meers (2003) described an "extensor carpi radialis longus" and an "extensor carpi radialis brevis" in reptiles such as *Iguana* and crocodylians. However, most authors argue that reptiles have a single "extensor carpi radialis," which corresponds to the extensor antebrachii et carpi radialis *sensu* this volume and is usually subdivided into three bundles in amphibians such as urodeles and reptiles such as turtles, crocodylians, and lepidosaurs; i.e., superficialis, profundus, and "intermedius" (e.g., Humphry 1872a,b; Walker 1973; Holmes 1977; Lewis 1989; Dilkes 1999; Abdala et al. 2008). Authors such as Humphry (1872a,b) and Lewis (1989) suggested that these three bundles might have given rise to the extensor carpi radialis (brevis + longus), brachioradialis, and supinator of mammals, respectively. However, in addition to these three bundles of the extensor antebrachii et carpi radialis, reptiles usually also have a muscle "supinator longus"/"tractor radii" *sensu* Holmes (1977), which is the probable homologue of the mammalian brachioradialis (see Tables 18.1 through 18.3 and the following). The evolution and homologies of these muscles are discussed in Chapter 17. It is however possible that the structure that has been often designated the intermedius head of the extensor antebrachii et carpi radialis in other nonmammalian tetrapods such as urodeles, for instance, corresponds to the structure that has given rise to the mammalian brachioradialis, as suggested by Humphry (1872a,b) and Lewis (1989). That is, it is possible, and even likely, that the intermedius head of taxa such as urodeles is not homologous to the intermedius head of reptiles such a crocodylians and turtles. In crocodylians, the extensor antebrachii et carpi radialis *sensu* this volume seems to include the extensor carpi radialis longus, the extensor carpi radialis brevis, and the "abductor radialis" *sensu* Meers (2003), although part of the extensor carpi radialis brevis *sensu* Meers (2003) (his pars ulnaris) might correspond to the abductor pollicis longus *sensu* this volume (see Tables 18.1 through 18.3). The crocodylian extensor carpi radialis longus, extensor carpi radialis brevis (or at least its "pars radialis"), and the abductor radialis *sensu* Meers (2003) might well correspond to the "pars superficialis, pars intermedia, and pars profunda of the extensor carpi radialis" *sensu* Holmes 1977 and, thus, to the structures that give rise to the distinct mammalian extensor carpi radialis longus, extensor carpi radialis brevis, and supinator muscles. The two former structures in crocodylians insert onto hand bones, as the mammalian extensor carpi radialis longus and extensor carpi radialis brevis usually do, while the latter, third structure does not reach the hand bones, inserting, distally, exclusively onto the forearm bones, as the mammalian supinator usually does. This hypothesis is supported by the designation by some authors of the "extensor carpi radialis profundus" *sensu* Holmes 1977 as "supinator" or "supinator brevis" (see, e.g., Walker 1973; see also Tables 18.1 through 18.3).

The supinator longus (tractor radii) *sensu* Holmes (1977), which in reptiles, such as turtles, is innervated by the "inferior brachial nerve" and the radial nerve (e.g., Haines 1939) clearly seems to correspond to the brachioradialis of mammals, because its origin on the humerus is more lateral and more proximal than that of the other derivatives of the "extensor antebrachii et carpi radialis anlage" (see, e.g., Figure 19 of Holmes [1977]). This idea is supported by the designation in the older literature of the mammalian brachioradialis as supinator longus and the reptilian tractor radii was often designated brachioradialis (see, e.g., Walker 1973; Chapter 17). Jollie (1962) suggested that the humeroradialis is present in crocodylians and *Sphenodon* and corresponds to the "tensor patagii" of birds and to the brachioradialis of mammals. Meers (2003) stated that the humeroradialis of crocodylians (Figure 18.13) is homologous to the "tensor propatagialis" of birds, but that this muscle is absent in other living reptiles, thus suggesting that the mammalian brachioradialis is homologous to the crocodylian muscle that he designated supinator. It is important to note that the overall configuration and function of the humeroradialis *sensu* Meers (2003) are somewhat similar to those of the mammalian brachioradialis, because the humeroradialis is ontogenetically derived from the dorsal (extensor) anlage but mainly acts as a flexor of the antebrachium (see, e.g., Meers 2003, Table 18.1, and Figure 18.13). However, regarding its innervation, the humeroradialis *sensu* Meers (2003) does not seem to be homologous to the mammalian brachioradialis, because it is innervated by the axillary nerve rather than the radial nerve. Moreover, the supinator *sensu* Meers (2003) also has an overall configuration and function that are similar to those of the mammalian brachioradialis (i.e., it is not only part of the extensor musculature but also acts mainly as a flexor of the antebrachium) and, unlike the humeroradialis, is mainly innervated by the radial nerve, as is the mammalian brachioradialis (see, e.g., Meers 2003). Therefore, the mammalian brachioradialis might well be homologous to the supinator, not the humeroradialis, *sensu* Meers (2003; see Chapter 17 for more details).

Haines (1939) correctly stated that the tractor radii (which, as explained earlier, very likely corresponds to the brachioradialis *sensu* this volume) is not present as a separate muscle in amphibians such as *Salamandra*; but, at the same time, he designated the intermedius head of the extensor antebrachii et carpi radialis of *Salamandra* as a "brachioradialis." This seems to support the hypothesis, proposed earlier, that the structure that is often designated as the intermedius head of the extensor antebrachii et carpi radialis in taxa such as urodeles is not directly homologous to the intermedius head of reptiles such as turtles and crocodylians. In any case, most authors argue that, even if the intermedius head of taxa such as urodeles corresponds to the structure that has given rise to the mammalian brachioradialis, this intermedius head is just a poorly differentiated part of the extensor antebrachii et carpi radialis, not a distinct, separate muscle like the mammalian brachioradialis (see, e.g., Humphry 1872a,b; Howell 1936c; Straus 1941a,b; Lewis 1989; Tables 18.1 through 18.3 and Chapter 17).

Regarding the ulnar dorsal (extensor) muscular complex, the anconeus (often designated "extensor antebrachii ulnaris") and the extensor carpi ulnaris do not seem to be present as independent muscles in *Ambystoma* and *Timon*. But several authors do describe an "anconeus" in amphibians such as *Salamandra* and various reptiles, such as *Sphenodon*, and some birds (see the following). However, we did not find a distinct, separate anconeus such as that found in mammals in the turtles and in the numerous lizards we dissected. Furthermore, Howell (1936b,c) does not describe an anconeus in urodeles such as *Necturus* and lizards such as *Iguana*, and Meers (2003) does not describe a distinct anconeus in crocodylians. Moreover, Haines (1939), who did describe an anconeus in amphibians such as *Salamandra*, *Triton*, and *Rana*, argued that the anconeus is rarely present as a separate, distinct muscle in urodeles and apparently in anurans, thus suggesting that the anconeus of reptiles and amphibians and the anconeus of mammals were independently acquired in the evolution of these clades, i.e., that they are not homologous to each other. As the anconeus that authors such as Haines (1939), Sullivan (1962), Jouffroy (1971), Jouffroy and Lessertisseur (1971), Holmes (1977), and Shellswell and Wolpert (1977) describe in some amphibians and reptiles, and the anconeus of mammals have a similar overall configuration and a similar innervation and derive from the same anlage (i.e., derive from the "extensor antebrachii et carpi ulnaris" anlage), this would be a clear illustration of convergent/parallel evolution. For instance, Sullivan (1962), Shellswell and Wolpert (1977), Meyers (1996), and Maxwell and Larsson (2007) described a muscle anconeus (or "ectepicondylo-ulnaris") in birds such as chickens, which connects the distal dorsal margin of the humerus to the proximal dorsal margin of the ulnar and ontogenetically derives from the extensor antebrachii et carpi ulnaris *sensu* this volume, as does the anconeus of mammals. If further studies, including a broader sampling of amphibian and reptilian taxa, do show that an anconeus is present in at least some members of all, or at least most of, the major extant clades of limbed amphibians and reptiles, this would probably indicate that in the LCA of tetrapods, the extensor antebrachii et carpi ulnaris was at least already partially differentiated into an antebrachial, proximal portion and a carpal, distal portion, which then gave rise to the anconeus and extensor carpi ulnaris *sensu* this volume, respectively.

The remaining muscles of the dorsal (extensor) layer of the forearm are the extensor digitorum, the extensores digitorum breves (often designated "short extensors of the digits"), and the abductor pollicis longus, which, in urodeles such as *Ambystoma* and *Taricha*, is possibly fused with the short extensor of digit 1, forming the abductor et extensor digit 1 (see Tables 18.1 through 18.3). Among crocodylians, Meers (2003) included, in the intrinsic extensors of the manus: five "extensores digitorum superficiales" that often attach to the distal phalanges of digits 1–5; six "extensores digitorum profundi" that often attach to the distal phalanges of these five digits; one "extensor pollicis superficialis et indicis proprius" attaching to the distal portions of digits 1 and 2; one "extensor metacarpi I" attaching to metacarpal I; and one "extensor metacarpi IV" attaching to metacarpal IV. All these 14 muscles seem to correspond to/derive from the extensores digitorum breves *sensu* this volume, except the extensor metacarpi I or possibly the "extensor digiti I superficialis," as one of these two latter structures might well correspond to the abductor pollicis longus *sensu* this volume (Tables 18.1 through 18.3). The "extensor digiti I superficialis" might correspond to the abductor pollicis longus because, as it is often the case with the latter muscle, it is the largest and most lateral dorsal (extensor) muscle of the hand (compare, e.g., Figure 13 of Meers [2003] with Figure 2 of Abdala and Moro [2006]). However, this extensor digiti I superficialis inserts onto the distal phalanx of digit 1 rather than onto the metacarpal I as the abductor pollicis longus of other reptiles often does. This might indicate that the abductor pollicis longus *sensu* this volume, if present in crocodylians, corresponds to the extensor metacarpi I *sensu* Meers (2003), because the latter structure inserts onto metacarpal I rather than onto the distal phalanx of digit I. However, the most likely hypothesis in view of the evidence is that all 14 muscles described by Meers (2003) are part of the extensores digitorum breves *sensu* this volume and that the author failed to describe the abductor pollicis longus. Authors such as Holmes (1977) clearly stated that the latter muscle is present in crocodylians and is similar to the abductor pollicis longus of other reptiles, mainly extending from the ulna to the carpal/metacarpal region. Meers (2003) did not describe a muscle with such a configuration in crocodylians: his extensor metacarpi I extends from the radial bone to the metacarpal I, and his extensor digiti I superficialis extends from the radial bone to the distal phalanx of digit I.

Another plausible hypothesis is that the abductor pollicis longus *sensu* this volume corresponds to the "extensor carpi radialis brevis pars ulnaris" *sensu* Meers (2003), because the latter structure is well developed, originates from the ulna, and inserts onto the carpal/metacarpal region (onto the radial bone according to Meers [2003]), as the abductor pollicis longus of other reptiles usually does. Notably, Russell and Bauer (2008) described a "superficial extensores digitores brevis complex" and an "interossei dorsales" complex in lepidosaurs. According to the authors, the former complex is subdivided into superficial and deep components (see Figure 18.5). The "extensores digitores brevis profundus" of Russell and Bauer (2008) correspond to the dorsometacarpales *sensu* this volume (see Tables 18.1 through 18.3).

The homologies of the hand muscles of tetrapods were discussed in Chapter 17. Examples of amphibian and reptilian hand muscles include the flexores breves superficiales, which are ventral (palmar, superficial) to the other muscles; the abductor pollicis brevis and abductor digiti minimi, which usually lie on the ventrolateral (radial) and ventromesial (ulnar) surface of the hand and abduct the most lateral (radial) and most medial (ulnar) digits, respectively; the lumbricales, which are deeper and are usually associated with the tendons of the flexor digitorum communis/longus, being often related to the extension and/or flexion of different parts of the digits; the contrahentes digitorum, which are deep to the lumbricales; the flexores breves profundi, which are usually deep to the contrahentes digitorum and which often insert onto both the radial and ulnar sides of the digits (note that in the present volume, each of the "bicipital muscles" that are often described in the literature as going to both sides of the same digit are considered to be two distinct flexores breves profundi muscles: see Chapter 17); the intermetacarpales, which are the deepest (most dorsal) muscles of the ventral (palmar), and the dorsometacarpales, which are part of the dorsal layer of the hand and are thus the most dorsal intrinsic muscles of the hand (the dorsometacarpales are not present as distinct muscles in mammals) (Tables 18.1 through 8.3).

The flexores breves superficiales are consistently present in limbed amphibians and reptiles (Tables 18.1 through 18.3). However, there is some confusion, in the literature, about the presence of these muscles in birds. Holmes (1977) suggested that the flexores breves superficiales are present in all major extant groups of reptiles, but Ribbing (1938) argued that these muscles are not present as a group in birds. We could not identify, in the chickens we dissected, muscles that clearly correspond to the flexores breves superficiales of other reptiles. But it is possible, and even likely, that the "flexor indicis" *sensu* Sullivan (1962) and Shellswell and Wolpert (1977) (which goes to digit 1—i.e., to digit 2 according to most embryologists—and corresponds to the "flexor alulae" or "flexor pollicis" or "flexor digiti II" or "flexor digiti secundi manus" or "adductor indicis" *sensu* Meyers 1996) and/or the "flexor digiti quarti" *sensu* Sullivan (1962) and Shellswell and Wolpert (1977) (which goes to digit 3—i.e., digit 4 according to most embryologists—and corresponds to the "flexor digiti minoris" or "flexor minimi digiti" or "flexor minimi digiti + flexor minimi digiti brevis" or "flexor digiti III" or "flexor digiti IV" or "flexor digiti quarti brevis + abductor digiti quarti proprius" or "flexor digiti quarti manus longus" or "flexor longus muscle of the fourth digit" *sensu* Meyers [1996]) are part of the flexores breves superficiales *sensu* this volume. However, we cannot completely discard the hypothesis that at least some of the latter muscles correspond, instead, to a part of the flexores breves profundi, if the "interossei ventralis" and "interossei dorsalis" *sensu* Sullivan (1962), Shellswell and Wolpert (1977), and Meyers (1996) correspond to the intermetacarpales and dorsometacarpales *sensu* this volume, respectively (see Tables 18.1 through 18.3). In amphibians, the flexores breves superficiales have a particular conformation, because they are often markedly reduced and mainly associated with the structure that is often designated "palmar aponeuroses" in the literature (e.g., Ecker 1889; Walthall and Ashley Ross 2006). It should be taken into account, however, that the name "palmar aponeurosis" is misleading, as this structure is not an aponeurosis, but a strong

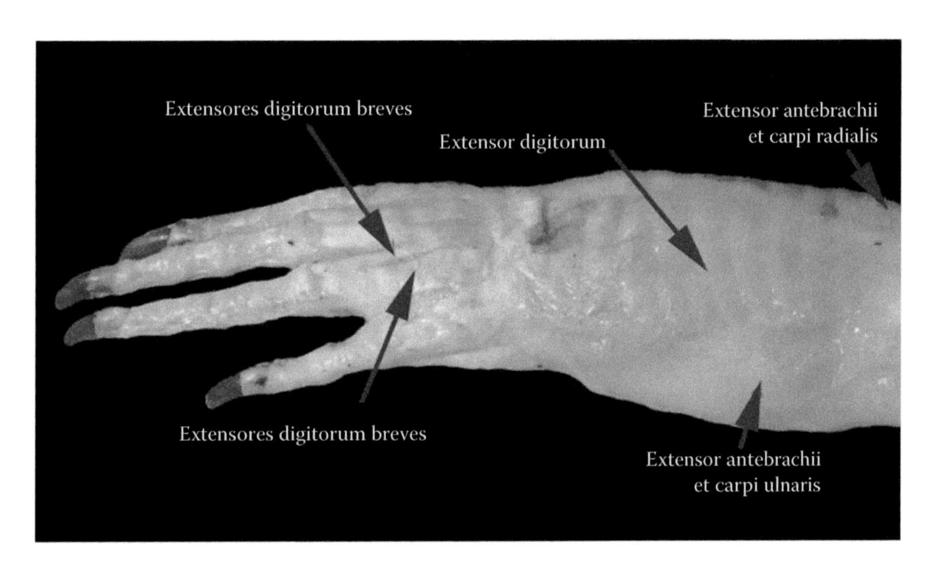

FIGURE 18.5 *Tupinambis meriane* (Reptilia, Lepidosauria): dorsal view of the dorsal (extensor) muscles of the forearm and hand.

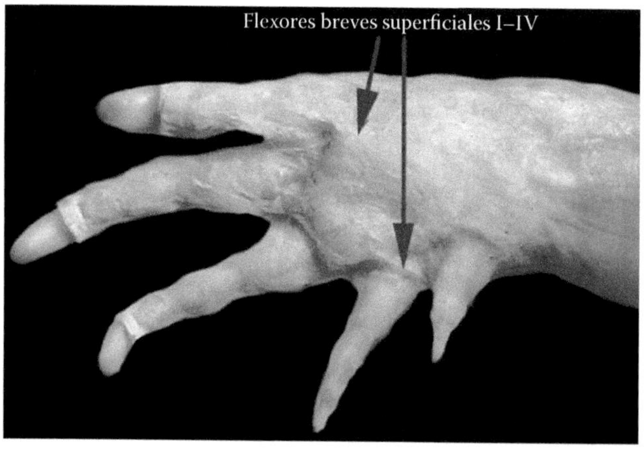

FIGURE 18.6 *Caiman latirostris* (Reptilia, Crocodylia): ventral view of the most superficial (ventral, or palmar) layer of the hand muscles.

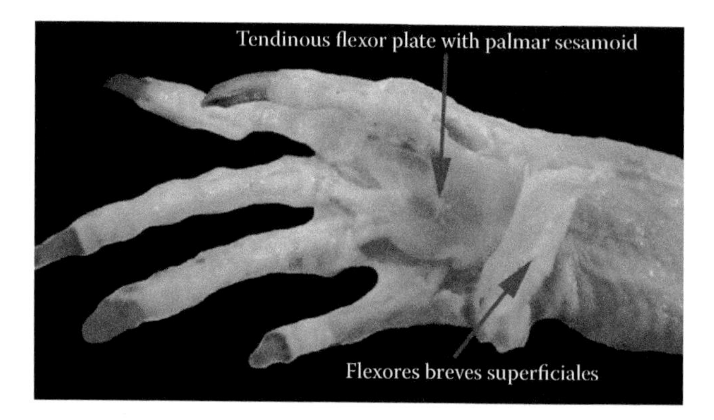

FIGURE 18.7 *Tupinambis meriane* (Reptilia, Lepidosauria): ventral view of the flexor plate after resection of the superficial layer of flexores breves superficiales.

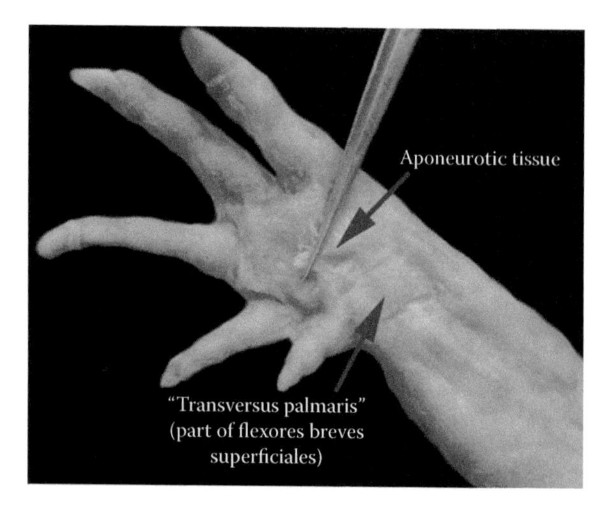

FIGURE 18.8 *Caiman latirostris* (Reptilia, Crocodylia): ventral view of the palm of the hand after resection of part of the aponeurotic tissues covering it.

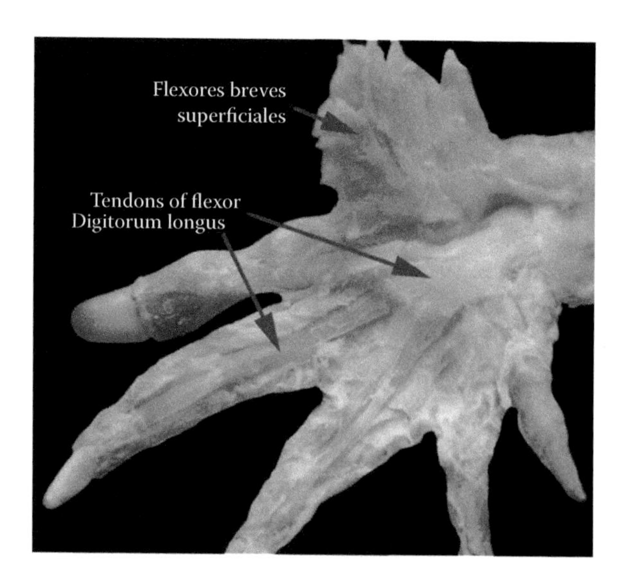

FIGURE 18.9 *Caiman latirostris* (Reptilia, Crocodylia): ventral view of the palm of the hand after resection of the layer of the flexores breves superficiales, showing that the expanded tendon of the flexor digitorum longus does not form a flexor plate.

tendon with a palmar sesamoid embedded in it. We found this structure in anurans such as *Bufo* and *Telmatobius* and called it flexor plate (Figure 18.4; see also Figures 18.5 through 18.10). However, in some anurans, this flexor plate might be very small (e.g., *Pseudis minutus*) or even completely absent (e.g., *Pseudis paradoxa*) (see, e.g., Manzano 1996). Some reptiles do have a "true palmar aponeurosis," that is, a superficial (ventral) structure that has a typical aponeurotic configuration and that is often related to the flexores breves superficiales (see, e.g., Haines 1950; Meers 2003; Abdala et al. 2008).

The lumbricales are usually present in anurans, turtles, lepidosaurs, and crocodylians, but absent in most urodeles and apparently in birds. In fact, our dissections indicate that birds such as chickens have no distinct lumbricales muscles such as those seen in other tetrapods. Authors such as Sullivan (1962), Shellswell and Wolpert (1977), and Meyers (1996) have not used the name "lumbricales" to describe any of the hand muscles of chickens and of other groups of Aves. However, they do describe a muscle ("abductor medius" *sensu* Sullivan [1962] and Shellswell and Wolpert [1977], which attaches to digit 2—i.e., digit 3 according to most embryologists—and corresponds to the "abductor digiti majoris" *sensu* Meyers [1996]) that is "applied to" the tendons of the flexor pollicis longus according to Sullivan (1962) and Shellswell and Wolpert (1977) and "covered by" these tendons according to Meyers (1996) and that might thus correspond to a part of the lumbricales *sensu* this volume. If this muscle is not a part of the lumbricales, it would probably correspond to a part of the intermetacarpales *sensu* this volume, because it seems to mainly abduct digit 2 (i.e., digit 3 according to most embryologists). Regarding the contrahentes digitorum, the flexores breves profundi, the intermetacarpales, and the dorsometacarpales, one or more muscles of each of these four muscular groups is always present in at least some urodeles,

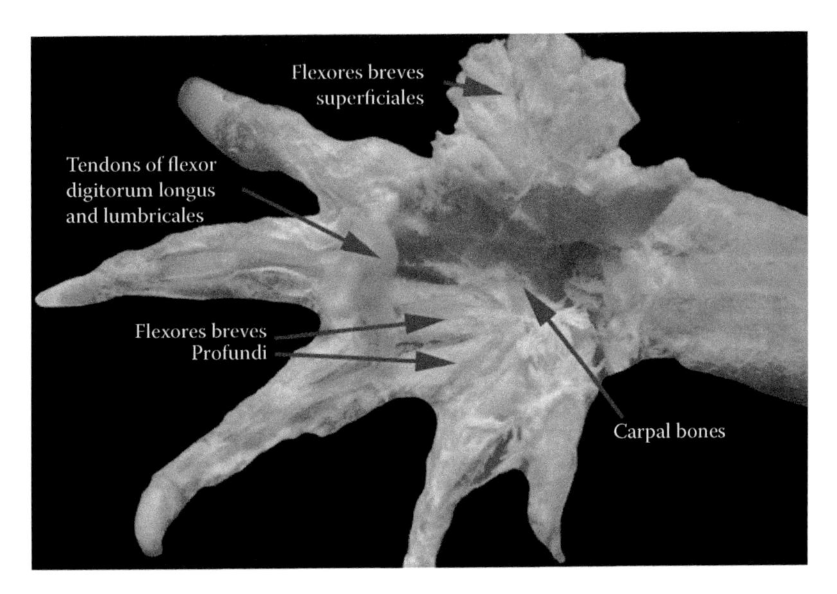

FIGURE 18.10 *Caiman latirostris* (Reptilia, Crocodylia): ventral view of the deep (dorsal) musculature of the palm of the hand after resection of the more superficial (ventral or palmar) layers.

anurans, lepidosaurs, turtles, crocodylians, and birds (Tables 18.1 through 18.3). In fact, although the dorsometacarpales do not seem to be present as distinct muscles in the *Ambystoma* specimens we dissected, and are not described in urodeles such as *Taricha* by authors such as Walthall and Ashley-Ross (2006), these muscles have been described in other urodeles. For instance, Straus (1941a,b) stated that urodeles such as *Salamandra* and *Cryptobranchus* do have dorsometacarpales (see his Table 1) and illustrated a *Necturus* specimen with "dorsometacarpales" in his Figure 1, although he explains that in this specific case, the dorsometacarpales of *Necturus* probably correspond to the "extensores digitorum breves" *sensu* other authors. Haines (1939) argued that anurans such as *Rana* and urodeles such as *Salamandra* clearly have both extensores digitorum breves and dorsometacarpales, so at least some amphibians do seem to have dorsometacarpales *sensu* this volume (see Tables 18.1 through 18.3). This subject was discussed in some detail in Chapter 17. Lastly, the abductor pollicis brevis and abductor digiti minimi are also consistently present in most major extant clades of limbed amphibians and reptiles, the exceptions being urodeles, which lack an abductor pollicis brevis, and birds, which seem to lack an abductor digiti minimi, as will be discussed in the following (Tables 18.1 through 18.3).

CHAMELEON LIMB MUSCLES, MACROEVOLUTION, AND PATHOLOGY

In this section, we compare and contrast the forelimb muscles of chameleons with those of a morphologically and behaviourally more generalized lizard and discuss the functional implications of derived musculoskeletal characteristics in the forelimbs of chameleons. In addition, we discuss the broader implications of the evolution and development of chameleons in terms of macroevolution and pathologies. This section is excerpted from a recent study published in the *Journal of Morphology* (Molnar et al. 2017b).

The lineage leading to modern Chamaeleonidae diverged from its sister taxon, the Agamidae, (forming the clade Acrodonta) ~126–120 million years ago (MYA) during the early Cretaceous (Okajima and Kumazawa 2010; Zheng and Wiens 2016). The family Chamaeleondiae is hypothesized to have evolved ~60–65 MYA during the Late Cretaceous/Early Tertiary (Okajima and Kumazawa 2010; Tolley et al. 2013). Chameleons are thus a relatively young clade of lizards that present a highly specialized suite of phenotypic traits associated with true arboreal locomotion, in contrast to the sprawling gait typical of lizards (Webb and Gans 1982; Peterson 1984; Higham and Jayne 2004; Fischer et al. 2010) and most tetrapods. Adaptations for life in the trees include an increased dependence on a projectile tongue for feeding, large eyes, and increased dependence on processing visual sensory information, a laterally compressed body, a prehensile tail, and zygodactyly of the autopodia (manus/"hand" and pes/"foot").

In chameleons, the ulnar and radial sides of each autopodium are widely separated from each other by a cleft. The syndactylous complexes formed by manual digits 1–3 and 4 and 5 and pedal digits 1 and 2 and 3–5 are often designated as "superdigits" because each constitutes a single functional unit. The cleft, associated with other changes in the wrist and ankle, allows the two superdigits of each autopod to face in opposite directions, resulting in opposable autopodia that are related to the characteristic chameleon mode of autopodial grasping (Diaz and Trainor 2015; for recent reviews, see Diaz et al. 2015a). These features represent an important morphological trade-off that has led to the reduced speed of locomotion; a decreased dependence on olfaction, audition, the use of the snout for catching prey; and potentially a decrease in gustation (taste) (reviewed by Tolley and Herrel [2013]).

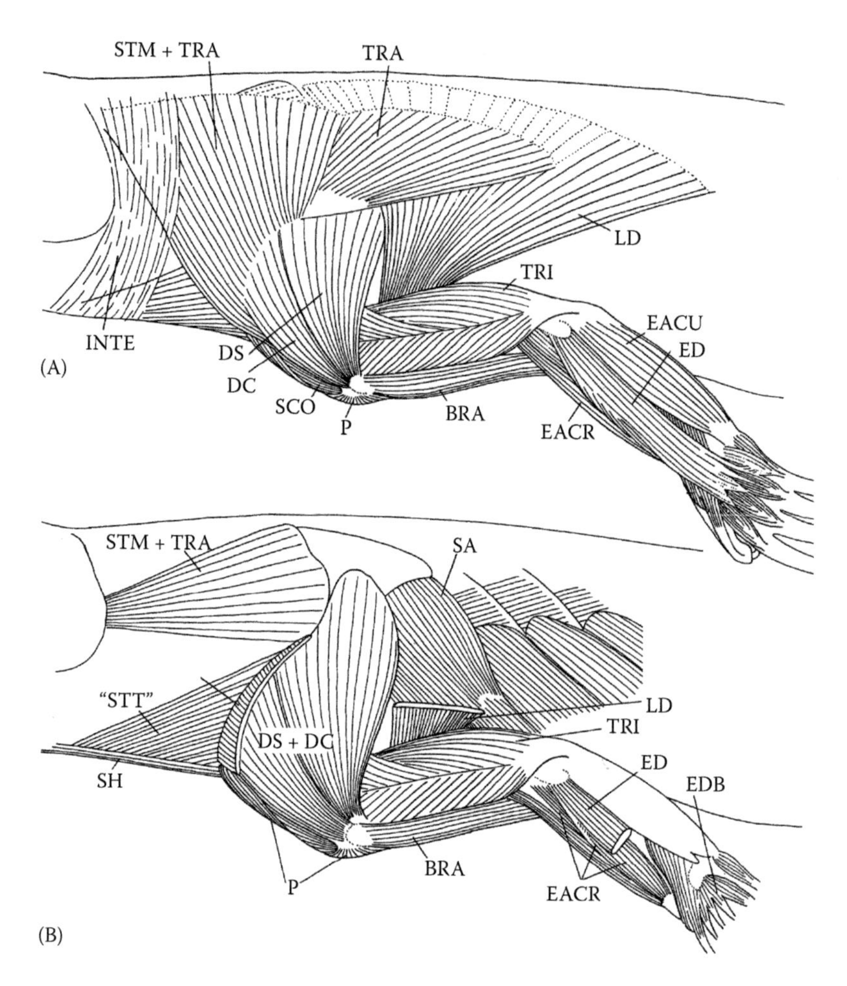

FIGURE 18.11 *Iguana* sp. (Reptilia, Lepidosauria). (A) Lateral view of the musculature of the pectoral girdle, arm, forearm, and hand; (B) same view; the trapezius, the latissimus dorsi, and the superficial layer of the forearm muscles were cut to show the deeper muscles (anterior is toward the left of the figure) (the nomenclature of the structures illustrated follows that of the present work). (Modified from Jollie, M., *Chordate Morphology*, Reinhold, New York, 1962.) BRA, brachialis; DC, deltoideus clavicularis; DS, deltoideus scapularis; EACR, extensor antebrachii et carpi radialis; EACU, extensor antebrachii et carpi ulnaris; ED, extensor digitorum; EDB, extensores digitorum breves; INTE, interhyoideus (hyoid muscle); LS, levator scapulae; LD, latissimus dorsi; P, pectoralis; SA, serratus anterior; SCO, supracoracoideus; SH, sternohyoideus (hypobranchial muscle); STM, sternocleidomastoideus (branchial muscle); "STT," sternothyroideus (hypobranchial muscle); TRA, trapezius (branchial muscle); TRI, triceps brachii.

Because of these unusual traits, chameleons have recently been adopted as model organisms for reptile development and the study of macroevolution. These lizards represent a good model system to study reptile embryonic development because of their well-known husbandry requirements and slow development (Diaz and Trainor 2015; Diaz et al. 2015a,b; Stower et al. 2015). Thus, chameleons allow us to study not only the skeletal development of the autopodia, but also the developmental integration of soft and hard tissues of the limb more broadly. These developmental studies may help us to better understand how changes in locomotor mode and life history lead to anatomical changes (Abdala et al. 2009). Chameleons have also recently been adopted as models for linking evolution, development, and pathology; for instance, development of the autopodium in chameleons has been studied as a possible model for the genetic and developmental processes that lead to split hand/foot and syndactyly in humans (Diogo et al. 2016a). If the distinctive autopodia of chameleons arise *via* deviation of "normal" development, they could also be seen as an example of an evolutionary teratology *sensu* Guinard (2015), or even as "hopeful monsters" *sensu* Goldschmidt (1940) (Diogo et al. 2017b).

However, these studies were mainly based on the skeleton, and the soft-tissue anatomy of chameleons—particularly that of the autopodia—is poorly known. Here, we describe the adult musculature of the forelimbs and hindlimbs of *Chamaeleo calyptratus* and compare its muscles to those of other chameleons and other lizards. We also discuss the locomotor implications of the distinctive features of chameleon limb muscle anatomy, particularly for grasping, and the striking similarities between the muscles of the distal regions of the forelimb and hindlimb. This descriptive work will serve as a basis for better understanding chameleon anatomy and evolution and to pave the way for linking anatomical studies with developmental and experimental studies on these new model organisms, including the mechanisms that lead to phenotypes

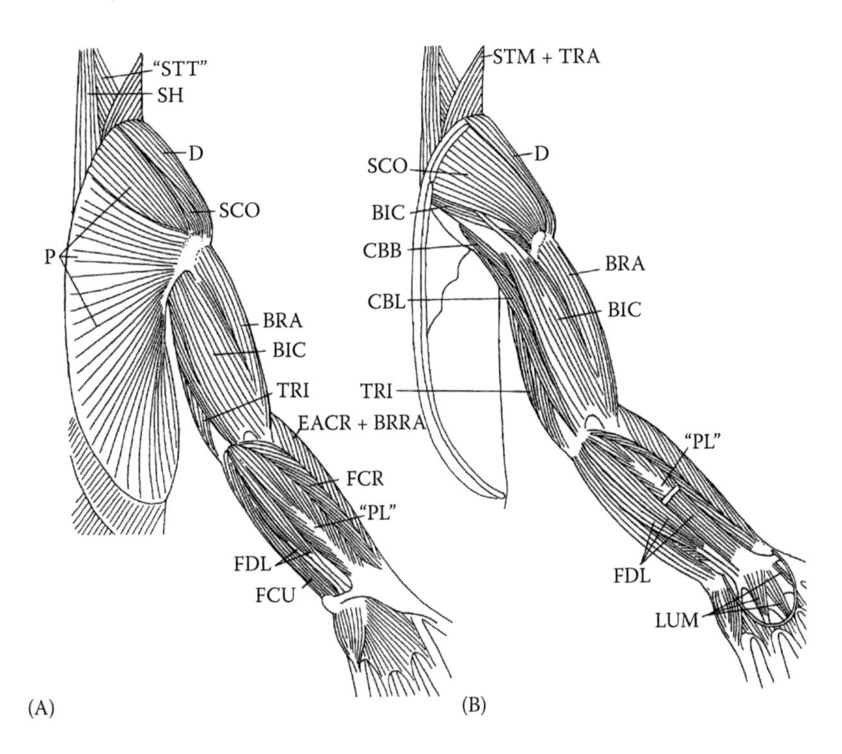

FIGURE 18.12 *Iguana* sp. (Reptilia, Lepidosauria). (A) Ventral view of the musculature of the pectoral girdle, arm, forearm, and hand; (B) same view, but some superficial muscles (e.g., pectoralis, palmaris longus) were cut to show the deeper muscles (anterior is toward the top of the figure) (the nomenclature of the structures illustrated follows that of the present work). (Modified from Jollie, M., *Chordate Morphology*, Reinhold, New York, 1962.) BIC, biceps brachii; BRA, brachialis; BRRA, brachioradialis; CBB, CBL, brevis and longus sections of the coracobrachialis; D, deltoideus; EACR, extensor antebrachii et carpi radialis; FCR, flexor carpi radialis; FCU, flexor carpi ulnaris; FDL, flexor digitorum longus; LUM, lumbricales; P, pectoralis; PL, "palmaris longus"; SCO, supracoracoideus; SH, sternohyoideus (hypobranchial muscle); STM, sternocleidomastoideus (branchial muscle); "STT," sternothyroideus (hypobranchial muscle); TRA, trapezius (branchial muscle); TRI, triceps brachii.

such as syndactyly and zygodactyly, which resemble human congenital malformations.

The forelimb muscle anatomy of *C. calyptratus* is shown in Figure 18.14. The dorsal shoulder muscles in chameleons (exemplified by *C. calyptratus*) are highly modified compared to those of most other lizards (exemplified by *Aspidoscelis uniparens*), and these modifications extend to the muscles anchoring the girdle to the body wall. The serratus anterior in *C. calyptratus* has only three bellies compared with the six described by Fisher and Tanner (1970) in teiid lizards such as *A. uniparens* (however, we only observed two bellies of this muscle in our specimens). The latissimus dorsi lies posterior to, rather than deep to, the reduced trapezius, and its area of origin includes not only the vertebrae but also several thoracic ribs as well. The costocoracoideus in *C. calyptratus* is reduced to a membrane between the first rib and the shoulder girdle. Because adult chameleons normally lack a clavicle (Skinner 1958), the shoulder muscles that attach to the clavicle in the lizard *A. uniparens*—the levator claviculae and deltoideus acromialis et clavicularis—usually attach to the coracoid and/or sternum in chameleons. The levator scapulae in *C. calyptratus* and *Trioceros melleri* is fused with the levator claviculae and originates from both the atlas and basioccipital, whereas it originates from the atlas only in most other lizards (Russell and Bauer 2008) and from the basioccipital only according to Mivart's (1870) description of *C. parsonii*. Furthermore, as pointed out by Peterson (1973), this muscle has longer, more diagonal fibers in arboreal lizards such as chameleons and shorter, more horizontal fibers in terrestrial lizards such as *Agama* and *Aspidoscelis*.

In *C. calyptratus*, the deltoideus scapularis is divided into two parts—posterior/superficial and anterior/deep—which insert adjacent to each other onto the humerus. Likewise, the deltoideus acromialis et clavicularis has dorsal and ventral parts, but they are only separable at their origins. Peterson (1973) noted that these muscles tend to insert more proximally on the humerus in arboreal lizards than terrestrial ones, and the specimens we examined also follow this trend. The sternocoracoideus in *C. calyptratus* has a single internal head that originates from the internal sternum, but in most other lizards, it has both internal and external heads and originates from both sternum and ribs (Russell and Bauer 2008). The scapulohumeralis anterior is present in *C. calyptratus* and *T. melleri*, *contra* Mivart's (1870) description of *C. parsonii*, originating from the lateral scapula just dorsal to the glenoid and passing underneath the scapulohumeral ligament to insert on the proximal humerus between the medial and lateral heads of triceps. This muscle has a similar configuration in *A. uniparens*. The configuration of the *triceps* in chameleons is also different from that in other lizards. In chameleons, the coracoid head of triceps is either absent or completely fused with other heads; the single long head of the triceps, which originates from the posterior scapula just above the glenoid and the scapulohumeral ligament, corresponds mainly to the

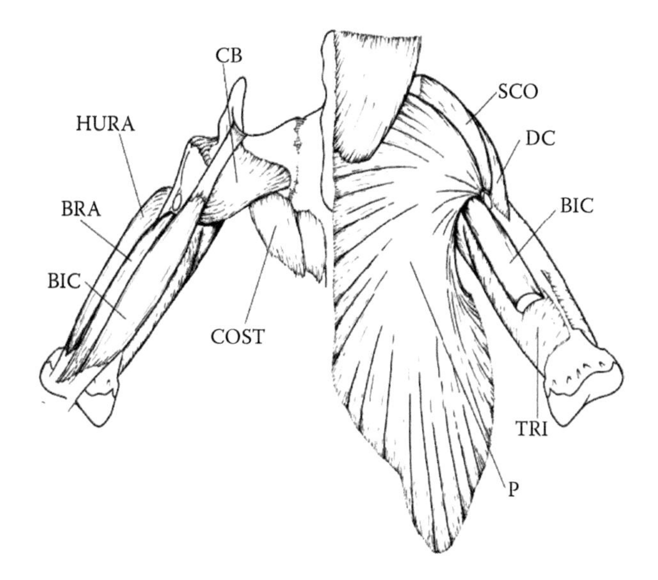

FIGURE 18.13 *Alligator mississippiensis* (Reptilia, Crocodylia): ventral view of the musculature of the pectoral girdle and arm, with several muscles removed on each side to facilitate viewing the deeper musculature: the muscles removed were the brachioradialis and humeroradialis, on the left side, and the deltoideus, the pectoralis, and the supracoracoideus, on the right side (anterior is toward the top of the figure) (the nomenclature of the structures illustrated follows that of the present work). (Modified from Meers, M. B., . *Anat Rec*, 274A, 891–916, 2003.) BIC, biceps brachii; BRA, brachialis; COST, costocoracoideus; CB, coracobrachialis; DC, deltoideus clavicularis; HUA, humeroradialis; P, pectoralis; SCO, supracoracoideus; TRI, triceps brachii.

scapular head of other lizards (Peterson 1973). The chameleons we studied also possess a third humeral head of triceps (triceps internus *sensu* Mivart [1870]) that originates from the humerus just distal to the insertion of the latissimus dorsi.

The ventral shoulder muscles are similar between *C. calyptratus* and *A. uniparens*, but the ventral arm muscles are noticeably different. The origin and insertion of the biceps brachii in *C. calyptratus* are the same as they are in *A. uniparens*, but its architecture is different. In *A. uniparens*, the biceps brachii has a single belly that inserts on the radius. However, the probable ancestral condition in lepidosaurs is a fleshy origin and an intermediate tendon that divides the muscle into proximal and distal bellies (unlike the condition in mammals, which often has medial and lateral heads that are separate proximally) (Russell and Bauer 2008). As in *A. uniparens*, the biceps brachii in *C. calyptratus* and *T. melleri* has a tendinous origin and lacks an intermediate tendon, but it has distinct medial and lateral bellies that are separate distally, near the insertion of the muscle onto the forearm bones. The brachialis, which blends distally with the biceps brachii, has distinct superficial and deep layers in *C. calyptratus* but only a single layer in *A. uniparens*. Two layers of the brachialis were also described in *Gekko* (Lecuru-Renous 1968).

Many of the muscles in the dorsal forearm of *C. calyptratus* are divided into multiple distinct bellies. The extensor carpi radialis and brachioradialis are fused in *A. uniparens* and in many other lizards as well (Russell and Bauer 2008), but in *C. calyptratus*, they are well separated: the former muscle arises from the humerus as usual but blends distally with the abductor pollicis longus. Also, in *C. calyptratus*, the extensor antebrachii et carpi ulnaris lacks an individual origin, consisting of a second belly that splits from the flexor carpi ulnaris in the distal forearm and inserts onto metacarpal V. The extensor digitorum in *A. uniparens* consists of a single belly that originates from the ulnar epicondyle of the humerus and inserts onto metacarpals II-V. In *C. calyptratus*, this muscle has two distinct parts (extensor carpi radialis longior and brevior *sensu* Mivart 1870) that originate together from the medial condyle of the humerus and split in the proximal forearm, where they are connected to the shaft of the radius. Both parts pass beneath an extensor retinaculum before inserting separately onto metacarpals III and IV. Russell and Bauer (2008) referred to an additional insertion onto metacarpal V in chameleons, but this was not the case in the chameleons dissected by us (*T. melleri* and *C. calyptratus*) or by Mivart (1870) (*C. parsonii*).

Like its counterpart on the dorsal forearm, the flexor digitorum longus consists of two distinct bellies (flexor pollicis longus and flexor profundus digitorum *sensu* Mivart 1870) whose tendons split and exchange slips in the wrist. In *C. calyptratus*, the tendons from more radial belly (part 1) insert on the distal phalanges of digits 2–5, while those from the more ulnar belly (part 2) insert on digits 1–4. In *C. parsonii*, however, the more radial belly serves digits 4 and 5 only (Mivart 1870), and the same pattern appeared to be present in *T. melleri*. The more radial belly is fused with the flexor carpi ulnaris. It originates from the radial epicondyle, the origin tendon of the pronator accessorius (Mivart [1870] did not describe this connection in *C. parsonii*, but it was present in both chameleon species we dissected) and the ulnar shaft. The more ulnar belly originates from the proximal ulna. The flexor carpi radialis arises *via* two tendons that surround the origin of the flexor digitorum longus in *C. calyptratus* and in *C. parsonii* according to Mivart (1870), but in both forelimbs of our specimen of *T. melleri* (and in *A. uniparens*), the ulnar tendon was absent. In *A. uniparens*, a palmar head joins the tendon of flexor digitorum longus to digit 2, but we did not observe this head in either chameleon we dissected.

The flexor carpi ulnaris has multiple heads of origin in *C. calyptratus*: in addition to the normal origin from the medial epicondyle of the humerus, it also attaches *via* a long tendon to the elbow joint capsule and fleshily to the posterior ulna. In addition to inserting on the pisiform, as it does in *A. uniparens*, the muscle also sends a slip to the flexor digitorum longus and attaches to the flexor retinaculum. The pronator accessorius in *C. calyptratus* inserts not only onto the radiale, as it does in *A. uniparens* and in many other lizards (Russell and Bauer 2008), but also extensively onto the distal 2/3 of the radius.

Five extensores digitorum breves are present in *C. calyptratus*, as in other lizards, but several have multiple divisions that could be counted separately. For example, Mivart (1870) describes three muscles inserting on digit 5, but as they are fused with each other, we count them as divisions of a single short extensor to digit 5. Mivart (1870) described the extensores

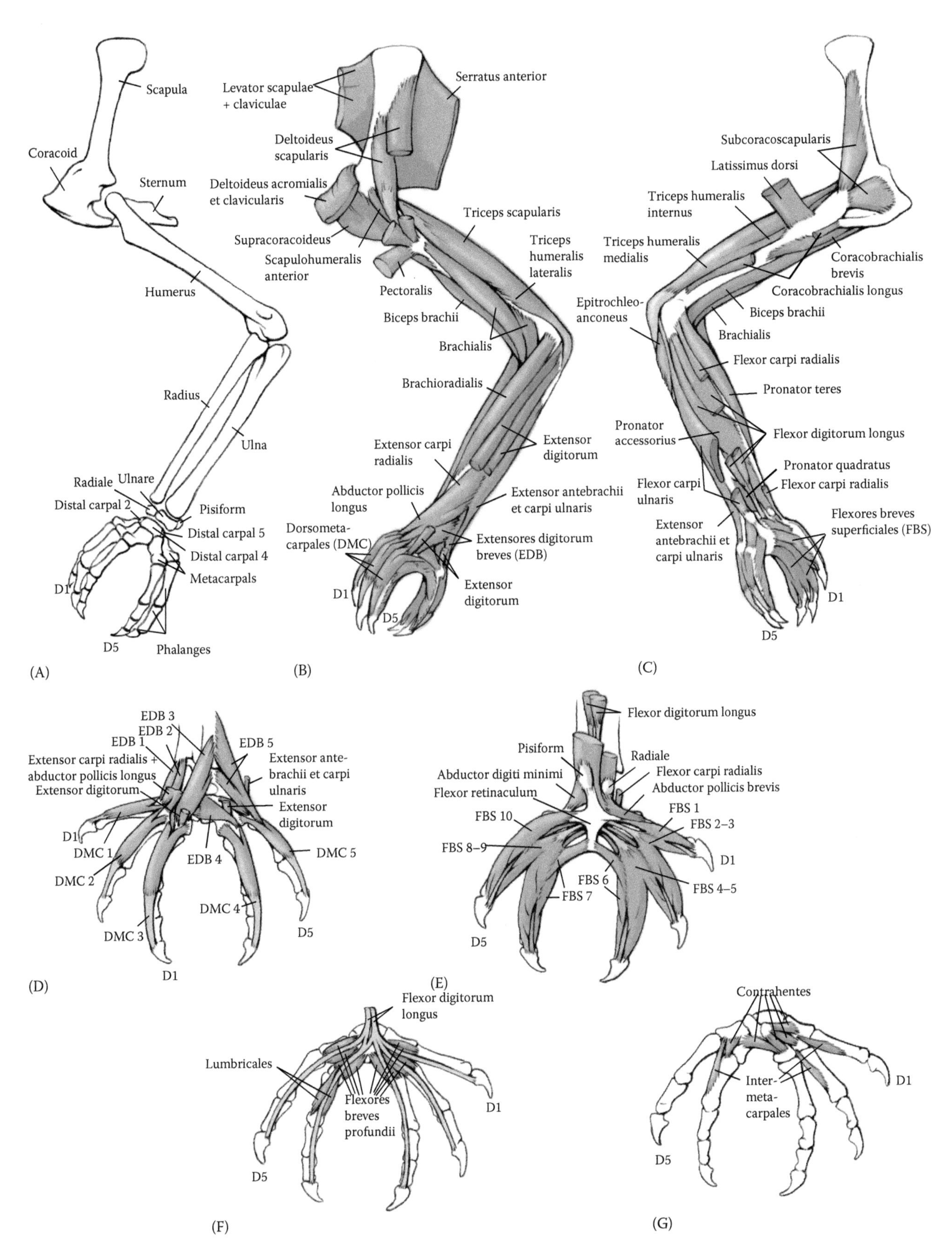

FIGURE 18.14 *Trioceros melleri* (Reptilia, Lepidosauria): left upper limb. (Modified from Molnar, J. et al., *J Morphol*, 278, 1241–1261, 2017.). (A) Skeleton in lateral view. (B) Muscles in lateral view. (C) Muscles in medial view. (D) Dorsal muscles of manus. (E) Superficial ventral muscles of manus. (F) Intermediate muscles of manus. (G) Deep muscles of manus. D, digit.

digitorum breves (his "extensores III-IX") as inserting on the metacarpals, but in our specimens, they inserted mainly onto the proximal phalanges, as in *A. uniparens*, between the two heads of the dorsometacarpales of the respective digits. The exception is the deep portion of the muscle to digit 5, which also inserts onto metacarpal V. The muscle to digit 1 (extensor III *sensu* Mivart 1870) originates from the styloid process of the radius. The muscles to digits 2, 3, and 5 (extensores IV, V, and VII+VIII+IX *sensu* Mivart [1870]) originate *via* three heads each from the distal ulna, ulnare, and metacarpals III–IV. The muscle to digit 4 (extensor VI *sensu* Mivart [1870]) solely originates from metacarpals III–IV. The abductor pollicis longus in *C. calyptratus* originates from the radius and ulna rather than from the ulna and ulnare as in *A. uniparens*. The intermetacarpales are present as usual, except that the muscles between metacarpals III–IV (where the manus is cleft) and IV–V are absent in *C. calyptratus* (the latter muscle is present in *T. melleri*).

The normal complement of 10 flexores breves superficiales are present in *C. calyptratus*, but their configuration is somewhat different from that in *A. uniparens* (Figure 18.14). First, rather than inserting onto the tendons of the *flexor digitorum longus*, these muscles directly insert onto the radial and ulnar aspects of proximal, middle, and distal phalanges. Second, the flexors that lie between syndactylous digits (2 and 3, 4 and 5, and 8 and 9) are fused to each other and arise *via* shared tendons and split at the level of the proximal phalanges. Third, the flexor retinaculum from which the flexores breves superficiales originate (annular ligament *sensu* Russell and Bauer 2008) extends much further distally and dorsally than in *A. uniparens* or other lizards, forming a structure that resembles an extensive V-shaped palmar aponeurosis. The retinaculum is connected to the ligaments between metacarpals III–IV on the dorsum of the manus and divides the palm into the medial and lateral portions related to the zygodactyl and opposable manus of chameleons. The two flexores breves superficiales (FBS) on either side of the cleft between digits III and IV (FBS 6 and 7) originate from the proximal extreme of this palmar aponeurosis, so their proximal portions are obliquely oriented—almost transversely—to the axis of the limb. The functional implications of this arrangement will be considered in the following.

While five lumbricales are present in *A. uniparens*, originating from the insertion tendons of flexor digitorum longus to digits 2–5 and inserting on the respective digits, only two or three lumbricales appear to be present in chameleons. In *C. calyptratus*, two muscles are present, originating from the tendons to digits 3 and 4 and inserting onto the ulnar aspects of the proximal phalanges of the respective digits. In *T. melleri*, both *lumbricales* were associated with digit 4, and in *C. parsonii*, an additional *lumbrical* inserting on the radial aspect of digit 3 was described (Mivart 1870). Mivart (1870) did not describe contrahentes muscles in *C. parsonii* (probably because he did not focus on the autopodia), but these muscles were present in all the chameleons we dissected. Their configuration is very similar to that of other lizards and tetrapods in general: they mainly arise from the *contrahens* fascia and insert onto the proximal phalanges. However, in the chameleons we dissected, the contrahens to digit 4 originated from metacarpal V, and in the place of a contrahens to digit 5 was a second contrahens to digit 3, also originating from metacarpal V.

The overall appendicular muscle anatomy of chameleons, particularly in terms of the number and identity of the muscles, is surprisingly conservative considering the remarkable structural and functional modifications of the limb skeleton. The most unusual soft tissue feature that we found in the limbs of chameleons was the configuration of the flexor retinacula and the short superficial digital flexors attached to them (Figure 18.14). However, even in this unusual configuration, no muscles commonly present in other lizards were absent and no additional muscles were present. In fact, the number and identities of the autopodial muscles in chameleons are surprisingly plesiomorphic, not only among squamates, but also among reptiles and even tetrapods in general.

Chameleons are specialized for arboreal locomotion, and many of their locomotor adaptations parallel those of mammals (Fischer et al. 2010). The functional morphology of the chameleon shoulder was characterized by Peterson (1973). In terms of kinematics, the shoulder of chameleons differs from that of other lizards in two main ways: (a) increased anterior–posterior movement of the pectoral girdle relative to the body wall and (b) greater anterior–posterior arc of movement of the limb, primarily achieved through increased protraction of the humerus (Peterson 1984). These differences are reflected in muscle anatomy as well; for example, the serratus anterior has relatively longer bellies in chameleons, which may facilitate the movement of the shoulder girdle on the body wall (Peterson 1973). Other differences in muscle anatomy, including the loss or fusion of the coracoid head of the triceps and of an associated coracoid arm of the sternoscapular ligament and the more proximal location of the pectoralis and deltoideus insertions, have been interpreted as adaptations for greater shoulder mobility (Peterson 1973; Russell and Bauer 2008). The more proximal insertion of the glenohumeral muscles translates into greater shoulder mobility because the same amount of muscle shortening produces a larger arc of movement of the humerus, at the expense of leverage (Peterson 1973). Likewise, the decreased role of humeral rotation due to the more sagittal orientation of the limbs in arboreal forms such as chameleons is reflected in the insertion of muscles such as deltoideus scapularis closer to the long axis of the humerus (Peterson 1973).

One of the most important adaptations for arboreal locomotion in both reptiles and mammals is prehensile appendages (Fischer et al. 2010). Several adaptations of the autopodial musculature of chameleons would appear to enhance grasping ability. For example, due to the distal and dorsal expansions of the flexor retinaculum from which they originate, the orientation of the flexores breves superficiales that insert on the digits adjacent to the cleft (i.e., 3 and 4 in the manus; 2 and 3 in the pes) is more transverse than proximodistal (Figure 18.14). Without changing the number of muscles or their main attachments, this configuration would allow muscles that in most lizards act as flexors to instead adduct and thus oppose the two superdigits of each autopodium. This change would presumably enhance the animal's prehensile ability (fittingly, Mivart [1870] designated these muscles "adductor digiti terti" and "adductor digit quarti"). The relatively large cross-sectional area of these muscles and their insertion directly onto the phalanges may also increase grasping ability. Not surprisingly, chameleons can generate substantially greater grasping forces than other lizards (Abdala et al. 2009). In contrast, the muscles lying between the individual digits within each superdigit are often either fused with other muscles (e.g., flexores digitorum breves profundi and extensores digitorum breves) or reduced in size (e.g., intermetatarsales). In fact, the reduction of the intermetacarpales and intermetatarsales is the only change in number of muscles clearly related to a particular functional feature in chameleons (cleft and related zygodactyly). While most pentadactyl tetrapods have four of these muscles in each autopod occupying the spaces between the five digits, in all chameleon taxa in which the myology has been analyzed, so far, the muscles between the digits affected by the cleft are lost, and there are normally only three intermetacarpales/intermetatarsales in each autopodium.

In summary, the most obvious musculoskeletal features related to the peculiar opposable autopodia of chameleons (Figure 18.14) are (a) syndactyly and zygodactyly of the autopod; (b) shortened metacarpals and modification of the carpal and tarsal bones; (c) peculiar broad, V-shaped plantar and palmar aponeuroses between the digits separated by a cleft; (d) oblique orientation of the superficial short flexors originating from these aponeuroses, which probably allows these muscles to act as powerful adductors of the two superdigits leading to their opposition; (e) absence of intermetacarpales and intermetatarsales between the digits separated by a cleft; and (f) well-developed abductor digiti minimi and abductor pollicis/hallucis brevis muscles, which could act as powerful abductors of the superdigits, leading to their reposition.

GENERAL REMARKS

The case study of chameleons exemplifies how comparative anatomical studies can shed light on the evolution of morphological and functional innovations, including phenotypes such as syndactyly and cleft autopodium that closely resemble congenital malformations of the human limb. By doing so, these studies also pave the way for developmental experimental works to investigate the mechanisms that lead to those phenotypes. While many studies have described the anatomical diversity of chameleons (e.g., Mivart 1870; Anderson and Higham 2014), few have focused on the development and morphogenesis of their unique anatomical traits (e.g., Rieppel 1993; Diaz and Trainor 2015). Given that there are more than 200 species of chameleons which greatly vary in body size, complement of wrist/ankle bones, and limb proportions, our understanding of the musculoskeletal system of the limbs and girdles in chameleons is limited. In addition to a detailed comparative study of the soft tissues of species of all chameleon genera, future work on chameleon development and genetics would help explain the processes behind their remarkable anatomy and evolutionary history. In a broader context, the same type of detailed study is also needed for each of the major groups we show in Tables 18.1 through 18.3 and should ideally be performed for each major extant chordate taxon.

19 Hindlimb Muscles of Tetrapods and More Insights on Pectoral–Pelvic Nonserial Homology

In the preceding chapters, we have discussed and illustrated in detail the muscular anatomy of the pelvic appendages of fishes and salamanders (see, e.g., Figures 16.2 through 16.4). In Chapter 19, we summarize the major differences between, and hypotheses of homology within, the hindlimb muscles of tetrapods. In addition to the information provided by Diogo and Molnar (2014), which is the main basis for this chapter, we also include information from our more recent works on frogs (Diogo and Ziermann 2014), marsupials (Diogo et al. 2016a), chameleons (Molnar et al. 2017a), and primates, including humans and our closest relatives, the common chimpanzees and bonobos (Diogo et al. 2017a). Therefore, Chapter 19 includes data about the evolution and homologies of all hindlimb muscles of representatives of each major tetrapod group, i.e., urodeles (Figures 16.2 through 16.4), anurans (Figure 19.1), turtles, lepidosaurs (Figure 19.2), crocodylians, birds, monotremes, marsupials (Figure 19.3), and placentals, including our own species (Figure 19.4; Tables 19.1 through 19.13). Namely, Tables 19.1 through 19.4 compare the hindlimb muscles of the following key taxa, chosen from among the many tetrapod taxa dissected by the author and colleagues (see the following): the salamander *Ambystoma mexicanum*, the lizard *Timon lepidus*, the marsupial *Didelphis virginiana*, the placental rodent *Rattus norvegicus*, and modern humans *Homo sapiens*. Tables 19.5 through 19.8 compare the hindlimb muscles of *Ambystoma* and the frog *Xenopus laevis*, while Tables 19.9 through 19.12 compare the turtle *Trachemys scripta*, the lizard *Timon sp.*, the crocodilian *Alligator mississippiensis*, and the bird *Gallus gallus*.

In addition, we provide Tables 19.14 through 19.22, which summarize the "topological correspondences" between forelimb and hindlimb muscles of representatives of all major tetrapod taxa, specifically the leg vs. forearm and foot vs. hand (no apparent topological correspondences were found in the pelvic and thigh vs. pectoral and arm muscles: see the following). We opted to use the simplest, most conservative and thus least subjective method available to compare these muscles. This method, described in detail by Diogo et al. (2013), uses the historical criteria for "correspondence" between parts of an organism, i.e., "positional or topographical similarity" *sensu* Owen (1866) or "morphological congruence" *sensu* Shubin and Alberch (1986). Anatomical characteristics of each muscle (distal attachments, proximal attachments, number of divisions, relations to other hard and soft-tissue structures, orientation of fibers, and number of joints crossed) were analyzed to determine which hindlimb muscles shown in Tables 19.1 through 19.12 have clear, direct (one to one) "topological equivalents" in the forelimb of the same taxa (Tables 19.14 through 19.22). This method also allows meaningful comparison between taxa, the results of which are given in Tables 19.14 through 19.22, including new information about the evolution and homologies of all hindlimb muscles in the taxa shown in Tables 19.1 through 19.12. Nontopological data, which might suggest more "speculative/overdone" and/or "obscure/less apparent" cases of "correspondence" (e.g., ontogeny of wildtype animals, development of transgenic animals, gene expression), are discussed in relevant sections of the text and in other tables. For more details about the methodology and illustrative examples of how it was applied, see Diogo et al. (2013).

Adoption of a unifying nomenclature for hindlimb muscles across the whole tetrapod clade (Tables 19.1 through 19.22) may be crucial to solve some major comparative problems in tetrapod evolution and anatomy. For instance, classical authors such as Ecker (1864) compared human muscles with those of frogs and birds without taking into account the configuration found in anatomically more plesiomorphic reptiles and amphibians, thus erroneously using the same names for nonhomologous muscles. Unfortunately, many of these names are still used in current textbooks, ontologies, and specialized papers. An illustrative example is Přikryl et al.'s (2009) study, which is otherwise excellent but uses names of human muscles to designate several nonhomologous pelvic–thigh muscles of frogs. Only one pelvic–thigh muscle that was present in the most recent common ancestor of amphibians and amniotes seems to be homologous with an individual human pelvic-thigh muscle (the gracilis; Tables 19.1 and 19.5). This confusing nomenclature is a major problem not only for comparative anatomy in general, but also for amphibian comparative anatomy and broader evolutionary biology because different names are used to designate homologous muscles in frogs and salamanders (see, e.g., Table 2 of Přikryl et al. [2009]). In fact, as recognized by Přikryl et al. (2009), many researchers still follow anuran and urodelan muscle nomenclatures published several decades ago (see, e.g., Gaupp 1896 for anurans and Francis 1934 for urodeles), when many authors did not consider frogs and salamanders to be closely related to each other (see, e.g., Edgeworth 1935).

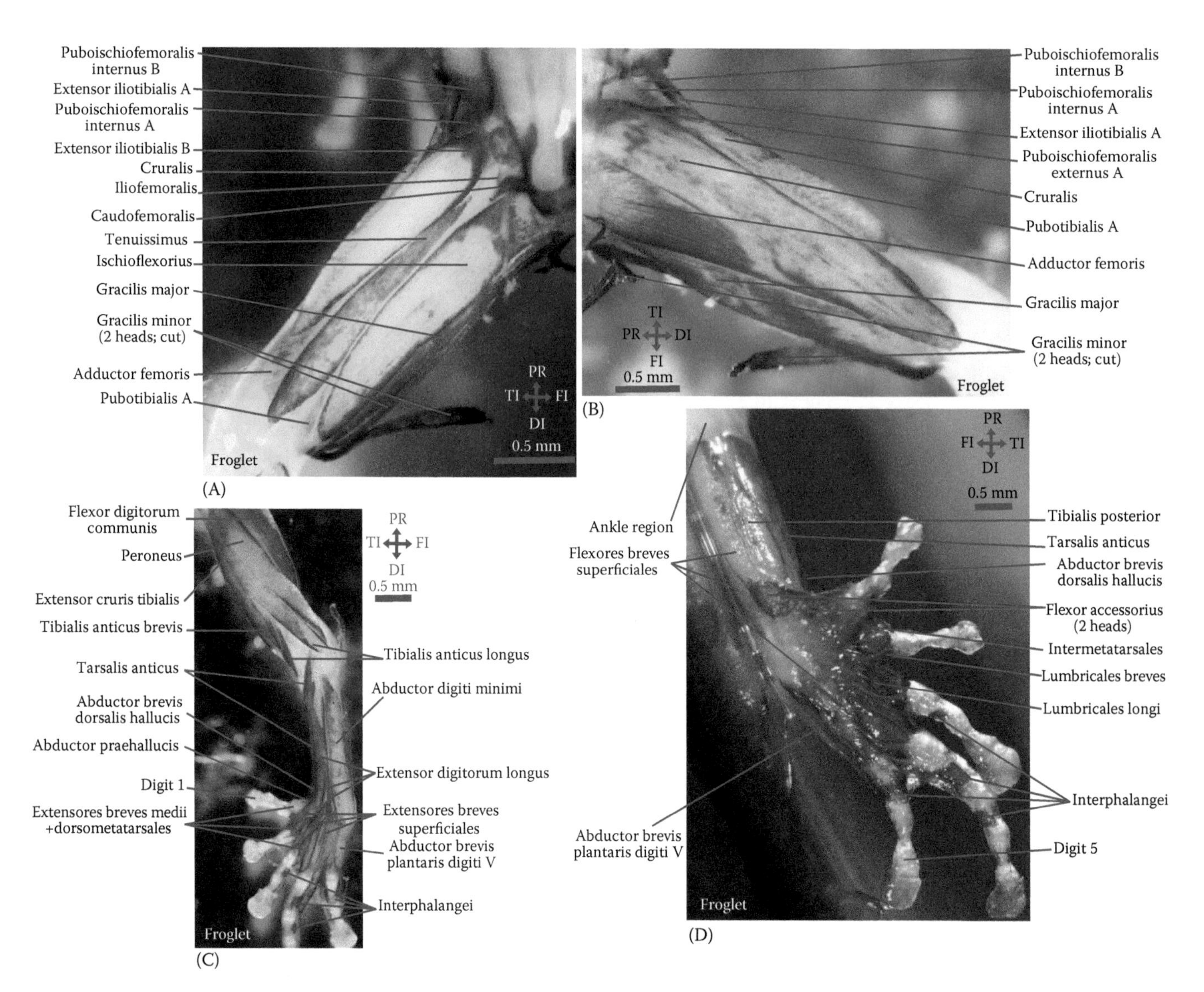

FIGURE 19.1 *E. coqui* (Amphibia, Anura): froglet (not adult) stage. (Modified from Diogo, R., and Ziermann, J. M., *J Exp Zool B (Mol Dev Evol)*, 322, 86–105, 2014.) (A) Dorsal view of the left pelvic and thigh muscles. (B) Ventral view of the left pelvic and thigh muscles. (C) Dorsal view of the left leg and foot muscles. (D) Ventral view of the left leg and foot muscles.

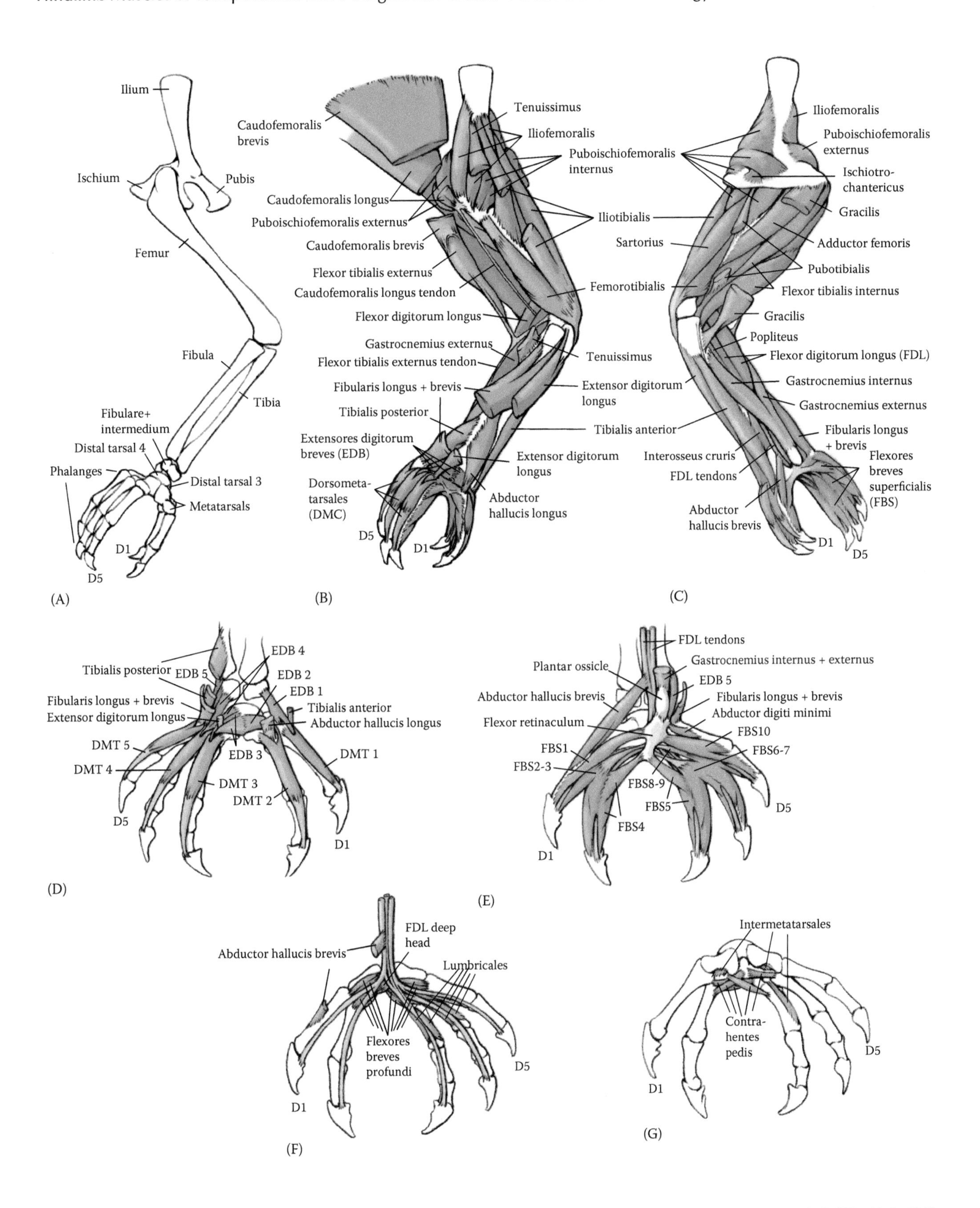

FIGURE 19.2 *Trioceros melleri* (Reptilia, Lepidosauria): left hindlimb. (Modified from Molnar, J. et al., *J Morphol*, 278, 1241–1261, 2017.) (A) Skeleton in lateral view. (B) Muscles in lateral view. (C) Muscles in medial view. (D) Dorsal muscles of pes. (E) Superficial ventral muscles of pes. (F) Intermediate muscles of pes. (G) Deep muscles of pes. D, digit.

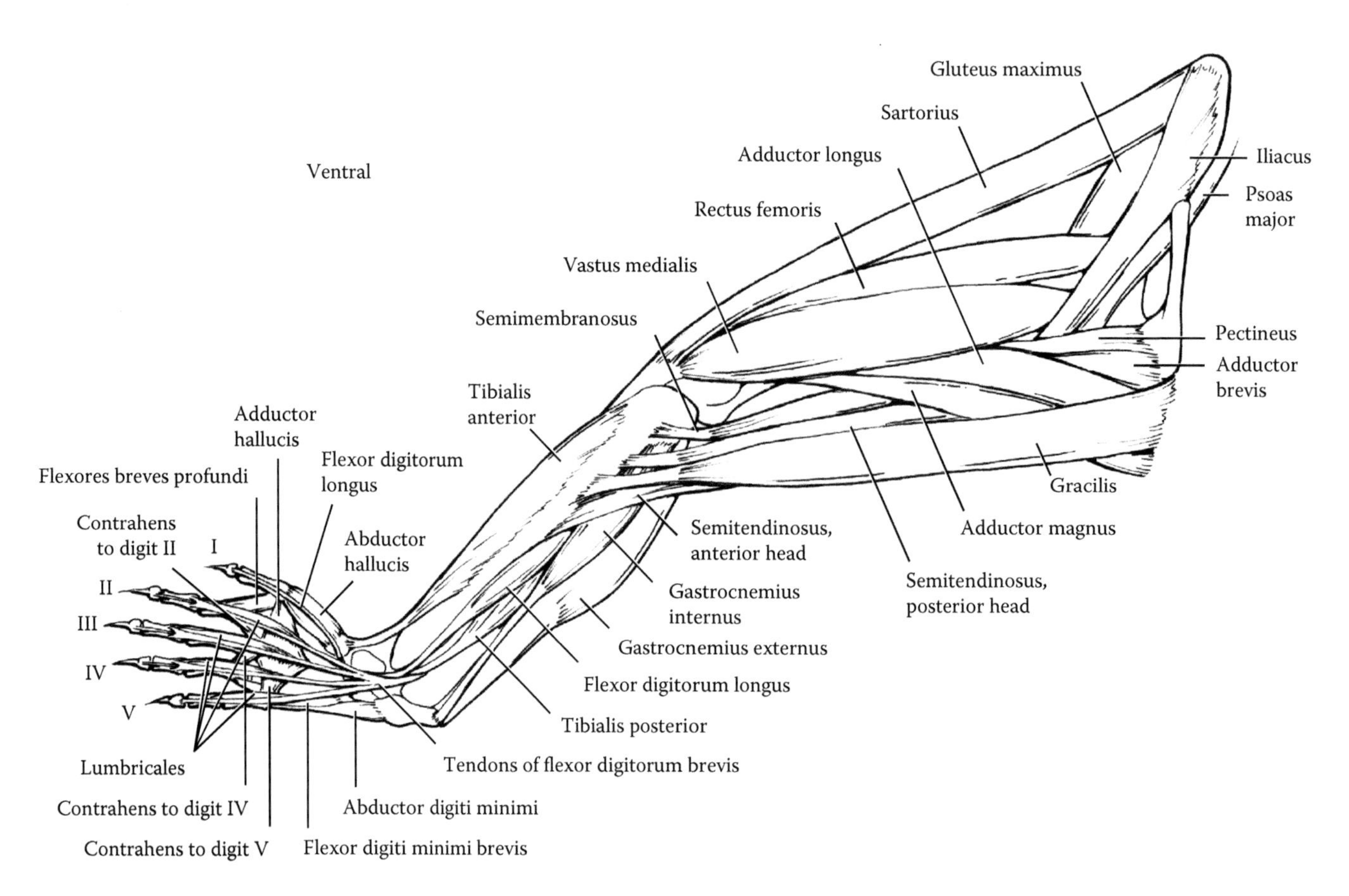

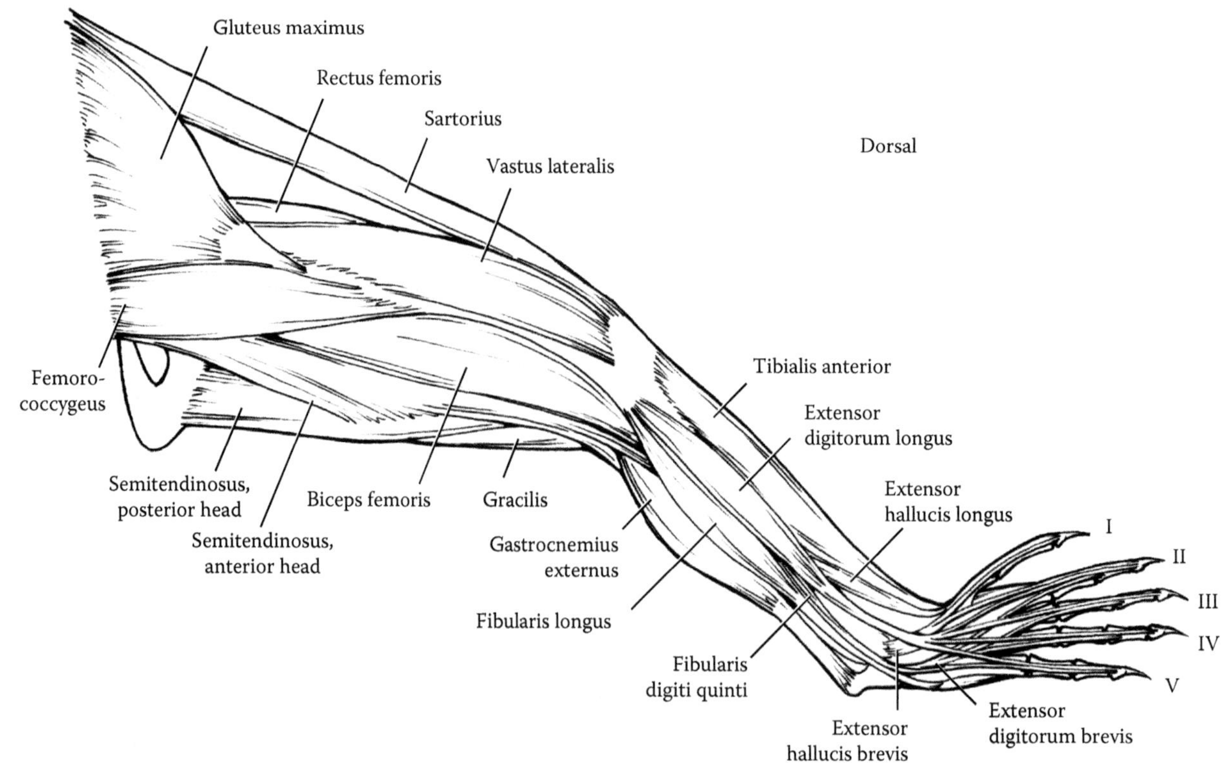

FIGURE 19.3 *D. virginiana* (Mammalia, Marsupialia). Schematic drawing by J. Molnar of hindlimb muscles.

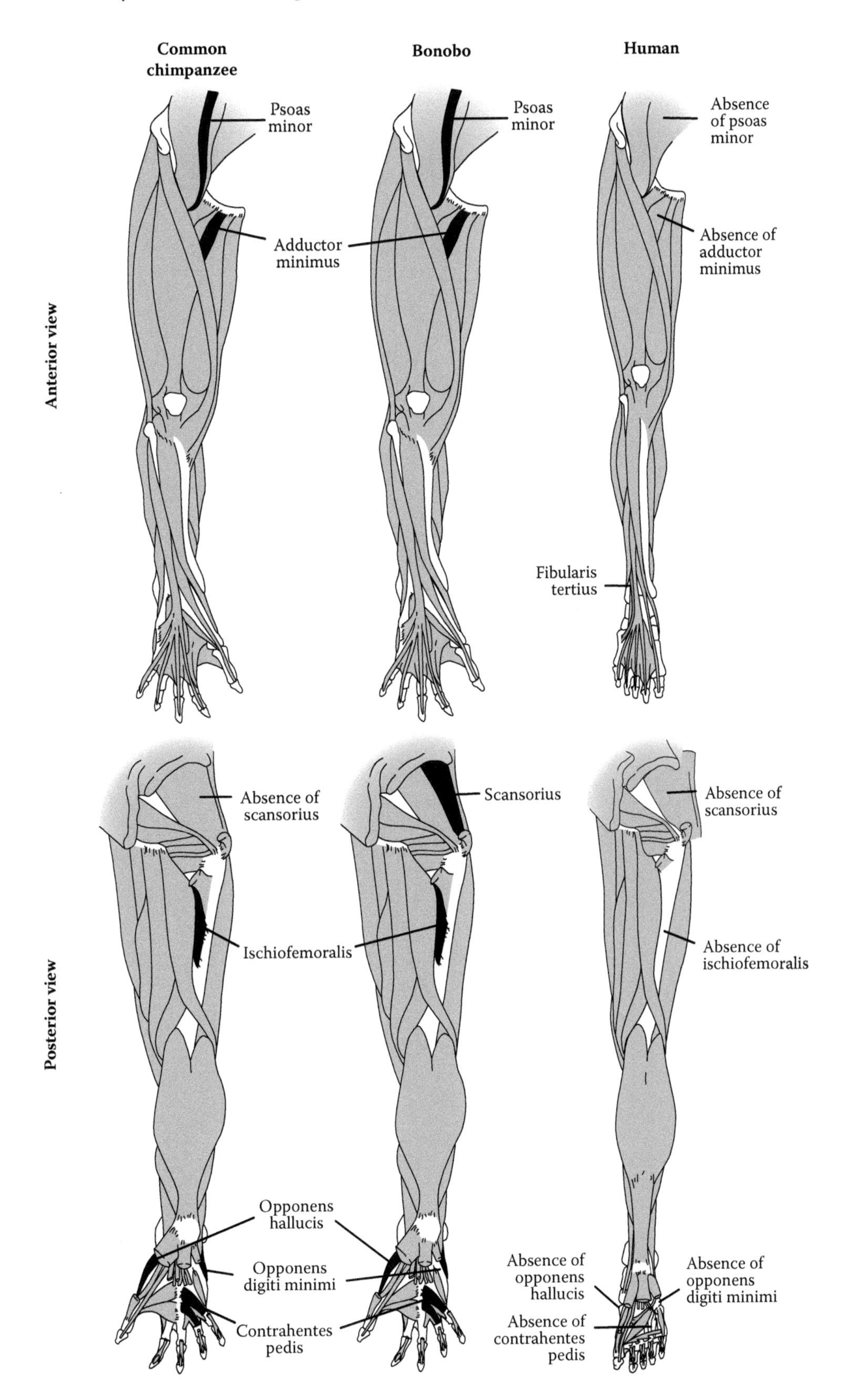

FIGURE 19.4 Differences between hindlimb muscles of common chimps, bonobos, and humans (drawing by Julia Molnar). (Modified from Diogo, R. et al., *Sci Rep*, 7, 608, 2017.) The only consistent difference between bonobos (*center*) and common chimps (*left*) concerning the presence/absence of muscles (shown in colors in the common chimps figure) is that the latter usually lack the scansorius, as is the case in humans. In contrast, there are many differences between bonobos and humans (right) concerning the presence/absence of muscles (shown in colors and/or with labels in the human figure; muscles present in chimps and not in humans are shown in black, in chimps).

TABLE 19.1
Hypotheses Regarding the Evolution and Homologies of the Pelvic/Thigh Muscles of Adults of Tetrapod Taxa

		Ambystoma (12 m.) [*see notes 1 below]	*Timon lepidus* (16 m.) [*see notes 2 below]	*Ornithorhynchus* (21 m.)	*LCA mar.+pla.* (30 m.)	*Didelphis* (27 m.)	*Rattus* (27 m.) [*see notes 4 below]	*Homo* (26 m.) [*see notes 5 below]
Axial		---	Quadratus lumborum	Quadratus lumborum	Quadratus lumborum	Quadratus lumborum	Quadratus lumborum	Quadratus lumborum
		---	---	Psoas minor	Psoas minor	Psoas minor	Psoas minor	---
Appendicular: Dorsal mass	Posterior	Iliofemoralis	Iliofemoralis	Gluteus medius	Gluteus medius	Gluteus medius	Gluteus medius (a+ p h.)	Gluteus medius (has a single h.)
		---	---	--- (absent monot.)	Piriformis	Piriformis	Piriformis	Piriformis
		---	---	--- (present other monot.)	Gluteus minimus	Gluteus minimus	Gluteus minimus (includes 'scansorius' h.)	Gluteus minimus
		---	---	--- (absent monot.)	Scansorius	Scansorius	--- (yes early ontogeny, then fused with gluteus minimus)	---
		---	---	--- (absent monot.)	---	--- (ab. mar.)	Tensor fasciae latae	Tensor fasciae latae
		---	---	--- (absent monot.)	Femorococcygeus	Femorococcygeus	--- (yes early ontogeny, then fused with biceps femoris)	---(fem.coc. apparentlyfused with gl.ma.)
		Extensor iliotibialis ('iliotibialis'; a.+p.h.)	Iliotibialis	Gluteus maximus	Gluteus maximus	Gluteus maximus	Gluteus maximus	Gluteus maximus
		---	---	Rectus femoris	Rectus femoris	Rectus femoris	Rectus femoris	Rectus femoris
		Tenuissimus ('iliofibularis')	Tenuissimus ('iliofibularis')	--- (absent monot.)	Tenuissimus	--- (but psom.)	--- (tenuissimus in early ontogeny, then disappears)	---
		---	Femorotibialis (3 h. in some reptiles)	Vastus lateralis	Vastus lateralis	Vastus lateralis	Vastus lateralis	Vastus lateralis
		---	---	--- (absent monot.)	---	---(apparently ab. mar.)	Vastus intermedius	Vastus intermedius
		---	---	Vastus medialis	Vastus medialis	Vastus medialis	Vastus medialis	Vastus medialis
	*	---	Sartorius ('ambiens')	Sartorius	Sartorius	Sartorius	---(yes early ontogeny, then apparently fused with gl.ma./te.fa.la.)	Sartorius
	Ante-	Puboischiofemoralis internus	Puboischiofemoralis internus (3 h.)	Iliopsoas	Psoas major	Psoas major	Psoas major	Psoas major
		---	---	---	Iliacus	Iliacus	Iliacus	Iliacus
#	#	---	---	Pectineus	Pectineus	Pectineus	Pectineus	Pectineus
Appendicular: Ventral mass	Adductors	Adductor femoris ('pubofemoralis')	Adductor femoris	Adductor magnus	Adductor magnus	Adductor magnus	Adductor magnus (includes adductor minimus h.)	Adductor magnus (includes ad. mi. h.)
		---	---	Adductor brevis	Adductor brevis	Adductor brevis	Adductor brevis	Adductor brevis
		Pubotibialis	Pubotibialis	Adductor longus	Adductor longus	Adductor longus	Adductor longus	Adductor longus
		Gracilis ('puboischiotibialis,' prox. and dis. h.)	Gracilis ('puboischiotibialis'; 3 h.)	Gracilis (a+p h.)	Gracilis	Gracilis	Gracilis (anterior and posterior h.)	Gracilis (has a single h.)
		---	---	Obturator externus	Obturator externus	Obturator externus	Obturator externus	Obturator externus
	Hamstrings	Femorofibularis	---	---	---	---	---	---
		Ischioflexorius	Flexor tibialis internus (2 h.)	Semimembranosus (a+ph.)	Semimembranosus	Semimembranosus	Semimembranosus	Semimembranosus
		---	---	--- (absent monot.)	Presemimembranosus	--- (but psom.)	--- (yes early ontogeny, then disappears)	---
		---	---	Biceps femoris	Biceps femoris	Biceps femoris	Biceps femoris (anterior and posterior h.)	Biceps femoris (short and long h.)
		---	Fl. tib. ext. (seems to include fl.cr.ta.ti.)	Semitendinosus(a+p h.)	Semitendinosus (a+p h.)	Semitendinosus (a+p h.)	Semitendinosus (a+p h.; a. h often designated 'crurococcygeus')	Semitendinosus (has a single h.)
		Caudofemoralis	Caudofemoralis longus	Caudofemoralis	Caudofemoralis	--- (but psom.)	Caudofemoralis	---
		---	Caudofemoralis brevis	---	---	---	---	---
	Ischiotro-	Puboischiofemoralis externus	Puboischiofemoralis externus (2 major h.)	Quadratus femoris	Quadratus femoris	Quadratus femoris	Quadratus femoris	Quadratus femoris
		Ischiotrochantericus ('ischiofemoralis')	Ischiotrochantericus	Ischiotrochantericus	Obturator internus	Obturator internus	Obturator internus	Obturator internus
		---	---	---	Gemellus superior	Gemellus superior	Gemellus superior	Gemellus superior
		---	---	---	Gemellus inferior	Gemellus inferior	Gemellus inferior	Gemellus inferior

Source: Diogo, R. et al., *Anat Rec*, 299, 1224–1255, 2016.

Note: See footnotes of tables in preceding chapters. *, Features of both anterior and posterior masses of dorsal mass; #, features of both ventral and dorsal masses. a+p, anterior + posterior; ab. mar. absent in marsupials; ad. adductor; ad. mi. adductor minimus; ant. anterior; dis. distal; fem.coc. femorococcygeus; fl.cr.ta.ti, flexor cruris et tarsi tibialis; fl.tib.ext. flexor tibialis externus; gl.ma. gluteus maximus; h. head(s); m. muscles; mar. + plac. marsupials + placentals; monot. monotremes; pos. posterior; pro. proximal; psom. present some other marsupials); te.fa.la. tensor fasciae latae.

(*Continued*)

TABLE 19.1 (CONTINUED)
Hypotheses Regarding the Evolution and Homologies of the Pelvic/Thigh Muscles of Adults of Tetrapod Taxa

[1A] Kardong (2002) states that the "ilioextensorius," "part of puboischiofemoralis externus," and iliotibialis of salamanders give rise to, respectively, the femorotibialis, iliotibialis, and ambiens of reptiles; however, as noted by Gadow (1882) and reported by Ashley-Ross (1992) and Walthall and Ashley-Ross (2006), in salamanders, there is usually a single muscle extensor iliotibialis in this region (which however often has anterior and posterior heads that are often named, respectively, as iliotibialis and ilioextensorius in urodeles: see, e.g., Appleton 1928), which thus seems to correspond to the femorotibialis, iliotibialis, and ambiens of lizards, as defended by authors such as Gadow 1882 and Walker 1954. Our dissections and comparisons between amphibians and reptiles strongly support this idea.

[1B] At first sight, the "pubofemoralis" *sensu* Ashley-Ross (1992) could seems to correspond to the ambiens of reptiles and thus the sartorius of mammals, because in a dorsal view is just deep and anterior to the iliotibialis and at least partially deep to the puboischiofemoralis internus (e.g., Ashley-Ross 1992; fig. 3A), as is usually the ambiens of reptiles (e.g., Figure 1.38A of Russell and Bauer [2008]); however, reptiles may have both an ambiens and a pubifemoralis, although these muscles are often closely related and sometimes the two names are even used as synonyms (see, e.g., Russell and Bauer 2008); our dissections and comparisons between amphibians and reptiles clearly support the idea that the pubifemoralis of amphibians such as salamanders corresponds instead to the adductor femoris of reptiles such as lizards; this is because both the amphibian pubofemoralis and the reptilian adductor femoris are deep ventral muscle that lie just deep (dorsal) to the pubotibialis and that extend from the ventral surface of the pelvic girdle to the ventral surface of the femur, while the reptilian ambiens is a dorsal muscle (see, e.g., Romer 1942 and Lance-Jones 1979); moreover, if one compares the ventral pelvic/thigh muscles of reptiles such as lizards and amphibians such as salamanders, almost all homologies are straightforward and somewhat consensual, with two exceptions: the pubifemoralis and ischiofemoralis of amphibians and the adductor femoris and ischiotrochantericus of reptiles; only a few authors have suggested that the amphibian ischiofemoralis might correspond to the adductor femoris (e.g., Gadow 1882), most authors agreeing that it corresponds instead to the reptilian ischiotrochantericus, as argued here; for instance, Walker (1954: 94) defends the latter hypothesis and states that the adductor femoris of *Necturus* is deeply blended with the puboischiofemoralis externus, as is the case with the pubofemoralis (see, e.g., Ashley-Ross 1992; our dissections), thus supporting the two homologies defended in the present work; Kardong (2002: Table 10.2) also clearly states that salamanders have a muscle that is directly homologous to the adductor femoris of reptiles and that this muscle is topologically related to the pubotibialis, as is the amphibian pubofemoralis (see preceding text); Appleton (1928: 434) also states that the amphibian ischiofemoralis corresponds to the reptilian ischiotrochantericus, and the pubifemoralis–adductor femoris synonymy was also argued by Francis (1934: 106); our dissections and comparisons confirm that both the latter muscles are ventral muscles that originate mainly from the posterior portion of the pelvic girdle and that mainly extend parallel to the anteroposterior axis of the girdle to attach onto the proximal region of the femur (compare, e.g., Figure 1.38 of Russell and Bauer 2008 to Figure 6A of Ashley-Ross 1992).

[1C] As noted by authors such as McMurrich (1905) and Bardeen (1906) in urodeles such as *Ambystoma*, the leg muscle flexor cruris et tarsi tibialis (*sensu* the present work) seems to correspond to the distal (leg) portion of the ischioflexorius. Most authors agree that the ischioflexorius of salamanders corresponds to both the flexor tibialis internus and flexor tibialis externus of lizards (e.g., Walker 1954 and Kardong 2002), but some authors suggested that part of these two lizards muscles (e.g., of the flexor tibialis externus according to Gadow [1882]) might instead derive from more caudal musculature such as the caudalipuboischiotibialis (e.g., Mivart 1869 and Gadow 1882); the latter muscle is however often regarded as part of the caudal–pelvic girdle musculature, although unlike caudal–pelvic muscles such as the ischiocaudalis that connect the pelvis to the tail, the caudalipuboischiotibialis mainly connects the tail to the puboischiotibialis muscle, thus being probably related not only to flexion of the tail but also to limb retraction (e.g., Walthall and Ashley-Ross 2006).

[2A] Lance-Jones (1979) stated that authors such as Haines (1934) and Howell (1938) argued that the iliofemoralis and at least part of the iliotibialis of lizards correspond to, respectively, the deep and superficial gluteal muscles of mammals and that her embryological comparisons support this idea, because early in development of mice, the deep and superficial gluteal muscles are positioned, respectively, deep and superficial to the fibular nerve trunk, like the embryonic reptilian iliotibialis and iliofemoralis. The deep anlage in the mouse, i.e., the gluteus medius, gluteus minimus, and piriformis, is homologous to the lizard iliofemoralis, because in both groups, this anlage has an acetabular bony origin, a proximal insertion, and a position posterior to the fibular nerve trunk. Although some authors (e.g., Pearson 1926) have homologized the piriformis with reptilian, ventrally derived caudofemoralis, no embryonic association with ventral anlagen was ever noted in the mouse. The tensor fasciae latae appears to cleave from the anterior tip of the deep gluteal mass and become secondarily associated with the gluteus maximus, although these features indicate that this muscle also derives from the iliofemoralis, and the origin of this muscle from the superficial mass cannot be completely excluded as its differentiation closely follows the separation of deep and superficial gluteal layers.

[2B] Lance-Jones (1979) stated that although the reptilian iliotibialis is difficult to homologize with mammalian muscles, it shares proximal features with the superficial gluteals and distal features with the rectus femoris in the adult condition (e.g., Romer 1922 and Appleton 1928). In early lizard ontogeny, the iliotibialis anlage is a massive sheet of mesenchyme that originates from the acetabular region and covers the dorsal surface of the thigh superficial to the femoral nerve trunk. A similar condensation is found in E12 mice, but in mice, the cleavage divides this mass into anterior and posterior parts, the anterior part maintaining its acetabular origin and distal insertion and giving rise to the rectus femoris. The posterior part gives rise to the femorococcygeus and gluteus maximus. Both the anterior and posterior parts of the mammalian lateral mass and the reptilian anterior and posterior iliotibialis are also comparably positioned with respect to the femoral and fibular nerve trunks.

(*Continued*)

TABLE 19.1 (CONTINUED)
Hypotheses Regarding the Evolution and Homologies of the Pelvic/Thigh Muscles of Adults of Tetrapod Taxa

[2C] Lance-Jones (1979) argued that the tenuissimus of mammals is homologous to the iliofibularis of reptiles, both muscles splitting off from the posterior dorsal mass of the pelvis/thigh very early in ontogeny, lying at the border of the dorsal and ventral masses postaxially and innervated by a distal branch of the fibular nerve that separates them from the rest of the postaxial dorsal mass. The position of the tenuissimus deep to the biceps femoris in adult mammals has prompted doubt about this homology in the past, but such a deep position is clearly derived because the tenuissimus of mice is superficial until E13.

[2D] Lance-Jones (1979) stated that the lizard femorotibialis and mammalian vasti muscles are clearly homologous, both extending from the femur and being separated from more superficial extensors by the distal femoral nerve trunk; some reptiles, e.g. chickens, have a femorotibialis divided into "externus," "medialis," and "internus" heads that seem to correspond, respectively, to the vastus lateralis, intermedius, and medialis of mammals, although these heads are deeply blended to each other and do not form separate muscles as in mammals (see, e.g., Schroeter and Tosney 1991).

[2E] The adult reptilian ambiens and the mammalian sartorius are quite morphologically different, especially with respect to their bony origins (Vaughn 1956). However, according to Lance-Jones (1979), the vestigial sartorius in mice shares several developmental features with the ambiens of lizards, e.g., both these muscles are derived from the preaxial dorsal mass, show a persistent proximal connection to the iliopsoas/puboischiofemoralis internus, and lie anterodorsal to the femoral and saphenous nerves and vessels. The clear and early separation of the sartorius from the gluteals and quadriceps proximally and its lack of any association with the ventral musculature argue against homology with either the iliotibialis (e.g., Howell 1938) or an adductor (e.g., Haines 1934). Our dissections of lizards clearly support the idea that the ambiens is homologous to the mammalian sartorius, because by applying a medial rotation of the femur in lizards as the one seeing during the ontogeny of mammals, the origin of the muscle becomes anteroproximolateral and the fibers of the muscle run distolaterally to an anterodistomedial insertion, exactly as in mammals.

[2F] Lance-Jones (1979) stated that in lizards, the puboischiofemoralis internus derives from the anterior dorsal anlage, together with the ambiens and corresponds to the iliopsoas (iliacus + psoas major) and the dorsal portion of the pectineus of mice, based on adult topology (Romer 1922; Howell 1938) and ontogeny. For instance, at early development, both the mouse iliopsoas and the lizard puboischiofemoralis internus lie adjacent to the pubis and iliac acetabular regions and anterior to the femoral nerve trunk, insert laterally on the femur, and are connected distally to the anlage of the quadriceps/triceps group. At later stages the puboischiofemoralis internus of lizards migrates over the dorsal surface of the pubis and into the interior of the girdle, the iliopsoas of mice migrating anteriorly onto the iliac blade and trunk. The dorsal portion of the pectineus of mammals is often homologized with part of the puboischiofemoralis of lizards (e.g., Leche 1886; Romer 1922) and the embryonic derivation of the dorsal part of the pectineus from the anterior and proximal dorsal muscle mass in mice, opossums, and humans. The ventral portion of the pectineus has been homologized with an adductor (e.g., Leche 1886; Howell 1938) or the pubotibialis (e.g., Vaughn 1956) of reptiles and Lance-Jones's ontogenetic comparisons support the idea that both the adductor longus and ventral portion of the pectineus of mammals correspond to the pubotibialis of lizards. This is because in lizard embryos, the pubotibialis arises from the superficial anterior part of the ventral muscle mass and, according to her, the homology of the ventral pectineus portion with the adductor femoris or puboischiofemoralis of lizards is excluded because of the position of the two latter muscles deep to the nerve to the gracilis. Further evidence that indicates a homology between a part of the pubotibialis and the adductor longus include the identical positions of the two structures with respect to the branch of the obturator nerve to the gracilis and the origin of both structures from the anterior or adductor mass; one must assume that the tibial insertion was lost during the transition from reptiles to mammals. Our dissections and comparisons support the idea that the puboischiofemoralis internus gave rise to the pectineus, iliacus, and psoas major, and one could hypothesize, based on the topology of its bundles, that parts 3, 2, and 1 directly correspond to a part/the totality of the pectineus (the more posterior/medial in mammals), psoas major, and iliacus (the more anterior/lateral in mammals).

[2G] Lance-Jones (1979) stated that the adductor femoris in lizards is usually a stout mass that arises deep to the puboischiotibialis and is fused with the flexors on their inner surface, the "adductor brevis" (main body of adductor magnus *sensu* the present work) and adductor magnus ("adductor minimus" head of adductor magnus *sensu* the present work) developing in a similar position in the mouse; they arise initially as superficial extensions of the proximal anlagen and lie posterior to the major obturator trunk, the "adductor brevis" (main body of adductor magnus *sensu* the present work) having an insertion that extends further distally than that of the adductor femoris, but this develops very late in the mouse. However, because the adductor brevis of the mouse, *sensu* the present work, apparently corresponds to either part of the "adductor longus" and/or of the "adductor magnus" *sensu* Lance-Jones (1979), it is not clear if the adductor brevis really derives from the adductor femoris or from the pubotibialis (the muscle giving rise to the adductor longus *sensu* Lance-Jones).

[2H] Lance-Jones (1979) stated that the "puboischiotibialis" or lizards and the gracilis of mammals such as mice arise superficial to both the deep anterior part of the ventral muscle mass and the deep flexors and are homologous; both separate from the rest of the ventral mass early, have slips that migrate anteriorly during development, and are innervated by branches of the obturator that pierce homologous deep adductor muscles.

(*Continued*)

TABLE 19.1 (CONTINUED)
Hypotheses Regarding the Evolution and Homologies of the Pelvic/Thigh Muscles of Adults of Tetrapod Taxa

[2I] Lance-Jones (1979) stated that the deep hamstring or flexor region of the ventral mass gives rise to the flexor tibialis internus complex in lizards and to the ventral head of the semitendinosus, biceps, and semimembranosus in mice. Apart other clear similarities in ontogeny, both the flexor tibialis internus and the corresponding anlage in the mice divide early in development into two parts separated by a branch of the hamstring nerve, the more superficial part giving rise to the flexor tibialis internus 2 of lizards and to the biceps and ventral semitendinosus of mammals and being often blended to the caudofemoralis/dorsal semitendinosus of mammals and the caudofemoralis/flexor tibialis externus of reptiles, the deeper portion giving rise to the flexor internus 1 of lizards and to the semimembranosus of mammals, being often blended to the adductors in both mammals and reptiles.

[2J] Lance-Jones (1979) stated that because of its adult lateral position, the biceps femoris has frequently be confused with gluteal muscles (e.g., Greene 1935) or homologized with dorsally derived reptilian muscles (e.g., Gregory and Camp 1918; Romer 1922; Haines 1935b). However, it clearly originates in mouse in association with the ventral musculature, migrating laterally or dorsally as ontogeny progresses. The mammalian semitendinosus is usually described as a single muscle with ventral (ischial) and dorsal heads of origin. Historically, however, the dorsal head has been treated separately and termed the "flexor cruris caput dorsale" (Appleton 1928), the "caudal head of the biceps" (Cunningham 1881), or the crurococcygeus (Romer 1942). When present, the semitendinosus is distinguishable from the other hamstrings by its caudal superficial origin and its course dorsal to the caudofemoralis and the perinaealis and hamstring nerves. Its frequent large size and caudal origin led Romer (1922) to homologize the dorsal head of the semitendinosus with a part of the reptilian caudofemoralis, but its position with respect to nerves led Appleton (1928) to homologize it with the flexor tibialis externus. As the caudofemoralis and dorsal head of the semitendinosus are closely associated during ontogeny, it is difficult to reject an origin from the reptilian caudofemoralis or from the reptilian flexor tibialis externus. However, the posterior ontogenetic origin of the dorsal part of the semitendinosus and its extensive distal association with the other flexors suggest its homology with the flexor tibialis externus, because in early ontogeny, both structures extend down the posterior surface of the thigh, later establishing more medial insertions on the shank.

[2K] Lance-Jones (1979) stated that the proximal superficial hamstring region gives rise to the caudofemoralis and dorsal head of the semitendinosus in mammals and to the caudofemoralis and flexor tibialis externus in reptiles. Although the reptilian caudofemoralis have been homologized with the mammalian caudofemoralis on the basis of similar relationships to nerves (Appleton 1928), they form a large complex of muscles that extends from the tail region to the femur as the major retractors of the thigh. Further, the bulk of the reptilian caudofemoralis inserts proximally, and the mammalian caudofemoralis, distally, and therefore, some researchers suggested that the caudofemoralis of reptiles correspond to caudal muscles (e.g., Haines 1935b) or instead to proximally inserting muscles such as the piriformis in mammals (e.g., Pearson 1926). However, both the reptilian caudofemoralis and the mammalian caudofemoralis form from the anterior part of the superficial hamstring mass, show medial connections to deep flexors and adductors at early stages, and are closely ontogenetically related with the ischiotrochanteric anlagen. At intermediate and late stages, the reptilian caudifemorales expand, and their origin migrates dorsocaudally, while the mammalian caudofemoralis has an origin that migrates only slightly dorsally and an insertion that migrates distally.

[2L] Lance-Jones (1979) stated that the mammalian obturator externus shares the following characteristics with the anterior portion of the puboischiofemoralis externus of reptiles: (a) an early posterior fusion with the ischiotrochanteric mass, (b) an anterior origin from the foramen and the pubis, and (c) an insertion early in development near the anteromedial trochanter. The anlage of the mammalian quadratus femoris is large and originating at early stages from the lateral surface of the ascending ramus of the ischium. This anlage is closely associated with the adductor magnus and obturator externus but shows only a very tenuous association with the caudofemoralis at early cleavage stages; these characteristics are very similar to those displayed by the posterior part of the reptilian puboischiofemoralis externus and support the homology of these two structures.

[2M] Lance-Jones (1979) stated that there is little question that the mammalian obturator internus and gemelli are homologous to the reptilian ischiotrochantericus, for they are morphologically quite similar in the adult condition and share ontogenetic features such as an early association with the caudifemorales/caudofemoralis, an initial lateral bony origin that later migrates to the pelvic surface of the girdle and a persistent connection to the deep adductors.

[3A] According to Lance-Jones (1979), the scansorius is present in 12- and 13-day-old mouse embryos (E12 and E13), being developed from the anlage of the iliopsoas (not of the tensor fasciae latae/gluteal muscles) and then fusing with the gluteus minimus, which is thus divided at its origin by the passage of the nerve to the tensor fasciae latae into a gluteus minimus proper originating from the posterior part of the gluteal fossa and into a scansorius head originating from the lateral crest of the ilium and the iliac fossa.

[3B] Also according to Lance-Jones (1979), the femorococcygeus can be recognized at E13.5 mice but is then integrated in the biceps femoris; our dissections of rats and mice indicate that it probably corresponds to the anterior head of the biceps femoris of adult rats and mice.

(Continued)

TABLE 19.1 (CONTINUED)
Hypotheses Regarding the Evolution and Homologies of the Pelvic/Thigh Muscles of Adults of Tetrapod Taxa

[3C] According to Greene (1935), in rats, the biceps femoris has an anterior head that corresponds to the short head of humans originating from the last sacral and first caudal vertebrae with the caudofemoralis, a posterior head originating from the sciatic tuber anterior to the accessory head, and an accessory head originating from the sciatic tuber with the semimembranosus, the two latter heads forming the long head of the biceps of humans and all three heads attaching together onto the distal femur and proximal 2/3 of the tibia. However, in both mice and rats, we found a biceps divided into the anterior head *sensu* Greene (1935) (which likely corresponds to the femorococcygeus, as suggested by Lance-Jones [1979] and by the fact that this head is innervated by the inferior gluteal nerve as reported by Greene 1935; see notes in the following) and a posterior head including the posterior and accessory heads *sensu* Greene (1935). (NB: authors such as Walker and Homberger [1997] refer to two heads as observed by us.) A major problem with Greene's (1935) hypothesis of homology between the long head of humans and the anterior head of rats is that the anterior head of rats crosses three joints (between axis and pelvis, pelvis and thigh, and thigh and leg), while the short head of humans only crosses one joint (between the thigh and leg; NB: both the posterior head of rats and the long head of humans cross two joints because they originate from the pelvis and insert onto the leg). In fact, the homology hypothesis of Greene (1935) was questioned by Lance-Jones (1979), who states that the mouse muscle tenuissimus that derives from the gluteal anlage and that, in E18 lies on the distal portion of the femur disappearing later in ontogeny, corresponds to the short head of the biceps femoris of humans. The topology of the tenuissimus is more similar to that of the short head of the biceps femoris of humans and its innervation as well because the tenuissimus is innervated by the fibular nerve and the short head of humans is innervated by the common fibular division of the sciatic nerve, but if Lance-Jones' (1979) homology hypothesis is right, it would mean that the short head derives from the gluteal anlage as does the tenuissimus of rats (and not from the hamstring anlage, as does the long head of the biceps femoris of humans). In the 0-day-old (P0) mice we dissected, there was, in the region where Lance-Jones (1979) reported a tenuissimus, what clearly seemed to be muscle fibers connected with the semitendinosus and/or the posterior head of the biceps femoris, extending dorsoventrally as shown in Figure 10 of Lance-Jones (1979), surrounded by some fatty/loose connective tissue. In the adult rats and 6-week-old mice we dissected, there were no muscle fibers on this region, just fatty/loose connective tissue. This thus confirms the observations of Lance-Jones according to which the muscle starts to degenerate after E18 but only disappears completely after the first week after birth and shows how a muscle can be progressively replaced by what seems to be fatty/loose connective tissue until there are no more muscle fibers left.

[3D] Greene (1935) stated that in adult rats, the rectus femoris has well-differentiated anterior and posterior heads, but our dissections of rats and mice indicated that this muscle is mainly undivided, as reported by Walker and Homberger (1997).

[3E] Distally some fibers of the vastus intermedius are somewhat distinct from those of its main body, so the former fibers may correspond to the ones that are sometimes considered to form the "articularis genus" of other mammalian taxa, which has a similar topology to the upper limb muscle anconeus of most mammals (which is however apparently a forearm muscle derived from the extensor carpi ulnaris) and to the posterior head of the triceps brachii of primates such as the strepsirrhines.

[3F] Lance-Jones (1979) stated that the mice sartorius is ontogenetically associated with, and perhaps a link between, the anterior dorsal mass (particularly the iliacus) and the posterior dorsal mass (particularly the quadriceps) and then disappears early in the ontogeny (after E13; according to Greene 1935, in adult rats, it is fused with the gluteus maximus and/or tensor fasciae latae), probably corresponding to the ambiens of reptiles and not to the structure that is sometimes designated sartorius in, e.g., chickens.

[3G] Lance-Jones (1979) also stated that in mice, the dorsal portion of the pectineus derives from the anterior ventral anlage together with the iliacus and the psoas major while the ventral portion of the pectineus derives from the adductor anlage together with the adductor longus. According to her, the double ontogenetic origin of the pectineus is "ancestral," although it is not clear if for rodents or for mammals as a whole.

[3H] The adductor brevis *sensu* Greene (1935) corresponds to the adductor magnus of Lance-Jones (1979), because it is deep to (from a medial view; i.e., more lateral than) the adductor magnus *sensu* Greene (1935; adductor brevis *sensu* Lance-Jones 1979). However, our dissections of rats and particularly mice clearly support a much more parsimonious hypothesis, i.e., that the muscles of these rodents are basically the same as seen in modern humans; i.e., particularly in mice, the fleshy head that gave rise to the well-defined tendon of the adductor longus is thinner than is normally shown in illustrations of rats and mice, and just deep/posterior to the adductor longus, there is a relatively well-separated structure that attaches onto the femur at about the same level as the adductor longus, just posterior to the attachment of this muscle. This structure is thus topologically very similar to the adductor brevis of modern humans, although it is thinner than the muscle of humans and a little bit more blended to the adductor magnus, and that is why it was probably not identified as a separate muscle in rats by Greene (1935), Walker and Homnerger (1997), and Wingerd (1998) and not identified in the mice embryos analyzed by Lance-Jones (1979). Then, more posteriorly, lies the adductor magnus, which is also topologically similar to that of humans, including a main body (which corresponds to the adductor brevis of Lance-Jones [1979] and to the adductor magnus *sensu* Greene [1935]) and an adductor minimus head (which corresponds to the adductor magnus of Lance-Jones [1979] and to the adductor brevis *sensu* Greene [1935]) that is posterior and proximal to the main body of the muscle, exactly as it often happens in modern humans (see, e.g., plate 463 of Netter 1989).

[3I] According to Lance-Jones (1979) stated that a gracilis formed by a single head is the plesiomorphic condition for mammals; i.e., that the presence of two heads as in e.g. mice is a derived condition.

(*Continued*)

TABLE 19.1 (CONTINUED)
Hypotheses Regarding the Evolution and Homologies of the Pelvic/Thigh Muscles of Adults of Tetrapod Taxa

[3J] According to Lance-Jones (1979), at E13 mice, a slip connects the caudofemoralis and semimembranosus in the proximal thigh region and then appears to fuse with the semimembranosus, thus suggesting that it is a presemimembranosus, and at E14, this presemimembranosus fuses dorsally with the caudofemoralis and then later in ontogeny disappears. According to her, the presemimembranosus is homologous to the hamstring (sciatic) part of the adductor magnus of humans and is present as a distinct muscle in other mammals, and the presence of both a caudofemoralis and a presemimembranosus is plesiomorphic according to her, this being a further case of parallelism between ontogeny and phylogeny.

[3K] According to Lance-Jones (1979), during mouse ontogeny, the posterior (her "ventral") and anterior (her "dorsal") heads of the semitendinosus have more affinity with the biceps femoris and the caudofemoralis, respectively, and the absence of the anterior (her dorsal) head in, e.g., humans and of the posterior (her ventral) head in some rodents is a derived condition.

[3L] The "cranial head of the semimembranosus" *sensu* authors such as Walker and Homberger (1997) corresponds to the caudofemoralis *sensu* Greene (1935), Lance-Jones (1979), and the present work.

[4] In addition to the major differences shown directly in the table, the major differences between adult humans and adult rats are (a) the gluteus maximus of humans is innervated by the inferior gluteal nerve, while that of rats is by this nerve and the inferior branch of the superior gluteal nerve (Greene 1935); (b) the piriformis and obturator internus are innervated by different nerves, but in rats, there is a branch of the lumbosacral plexus that innervates both these muscles (Greene 1935); (c) in humans, the quadratus femoris is innervated by a special nerve from the sacral plexus, while in rats, it is innervated by the posterior division of the femoral nerve (Greene 1935); (d) in humans, the psoas major and iliacus are innervated by L2 and L3 through the femoral nerve, while in rats, these muscles are said to be innervated directly from L2 and L3 of the lumbar plexus (Greene 1935); (e) the pectineus of humans is innervated by the femoral and obturator nerves, while in rats, it is only by the femoral nerve (Greene 1935); (f) in humans, the adductor magnus is innervated by the sciatic and obturator nerves, while in rats, it is only by the obturator nerve (Greene 1935); (g) in humans, the long and short heads of the biceps femoris are innervated by, respectively, the tibial and fibular divisions of the sciatic nerve, while in rats, the anterior head is by the inferior gluteal nerve and the posterior head (which corresponds to the posterior + accessory heads *sensu* Greene [1935]) is by the tibial division of sciatic nerve (Greene [1935]); (h) in humans, the semitendinosus is by the tibial division of the sciatic nerve, while in rats, the anterior ("principal") head is by the tibial division of the sciatic nerve and the posterior ("accessory") head is by a branch of the lumbosacral plexus (Greene 1935).

TABLE 19.2
Hypotheses on the Evolution/Homologies of Ventral/Flexor Leg Muscles of Adults of Representative Tetrapod Taxa

	Ambystoma (6 m.)	*Timon* (6 m.) [*see notes 2 below]	*Ornithorhynchus* (6 m.)	*LCA mar. + pla.* (8 m.)	*Didelphis* (7 m.)	*Rattus* (7 m.) [*see notes 4 below]	*Homo* (8 m.)
Long flexors	---(fle.cr. ta.ti. seems included in ischioflexorius)	---(fle.cr. ta.ti. seems included in fl.ti.ex. and ga.in)	---	---	---	---(fle.cr. ta.ti. seems included in ha.fa. and ga.)	---
	---	Gastrocnemius internus ('femorotibial')	Gastrocnemius internus	Gastrocnemius internus	Gastrocnemius internus	Gastrocnemius (medial and lateral h.)	Gastrocnemius (med.and lat. h.)
	---	Gastrocnemius externus ('femoral'; sup. and deep h.)	Gastrocnemius externus	Gastrocnemius externus	Gastrocnemius externus	Soleus	Soleus
	---	---	Plantaris	Plantaris	Plantaris	Plantaris	Plantaris
	Fle. dig. com. (d.1-5; includes fl.c.ta.fi.)	Fle. digitorum longus (d. 1-5)	Fle. digitorum longus (d. 1-5)	Fle. digitorum longus (d. 1-5)	Fle. digitorum longus (d. 1-5)	Fle. digitorum longus (d. 1-5)	Fle. digitorum longus (d. 2-5)
	---	---	---	---	---	---	Fle. hallucis longus (d. 1)
	Fle. accessorius medialis	---('fle. accessorius' is distal h. of fle. digitorum longus)	Quadratus plantae	Quadratus plantae	Quadratus plantae	Quadratus plantae (or derived from foot m.?)	Quadratus plantae (med. and lat. h.)
	Fle. accessorius lateralis	---	---	---	---	---	---
	Contrahentium caput longum	---	---	---	---	---	---
	---	Popliteus	Popliteus	Popliteus	---	Popliteus	Popliteus
	Interosseus cruris	Interosseus cruris	---(but present in Echidnas)	Interosseus cruris	Interosseus cruris	---	---
	Tibialis posterior ('pronator profundus')	Tibialis posterior ('pronator profundus')	Tibialis posterior	Tibialis posterior	Tibialis posterior	Tibialis posterior	Tibialis posterior

Note: See footnote of Table 19.1. d. digits; fle. flexor; fle.cr.ta.fi. flexor cruris tarsi fibularis; fle.cr.ta.ti. flexor cruris tarsi tibialis; fle.dig.com. flexor digitorum communis; fle.ti.ex. flexor tibialis externus; ga. gastrocnemius; ga.in. gastrocnemius internus; h. head(s); ha.fa. fascial insertion of hamstring muscles; Inc. includes; mar.+pla. marsupials + placentals; sup. superficial.

[1A] In lizards such as *Timon*, the abductor et extensor digit 1 apparently includes both an abductor and a part of the short extensor anlage to digit 1, the extensores digitorum breves including also a part of the short extensor anlage to digit 1 and short extensors to digits 2–5; in mammals, the short extensor to digit 1 is apparently not fused to an abductor and often forms a distinct muscle extensor hallucis brevis, while the short extensors to the other digits might be included into the extensor digitorum brevis complex or instead also form distinct muscles (e.g., the fibularis digiti quarti to digit 4 and the fibularis digiti quinti to digit 5 in mice), or alternatively be absent (e.g., the short extensor to digit 5 seems to be absent in humans).

(Continued)

TABLE 19.2 (CONTINUED)
Hypotheses on the Evolution/Homologies of Ventral/Flexor Leg Muscles of Adults of Representative Tetrapod Taxa

[2A] Most authors agree that rats have no fibularis tertius; i.e., in addition to the fibularis brevis and fibularis longus (which are present in humans), rats have only a fibularis digiti quinti and a fibularis digiti quarti. These two latter muscles are often considered to form a group with the fibularis longus and brevis, but according to authors such as Greene (1935), they are innervated by the deep fibular nerve, which innervates extensor hindlimb muscles such as the extensor digitorum brevis and does not innervate the fibularis longus and fibularis brevis (these muscles are instead innervated by the superficial fibular nerve). This fact, together with the fact that in rats, the extensor indicis of the forelimb is very similar to the extensor digitorum brevis of the hindlimb because both these muscles go to digits 2 and 3 and that the extensor pollicis longus is very similar to the extensor hallucis longus because both these muscles go to digit 1, strongly suggests that in these rodents, the fibularis digiti quarti and fibularis digiti quinti are the short extensors to digits 4 and 5 of the hindlimb, respectively, i.e., that they "correspond" to the extensor digiti quarti and extensor digiti quinti of the forelimb. This would imply that the fibularis digiti quinti and fibularis digiti quarti of rats lost the contact with the digital phalanges, because they insert, respectively, onto the distal ends of metatarsals V and IV. In fact, in the mice, we dissected and various other primates including primate taxa; the fibularis digiti quinti and/or fibularis digiti quarti do insert directly onto the extensor expansions of the digits, so for those cases, one would not even need to explain any major change. So, this is clearly more parsimonious than accepting the following changes to contradict that the fibularis digiti quarti and fibularis digiti quinti are not short extensors of the digits, e.g. in rats, (a) the short extensor of digit 4 of the hindlimb would be lost; (b) the short extensor of digit 5 of the hindlimb would also be lost, while in the meanwhile, two new muscles precisely going do digits 4 and 5 would be formed; i.e., (c) the fibularis digiti quarti would be formed from the lateral mass of leg muscles and (d) the fibularis digiti quinti would be formed from the lateral mass of leg muscles; and moreover (e) the innervation of these two muscles, which should then be by the superficial fibular nerve (which innervates the lateral muscles of the leg) was changed, the two muscles now being innervated by the deep fibular nerve that innervates the short extensors of the leg/foot. Be that as it may, contrary to the correspondence in rats of the extensor indicis of the forelimb with the extensor digitorum brevis of the hindlimb (both muscles are clearly short extensors going to the phalanges of digits 2 and 3), the potential correspondence of the fibularis digiti quinti and fibularis digiti quarti of the hindlimb with the extensor digiti minimi and extensor digiti quarti of the forelimb in rats is a much less obvious one and can only be suggested after a much more extensive analysis of the data (including innervation and information about other muscles) because the two former muscles are clearly located in the lateral and not the extensor compartment of the leg and moreover do not go to the phalanges of the digits, as the extensor digiti minimi and extensor digiti quarti do, but instead to the metatarsals. Therefore, following the methodology outlined in the Materials and Methods section of Diogo et al. (2012), these potential correspondences are not directly shown in the tables of this chapter. A different, but related question concerns the correspondence of the "fibularis tertius" of humans, which is innervated by the deep fibular nerve and goes to metatarsal V in humans and could thus be seen as a remaining of the short extensor of the digit 5 of the hindlimb and thus of the fibularis digiti quinti of rats and thus to the extensor digiti minimi of the forelimb as suggested by Quain et al. (1894). Some authors have also supported such a homology, because they noted that in the human upper limb the short extensor to digit 5 (extensor digiti minimi) is also superficial and deeply blended to the extensor digitorum, so it would make sense that the fibularis tertius is a short extensor of digit 5 deeply blended to the extensor digitorum longus in the lower limb. However, there are four main points contradicting this idea: (a) lizards often have both a fibularis digiti quinti and a fibularis tertius indicating that these muscles cannot be homologous to each other; (b) developmental studies indicate that the fibularis tertius derives from the anlage of the extensor digitorum longus and not of the extensor digitorum brevis of the hindlimb (e.g., Lewis 1910); (c) to consider that humans retain a short extensor of digit 5 as a fibularis tertius would require numerous secondary losses as other primates do normally do not have a fibularis tertius; (d) Lewis (1989), based on his studies of primates, also supports the idea that the fibularis tertius is a peculiar, derived muscle that only consistently appeared during human evolution and not a short extensor to digit 5 of the foot, a view also recently supported by our recent studies of common chimpanzees and bonobos, in which the fibularis tertius can be present as a variation (Diogo et al. 2017a).

TABLE 19.3

Hypotheses Regarding the Evolution and Homologies of the Foot Muscles of Adults of Representative Tetrapod Taxa

	Ambystoma (32 m.) [*see notes 1 below]	*Timon* (25 m.) [*see notes 2 below]	*Ornithorhynchus* (21 m.)	*LCA mar.+ pla.* (25 m.)	*Didelphis* (25 m.)	*Rattus* (22 m.) [*see notes 4 below]	*Homo* (19 m.) [*see notes 5 below]
Intrinsic foot muscles	Fles. breves superficiales (5 m. d.1-5)	Fles. breves superficiales (3 m. d.1-3)	Fle. digitorum brevis (d.2-4)	Fle. digitorum brevis (d.2-5)	Fle. digitorum brevis (d.2-5)	Fle. digitorum brevis (d.2-4)	Flexor digitorum brevis (d.2-5)
	—	Lumbricales (2 m. d.3-4)	—	Lumbricales (4 m. to d.2-5)	Lumbricales (4 m.. to d.2-5)	Lumbricales (4 m. to d.2-5)	Lumbricales (4 m. to d.2-5)
	Abductor digiti minimi (me. 5)	Abductor digiti minimi (me. 5)	Abductor digiti minimi (d.5)	Abductor digiti minimi (d.5)	Abductor digitiminimi (d.5)	Abductor digiti minimi (d.5)	Abductor digiti minimi (d.5)
	—	—	—	Abductor hallucis (d.1)	Abductor hallucis (d.1)	— Apparently fused with fle. hallucis brevis)	Abductor hallucis (d.1)
	—	—	—	—	—	—	Adductor hallucis accessorius (d.1)
	Contrahentes pedis (5 m. to d. 1-5)	Contrahentes pedis (5 m. d.1-5)	Contrahentes pedis (5 m. to d.1-5)	Contrahentes pedis (4 m.; d.1, 2, 4, 5)	Contrahentes pedis (4 m.to d.1, 2, 4, 5)	Contrahentes pedis (2 m. to d.2, 5)	—
	—	—	—	—	—	—	Adductor hallucis (obli. and transverse h.)
	—	—	Fle. digiti minimi brevis (=f.b.p.10, to d.5)	Fle. digiti minimi brevis (to d.5)	Fle. digiti minimi brevis (2h. to d.5)	Fle. digiti minimi brevis (d.5)	Fle. digiti minimi brevis (d.5)
	—	—	Fle. brevis profundus 2 (fibular side d.1)	Flex. brevis profundus 2 (d.1)	Fle. brevis profundus 2 (d.1)	Fle. brevis profundus 2 (d.1)	Fle. brevis profundus 2 (d.1)
	—	—	Fle. hallucis brevis (=f.b.p.1, tibial side d.1)	Fle. hallucis brevis (d.1)	Fle. hallucis brevis (=f.b.p.1, d.1)	Fle. hallucis brevis (d.1)	Fle. hallucis brevis (d.1)
	Fles. digitorum minimi (4 m. d.2-5)	—	—	—	—	—	—
	Interphalangeus digiti 3 (1 m. d.3)	—	—	—	—	—	—
	Interphalangei digiti 4 (2 m. d.4)	—	—	—	—	—	—
	Fles. breves profundi (10 m. d.1-5)	Fles. breves profundi (10 m. d.1-5)	Fles. breves profundi (3-9; 7 m. to d.2-5)	Fles. breves profundi (3-9; 7 m. to d.2-5)	Fles. breves profundi (3-9; 7 m. to d.2-5)	Fles. breves profundi (3-9, 7 m. to d.2-5)	Int. plantares (3 f.b.p. to d.3,4,5, =3 m.)
	—	—	—	—	—	—	Int. dorsales (4 f.b.p. + 4 interm. =4 m.)
	Intermetatarsales (4 m. between me.1-5)	Intermetatarsales (4 m. between m.1-5)	Intermetatarsales (4 m. between d.1-5)	Intermetatarsales (4 m. between d.1-5)	Intermetatarsales (4 m. between d.1-5)	Intermetatarsales (4 m. between d.1-5)	—

Note: See footnote of Table 19.1. d. digits; di. digitorum; fle. flexor; fles. flexores; f.b.p. flexores breves profundi; h. heads; int, interossei; interm. intermetatarsales; m. muscle(s); mar. + pla. marsupials and placentals; me. metatarsal(s); obli. oblique.

[1A] Most authors consider that the femorofibularis is a thigh muscle and that at least part of the gastrocnemius of reptiles has to derive from the flexor digitorum communis of plesiomorphic tetrapods; for instance, some authors stated that the flexor digitorum communis of salamanders corresponds to both the gastrocnemius externus plus the flexor digitorum longus of lizards (page 385 of Russell and Bauer [2008]). This is mainly the hypothesis followed in the present work. However, it cannot be discarded that the femorofibularis os salamanders is part of the flexor musculature of the leg, corresponding to the extensor cruris et tarsi tibialis and/or extensor cruris et tarsi fibularis of the extensor musculature of the leg and thus to the gastrocnemius complex (internus and/or externus) of reptiles; Howell (1938) described a very similar muscle in birds, which he named "femorocruralis" and is innervated by the tibial nerve (and not by the obturator nerve as suggested by Gadow [1882]); according to Howell (1938), this avian muscle seems to be the result of a proximal migration of part of the gastrocnemius complex, thus providing some support to the idea that the femorofibularis of amphibians might correspond to at least part of the gastrocnemius complex of reptiles. However, as noted by authors such as McMurrich (1905) and Bardeen (1906) in urodeles such as *Ambystoma*, the leg muscle flexor cruris et tarsi tibialis (*sensu* the present work) seems to correspond to the distal (leg) portion of the ischioflexorius, and the flexor cruris et tarsi fibularis, to the femoral head of the flexor digitorum communis. If this is so, in urodeles, the configuration of the ventral muscles of the forearm is very similar to that of the ventral muscles of the leg, including the presence of a flexor antebrachii et carpi radialis/flexor cruris et tarsi tibialis anlage and of a flexor antebrachii et carpi ulnaris/flexor cruris et tarsi fibularis anlage, the main difference being that in the forearm, the muscle/muscles derived from each anlage are present as distinct muscles, while in the leg, they are fused with the thigh muscle ischioflexorius and with the leg muscle flexor digitorum communis, respectively. Therefore, as explained earlier, we mainly follow here the consensual view that the femorofibularis is a thigh muscle, particularly because it is exclusively innervated by the sciatic nerve (e.g., Francis 1934) and because there is no such muscle or any apparent derivative of it in the leg of anurans, strongly supporting the idea that this is a peculiar, derived thigh muscle of urodeles and not an important part of the tetrapod leg bodyplan that gave rise to amniote leg muscles.

(*Continued*)

TABLE 19.3 (CONTINUED)
Hypotheses Regarding the Evolution and Homologies of the Foot Muscles of Adults of Representative Tetrapod Taxa

1B As explained by authors such as McMurrich (1905), the contrahentium caput longum is apparently completely absent in amniotes, although it cannot be discarded that part of it is fused with the flexor digitorum longus (or, much less likely, to deeper muscles such as the tibialis posterior/interosseous cruris).

2A According to Bardeen (1906), in lizards, the leg muscle flexor cruris et tarsi tibialis (*sensu* the present work) usually corresponds to the long distal tendon of the thigh muscle flexor tibialis externus, although it cannot be completely discarded that it also/instead corresponds to part of the leg muscle gastrocnemius internus, as proposed by McMurrich (1905).

2B McMurrich (1905) and Ribbing (1909) named the distal head of the flexor digitorum longus of lizards and *Sphenodon* (*sensu*, e.g., Russell and Bauer [2008]) as the "plantaris profundus accessorius 1" and the "flexor accessorius," respectively. McMurrich (1905) suggested that this structure derives from the pronator profundus, i.e., from the tibialis posterior *sensu* the present work. However, as Ribbing's (1909) name indicates, the distal head of the flexor digitorum longus of lizards is very similar and seems to correspond to the flexor accessorius complex of amphibians, being mainly situated on the tarsal region and running distomedially to attach onto the plantar aponeurosis and/or the structures associated with it (see, e.g., Figures 1.45 and 1.48 of Russell and Bauer [2008]). Ribbing (1909) also used the name "flexor accessorius lateralis" to designate the "tarsal part of the fibular head" of the flexor digitorum longus *sensu* Walker (1973). Haughton (1868) in turn used the name "flexori longus accessorius" to designate the distal part of the flexor digitorum longus of crocodylians such as alligators. That is, the flexor accessorius complex is a distinct structure in amphibians and apparently corresponds to the distal bundle of the flexor digitorum longus of reptiles such as lizards, *Sphenodon*, turtles, and crocodylians and to the distinct muscle quadratus plantae of mammals such monotremes, rats, and humans. It is however not clear if the LCA of amphibians and amniotes had a flexor accessorius complex divided into medial and lateral parts, as is usually the case in urodeles and often the case in some mammalian taxa (e.g., Lewis 1989) or if it was mainly an undivided structure, as is usually the case in anurans and in turtles and squamates (where it is an undivided bundle of the flexor digitorum longus) and in mammals such as humans, for instance.

2C Rewcastle's (1977) proposition that the popliteus is derived from the anlage of the interosseous cruris, together with the fact that in reptiles such as the lizard *Timon*, the muscle apparently usually originates from the tibia only and not the femur, as well as that the forearm of lizards usually has a pronator accessorius that runs from the radius to the fibula, and indicates that the popliteus derives from the interosseous cruris. In fact, according to Walker (1973), some turtles have "proximosuperficial" and "distodeep" portions of the interosseous cruris, which are apparently similar, respectively, to the popliteus and interosseous cruris of other reptiles such as lizards. Developmental studies of humans have also supported this hypothesis, because the popliteus does not derive from the gastrocnemius anlage but instead from a deeper anlage (e.g., Bardeen 1906). Therefore, this shows a further example of homoplasy between the hind and forelimbs, because the mammalian pronator teres of the upper limb derives from the flexor antebrachii et carpi radialis anlage and not from the pronator quadratus anlage; the pronator teres is a superficial muscle of the forearm in contrast to the pronator quadratus which is a very deep forearm muscle, while in the leg, the popliteus is a very deep muscle. That is, in reptiles such as lizards, the popliteus is ontogenetically more similar to the forelimb muscle pronator accessorius than to the forelimb muscle pronator teres.

3A Kardong's (2002) Table 10.2 states that the femorotibial or "internal" head of the gastrocnemius gave rise to the mammalian gastrocnemius medialis and flexor hallucis longus, but it instead very likely gave rise to the mammalian gastrocnemius medialis, while the femoral or "external" head gave rise to the gastrocnemius lateralis, soleus, and plantaris. Our dissections of mice and rats support the idea that the soleus and plantaris derived from the lateral part of the gastrocnemius complex as proposed in the developmental works such as Bardeen (1906), because topologically in rats, these muscles lie more laterally (on the fibular side) than in humans. According to Bardeen (1906), in mammals, the leg muscle flexor cruris et tarsi tibialis (*sensu* the present work) usually corresponds to the fascial distal insertion of the thigh hamstring muscles, although it cannot be completely discarded that it also/instead corresponds to part of the leg muscle gastrocnemius, as proposed by McMurrich (1905).

(*Continued*)

TABLE 19.3 (CONTINUED)
Hypotheses Regarding the Evolution and Homologies of the Foot Muscles of Adults of Representative Tetrapod Taxa

[3B] Our dissections of mice and rats show that the quadratus plantae of these mammals shares more similarities with the palmaris brevis of the hand than it does in humans, because the quadratus plantae of mice and rats also lies deep (less plantar) to the flexor digitorum brevis but is quite superficial and mainly lies on the fibular side of the foot (in humans the muscle is broader and less fibular), in the same way as the palmaris brevis lies mostly on the ulnar side of the hand; this could seem to indicate that as the palmaris brevis of the hand derives from the flexores breves superficiales, the quadratus plantae could derive from the flexores breves superficiales group or from other foot muscles. However, it is commonly accepted that the quadratus plantae is instead a ventral (flexor) leg muscle that is closely related to the flexor digitorum/hallucis longus (e.g., Quain et al. 1894; McMurrich 1905; Bardeen 1906; Schaeffer 1941). In fact, this muscle seems to correspond specifically to the ventral leg muscle complex formed by the flexor accessorius lateralis and flexor accessorius medialis of urodeles, because (a) during human ontogeny, the quadratus plantae clearly starts to develop near the flexor digitorum/hallucis longus and then migrates distally; i.e., it is very likely a leg muscle that migrates to the foot (e.g., Bardeen 1906; Cihak 1972); (b) the flexor accessorius complex of plesiomorphic tetrapods such as urodeles mainly runs distomedially (to the tibial side) to insert onto the plantar aponeurosis formed by the flexor digitorum communis, and the quadratus plantae of mammals mainly runs distomedially to insert onto the tendons of the flexor digitorum longus, which derives from the flexor digitorum communis of amphibians; (c) in amphibians such as urodeles, the flexor accessorius complex is usually differentiated into a muscle flexor accessorius medialis and a muscle flexor accessorius lateralis, and the quadratus plantae of mammals, which is often designated "flexor accessorius" in the literature, is often differentiated into similar medial and lateral bundles/muscles (e.g., Lewis 1989); and (d) Schroeder et al. (2014) state that medial head of quadratus plantae of humans is unique and is derived from part of the flexor hallucis longus of other primates, reinforcing the idea that the whole quadratus plantae is derived from the leg muscles and is therefore not an intrinsic foot muscle.

[4A] There is much confusion in the literature about the origin of the flexor digitorum longus, flexor hallucis longus, and tibialis posterior of mammals, and interestingly, this confusion is somewhat similar to the confusion concerning the flexor digitorum profundus and flexor hallucis longus of the upper limb and illustrates what is perhaps the most serious common problem in comparative myology: the tendency to consider homologous muscles of different animals as nonhomologous, "special" muscles. Greene (1935) reports a "flexor hallucis longus"/"flexor fibularis" running from the fibula, interosseous membrane, and tibia to the distal phalanx of digits 1–5 and a "flexor digitorum longus"/"flexor tibialis" extending from the tibia and fibula to the terminal phalanx of digits 1–5 together (i.e., fused) with the tendons of the flexor hallucis longus/flexor fibularis, which is designated flexor hallucis longus because it is more fibular than the flexor digitorum longus, as in humans, the belly of the flexor hallucis longus is more fibular than the belly of the flexor digitorum longus. However, unlike humans, in rats and mice, there is no distinct flexor hallucis longus going only to the hallux; i.e., the configuration of these two muscle structures is very similar to the heads of the flexor digitorum profundus of the rat forearm, which sends tendons to digits 1-5 of the hand; i.e., in rats, there is no distinct flexor pollicis longus extending to the hand only. Therefore, the two structures reported by Greene (1935) and most authors as two different muscles correspond to the fibular and tibial heads of the flexor digitorum longus *sensu* the present work; our dissections confirmed that the fleshy bellies of these two heads were blended to each other as well. The homology of the tibialis posterior of mammals is also straightforward: in amphibians and reptiles, there is a deep muscle pronator profundus that usually runs mainly distomedially from the fibula to the tibiale and/or metatarsal I; in phylogenetically plesiomorphic mammals such as marsupials, the deep muscle tibialis posterior runs mainly distomedially from the fibula to the navicular tuberosity, which represents the tibiale of phylogenetically plesiomorphic tetrapods. The homology of the pronator profundus of nonmammalian tetrapods and of the tibialis posterior of mammals is thus clear and was pointed out more than 14 decades ago by authors such as Owen (1866; see also, e.g., Schaeffer 1941). However, in more recent textbooks, it is often stated that lizards and other reptiles do not have a tibialis posterior and that the tibialis posterior of mammals corresponds to part of the gastrocnemius complex of lizards (e.g., Kardong 2002, which is otherwise an excellent book). Also, authors such as Lewis (1989), who wrote a monograph that is also otherwise excellent, continue to accept the confusing statements of authors such as McMurrich (1905) and Bardeen (1906) that the flexor digitorum longus and the tibialis posterior are phylogenetically and ontogenetically more related to each other than to the flexor hallucis longus.

TABLE 19.4
Hypotheses Regarding the Evolution and Homologies of the Dorsal/Extensor Muscles of Adults of Representative Tetrapod Taxa

	Ambystoma (9 m.) [* see notes 1 below]	*Timon* (9 m.)	*Ornithorhynchus* (6 m.)	*LCA mar.+ pla.* (9 m.)	*Didelphis* (8 m.)	*Rattus* (8 m.) [* see notes 4 below]	*Homo* (8 m.) [* see notes 5 below]
Long extensors	---	---	Ex. hallucis longus (d. 1)	Ex. hallucis longus (d. 1)	Ex. hallucis longus (d. 1)	Ex. hallucis longus (d. 1)	Ex. hallucis longus (d. 1)
	Ex. digito. longus (me. 1-5)	Ex. digito. longus (me. 2-3)	Ex. digito. longus (d. 2-5)	Ex. digito. longus (d. 2-5)	Ex. digito. longus (d. 2-5)	Ex. digito. longus (d. 2-5)	Ex. digito. longus (d. 2-5)
	---	---	---	---	---	---	Fibularis tertius (me. 5)
	Ex. tarsi tibialis	Tibialis anterior	Tibialis anterior	Tibialis anterior	Tibialis anterior	Tibialis anterior	Tibialis anterior
	Ex. cruris tibialis	---	---	---	---	---	---
	Ex. cruris et tarsi fibularis	Fibularis longus	Fibularis longus	Fibularis longus	Fibularis longus	Fibularis longus	Fibularis longus
	---	Fibularis brevis	--- (apparently not separate m. in monot.)	Fibularis brevis	Fibularis brevis	Fibularis brevis	Fibularis brevis
Short extensors	---	---	---	Fibularis digiti quarti (me.4/d.4)	--- (distinct some marsupials)	Fibularis digiti quarti (me.4 rats; d.4 mice)	---
	---	---	Fibularis digiti quinti (short ex. d. 5; me.5/d.5)	Fibularis digiti quinti (me.5/d.5)	Fibularis digiti quinti (d.5)	Fibularis digiti quinti (me.4 rats; d.5 mice)	---
	Exs. digito. breves (4m. to d.2-5)	Exs. digito. breves (4 m. d. 2-5 + part short ex. d.1)	Ex. digito. brevis (1 m. to d. 1-4 or 1-5?)	Ex. digito. brevis (1 m. to d.2-3)	Ex. digito. brevis (1m.; d.2-4)	Ex. digito. brevis (1 m. to d. 2-3)	Ex. digito. brevis (1 m. d.2-4)
	---	---	---	Ex. hallucis brevis (d.1)	Ex. hallucis brevis (d.1)	---	Ex. hallucis brevis (d.1)
	Abd. et ex. digiti 1 (d.1; abd.+ short ex. d.1)	Abd. et ex. digiti 1 (d.1; abd. + part short ex. d.1)	---	---	---	---	---

Note: See footnote of Table 19.1. abd. abductor; d. digits; digito, digitorum; ex. extensor; exs. extensores; h. heads; m. muscle(s); mar. + pla. marsupials and placentals; me. metatarsal(s); monot. monotremes.

1A It should be noted that our recent studies of *Ambystoma mexicanum* (Diogo and Tanaka 2012) confirmed that the members of this species usually have a muscle interphalangeus digiti 3 in the hand connecting the metacarpophalangeal and first interphalangeal joints of digit 3, as reported by, e.g., Grim and Carlson (1974a,b). This muscle is very similar to the muscle interphalangeus digiti 3 of the foot, which connects the metatarsophalangeal and first interphalangeal joints of digit 3; our dissections and comparisons revealed that unlike the hand, in the members of this species, digit 4 of the foot has a similar muscle interphalangeus as well as a more distal muscle interphalangeus apparently connecting the first and second interphalangeal joints of this digit.

2A Greene (1935) described a single contrahens muscle in rats, from the contrahens fascia to the sesamoid bone of the fibular side of digit 2 ("adductor indicis"), but there is clearly also a contrahens from the contrahens fascia to the tibial side of digit 5 in the mice and rats we dissected; many other rodents also have contrahentes to digits 2 and 5 of the foot (e.g., Rocha-Barbosa et al. 2007).

2B Greene (1935) described interossei to digits 1 (1), 5 (1) and 2, 3 and 4 (2 each), in a total of 8 muscles; as he did in the descriptions of the hand muscles, he clearly confused the identity of these foot muscles, because as in the hand, our dissections of the foot indicate that rats have a full series of flexores breves profundi with 10 muscles in total, one to each side of each digit, and of intermetatarsales with 4 muscles in total, which mainly abduct the digits. All these muscles usually originate from the tarsal and/or metatarsal region, and the flexores breves profundi 1, 2, and 10 correspond to the "medial head of the flexor hallucis brevis," "lateral head of flexor hallucis brevis," and "flexor digiti minimi brevis" *sensu* Greene (1935), the other 7 flexores breves profundi thus corresponding to part of the interossei *sensu* Greene (1935). Our dissections of mice and rats confirmed mice that the 10 flexores breves profundi, and the four intermetatarsales are present as distinct muscles, i.e., more separated from each other than in humans, the axis i.e., the digit having 2 intermetatarsales, being apparently digit 4 as is the case of the foot of primates such as lemurs, not digit 2 as in the human foot.

3A As explained by Diogo et al. (2012), in the human foot, the adductor pollicis accessorius ("volaris primus of Henle": Bello-Hellegouarch 2012) is mirrored by the "interosseous plantaris hallucis" *sensu* Cihak (1972), which as the "volaris primus of Henle" is probably derived from the contrahentes layer and, namely, from the adductor hallucis in this specific case, but could alternatively be derived from the flexores breves profundi layer. It should also be noted that both the transverse and oblique heads of the adductor pollicis are mainly ontogenetically derived from the anlage of contrahens 1, while in the foot, the oblique head of the adductor hallucis is derived from the anlage of contrahens 1, but the transverse head is derived from a different, neomorphic anlage of the contrahens layer (e.g., Cihak 1972).

3B According to the developmental study of Cihak (1972), unlike the opponens hallucis (which is absent at all stages of normal human ontogeny, although may be present as anomaly/variation in some humans), the opponens digiti minimi of the foot is always present as a distinct structure in early normal human ontogeny, just next to the flexor digiti minimi brevis, but then becomes unrecognizable as a distinct structure, in later ontogenetic stages.

3C As explained by Cihak (1972), as some dorsal interossei of the human hand apparently include part of the dorsometacarpales, which are originally part of the dorsal/extensor muscles of the forearm, some dorsal interossei of the human foot apparently also include part of the dorsometacarpales, which are originally part of the dorsal/extensor muscles of the leg (the dorsometacarpales are present as distinct muscles in adults of tetrapod taxa such as the Anura, for instance).

TABLE 19.5

Diagram Illustrating Hypotheses Regarding the Homologies of the Appendicular Pelvic and Thigh Muscles of Adults of Representative Urodele and Anuran Amphibian Taxa

Lower limb m.		***Ambystoma mexicanum* [12 pelvic and thigh muscles]**	***Xenopus laevis* [17 pelvic and thigh muscles][1D]**
Dorsal mass	Posterior	Iliofemoralis	Iliofemoralis
		Tenuissimus ('iliofibularis')	Tenuissimus ('iliofibularis')
		Extensor iliotibialis ('iliotibialis'; anterior and posterior heads)	Extensor iliotibialis A ('tensor fasciae latae')
		---	Extensor iliotibialis B ('**gluteus maximus**')
		---	Cruralis
	Ante-rior	Puboischiofemoralis internus	Puboischiofemoralisinternus A ('iliacus internus')
		---	Puboischiofemoralis internus B ('iliacus externus'; 3 heads)
Ventral mass	Adductors	Adductor femoris ('pubofemoralis')	Adductor femoris ('**adductor magnus**'; 2 heads)[1A]
		---	---(but pubotibialis B; i.e.,'semitendinosus,' present in other anurans)[1A]
		Pubotibialis	Pubotibialis A ('sartorius'; 2 heads)[1A]
		Gracilis ('puboischiotibialis,' proximal and distal heads)	**Gracilis major**[1B]
		---	**Gracilis minor**[1B]
	Ham-strings	Ischioflexorius (seems to include part/totality of fl.cr.ta.ti.)	Ischioflexorius ('**semimembranosus**'; 2 heads; fl.cr.ta.ti. seems to correspond to cruroastragalus)[1B]
		Femorofibularis	---
		Caudofemoralis	Caudofemoralis ('piriformis'; may be reduced or absent in some *Xenopus* specimens
	Ischiotro-chantericus	Puboischiofemoralis externus	Puboischiofemoralis externus A ('pectineus')
		---	---(but puboischiofemoralis externus B; i.e.,'adductor longus,' present in other anurans)
		Ischiotrochantericus ('ischiofemoralis')	Ischiotrochantericus A ('**gemellus**')[1C]
		---	Ischiotrochantericus B ('obturator externus')[1C]
		---	---(but ischiotrochantericus C; i.e.,'quadratus femoris,' present in other anurans)[1C]
		---	Ischiotrochantericus D ('**obturator internus**')[1C]

Note: NB: axial pelvic muscles such as the psoas minor and quadratus lumborum do not seem to be differentiated in amphibians. The muscles that are usually named "tensor fasciae latae," gluteus maximus, gemellus, obturator externus, quadratus femoris, obturator internus, pectineus," adductor longus, "iliacus," "sartorius," "semitendinosus," and "adductor magnus" in anurans are clearly not homologous to the muscles designated with the same name in humans, because the LCA of amphibians and amniotes did not have any of these muscles, hence, we use a different name in anurans. There are however a few cases (highlighted in bold) in which although the muscle that is designated with a certain name in anurans does not directly correspond to the muscle designated with the same name in humans, both the anuran and human muscles derive from a similar anlage/ancestral muscle: the anuran gluteus maximus and the human gluteus maximus derive from the extensor iliotibialis; the anuran adductor magnus and the human adductor magnus derive from the adductor femoris; the anuran gracilis major, and gracilis minor, and the human gracilis derive from the gracilis; the anuran semimembranosus and the human semimembranosus derive from the ischioflexorius; and the anuran gemellus and obturator internus and the human gemelli and obturator internus derive from the ischiotrochantericus. We prefer to use A, B, C, and D to designate the anuran muscles, instead of I, II, III, and IV, because names such as "puboischiofemoralis internus I" or "puboischiofemoralis internus II" are often used to designate the heads of the muscle puboischiofemoralis internus of reptiles such as lizards (for more details, see text). fl.cr.ta.ti, flexor cruris et tarsi tibialis.

[1A] The adductor magnus of anurans clearly seems to correspond to the adductor femoris (i.e., the pubofemoralis) of salamanders, although the hypothesis of Noble (1922) that it also receives a contribution from the pubotibialis of salamanders cannot be completely discarded.

[1B] We agree with Přikryl et al. (2009) that the anuran gracilis major and minor seem to correspond to the gracilis of urodeles, but we agree with Noble (1922) that the anuran semimembranosus mainly seems to correspond to the ischioflexorius of urodeles and, namely, to its proximal (thigh) portion (the distal portion of the ischioflexorius of urodeles apparently corresponds to the flexor cruris et tarsi tibialis). Přikryl et al. (2009) suggested that the anuran semimembranosus derives from the urodele gracilis, but then in their text and their Table 2, they do not explain to which anuran muscle/muscles corresponds the urodele ischioflexorius. They also seem to have forgotten/neglected the urodele pubofemoralis in their Table 2 because they do not list that muscle, which is also clearly part of the *Bauplan* of urodeles as can be seen in their own figures and descriptions of these amphibians. However, the hypothesis of Noble (1922) that the anuran gracilis major and minor might also receive a contribution from the ischioflexorius and the hypothesis of Přikryl et al. (2009) that the "anuran" semimembranosus might receive a contribution from the gracilis of salamanders cannot be completely discarded.

[1C] We agree with Přikryl et al. (2009) that the ischiotrochantericus of axolotls correspond to the "quadratus femoris," gemellus, and "obturator externus" of anurans and, possibly, the "obturator internus" of anurans as proposed by Noble (1922).

[1D] In general, the ontogenetic study of Dunlap (1966) supports the hypothesis of homology and evolution of the anuran pelvic and thigh muscles shown in this table, with two major exceptions: the tensor fasciae latae shares an anlage with the pectineus and abductor longus and not with the cruralis and gluteus maximus, and the adductor magnus develops from the anlage that gives rise to the gemellus, obturator externus, quadratus femoris, and obturator internus.

TABLE 19.6
Diagram Illustrating Hypotheses Regarding the Homologies of the Dorsal/Extensor Leg Muscles of Adults of Representative Urodele and Anuran Amphibian Taxa

	Ambystoma mexicanum [9 dorsal/extensor leg muscles]	*Xenopus laevis* [16 dorsal/extensor leg muscles]
Long extensors	Extensor digitorum longus ('extensor digitorum communis'; m.I-V)	Extensor digitorum longus (to extensores digitorum breves of digits 3 and 4)
	---	Tarsalis anticus[1A]
	Extensor tarsi tibialis	**Tibialis anticus brevis**[1A]
	Extensor cruris tibialis	Extensor cruris tibialis ('extensor cruris brevis')[1A]
	Extensor cruris et tarsi fibularis	**Peroneus**[1A]
	---	Tibialis anticus longus[1A]
Short ext.	---	Dorsometatarsales ('extensores breves profundi'; 5 muscles to d.1-5)[1B]
	Extensores digitorum breves (4 muscles to d.2-5)	Extensores digitorum breves ('e.b.s.+e.b.m.'; 4 muscles to d.1-4)[1B]
	Abductor et extensor digiti 1 (d.1; abductor + part short extensor d.1)	Abductor brevis dorsalis hallucis (m.I)

Note: As can be seen in the names highlighted in bold, there are some cases in which although the muscle that is designated with a certain name in anurans does not directly correspond to the muscle designated with a similar name in humans (but not the same; hence, we can use the commonly used anuran names), both the anuran and human muscles derive from a similar anlage/ancestral muscle: the anuran "tibialis anticus brevis" and the human tibialis anterior (anticus) derive from the extensor tarsi tibialis, and the anuran "peroneus" and the human fibularis (peroneus) longus/brevis derive from the extensor cruris et tarsi fibularis (NB: regarding the extensor digitorum longus and the extensores digitorum breves of anurans, they mainly correspond directly to the extensor digitorum longus and extensor digitorum brevis of humans, because they were present in the LCA of tetrapods). d., digits; m, metatarsals (attachments are only shown for muscles that often insert onto digits). "e.b.s.+e.b.m.+e.b.p.," "extensores breves superficiales" plus "extensores breves medii" *sensu* Dunlap (1960); ext., extensors.

[1A] The peroneus of anurans clearly seems to correspond to part or the totality of the extensor cruris et tarsi fibularis of urodeles, as this muscle lies on the fibular side of the dorsal leg and inserts on the fibularis. Regarding the tibialis anticus longus, the name might suggest that it corresponds to part of the extensor cruris et tarsi tibialis anlage and not of the extensor cruris et tarsi fibularis anlage as proposed in the present work. However, the hypothesis defended in the present work is supported by both topological and ontogenetic evidence because the tibialis anticus longus mainly lies on the fibular side of the leg and is usually partially inserted onto the fibulare and fused with the peroneus and, moreover, seems to develop together with the peroneus from the same anlage (Dunlap 1966). Some authors refer to the tibialis anticus longus as a head of the peroneus (see, e.g., Ecker 1864). If the tibialis anticus longus and peroneus of frogs derive from the same (fibular) anlage, then they are somewhat similar to the fibularis brevis and fibularis longus of amniotes, although it is not clear if this is due to a parallelism or instead to a true homology; i.e., if the LCA of tetrapods already had these two separated muscles. As noted by Dunlap (1960, 1966), the tibialis anticus brevis and tarsalis anticus apparently derive from the same anlage (together with the extensor digitorum communis) and, in some anurans, the configuration and attachments of these two muscles thus supporting the idea that they derive from the same muscle mass.

[1B] The extensores breves profundi *sensu* Dunlap (1960) clearly derive from the extensores digitorum breves, because they are the ones that send tendons to the distal phalanges of the digits, as do the extensores digitorum breves of the hindlimb and of the forelimb of urodeles. However, unlike urodeles, in anurans the extensores breves profundi are separate, distinct muscles, as reported by, e.g., Dunlap (1960), thus constituting the dorsometatarsales series of muscle *sensu* the present work, as the extensores breves profundi of the anuran forelimb constitute the dorsometacarpales series. The abductor brevis dorsalis hallucis corresponds to the abductor part of the abductor et extensor digit 1 of urodeles, but as noted by Dunlap (1960), it is usually present as a distinct muscle in anurans, although in some taxa, it might be deeply blended with the extensor digiti brevis 1 or even be absent as a distinct muscle.

EVOLUTION AND HOMOLOGIES OF HINDLIMB MUSCLES, WITH SPECIAL ATTENTION TO MAMMALS

Table 19.13 provides a detailed, textual account of the hindlimb muscles of all major groups of tetrapods, which complements the data summarized in Tables 19.1 through 19.12. Therefore, in this section, we will not repeat the detailed information provided in these tables, but will instead discuss two main subjects: the origin of the tetrapod hindlimb muscles and the implications of recent data on marsupials for our understanding of mammalian evolution.

The transitions from the muscles of the pelvic appendage of fishes to the hindlimb of tetrapods have already been discussed in detail in Chapter 16. However, we will add some comments about Gadow's (1882) ideas, which, despite the detailed nature of his analysis, are often neglected in such discussions. For instance, Gadow (1882) argued that the evolution of the extensor iliotibialis illustrates his "case study 1"; i.e., the splitting of proximal and distal muscle masses from within the origin of the individualized tetrapod limb muscles. According to him, first the "obliquus externus abdominis" probably contacted the pelvic girdle, and then the extensor iliotibialis mass started to form, as seen in urodeles such as *Cryptobranchus* where the "extensor iliotibialis" scarcely takes its origin from the pelvic girdle and is mainly a continuation of part of the obliquus externus abdominis. According to Gadow, a similar condition is seen in the dipnoan fish *Neoceratodus*, in which the "externus obliquus" anlage has a proximal part attached to the girdle and a more distal part separated from the former by a tendinous intersection and corresponding to the extensor iliotibialis of amphibians. A similar scenario was described for the puboischiotibialis, which in urodeles such as *Cryptobranchus* is continuous with

TABLE 19.7
Diagram Illustrating Hypotheses Regarding the Homologies of the Ventral/Flexor Leg Muscles of Adults of Representative Urodele and Anuran Amphibian Taxa

	Ambystoma mexicanum [6 ventral/flexor leg muscles]	*Xenopus laevis* [6 ventral/flexor leg muscles]
Long flexors	Flexor digitorum communis (d.1-5; includes fl.cr.ta.fi.)	Flexor digitorum communis ('**plantaris longus**'; directly only to d.1-2)[1A]
	--- (fl.cr. ta.ti. seems to be included in ischioflexorius)	Cruroastragalus ('tibialis posticus'; seems to correspond to fl.cr.ta.ti.)[1A]
	Flexor accessorius medialis	Flexor accessorius ('transversus plantae proximalis et distalis')[1A]
	Flexor accessorius lateralis	---
	Contrahentium caput longum	Contrahentium caput longum ('tarsalis posticus'; derived also/instead from f.b.s?)[1A]
	Interosseus cruris	Interosseus cruris ('intertarsalis')[1A]
	Tibialis posterior ('pronator profundus')	Tibialis posterior ('plantaris profundus'; or 'palmaris profundus' derived instead from f.b.s?)[1A]

Note: As can be seen in the names highlighted in bold, there are some cases in which although the muscle that is designated with a certain name in anurans does not directly correspond to the muscle designated with a similar name in humans (but not the same name in the case of the plantaris longus; hence, we can use this commonly used anuran name), both the anuran and human muscles derive from a similar anlage/ancestral muscle: the anuran plantaris longus and the human plantaris derive from the flexor digitorum communis (NB: the anuran tibialis posticus and the human tibialis posterior do not seem to be homologous, as shown in this table). d., digits; f.b.s., flexores breves superficiales; fl.cr.ta.fi., flexor cruris et tarsi fibularis; fl.cr.ta.ti., flexor cruris et tarsi tibialis.

[1A] The presence of a single leg bone (cruris) and the remarkable proximodistal elongation of the tibiale and fibulare bones in anurans seems to have led to a functional convergence between these bones and the tibia and fibula of tetrapods such as urodeles; an illustrative example supporting this idea is the fact that the intertarsalis muscle of anurans, which clearly seems to mainly correspond to the interosseous cruris of urodeles, fills the space between the tibiale and the fibulare, while in urodeles, it fills the space between the tibia and fibula. Another example is that the anuran "tarsalis posticus" probably corresponds to the contrahentium caput longum of urodeles, because the anuran muscle is ventral (superficial) to the intertarsalis and develops from the same anlage that gives rise to intertarsalis (Dunlap 1966). Dunlap (1966) stated that this anlage is closely related to the anlage of the flexores breves superficiales and gives rise to the plantaris profundus. Authors such as Gaupp (1896) and Perrin (1892) suggested that the plantaris profundus probably corresponds to the most tibial part of the flexores breves superficiales of urodeles because (a) it is a relatively superficial, tibial muscle that is related to the plantar aponeurosis, which normally sends tendons to digits 1 and 2 in anurans; (b) in anurans, there are usually no separate flexores breves superficiales to the foot digits 1 and 2 (e.g., Dunlap 1960); and (c) in anurans, and particularly in the most plesiomorphic anuran taxa, the plantaris profundus is deeply fused with the flexores breves superficiales (e.g., Dunlap 1960). However, our dissections and comparisons, as well as the ontogenetic data published by Dunlap (1966) indicate that the plantaris profundus of anurans corresponds to the pronator profundus of urodeles, i.e., to the tibialis posterior *sensu* the present work. This is because (a) Dunlap (1966) stated that in anurans, the tarsalis posticus, the intertarsalis, and the plantaris profundus derive from the same ontogenetic anlage, as normally do the interosseous cruris and the pronator profundus (i.e., the tibialis posterior), which are closely related to the contrahentium caput longum; in adult anurans, the tarsalis posticus, the intertarsalis, and the plantaris profundus are topologically associated because they all lie on the tarsal region, as does the "transversus plantae proximalis et distalis," which is apparently also a deep flexor muscle of the leg originally. The correspondence of this transversus plantae proximalis et distalis and the flexor accessorius medialis/lateralis of the hindlimb of urodeles seems clear, because the latter usually also originates at least partially from the fibulare and inserts onto the dorsal surface of the plantar fascia, also extending in a fibulotibial direction; in derived anurans, the muscle is often divided into proximal and distal bundles that resemble the flexor accessorius medialis and flexor accessorius lateralis of the urodele hindlimb, respectively. However, according to Dunlap (1960), in anurans, the "transversus plantae proximalis and distalis" is plesiomorphically not divided into distal and proximal bundles, so these bundles are not directly homologous to the two muscles of urodeles because they seemed to be absent in the LCA of anurans; the undivided muscle of plesiomorphic anurans thus clearly seems to correspond to the usually also undivided flexor accessorius of the anuran forelimb. In summary, it would be difficult to accept that the so-called tibialis posticus of anurans really corresponds to the tibialis posterior *sensu* the present work, i.e., to the pronator profundus of urodeles, because that would mean that in anurans, the pronator profundus is well separated, both ontogenetically and topologically, from all the other original deep flexors of the leg (i.e., the contrahentium caput longum, the flexor accessorius and the interosseous cruris). The structure that is often named "tibialis posticus" in anurans seems to therefore correspond to the flexor cruris et tarsi tibialis, which in urodeles very likely corresponds to the distal (leg) part of the ischioflexorius, as pointed out by authors such as McMurrich (1905) and Bardeen (1906).

the external obliquus and which has proximal and distal heads divided by a clear tendinous intersection in several urodeles, including axolotls. Gadow also described in *Cryptobranchus* blending between pelvic/thigh muscles and leg/foot muscles, reminding us that the flexors and tensors of the toes are simply distal parts of the original muscles. That is, the development of the joints in the hindlimb (e.g., knee joint, ankle joint) in early tetrapods must have caused the long extensor and flexor muscles passing over the knee to develop a tendon, which led to their partition, as did the tendinous intersections seen in the pelvic region. Gadow argued that even some reptiles still have features showing the existence of such long muscles; e.g., the sartorius ("ambiens") of crocodylians often passes right over the knee and forms one of the heads of the digital flexors. In other reptiles, the tendon of this muscle is often inserted on the knee and the heads of the digital flexors arise only from the tibia and fibula, and the part of the tendon connecting the "ambiens" with the muscles of the toes has disappeared. This case study illustrates how an original extensor muscle extending from the pelvic girdle to the toes might have split into

TABLE 19.8
Diagram Illustrating Hypotheses Regarding the Homologies of the Foot Muscles of Adults of Representative Urodele and Anuran Amphibian Taxa

	Ambystoma mexicanum [28 foot muscles]	*Xenopus laevis* [30 foot muscles]
Intrinsic foot muscles	Flexores breves superficiales (5 muscles to d.1-5)	Flexores breves superficiales ('**flexor digitorum brevis superficialis**'; 3 muscles to d. 3-5; also gave rise to 'pl.pr.' and/or 'ta.po.'?)
	---	Lumbricales ('lumbricales breves' plus 'lumbricales longi' and 'lumbricalis longissimus IV'; 8 muscles to d.2-5)[1A]
	---	--- (but abductor praehallucis present in other anurans, going to prehallux)[1A]
	---	Abductor brevis plantaris hallucis (1 muscle, 1 muscle to m.I; absent in *E. coqui*)[1A]
	Abductor digiti minimi (1 muscle to m.V)	Abductor digiti minimi ('abductor brevis dorsalis digiti V'; 1 muscle to m.V)[1B]
	Contrahentes pedis (5 muscles to d.1-5)	Contrahentes pedis ('contrahentes digitorum'; 3 muscles to d.1, 2 and to m.V)[1B]
	---	--- (but 'opponens hallucis' present in other aurans, going to m.I)[1B]
	Flexores digitorum minimi (4 muscles to d.2-5)	Flexores digitorum minimi ('flexores teretes digitorum'; 2 muscles to d.2-5)
	Interphalangeus digiti 3 (1 muscle to d.3)	Interphalangeus digiti 3 (1 muscle to d.3)
	Interphalangei digiti 4 (2 muscles to d.4)	Interphalangei digiti 4 (2 muscles to d.4)
	---	Interphalangeus digiti 5 (1 muscle to d.5)
	Flexores breves profundi (10 muscles to d.1-5)	Flexores breves profundi ('flexores ossi metatarsales'; 2 muscles to m.II and III)
	---	Abductor brevis plantaris digiti V (1 muscle to m.V)[1C]
	---	Abductor proprius digiti IV (1 muscle to m.IV)[1C]
	Intermetatarsales (4 muscles metatarsals 1-5)	Intermetatarsales ('transversi metatarsi'; 4 muscles connecting metatarsals 1-5)

Note: As can be seen in the name highlighted in bold, there are some cases in which although the muscle that is designated with a certain name in anurans does not directly correspond to the muscle designated with a similar name in humans, both the anuran and human muscles derive from a similar anlage/ancestral muscle: the anuran flexor digitorum brevis superficialis and the human flexor digitorum brevis derived from the flexores breves superficiales (NB: It is not clear if the lumbricales were present, or not, in the LCA of tetrapods because they are usually present in anurans and amniotes but not in urodeles: if they present in the LCA of tetrapods, then the anuran lumbricales would directly correspond to the human lumbricales; the abductor digiti minimi of anurans clearly seems to directly correspond to the abductor digiti minimi of the human foot). d., digits; m, metatarsals. pl.pr., plantaris profundus; ta.po., tarsalis posticus.

[1A] Regarding the plantaris profundus, see preceding text; regarding the abductor prehallucis and abductor brevis plantaris hallucis, these muscles seem to derive from the same anlage because they are deeply blended in various anurans (e.g., Dunlap 1960), and according to Ribbing (1938), these muscles derive from the lumbricales, but if cannot be discarded that they also/instead receive a contribution from the flexores breves superficiales.

[1B] All the contrahentes, i.e., to all five digits, are present in plesiomorphic anuran taxa such as *Leiopelma* (Dunlap 1960). Regarding the opponens hallucis, which is present in most anurans but not in *Xenopus*, Dunlap (1960) suggested that it derived from the contrahentes, but the hypothesis of authors such as Perrin (1892) that it derived instead from the flexores breves profundi cannot be discarded particularly because there is no flexor brevis profundus directly going to metatarsal I.

[1C] We agree with Gaupp (1896), who stated that the abductor proprius digiti IV derives from the "flexor ossis metatarsi IV" due to the topology of the two muscles and the fact that they are innervated by the same nerve; that is, as the abductor brevis plantaris digiti V, the abductor proprius digiti IV seems to derive from the flexores breves profundi sensu the present work, although it cannot be discarded that is also/instead receives a contribution from the intermetatarsales.

three: the iliotibialis originating from the ilium, the femorotibialis originating from the femur, and the leg muscle extensor digitorum. In this way, the animal would be able to use a similar amount of muscle mass to perform three very specialized actions, without one movement interfering with the other.

We propose that the reason the ventral/flexor muscles were mainly involved in the splitting of muscles within the joints of the autopodium—meaning that all the intrinsic muscles of the autopodium (plantar/palmar) are ventral muscles (e.g., innervated only by median/ulnar nerve in the hand)—is that the ventral surface of the autopodium is in direct contact with the ground and thus in physical contact with, e.g., sources of food. We also propose that the muscles of the distal regions of the fore- and hindlimbs (i.e., foot/hand and leg/forearm) are so similar to each other because they are formed by a completely new distal extension of the existing muscles to neomorphic regions of the body. These new muscles were subject to very similar topological, functional, and developmental constraints, resulting in strikingly similar homoplastic muscles in the distal regions of the fore- and hindlimb. This type of proximodistal division and splitting of the muscles/muscle groups in the joint regions has been demonstrated in more recent developmental studies; e.g., in the developmental work of Kardon (1998), as noted in Chapter 16. This study of muscle and tendon development in the chicken hindlimb showed that early in the ontogeny of this limb, the pairs (dorsal and ventral) of tendon primordia appear in a generally proximal to distal sequence, namely, in the knee region, the intertarsal region and then at the metatarsophalangeal and interphalangeal joints (Figure 16.6). The thigh, leg, and foot muscles differentiate between these three pairs of tendon primordia (Figure 16.6). This seems to be a further example of parallelism between ontogeny and phylogeny; during evolution, the tendons are thought to have arrived after the muscles at joint regions as explained earlier. Kardon (1998) stated that in development, the muscle cell precursors populate the limb as the early bud forms but do not begin to form myotubules until just after the proximal dorsal or ventral intermediate tendon primordia can be seen. However, phenotypically, the actual formation of muscle is usually seen before the actual

TABLE 19.9

Diagram Illustrating Hypotheses Regarding the Homologies of the Axial and Appendicular Pelvic and Thigh Muscles of Adults of Representative Reptilian Taxa

		Trachemys scripta [15 pelvic and thigh muscles]	*Timon lepidus* [16 pelvic and thigh muscles]	*Alligator mississippiensis* [15 pelvic and thigh muscles]	*Gallus gallus* [18 pelvic and thigh muscles]
Axial		Quadratus lumborum	Quadratus lumborum	Quadratus lumborum	Quadratus lumborum
Dorsal mass	Posterior	Iliofemoralis	Iliofemoralis	Iliofemoralis	Iliotrochantericus externus ('**gluteus med. et min.**')
		---	---	---	Iliotrochantericus caudalis
		---	---	---	Iliotrochantericus cranialis
		---	---	---	Iliotrochantericus medius
		Iliotibialis	Iliotibialis	Iliotibialis (3 heads)	Iliotibialis ('iliotibialis lateralis')[1A]
		Tenuissimus ('iliofibularis')	Tenuissimus ('iliofibularis')	Tenuissimus ('iliofibularis')	Tenuissimus ('iliofibularis part of **biceps**')
		---	---	---	Iliotibialis cranialis ('sartorius')[1A]
		Femorotibialis (3 heads)	Femorotibialis (3 heads in some lizards)	Femorotibialis (2 heads)	Femorotibialis (3 heads)
	*	Sartorius ('ambiens')	Sartorius ('ambiens')	Sartorius ('ambiens'; 2 heads)	Sartorius ('ambiens')
	An	Puboischiofemoralis internus (2 heads)	Puboischiofemoralis internus (3 heads)	Puboischiofemoralis internus (2 heads)	Puboischiofemoralis internus ('**iliacus**'; 3 heads)
Ventral mass	Adductors	Adductor femoris ('ischiofemoralis')	Adductor femoris	Adductor femoris (2 heads)	Adductor femoris ('puboischiofemoralis', '**adductor**'; 2 heads)
		Pubotibialis	Pubotibialis	---	---
		Gracilis ('puboischiotibialis')	Gracilis ('puboischiotibialis'; 3 heads)	Gracilis ('puboischiotibialis')	---
		Flexor tibialis internus (2 heads)	Flexor tibialis internus (2 heads)	Flexor tibialis internus (4 heads)	Flexor tibialis internus ('ischioflexorius'; '**semimembranosus**', 'f.c.m.')[1B]
		---	---	---	Flexor cruris lateralis accessorius ('accessorius')[1B]
		Flexor tibialis externus (might include fl.cr.ta.ti)	Flexor tibialis externus (seems to include fl.cr.ta.ti.)	Flexor tibialis externus (might include fl.cr.ta.ti)	Flexor tibialis externus ('caudilioflexorius', '**semitendinosus**', 'f.c.l.')[1B]
		Caudofemoralis ('caudi-iliofemoralis')	Caudofemoralis longus	Caudofemoralis longus	Caudofemoralis ('piriformis'; 2 heads)
		---	Caudofemoralis brevis	Caudofemoralis brevis	---
Ischio troch.		Puboischiofemoralis externus (2 major heads)	Puboischiofemoralis externus (2 major heads)	Puboischiofemoralis externus (3 major heads)	Puboischiofemoralis externus ('**obturator**'; 2 heads)
		Ischiotrochantericus	Ischiotrochantericus	Ischiotrochantericus	Ischiotrochantericus ('ischiofemoralis')

Note: Features of both anterior and posterior masses of dorsal mass. The muscle that is usually named piriformis in birds is not homologous to the muscle designated with the same name in humans, because the LCA of reptiles and mammals did not have a distinct muscle piriformis. There are however a few cases (highlighted in bold) in which although the muscle that is designated with a certain name in birds does not directly correspond to the muscle designated with the same name in humans, both the avian and human muscles derive from a similar anlage/ancestral muscle: the avian "gluteus medius" and "gluteus minimus," and the human gluteus medius and gluteus minimus derive from the iliofemoralis; the avian obturator and the human obturator externus derive from the puboischiofemoralis externus; the avian "iliacus" and the human iliacus derive from the puboischiofemoralis internus; the avian adductor and human adductor brevis and at least part of the adductor magnus derive from the adductor femoris, part of the avian "biceps" and of the human biceps apparently derive from the tenuissimus; and at least part of the avian semimembranosus and "semitendinosus" and of the human semimembranosus and human semitendinosus derive, respectively, from the flexor tibialis internus and externus. An., anterior; f.c.l., flexor cruris lateralis; f.c.m., flexor cruris medialis; fl.cr.ta.ti., flexor cruris tarsi tibialis; Ischiotroch., ischiotrochantericus; med. et min., medius et minimus. A line with two circles means that the muscle is the ontogenetic result of the fusion of two different anlagen.

[1A] According to Lance-Jones (1979) during ontogeny, the superficial/anterior dorsal mass or iliotibialis of birds expands anteriorly to form the iliotibialis cranialis (sartorius), the topology of which resembles the sartorius of mammals. However, the development of the iliotibialis cranialis instead suggests a homology with either the mammalian rectus femoris or part of the superficial gluteals, all being superficial dorsal mass derivatives. The extreme anterior position of the avian sartorius makes the homology with the rectus femoris most likely, both arising just superficial and anterior to the femoral nerve trunk and vasti/femorotibialis muscles. However, one cannot completely rule out a partial homology of the avian sartorius and the mammalian sartorius, because Lance-Jones (1979) shows how the superficial dorsal mass occasionally contributes a slip of mesenchyme to the mouse sartorius as well.

[1B] According to Lance-Jones (1979) the ischioflexorius and caudilioflexorius of birds seem to correspond, respectively, to heads 1 and 2 of the flexor tibialis internus of lizards, and the avian "accessory muscle" (flexor cruris lateralis accessorius *sensu* the present work) is probably either a slip of the gastrocnemius that has migrated proximally or a distal slip of the caudofemoralis. However, more recent and extensive comparative works between birds and other reptiles such as Dilkes (1999) and Fechner (2009) point out that the flexor tibialis externus of crocodylians mainly corresponds to the "flexor cruris lateralis" (semitendinosus or caudilioflexorius) of birds, while the flexor tibialis internus of crocodylians corresponds mainly to the flexor cruris medialis (semimembranosus or ischioflexorius) of birds.

TABLE 19.10
Diagram Illustrating Hypotheses Regarding the Homologies of the Dorsal/Extensor Leg Muscles of Adults of Representative Reptilian Taxa

	Trachemys scripta [11 dorsal/extensor leg muscles]	*Timon lepidus* [9 dorsal/extensor leg muscles]	*Alligator mississippiensis* [8 dorsal/extensor leg muscles]	*Gallus gallus* [8 dorsal/extensor leg muscles]
Long extensors	Extensor digitorum longus (m. I-IV and d.1)	Ex. digitorum longus (m.II-III, not to digits)	Extensor digitorum longus (m. I-III, sometimes also to m.IV)	Ex. digitorum longus (to d. 2-4)
	Tibialis anterior	Tibialis anterior	Tibialis anterior	Tibialis anterior ('tibialis cranialis')
	Fibularis longus (m.V and d.5)	Fibularis longus	Fibularis longus	Fibularis longus
	Fibularis brevis	Fibularis brevis	Fibularis brevis	Fibularis brevis
Short extensors	Extensores digitorum breves (3 muscles to d.2-4)	Extensores digitorum breves(4 muscles to d.2-5 + part short ex. d.1)	Extensores digitorum breves (3 muscles to d. 2-4)	Extensores digitorum breves (3 muscles to d. 2-4)
	Dorsometarsales ('interossei dorsales'; 3 muscles to d.2-4)	---	---	---
	Ab. et ex. digiti 1 ('ex.h.p.+ab.h.'; d.1; ab. + short ex. d.1)[1A]	Ab. et ex. digiti 1(to d.1; ab. + part short ex. d.1)	Ab. et ex. digiti 1('ex.h.l.'; d.1; ab. + short ex. d.1)	Ab. et ex. digiti 1('ex.h.l.'; d.1; ab. + short ex. d.1)

Note: The extensor digitorum longus, tibialis anterior, fibularis brevis, fibularis longus, and extensores digitorum breves of birds clearly correspond to the muscles designated by similar names in humans. However, the "extensor hallucis longus" in birds is not homologous to the muscle designated with the same name in humans, because the LCA of reptiles and mammals did not have a distinct extensor hallucis longus, and moreover, the bird muscle corresponds to the abductor et extensor digiti 1 *sensu* the present work, while the human extensor hallucis longus derives from the extensor digitorum longus. ab., abductor; ab.h, abductor hallucis; d., digits; ex., extensor; ex.h.l., extensor hallucis longus; ex.h.p., extensor hallucis proprius; m, metatarsals (attachments are only shown for muscles that often insert onto digits).

[1A] In *Trachemys scripta* and various other turtles, the "extensor hallucis proprius" runs from the distal fibula to the proximal phalanx of digit 1, thus corresponding to the abductor part of the abductor et extensor digiti 1 *sensu* the present work, while the abductor hallucis runs from the tarsal and metatarsal regions to the distal phalanx of digit 1, thus corresponding to the extensor part of the abductor et extensor digiti 1 *sensu* the present work, because these two structures are blended as recognized by Walker (1973).

TABLE 19.11

Diagram Illustrating Hypotheses Regarding the Homologies of the Ventral/Flexor Leg Muscles of Adults of Representative Reptilian Taxa

	Trachemys scripta [5 ventral/flexor leg muscles]	*Timon lepidus* [6 ventral/flexor leg muscles]	*Alligator mississippiensis* [7 ventral/flexor leg muscles]	*Gallus gallus* [4 ventral/flexor leg muscles]
Long flexors	--- (fl.cr. ta.ti. seems included in f.ti.ex. and/or ga.in.)	--- (fl.cr. ta.ti. seems included in fl.ti.ex. and ga.in)	--- (fl.cr. ta.ti. seems included in f.ti.ex. and/or ga.in.)	--- (fl.cr. ta.ti. seems included in ga.co. or lost)
	---	---	Gastrocnemius plantaris ('**plantaris**')[2A]	Gastrocnemius plantaris ('**plantaris**')[3A]
	Gastrocnemius internus ('gastrocnemius'; 2 heads)	Gastrocnemius internus ('femorotibial')	Gastrocnemius internus	Gastrocnemius (3 heads)
	Gastrocnemius externus ('f.d.l. sublimis')	Gastrocnemius externus ('femoral'; sup. and deep heads)	Gastrocnemius externus	---
	Flexor digitorum communis (to d.1-4)	Flexor digitorum longus (to d.1-5)	Flexor digitorum longus (d.1-4)	Flexor digitorum longus ('f.d.l.'+'**f.h.l.**'; to d. 1-4)
	--- (but some turtles have a popliteus)[1A]	Popliteus	Popliteus	Popliteus
	Interosseus cruris[1A]	Interosseus cruris	Interosseus cruris	---
	Tibialis posterior ('pronator profundus')[1A]	Tibialis posterior ('pronator profundus')	Tibialis posterior ('pronator profundus')	---

Note: The flexor digitorum longus, popliteus, and gastrocnemius of humans mainly correspond to at least part of the muscles designated by similar names in birds. However, the muscle that are usually named plantaris in birds is not homologous to the muscle designated with the same name in humans, because the LCA of reptiles and mammals did not have a distinct plantaris muscle; hence, we use a different name in birds. But, although the plantaris of birds does not directly correspond to the plantaris of humans, both the avian and human muscles derive from a similar anlage/ancestral muscle, i.e., from the gastrocnemius complex; hence, the avian muscle is highlighted in bold. Regarding the avian flexor hallucis longus, although this structure and the extensor hallucis longus of humans are both derived from the flexor digitorum longus, the avian structure is simply a bundle of the flexor digitorum longus and not a separate muscle; the LCA of reptiles and mammals did not have a distinct flexor hallucis longus. d., digits; fl.cr.ta.ti., flexor cruris tarsi tibialis; f.d.l., flexor digitorum longus; f.h.l., flexor hallucis longus; fl.ti.ex., flexor tibialis externus; ga.co., gastrocnemius complex; ga.in., gastrocnemius internus; sup. superficial.

[1A] The "popliteus" *sensu* Walker (1973) connects the tibia and fibula and is deep to the tibialis posterior, so it clearly corresponds to the interosseous cruris *sensu* the present work. According to Walker (1973), some turtles have however "proximosuperficial" and "distodeep" portions, which are apparently similar, respectively, to the popliteus and interosseous cruris of other reptiles such as lizards, so this indicates that the popliteus does not derive from the flexor digitorum communis/gastrocnemius anlage, but instead from the interosseous cruris anlage.

[2A,3A] The plantaris of birds and crocodylians is similar to the plantaris of mammals such as humans because it lies deep to and derives from the gastrocnemius complex, but it is almost surely the result of an evolutionary parallelism because the LCA of amniotes and of reptiles apparently did not have a muscle plantaris. The plantaris of birds is functionally and topologically different from that of mammals, e.g., not inserting onto the calcaneal region or with the flexor digitorum brevis and not originating from the femur as it typically does in various mammals but instead from the proximal tibia; i.e., it is a much shorter (proximodistally) muscle that is very short in most birds, as noted by Vanden Berge (1975), hence, the use of the name plantaris gastrocnemius plantaris in the present work. However, the plantaris of crocodylians such as *Alligator* is a long muscle that runs from the distal end of the thigh muscles to both the calcaneal region and the flexor digitorum brevis, as usually does the plantaris of mammals, so it is possible that the LCA of birds and crocodylians did have a condition that is quite similar to that of the plantaris of mammals.

TABLE 19.12

Diagram Illustrating Hypotheses Regarding the Homologies of the Foot Muscles of Adults of Representative Reptilian Taxa

	Trachemys scripta (20 Foot Muscles)	*Timon lepidus* (25 Foot Muscles)	*Alligator mississippiensis* (At Least 21 Foot Muscles)	*Gallus gallus* (10 Foot Muscles)
Intrinsic Foot Muscles	Flexores breves sup.("fl.di.com.sub."; 4 muscles to d. 1-4)	Flexores breves sup. (3 muscles to d. 1-3)	Flexores breves sup. (4 muscles to d. 1-4)	Flexores breves sup. ("f.p.p.II,III + f.p.II,III,IV"; 5 muscles to d. 2-4)
	Lumbricales (4 muscles to d. 2-5)	Lumbricales (2 muscles, to d. 3-4)	Lumbricales (4 muscles, to d. 1-4)	Lumbricales (2 muscles, to d. 3-4)
	–	Abductor digiti minimi (1 muscle to m. V)	Abductor digiti minimi (mainly to m. IV)	Abductor digiti minimi ("ab.di.IV"; 1 muscle to d. 4)
	Contrahentes pedis (p. "i.p"; 4 muscles to d. 1-4)[1A]	Contrahentes pedis (5 muscles to d. 1-5)	Contrahentes pedis (atleast 1 muscle to d. 1)	Contrahentes pedis ("adductor digiti II"; 1 muscle to d. 2)
	Flexores breves profundi (p. "i.p"; 4 muscles to d. 1-4)[1A]	Flexores breves profundi (10 muscles to d. 1-5)	Flexores breves profundi (apparently 8 muscles tod. 1-4)	Flexores breves profundi ("f.h.b." 1 muscle to d. 1)
	Intermetatarsales (p. "i.p"; 4 connecting m. 1-5)[1A]	Intermetatarsales (4 connecting m. 1-5)	Intermetatarsales (apparently 3 muscles connecting m. 1-4)	–

Note: The flexores breves superficiales, lumbricals, and abductor digiti minimi of birds clearly correspond to the muscles designated by similar names in humans. Regarding the avian "flexor hallucis brevis," this muscle corresponds to the first flexor brevis profundus, so it does correspond to the flexor hallucis brevis of humans *sensu* the present work (i.e., to the structure usually designated "medial head of the flexor hallucis longus" in human anatomy). However, in the LCA of tetrapods and of amniotes, this first flexor brevis profundi was clearly part of a group of muscles, the flexores breves profundi, and was not a distinct, separate, peculiar muscle as is the case in birds and humans, so the bird and human condition is the result of an evolutionary homoplasy (hence, the bird muscle is highlighted in bold). ab.di.IV, abductor digiti IV; d., digits; fl.di.com.sub., flexor digitorum communis sublimis; f.h.b., flexor hallucis brevis; f.p.p.II,III + f.p. II,III,IV; flexor perforans et perforatus digiti II and III plus flexor perforatus digiti II, III and IV; m, metatarsals; p."i.p.," part of "interossei plantares" *sensu* Walker [1973]; sup., superficiales.

[1A] Walker (1973) states that in turtles such as *Trachemys scripta*, the "interossei plantares" extend from tarsals and metatarsals to the fibular side of the proximal phalanges of digits 1-4; his descriptions and particularly his illustrations (see, e.g., his Figures 25 and 30), as well as the fact that he recognized that his interossei plantares correspond to the flexores breves profundi, contrahentes, and "interdigitales" *sensu* Ribbing (1938), indicate that his interossei plantares do include the flexores breves profundi (thus probably 4 muscles to digits 1-4), contrahentes (thus probably 4 muscles to digits 1-4), and intermetatarsales (thus probably 4 muscles connecting the metatarsals 1-5; i.e., inserting onto metatarsals 1-4) *sensu* the present work.

TABLE 19.13
Hindlimb Muscles of Adults of Representative Tetrapod Taxa

Amphibia (Caudata): *Ambystoma mexicanum* (Axolotl)	Amphibia (Anura): *X. laevis* (African Clawed Frog)	Reptilia (Testudines): *Trachemys scripta* (Pond Slider)	Reptilia (Lepidosauria): *Timon lepidus* (Ocellated Lizard)	Reptilia (Crocodylia): *Alligator mississippiensis* (American Alligator)	Reptilia (Aves): *Gallus domesticus* (Chicken)	Mammalia (Monotremata): *Ornithorhynchus anatinus* (Platypus)	Mammalia (Marsupialia): *D. virginiana* (Virginia Opossum)	Mammalia (Placentalia): *Rattus norvegicus* (Brown Rat)	Mammalia (Placentalia): *Homo sapiens* (Modern Human)
Caudalipuboischiotibialis [Appleton [1928] considers that the caudalipuboischiotibialis of urodeles is also found in frogs and that this muscle is derived from the same anlage giving rise to the caudofemoralis, i.e., that it is an appendicular muscle; this muscle is however often regarded as part of the caudal–pelvic girdle musculature, although unlike caudal–pelvic muscles such as the ischiocaudalis that connect the pelvis to the tail, in amphibians such as the axolotl, the caudalipuboischiotibialis mainly connects the tail (caudal/postsacral vertebrae) to the puboischiotibialis muscle, thus being probably related not only to flexion of the tail but also to limb retraction (e.g., Walthall and Ashley-Ross 2006), and some authors thus consider that it might be an hamstring muscle (e.g., Mivart 1869, Gadow 1882); in urodeles, it is innervated by pudendal nerve (Francis 1934)]	– [only present in one of the most plesiomorphic anuran genera, *Ascaphus*, according to Dunlap [1960] and *Leiopelma* from the same family according to Přikryl et al. [2009]]	–	–	– [see following cell]	–	–	–	–	–

(Continued)

TABLE 19.13 (CONTINUED)
Hindlimb Muscles of Adults of Representative Tetrapod Taxa

Amphibia (Caudata): *Ambystoma mexicanum* (Axolotl)	Amphibia (Anura): *X. laevis* (African Clawed Frog)	Reptilia (Testudines): *Trachemys scripta* (Pond Slider)	Reptilia (Lepidosauria): *Timon lepidus* (Ocellated Lizard)	Reptilia (Crocodylia): *Alligator mississippiensis* (American Alligator)	Reptilia (Aves): *Gallus domesticus* (Chicken)	Mammalia (Monotremata): *Ornithorhynchus anatinus* (Platypus)	Mammalia (Marsupialia): *D. virginiana* (Virginia Opossum)	Mammalia (Placentalia): *Rattus norvegicus* (Brown Rat)	Mammalia (Placentalia): *Homo sapiens* (Modern Human)
Ischiocaudalis [in axolotls it extends from the tail (caudal/postsacral vertebrae) to the puboischiac plate; in urodeles, it is innervated by the pudendal nerve (Francis 1934)]	– [seems to be absent in anurans, including *Ascaphus*, which has a vestigial tail; at least it was not described by Dunlap [1960]]	**Ischiocaudalis** ("ischiococcygeus") [Walker [1973] described four muscles in *Trachemys scripta* that according to him are axial muscles attached to the pelvic appendage and, namely, the pelvic girdle: a testoiliacus from the shell to the ilium, a rectus abdominus from the plastron to the pubic process, and an ischiococcygeus (= ischiocaudalis) and a pubococcygeus from the pubic and ischial symphysis to the sacral and caudal vertebrae]	**Ischiocaudalis** (ischiococcygeus)	**Ischiocaudalis** (ischiococcygeus) [Romer [1923] described five muscles in *A. mississippiensis* that according to him are axial muscles attached to the pelvic appendage and, namely, the pelvic girdle: obliquus abdominus externus, obliquus abdominis internus, transversis abdominis, rectus abdominis, and ischiococcygeus (or "ilio-ischio-caudalis"); the latter muscle is subdivided into "iliocaudalis" (iliococcygeus in squamates and iliococcygeus + depressor coccygeus in birds) and "ischiocaudalis" (ischiococcygeus in squamates and birds), which probably correspond to the turtle muscles ischiococcygeus and iliococcygeus: see cells on the left]	**Ischiocaudalis** ("ischiococcygeus" + depressor coccygeus) [see crocodylians, on the left; no ischiocaudalis is listed by Vanden Berge [1975] in chicken, who just described a pubocaudalis internus and externus, but see on the left]	**Ischiocaudalis** (ischiococcygeus *sensu* Pearson 1926) [from caudal vertebrae to ischium; also present in echidna according to Gregory and Camp [1918], being designated "flexor lateralis of the tail"]	**Ischiocaudalis** (ischiococcygeus *sensu* Coues 1872) [from ischial part of pelvic bone to five or six first caudal vertebrae]	**Ischiocaudalis** [in mammals such as cats, the ischiocaudalis is often designated abductor caudae internus, according to Gregory and Camp [1918]]	–
? [no information available for this species]	? [no information available for this species]	**Pubococcygeus** [see preceding cell]	? [no information available for this species]	? [no information available for this species]	? [no information available for this species]	? [no information available for this species]	**Pubococcygeus** [as described by Coues [1872], extends from pelvic bone, mainly from pubes opposite the articulation of the marsupial bone, to the "second V-shaped bone" (*sensu* Coues [1872]), being fused to the iliocaudalis]	**Pubococcygeus**	–

(Continued)

TABLE 19.13 (CONTINUED)
Hindlimb Muscles of Adults of Representative Tetrapod Taxa

Amphibia (Caudata): *Ambystoma mexicanum* (Axolotl)	Amphibia (Anura): *X. laevis* (African Clawed Frog)	Reptilia (Testudines): *Trachemys scripta* (Pond Slider)	Reptilia (Lepidosauria): *Timon lepidus* (Ocellated Lizard)	Reptilia (Crocodylia): *Alligator mississippiensis* (American Alligator)	Reptilia (Aves): *Gallus domesticus* (Chicken)	Mammalia (Monotremata): *Ornithorhynchus anatinus* (Platypus)	Mammalia (Marsupialia): *D. virginiana* (Virginia Opossum)	Mammalia (Placentalia): *Rattus norvegicus* (Brown Rat)	Mammalia (Placentalia): *Homo sapiens* (Modern Human)
Iliocaudalis [really present in axolotls?; Ashley-Ross [1992] did not describe it in her *Ambystoma* papers, but Walthall and Ashley-Ross [2006] describe it in *Taricha*, extending from the caudal vertebrae to the ilium, being parallel and blended to the epaxial muscles; in urodeles, it is innervated by the dorsal and ventral rami of spinal nerves (Francis 1934)]	– [seems to be absent in anurans, including *Ascaphus* which has a vestigial tail, at least it was not described by Dunlap [1960]]	**Iliocaudalis** (iliococcygeus) [see ischiocaudalis]	**Iliocaudalis** (iliococcygeus) [see ischiocaudalis]	**Iliocaudalis** (iliococcygeus) [see ischiocaudalis]	**Iliocaudalis** (iliococcygeus) [no iliocaudalis listed in Avian Nomina nor described by Vanden Berge [1975] in chicken, who just described a pubocaudalis internus and externus, but see ischiocaudalis]	**Iliocaudalis** [also present in echidna according to Gregory and Camp [1918], being designated extensor lateralis of the tail]	**Iliocaudalis** (iliococcygeus *sensu* Coues [1872]) [from pelvis to caudal vertebrae, being fused to the pubococcygeus]	**Iliocaudalis** [in mammals such as cats, the iliocaudalis is often designated intertransversarii caudae, according to Gregory and Camp [1918]]	–
– [according to Gadow [1882], the quadratus lumborum and the psoas minor are not present as separate muscles in amphibians, nor in fish]	– [see on the left]	**Quadratus lumborum** [apparently present in turtles; at least a similar muscle has been described by various authors, see, e.g., Gadow [1882] and Gregory and Camp [1918]]	**Quadratus lumborum** [present in lepidosaurs, including *Sphenodon*, according to Gregory and Camp [1918], corresponding to both the quadratus lumborum and the psoas minor of mammals]	**Quadratus lumborum** [present in crocodylians according to Gregory and Camp [1918], corresponding to both the quadratus lumborum and the psoas minor of mammals]	**Quadratus lumborum** [apparently present in birds; at least a similar muscle has been described by various authors, see, e.g., Gadow [1882] and Gregory and Camp [1918]]	**Quadratus lumborum** [from ribs and vertebrae to ilium; Gregory and Camp [1918] confirmed its presence in monotremes]	**Quadratus lumborum** [present according to Coues [1872]; apparently from ribs and vertebrae to pelvis]	**Quadratus lumborum** [from vertebrae and ribs to hip bone (ischium and acetabulum)]	**Quadratus lumborum** [from ribs and vertebrae to pelvis]
–	–	–	–	–	– [only present in mammals, apparently; see quadratus lumborum mentioned earlier]	**Psoas minor** [from last three thoracic vertebrae and first lumbar vertebra and last two ribs to ilium; Gregory and Camp [1918] confirmed its presence in monotremes]	**Psoas minor** (psoas parvus *sensu* Coues [1872] and Stein [1981]) [from lumbar vertebrae to pubis, according to Coues [1872], and to ischium, according to Stein [1981]; NB: according to Stein [1981], in some marsupials such as *Chironectes*—but not in *Didelphis*—there is also a "psoas tertius" running from lumbar vertebrae to the femur]	**Psoas minor** [rat: second to sixth lumbar vertebrae to ilium; mouse: from vertebrae to femur according to Charles et al. [2016]]	– [present as a variant in humans, in about 40% of the cases]
			–				–	– [seems to be absent in rats/mice]	–

(Continued)

TABLE 19.13 (CONTINUED)
Hindlimb Muscles of Adults of Representative Tetrapod Taxa

Amphibia (Caudata): *Ambystoma mexicanum* (Axolotl)	Amphibia (Anura): *X. laevis* (African Clawed Frog)	Reptilia (Testudines): *Trachemys scripta* (Pond Slider)	Reptilia (Lepidosauria): *Timon lepidus* (Ocellated Lizard)	Reptilia (Crocodylia): *Alligator mississippiensis* (American Alligator)	Reptilia (Aves): *Gallus domesticus* (Chicken)	Mammalia (Monotremata): *Ornithorhynchus anatinus* (Platypus)	Mammalia (Marsupialia): *D. virginiana* (Virginia Opossum)	Mammalia (Placentalia): *Rattus norvegicus* (Brown Rat)	Mammalia (Placentalia): *Homo sapiens* (Modern Human)
Extensor iliotibialis (iliotibialis or ilioextensorius *sensu* Kardong [2002]) [in urodeles, it is usually innervated by the n. extensorius (Francis 1934); as in *A. tigrinum* (Ashley-Ross 1992); in the axolotls we dissected, it is divided into anterior (iliotibialis) and posterior (ilioextensorius) heads and originates from the ilium to a wide tendon inserting onto the crista tibialis and blending with the distal portions of the extensor cruris tibialis and the extensor digitorum longus]	**Extensor iliotibialis A** (tensor fasciae latae *sensu* Dunlap [1960]) [the muscle is clearly not homologous to the tensor fasciae latae of humans, because the LCA of amphibians and amniotes did not have the latter muscle; hence, we use a different name; usually innervated in anurans by the femoral nerve of the plexus cruralis (Noble 1922); in the *Rana pipiens*, and *E. coqui* specimens we dissected, the tensor fasciae latae is less developed than in *X. laevis*; it usually extends from the girdle to the cruralis and "gluteus," as reported by Dunlap [1960]; Appleton [1928] suggested that the extensor iliotibialis of urodeles such as axolotls (*sensu* the present work) corresponds to the iliacus internus, iliacus externus, tensor fasciae latae, gluteus, and cruralis of anurans, but subscribe to Noble's [1922] and Přikryl's [2009] hypothesis that it corresponds only to the tensor fasciae latae, gluteus, and cruralis of anurans; Noble [1922] states that in plesiomorphic anurans such as *Ascaphus*, the tensor fasciae latae is a large muscle and that some salamanders have an anterior bundle of the extensor iliotibialis that is very similar to the cruralis of anurans]	**Iliotibialis** (gluteus maximus *sensu* Ogushi [1913]) [Walker [1973] stated that in *Trachemys scripta*, the iliotibialis runs from the ilium (two heads of origin) to the tibia; Appleton [1928] suggests that the extensor iliotibialis of axolotls corresponds to both the gluteus maximus and "caput longum of the vastus femoris" of turtles *sensu* Gadow [1882]; according to Walker [1973], the "caput longum" *sensu* Ogushi [1913] is however only present in the turtle family Trionynchidae, being designated by him as "musculus triceps femoris, caput ischiadicum"]	**Iliotibialis** [Russell and Bauer [2008] stated that the iliotibialis is often treated as a component of a functional "quadriceps femoris," which is clearly not homologous to the quadriceps femoris of mammals, consisting of, e.g., femorotibialis, iliofemoralis and sartorius (ambiens); in lizards, the iliotibialis usually extends *via* a broad aponeurosis from the lateral surface of the ilium to the cnemial crest of the tibia in common with the tendon of the femorotibialis; innervation in lizards: doubly, by femoral/obturator and/or tibial/fibular plexus (Romer 1942); as reported by Russell and Bauer [2008] in *Iguana*, in the *Timon* specimens we dissected, the iliotibialis extends from a broad aponeurosis from the lateral surface of the ilium to the cnemial crest of the tibia in common with the tendon of the femorotibialis]	**Iliotibialis** (*sensu* Romer [1923]) [in *A. mississippiensis*, the iliotibialis has 3 heads and runs from the ilium to the tibia (Romer 1923); Appleton [1928] suggests that the extensor iliotibialis of axolotls corresponds to both the extensor iliotibialis and the "iliofibularis II" of crocodylians, *sensu* Ogushi [1913]; innervation crural and sacral nerves (Romer 1923)]	**Iliotibialis** (iliotibialis lateralis *sensu* Schroeter and Tosney [1991]) [in adult chicken, the "iliotibialis lateralis" runs from iliac crest to patellar tendon to the tibiotarsus (Vanden Berge 1975)]	**Gluteus maximus** (gluteus superficialis *sensu* Gambaryan et al. [2002]) [from sacral vertebrae 2, 3, and 4 and last sacral vertebra to proximal femur; innervation: inferior gluteal nerve (Appleton 1928)]	**Gluteus maximus** [from two sacral and first three caudal vertebrae to femur]	**Gluteus maximus** (gluteus superficialis *sensu* Walker and Homberger [1997]) [rat: from ilium and vertebrae to femur, blended with tensor fasciae latae; mouse according to Charles et al. [2016]: from iliac crest to proximal femur; Lance-Jones [1979]: in *Mus*, the gluteus maximus derives from the posterior dorsal anlage, together with the quadriceps, the tenuissimus, gluteus medius, gluteus minimus, piriformis, tensor fasciae latae, and femoro-coccygeus, having more affinity with the femorococcygeus, which can be recognized at 13.5 days but (a) is then fused with/integrated in the biceps femoris (NB: it seems to correspond to the anterior head of the biceps femoris of the adult *Rattus*) and (b) migrates to insert onto the patella and to originate from the vertebrae (originally the gluteus maximus and femorococcygeus do not attach onto vertebrae, so they are "superficial"/"appendicular" muscles according to the "in–out" model); innervation: inferior gluteal nerve and inferior branch of the superior gluteal nerve (Greene 1935); as	**Gluteus maximus**

(Continued)

TABLE 19.13 (CONTINUED)
Hindlimb Muscles of Adults of Representative Tetrapod Taxa

Amphibia (Caudata): *Ambystoma mexicanum* (Axolotl)	Amphibia (Anura): *X. laevis* (African Clawed Frog)	Reptilia (Testudines): *Trachemys scripta* (Pond Slider)	Reptilia (Lepidosauria): *Timon lepidus* (Ocellated Lizard)	Reptilia (Crocodylia): *Alligator mississippiensis* (American Alligator)	Reptilia (Aves): *Gallus domesticus* (Chicken)	Mammalia (Monotremata): *Ornithorhynchus anatinus* (Platypus)	Mammalia (Marsupialia): *D. virginiana* (Virginia Opossum)	Mammalia (Placentalia): *Rattus norvegicus* (Brown Rat)	Mammalia (Placentalia): *Homo sapiens* (Modern Human)
								described by Greene [1935], in the rats we dissected, this is a thin muscle from fascia from the dorsal border of the ilium and from vertebrae to the third trochanter (which clearly seems to correspond to the gluteal tuberosity of humans and to the deltoid tuberosity of the upper limb of humans and particularly rats) by a tendon; the muscle is blended anteriorly with the tensor fasciae latae and posteriorly covered by the biceps femoris]	
–	**Extensor iliotibialis B** (gluteus *sensu* Noble [1922]; gluteus magnus *sensu* Dunlap [1960]) [the muscle is not homologous to the gluteus maximus/"magnus" of humans, because the LCA of amphibians and amniotes did not have the latter muscle; hence, we use a different name; innervation: femoral nerve of plexus sacralis (Noble 1922); as reported by Dunlap [1960] in the *R. pipiens*, *X. laevis*, and *E. coqui* specimens we dissected, the gluteus magnus extends from the ilium to the "cruralis"; see extensor iliotibialis A mentioned earlier]	–	–	–	–	–	–	–	–

(Continued)

TABLE 19.13 (CONTINUED)
Hindlimb Muscles of Adults of Representative Tetrapod Taxa

Amphibia (Caudata): *Ambystoma mexicanum* (Axolotl)	**Amphibia (Anura): *X. laevis* (African Clawed Frog)**	**Reptilia (Testudines): *Trachemys scripta* (Pond Slider)**	**Reptilia (Lepidosauria): *Timon lepidus* (Ocellated Lizard)**	**Reptilia (Crocodylia): *Alligator mississippiensis* (American Alligator)**	**Reptilia (Aves): *Gallus domesticus* (Chicken)**	**Mammalia (Monotremata): *Ornithorhynchus anatinus* (Platypus)**	**Mammalia (Marsupialia): *D. virginiana* (Virginia Opossum)**	**Mammalia (Placentalia): *Rattus norvegicus* (Brown Rat)**	**Mammalia (Placentalia): *Homo sapiens* (Modern Human)**
–	**Cruralis** [innervation: femoral nerve of plexus sacralis (Noble 1922); as reported by Dunlap [1960] in the *R. pipiens*, *X. laevis*, and *E. coqui* specimens we dissected, the cruralis runs from the girdle to the distal femur and proximal cruris; see extensor iliotibialis A mentioned earlier]	–	–	–	–	–	–	–	–
–	–	–	– [sartorius of birds corresponds to the anterior portion of the iliotibialis of other reptiles and apparently to the rectus femoris of mammals, although it can also contribute to part of the sartorius of mammals; see on the right]	–	**Iliotibialis cranialis** (sartorius of birds *sensu* Romer [1927], George and Berger [1966], and Lance-Jones [1979]) [muscle only differentiated in birds, corresponding to the anterior portion of the iliotibialis of other reptiles and apparently with the rectus femoris of mammals, although it can also contribute to part of the sartorius of mammals: see iliotibialis mentioned earlier; Vanden Berge 1975: in adult chicken, the "iliotibialis cranialis" runs from ilium to patella or directly to tibiotarsus; innervation: femoral nerve (Vanden Berge 1975); see iliotibialis]	–	–	–	–

(Continued)

TABLE 19.13 (CONTINUED)
Hindlimb Muscles of Adults of Representative Tetrapod Taxa

Amphibia (Caudata): *Ambystoma mexicanum* (Axolotl)	Amphibia (Anura): *X. laevis* (African Clawed Frog)	Reptilia (Testudines): *Trachemys scripta* (Pond Slider)	Reptilia (Lepidosauria): *Timon lepidus* (Ocellated Lizard)	Reptilia (Crocodylia): *Alligator mississippiensis* (American Alligator)	Reptilia (Aves): *Gallus domesticus* (Chicken)	Mammalia (Monotremata): *Ornithorhynchus anatinus* (Platypus)	Mammalia (Marsupialia): *D. virginiana* (Virginia Opossum)	Mammalia (Placentalia): *Rattus norvegicus* (Brown Rat)	Mammalia (Placentalia): *Homo sapiens* (Modern Human)
–	–	–	– [see iliotibialis]	–	– [see iliotibialis]	– [femorococcygeus seem to be absent as distinct muscle in *Ornithorhynchus* according to Gregory and Camp [1918]; also not described by Pearson [1926] nor in the detailed study of Gambaryan et al. [2002]]	**Femorococcygeus** [from third and fourth caudal vertebrae to femur, being usually deeply blended with gluteus maximus as stated by Coues [1872]]	– [the femorococcygeus is not described by Greene [1935] and Walker and Homberger [1997], but it was described by Lance-Jones [1979] in mice embryos, developing from the same anlage that gives rise to the gluteus maximus but then apparently fusing with the biceps femoris, probably corresponding to the anterior head of the biceps femoris: see gluteus maximus mentioned earlier]	–

(Continued)

TABLE 19.13 (CONTINUED)
Hindlimb Muscles of Adults of Representative Tetrapod Taxa

Amphibia (Caudata): *Ambystoma mexicanum* (Axolotl)	Amphibia (Anura): *X. laevis* (African Clawed Frog)	Reptilia (Testudines): *Trachemys scripta* (Pond Slider)	Reptilia (Lepidosauria): *Timon lepidus* (Ocellated Lizard)	Reptilia (Crocodylia): *Alligator mississippiensis* (American Alligator)	Reptilia (Aves): *Gallus domesticus* (Chicken)	Mammalia (Monotremata): *Ornithorhynchus anatinus* (Platypus)	Mammalia (Marsupialia): *D. virginiana* (Virginia Opossum)	Mammalia (Placentalia): *Rattus norvegicus* (Brown Rat)	Mammalia (Placentalia): *Homo sapiens* (Modern Human)
Iliofemoralis [innervation: n. extensorius (Francis 1934); as in *A. tigrinum* (Ashley-Ross 1992), in the axolotls we dissected, it extends from the ilium and lateral border of puboischiac plate to the posterior surface of the femur and femoral trochanter]	**Iliofemoralis** [innervation: ischiadicus nerve of plexus sacralis (Noble 1922); as reported by Dunlap [1960] in the *R. pipiens*, *X. laevis*, and *E. coqui* specimens we dissected, the iliofemoralis runs from the girdle to the femur]	**Iliofemoralis** (testofemoralis *sensu* Ogushi [1913]) [Walker [1973]: in *Trachemys scripta*, the iliofemoralis runs from the ilium, first sacral and one or two presacral vertebrae and adjacent parts of the carapace, to the trochanter major of the femur]	**Iliofemoralis** [Appleton 1928: contrary to Lance-Jones [1979] and to the idea defended in the present work considered that not only the tensor fasciae but also the gluteus maximus and the femorococcygeus derive from the reptilian iliofemoralis and not from the iliotibialis; innervation: doubly, by femoral/ obturator and/or tibial/ fibular plexus (Romer 1942); as described by Russell and Bauer [2008], for *Iguana*, in the *Timon* specimens we dissected, the iliofemoralis runs from ilium to the femur, namely, to the distal portion of the femur, being blended distally with the iliotibialis and the sartorius (ambiens)]	**Iliofemoralis** (*sensu* Romer [1923]) [Romer [1923]: in *A. mississippiensis*, the iliofemoralis runs from the ilium to the femur; innervation: crural and sacral nerves (Romer 1923)]	**Iliotrochantericus externus** (gluteus medius et minimus or iliofemoralis externus *sensu* George and Berger [1966] and Lance-Jones [1979]) [corresponds only to a part of the iliofemoralis of lizards, the other part of the latter muscle corresponding to the iliotrochanterici of birds; that is why it makes sense to use the name iliofemoralis externus for birds; Vanden Berge 1975: in adult chicken, the iliotrochantericus externus runs from ilium to femur; Lance-Jones 1979: the deep dorsal mass of the chick originates from the iliac acetabular region and inserts on the posterior and lateral aspects of the femoral head, giving rise to small slips, i.e., the iliofemoralis externus and the iliotrochanterici, which share features with the deep gluteal muscles of mammals; however, the innervation of the deep gluteals of mammals is by the sciatic nerve only in mouse, while the innervation of the iliotrochanterici and iliofemoralis externus of chicken is by both sciatic and femoral nerves; however according to Vanden Berge [1975], innervation is only by ischiatic nerve in chickens]	**Gluteus medius** [from hip bone—tuber coxae—and soft tissues to proximal femur; see also Gregory and Camp [1918]]	**Gluteus medius** [from ilium to femur, being deeply blended with gluteus minimus, gluteus maximus, and piriformis]	**Gluteus medius** [rat: from ilium and sacrum to femur, being blended with piriformis; it has anterior head + posterior head; see preceding text; innervation: superior gluteal nerve (Greene 1935); as described by Greene [1935], in the mice we dissected, this large muscle extends from fascia of sacrum and the dorsal and ventral borders of the ilium to the greater trochanter, being completely covered by the gluteus maximus; it has an anterior head and a posterior head, which is deeply blended with the piriformis, see, e.g., Figure 2-7 of Walker and Homberger [1997]]	**Gluteus medius**

(Continued)

TABLE 19.13 (CONTINUED)
Hindlimb Muscles of Adults of Representative Tetrapod Taxa

Amphibia (Caudata): *Ambystoma mexicanum* (Axolotl)	Amphibia (Anura): *X. laevis* (African Clawed Frog)	Reptilia (Testudines): *Trachemys scripta* (Pond Slider)	Reptilia (Lepidosauria): *Timon lepidus* (Ocellated Lizard)	Reptilia (Crocodylia): *Alligator mississippiensis* (American Alligator)	Reptilia (Aves): *Gallus domesticus* (Chicken)	Mammalia (Monotremata): *Ornithorhynchus anatinus* (Platypus)	Mammalia (Marsupialia): *D. virginiana* (Virginia Opossum)	Mammalia (Placentalia): *Rattus norvegicus* (Brown Rat)	Mammalia (Placentalia): *Homo sapiens* (Modern Human)
–	–	–	–	–	**Iliotrochantericus caudalis** (head III of Romer [1927]) [Vanden Berge 1975: in adult chicken, the iliotrochantericus caudalis extends from ilium to femur; Lance-Jones 1979: see iliotrochantericus externus mentioned earlier; innervation: sciatic and femoral nerves (Lance-Jones 1979), but only femoral nerve (cranial gluteal nerve) for Vanden Berge [1975]]	–	–	–	–
–	–	–	–	–	**Iliotrochantericus cranialis** (head I of Romer [1927]) [Vanden Berge 1975: in adult chicken, the iliotrochantericus cranialis runs from ilium to femur; Lance-Jones 1979: see iliotrochantericus externus mentioned earlier; innervation: sciatic and femoral nerves (Lance-Jones [1979]), but only femoral nerve (cranial gluteal nerve) for Vanden Berge [1975]]	–	–	–	–
–	–	–	–	–	**Iliotrochantericus medius** (head II of Romer [1927]) [Vanden Berge [1975]: in adult chicken, the iliotrochantericus medius extends from ilium to femur; Lance-Jones 1979: see iliotrochantericus externus mentioned earlier; innervation: sciatic and femoral nerves (Lance-Jones 1979), but only femoral nerve (cranial gluteal nerve) for Vanden Berge [1975]]	–	–	–	–

(Continued)

TABLE 19.13 (CONTINUED)
Hindlimb Muscles of Adults of Representative Tetrapod Taxa

Amphibia (Caudata): *Ambystoma mexicanum* (Axolotl)	Amphibia (Anura): *X. laevis* (African Clawed Frog)	Reptilia (Testudines): *Trachemys scripta* (Pond Slider)	Reptilia (Lepidosauria): *Timon lepidus* (Ocellated Lizard)	Reptilia (Crocodylia): *Alligator mississippiensis* (American Alligator)	Reptilia (Aves): *Gallus domesticus* (Chicken)	Mammalia (Monotremata): *Ornithorhynchus anatinus* (Platypus)	Mammalia (Marsupialia): *D. virginiana* (Virginia Opossum)	Mammalia (Placentalia): *Rattus norvegicus* (Brown Rat)	Mammalia (Placentalia): *Homo sapiens* (Modern Human)
–	–	–	–	–	–	– [present in *Ornithorhynchus* according to Gregory and Camp [1918], but most authors state that it is not present as a distinct muscle in platypus, being apparently fused with the gluteus medius; this was confirmed by Gambaryan et al. [2002], who stated that the gluteus minimus is however present in other extant monotremes such as *Trachyglossus*]	**Gluteus minimus** [from ilium to femur]	**Gluteus minimus** (gluteus minimus proper plus scansorius *sensu* Lance-Jones [1979]; gluteus profundus *sensu* Walker and Homberger [1997]) [rat: from ilium to femur; see [preceding text; innervation: superior gluteal nerve (Greene 1935); as described by Greene [1935], in the mice we dissected, it is a well-developed muscle extending from both the dorsal and lateral borders of the ilium to the greater trochanter, being completely covered by the gluteus medius]	**Gluteus minimus**
–	–	–	–	–	–	– [Appleton [1928] stated that the monotreme gluteus maximus is solely innervated by the inferior gluteal nerve, and there is no separate tensor fasciae latae; see also cell on the right]	– [the tensor fasciae latae is likely a derived feature of placentals, because it is apparently absent in monotremes and marsupials: see, e.g., Bardeen 1906, Gregory and Camp 1918, and Appleton 1928]	**Tensor fasciae latae** [rat: from ilium to tibia, being blended with gluteus maximus; mouse: from ilium to fibula according to Charles et al. [2016]; see preceding text; innervation: superior gluteal nerve (Greene 1935); as described by Greene [1935], in the mice we dissected, it is a thin muscle extending from iliac crest to fascia latae, being blended with the gluteus maximus]	**Tensor fasciae latae** (tensor fasciae femoris *sensu* Osman [1955], often also designated tensor vaginae femoris)

(Continued)

TABLE 19.13 (CONTINUED)
Hindlimb Muscles of Adults of Representative Tetrapod Taxa

Amphibia (Caudata): *Ambystoma mexicanum* (Axolotl)	Amphibia (Anura): *X. laevis* (African Clawed Frog)	Reptilia (Testudines): *Trachemys scripta* (Pond Slider)	Reptilia (Lepidosauria): *Timon lepidus* (Ocellated Lizard)	Reptilia (Crocodylia): *Alligator mississippiensis* (American Alligator)	Reptilia (Aves): *Gallus domesticus* (Chicken)	Mammalia (Monotremata): *Ornithorhynchus anatinus* (Platypus)	Mammalia (Marsupialia): *D. virginiana* (Virginia Opossum)	Mammalia (Placentalia): *Rattus norvegicus* (Brown Rat)	Mammalia (Placentalia): *Homo sapiens* (Modern Human)
–	–	–	–	–	–	– [not present in monotremes: see, e.g., Pearson [1926] and Gambaryan et al. [2002]]	**Scansorius** (iliofemoralis *sensu* Coues [1872] and Stein [1981]) [small, very slender fasciculus extending from acetabulum of hip bone to lesser trochanter of femur, lying in the same position as the iliofemoral ligament of humans, and corresponding, according to Coues [1872], to the scansorius of nonhuman primates; Stein [1981] could not find this muscle in *D. virginiana*, contrary to Coues [1872]]	– [not described by Greene [1935] and Walker and Homberger [1997] and not found by us; reported by Lance-Jones [1979] only in embryo mice, being part of the gluteus minimus]	–
Ischiotrochantericus (ischiofemoralis *sensu* Gadow [1882], Walker [1954], and Ashley-Ross [1992]) [innervation: ischiadicus ventralis (Francis 1934); as in *Ambystoma tigrinum* (Ashley-Ross 1992), in our axolotls, the ischiofemoralis is blended with the puboischiofemoralis externus and extends from the lateral and dorsal surfaces of the puboischiac plate to the tuberosity of the posterior side of the head of the femur through a strong, stout tendon; see adductor femoris]	**Ischiotrochantericus A** (gemellus *sensu* Noble [1922] and Dunlap [1960]) [the muscle is not homologous to any gemellus of humans, because the LCA of amphibians and amniotes did not have the latter muscle; hence, we use a different name; innervation: ischiadicus nerve of plexus sacralis (Noble 1922); as reported by Dunlap [1960], in the *R. pipiens*, *X. laevis*, and *E. coqui* specimens we dissected, the gemellus extends from the ischium to the proximal femur]	**Ischiotrochantericus** (quadratus femoris *sensu* Ogushi [1913]) [Walker 1973: in *Trachemys scripta*, the ischiotrochantericus extends from the ilium, ischium and pubis and membrane covering the thyroid fenestra, to the trochanter major and adjacent part of the intertrochanteric fossa of the femur]	**Ischiotrochantericus** ("puboischiofemoralis posterior" *sensu* Gregory and Camp [1928]) [innervation: tibial/fibular plexus (Romer 1942); as described by Russell and Bauer [2008] for *Iguana* and other lizards, in our *Timon* specimens, the ischiotrochantericus extends from the ischial tuberosity/posterior surface of the ischium, to the proximal femur, being mainly a cuff muscle that assists in retraction of the hindlimb]	**Ischiotrochantericus** (*sensu* Romer [1923]; part of puboischiofemoralis posterior *sensu* Gadow [1882]) [Romer 1923: in *A. mississippiensis*, the ischiotrochantericus extends from the ischium to the proximal femur; innervation: ischiatic nerve (Romer 1923)]	**Ischiotrochantericus** (ischiofemoralis *sensu* Romer [1927], George and Berger [1966], Lance-Jones [1979], and Schroeter and Tosney [1991]) [Vanden Berge 1975: in adult chicken, the ischiofemoralis extends from ischium to trochanteric ridge of femur; Lance-Jones 1979: the ischiotrochantericus (ischio-femoralis) of birds and the ischio-trochantericus of mammals ontogenetically originate from the anterodorsal part of the ventral mass and clearly seem to be homologous structures; Vanden Berge 1975: in adult chicken, the iliacus extends from ilium to proximal femur; innervation: ischiofemoral nerve, branch of sacral plexus (Vanden Berge 1975)]	**Ischiotrochantericus** (gemellus *sensu* Gambaryan et al. [2002]) [from ischium to proximal femur; the obturator internus is present in monotremes according to Appleton [1928] and in *Ornithorhynchus* but not in echidna according to Gregory and Camp [1918]; however, as stated by Jouffroy and Lessertisseur 1971, the structure designated "obturator internus" or as "gemellus" by some authors seems to correspond simply to an undifferentiated ischiotrochantericus]	**Obturator internus** [from hip bone—ischium and pubis—to femur, being fused with the gemellus superior and with the gemellus inferior, distally]	**Obturator internus** [rat: from ischium to femur; mouse according to Charles et al. [2016]: from pubic tubercle to greater trochanter; Lance-Jones 1979: in *Mus*, the gemellus superior, gemellus inferior, obturator internus, and quadratus femoris derive from the ischiotrochanteric ventral anlage, the gemelli having more affinity obturator internus; innervation: special nerve from sacral plexus (Greene 1935); as described by Greene [1935], in our rats, it runs from medial surface of ischium to the trochanteric fossa of the femur through a long tendon; it does not cover the obturator membrane]	**Obturator internus**

(*Continued*)

TABLE 19.13 (CONTINUED)
Hindlimb Muscles of Adults of Representative Tetrapod Taxa

Amphibia (Caudata): *Ambystoma mexicanum* (Axolotl)	Amphibia (Anura): *X. laevis* (African Clawed Frog)	Reptilia (Testudines): *Trachemys scripta* (Pond Slider)	Reptilia (Lepidosauria): *Timon lepidus* (Ocellated Lizard)	Reptilia (Crocodylia): *Alligator mississippiensis* (American Alligator)	Reptilia (Aves): *Gallus domesticus* (Chicken)	Mammalia (Monotremata): *Ornithorhynchus anatinus* (Platypus)	Mammalia (Marsupialia): *D. virginiana* (Virginia Opossum)	Mammalia (Placentalia): *Rattus norvegicus* (Brown Rat)	Mammalia (Placentalia): *Homo sapiens* (Modern Human)
–	**Ischiotrochantericus B** (obturator externus *sensu* Noble [1922] and Dunlap [1960]) [the muscle is not homologous to the obturator externus of humans, because the LCA of amphibians and amniotes did not have the latter muscle; hence, we use a different name; innervation: obturatorius nerve of plexus cruralis (Noble 1922) as reported by Dunlap [1960] in the *R. pipiens*, *X. laevis*, and *E. coqui* specimens we dissected, the obturator externus extends from the pubis to the femur; see ischiotrochantericus A]	–	–	–	–	–	–	–	–

(Continued)

TABLE 19.13 (CONTINUED)
Hindlimb Muscles of Adults of Representative Tetrapod Taxa

Amphibia (Caudata): *Ambystoma mexicanum* (Axolotl)	Amphibia (Anura): *X. laevis* (African Clawed Frog)	Reptilia (Testudines): *Trachemys scripta* (Pond Slider)	Reptilia (Lepidosauria): *Timon lepidus* (Ocellated Lizard)	Reptilia (Crocodylia): *Alligator mississippiensis* (American Alligator)	Reptilia (Aves): *Gallus domesticus* (Chicken)	Mammalia (Monotremata): *Ornithorhynchus anatinus* (Platypus)	Mammalia (Marsupialia): *D. virginiana* (Virginia Opossum)	Mammalia (Placentalia): *Rattus norvegicus* (Brown Rat)	Mammalia (Placentalia): *Homo sapiens* (Modern Human)
–	**Ischiotrochantericus C** (quadratus femoris *sensu* Noble [1922] and Dunlap [1960]) [the muscle is not homologous to the quadratus femoris of humans, because the LCA of amphibians and amniotes did not have the latter muscle; hence, we use a different name; innervation: obturatorius nerve of plexus cruralis (Noble 1922) as reported by Dunlap [1960], in the *Rana* and *Eleutherodactylus* specimens we dissected, the quadratus femur runs from ischium to proximal femur, but in *Xenopus*, it is fused with the obturator externus, i.e., not really present as a distinct muscle, being probably undifferentiated from the obturator externus, as it is in plesiomorphic anurans such as *Ascaphus*; see ischiotrochantericus A]	–	–	–	–	–	–	–	–

(Continued)

TABLE 19.13 (CONTINUED)
Hindlimb Muscles of Adults of Representative Tetrapod Taxa

Amphibia (Caudata): *Ambystoma mexicanum* (Axolotl)	**Amphibia (Anura): *X. laevis* (African Clawed Frog)**	**Reptilia (Testudines): *Trachemys scripta* (Pond Slider)**	**Reptilia (Lepidosauria): *Timon lepidus* (Ocellated Lizard)**	**Reptilia (Crocodylia): *Alligator mississippiensis* (American Alligator)**	**Reptilia (Aves): *Gallus domesticus* (Chicken)**	**Mammalia (Monotremata): *Ornithorhynchus anatinus* (Platypus)**	**Mammalia (Marsupialia): *D. virginiana* (Virginia Opossum)**	**Mammalia (Placentalia): *Rattus norvegicus* (Brown Rat)**	**Mammalia (Placentalia): *Homo sapiens* (Modern Human)**
–	**Ischiotrochantericus D** (obturator internus *sensu* Noble 1922 and Dunlap 1960) [the muscle is not homologous to the obturator internus of humans, because the LCA of amphibians and amniotes did not have the latter muscle, hence we use a different name; innervation: ischiadicus nerve of plexus sacralis (Noble 1922); as reported by Dunlap 1960 in the *R. pipiens*, *X. laevis*, and *E. coqui* specimens we dissected the "obturator internus" extends from the girdle to the proximal femur; see ischiotrochantericus A]	–	–	–	–	–	–	–	–
–	–	–	–	–	–	–	**Gemellus superior** [from ischium to femur]	**Gemellus superior** [rat: from ischium to femur, being blended with obturator internus; mouse: according to Charles et al. [2016], there is only one gemellus, from ischial spine to greater trochanter, superior to obturator internus; see preceding text; innervation: sacral plexus (Greene 1935); as described by Greene [1935], in our rats, this muscle is situated anteriorly to the tendon of the obturator internus and extends from the anterior part of the dorsal border of the ischium to the trochanteric fossa trough the tendon of the obturator internus]	**Gemellus superior**

(Continued)

TABLE 19.13 (CONTINUED)
Hindlimb Muscles of Adults of Representative Tetrapod Taxa

Amphibia (Caudata): *Ambystoma mexicanum* (Axolotl)	Amphibia (Anura): *X. laevis* (African Clawed Frog)	Reptilia (Testudines): *Trachemys scripta* (Pond Slider)	Reptilia (Lepidosauria): *Timon lepidus* (Ocellated Lizard)	Reptilia (Crocodylia): *Alligator mississippiensis* (American Alligator)	Reptilia (Aves): *Gallus domesticus* (Chicken)	Mammalia (Monotremata): *Ornithorhynchus anatinus* (Platypus)	Mammalia (Marsupialia): *D. virginiana* (Virginia Opossum)	Mammalia (Placentalia): *Rattus norvegicus* (Brown Rat)	Mammalia (Placentalia): *Homo sapiens* (Modern Human)
–	–	–	–	–	–	–	**Gemellus inferior** [from ischium to femur]	**Gemellus inferior** [rat: from ischium to femur, being blended with obturator internus; mouse: see preceding cell; see preceding text; innervation: sacral plexus (Greene 1935); as described by Greene [1935], in our rats, this muscle is situated posteriorly to the tendon of the obturator internus and extends from the posterodorsal margin of the ischium to the trochanteric fossa trough the tendon of the obturator internus]	**Gemellus inferior**
Puboischiofemoralis externus [innervation: obturator and ischiadicus ventralis (Francis 1934); as in *Ambystoma tigrinum* (Ashley-Ross 1992), in our axolotls, it extends from the ventral midline and anterior border of the puboischiac plate as well as pubis and ischium to the femoral crest and to the femoral shaft distal to this crest, as well as to the femoral trochanter]	**Puboischiofemoralis externus A** (pectineus *sensu* Noble [1922] and Dunlap [1960]) [the muscle is not homologous to the obturator pectineus of humans, because the LCA of amphibians and amniotes did not have the latter muscle; hence, we use a different name; innervation: femoralis nerve of plexus cruralis (Noble 1922); as reported by Dunlap [1960] in the *R. pipiens*, *X. laevis*, and *E. coqui* specimens we dissected, the pectineus runs from the ilium and pubis to the femur]	**Puboischiofemoralis externus** (obturatorius externus *sensu* Ogushi [1913]) [Walker 1973: in *Trachemys scripta*, the puboischiofemoralis externus extends from the pubis and epipubic cartilage (anterior head) and the ischium and membrane, enclosing the thyroid fenestra (posterior head) to the trochanter minor of the femur]	**Puboischiofemoralis externus** (includes iliacus, pectineus, obturator internus, obturator externus, and adductor brevis of various authors: see Russell and Bauer [2008]) [the quadratus femoris derives from the posterior part of the puboischiofemoralis externus according to Lance-Jones [1979]; innervation: doubly, by femoral/obturator and/or tibial/fibular plexus (Romer 1942); as reported by Russell and Bauer [2008] in *Iguana* and other lizards, in our *Timon* specimens, the puboischiofemoralis externus is mainly divided into an anterior group and a posterior group extending from the pubis and ischium to the intertrochanteric fossa and internal trochanter of the femur,	**Puboischiofemoralis externus** (*sensu* Romer [1923]) [Romer 1923: in *A. mississippiensis*, the puboischiofemoralis externus has 3 heads and extends from the pubis, ischium, and last abdominal rib to proximal femur; innervation: at least by obturator nerve (Romer 1923)]	**Puboischiofemoralis externus** (anterior puboischiofemoralis externus or obturators *sensu* Romer [1927], George and Berger [1966], Lance-Jones [1979], and Schroeter and Tosney [1991]) [lateral and medial heads: Romer 1927, George and Berger 1966, Schroeter and Tosney 1991; Vanden Berge 1975: in adult chicken the "obturator" runs from ischium, pubis, and ilium to proximal femur; Lance-Jones 1979: the first visible indentations of the ventral mass of muscles divide anterior anlagen innervated by the obturator from posterior sciatically innervated anlagen, the anterior and posterior masses showing further signs of division into proximally and distally	**Obturator externus** [from hip bone to proximal femur]	**Obturator externus** [from hip bone—pubis and ischium—to femur]	**Obturator externus** [rat: from margin of obturator foramen to femur; mouse according to Charles et al. [2016]: from pubic ramus to greater trochanter; Lance-Jones 1979: in *Mus*, the obturator externus, adductor longus, adductor magnus, adductor brevis, and gracilis derive from the adductor ventral anlage, the obturator externus more affinity with the adductor longus and the structure she designates as adductor magnus (adductor minimus head of the adductor magnus *sensu* the present work); so the adult innervation and topology and the development of the obturator externus and of the obturator internus are markedly different; innervation: posterior	**Obturator externus**

(*Continued*)

TABLE 19.13 (CONTINUED)
Hindlimb Muscles of Adults of Representative Tetrapod Taxa

Amphibia (Caudata): *Ambystoma mexicanum* (Axolotl)	Amphibia (Anura): *X. laevis* (African Clawed Frog)	Reptilia (Testudines): *Trachemys scripta* (Pond Slider)	Reptilia (Lepidosauria): *Timon lepidus* (Ocellated Lizard)	Reptilia (Crocodylia): *Alligator mississippiensis* (American Alligator)	Reptilia (Aves): *Gallus domesticus* (Chicken)	Mammalia (Monotremata): *Ornithorhynchus anatinus* (Platypus)	Mammalia (Marsupialia): *D. virginiana* (Virginia Opossum)	Mammalia (Placentalia): *Rattus norvegicus* (Brown Rat)	Mammalia (Placentalia): *Homo sapiens* (Modern Human)
			as also described by Romer [1942], being mainly a protractor and rotator of the femur]		inserting groups; three muscles are derived from the anterior part of the ventral muscle mass in chicks: one proximally inserting obturator or "anterior puboischio-femoralis externus" (included in the pubo-ischiofemoralis externus *sensu* the present work) and two distally inserting adductors or "puboischio-femoralis" (adductor femoris *sensu* the present work); ontogenetically, they are respectively homologous to the obturator externus and to the adductor magnus (adductor minimus head of the adductor magnus *sensu* the present work) and adductor brevis (main body of adductor magnus *sensu* the present work) of mice; both the puboischiofemoralis externus of birds and the obturator externus of mammals originate from the deepest part of the preaxial ventral musculature opposite the acetabular region of the pubis; the quadratus femoris of mammals is homologous to the posterior part of the puboischiofemoralis externus of lizards, a structure that is not present in birds; innervation: obturator nerve (Vanden Berge 1975)]			division of obturator nerve (Greene 1935); as described by Greene [1935], in our rats this thin muscle extends from the obturator membrane and the bony margin of the obturator foramen, to the trochanteric fossa between quadratus femoris and obturator internus]	

(Continued)

TABLE 19.13 (CONTINUED)
Hindlimb Muscles of Adults of Representative Tetrapod Taxa

Amphibia (Caudata): *Ambystoma mexicanum* (Axolotl)	Amphibia (Anura): *X. laevis* (African Clawed Frog)	Reptilia (Testudines): *Trachemys scripta* (Pond Slider)	Reptilia (Lepidosauria): *Timon lepidus* (Ocellated Lizard)	Reptilia (Crocodylia): *Alligator mississippiensis* (American Alligator)	Reptilia (Aves): *Gallus domesticus* (Chicken)	Mammalia (Monotremata): *Ornithorhynchus anatinus* (Platypus)	Mammalia (Marsupialia): *D. virginiana* (Virginia Opossum)	Mammalia (Placentalia): *Rattus norvegicus* (Brown Rat)	Mammalia (Placentalia): *Homo sapiens* (Modern Human)
–	– (adductor longus *sensu* Noble [1922] and Dunlap [1960]) [the muscle is not homologous to the adductor longus pectineus of humans, because the LCA of amphibians and amniotes did not have the latter muscle; hence, we use a different name; innervation: femoralis nerve of plexus cruralis (Noble 1922); as reported by Dunlap [1960], in the *R.* specimens we dissected, the adductor longus extends from the girdle to the adductor magnus and to the knee region; in the *Xenopus* and *Eleutherodactylus* specimens we dissected, the muscle is not present as a distinct structure, being probably fused with the pectineus, as it is in plesiomorphic anurans such as *Ascaphus*]	–	–	–	–	–	–	–	–

(Continued)

TABLE 19.13 (CONTINUED)
Hindlimb Muscles of Adults of Representative Tetrapod Taxa

Amphibia (Caudata): *Ambystoma mexicanum* (Axolotl)	Amphibia (Anura): *X. laevis* (African Clawed Frog)	Reptilia (Testudines): *Trachemys scripta* (Pond Slider)	Reptilia (Lepidosauria): *Timon lepidus* (Ocellated Lizard)	Reptilia (Crocodylia): *Alligator mississippiensis* (American Alligator)	Reptilia (Aves): *Gallus domesticus* (Chicken)	Mammalia (Monotremata): *Ornithorhynchus anatinus* (Platypus)	Mammalia (Marsupialia): *D. virginiana* (Virginia Opossum)	Mammalia (Placentalia): *Rattus norvegicus* (Brown Rat)	Mammalia (Placentalia): *Homo sapiens* (Modern Human)
–	–	–	–	–	–	– [piriformis absent in monotremes according to Gambaryan et al. [2002], who stated that the piriformis *sensu* authors such as Coues [1970] corresponds to the gluteus maximus; the piriformis *sensu* Pearson [1926] seems to correspond to the caudofemoralis]	**Piriformis** [from second sacral and first caudal vertebrae to femur; Appleton [1928] stated that the muscle that is often designated piriformis in marsupials corresponds to the caudofemoralis and not to the piriformis, of other mammals, but authors such as Stein [1981] consider the marsupial piriformis homologous to the placental piriformis, and the attachments of the former are similar to the latter, so there is no reason to think that the piriformis was lost in marsupials such as the opossum and that at the same time the insertion of the caudofemoralis became more proximal insertion (two evolutionary steps); it is more parsimonious to assume that the caudofemoralis muscle is absent in the opossum (one evolutionary step), considering that the crurococcygeus of opossum seems to correspond to the dorsal/anterior head of the semitendinosus rather than to caudo-femoralis, see semitendinosus below; the loss of a caudofemoralis in marsupials such as the opossum is supported by dissections of other marsupials that do have this muscle, such as Osgood 1921, who stated that *Didelphis* species do not have a caudofemoralis]	**Piriformis** [rat: from sacrum to femur, being blended with gluteus medius; mouse not described by Charles et al. [2016]; see preceding text; Walker and Homberger 1997: the piriformis is usually completely fused to the gluteus medius in rats; innervation: branch of lumbosacral plexus, which also innervates the obturator internus (Greene 1935); as described by Greene [1935], in our rats, the piriformis extends from the sacrum to the greater trochanter of the femur, being deeply blended to the gluteus medius but still recognizable as a distinct muscle]	**Piriformis**

(*Continued*)

TABLE 19.13 (CONTINUED)
Hindlimb Muscles of Adults of Representative Tetrapod Taxa

Amphibia (Caudata): *Ambystoma mexicanum* (Axolotl)	Amphibia (Anura): *X. laevis* (African Clawed Frog)	Reptilia (Testudines): *Trachemys scripta* (Pond Slider)	Reptilia (Lepidosauria): *Timon lepidus* (Ocellated Lizard)	Reptilia (Crocodylia): *Alligator mississippiensis* (American Alligator)	Reptilia (Aves): *Gallus domesticus* (Chicken)	Mammalia (Monotremata): *Ornithorhynchus anatinus* (Platypus)	Mammalia (Marsupialia): *D. virginiana* (Virginia Opossum)	Mammalia (Placentalia): *Rattus norvegicus* (Brown Rat)	Mammalia (Placentalia): *Homo sapiens* (Modern Human)
–	–	–	– [difficult to discern if it is really absent in *Timon*; a distinct quadratus femoris was described by Mivart [1870] in chameleons, as part of the puboischiofemoralis externus]	–	– [the posterior part of the puboischiofemoralis externus of lizards, which corresponds to the quadratus femoris of mammals, is absent in chicken; see puboischiofemoralis externus]	**Quadratus femoris** [from ischium to proximal femur]	**Quadratus femoris** [from ischium to femur]	**Quadratus femoris** [rat: from ischium to femur; mouse according to Charles et al. [2016]: from pubic tubercle to midfemur; see preceding text; innervation: posterior division of femoral nerve (Greene 1935); as described by Greene [1935], the quadratus femoris extends from the posterior border of the ischium to the lesser trochanter]	**Quadratus femoris**
Puboischiofemoralis internus [includes "pubo-extensorius" *sensu* Francis [1934], who stated that the pubo-extensorius is present in *Salamandra*, extending from pubis to the knee region and being innervated by the femoral nerve but deeply blended with the puboischiofemoralis internus; innervation: femoral and small contribution from obturator (Francis 1934); as in *Ambystoma tigrinum* (Ashley-Ross 1992), in our axolotls, it extends from the anterior portion of the dorsal midline of the pubo-ischiac plate, ypsiloid cartilage and pubis to the anterior, dorsal and ventral faces of the femur]	**Puboischiofemoralis internus A** (iliacus internus *sensu* Noble [1922] and Dunlap [1960]) [the muscle is not homologous to the iliacus nor to an iliacus bundle of humans, because the LCA of amphibians and amniotes did not have an iliacus; thus, we use a different name; innervation: femoralis nerve of plexus cruralis (Noble 1922); as reported by Dunlap [1960], in the *R. pipiens*, *X. laevis*, and *E. coqui* specimens we dissected, the iliacus internus extends from the ilium to the dorsal and medial region of the femur]	**Puboischiofemoralis internus** (obturatorius internus plus iliacus *sensu* Ogushi [1913]) [Walker 1973: in *Trachemys scripta*, the puboischiofemoralis internus extends from pubis, epipubic cartilage (anteroventral head, which apparently corresponds to the head 1 of lizards), the first sacral rib and vertebra, and two presacral vertebrae (posterodorsal head, which apparently corresponds to head 2 of lizards) to the femoral shaft and trochanter minor]	**Puboischiofemoralis internus** [Russell and Bauer 2008: the puboischiofemoralis internus of lizards protracts (moves anteriorly) the femur, corresponds to the deltoideus scapularis of the lizard forelimb, and is often divided into three parts, part 3 (ventral) being the more ventral and anterior, lying ventrally to the sartorius (ambiens) and extending from the anterior rim of the pubic plate to the femoral trochanter, part 2 (medial) mainly originating with part 3 and inserting with part 1 and part 1 (dorsal) extending from the ischial symphysis, main body of ischium and posterior pubis, to the femoral trochanter (together with part 3) and to the posterodorsal margin of the femur at its proximal end (together with part 2);	**Puboischiofemoralis internus** (*sensu* Romer [1923]) [Romer [1923]: in *A. mississippiensis*, the puboischiofemoralis internus has 2 heads and extends from six lumbar vertebrae, sacral ribs, ilium, and ischium to proximal femur; innervation: crural nerve (Romer 1923)]	**Puboischiofemoralis internus** (iliofemoralis internus or iliacus *sensu* Romer [1927], George and Berger [1966], Schroeter and Tosney [1991], and Lance-Jones [1979]) [Vanden Berge 1975: in adult chicken, the iliacus extends from ilium to proximal femur; Lance-Jones 1979: the "iliofemoralis internus" or iliacus of the chick (puboischiofemoralis internus *sensu* the present work) is mainly homologous to the mammalian iliopsoas, both originating from the anteroproximal part of the dorsal mass, inserting proximally on the anteromedial aspect of the femur, and showing an early ontogenetic fusion with a small mass which lies superficial and anterior to them; the adult mammalian iliopsoas is a much larger muscle,	**Iliopsoas** [iliacus and psoas major not differentiated, the iliopsoas running from lumbar vertebrae and first sacral vertebra and ilium to proximal femur: see, e.g., Gambaryan et al. 2002]	**Iliacus** [from hip bon—ilium—to femur, being deeply blended with the psoas major]	**Iliacus** [rat: from L5 and L6 to femur, being blended with psoas major; mouse: according to Charles et al. [2016]: from iliac crest to less trochanter; Lance-Jones 1979: in *Mus*, the iliacus derives from the anterior dorsal anlage, together with the psoas major, forming the iliopsoas, the sartorius being ontogenetically associated with, and perhaps a link between, the anterior dorsal mass and the posterior dorsal mass, particularly the quadriceps (originally, the iliacus does not attach onto vertebrae, so it is a superficial/ appendicular muscle according to the in–out model); innervation: lumbar plexus, namely, L2 and L3 (Greene 1935); as described by Greene [1935], in our rats, the iliacus runs from L5 and L6 to the lesser trochanter together with the psoas major]	**Iliacus**

(*Continued*)

TABLE 19.13 (CONTINUED)
Hindlimb Muscles of Adults of Representative Tetrapod Taxa

Amphibia (Caudata): *Ambystoma mexicanum* (Axolotl)	Amphibia (Anura): *X. laevis* (African Clawed Frog)	Reptilia (Testudines): *Trachemys scripta* (Pond Slider)	Reptilia (Lepidosauria): *Timon lepidus* (Ocellated Lizard)	Reptilia (Crocodylia): *Alligator mississippiensis* (American Alligator)	Reptilia (Aves): *Gallus domesticus* (Chicken)	Mammalia (Monotremata): *Ornithorhynchus anatinus* (Platypus)	Mammalia (Marsupialia): *D. virginiana* (Virginia Opossum)	Mammalia (Placentalia): *Rattus norvegicus* (Brown Rat)	Mammalia (Placentalia): *Homo sapiens* (Modern Human)
			innervation: by femoral/obturator plexus (Romer 1942); as described by Russell and Bauer [2008] for *Iguana*, in our *Timon* specimens, the puboischiofemoralis internus is divided into three parts, parts 3 and 2, as described earlier, and part 1 (dorsal) running from the ischium and pubis to the region of the femoral trochanter (together with part 3) and to the posterodorsal margin of the femur at its proximal end (together with part 2); our dissections and comparisons support the idea that this muscle gave rise to the pectineus, iliacus, and psoas major, and one could hypothesize, based on the topology of its bundles, that parts 3, 2, and 1 directly correspond to the pectineus (the more posterior/medial in mammals), psoas major, and iliacus (the more anterior/lateral in mammals)]		but it is thought that parts of the puboischiofemoralis internus were lost from the transitions leading to the origin of birds; the pectineus is a mammalian neomorph and thus lacks an avian homologue; innervation: femoral nerves (Vanden Berge 1975)]				

(Continued)

TABLE 19.13 (CONTINUED)
Hindlimb Muscles of Adults of Representative Tetrapod Taxa

Amphibia (Caudata): *Ambystoma mexicanum* (Axolotl)	Amphibia (Anura): *X. laevis* (African Clawed Frog)	Reptilia (Testudines): *Trachemys scripta* (Pond Slider)	Reptilia (Lepidosauria): *Timon lepidus* (Ocellated Lizard)	Reptilia (Crocodylia): *Alligator mississippiensis* (American Alligator)	Reptilia (Aves): *Gallus domesticus* (Chicken)	Mammalia (Monotremata): *Ornithorhynchus anatinus* (Platypus)	Mammalia (Marsupialia): *D. virginiana* (Virginia Opossum)	Mammalia (Placentalia): *Rattus norvegicus* (Brown Rat)	Mammalia (Placentalia): *Homo sapiens* (Modern Human)
–	**Puboischiofemoralis internus B** (iliacus externus *sensu* Noble [1922] and Dunlap [1960]) [the muscle is not homologous to the iliacus nor to a iliacus bundle of humans, because the LCA of amphibians and amniotes did not have a iliacus; hence, we use a different name; innervation: femoralis nerve of plexus cruralis (Noble 1922); as reported by Dunlap [1960], in the *R. pipiens*, *X. laevis,* and *E. coqui* specimens we dissected the, iliacus externus extends from the ilium to the posterodorsal surface of the caput femoris]	–	–	–	–	–	–	–	–
–	–	–	– [corresponds directly to puboischiofemoralis internus 2?]	–	– [corresponds to part of the puboischiofemoralis internus of chicks?]	– [see preceding cell]	**Psoas major** [from last three lumbar and first two sacral vertebrae to femur]	**Psoas major** [rat: from lumbar vertebrae to femur; mouse according to Charles et al. [2016]: from lower vertebrate to lesser trochanter; see preceding text; innervation: L2 and L3 through the femoral nerve (Greene 1935); as described by Greene [1935], in our rats, the psoas major extends from all lumbar vertebrae to the lesser trochanter together with the iliacus]	**Psoas major** (psoas magnus *sensu* Woollard [1925] and Osman [1955])
–	–	–	–	–	–	–	–	–	– [at least sometimes present in humans, as a variation]

(Continued)

TABLE 19.13 (CONTINUED)
Hindlimb Muscles of Adults of Representative Tetrapod Taxa

Amphibia (Caudata): *Ambystoma mexicanum* (Axolotl)	Amphibia (Anura): *X. laevis* (African Clawed Frog)	Reptilia (Testudines): *Trachemys scripta* (Pond Slider)	Reptilia (Lepidosauria): *Timon lepidus* (Ocellated Lizard)	Reptilia (Crocodylia): *Alligator mississippiensis* (American Alligator)	Reptilia (Aves): *Gallus domesticus* (Chicken)	Mammalia (Monotremata): *Ornithorhynchus anatinus* (Platypus)	Mammalia (Marsupialia): *D. virginiana* (Virginia Opossum)	Mammalia (Placentalia): *Rattus norvegicus* (Brown Rat)	Mammalia (Placentalia): *Homo sapiens* (Modern Human)
–	–	–	– [see iliotibialis]	–	– [see iliotibialis]	**Rectus femoris** [from ilium to patella and thus indirectly to tibia, being deeply blended distally with vastus lateralis and with vastus medialis]	**Rectus femoris** [from ilium to patella and to tibia, being deeply blended distally with vastus lateralis and with vastus medialis]	**Rectus femoris** [rat: from the anterior inferior iliac spine and border of the acetabulum to the patella and then the tuberosity of the tibia; mouse: according to Charles et al. [2016], from ilium to patella; see preceding text; innervation: femoral nerve (Greene 1935); as described by Greene [1935], in our rats, this muscle extends from the anterior inferior iliac spine and border of the acetabulum to the patella and then the tuberosity of the tibia; we could not see well-differentiated anterior and posterior heads of the rectus femoris as described by Greene [1935], i.e., seemed to be a single bundle as reported by Walker and Homberger [1997]]	**Rectus femoris**

(*Continued*)

TABLE 19.13 (CONTINUED)
Hindlimb Muscles of Adults of Representative Tetrapod Taxa

Amphibia (Caudata): *Ambystoma mexicanum* (Axolotl)	Amphibia (Anura): *X. laevis* (African Clawed Frog)	Reptilia (Testudines): *Trachemys scripta* (Pond Slider)	Reptilia (Lepidosauria): *Timon lepidus* (Ocellated Lizard)	Reptilia (Crocodylia): *Alligator mississippiensis* (American Alligator)	Reptilia (Aves): *Gallus domesticus* (Chicken)	Mammalia (Monotremata): *Ornithorhynchus anatinus* (Platypus)	Mammalia (Marsupialia): *D. virginiana* (Virginia Opossum)	Mammalia (Placentalia): *Rattus norvegicus* (Brown Rat)	Mammalia (Placentalia): *Homo sapiens* (Modern Human)
–	–	**Femorotibialis** (includes at least the caput internum and caput externum of vastus femoris *sensu* Ogushi [1913]) [Walker [1973]: in *Trachemys scripta*, the femorotibialis extends from the dorsal surface of the femur ("vastus medialis" *sensu* Walker [1973]), trochanter minor, anteroventral surface of the femur ("vastus internus" *sensu* Walker [1973]), trochanter major and posteroventral surface of the femur ("vastus externus" *sensu* Walker 1973) to tibia *via* the patellar tendon]	**Femorotibialis** [Russell and Bauer 2008: in lizards, the femorotibialis usually originates from the entire length of the femoral shaft and diverges posteriorly into two masses that subsequently converge onto the insertion tendon of the iliotibialis, the muscle being therefore often subdivided into external and internal and sometimes cruraeus, heads; innervation: by femoral/obturator plexus, namely, the femoral nerve; as described by Russell and Bauer [2008], in *Iguana*, in our *Timon* specimens, the femorotibialis extends from the entire length of the femoral shaft to the tendon of the iliotibialis; the femorotibialis is not divided into clear bundles]	**Femorotibialis** (*sensu* Romer [1923]) [Romer 1923: in *A. mississippiensis*, the femorotibialis extends from femur to the tibia]	**Femorotibialis** (external, medial and internal heads: George and Berger 1966, Schroeter and Tosney 1991) [Vanden Berge 1975: in adult chicken, the femorotibialis extends from femur to patellar crest of tibiotarsus; the medial, external, and internal heads being deeply blended to each other; Lance-Jones 1979: the bulk of the chick "triceps" complex consists of two large lateral muscles, one from ilium, the iliotibialis, and one from the femur, the femorotibialis; the femorotibialis ontogenetically arises deep to the femoral nerve trunk and is fused posteriorly and distally with the iliotibialis; the vasti of mammals also lie deep to the femoral nerve trunk at early stages, are fused with the more superficial rectus anlage, and originate from the femoral shaft, such features thus supporting the main homology of the femorotibialis and vasti; Schroeter and Tosney 1991: chickens have a femorotibialis divided into externus, medialis, and internus heads that seem to correspond, respectively, to the vastus lateralis, intermedius and medialis of mammals, although these heads are deeply blended to each other and do not form separate muscles as in mammals; innervation: femoral nerve (Vanden Berge 1975)]	– [absent as a distinct muscle in monotremes, see e.g. Gambaryan et al. 2002]	– [according to Coues [1872] and Stein [1981], there is no distinct vastus intermedius in *D. virginiana*, the quadriceps femoris complex being therefore instead a triceps complex; as stated by Warburton [2003], the vastus intermedius (crureus) seems to be absent as a separate, well-differentiated muscle in marsupials, being instead deeply blended with the other parts of the quadriceps femoris: see also, e.g., Thompson and Hillier 1905 and Osgood 1921]	**Vastus intermedius** [rat: from the whole length of the extensor surface of the shaft of the femur to the patella and then the tuberosity of the tibia; mouse: according to Charles et al.[2016]: from proximal femur to patella; see preceding text; innervation: femoral nerve (Greene 1935); as described by Greene [1935], in our rats, this muscle extends from the whole length of the extensor surface of the shaft of the femur to the patella and then the tuberosity of the tibia; distally some of its fibers are somewhat distinct from those of its main body, so perhaps those fibers are the ones that are sometimes considered as "articularis genus" in other taxa, and this could correspond to either the anconeus of most mammals or the triceps posterior of primates such as strepsirrhines]	**Vastus intermedius** (crureus *sensu* Osman [1955])

(Continued)

TABLE 19.13 (CONTINUED)
Hindlimb Muscles of Adults of Representative Tetrapod Taxa

Amphibia (Caudata): *Ambystoma mexicanum* (Axolotl)	Amphibia (Anura): *X. laevis* (African Clawed Frog)	Reptilia (Testudines): *Trachemys scripta* (Pond Slider)	Reptilia (Lepidosauria): *Timon lepidus* (Ocellated Lizard)	Reptilia (Crocodylia): *Alligator mississippiensis* (American Alligator)	Reptilia (Aves): *Gallus domesticus* (Chicken)	Mammalia (Monotremata): *Ornithorhynchus anatinus* (Platypus)	Mammalia (Marsupialia): *D. virginiana* (Virginia Opossum)	Mammalia (Placentalia): *Rattus norvegicus* (Brown Rat)	Mammalia (Placentalia): *Homo sapiens* (Modern Human)
–	–	–	–	–	–	**Vastus lateralis** [from femur to patella and indirectly to tibia, being deeply blended with vastus medialis]	**Vastus lateralis** [from femur to patella and tibia, being deeply blended with vastus medialis]	**Vastus lateralis** [rat: from the greater trochanter and third trochanter of the femur to the patella and then the tuberosity of the tibia; mouse: according to Charles et al. [2016], from proximal femur to patella; see preceding text; innervation: femoral nerve (Greene 1935); as described by Greene [1935], in our rats, this muscle extends from the greater trochanter and third trochanter (= gluteal tuberosity, I think) of the femur to the patella and then the tuberosity of the tibia]	**Vastus lateralis** (vastus externus *sensu* Allen [1897] and Woollard [1925])
–	–	–	–	–	–	**Vastus medialis** [from femur to patella and indirectly to tibia]	**Vastus medialis** [from femur to patella and tibia]	**Vastus medialis** [rat: from the neck and proximal end of the shaft of the femur to the patella and then the tuberosity of the tibia; mouse: according to Charles et al. [2016], from proximal femur to patella; see preceding text; innervation: femoral nerve (Greene 1935); as described by Greene [1935], in our rats, this muscle extends from the neck and proximal end of the shaft of the femur to the patella and then the tuberosity of the tibia]	**Vastus medialis** (vastus internus *sensu* Allen [1897])

(Continued)

TABLE 19.13 (CONTINUED)
Hindlimb Muscles of Adults of Representative Tetrapod Taxa

Amphibia (Caudata): *Ambystoma mexicanum* (Axolotl)	Amphibia (Anura): *X. laevis* (African Clawed Frog)	Reptilia (Testudines): *Trachemys scripta* (Pond Slider)	Reptilia (Lepidosauria): *Timon lepidus* (Ocellated Lizard)	Reptilia (Crocodylia): *Alligator mississippiensis* (American Alligator)	Reptilia (Aves): *Gallus domesticus* (Chicken)	Mammalia (Monotremata): *Ornithorhynchus anatinus* (Platypus)	Mammalia (Marsupialia): *D. virginiana* (Virginia Opossum)	Mammalia (Placentalia): *Rattus norvegicus* (Brown Rat)	Mammalia (Placentalia): *Homo sapiens* (Modern Human)
–	–	**Sartorius** (caput rectum of vastus femoris *sensu* Ogushi [1913]; ambiens *sensu* Walker [1973]) [Walker 1973: in *Trachemys scripta*, the ambiens extends from the pubic process and puboischiadic ligament to the tibia *via* the patellar tendon]	**Sartorius** (ambiens *sensu* e.g. Russell and Bauer [2008]) [as the ambiens seems to be directly homologous to the sartorius of mammals, we prefer to use the name sartorius, in order to use the same name directly designating homologous muscles in different taxa; Russell and Bauer 2008: in lizards, the sartorius (ambiens) usually extends from the pubis, near or contacting the acetabulum, to the region of insertion of the iliotibialis, i.e., extending to the proximal tibia; innervation: by femoral/obturator plexus, namely, the femoral nerve (Romer 1942); as described by Russell and Bauer [2008] in *Iguana*, in our *Timon* specimens, the sartorius (ambiens) usually extends from the pubis to the region of insertion of the iliotibialis, i.e., extending to the proximal tibia; our dissections of lizards clearly support the idea that the ambiens is homologous to the mammalian sartorius, because by applying a medial rotation of the femur in lizards as the one seeing during the ontogeny of mammals, the origin of the muscle becomes anteroproximolateral, the muscle running distolaterally and the insertion being anterodistomedial, exactly as in mammals]	**Sartorius** (ambiens *sensu* Romer [1923]) [Romer 1923: in *A. mississippiensis*, the ambiens has two heads and extends from the ilium, pubis, and ischium to the iliotibialis and gastrocnemius muscles]	**Sartorius** (ambiens *sensu* Romer [1927], George and Berger [1966], Lance-Jones [1979], and Schroeter and Tosney [1991]) [NB: it does not correspond to the structure that is often named sartorius in chicken, which corresponds instead to part of the iliotibialis: see iliotibialis in the following; Vanden Berge 1975: in adult chicken, the ambiens extends from ilium to fibular head and to flexor musculature of the leg; Lance-Jones 1979: the small ambiens of the chick corresponds to the sartorius of mammals, both ontogenetically arising from the most anterior aspect of the triceps and being separable from the latter by the insertion of the iliopsoas/puboischiofemoralis internus, both also originating early in ontogeny near the pectineal process and fusing distally with the triceps/quadriceps; innervation: femoral nerve (Vanden Berge 1975)]	**Sartorius** [not described in platypus by Gambaryan et al. [2002], but generally considered to be present in monotremes, extending from the pubis (and not ilium as in therians) to tibia: see, e.g., Gregory and Camp 1918, Pearsons 1926, and Jouffroy and Lessertisseur 1971]	**Sartorius** [from ilium to patella and tibia; it is however not fused with the vasti nor with the rectus femoris]	– [not present in adult rats nor in adult mice, but apparently normally present in mice embryos (E12 to E13) according to Lance-Jones [1979]; Greene [1935] states that the sartorius is usually not present as a distinct muscle in rats because it is fused with the gluteus maximus and/or the tensor fasciae latae; Walker and Homberger [1997] do not describe a sartorius in rats]	**Sartorius**

(*Continued*)

TABLE 19.13 (CONTINUED)
Hindlimb Muscles of Adults of Representative Tetrapod Taxa

Amphibia (Caudata): *Ambystoma mexicanum* (Axolotl)	Amphibia (Anura): *X. laevis* (African Clawed Frog)	Reptilia (Testudines): *Trachemys scripta* (Pond Slider)	Reptilia (Lepidosauria): *Timon lepidus* (Ocellated Lizard)	Reptilia (Crocodylia): *Alligator mississippiensis* (American Alligator)	Reptilia (Aves): *Gallus domesticus* (Chicken)	Mammalia (Monotremata): *Ornithorhynchus anatinus* (Platypus)	Mammalia (Marsupialia): *D. virginiana* (Virginia Opossum)	Mammalia (Placentalia): *Rattus norvegicus* (Brown Rat)	Mammalia (Placentalia): *Homo sapiens* (Modern Human)
–	–	–	– [directly corresponds to puboischiofemoralis internus 1 head of lizard?]	–	– [corresponds to a part of the puboischiofemoralis internus of chicks (dorsal pectineus), the muscle including the part corresponding to the ventral pectineus being apparently absent in chicks (i.e., pubotibialis)]	**Pectineus** [from hip bone—apparently pubis—and epipubic bone ("marsupial bone," although it is present in monotremes) to femur]	**Pectineus** [from hip bone—ischium, according to Stein [1981], but Coues [1872] states that it is from a distinct bone, the epipubic bone (marsupial bone)—to femur]	**Pectineus** [rat: from hip bone—the pubic arch and the iliopectineal tubercle—to the medial ridge of the femoral shaft just proximal to the insertion of the adductor longus; dorsal pectineus + ventral pectineus; mouse: according to Charles et al. [2016]: from pubic ramus to proximal femur; Lance-Jones 1979: in *Mus*, the pectineus develops from two different anlagen; i.e., from the ventral and dorsal anlage, the dorsal pectineus deriving from the anterior ventral anlage together with the iliacus and the psoas major and the ventral pectineus deriving from the adductor anlage together with the adductor longus; according to her, the double ontogenetic origin of the pectineus is ancestral, although it is not clear if for rodents or for mammals as a whole; innervation: femoral nerve (Greene 1935); as described by Greene [1935], in our rats, the pectineus extends from the pubic arch and the iliopectineal tubercle to the medial ridge of the shaft just proximal to the insertion of the adductor longus]	**Pectineus**

(*Continued*)

TABLE 19.13 (CONTINUED)
Hindlimb Muscles of Adults of Representative Tetrapod Taxa

Amphibia (Caudata): *Ambystoma mexicanum* (Axolotl)	Amphibia (Anura): *X. laevis* (African Clawed Frog)	Reptilia (Testudines): *Trachemys scripta* (Pond Slider)	Reptilia (Lepidosauria): *Timon lepidus* (Ocellated Lizard)	Reptilia (Crocodylia): *Alligator mississippiensis* (American Alligator)	Reptilia (Aves): *Gallus domesticus* (Chicken)	Mammalia (Monotremata): *Ornithorhynchus anatinus* (Platypus)	Mammalia (Marsupialia): *D. virginiana* (Virginia Opossum)	Mammalia (Placentalia): *Rattus norvegicus* (Brown Rat)	Mammalia (Placentalia): *Homo sapiens* (Modern Human)
–	–	–	– [see adductor femoris in the following]	–	– [see adductor femoris in the following]	**Adductor brevis** [from pubis and ischium to femur, lying between adductor magnus and adductor longus]	**Adductor brevis** (adductor parvus *sensu* Coues [1872] and Stein [1981]) [from pubis to femur; note that this muscle is deep to the adductor longus, lying between it and the adductor magnus, so the condition is similar to that of humans, unlike rats/mice, see cells on the right]	**Adductor brevis** (part of adductor magnus and/or adductor magnus *sensu* Lance-Jones [1979]) [rat: from ischium and pubis to femur; mouse: according to Charles et al. [2016], from pubic ramus to distal femur; Lance-Jones 1979: see preceding cells; innervation: obturator nerve (Greene 1935)]	**Adductor brevis**
Pubotibialis [innervation: ischiadicus ventralis (Francis 1934); as in *A. tigrinum* (Ashley-Ross 1992), in our axolotls, it extends from the anterolateral border of the puboischiac plate to the proximal tibia, being blended with the gracilis (puboischiotibialis)]	**Pubotibialis A** (sartorius *sensu* Noble [1922] and Dunlap [1960]) [the muscle is not homologous to the sartorius of humans, because the LCA of amphibians and amniotes did not have a Sartorius; hence, we use a different name; innervation: ischiadic nerve of plexus cruralis (Noble [1922]); as reported by Dunlap [1960], in the *R.* and *Eleutherodactylus* specimens we dissected, the sartorius extends from the ilium and pubis to the cruralis, "gracilis" and semitendinosus; in *Xenopus*, the sartorius and semitendinosus are fused into a single muscle that however as two heads corresponding to these structures]	**Pubotibialis** (sartorius *sensu* Ogushi [1913]) [Walker 1973: in *Trachemys scripta*, the pubotibialis extends from the puboischiadic ligament to the tibia]	**Pubotibialis** [Lance-Jones 1979: see puboischiofemoralis internus; innervation: obturator nerve (Lance-Jones 1979; Russell and Bauer 2008); as reported by Russell and Bauer [2008], in *Iguana* and other lizards, in our *Timon* specimens, the pubotibialis extends from the pubis to the tibia, having mainly a single bundle as described by Romer [1942] and not two as noted by Russell and Bauer [2008] for *Iguana* and *Sphenodon*]	– [pubotibialis was lost in crocodylians according to Romer [1923]]	– [Lance-Jones 1979: the pubotibialis was lost during the transitions leading to birds]	**Adductor longus** [from pubis to femur]	**Adductor longus** [from pubis and ischium to femur]	**Adductor longus** [rat: from pubis to femur; mouse: according to Charles et al. [2016]: from pubic ramus to medial femoral condyle; Lance-Jones 1979: see preceding cells; innervation: obturator nerve (Greene 1935); as described by Greene [1935], the adductor longus extends from the pubis to the shaft of the femur]	**Adductor longus**

(*Continued*)

TABLE 19.13 (CONTINUED)
Hindlimb Muscles of Adults of Representative Tetrapod Taxa

Amphibia (Caudata): *Ambystoma mexicanum* (Axolotl)	Amphibia (Anura): *X. laevis* (African Clawed Frog)	Reptilia (Testudines): *Trachemys scripta* (Pond Slider)	Reptilia (Lepidosauria): *Timon lepidus* (Ocellated Lizard)	Reptilia (Crocodylia): *Alligator mississippiensis* (American Alligator)	Reptilia (Aves): *Gallus domesticus* (Chicken)	Mammalia (Monotremata): *Ornithorhynchus anatinus* (Platypus)	Mammalia (Marsupialia): *D. virginiana* (Virginia Opossum)	Mammalia (Placentalia): *Rattus norvegicus* (Brown Rat)	Mammalia (Placentalia): *Homo sapiens* (Modern Human)
–	– (semitendinosus *sensu* Noble [1922] and Dunlap [1960]) [the muscle is not homologous to the semitendinosus of humans, because the LCA of amphibians and amniotes did not have a semitendinosus; hence, we use a different name; innervation: ischiadic nerve of plexus cruralis (Noble 1922); as reported by Dunlap [1960], in *R.* and *Eleutherodactylus*, the semitendinosus extends from the pelvic rim to the proximal leg (cruris); in *Xenopus*, the sartorius and semitendinosus are fused into a single muscle that however as two heads corresponding to these structures]	–	–	–	–	–	–	–	–

(*Continued*)

TABLE 19.13 (CONTINUED)
Hindlimb Muscles of Adults of Representative Tetrapod Taxa

Amphibia (Caudata): *Ambystoma mexicanum* (Axolotl)	Amphibia (Anura): *X. laevis* (African Clawed Frog)	Reptilia (Testudines): *Trachemys scripta* (Pond Slider)	Reptilia (Lepidosauria): *Timon lepidus* (Ocellated Lizard)	Reptilia (Crocodylia): *Alligator mississippiensis* (American Alligator)	Reptilia (Aves): *Gallus domesticus* (Chicken)	Mammalia (Monotremata): *Ornithorhynchus anatinus* (Platypus)	Mammalia (Marsupialia): *D. virginiana* (Virginia Opossum)	Mammalia (Placentalia): *Rattus norvegicus* (Brown Rat)	Mammalia (Placentalia): *Homo sapiens* (Modern Human)
Adductor femoris (pubofemoralis *sensu* Ashley-Ross 1992) [innervation: ischiadicus ventralis (Francis 1934); as in *A. tigrinum* (Ashley-Ross 1992), the pubifemoralis extends from the anterior and ventral surface of the puboischiac plate and inserts on the ventral face of the distal half of the femur]	**Adductor femoris** (adductor magnus *sensu* Noble [1922] and Dunlap [1960]) [the muscle is not homologous to the adductor magnus of humans, because the LCA of amphibians and amniotes did not have an adductor magnus; hence, we use a different name; innervation: ischiadic nerve of plexus cruralis (Noble 1922); as reported by Dunlap [1960], in the *R. pipiens*, *X. laevis*, and *E. coqui* specimens we dissected, the adductor magnus originates by two heads from the pelvis to the femur, the ventral head fusing to the dorsal head in the three genera and with an accessory head of the semitendinosus in *R.* and *Eleutherodactylus* but not in *Xenopus* where such an accessory head is absent]	**Adductor femoris** [Walker 1973: in *Trachemys scripta*, the adductor femoris (often designated ischiofemoralis) extends from the ischium and puboischiadic ligament to the proximal third of the femoral shaft]	**Adductor femoris** [innervation: by femoral/obturator and tibial/fibular plexus (Romer 1942); as reported by Russell and Bauer [2008], in *Iguana* and other lizards, the adductor femoris extends from the puboischiadic ligament to the middle 1/3 of the femur, being mainly a protractor and adductor of the femur; because the adductor brevis of the mouse, *sensu* the present work, apparently corresponds to either a part of the adductor longus or of the adductor magnus *sensu* Lance-Jones [1979], it is not clear if the adductor brevis really derives from the adductor femoris or from the pubotibialis (the muscle giving rise to the adductor longus *sensu* Lance-Jones [1979])]	**Adductor femoris** (*sensu* Romer [1923]) [Romer 1923: in *A. mississippiensis*, the adductor femoris has two heads and extends from the ischium to the femur; innervation: obturator and sacral nerves (Romer 1923)]	**Adductor femoris** (puboischiofemoralis externus and internus *sensu* Romer [1927]; puboischiofemoralis or adductors *sensu* George and Berger [1966], Lance-Jones [1979], Schroeter and Tosney [1991]) [pars lateralis and pars medialis, corresponding, respectively, to the adductor brevis and adductor longus *sensu* George and Berger [1966]; Vanden Berge 1975: in adult chicken, the puboischiofemoralis has a pars externus and a pars internus, running from ischium and pubis to distal femur and proximal portion of gastrocnemius; Lance-Jones 1979: the adductor magnus (adductor minimus head of the adductor magnus *sensu* the present work) and adductor brevis (main body of adductor magnus *sensu* the present work) of mice correspond to the adductor femoris (puboischiofemoralis or adductors) of birds, the anlagen in both birds and mammals originating superficially from both the ischium and pubis and being closely associated with the flexors at their posterior and dorsal borders; innervation: obturator nerve (Vanden Berge 1975)]	**Adductor magnus** [from pubis and ischium to femur, being blended with adductor brevis, semimembranosus and adductor longus]	**Adductor magnus** [from pubis and ischium to femur]	**Adductor magnus** (adductor brevis plus adductor magnus *sensu* Greene [1935], Walker and Homnerger [1997], and Wingerd [1998], as well as Lance-Jones [1979]) [rat: from pubis and ischium to femur—adductor minimus head plus main body including a pars genicularis and a pars femoris; mouse: according to Charles et al. [2016], from pubic ramus to distal third of femur; Lance-Jones 1979: see preceding cells; innervation: obturator nerve (Greene 1935; Lance-Jones [1979], who states that a single innervation by the obturator is seen in most mammals, although some "insectivores" and marsupials have a obturator plus sciatic innervation)]	**Adductor magnus**

(*Continued*)

TABLE 19.13 (CONTINUED)
Hindlimb Muscles of Adults of Representative Tetrapod Taxa

Amphibia (Caudata): *Ambystoma mexicanum* (Axolotl)	Amphibia (Anura): *X. laevis* (African Clawed Frog)	Reptilia (Testudines): *Trachemys scripta* (Pond Slider)	Reptilia (Lepidosauria): *Timon lepidus* (Ocellated Lizard)	Reptilia (Crocodylia): *Alligator mississippiensis* (American Alligator)	Reptilia (Aves): *Gallus domesticus* (Chicken)	Mammalia (Monotremata): *Ornithorhynchus anatinus* (Platypus)	Mammalia (Marsupialia): *D. virginiana* (Virginia Opossum)	Mammalia (Placentalia): *Rattus norvegicus* (Brown Rat)	Mammalia (Placentalia): *Homo sapiens* (Modern Human)
–	–	–	–	–	–	– [adductor minimus not present as separate muscle: see, e.g., Gambaryan et al. 2002]	– [adductor minimus not present as separate muscle: see, e.g., Coues 1872 and Stein 1981]	– [not described by Greene [1935], Lance-Jones [1979], Walker and Homberger [1997], James et al. [2016], and in this work; we are considering this structure simply as a head of the adductor magnus, which is present in mice and rats: see adductor brevis]	– [the adductor minimus is not really present as a distinct muscle in humans, although it is recognized as a distinct muscle in Wikipedia: here it is instead seen as a head of the adductor magnus, as shown in most anatomical atlases, see e.g. plate 463 of Netter [1989]]
Gracilis (puboischiotibialis *sensu*, e.g., Ashley-Ross [1992]) [innervation: ischiadicus ventralis (Francis 1934); as in *A. tigrinum* (Ashley-Ross 1992), it has proximal and distal heads and extends from the ventral midline of the puboischiac plate to the proximal two-thirds of the anteromedial face of the tibia]	**Gracilis major** [innervation: ischiadic nerve of plexus sacralis (Noble 1922); as reported by Dunlap [1960], in the *R. pipiens*, *X. laevis*, and *E. coqui* specimens we dissected, the gracilis major extends from the pelvis to the knee aponeurosis and the proximal cruris (leg)]	**Gracilis** (sartorius *sensu* Ogushi [1913]; puboischiotibialis *sensu* Walker [1973]) [Walker 1973: in *Trachemys scripta*, the puboischiotibialis is a vestigial muscle (it is absent in numerous other turtles) extending from the puboischiadic ligament to the tibia]	**Gracilis** (puboischiotibialis *sensu*, e.g., Russell and Bauer [2008]) [Russell and Bauer (2008): the gracilis (puboischiotibialis) of lizards is often divided into three heads, although in adults, some of these heads may be fused and thus two heads or only one head may be present; this is the most ventral muscle of the thigh and aponeurotically originates with the pubotibialis and from the ischiopubic ligament, inserting onto the proximomesial aspect of the tibial shaft; the gracilis (puboischiotibialis) is not only mainly a flexor of the leg, but may also help adduct the thigh; innervation: obturator nerve (Lance-Jones 1979); obturator and sacral components (Romer 1942); in *Timon* specimens, the gracilis (puboischiotibialis) has	**Gracilis** (puboischiotibialis *sensu* Romer [1923]) [Romer 1923: in *A. mississippiensis*, the puboischiotibialis extends from the ischium to the tibia; innervation: obturator nerve (Romer 1923)]	– [Lance-Jones 1979: the gracilis (puboischiotibialis) was lost during the transitions leading to birds]	**Gracilis** [according to Gambaryan et al. [2002], it is divided into a "gracilis posterior" from epipubis and ischium to tibia, which is blended with the semitendinosus, and a "gracilis anterior"; extends from hip bone to tibia and indirectly to patella]	**Gracilis** [from pubis to tibia]	**Gracilis** (gracilis anticus plus gracilis posticus *sensu* Greene [1935]; gracilis cranialis plus gracilis caudalis *sensu* Walker and Homberger [1997]) [rat: there is a "gracilis anticus" (anterior head of the gracilis *sensu* the present work) extending from the pubic symphysis to the crest and medial border of the tibia and more lateral "gracilis posticus" (posterior head of the gracilis *sensu* the present work) extending from the ischium to the tibial tuberosity; mouse: according to Charles et al. [2016], anticus from pubis to proximal tibia, posticus from pubis to proximal tibia; see preceding text; innervation: anterior division of obturator nerve (Greene 1935; Lance-Jones 1979); as described by Greene [1935], there is a gracilis anticus	**Gracilis**

(Continued)

TABLE 19.13 (CONTINUED)
Hindlimb Muscles of Adults of Representative Tetrapod Taxa

Amphibia (Caudata): *Ambystoma mexicanum* (Axolotl)	Amphibia (Anura): *X. laevis* (African Clawed Frog)	Reptilia (Testudines): *Trachemys scripta* (Pond Slider)	Reptilia (Lepidosauria): *Timon lepidus* (Ocellated Lizard)	Reptilia (Crocodylia): *Alligator mississippiensis* (American Alligator)	Reptilia (Aves): *Gallus domesticus* (Chicken)	Mammalia (Monotremata): *Ornithorhynchus anatinus* (Platypus)	Mammalia (Marsupialia): *D. virginiana* (Virginia Opossum)	Mammalia (Placentalia): *Rattus norvegicus* (Brown Rat)	Mammalia (Placentalia): *Homo sapiens* (Modern Human)
			3 heads exactly as shown by Romer [1942], extending from the region surrounding the ischiopubic ligament to the proximomesial aspect of the tibial shaft, near the insertion of the flexor tibialis internus; the muscle clearly seems to correspond to the gracilis of rats and mice]					(anterior head of the gracilis *sensu* the present work) extending from the pubic symphysis to the crest and medial border of the tibia and more lateral gracilis posticus (posterior head of the gracilis *sensu* the present work) extending from the ischium to the tibial tuberosity; these two structures are distinct but lie very closely together and are connected by connective tissue, so they seem to be simply two heads of the muscle gracilis]	
–	**Gracilis minor** [innervation: ischiadic nerve of plexus sacralis (Noble 1922); as reported by Dunlap [1960], in the *R. pipiens*, *X. laevis*, and *E. coqui* specimens we dissected, the gracilis minor extends from the pelvis to the knee aponeurosis and the proximal cruris (leg) together with the gracilis major and is attached to the skin, and only in *E. coqui*, the muscle has two heads, one originating from the pelvis and the other from the skin; see gracilis major]	–	–	–	–	–	–	–	–

(Continued)

TABLE 19.13 (CONTINUED)
Hindlimb Muscles of Adults of Representative Tetrapod Taxa

Amphibia (Caudata): *Ambystoma mexicanum* (Axolotl)	**Amphibia (Anura): *X. laevis* (African Clawed Frog)**	**Reptilia (Testudines): *Trachemys scripta* (Pond Slider)**	**Reptilia (Lepidosauria): *Timon lepidus* (Ocellated Lizard)**	**Reptilia (Crocodylia): *Alligator mississippiensis* (American Alligator)**	**Reptilia (Aves): *Gallus domesticus* (Chicken)**	**Mammalia (Monotremata): *Ornithorhynchus anatinus* (Platypus)**	**Mammalia (Marsupialia): *D. virginiana* (Virginia Opossum)**	**Mammalia (Placentalia): *Rattus norvegicus* (Brown Rat)**	**Mammalia (Placentalia): *Homo sapiens* (Modern Human)**
–	–	–	–	–	–	**Biceps femoris** [from ischium to patella and tibia (not fibula)]	**Biceps femoris** [single head from ischium to fibula, being blended with the semitendinosus; the short head of the biceps femoris is therefore apparently absent in *D. virginiana*, as stated by Coues [1872]]	**Biceps femoris** (biceps femoris plus femorococcygeus *sensu* Lance-Jones 1979) [anterior, posterior, and accessory heads according to Greene [1935]; anterior and posterior heads according to our dissections and to Walker and Homberger [1997]; rat: from the last sacral and first caudal vertebrae, blended with the caudofemoralis, as well as from sciatic tuberosity, blended with semimembranosus, to distal femur and proximal 2/3 of the tibia; mouse: anterior head from ischial tuberosity to lateral femoral condyle, posterior head from ischial tuberosity to head of fibula and adjacent fascia, per Charles et al. [2016]; Lance-Jones 1979: in *Mus*, the biceps femoris, semitendinosus, semimembranosus, caudofemoralis and presemimembranosus derive from the hamstring ventral anlage, the biceps femoris (she only described a single head) having more affinity with the ventral head of the semitendinosus (originally, the biceps femoris does not attach onto vertebrae, so it is a superficial/appendicular muscle according to the in–out model); innervation: anterior head by inferior gluteal nerve and posterior and accessory heads by tibial division of sciatic nerve (Greene 1935)]	**Biceps femoris** [caput longum and caput breve]

(Continued)

TABLE 19.13 (CONTINUED)
Hindlimb Muscles of Adults of Representative Tetrapod Taxa

Amphibia (Caudata): *Ambystoma mexicanum* (Axolotl)	Amphibia (Anura): *X. laevis* (African Clawed Frog)	Reptilia (Testudines): *Trachemys scripta* (Pond Slider)	Reptilia (Lepidosauria): *Timon lepidus* (Ocellated Lizard)	Reptilia (Crocodylia): *Alligator mississippiensis* (American Alligator)	Reptilia (Aves): *Gallus domesticus* (Chicken)	Mammalia (Monotremata): *Ornithorhynchus anatinus* (Platypus)	Mammalia (Marsupialia): *D. virginiana* (Virginia Opossum)	Mammalia (Placentalia): *Rattus norvegicus* (Brown Rat)	Mammalia (Placentalia): *Homo sapiens* (Modern Human)
–	–	–	–	–	–	–	–	–	**"Annectant Mass"** (piriformis *sensu* Burmeister [1846])
Tenuissimus (iliofibularis *sensu* Ashley-Ross 1992) [Gadow 1882: supported the idea that the iliofibularis became part of the flexor musculature during amniote evolution; innervation: n. extensorius (Francis 1934); as in *A. tigrinum* (Ashley-Ross 1992), the tenuissimus (iliofibularis) extends from the tendon of the extensor iliotibialis to the posterior border of the fibula]	**Tenuissimus** (iliofibularis *sensu* Noble [1922], Appleton [1928], and Dunlap [1960]) [innervation: fibular nerve of plexus sacralis (Noble 1922); as reported by Dunlap [1960], in the *R. pipiens*, *X. laevis*, and *E. coqui* specimens we dissected, the iliofibularis extends from the pelvis to the aponeurosis of the knee]	**Tenuissimus** (gluteus minimus *sensu* Ogushi [1913]; iliofibularis *sensu* Appleton [1928] and Walker [1973]) [Walker 1973: in *Trachemys scripta*, the iliofibularis runs from the iliac crest to the fibula]	**Tenuissimus** (iliofibularis *sensu* Russell and Bauer [2008]) [Russell and Bauer 2008: in almost all saurian taxa, the iliofibularis (tenuissimus *sensu* the present work) extends from the posteroventral margin of the ilium to the proximal fibular shaft, through a tendon that passes between the "peroneus longus" and the "femoral gastrocnemius," a major exception being *Sphenodon* where the iliofibularis originates not from the ilium but more posteriorly from the pygal vertebrae; Lance-Jones 1979: tenuissimus of mammals is homologous to the iliofibularis of reptiles, both muscles splitting off from the posterior dorsal mass of the pelvis/thigh very early in ontogeny, lying at the border of the dorsal and ventral masses postaxially, and innervated by a distal branch of the fibular nerve that separates them from the rest of the postaxial dorsal mass; the position of the tenuissimus deep to the biceps femoris in adult mammals has prompted doubt about this homology in the	**Tenuissimus** (iliofibularis I *sensu* Gadow [1882]; iliofibularis *sensu* Romer [1923] and Appleton [1928]) [Romer 1923: in *A. mississippiensis*, the iliofibularis extends from the ilium to the fibula]	**Tenuissimus** (iliofibularis or part of biceps *sensu* Romer [1927], George and Berger [1966], Lance-Jones [1979], and Schroeter and Tosney [1991]) [Vanden Berge 1975: in adult chicken, the iliofibularis extends from ilium to fibula; Lance-Jones 1979: the posterior part of the superficial dorsal muscle mass in the chick becomes distinct early as the iliofibularis (biceps), which lies superficial to the fibular nerve and deep to the posterior iliotibialis, thus sharing characteristics with the mammalian tenuissimus and thus with the short head of the biceps femoris of humans; innervation: ischiatic nerve (Vanden Berge 1975)]	– [not present in monotremes according to, e.g., Gregory and Camp 1918, Pearson 1926, Appleton 1928, and Gambaryan et al. 2002; as the muscle is apparently present in amphibians, reptiles and therians, its absence/undifferentiation is apparently a derived feature]	– [but tenuissimus ("bicipiti accessorius") present in some other marsupials, such as marsupial moles: see, e.g., Warburton 2003]	– [the tenuissimus is not present in adult mice nor adult rats, but present and innervated by the fibular nerve early in E12.5 mice extending from the posterolateral surface of the thigh to a region deep to the biceps femoris just above the knee region; then the tenuissimus loses innervation at E18, and then at P10 (10 days old), it is completely absent according to Lance-Jones [1979], who considers that the tenuissimus corresponds to the short head of the biceps femoris of humans and that the absence of this muscle or its position deep to the main body of the biceps as is the case in humans is a derived condition (see biceps femoris mentioned earlier)]	–

(*Continued*)

TABLE 19.13 (CONTINUED)
Hindlimb Muscles of Adults of Representative Tetrapod Taxa

Amphibia (Caudata): *Ambystoma mexicanum* (Axolotl)	Amphibia (Anura): *X. laevis* (African Clawed Frog)	Reptilia (Testudines): *Trachemys scripta* (Pond Slider)	Reptilia (Lepidosauria): *Timon lepidus* (Ocellated Lizard)	Reptilia (Crocodylia): *Alligator mississippiensis* (American Alligator)	Reptilia (Aves): *Gallus domesticus* (Chicken)	Mammalia (Monotremata): *Ornithorhynchus anatinus* (Platypus)	Mammalia (Marsupialia): *D. virginiana* (Virginia Opossum)	Mammalia (Placentalia): *Rattus norvegicus* (Brown Rat)	Mammalia (Placentalia): *Homo sapiens* (Modern Human)
			past, but such a deep position is clearly derived because the tenuissimus of mice is superficial until E13; innervation: distal branch of fibular nerve (Lance-Jones 1979); as described by Russell and Bauer [2008] in *Iguana*, the iliofibularis (tenuissimus *sensu* the present work) extends from the posteroventral margin of the ilium to the proximal fibular shaft, through a tendon that passes between the peroneus longus and the femoral gastrocnemius]						
Femorofibularis [Appleton 1928: considers that the femorofibularis is derived thigh muscle only found in urodeles, not on amniotes nor in anurans; innervation: sciaticus (Francis 1934); as in *Ambystoma tigrinum* (Ashley-Ross 1992), it extends from the posteroventral border of the femur, at a point approximately halfway to the knee, to the posterolateral border of the fibula between the insertion of the extensor cruris tibialis and the fibular portion of origin of the flexor digitorum communis]	–	–	–	–	–	–	–	–	–

(*Continued*)

TABLE 19.13 (CONTINUED)
Hindlimb Muscles of Adults of Representative Tetrapod Taxa

Amphibia (Caudata): *Ambystoma mexicanum* (Axolotl)	Amphibia (Anura): *X. laevis* (African Clawed Frog)	Reptilia (Testudines): *Trachemys scripta* (Pond Slider)	Reptilia (Lepidosauria): *Timon lepidus* (Ocellated Lizard)	Reptilia (Crocodylia): *Alligator mississippiensis* (American Alligator)	Reptilia (Aves): *Gallus domesticus* (Chicken)	Mammalia (Monotremata): *Ornithorhynchus anatinus* (Platypus)	Mammalia (Marsupialia): *D. virginiana* (Virginia Opossum)	Mammalia (Placentalia): *Rattus norvegicus* (Brown Rat)	Mammalia (Placentalia): *Homo sapiens* (Modern Human)
Ischioflexorius (ischioflexorius plus plantaris superficialis medialis *sensu* McMurrich 1905) [apparently includes flexor cruris et tarsi tibialis, see notes of Table 19.1 as well as flexor cruris et tarsi tibialis row in the following; innervation: pars propria by ischiadicus ventralis and pars plantaris by sciaticus (Francis 1934); as in *A. tigrinum* (Ashley-Ross 1992), the ischioflexorius extends from the posterolateral corner of the puboischiac plate to the plantar aponeurosis, the muscle being divided into proximal and distal portions divided by a tendinous plate at the level of one-third to one-half the distance between the origin and insertion of the muscle]	**Ischioflexorius** (semimembranosus *sensu* Dunlap [1960]) [innervation: ischiadic nerve of plexus sacralis (Noble 1922); as reported by Dunlap [1960] in the *R. pipiens*, *X. laevis*, and *E. coqui* specimens we dissected, the semimembranosus extends from the pelvis to the distal femur and proximal cruris (leg); see gracilis major]	**Flexor tibialis internus** (ischiococcygeotibialis *sensu* Ogushi [1913]) [Walker 1973: in *Trachemys scripta*, the flexor tibialis internus extends from the ischium, puboischiadic ligament (anteroventral head), and last sacral and first two postsacral vertebrae (posterodorsal head) to the tibia; see flexor cruris et tarsi tibialis]	**Flexor tibialis internus** [Russell and Bauer 2008: this muscle is usually divided in lizards into a posterior portion (portion 1) and an anterior portion (portion 2), portion 2 being in turn divided into superficial (from perineal region and the anterior aspect of the iliotibial/fibular ligament to the mesial aspect of the proximal end of the tibial shaft and to the proximolateral tibia through two distinct slips) and deep (from posteroventral margin of ischium to the proximolateral tibia) portions; portion 2 runs from the iliotibial/ fibular ligament to the insertion of part 3 of the gracilis (puboischiotibialis) onto the proximomesial tibia; the developmental association between the caudifemorales and the flexor tibialis internus is clearly also seen in various adult lizards and is particularly evident in *Sphenodon* adults, in which there is no iliotibial/fibular ligament, and thus, the posterior component (part 1) of the flexor tibialis internus arises from the ventral surface of the transverse processes of the six anterior caudal vertebrae; functionally, the flexor tibialis	**Flexor tibialis internus** (*sensu* Romer 1923) [Romer 1923: in *A. mississippiensis*, the flexor tibialis internus mainly extends from the ischium region to the tibia and has 4 heads; innervation: obturator, sacral and ischiadic nerves (Romer 1923); see flexor cruris et tarsi tibialis]	**Flexor tibialis internus** (semimembranosus or ischioflexorius or flexor cruris medialis *sensu* George and Berger [1966], Lance-Jones [1979], and Dilkes [1999]) [Vanden Berge 1975: in adult chicken, the "flexor cruris medialis" extends from ischium and adjacent pubis to proximal end of tibiotarsus; Lance-Jones 1979: the long flexors of the chick include the ischioflexorius and the caudilioflexorius, while in the mouse, they include the semimembranosus, biceps, and ventral semitendinosus; the ischioflexorius of birds clearly seems to correspond to the semimembranosus of mice (but only to the head 1 of the flexor tibialis internus) because they arise from the deep posterior ventral mass between the adductors and more posterior flexors; the biceps and ventral semitendinosus arise more superficially and posteriorly like the caudilioflexorius, which inserts *via* the accessory muscle on the femur, whereas the mammalian muscles insert on the lateral and medial aspects of the shank; the accessory muscle has no homologue being	**Semimembranosus** [according to Gambaryan et al. [2002], it is divided into a "semimembranosus posterior" from ischium to tibia and a "semimembranosus anterior" from ischium to tibia as well; the semimembranosus is blended with the semitendinosus according to Pearson [1926]]	**Semimembranosus** [from ischium to tibia]	**Semimembranosus** (caudal head of semimembranosus *sensu* Walker and Homberger [1997]) [rat: from hip bone—sciatic tuber—to the ridge and medial surface of the tibia and into the medial fabella; mouse: according to Charles et al. [2016]: from ischial tuberosity to proximal tibia; see preceding text; innervation: tibial division of sciatic nerve (Greene 1935); as described by Greene [1935], the semimembranosus extends from the sciatic tuber to the ridge and medial surface of the tibia and into the medial fabella]	**Semimembranosus**

(*Continued*)

TABLE 19.13 (CONTINUED)
Hindlimb Muscles of Adults of Representative Tetrapod Taxa

Amphibia (Caudata): *Ambystoma mexicanum* (Axolotl)	Amphibia (Anura): *X. laevis* (African Clawed Frog)	Reptilia (Testudines): *Trachemys scripta* (Pond Slider)	Reptilia (Lepidosauria): *Timon lepidus* (Ocellated Lizard)	Reptilia (Crocodylia): *Alligator mississippiensis* (American Alligator)	Reptilia (Aves): *Gallus domesticus* (Chicken)	Mammalia (Monotremata): *Ornithorhynchus anatinus* (Platypus)	Mammalia (Marsupialia): *D. virginiana* (Virginia Opossum)	Mammalia (Placentalia): *Rattus norvegicus* (Brown Rat)	Mammalia (Placentalia): *Homo sapiens* (Modern Human)
			internus is in general a flexor of the leg and may help adduct the thigh; innervation: by tibial/fibular plexus in the lacertid *Podarcis muralis*, but sometimes also by obturator nerve in other reptiles (Romer 1942); similar to description of Russell and Bauer 2008 for *Iguana*, there were clearly two main portions of the muscle, 1 and 2, as also reported by Romer [1942]; see flexor cruris et tarsi tibialis]		perhaps either a slip of the gastrocnemius that has migrated proximally (Romer 1927, 1942) or a distal slip of the coccygeofemoralis (Appleton 1928); innervation: ischiatic nerve (Vanden Berge 1975)]				
–	–	–	–	–	**Flexor cruris lateralis accessorius** (accessorius semitendinosi *sensu* George and Berger [1966]; accessorius Lance-Jones 1979; flexor cruris lateralis pars accessoria *sensu* Schroeter and Tosney [1991]) [Vanden Berge 1975: in adult chicken, the "flexor cruris lateralis pars accessorius" extends from ilium and caudal pelvis to distal half of femur and popliteal area; Lance-Jones 1979: see ischioflexorius mentioned earlier; innervation: ischiatic nerve (Vanden Berge 1975); see flexor tibialis internus]	–	–	–	–

(*Continued*)

TABLE 19.13 (CONTINUED)
Hindlimb Muscles of Adults of Representative Tetrapod Taxa

Amphibia (Caudata): *Ambystoma mexicanum* (Axolotl)	Amphibia (Anura): *X. laevis* (African Clawed Frog)	Reptilia (Testudines): *Trachemys scripta* (Pond Slider)	Reptilia (Lepidosauria): *Timon lepidus* (Ocellated Lizard)	Reptilia (Crocodylia): *Alligator mississippiensis* (American Alligator)	Reptilia (Aves): *Gallus domesticus* (Chicken)	Mammalia (Monotremata): *Ornithorhynchus anatinus* (Platypus)	Mammalia (Marsupialia): *D. virginiana* (Virginia Opossum)	Mammalia (Placentalia): *Rattus norvegicus* (Brown Rat)	Mammalia (Placentalia): *Homo sapiens* (Modern Human)
–	–	–	–	–	–	– [not described by authors such as Pearson [1926] and Gambaryan et al. [2002]]	– [presemimembranosus not present in *D. virginiana* and apparently absent in other marsupials (see, e.g., Osgood 1921), but has been described in some others, such as in marsupial moles by Thompson and Hillier [1905] and Warburton [2003]; presemimembranosus of mammals comes from caudofemoralis or from flexor tibialis internus of reptiles?]	– [the "presemimembranosus" is not present in adult mice and adult rats, but is present at early ontogenetic stages of karyotypically mice according to Lance-Jones [1979], coming from the anlage of the caudofemoralis which is superficial to the presemimembranosus]	–

(*Continued*)

TABLE 19.13 (CONTINUED)
Hindlimb Muscles of Adults of Representative Tetrapod Taxa

Amphibia (Caudata): *Ambystoma mexicanum* (Axolotl)	Amphibia (Anura): *X. laevis* (African Clawed Frog)	Reptilia (Testudines): *Trachemys scripta* (Pond Slider)	Reptilia (Lepidosauria): *Timon lepidus* (Ocellated Lizard)	Reptilia (Crocodylia): *Alligator mississippiensis* (American Alligator)	Reptilia (Aves): *Gallus domesticus* (Chicken)	Mammalia (Monotremata): *Ornithorhynchus anatinus* (Platypus)	Mammalia (Marsupialia): *D. virginiana* (Virginia Opossum)	Mammalia (Placentalia): *Rattus norvegicus* (Brown Rat)	Mammalia (Placentalia): *Homo sapiens* (Modern Human)
Caudofemoralis [Appleton 1928: many authors state that the piriformis of mammals derives from the caudofemoralis of other tetrapods, but this is based only on some topological similarity because the detailed analysis of the nerves and overall configuration of the muscles clearly shows that this hypothesis is wrong; innervation: pudendal (Francis 1934); as in *A. tigrinum* (Ashley-Ross 1992), it extends from the caudal vertebrae to the femur, just posterior to the femoral trochanter]	**Caudofemoralis** (piriformis *sensu* Noble [1922] and Dunlap [1960]) [innervation: ischiadic nerve of plexus sacralis (Noble 1922); as reported by Dunlap [1960], in the *R. pipiens*, *X. laevis*, and *E. coqui* specimens we dissected, the piriformis extends from the coccyx to the proximal femur]	**Caudofemoralis** (piriformis *sensu* Ogushi [1913]; caudiiliofemoralis *sensu* Walker [1973]) [Walker 1973: in *Trachemys scripta*, the caudiliofemoralis extends from the sacral vertebrae, ribs, and iliac crest, and two postsacral vertebrae to the tibia; Walker [1973] suggested that the caudiiliofemoralis of turtles might correspond only to the caudofemoralis brevis of lizards, i.e., that the caudofemoralis longus might be absent in turtles]	**Caudofemoralis longus** [Romer 1942: the caudofemoralis longus originally in evolution probably attached to connective tissue connected to the flexor tibialis internus, and then its tendon "took a ride" with it to now attach onto the tibia; Russell and Bauer 2008: in lizards, the caudofemoralis longus usually extends from caudal vertebrae to the femoral trochanter, also sending a long tendon to the knee joint, being mainly not only a retractor of the thigh but also a flexor of the leg; innervation: by tibial/fibular plexus (Romer 1942); as described by Russell and Bauer 2008 for *Iguana*, the caudofemoralis longus extends from caudal vertebrae to the femoral trochanter, also sending a long tendon to the knee joint; the evolutionary hypothesis proposed by Romer [1942] for the distal attachment of the caudofemoralis longus onto the tibia seems very much like with what happened with the procoracohumeralis, which is a pectoral muscle that probably took a ride with the coracobrachialis and/or humeroantebrachialis to attach to the forearm and then apparently contributed to the	**Caudofemoralis longus** ("coccygeofemoralis longus" *sensu* Romer [1923]) [Romer 1923: in *A. mississippiensis*, the caudofemoralis longus extends from the caudal vertebrae to the femur and then sends a long tendon to the fibula, as it usually does in lizards]	**Caudofemoralis** (piriformis or coccygeofemoralis *sensu* George and Berger [1966] and Lance-Jones [1979]; caudiliofemoralis *sensu* Vanden Berge [1975]) [apparently corresponds to the caudofemoralis longus and caudofemoralis brevis of lizards; Schroeter and Tosney [1991] divide this muscle in a pars caudofemoralis and a pars iliofemoralis; Vanden Berge 1975: in adult chicken, the caudiliofemoralis extends from cruciate aponeurosis of depressor caudae to femur (pars caudofemoralis) and from ilium and sometimes ischium to femur (pars iliofemoralis); Lance-Jones 1979: the early development of the posterior ventral mass of the chick differs from that of the mouse in that it includes a large, caudally extending portion, the caudofemoralis or coccygeofemoralis or piriformis anlage; this anlage is fused anteriorly with a short proximally inserting mass, the ischiotrochantericus or ischiofemoralis and ventrally with the flexorius mass; the caudofemoralis and dorsal head of the	**Caudofemoralis** (piriformis *sensu* Pearson [1926]) [from first caudal vertebra to femur; Appleton [1928] stated that in monotremes, the caudofemoralis inserts often near the middle (proximodistally) of the femur; according to him, Gregory and Camp 1918 failed to recognize the homology between the caudofemoralis of echidna and the so-called piriformis of *Ornithorhynchus*]	– [Appleton [1928] stated that in marsupials, the caudofemoralis often inserts onto the proximal femur, hence, the use of the name piriformis to designate this muscle in these mammals; however, the piriformis muscle described in the opossum by authors such as Coues [1872] and Stein [1981] seems to correspond to the piriformis of placental mammals, and thus the caudofemoralis seems to be absent in the opossum; see piriformis mentioned earlier]	**Caudofemoralis** (cranial head of semimembranosus *sensu* Walker and Homberger [1997]) [rat: from the posterior sacral and first caudal vertebrae to the distal portion of the dorsal (flexor) margin of the femur; mouse: according to Charles et al. [2016], from ischial tuberosity to medial femoral condyle; Appleton 1928: in placentals, the caudofemoralis often inserts onto the distal portion of the femur; see preceding text; innervation: tibial division of sciatic nerve (Greene 1935); as described by Greene [1935], the caudofemoralis extends from the posterior sacral and first caudal vertebrae to the distal portion of the dorsal (flexor) margin of the femur]	–

(Continued)

TABLE 19.13 (CONTINUED)
Hindlimb Muscles of Adults of Representative Tetrapod Taxa

Amphibia (Caudata): *Ambystoma mexicanum* (Axolotl)	Amphibia (Anura): *X. laevis* (African Clawed Frog)	Reptilia (Testudines): *Trachemys scripta* (Pond Slider)	Reptilia (Lepidosauria): *Timon lepidus* (Ocellated Lizard)	Reptilia (Crocodylia): *Alligator mississippiensis* (American Alligator)	Reptilia (Aves): *Gallus domesticus* (Chicken)	Mammalia (Monotremata): *Ornithorhynchus anatinus* (Platypus)	Mammalia (Marsupialia): *D. virginiana* (Virginia Opossum)	Mammalia (Placentalia): *Rattus norvegicus* (Brown Rat)	Mammalia (Placentalia): *Homo sapiens* (Modern Human)
			biceps brachii, as the caudofemoralis longus may contribute to some flexors of the leg; according to Russell and Bauer [2008], the original origin of the flexor tibial muscles was from the caudal vertebrae, together with the caudofemoralis, a condition more similar to that found in *Sphenodon* and particularly similar to adult urodeles when these two muscle masses form a continuum; one could propose a correspondence between the bundle of the caudofemoralis sending the long tendon to the knee region and the bundle of the procoracohumeralis (i.e., the coracoradialis) sending a long tendon to the elbow joint, because the former clearly forms part of the hamstrings (Lance-Jones 1979), while the latter clearly seems to form part of the arm flexor musculature and, namely, to contribute to the biceps brachii, so one would be talking about similar topological and function conditions, but of different ontogenetic origins, because the caudofemoralis comes from the ventral musculature of the pelvis/thigh, while the procoracohumeralis comes from the dorsal musculature of the pectoral girdle/arm]		semitendinosus of the mouse ontogenetically originate from the posterodorsal portion of the ventral muscle mass; they also show similar topographical relationships including deep proximal and distal insertions; therefore, although they are significantly smaller than the caudofemoralis of birds, their homology is suggested, the muscle of birds thus corresponding to the caudofemoralis longus and brevis plus at least part of the flexor tibialis externus of lizards; innervation: caudal gluteal nerve, branch of sacral plexus (Vanden Berge 1975); see flexor tibialis internus and flexor cruris et tarsi tibialis]				

(Continued)

TABLE 19.13 (CONTINUED)
Hindlimb Muscles of Adults of Representative Tetrapod Taxa

Amphibia (Caudata): *Ambystoma mexicanum* (Axolotl)	Amphibia (Anura): *X. laevis* (African Clawed Frog)	Reptilia (Testudines): *Trachemys scripta* (Pond Slider)	Reptilia (Lepidosauria): *Timon lepidus* (Ocellated Lizard)	Reptilia (Crocodylia): *Alligator mississippiensis* (American Alligator)	Reptilia (Aves): *Gallus domesticus* (Chicken)	Mammalia (Monotremata): *Ornithorhynchus anatinus* (Platypus)	Mammalia (Marsupialia): *D. virginiana* (Virginia Opossum)	Mammalia (Placentalia): *Rattus norvegicus* (Brown Rat)	Mammalia (Placentalia): *Homo sapiens* (Modern Human)
–	–	–	**Caudofemoralis brevis** [Lance-Jones 1979: see caudofemoralis longus mentioned earlier; as described by Russell and Bauer [2008], for *Iguana* and other lizards, the caudofemoralis brevis extends from pygal vertebrae to the femoral trochanter]	**Caudofemoralis brevis** (“coccygeofemoralis brevis” *sensu* Romer [1923]) [Romer 1923: in *A. mississippiensis*, the caudofemoralis brevis extends from the last sacral and first caudal vertebrae to the femur]	– [see caudofemoralis mentioned earlier]	–	–	–	–
–	–	**Flexor tibialis externus** [in turtles such as *Trachemys*, the flexor cruris et tarsi tibialis is apparently included in the flexor tibialis externus part that inserts onto the gastrocnemius internus and/or included in the latter muscle (semitendinosus and coccygeofibularis *sensu* Ogushi [1913]); Walker 1973: in *Trachemys scripta*, the flexor tibialis externus extends from the iliac crest (dorsal head) and ischium (ventral head) to the tibia and to the internal head of the gastrocnemius]	**Flexor tibialis externus** [Russell and Bauer 2008: in lizards, the flexor tibialis externus often has a single belly extending from the iliotibial/fibular ligament (in common with the flexor tibialis internus 1) to the proximolateral tibial (both fleshy and tendinously) and sends a long, slender tendon that follows the lateral border of the femorotibial gastrocnemius, receives fibers from it, and contributes to the plantar aponeurosis, this connection being a relic of the early developmental stages in which thigh and crural muscles are part of a single, undifferentiated mass; the flexor tibialis externus is a flexor of the thig,h but its distal extension also enables the muscle to contribute to pedal plantar flexion; innervation: by tibial/fibular plexus (Romer 1942); similar to description of Russell and Bauer [2008] for *Iguana*]	**Flexor tibialis externus** (*sensu* Romer [1923]) [in crocodylians such as *Alligator*, the flexor cruris et tarsi tibialis is apparently included in the flexor tibialis externus part that inserts onto the gastrocnemius internus and/or included in the latter muscle; Romer 1923: in *A. mississippiensis* the flexor tibialis externus extends from the iliac blade to the gastrocnemius; innervation: sacral nerve (Romer 1923)]	**Flexor tibialis externus** (semitendinosus *sensu* George and Berger [1966]; flexor cruris lateralis *sensu* Vanden Berge [1975]; semitendinosus or caudilioflexorius *sensu* Lance-Jones [1979]; flexor cruris lateralis pars pelvica *sensu* Schroeter and Tosney [1991]) [Vanden Berge 1975: in adult chicken, the “flexor cruris lateralis” extends from ilium and caudal pelvis to proximal tibia; Lance-Jones 1979: see ischioflexorius mentioned earlier; innervation: ischiatic nerve (Vanden Berge 1975); see flexor tibialis internus]	**Semitendinosus** [anterior head that corresponds to the crurococcygeus *sensu* Pearson [1926] and to the “dorsal” head of Lance-Jones [1975] and to “semitendinosus pars vertebralis” *sensu* Gambaryan et al. [2002]; extending from caudal and lumbar vertebrae to tibia, extending also all the way to attach onto metatarsal V and metatarsal I + posterior head that corresponds to the “ventral” head of Lance-Jones [1975] and to the semitendinosus *sensu* Pearson [1926] and to “semitendinosus pars ischiadicus” *sensu* Gambaryan et al. 2002, extending from ischium to tibia]	**Semitendinosus** [anterior head (corresponds to the “crurococcygeus muscle” *sensu* Coues 1872 and Stein 1981 and to the “dorsal” head of Lance-Jones 1975) + posterior head (corresponds to the “ventral” head of Lance-Jones [1975] and to the “semitendinosus” *sensu* Coues [1872] and Stein [1981], because, as stated by the latter authors, even if the two heads are clearly differentiated in *D. virginiana*, they are still deeply blended proximally); the posterior/ventral/ semitendinosus head connects the ischium to the tibia as described by Coues [1872] (not the femur, as stated by Stein [1981]), while the anterior/dorsal/ crurococcygeus head connects the third coccygeal vertebra to the tibia as described by Coues [1872] (not the tibia + fibula as described by Stein 1981)]	**Semitendinosus** (anterior head that corresponds to the “principal” head of Greene [1935]and to the dorsal head of Lance-Jones [1975] and + posterior head that corresponds to the accessory head of Greene [1935] and to the ventral head of Lance-Jones [1975]) [rat: the semitendinosus has an anterior (his principal) head from the hip bone—sciatic tuber—and a posterior (his accessory) head from the last sacral and first two caudal vertebrae, the two heads inserting together onto the medial side of the leg, including the tuberosity of the tibia; mouse: according to Charles et al. [2016], from ischial tuberosity to proximal tibia; see preceding text; innervation: anterior (his principal) head by tibial division of sciatic nerve while posterior (his accessory) head is by a branch of the lumbosacral plexus (Greene 1935)]	**Semitendinosus**

(*Continued*)

TABLE 19.13 (CONTINUED)
Hindlimb Muscles of Adults of Representative Tetrapod Taxa

Amphibia (Caudata): *Ambystoma mexicanum* (Axolotl)	Amphibia (Anura): *X. laevis* (African Clawed Frog)	Reptilia (Testudines): *Trachemys scripta* (Pond Slider)	Reptilia (Lepidosauria): *Timon lepidus* (Ocellated Lizard)	Reptilia (Crocodylia): *Alligator mississippiensis* (American Alligator)	Reptilia (Aves): *Gallus domesticus* (Chicken)	Mammalia (Monotremata): *Ornithorhynchus anatinus* (Platypus)	Mammalia (Marsupialia): *D. virginiana* (Virginia Opossum)	Mammalia (Placentalia): *Rattus norvegicus* (Brown Rat)	Mammalia (Placentalia): *Homo sapiens* (Modern Human)
Extensor digitorum longus (extensor digitorum communis *sensu* Ashley-Ross 1992) [innervation: dorsales pedis, from fibular nerve (Francis 1934); as in *A. tigrinum* (Ashley-Ross 1992), it extends from the femoral condyles to the proximal end of metatarsals I–V]	**Extensor digitorum longus** ("extensor longus digiti IV" *sensu* Gaupp [1896]; "extensor digitorum communis longus" *sensu* Dunlap [1960]) [as reported by Dunlap [1960] in the *R. pipiens*, *X. laevis*, and *E. coqui* specimens we dissected, the extensor digitorum communis longus extends from the distal extremity of the leg (cruris) to the short extensors of digits 3 and 4 in *X. laevis* and sometimes of digit 5 in *E. coqui* and only of digit 4 in *R. pipiens* (but also goes to digit 3 in other *R.* species)]	**Extensor digitorum longus** (extensor digitorum communis *sensu* Walker [1973]) [Walker 1973: in *Trachemys scripta*, the extensor digitorum longus extends from the distal femur to the distal phalanx of digit 1 and to metatarsals I–IV]	**Extensor digitorum longus** [as described by Russell and Bauer [2008] for *Iguana* and other lizards and *Sphenodon*, it extends from dorsomedial aspect of the lateral femoral epicondyle to metatarsals II and III, thus being a pedal dorsiflexor and not an extensor of the digits]	**Extensor digitorum longus** [Haughton 1868: in crocodylians, it usually connects to metatarsals I–III following fusion with the tibialis anterior; Schaeffer 1941: in *Alligator*, it extends from femur to metatarsals II and III; Fechner 2009: in crocodylians, including *Alligator*, it usually attaches to metatarsals I–V]	**Extensor digitorum longus** [Vanden Berge 1975: in adult chicken, the extensor digitorum longus extends from proximal tibia to distal phalanges of digits 2, 3, and 4; innervation: common fibular nerve (Vanden Berge 1975)]	**Extensor digitorum longus** (extensor digitorum superficialis *sensu* Gambaryan et al. [2002]) [from fibula to metatarsal II and distal phalanges of digits 2–5, as described by Gambaryan et al. [2002]; see main text]	**Extensor digitorum longus** [fibula to distal phalanges of digits 2–5, being deeply blended with fibularis brevis]	**Extensor digitorum longus** [rat: from the lateral epicondyle of the femur to the base of the distal phalanges of digits 2–5; mouse: from proximal fibula to distal phalanges of digits 2–5 according to Charles et al. [2015]; innervation: deep fibular nerve (Greene 1935)]	**Extensor digitorum longus** (extensor longus digitorum *sensu* Woollard 1925)
–	–	–	–	–	–	**Extensor hallucis longus** [fibula to distal phalanx of digit 1]	**Extensor hallucis longus** [fibula to distal phalanx of digit 1, being deeply blended with extensor digitorum brevis]	**Extensor hallucis longus** (extensor hallucis proprius *sensu* Greene 1935) [rat: as described by Greene [1935], the extensor hallucis longus extends from the fibula and interosseous membrane to the distal phalanx of the thumb; mouse: extends from tibia to distal phalanx of digit 1 according to Charles et al. [2015]; innervation: deep fibular nerve (Greene 1935)]	**Extensor hallucis longus** (extensor hallucis, extensor longus hallucis *sensu* Woollard 1925)

(Continued)

TABLE 19.13 (CONTINUED)
Hindlimb Muscles of Adults of Representative Tetrapod Taxa

Amphibia (Caudata): *Ambystoma mexicanum* (Axolotl)	**Amphibia (Anura): *X. laevis* (African Clawed Frog)**	**Reptilia (Testudines): *Trachemys scripta* (Pond Slider)**	**Reptilia (Lepidosauria): *Timon lepidus* (Ocellated Lizard)**	**Reptilia (Crocodylia): *Alligator mississippiensis* (American Alligator)**	**Reptilia (Aves): *Gallus domesticus* (Chicken)**	**Mammalia (Monotremata): *Ornithorhynchus anatinus* (Platypus)**	**Mammalia (Marsupialia): *D. virginiana* (Virginia Opossum)**	**Mammalia (Placentalia): *Rattus norvegicus* (Brown Rat)**	**Mammalia (Placentalia): *Homo sapiens* (Modern Human)**
–	–	–	– [not present in *Timon*, nor in *Iguana* according to Russell and Bauer [2008]]	–	–	–	– [see fibularis digiti quinti in the following]	– [really not present? Charles et al. [2016] do describe a fibularis tertius in addition to the brevis, longus, digiti quinti, and digit quarti, so it might be present in mice; however, according to them, it extends from proximal fibula to cuboid, what seems a strange attachment for a true fibularis tertius; moreover, the fibularis tertius has never been described in rats according to our literature review and is apparently absent in plesiomorphic primates, so seems to have been acquired later in humans evolution, namely, from the extensor digitorum longus, being only present as a variation in a few nonhuman primates such as bonobos: see, e.g., Diogo et al. 2017b; see also notes about fibularis digiti quinti in the following]	**Fibularis tertius** [see notes about fibularis digiti quinti in the following]

(Continued)

TABLE 19.13 (CONTINUED)
Hindlimb Muscles of Adults of Representative Tetrapod Taxa

Amphibia (Caudata): *Ambystoma mexicanum* (Axolotl)	Amphibia (Anura): *X. laevis* (African Clawed Frog)	Reptilia (Testudines): *Trachemys scripta* (Pond Slider)	Reptilia (Lepidosauria): *Timon lepidus* (Ocellated Lizard)	Reptilia (Crocodylia): *Alligator mississippiensis* (American Alligator)	Reptilia (Aves): *Gallus domesticus* (Chicken)	Mammalia (Monotremata): *Ornithorhynchus anatinus* (Platypus)	Mammalia (Marsupialia): *D. virginiana* (Virginia Opossum)	Mammalia (Placentalia): *Rattus norvegicus* (Brown Rat)	Mammalia (Placentalia): *Homo sapiens* (Modern Human)
Extensor tarsi tibialis [innervation: dorsales pedis intermedius, from fibular nerve (Francis 1934); as in *A. tigrinum* (Ashley-Ross 1992), it extends from the tibial epicondyle condyle of the femur to the tibiale bone]	**Tibialis anticus brevis** [as reported by Dunlap [1960], in the *R. pipiens*, *X. laevis,* and *E. coqui* specimens we dissected, the tibialis anticus brevis extends from the cruris to the tibiale; as also noted by Dunlap [1960, 1966], the tibialis anticus brevis and tarsalis anticus apparently derive from the same anlage (together with the extensor digitorum communis) and, in some anurans, the configuration and attachments of these two muscles thus supporting the idea that they derive from the same muscle mass]	**Tibialis anterior** [Walker 1973: in *Trachemys scripta*, the tibialis anterior extends from the tibia to metatarsal I]	**Tibialis anterior** [as described by Russell and Bauer [2008] for *Iguana* and other lizards and *Sphenodon*, it extends from the most proximal point of the anteromesial aspect of the tibial head, to metatarsal I, the muscle being mainly a pedal dorsiflexor]	**Tibialis anterior** [Schaeffer 1941: in *Alligator*, it extends from tibia to metatarsal I; Dilkes 1999; Fechner 2009: in *Alligator*, it extends from tibia to metatarsals I, II and III]	**Tibialis anterior** (tibialis cranialis *sensu* Vanden Berge [1975]) [Vanden Berge 1975: in adult chicken, it extends from distal femur and proximal tibia to tarsometatarsal region of the foot; innervation: common fibular nerve (Vanden Berge 1975)]	**Tibialis anterior** (tibialis cranialis *sensu* Gambaryan et al. [2002]) [from tibia, fibula and patella to medial cuneiform]	**Tibialis anterior** (flexor tarsi tibialis *sensu* Coues [1872]) [from tibia to medial cuneiform]	**Tibialis anterior** (tibialis cranialis *sensu* Walker and Homberger [1997]) [rat: as described by Greene [1935], the tibialis anterior extends from the tibia to the first cuneiform and to the proximal end of metatarsal I; mouse: according to Charles et al. [2016], from proximal tibia to medial cuneiform and metatarsal I; innervation: deep fibular nerve (Greene 1935)]	**Tibialis anterior** (tibialis anticus *sensu* Woollard [1925])
–	**Tarsalis anticus** [as reported by Dunlap [1960], in the *R. pipiens*, *X. laevis,* and *E. coqui* specimens we dissected, the tarsalis anticus extends from the distal cruris to the tibiale; see tibialis anticus brevis]	–	–	–	–	–	–	–	–
Extensor cruris tibialis [innervation: dorsales pedis intermedius and dorsalis pedis tibialis (n. saphenous) (Francis 1934); as in *A. tigrinum* (Ashley-Ross 1992), it extends from the tibial epicondyle of the femur to the anteroventral and anterodorsal margins of the tibia]	**Extensor cruris tibialis** (extensor cruris brevis *sensu* Dunlap [1960]) [as reported by Dunlap [1960], in the *R. pipiens*, *X. laevis,* and *E. coqui* specimens we dissected, the extensor cruris brevis extends from the medial condyle of the femur to the cruris]	–	–	–	–	–	–	–	–

(*Continued*)

TABLE 19.13 (CONTINUED)
Hindlimb Muscles of Adults of Representative Tetrapod Taxa

Amphibia (Caudata): *Ambystoma mexicanum* (Axolotl)	Amphibia (Anura): *X. laevis* (African Clawed Frog)	Reptilia (Testudines): *Trachemys scripta* (Pond Slider)	Reptilia (Lepidosauria): *Timon lepidus* (Ocellated Lizard)	Reptilia (Crocodylia): *Alligator mississippiensis* (American Alligator)	Reptilia (Aves): *Gallus domesticus* (Chicken)	Mammalia (Monotremata): *Ornithorhynchus anatinus* (Platypus)	Mammalia (Marsupialia): *D. virginiana* (Virginia Opossum)	Mammalia (Placentalia): *Rattus norvegicus* (Brown Rat)	Mammalia (Placentalia): *Homo sapiens* (Modern Human)
Extensor cruris et tarsi fibularis [innervation: fibular nerve (Francis 1934); as in *A. tigrinum* (Ashley-Ross 1992), it extends from the femoral condyles to the posterodorsal face of the fibula (the posterior part of the muscle) and the fibulare (the anterior part of the muscle)]	**Peroneus** [as reported by Dunlap [1960], in the *R. pipiens*, *X. laevis*, and *E. coqui* specimens we dissected, the peroneus extends from the region of the knee joint to the distal cruris and fibulare]	**Fibularis brevis** (peroneus brevis *sensu* Ogushi [1913]; peroneus posterior *sensu* Walker [1973]) [Walker 1973: in *Trachemys scripta*, the fibularis brevis extends from the distal fibula to metatarsal V]	**Fibularis brevis** (peroneus brevis *sensu* Russell and Bauer [2008]) from the fibular shaft to metatarsal V, being a dorsiflexor of the foot; unlike *Iguana*, in *Sphenodon*, the fibularis longus and brevis are mainly fused and originate from the fibula only, and this is the plesiomorphic condition for lepidosaurs according to Russell and Bauer [2008], the condition in *Timon* being different because the two muscles are blended, but they retain the characteristic origins and insertions of the longus and brevis of lizards such as *Iguana*]	**Fibularis brevis** [Haughton 1868: in *A. mississippiensis*, it connects to cuboid and metatarsal V; Schaeffer 1941: in *Alligator*, it extends from fibula to metatarsal V and calcaneum; Fechner 2009: in crocodylians, including *Alligator*, it extends from fibula to metatarsal V]	**Fibularis brevis** [Vanden Berge 1975: in adult chicken, it extends from tibia and fibula to plantar side of tarsometatarsal region of foot; innervation: deep fibular nerves (Vanden Berge 1975)]	– [not present as a distinct muscle in monotremes according to Gambaryan et al. [2002]]	**Fibularis brevis** [fibula to metatarsal V, being deeply blended with fibularis longus]	**Fibularis brevis** (peroneus brevis *sensu* Greene [1935]) [rat: as described by Greene [1935], the fibularis brevis extends from the fibula and interosseous membrane to metatarsal V, passing under the lateral malleolus near the tendons of the other three fibular muscles; mouse: according to Charles et al. [2016], from fibula to base of metatarsal V; innervation: superficial fibular nerve (Greene 1935)]	**Fibularis brevis** (peroneus brevis *sensu* Woollard 1925, 1955, Lewsi 1989; peronaeus longus *sensu* Osman 1955)
–	**Tibialis anticus longus** [as reported by Dunlap [1960], in the *R. pipiens*, *X. laevis*, and *E. coqui* specimens we dissected, the tibialis anticus longus extends from the medial condyle of the femur to the fibulare (its medial head) and to the tibiale (its lateral head), the muscle being often fused with the peroneus; see peroneus]	–	–	–	–	–	–	–	–

(*Continued*)

TABLE 19.13 (CONTINUED)
Hindlimb Muscles of Adults of Representative Tetrapod Taxa

Amphibia (Caudata): *Ambystoma mexicanum* (Axolotl)	Amphibia (Anura): *X. laevis* (African Clawed Frog)	Reptilia (Testudines): *Trachemys scripta* (Pond Slider)	Reptilia (Lepidosauria): *Timon lepidus* (Ocellated Lizard)	Reptilia (Crocodylia): *Alligator mississippiensis* (American Alligator)	Reptilia (Aves): *Gallus domesticus* (Chicken)	Mammalia (Monotremata): *Ornithorhynchus anatinus* (Platypus)	Mammalia (Marsupialia): *D. virginiana* (Virginia Opossum)	Mammalia (Placentalia): *Rattus norvegicus* (Brown Rat)	Mammalia (Placentalia): *Homo sapiens* (Modern Human)
–	–	**Fibularis longus** (peroneus longus *sensu* Ogushi [1913]; peroneus anterior *sensu* Walker [1973]) [Walker 1973: in *Trachemys scripta*, the fibularis longus extends from the distal half of the fibula to metatarsal V and digit 5]	**Fibularis longus** (peroneus longus *sensu* Russell and Bauer [2008]) [from the lateral femoral epicondyle to metatarsal V, being a dorsiflexor of the foot; unlike *Iguana*, in *Sphenodon*, the fibularis longus and brevis are mainly fused and originate from the fibula only, and this is the plesiomorphic condition for lepidosaurs according to Russell and Bauer [2008], the condition in *Timon* being different because the two muscles are blended, but they retain the characteristic origins and insertions of the longus and brevis of lizards such as *Iguana*]	**Fibularis longus** [Schaeffer 1941: in *Alligator*, it mainly extends from femur/patellar tendon to tuber; Fechner 2009: in crocodylians, including *Alligator*, it extends from fibula to calcaneum]	**Fibularis longus** [Vanden Berge 1975: in adult chicken, it extends from patellar tendon, fibula and tibia to tibial cartilage and tendon of flexor digitorum longus to digit 3; innervation: common fibular nerve (Vanden Berge 1975)]	**Fibularis longus** [fibula to metatarsal I]	**Fibularis longus** [fibula to metatarsal I]	**Fibularis longus** (peroneus longus *sensu* Greene [1935]) [rat: as described by Greene [1935], the fibularis longus extends from the fibula and tibia to metatarsal I and first cuneiform, passing under the lateral malleolus near the tendons of the other three fibular muscles; mouse: from fibula to first (medial) cuneiform according to Charles et al. [2016]; innervation: superficial fibular nerve (Greene 1935)]	**Fibularis longus** (peroneus primus *sensu* Burmeister [1846]; peroneus longus *sensu* Allen [1897], Woollard [1925, 1955], and Osman 1955)
–	–	–	– [not present in *Timon*, nor in *Iguana* according to Russell and Bauer [2008]; i.e., this structure seems to just correspond to the short extensor of digit 4]	–	–	–	– [fibularis digiti quarti not present as distinct muscle in *Didelphis*, confirming that this is simply a short extensor to digit 4, because in *Didelphis*, the short extensor to digit 4 is fused with the main body of the extensor digitorum brevis, this muscle thus attaches to digits 2–4; Stein [1981] uses the name extensor digitorum brevis in *Didelphis*, while he uses the names "fibularis digiti quinti" and fibularis digiti quarti in *Chironectes*, alternatively]	**Fibularis digiti quarti** (peroneus digiti quarti *sensu* Greene [1935]) [rat: as described by Greene [1935], in rats, the fibularis digiti quarti runs from the fibula to the distal end of metatarsal IV, passing under the lateral malleolus near the tendons of the other three fibular muscles, but in mice, it attaches to proximal phalanx of digit 4, clearly supporting the hypothesis that this is a short extensor muscle; i.e., probably the short extensor to digit 4; mouse: from fibula to distal phalanx of digit 4 according to Charles et al. [2015]; innervation: deep fibular nerve (Greene 1935)]	–

(Continued)

TABLE 19.13 (CONTINUED)
Hindlimb Muscles of Adults of Representative Tetrapod Taxa

Amphibia (Caudata): *Ambystoma mexicanum* (Axolotl)	Amphibia (Anura): *X. laevis* (African Clawed Frog)	Reptilia (Testudines): *Trachemys scripta* (Pond Slider)	Reptilia (Lepidosauria): *Timon lepidus* (Ocellated Lizard)	Reptilia (Crocodylia): *Alligator mississippiensis* (American Alligator)	Reptilia (Aves): *Gallus domesticus* (Chicken)	Mammalia (Monotremata): *Ornithorhynchus anatinus* (Platypus)	Mammalia (Marsupialia): *D. virginiana* (Virginia Opossum)	Mammalia (Placentalia): *Rattus norvegicus* (Brown Rat)	Mammalia (Placentalia): *Homo sapiens* (Modern Human)
–	–	–	– [not present in *Timon*, nor in *Iguana* according to Russell and Bauer [2008]; i.e., this structure seems to just correspond to the short extensor of digit 5]	–	–	**Fibularis digiti quinti** [from fibula to proximal and distal phalanges of digit 5, so it clearly corresponds to the short extensor of digit 5, as in therians such as rats and opossums, and not to the fibularis tertius, which is a derived structure of placentals that often attaches instead onto metatarsal V]	**Fibularis digiti quinti** (peroneus tertius *sensu* Coues [1872] and Stein [1981]) [from fibula to distal phalanx of digit 5, so it clearly corresponds to the short extensor of digit 5, as in rats, and not to the fibularis tertius, which is a derived structure of placentals that often attaches instead onto metatarsal V]	**Fibularis digiti quinti** (peroneus digiti quinti *sensu* Greene [1935]) [rat: as described by Greene [1935] in rats, the fibularis digiti quinti extends from the fibula to the distal end of metatarsal 5, passing under the lateral malleolus near the tendons of the other three fibular muscles, but in mice, it extends to the proximal end of the attaches phalanx of digit 5, clearly supporting the hypothesis that this is a short extensor muscle, probably the one going to digit 5; mouse: from fibula to distal phalanx of digit 5 according to Charles et al. [2015], being different from their "peroneus tertius," see preceding text; innervation: deep fibular nerve (Greene 1935)]	–

(*Continued*)

TABLE 19.13 (CONTINUED)
Hindlimb Muscles of Adults of Representative Tetrapod Taxa

Amphibia (Caudata): *Ambystoma mexicanum* (Axolotl)	Amphibia (Anura): *X. laevis* (African Clawed Frog)	Reptilia (Testudines): *Trachemys scripta* (Pond Slider)	Reptilia (Lepidosauria): *Timon lepidus* (Ocellated Lizard)	Reptilia (Crocodylia): *Alligator mississippiensis* (American Alligator)	Reptilia (Aves): *Gallus domesticus* (Chicken)	Mammalia (Monotremata): *Ornithorhynchus anatinus* (Platypus)	Mammalia (Marsupialia): *D. virginiana* (Virginia Opossum)	Mammalia (Placentalia): *Rattus norvegicus* (Brown Rat)	Mammalia (Placentalia): *Homo sapiens* (Modern Human)
Extensores digitorum breves [in addition to the "extensores digitorum breves superficiales," Francis [1934] also reported "extensores digitorum breves profundi" in *Salamandra*, also innervated by nn. dorsales pedis intermedius and fibularis, connecting the tarsal region to the dorsal side of the tendons of the extensores digitorum breves superficiales at the distal end of the metatarsal bones; that is, as described by authors such as Francis [1934], these bundles originate from different bones but are then fused distally, and each whole muscle formed by the two bundles sends a single tendon to the respective digit, as it also happens in the urodele hand, so there are no distinct, separate muscles dorsometatarsales as is the case in, e.g., anurans; innervation: dorsales pedis intermedius and fibular nerve (Francis 1934); as in *T. torosa* (Walthall and Ashley-Ross 2006), there are four muscles extending from the distal tarsal bones to the dorsal surface of the proximal end of the distal phalanx of each digit *via* a long tendon]	**Extensores digitorum breves** ("extensores breves superficiales and medii" of Dunlap [1960]) [as reported by Dunlap [1960], in the *R. pipiens*, *X. laevis*, and *E. coqui* specimens we dissected, the "extensores breves superficiales" mainly extend from the fibulare to the proximal phalanx of digit 1–4 except in *Rana* where it also attaches to digit 5; and the "extensores breves medii" mainly extend from the fused distal extremities of the tibiale and fibulare to the proximal phalanx of digits 1–3, except in *Eleutherodactylus* where it also connects to digit 4; see dorsometatarsales in the following]	**Extensores digitorum breves** (extensor digitorum communis brevis and apparently profundus *sensu* Ogushi [1913], according to Walker [1973]) [Walker 1973: in *Trachemys scripta* the extensores digitorum breves extend from the tarsals to the distal phalanges of digits 2–4]	**Extensores digitorum breves** [in *Timon*, they extend from tarsal/metatarsal region, sending a bundle to digit 1 and two or three bundles to each of the other digits, thus including all short extensors except apparently part of the extensor to digit 1, which is apparently included into the abductor et extensor digit 1; in *Sphenodon*, the muscle also connects to digit 1 only, while in *Iguana*, the condition is similar, but the latter muscle apparently also connects to digit 2 thus also including part of the short extensor to digit 2 ("adductor et extensor hallucis et indicis" *sensu* Russell and Bauer [2008]); as stated by Russell and Bauer [2008], contrary to what happens to the extensores digitorum breves of the lizard forelimb, only a few authors have divided the extensores digitorum breves of the lizard hindlimb into a superficial and a deep layer, and those few authors did not consider the deep layer as separate, distinct muscles dorsometatarsales (in the forelimb, the structures of the deep layer are often considered to be distinct muscles; i.e., the dorsometacarpales; however in chameleons, the dorsometatarsales are clearly present and exactly as the dorsometacarpales of the hand, as we confirmed in our dissections)]	**Extensores digitorum breves** [Haughton 1868: in *A. mississippiensis*, it attaches to all four digits; Schaeffer 1941: in *Alligator*, they mainly extend from the tarsal region to the distal phalanges of digits; Fechner 2009: in crocodylians, including *Alligator*, they usually attach to digits 2–4]	**Extensores digitorum breves** (abductor/extensor digit II, extensor brevis or proprius digit III and extensor brevis digit IV *sensu* Vanden Berge [1975]) [Vanden Berge 1975: in adult chicken, there are three separate short muscles, from the tarsometatarsal region to proximal phalanges of digit 2 ("abductor/extensor digit II"), digit 3 ("extensor brevis/proprius digit III"), and digit 4 ("extensor brevis digit IV"); innervation: deep fibular nerve for short extensor of digit 2 and superficial fibular nerve for those of digits 3 and 4 (Vanden Berge 1975)]	**Extensor digitorum brevis** (extensor digitorum profundus *sensu* Gambaryan et al. [2002]) [from fibula to metatarsal V, distal phalanges of digits 2–4, and apparently proximal and distal phalanx of digit 5: see extensor digitorum longus mentioned earlier]	**Extensor digitorum brevis** [from fibula to middle phalanges of digits 2–4; as a variation the tendon to digit 4 can send a thin branch to digit 5 as well: see, e.g., Coues 1872; the condition seen in *Didelphis* is also often seen in other marsupials: see, e.g., Osgood 1921]	**Extensor digitorum brevis** [rat: as described by Greene [1935] the extensor digitorum brevis extends from the lateral process of the calcaneus to the base of the middle phalanx of digits 2 and 3; innervation: deep fibular nerve (Greene 1935)]	**Extensor digitorum brevis** (extensor brevis digitorum pedis *sensu* Allen [1897]; extensor brevis digitorum *sensu* Woollard [1925] and Osman [1955])

(*Continued*)

TABLE 19.13 (CONTINUED)
Hindlimb Muscles of Adults of Representative Tetrapod Taxa

Amphibia (Caudata): *Ambystoma mexicanum* (Axolotl)	Amphibia (Anura): *X. laevis* (African Clawed Frog)	Reptilia (Testudines): *Trachemys scripta* (Pond Slider)	Reptilia (Lepidosauria): *Timon lepidus* (Ocellated Lizard)	Reptilia (Crocodylia): *Alligator mississippiensis* (American Alligator)	Reptilia (Aves): *Gallus domesticus* (Chicken)	Mammalia (Monotremata): *Ornithorhynchus anatinus* (Platypus)	Mammalia (Marsupialia): *D. virginiana* (Virginia Opossum)	Mammalia (Placentalia): *Rattus norvegicus* (Brown Rat)	Mammalia (Placentalia): *Homo sapiens* (Modern Human)
–	**Dorsometatarsales** (extensores breves profundi of Dunlap [1960]) [as reported by Dunlap [1960], in the *R. pipiens*, *X. laevis*, and *E. coqui* specimens we dissected, the extensores breves profundi mainly extend from the metatarsales to the distal phalanges of digits 1–5]	**Dorsometatarsales** (extensor hallucis brevis + metatarso-phalangis dorsalis *sensu* Ogushi [1913]; interossei dorsales *sensu* Walker [1973]) [Walker 1973: in *Trachemys scripta*, the dorsal interossei extend from the dorsal surface of the metatarsals of digits 1–4 and adjacent tarsals, to the distal phalanges of digits 1–4, thus clearly correspond to the dorsometatarsales and not to the intermetatarsales, *sensu* the present work]	– [but present as distinct muscles in lizards such as chameleons; see extensores digitorum breves]	– [Haughton 1868: in *A. mississippiensis*, these do not seem to be present as distinct muscles]	–	–	–	–	– [but present early in human ontogeny; see notes about interossei dorsales]
–	–	–	–	–	–	–	**Extensor hallucis brevis** (extensor brevis pollicis *sensu* Coues [1872]) [from fibula—lateral malleolus—to distal phalanx of digit 1; as a variation, it can attach, in some cases, on both digits 1 and 2: see, e.g., Coues 1872 and Stein 1981]	– [see preceding cells]	**Extensor hallucis brevis** [some authors consider that the in humans the extensor hallucis brevis is just a head of the extensor digitorum brevis, but following authors such as Netter (2006) and our own dissections, we consider that it is a distinct muscle]

(Continued)

TABLE 19.13 (CONTINUED)
Hindlimb Muscles of Adults of Representative Tetrapod Taxa

Amphibia (Caudata): *Ambystoma mexicanum* (Axolotl)	Amphibia (Anura): *X. laevis* (African Clawed Frog)	Reptilia (Testudines): *Trachemys scripta* (Pond Slider)	Reptilia (Lepidosauria): *Timon lepidus* (Ocellated Lizard)	Reptilia (Crocodylia): *Alligator mississippiensis* (American Alligator)	Reptilia (Aves): *Gallus domesticus* (Chicken)	Mammalia (Monotremata): *Ornithorhynchus anatinus* (Platypus)	Mammalia (Marsupialia): *D. virginiana* (Virginia Opossum)	Mammalia (Placentalia): *Rattus norvegicus* (Brown Rat)	Mammalia (Placentalia): *Homo sapiens* (Modern Human)
Abductor et extensor digiti 1 [innervation: dorsales pedis intermedium and fibular nerve (Francis 1934); as in *T. torosa* (Walthall and Ashley-Ross 2006), this muscle includes the abductor plus the short extensor to digit 1, thus connecting to metatarsal I and to the distal phalanx of digit 1, respectively]	**Abductor brevis dorsalis hallucis** [as reported by Dunlap [1960], in the *R. pipiens*, *X. laevis*, and *E. coqui* specimens we dissected, the abductor brevis dorsalis hallucis extends from the prehallux and centrale to the dorsomedial border of metatarsal I, being blended with the extensor brevis superficialis of digit 1 in *E. coqui*; this muscle clearly corresponds to the abductor part of the abductor et extensor digit 1 of urodeles, but is usually present as a distinct muscle in anurans, although in some taxa, it might be deeply blended with the short extensor to digit 1 or even be absent as a distinct muscle]	**Abductor et extensor digiti 1** (extensor hallucis longus and adductor hallucis dorsalis plus abductor hallucis *sensu* Ogushi [1913]; extensor hallucis proprius and abductor hallucis *sensu* Walker [1973])	**Abductor et extensor digiti 1** [from distal end of fibula and astragalocalcaneum to distal phalanx of digit 1; in *Iguana*, the condition is similar, but the muscle apparently also attaches to digit 2, thus also including a part of the short extensor to digit 2 ("adductor et extensor hallucis et indicis" *sensu* Russell and Bauer [2008]); our work on chameleons might support that part of the extensor brevis 1 is included in the abductor et extensor digit 1, see Molnar et al. 2017a]	**Abductor et extensor digiti 1** (extensor hallucis longus *sensu* Fechner [2009]) [Schaeffer 1941: in *Alligator*, it extends from fibula and fibulare to digit 1; Fechner 2009: in crocodylians, including *Alligator*, it usually extends from fibula to metatarsal I and to digit 1]	**Abductor et extensor digiti 1** (extensor hallucis longus *sensu* Vanden Berge [1975]) [Vanden Berge 1975: in adult chicken, it extends from tarsometatarsal region to distal phalanx of digit 1; innervation: deep fibular nerve (Vanden Berge 1975); because the muscle connects to the distal phalanx of digit 1, and there is no separate short extensor to digit 1, as in lizards, it probably includes the abductor plus at least part or the totality of the short extensor to digit 1, as it does in lizards as well]	–	–	–	–
–	–	**Gastrocnemius internus** (gastrocnemius *sensu* Ogushi [1913]) [Walker 1973: in *Trachemys scripta*, the gastrocnemius internus has two heads and extends from the tibia to the plantar aponeurosis]	**Gastrocnemius internus** (femorotibial gastrocnemius *sensu* Russell and Bauer [2008]) [from various tendons from the proximal tibia and surrounding regions to plantar aponeurosis, the flexor digitorum brevis, astragalocalcaneum, some tarsals, and digit 5 and surrounding muscles, being a very complex muscle; in *Iguana*, the condition is similar, but the muscle originates from the distal femur, knee joint, and proximal tibia, and in *Sphenodon*, the origin is apparently only from the tibia as in *Timon* (Russell and Bauer 2008)]	**Gastrocnemius internus** [Haughton 1868: in *A. mississippiensis*, it extends from tibia and femur to outer tarsal bone; Schaeffer 1941: in *Alligator*, it mainly extends from tibia to tuber and plantar aponeurosis]	**Gastrocnemius** (gastrocnemius intermedius, gastrocnemius internus, and gastrocnemius externus *sensu* Kardon [1998]) [Vanden Berge 1975: in adult chicken, the gastrocnemius has a pars interna, a pars externa, and a pars media, being however apparently a single, complex muscle as in most mammals, extending from the distal femur and proximal tibia to the tarsometatarsal region of the foot; innervation: medial and lateral tibia nerves (Vanden Berge 1975)]	**Gastrocnemius internus** [medial condyle of femur and tibia to calcaneus *via* Achilles tendon, being deeply blended with gastrocnemius externus]	**Gastrocnemius internus** [femur to calcaneus]	**Gastrocnemius** (part of triceps surae *sensu* Greene [1935]) [rat: as described by Greene [1935], the gastrocnemius extends from the medial epicondyle of the femur and medial fabella (medial sesamoid bone) and from the lateral epicondyle of the femur and lateral fabella (lateral sesamoid bone) to the calcaneus—tuber calcanei—*via* the calcaneal tendon; mouse: according to Charles et al. [2016], medial head extends from medial supracondylar ridge of femur to calcaneus and lateral head extends from lateral supracondylar ridge of femur to calcaneus; innervation: tibial nerve (Greene 1935)]	**Gastrocnemius** [medial head + lateral head]

(*Continued*)

TABLE 19.13 (CONTINUED)
Hindlimb Muscles of Adults of Representative Tetrapod Taxa

Amphibia (Caudata): *Ambystoma mexicanum* (Axolotl)	Amphibia (Anura): *X. laevis* (African Clawed Frog)	Reptilia (Testudines): *Trachemys scripta* (Pond Slider)	Reptilia (Lepidosauria): *Timon lepidus* (Ocellated Lizard)	Reptilia (Crocodylia): *Alligator mississippiensis* (American Alligator)	Reptilia (Aves): *Gallus domesticus* (Chicken)	Mammalia (Monotremata): *Ornithorhynchus anatinus* (Platypus)	Mammalia (Marsupialia): *D. virginiana* (Virginia Opossum)	Mammalia (Placentalia): *Rattus norvegicus* (Brown Rat)	Mammalia (Placentalia): *Homo sapiens* (Modern Human)
–	–	**Gastrocnemius externus** (flexor digitorum longus sublimis *sensu* Ogushi [1913]) [Walker 1973: in *Trachemys scripta*, the gastrocnemius externus extends from the femur to the plantar aponeurosis]	**Gastrocnemius externus** (femoral gastrocnemius *sensu* Russell and Bauer [2008]) [has deep and superficial heads and extends from femur to mainly plantar aponeurosis, metatarsal V, digits 2–4, and posterior portion of flexor digitorum brevis; in *Iguana* and *Sphenodon*, the condition is similar (Russell and Bauer 2008)]	**Gastrocnemius externus** [Haughton 1868: in *A. mississippiensis*, it extends from thigh muscles to outer tarsal bone; Schaeffer 1941: in *Alligator*, it mainly extends from femur to tuber]	–	**Gastrocnemius externus** [fibula to calcaneus *via* Achilles tendon]	**Gastrocnemius externus** (gastrocnemius externus et soleus *sensu* Coues [1872] and Stein [1981]) [extends from femur (a part corresponding to lateral head of gastrocnemius of placental mammals) and fibula (a part corresponding to soleus of gastrocnemius of placental mammals) to calcaneus, being well separated from the gastrocnemius internus; as stated by Coues [1872] and Stein [1981], in *D. virginiana* the soleus is not differentiated from the gastrocnemius externus, the soleus being mainly a placental feature according to Lewis [1989]]	–	–
–	–	–	–	**Gastrocnemius plantaris** (plantaris *sensu* Haughton [1868]) [Haughton 1868: in *A. mississippiensis*, it extends from thigh muscles to the calcaneum and to the flexor digitorum brevis]	**Gastrocnemius plantaris** (plantaris *sensu* Vanden Berge [1975]) [Vanden Berge 1975: in adult chicken, the plantaris extends from proximal tibia *via* a small belly to a slender tendon lying deep to the gastrocnemius and attaching onto the proximomedial end of the tibia cartilage; innervation: medial tibial nerve (Vanden Berge 1975); see notes on the left]	–	–	–	–

(Continued)

TABLE 19.13 (CONTINUED)
Hindlimb Muscles of Adults of Representative Tetrapod Taxa

Amphibia (Caudata): ***Ambystoma mexicanum*** **(Axolotl)**	**Amphibia (Anura):** ***X. laevis*** **(African Clawed Frog)**	**Reptilia (Testudines):** ***Trachemys scripta*** **(Pond Slider)**	**Reptilia (Lepidosauria):** ***Timon lepidus*** **(Ocellated Lizard)**	**Reptilia (Crocodylia):** ***Alligator mississippiensis*** **(American Alligator)**	**Reptilia (Aves):** ***Gallus domesticus*** **(Chicken)**	**Mammalia (Monotremata):** ***Ornithorhynchus anatinus*** **(Platypus)**	**Mammalia (Marsupialia):** ***D. virginiana*** **(Virginia Opossum)**	**Mammalia (Placentalia):** ***Rattus norvegicus*** **(Brown Rat)**	**Mammalia (Placentalia):** ***Homo sapiens*** **(Modern Human)**
–	–	–	–	–	– [but see preceding cells]	**Plantaris** [from fibula to plantar aponeurosis; see main text]	**Plantaris** (tensor fasciae plantaris *sensu* Coues [1872]) [from fibula (not femur as in, e.g., humans) to calcaneus, being blended with the gastrocnemius externus]	**Plantaris** [rat: as described by Greene [1935], the plantaris is much broader than in modern humans and extends from the lateral epicondyle of the femur, lateral fabella, and medial border of the head of the fibula to the region of the tuber calcanei, passing over this structure to become blended with the flexor digitorum brevis; mouse: from femur to calcaneus, according to Charles et al. [2016]; our dissections of mice support the idea that the soleus, and likely the plantaris, could correspond to a part of the flexor carpi ulnaris of the upper limb, because they mainly lie on the lateral side of the leg; innervation: tibial nerve (Greene 1935)]	**Plantaris**
–	–	–	–	–	–	–	– [soleus not present as a distinct muscle in marsupials according to McMurrich [1905] and Lewis [1989]; see gastrocnemius externus mentioned earlier]	**Soleus** [rat: as described by Greene [1935], the soleus extends from the head of the fibula to the tuber calcanei *via* the calcaneal tendon; our dissections of mice support the idea that the soleus, and likely the plantaris, could correspond to a part of the flexor carpi ulnaris of the upper limb, because they mainly lie on the lateral side of the leg; mouse: according to Charles et al. [2016], from upper and mid part of fibula to calcaneus; tibial nerve (Greene 1935)]	**Soleus**

(Continued)

TABLE 19.13 (CONTINUED)
Hindlimb Muscles of Adults of Representative Tetrapod Taxa

Amphibia (Caudata): *Ambystoma mexicanum* (Axolotl)	Amphibia (Anura): *X. laevis* (African Clawed Frog)	Reptilia (Testudines): *Trachemys scripta* (Pond Slider)	Reptilia (Lepidosauria): *Timon lepidus* (Ocellated Lizard)	Reptilia (Crocodylia): *Alligator mississippiensis* (American Alligator)	Reptilia (Aves): *Gallus domesticus* (Chicken)	Mammalia (Monotremata): *Ornithorhynchus anatinus* (Platypus)	Mammalia (Marsupialia): *D. virginiana* (Virginia Opossum)	Mammalia (Placentalia): *Rattus norvegicus* (Brown Rat)	Mammalia (Placentalia): *Homo sapiens* (Modern Human)
Flexor digitorum communis (flexor primordialis communis *sensu* Francis [1934]; plantaris superficialis lateralis plus plantaris profundus III *sensu* McMurrich [1905]) [Liem 1977: the flexor digitorum communis of salamanders corresponds to both the gastrocnemius externus and the flexor digitorum longus of lizards (page 385 of Russell and Bauer [2008]); innervation: fibular nerve (Francis 1934) as in *T. torosa* (Walthall and Ashley-Ross 2006), the muscle extends from fibular condyle of the femur to the distal phalanges of digits 1–5]	**Flexor digitorum communis** (plantaris longus *sensu* Dunlap [1960]) [as reported by Dunlap [1960] in the *R. pipiens*, *X. laevis*, and *E. coqui* specimens we dissected, the plantaris longus extends from the knee region to the palmar aponeurosis (and thus indirectly to digits 1 and 2), distal cruris, and proximal tarsus; the plantaris longus of anurans mainly corresponds to the flexor digitorum communis of urodeles]	**Flexor digitorum longus** (flexor digitorum longus profundus *sensu* Ogushi [1913]) [Walker 1973: in *T. scripta*, the flexor digitorum longus extends from the femur, fibula, and tarsal region to the distal phalanges of digits 1–4]	**Flexor digitorum longus** [contrary to what was described for *Iguana* (i.e., muscle only attaching to digits 1–4) in *Timon*, the muscle connects to digits 1–5; as in Iguana and *Sphenodon*, it is associated with two lumbricales attaching to digits 3 and 4 and mainly originates from the femur, fibula, and astragalocalcaneum (Russell and Bauer 2008)]	**Flexor digitorum longus** (flexor hallucis longus plus flexor digitorum longus *sensu* Haughton [1868] and Fechner [2009]) [Haughton 1868: in *A. mississippiensis*, it attaches to digits 1–3 (2–4 *sensu* the present work) only; Schaeffer 1941: in *Alligator*, it mainly extends from fibula and femur to the digits; Fechner 2009: in crocodylians, the flexor hallucis longus connects to digit 1 and the flexor digitorum longus attaches to digits 2–4]	**Flexor digitorum longus** (flexor digitorum longus plus flexor hallucis longus *sensu* Vanden Berge [1975]) [Vanden Berge 1975: in adult chicken, the flexor digitorum longus extends from tibia and fibula to the distal phalanx of digits 2–4, while the flexor hallucis longus extends from the popliteal region to the distal phalanx of digit 1, the two structures and their distal tendons being deeply blended and forming a single muscle complex, which corresponds to the flexor digitorum longus *sensu* the present work; i.e., there is no distinct, separate flexor hallucis longus as in, e.g., humans; innervation: deep medial tibial nerve (Vanden Berge 1975)]	**Flexor digitorum longus** (flexor digitorum fibularis *sensu* Gambaryan et al. [2002]) [fibula to distal phalanges of digits 1–5]	**Flexor digitorum longus** (flexor hallucis longus + flexor digitorum longus *sensu* Coues [1872] and Stein [1981]) [fibula and tibia to distal phalanx of digits 1–5, being deeply blended with the quadratus plantae, interosseous cruris, tibialis anterior, flexor digitorum brevis, and the four lumbricales; as noted by McMurrich [1905] and Lewis [1989], Coues [1872], and Stein [1981] erroneously stated that the flexor hallucis longus part of the flexor digitorum longus *sensu* the present work did not attach onto digit 1 because they erroneously stated that the "flexor brevis hallucis obliquus" *sensu* Stein [1981] and "flexor brevis pollicis obliquus" *sensu* Coues [1872] were the muscles sending the tendon to this digit, when the muscle corresponds instead to the quadratus plantae (accessorius), and therefore, the tendon to the distal phalanx described by Coues [1872] and Stein [1981] is the tendon of the flexor digitorum longus to digit 1, to which the quadratus plantae is attached to, as usual in other mammals]	**Flexor digitorum longus** (flexor hallucis longus or flexor fibularis + flexor digitorum longus or flexor tibialis *sensu* Greene [1935]) [rat: tibial head + fibular head to distal phalanges of digits 1–5; mouse: from tibia to distal phalanges of digits 1–5 according to Charles et al. [2016]; innervation: tibial nerve (Greene 1935) → similar to humans; there is much confusion in the literature about the flexor digitorum longus of mammals, and interestingly, this confusion is similar to the confusion concerning the flexor digitorum profundus of the upper limb; this is illustrated by Greene's (1935) work, who reports a flexor hallucis longus/flexor fibularis extending from the fibula, interosseous membrane and tibia to the distal phalanx of digits 1–5, and a flexor digitorum longus/flexor tibialis extending from the tibia and fibula to the terminal phalanx of digits 1–5 together (i.e., fused) with the tendons of the flexor hallucis longus/flexor fibularis, which is designated flexor hallucis longus because it is more fibular than the flexor digitorum longus, as in humans the belly of the flexor hallucis longus is	**Flexor digitorum longus** (flexor communitis digitorum *sensu* Allen [1897]; flexor longus digitorum *sensu* Woollard [1925]; flexor digitorum tibialis or medialis *sensu* Osman [1955])

(Continued)

TABLE 19.13 (CONTINUED)
Hindlimb Muscles of Adults of Representative Tetrapod Taxa

Amphibia (Caudata): *Ambystoma mexicanum* (Axolotl)	Amphibia (Anura): *X. laevis* (African Clawed Frog)	Reptilia (Testudines): *Trachemys scripta* (Pond Slider)	Reptilia (Lepidosauria): *Timon lepidus* (Ocellated Lizard)	Reptilia (Crocodylia): *Alligator mississippiensis* (American Alligator)	Reptilia (Aves): *Gallus domesticus* (Chicken)	Mammalia (Monotremata): *Ornithorhynchus anatinus* (Platypus)	Mammalia (Marsupialia): *D. virginiana* (Virginia Opossum)	Mammalia (Placentalia): *Rattus norvegicus* (Brown Rat)	Mammalia (Placentalia): *Homo sapiens* (Modern Human)
								more fibular than the belly of the flexor digitorum longus; however, unlike humans, in rats and mice, there is no distinct flexor hallucis longus attaching only to the hallux; i.e., the configuration of these two muscle structures is very similar to the heads of the flexor digitorum profundus of the rat forearm, which sends tendons to digits 1–5 of the hand; i.e., in rats, there is no distinct flexor pollicis longus going to the hand only; therefore, the two structures reported by Greene (1935) and most authors as two different muscles correspond to the fibular and tibial heads of the flexor digitorum longus *sensu* the present work; our dissections confirmed that the fleshy bellies of these two heads were blended to each other as well]	
– [in urodeles such as *Ambystoma*, the flexor cruris et tarsi tibialis (plantaris superficialis medialis *sensu* McMurrich [1905]) likely corresponds to the distal (leg) portion of the ischioflexorius, as explained by authors such as McMurrich [1905] and Bardeen [1906]]	**Cruroastragalus** (tibialis posticus *sensu* Dunlap [1960]) [as reported by Dunlap [1960], in the *R. pipiens*, *X. laevis*, and *E. coqui* specimens we dissected, the tibialis posticus extends from the cruris to the tibiale]	– [in turtles such as *Trachemys*, the flexor cruris et tarsi tibialis is apparently included in the flexor tibialis externus tarsi that inserts onto the gastrocnemius internus and/or included in the latter muscle]	– [as noted by Bardeen 1906, in lizards such as *Timon*, the flexor cruris et tarsi tibialis apparently corresponds to the long crural tendon of the flexor tibialis externus that follows the border of the gastrocnemius internus, receives fibers from it, and contributes to plantar aponeurosis with it, so a part of the muscle may also be included in the gastrocnemius internus]	– [in crocodylians such as *Alligator* ,the flexor cruris et tarsi tibialis is apparently included in the flexor tibialis externus part that inserts onto the gastrocnemius internus and/or included in the latter muscle]	– [in birds such as *Gallus*, the flexor cruris et tarsi tibialis is apparently completely lost or included in the gastrocnemius complex because none of the "hamstrings" is directly going to the distal leg/foot]	–	– [according to Bardeen [1906], in marsupials and in mammals, including humans, the flexor cruris et tarsi tibialis is usually represented only by the fascial insertion of the hamstring muscles]	– [see on the left]	– [see on the left]

(Continued)

TABLE 19.13 (CONTINUED)
Hindlimb Muscles of Adults of Representative Tetrapod Taxa

Amphibia (Caudata): *Ambystoma mexicanum* (Axolotl)	**Amphibia (Anura): *X. laevis* (African Clawed Frog)**	**Reptilia (Testudines): *Trachemys scripta* (Pond Slider)**	**Reptilia (Lepidosauria): *Timon lepidus* (Ocellated Lizard)**	**Reptilia (Crocodylia): *Alligator mississippiensis* (American Alligator)**	**Reptilia (Aves): *Gallus domesticus* (Chicken)**	**Mammalia (Monotremata): *Ornithorhynchus anatinus* (Platypus)**	**Mammalia (Marsupialia): *D. virginiana* (Virginia Opossum)**	**Mammalia (Placentalia): *Rattus norvegicus* (Brown Rat)**	**Mammalia (Placentalia): *Homo sapiens* (Modern Human)**
– [in urodeles such as *Ambystoma*, the flexor cruris et tarsi fibularis (plantaris superficialis lateralis *sensu* McMurrich [1905]) likely corresponds to the femoral head of the flexor digitorum communis, as explained by authors such as McMurrich [1905] and Bardeen [1906]]	– [in anurans such as *Xenopus*, the flexor cruris et tarsi fibularis likely corresponds to the femoral head of the flexor digitorum communis, as explained by authors such as McMurrich [1905] and Bardeen [1906]]	– [in amniotes, the flexor cruris et tarsi fibularis likely corresponds to a part of the gastrocnemius complex and/or its derivatives, as explained by authors such as McMurrich [1905] and Bardeen [1906]]	– [see cells on the left]	– [see cells on the left]	– [see cells on the left]	– [see cells on the left]	– [see cells on the left]	– [see cells on the left]	– [see cells on the left]
–	–	–	–	–	–	–	– [the flexor hallucis longus is not present as a distinct muscle; see flexor digitorum longus]	– [see on the left]	**Flexor hallucis longus** (flexor longus pollicis pedis *sensu* Allen [1897]; flexor longus hallucis *sensu* Woollard [1925]; flexor digitorum fibularis or lateralis *sensu* Osman [1955])
–	–	– [but popliteus is present in some turtles; see interosseous cruris in the following]	**Popliteus** [Rewcastle 1977: the popliteus is derived from the anlage of the interosseous cruris; extends from proximal tibia (not the whole length of tibia as reported for *Lacerta* by Russell and Bauer [2008]) to proximal fibula; in *Sphenodon*, it is a broad muscle originating from almost all the length of the tibia, while in *Iguana*, it originates only from the proximal tibia (Russell and Bauer 2008)]	**Popliteus** [Haughton 1868: in *A. mississippiensis*, there is no popliteus; Fechner 2009: in crocodylians, the popliteus originates from tibia and fibula, dorsally, and inserts onto the interosseous cruris]	**Popliteus** [Vanden Berge 1975: in adult chicken, the popliteus extends from proximal tibia and fibula to the head of the fibula; innervation: medial tibial nerve (Vanden Berge 1975)]	**Popliteus** [from fibula to tibia]	– [the popliteus is not present as a separate muscle, being instead a part of the interosseous cruris, see the following; this seems to be usually the case in other marsupials as well: e.g., Bardeen 1906]	**Popliteus** [rat: as described by Greene [1935], the popliteus extends from the lateral epicondyle of the femur to the proximal third of the medial surface of the tibia; mouse: according to Charles et al. [2016], extends from lateral femoral condyle to medial tibial condyle; innervation: tibial nerve (Greene 1935)]	**Popliteus**

(Continued)

TABLE 19.13 (CONTINUED)
Hindlimb Muscles of Adults of Representative Tetrapod Taxa

Amphibia (Caudata): *Ambystoma mexicanum* (Axolotl)	Amphibia (Anura): *X. laevis* (African Clawed Frog)	Reptilia (Testudines): *Trachemys scripta* (Pond Slider)	Reptilia (Lepidosauria): *Timon lepidus* (Ocellated Lizard)	Reptilia (Crocodylia): *Alligator mississippiensis* (American Alligator)	Reptilia (Aves): *Gallus domesticus* (Chicken)	Mammalia (Monotremata): *Ornithorhynchus anatinus* (Platypus)	Mammalia (Marsupialia): *D. virginiana* (Virginia Opossum)	Mammalia (Placentalia): *Rattus norvegicus* (Brown Rat)	Mammalia (Placentalia): *Homo sapiens* (Modern Human)
Flexor accessorius medialis (apparently includes the plantaris profundus III minor, as noted by Francis [1914]) [innervation: ramus medianus of fibular nerve (Francis 1934); as in *T. torosa* (Walthall and Ashley-Ross 2006), it extends from the distal region of the fibula, the fibulare and intermedium to the plantar fascia, probably helping in pronating the foot]	**Flexor accessorius** ("transversus plantae proximalis et distalis" *sensu* Dunlap [1960]) [as reported by Dunlap [1960], in the *R. pipiens*, *X. laevis,* and *E. coqui* specimens we dissected, the transversus plantae proximalis et distalis extends from the fibulare to the dorsal surface of the plantar aponeurosis and the tibiale, the muscle being not divided into distal and proximal bundles in *Xenopus* (as well as in the more plesiomorphic anuran taxa) and being divided into these two bundles in *Rana* and *Eleutherodactylus*; see cruroastragalus]	– [flexor accessorius complex corresponds to distal head of flexor digitorum longus of turtles, see cell on the right]	– [flexor accessorius complex corresponds to distal head of flexor digitorum longus of lizards and *Sphenodon*, see the following]	– [flexor accessorius complex probably corresponds to the distal part of the flexor digitorum longus of crocodylians, see cell on the left]	– [flexor accessorius complex probably corresponds to part of flexor digitorum longus, but no distinct distal/accessorius head of this muscle was described by Vanden Berge [1975]]	**Quadratus plantae** (flexor accessorius *sensu* McMurrich [1905] and Lewis [1989]) [from calcaneus to tendon of flexor digitorum longus]	**Quadratus plantae** (flexor accessorius *sensu* McMurrich [1905]; flexor brevis pollicis obliquus *sensu* Coues [1872]; flexor hallucis obliquus *sensu* Stein [1981]; accessorius *sensu* Lewis [1989]) [from calcaneus to tendon of flexor digitorum longus: see flexor digitorum longus]	**Quadratus plantae** [as described by Greene [1935], the quadratus plantae extends from the calcaneus to the tendon of the flexor digitorum longus; innervation: lateral plantar nerve (Greene 1935)]	**Quadratus plantae** (flexor accessorius *sensu* McMurrich [1905] and Osman [1955])
Flexor accessorius lateralis (plantaris profundus II *sensu* McMurrich [1905]) [innervation: ramus medianus of fibular nerve (Francis 1934); as in *T. torosa* (Walthall and Ashley-Ross 2006), it extends from the fibulare to the plantar fascia, probably helping to pronate the foot]	– **[see flexor accessorius mentioned earlier]**	–	–	–	–	–	–	–	–

(Continued)

TABLE 19.13 (CONTINUED)
Hindlimb Muscles of Adults of Representative Tetrapod Taxa

Amphibia (Caudata): *Ambystoma mexicanum* (Axolotl)	Amphibia (Anura): *X. laevis* (African Clawed Frog)	Reptilia (Testudines): *Trachemys scripta* (Pond Slider)	Reptilia (Lepidosauria): *Timon lepidus* (Ocellated Lizard)	Reptilia (Crocodylia): *Alligator mississippiensis* (American Alligator)	Reptilia (Aves): *Gallus domesticus* (Chicken)	Mammalia (Monotremata): *Ornithorhynchus anatinus* (Platypus)	Mammalia (Marsupialia): *D. virginiana* (Virginia Opossum)	Mammalia (Placentalia): *Rattus norvegicus* (Brown Rat)	Mammalia (Placentalia): *Homo sapiens* (Modern Human)
Contrahentium caput longum (fibulotarsalis *sensu* McMurrich [1905]; caput longum musculorum contrahentium *sensu* Walthall and Ashley-Ross [2006]) [innervation: ramus medianus of fibular nerve (Francis 1934); AMP4-1: As in *T. torosa* (Walthall and Ashley-Ross 2006), it lies deep to the plantar fascia and extends from the distal portion of the fibula to the distal tarsal bones, being in contact with the contrahentes pedis and probably helping to flex (depress) the tarsus]	**Contrahentium caput longum** (tarsalis posticus *sensu* Dunlap [1960]) [as reported by Dunlap [1960], in the *R. pipiens*, *X. laevis,* and *E. coqui* specimens we dissected, the tarsalis posticus extends from the ligamentum calcanei together with the plantaris profundus to the tibiale; see cruroastragalus]	–	–	–	–	–	–	–	–
Interosseous cruris [innervation: n. interosseous (Francis 1934); as in *T. torosa* (Walthall and Ashley-Ross 2006), it extends from the proximal part of the fibula to the distal portion of the tibia, probably helping to stabilize the two long bones of the crus (and in pronation, in our opinion)]	**Interosseous cruris** (intertarsalis) [as reported by Dunlap [1960], in the *R. pipiens*, *X. laevis,* and *E. coqui* specimens we dissected, the intertarsalis extends from the tibiale and fibulare to the os centrale and, only in *Xenopus*, to the prehallux and the base of metatarsal I; see cruroastragalus]	**Interosseous cruris** (popliteus *sensu* Ogushi [1913] and Walker [1973]) [Walker 1973: in *Trachemys scripta*, the popliteus is an undivided muscle extending from the tibia to the fibula]	**Interosseous cruris** [as in *Iguana* (Russell and Bauer 2008), I found an interosseous cruris deep to the pronator profundus, connecting the distal tibia to the distal fibula]	**Interosseous cruris** [Fechner 2009: in crocodylians, the interosseous cruris is present]	– [not described by Kardon [1998] and particularly by Vanden Berge [1975], who described all muscles in detail, although this muscle is many times neglected even in detailed descriptions of other taxa; Fechner 2009: in Aves interosseous cruris is absent]	– [according to Gambaryan et al. [2002], the interosseous cruris (peronotibialis) is present in echidnas (which thus have both a popliteus and an interosseous cruris, as do many reptiles) but not in platypus]	**Interosseous cruris** (pronator tibiae *sensu* McMurrich [1905]; rotator fibulae or popliteus *sensu* Lewis [1989] and Coues [1872]) [from tibia to fibula; as stated by Stein [1981], in *D. virginiana*, the popliteus is part of the larger interosseous cruris, which basically occupies the whole proximodistal space between the tibia and fibula, in a condition very similar to that found in many nonmammalian tetrapods]	–	– [rotator fibulae as possible human variation]

(Continued)

TABLE 19.13 (CONTINUED)
Hindlimb Muscles of Adults of Representative Tetrapod Taxa

Amphibia (Caudata): *Ambystoma mexicanum* (Axolotl)	Amphibia (Anura): *X. laevis* (African Clawed Frog)	Reptilia (Testudines): *Trachemys scripta* (Pond Slider)	Reptilia (Lepidosauria): *Timon lepidus* (Ocellated Lizard)	Reptilia (Crocodylia): *Alligator mississippiensis* (American Alligator)	Reptilia (Aves): *Gallus domesticus* (Chicken)	Mammalia (Monotremata): *Ornithorhynchus anatinus* (Platypus)	Mammalia (Marsupialia): *D. virginiana* (Virginia Opossum)	Mammalia (Placentalia): *Rattus norvegicus* (Brown Rat)	Mammalia (Placentalia): *Homo sapiens* (Modern Human)
Tibialis posterior (pronator profundus) [innervation: n. interosseous (Francis 1934); as in *T. torosa* (Walthall and Ashley-Ross 2006), it extends from the medial part of the fibula to the distal portion of the tibia, the tibiale and the base of metatarsal I, probably helping in pronating the foot]	**Tibialis posterior** (plantaris profundus) [as reported by Dunlap [1960], in the *R. pipiens*, *X. laevis*, and *E. coqui* specimens we dissected, the plantaris profundus extends from the ligamentum calcanei to the dorsal surface of the aponeurosis plantaris, and only in *Xenopus*, to the prehallux; see contrahentium caput longum, and cruroastragalus]	**Tibialis posterior** (pronator profundus; flexor hallucis longus *sensu* Ogushi [1913]) [Walker 1973: in *Trachemys scripta*, the pronator profundus extends from the fibula to the tarsal region and to metatarsal I]	**Tibialis posterior** (pronator profundus) [as in *Iguana* (Russell and Bauer 2008), extends from fibular shaft to tarsometatarsal joints 1–3 and surrounding structures, i.e., tarsals and metatarsals]	**Tibialis posterior** (pronator profundus *sensu* Fechner [2009]) [Haughton 1868: in *A. mississippiensis*, it attaches to metatarsals I and II; Schaeffer 1941: in *Alligator*, it mainly extends from fibula to metatarsal I; Fechner 2009: in crocodylians, it extends from fibula and tibia to metatarsals I and II]	– [not described by Kardon [1998] and particularly by Vanden Berge [1975], who described all muscles in detail, although this muscle is many times neglected even in detailed descriptions of other taxa]	**Tibialis posterior** (tibialis caudalis *sensu* Gambaryan et al. [2002]) [from tibia to calcaneus]	**Tibialis posterior** [from tibia and fibula to navicular bone]	**Tibialis posterior** (tibialis caudalis *sensu* Walker and Homberger [1997]) [rat: as described by Greene [1935], the tibialis posterior extends from the tibia and interosseous membrane and fibula to the navicular and the first cuneiform; mouse: extends from tibia to navicular according to Charles et al. [2015]; innervation: tibial nerve (Greene 1935)]	**Tibialis posterior** (tibialis posticus *sensu* Allen [1897] and Woollard [1925])
Flexores digitorum minimi (*sensu* e.g., Francis [1934], which correspond to the "median slips of flexor brevis profundi" *sensu* McMurrich [1904]) [innervation: plantar nerve loop (Francis 1934); we could not discern how many muscles in axolotls, but according to Francis [1934], in *Salamandra*, there are four muscles extending from the metatarsals to the base of the proximal phalanx of digits 2–5, deep to the contrahentes]	**Flexores digitorum minimi** ("flexores teretes digitorum" *sensu* Dunlap [1960]) [as reported by Dunlap [1960], in the *R. pipiens*, *X. laevis*, and *E. coqui* specimens we dissected, the flexores teretes digitorum, which might be either dorsal or ventral to the transversi metatarsi (metatarsales *sensu* the present work) mainly extend from the ventral margin of the metatarsals to the ventral base of the proximal phalanx of digits 2 and 4 in *Xenopus* and 2–5 in the other two genera]	–	–	–	–	–	–	–	–
Interphalangeus digiti 3 [Francis describes two muscles interphalangei in digit 3 and two in digit 4 in *Salamandra*, but we analyzed this in detail in green fluorescence protein (GFP) axolotls and only found one muscle, connecting the metatarsophalangeal and first interphalangeal joints of digit 3; innervation: plantar nerve loop (Francis 1934)]	**Interphalangeus digiti 3** ("interphalangealis digiti III" *sensu* Dunlap [1960]) [as reported by Dunlap [1960], in the *R. pipiens*, *X. laevis*, and *E. coqui* specimens we dissected, the interphalangeus digiti 3 extends from the proximal phalanx of digit 3 to the second phalanx of this digit; in *Xenopus*, the interphalangei also originate from metatarsals]	–	–	–	–	–	–	–	–

(Continued)

TABLE 19.13 (CONTINUED)
Hindlimb Muscles of Adults of Representative Tetrapod Taxa

Amphibia (Caudata): ***Ambystoma mexicanum*** **(Axolotl)**	**Amphibia (Anura):** ***X. laevis*** **(African Clawed Frog)**	**Reptilia (Testudines):** ***Trachemys scripta*** **(Pond Slider)**	**Reptilia (Lepidosauria):** ***Timon lepidus*** **(Ocellated Lizard)**	**Reptilia (Crocodylia):** ***Alligator mississippiensis*** **(American Alligator)**	**Reptilia (Aves):** ***Gallus domesticus*** **(Chicken)**	**Mammalia (Monotremata):** ***Ornithorhynchus anatinus*** **(Platypus)**	**Mammalia (Marsupialia):** ***D. virginiana*** **(Virginia Opossum)**	**Mammalia (Placentalia):** ***Rattus norvegicus*** **(Brown Rat)**	**Mammalia (Placentalia):** ***Homo sapiens*** **(Modern Human)**
Interphalangei digiti 4 [see preceding cell; innervation: plantar nerve loop (Francis 1934); two muscles, one similar to the interphalangeus digiti 3, connecting the metatarsophalangeal and first interphalangeal joints of digit 4, and the other connecting the first and second interphalangeal joints of this digit 4]	**Interphalangei digiti 4** (“interphalangealis digiti IV” plus “interphalangealis distalis digit IV” *sensu* Dunlap [1960]) [as reported by Dunlap [1960], in the *R. pipiens*, *X. laevis*, and *E. coqui* specimens we dissected, there are two muscles interphalangei digiti 4, one extending from the proximal phalanx of digit 4 to the second phalanx of this digit, and the other from the second phalanx to the third phalanx of this digit; in *Xenopus*, the interphalangei also originate from metatarsals]	–	–	–	–	–	–	–	–
–	**Interphalangeus digiti 5** (“interphalangealis digiti V” *sensu* Dunlap [1960]) [as reported by Dunlap [1960], in the *R. pipiens*, *X. laevis*, and *E. coqui* specimens we dissected, the interphalangeus digiti 5 extends from the proximal phalanx of digit 5 to the second phalanx of this digit; in *Xenopus*, the interphalangei also originate from metatarsals]	–	–	–	–	–	–	–	–

(*Continued*)

TABLE 19.13 (CONTINUED)
Hindlimb Muscles of Adults of Representative Tetrapod Taxa

Amphibia (Caudata): *Ambystoma mexicanum* (Axolotl)	**Amphibia (Anura): *X. laevis* (African Clawed Frog)**	**Reptilia (Testudines): *Trachemys scripta* (Pond Slider)**	**Reptilia (Lepidosauria): *Timon lepidus* (Ocellated Lizard)**	**Reptilia (Crocodylia): *Alligator mississippiensis* (American Alligator)**	**Reptilia (Aves): *Gallus domesticus* (Chicken)**	**Mammalia (Monotremata): *Ornithorhynchus anatinus* (Platypus)**	**Mammalia (Marsupialia): *D. virginiana* (Virginia Opossum)**	**Mammalia (Placentalia): *Rattus norvegicus* (Brown Rat)**	**Mammalia (Placentalia): *Homo sapiens* (Modern Human)**
Flexores breves profundi [innervation: plantar nerve loop (Francis 1934); as in *A. tigrinum* (Ashley-Ross 1992), the flexores breves profundi extend from the carpal/metacarpal region to each side of digits 1-5, in a total of 10 muscles]	**Flexores breves profundi** ("flexores ossi metatarsales" *sensu* Dunlap [1960]) [as reported by Dunlap [1960], in the *R. pipiens*, *X. laevis,* and *E. coqui* specimens we dissected, the flexores ossi metatarsales mainly extend from the fibulare to the metatarsals II–IV, with the exception of *Xenopus* where the muscle to metatarsal IV is absent; see opponens hallucis and cruroastragalus]	**Flexores breves profundi** (*sensu* Ribbing [1938]; i.e., part of interossei plantares *sensu* Walker [1973]) [four muscles to digits 1–4]	**Flexores breves profundi** (interossei plantares + flexor hallucis *sensu* Russell and Bauer [2008]) [from metatarsals to digits 1–5, one for each side of each digit, including the so-called flexor hallucis to the tibial side of digit 1]	**Flexores breves profundi** [apparently eight muscles to digits 1–4, according to Suzuki et al. [2011]]	**Flexores breves profundi** (flexor hallucis brevis *sensu* Vanden Berge [1975]) [Vanden Berge 1975: in adult chicken, the flexor hallucis brevis extends from plantar metatarsal groove to the proximal phalanx of digit 1, this being thus the only flexor brevis profundus present in the foot; innervation: lateral plantar nerve (Vanden Berge 1975)]	**Flexores breves profundi** [number 1 corresponds to "medial head of flexor hallucis brevis" *sensu* Coues [1870] and to "abductor hallucis brevis" *sensu* Gambaryan et al. [2002] and is described under flexor hallucis brevis below; number 2 corresponds to "lateral head of flexor hallucis brevis" *sensu* Coues [1872] and to "flexor hallucis brevis" *sensu* Gambaryan et al. [2002] and extends from metatarsal I to lateral side of proximal phalanx of digit 1; numbers 3, 4, 5, 6, 7, and 8 correspond to "interossei plantares" *sensu* Coues [1870], connecting each metatarsal (II, III, and IV) to the medial and lateral sides of the proximal phalanx of the respective digit; number 9 corresponds to "medial head of flexor brevis digiti minimi" *sensu* Coues [1872] and extends from metatarsal V to lateral side of proximal phalanx of digit 5; number 10 is described in flexor digiti minimi brevis in the following]	**Flexores breves profundi** [number 1 corresponds to "superficial or medial head of flexor hallucis brevis" and is described under flexor hallucis brevis below; number 2 corresponds to "lateral head of flexor hallucis brevis" and extends from medial cuneiform to proximal phalanx of digit 1; numbers 3 and 4 (often designated "flexor brevis indicis") extend from intermediate cuneiform to proximal phalanx of digit 2; numbers 5 and 6 (often designated "flexor brevis medii") extend from lateral cuneiform to proximal phalanx of digit 3; numbers 7 and 8 (often designated "flexor brevis annularis") extend from cuboid to proximal phalanx of digit 4; number 9 extends from metatarsal V and cuboid to proximal phalanx of digit 5; number 10 is described in flexor digiti minimi brevis in the following]	**Flexores breves profundi**, [number 1 corresponds to "superficial or tibial head of flexor hallucis brevis" and is described under flexor hallucis brevis in the following; number 10 corresponds to flexor digiti minimi brevis and is described in the following; number 2 corresponds to "lateral head of flexor hallucis brevis," extends from navicular to proximal phalanx of digit 1; numbers 3 and 4 extend from navicular to proximal phalanx of digit 2; numbers 5 and 6 extend from cuboid to proximal phalanx of digit 3; numbers 7 and 8 extend from cuboid to proximal phalanx of digit 4; number 9 extends from cuboid to proximal phalanx of digit 5; innervation: lateral plantar nerve (Greene 1935)]	–

(*Continued*)

TABLE 19.13 (CONTINUED)
Hindlimb Muscles of Adults of Representative Tetrapod Taxa

Amphibia (Caudata): *Ambystoma mexicanum* (Axolotl)	Amphibia (Anura): *X. laevis* (African Clawed Frog)	Reptilia (Testudines): *Trachemys scripta* (Pond Slider)	Reptilia (Lepidosauria): *Timon lepidus* (Ocellated Lizard)	Reptilia (Crocodylia): *Alligator mississippiensis* (American Alligator)	Reptilia (Aves): *Gallus domesticus* (Chicken)	Mammalia (Monotremata): *Ornithorhynchus anatinus* (Platypus)	Mammalia (Marsupialia): *D. virginiana* (Virginia Opossum)	Mammalia (Placentalia): *Rattus norvegicus* (Brown Rat)	Mammalia (Placentalia): *Homo sapiens* (Modern Human)
–	**Abductor brevis plantaris digiti V** [as reported by Dunlap [1960], in the *R. pipiens*, *X. laevis,* and *E. coqui* specimens we dissected, the abductor brevis plantaris digiti V extends from fibulare to the lateroventral border of metatarsal V; clearly seems to correspond to the most fibular flexor brevis profundus]	–	–	–	–	–	–	–	–
–	**Abductor proprius digiti IV** [as reported by Dunlap [1960], in the *R. pipiens*, *X. laevis,* and *E. coqui* specimens we dissected, the abductor proprius digiti IV extends from metatarsal V to metatarsal IV]	–	–	–	–	–	–	–	–
Intermetatarsales [innervation: plantar nerve loop (Francis 1934); as in *T. torosa* (Walthall and Ashley-Ross 2006), there are four muscles connecting the metatarsals I–V, probably helping in adducting the metatarsals and the attached digits]	**Intermetatarsales** (transversi metatarsi *sensu* Dunlap [1960]) [as reported by Dunlap [1960], in the *R. pipiens*, *X. laevis,* and *Eleutherodactylus coqui* specimens we dissected, the four transversi metatarsi extend inbetween metatarsal I–V]	**Intermetatarsales** (interdigitales *sensu* Ribbing [1938], i.e., part of interossei plantares *sensu* Walker [1973]) [see flexores breves profundi]	**Intermetatarsales** (interossei dorsales *sensu* Russell and Bauer [2008]) [four muscles between metatarsals I–V]	**Intermetatarsales** [apparently three muscles connecting metatarsals I–IV according to Gadow [1882]; although Suzuki et al. [2011] seem to suggest that there are only two muscles in crocodylians]	– [Vanden Berge 1975: in adult chicken, there are no intermetacarpales]	**Intermetatarsales** (interossei dorsales *sensu* Coues [1870] and Gambaryan et al. [2002]) [number 1 is vestigial, just a tendon, extending from metatarsal II to lateral side of proximal phalanx of digit 1, being blended with the adductor hallucis; number 2 extends from metatarsals II and III to proximal phalanx of digit 2; number 3, from metatarsal IV to proximal phalanx of digit 3; number 4, from metatarsal IV to proximal phalanx of digit 4; see, e.g., Gambaryan et al. 2002]	**Intermetatarsales** [number 1 extends from metatarsal I to medial side of proximal phalanx of digit 2; number 2, from metatarsals II and III to medial side of proximal phalanx of digit 3; number 3, from metatarsals III and IV to lateral side of proximal phalanx of digit 3 and to medial side of proximal phalanx of digit 4; number 4, from metatarsals IV and V to lateral side of proximal phalanx of digit 4 and to medial side of proximal phalanx of digit 5; see, e.g., Figure 16.1 of Lewis [1989]]	**Intermetatarsales** [number 1 extends from metatarsal I to tibial side of proximal phalanx of digit 2; number 2 from metatarsals II and III to tibial side of proximal phalanx of digit 3; number 3, from metatarsals III and IV to tibial side of proximal phalanx of digit 4; number 4, from metatarsals IV and V to fibular side of proximal phalanx of digit 4, so axis is apparently digit 4, and not digit 2 as in the human foot; innervation: lateral plantar nerve (Greene 1935); see flexores breves profundi mentioned earlier]	–

(Continued)

TABLE 19.13 (CONTINUED)
Hindlimb Muscles of Adults of Representative Tetrapod Taxa

Amphibia (Caudata): *Ambystoma mexicanum* (Axolotl)	Amphibia (Anura): *X. laevis* (African Clawed Frog)	Reptilia (Testudines): *Trachemys scripta* (Pond Slider)	Reptilia (Lepidosauria): *Timon lepidus* (Ocellated Lizard)	Reptilia (Crocodylia): *Alligator mississippiensis* (American Alligator)	Reptilia (Aves): *Gallus domesticus* (Chicken)	Mammalia (Monotremata): *Ornithorhynchus anatinus* (Platypus)	Mammalia (Marsupialia): *D. virginiana* (Virginia Opossum)	Mammalia (Placentalia): *Rattus norvegicus* (Brown Rat)	Mammalia (Placentalia): *Homo sapiens* (Modern Human)
Abductor digiti minimi (abductor digiti V *sensu* Walthall and Ashley-Ross [2006]) [innervation: fibular nerve (Francis 1934); as in *T. torosa* (Walthall and Ashley-Ross 2006), it extends from the distal end of the fibula to the fibulare, basale V and the base of metatarsal V, not extending to the phalanges of digit 5, probably helping in abducting metatarsal V]	**Abductor digiti minimi** ("abductor brevis dorsalis digiti V" *sensu* Dunlap [1960]) [as reported by Dunlap [1960], in the *R. pipiens*, *Xenopus laevis*, and *E. coqui* specimens we dissected, the "abductor brevis dorsalis digiti V" mainly extends from the fibulare to the proximal portion of metatarsal V]	– [absent according to Walker [1973]]	**Abductor digiti minimi** (abductor digiti quinti *sensu* Russell and Bauer [2008]) [extends from astragalocalcaneum to proximal end of metatarsal V, as is usually the case in other lepidosaurs (Russell and Bauer 2008)]	**Abductor digiti minimi** [mainly to metatarsal IV according to Gadow [1882]]	**Abductor digiti minimi** (abductor digiti IV *sensu* Vanden Berge [1975]) [Vanden Berge 1975: in adult chicken, it extends from tarsometatarsal region to proximal phalanx of digit 4; innervation: lateral plantar nerve (Vanden Berge 1975)]	**Abductor digiti minimi** [extends from calcaneus to lateral side of proximal and middle phalanges of digit 5 and of metatarsal V, being blended with the semitendinosus]	**Abductor digiti minimi** [extends from calcaneus to proximal phalanx of digit 5; authors often refer to two heads of the abductor digiti minimi in marsupials (see, e.g., Figure 16.1 of Lewis [1989]), but in reality, what they designate as the more superficial head is similar to the abductor digiti minimi of, e.g., humans, while what they designate as the more deep head (or often as "calcaneo-metatarsales" muscle, *sensu* Coues [1872] and Stein [1981]) is similar to the opponens digiti minimi of, e.g., apes because it attaches to metatarsal V, not to the proximal phalanx of digit 5; therefore the "deep head" might well correspond instead to a deeper bundle of the flexor brevis profundus 10 (i.e., flexor digiti minimi brevis) that resembles (homoplastically) the opponens digiti minimi, as apparently also happens in the hand of *D. virginiana*]	**Abductor digiti minimi** (to digit 5 but does not reach its phalanges; abductor digiti quinti *sensu* Greene [1935]) [as described by Greene [1935], the abductor digiti minimi runs from the calcaneus to the tuberosity of the metatarsal 5; i.e., it does not reach the phalanges of digit 5; innervation: Lateral plantar nerve (Greene 1935)]	**Abductor digiti minimi** (to digit 5; abductor minimi digiti *sensu* Allen [1897] and Woollard [1925]; abductor quinti digiti *sensu* Osman [1955])
–	–	–	– [but abductor hallucis brevis present in lizards such as chameleons, as described by Mivart [1870] (flexor brevis hallucis) and confirmed by my dissections]	–	–	– [not present as a distinct muscle, see flexores breves profundi mentioned earlier]	**Abductor hallucis** [from navicular bone to proximal phalanx of digit 1]	– [the abductor hallucis is not present as a distinct muscle, it clearly seems to be fused with the "medial head of the flexor hallucis brevis," see flexor hallucis brevis in the following]	**Abductor hallucis**

(*Continued*)

TABLE 19.13 (CONTINUED)
Hindlimb Muscles of Adults of Representative Tetrapod Taxa

Amphibia (Caudata): *Ambystoma mexicanum* (Axolotl)	Amphibia (Anura): *X. laevis* (African Clawed Frog)	Reptilia (Testudines): *Trachemys scripta* (Pond Slider)	Reptilia (Lepidosauria): *Timon lepidus* (Ocellated Lizard)	Reptilia (Crocodylia): *Alligator mississippiensis* (American Alligator)	Reptilia (Aves): *Gallus domesticus* (Chicken)	Mammalia (Monotremata): *Ornithorhynchus anatinus* (Platypus)	Mammalia (Marsupialia): *D. virginiana* (Virginia Opossum)	Mammalia (Placentalia): *Rattus norvegicus* (Brown Rat)	Mammalia (Placentalia): *Homo sapiens* (Modern Human)
–	–	–	– [topologically the abductor digiti minimi of lizards extends from the calcaneus to the metatarsal V, but we think it is homologous to the abductor digiti minimi of humans; i.e., it is not a distinct, nonhomologous "abductor metatarsi quinti"]	–	–	–	– [really absent? see abductor digiti minimi and flexor digiti minimi brevis; there is an abductor metatarsi quinti, but we think it is part of the flexor brevis profundus 10, as is the case in the hand]	– [topologically the abductor digiti minimi of rats goes from the calcaneus to the metatarsal V, but we think it is homologous to the abductor digiti minimi of humans; i.e., it is not a distinct, nonhomologous abductor metatarsi quinti]	– [designated in other primates as abductor ossis metacarpi quinti *sensu* Woollard [1925], abductor ossis metatarsi quinti *sensu* Osman [1955], and abductor ossis metatarsi or Wood's muscle *sensu* Osman [1955]]
Flexores breves superficiales (flexores digitorum breves *sensu* Walthall and Ashley-Ross 2006) [McMurrich [1906] states that urodeles, as most tetrapods, and lizards almost always have a superficial and a deep layer of flexores breves superficiales; innervation: plantar nerve loop (Francis 1934); as in *T. torosa* (Walthall and Ashley-Ross 2006), there are five muscles extending from dorsal side of the plantar fascia to metatarsals I-V and digits 2–4 (the first muscle only goes to metatarsal I, not to digit 1), probably helping in flexing the digits]	**Flexores breves superficiales** [as reported by Dunlap [1960], in the *R. pipiens*, *X. laevis*, and *E. coqui* specimens we dissected, the flexores breves superficiales extend from the ligamentum calcanei to the plantar aponeurosis and then giving rise to the superficial flexor tendons of digits 3–5; see contrahentium caput longum]	**Flexores breves superficiales** (flexor digitorum communis sublimis *sensu* Ogushi 1938 and Walker 1973) [Walker 1973: in *Trachemys*, *scripta* the flexores breves superficiales connect the flexor plate of the flexor digitorum longus to the tibial and fibular side of the proximal phalanges of digits 1–4, and that is why they are sometimes designated the perforatus group of muscles]	**Flexores breves superficiales** (flexores digitores breves *sensu* Russell and Bauer 2008) [McMurrich 1906: states that lizards, as most tetrapods, almost always have a superficial and a deep layer of flexores breves superficiales; extends from aponeurosis of the deep head of the gastrocnemius externus to digits 1–3, as is usually the case *Sphenodon* (Russell and Bauer 2008)]	**Flexores breves superficiales** (flexores digitorum brevis *sensu* Haughton 1868) [Haughton 1868: in *A. mississippiensis*, the flexor digitorum brevis; consists of three muscles, i.e., a "flexor 2 perforatus" extends from the calcaneum and plantaris to the penultimate phalanx of digit 2, a "flexor 3 perforatus" to the penultimate phalanx of digit 3, and a "flexor 4 perforans" to the distal phalanx of digit 4; Fechner 2009: in crocodylians, including *Alligator*, the flexores breves superficiales usually attach to digits 1–4; four muscles to digits 1–4 according to Suzuki et al. 2011]	**Flexores breves superficiales** (flexor perforans et perforatus digiti II and III + flexor perforatus digiti II, III and IV *sensu* Vanden Berge [1975]) [Vanden Berge 1975: in adult chicken, there is a superficial layer ("flexor perforatus digiti II, III, and IV") extending from femur, patella, proximal tibia, and associated structures to the proximal phalanges of digits 2, 3, and 4 and sending tendons to the proximal, middle, and distal interphalangeal joints of digit 4; these three muscles clearly correspond to flexores breves superficiales to digits 2–4, and their tendons are perforated by both the tendons of the intermediate layer and the tendons of the flexor digitorum longus; the intermediate layer ("flexor perforans and perforatus digiti II and III") extends from the distal femur, patellar tendon, fibula, and tibia to the proximal interphalangeal joint of digit 2 and the middle interphalangeal joint of digit 3; the tendons of these two muscles perforating those of the	**Flexor digitorum brevis** (flexor digiti quarti brevis + flexor digitorum intermedii *sensu* Gambaryan et al. [2002]) [extends from calcaneus to proximal phalanx of digits 4 ("flexor digiti quarti brevis") and from tendon of flexor digitorum longus to proximal phalanges of digits 2 and 3 ("flexor digitorum intermedii")]	**Flexor digitorum brevis** [extends from tendon of flexor digitorum longus to middle phalanges of digits 2–5, the configuration being thus similar to humans, with the main difference that the muscle does not originate from calcaneus, but instead from the tendon of flexor digitorum longus; this resembles the configuration seen in the upper limb, where the flexor digitorum profundus and flexor digitorum superficialis are also more blended to each other than they are in humans]	**Flexor digitorum brevis** [as described by Greene [1935], the flexor digitorum brevis extends from the distal tendon of the plantaris muscle to the second phalanges of digits 2–4, the tendons to each of these digits being bifurcated and the tendons of the flexor digitorum longus for digits 2–4 passing between the tendons of the flexor digitorum brevis; innervation: medial plantar nerve (from tibial nerve) (Greene 1935)]	**Flexor digitorum brevis** (flexor brevis digitorum *sensu* Woollard [1925] and Osman [1955]; flexor perforates *sensu* Osman [1955]) [McMurrich [1906] states that humans, as most tetrapods, have a superficial and a deep layer of flexores breves superficiales, but that these muscles become fused early in human ontogeny]

(Continued)

TABLE 19.13 (CONTINUED)
Hindlimb Muscles of Adults of Representative Tetrapod Taxa

Amphibia (Caudata): *Ambystoma mexicanum* (Axolotl)	Amphibia (Anura): *X. laevis* (African Clawed Frog)	Reptilia (Testudines): *Trachemys scripta* (Pond Slider)	Reptilia (Lepidosauria): *Timon lepidus* (Ocellated Lizard)	Reptilia (Crocodylia): *Alligator mississippiensis* (American Alligator)	Reptilia (Aves): *Gallus domesticus* (Chicken)	Mammalia (Monotremata): *Ornithorhynchus anatinus* (Platypus)	Mammalia (Marsupialia): *D. virginiana* (Virginia Opossum)	Mammalia (Placentalia): *Rattus norvegicus* (Brown Rat)	Mammalia (Placentalia): *Homo sapiens* (Modern Human)
					superficial layer and being at the same time perforated by the tendons of the flexor digitorum longus; innervation: lateral plantar nerve (Vanden Berge 1975); these two intermediate muscles mainly have not only characteristics of flexores breves superficiales (superficial to flexor digitorum longus and perforated by its tendons and usually deeply fused with the superficial layer) but also a few features of the flexor digitorum longus (perforating the superficial layer and being sometimes blended with the flexor digitorum longus as well); therefore, they very likely include a part of the flexores breves profundi; but they may also include a part of the flexor digitorum longus; be that as it may, it is sure that the at least the superficial muscle includes flexores breves superficiales; i.e., in the chicken hindlimb, these muscles have migrated proximally, as did the muscles of the forelimb of chickens, thus showing a case of parallelism and apparently integration between the two limbs in the same taxon and a case of parallelism with the forelimb (but not the hindlimb) of mammals, where the flexores breves superficiales also migrated proximally to form the flexor digitorum superficiales of these tetrapods]				

(Continued)

TABLE 19.13 (CONTINUED)
Hindlimb Muscles of Adults of Representative Tetrapod Taxa

Amphibia (Caudata): *Ambystoma mexicanum* (Axolotl)	Amphibia (Anura): *X. laevis* (African Clawed Frog)	Reptilia (Testudines): *Trachemys scripta* (Pond Slider)	Reptilia (Lepidosauria): *Timon lepidus* (Ocellated Lizard)	Reptilia (Crocodylia): *Alligator mississippiensis* (American Alligator)	Reptilia (Aves): *Gallus domesticus* (Chicken)	Mammalia (Monotremata): *Ornithorhynchus anatinus* (Platypus)	Mammalia (Marsupialia): *D. virginiana* (Virginia Opossum)	Mammalia (Placentalia): *Rattus norvegicus* (Brown Rat)	Mammalia (Placentalia): *Homo sapiens* (Modern Human)
–	**Lumbricales** ("lumbricales breves" plus "lumbricales longi" plus "lumbricalis longissimus digiti IV" *sensu* Dunlap [1960]) [as reported by Dunlap [1960], in the *R. pipiens*, *X. laevis,* and *E. coqui* specimens we dissected, the lumbricales breves mainly extend from the aponeurosis plantaris to the metatarsophalangeal joint of digits 1–5, usually by lateral and medial muscles to each digit, but *Xenopus* does not have any head of a "lumbricalis brevis hallucis" attaching to digit 1, and *Rana* and *Eleutherodactylus* only have a muscle connecting to digit 2 and one to digit 3; in the three taxa, the lumbricales longi mainly extend from the aponeurosis plantaris to the second phalanx of digit 3, the interphalangeal joint capsule of digit 4, the second phalanx of digit 4 (this is usually considered to be a separate lumbricalis longissimus digiti IV) and the proximal interphalangeal joint of digit 5]	**Lumbricales** (*sensu* Walker [1973]) [Walker 1973: in *Trachemys scripta*, the lumbricales extend from the dorsal surface of the flexor plate to the tibial side of the penultimate phalanx of digits 2–5]	**Lumbricales** [as in *Iguana* and *Sphenodon,* there are two lumbricales going to digits 3 and 4 (Russell and Bauer 2008)]	**Lumbricales** [apparently four muscles, some of them with more than one bundle, according to Suzuki et al. 2011]	**Lumbricales** [Vanden Berge 1975: in adult chicken, there are two lumbricales extendng from the tendons of the flexor digitorum longus to the fibrocartilaginous joint pulley of digits 3 and 4; innervation: lateral plantar nerve (Vanden Berge 1975)]	– [seem to be absent as distinct muscles in monotremes: see, e.g., Coues 1870 and Gambaryan et al. 2002]	**Lumbricales** [four muscles extending from tendons of flexor digitorum longus to middle phalanges of digits 2–5, respectively]	**Lumbricales** (abductor digiti quinti *sensu* Greene [1935]) [Greene [1935] describes more than four structures, but we clearly found four lumbricales extending from the tendons of the flexor digitorum longus to the tibial sides of the proximal phalanx of digits 2–5 and then to soft tissues; i.e., the extensor expansions of these digits; innervation: lateral plantar nerve to lumbricales 2–4 and medial plantar nerve to lumbrical 1 (Greene [1935])]	Lumbricales

(Continued)

TABLE 19.13 (CONTINUED)
Hindlimb Muscles of Adults of Representative Tetrapod Taxa

Amphibia (Caudata): *Ambystoma mexicanum* (Axolotl)	Amphibia (Anura): *X. laevis* (African Clawed Frog)	Reptilia (Testudines): *Trachemys scripta* (Pond Slider)	Reptilia (Lepidosauria): *Timon lepidus* (Ocellated Lizard)	Reptilia (Crocodylia): *Alligator mississippiensis* (American Alligator)	Reptilia (Aves): *Gallus domesticus* (Chicken)	Mammalia (Monotremata): *Ornithorhynchus anatinus* (Platypus)	Mammalia (Marsupialia): *D. virginiana* (Virginia Opossum)	Mammalia (Placentalia): *Rattus norvegicus* (Brown Rat)	Mammalia (Placentalia): *Homo sapiens* (Modern Human)
–	– [as reported by Dunlap [1960], in *Xenopus*, the abductor praehallucis is absent, but in *Rana* and *Eleutherodactylus*, it extends from the aponeurosis plantaris to the prehallux]	–	–	–	–	–	–	–	–
–	**Abductor brevis plantaris hallucis** [as reported by Dunlap [1960], in *Rana* and *Xenopus*, the abductor brevis plantaris hallucis extends from the aponeurosis plantaris to metatarsal I, but in *Eleutherodactylus*, the muscle is absent; see abductor praehallucis; this muscle was probably plesiomorphically absent in anurans, as it is apparently absent in the most plesiomorphic anuran genera (e.g., Dunlap 1960)]	–	–	–	–	–	–	–	–
–	–	–	–	–	–	–	–	– [completely absent]	**Adductor hallucis** [caput obliquum and caput transversum]
–	–	–	–	–	–	–	–	–	**Adductor hallucis accessorius** (interosseous plantaris hallucis) [not clear if it is usually present, or not, in modern humans: see adductor hallucis]

(*Continued*)

TABLE 19.13 (CONTINUED)
Hindlimb Muscles of Adults of Representative Tetrapod Taxa

Amphibia (Caudata): *Ambystoma mexicanum* (Axolotl)	Amphibia (Anura): *X. laevis* (African Clawed Frog)	Reptilia (Testudines): *Trachemys scripta* (Pond Slider)	Reptilia (Lepidosauria): *Timon lepidus* (Ocellated Lizard)	Reptilia (Crocodylia): *Alligator mississippiensis* (American Alligator)	Reptilia (Aves): *Gallus domesticus* (Chicken)	Mammalia (Monotremata): *Ornithorhynchus anatinus* (Platypus)	Mammalia (Marsupialia): *D. virginiana* (Virginia Opossum)	Mammalia (Placentalia): *Rattus norvegicus* (Brown Rat)	Mammalia (Placentalia): *Homo sapiens* (Modern Human)
Contrahentes pedis (contrahentes digitorum *sensu* Walthall and Ashley-Ross [2006]) [innervation: plantar nerve loop (Francis 1934); as in *T. torosa* (Walthall and Ashley-Ross 2006), there are five muscles extending from the tendon of the contrahentium caput longum and tarsal bones to the proximal phalanx of digits 1–5]	**Contrahentes pedis** [as reported by Dunlap [1960], in *Xenopus*, the contrahentes mainly extend from the tarsal region to the metatarsophalangeal joints of digits 1 and 2 and to the metatarsal V; in *Rana* and *Eleutherodactylus*, only the muscle to digit 1 is present; all the contrahentes, i.e., to all five digits, are present in plesiomorphic anuran taxa such as *Leiopelma* (Dunlap 1960); see also opponens hallucis in the following]	**Contrahentes pedis** (contrahentes digitorum *sensu* Ribbing [1938]; i.e., part of interossei plantares *sensu* Walker [1973]) [see flexores breves profundi]	**Contrahentes pedis** (contrahentes plus adductor digiti quinti *sensu* Russell and Bauer [2008]) [mainly extending from metatarsals to digits 1–5, as is usually the case in lepidosaurs including *Sphenodon* which has muscles to at least digits 1–4 (Russell and Bauer 2008)]	**Contrahentes pedis** [at least one muscle to digit 1 according to Suzuki et al. [2011]; Gadow [1882] suggests that there are five muscles to digits 1–4, but it is not sure that these are really contrahentes]	**Contrahentes pedis** (adductor digiti II *sensu* Vanden Berge [1975]) [Vanden Berge 1975: in adult chicken, the "adductor digiti II" extends from tarsometatarsal region to proximal phalanx of digit 2, being the only contrahens of the foot; innervation: lateral plantar nerve (Vanden Berge 1975)]	**Contrahentes pedis** (adductor hallucis brevis + contrahentes *sensu* Gambaryan et al. [2002]) [contrahens 1 extends from contrahens aponeurosis to lateral side of proximal phalanx of digit 1; contrahens 2 extends from contrahens aponeurosis to medial side of metatarsal II and proximal and middle phalanges of digit 2; contrahens 3 extends from contrahens aponeurosis to medial side of metatarsal III and proximal and middle phalanges of digit 3; contrahens 4 extends from contrahens aponeurosis to medial side of metatarsal IV and proximal and middle phalanges of digit 4; contrahens 5 extends from contrahens aponeurosis to medial side of metatarsal V and proximal and middle phalanges of digit 5]	**Contrahentes pedis** [that of digit 1 extends from metatarsal III to proximal phalanx of digit 1; that of digit 2 extends from metatarsal III to proximal phalanx of digit 2; that of digit 4 extends from metatarsal III to proximal phalanx of digit 4; that of digit 5 extends from metatarsal III to proximal phalanx of digit 5]	**Contrahentes pedis** (includes adductor indicis *sensu* Greene [1935]) [innervation: medial plantar nerve (from tibial nerve) (Greene 1935)]	– [contrahentes pedis absent as a group, only the adductor hallucis and the adductor hallucis accessorius to digit 1 are present]
–	–	–	–	–	–	**Flexor digiti minimi brevis** (abductor metatarsi V *sensu* Gambaryan et al. [2002]; lateral head of flexor brevis quinti digiti *sensu* Coues [1870]) [from calcaneus to lateral side of proximal phalanx of digit 5 and of metatarsal 5, being blended with the semitendinosus]	**Flexor digiti minimi brevis** (includes calcaneo-metatarsales muscle *sensu* Coues [1872] and Stein [1981] and "abductor ossis metatarsi quinti digiti" *sensu* McMurrich [1906]) [extends from calcaneus and cuboid to base of metatarsal V and to proximal phalanx of digit 5; as noted earlier, the calcaneo-metatarsales muscle *sensu* Coues [1872] and Stein [1981] and "abductor ossis metatarsi quinti digiti" muscle *sensu* McMurrich [1906] seem to correspond to a deep head of the flexor brevis profundus 10]	**Flexor digiti minimi brevis** (flexor digiti quinti brevis *sensu* Greene [1935]) [as described by Greene [1935], the flexor digiti minimi brevis extends from the cuboid to the fibular side of the proximal phalanx of digit 5; innervation: lateral plantar nerve (Greene 1935)]	**Flexor digiti minimi brevis** (flexor brevis minimi digiti *sensu* Murie and Mivart [1872]; flexor digiti quinti brevis *sensu* Osman [1955])

(Continued)

TABLE 19.13 (CONTINUED)
Hindlimb Muscles of Adults of Representative Tetrapod Taxa

Amphibia (Caudata): *Ambystoma mexicanum* (Axolotl)	Amphibia (Anura): *X. laevis* (African Clawed Frog)	Reptilia (Testudines): *Trachemys scripta* (Pond Slider)	Reptilia (Lepidosauria): *Timon lepidus* (Ocellated Lizard)	Reptilia (Crocodylia): *Alligator mississippiensis* (American Alligator)	Reptilia (Aves): *Gallus domesticus* (Chicken)	Mammalia (Monotremata): *Ornithorhynchus anatinus* (Platypus)	Mammalia (Marsupialia): *D. virginiana* (Virginia Opossum)	Mammalia (Placentalia): *Rattus norvegicus* (Brown Rat)	Mammalia (Placentalia): *Homo sapiens* (Modern Human)
–	–	–	–	–	–	**Flexor hallucis brevis** (medial head of flexor hallucis brevis *sensu* Coues [1870]) [extends from metatarsal I to medial side of proximal phalanx of digit 1]	**Flexor hallucis brevis** (medial head of flexor hallucis brevis *sensu* Stein [1981], corresponding to flexor brevis profundus 1 *sensu* the present work) [extends from medial cuneiform to proximal phalanx of digit 1]	**Flexor hallucis brevis** (medial head of flexor hallucis brevis *sensu* Greene [1935], corresponding to flexor brevis profundus 1 *sensu* the present work) [as described by Greene [1935], the flexor hallucis brevis (which corresponds only to the "medial head of the flexor hallucis brevis" *sensu* Greene [1935], the "lateral head of the flexor hallucis brevis" *sensu* Greene corresponding to the flexor brevis profundus 2 and thus being included in the flexores breves profundi mentioned earlier) extends from the navicular to the tibial side of the proximal phalanx of digit 1; innervation: medial plantar nerve (Greene 1935)]	**Flexor hallucis brevis** (part of flexor brevis hallucis *sensu* Murie and Mivart [1872], of flexor brevis hallucis superficialis *sensu* Woollard [1925], and of flexor brevis superficialis *sensu* Osman 1955) [caput mediale only, see preceding text]

(*Continued*)

TABLE 19.13 (CONTINUED)
Hindlimb Muscles of Adults of Representative Tetrapod Taxa

Amphibia (Caudata): *Ambystoma mexicanum* (Axolotl)	**Amphibia (Anura): *X. laevis* (African Clawed Frog)**	**Reptilia (Testudines): *Trachemys scripta* (Pond Slider)**	**Reptilia (Lepidosauria): *Timon lepidus* (Ocellated Lizard)**	**Reptilia (Crocodylia): *Alligator mississippiensis* (American Alligator)**	**Reptilia (Aves): *Gallus domesticus* (Chicken)**	**Mammalia (Monotremata): *Ornithorhynchus anatinus* (Platypus)**	**Mammalia (Marsupialia): *D. virginiana* (Virginia Opossum)**	**Mammalia (Placentalia): *Rattus norvegicus* (Brown Rat)**	**Mammalia (Placentalia): *Homo sapiens* (Modern Human)**
–	– (opponens hallucis *sensu* Dunlap [1960]) [as reported by Dunlap [1960], in *Xenopus*, the opponens hallucis is absent, but in the two other genera, it extends from the tarsal region to the metatarsal I; regarding the opponens hallucis, which is present in most anurans but not in *Xenopus*, Dunlap [1960] suggested that it derived from the contrahentes, but the hypothesis of authors such as Perrin [1892] that it derived instead from the flexores breves profundi cannot be discarded, particularly because there is no flexor brevis profundus going directly to metatarsal I]	–	–	–	–	–	–	–	– [opponens hallucis is absent at all stages of normal human ontogeny, although it may be present as anomaly/variation in some humans, see notes about opponens digiti minimi]
–	–	–	–	–	–	–	– [see flexor digiti minimi brevis mentioned earlier; seems to be homoplastically present, e.g., in *D. virginiana*, being part of the flexor digiti minimi brevis]	–	– (opponens minimi digiti pedis *sensu* Murie and Mivart [1872]) [opponens digiti minimi is usually absent in humans]
–	–	–	–	–	–	–	–	–	**Interossei plantares** (plantar interossei *sensu* Allen [1897])
–	–	–	–	–	–	– [see flexores breves profundi and intermetatarsales]	– [see flexores breves profundi and intermetatarsales]	– [see flexores breves profundi and intermetatarsales]	**Interossei dorsales** (dorsal interossei *sensu* Allen [1897])

Note: The nomenclature of the muscles shown in bold follows that of the text; in order to facilitate comparisons, in some cases, names often used by other authors to designate a certain muscle/bundle are given in front of that muscle/bundle, between round brackets; additional comments are given between square brackets. Data compiled from evidence provided by our own dissections and comparisons and by an overview of the literature (see text, Tables 19.1 through 19.22, and Figures 16.2 through 16.4 and 19.1 through 19.4).

TABLE 19.14

Diagram Illustrating the "Clear Topological Correspondences" between the Dorsal Leg and Forearm Muscles of Adults of Representative Tetrapod Taxa

	Ambystoma mexicanum **[6 dorsal/extensor leg muscles/muscle groups: 6 (100%) seem to 'correspond' directly to forelimb muscles/muscle groups]**	***Timon lepidus*** **[6 dorsal/extensor leg muscles/muscle groups: 4 (67%) seem to 'correspond' directly to forelimb muscles/muscle groups]**	***Rattus norvegicus*** **[8 dorsal/extensor leg muscles/muscle groups: 3 (38%) seem to 'correspond' directly to forelimb muscles/muscle groups; a remarkable example of integration/homoplasy is the fact that the short extensors to digits 2 and 3 are fused to form an undivided muscle; i.e., the forelimb muscle extensor indicis and the hindlimb muscle extensor digitorum brevis (NB: the short extensors of digits 4 and 5 are present in both the forelimb and the hindlimb but topologically there are very different and thus do not 'clearly correspond'; i.e., the fibularis digiti quarti and fibularis digi quinti of the hindlimb are very different from the extensor digiti minimi and extensor digiti quarti of the forelimb]**	***Homo sapiens*** **[8 dorsal/extensor leg muscles/muscle groups: 3 (38%) seem to 'correspond' directly to forelimb muscles/muscle groups]**
Long extensors	---	---	Extensor hallucis longus (d. 1) *[Extensor pollicis longus]*	Extensor hallucis longus (d. 1) *[Extensor pollicis longus]*
	Extensor digitorum longus ('extensor digitorum communis'; m.I-V) [Extensor digitorum]	Extensor digitorum longus (m.II-III, not to digits) [Extensor digitorum]	Extensor digitorum longus (d. 2-5) [Extensor digitorum]	Extensor digitorum longus (d.2-5) [Extensor digitorum]
	---	---	---	Fibularis tertius (m.V)
	Extensor tarsi tibialis [Extensor carpi radialis]	Tibialis anterior [Extensor antebrachii et carpi radialis]	Tibialis anterior	Tibialis anterior
	Extensor cruris tibialis [Supinator]	---	---	---
	Extensor cruris et tarsi fibularis [Extensor antebrachii et carpi ulnaris]	Fibularis longus	Fibularis longus	Fibularis longus
	---	Fibularis brevis	Fibularis brevis	Fibularis brevis
Short extensors	---	---	Fibularis digiti quarti (m.IV in rats and d.4 in mice)	---
	---	---	Fibularis digiti quinti (m.V in rats and d.5 in mice)	---
	Extensores digitorum breves (4 muscles to d.2-5) [Extensores digitorum breves]	Extensores digitorum breves (4 muscles to d.2-5 + part short ex. d.1) [Extensores digitorum breves]	Extensor digitorum brevis (2 muscles to d.2-3) [Extensor indicis]	Extensor digitorum brevis (3 muscles to d.2-4)
	---	---	--- (but ex.ha.br. found in various non-primate mammals)	Extensor hallucis brevis (d.1; short ext. d.1) *[Extensor pollicis brevis]*
	Ab. et ex. digiti 1 (d.1; ab. + part short ex. d.1) [Ab. et ex. digiti 1]	Ab. et ex. digiti 1 (d.1; ab. + part short ex. d.1) [Abductor pollicis longus]	---	---

Note: Data compiled from evidence provided by our own dissections and comparisons and from a review of the literature. The black arrows indicate hypotheses that are most strongly supported by the evidence available; the gray arrows indicate alternative hypotheses that are supported by some data but, overall, are not as strongly supported by the evidence as are the hypotheses indicated by the black arrows. Below the names of those hindlimb muscles/muscle groups that have clear, direct topological forelimb "equivalents" in the same taxon, we provide (in bold, between square brackets) the names of those forelimb equivalents. In those cases in which those equivalents clearly have a different ontogeny, these are shown in italics. ab., abductor; d., digits; ex., extensor; m, metatarsals (attachments are in general only shown for muscles that often insert onto digits).

TABLE 19.15
Diagram Illustrating the Clear Topological Correspondences between the Ventral Leg and Forearm Muscles of Adults of Representative Tetrapod Taxa

	Ambystoma mexicanum [6 ventral/flexor leg muscles/muscle groups: 6 (100%) seem to 'correspond' directly to forelimb muscles/muscle groups]	*Timon lepidus* [6 ventral/flexor leg muscles/muscle groups: 4 (67%) seem to 'correspond' directly to forelimb muscles/muscle groups]	*Rattus norvegicus* [7 ventral/flexor leg muscles/muscle groups: 4 (57%) seem to 'correspond' directly to forelimb muscles/muscle groups]	*Homo sapiens* [8 ventral/flexor leg muscles/muscle groups: 5 (62%) seem to 'correspond' directly to forelimb muscles/muscle groups; 2 further examples of remarkable homoplasy, the popliteus and the pronator teres of mammals, they are said to correspond to each other but clearly do not correspond because the popliteus derived from the interosseus cruris and is very similar to the pronator accessorius of the forearm of lizards; the flexor carpi radialis and the tibialis posterior are also said to 'correspond' to each other but being ontogenetically very different because the tibialis posterior corresponds to deep muscle 'pronator profundus' of amphibians and reptiles, which is ontogenetically similar to the palmaris profundus 1 of the forelimb as can be also topologically seen in amphibians; so there are two interesting cases where a 'clear topological correspondence' seen in certain taxa gets less clear in other taxa where there is another apparently 'clear topological correspondence' (in this case mainly due to the absence of palmaris profundus 1 and pronator accessorius in the forearm of mammals, the tibialis posterior and the popliteus apparently 'correspond' to the flexor carpi radialis and pronator teres of the mammalian forearm, respectively]
Long flexors	--- (fl.cr. ta.ti. seems included in ischioflexorius)	--- (fl.cr. ta.ti. seems included in fl.ti.ex. and ga.in)	--- (fl.cr. ta.ti. seems included in ha.fa. and ga.)	--- (fl.cr. ta.ti. seems included in ha.fa. and ga.)
	---	Gastrocnemius internus ('femorotibial')	Gastrocnemius (medial and lateral heads)	Gastrocnemius (medial and lateral heads)
	---	Gastrocnemius externus ('femoral'; sup. and deep heads)	Plantaris *[Palmaris longus]*	Plantaris *[Palmaris longus]*
	---	---	Soleus	Soleus
	Fl. dig. com. (d.1-5; includes fl.c.ta.fi.) [Flexor digitorum communis]	Flexor digitorum longus (d. 1-5) [Flexor digitorum longus]	Flexor digitorum longus (d. 1-5) [Flexor digitorum profundus]	Flexor digitorum longus (d. 2-5) [Flexor digitorum profundus]
	---	---	---	Flexor hallucis longus (d. 1) [Flexor pollicis longus]
	Flexor accessorius medialis [Flexor accessorius medialis]	--- ('flexor accessorius' is distal head of flexor digitorum longus)	Quadratus plantae	Quadratus plantae
	Flexor accessorius lateralis [Flexor accessorius lateralis]	---	---	---
	Contrahentium caput longum [Contrahentium caput longum]	---	---	---
		Popliteus [Pronator accessorius]	Popliteus *[Pronator teres]*	Popliteus *[Pronator teres]*
	Interosseus cruris [Pronator quadratus]	Interosseus cruris [Pronator quadratus]	---	---
	Tibialis posterior ('pronator profundus') [Palmaris profundus 1]	Tibialis posterior ('pronator profundus') [Palmaris profundus 1]	Tibialis posterior *[Flexor carpi radialis]*	Tibialis posterior *[Flexor carpi radialis]*

Note: Data compiled from evidence provided by our own dissections and comparisons and from a review of the literature. The black arrows indicate hypotheses that are most strongly supported by the evidence available; the gray arrows indicate alternative hypotheses that are supported by some data but, overall, are not as strongly supported by the evidence as are the hypotheses indicated by the black arrows. Below the names of those hindlimb muscles/muscle groups that have clear, direct topological forelimb equivalents in the same taxon, we provide (in bold, between square brackets) the names of those forelimb equivalents. In those cases in which those equivalents clearly have a different ontogeny, these are shown in italics. d., digits; fl.cr.ta.fi., flexor cruris tarsi fibularis; fl.cr.ta.ti., flexor cruris tarsi tibialis; fl.dig. com., flexor digitorum communis; fl. ti.ex., flexor tibialis externus; ga., gastrocnemius; ga.in., gastrocnemius internus; ha.fa., fasical insertion of hamstring muscles; sup. superficial.

TABLE 19.16
Diagram Illustrating the Clear Topological Correspondences between the Hand and Foot Muscles of Adults of Representative Tetrapod Taxa

	Ambystoma mexicanum **[7 foot muscles/muscle groups: 7 (100%) seem to 'correspond' directly to forelimb muscles/muscle groups]**	***Timon lepidus*** **[6 foot muscles/muscle groups: 6 (100%) seem to 'correspond' directly to forelimb muscles/muscle groups]**	***Rattus norvegicus*** **[9 foot muscles/muscle groups: 9 (100%) seem to 'correspond' directly to forelimb muscles/muscle groups]**	***Homo sapiens*** **[11 foot muscles/muscle groups: 11 (100%) seem to 'correspond' directly to forelimb muscles/muscle groups; further examples of parallelism are that in the human foot, the adductor pollicis accessorius ('volaris primus of Henle': Bello-Hellegouarch 2012) is mirrored by the 'interosseus plantaris hallucis' *sensu* Cihak (1972), and that both the transverse and oblique heads of the adductor pollicis are mainly derived ontogenetically from the anlage of contrahens 1, while in the foot the oblique head of the adductor hallucis is derived from the anlage of contrahens 1 but the transverse head is derived from a different, neomorphic anlage of the contrahens layer (e.g., Cihak 1972)]**
Intrinsic foot muscles	Flexores breves superficiales (5 muscles to d.1-5) [Flexores breves superficiales]	Flexores breves superficiales (3 muscles to d.1-3) [Flexores breves superficiales]	Flexor digitorum brevis (d.2-4) [Flexor digitorum superficialis]	Flexor digitorum brevis (d.2-5) [Flexor digitorum superficialis]
	---	Lumbricales (2 muscles, to d.3-4) [Lumbricales]	Lumbricales (4 muscles, to d.2-4) [Lumbricales]	Lumbricales (4 muscles to d.2-4) [Lumbricales]
	Abductor digiti minimi (m.V) [Abductor digiti minimi]	**Abductor digiti minimi (m.V)** [Abductor digiti minimi]	**Abductor digiti minimi (d.5)** [Abductor digiti minimi]	**Abductor digiti minimi (d.5)** [Abductor digiti minimi]
	---	---	--- (*abductor hallucis* apparently fused with flexor hallucis brevis)	Abductor hallucis (d.1) [Abductor pollicis brevis]
	---	---	---	Adductor hallucis accessorius (interosseus plantaris hallucis) (d.1) [Adductor pollicis accessorius, or 'volaris primusof Henle']
	Contrahentes pedis (5 muscles to d.1-5) [Contrahentes digitorum]	Contrahentes pedis (5 muscles to d.1-5) [Contrahentes digitorum]	Contrahentes pedis (2 muscles, to d.2,5) [Contrahentes digitorum]	---
	---	---	---	Adductor hallucis (oblique and transverse heads) [*Adductor pollicis*]
	---	---	Flexor digiti minimi brevis (= f.b.p. 10, to d.5) [Flexor digiti minimi]	Flexor digiti minimi brevis (= f.b.p. 10, to d.5) [Flexor digiti minimi]
	---	---	Flexor brevis profundus 2 (to fibular side of d.1) [Flexor brevis profundus 2]	Flexor brevis profundus 2 (fibular side of d.1, = deep head fpb) [Flexor brevis profundus 2, = lateral head fhb]
	Flexores digitorum minimi (4 muscles to d.2-5) [Flexores digitorum minimi]			
	---	---	Flexor hallucis brevis (= f.b.p. 1, to tibial side of d.1) [Flexor pollicis brevis]	Flexor hallucis brevis (= f.b.p. 1, tibial side of d.1) [Flexor pollicis brevis]
	Interphalangei digiti 4 (2 muscles to d.4) [Interphalangei]	---	---	---
	Flexores breves profundi (10 muscles to d.1-5) [Flexores breves profundi]	Flexores breves profundi (10 muscles to d.1-5) [Flexores breves profundi]	Flexores breves profundi (3,4,5,6,7,8,9; i.e.,7 muscles to d.2-5) [Flexores breves profundi]	Interossei plantares (3 f.b.p. muscles to d.3,4,5) [Interossei palmares]
	---	---	---	Interossei dorsales (4 f.b.p. muscles + 4 interm. = 4 muscles, to d.2-4) [Interossei dorsales]
	Intermetatarsales (4 muscles connecting m.1-5) [Intermetacarpales]	Intermetatarsales (4 muscles connecting m.1-5) [Intermetacarpales]	Intermetatarsales (4 muscles connecting d.1-5) [Intermetacarpales]	---

Note: Data compiled from evidence provided by our own dissections and comparisons and from a review of the literature. The black arrows indicate hypotheses that are most strongly supported by the evidence available; the gray arrows indicate alternative hypotheses that are supported by some data but, overall, are not as strongly supported by the evidence as are the hypotheses indicated by the black arrows. Below the names of those hindlimb muscles/muscle groups that have clear, direct topological forelimb equivalents in the same taxon, we provide (in bold, between square brackets) the names of those forelimb equivalents. In those cases in which those equivalents clearly have a different ontogeny, these are shown in italics. The muscles shown in bold are stressing the fact that the LCA of tetrapods probably only had a muscle, at the maximum, that directly corresponds to an individual muscle present in modern humans, i.e., the adductor digiti minimi. d., digits; sup. superficial. d., digits; f.b.p., flexores breves pronfudi; interm., intermetatarsales; m, metatarsal.

TABLE 19.17
Diagram Illustrating the Clear Topological Correspondences between the Dorsal Leg and Forearm Muscles of Adults of Representative Urodele and Anuran Taxa

	Ambystoma mexicanum [6 dorsal/extensor leg muscles/muscle groups: 6 (100%) seem to 'correspond' directly to forelimb muscles/muscle groups]	*Xenopus laevis* [9 dorsal/extensor leg muscles/muscle groups: 3 (33%) seem to 'correspond' directly to forelimb muscles/muscle groups; i.e. many of the 'clear topological correspondences' have been lost and the only muscles/muscle groups that can be easily 'matched' to each other are the extensor digitorum, the dorsometacarpales and the extensores digitorum breves; one of the reasons leading to the topological difference between the dorsal leg and forearm arm muscles of anurans concerns the fact that in the hindlimb the highly elongated tibiale and fibulare somehow functionally replaces the ulna and radius and thus some leg muscles moved distally to the tarsal region; however, there are precisely 8 muscles/muscle groups in the dorsal side of the forearm and moreover the new two muscles in the forearm derive from the muscles that 'correspond' to the leg muscles that also gave rise to the two new muscles of the leg; i.e., the brachioradialis derives from the extensor carpi radialis and the anconeus derives from the extensor antebrachii et carpi ulnaris, in a further remarkable example of integration/homoplasy; another example is the fact that both both in the anuran forelimb and the anuran hindlimb there is a dorsometatarsales/metacarpales series and there is no abductor et extensor 1 because the short extensor to digit 1 is a separate muscle]
Long extensors	Extensor digitorum longus ('extensor digitorum communis'; m.I-V) [Extensor digitorum]	Extensor digitorum longus (to short extensors of digits 3 and 4) [Extensor digitorum]
	---	Tarsalis anticus
	Extensor tarsi tibialis [Extensor carpi radialis]	**Tibialis anticus brevis**
	Extensor cruris tibialis [Supinator]	Extensor cruris tibialis ('extensor cruris brevis')
	Extensor cruris et tarsi fibularis [Extensor antebrachii et carpi ulnaris]	**Peroneus**
	---	Tibialis anticus longus
Short extensors	---	Dorsometatarsales [Dorsometacarpales]
	Extensores digitorum breves (4 muscles to d.2-5) [Extensores digitorum breves]	Extensores digitorum breves ('e.b.s.+e.b.m.+e.b.p.'; 5 muscles to d.1-5) [Extensores digitorum breves]
	Abductor et extensor digiti 1 (d.1; abductor + part short extensor d.1) [Abductor et extensor digiti 1]	Abductor brevis dorsalis hallucis (m.I)

Note: Data compiled from evidence provided by our own dissections and comparisons and from a review of the literature. The black arrows indicate hypotheses that are most strongly supported by the evidence available; the gray arrows indicate alternative hypotheses that are supported by some data but, overall, are not as strongly supported by the evidence as are the hypotheses indicated by the black arrows. Below the names of those hindlimb muscles/muscle groups that have clear, direct topological forelimb equivalents in the same taxon, we provide (in bold, between square brackets) the names of those forelimb equivalents. In those cases in which those equivalents clearly have a different ontogeny, these are shown in italics. As can be seen in the names highlighted in bold, there are some cases in which although the muscle that is designated with a certain name in anurans does not directly correspond to the muscle designated with a similar name in humans (but not the same; hence, we can use the commonly used anuran names), both the anuran and human muscles derive from a similar anlage/ancestral muscle: the anuran "tibialis anticus brevis" and the human tibialis anterior (anticus) derive from the extensor tarsi tibialis, and the anuran peroneus and the human fibularis (peroneus) longus/brevis derive from the extensor cruris et tarsi fibularis (NB: regarding the extensor digitorum longus and the extensores digitorum breves of anurans, they mainly correspond directly to the extensor digitorum longus and extensor digitorum brevis of humans, because they were present in the LCA of tetrapods). d., digits; m, metatarsals (attachments are only shown for muscles that often insert onto digits). e.b.s.+e.b.m.+e.b.p., "extensores breves superficiales plus extensores breves medii" *sensu* Dunlap (1960); ext., extensors.

formation of tendon in limb development, as explained in the studies of Schweitzer et al. (2010) and Manzano et al. (2012).

Within the more proximal regions of the limbs, i.e., girdles and thigh/arm, the evolution of muscles has much more to do with "case study 2" and "case study 3" by Gadow (1882). These case studies concern dorsoventral (deep/superficial: e.g., the ventral mass gives rise to the deep iliofemoralis and the superficial extensor iliotibialis of amphibians) or anteroposterior (longitudinal: e.g., the ischioflexorius gives rise to the flexor tibialis internus and flexor tibialis externus of reptiles) differentiation of muscles that were originally, and continue to be, associated with those existing, very different parts of the body (pelvic and pectoral regions). Moreover, the pectoral and pelvic regions perform very different functions in terrestrial locomotion and consequently experience very different evolutionary torsions and rotations (see, e.g., Figure 1.1

TABLE 19.18
Diagram Illustrating the Clear Topological Correspondences between the Ventral Leg and Forearm Muscles of Adults of Representative Urodele and Anuran Taxa

	Ambystoma mexicanum [6 ventral/flexor leg muscles/muscle groups: 6 (100%) seem to 'correspond' directly to forelimb muscles/muscle groups]	*Xenopus laevis* [6 ventral/flexor leg muscles/muscle groups: 1 (17%) seem to 'correspond' directly to forelimb muscles/muscle groups; unlike the dorsal leg and forearm muscles, the number of ventral leg and forearm muscles is not the same; i.e., there are 8 ventral forearm muscles and only 6 ventral leg muscles in anurans; i.e. many of the 'clear topological correspondences' have been lost and the only muscles/muscle groups that can be easily 'matched' to each other are the extensor digitorum, the dorsometacarpales and the extensores digitorum breves; one of the reasons leading to the topological difference between the ventral leg and forearm arm muscles of anurans concerns the fact that in the hindlimb the highly elongated tibiale and fibulare somehow functionally replaces the ulna and radius and thus some leg muscles moved distally to the tarsal region; however, there is still an examples of homoplasy/integration, which is the presence if a single flexor accessorius in both the forelimb and hindlimb as the apparently plesiomorphic condition for anurans]
Long flexors	Flexor digitorum communis (d.1-5; includes fl.cr.ta.fi.) [Flexor digitorum communis]	Flexor digitorum communis (**'plantaris longus'**; directly only to d.1-2) [Flexor digitorum communis]
	--- (fl.cr. ta.ti. seems to be included in ischioflexorius)	Cruroastragalus ('tibialis posticus'; seems to correspond to fl.cr.ta.ti.)
	Flexor accessorius medialis [Flexor acessorius medialis] →	Flexor accessorius ('transversus plantae proximalis et distalis')
	Flexor accessorius lateralis [Flexor acessorius lateralis] ↗	---
	Contrahentium caput longum [Contrahentium caput longum]	Contrahentium caput longum ('tarsalis posticus'; derived also/instead from f.b.s?)
	Interosseus cruris [Pronator quadratus]	Interosseus cruris ('intertarsalis')
	Tibialis posterior ('pronator profundus') [Palmaris profundus 1]	Tibialis posterior ('plantaris profundus'; or 'palmaris profundus' derived instead from f.b.s?)

Note: Data compiled from evidence provided by our own dissections and comparisons and from a review of the literature. The black arrows indicate hypotheses that are most strongly supported by the evidence available; the gray arrows indicate alternative hypotheses that are supported by some data but, overall, are not as strongly supported by the evidence as are the hypotheses indicated by the black arrows. Below the names of those hindlimb muscles/muscle groups that have clear, direct topological forelimb "equivalents" in the same taxon, we provide (in bold, between square brackets) the names of those forelimb equivalents. In those cases in which those equivalents clearly have a different ontogeny, these are shown in italics. As can be seen in the names highlighted in bold, there are some cases in which although the muscle that is designated with a certain name in anurans does not directly correspond to the muscle designated with a similar name in humans (but not the same name in the case of the plantaris longus; hence, we can use this commonly used anuran name), both the anuran and human muscles derive from a similar anlage/ancestral muscle: the anuran plantaris longus and the human plantaris derive from the flexor digitorum communis (NB: the anuran tibialis posticus and the human tibialis posterior do not seem to be homologous, as shown in this table). d., digits; f.b.s., flexores breves superficiales; fl.cr.ta.fi., flexor cruris et tarsi fibularis; fl.cr.ta.ti., flexor cruris et tarsi tibialis.

of Russell and Bauer (2008); for instance, the hindlimb is the major player in forward movement). In this case, major phylogenetic and functional constraints prevent the similar ontogenetic constraints and genetic pathways from resulting in such a striking homoplastic similarity as that seen in the more distal regions of the fore and hindlimbs. It is also easier to understand why more proximal muscles tend to have a more mixed innervation than more distal muscles (e.g., the extensor forearm muscles are exclusively innervated by the radial nerve). It is also easier to understand why there is a trend toward less mixing of nerves in the proximal regions during the evolutionary history of tetrapods, e.g., about 27% in urodeles, 25% in lizards, and 5% in humans (Gadow 1882). That is, according to this idea, the proximal muscles were plesiomorphically innervated by multiple nerves because they occupied broad anteroposterior and dorsoventral areas, and then during the splitting, the parts lying in different areas gave rise to new muscles that were innervated by a single nerve. For instance, the ventral mass gave rise, in humans, to flexor hamstring (sciatic nerve), adductor (obturator nerve), and ischiotrochanteric (gemelli, quadratus femoris, and obturator internus: sacral plexus) groups. Older and unfortunately much neglected theories such as those of Gadow (1882) can thus now be reinterpreted in light of new developmental and phylogenetic information. For instance, some of the homologies and theories proposed by Gadow and other authors proved to be incorrect in face of developmental studies of muscle development (see, e.g., Lance-Jones 1979) and phylogeny. Importantly, the condition found in amphibians such as *Cryptobranchus* cannot be seen as the ancestral condition for tetrapods because the last common ancestor (LCA) (crown) tetrapods must have already possessed most of the differentiated muscles present in amphibians such as *Ambystoma*, because these muscles are also present in amniotes. Instead, the condition found in amphibians such as *Cryptobranchus* is probably due to heterochrony leading to less muscle differentiation than seen in the

TABLE 19.19
Diagram Illustrating the Clear Topological Correspondences between the Foot and Hand Muscles of Adults of Representative Urodele and Anuran Taxa

	Ambystoma mexicanum **[7 foot muscles/muscle groups: 7 (100%) seem to 'correspond' directly to forelimb muscles/muscle groups]**	***Xenopus laevis*** **[11 foot muscles/muscle groups: 8 (73%) seem to 'correspond' directly to forelimb muscles/muscle groups [further examples of homoplasy/integration is the fact that the lumbricales have both 'breves' and 'longus' layers in the anuran hand and foot and this is apparently a derived feature among tetrapods i.e. these bundles were apparently acquired independently in the foot and in the hand during anuran evolutionary history]**
Intrinsic foot muscles	Flexores breves superficiales (5 muscles to d.1-5) [Flexores breves superficiales]	Flexores breves superficiales ('**flexor digitorum brevis superficialis**'; 3 muscles to d. 3-5; also gave rise to 'pl.pr.' and/or 'ta.po.'?) [Flexores breves superficiales]
	---	Lumbricales ('lumbricales breves' plus 'lumbricales longi' and 'lumbricalis longissimus IV'; 8 muscles to d.2-5) [Lumbricales]
	---	--- (but abductor praehallucis present in other anurans, going to prehallux)
	---	Abductor brevis plantaris hallucis (1 muscle, 1 muscle to m.I)
	Abductor digiti minimi (1 muscle to m.V) [Abductor digiti minimi]	Abductor digiti minimi ('abductor brevis dorsalis digiti V'; 1 muscle to m.V) [Abductor digiti minimi]
	Contrahentes pedis (5 muscles to d.1-5) [Contrahentes digitorum]	Contrahentes pedis ('contrahentes digitorum'; 3 muscles to d.1, 2 and to m.V) [Contrahentes digitorum]
	---	--- (but 'opponens hallucis' present in other aurans, going to m.I)
	Flexores digitorum minimi (4 muscles to d.2-5) [Flexores digitorum minimi]	Flexores digitorum minimi ('flexores teretes digitorum'; 2 muscles to d.2-5) [Flexores digitorum minimi]
	Interphalangei (3 muscles to d.3-4) [Interphalangei]	Interphalangei (4 muscle to d.3-5) [Interphalangei]
	Flexores breves profundi (10 muscles to d.1-5) [Flexores breves profundi]	Flexores breves profundi ('flexores ossi metatarsales'; 2 muscles to m.II and III) [Flexores breves profundi]
	---	Abductor brevis plantaris digiti V (1 muscle to m.V)
	---	Abductor proprius digiti IV (1 muscle to m.IV)
	Intermetatarsales (4 muscles metatarsals 1-5) [Intermetacarpales]	Intermetatarsales ('transversi metatarsi'; 4 muscles connecting metatarsals 1-5) [Intermetacarpales]

Note: Data compiled from evidence provided by our own dissections and comparisons and from a review of the literature. The black arrows indicate hypotheses that are most strongly supported by the evidence available; the gray arrows indicate alternative hypotheses that are supported by some data but, overall, are not as strongly supported by the evidence as are the hypotheses indicated by the black arrows. Below the names of those hindlimb muscles/muscle groups that have clear, direct topological forelimb equivalents in the same taxon, we provide (in bold, between square brackets) the names of those forelimb equivalents. In those cases in which those equivalents clearly have a different ontogeny, these are shown in italics. As can be seen in the names highlighted in bold, there are some cases in which although the muscle that is designated with a certain name in anurans does not directly correspond to the muscle designated with a similar name in humans, both the anuran and human muscles derive from a similar anlage/ancestral muscle: the anuran flexor digitorum brevis superficialis and the human flexor digitorum brevis derived from the flexores breves superficiales (NB: it is not clear if the lumbricales were present, or not, in the LCA of tetrapods because they are usually present in anurans and amniotes but not in urodeles: if they present in the LCA of tetrapods, then the anuran lumbricales would directly correspond to the human lumbricales; the abductor digiti minimi of anurans clearly seems to directly correspond to the abductor digiti minimi of the human foot). d., digits; m, metatarsals. pl.pr., plantaris profundus; ta.po., tarsalis posticus.

TABLE 19.20

Diagram Illustrating the Clear Topological Correspondences between the Dorsal/Extensor Leg and Forearm Muscles of Adults of Representative Reptilian Taxa

	Trachemys scripta [7 dorsal/extensor leg muscles/muscle groups: 5 (71%) seem to 'correspond' directly to forelimb muscles/muscle groups]	*Timon lepidus* [6 dorsal/extensor leg muscles/muscle groups: 4 (67%) seem to 'correspond' directly to forelimb muscles/muscle groups]	*Alligator mississippiensis* [6 dorsal/extensor leg muscles/muscle groups: 4 (67%) seem to 'correspond' directly to forelimb muscles/muscle groups]	*Gallus gallus* [6 dorsal/extensor leg muscles/muscle groups: 4 (67%) seem to 'correspond' directly to forelimb muscles/muscle groups]
Long extensors	Extensor digitorum longus (m.I-IV and d.1) [Extensor digitorum]	Ex. digitorum longus (m.II-III, not to digits) [Extensor digitorum]	Extensor digitorum longus (m.I-III, sometimes also to m.IV) [Extensor digitorum]	Ex. digitorum longus (to d. 2-4) [Extensor digitorum]
	Tibialis anterior [Extensores antebrachii et carpi radialis]	Tibialis anterior [Extensores antebrachii et carpi radialis]	Tibialis anterior [Extensores antebrachii et carpi radialis]	Tibialis anterior ('tibialis cranialis') [Extensores antebrachii et carpi radialis]
	Fibularis longus (m.V and d.5)	Fibularis longus	Fibularis longus	Fibularis longus
	Fibularis brevis	Fibularis brevis	Fibularis brevis	Fibularis brevis
Short extensors	Extensores digitorum breves (3 muscles to d.2-4) [Extensores digitorum breves]	Extensores digitorum breves (4 muscles to d.2-5 + part short ex. d.1) [Extensores digitorum breves]	Extensores digitorum breves (3 muscles to d.2-4) [Extensores digitorum breves]	Extensores digitorum breves (3 muscles to d.2-4) [Extensores digitorum breves]
	Dorsometarsales ('interossei dorsales'; 3 muscles to d.2-4) [Dorsometacarpales]	---	---	---
	Ab. et ex. digiti 1 ('ex.h.p.+ab.h.'; d.1; ab. + short ex. d.1) [Abductor pollicis longus]	Ab. et ex. digiti 1 (to d.1; ab. + part short ex. d.1) [Abductor pollicis longus]	Ab. et ex. digiti 1 ('ex.h.l.'; d.1; ab. + short ex. d.1) [Abductor pollicis longus]	Ab. et ex. digiti 1 ('ex.h.l.'; d.1; ab. + short ex. d.1) [Abductor pollicis longus]

Note: Data compiled from evidence provided by our own dissections and comparisons and from a review of the literature. The black arrows indicate hypotheses that are most strongly supported by the evidence available; the gray arrows indicate alternative hypotheses that are supported by some data but, overall, are not as strongly supported by the evidence as are the hypotheses indicated by the black arrows. Below the names of those hindlimb muscles/muscle groups that have clear, direct topological forelimb equivalents in the same taxon, we provide (in bold, between square brackets) the names of those forelimb equivalents. In those cases in which those equivalents clearly have a different ontogeny, these are shown in italics. The extensor digitorum longus, tibialis anterior, fibularis brevis, fibularis longus, and extensores digitorum breves of birds clearly correspond to the muscles designated by similar names in humans. However, the extensor hallucis longus in birds is not homologous to the muscle designated with the same name in humans, because the LCA of reptiles and mammals did not have a distinct extensor hallucis longus, and moreover, the bird muscle corresponds to the abductor et extensor digiti 1 *sensu* the present work, while the human extensor hallucis longus derives from the extensor digitorum longus. ab., abductor; ab.h, abductor hallucis; d., digits; ex., extensor; ex.h.l., extensor hallucis longus; ex.h.p., extensor hallucis proprius; m, metatarsals (attachments are only shown for muscles that often insert onto digits).

TABLE 19.21

Diagram Illustrating the Clear Topological Correspondences between the Ventral/Flexor Leg and Forearm Muscles of Adults of Representative Reptilian Taxa

	Trachemys scripta [5 ventral/flexor leg muscles/muscle groups: 3 (60%) seem to 'correspond' directly to forelimb muscles/muscle groups]	*Timon lepidus* [6 ventral/flexor leg muscles/muscle groups: 4 (67%) seem to 'correspond' directly to forelimb muscles/muscle groups]	*Alligator mississippiensis* [7 ventral/flexor leg muscles/muscle groups: 4 (57%) seem to 'correspond' directly to forelimb muscles/muscle groups; two further examples of remarkable homoplasy, the popliteus/pronator teres and the tibialis posterior/flexor carpi radialis of crocodylians, as those of mammals, are said to correspond to each other but they do not, so this is also an interesting case where a 'clear topological correspondence' seen in certain taxa gets less clear in other tax a where there is another apparently 'clear topological correspondence' (in this case mainly due to the absence of pronator accessorius and palmaris profundus 1 in the forearm of crocodylians, the popliteus and tibialis posterior apparently 'corresponding' to the pronator teres and flexor carpi radialis of the crocodylian forearm, respectively]	*Gallus gallus* [4 ventral/flexor leg muscles/muscle groups: 2 (50%) seem to 'correspond' directly to forelimb muscles/muscle groups; so there is a trend to reduce the number of 'clear topological equivalents' in the transitions to crocodylians and then to birds, which mainly use the limbs for very different functions; i.e., the forelimbs to fly; so contrary to what happens in humans, this is an expected case where functional divergence is associated with anatomical divergence; the increase from turtles to lizards is not really unexpected, because although turtles are phylogenetically more plesiomorphic, they are highly derived in some anatomical regions, contrary to what happens with, e.g., salamanders within the tetrapod clade; moreover some turtles have a popliteus, so this muscle was very likely present in the LCA of turtles and other reptiles and thus there is no new muscle in reptiles that 'gained' a 'clear topological equivalent' in the other limb]
Long flexors	--- (fl.cr. ta.ti. seems included in f.ti.ex. and/or ga.in.)	--- (fl.cr. ta.ti. seems included in fl.ti.ex. and ga.in)	--- (fl.cr. ta.ti. seems included in f.ti.ex. and/or ga.in.)	--- (fl.cr. ta.ti. seems included in ga.co. or lost)
	---	---	Gastrocnemius plantaris ('**plantaris**')	Gastrocnemius plantaris ('**plantaris**')
	Gastrocnemius internus ('gastrocnemius'; 2 heads)	Gastrocnemius internus ('femorotibial')	Gastrocnemius internus	Gastrocnemius (3 heads)
	Gastrocnemius externus ('f.d.l. sublimis')	Gastrocnemius externus ('femoral'; sup. and deep heads)	Gastrocnemius externus	---
	Flexor digitorum communis (to d.1-4) [Flexor digitorum longus]	Flexor digitorum longus (to d. 1-5) [Flexor digitorum longus]	Flexor digitorum longus (d. 1-4) [Flexor digitorum longus]	Flexor digitorum longus ('f.d.l.'+'**f.h.l.**'; to d. 1-4) [Flexor digitorum longus]
	--- (but some turtles have a popliteus)	Popliteus [Pronator accessorius]	Popliteus *[Pronator teres]*	Popliteus *[Pronator teres]*
	Interosseus cruris [Pronator quadratus]	Interosseus cruris [Pronator quadratus]	Interosseus cruris [Pronator quadratus]	---
	Tibialis posterior ('pronator profundus') [Palmaris profundus 1]	Tibialis posterior ('pronator profundus') [Palmaris profundus 1]	Tibialis posterior ('pronator profundus') *[Flexor carpi radialis]*	---

Note: Data compiled from evidence provided by our own dissections and comparisons and from a review of the literature. The black arrows indicate hypotheses that are most strongly supported by the evidence available; the gray arrows indicate alternative hypotheses that are supported by some data but, overall, are not as strongly supported by the evidence as are the hypotheses indicated by the black arrows. Below the names of those hindlimb muscles/muscle groups that have clear, direct topological forelimb equivalents in the same taxon, we provide (in bold, between square brackets) the names of those forelimb equivalents. In those cases in which those equivalents clearly have a different ontogeny, these are shown in italics. The flexor digitorum longus, popliteus, and gastrocnemius of humans mainly correspond to at least part of the muscles designated by similar names in birds. However, the muscle that are usually named plantaris in birds is not homologous to the muscle designated with the same name in humans, because the last LCA of reptiles and mammals did not have a distinct plantaris muscle; hence, we use a different name in birds. But, although the plantaris of birds does not directly correspond to the plantaris of humans, both the avian and human muscles derive from a similar anlage/ancestral muscle, i.e., from the gastrocnemius complex; hence, the avian muscle is highlighted in bold. Regarding the avian flexor hallucis longus, although this structure and the extensor hallucis longus of humans are both derived from the flexor digitorum longus, the avian structure is simply a bundle of the flexor digitorum longus and not a separate muscle; the LCA of reptiles and mammals did not have a distinct flexor hallucis longus. d., digits; fl.cr.ta.ti., flexor cruris tarsi tibialis; f.d.l., flexor digitorum longus; f.h.l., flexor hallucis longus; fl.ti.ex., flexor tibialis externus; ga.co., gastrocnemius complex; ga.in., gastrocnemius internus; sup. superficial.

TABLE 19.22

Diagram Illustrating the Clear Topological Correspondences between the Foot and Hand Muscles of Adults of Representative Reptilian Taxa

	Trachemys scripta **[5 foot muscles/muscle groups: 5 (100%) seem to 'correspond' directly to forelimb muscles/muscle groups]**	***Timon lepidus*** **[6 foot muscles/muscle groups: 6 (100%) seem to 'correspond' directly to forelimb muscles/muscle groups]**	***Alligator mississippiensis*** **[6 foot muscles/muscle groups: 6 (100%) seem to 'correspond' directly to forelimb muscles/muscle groups]**	***Gallus gallus*** **[5 foot muscles/muscle groups: 3 (60%) seem to 'correspond' directly to forelimb muscles/muscle groups; a further remarkable case of homoplasy/integration is the fact that both in the forelimb and the hindlimb of birds the flexores breves superficiales have migrated proximally and are now located in the forearm and leg regions respectively; this is also a case of parallelism with the forelimb (but not the hindlimb) of mammals, where the flexores breves superficiales also migrated proximally to form the flexor digitorum superficiales of these tetrapods]**
Intrinsic foot muscles	Flexores breves sup. ('fl.di.com.sub.'; 4 muscles to d.1-4) [Flexores breves superficiales]	Flexores breves sup. (3 muscles to d.1-3) [Flexores breves superficiales]	Flexores breves sup. (4 muscles to d.1-4) [Flexores breves superficiales]	Flexores breves sup. ('f.p.p.II,III + f.p.II,III,IV'; 5 muscles to d.2-4) [Flexores breves superficiales]
	Lumbricales (4 muscles to d.2-5) [Lumbricales]	Lumbricales (2 muscles to d.3-4) [Lumbricales]	Lumbricales (4 muscles, to d.1-4) [Lumbricales]	Lumbricales (2 muscles, to d.3-4)
	---	Abductor digiti minimi (1 muscle to m.V) [Abductor digiti minimi]	Abductor digiti minimi (mainly to m.IV) [Abductor digiti minimi]	Abductor digiti minimi ('ab.di.IV'; 1 muscle to d.4)
	Contrahentes pedis (p.'i.p'; 4 muscles to d.1-4) [Contrahentes digitorum]	Contrahentes pedis (5 muscles to d.1-5) [Contrahentes digitorum]	Contrahentes pedis (at least 1 muscle to d.1) [Contrahentes digitorum]	Contrahentes pedis ('adductor digiti II'; 1 muscles to d.2) [Contrahentes digitorum]
	Flexores breves profundi (p.'i.p';; 4 muscles to d.1-4) [Flexores breves profundi]	Flexores breves profundi (10 muscles to d.1-5) [Flexores breves profundi]	Flexores breves profundi (apparently 8 muscles to d.1-4) [Flexores breves profundi]	Flexores breves profundi ('**f.h.b.**'; 1 muscle to d.1) [Flexores breves profundi]
	Intermetatarsales (p.'i.p'; 4 muscles connecting m.1-5) [Intermetatarsales]	Intermetatarsales (4 muscles connecting m.1-5) [Intermetatarsales]	Intermetatarsales (apparently 3 muscles connecting m.1-4) [Intermetatarsales]	---

Note: Data compiled from evidence provided by our own dissections and comparisons and from a review of the literature. The black arrows indicate hypotheses that are most strongly supported by the evidence available; the gray arrows indicate alternative hypotheses that are supported by some data but, overall, are not as strongly supported by the evidence as are the hypotheses indicated by the black arrows. Below the names of those hindlimb muscles/muscle groups that have clear, direct topological forelimb equivalents in the same taxon, we provide (in bold, between square brackets) the names of those forelimb equivalents. In those cases in which those equivalents clearly have a different ontogeny, these are shown in italics. The flexores breves superficiales, lumbricales, and abductor digiti minimi of birds clearly correspond to the muscles designated by similar names in humans. Regarding the avian flexor hallucis brevis, this muscle corresponds to the first flexor brevis profundus, so it does correspond to the flexor hallucis brevis of humans *sensu* the present work (i.e., to the structure usually designated "medial head of the flexor hallucis longus" in human anatomy). However, in the LCA of tetrapods and of amniotes, this first flexor brevis profundi was clearly a part of a group of muscles, the flexores breves profundi, and was not a distinct, separate, peculiar muscle as is the case in birds and humans, so the bird and human condition is the result of an evolutionary homoplasy (hence, the bird muscle is highlighted in bold). ab.di.IV, abductor digiti IV; d., digits; fl.di.com.sub., flexor digitorum communis sublimis; f.h.b., flexor hallucis brevis; f.p.p.II,III + f.p. II,III,IV; flexor perforans et perforatus digiti II and III plus flexor perforatus digiti II, III and IV; m, metatarsals; p."i.p.," part of "interossei plantares" *sensu* Walker (1973); sup., superficiales.

adult forms of the LCA of tetrapods. Less muscle differentiation is also seen in adult members of derived amniotes with limb and/or digit reduction (see, e.g., Abdala et al. 2015).

The new information we recently obtained for marsupials (Diogo et al. 2016a) is the basis for this section. The hindlimb muscles present in placentals such as rats/mice and humans (Figure 19.4) are, with very few exceptions, also present in marsupials such as opossums (Figure 19.3; Tables 19.1 through 19.4 and 19.13). This fact runs contrary to previous ideas of "scala naturae" from marsupials to placentals, culminating in humans (see Diogo et al. 2015d,e; 2016c), and it means that the homologies between marsupials and placentals are not particularly difficult to accept. Instead, the main difficulty lies in comparing the muscles of these two groups with those of monotremes, and such comparisons will thus be the main focus of this section. One of the few major controversies about the homologies between the hindlimb muscles of marsupials and placentals concerns the piriformis. Appleton (1928) stated that the muscle often designated "piriformis" in marsupials corresponds to the caudofemoralis rather than the piriformis of other mammals, but Stein (1981) considered the marsupial piriformis to be homologous with the placental piriformis. Our dissections and comparisons show that the attachments of the marsupial piriformis are similar to those of the placental piriformis (see Tables 19.1 through 19.4 and 19.13). Considering that the "crurococcygeus" of opossums corresponds to the dorsal/anterior head of the semitendinosus rather than to a true caudofemoralis, it seems that the caudofemoralis was secondarily lost in the opossum (Tables 19.1 through 19.4 and 19.13; NB: the caudofemoralis does not correspond to the femorococcygeus, as both these muscles are present in other marsupials and in early developmental stages of placentals such as mice: Lance-Jones 1979). Furthermore, some marsupials have a piriformis, a caudofemoralis, and a crurococcygeus (e.g., Osgood 1921, who also stated that the caudofemoralis is not present in *Didelphis* species).

Regarding the origin of and homologies between the hindlimb muscles in monotremes, therians, and other tetrapods, we agree with the homologies proposed by Gambaryan et al. (2002), with the following five major exceptions. First, the sartorius was not described in platypus by Gambaryan et al. (2002), but it is usually considered to be present in extant monotreme species. It extends from the pubis (and not the ilium as is usually the case in therians) to the tibia (Tables 19.1 through 19.4 and 19.13; see, e.g., Gregory and Camp, 1918; Pearson 1926; and Jouffroy and Lessertisseur, 1971).

Second, the "obturator internus" is present in monotremes according to Appleton (1928) and in *Ornithorhynchus* but not in echidna according to Gregory and Camp (1918). However, as stated by Jouffroy and Lessertisseur (1971), the structure designated obturator internus or "gemellus" (e.g., Gambaryan et al. 2015) in monotremes seems to simply correspond to an undifferentiated ischiotrochantericus, meaning that only one muscle derived from this group is present in the adult platypus (Tables 19.1 through 19.4 and 19.13).

Third, according to Gambaryan et al. (2002), the "plantaris" described in monotremes by authors such as Coues (1870) does not correspond to the therian plantaris, but instead to the flexor hallucis longus ("flexor digitorum tibialis"). According to Gambaryan et al. (2002), in platypus, the plantaris extends from the fibula to the plantar fascia and then to a sesamoid lying on the tibial side of the tarsal region, not the Achilles tendon, and it is broader than the plantaris of many therians. However, the authors did not employ a rigorous comparative anatomical framework, and in particular, they did not draw comparisons with marsupials. Coues (1870, 1872), who did compare the opossum with platypus, designated the platypus muscle plantaris because it closely resembles the plantaris of opossums in, e.g., its origin from the fibula and topological relationship with the gastrocnemius externus. Moreover, unlike the plantaris, which was clearly present in the LCA of placentals and marsupials, the flexor hallucis longus is consistently present as a well-separated muscle only in some placental clades (Tables 19.1 through 19.4 and 19.13). Further (as recognized by Gambaryan et al. [2002]), in platypus the flexor digitorum longus goes to digits 1–5. Therefore, it is extremely unlikely that platypus would also have a separate flexor hallucis longus because this muscle is derived from the part of the flexor digitorum longus going to digit 1 (Tables 19.1 through 19.4 and 19.13). Finally, the distal attachment of the plantaris onto the plantar aponeurosis is not surprising because the name plantaris refers to the "topological equivalence" of this muscle with the palmaris longus in the forelimb, which usually inserts mainly on the palmar aponeurosis.

Fourth, Gambaryan et al. (2002) did not describe a flexor digitorum brevis in monotremes, but this muscle is clearly present in these mammals (e.g., Coues 1870) (see Tables 19.1 through 19.4 and 19.13). Instead, they divide the flexor digitorum brevis into two muscles: the part attaching to digits 2 and 3 and the part attaching to digit 4. They explain that the first part ("flexor digitorum intermedii") originates from the tendon of the flexor digitorum longus rather than from the calcaneus, which is the origin of the flexor digitorum brevis in most therians. Accurately identifying it as one of the short flexors of the foot, they named the second part, which does originate from the calcaneus, "flexor digiti quarti brevis." However, had the authors compared platypus with other marsupials such as *D. virginiana*, they would have found that the entire flexor digitorum brevis originates from the tendon of the flexor digitorum longus. Therefore, the condition seen in platypus, in which the muscle originates partly from the calcaneus and partly from the tendons of the flexor digitorum longus, is intermediate between the condition seen in opossums and that seen in many placentals.

Fifth, Gambaryan et al. (2002) misinterpreted the homologies of some foot muscles. Once again, their errors resulted from the lack of comparison with other mammals, particularly marsupials. For example, they stated that the flexor digiti minimi brevis is absent in platypus, but they described an "abductor metatarsi V" and an "abductor hallucis brevis" in this taxon. Their description of the flexores breves profundi and of the intermetacarpales was extremely confusing, and they erroneously used the names "interossei plantares" and "interossei dorsales," respectively, for these muscles.

However, the homologies of these foot muscles had already been resolved prior to Gambaryan et al.'s (2002) study. Earlier homology hypotheses (e.g., Coues 1870, Jouffroy and Lessertisseur 1971) were very simple and are well supported by recent work, including the present study: the foot muscles found in platypus are essentially the same as those found in *D. virginiana*, the only exception being that platypus lacks a distinct abductor hallucis brevis (Tables 19.1 and 19.13).

Despite their lack of an in-depth comparative anatomical framework, the work by Gambaryan et al. (2002) is extremely valuable because it describes all the hindlimb muscles of monotremes and resolves some points of controversy. For instance, enormous confusion has surrounded the homologies of the extensor muscles of the leg in monotremes. However, taking into account the configuration of muscles, particularly their distal attachments, in phylogenetically basal (see Diogo et al. 2015d) therians such as opossums, it is clear that the patterns in monotremes are very similar to those found in many therians. Our homology hypotheses largely agree with those proposed by Gambaryan et al. (2002); their "extensor digitorum superficialis" and "extensor digitorum profundus" correspond to the extensor digitorum longus and extensor digitorum brevis of the present work (Tables 19.1 through 19.4 and 19.13). Thus, the muscle that inserts onto the distal phalanx of digit 5 corresponds to the fibularis digiti quinti of therians, which is very likely a short extensor to digit 5. That is why in therians, the extensor digitorum brevis, formed by the short extensors, normally does not include a tendon to digit 5; this condition is, e.g., seen in the opossum. It is not clear whether the extensor digitorum brevis in platypus sends a tendon to digit 5. Gambaryan et al. (2002) state on page 18 of their work that it does, but then on page 19, they state that it only sends tendons to digits 1–4 and that the individual tendon to digit 5 is the tendon of the fibularis digiti quinti, as is the case in many therians. If the extensor digitorum brevis does send a tendon to digit 5, a part of the short extensor to digit 5 might be included in this muscle in platypus, or, alternatively, the short extensor to digit 4 might send a tendon to digit 5; as such, evolutionary changes are not uncommon in tetrapods.

Several pieces of evidence suggest that the "superficial extensor" of platypus corresponds to the extensor digitorum longus of other mammals. For one thing, this muscle does not go to digit 1; the portion attaching to digit 1 is differentiated into an extensor hallucis longus, as is the case in other mammals. Also, the tendons of both this superficial extensor to digits 2–5 and of the extensor hallucis longus to digit 1 pass superficial to the tendons of the "extensor profundus" in platypus. Likewise, in other mammals, the tendons of the extensor digitorum longus are superficial to the tendons of the short extensors (which form the extensor digitorum brevis and extensor hallucis brevis). In fact, the overall number of extensor muscles and the pattern of their distal attachments in platypus is almost the same as that found in *D. virginiana*, with only two differences (Tables 19.1 through 19.4 and 19.13): (a) the extensor hallucis brevis is not differentiated as a separate muscle in monotremes and (b) the proximal attachment of the short extensors (i.e., of the extensor digitorum brevis) migrated proximally, all the way to the proximal leg. The latter configuration is also found in the hindlimb of therians (fibularis digiti quinti and fibularis digiti quarti) and in the short forelimb flexors of therians (flexores breves superficiales), which are integrated into the long forearm muscle flexor digitorum superficialis (see Tables 19.1 through 19.4 and 19.13).

Our results thus allow us to provide a very detailed list of hindlimb muscle synapomorphies shared by extant marsupials, extant therians, and extant mammals as a whole. Based on this comparison, extant mammals share 18 muscle synapomorphies for the hindlimb. Most of the synapomorphic changes in the hindlimb leading to the LCA of extant mammals occurred in the proximal region, supporting this hypothesis. There are 16 synapomorphic changes in the proximal (girdle/thigh) hindlimb region—differentiation of psoas minor, gluteus medius, gluteus minimus, gluteus maximus, rectus femoris, vastus lateralis, vastus medialis, iliopsoas, pectineus, adductor magnus, adductor longus, obturator externus, semimembranosus, biceps femoris, semitendinosus, and quadratus femoris. In contrast, there are only two changes in the distal (leg/foot) hindlimb muscles: differentiation of plantaris and extensor hallucis longus (Tables 19.1 through 19.4 and 19.13).

There are 9 synapomorphic changes from the LCA of extant mammals to the LCA of extant therians: all of them concern the proximal (girdle/thigh) hindlimb muscles (differentiation of piriformis, scansorius, femorococcygeus, tenuissimus, psoas major, iliacus, obturator internus, gemellus superior, and gemellus inferior) (Tables 19.1 through 19.4 and 19.13). Therefore, there were only half as many changes in the hindlimb in the transitions leading to therians as in those leading to mammals (9 vs. 18). Lastly, there are 4 synapomorphies of extant placentals: two in the proximal (girdle/thigh) hindlimb (differentiation of tensor fasciae latae and vastus intermedius) and two in the distal (leg/foot) hindlimb (differentiation of soleus; loss of interosseous cruris).

COMPARISON BETWEEN THE TETRAPOD HINDLIMB AND FORELIMB MUSCLES

This section, as well as Tables 19.14 through 19.22 that go with it, is an updated version of the study by Diogo and Molnar (2014), which show the topological similarities between the tetrapod forelimb and hindlimb muscles. In particular, we observed some remarkable cases of topological similarity between the forelimb and hindlimb muscles in specific tetrapod taxa, which are discussed in the following.

For instance, Table 19.14 shows a notable example of forelimb–hindlimb integration due to homoplasy: in rats, the short extensors of digits 2 and 3 are fused to form an undivided muscle in both pairs of limbs; i.e., extensor indicis in the forelimb and extensor digitorum brevis in the hindlimb. As shown in this table, these fusions, and the resulting similarity between the hindlimb extensor digitorum brevis and the forelimb extensor indicis in rats, were clearly acquired independently in each limb and constitute an example of evolutionary parallelism/convergence (as opposed to serial homology) between the forelimb and hindlimb.

Table 19.15 shows two further examples of forelimb–hindlimb homoplasy. The hindlimb muscle popliteus and the forelimb muscle pronator teres of mammals are often said to "topologically correspond" to each other (e.g., Quain et al. 1894; Diogo et al. 2013). However, the popliteus evolutionarily derives from the interosseous cruris and is very similar, topologically, to the pronator accessorius of the forearm of lizards (Table 19.15). The forelimb muscle flexor carpi radialis and the hindlimb muscle tibialis posterior are also said to be topological equivalents (e.g., Quain et al. 1894; Diogo et al. 2013). However, the tibialis posterior corresponds to a deep leg muscle in amphibians and reptiles often designated "pronator profundus," which is ontogenetically and topologically similar to the palmaris profundus 1 of the forelimb in adult amphibians. In both cases, topological correspondence between a certain hindlimb muscle and a certain forelimb muscle in phylogenetically more plesiomorphic taxa becomes less perceptible in more derived taxa; however, in the latter taxa, there is an apparent topological correspondence between the original hindlimb muscle and a different forelimb muscle. The muscles tibialis posterior and popliteus of the mammalian leg seem to "correspond," respectively, to the flexor carpi radialis and pronator teres of the mammalian forearm due to the absence of the hindlimb muscle palmaris profundus 1 and the forelimb muscle pronator accessorius in mammals (Table 19.15). Interestingly, the configuration of these muscles in mammals is mirrored in phylogenetically derived reptilian groups such as crocodylians, in which the hindlimb muscles popliteus and tibialis posterior are also remarkably similar topologically to the pronator teres and flexor carpi radialis, respectively (Table 19.21). These examples, which clearly illustrate the highly complex and homoplastic evolutionary history of the structures of the tetrapod forelimb and hindlimb, have been previously recognized by some researchers in their efforts to present evidence in support of the forelimb–hindlimb serial homology hypothesis. For instance, Humphry (1872b) stated that the mammalian hindlimb muscle tibialis posterior seems to topologically correspond to the mammalian forelimb muscle pronator teres but "corresponds ontogenetically" to the forelimb muscle palmaris profundus 1 of other tetrapods. He wrote that such examples "furnish a good illustration of the way in which certain nearly corresponding muscles in the two limbs are differently segmented from the parent mass; they afford further evidence that homological rule is not so rigidly followed in development as we may be disposed to think, and that we must not be too severe in our attempts to institute homeological comparison" (Humphry 1872b: 175).

Additional examples of forelimb–hindlimb homoplasy are seen in the human foot and hand (Table 19.16).The phylogenetically derived human hand muscle adductor pollicis accessorius ("volaris primus of Henle": Bello-Hellegouarch 2013) is mirrored by the likewise derived foot muscle "interosseous plantaris hallucis" *sensu* Cihak (1972). The transverse and oblique heads of the adductor pollicis and the oblique head of the adductor hallucis develop from the anlage of contrahens 1, but the transverse head of the adductor hallucis develops from a different, neomorphic anlage of the contrahens layer (e.g., Cihak 1972).

Table 19.17 shows that many of the apparent topological correspondences seen in anatomically plesiomorphic tetrapods such as urodeles have been lost in anurans, in which the only dorsal forearm/leg muscles or muscle groups with apparent topological equivalents in the other limb are the extensor digitorum/extensor digitorum longus, the dorsometacarpales/dorsometatarsales, and the extensores digitorum breves/extensores digitorum breves. One reason for the marked topological difference between the dorsal leg and forearm muscles is that in anurans, the greatly elongated tarsal bones tibiale and fibulare functionally partially replace the tibia and fibula of other tetrapods. Consequently, some leg muscles have migrated distally to the tarsal region (NB: developmental evidence suggests that the transformation of the anuran tibiale and fibulare represents a distal shift in the zeugo-autopodial border; i.e., these bones have acquired a zeugopodial [leg] identity; e.g., Wagner and Chiu 2003). Despite these transformations, the dorsal side of the forearm and the leg have the same number of muscles/muscle groups (eight; Table 19.16). Moreover, although the apomorphic (uniquely derived) dorsal muscles of the anuran forearm and leg are not topologically similar, they evolutionarily derive from "topologically corresponding" muscles. The apomorphic hindlimb muscle tibialis anticus brevis derives from the extensor tarsi tibialis (Table 19.17), and the apomorphic forelimb muscle brachioradialis derives from the extensor carpi radialis (Diogo and Abdala 2010). The apomorphic hindlimb muscle peroneus derives from the extensor cruris et tarsi fibularis (Table 19.17), and the apomorphic forelimb muscle anconeus derives from the extensor antebrachii et carpi ulnaris (Diogo and Abdala 2010). In yet another example of forelimb–hindlimb integration and evolutionary convergence/parallelism, both the anuran forelimb and hindlimb have dorsometatarsales/metacarpales series and lack an abductor et extensor digiti 1 (because the short extensor to digit 1 is a completely separate muscle in both of these limbs).

As in the dorsal forearm/leg, many apparent "topological correspondences" between the ventral forearm/leg muscles have been lost in anurans (Table 19.18), but unlike the former case, the number of ventral forearm muscles/muscle groups in anurans (eight; Diogo and Abdala 2010) is not equal to the number of ventral leg muscles/muscle groups (six; Table 19.18). However, there is a single flexor accessorius in both the forelimb and hindlimb, constituting an example of homoplasy/integration that seems to be plesiomorphic for frogs (Table 19.18; NB: because of the dramatic transformations of the anuran hindlimb, in frogs such as *Xenopus*, the leg muscle flexor accessorius has migrated distally and lies in the foot region; therefore, it is topologically very different from the forearm muscle flexor accessorius and is not shown in Table 19.18). Examples of integration/homoplasy are also found in the foot/hand, including the presence in both autopodia of "breves" and "longus" layers of lumbricales, which were likely independently acquired during anuran evolution (Table 19.19).

Unsurprisingly, considering their marked functional differences, in birds, there are fewer apparent topological equivalents between the leg–foot and forearm–hand (Table 19.21). On the contrary, the very different functions of the forelimb and hindlimb in humans are not associated with marked divergence in muscle anatomy (see Diogo et al. 2013). However, the forelimb and hindlimb of birds provide an additional example of integration/homoplasy, despite their functional and anatomical divergence: in both limbs, the flexores breves superficiales have migrated proximally to the forearm and leg regions, respectively (e.g., Vanden Berge 1975; Diogo and Abdala 2010; this book).

Combined with the data provided in Tables 19.14 through 19.22, the preceding examples of hindlimb muscles that mirror the evolution of the forelimb muscles in each major tetrapod clade and in Tetrapoda as a whole strongly support the idea that a substantial number of forelimb–hindlimb similarities found in these vertebrates result from parallelism/convergence (homoplasy) *during the evolutionary history of these clades*, not from serial homology (i.e., *ancestral* duplication of the structures of the paired appendages). As explained in the preceding chapters, Diogo et al. (2013) provided an extensive review of several lines of evidence—published by other researchers—that contradict arguments often used to support the forelimb-hindlimb serial homology hypothesis. Curiously, those other researchers did not question or contradict the serial homology hypothesis, which continues to be treated as dogma by many.

As explained earlier, a major weakness of the "paired appendages serial homology followed by functional/anatomical divergence" hypothesis (Figure 12.3) comes from such an integrative analysis combined with the first detailed examination and comparison of the soft tissues of the paired appendages in representatives of all major extant gnathostome clades (Figure 12.4). As noted in Figure 12.4, the pectoral and pelvic appendages of extant plesiomorphic gnathostome fish have largely undifferentiated adductor and abductor muscle masses, with only a few other muscles being differentiated. Therefore, all the numerous and (as explained earlier) in many cases strikingly similar muscles, as well as many bones (e.g., Don et al. 2013), of the tetrapod forelimb and hindlimb were undoubtedly acquired independently during the evolutionary transitions between early gnathostomes and tetrapods. Thus, under the historical (phylogenetic) definition of homology, these structures cannot be considered serial homologues. It could be argued that they are homologous under the morphological or developmental definitions (see Wagner 1994), but this argument is contradicted by the analysis of the evolution of the soft tissues of the gnathostome paired appendages, particularly the muscles, as noted in the preceding paragraphs.

As explained earlier, the ancestral morphological or developmental serial homology of the forelimb and hindlimb implies that they were originally similar and then diverged anatomically/functionally (Figure 12.3). On the contrary, the hard tissues of the pectoral and pelvic appendages plesiomorphically were anatomically very different (Zhu et al. 2012), and both the hard and soft tissues of the proximal region of the pectoral and pelvic appendages (particularly the girdles) have remained markedly different in all gnathostome clades (Coates and Cohn 1998; Diogo et al. 2013; Figure 12.4; Tables 19.14 through 19.22). For instance, among all tetrapod clades listed in these tables, including anatomically plesiomorphic taxa such as urodeles, not a single pelvic–thigh muscle has a clear topological equivalent in the pectoral region and arm. This lack of equivalence seems to be due to a strong phylogenetic constraint; i.e., the pelvic and pectoral girdles of fish plesiomorphically were markedly different anatomically, and consequently, the girdles of tetrapods are also quite different. However, the derived distal regions of the tetrapod forelimb and hindlimb, particularly the autopodium (hand/foot), have at least some new bones and a very different developmental plan from the pectoral and pelvic appendages in fish (NB: even if digits are derived from distal fin rays as suggested by, e.g., Johanson et al. 2007 [see also, e.g., Davis et al. 2007], it is broadly accepted that at least some tetrapod wrist/ankle bones are neomorphic structures; e.g., Don et al. 2012). Therefore, the evolution of the autopodium represents a major "evolutionary novelty" and is subject to fewer phylogenetic constraints than the limb girdles, and the developmental constraints/factors resulting from further *derived* (e.g., Roth 1994) cooption of some similar genes in the ontogeny of the forelimb and hindlimb led to greater similarity between the distal regions of these limbs in basal tetrapods ("similarity bottleneck": Figure 12.4). The detailed data on the tetrapod hindlimb and forelimb muscles summarized in the preceding paragraphs supports this hypothesis and complements the information about the forearm–hand and leg–foot muscle similarities provided by Diogo et al. (2013). In particular, these data strongly support the idea that evolutionary divergence and convergence/parallelism have both played crucial, sometimes counterbalancing roles in the evolution of the tetrapod forelimb and hindlimb.

Regarding the strong similarity between the leg-foot and forearm-hand muscles in urodeles ("similarity bottleneck": Figure 12.4) and its bearing on the forelimb–hindlimb homology hypothesis, it should be emphasized that the developmental changes associated with the "fins–limbs transition" that led to this similarity consist of phylogenetically *derived* cooption of a few similar genes ("genetic piracy" *sensu* Roth 1994; see also the review of Pavlicev and Wagner 2012). As noted in Chapter 12, it is now broadly accepted that the independent cooption of similar *ancestral* genes to form structures such as complex eyes in vertebrate and nonvertebrate animals ("deep homology") is a case of evolutionary parallelism and thus of homoplasy (e.g., Willmer 2003). Therefore, it would be very difficult to argue that the *derived* cooption of a few genes represents a case of forelimb–hindlimb serial homology under the developmental concept of homology. This logic was recognized by Tabin (1992), who stated that the structures of the forelimb and hindlimb of tetrapods independently evolved from the pectoral and pelvic fins, respectively, of fish. Tabin argued that this independent evolution is reflected in the significant differences between the forelimb and hindlimb patterns (despite their general similarity) in fish prior to the evolution of limbs, as noted by Rackoff (1980). According to Tabin (1992), the similarities between the tetrapod forelimb and hindlimb may be a direct consequence of the evolution of the limb buds by reorienting the expression of similar genes (e.g., *Hox-1* and *Hox-4*) along orthogonal axes, which have an effect on downstream target genes.

In fact, the musculature of the leg–foot in the first tetrapods might have been even more similar to the musculature of the forearm–hand than it is in anatomically plesiomorphic extant tetrapods such as salamanders. According to the comparative studies by authors such as McMurrich (1905) and Bardeen (1906) on extant salamanders such as *Ambystoma,* the distal (leg) portion of the ischioflexorius seems to correspond to an ancestral muscle flexor cruris et tarsi tibialis, while the femoral head of the flexor digitorum communis seems to correspond to an ancestral muscle flexor cruris et tarsi fibularis. If these correspondences are valid, all eight ventral forearm muscles of the first tetrapods had a clear topological equivalent in the leg: the flexor antebrachii et carpi radialis and the flexor antebrachii et carpi ulnaris "corresponded" to the flexor cruris et tarsi tibialis and to the flexor cruris et tarsi fibularis, respectively. Therefore, the first tetrapods would have more leg–foot muscles/muscle groups with "topological equivalents" in the forearm-hand than those found in extant salamanders (NB: because the flexor cruris et tarsi tibialis and flexor cruris et tarsi fibularis are not present as separate structures in extant salamanders, only six rather than eight of the individual salamander leg muscles have apparent topological equivalents in the forearm; Table 19.15). More importantly, the particularly strong ontogenetic factors/constraints, as well as the functional/topological factors/constraints (see the following), involved in the fins–limbs transition may thus have produced dorsal forearm musculature in early tetrapods that was similar not only to the dorsal leg musculature, but also to the ventral forearm and leg musculature (dorsoventral symmetry).

As noted in Chapter 12, even the cooption of similar genes during the fins–limbs transition is not sufficient to explain the strong similarity between the leg–foot and forearm–hand muscles of phylogenetically and anatomically derived tetrapods such as our bipedal species, *H. sapiens* (Figure 12.4). Many of the forelimb–hindlimb muscles/muscle groups in modern humans with apparent topological equivalents were undoubtedly acquired independently during the evolutionary history of primates (Diogo and Wood 2012a; Diogo et al. 2013). Such similarity bottlenecks have also occurred during the evolutionary history of other derived tetrapod and even nontetrapod clades; e.g., the pelvic appendicular muscles of plesiomorphic teleosts were derived independently from the pectoral appendicular muscles of these fish but are topologically very similar to them (e.g., Winterbottom 1974). Additional examples of homoplasy between forelimb and hindlimb bones can be seen in horses and *Plesiosaurus* (e.g., Owen 1849). In addition to ontogenetic constraints, the similarity bottlenecks leading to such derived clades (Figure 12.4) are clearly due to topological and functional constraints. For instance, the only functional configuration for an abductor hallucis/pollicis brevis is a muscle that lies tibial/radial to the first digit of the foot/hand and inserts onto the tibial/radial side of this digit. Both muscles were homoplastically acquired during tetrapod evolution (Table 19.16).

That topologically similar leg–foot and forearm–hand muscles in such taxa do not develop from similar anlagen (Tables 19.14 through 19.16) is a crucial point in support of the idea that ontogenetic constraints are not sufficient to explain the similarity bottlenecks leading to derived taxa such as modern humans (Figure 12.4). For instance, the extensor pollicis longus and extensor hallucis longus in modern humans are remarkably similar topologically, but the former derives from the anlage of the short extensors of the hand, while the latter derives from the anlage of the long extensors of the leg (Diogo and Wood 2012a; Table 19.16). Moreover, many of these muscles, or their subdivisions, formed at different geological times and/or phylogenetic nodes. For example, the adductor hallucis and adductor pollicis in modern humans are particularly similar to each other because they have well-differentiated transverse and oblique heads. However, unlike the adductor hallucis heads, which are well differentiated in phylogenetically plesiomorphic primates such as lemurs, the adductor pollicis heads only became well differentiated in the much more derived node leading to catarrhines (old world monkeys + hominoids; for more examples, see Diogo et al. 2013). In addition to those described above, many examples of similar forelimb-hindlimb muscles derived from different anlagen and/or appearing at different geological times referred to earlier are shown in Tables 19.14 through 19.22.

GENERAL REMARKS

In summary, evolutionary divergences in the history of tetrapods have produced differences between the forelimb and hindlimb musculatures that are not seen in anatomically plesiomorphic tetrapods such as urodeles. However, there are also cases of evolutionary parallelism/convergence leading to subsequent similarity bottlenecks between the forelimb and hindlimb in more derived taxa. The striking similarity between many pectoral and pelvic muscles found in gnathostome taxa such as modern humans and teleost fishes is undoubtedly the result of homoplasy due to a complex interplay between ontogenetic, topological, functional, and even phylogenetic (Diogo et al. 2013) constraints/factors, not of serial homology. We hope that the new data and the discussion provided in this book will lead to deconstruction of the serial homology hypothesis and promote more integrative studies on the puzzling and fascinating evolutionary history of the paired appendages in all its complexity, including the importance of homoplasy and constraints, under a new paradigm.

20 Development of Limb Muscles in Tetrapods

In Chapter 20, we use the neotenous axolotl *Ambystoma mexicanum* (Amphibia: Urodela) as a case study to illustrate the development of limb muscles in tetrapods, i.e., this chapter is mainly based on the study by Diogo and Tanaka (2014). One of the main reasons for this choice of taxon is that the axolotl is increasingly becoming one of the most popular model organisms in evolutionary, developmental, and regenerative studies, being a particularly powerful model of regeneration (e.g., Carlson 2007; Kragl et al. 2009; Nacu and Tanaka 2011; Stocum and Cameron 2011; Boisvert et al. 2013; Ziermann and Diogo 2013; Sefton et al. 2015). Another reason is that previous studies providing details about the development of axolotl limb muscles, such as Grim and Carlson's (1974a,b) ontogenetic work of the forearm and hand muscles and Boisvert et al.'s (2013) study on the ontogeny of the pelvic muscles, as well as those performed by ourselves and our colleagues (Diogo and Tanaka 2014), show that axolotls are among the best living models to study the basal configuration and developmental patterns of these muscles in tetrapods. We first provide here a detailed morphological description of the ontogeny of the axolotl pectoral, arm, forearm, hand, pelvic, thigh, leg, and foot muscles, which is based on analyses of transgenic animals that express green fluorescence protein (GFP) in muscle fibers (Diogo and Tanaka 2014); details about the methodology used for this analysis are given in that paper. We then discuss broad developmental and evolutionary topics such as the evolution and morphogenesis of tetrapod limbs and come back to subject concerning the supposed serial homology between the fore and hindlimbs.

DEVELOPMENT OF PECTORAL AND ARM MUSCLES

The pectoral and arm muscles seen in the pictures obtained from imaging the developing axolotls we analyzed are the procoracohumeralis, pectoralis, coracobrachialis, supracoracoideus, humeroantebrachialis, latissimus dorsi, deltoideus scapularis, and triceps brachii. All these muscles are appendicular muscles *sensu* the present book (see Chapters 16 and 18 for a list of all limb muscles, and their subgroups, found in adult axolotls). That is, none of the two adult axolotl axial pectoral muscles *sensu* the present book (i.e., serratus anterior and levator scapulae) could be clearly seen in the forelimb developmental stages we examined, i.e., from stage 46 to stage 54. This is likely because the axial pectoral muscles are often deep muscles superficially covered by the appendicular pectoral muscles (Figure 20.1; see also preceding chapters and, e.g., Valasek et al. 2011) and not due to the absence of all the axial pectoral muscles in these stages of development (e.g., these muscles appear early in the ontogeny of the amphibian urodele *Necturus*: see, e.g., Chen 1935 and the following discussion). Furthermore, the muscle subcoracoscapularis, which is an appendicular but very deep pectoral muscle, could not be seen in the pictures of the developing axolotls we obtained.

In adults, the procoracohumeralis, supracoracoideus, coracoradialis, and pectoralis cover the ventral surface of the pectoral region, the procoracohumeralis and pectoralis being respectively the most anterior and posterior structures within these muscles (Figure 20.1). The pectoralis (Figure 20.1) originates from the coracoid near the midline and from the rectus abdominis and sternum and inserts onto the proximal portion of the humerus. The supracoracoideus (or "coracohumeralis"; Figure 20.1) extends from the superficial ventral surface of the coracoid to the proximal portion of humerus. The procoracohumeralis (or "procoracohumeralis longus"; Figure 20.1) extends from the procoracoid cartilage to the proximal portion of the humerus. The coracoradialis (or "sternoradialis") has a well-developed fleshy belly that originates from the ventral surface of the coracoid deep (dorsal) to the superficial (ventral) layers of the pectoralis and supracoracoideus and that gives rise to a long and thin tendon that inserts onto the proximal portion of the radius. The humeroantebrachialis and coracobrachialis lie respectively on the radial and ulnar sides of the ventral surface of the arm (Figures 20.1 and 20.2). The humeroantebrachialis (or "brachialis"; Figures 20.1 and 20.2) connects the proximal portions of the humerus and radius. The coracobrachialis is differentiated into two bundles that originate from the posteroventral and posterolateral margin of the coracoid: the coracobrachialis longus inserts onto the distal humerus and the elbow joint, while the coracobrachialis brevis inserts onto the medial side of approximately the proximal 1/3 of the humerus. The deltoideus scapularis and latissimus dorsi lie on the dorsal side of the pectoral region, while the triceps brachii mainly lies on the surface of the arm (Figures 20.1 and 20.2). The deltoideus scapularis (or "dorsalis scapulae"; Figure 20.1) extends from the lateral and dorsal surfaces of the suprascapular cartilage to the ventral margin of the proximal portion of the humerus. The latissimus dorsi (Figure 20.1) is divided into a broader, posterior bundle, and a smaller, anterior bundle, which extend from the dorsal fascia and insert onto the proximal portion of the humerus. The triceps brachii (sometimes erroneously designated "anconeus"; Figures 20.1 and 20.2) is divided into four bundles, all being blended distally to insert onto the olecranon process of the ulna: the triceps coracoideus originates from a prominent posterior process of the coracoid and is medial to the triceps scapularis medialis, which originates from the scapula and

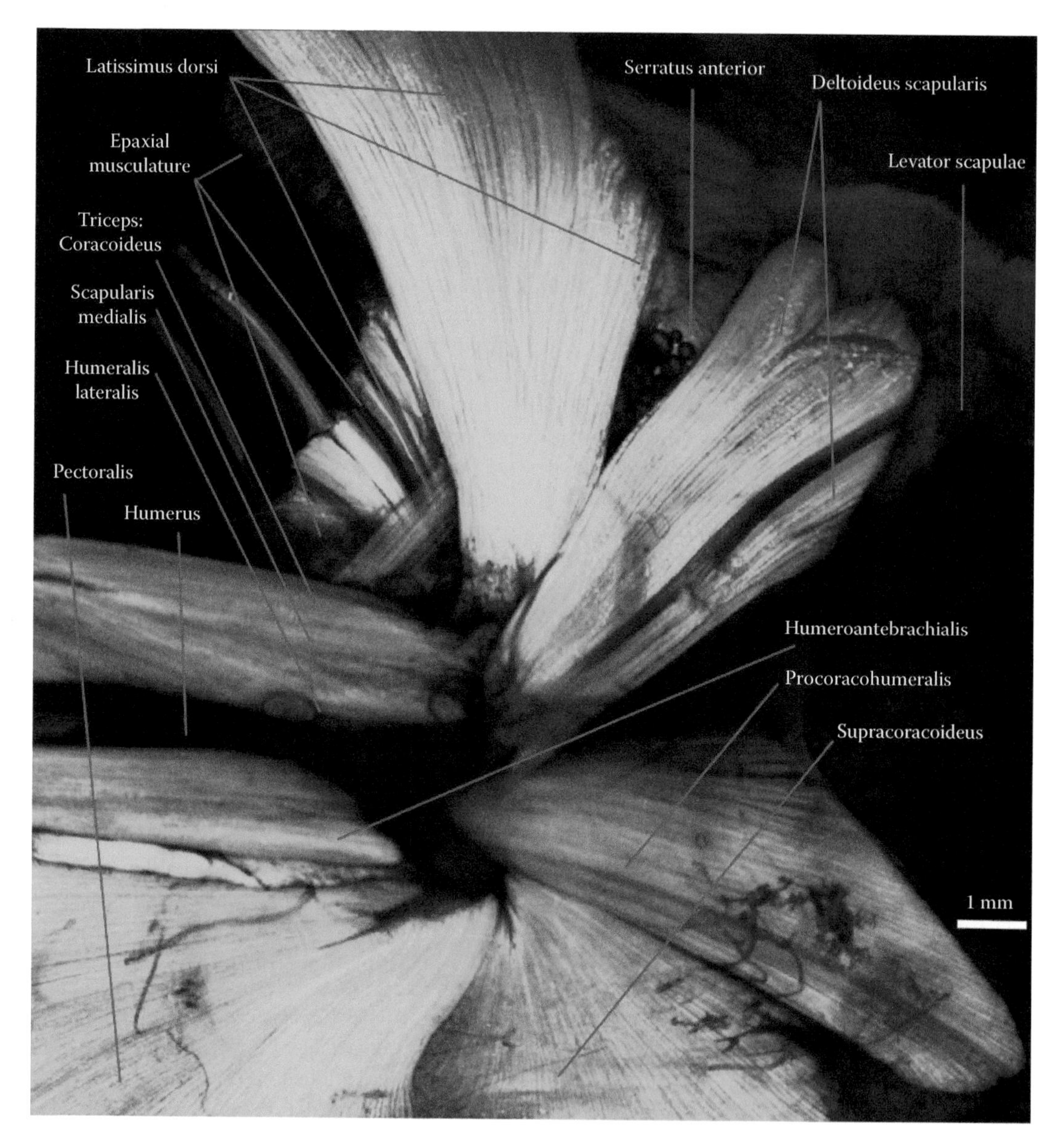

FIGURE 20.1 *Ambystoma mexicanum* (Amphibia, Caudata): adult specimen CRTD AM11. Lateral view of the left pectoral and arm muscles (anterior is to the right, dorsal to the top); the protractor pectoralis (not a pectoral nor a forelimb muscle) was removed (horizontal flop done with the LAS-AF software, V 2.6).

from connective tissue surrounding the shoulder joint capsule; the triceps humeralis lateralis originates from the lateral surface of the proximal portion of the humerus, while the triceps humeralis medialis originates from about 1/3 of the proximodistal length of the humerus.

Regarding the development of the pectoral and arm structures, the first appearance of muscles can be seen early at stage 46, in which the anlage of the pectoralis (and possibly the coracobrachialis) and the anlage of the supracoracoideus + coracoradialis and probably of the procoracohumeralis are seen in a ventral view, while the anlagen of the deltoideus scapularis + latissimus dorsi are seen in a dorsal view. Importantly, all these muscle anlagen lie near the distal region of the arm. In middle of stage 46, the pectoralis, coracobrachialis, supracoracoideus + coracoradialis, procoracohumeralis, deltoideus scapularis, and latissimus dorsi are differentiated and more elongated proximodistally (Figure 20.3A). All these muscles become considerably broader at later stages (e.g., Figures 20.3B and C and 20.4A), and, with the exception of the coracobrachialis, they all extend proximally toward the midline of the body to cover a significant part of the dorsal and ventral sides of the thoracic region, as they do in adults (Figure 20.1). The coracobrachialis brevis and coracobrachialis longus become differentiated in the middle of stage 47, i.e., at the stage in which the humeroantebrachialis (Figure 20.3C) and triceps brachii become clearly visible. At later stages, these four arm structures extend distally (e.g., Figure 20.4A and B), and the bundles of the triceps brachii become clearly differentiated in late stage 47 (e.g., Figure 20.4B); at stage

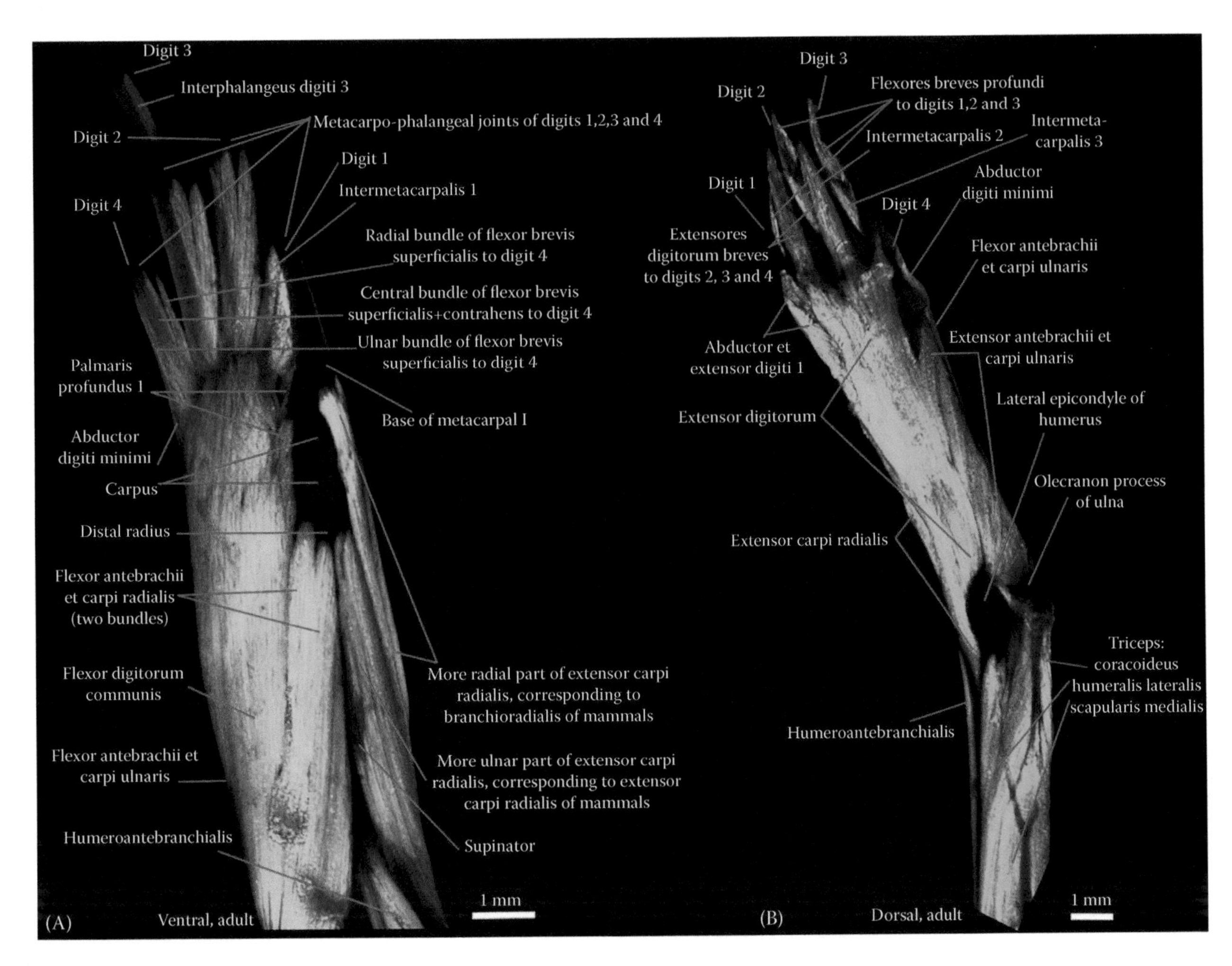

FIGURE 20.2 *Ambystoma mexicanum* (Amphibia, Caudata): adult specimen CRTD AM11. (A) Ventral view of the forearm and hand muscles of the left limb of CRTD AM11 (radial to the right; distal to the top; horizontal flop automatically done with the LAS-AF software, V 2.6). (B) Dorsal view of the forelimb muscles of the left limb of CRTD AM11 (radial to the left; distal to the top; horizontal flop done with the LAS-AF software, V 2.6).

50, the overall configuration and attachments of these four muscles are similar to those seen in adults (Figure 20.5A and B).

DEVELOPMENT OF VENTRAL/FLEXOR FOREARM MUSCLES

In adults, the palmaris profundus 1 (or "pronator profundus"; Figure 20.2A) is deep to the contrahentium caput longum and extends from the ulna, intermedium, and ulnare to the carpus and base of metacarpal I. The pronator quadratus (or "interosseous antebrachii") lies deep to the palmaris profundus 1 and thus to the other ventral forearm muscles, running from the ulna to the radius. The contrahentium caput longum (or "ulnocarpalis") lies deep (dorsal) to the flexor antebrachii et carpi ulnaris and to the flexor digitorum communis and originates from the proximal portion of the ulna and inserts onto the carpus through a tendon, being continuous with the contrahentes digitorum. The flexor accessorius lateralis (or "palmaris profundus 3") extends from the distal portion of the ulna and the ulnare to the dorsal surface of the palmar fascia, passing superficially (ventrally) to the contrahentium caput longum. The flexor accessorius medialis (or "palmaris profundus 2") connects the distal portion of the ulna and the intermedium to the dorsal surface of the palmar fascia, originating deep (dorsal) to the contrahentium caput longum. The flexor digitorum communis (or "palmaris superficialis"; Figures 20.2A) is the most superficial (ventral) muscle of the ventral forearm, originating from the medial epicondyle of the humerus, and sending a thin tendon to the distal phalanx of each of the four digits. The flexor antebrachii et carpi ulnaris (Figure 20.2A) lies on the ulnar side of, and is partially covered ventrally by, the flexor digitorum communis, running from the medial epicondyle of the humerus to the ulna and ulnare. The flexor antebrachii et carpi radialis (Figure 20.2A) is divided into a longer, more superficial (ventral) and proximodistally oriented bundle that runs from the medial epicondyle of the humerus to the radius and radiale and a shorter, deeper (dorsal), and more oblique bundle that runs from the medial epicondyle of the humerus to about half of the length of the radius, but that instead of inserting onto the radius directly, is mainly associated distally with the longer bundle.

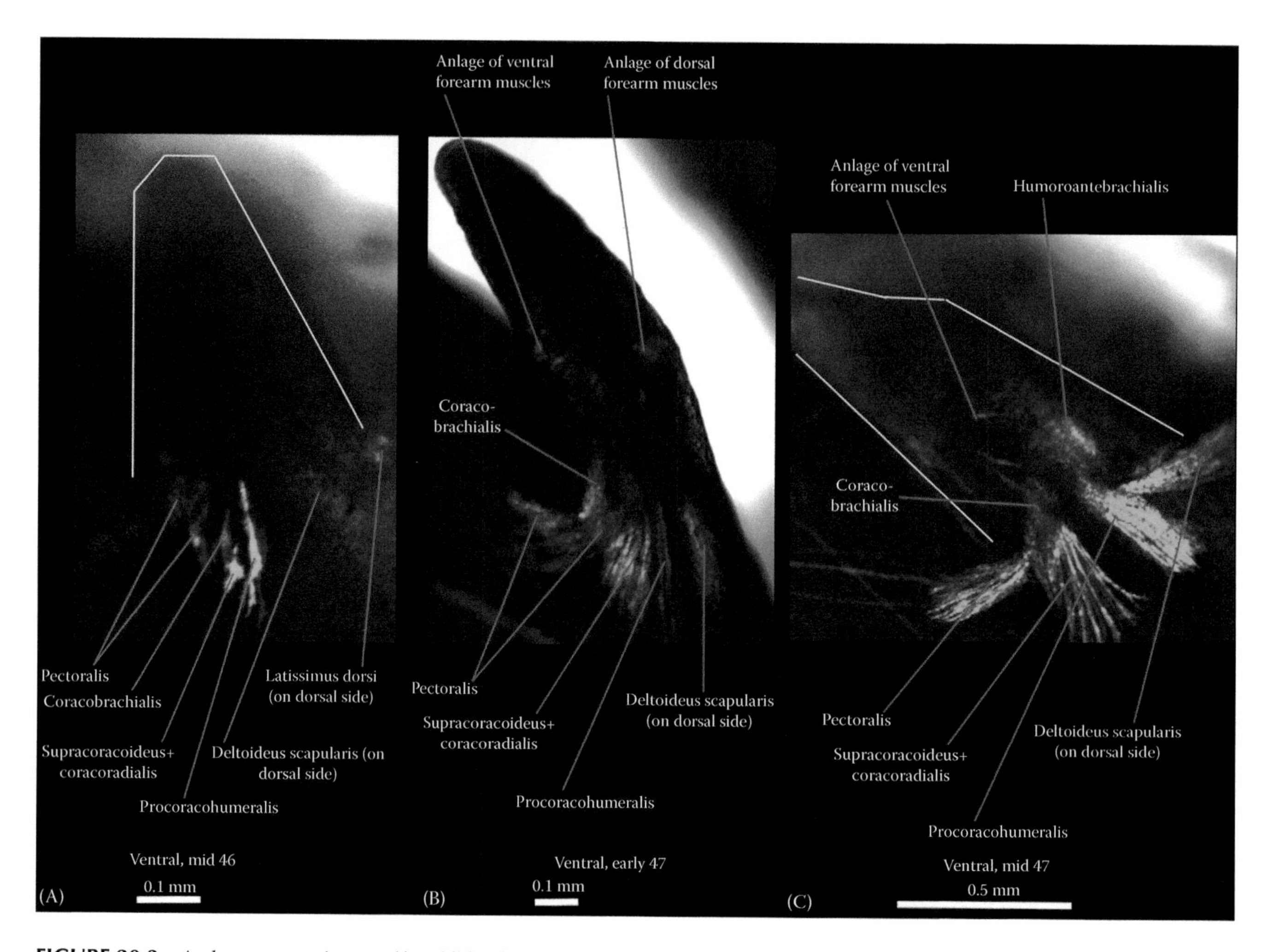

FIGURE 20.3 *Ambystoma mexicanum* (Amphibia, Caudata): developmental stages mid-46 and early and mid 47. (A) Ventral view of the muscles of the right forelimb of CRTD AM126 at stage mid-46 (radial to the right; distal to the top). (B) Ventral view of the muscles of the right forelimb of CRTD AM127 at stage early 47 (radial to the right; distal to the top). (C) Ventral view of the muscles of the right forelimb of CRTD AM128 at stage mid-47 (radial to the right; distal to the top). In this figure and the next figures, yellow lines are sometimes used to help show the limits of the developing limbs, when these limits cannot be clearly seen on the original picture.

Concerning development, the first appearance of an anlage—still undifferentiated—of ventral forearm muscles is early in stage 47 (Figure 20.3B). Late in stage 47, the flexor antebrachii et carpi radialis is a separate muscle, but at this stage, and even early in stage 48, the flexor antebrachii et carpi ulnaris is apparently still undifferentiated from the anlage of the flexor digitorum communis or is just a small muscle covered ventrally by the latter muscle (Figure 20.4A). Early in stage 48, the flexor antebrachii et carpi radialis starts to differentiate into the two heads that are found in adults (Figure 20.4A; see also Figures 20.5A, 20.6A, 20.7A, and 20.2A). The flexor antebrachii et carpi ulnaris, as well as the deeper muscles palmaris profundus 1, flexor accessorius lateralis, flexor accessorius medialis, and contrahentium caput longum, are clearly visible at stage 50 (Figure 20.5A; NB: in the images of the developing forelimb, the deeper pronator quadratus could not be seen). Late in stage 52, the overall configuration and attachments of these and other ventral forearm muscles are basically similar to those found in adults (Figure 20.6A).

DEVELOPMENT OF DORSAL/EXTENSOR FOREARM MUSCLES

In adults the extensor carpi radialis (Figure 20.2B) is divided into two bundles: the more radial bundle extends from the distal 1/2 of the radial side of the humerus to the radial side of the distal radius, carpus, and base of metacarpal I; the more ulnar bundle extends from the lateral epicondyle of the humerus to the radial side of the distal radius and radiale. The supinator (or "extensor antebrachii radialis") originates from the lateral epicondyle of the humerus and is distally separated from the extensor carpi radialis by the whole ventral side of the radius, because it inserts onto the medial surface of the distal 1/3 of the radius. The extensor antebrachii et carpi ulnaris (Figure 20.2B) extends from the lateral epicondyle of the humerus to the ulna and ulnare. The extensor digitorum (or "humer odorsalis"/"humerometacarpalis"; Figure 20.2B) is the most superficial (dorsal) of the dorsal forearm muscles, originating from the lateral epicondyle of the humerus and sending thin tendons to the metacarpals I–IV. There are three extensores

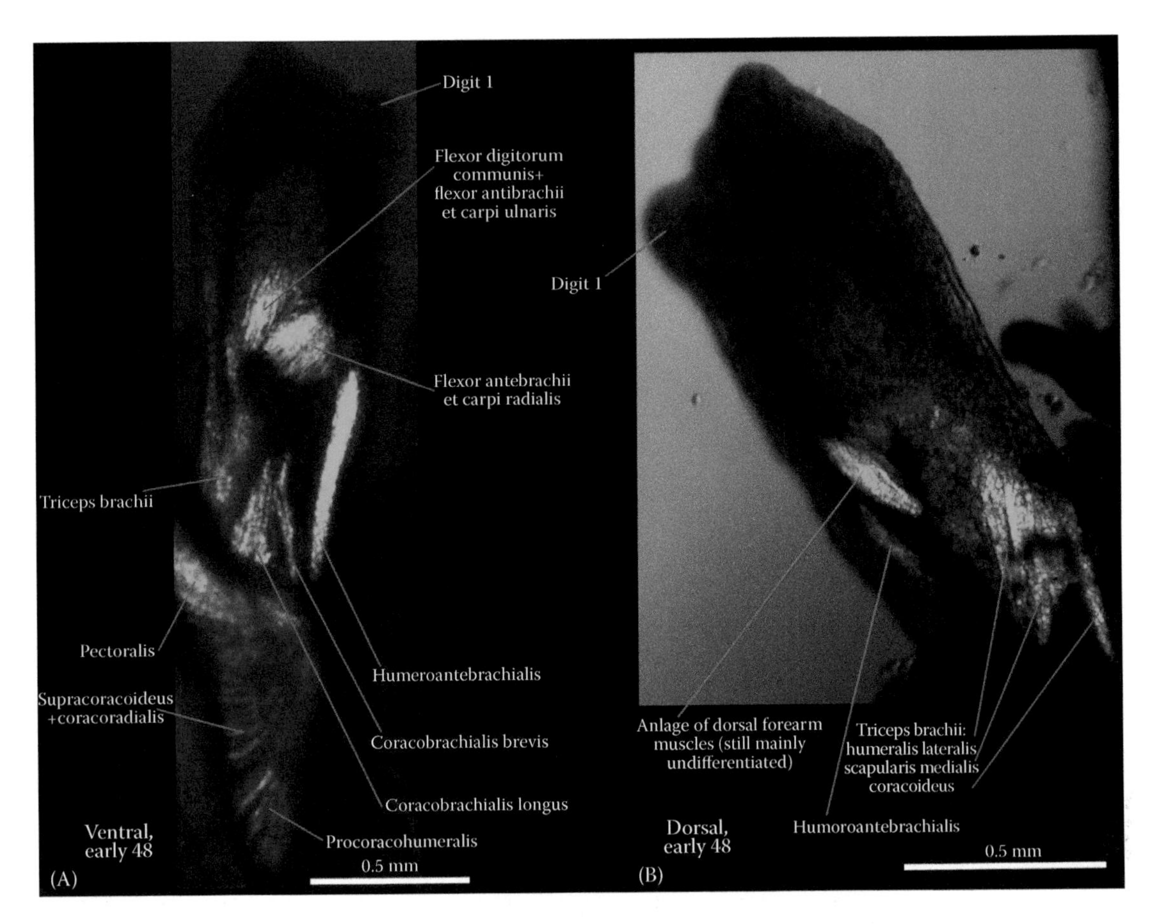

FIGURE 20.4 *Ambystoma mexicanum* (Amphibia, Caudata): developmental stage early 48, CRTD AM 129. Ventral ((A) radial to the right; distal to the top) and ventral ((B) radial to the left; distal to the top) views of the muscles of the right forelimb.

digitorum breves (Figure 20.2B), which lie deep (ventral) to the main body of the extensor digitorum, originate from carpal bones, and send thin tendons to the distal phalanx of digits 2–4. The abductor et extensor digiti 1 (or "supinator manus"; Figure 20.2B) originates from carpal bones and from the distal portions of the radius and ulna and has two bundles: the most radial bundle inserts onto the radial side of the base of metacarpal I and corresponds to the abductor part of this muscle; the most ulnar bundle sends a thin tendon to the distal phalanx of digit 1 and corresponds to the extensor breves of digit 1.

Concerning the development, the first appearance of an anlage—still undifferentiated—of dorsal forearm muscles is early in stage 47 (Figure 20.3B). During the time from the middle of stage 47 to early stage 48, this anlage becomes broader but remains mainly undifferentiated (Figure 20.4B). The abductor et extensor digit 1, extensor digitorum, supinator, extensor carpi radialis, and the extensor antebrachii et carpi ulnaris are present as separate muscles at stage 50 (Figure 20.5B). In the middle of stage 52 the short extensor of digit 2 is present and the short extensor of digit 3 is starting to form, the "brachioradialis" and "extensor carpi radialis" bundles of the extensor carpi radialis are differentiated; the short extensor of digit 4 does not start to form until late during stage 53 (Figures 20.6B and 20.7B). At later stages, the overall configuration and attachments of the dorsal forearm muscles are similar to those found in adults.

DEVELOPMENT OF HAND MUSCLES

In adults, the flexores breves superficiales (Figure 20.2A) are the most ventral intrinsic muscles of the hand, extending from the aponeurosis and tendons of the flexor digitorum communis and the palmar fascia to the metacarpophalangeal joint of each digit; all digits receive an ulnar, a radial, and a central bundle from these muscles, the latter bundle being blended with the contrahens going to the same digit. The contrahentes digitorum (or "flexores breves medii"; Figure 20.2A) constitute the second layer of intrinsic hand muscles and run from the carpus to the metacarpophalangeal joints of digits 1–4. The flexores breves profundi form the third layer, running from the carpals at the base of their respective digits to the distal portion of the metacarpals I–IV, being mainly attached to the ulnar and radial sides of each metacarpal. The flexores digitorum minimi are two short, thin muscles that extend from the central part of the ventral surface of metacarpals I and IV to the metacarpophalangeal joints of digits 1 and 4, respectively, together with the flexores breves superficiales and the contrahentes digitorum. The interphalangeus digiti 3 (Figure 20.2A) connects the ventral margin of the base of the proximal phalanx of digit 3 to the proximal interphalangeal joint of this digit. The abductor digiti minimi (or "extensor lateralis digiti IV"; Figure 20.2A) extends from the ulnare to the ulnar side of the base of metacarpal IV, being proximally blended

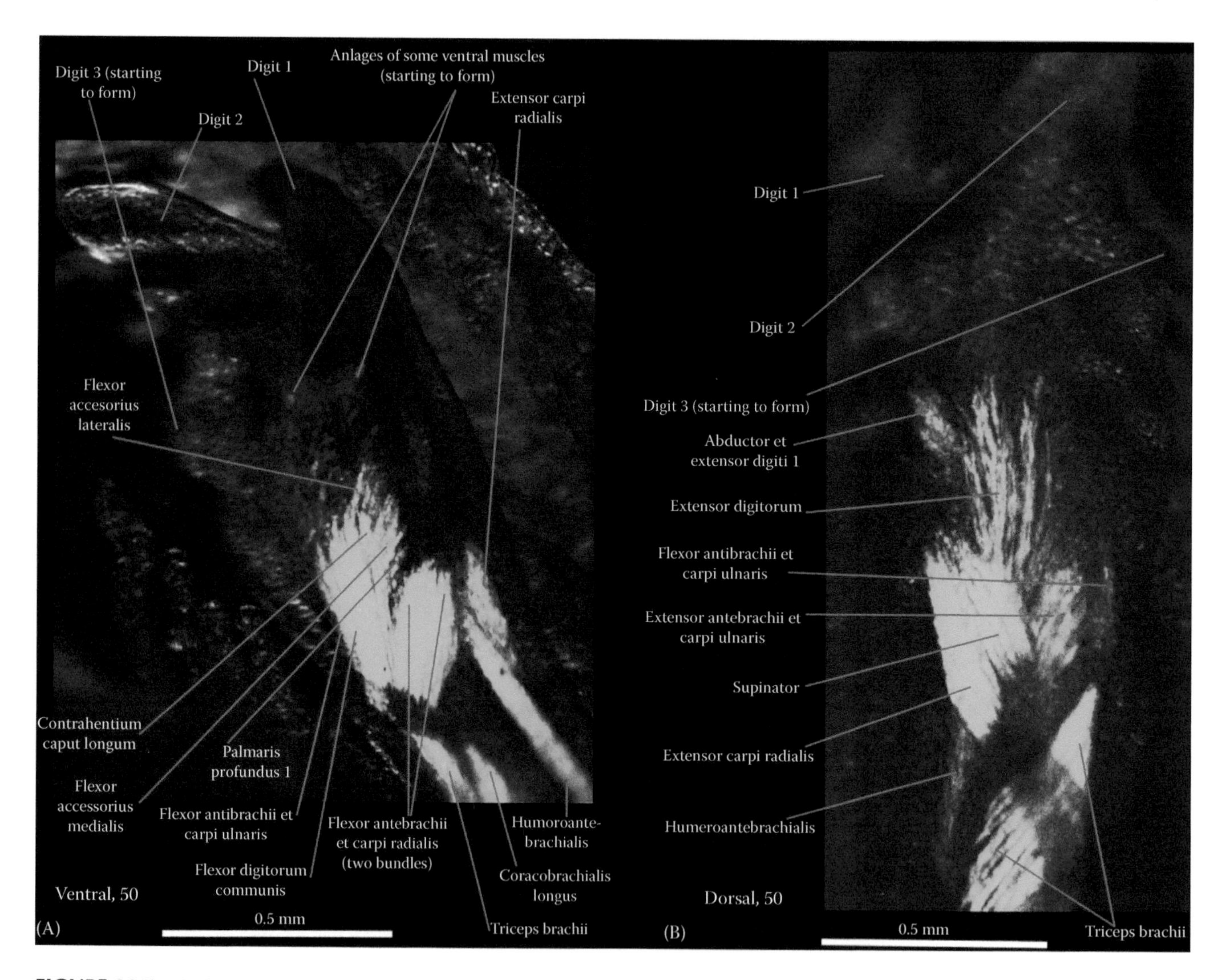

FIGURE 20.5 *Ambystoma mexicanum* (Amphibia, Caudata): developmental stage 50, CRTD AM 132. Ventral ((A) radial to the right; distal to the top) and ventral ((B) radial to the left; distal to the top) views of the muscles of the right forelimb.

to the tendon of the flexor carpi ulnaris and distally to the ulnar bundle of the flexor brevis profundus of digit 4. There are three intermetacarpales (Figure 20.2A), which extend distally and radially from the radial sides of metacarpals II–IV to the ulnar sides of metacarpals I, II, and IV, respectively.

During development, the first appearance of anlagen—still undifferentiated—of hand muscles is seen at stage 50 (Figure 20.5A). From the middle of stage 52 on, the intermetacarpalis 1 and the contrahentes and flexores breves superficiales of digits 1 and 2 are clearly visible, and the intermetacarpalis 2 and the contrahentes and flexores breves superficiales of digit 3 are visible in late stage 52 (Figure 20.6A). Early in stage 53, the intermetacarpalis 3 and the anlage of the other muscles of digit 4 are starting to form. Late in stage 53, the contrahens and flexores breves superficiales of digit 4 are differentiated and elongated, as is the abductor digiti minimi (Figure 20.7A), and the interphalangeus of digit 3 is just starting to form. The flexores breves profundi are clearly visible early in stage 54, and the interphalangeus digit 3 becomes more elongated proximodistally. From stage 55, on the configuration of the hand muscles is essentially similar to that seen in adults (NB: the deep flexores digitorum minimi could not be seen in the images of the developing forelimb we obtained).

DEVELOPMENT OF PELVIC AND THIGH MUSCLES

The ventral pelvic/thigh muscles that were visible in the images of the developing hindlimbs we obtained are the pubotibialis, femorofibularis, gracilis, ischioflexorius, and caudofemoralis. Among these muscles, the gracilis (or "puboischiotibialis") is the most ventral thigh muscle in adults (Figure 20.8A). This muscle has proximal and distal heads (only the latter head can be seen in Figure 20.8A) and extends from the ventral midline of the puboischiac plate to the proximal two-thirds of the anteromedial face of the tibia. On the fibular side of this muscle lies the ischioflexorius (Figure 20.8A), which connects the posterolateral corner of the puboischiac plate to the plantar aponeurosis and is divided into proximal and distal portions separated by a tendinous plate at the level of one-third to one-half the distance between the origin and insertion of the muscle. On the tibial side of the gracilis lies the pubotibialis (Figure 20.8A), which connects the anterolateral

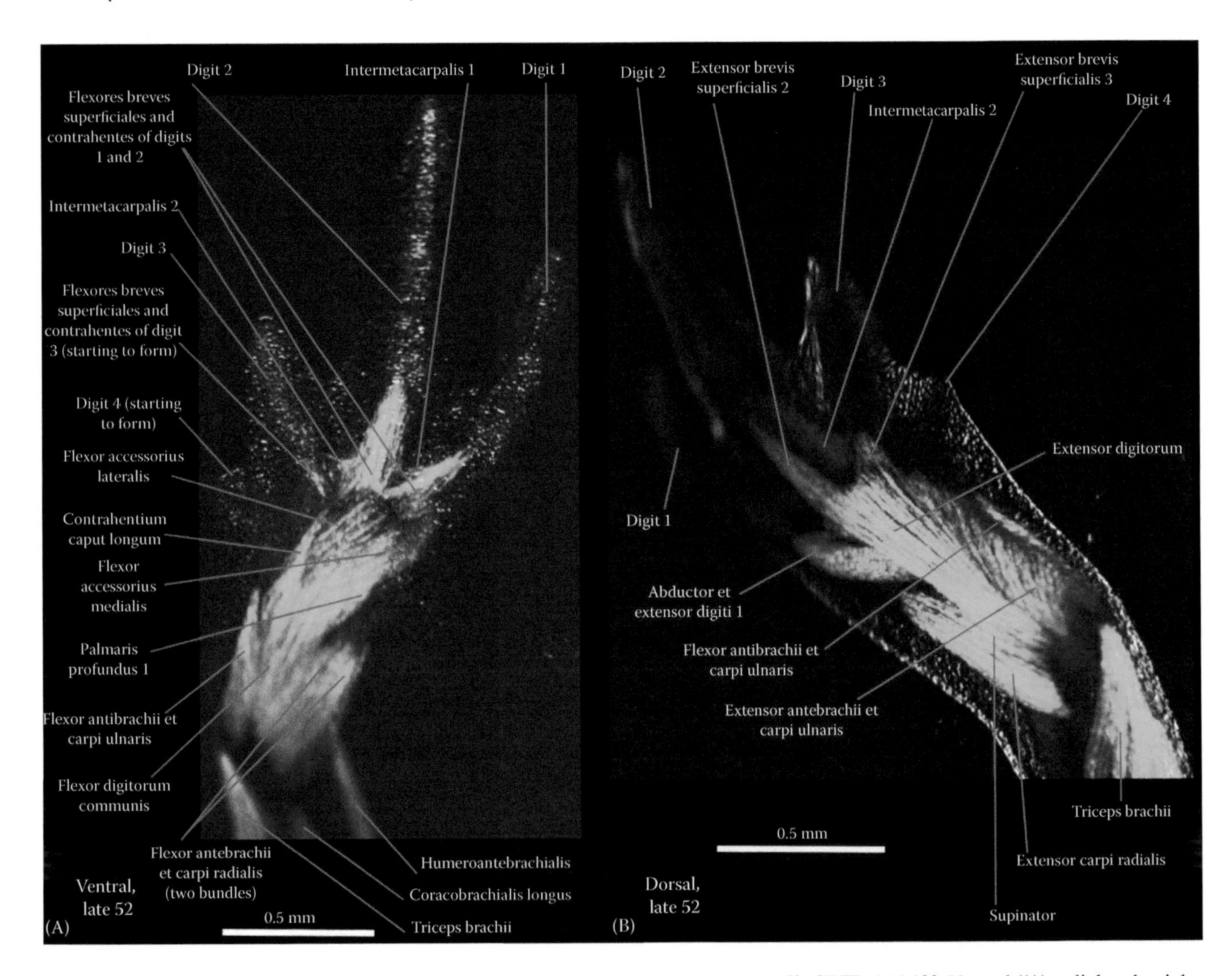

FIGURE 20.6 *Ambystoma mexicanum* (Amphibia, Caudata): developmental stage late 52, CRTD AM 132. Ventral ((A) radial to the right; distal to the top) and ventral ((B) radial to the left; distal to the top) views of the muscles of the right forelimb.

border of the puboischiac plate to the proximal tibia. The femorofibularis lies on the deep ventral side of the thigh and extends from the posteroventral border of the femur, at a point approximately halfway to the knee, to the posterolateral border of the fibula between the insertion of the extensor cruris tibialis and the fibular portion of origin of the flexor digitorum communis. The caudofemoralis attaches onto the caudal vertebrae and onto the femur (Figure 20.8B). The dorsal pelvic/thigh muscles that were visible in the images of the developing hindlimbs we obtained are the extensor iliotibialis, tenuissimus, and puboischiofemoralis internus. Among these muscles, the extensor iliotibialis (or "iliotibialis") is the most dorsal thigh muscle in adults (Figure 20.8B). This muscle is divided into anterior (iliotibialis) and posterior ("ilioextensorius") heads and extends from the ilium to a wide tendon inserting onto the crista tibialis and blending with the distal portions of the extensor cruris tibialis and the extensor digitorum longus. On the fibular side of the extensor iliotibialis lies the tenuissimus (often also designated "iliofibularis") (Figure 20.8B), which connects the tendon of the extensor iliotibialis to the posterior border of the fibula. On the tibial side of the extensor iliotibialis lies the puboischiofemoralis internus (Figure 20.8B), which connects the anterior portion of the dorsal midline of the puboischiac plate, ypsiloid cartilage, and pubis to the femur.

During development of the hindlimb, the first appearance of pelvic and thigh muscles is seen early in stage 54, in which the caudofemoralis and the dorsal muscles tenuissimus, extensor iliotibialis (apparently with only one head), and puboischiofemoralis internus are present but still very small. In the middle of stage 54, the two heads of the iliotibialis seem to be differentiated, and early in stage 55, all these three dorsal muscles become more elongated (Figure 20.9A and C). Regarding the ventral pelvic/thigh muscles, in the middle of stage 54, at least one anlage of these muscles was present, but it is difficult to discern if some muscles are already differentiated or not. Late in stage 55, the ventral muscles ischioflexorius, gracilis, and pubotibialis are clearly differentiated (Figure 20.9B). Late in stage 56, the configuration and attachments of all the analyzed pelvic/thigh muscles, including the femorofibularis, are basically similar to those found in adults (Figure 20.10A and B).

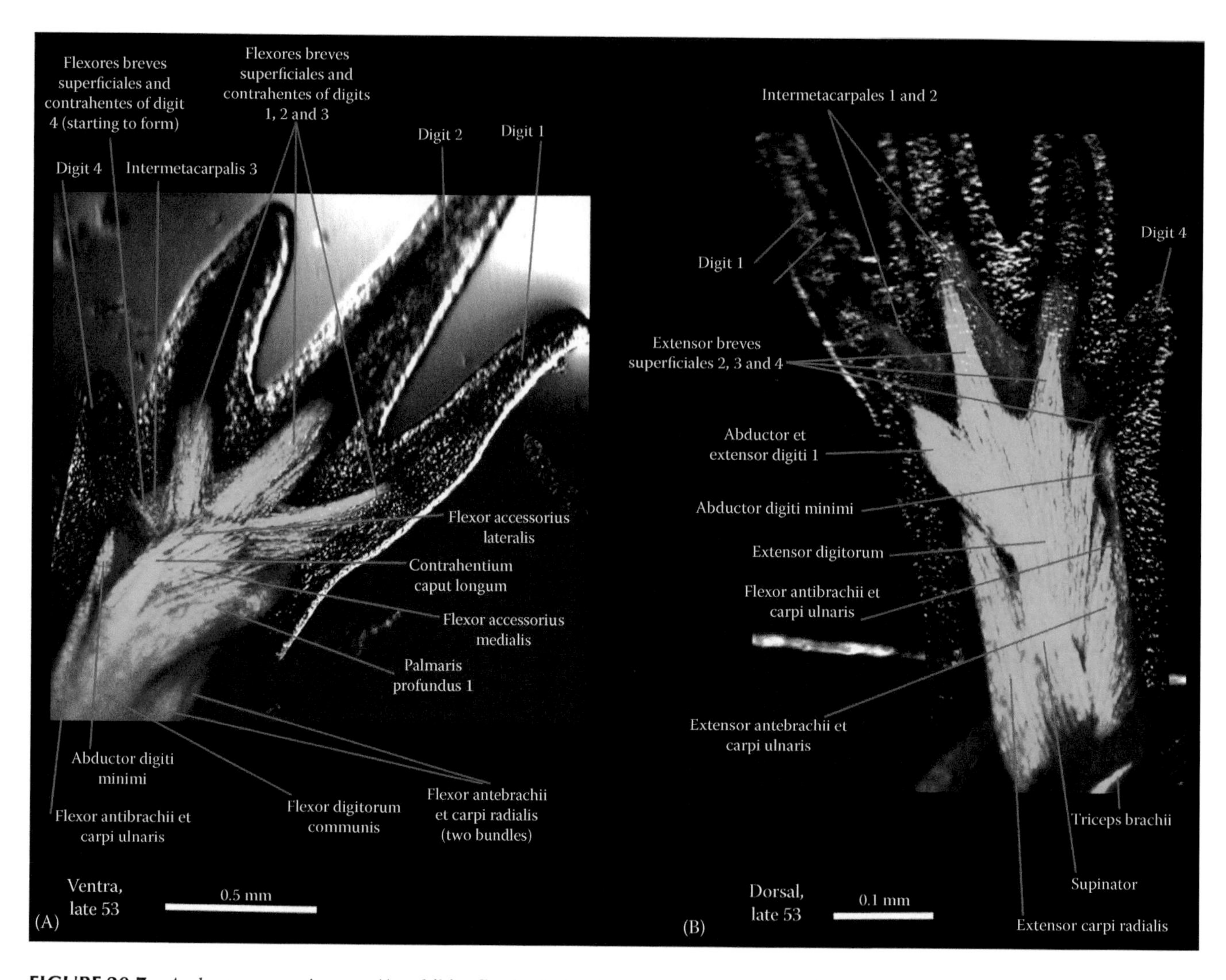

FIGURE 20.7 *Ambystoma mexicanum* (Amphibia, Caudata): developmental stage late 53, CRTD AM 134. Ventral ((A) radial to the right; distal to the top) and ventral ((B) radial to the left; distal to the top) views of the muscles of the right forelimb.

DEVELOPMENT OF VENTRAL/FLEXOR LEG MUSCLES

The ventral (flexor) leg muscles found in adults are the flexor digitorum communis, flexor accessorius medialis, flexor accessorius lateralis, contrahentium caput longum, interosseous cruris, and tibialis posterior. The most ventral of these muscles is the flexor digitorum communis (Figure 20.8A), which originates from the fibular condyle of the femur and sends a broad tendon to the distal phalanges of digits 1–5. The deeper ventral leg muscles can be seen in Figure 20.8A because this broad tendon appears as transparent in the figure. The tibialis posterior (or "pronator profundus") is the most tibial of these deeper muscles (Figure 20.8A), extending from the medial part of the fibula to the distal portion of the tibia, the tibiale, and the base of metatarsal I. On the fibular side of the tibialis posterior lies the flexor accessorius medialis (Figure 20.8A), which extends from the distal region of the fibula, the fibulare, and intermedium to the plantar fascia, then the contrahentium caput longum (Figure 20.8A), which lies deep to the plantar fascia and connects the distal portion of the fibula to the distal tarsal bones and the contrahentes, and then the flexor accessorius lateralis (Figure 20.8A), which extends from the fibulare to the plantar fascia. The interosseous cruris is the deeper ventral leg muscle, connecting the proximal part of the fibula to the distal portion of the tibia.

During development, ventral leg muscles first appear late in stage 55, in which the flexor digitorum communis is present (Figure 20.9B). Early in stage 56, deep (dorsal) to the flexor digitorum communis, a well-developed muscle anlage is visible but does not yet seem to be differentiated into the flexor accessorius lateralis, flexor accessorius medialis, contrahentium caput longum, interosseous cruris, and tibialis posterior muscles. These deep muscles become differentiated in the middle of stage 56 (Figure 20.11A), and late in stage 56, the overall configuration and attachments of the ventral leg muscles are essentially similar to those found in adults (Figure 20.10A).

DEVELOPMENT OF DORSAL/EXTENSOR LEG MUSCLES

The dorsal (extensor) leg muscles found in adults are the extensor digitorum longus, extensor tarsi tibialis, extensor

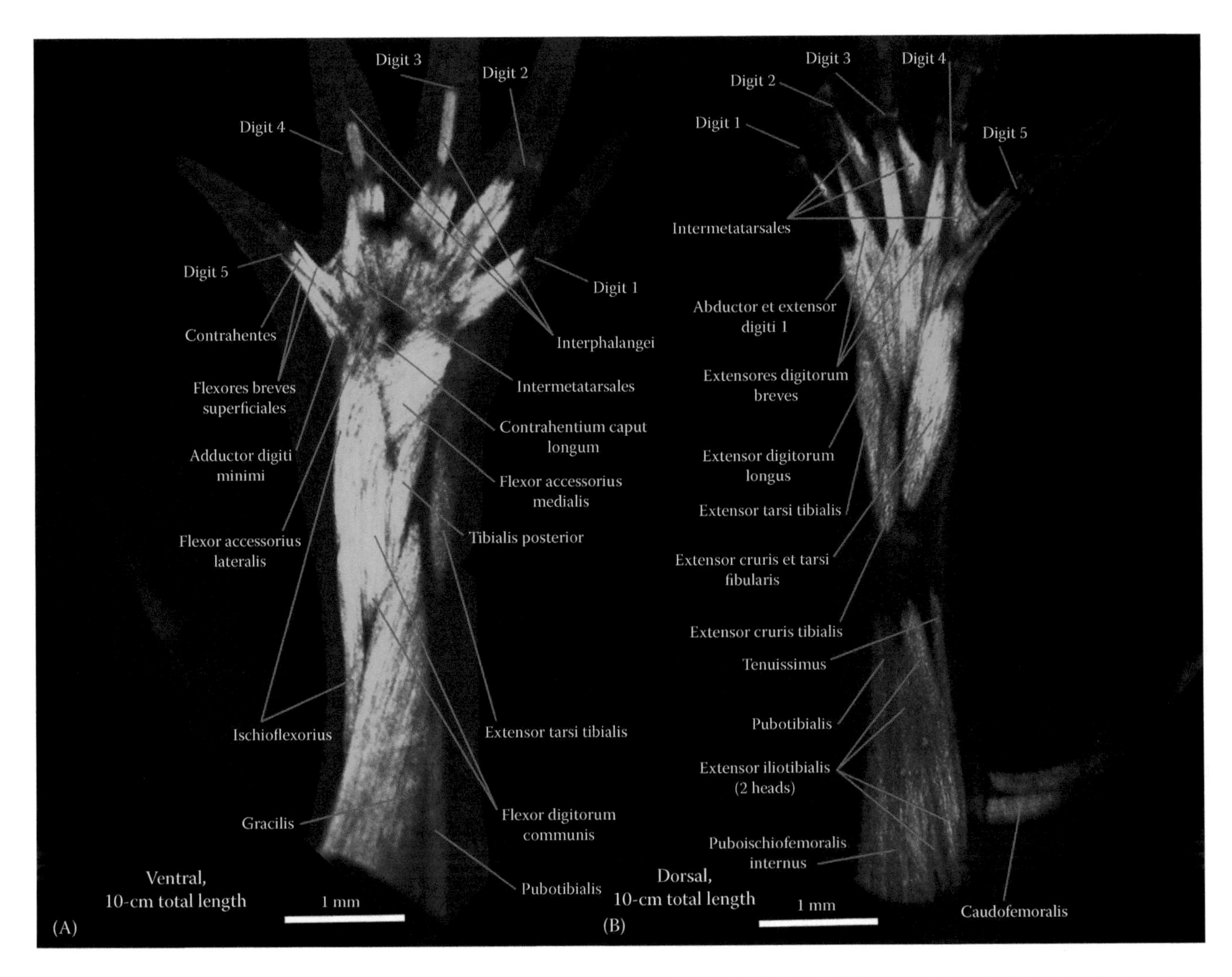

FIGURE 20.8 *Ambystoma mexicanum* (Amphibia, Caudata): specimen CRTD AM125 with 10 cm total length. (A) Ventral view of the right hindlimb muscles (tibial is to the right and distal to the top). (B) Dorsal view of the right hindlimb muscles (tibial is to the left and distal to the top).

cruris tibialis, extensor cruris et tarsi fibularis, extensores digitorum breves, and abductor et extensor digiti 1. The most dorsal of these muscles is the extensor digitorum longus (or "extensor digitorum communis") (Figure 20.8B), which extends from the femoral condyles to the proximal end of metatarsals I–V. On the tibial side of this muscle lies the extensor cruris tibialis (Figure 20.8B), extending from the tibial epicondyle of the femur to the anteroventral and anterodorsal margins of the tibia, and then the extensor tarsi tibialis (Figure 20.8B), extending from the tibial epicondyle condyle of the femur to the tibiale bone. On the fibular side of the extensor digitorum longus lies the extensor cruris et tarsi fibularis (Figure 20.8B), connecting the femoral condyles to the posterodorsal face of the fibula (the posterior part of the muscle) and the fibulare (the anterior part of the muscle). Distal to these four muscles lie the four extensores digitorum breves (Figure 20.8B), which connect the distal tarsal bones to the dorsal surface of the proximal end of the distal phalanx of digits 2–5 through a long tendon. The extensor digitorum brevis of digit 1 is fused with an abductor muscle to form the abductor et extensor digiti 1 muscle (Figure 20.8B), which extends from the distal tarsal bones to the metatarsal I and distal phalanx of digit 1.

With respect to development, the first appearance of dorsal leg muscles is early in stage 55, in which the extensor digitorum longus, extensor cruris et tarsi fibularis, and an undifferentiated extensor cruris tibialis + extensor tarsi tibialis are present (Figure 20.9A). Late in stage 55, the extensor cruris tibialis and extensor tarsi tibialis start to differentiate (Figure 20.9C). Early during stage 56, the abductor et extensor digiti 1 is present and the short extensors of digits 2 and 3 start to be clearly visible, the short extensor of digit 4 starting to be clearly visible at stage mid-56 (Figure 20.11B). At stage late 56, the short extensor of digit 5 starts to be visible (Figure 20.10B), and at stage early 57, the overall configuration and attachments of the dorsal leg muscles are essentially similar to those seen in adults (Figure 20.12B).

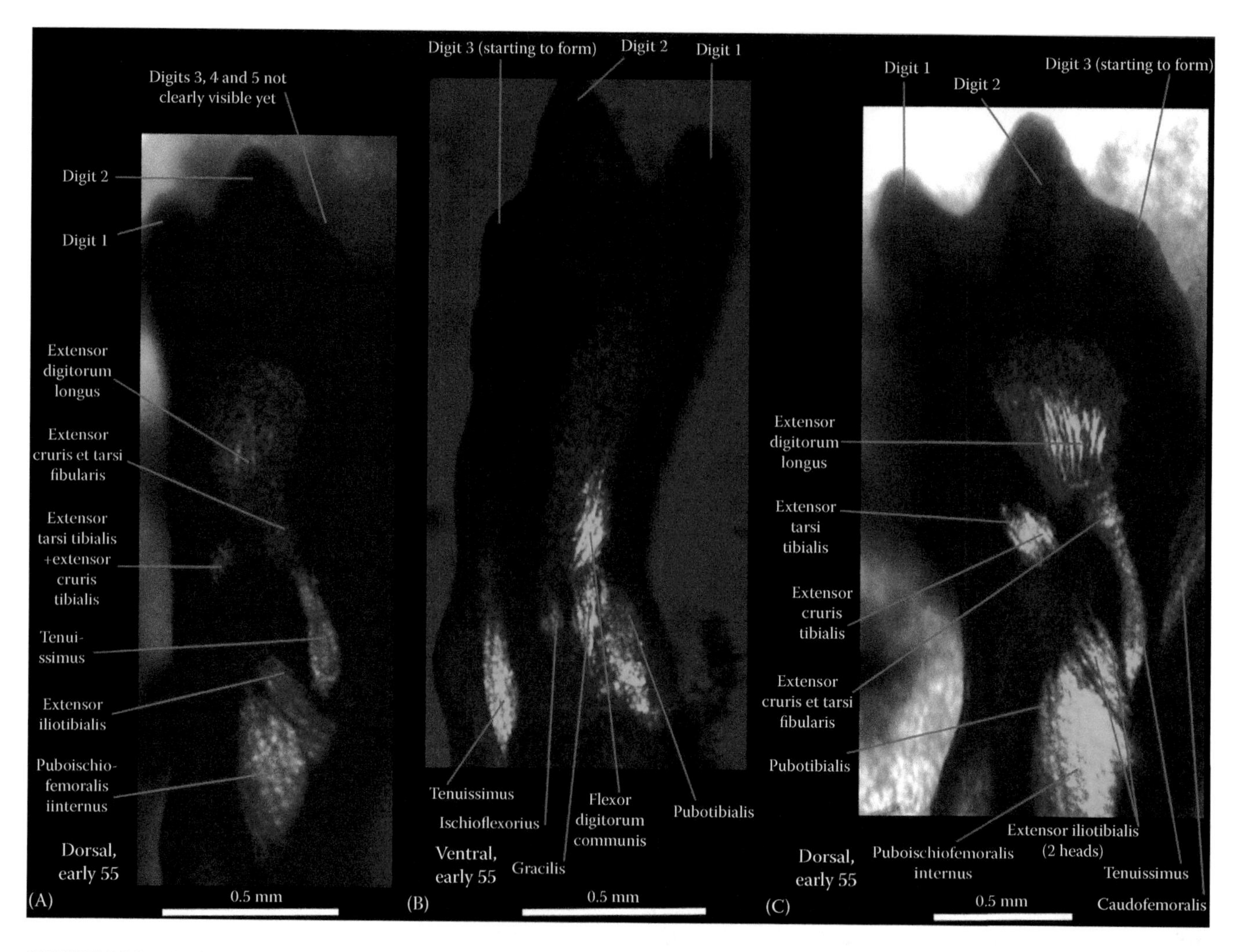

FIGURE 20.9 *Ambystoma mexicanum* (Amphibia, Caudata): developmental stages early and late 55, CRTD AM116. (A) Dorsal view of the muscles of the right hindlimb at early stage 55 (tibial is to the left and distal to the top). (B and C) Ventral ((B) tibial is to the right and distal to the top) and dorsal ((C) tibial is to the left and distal to the top) views of the muscles of the right hindlimb at late stage 55.

DEVELOPMENT OF FOOT MUSCLES

The intrinsic foot muscles found in adults are the flexores breves superficiales, flexores breves profundi, abductor digiti minimi, contrahentes pedis, flexores digitorum minimi, interphalangei, and intermetatarsales. The most ventral of these muscles are the flexores breves superficiales (Figure 20.8A), which extend from the dorsal side of the plantar fascia to metatarsals I–V and digits 2–4 (the first muscle only goes to metacarpal I, not to digit 1), and the abductor digiti minimi (Figure 20.8A), which extends from the distal end of the fibula to the fibulare, basale V, and the base of metatarsal V. The five contrahentes pedis are deep to, and lie between, the flexores breves superficiales (Figure 20.8A), connecting the tendon of the contrahentium caput longum and tarsal bones to the proximal phalanx of digits 1–5. Deep to these muscles lie the flexores breves profundi, extending from the carpal/metacarpal region to each side of digits 1–5, and then the flexores digitorum minimi, which are deep and small muscles extending from the metatarsals to the ventral side of the base of the proximal phalanx of digits 2–5. The four intermetatarsales connect the metatarsals of digits 1–5 (Figure 20.8A and B). The interphalangei are the most distal foot muscles (Figure 20.8A). Digit 3 has one interphalangeus connecting the metatarsophalangeal and first interphalangeal joints of this digit. Digit 4 has two muscles, one similar to the interphalangeus digiti 3, connecting the metatarsophalangeal and first interphalangeal joints of digit 4, and the other connecting the first and second interphalangeal joints of this digit 4.

Concerning development, the first appearance of intrinsic foot muscles is early in stage 56, in which an undifferentiated muscle anlage can be seen in the proximal and central region of the foot. In the middle of stage 56, the contrahentes and flexores breves superficiales of digits 1 and 2 are differentiated, as is the intermetatarsalis 1 lying between these two digits (Figure 20.11A). Late in stage 56, the flexores breves superficiales, flexores breves profundi, and contrahentes at digits 1–4 are formed, as are the intermetatarsales 1–4 and the abductor digiti minimi, and the flexores breves superficiales and the contrahens of digit 5 are starting to form (Figure 20.10A). Early in stage 57, the interphalangeus of digit 3 is clearly visible, and the proximal interphalangeus of digit 4 is starting to form (Figure 20.12A). Late in stage 57, the distal

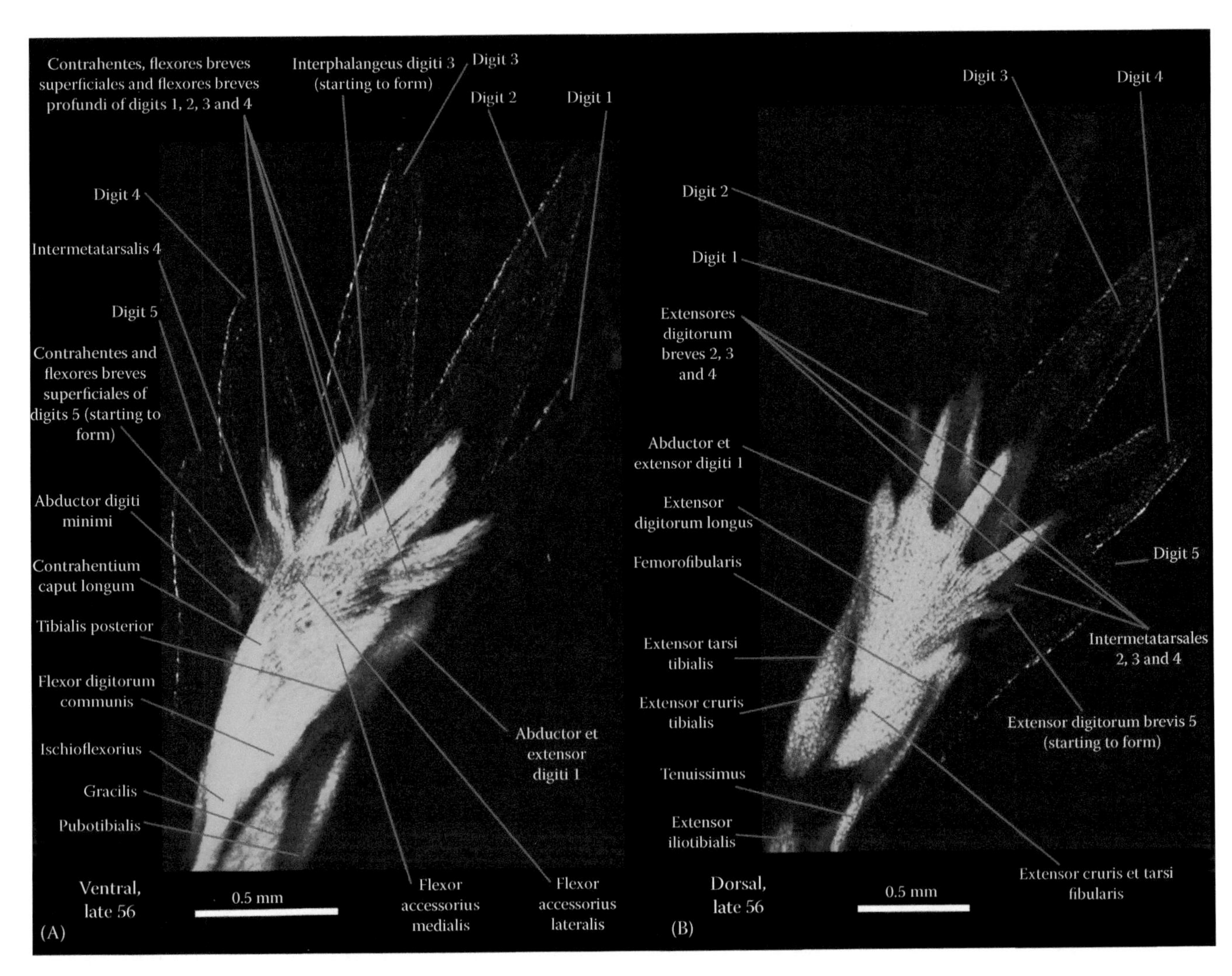

FIGURE 20.10 *Ambystoma mexicanum* (Amphibia, Caudata): developmental stage late 56, CRTD AM112. Ventral ((A) tibial is to the right and distal to the top) and dorsal ((B) tibial is to the left and distal to the top) views of the muscles of the right hindlimb.

interphalangeus of digit 4 starts to be visible, and at later stages, the overall configuration and attachments of all foot muscles are thus similar to those found in adults.

MORPHOGENESIS AND MYOLOGICAL PATTERNS

In general, our observations of the development of the axolotl fore- and hindlimb muscles agree with the scant detailed information available in the literature on the ontogeny of these muscles in axolotls and other urodeles. For instance, Grim and Carlson (1974a,b) also reported that the long flexors and extensors of the forearm form before muscles such as the abductor et extensor digiti 1, which in turn forms before most intrinsic hand muscles (Figures 20.3 through 20.7). Our observations also support Chen's (1935) statements (concerning his ontogenetic study of the aquatic salamander *Necturus*) that the latissimus dorsi, deltoideus scapularis, and triceps brachii develop from dorsal embryonic anlagen while the pectoralis, supracoracoideus, coracoradialis, procoracohumeralis, coracobrachialis, and humeroantebrachialis develop from ventral embryonic anlagen (Figures 20.3 through 20.5). Both Chen's and our ontogenetic studies of salamanders thus contradict the idea, advanced by authors such as Howell (1935) and mainly based on comparisons of adult morphology that the procoracohumeralis is a dorsal pectoral muscle. Interestingly, in the study of Diogo and Tanaka (2014), the coracobrachialis seems to develop from the anlage of, or from an anlage that lies very near the anlage of, the pectoralis (Figure 20.3). One difference between our observations in axolotls and Chen's (1935) observations in *Necturus* is that in our study, the coracobrachialis and humeroantebrachialis appear at different stages (respectively in the middle of the stages 46 and 47), while in Chen's study, the two muscles appear at the same stage (17.5 mm). Like Chen's (1935) study of *Necturus*, our study showed that the long forearm extensors form before the short extensors and that the superficial forearm flexors form before the deep forearm flexors (i.e., flexor accessorius medialis and lateralis, contrahentium caput longum, palmaris profundus 1, and pronator quadratus).

In general, our observations also agree with Boisvert et al.'s (2013) descriptions of the development of the axolotl pelvic/thigh muscles, but some of the captions of the figures provided by those authors seem to be wrong. For instance,

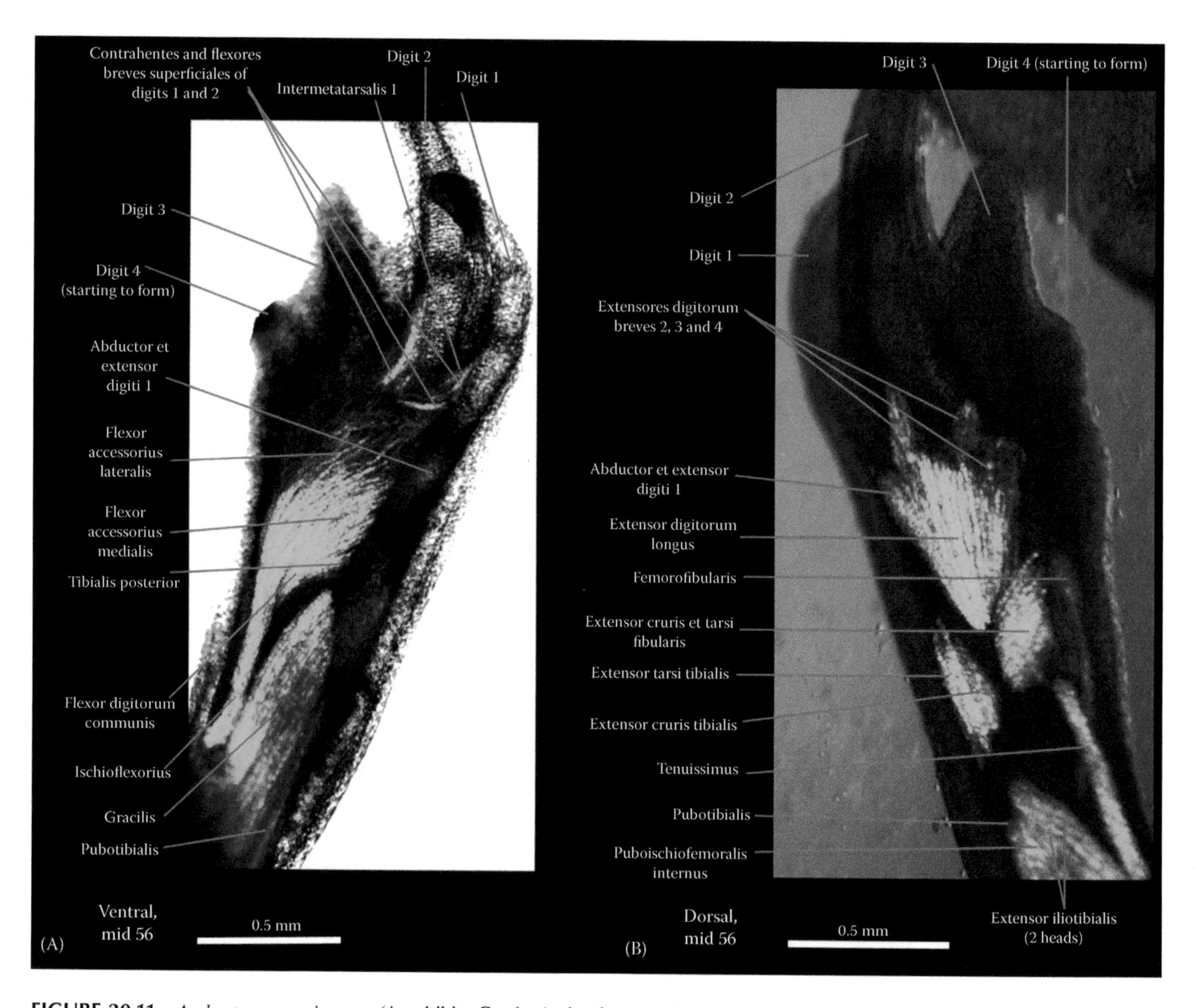

FIGURE 20.11 *Ambystoma mexicanum* (Amphibia, Caudata): developmental stage mid 56, CRTD AM111. Ventral ((A) tibial is to the right and distal to the top) and dorsal ((B) tibial is to the left and distal to the top) views of the muscles of the right hindlimb.

their Figure 4BL is an external (lateral) view of the hindlimb and can therefore only mainly show dorsal muscles (ventral hindlimb muscles usually have to be seen from a medial/ventral view). However, in their caption, they wrote ischioflexorius, which is one of the most ventral pelvic/thigh muscles. In fact, our ontogenetic observations support the idea that the extensor iliotibialis, puboischiofemoralis internus, and tenuissimus develop from dorsal, and that the gracilis, pubotibialis and ischioflexorius develop from ventral, pelvic/thigh anlagen, as shown in Chapter 19. Still concerning muscles associated with the girdles, our observations support Valasek et al.'s (2011) "in–out" mechanistic hypothesis, which was based on a review of ontogenetic data from various tetrapod taxa. According to those authors the superficial girdle muscles ("appendicular pectoral muscles" *sensu* the present book) develop by an in–out mechanism whereby the migration of myogenic cells from the somites into the limb bud is followed by their extension from the proximal limb bud out onto the thorax; the deep girdle muscles ("axial pectoral muscles" *sensu* the present book) are induced by the forelimb field that promotes myotomal extension directly from the somites. As described earlier, that the axial pectoral muscles are often deep girdle muscles probably explains why the levator scapulae and serratus anterior could not be seen in the pictures we obtained of the developing muscles of axolotls from stages 46 to 54. Like our ontogenetic study of frog limbs (Diogo and Ziermann 2014), the present study of axolotls also supports Valasek et al.'s (2011) in–out mechanism because at earlier ontogenetic stages, the appendicular pectoral muscles start to develop far from the midline, at the level of the proximal region of the arm. Only later in development do they extend medially to cover a substantial part of the ventral and dorsal surfaces of the thoracic region (compare, e.g., Figure 20.1 with Figure 20.3).

The regeneration of urodele forelimbs is a classic case study for the investigation of the morphogenesis of both hard

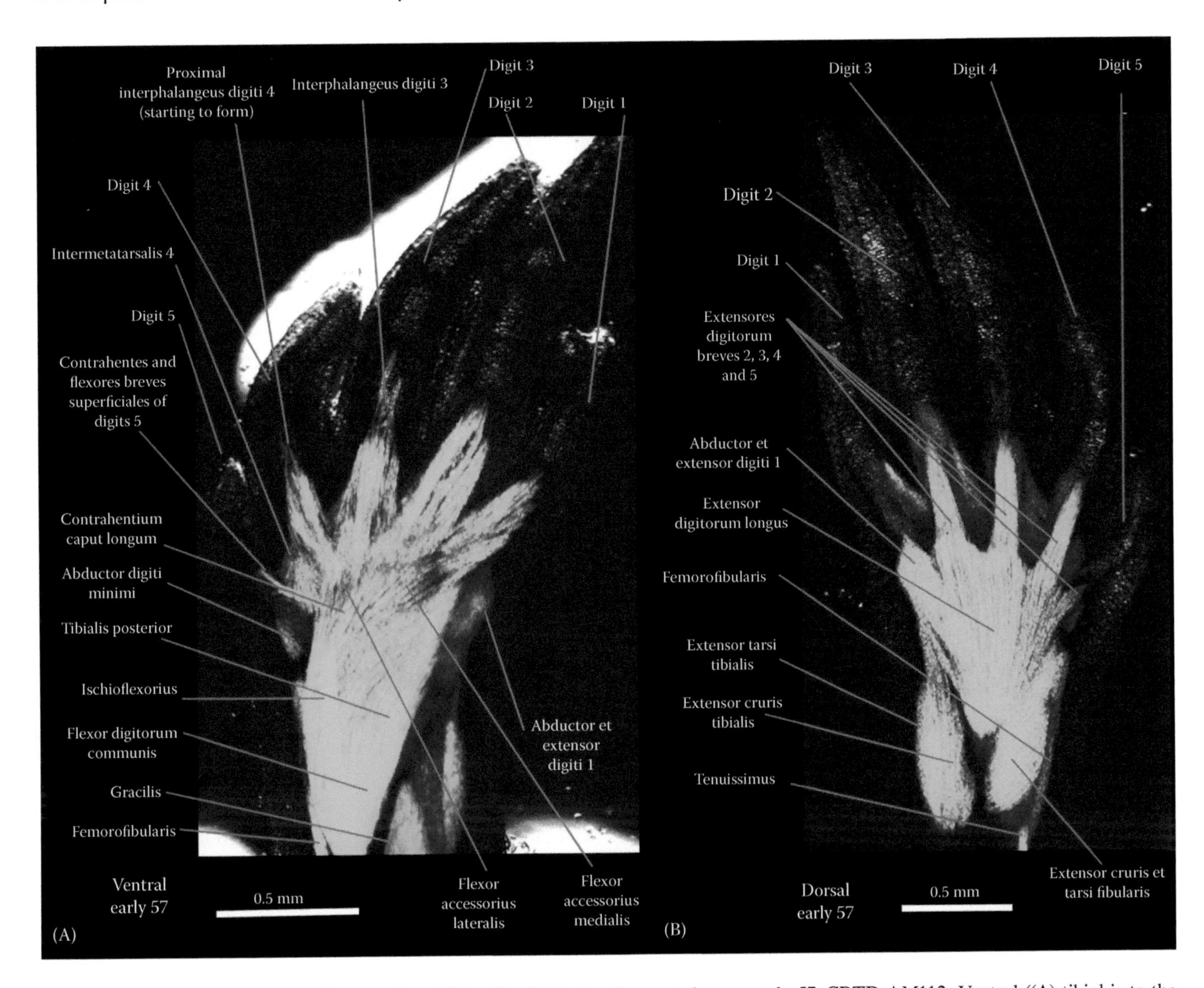

FIGURE 20.12 *Ambystoma mexicanum* (Amphibia, Caudata): developmental stage early 57, CRTD AM112. Ventral ((A) tibial is to the right and distal to the top) and dorsal ((B) tibial is to the left and distal to the top) views of the muscles of the right hindlimb.

and soft tissues (see, e.g., Carlson 2007). In the study of axolotl forelimb regeneration (Diogo et al. 2014b), the tempo and mode of the morphological events observed during regeneration were similar to those reported by other authors. The formation and differentiation of the muscles followed a proximodistal and a radioulnar gradient (e.g., Grim and Carlson 1974a,b). However, in addition to these two morphogenetic gradients described in the literature, the results of Diogo et al.'s (2014b) study indicated that there is also a marked ventrodorsal gradient during the regeneration of at least some axolotl forearm muscles. This contrasts with the results of Diogo et al.'s (2014c) study of axolotl hindlimb regeneration, in which we found proximodistal and tibiofibular morphogenetic gradients but not a ventrodorsal gradient.

The ontogeny of axolotl limb muscles reveals a similar pattern of gradients. The proximodistal gradient is clearly seen in Figures 20.3 through 20.7 and Figures 20.9 through 20.12: in the forelimb, the pectoral/arm muscles form first, followed by the forearm muscles and then by the hand muscles; in the hindlimb, the pelvic/thigh muscles are the first to form, followed by the leg and then the foot muscles. Radioulnar and tibiofibular gradients can be seen in the zeugopodium (forearm/leg). For example, early in stage 48, the flexor antebrachii et carpi radialis is differentiated and well developed, but the flexor antebrachii et carpi ulnaris still appears to be undifferentiated from the anlage of the flexor digitorum communis or is just a very small muscle covered ventrally by the latter muscle (Figure 20.4A). These gradients are particularly evident in the region of the autopodia (hand/foot); e.g., the first intrinsic hand/foot muscles and the short extensors to form and differentiate are those of the most radial/tibial digits, while the last ones to form and differentiate are those of the most ulnar/fibular digit (digit 4 in the hand; digit 5 in the foot) (Figures 20.5 through 20.7 and 20.10 through 20.12). The ventrodorsal gradient seen in forelimb ontogeny is clearly illustrated early in stage 48, in which the anlage of the dorsal forearm muscles is still relatively thin and not differentiated, while the ventral forearm muscles flexor digitorum communis and flexor antebrachii et carpi radialis are already differentiated and well developed (Figure 20.4A and B).

The presence of a ventrodorsal gradient in the regeneration and ontogeny of the forelimb, but not of the hindlimb, muscles might represent a genuine difference between the forelimbs and hindlimbs. That is, this could potentially be added to the list of differences between the tetrapod forelimbs and hindlimbs provided by Diogo et al. (2013), who proposed that these limbs are not serial homologues but instead the result of homoplasy (convergence/parallelism: see Chapters 12, 15, 16, and 19 as well as the following section). However, a dorsoventral gradient was reported in Kardon's (1998) study of the ontogeny of the hindlimb muscles of chickens. Moreover, our recent ontogenetic study of the frog *Eleutherodactylus coqui* (Diogo and Ziermann 2014) has shown that both limbs display a proximodistal muscle morphogenetic gradient and that the hindlimb displays a dorsoventral gradient, as reported in chickens by Kardon (1998). Also, instead of a radioulnar/tibiofibular gradient as seen in the regeneration and ontogeny of the musculature of both limbs of axolotls, there is mainly an ulnoradial/fibulotibial gradient in the ontogeny of the musculature of the fore- and hindlimbs of this frog. The ulnoradial muscle morphogenetic gradient observed in frogs is thus more similar to the ulnoradial gradient seen during the ontogenesis of the limb skeletal structures in other nonurodele tetrapod groups and during the ontogenesis of limb muscles seen in at least some of these groups (e.g., chickens: see, e.g., Carlson 2007).

The existence of different morphogenetic gradients of muscle formation and differentiation in the same limbs of different taxa and in different limbs of the same taxon suggests remarkable plasticity in the morphogenesis of tetrapod limbs. This plasticity makes it difficult to discern and discuss plesiomorphic states and general morphogenetic patterns in tetrapods. For instance, one hypothesis is that a general radioulnar/tibiofibular gradient might help explain why there are, for instance, more radial/tibial than ulnar/fibular muscles in tetrapod taxa from all major groups (amphibians, reptiles and mammals). That is, the radial musculature would become more differentiated than the ulnar musculature before reaching adulthood. For instance, in the axolotl hindlimb, there are two tibial dorsal long extensors (extensor cruris tibialis and extensor tarsi tibialis) and only one fibular long extensor (extensor cruris et tarsi tibialis) (Figure 20.8). A similar pattern is seen in the adult axolotl forearm, in which there are two radial dorsal long extensors (supinator and extensor carpi radialis, which is in turn divided into two bundles) and only one ulnar long extensor (extensor antebrachii et carpi ulnaris). A similar pattern is seen in the adult human forearm, which has four radial long extensors (supinator, brachioradialis, extensor carpi radialis longus, and extensor carpi radialis brevis) versus two ulnar long extensors (anconeus and extensor carpi ulnaris) (Diogo and Wood 2012a), and in the adult forelimb autopodia of humans and birds such as chickens, in which the most radial digit (thumb in humans) has more muscles associated with it than the most ulnar digit does.

However, in amphibians such as frogs and amniotes such as chickens, there is instead an ulnoradial muscle and skeletal morphogenetic gradient. It is possible that both radioulnar/tibiofibular and proximodistal gradients in both limbs represent the plesiomorphic condition for tetrapods and that this condition is explained by dependence of muscle patterning upon the patterning of connective tissue. In both axolotl development and regeneration, as well as in the ontogeny of other tetrapods, many markers and patterning genes have been extensively implicated in patterning of limb connective tissue and are upregulated in proximodistal (e.g., *Hox*, *FGF*s, *RA*) and radioulnar and/or tibiofibular fashion (e.g., *Shh*) (e.g., Gilbert 2006; Carlson 2007). Therefore, these molecules may either influence muscle patterning directly or *via* patterning of connective tissue, which might explain, at least in part, the general presence of more radial/tibial than ulnar/fibular muscles in tetrapods. In fact, some early amphibian fossils (that are likely phylogenetically more basal than the last common ancestor of extant urodeles and anurans) seem to show a mainly preaxial (radioulnar) sequence of digit development; this evidence also raises questions about whether or not the postaxial (ulnoradial) sequence of digit development seen in extant anurans and amniotes represents the plesiomorphic condition for tetrapods (see, e.g., Fröbisch and Shubin 2011).

FORE–HINDLIMB ENIGMA AND THE ANCESTRAL *BAUPLAN* OF TETRAPODS

Based on an extensive anatomical comparison of the adult muscles of the pectoral and pelvic appendages of numerous vertebrate taxa and on a review of other lines of evidence available (e.g., from paleontology, functional morphology, evo–devo, and genetics), Diogo et al. (2013), Diogo and Molnar (2014), and Miyashita and Diogo (2016), among others, argued that contrary to the commonly accepted dogma, there is no serial homology between these appendages. Instead, the enigmatic similarity between many forelimb and hindlimb structures ("fore–hindlimb enigma"), including muscles, was acquired during the "fins–limbs transition" through *derived* cooption of some similar genes for the development of the more distal parts of both the forelimb and the hindlimb ("genetic piracy" *sensu* Roth [1994]; see also the recent review of Pavlicev and Wagner [2012]). Therefore, while it is possible that the fore- and hindlimbs display "deep homology" in the developmental sense (i.e., shared gene regulatory circuits: Shubin et al. 2009), they are not homologous in the morphological or phylogenetic sense used in this book.

This subject was also discussed by Diogo and Ziermann (2014) in the context of our ontogenetic study of frog limbs. Both in the larvae and the froglets/adults of *Eleutherodactylus coqui*, there is a marked similarity between many forearm/hand and leg/foot muscles. The similarity is much more noticeable in axolotls, in which all the leg/foot and forearm/hand muscles have a clear "topological equivalent" in the other limb, with the exception of the flexor antebrachii et carpi radialis and flexor antebrachii et carpi ulnaris (see Chapter 19). The limb musculature of urodeles such as axolotls is much more similar to the limb musculature that was plesiomorphically present in tetrapods than is the musculature of frogs, as noted earlier. Therefore, the striking similarity between the zeugopodial and autopodial musculature of the fore- and hindlimb of axolotls indicates that this similarity was very likely acquired in the transitions from fins to limbs, supporting Diogo et al.'s (2013) hypothesis of nonserial homology. This hypothesis is also supported by the lack of clear similarity or correspondence between any pelvic/thigh and any pectoral/arm muscles, even in anatomically plesiomorphic tetrapods such as salamanders (compare, e.g., Figures 20.1 and 20.2 with Figure 20.8). According to Diogo et al. (2013), this is due to a phylogenetic constraint; i.e., the musculature of the girdles is phylogenetically very ancient (unlike the zeugopodial and autopodial musculature, which was only acquired in the node leading to tetrapods), and the pelvic and pectoral muscles were very different from the beginning. The muscles of the pectoral girdle are extremely different from the muscles of the pelvic girdle in not only tetrapods but also in nontetrapod gnathostomes, reinforcing the idea that the pelvic and pectoral appendages are not serial homologues (Diogo et al. 2013).

The ontogenetic observations provided in Chapter 20 are very important in this context. One possible defense of the serial homology hypothesis is the adult configuration found in tetrapods is highly modified from that seen at earlier stages of development, and the latter might provide more evidence of similarities between the muscles of the two girdles. However, the data presented in Chapter 20 show that even in anatomically plesiomorphic tetrapods such as salamanders, the muscles of the two girdles are markedly different from each other from early ontogenetic stages. Such a marked difference was also found in our recent study of frog limb development (Diogo and Ziermann 2014).

Defenders of the serial homology hypothesis still might argue that the evidence may be interpreted as a change of the developmental mechanisms during evolutionary history, i.e., that serially homologous muscles now develop in completely different ways. For example, despite the extreme anatomical divergence in the adult forelimb of tetrapods between the wings of birds and the pectoral fins of dolphins, there is no doubt that the forelimbs of different tetrapod taxa are homologous to each other. However, as explained by Diogo and Ziermann (2014), there is a clearly recognizable pattern/*Bauplan* in the development of the muscles of the forelimbs of reptiles and mammals, despite some developmental differences (e.g., heterochronic changes where some muscles develop before others do, in different taxa). Likewise, if the hard and soft tissues of the pectoral and pelvic girdle in fish and tetrapods are serially homologous, recognizable common features and *Bauplan* in their ontogeny would be expected. However, so far no researcher has been able to find any clear resemblances between either the adult configuration or the ontogeny of any single pectoral and pelvic muscle in any tetrapod taxon. As explained earlier, we could not find any resemblance within our direct comparisons between the development of the forelimbs and that of the hindlimbs of the very same, anatomically plesiomorphic, tetrapod species (*Ambystoma mexicanum*).

These new insights into the serial homology versus homoplasy of the structures of the tetrapod limbs and the new data obtained in the present work regarding the development of the limb muscles of the myologically plesiomorphic axolotls provide a more comprehensive picture of the ancestral *Bauplan* of tetrapods. It is now clear that the pelvic/thigh muscles of the first tetrapods were very topologically different from the pectoral/arm muscles, as are/were the pectoral and pelvic girdle muscles in the vast majority of fish. Although the humerus and femur clearly topologically correspond to each other, these structures very likely appeared in fish long before the origin of tetrapods, and moreover, the thigh/arm muscles topologically originate from the pelvic/pectoral girdles, which were anatomically very different in the fish that gave rise to tetrapods (e.g., Coates and Cohn 1998; Diogo et al. 2013). This explains why not only the pelvic/pectoral, but also the thigh/arm muscles are topologically very different. In contrast, the forearm/hand and leg/foot muscles are very similar due to a derived cooption, during the fin–limb transition, of some similar genes for the development of the zeugopodia and autopodia, which are mainly de novo structures, as explained earlier.

But how similar were the zeugopodial/autopodial muscles of the two limbs in the first tetrapods? As explained earlier, in myologically plesiomorphic tetrapods such as salamanders all the leg/foot and forearm/hand muscles have a clear topological equivalent in the other limb, with the exception of the flexor antebrachii et carpi radialis and flexor antebrachii et carpi ulnaris. However, there is some evidence indicating that in the first tetrapods, the musculature of the leg–foot might have been even more similar to the musculature of the forearm–hand than is the case in these anatomically plesiomorphic extant tetrapods. For instance, the comparative studies of McMurrich (1905) and Bardeen (1906) suggested that in extant salamanders, the distal (leg) portion of the ischioflexorius corresponds to an ancestral muscle flexor cruris et tarsi tibialis, while the femoral head of the flexor digitorum communis corresponds to an ancestral muscle flexor cruris et tarsi fibularis. These and most other authors agree that the femorofibularis is a thigh muscle, as we do in the present book (see Chapter 19). However, Howell (1936b) described a "femorocruralis" muscle in birds that is very similar to the femorofibularis of salamanders and that is innervated by the tibial nerve (and not by the obturator nerve as suggested by Gadow [1882]). According to Howell (1936b), this avian muscle seems to be the result of a proximal migration of part of the gastrocnemius complex, thus providing some support for the idea that the femorofibularis of amphibians might correspond to at least part of the gastrocnemius complex of reptiles and that it is a leg muscle rather than a thigh muscle. Alternatively, the avian femorocruralis might correspond to part of, or the whole, distal (leg) portion of the ischioflexorius that has migrated proximally.

The study of axolotl limb muscles sheds light on this question because the development and topology of the femorofibularis, including its fibular position in the ventral side of the leg and the position of its distal portion near and just proximal to the abductor digiti minimi, indicate that it may well correspond topologically to the flexor antebrachii et carpi ulnaris of the forearm (e.g., Figures 20.10B, 20.11B, and 20.12B compare with Figures 20.5B, 20.6B, and 20.7B). That is, our study supports the idea that the distal (leg) portion of the ischioflexorius corresponds to an ancestral muscle flexor cruris et tarsi tibialis and points out that the femorofibularis might correspond to an ancestral muscle flexor cruris et tarsi fibularis. However, we have to stress here that the more consensual view is that the salamander femorofibularis is a thigh muscle, particularly because it is said to be innervated by the sciatic nerve (e.g., Francis 1934) and because there is no such muscle or any apparent derivative of it in the leg of anurans (see Chapter 19), thus supporting the idea that this is a peculiar, derived thigh muscle of urodeles and not an important part of the tetrapod leg *Bauplan* that gave rise to amniote leg muscles. But it should also be noted that some previous authors, such as Humphry (1872a,b), have also proposed that the femorofibularis is a leg muscle that might have given rise to at least part of the short head of the biceps femoris of mammals.

If the femorofibularis and/or by part of the flexor digitorum communis represents a flexor cruris et tarsi fibularis (as suggested by McMurrich 1905 and Bardeen 1906), the ancestral *Bauplan* of the zeugopodial and autopodial musculature of the pectoral and pelvic limbs is even more similar. Thus, due to the *derived* (homoplastic) cooption of similar genes for the development of the fore- and hindlimb zeugopodia and autopodia in the fins–limbs transition, in the first tetrapods, all the forearm/hand muscles probably had a topological "equivalent" in the hindlimb. This similarity is still present, to a lesser degree, in myologically plesiomorphic tetrapods such as salamanders, which have flexor antebrachii et carpi radialis/ulnaris anlagen and seem to have flexor cruris et tarsi tibialis/fibularis anlagen. However, the similarity is not as evident in salamander adult morphology because in the forearm, the derivatives of each anlage are present as distinct muscles, while in the leg, the flexor cruris et tarsi tibialis is fused with the thigh portion of the muscle ischioflexorius, and the flexor cruris et tarsi fibularis either migrated slightly proximally (forming the femorofibularis) or became fused with the flexor digitorum communis.

These comparisons indicate that in the first tetrapods, the dorsal forearm musculature was strikingly similar not only to the dorsal leg musculature but also to the ventral forearm and to the ventral leg musculature (dorso-ventral symmetry). The forelimb has extensor antebrachii et carpi radialis/ulnaris anlagen and the hindlimb has extensor cruris et tarsi tibialis/fibularis anlagen, meaning that the first tetrapods probably had a very similar configuration in the eight sides of the four zeugopodia (i.e., dorsal and ventral sides of the four limbs), with a recognizable central muscle (extensor digitorum (longus)/flexor digitorum communis) surrounded by radial/tibial and ulnar/fibular muscles (derived by the flexor/extensor antebrachii/cruris et carpi/tarsi radialis/tibialis/ulnaris/fibularis anlagen). However, there probably were not 16 symmetrical parts within the four adult zeugopodia of these first tetrapods (i.e., radioulnar and tibiofibular symmetry as well) because the flexor/extensor antebrachii/cruris et carpi/tarsi radialis/tibialis anlagen usually give rise to more adult muscles and bundles than do the flexor/extensor antebrachii/cruris et carpi/tarsi ulnaris/fibularis anlagen. Interestingly, our ontogenetic study of axolotls revealed that the dorsoventral symmetry of the zeugopodia might concern not only the central, the radial/tibial and ulnar/fibular anlagen, but also the anlagen of the other, shorter muscles of the zeugopodia. During the development of the axolotl muscles, e.g., in stage 50 (Figure 20.5), the morphogenesis of, for instance, the forearm abductor et extensor digiti 1 is very similar to that of the palmaris profundus 1; both muscles form distally and deep to the central extensor/flexor (extensor digitorum/flexor digitorum communis) and run distally and radially to reach the region of the thumb (digit 1). In this sense, the short and deep dorsal forearm muscles extensor digitorum breves 2, 3, and 4 (which are part of the group that includes the abductor et extensor digiti 1) might topologically correspond to the short and deep ventral forearm muscles flexor accessorius medialis, contrahentium caput longum, and flexor accessorius lateralis (which are part of the group that includes the palmaris profundus 1). However, further studies, including (ideally) mechanistic developmental studies of a wide range of tetrapods, are clearly needed to test this hypothesis.

GENERAL REMARKS

The more we know about the comparative anatomy and developmental biology of tetrapod limbs, the better we understand the origin and ancestral *Bauplan* of the tetrapod limbs, and the more strongly the fore–hindlimb serial homology hypothesis is contradicted (NB: for recent discussion on our use of the term *Bauplan*, see Diogo et al. 2015d, 2015e, 2016c; Diogo 2017). However, much remains to be done, and it is striking that some essential aspects of tetrapod limb morphogenesis (e.g., the occurrence of a ventrodorsal morphogenetic gradient in axolotl forelimb muscle formation and differentiation) are only now being discovered and reported. We plan to continue studying the ontogeny of tetrapod limbs; specifically, after studying frog and axolotl development, we plan to analyze the ontogeny of limb muscles in reptiles and mammals and to undertake mechanistic developmental studies in all tetrapod groups. For example, we are now starting a project together with other colleagues (e.g., Raul Diaz) on the development and genetics of limb development of chameleons, which are proving to be a very good model organism to study not only macroevolution but also human birth defects (e.g., Diogo et al. 2017b; Molnar et al. 2017a).

However, a major goal of this chapter, and of this book in general, is to stimulate other researchers to investigate the comparative anatomy, evolution, and development of muscles, not only in tetrapods but also in chordates as a whole, and if possible in other organisms. Only a multidisciplinary, comprehensive, collaborative approach can help solve the crucial and fascinating questions that remain to be answered about the origin, evolution, development, and pathologies of muscles. For us, this is the end of a long journey through dissection rooms, natural history museums, hospitals, and libraries. But it is also a new beginning and an opportunity to explore new ideas and new lines of research that, either inside or outside of any of these spaces, will surely continue to pursue the most puzzling, charming, and fascinating subjects among natural sciences and, in our opinion, among sciences as a whole: the evolution and normal and abnormal development of biological organisms.

References

Abdala V, Diogo R (2010) Comparative anatomy, homologies and evolution of the pectoral and forelimb musculature of tetrapods with special attention to extant limbed amphibians and reptiles: The pectoral and forelimb musculature of tetrapods. *J Anat* 217, 536–573.

Abdala V, Grizante M, Diogo R, Molnar J, Kohlsdorf T (2015) Musculoskeletal anatomical changes that accompany limb reduction in lizards. *J Morphol* 276, 1290–1310.

Abdala V, Manzano AS, Herrel A (2008) The distal forelimb musculature in aquatic and terrestrial turtles: Phylogeny or environmental constraints? *J Anat* 213, 159–172.

Abdala V, Manzano AS, Nieto L, Diogo R (2010) Comparative myology of Leiosauridae (Squamata) and its bearing on their phylogenetic relationships. *Belg J Zool* 139, 109–123.

Abdala V, Manzano AS, Tulli MJ, Herrel A (2009) The tendinous patterns in the palmar surface of the lizard manus: Functional consequences for grasping ability. *Anat Rec Adv Integr Anat Evol Biol* 292, 842–853.

Abdala V, Moro S (1996) Cranial musculature of South American Gekkonidae. *J Morphol* 229, 59–70.

Abdala V, Moro S (2003) A cladistic analysis of ten lizard families (Reptilia: Squamata) based on cranial musculature. *Russ J Herpetol* 10, 53–78.

Abdala V, Moro S (2006) Comparative myology of the forelimb of *Liolaemus* sand lizards (Liolaemidae). *Acta Zool* 87, 1–12.

Abramowitz I (1955) On the existence of a palmar interosseous muscle in the thumb with particular reference to the Bantle-speaking Negro. *South African J Sci* 51, 270–276.

Adamicka P, Ahnelt H (1992) Two jaw articulations in *Latimeria chalumnae* (Actinistia, Coelacanthidae). *Zool Jb Anat* 122, 107–112.

Adams LA (1919) A memoir on the phylogeny of the jaw muscles in recent and fossils vertebrates. *Annals NY Acad Sci* 28, 51–166.

Adams SB, Wheeler JFG, Edgeworth FH (1929) On the innervation of the platysma and the mandibulo-auricularis. *J Anat* 63, 242–252.

Adriaens D, Decleyre D, Verraes W (1993) Morphology of the pectoral girdle in *Pomatoschistus lozanoi* De Buen, 1923 (Gobiidae), in relation to pectoral fin adduction. *Belg J Zool* 123, 135–157.

Agathon A, Thisse C, Thisse B (2003) The molecular nature of the zebrafish tail organizer. *Nature* 424, 448–452.

Agnarsson I, Coddington JA (2007) Quantitative tests of primary homology. *Cladistics* 23, 1–11.

Ahlberg PE (1989) Paired fin skeletons and relationships of the fossil group Porolepiformes (Osteichthyes: Sarcopterygii). *Zool J Lin Soc* 96, 119–166.

Ahlberg PE (2011) Humeral homology and the origin of the tetrapod elbow: A reinterpretation of the enigmatic specimens ANSP 21350 and GSM 104536. *Spec Pap Palaeontol* 86, 17–29.

Albrecht PE (1876) *Beitrag zur Morphologie des M. omo-hyoides und der ventralen inneren Interbranchialmusculatur in der Reihe der Wirbelthiere.* PhD Dissertation, Universität zu Kiel, Kiel.

Alcock R (1898) The peripheral distribution of the cranial nerves of ammocoetes. *J Anat Physiol* 33, 131–154.

Alexander R McN (1973) Jaw mechanisms of the coelacanth *Latimeria*. *Copeia* 1973, 156–158.

Allen WL, Stevens M, Higham JP (2014) Character displacement of Cercopithecini primate visual signals. *Nat Commun* 5, 4266.

Allis EP (1897) The cranial muscles and cranial nerves of *Amia calva*. *J Morphol* 12, 487–737.

Allis EP (1917) The prechordal portion of the chondrocranium of *Chimaera colliei*. *Proc Zool Soc Lond* 1917, 105–143.

Allis EP (1919) The lips and the nasal apertures in the gnathostome fishes. *J Morphol* 32, 123–221.

Allis EP (1922) The cranial anatomy of *Polypterus*, with special reference to *Polypterus bichir*. *J Anat* 56, 180–294.

Allis EP (1923) The cranial anatomy of *Chlamydoselachus anguineus*. *Acta Zool* 4, 162–219.

Allis EP (1931) Concerning the mouth opening and certain features of the visceral endoskeleton of *Cephalaspis*. *J Anat* 65, 509–527.

Anderson CV, Higham TE (2014) Chameleon anatomy. In *The Biology of Chameleons* (eds. Tolley KA, Herrel, A), pp. 7–55. University of California Press, Berkeley, CA.

Anderson PSL (2008) Cranial muscle homology across modern gnathostomes. *Biol J Linn Soc* 94, 195–216.

Anderson RJ (1880) A variety of the mylopharyngeus and other unusual muscular abnormalities. *J Anat Physiol* 14, 357–359.

Anderson RJ (1881) The morphology of the muscles of the tongue and pharynx. *J Anat Physiol* 15, 382–391.

Andrew RJ (1963) Evolution of facial expression. *Science* 142, 1034–1041.

Ang S-L, Conlon RA, Jin O, Rossant J (1994) Positive and negative signals from mesoderm regulate the expression of mouse *Otx2* in ectoderm explants. *Development* 120, 2979–2989.

Anthwal N, Joshi L, Tucker AS (2013) Evolution of the mammalian middle ear and jaw: Adaptations and novel structures. *J Anat* 222, 147–160.

Anthony J (1980) Évocation des travaux français sur *Latimeria* notamment depuis 1972. *Proc R Soc Lond B* 208, 349–367.

Appleton AB (1928) The muscles and nerves of the post-axial region of the tetrapod thigh. *J Anat* 62, 364–438.

Ashley-Ross MA (1992) The comparative myology of the thigh and crus in the salamanders *Ambystoma tigrinum* and *Dicamptodon tenebrosus* (1992). *J Morphol* 211,147–163.

Aziz MA (1981) Possible "atavistic" structures in human aneuploids. *Am J Phys Anthropol* 54, 347–353.

Aziz MA, Cowie RJ, Skinner CE, Abdi TS, Orzame G (1998) Are the two heads of the human lateral pterygoid separate muscles? A perspective based on their nerve supply. *J Orofac Pain* 2, 226–239.

Balfour FM (1878) The development of the elasmobranchial fishes. *J Anat Physiol* 11, 405–706.

Balfour FM (1881). On the development of the skeleton of the paired fins of Elasmobranchii, considered in relation to its bearings on the nature of the limbs of the vertebrata. *Proc Zool Soc Lond* 1881, 656–671.

Ballard WW, Mellinger J, Lechenault H (1993) A series of normal stages for development of Scyliorhinus canicula, the lesser spotted dogfish (Chondrichthyes: Scyliorhinidae). *J Exp Zool* 267, 318–336.

Bardeen CR (1906) Development and variation of the nerves and the musculature of the inferior extremity and of the neighboring regions of the trunk in man. *Am J Anat* 6, 259–390.

Barghusen HR (1968) The lower jaw of cynodonts (Reptilia, Therapsida) and the evolutionary origin of mammal-like adductor jaw musculature. *Postilla* 116, 1–49.

Barghusen HR (1986) On the evolutionary origin of the therian tensor veli palatini and tensor tympani muscles. In *The Ecology and Niology of Mammal-like Reptiles* (eds. Hotton N, MacLean PD, Roth JJ, Roth EC), pp. 253–262. Smithsonian Institution Press, Washington, DC.

Barrow JR, Capecchi MR (1999) Compensatory defects associated with mutations in *Hoxa1* restore normal palatogenesis to *Hoxa2* mutants. *Development* 126, 5011–5026.

Bartsch P (1992) On the constructional anatomy of the jaw suspension and the cranial base in the larva of *Neoceratodus forsteri* (Krefft 1870). *Zool Jb Anat* 122, 113–127.

Bartsch P (1993) Development of the snout in the Australian lungfish *Neoceratodus forsteri* (Krefft, 1870), with special reference to cranial nerves. *Acta Zool (Stockh)* 74, 15–29.

Bartsch P (1994) Development of the cranium of *Neoceratodus forsteri*, with a discussion of the suspensorium and the opercular apparatus in Dipnoi. *Zoomorphology* 114, 1–31.

Bauer WJ (1992) A contribution to the morphology of the m. interhyoideus posterior (VII) of urodele Amphibia. *Zool Jb Anat* 122, 129–139.

Bauer WJ (1997) A contribution to the morphology of visceral jaw-opening muscles of urodeles (Amphibia: Caudata). *J Morphol* 233, 77–97.

Baumel JJ, King AS, Lucas AM, Breazile JE, Evans HE (1979) *Nomina Anatomica Avium*. New York: Academic Press.

Baur G (1886) Osteologische Notizen über Reptilien. *Zool Anz* 685–690.

Bello-Hellegouarch G, Aziz MA, Ferrero EM, Kern M, Francis N, Diogo R (2013) Pollical palmar interosseous muscle (*musculus adductor pollicis accessorius*): Attachments, innervation, variations, phylogeny, and implications for human evolution and medicine. *J Morphol* 274, 275–293.

Bemis WE (1984) Paedomorphosis and the evolution of Dipnoi. *Paleobiology* 10, 293–307.

Bemis WE (1986) Feeding mechanisms of living Dipnoi: Anatomy and function. *J Morphol*, suppl. 1, 249–275.

Bemis WE, Findeis EK, Grande L (1997) An overview of Acipenseriformes. *Environ Biol Fish* 48, 25–71.

Bemis WE, Findeis EK, Grande L (2002) An overview of Acipenseriformes. In *Sturgeon Biodiversity and Conservation* (eds. Birstein VJ, Waldman JR, Bemis WE), pp. 25–71. Kluwer Academic Publishers, Dordrecht.

Bemis WE, Lauder CV (1986) Morphology and function of the feeding apparatus of the lungfish, *Lepidosiren paradoxa* (Dipnoi). *J Morphol* 187, 81–108.

Bemis WE, Schwenk K, Wake MH (1983) Morphology and function of the feeding apparatus in *Dermophis mexicanus*. *Zool J Linn Soc* 77, 75–96.

Bird NC, Mabee PM (2003) Developmental morphology of the axial skeleton of the zebrafish, *Danio rerio* (Ostariophysi: Cyprinidae). *Dev Dyn* 228, 337–357.

Birou G, Garcier JM, Guillot M, Vanneuville G, Chazal J (1991) A study of the lateral pterygoid muscle: Anatomic sections and CT appearances. *Surg Radiol Anat* 13, 307–311.

Birstein VJ, Doukakis P, DeSalle R (2002) Molecular phylogeny of Acipenseridae: Nonmonophyly of Scaphirhynchinae. *Copeia* 2002, 287–301.

Bischoff TLW (1840) Description anatomique du *Lepidosiren paradoxa*. *Ann Sci Nat* 14, 116–159.

Bishop PJ (2014) The humerus of *Ossinodus pueri*, a stem tetrapod from the Carboniferous of Gondwana, and the early evolution of the tetrapod forelimb. *Aust J Palaeontol* 38, 209–238.

Blair JE, Hedges SB (2005) Molecular phylogeny and divergence times of deuterostome animals. *Mol Biol Evol* 22: 2275–2284.

Blake RW (1979) The mechanics of labriform locomotion I: Labriform locomotion in the angelfish (*Pterophyllum eimekei*): An analysis of the power stroke. *J Exp Biol* 82, 255–271.

Boas JEV, Paulli S (1908) *The Elephant Head: Studies in the Comparative Anatomy of the Organs of the Head of the Indian Elephant and Other Mammals, 1: The Facial Muscles and the Proboscis*. Gustav Fischer, Jena.

Boas JEV, Paulli, S (1925) *The Elephant's Head: Studies in the Comparative Anatomy of the Organs of the Head of the Indian Elephant and Other Mammals*, 2. Copenhagen: Folio: Gustav Fisher, Jena.

Böck P (1989) *Romeis Mikroskopische Technik*. Urban and Schwarzenberg, München.

Boisvert CA, Joss JM, Ahlberg PE (2013) Comparative pelvic development of the axolotl (*Ambystoma mexicanum*) and the Australian lungfish (*Neoceratodus forsteri*): Conservation and innovation across the fish-tetrapod transition. *Evodevo* 4, 3.

Boisvert CA, Mark-Kurik E, Ahlberg PE (2008) The pectoral fin of *Panderichthys* and the origin of digits. *Nature* 456, 636–638.

Bone Q (1960) The central nervous system in amphioxus. *J Comp Neurol* 115, 27–64.

Bonner JT (1988) *The Evolution of Complexity by Means of Natural Selection*. Princeton University Press, Princeton, NJ.

Boord RL, Sperry DG (1991) Topography and nerve supply of the cucullaris (trapezius) of skates. *J Morphol* 207, 165–172.

Borden WC (1998) Phylogeny of the unicornfishes (*Naso*, Acanthuridae) based on soft anatomy. *Copeia* 1998, 104–113.

Borden WC (1999) Comparative myology of the unicornfishes, *Naso* (Acanthuridae, Percomorpha), with implications for phylogenetic analysis. *J Morphol* 239, 191–224.

Bordzilovskaya NP, Dettlaff TA, Duhon ST, Malacinski GM (1989) Developmental-stage series of axolotl embryos. In *Developmental Biology of the Axolotl* (eds. Armstrong JB, Malacinski GM). Oxford University Press, New York.

Boruel X, Noden DM (2004) Normal and aberrant craniofacial myogenesis by grafted trunk somitic and segmental plate mesoderm. *Development* 131, 3967–3980.

Boto L, Doadrio I, Diogo R (2009) Prebiotic world, macroevolution, and Darwin's theory: A new insight. *Biol and Philos* 24, 119–128.

Brandell BR (1965) Innervation of the hand muscles of *Didelphis Marsupialis Virginiana*, Kerr. *J Morphol* 116, 133–139.

Brand-Saberi B, ed. (2002) *Vertebrate Myogenesis*. Springer-Verlag, Berlin.

Braus H (1941) Die Muskeln und Nerven der Ceratodus-Vösse. *Semons Zool Forschungsreisen* 1, 139–300.

Brinkmann H, Venkatesh B, Brenner S, Meyer A (2004) Nuclear protein-coding genes support lungfish and not the coelacanth as the closest living relatives of land vertebrates. *PNAS* 101, 4900–4905.

Brock GT (1938) The cranial muscles of the *Gecko*—A general account with a comparison of muscles in other gnathostomes. *Proc Zool Soc Lond (Ser B)* 108, 735–761.

Bronner ME, LeDouarin NM (2012) Development and evolution of the neural crest: An overview. *Dev Biol* 366, 2–9.

Brooks HSJ (1886a) On the morphology of the intrinsic muscles of the little finger, with some observations on the ulnar head of the short flexor of the thumb. *J Anat Physiol* 20, 644–661.

Brooks HSJ (1886b) Variations in the nerve supply of the flexor brevis pollicis muscle. *J Anat Physiol* 20, 641–644.

Brooks HSJ (1887) On the short muscles of the pollex and hallux of the anthropoid apes, with special reference to the opponens hallucis. *J Anat Physiol* 22, 78–95.

Brooks HSJ (1889) Morphology of the muscles on the extensor aspect of the middle and distal segments of the limbs: With an account of the various paths which are adopted by the nerve-trunks in these segments. *Stud Mus Zool Univ Coll Dundee* 1, 1–17.

Brousseau RA (1976a) The pectoral anatomy of selected Ostariophysi: I. The Characiniformes. *J Morphol* 148, 89–135.

Brousseau RA (1976b) The pectoral anatomy of selected Ostariophysi: II. The Cypriniformes and Siluriformes. *J Morphol* 150, 79–115.

Bryant MD (1945) Phylogeny of Nearctic Sciuridae. *Am Midland Nat* 33, 257–390.

Bunnell S (1942) Surgery of the intrinsic muscles of the hand other than those producing opposition of the thumb. *J Bone Joint Surg Am* 24, 1–31.

Burke AC, Feduccia A (1997) Developmental patterns and the identification of homologies in the avian hand. *Science* 278, 666–668.

Burrows AM (2008) The facial expression musculature in primates and its evolutionary significance. *Bioessays* 30, 212–225.

Burrows A, Diogo R, Waller BM, Bonar CJ, Liebal K (2011) Morphology of facial expression in a monogamous ape: Evaluating the relative influences of ecological and phylogenetic factors in hylobatids. *Anat Rec* 294, 645–663.

Burrows AM, Smith TD (2003) Muscles of facial expression in *Otolemur* with a comparison to Lemuroidea. *Anat Rec* 274, 827–836.

Burrows AM, Waller BM, Parr LA, Bonar CJ (2006) Muscles of facial expression in the chimpanzee (*Pan troglodytes*): Descriptive, comparative and phylogenetic contexts. *J Anat* 208, 153–167.

Burton TC (1998) Pointing the way: The distribution and evolution of some characters of the finger muscles of frogs. *Am Mus Nov* 3229, 1–13.

Burtt EH (1986) An analysis of physical, physiological, and optical aspects of avian coloration with emphasis on wood-warblers. *Ornithol Monogr.*

Burtt Jr. EH, Ichida JM (2004) Gloger's rule, feather-degrading bacteria, and color variation among song sparrows. *The Condor* 106, 681–686.

Busbey AB (1989) Form and function of the feeding apparatus of *Alligator mississippiensis. J Morphol* 202, 99–127.

Butler AB, Hodos W (2005) *Comparative Vertebrate Neuroanatomy: Evolution and Adaptation.* Wiley Interscience, Hoboken, NJ.

Cameron CB (2002) The anatomy, life habits, and later development of a new species of enteropneust, *Harrimania planktophilus* (Hemichordata: Harrimaniidae) from Barkley Sound. *Biol Bull* 202, 182–191.

Camp CL (1923) Classification of the lizards. *Bull Amer Mus Nat Hist* 48, 289–481.

Campbell B (1939) The comparative anatomy of the dorsal interosseous muscles. *Anat Rec* 73,115–125.

Campbell KSW, Barwick RE (1986) Paleozoic lungfishes—A review. *J Morphol*, suppl 1, 93–131.

Candiani S (2012) Focus on miRNAs evolution: A perspective from amphioxus. *Brief Func Genomics*, els004.

Cannatella DC (1999) 4. Architecture: Cranial and axial musculoskeleton. In *Tadpoles—The Biology of Anuran Larvae* (eds. McDiarmid RW, Altig R). University of Chicago Press, Chicago, IL.

Carlson BM (2007) *Principles of Regenerative Biology.* Elsevier, Amsterdam.

Caro T (2005) The adaptive significance of coloration in mammals. *Bioscience* 55, 125–136.

Caron J-B, Morris SC, Cameron CB (2013) Tubicolous enteropneusts from the Cambrian period. *Nature* 495, 503–506.

Carroll AM, Wainwright PC (2003) Functional morphology of feeding in the sturgeon, *Scaphirhyncus albus. J Morphol* 256, 270–284.

Carroll RL (2007) The Paleozoic ancestry of salamanders, frogs and caecilians. *Zool J Linn Soc* 150, 1–140.

Carroll RL, Holmes R (1980) The skull and jaw musculature as guides to the ancestry of salamanders. *Zool J Linn Soc Lond* 68, 1–40.

Carroll SB (2005) *The New Science of Evo–Devo: Endless Forms Most Beautiful.* W. W. Norton and Company, New York.

Carroll SB, Grenier JK, Weathrbee SD (2005) *From DNA to Diversity—Molecular Genetics and the Evolution of Animal Design* (2nd Ed.). Blackwell Science, Malden, MA.

Carvajal JJ, Cox D, Summerbell D, Rigby PW (2001) A BAC transgenic analysis of the Mrf4/Myf5 locus reveals interdigitated elements that control activation and maintenance of gene expression during muscle development. *Development* 128, 1857–1868.

Cerny R, Cattell M, Sauka-Spengler T, Bronner-Fraser M, Yu F, Meulemans Medeiros D (2010) Evidence for the prepattern/cooption model of vertebrate jaw evolution. *PNAS* 107, 17262–17267.

Cerny R, Lwigale P, Ericsson R, Meulemans D, Epperlein HH, Bronner-Fraser M (2004) Developmental origins and evolution of jaws: New interpretation of "maxillary" and "mandibular." *Dev Biol* 276, 225–36.

Chanoine C, Hardy S (2003) *Xenopus* muscle development: From primary to secondary myogenesis. *Dev Dyn* 226, 12–23.

Chapman HC (1902) Observations upon *Galeopithecus volans. Proc Acad Nat Sci Philad* 54, 241–254.

Chen H-K (1935) The development of the pectoral limb of *Necturus maculosus. Illi Biol Monogr* 14, 1–71.

Cheng C-C (1955) The development of the shoulder region of the opossum, *Didelphis virginiana*, with special reference to musculature. *J Morphol* 97, 415–471.

Chew KY, Shaw G, Yu H, Pask AJ, Renfree MB (2014) Heterochrony in the regulation of the developing marsupial limb. *Dev Dyn* 243, 324–338.

Chi TK (1937) An unusual variation of the pharyngeal muscles. *J Anat* 72, 134–135.

Cihak R (1972) Ontogenesis of the skeleton and intrinsic muscles of the human hand and foot. *Adv Anat Embryol Cell Biol* 46, 1–194.

Clark AJ, Maravilla EJ, Summers AP (2010) A soft origin for a forceful bite: Motorpatterns of the feeding musculature in Atlantic hagfish, *Myxine glutinosa. Zoology* 113, 259–268.

Clark AJ, Summers AP (2007) Morphology and kinematics of feeding in hagfish: Possible functional advantages of jaws. *J Exp Biol* 210, 3897–3909.

Coates MI, Cohn MJ (1998) Fins, limbs, and tails: Outgrowths and axial patterning in vertebrate evolution. *BioEssays* 20, 371–381.

Coates MI, Jeffery JE, Ruta M (2002) Fins to limbs: What the fossils say. *Evol Dev* 4, 390–401.

Coates MI, Ruta M, Friedman M (2008) Ever since Owen: Changing perspectives on the early evolution of tetrapods. *Ann Rev Ecol Evol Syst* 39, 571–592.

Coghill GE (1902) The cranial nerves of *Amblystoma tigrinum. J Comp Neurol* 12, 205–289.

Cohen SR, Chen L, Trotman CA, Burdi AR (1993) Soft-palate myogenesis: A developmental field paradigm. *Cleft Palate Craniofac J* 30, 441–446.

Cohen SR, Chen LL, Burdi AR, Trotman CA (1994) Patterns of abnormal myogenesis in human cleft palates. *Cleft Palate Craniofac J* 31, 345–350.

Cole FJ (1896) On the cranial nerves of *Chimaera monstrosa* (Linn. 1754); with a discussion of the lateral line system and the morphology of the chorda tympani. *Trans R Soc Edinb* 38, 631–680.

Cole FJ (1907) A monograph on the general morphology of the myxinoid fishes, based on a study of Myxine: Part II. The anatomy of the muscles. *Trans R Soc Edinb* 45, 683–757.

Cole NJ, Hall TE, Don EK, Berger S, Boisvert AC, Nyet C, Ericsson R, Joss J, Gurevich DB, Currie PD (2011) Development and evolution of the muscles of the pelvic fin. *PLoS Biol* 9, e1001168.

Collard M, Wood B (2000) How reliable are human phylogenetic hypotheses? *PNAS* 97, 5003–5006.

Conrad JL (2008) Phylogeny and systematics of Squamata (Reptilia) based on morphology. *Bull Am Mus Nat Hist* 310, 1–182.

Consi TR, Seifert PA, Triantafyllou MS, Edelman ER (2001) The dorsal fin engine of the seahorse (*Hippocampus sp.*). *J Morphol* 248, 80–97.

Coues E (1870) On the myology of the *Ornithorhynchus. Proc Essex Inst* 6, 127–173.

Coues E (1872) On the osteology and myology of *Didelphys virginiana. Mem Boston Soc Nat Hist* 2, 41–154.

Couly GF, Coltey PM, LeDouarin NM (1992) The developmental fate of the cephalic mesoderm in quail-chick chimeras. *Development* 114, 1–15.

Crelin ES (1987) *The Human Vocal Tract: Anatomy, Function, Development, and Evolution*. Vantage Press, New York.

Crompton AW (1963) The evolution of the mammalian jaw. *Evolution* 17, 431–439.

Crompton AW, Parker P (1978) Evolution of the mammalian masticatory apparatus: The fossil record shows how mammals evolved both complex chewing mechanisms and an effective middle ear, two structures that distinguish them from reptiles. *Am Sci* 66, 192–201.

Cubbage CC, Mabee PM (1996) Development of the cranium and paired fins in the zebrafish *Danio rerio* (Ostariophysi, Cyprinidae). *J Morphol* 229, 121–160.

Cunningham DJ (1881) The nerves of the hind-limb of *Thylacinus harrisii* and *Phalangista maculata. J Anat Phys* 15, 265–277.

Cuvier G, Laurillard L (1849) Recueil de planches de myologie. *Anatomie comparée* 3, Paris.

Czajkowski MT, Rassek C, Lenhard DC, Bröhl D, Birchmeier C (2014) Divergent and conserved roles of Dll1 signaling in development of craniofacial and trunk muscle. *Dev Biol* 395, 307–316.

Damas H (1935) Contribution a l'etude de la metamorphose de la tete de la lamproie. *Arch Biol* 46, 171–227.

Danforth CH (1913) The myology of *Polyodon. J Morphol* 24, 107–146.

Daniel JF (1928) The elasmobranch fishes (2nd. Ed.). University of California Press, Berkeley, CA.

Darwin C (1871) *The Descent of Man, and Selection in Relation to Sex.* J. Murray, London.

Davis CA, Holmyard DP, Millen KJ, Joyner AL (1991) Examining pattern formation in mouse, chicken and frog embryos with an En-specific antiserum. *Development* 111, 287–298.

Davis MC, Dahn RD, Shubin NH (2007) An autopodial-like pattern of *Hox* expression in the fins of a basal actinopterygian fish. *Nature* 447, 473–476.

Dawkins R (2004) *The ancestor's tale—A pilgrimage to the dawn of life.* Houghton Mifflin, Boston, MA.

De Beer GR (1940) *Embryos and Ancestors.* Clarendon Press Oxford, UK.

De Beer GR (1959) Part 1, Chapter 3: *Petromyzon*, a chordate with a skull, heart, and kidney. In *Vertebrate Zoology—An Introdcution to the Comparative Anatomy and Evolution of Chordate Animals.* (8th Ed.; 2nd reprint), pp. 17–30.

De Pinna MCC (1991) Concepts and tests of homology in the cladistic paradigm. *Cladistics* 7, 367–394.

Delarbre C, Gallut C, Barriel V, Janvier P, Gachelin G (2002) Complete mitochondrial DNA of the hagfish, *Eptatretus burgeri*: The comparative analysis of mitochondrial DNA sequences strongly supports the cyclostomes monophyly. *Mol Phylogenet Evol* 22, 184–192.

Delsuc F, Brinkmann H, Chourrout D, Philippe H (2006) Tunicates and not cephalochordates are the closest living relatives of vertebrates. *Nature* 439, 965–968.

Dennell R (1950) Note on the feeding of amphioxus (*Branchiostoma bermudæ*). *P Roy Soc Edinb B*, 64, 229–234.

Dent JA, Polson AG, Klymkowsky MW (1989) A whole-mount immunocytochemical analysis of the expression of the intermediate filament protein vimentin in *Xenopus. Development* 105, 61–74.

Depew MJ, Lufkin T, Rubenstein JL. (2001) Specification of jaw subdivisions by Dlx genes. *Science* 298, 381–385.

Diaz RE, Anderson CV, Baumann DP, Kupronis R, Jewell D, Piraquive C, Kupronis J, Winter K, Bertocchini F, Trainor PA (2015a) The veiled chameleon (*Chamaeleo calyptratus* Duméril and Duméril 1851): A model for studying reptile body plan development and evolution. *Cold Spring Harb Protoc* 2015, 889–894.

Diaz RE, Anderson CV, Baumann DP, Kupronis R, Jewell D, Piraquive C, Kupronis, .J, Winter K, Greek F, Trainor PA (2015b) Captive care, raising, and breeding of the veiled chameleon *(Chamaeleo calyptratus). Cold Spring Harb Protoc* 943–949.

Diaz RE, Trainor PA (2015) Hand/foot splitting and the "re-evolution" of mesopodial skeletal elements during the evolution and radiation of chameleons. *BMC Evol Biol* 15, 1–25.

Didier DA (1987) *Myology of the Pectoral, Branchial, and Jaw Regions of the Ratfish Hydrolagus colliei (Holocephali).* Honors Projects, paper 46. Illionois Wesleyan University, Bloomington, IL.

Didier DA (1995) Phylogenetic systematics of extant chimaeroid fishes (Holocephali, chimaeroidei). *Am Mus Novit* 3119, 86.

Dilkes DW (1999) Appendicular myology of the hadrosaurian dinosaur *Maiasaura peeblesorum* from the Late Cretaceous (Campanian) of Montana. *Trans R Soc Edin (Earth Sci)* 90, 87–125.

Diogo R (2004a) *Morphological Evolution, Aptations, Homoplasies, Constraints, and Evolutionary Trends: Catfishes as a Case Study on General Phylogeny and Macroevolution.* Science Publishers, Enfield, NH.

Diogo R (2004b) Muscles versus bones: Catfishes as a case study for an analysis on the contribution of myological and osteological structures in phylogenetic reconstructions. *Anim Biol* 54, 373–391.

Diogo R (2005a) Evolutionary convergences and parallelisms: Their theoretical differences and the difficulty of discriminating them in a practical phylogenetic context. *Biol Philos* 20, 735–744.

Diogo R (2005b) Osteology and myology of the cephalic region and pectoral girdle of *Pimelodus blochii*, comparison with other pimelodines, and comments on the synapomorphies and phylogenetic relationships of the Pimelodinae (Ostariophysi: Siluriformes). *Eur J Morphol* 42, 115–126.

Diogo R (2007) *On the Origin and Evolution of Higher Clades: Osteology, Myology, Phylogeny and Macroevolution of Bony Fishes and the Rise of Tetrapods*. Science Publishers, Enfield, NH.

Diogo R (2008a) Comparative anatomy, homologies and evolution of the mandibular, hyoid and hypobranchial muscles of bony fish and tetrapods: A new insight. *Anim Biol* 58, 123–172.

Diogo R (2008b) *The Origin of Higher Clades: Osteology, Myology, Phylogeny and Evolution of Bony Fishes and the Rise of Tetrapods*. Science Publishers, Enfield, NH.

Diogo R (2009) The head musculature of the *Philippine colugo* (Dermoptera: *Cynocephalus volans*), with a comparison to tree-shrews, primates and other mammals. *J Morphol* 270, 14–51.

Diogo R (2017) *Evolution Driven by Organismal Behavior: A Unifying View of Life, Function, Form, Mismatches and Trends*. Springer, New York.

Diogo R, Abdala V (2007) Comparative anatomy, homologies and evolution of the pectoral muscles of bony fish and tetrapods: A new insight. *J Morphol* 268, 504–517.

Diogo R, Abdala V (2010) *Muscles of Vertebrates: Comparative Anatomy, Evolution, Homologies and Development*. Taylor & Francis, Oxford, UK.

Diogo R, Abdala V, Aziz MA, Lonergan N, Wood BA (2009b) From fish to modern humans—Comparative anatomy, homologies and evolution of the pectoral and forelimb musculature. *J Anat* 214, 694–716.

Diogo R, Abdala V, Lonergan N, Wood BA (2008a) From fish to modern humans—Comparative anatomy, homologies and evolution of the head and neck musculature. *J Anat* 213, 391–424.

Diogo R, Bello-Hellegouarch G, Kohlsdorf T, Esteve-Altava B, Molnar J (2016a) Comparative myology and evolution of marsupials and other vertebrates, with notes on complexity, Bauplan, and "Scala Naturae." *Anat Rec* 299, 1224–1255.

Diogo R, Chardon M (2000) Homologies between different adductor mandibulae sections of teleostean fishes, with a special regard to catfishes (Teleostei: Siluriformes). *J Morphol* 243, 193–208.

Diogo R, Chardon M, Vandewalle P (2001a) Osteology and myology of the cephalic region and pectoral girdle of *Bunocephalus knerii*, and a discussion on the phylogenetic relationships of the Aspredinidae (Teleostei: Siluriformes). *Neth J Zool* 51, 457–481.

Diogo R, Chardon M, Vandewalle P (2002) Osteology and myology of the cephalic region and pectoral girdle of the Chinese catfish *Cranoglanis bouderius*, with a discussion on the autapomorphies and phylogenetic relationships of the Cranoglanididae (Teleostei: Siluriformes). *J Morphol* 253, 229–242.

Diogo R, Chardon M, Vandewalle P (2003) Osteology and myology of the cephalic region and pectoral girdle of *Erethistes pusillus*, comparison with other erethistids, and discussion on the autapomorphies and phylogenetic relationships of the Erethistidae (Teleostei: Siluriformes). *J Fish Biol* 63, 1160–1176.

Diogo R, Chardon M, Vandewalle P (2004) Osteology and myology of the cephalic region and pectoral girdle of *Batrachoglanis raninus*, with a discussion on the synapomorphies and phylogenetic relationships of the Pseudopimelodinae and the Pimelodidae (Teleostei: Siluriformes). *Anim Biol* 54, 261–280.

Diogo R, Chardon M, Vandewalle P (2006) Osteology and myology of the cephalic region and pectoral girdle of *Cetopsis coecutiens*, comparison with other cetopsids, and discussion on the synapomorphies and phylogenetic position of the Cetopsidae (Teleostei: Siluriformes). *Belg J Zool* 136, 3–13.

Diogo R, Doadrio I, Vandewalle P (2008b) Teleostean phylogeny based on osteological and myological characters. *Int J Morphol* 26, 463–522.

Diogo R, Esteve-Altava B, Smith C, Boughner JC, Rasskin-Gutman D (2015a) Anatomical network comparison of human upper and lower, newborn and adult, and normal and abnormal limbs, with notes on development, pathology and limb serial homology vs. homoplasy. *PLoS ONE* 10, e0140030.

Diogo R, Guinard G, Diaz RE Jr. (2017b) Dinosaurs, chameleons, humans and evo-devo-path: Linking Étienne Geoffroy's teratology, Waddington's homeorhesis, Alberch's logic of "monsters," and Goldschmidt hopeful "monsters." *J Exp Zool B (Mol Dev Evol)* 328, 207–229.

Diogo R, Hinits Y, Hughes SM (2008c) Development of mandibular, hyoid and hypobranchial muscles in the zebrafish: Homologies and evolution of these muscles within bony fishes and tetrapods. *BMC Dev Biol* 8, 24–46.

Diogo R, Johnston P, Molnar J, Esteve-Altava B (2016c) Characteristic tetrapod musculoskeleletal limb phenotype emerged more than 400 MYA in basal lobe-finned fishes. *Sci Rep* 6, 1–9.

Diogo R, Kelly R, Christiaen L, Levine M, Ziermann JM, Molnar J, Noden D, Tzahor E (2015b) A new heart for a new head in vertebrate cardiopharyngeal evolution. *Nature* 520, 466–473.

Diogo R, Linde-Medina M, Abdala V, Ashley-Ross MA (2013) New, puzzling insights from comparative myological studies on the old and unsolved forelimb/hindlimb enigma. *Biol Rev* 88, 196–214.

Diogo R, Molnar J (2014) Comparative anatomy, evolution and homologies of the tetrapod hindlimb muscles, comparisons with forelimb muscles, and deconstruction of the forelimb-hindlimb serial homology hypothesis. *Anat Rec Hoboken* 297, 1047–1075.

Diogo R, Molnar J, Smith TD (2014c) The anatomy and ontogeny of the head, neck, pectoral, and upper limb muscles of *Lemur catta* and *Propithecus coquereli* (primates): Discussion on the parallelism between ontogeny and phylogeny and implications for evolutionary and developmental biology. *Anat Rec* 297, 1435–1453.

Diogo R, Molnar J, Wood BA (2017a) Bonobo anatomy reveals stasis and mosaicism in chimpanzee evolution, and supports bonobos as the most appropriate extant model for the common ancestor of chimpanzees and humans. *Sci Rep* 7, 608.

Diogo R, Murawala P, Tanaka EM (2014a) Is salamander hindlimb regeneration similar to that of the forelimb? Anatomical and morphogenetic analysis of hindlimb muscle regeneration in GFP-transgenic axolotls as a basis for regenerative and developmental studies. *J Anat* 224, 459–468.

Diogo R, Nacu E, Tanaka EM (2014b) Is salamander limb regeneration perfect? Anatomical and morphogenetic analysis of forelimb muscle regeneration in GFP-transgenic axolotls as a basis for regenerative, development and evolutionary studies. *Anat Rec* 297, 1076–1089.

Diogo R, Noden D, Smith CM, Molnar JA, Boughner J, Barrocas C, Bruno J (2016b) Understanding human anatomy and pathology: An evolutionary and developmental guide for medical students. Taylor and Francis (Oxford, UK).

Diogo R, Oliveira C, Chardon M (2001b) On the osteology and myology of catfish pectoral girdle, with a reflection on catfish (Teleostei: Siluriformes) plesiomorphies. *J Morphol* 249, 100–125.

Diogo R, Richmond BG, Wood B (2012) Evolution and homologies of modern human hand and forearm muscles, with notes on thumb movements and tool use. *J Hum Evol* 63: 64–78.

Diogo R, Santana SE (2017) Evolution of facial musculature and relationships with facial color patterns, mobility, social group size, development, birth defects and assymetric use of facial expressions. In *The Science of Facial Expression.* (eds. Russel R, Dols JMF), pp. 133–152. Oxford University Press, Oxford, UK.

Diogo R, Smith C, Ziermann JM (2015c) Evolutionary Developmental Pathology and Anthropology: A new area linking development, comparative anatomy, human evolution, morphological variations and defects, and medicine. *Dev Dyn* 244, 1357–1374.

Diogo R, Tanaka EM (2012) Anatomy of the pectoral and forelimb muscles of wildtype and green fluorescent protein-transgenic axolotls and comparison with other tetrapods including humans: A basis for regenerative, evolutionary and developmental studies. *J Anat* 221, 622–635.

Diogo R, Tanaka EM (2014) Development of fore- and hindlimb muscles in GFP-transgenic axolotls: Morphogenesis, the tetrapod Bauplan, and new insights on the forelimb–hindlimb enigma. *J Exp Zool B (Mol Dev Evol)* 322, 106–127.

Diogo R, Wood B (2011) Soft-tissue anatomy of the primates: Phylogenetic analyses based on the muscles of the head, neck, pectoral region and upper limb, with notes on the evolution of these muscles. *J Anat* 219, 273–359.

Diogo R, Wood BA (2009) Comparative anatomy and evolution of the pectoral and forelimb musculature of primates: A new insight. *Am J Phys Anthropol, Meeting Suppl* 48, 119.

Diogo R, Wood BA (2012a) *Comparative Anatomy and Phylogeny of Primate Muscles and Human Evolution.* Taylor & Francis, Oxford, UK.

Diogo R, Wood BA (2012b) Violation of Dollo's law: Evidence of muscle reversions in primate phylogeny and their implications for the understanding of the ontogeny, evolution and anatomical variations of modern humans. *Evolution* 66, 3267–3276.

Diogo R, Wood BA (2013) The broader evolutionary lessons to be learned from a comparative and phylogenetic analysis of primate muscle morphology. *Biol Rev 88*, 988–1001.

Diogo R, Wood BA, Aziz MA, Burrows A (2009a). On the origin, homologies and evolution of primate facial muscles, with a particular focus on hominoids and a suggested unifying nomenclature for the facial muscles of the Mammalia. *J Anat* 215, 300–319.

Diogo R, Ziermann JM (2014) Development of fore- and hindlimb muscles in frogs: Morphogenesis, homeotic transformations, digit reduction, and the forelimb-hindlimb enigma. *J Exp Zool B (Mol Dev Evol)* 322, 86–105.

Diogo R, Ziermann JM (2015a) Development, metamorphosis, morphology, and diversity: The evolution of chordate muscles and the origin of vertebrates. *Dev Dyn* 244, 1046–1057.

Diogo R, Ziermann JM (2015b) Muscles of chondrichthyan paired appendages: Comparison with osteichthyans, deconstruction of the fore–hindlimb serial homology dogma, and new insights on the evolution of the vertebrate neck. *Anat Rec* 298, 513–530.

Diogo R, Ziermann JM, Linde-Medina M (2015d) Is evolutionary biology becoming too politically correct? A reflection on the scala naturae, phylogenetically basal clades, anatomically plesiomorphic taxa, and 'lower' animals. *Biol Rev* 90, 502–521.

Diogo R, Ziermann JM, Linde-Medina M (2015e) Specialize or risk disappearance—Empirical evidence of anisomerism based on comparative and developmental studies of gnathostome head and limb musculature. *Biol Rev* 90, 964–978.

Dobson SD (2009a) Socioecological correlates of facial mobility in nonhuman anthropoids. *Am J Phys Anthropol* 139, 413–420.

Dobson SD (2009b). Allometry of facial mobility in anthropoid primates: Implications for the evolution of facial expression. *Am J Phys Anthropol* 138, 70–81.

Don EK, Currie PD, Cole NJ (2013) The evolutionary history of the development of the pelvic fin/hindlimb. *J Anat* 222, 114–133.

Donoghue PJC, Smith MP (2001) The anatomy of *Turinia pagei* (Powrie) and the phylogenetic status of the Thelodonti. *Trans R Soc Edinb Earth Sci* 92, 15–37.

Drucker E, Jensen J (1996a) Pectoral fin locomotion in the striped surfperch: I. Kinematic effects of swimming speed and body size. *J Exp Biol* 199, 2235–2242.

Drucker E, Jensen J (1996b) Pectoral fin locomotion in the striped surfperch: II. Scaling swimming kinematics and performance at a gait transition. *J Exp Biol* 199, 2243–2252.

Drucker EG, Lauder GV (2003) Function of pectoral fins in rainbow trout: Behavioral repertoire and hydrodynamic forces. *J Exp Biol* 206, 813–826.

Duboc V, Logan MP (2011) *Pitx1* is necessary for normal initiation of hindlimb outgrowth through regulation of *Tbx4* expression and shapes hindlimb morphologies *via* targeted growth control. *Development* 138, 5301–5309.

DuBrul EL (1958) *Evolution of the speech apparatus.* Thomas, Springfield, IL.

Duellman WE, Trueb L (1986) *The biology of amphibians.* McGraw-Hill Book Company, New York.

Duméril C, Duméril AHA (1851) *Catalogue méthodique de la collection des reptiles* (Gide et Baudry).

Dunlap DG (1960) The comparative myology of the pelvic appendage in the salientia. *J Morphol* 106, 1–76.

Dunlap DG (1966) The development of the musculature of the hindlimb in the frog, *Rana pipiens. J Morphol* 119, 241–258.

Durland JL, Sferlazzo M, Logan M, and Burke AC (2008) Visualizing the lateral somitic frontier in the Prx1Cre transgenic mouse. *J Anat* 212, 590–602.

Dutel H, Herrel A, Clément G, Herbin M (2015) Redescription of the hyoid apparatus and associated musculature in the extant coelacanth *Latimeria chalumnae*: Functional implications for feeding. *Anat Rec* 298, 579–601.

Eble GJ (2004) The macroevolution of phenotypic integration. In *Phenotypic Integration. Studying the Ecology and Evolution of Complex Phenotypes* (ed. Pigliucci M). Oxford University Press, Oxford, UK.

Eble GJ (2005) Morphological modularity and macroevolution: Conceptual and empirical aspects. In *Modularity. Understanding the Development and Evolution of Natural Complex Systems* (eds. Callebaut W, Rasskin-Gutman D). MIT Press, Cambridge, MA.

Ecker A (1864) *Die Anatomie des Frosches: Ein Handbuch für Physiologen, Ärzte und Studirende.* F. Vieweg und Sohn, Braunschweig.

Ecker A (1889) *Anatomy of the Frog.* Clarendon Press, Oxford, UK.

Edgeworth FH (1902) The development of the head muscles in *Scyllium canicula. J Anat Physiol* 37, 73–88.

Edgeworth FH (1911) On the morphology of the cranial muscles in some vertebrates. *Q J Micr Sci N S* 56, 167–316.

Edgeworth FH (1923) On the development of the hypobranchial, branchial and laryngeal muscles of *Ceratodus*, with a note on the development of the quadrate and epihyal. *Q J Micr Sci N S* 67, 325–368.

Edgeworth FH (1926a) On the hyomandibula of Selachii, Teleostomi and *Ceratodus. J Anat Physiol* 60, 173–193.

Edgeworth FH (1926b) On the development of the coraco-branchialis and cucullaris in *Scyllium canicula. J Anat Physiol* 60, 298–308.

Edgeworth FH (1926c) On the development of the cranial muscles in *Protopterus* and *Lepidosiren*. *Trans R Soc Edinb* 54, 719–734.

Edgeworth FH (1928) The development of some of the cranial muscles of ganoid fishes. *Philos Trans R Soc Lond (Biol)* 217, 39–89.

Edgeworth FH (1935) *The cranial muscles of vertebrates*. Cambridge University Press, Cambridge, London.

El Haddioui AL, Zouaoui A, Bravetti P, Gaudy JF (2005) Functional anatomy of the human lateral pterygoid muscle. *Surg Radiol Anat* 27, 271–286.

Elinson RP (2013) Metamorphosis in a frog that does not have a tadpole. *Curr Top Dev Biol* 103, 259–76.

Elinson RP, Fang H (1998) Secondary coverage of the yolk sac by the body wall in the direct developing frog, *Eleutherodactylus coqui*: An unusual process for amphibian embryos. *Dev Genes Evol* 208, 457–466.

Ellies DL, Tucker AS, Lumsden A (2002) Apoptosis of premigratory neural crest cells in rhombomeres 3 and 5: Consequences for patterning of the branchial region. *Dev Biol* 251, 118–128.

Elzanowski A (1987) Cranial and eyelid muscles and ligaments of the tinamous (Aves: Tinamiformes). *Zool Jb Anat* 116, 63–118.

Engels WL (1938) Tongue musculature of passerine birds. *Auk* 55, 642–650.

Epperlein HH, Khattak S, Knapp D, Tanaka EM, Malashichev Y (2012) Neural crest does not contribute to the neck and shoulder in the Axolotl (*Ambystoma mexicanum*). *PLoS ONE* 7, e52244.

Ericsson R, Cerny R, Falck P, Olsson L (2004) Role of cranial neural crest cells in visceral arch muscle positioning and morphogenesis in the Mexican axolotl (*Ambystoma mexicanum*). *Dev Dynam* 231, 237–247.

Ericsson R, Joss J, Olsson L (2010) Early head development in the Australian lungfish, *Neoceratodus forsteri*. In *The Biology of Lungfishes* (eds. Jørgensen JM, Joss J), pp. 149–168. Taylor & Francis, London.

Ericsson R, Knight R, Johanson Z (2013) Evolution and development of the vertebrate neck. *J Anat* 222, 67–78.

Ericsson R, Olsson L (2004) Patterns of spatial and temporal visceral arch muscle development in the Mexican axolotl (*Ambystoma mexicanum*). *J Morphol* 261, 131–140.

Fechner R (2009) *Morphofunctional Evolution of the Pelvic Girdle and Hindlimb of Dinosauromorpha on the Lineage to Sauropoda*. PhD Thesis, Ludwigs Maximilians Universität, München.

Findeis EK (1997) Osteology and phylogenetic relationships of recent sturgeons. *Environ Biol Fish* 48, 73–126.

Fischer MS, Krause C, Lilje KE (2010) Evolution of chameleon locomotion, or how to become arboreal as a reptile. *Zoology* 113, 67–74.

Fisher DL, Tanner WW (1970) Osteological and myological comparisons of the head and thorax regions of *Cnemidophorus tigris septentrionalis* Burger and *Ameiva undulata parva* Barbour and Nobel (Family Teiidae). *Brigh Young Univ Sci Bull Biol Ser* 11, 1.

Flammang BE (2009) Functional morphology of the radialis muscle in shark tails. *J Morphol* 271, 340–52.

Flammang BE (2014) The fish tail as a derivation from axial musculoskeletal anatomy: An integrative analysis of functional morphology. *Zoology* 117, 86–92.

Flammang BE, Lauder GV (2009) Caudal fin shape modulation and control during acceleration, braking and backing maneuvers in bluegill sunfish, *Lepomis macrochirus*. *J Exp Biol* 212, 277–286.

Flood PR (1966) A peculiar mode of muscular innervation in amphioxus: Light and electron microscopic studies of the so-called ventral roots. *J Comp Neurol* 126, 181–217.

Forey PL (1986) Relationships of lungfishes. *J Morphol*, suppl. 1, 75–91.

Forster A (1917) Die mm. contrahentes und interossei manus in der Säugetierreihe und beim Menschen. *Arch Anat Physiol Anat Abt* 1916, 101–378.

Fox H (1959) A study of the development of the head and pharynx of the larval urodele *Hynobius* and its bearing on the evolution of the vertebrate head. *Philos Trans R Soc Lond B Biol Sci* 242, 151–204.

Francis ETB (1934) *The anatomy of the salamander*. Clarendon Press, Oxford.

Frazzetta TH (1962) A functional consideration of cranial kinesis in lizards. *J Morphol* 111, 287–320.

Freitas R, Zhang G, Cohn MJ (2006) Evidence that mechanisms of fin development evolved in the midline of early vertebrates. *Nature* 442, 1033–1037.

Fritzsch B, Northcutt RG (1993) Cranial and spinal nerve organization in Amphioxus and Lampreys: Evidence for an ancestral craniate pattern. *Cells Tissues Organs* 148, 96–109.

Fröbisch NB, Shubin NH (2011) Salamander limb development: Integrating genes, morphology, and fossils. *Dev Dyn* 240, 1087–1099.

Fujimoto S, Oisi Y, Kuraku S, Ota KG, Kuratani S (2013) Nonparsimonious evolution of hagfish Dlx genes. *BMC Evol Biol* 13, 15.

Fürbringer M (1874) Zur vergleichenden Anatomie der Schultermuskeln: 2. Teil. *Jena Zeitschr Naturwiss* 8, 175–280.

Fürbringer M (1876) Zur vergleichenden Anatomie der Schultermuskeln: 3. Teil, Capitel IV, Saurier und Crocodile. *Gegenbaurs Morphol Jahrb* 1, 636–816.

Fürbringer P (1875) Untersuchungen zur vergleichenden Anatomie der Muskulatur des Kopfskelettes der Cyclostomen. *Jenaer Zeitschriften* 9, 1–93.

Furlong RF, Holland PWH (2002) Bayesian phylogenetic analysis supports monophyly of Ambulacraria and of cyclostomes. *Zool Sci* 19, 593–599.

Gadow H (1882) Observations in comparative myology. *J Anat Physiol* 16, 493–514.

Gai Z, Donoghue PC, Zhu M, Janvier P, Stampanoni M (2011) Fossil jawless fish from China foreshadows early jawed vertebrate anatomy. *Nature* 476, 324–327.

Galis F, Kundrát M, Metz JAJ (2005). *Hox* genes, digiti identities and the theropod/bird transition. *J Exp Zool B Mol Dev Evol* 304, 198–205.

Galis F, Kundrát M, Sinervo B (2003). An old controversy solved: Bird embryos have five fingers. *Trends Ecol Evol* 18, 7–9.

Gambaryan PP, Aristov AA, Dixon JM, Zubtsova GY (2002) Peculiarities of the hind limb musculature in monotremes: An anatomical description and functional approach. *Russian J Theriol* 1, 1–36.

Gambaryan PP, Kuznetsov AN, Panyutina AA, Gerasimov SV (2015) Shoulder girdle and forelimb myology of extant Monotremata. *Russian J Theriol* 14, 1–56.

Gans C (1989) Stages in the origin of vertebrates: Analysis by means of scenarios. *Biol Rev Camb Philos Soc* 64, 221–268.

Gans C, Northcutt RG (1983) Neural crest and the origin of vertebrates: A new head. *Science* 220, 268–274.

Garcia-Fernàndez J, Benito-Gutiérrez É (2009) It's a long way from amphioxus: Descendants of the earliest chordate. *Bioessays* 31, 665–675.

Garcia-Fernàndez J, Holland PWH (1994) Archetypal organization of the amphioxus *Hox* gen cluster. *Nature* 370, 563–566.

Gardner CA, Barald KF (1992) Expression patterns of engrailed-like proteins in the chick embryo. *Dev Dyn* 193, 370–388.

Garstang W (1928) Memoirs: The morphology of Tunicata, and its bearing on the phylogeny of the Chordata. *Q J Micro Sci* 2, 51–187.

Gasser RF (1967) The development of the facial muscles in man. *Am J Anat* 120, 357–376.

Gaunt AS, Gans C (1969) Mechanisms of respiration in the snapping turtle, *Chelydra serpentina* (Linné). *J Morphol* 128, 195–228.

Gaupp E (1896) A. *Ecker's und R. Wiedersheim's Anatomie des Frosches I* Friedrich Vieweg und Sohn, Braunschweig.

Gaupp E (1899) *Ecker's and R. Wiedersheim's Anatomie des Frosches, II.* Braunshweig: Vieweg.

Gaupp E (1912) Die Reichertsche Theorie. (Hammer-, Amboss- und Kerferfrage). *Arch Anat Physiol Suppl*, 1–416.

Gauthier J, Kluge AG, Rowe T (1988) Amniote phylogeny and the importance of fossils. *Cladistics* 4, 105–209.

Gee H (1996) *Before the Backbone: Views on the Origin of the Vertebrates.* Springer, New York.

Gee H (2001) Deuterostome phylogeny: The context for the origin and evolution of chordates. In *Major Events in Early Vertebrate Evolution: Palaeontology, Phylogeny, Genetics and Development* (ed. Ahlberg PE). Taylor and Francis, London.

Gegenbaur C (1859) *Grundzüge der Vergleichenden Anatomie.* Wilhelm Engelmann, Leipzig.

Gegenbaur C (1872) *Untersuchungen zur Vergleichenden Anatomie der Wirbelthiere. III. Das Kopfskelet der Selachier.* Engelmann, Leipzig.

Gegenbaur C (1878) *Elements of Comparative Anatomy.* Macmillan and Co., London.

Gegenbaur C (1888) The metamerism of the head and the vertebrate theory of the skull. *Brain* 10, 559–565.

George RM (1977) The limb musculature of the Tupaiidae. *Primates* 18, 1–34.

Gess RW, Coates MI, Rubidge BS (2006) A lamprey from the devonian of South America. *Nature* 443, 981–984.

Ghetie V, Chitescu ST, Cotofan V, Hillebrand A (1981) *Atlas de anatomía de las aves doméstica*s: Editorial Acribia, Madrid.

Gibbs S (1999) *Comparative Soft Tissue Morphology of the Extant Hominoidea, including Man.* PhD Thesis, University of Liverpool, Liverpool.

Gibbs S, Collard M, Wood BA (2000) Soft-tissue characters in higher primate phylogenetics. *PNAS* 97, 11130–11132.

Gibbs S, Collard M, Wood BA (2002) Soft-tissue anatomy of the extant hominoids: A review and phylogenetic analysis. *J Anat* 200, 3–49.

Gilbert SF (2006) *Developmental Biology* (8th Ed.). Sinauer, Sunderland, MA.

Gill T (1895) The lowest of the vertebrates and their origin. *Science* 1, 645–649.

Gillis JA, Dahn RD, Shubin NH (2009) Shared developmental mechanisms pattern the gill arch and paired fin skeletons in vertebrates. *PNAS* 106, 5720–5724.

Gillis JA, Fritzenwanker JH, Lowe CJ (2012) A stem-deuterostome origin of the vertebrate pharyngeal transcriptional network. *Proc Biol Sci B* 279, 237–246.

Gillis JA, Hall BK (2016) A shared role for sonic hedgehog signalling in patterning chondrichthyan gill arch appendages and tetrapod limbs. *Development* 143, 1313–1317.

Gillis JA, Modrell MS, Baker CV (2013) Developmental evidence for serial homology of the vertebrate jaw and gill arch skeleton. *Nat Commun* 4, 1436.

Goldschmidt R (1940) *The Material Basis of Evolution.* Yale University Press, New Haven, CT.

González-Isáis M (2003) Anatomical comparison of the cephalic musculature of some members of the superfamily Myliobatoidea (Chondrichthyes): Implications for evolutionary understanding. *Nat Rec A Discov Mol Cell Evol Biol* 271, 259–272.

González-Isáis M, Domínguez HMM (2004) Comparative anatomy of the superfamily Myliobatoidea (Chondrichthyes) with some comments on phylogeny. *J Morphol* 262, 517–535.

Goodrich ES (1930) *Studies on the Structure and Development of Vertebrates.* Macmillan, London.

Goodrich ES (1958) *Studies on the Structure and Development of Vertebrates.* Dowwer Publications, New York.

Gopalakrishnan S, Comai G, Sambasivan R, Francou A, Kelly RG, Tajbakhsh S (2015) A cranial mesoderm origin for esophagus striated muscles. *Dev Cell* 34, 694–704.

Göppert E (1937) Kehlkopf und Trachea. In *Handbuch der Vergleichenden, Anatomie der Wirbeltiere, Vol. 3* (eds. Bolk L, Göppert E, Kallius E, Lubosch W), pp. 797–866. Urban and Schwarzenberg, Berlin.

Gorniak GC (1985) Trends in the action of mammalian masticatory muscles. *Am Zool* 25, 331–337.

Gosline WA (1971) *Functional Morphology and Classification of Teleostean Fishes.* University Press of Hawaii, Honolulu, HI.

Gosline WA (1989) Two patterns of differentiation in the jaw musculature of teleostean fishes. *J Zool (Lond)* 218, 649–661.

Goswami A, Binder WJ, Meachen J, O'Keefe FR (2015) The fossil record of phenotypic integration and modularity: A deep-time perspective on developmental and evolutionary dynamics. *PNAS* 112, 4891–4896.

Goswami A, Polly PD, Mock OB, Sánchez-Villagra MR (2012) Shape, variance and integration during craniogenesis: Contrasting marsupial and placental mammals. *J Evol Biology* 25, 862–72.

Goto T, Nishida K, Nakaya K (1999) Internal morphology and function of paired fins in the epaulette shark, *Hemiscyllium ocellatum. Ichthyol Res* 46, 281–287.

Gould SJ (1977) *Ontogeny and Phylogeny.* Harvard University Press, Cambridge, MA.

Gould SJ (2002) *The Structure of Evolutionary Theory*: Belknap Press, Cambridge, MA.

Gräfenberg E (1906) Die Entwicklung der Knochen, Muskeln und Nerven der Hand und der für die Bewegungen der Hand bestimmten Muskeln des Unterarmes. *Anat H* 30, 7–154.

Graham A (2003) Development of the pharyngeal arches. *Am J Med Genet* 119A, 251–256.

Graham A, Shimeld SM (2013) The origin and evolution of the ectodermal placodes. *J Anat* 222, 32–40.

Grandel H, Schulte-Merker S (1998) The development of the paired fins in the zebrafish (*Danio rerio*). *Mech Dev* 79, 99–120.

Grant RA, Haidarliu S, Kennerley NJ, Prescott TJ (2013) The evolution of active vibrissal sensing in mammals: Evidence from vibrissal musculature and function in the marsupial opossum *Monodelphis domestica. J Exp Biol* 216, 3483–3494.

Gray JE (1865) Revision of the genera and species of Chamaeleonidae, with the description of some new species. *Proc Zool Soc Lond* 1864, 465–479.

Green SA, Bronner ME (2014) The lamprey: A jawless vertebrate model system for examining origin of the neural crest and other vertebrate traits. *Differentiation* 87, 44–51.

Greene CW, Greene CH (1914) *The Skeletal Musculature of the King Salmon.* US Government Printing Office, Washington, DC.

Greene EC (1935) *Anatomy of the Rat.* Hafner Publishing Co, New York.

Greenwood PH (1971) Hyoid and ventral gill arch musculature in osteoglossomorph fishes. *Bull Brit Mus Nat Hist (Zool.)* 22, 1–55.
Greenwood PH (1977) Notes on the anatomy and classification of elopomorph fishes. *Bull Br Mus Nat Hist (Zool.)* 32, 65–103.
Greenwood PH, Thomson KS (1960) The pectoral anatomy of *Pantodon buchholzi* Peters (a freshwater flying fish) and the related Osteoglossidae. *Proc Zool Soc Lond* 135, 283–301.
Gregory WK (1935) On the evolution of the skulls of vertebrates with special reference to heritable changes in proportional diameters (anisomerism). *PNAS* 21, 1–8.
Gregory WK, Camp CL (1918) Studies in comparative myology and osteology. *Bull Am Mus Nat Hist* 38, 447–564.
Gregory WK, Raven HC (1941) Paired fins and girdles in ostracoderms, placoderms, and other primitive fishes. *Ann NY Acad Sci* 42, 275–291.
Grim M (1971) Development of the primordia of the latissimus dorsi muscle of chicken. *Folia Morphol (Prague)* 19, 252–258.
Grim M, Carlson BM (1974a) A comparison of morphogenesis of muscles of the forearm and hand during ontogenesis and regeneration in the axolotl (*Ambystoma mexicanum*): I. Anatomical description of muscles of the forearm and head. *Z Anat Entwickl-Gesch* 145, 137–148.
Grim M, Carlson BM (1974b) A comparison of morphogenesis of muscles of the forearm and hand during ontogenesis and regeneration in the axolotl (*Ambystoma mexicanum*): II. The development of muscular pattern in the embryonic and regenerating limb. *Z Anat Entwickl-Gesch* 145, 149–167.
Grimaldi A, Parada C, Chai Y (2015) A comprehensive study of soft palate development in mice. *PLoS ONE* 10, e0145018.
Grom K, Pasicka E, Tarnawski K (2016) Comparative anatomy of pectoral girdle and pectoral fin in Russian sturgeon and American paddlefish. *Folia Morphol* 75, 173–178.
Gudo M, Homberger DG (2002) The functional morphology of the pectoral fin girdle of the Spiny Dogfish (*Squalus acanthias*): Implications for the evolutionary history of the pectoral girdle of vertebrates. *Senckenbergiana lethaca* 82, 241–252.
Guinard G (2015) Introduction to evolutionary teratology, with an application to the forelimbs of Tyrannosauridae and Carnotaurinae (Dinosauria: Theropoda). *Evol Biol* 42, 20–41.
Gunnell GF, Simmons NB (2005) Fossil evidence and the origin of bats. *J Mammal Evol* 12, 209–246.
Haas A (1996a) Das larvale Cranium von *Gastrotheca riobambae* und seine Metamorphose (Amphibia, Anura, Hylidae). *Verh Naturwiss Ver Hamburg* 36, 33–162.
Haas A (1996b) The larval hyobranchial apparatus of discoglossoid frogs: Its structure and bearing on the systematics of the Anura (Amphibia: Anura). *J Zool Syst Evol Res* 35, 179–197.
Haas A (2001) The mandibular arch musculature of anuran tadpoles with comments on the homologies of amphibian jaw muscles. *J Morphol* 247, 1–33.
Haas A (2003) Phylogeny of frogs as inferred from primarily larval characters (Amphibia: Anura). *Cladistics* 19, 23–89.
Haas G (1973) Muscles of the jaws and associated structures in the Rhynchocephalia and Squamata. In *Biology of the Reptilia, Vol. 14* (eds. Gans C, Parsons TS), pp. 285–490. Academic Press, New York.
Hacker A, Guthrie S (1998) A distinct developmental program for the cranial paraxial mesoderm in the chick embryo. *Development* 125, 3461–3472.
Hadjiaghai O, Ladich F (2015) Sex-specific differences in agonistic behaviour, sound production and auditory sensitivity in the callichthyid armoured catfish *Megalechis thoracata. PLOS ONE* 10, e0121219.
Haidarliu S, Golomb D, Kleinfeld D, Ahissar E (2012) Dorsorostral snout muscles in the rat subserve coordinated movement for whisking and sniffing. *Anat Rec* 295, 1181–1191.
Haines RW (1934) The homologies of the flexor and adductor muscles of the thigh. *J Morphol* 56, 21–49.
Haines RW (1935) Some muscular changes in the tail and thigh of reptiles and mammals. *J Morphol* 58, 355–383.
Haines RW (1939) A revision of the extensor muscles of the forearm in tetrapods. *J Anat* 73, 211–233.
Haines RW (1946) A revision of the movements of the forearm in tetrapods. *J Anat* 80, 1–11.
Haines RW (1950) The flexor muscles of the forearm and hand in lizards and mammals. *J Anat* 84, 13–29.
Haines RW (1951) The extensor apparatus of the finger. *J Anat* 85, 251–259.
Haines RW (1952) The shoulder joint of lizards and the primitive reptilian shoulder mechanism. *J Anat* 86, 412–422.
Haines RW (1955) The anatomy of the hand of certain insectivores. *Proc Zool Soc Lond* 125, 761–777.
Haines RW (1958) Arboreal or terrestrial ancestry of placental mammals. *Q Rev Biol* 33, 1–23.
Hale ME (2006) Pectoral fin coordination and gait transitions in steadily swimming juvenile reef fishes. *J Exp Biol* 209, 3708–3718.
Hall BK (1984) Developmental mechanisms underlying the formation of atavisms. *Biol Rev* 59, 89–124.
Hall BK (2007) Homoplasy and homology: Dichotomy or continuum? *J Hum Evol* 52, 473–479.
Hall BK (2013) Homology, homoplasy, novelty, and behavior. *Dev Psychobiol* 55, 4–12.
Hall BK, Gillis JA (2013) Incremental evolution of the neural crest, neural crest cells and neural crest-derived skeletal tissues. *J Anat* 222, 19–31.
Haninec P, Tomás R, Kaiser R and Cihak R (2009) Development and clinical significance of the musculus dorsoepitrochlearis in men. *Clin Anat* 22, 481–488.
Hanken J, Thorogood P (1993) Evolution and development of the vertebrate skull: The role of pattern formation. *Trends Ecol Evol* 8, 9–15.
Harel I, Nathan E, Tirosh-Finkel L, Zigdon H, Guimarães-Camboa N, Evans SM, Tzahor E (2009) Distinct origins and genetic programs of head muscle satellite cells. *Dev Cell* 16, 822–832.
Hardisty MW (2006) *Lampreys: Life without Jaws.* Forrest Text, Ceredigion.
Harel I, Tzahor E (2013) Head muscle development. In: *Craniofacial Muscles, a New Framework for Understanding the Effector Side of Craniofacial Muscle Control* (eds. McLoon LK, Andrade FH). Springer, New York.
Harris JE (1938) The role of the fins in the equilibrium of the swimming fish. *J Exp Biol* 15, 32–47.
Harrison DFN (1995) *The anatomy and physiology of the mammalian larynx.* Cambridge University Press, Cambridge, UK.
Hartman SE (1988) A cladistic analysis of hominoid molars. *J Human Evol* 17, 489–502.
Hatta K, Bremiller R, Westerfield M, Kimmel CB (1991) Diversity of expression of engrailed-like antigens in zebrafish. *Development* 112, 821–832.
Hatta K, Schilling TF, Bremiller R, Kimmel CB (1990) Specification of jaw muscle identity in zebrafish: Correlation with engrailed-homeoprotein expression. *Science* 250, 802–805.
Haughton S (1868) On the muscular anatomy of the alligator. *J Nat Hist Ser* 1, 582–592.
Hedges SB (2012) Amniote phylogeny and the position of turtles. *BMC Biology* 10, 64.

Heimberg AM, Cowper-Sal·lari R, Sémon M, Donoghue PCJ, Peterson KJ (2010) microRNAs reveal the interrelationships of hagfish, lampreys, and gnathostomes and the nature of the ancestral vertebrate. *PNAS* 107, 19379–19383.

Helms JA, Cordero D, Tapadia MD (2005) New insights into craniofacial morphogenesis. *Development* 132, 851–61.

Henkel-Kopleck A, Schmidt HM (2000) Das Ligamentum metacarpale pollicis. *Handchir Mikrochir Plast Chir* 32, 1–8.

Hernandez LP, Barresi MJF, Devoto SH (2002) Functional morphology and developmental biology of zebrafish: Reciprocal illumination from an unlikely couple. *Integr Comp Biol* 42, 222–231.

Hernandez LP, Patterson SE, Devoto SH (2005) The development of muscle fiber type identity in zebrafish cranial muscles. *Anat Embryol* 209, 323–334.

Herrel A, Aerts P, Fret J, De Vree F (1999) Morphology of the feeding system in agamid lizards: Ecological correlates. *Anat Rec* 254, 496–507.

Herrel A, Canbek M, Özelmas Ü, Uyanoglu M, Karakaya M (2005) Comparative functional analysis of the hyolingual anatomy in lacertid lizards. *Anat Rec A* 284, 561–573.

Hetherington TE, Tugaoen JR (1990) Histochemical studies on the amphibian opercularis muscle (Amphibia). *Zoomorphology* 109, 273–279.

Heude É, Bouhali K, Kurihara Y, Kurihara H, Couly G, Janvier P, Levi G (2010) Jaw muscularization requires *Dlx* expression by cranial neural crest cells. *PNAS* 107, 11441–11446.

Heyland A, Hodin J, Reitzel AM (2005) Hormone signaling in evolution and development: A non-model system approach. *Bioessays* 27, 64–75.

Higashiyama H, Kuratani S (2014) On the maxillary nerve. *J Morphol* 275, 17–38.

Higham TE (2004). In vivo muscle activity in the hindlimb of the arboreal lizard, *Chamaeleo calyptratus*: General patterns and the effects of incline. *J Exp Biol* 207, 249–261.

Higham TE (2005) Constraints on starting and stopping: Behavior compensates for reduced pectoral fin area during braking of the bluegill sunfish *Lepomis macrochirus*. *J Exp Biol* 208, 4735–4746.

Higham TE, Jayne BC (2004) Locomotion of lizards on inclines and perches: Hindlimb kinematics of an arboreal specialist and a terrestrial generalist. *J Exp Biol* 207, 233–248.

Hiiemae K, Jenkins F (1969) The anatomy and internal architecture of the muscles of mastication in *Didelphis marsupialis*. *Postilla* 140, 1–49.

Hinits Y, Williams VC, Sweetman D, Donn TM, Ma TP, Moens CB, Hughes SM (2011) Defective cranial skeletal development, larval lethality and haploinsufficiency in MyoD mutant zebrafish. *Dev Biol* 358: 102–112.

Holland LZ (2014) Genomics, evolution and development of amphioxus and tunicates: The goldilocks principle. *J Exp Zool B (Mol Dev Evol)* 324B, 342–352.

Holland LZ, Holland ND (1998) Developmental gene expression in amphioxus: New insights into the evolutionary origin of vertebrate brain regions, neural crest, and rostrocaudal segmentation. *Am Zool* 38, 647–658.

Holland LZ, Holland ND (2001) Evolution of neural crest and placodes: Amphioxus as a model for the ancestral vertebrate? *J Anat* 199, 85–98.

Holland LZ, Holland ND, Gilland E (2008) *Amphioxus* and the evolution of head segmentation. *Integr Comp Biol* 48, 630–646.

Holland ND, Holland LZ, Holland PW (2015) Scenarios for the making of vertebrates. *Nature* 520, 450–455.

Holland ND, Holland LZ, Honma Y, Fudjii T (1993) *Engrailed* expression during development of a lamprey, *Lampetra japonica*: A possible clue to homologies between agnathan and gnathostome muscles of the mandibular arch. *Dev Growth Diff* 35, 153–160.

Holland ND, Venkatesh TV, Holland LZ, Jacobs DK, Bodmer R (2003) *AmphiNk2–tin*, an amphioxus homeobox gene expressed in myocardial progenitors: Insights into evolution of the vertebrate heart. *Dev Biol* 255, 128–137.

Holliday CM (2009) New insights into dinosaur jaw muscle anatomy. *Anat Rec* 292, 1246–1265.

Holliday CM, Witmer LM (2007) Archosaur adductor chamber evolution: Integration of musculoskeletal and topological criteria in jaw muscle homology. *J Morphol* 268, 457–484.

Holmes R (1977) The osteology and musculature of the pectoral limb of small captorhinids. *J Morphol* 152, 101–140.

Holmes W (1953) The atrial nervous system of amphioxus (*Branchiostoma*). *Q J Micro Sci* 3, 523–535.

Holmgren N (1946) On two embryos of *Myxine glutinosa*. *Acta Zooligica* XXVII: 1–91.

Horigome N, Myojin M, Ueki T, Hirano S, Aizawa S, Kuratani S (1999) Development of cephalic neural crest cells in embryos of *Lampetra japonica*, with special reference to the evolution of the jaw. *Dev Biol* 207, 287–308.

Horovitz I, Sánchez-Villagra MR (2003) A morphological analysis of marsupial mammal higher-level phylogenetic relationships. *Cladistics* 19, 181–212.

Horton P (1982) Diversity and systematic significance of anuran tongue musculature. *Copeia* 1982, 595–602.

House EL (1953) A myology of the pharyngeal region of the albino rat. *Anat Rec* 116, 363–381.

Howe DG, Bradford YM, Eagle A, Fashena D, Frazer K, Kalita P, Mani P et al. (2017) The zebrafish model organism database: New support for human disease models, mutation details, gene expression phenotypes and searching. *Nucleic Acids Res* 45, D758–D768.

Howell AB (1933a) Morphogenesis of the shoulder architecture: Part I—General considerations. *Q Rev Biol* 8, 247–259.

Howell AB (1933b) Morphogenesis of the shoulder architecture: Part II—Pisces. *Q Rev Biol* 8, 434–456.

Howell AB (1935) Morphogenesis of the shoulder architecture: Part III—Amphibia. *Q Rev Biol* 10, 397–431.

Howell AB (1936a) Morphogenesis of the archicteture of the hip and thigh. *J Morphol* 62, 177–218.

Howell AB (1936b) Morphogenesis of the shoulder architecture: Part IV—Reptilia. *Q Rev Biol* 11, 183–208.

Howell AB (1936c) Phylogeny of the distal musculature of the pectoral appendage. *J Morphol* 60, 287–315.

Howell AB (1936d) The phylogenetic arrangement of the muscular system. *Anat Rec* 66, 295–316.

Howell AB (1937a) Morphogenesis of the shoulder architecture: Part V—Monotremata. *Q Rev Biol* 12, 191–205.

Howell AB (1937b) Morphogenesis of the shoulder architecture: Part IV—Therian Mammalia. *Q Rev Biol* 12, 440–463.

Howell AB (1937c) The swimming mechanism of the platypus. *J Mammal* 18, 217–222.

Howell AB (1937d) The musculature of antebrachium and manus in the platypus. *Amer J Anat* 59, 425–432.

Howell AB (1938). Muscles of the avian hip and thigh. *Auk* 55, 71–81.

Howell AB, Straus WL (1933) The muscular system. In *The Anatomy of the Rhesus Monkey* (eds. Hartman CG, Straus WL), pp. 89–175. Williams and Wilkins Co, Baltimore, MD.

Hoyos JM (1999) The musculus depressor mandibulae complex in frogs, with special remarks on ranids. *Amphibia-Reptilia* 20, 265–278.

Huang AH, Riordan TJ, Pryce B, Weibel JL, Watson SS, Long F, Lefebvre V et al. (2015) Musculoskeletal integration at the wrist underlies the modular development of limb tendons. *Development* 142, 2431–2441.

Huang R, Zhi Q, Izpisua-Belmonte J-C, Christ B, Patel K (1999) Origin and development of the avian tongue muscles. *Anat Embryol* 200, 137–152.

Huber E (1930a) Evolution of facial musculature and cutaneous field of trigeminus—Part I. *Q Rev Biol* 5, 133–188.

Huber E (1930b) Evolution of facial musculature and cutaneous field of trigeminus—Part II. *Q Rev Biol* 5, 389–437.

Huber E (1931) *Evolution of Facial Musculature and Expression.* Johns Hopkins University Press, Baltimore, MD.

Hübler M, Molineaux AC, Keyte A, Schecker T, Sears KE (2013) Development of the marsupial shoulder girdle complex: A case study in *Monodelphis domestica. Evol Dev* 15, 18–27.

Hudson GE, Schreiweis DO, Wang SYC (1972) A numerical study of the wing and leg muscles of *Tinamous* (Tinamidae). *Northwest Sci* 46, 207–255.

Humphry GM (1871) The muscles and nerves of the *Cryptobranchus japonicus. J Anat Physiol* 6, 1–61.

Humphry GM (1872a) Observations in myology—The myology of Cryptobranch, *Lepidosiren*, dog-fish, *Ceratodus*, and *Pseudopus palasii*, with the nerves of *Cryptobranchus* and *Lepidorsiren* and the disposition of muscles in vertebrate animals. Macmillan and Co, Cambridge, MA.

Humphry GM (1872b) On the disposition of muscles in vertebrate animals. *J Anat Physiol* 6, 293–376.

Hunter JP, Jernvall J (1995) The hypocone as a key innovation in mammalian evolution. *PNAS* 92, 10718–10722.

Hunter MP, Prince VE (2002) Zebrafish Hox Paralogue group 2 genes function reduntantly as selector genes to pattern the second pharyngeal arch. *Dev Biol* 247, 367–389.

Huntington GS (1904) The derivation and significance of certain supernumerary muscles of the pectoral region. *J Anat Physiol* 39, 1–54.

Huxley TH (1876b) Contributions to morphology: Ichthyopsida No. 1—on *Ceratodus Forsteri*, with observations on the classification of fishes. *Proc Zool Soc (Lond)* 1876, 24–59.

Huxley TH (1876a) The nature of the craniofacial apparatus of *Petromyzon. J Anat Physiol* 10, 412–429.

Inagi K, Schultz E, Ford C (1998) An anatomic study of the rat larynx: Establishing the rat model for neuromuscular function. *Otolaryngol Head Neck Surg* 118, 74–81.

Iordansky NN (1964) The jaw muscles of the crocodiles and some relating structures of the crocodylian skull. *Anat Anz* 115, 256–280.

Iordansky NN (1992) Jaw muscles of the Urodela and Anura: Some features of development, functions, and homology. *Zool Jb Anat* 122, 225–232.

Iordansky NN (1996) Evolution of the musculature of the jaw apparatus in the Amphibia. *Adv Amph Res* 1, 3–26.

Iordansky NN (2000) Jaw muscles of the crocodiles: Structure, synonymy, and some implications on homology and functions. *Russ J Herpetol* 7, 41–50.

Iordansky NN (2004) The jaw apparatus of *Lialis* (Lacertilia, Pygopodidae): Patterns of intensification of cranial kinesis and the problem of the origin of snakes. *Entomol Rev*, suppl. 1, 84, 99–109.

Iordansky NN (2008) Intramandibular muscles and some problems of evolution of the jaw apparatus in vertebrates. *Zool Zhurnal* 87, 49–61.

Iordansky NN (2010) Pterygoideus muscles and other jaw adductors in amphibians and reptiles. *Biology Bulletin* 37, 905–914.

Janeka JE, Miller W, Pringle TH, Wiens F, Zitzmann A, Helgen KM, Springer MS, Murphy WJ (2007) Molecular and genomic data identify the closest living relative of Primates. *Science* 318, 792–794.

Janvier P (1996) *Early vertebrates* (Vol. 33). Clarendon Press, Oxford, UK.

Janvier P (2007) Homologies and evolutionary transitions in early vertebrate history. In: *Major Transitions in Vertebrate Evolution.* (eds. Anderson JS, Sues HD), pp 57–121. Indiana University Press, Bloomington, IN.

Jarvik E (1963) The composition of the intermandibular division of the head in fishes and tetrapods and the diphyletic origin of the tetrapod tongue. *Kungl Sven Veten Handl* 9, 1–74.

Jarvik E (1965) On the origin of girdles and paired fins. *Isr J Zool* 14, 141–172.

Jarvik E (1980) *Basic Structure and Evolution of Vertebrates.* Academic Press, London.

Jefferies RPS (1986) *The Ancestry of the Vertebrates.* British Museum (Natural History), London.

Jenkins F, Weijs W (1979) The functional anatomy of the shoulder in the Virginia opossum (*Didelphis virginiana*). *J Zool (London)* 188, 379–410.

Jernvall J (2000) Linking development with generation of novelty in mammalian teeth. *PNAS* 97, 2641–2645.

Jessen H (1972) Schultergürtel und Pectoralflosse bei Actinopterygiern. *Fossils and Strata* 1, 1–101.

Johanson Z (2003) Placoderm branchial and hypobranchial muscles and origins in jawed vertebrates. *J Vert Paleont* 23, 735–749.

Johanson Z, Joss J, Boisvert CA, Ericsson R, Sutija M, Ahlberg PE (2007) Fish fingers: Digit homologues in sarcopterygian fish fins. *J Exp Zoolog B Mol Dev Evol* 308B, 757–768.

Johnston P (2011) Cranial muscles of the anurans *Leiopelma hochstetteri* and *Ascaphus truei* and the homologies of the mandibular adductors in Lissamphibia and other gnathostomes. *J Morphol* 272, 1492–1512.

Jollie M (1962) *Chordate Morphology.* Reinhold, New York.

Jollie M (1982) Ventral branchial musculature and synapomorphies questioned. *Zool J Linn Soc* 75, 35–47.

Jones CL (1979) The morphogenesis of the thigh of the mouse with special reference to tetrapod muscle homologies. *J Morphol* 162, 275–310.

Jørgensen JM (1998) *The Biology of hagfishes* (1st ed.) Chapman and Hall, New York.

Jouffroy FK (1971) Musculature des membres. In *Traité de Zoologie, XVI: 3 (Mammifères)* (ed. Grassé PP), pp. 1–475. Masson et Cie, Paris.

Jouffroy FK, Lessertisseur J (1971) Particularités musculaires des Monotrémes—Musculature post-craniénne. In *Traité de Zoologie, XVI: 3 (Mammifères)* (ed. Grassé PP), pp. 679–836. Masson et Cie, Paris.

Jouffroy FK, Saban R (1971) Musculature peaucière. In *Traité de Zoologie, XVI: 3 (Mammifères).* (ed. Grassé PP), pp. 477–611. Masson et Cie, Paris.

Jude E, Johanson Z, Kearsley A, Friedman M (2014) Early evolution of the lungfish pectoral-fin endoskeleton: Evidence from the Middle Devonian (Givetian) *Pentlandia macroptera. Front. Earth Sci.* 2, 18.

Kaji T, Aizawa S, Uemura M, Yasui K (2001) Establishment of left-right asymmetric innervation in the lancelet oral region. *J Comp Neurol* 435, 394–405.

Kaji T, Keiji S, Artinger KB, Yasui K (2009) Dynamic modification of oral innervation during metamorphosis in *Branchiostoma belcheri*, the oriental lancelet. *Biol Bull* 217, 151–160.

Kamilar JM, Bradley BJ (2011) Interspecific variation in primate coat colour supports Gloger's rule. *J Biogeogr* 38, 2270–2277.

Kanagasuntheram R (1952–1954) Observations on the anatomy of the hoolock gibbon. *Ceylon J Sci Sect G* 5, 11–64+69–122.

Kaneff A (1979) Évolution morphologique des musculi extensores digitorum et abductor pollicis longus chez l'Homme: I. Introduction, méthodologie M extensor digitorum. *Gegenbaurs Morphol Jahrb* 125, 818–873.

Kaneff A (1980a) Évolution morphologique des musculi extensores digitorum et abductor pollicis longus chez l'Homme: II. Évolution morphologique des m. extensor digiti minimi, abductor pollicis longus, extensor pollicis brevis et extensor pollicis longus chez l'homme. *Gegenbaurs Morphol Jahrb* 126, 594–630.

Kaneff A (1980b) Évolution morphologique des musculi extensores digitorum et abductor pollicis longus chez l'Homme: III. Évolution morphologique du m. extensor indicis chez l'homme, conclusion générale sur l'évolution morphologique des musculi extensores digitorum et abductor pollicis longus chez l'homme. *Gegenbaurs Morphol Jahrb* 126, 774–815.

Kaplan N, Razy-Krajka F, Christiaen L (2015) Regulation and evolution of cardiopharyngeal cell identity and behavior: Insights from simple chordates. *Cur Opin Genet Dev* 32, 119–128.

Kardon G (1998) Muscle and tendon morphogenesis in the avian hind limb. *Development* 125, 4019–32.

Kardong KV (2002) *Vertebrates: Comparative anatomy, function, evolution*, 3rd ed. McGraw-Hill, New York.

Kardong KV (2011) *Vertebrates: Comparative Anatomy, Function, Evolution* (7th ed.) Boston, MA: Mcgraw-Hill.

Kardong KV, Zalisko EJ (1998) *Comparative Vertebrate Anatomy—A Laboratory Dissection Guide*. McGraw-Hill, New York.

Kaufman MH, Kaufman MH (1992) *The Atlas of Mouse Development*. Academic Press, San Diego.

Kelly EM, Sears KE (2011) Reduced phenotypic covariation in marsupial limbs and the implications for mammalian evolution. *Biol J Linn Soc* 102, 22–36.

Kemp TS (2005) *The Origin and Evolution of Mammals*. Oxford University Press, New York.

Kesteven HL (1933) The anatomy of the head of *Callorhynchus antarcticus*. *J Anat* 67, 443.

Kesteven HL (1942) The evolution of the skull and the cephalic muscles: Part I. The fishes. *Mem Aust Mus* 8, 1–132.

Kesteven HL (1942–1945) The evolution of the skull and the cephalic muscles. *Mem Austr Mus* 8, 1–361.

Keyte AL, Smith KK (2010) Developmental origins of precocial forelimbs in marsupial neonates. *Development* 137, 4283–4294.

Khattak S, Schuez M, Richter T, Knapp D, Haigo SL, Sandoval-Guzmán T, Hradlikova K, Duemmler A, Kerney R, Tanaka EM (2013) Germline transgenic methods for tracking cells and testing gene function during regeneration in the axolotl. *Stem Cell Reports* 1, 90–103.

Kimmel CB, Miller CT, Keynes RJ (2001) Neural crest patterning and the evolution of the jaw. *J Anat* 199, 105–119.

King HM, Hale ME (2014) Musculoskeletal morphology of the pelvis and pelvic fins in the lungfish *Protopterus annectens*. *J Morphol* 275, 431–441.

Kingdon J (1992) Facial patterns as signals and masks. Jones Ed. Camb. Encycl. *Hum Evol* 161–165.

Kingdon J (2007) Primate visual signals in noisy environments. *Folia Primatol (Basel)* 78, 389–404.

Kintner CR, Brockes JP (1985) Monoclonal antibodies identify blastemal cells derived from dedifferentiating muscle in newt limb regeneration. *Nature* 308, 67–69.

Kirschner MW, Gerhart JC (2005) *The Plaustibility of Life—Resolving Darwin's Dilemma*. Yale University Press, London.

Kishimoto H, Yamada S, Kanahashi T, Yoneyama A, Imai H, Matsuda T, Takeda T, Kawai K, Suzuki S (2016) Three-dimensional imaging of palatal muscles in the human embryo and fetus: Development of levator veli palatini and clinical importance of the lesser palatine nerve. *Dev Dyn* 245, 123–131

Kisia SM, Onyango DW (2005) *Muscular system of vertebrates*. Science Publishers, Enfield, NH.

Kladetsky J, Kobold H (1966) Das teres-minor problem, der nervus auxilliaris und die Hautrumpftmuskulatur bei *Tupaia glis* (Diard 1820). *Anat Anz* 119, 1–29.

Kleinteich T, Haas A (2007) Cranial musculature in the larva of the caecilian, *Ichthyophis kohtaoensis* (Lissamphibia: Gymnophiona). *J Morphol* 268, 74–88.

Kleinteich T, Haas A (2011) The hyal and ventral branchial muscles in caecilian and salamander larvae: Homologies and evolution. *J Morphol* 272, 598–613.

Knight RD, Mebus K, d'Angelo A, Yokoya K, Heanue T, Tubingen 2000 Screen Consortium, Rohel H (2011) Ret signalling integrates a craniofacial muscle module during development. *Development* 138, 2015–2024.

Knight RD, Mebus K, Roehl HH (2008) Mandibular arch muscle identity is regulated by a conserved molecular process during vertebrate development. *J Exp Zool (Mol Dev Evol)* 310B, 355–369.

Konstantinidis P, Warth P, Naumann B, Metscher M, Hilton EJ, Olsson L (2015) The developmental pattern of the musculature associated with the mandibular and hyoid arches in the long-nose gar, *Lepisosteus osseus* (Actinopterygii, Ginglymodi, Lepisosteiformes). *Copeia* 103, 920–932.

Köntges G, Lumsden A (1996) Rhombencephalic neural crest segmentation is preserved throughout craniofacial ontogeny. *Development* 122, 3229–3242.

Koop D, Chen J, Theodosiou M, Carvallo JE, Alvarez S, de Lera AR, Holland LZ, Schubert M (2014) Roles of retinoic acid and Tbx1/10 in pharyngeal segmentation: Amphioxus and the ancestral chordate condition. *EvoDevo* 5, 1.

Koop D, Holland LZ (2008) The basal chordate amphioxus as a simple model for elucidating developmental mechanisms in vertebrates. *Birth Def Res C: Embryo Today: Revs* 84, 175–187.

Kragl M, Knapp D, Nacu E, Khattak S, Maden M, Epperlein HH, Tanaka EM (2009) Cells keep a memory of their tissue origin during axolotl limb regeneration. *Nature* 460, 60–65.

Krieger J, Hett AK, Fuerst PA, Artyukhin E, Ludwig A (2008) The molecular phylogeny of the order Acipenseriformes revisited. *J Appl Ichthyol* 24, 36–45.

Kryukova NV (2017) Functional analysis of the musculo-skeletal system of the gill apparatus in *Heptranchias perlo* (Chondrichthyes: Hexanchidae). *J Morphol* 278, 1075–1090.

Kundrát M (2009). Primary chondrification foci in the wing basipodium of *Struthio camelus* with comments on interpretation of autopodial elements in Crocodilia and Aves. *J Exp Zool B Mol Dev Evol* 312, 30–41.

Kundrát M, Janacek J, Martin M (2009) Development of transient head cavities during early organogenesis of the Nile Crocodile (*Crocodylus niloticus*). *J Morphol* 270, 1069–1083.

Kuraku S, Hoshiyama D, Katoh K, Suga H, Miyata T (1999) Monophyly of lampreys and hagfishes supported by nuclear DNA-coded genes. *J Mol Evol* 49, 729–735.

Kuraku S, Kuratani S (2006) Time scale for cyclostome evolution inferred with a phylogenetic diagnosis of hagfish and lamprey cDNA sequences. *Zool Sci* 23, 1053–1064.

Kuraku S, Takio Y, Sugahara F, Takechi M, Kuratani S (2010) Evolution of oropharyngeal patterning mechanisms involving *Dlx* and *endothelins* in vertebrates. *Dev Biol* 341, 315–323.

Kuratani S (1997) Spatial distribution of postotic crest cells defines the head/trunk interface of the vertebrate body: Embryological interpretation of peripheral nerve morphology and evolution of the vertebrate head. *Anat Embryol* 195, 1–13.

Kuratani S (2004) Evolution of the vertebrate jaw: Comparative embryology and molecular developmental biology reveal the factors behind evolutionary novelty. *J Anat* 205, 335–347.

Kuratani S (2005a) Craniofacial development and the evolution of the vertebrates: The old problems on a new background. *Zoolog Sci* 22, 1–19.

Kuratani S (2005b) Cephalic neural crest cells and the evolution of craniofacial structures in vertebrates: Morphological and embryological significance of the premandibular-mandibular boundary. *Zoology* 108, 13–25.

Kuratani S (2005c) Developmental studies of the lamprey and hierarchical evolutionary steps towards the acquisition of the jaw. *J Anat* 207, 489–499.

Kuratani S (2008a) Evolutionary developmental studies of cyclostomes and the origin of the vertebrate neck. *Develop Growth Differ* 50, 189–194.

Kuratani S (2008b) Is the vertebrate head segmented? Evolutionary and developmental considerations. *Integr Comp Biol* 48, 647–657.

Kuratani S (2012) Evolution of the vertebrate jaw from developmental perspectives. *Evol Dev* 14, 76–92.

Kuratani S, Adachi N, Wada N, Oisi Y, Sugahara F (2013) Developmental and evolutionary significance of the mandibular arch and prechordal/premandibular cranium in vertebrates: Revising the heterotopy scenario of gnathostome jaw evolution. *J Anat* 222, 41–55.

Kuratani S, Horigome N, Hirano S (1999) Developmental morphology of the head mesoderm and reevaluation of segmental theories of the vertebrate head: Evidence from embryos of an Agnathan Vertebrate, *Lampetra japonica*. *Dev Biol* 210, 381–400.

Kuratani S, Kuraku S, amd Murakami Y (2002) Lamprey as an evo-devo model: Lessons from comparative embryology and molecular phylogenetics. *Genesis* 34, 175–183.

Kuratani S, Murakami Y, Nobusada Y, Kusakabe R, Hirano S (2004) Developmental fate of the mandibular mesoderm in the lamprey, *Lethenteron japonicum*: Comparative morphology and development of the gnathostome jaw with special reference to the nature of the trabecula cranii. *J Exp Zool (Mol Dev Evol)* 302B, 458–468.

Kuratani S, Nobusada Y, Horigome N, Shigetani Y (2001) Embryology of the lamprey and evolution of the vertebrate jaw: Insights from molecular and developmental perspectives. *Philos Trans R Soc Lond B Biol Sci* 356, 1615–1632.

Kuratani S, Ota KG (2008a) Hagfish (Cyclostomata, Vertebrata): Searching for the ancestral developmental plan of vertebrates. *BioEssays* 30, 167–172.

Kuratani S, Ota KG (2008b) Primitive versus derived traits in the developmental program of the vertebrate head: Views from cyclostome developmental studies. *J Exp Zool (Mol Dev Evol)* 310B, 294–314.

Kuratani S, Schilling T (2008) Head segmentation in vertebrates. *Integr Comp Biol* 48, 604–610.

Kuratani S, Ueki T, Aizawa S, Hirano S (1997) Peripheral development of cranial nerves in a cyclostome, *Lampetra japonica*: Morphological distribution of nerve branches in the vertebrate body plan. *J Comp Neurol* 384, 483–500.

Kusakabe R, Kuratani S (2005) Evolution and developmental. patterning of the vertebrate skeletal muscles: Perspectives from the lamprey. *Dev Dyn* 234, 824–834.

Kusakabe R, Kuraku S, Kuratani S (2011) Expression and interaction of muscle-related genes in the lamprey imply the evolutionary scenario for vertebrate skeletal muscle, in association with the acquisition of the neck and fins. *Dev Biol* 350, 217–227.

Ladich F, ed. (2006) *Communication in Fishes*. Science Publishers, Enfield, NH.

Lakjer T (1926) *Die Trigeminus-versorgte Kaumuskulatur der Sauropsiden*. CA Reitzel, Copenhagen.

Lance Jones C (1979) The morphogenesis of the thigh of the mouse with special reference to tetrapod muscle homologies. *J Morphol* 162, 275–310.

Lander KF (1918) The pectoralis minor: A morphological study. *J Anat* 52, 292–318.

Langworthy OR (1932) The panniculus carnosus and pouch musculature of the opossum, a marsupial. *J Mammal* 13, 241–251.

Larsen JH, Guthrie DJ (1975) The feeding system of terrestrial tiger salamanders (*Ambystoma tigrinum melanostictum* Baird). *J Morphol* 147, 137–154.

Lauder GV (1980a) Evolution of the feeding mechanisms in primitive actinopterygian fishes: A functional anatomical analysis of *Polypterus*, *Lepisosteus*, and *Amia*. *J Morphol* 163, 283–317.

Lauder GV (1980b) On the evolution of the jaw adductor musculature in primitive gnathostome fishes. *Breviora* 460, 1–10.

Lauder GV (1980c) The role of the hyoid apparatus in the feeding mechanism of the coelacanth *Latimeria chalumnae*. *Copeia* 1980, 1–9.

Lauder GV (1989) Caudal fin locomotion in ray-finned fishes: Historical and functional analyses. *Am Zool* 29, 85–102.

Lauder GV, Liem KF (1983) The evolution and interrelationships of the actinopterygian fishes. *Bull Mus Comp Zool* 150, 95–197.

Lauder GV, Reilly SM (1988) Functional design of the feeding mechanism in salamanders: Causal bases of ontogenetic changes in function. *J exp Biol* 134, 219–233.

Lauder GV, Shaffer HB (1985) Functional morphology of the feeding mechanism in aquatic ambystomatid salamanders. *J Morphol* 185, 297–326.

Lauder GV, Shaffer HB (1988) Ontogeny of functional design in tiger salamanders (*Ambystoma tigrinum*): Are motor patterns conserved during major morphological transformations? *J Morphol* 197, 249–268.

Lautenschlager S, Gill P, Luo ZX, Fagan MJ, Rayfield EJ (2016) Morphological evolution of the mammalian jaw adductor complex. *Biol Rev*, doi: 10.1111/brv.12314.

Lawson R (1965) The anatomy of *Hypogeophis rostratus* Cuvier (Amphibia: Apoda or Gymnophiona), II, the musculature. *Proc Univ Newcastle Phil Soc* 1, 52–63.

Lazier EL (1945) *Anatomy of the Dogfish*. Stanford University Press, Palo Alto, CA.

Le Douarin NM, Creuzet S, Couly G, Dupin E (2004) Neural crest plasticity and its limits. *Development* 131, 4637–4650.

Le Gros Clark WE (1924) The myology of the tree-shrew (*Tupaia minor*). *Proc Zool Soc Lond* 1924, 461–497.

Le Gros Clark WE (1926) On the anatomy of the pen-tailed tree-shrew (*Ptilocercus lowii*). *Proc Zool Soc Lond* 1926, 1178–1309.

Le Lièvre C, Le Douarin NM (1975) Mesenchymal derivatives of the neural crest: Analysis of chimaeric quail and chick embryos. *J Embryol Exp Morphol* 34, 125–154.

Leche W (1886) Über die Säugethiergattung *Galeopithecus*: Eine morphologische Untersuchung. *Kungl Svenska Vetenskap Akad Stockholm* 21, 3–92.

Lecuru-Renous S (1968) Myologie et innervation du membre antérieur des Lacertiliens (Éditions du Muséum).
Lee MS (2000) Soft anatomy, diffuse homoplasy, and the relationships of lizards and snakes. *Zool Scripta* 29, 101–130.
Leite-Castro J, Beviano V, Rodrigues P, Freitas R (2016) HoxA genes and the fin-to-limb transition in vertebrates. *J Dev Biol* 4, 10.
Lemaire P (2011) Evolutionary crossroads in developmental biology: The tunicates. *Development* 138, 2143–2152.
Lescroart F, Hamou W, Francou A, Théveniau-Ruissy M, Kelly RG, Buckingham M (2015) Clonal analysis reveals a common origin between nonsomite-derived neck muscles and heart myocardium. *PNAS* 112: 1446–1451.
Lescroart F, Kelly RG, Le Garrec JF, Nicolas JF, Meilhac SM, Buckingham M (2010) Clonal analysis reveals common lineage relationships between head muscles and second heart field derivatives in the mouse embryo. *Development* 137, 3269–3279.
Lescroart F, Meilhac SM (2012) Cell lineages, growth and repair of the mouse heart. *Results Probl Cell Differ* 55, 263–289.
Lewis OJ (1989) *Functional Morphology of the Evolving Hand and Foot.* Clarendon Press, Oxford, UK.
Lewis WH (1910) The development of the muscular system. In: *Manual of Embryology*, II. (eds. Keibel F, Mall FP). J. B. Lippincott & Co, Philadelphia, PA.
Li Z, Clarke JA (2015) New insight into the anatomy of the hyolingual apparatus of *Alligator mississippiensis* and implications for reconstructing feeding in extinct archosaurs. *J Anat* 227, 45–1.
Lieschke GJ, Currie PD (2007) Animal models of human disease: Zebrafish swim into view. *Nat Rev Genet* 8, 353–367.
Lightoller GS (1928a) The action of the m. mentalis in the expression of the emotion of distress. *J Anat* 62, 319–332.
Lightoller GS (1928b) The facial muscles of three orangutans and two cercopithecidae. *J Anat* 63, 19–81.
Lightoller GS (1934) The facial musculature of some lesser primates and a *Tupaia*. *Proc Zool Soc Lond* 1934, 259–309.
Lightoller GS (1939) V. Probable homologues: A study of the comparative anatomy of the mandibular and hyoid arches and their musculature—Part I. Comparative myology. *Trans Zool Soc Lond* 24, 349–382.
Lightoller GS (1940a) The comparative morphology of the platysma: A comparative study of the sphincter colli profundus and the trachelo-platysma. *J Anat* 74, 390–396.
Lightoller GS (1940b) The comparative morphology of the m. caninus. *J Anat* 74, 397–402.
Lightoller GS (1942) Matrices of the facialis musculature: Homologization of the musculature in monotremes with that of marsupials and placentals. *J Anat* 76, 258–269.
Lima FC, Leite AV, Santos ALQ, Sabec-Pereira DK, Araújo EG, Pereira KF (2016) Muscular anatomy of the pectoral and forelimb of *Caiman crocodilus crocodilus* (Linnaeus, 1758) (Crocodylia: Alligatoridae). *Ciênc Anim Bras* 17, 285–295.
Lindström T (1949) On the cranial nerves of the cyclostomes with special reference to n. trigeminus. *Acta Zoologica* 30: 315–458.
Lingham-Soliar T (2005) Dorsal fin in the white shark, *Carcharodon carcharias*: A dynamic stabilizer for fast swimming. *J Morphol* 263, 1–11.
Lipscomb DL, Farris JS, Kallersjo M, Tehler A (1998) Support, ribosomal sequences and the phylogeny of the eukaryotes. *Cladistics* 14, 303–338.
Livingstone I (2014) I.Livingstone@BIODIDAC.
Lours-Calet C, Alvares LE, El-Hanfy AS, Gandesha S, Walters EH, Sobreira DR, Wotton KR et al. (2014) Evolutionarily conserved morphogenetic movements at the vertebrate head-trunk interface coordinate the transport and assembly of hypopharyngeal structures. *Dev Biol* 390: 231–246.
Lowe CJ, Clarke DN, Medeiros DM, Rokhsar DS, Gerhart J (2015) The deuterostome context of chordate origins. *Nature* 520, 456–465.
Lubosch W (1914) Vergleischende anatomie der Kaumusculatur der Wirbeltiere, in fünf Teilen: 1—die Kausmukulatur der Amphibien. *Jen Z Naturwiss* 53, 51–188.
Lund R (1967) An analysis of the propulsive mechanism of fishes: With reference to some fossil actinopterygian. *Ann Carnegie Mus* 39, 195–218.
Luo Z-X, Yuan C-X, Meng Q-J, Ji Q (2011) A Jurassic eutherian mammal and divergence of marsupials and placentals. *Nature* 476, 442–445.
Luther A (1913) Über die vom *N. trigeminus* versorgte Muskulatur des Ganoiden and Dipneusten. *Acta Soc Scient Fenn* 41, 1–72.
Luther A (1914) Über die vom *N. trigeminus* versorgte Muskulatur der Amphibien, mit einem vergleichenden Ausblick über den Adductor mandibulae der Gnathostomen, und einem Beitrag zum Verständnis der Organisation der Anurenlarven. *Acta Soc Scient Fenn* 44, 1–151.
Luther A (1938) Die Visceralmuskulatur der Acranier, Cyclostomen, und Fische. A. Acranier, Cyclostomen, Selachier, Holocephalen, Ganoiden und Dipnoer. In *Handbuch der vergleichenden Anatomie der Wirbeltiere, Vol. 5* (eds. Bolk L, Goppert E, Kallius E, Lubosch W), pp. 468–542. Urban and Schwarzenberg, Berlin.
Lynn WG (1961) Types of amphibian metamorphosis. *Am Zool* 1, 151–161.
Lyson TR, Bhullar BAS, Bever GS, Joyce WG, de Queiroz K, Abzhanov A, Gauthier JA (2013) Homology of the enigmatic nuchal bone reveals novel reorganization of the shoulder girdle in the evolution of the turtle shell. *Evol Dev* 15, 317–325.
Mabee PM, Crotwell PL, Bird NC, Burke AC (2002) Evolution of median fin modules in the axial skeleton of fishes. *J Exp Zool* 294, 77–90.
Macalister A (1869) On the arrangement of the pronator muscles in the limbs of vertebrate animals. *J Anat Physiol* 3, 335–340.
Macalister A (1872) The myology of the Cheiroptera. *Phil Trans R Soc Lond* 162, 125–171.
Macesic LJ, Kajiura SM (2010) Comparative punting kinematics and pelvic fin musculature of benthic batoids. *J Morphol* 271, 1219–1228.
Mackenzie S, Walsh FS, Graham A (1998) Migration of hypoglossal myoblast precursors. *Dev Dyn* 213, 349–358.
Mahadevan NR, Horton AC, Gibson-Brown JJ (2004) Developmental expression of the amphioxus *Tbx1/10* gene illuminates the evolution of vertebrate branchial arches and sclerotome. *Dev Genes Evol* 214, 559–566.
Maia A, Wilga CA (2013a) Function of dorsal fins in bamboo shark during steady swimming. *Zoology* 116, 224–231.
Maia A, Wilga CA (2013b). Anatomy and muscle activity of the dorsal fins in bamboo sharks and spiny dogfish during turning maneuvers: Functional anatomy of shark dorsal fins. *J Morphol* 274, 1288–1298.
Maier W (2008) Epitensoric position of the chorda tympani in Anthropoidea: A new synapomorphic character, with remarks on the fissura glaseri in Primates. In: *Mammalian Evolutionary Morphology: A Tribute to Frederick S. Szalay* (eds. Sargis EJ, and Dagosto M). Springer, Dordrecht.

Maier W, van den Heever J, and Durand F (1996) New therapsid specimens and the origin of the secondary hard and soft palate of mammals. *J Zoolog Syst Evol Res* 34, 9–19.

Majtánová Z, Symonová R, Arias-Rodriguez L, Sallan L, Ráb P (2017). "Holostei versus halecostomi" problem: Insight from cytogenetics of ancient nonteleost actinopterygian fish, bowfin Amia calva. *J Exp Zool (Mol Dev Evol) 328B*, 620–628.

Mallatt J (1996) Ventilation and the origin of jawed vertebrates: A new mouth. *Zool J Linn Soc* 117, 329–404.

Mallatt J (1997) Shark pharyngeal muscles and early vertebrate evolution. *Acta Zool* 78, 279–294.

Mallatt J (2008) The origin of the vertebrate jaw: Neoclassical ideas versus newer, development-based ideas. *Zool Sci* 25, 990–998.

Mallatt J, Chen JY (2003) Fossil sister group of craniates: Predicted and found. *J Morph* 258, 1–31.

Mallatt J, Winchell CJ (2007) Ribosomal RNA genes and deuterostome phylogeny revisited: More cyclostomes, elasmobranchs, reptiles, and a brittle star. *Mol Phylogenet Evol* 43, 1005–1022.

Mancini G, Ferrari PF, Palagi E (2013) Rapid facial mimicry in geladas. *Sci Rep* 3, 1527.

Mansfield JH, Haller E, Holland ND, Brent AE (2015) Development of somites and their derivatives in amphioxus, and implications for the evolution of vertebrate somites. *EvoDevo* 6, 21.

Manzanares M, Nieto MA (2003) A celebration of the new head and an evaluation of the new mouth. *Neuron* 37, 895–898.

Manzano A (1996) *Análisis de la musculatura de la familia Pseudidae (Amphibia: Anura).* PhD Thesis, Universidad Nacional de Tucumán, Tucumán.

Manzano A, Abdala V, Herrel A (2008) Morphology and function of the forelimb in arboreal frogs: Specializations for grasping ability? *J Anat* 213, 296–307.

Manzano A, Abdala V, Ponssa ML, Soliz M (2013) Ontogeny and tissue differentiation of the pelvic girdle and hind limbs of anurans. *Acta Zool* 94, 420–436.

Manzano A, Moro S, Abdala V (2003) The depressor mandibulae muscle in Anura. *Alytes* 20, 93–131.

Marcucio RS, Noden DM (1999) Myotube heterogeneity in developing chick craniofacial skeletal muscles. *Dev Dyn* 214, 178–194.

Marinelli W, Strenger A (1954) *Vergleichende Anatomie und Morphologie der Wirbeltiere. I Lieferung. Lampetra fluviatilis* (L), pp. 1–80. Deuticke, Wien.

Marinelli W, Strenger A (1956) *Vergleichende Anatomie und Morphologie der Wirbeltiere. II Lieferung. Myxine glutinosa* (L), pp. 81–172. Deuticke, Wien.

Marinelli W, Strenger A (1959) *Vergleichende Anatomie und Morphologie der Wirbeltiere. III*, pp. 173–308. Deuticke, Wien.

Marion GE (1905) Mandibular and pharyngeal muscles of *Acanthias* and *Raia. Am Nat* 39, 891–924.

Marivaux L, Bocat L, Chaimanee Y, Jaeger JJ, Marandat B, Srisuk P, Tafforeau P, Yamee C, Welcomme JL (2006) Cynocephalid dermopterans from the Palaeogene of South Asia (Thailand, Myanmar and Pakistan): Systematic, evolutionary and palaeobiogeographic implications. *Zool Scripta* 35, 395–420.

Matsuoka T, Ahlberg PE, Kessaris N, Iannarelli P, Dennehy U, Richardson WD, McMahon AP, Koentges G (2005) Neural crest origins of the neck and shoulder. *Nature* 436, 347–355.

Maxwell EE, Larsson HCE (2007) Osteology and myology of the wing of the Emu (*Dromarius novaehollandiae*), and its bearing on the evolution of vestigial structures. *J Morphol* 268, 423–441.

Mazet F, Hutt JA, Milloz J, Millard J, Graham A, Shimeld SM (2005) Molecular evidence from *Ciona intestinalis* for the evolutionary origin of vertebrate cranial placodes. *Dev Biol* 282, 494–508.

McClearn D, Noden DM (1988) Ontogeny of architectural complexity in embryonic quail visceral arch muscles. *Amer J Anat* 183, 277–293.

McGonnell IM (2001) The evolution of the pectoral girdle. *J Anat* 199, 189–194.

McKitrick MC (1991) Phylogenetic analysis of avian hindlimb musculature. *Misc Publ Mus Zool Univ Mich* 179, 1–85.

McMurrich JP (1903a) The phylogeny of the forearm flexors. *Amer J Anat* 2, 177–209.

McMurrich JP (1903b) The phylogeny of the palmar musculature. *Amer J Anat* 2, 463–500.

McMurrich JP (1905) The phylogeny of the crural flexors. *Am J Anat* 4, 33–76.

McMurrich JP (1906) The phylogeny of the plantar musculature. *Am J Anat* 6, 407–437.

McNamara KJ (1986) A guide to the nomenclature of heterochrony. *J Paleo* 60, 4–13.

McNamara KJ (1990) The role of heterochrony in evolutionary trends. In *Evolutionary Trends* (ed. McNamara KJ). London.

McShea DW, Brandon RN (2010) *Biology's First Law: The Tendency for Diversity and Complexity to Increase in Evolutionary Systems.* University of Chicago Press, Chicago, IL.

Medeiros DM, Crump JG (2012) New perspectives on pharyngeal dorsoventral patterning in development and evolution of the vertebrate jaw. *Dev Biol* 371, 121–135.

Meers MB (2003) Crocodylian forelimb musculature and its relevance to Archosauria. *Anat Rec* 274A, 891–916.

Meyer A, Dolven SI (1992) Molecules, fossils, and the origin of tetrapods. *J Mol Evol* 35, 102–113.

Meyers RA (1996) Morphology of the antebrachial musculature in the American Kestrel, *Falco sparverius* (Aves). *Ann Anat* 178, 49–60.

Miller GB (2009) Ancestral patterns in bird limb development: A new look at Hampe's experiment. *J Evol Biol* 2, 31–47.

Millot J, Anthony J (1958) *Anatomie de Latimeria chalumnae—I, squelette, muscles, et formation de soutiens.* CNRS, Paris.

Minchin JEN, Williams VC, Hinits Y, Low S, Tandon P, Fan C-F, Rawls JF, Hughes SM (2013) Oesophageal and sternohyal muscle fibres are novel Pax3-dependent migratory somite derivatives essential for ingestion. *Development* 140, 2972–2984.

Minkoff EC, Mikkelsen P, Cunningham WA, Taylor KW (1979) The facial musculature of the opossum (*Didelphis virginiana*). *J Mammal* 60, 46–57.

Minot CS (1897) Cephalic homologie. A contribution to the determination of the ancestry of vertebrates. *Am Nat* 31, 927–943.

Mivart S (1869) Notes on the myology of *Menopoma alleghaniense. Proc Zool Soc Lond* 1869, 254–261.

Mivart S (1879) On the fins of Elasmobranchii. *Trans Zool Soc Lond* 10, 439–484.

Mivart SG (1870) On the myology of *Chamaeleon parsonii. Proc Zool Soc Lond*, 850–890.

Miyake T, Kumamoto M, Iwata M, Sato R, Okabe M, Koie H, Kumai N and (2016) The pectoral fin muscles of the coelacanth *Latimeria chalumnae*: Functional and evolutionary implications for the fin-to-limb transition and subsequent evolution of tetrapods. *Anat. Rec* 299, 1203–1223.

Miyake T, McEachran JD, Hall BK (1992) Edgeworth's legacy of cranial muscle development with an analysis of muscles in the ventral gill arch region of batoid fishes (Chondrichthyes: Batoidea). *J Morphol* 212, 213–256.
Miyashita T (2012) *Comparative Analysis of the Anatomy of the Myxinoidea and the Ancestry of Early Vertebrate Lineages*. Thesis. University of Alberta, Edmonton, AB.
Miyashita T (2015). Fishing for jaws in early vertebrate evolution: A new hypothesis of mandibular confinement. *Biol Rev* 91, 611–657.
Miyashita T, Diogo R (2016) Evolution of serial patterns in the vertebrate pharyngeal apparatus and paired appendages *via* assimilation of dissimilar units. *Frontiers Ecol Evol—Evo Devo* 4, 71.
Modrell MS, Hockman D, Uy B, Buckley D, Sauka-Spengler T, Bronner ME, Baker CV (2014) A fate-map for cranial sensory ganglia in the sea lamprey. *Dev Biol* 385, 405–416.
Molnar J, Diaz RE, Skorka T, Dagliyan G, Diogo R (2017a) Comparative musculoskeletal anatomy of chameleon limbs, with implications for the evolution of arboreal locomotion in lizards and for teratology. *J Morphol* 278, 1241–1261.
Molnar J, Johnston PS, Esteve-Altava B, Diogo R (2017b) Musculoskeletal anatomy of the pelvic fin of *Polypterus*: Implications for phylogenetic distribution and homology of pre- and postaxial pelvic appendicular muscles. *J Anat* 230, 532–541.
Montero R, Moro SA, Abdala V (2002) Cranial anatomy of *Euspondylus acutirostris* (Squamata: Gymnophthalmidae) and its placement in a modern phylogenetic hypothesis. *Russ J Herpetol* 9, 215–228.
Mootoosamy RC, Dietrich S (2002) Distinct regulatory cascades for head and trunk myogenesis *Development* 129, 573–583.
Moreno TR, Rocha RM (2008) Phylogeny of the Aplousobranchia (Tunicata: Ascidiacea). *Rev Brasil Zool* 25, 269–298.
Moro SA, Abdala V (1998) Cranial myology of some species of *Liolaemus* and *Phymaturus* (Squamata: Tropiduridae: Liolaeminae). *Amphibia–Reptilia* 19, 171–192.
Moro SA, Abdala V (2000) Cladistic analysis of Teiidae (Squamata) based on myological characters. *Russ J Herpetol* 7, 87–102.
Moro SA, Abdala V (2004) Análisis descriptivo de la miología flexora y extensora del miembro anterior de *Polychrus acutirostris* (Squamata, Polychrotidae). *Pap Avulsos Zool (São Paulo)* 44, 81–89.
Motta PJ, Hueter RE, Tricas TC, Summers AP, Huber DR, Lowry D, Mara KR, Matott MP, Whitenack LB, Wintzer AP (2008) Functional morphology of the feeding apparatus, feeding constraints, and suction performance in the nurse shark *Ginglymostoma cirratum*. *J Morphol* 269, 1041–1055.
Motta PJ, Tricas TC, Hueter RE, Summers AP (1997) Feeding mechanism and functional morphology of the jaws of the lemon shark *Negaprion brevirostris* (Chondrichthyes, Carcharhinidae). *J Exp Biol* 200, 2765–2780.
Moustakas JE, Smith KK, Hlusko LJ (2011) Evolution and development of the mammalian dentition: Insights from the marsupial *Monodelphis domesticus*. *Dev Dyn* 240, 232–239.
Moynihan M (1967) *Comparative Aspects of Communication in New World primates*. Aldine, Chicago, IL.
Müller W, Weber E (1998) Re-discovery of a supposedly lost muscle in palaeognathous birds and its phylogenetic implications. *Mitt Mus Nat Berl (Zool Reihe)* 74, 11–18.
Murie J, Mivart ST (1869) On the anatomy of the Lemuroidea. *Trans Zool Soc Lond* 7, 1–113.
Murray PF (1981) A unique jaw mechanism in the echidna, *Tachglossus aculeatus* (Monotremata). *Austral J Zool* 29, 1–5.
Myojin M, Ueki T, Sugahara F, Murakami Y, Shigetani Y, Aizawa S, Hirano S, Kuratani S (2001) Isolation of Dlx and Emx gene cognates in an agnathan species, *Lampetra japonica*, and their expression patterns during embryonic and larval development: Conserved and diversified regulatory patterns of homeobox genes in vertebrate head evolution. *J Exp Biol Zool* 291, 68–84.
Nacu E, Tanaka EM (2011) Limb regeneration: A new development? *Annu Rev Cell Dev Biol* 27, 409–440.
Nagashima H, Sugahara F, Takechi M, Ericsson R, Kawashima-Ohya Y, Narita Y, Kuratani S (2009) Evolution of the turtle body plan by the folding and creation of new muscle connections. *Science* 325, 193–196.
Nagashima H, Sugahara F, Watanabe K, Shibata M, Chiba A, Sato N (2016) Developmental origin of the clavicle, and its implications for the evolution of the neck and the paired appendages in vertebrates. *J Anat* 229, 536–548.
Nakamura T, Gehrke AR, Lemberg J, Szymaszek J, Shubin NH (2016) Digits and fin rays share common developmental histories. *Nature* 537, 225–228.
Near TJ (2009) Conflict and resolution between phylogenies inferred from molecular and phenotypic data sets from hagfish, lampreys, and gnathostomes. *J Exp Zool B Mol Dev Evol* 312B, 749–761.
Negus VE (1949) *The Comparative Anatomy and Physiology of the Larynx*. New York: Hafner Publishing Company.
Nelson JS (2006) *Fishes of the World* (4th ed). John Wiley and Sons, New York.
Netter FH (2006) *Atlas of human anatomy* (4th ed). Saunders, Philadelphia, PA.
Niederreither K, Vermot J, Le Roux I, Schuhbaur B, Chambon P, Dolle P (2003) The regional pattern of retinoic acid synthesis by RALDH2 is essential for the development of posterior pharyngeal arches and the enteric nervous system. *Development* 130, 2525–2534.
Noble GK (1922) The phylogeny of the Salientia: I. The osteology and the thigh musculature; their bearing on classfication and phylogeny. *Bull Amer Mus Nat Hist* 46, 1–87.
Noda M, Miyake T, Okabe M (2017) Development of cranial muscles in the actinopterygian fish Senegal bichir, *Polypterus senegalus* Cuvier, 1829. *J Morphol* 278, 450–463.
Noden DM (1983a) The embryonic origins of avian cephalic and cervical muscles and associated connective tissues. *Am J Anat* 168, 257–276.
Noden DM (1983b) The role of the neural crest in patterning of avian cranial skeletal, connective, and muscle tissues. *Dev Biol* 96, 144–165.
Noden DM (1984) Craniofacial development: New views on old problems. *Anat Rec* 208, 1–13.
Noden DM (1986) Patterning of avian craniofacial muscles. *Dev Biol* 116, 347–356.
Noden DM, Francis-West P (2006) The differentiation and morphogenesis of craniofacial muscles. *Dev Dyn* 235, 1194–1218.
Noden DM, Marcucio R, Borycki AG, Emerson CP (1999) Differentiation of avian craniofacial muscles: I. Patterns of early regulatory gene expression and myosin heavy chain synthesis. *Dev Dyn* 216, 96–112.
Noden DM, Schneider RA (2006) Neural crest cells and the community of plan for craniofacial development: Historical debates and current perspectives. In *Neural Crest Induction and Differentiation—Advances in Experimental Medicine and Biology* 589 (ed. Saint-Jeannet J), pp. 1–31. Landes Bioscience, Georgetown, TX.

Northcutt RG (2005) The new head hypothesis revisited. *J Exp Zool (Mol Dev Evol)* 304B, 274–297.
Northcutt RG (2008) Historical hypotheses regarding segmentation of the vertebrate head. *Integr Comp Biol* 48, 611–619.
Nulens R, Scott L, Herbin M (2011) An updated inventory of all known specimens of the coelacanth, *Latimeria* spp. *Smithiana Spec. Publ.* 3.
Nye HLD, Cameron JA, Chernoff EAG, Stocum DL (2003) Extending the table of stages of normal development of Axolotl: Limb development. *Dev Dyn* 226, 555–560.
O'Gorman S (2005) Second branchial arch lineages of the middle ear of wild-type and *Hoxa2* mutant mice. *Dev Dyn* 234, 124–131.
Oelrich TM (1956) The anatomy of the head of *Ctenosaura pectinata. Misc Pub Mus Zool Univ Michigan* 94, 1–122.
Oisi Y, Fujimoto S, Ota KG, Kuratani S (2015) On the peculiar morphology and development of the hypoglossal, glossopharyngeal and vagus nerves and hypobranchial muscles in the hagfish. *Zool Lett* 1, 1.
Oisi Y, Ota KG, Fujimoto S, Kuratani (2013b). Development of the chondrocranium in hagfishes, with special reference to the early evolution of vertebrates. *Zoo Sci* 30, 944–961.
Oisi Y, Ota KG, Kuraku S, Fujimoto S, Kuratani S (2013a) Craniofacial development of hagfishes and the evolution of vertebrates. *Nature* 493, 175–180.
Okajima Y, Kumazawa Y (2010) Mitochondrial genomes of acrodont lizards: Timing of gene rearrangements and phylogenetic and biogeographic implications. *BMC Evol Biol* 10, 141.
Oken L (1843) *Lehrbuch der Naturphilosophie.* Friedrich Schulthess, Zürich.
Oldham JC (1975) *A Comparative Study of the Appendicular Musculature of Iguanine Lizards.* University of Colorado, Denver, CO.
Olsson L, Ericsson R, Cerny R (2005) Vertebrate head development: Segmentation, novelties, and homology. *Theory Biosci* 124, 145–63.
Olsson L, Ericsson R, Falck P (2000) Neural crest contributions to cranial muscle fate and patterning in the Mexican axolotl (*Ambystoma mexicanum*). In *Regulatory Processes in Development: The Legacy of Sven Hörstadius* (eds. Olsson L, Jacobson C-O), pp. 159–166. Portland Press, London.
Olsson L, Falck P, Lopez K, Cobb J, Hanken J (2001) Cranial neural crest cells contribute to connective tissue in cranial muscles in the anuran amphibian, *Bombina orientalis. Dev Biol* 237, 354–367.
Ortolani A (1999) Spots, stripes, tail tips and dark eyes: Predicting the function of carnivore colour patterns using the comparative method. *Biol J Linn Soc* 67, 433–476.
Osgood WH (1921) A monographic study of the American marsupial *Caenolestes. Field Mus Nat Hist Zool Ser* 16, 1–162.
Osse JWM (1969) Functional morphology of the head of the perch (*Perca fluviatis* L.): An electromyographic study. *Neth J Zool* 19, 289–392.
Ota KG, Fujimoto S, Oisi Y, Kuratani S (2011) Identification of vertebra-like elements and their possible differentiation from sclerotomes in the hagfish. *Nat Commun* 2, 373.
Ota KG, Kuraku S, Kuratani S (2007) Hagfish embryology with reference to the evolution of the neural crest. *Nature* 446, 672–675.
Owen R (1841) Description of the *Lepidosiren annectens. Trans Linn Soc Lond* 18, 327–361.
Owen R (1849) *On the nature of limbs.* John Van Voorst, London.
Owen R (1866) On the anatomy of vertebrates. Longmans, Green and Co, London.
Palavecino P (2001) *Desarrollo de la musculatura mandibular e hioidea en Leptodactylinae del noroeste argentino.* PhD thesis, Universidad Nacional de Tucumán, Tucumán.
Parada C, Han D, Chai Y (2012) Molecular and cellular regulatory mechanisms of tongue myogenesis. *J Dent Res* 91, 528–35.
Parichy DM, Elizondo MR, Mills MG, Gordon TN, Engeszer RE (2009) Normal table of postembryonic zebrafish development: Staging by externally visible anatomy of the living fish. *Dev Dyn* 238, 2975–3015.
Paris M, Escriva H, Schubert M, Brunet F, Brtko J, Ciesielski F, Roecklin D et al. (2008) Amphioxus postembryonic development reveals the homology of chordate metamorphosis. *Curr Biol* 18, 825–830.
Paris M, Hillenweck A, Bertrand S, Delous G, Escriva H, Zalko D, Cravedi J-P, Laudet V (2010) Active metabolism of thyroid hormone during metamorphosis of amphioxus. *Integr Comp Biol* 50, 63–74.
Paris M, Laudet V (2008) The history of a developmental stage: metamorphosis in chordates. *Genesis* 46, 657–672.
Parsons FG (1898) The muscles of mammals, with special relation to human myology: A course of lectures delivered at the Royal College of Surgeons of England - lecture I, the muscles of the head and neck. *J Anat Physiol* 32, 428–450.
Paterson NF (1939) The head of *Xenopus laevis. Q J Microsc Sci* 81, 161–234.
Patterson C (1988) Homology in classical and molecular biology. *Mol Biol Evol* 5, 603–625.
Patterson SE, Mook LB, Devoto SH (2008) Growth in the larval zebrafish pectoral fin and trunk musculature. *Dev Dyn* 237, 307–315.
Pavlicev M, Wagner GP (2012) A model of developmental evolution: Selection, pleiotropy and compensation. *Trends Ecol Evol* 27, 316–22.
Pearson JS (1926) Pelvic and thigh muscles of *Ornithorhynchus. J Anat* 60, 152–163.
Perrin A (1892) Contributions a l'etude de la myologie comparee: Membre posterieur chez un certain nombre de batraciens et de sauriens. *Bull Sci Fr Belg* 24, 373–552.
Peterka HE (1936) A study of the myology and osteology of tree sciurids with regard to adaptation to arboreal, glissant and fossorial habits. *Trans Kansas Acad Sci* 39, 313–332.
Peterson JA (1973). *Adaptation for arboreal locomotion in the shoulder region of lizards.* University of Chicago, IL.
Peterson JA (1984) The locomotion of *Chamaeleo* (Reptilia: Sauria) with particular reference to the forelimb. *J Zool* 202, 1–42.
Pettersen GC, Koltis GG, White MJ (1979) An examination of the spectrum of anatomic defects and variations found in eight cases of trisomy 13. *Amer J Med Genet* 3, 183–210.
Phisalix M (1914) Anatomie comparée de la tête et de l'appareil venimeux chez les serpents. *Ann Sci Nat (Ser 9)* 19, 1–114.
Piatt J (1938) Morphogenesis of the cranial muscles of *Ambystoma punctatum. J Morphol* 63, 531–587.
Piatt J (1939) Correct terminology in salamander myology. I. Intrinsic gill musculature. *Copeia* 4, 220–224.
Piekarski N, Olsson L (2007) Muscular derivatives of the cranialmost somites revealed by long-term fate mapping in the Mexican axolotl (*Ambystoma mexicanum*). *Evol Dev* 9, 566–578.
Piekarski N, Olsson L (2011) A somitic contribution to the pectoral girdle in the axolotl revealed by long-term fate mapping. *Evol Dev* 13, 47–57.
Pierce SE, Clack JA, Hutchinson JR (2012) Three-dimensional limb joint mobility in the early tetrapod *Ichthyostega. Nature* 486, 523–526.

Piotrowski T, Nüsslein-Volhard C (2000) The endoderm plays an important role in patterning the segmented pharyngeal region in zebrafish (*Danio rerio*). *Dev Biol* 225, 339–356.

Pollard HB (1892) On the anatomy and phylogenetic position of *Polypterus*. *Zool Jb* 5, 387–428.

Pough FH, Heiser JB, McFarland WN (1996) *Vertebrate Life* (4th Ed.). Prentice-Hall, Upper Saddle River, NJ.

Powell V., Esteve-Altava B, Molnar J, Villmoare B, Diogo R (2018) Primate modularity and evolution: First anatomical network analysis of primate head and neck musculoskeletal system. *Nature Sci Rep.*

Presley R, Horder T, Slipka J (1996) Lancelet development as evidence of ancestral chordate structure. *Isr J Zool* 42, 97–116.

Preuschoft S (2000) Primate faces and facial expressions. *Soc Res* 67, 245–271.

Preuschoft S, Van Hooff J (1997) The social function of "smile" and "laughter": Variations across primate species and societies. In: *Nonverbal Communication: Where Nature Meets Culture.* (eds. Segerstrale U, Molnar P). Erlbaum, Hillsdale, NJ.

Přikryl T, Aerts P, Havelková P, Herrl A, Rocek Z (2009) Pelvic and thigh musculature in frogs (Anura) and origin of anuran jumping locomotion. *J Anat* 214, 100–139.

Prunotto C, Crepaldi T, Forni PE, Leraci A, Kelly RG, Tajbakhsh S, Buckingham M, Ponzetto C (2004) Analysis of Mlc-lacZ Met mutants highlights the essential function of Met for migratory precursors of hypaxial muscles and reveals a role for Met in the development of hyoid arch-derived facial muscles. *Dev Dyn* 231, 582–591.

Putnam NH, Butts T, Ferrier DE, Furlong RF, Hellsten U, Kawashima T, Robinson-Rechavi M et al. (2008) The amphioxus genome and the evolution of the chordate karyotype. *Nature* 453, 1064–1071.

Quain J, Sharpey-Schäfer EA, Thane JD, Symington J (1894) Elements of Anatomy: General anatomy or histology (10th ed). Longmans, Green and Co, London.

Quint E, Smith A, Avaron F, Laforest L, Miles J, Gaffield W, Akimenko M-A (2002) Bone patterning is altered in the regenerating zebrafish caudal fin after ectopic expression of *sonic hedgehog* and *bmp2b* or exposure to cyclopamine. *PNAS* 99, 8713–8718.

Rackoff JS (1980) The origin of the tetrapod limb and the ancestry of vertebrates. In: *The Terrestrial Environment and the Origin of Land Vertebrates* (ed. Pachen AL). Academic Press, London.

Raikow RJ, Swierczewski EV (1975) Functional anatomy and sexual dimorphism of the cephalic clasper in the Pacific ratfish (*Chimaera collei*). *J Morphol* 145, 435–439.

Ramaswami LS (1942) The segmentation of the head of *Ichthyophius glutinosus* (Linné). *Proc Zool Soc Lond* 112, 105–112.

Razy-Krajka F, Lam K, Wang W, Stolfi A, Joly M, Bonneau R, Christiaen L (2014) Collier/OLF/EBF-dependent transcriptional dynamics control pharyngeal muscle specification from primed cardiopharyngeal progenitors. *Dev Cell* 29, 263–76.

Reichert KB (1937) Über die Visceralbogen der Wirbelthiere im Allgemeinen und deren Metamorphosen bei den Vögeln und Säugethieren. *Arch Anat Physiol Wissensch Med* 1837, 120–220.

Reilly S, Lauder GV (1989) Kinetics of tongue projection in *Ambystoma tigrinum*: Quantitative kinematics, muscle function and evolutionary hypotheses. *J Morphol* 199, 223–243.

Reilly S, Lauder GV (1990) The evolution of tetrapod feeding behavior: Kinematic homologies in prey transport. *Evolution* 44, 1542–1557.

Reilly S, Lauder GV (1991) Prey transport in the tiger salamander: Quantitative electromyography and muscle function in tetrapods. *J Exp Zool* 260, 1–17.

Reiss KZ (2001) Using phylogenies to study convergence: The case of the ant-eating mammals. *Amer Zool* 41, 507–525.

Reno PL, Kjosness KM, Hines JE (2016) The role of *Hox* in pisiform and calcaneus growth plate formation and the nature of the zeugopod/autopod boundary. *J Exp Zoolog B Mol Dev Evol* 326, 303–321.

Rensch B (1938) Some problems of geographical variation and species-formation. *Proc Linn Soc Lond* 150, 275–285.

Rewcastle SC (1977) *The structure and function of the crus and pes in extant Lacertilia.* Ph.D. Thesis, University of London, London.

Ribbing L (1907) Die distale Armmuskulatur der Amphibien, Reptilien und Säugetiere. *Zool Jb* 23, 587–680.

Ribbing L (1909) Die Unterschenkel- und Fussmuskulatur der Tetrapoden und ihr Verhalten zu der entsprechenden Armund Handmuskulatur. Lunds Univ. Årsskrift N F Afd 5, 1–158.

Ribbing L (1938) Die Muskeln und Nerven der Extremitäten. In *Handbuch der Vergleichenden Anatomie der Wirbeltiere Urban and Schwarzenberg* (eds. Bolk L, Göppert E, Kallius E, Lubosch W), pp. 543–656. Asher and Co, Amsterdam.

Richards RJ (2002) *The romantic conception of life: Science and philosophy in the age of Goethe.* University of Chicago Press, Chicago, IL.

Rieppel O (1980) The trigeminal jaw adductor musculature of *Tupinambis*, with comments on the phylogenetic relationships of the Teiidae (Reptilia: Lacertilia). *Zool J Linn Soc* 69, 1–29.

Rieppel O (1981) The skull and jaw adductor musculature in chamaleons. *Rev Suisse Zool* 88, 433–445.

Rieppel O (1984) The structure of the skull and jaw adductor musculature in the Gekkota, with comments on the phylogenetic relationships of the Xantusiidae (Reptilia: Lacertilia). *Zool J Linn Soc* 82, 291–318.

Rieppel O (1990) The development of the jaw adductor musculature in the turtle *Chelydra serpentina*. *Zool J Linn Soc* 98, 27–62.

Rieppel O (1993) Studies on skeleton formation in reptiles: II. *Chamaeleo hoehnelii* (Squamata: Chamaeleoninae), with comments on the homology of carpal and tarsal bones. *Herpetologica* 49, 66–78.

Rinon A, Lazar S, Marshall H, Büchmann-Møller S, Neufeld A, Elhanany-Tamir H, Taketo MM, Sommer L, Krumlauf R, Tzahor E (2007) Cranial neural crest cells regulate head muscle patterning and differentiation during vertebrate embryogenesis. *Development* 134, 3065–3075.

Roberts TDM (1950) The respiratory movements of the lamprey (*Lampetra fluviatilis*). *Proc R Soc Edinb (Biol)* 64, 235–252.

Rocha-Barbosa O, Loguercio MFC, Renous S, Gasc J-P (2007) Comparative study on the forefoot and hindfoot intrinsic muscles of some Cavioidea rodents (Mammalia, Rodentia). *Zoology* 110, 58–65.

Rodríguez-Vázquez JF, Sakiyama K, Abe H, Amano O, Murakami G (2016) Fetal tendinous connection between the tensor tympani and tensor veli palatini muscles: A single digastric muscle acting for morphogenesis of the cranial base. *Anat Rec* 299, 474–483.

Romer AS (1922) The locomotor apparatus of certain primitive and mammal-like reptiles. *Bull Am Mus Nat Hist* 46, 517–606.

Romer AS (1924) Pectoral limb musculature and shoulder-girdle structure in fish and tetrapods. *Anat Rec* 27, 119–143.

Romer AS (1927) The development of the thigh musculature of the chick twelve figures. *J Morphol* 43, 347–385.

Romer AS (1942) The development of tetrapod limb musculature-the thigh of Lacerta. *J Morphol* 71, 251–298.
Romer AS (1944) The development of tetrapod limb musculature—The shoulder region of *Lacerta. J Morphol* 74, 1–41.
Rose CS (2005) Integrating ecology and developmental biology to explain the timing of frog metamorphosis. *Trends Ecol Evol* 20, 129–135.
Rosen DE, Forey PL, Gardiner BG, Patterson C (1981) Lungfishes, tetrapods, paleontology and plesiomorphy. *Bull Am Mus Nat Hist* 167, 159–276.
Roth LV (1994) Within and between organisms: Replicators, lineages, and homologues. In: Homology: *The hierarchical basis of comparative biology* (Hall BK, ed), pp. 301–337. Academic Press, San Diego, CA.
Ruge G (1885) Über die Gesichtsmuskulatur der Halbaffen. *Morph Jb* 11, 243–315.
Ruge G (1897) *Über das peripherische Gebiet des nervus facialis bei Wirbelthieren.* Festschr f Gegenbaur, 3, Leipzig.
Ruge G (1910) Verbindungen des Platysma mit der tiefen Muskulatur des Halses beim Menschen. *Morph Jb* 41, 708–724.
Russel ES (1916) Form and function: A contribution to the history of animal morphology. J. Murray, London.
Russell AP (1988) Limb muscles in relation to lizard systematics: A reappraisal. In *Phylogenetic Relationships of Lizard Families: Essays Commemorating Charles L. Camp* (eds. Estes R, Pregill G), pp. 119–281. Stanford University Press, Standford, CA.
Russell AP, Bauer AM (2008) The appendicular locomotor apparatus of *Sphenodon* and normal-limbed squamates. In *Biology of Reptilia 21, The Skull and Appendicular Locomotor Apparatus of Lepidosauria* (eds. Gans C, Gaunt AS, Adler K), pp. 1–465. Society for the study of Amphibians and Reptilian, New York.
Ryan JM (1986) Comparative morphology and evolution of cheek pouches in rodents. *J Morphol* 190, 27–42.
Ryan JM (1989) Evolution of cheek pouches in African pouched rats (Rodentia: Cricetomyinae). *J Mammal* 70, 267–274.
Saban R (1968) Musculature de la tête. In *Traité de Zoologie, XVI: 3 (Mammifères)* (ed. Grassé PP), pp. 229–472. Masson et Cie, Paris.
Saban R (1971) Particularités musculaires des Monotrémes - muscu lature de la tête. In *Traité de Zoologie, XVI: 3 (Mammifères)* (ed. Grassé PP), pp. 681–732. Masson et Cie, Paris.
Sambasivan R, Kuratani S, Tajbakhsh S (2011) An eye on the head: The development and evolution of craniofacial muscles. *Development* 138, 2401–2415.
Sanchez S, Dupret V, Tafforeau P, Trinajstic KM, Ryll B, Gouttenoire PJ, Wretman L, Zylberberg L, Peyrin F, Ahlberg PE (2013) 3D microstructural architecture of muscle attachments in extant and fossil vertebrates revealed by synchrotron microtomography. *PLoS ONE* 8, e56992.
Sánchez-Villagra MR, Gemballa S, Nummela S, Smith KK, Maier W (2002) Ontogenetic and phylogenetic transformations of the ear ossicles in marsupial mammals. *J Morphol* 251, 219–238.
Sanger TJ, McCune AMY (2002) Comparative osteology of the *Danio* (Cyprinidae: Ostariophysi) axial skeleton with comments on *Danio* relationships based on molecules and morphology. *Zool J Linn Soc* 135, 529–546.
Santagati F, Rijli FM (2003) Cranial neural crest and the building of the vertebrate head. *Nat Rev Neurosci* 4, 806–18.
Santana SE, Alfaro JL, Noonan A, Alfaro ME (2013) Adaptive response to sociality and ecology drives the diversification of facial colour patterns in catarrhines. *Nat Commun* 4, 2765.
Santana SE, Dobson SD, Diogo R (2014) Plain faces are more expressive: Comparative study of facial colour, mobility and musculature in primates. *Biol Lett* 10.
Santana SE, Dobson SD, Diogo R (2014) Plain faces are more expressive: Comparative study of facial color, mobility and musculature in primates. *Biol Let* 10, 20140275.
Santana SE, Lynch Alfaro J, Alfaro ME (2012) Adaptive evolution of facial colour patterns in Neotropical primates. *Proc R Soc B Biol Sci* 279, 2204–2211.
Sargis EJ (2002a) The postcranial morphology of *Ptilocercus lowii* (Scandentia, Tupaiidae): An analysis of primatomorphan and volitantian characters. *J Mammal Evol* 9, 137–160.
Sargis EJ (2002b) Functional morphology of the forelimb of tupaiids (Mammalia, Scandentia) and its phylogenetic implications. *J Morphol* 253, 10–42.
Sargis EJ (2004) New views on tree shrews: The role of tupaiids in primate supraordinal relationships. *Evol Anthropol* 13, 56–66.
Sasaki CT (2006) Review: Part 1. Oral cavity, pharynx and esophagus: Anatomy and development and physiology of the larynx. *GI Motility online.*
Sasakura Y, Kanda M, Ikeda T, Horie T, Kawai N, Ogura Y, Yoshida R, Hozumi A, Satoh N, Fujiwara S (2012) Retinoic acid-driven *Hox1* is required in the epidermis for forming the otic/atrial placodes during ascidian metamorphosis. *Development* 139, 2156–2160.
Sauka-Spengler T, Le Mentec C, Lepage M, Mazan S (2002) Embryonic expression of *Tbx1*, a DiGeorge syndrome candidate gene, in the lamprey Lampetrafluviatilis. *GEP* 2, 99–103.
Saxena A, Cooper KL (2016) Fin to limb within our grasp. *Nature* 537, 176–177.
Saxena SC, Chandy M (1966) The pelvic girdle and fin in certain Indian hill stream fishes. *J Zool* 148, 167–190.
Schaeffer B (1941) The morphological and functional evolution of the tarsus in amphibians and reptiles. *Bull AMNH* 78, 395–472.
Schaeffer JP (1929) Some problems in genesis and development with special reference to the human palate. *Int J Orthod* 15, 291–310.
Schilling N (2011) Evolution of the axial system in craniates: Morphology and function of the perivertebral musculature. *Front Zool* 8, 1–19.
Schilling TF (2002) The morphology of larval and adult zebrafish. In *Zebrafish: A practical approach* (eds. Nüsslein-Volhard C, Dahm R), pp. 59–94. Oxford University Press, New York.
Schilling TF, Kimmel CB (1997) Musculoskeletal patterning in the pharyngeal segments of the zebrafish. *Development* 124, 2945–2960.
Schilling TF, Piotrowski T, Grandel H, Brand M, Heisenberg CP, Jiang YJ, Beuchle D et al. (1996) Jaw and branchial arch mutants in zebrafish I: Branchial arches. *Development* 123, 329–344.
Schlosser G, Roth G (1995) Nerves in tadpoles of *Discoglossus pictus:* Distribution of cranial and rostral spinal nerves in tadpoles of the frog *Discoglossus pictus* (Discoglossidae). *J Morphol* 226, 189–212.
Schmidt J, Piekarski N, Olsson L (2013) Cranial muscles in amphibians: Development, novelties and the role of cranial neural crest cells. *J Anat* 222, 134–146.
Schmidt KL, Cohn JF (2001) Human facial expressions as adaptations: Evolutionary questions in facial expression research. *Yearb Phys Anthropol* 44, 3–24.
Schneider H, Sulner B (2006) Innervation of dorsal and caudal fin muscles in adult zebrafish *Danio rerio. J Comp Neurol* 497, 702–716.

Schneider I, Aneas I, Gehrke AR, Dahn RD, Nobrega MA, Shubin NH (2011) Appendage expression driven by the Hoxd Global Control Region is an ancient gnathostome feature. *PNAS* 108, 12782–12786.

Schroeter S, Tosney KW (1991) Spatial and temporal patterns of muscle cleavage in the chick thigh and their value as criteria for homology. *Am J Anat* 191, 325–350.

Schubert M, Yu J-K, Holland ND, Escriva H, Laudet V, Holland LZ (2005) Retinoic acid signaling acts *via* Hox1 to establish the posterior limit of the pharynx in the chordate amphioxus. *Development* 132, 61–73.

Schumacher GH (1961) *Funktionelle morphologie der kaumuskulatur.* VEB Gustav Fischer Verlag, Jena.

Schumacher GH (1973) The head muscles and hyolaryngeal skeleton of turtles and crocodilians. In *Biology of the Reptilia, Vol. 14* (eds. Gans C, Parsons TS), pp. 101–199. Academic Press, New York.

Schweitzer R, Zelzer E, Volk T (2010) Connecting muscles to tendons: tendons and musculoskeletal development in flies and vertebrates. *Development* 137, 2807–2817.

Schwenk K (2001) Extrinsic vs. intrinsic lingual muscles: A false dichotomy? *Bull Mus Comp Zool (Harvard Univ)* 156, 219–235.

Sears KE (2004) Constraints on the morphological evolution of marsupial shoulder girdles. *Evolution* 58, 2353–2370.

Sears KE, Capellini TD, Diogo R (2015a) On the serial homology of the pectoral and pelvic girdles of tetrapods. *Evolution* 69, 2543–2555.

Sears KE, Maier JA, Rivas-Astroza M, Poe R, Zhong S, Kosog K, Marcot JD, Behringer RR, Cretekos CJ, Rasweiler JJ, Rapti Z (2015b) The relationship between gene network structure and expression variation among individuals and species. *PLoS Genet.* 11, e1005398.

Sedlmayr JC (2002) *Anatomy, evolution and functional significance of the cephalic vasculature in Archosauria.* PhD Thesis, Ohio University, Ohio.

Sedra SN (1949) The metamorphosis of the jaws and their muscles in the toad, *Bufo regularis* Reuss, correlated with the changes in the animal's feeding habits. *Proc Zool Soc Lond* 120, 405–449.

Sedra SN, Michael MI (1957) The development of the skull, visceral arches, larynx and visceral muscles of the South African Clawed Toad, *Xenopus laevis* (Daudin) during the process of metamorphosis (from Stage 55 to Stage 66). *Verh Koninklijke Nederlandse Akad Wetenschappen Afd Natuurkunde* 51, 3–80.

Sefton EM, Bhullar BAS, Mohaddes Z, Hanken J (2016) Evolution of the head-trunk interface in tetrapod vertebrates. *eLife* 5, e09972.

Sefton EM, Piekarski N, Hanken J (2015) Dual embryonic origin and patterning of the pharyngeal skeleton in the axolotl (*Ambystoma mexicanum*). *Evol Dev* 17, 175–84.

Seiler R (1971a) A comparison between the facial muscles of Catarrhini with long and short muzzles. *Proc. 3rd Int. Congr. Primat., Zürich 1970, vol. 1*, pp. 157–162. Karger, Basel.

Seiler R (1971b) Facial musculature and its influence on the facial bones of catarrhinous Primates. I. *Morphol Jb* 116, 122–142.

Seiler R (1971c) Facial musculature and its influence on the facial bones of catarrhine Primates. II. *Morphol Jb* 116, 147–185.

Seiler R (1971d) Facial musculature and its influence on the facial bones of catarrhine Primates. III. *Morphol Jb* 116, 347–376.

Seiler R (1971e) Facial musculature and its influence on the facial bones of catarrhine Primates. IV. *Morphol Jb* 116, 456–481.

Seiler R (1974a) Muscles of the external ear and their function in man, chimpanzees and *Macaca. Morphol Jb* 120, 78–122.

Seiler R (1974b) Particularities in facial muscles of *Daubentonia madagascariensis* (Gmelin 1788). *Folia Primatol* 22, 81–96.

Seiler R (1975) On the facial muscles in *Perodicticus potto* and *Nycticebus coucang. Folia Primatol* 23, 275–289.

Seiler R (1979) Criteria of the homology and phylogeny of facial muscles in primates including man. I. Prosimia and Platyrrhina. *Morphol Jb* 125, 191–217.

Seiler R (1980) Ontogenesis of facial muscles in primates. *Morphol Jb* 126, 841–864.

Setchell JM (2005) Do female mandrills prefer brightly colored males? *Int J Primatol* 26, 715–735.

Setchell JM, Wickings EJ, Knapp LA, Jean Wickings E (2006) Signal content of red facial coloration in female mandrills (*Mandrillus sphinx*). *Proc R Soc B Biol Sci* 273, 2395–2400.

Sewertzoff AN (1928) The head skeleton and muscles of *Acipenser ruthenus. Acta Zool-Stockholm* 9, 193–319.

Shann CEW (1920) The comparative myology of the shoulder girdle and pectoral fin of fishes. *Trans R Soc Edinb* 52, 531–570.

Shann EP (1924) Further observations on the myology of the pectoral region in fishes. *Proc Zool Soc London* 1924, 195–215.

Shann EW (2009) Further observations on the myology of the pectoral region in fishes. *Proc Zool Soc Lond* 94, 195–215.

Shattock SG (1882) A Kerato-thyro-hyoid muscle as a variation in human anatomy. *J Anat Physiol* 171, 124–125.

Shearman RM, Burke AC (2009) The lateral somitic frontier in ontogeny and phylogeny. *J Exp Biol (Mol Dev Evol)* 312B, 602–613.

Shellswell GB, Wolpert L (1977) The pattern of muscle and tendon development in the chick wing. In *Vertebrate Limb and Somite Morphogenesis* (eds. Ede D, Hinchcliffe R, Balls M), pp. 71–86. Cambridge University Press, Cambridge, MA.

Sherwood CC, Hof PR, Holloway RL, Semendeferi K, Gannon PJ, Frahm HD, Zilles K (2005) Evolution of the brainstem orofacial motor system in primates: A comparative study of trigeminal, facial, and hypoglossal nuclei. *J Hum Evol* 48, 45–84.

Shibata S (1959) On the facial musculature of *Macacus cyclopsis. OFAJ* 34, 159–176.

Shigetani Y, Sugahara F, Kuratani S (2005) A new evolutionary scenario for the vertebrate jaw. *Bioessays* 27, 331–338.

Shimeld SM, Holland PW (2000) Vertebrate innovations. *PNAS* 97, 4449–4452.

Shoshani J, Groves CP, Simons EL, Gunnell GF (1996) Primate phylogeny: Morphological vs molecular results. *Mol Phylogenet Evol* 5, 102–154.

Shrivastava RK (1962a) The deltoid musculature of the Marsupialia. *Amer Midland Nat* 67, 305–320.

Shrivastava RK (1962b) The deltoid musculature of the Monotremata. *Amer Midland Nat* 67, 434–440.

Shubin N, Tabin C, Carrol S (1997) Fossils, genes and the evolution of animal limbs. *Nature* 388, 639–648.

Shubin N, Tabin C, Carroll S (2009) Deep homology and the origins of evolutionary novelty. *Nature* 457, 818–823.

Shubin NH (2004) The early evolution of the tetrapod humerus. *Science* 304, 90–93.

Shubin NH, Alberch P (1986) A morphogenetic approach to the origin and basic organisation of the tetrapod limb. *Evol Biol* 1, 319–387.

Shubin NH, Daeschler EB, Coates MI (2004) The early evolution of the tetrapod humerus. *Science* 304, 90–93.

Shubin NH, Daeschler EB, Jenkins FA (2006) The pectoral fin of *Tiktaalik roseae* and the origin of the tetrapod limb. *Nature* 440, 764–771.

Silcox MT, Sargis EJ, Bloch JI, Boyer DM (2007) Primate origins and supraordinal relationships: Morphological evidence. In *Handbook of Palaeoanthropology, Vol. 2: Primate Evolution and Human Origins* (eds. Henke W, Tattersall I), pp- 831–859. Springer Verlag, Heidelberg.

Simões-Costa MS, Vasconcelos M, Sampaio AC, Cravo RM, Linhares VL, Hochgreb T, Yan CY, Davidson B, Xavier-Neto J (2005) The evolutionary origin of cardiac chambers. *Dev Biol* 277, 1–15.

Siomava N, Diogo R (in press) Comparative anatomy of zebrafish paired and medial fin muscles: Basis for functional, developmental, and macroevolutionary studies. *J Anat.*

Siomava N, Shkil F, Voronezhskaya E, Diogo R (in press) Development of zebrafish paired and medial fin musculature: Basis for comparative, developmental, and macroevolutionary studies. *Sci Rep.*

Skinner JH (1958) Ontogeny of the breast-shoulder apparatus of the South African lacertilian, *Microsaura pumila pumila* (Daudin). Stellenbosch University, Stellenbosch.

Smith KK (1988) Form and function of the tongue in agamid lizards with comments on its phylogenetic significance. *J Morphol* 196, 157–171.

Smith KK (1992) The evolution of the mammalian pharynx. *Zool J Linn Soc* 104, 313–349.

Smith KK (1994) Development of craniofacial musculature in *Monodelphis domestica* (Marsupialia: Didelphidae). *J Morphol* 222, 149–173.

Smith KK (2001) Early development of the neural plate: Neural crest and facial region in marsupials. *J Anat* 199, 121–131.

Smith KK (2006) Craniofacial development in marsupial mammals: Developmental origins of evolutionary change. *Dev Dyn* 235, 1181–1193.

Soares MC, Carvalho MR (2013a) Mandibular and hyoid muscles of galeomorph sharks (Chondrichthyes: Elasmobranchii), with remarks on their phylogenetic intrarelationships. *J Morphol* 274, 1111–1123.

Soares MC, de Carvalho MR (2013b) Comparative myology of the mandibular and hyoid arches of sharks of the order Hexanchiformes and their bearing on its monophyly and phylogenetic relationships (Chondrichthyes: Elasmobranchii). *J Morphol* 274: 203–214.

Sokol OT (1977) The free swimming *Pipa* larvae, with a review of pipoid larvae and pipid phylogeny (Anura: Pipidae). *J Morphol* 154, 357–425.

Sokoloff AJ (2000) Localization and contractile properties of intrinsic longitudinal motor units of the rat tongue. *J Neurophysiol* 84, 827–835.

Sokoloff AJ, Deacon TW (1992) Musculotopic organization of the hypoglossal nucleus in the cynomolgus monkey, *Macaca fascicularis. J Comp Neurol* 324, 81–93.

Song J, Boord RL (1993) Motor components of the trigeminal nerve and organization of the mandibular arch muscles in vertebrates: Phylogenetically conservative patterns and their ontogenetic basis. *Acta Anatomica* 148, 139–149.

Sordino P, van der Hoeven F, Duboule D (1995) *Hox* gene expression in teleost fins and the origin of vertebrate digits. *Nature* 375, 678–681.

Sprague JA (1943) The hyoid region of placental mammals with special reference to the bats. *Am J Anat* 72, 385–472.

Sprague JA (1944a) The hyoid region in the Insectivora. *Am J Anat* 74, 175–216.

Sprague JA (1944b) The innervation of the pharynx in the rhesus monkey, and the formation of the pharyngeal plexus in primates. *Anat Rec* 90, 197–208.

Sprague JM (1942) The hyoid apparatus of *Neotoma. J Mammal* 23, 405–411.

Stafford BJ, Szalay FS (2000) Craniodental functional morphology and taxonomy of dermopterans. *J Mammal* 81, 360–385.

Stahl FW (1995) George Streisinger (1927–1984). A Biographical Memoir. In: *Biographic Memoirs.* pp. 353–361. National Academies Press, Washington, DC.

Starck D, Schneider R (1960) Respirationsorgane. In *Primatologia III/2* (eds. Hofer H, Schultz AH, Starck D), pp. 423–587. S. Karger, Basel.

Starret P (1968) The phylogenetic significance of the jaw musculature in anuran amphibians. PhD thesis, University of Michigan, Ann Arbor, MI.

Stein B (1981) Comparative limb myology of two opossums, *Didelphis* and *Chironectes. J Morphol* 169, 113–140.

Stevens M, Merilaita S (2009) Animal camouflage: Current issues and new perspectives. *Philos Trans R Soc B Biol Sci* 364, 423–427.

Stewart TA, Hale ME (2013) First description of a musculoskeletal linkage in an adipose fin: Innovations for active control in a primitively passive appendage. *Proc R Soc Lond B Biol Sci* 280, 20122159.

Stiassny MLJ (2000) Gross functional anatomy: Muscular system. In *The handbook of experimental animals* (eds. Bullock G, Bunton TE), pp. 119–128. Academic Press, London.

Stiassny MLJ, Wiley EO, Johnson GD, De Carvalho MR (2004) Gnathostome fishes. In *Assembling the tree of life* (eds. Donaghue MJ, Cracraft J), pp. 410–429. Oxford University Press, New York.

Stock DW, Whitt GS (1992) Evidence from 18S ribosomal RNA sequences that lampreys and hagfishes form a natural group. *Science* 257, 787–789.

Stocum DL, Cameron JA (2011) Looking proximally and distally: 100 years of limb regeneration and beyond. *Dev Dyn* 240, 943–968.

Stokes M, Holland N (1995) Ciliary hovering in larval lancelets (= Amphioxus). *Biol Bull* 188, 231–233.

Stolfi A, Gainous TB, Young JJ, Mori A, Levine M, Christiaen L (2010) Early chordate origins of the vertebrate second heart field. *Science* 329, 565–568.

Stower MJ, Diaz RE, Fernandez LC, Crother MW, Crother B, Marco A, Trainor PA, Srinivas S, Bertocchini F (2015) Bi-modal strategy of gastrulation in reptiles. *Dev Dyn* 244, 1144–1157.

Straus WL (1941a) The phylogeny of the human forearm extensors. *Hum Biol* 13, 23–50.

Straus WL (1941b) The phylogeny of the human forearm extensors (concluded). *Hum Biol* 13, 203–238.

Straus WL (1942) The homologies of the forearm flexors: Urodeles, lizards, mammals. *Am J Anat* 70, 281–316.

Straus WL, Howell AB (1936) The spinal accessory nerve and its musculature. *Q Rev Biol* 11, 387–405.

Streisinger G, Walker C, Dower N, Knauber D, Singer F (1981) Production of clones of homozygous diploid zebra fish (*Brachydanio rerio*). *Nature* 291, 293–296.

Sugahara F, Murakami Y, Adachi N, Kuratani S (2013) Evolution of the regionalization and patterning of the vertebrate telencephalon: What can we learn from cyclostomes? *Curr Opin Genet Dev* 23: 475–483.

Sullivan GE (1962) Anatomy and embryology of the wing musculature of the domestic fowl (*Gallus*). *Aust J Zool* 10, 458–518.

Sullivan GE (1967) Abnormalities of the muscular anatomy in the shoulder region of paralysed chick embryos. *Aust J Zool* 15, 911–40.

Sullivan WE, Osgood CW (1927) The musculature of the superior extremity of the orang-utan, *Simia satyrus*. *Anat Rec* 35, 193–239.

Surlykke A, Miller LA, Møhl B, Andersen BB, Christensen-Dalsgaard J, and Buhl Jørgensen M (1993) Echolocation in two very small bats from Thailand *Craseonycteris thonglongyai* and *Myotis siligorensis*. *Behav Ecol Sociobiol* 33, 1–12.

Susman RL, Nyati L, Jassal MS (1999) Observations on the pollical palmar interosseus muscle (of Henle). *Anat Rec* 254, 159–165.

Suzuki D, Murakami G, Minoura N (2003) Crocodilian bone-tendon and bone-ligament interfaces. *Ann Anat* 185, 425–33.

Symington J (1898) The marsupial larynx. *J Anat Physiol* 33, 31–49.

Tabin CJ (1992) Why we have (only) five fingers per hand: Hox genes and the evolution of paired limbs. *Development* 116, 289–296.

Tada M, and Kuratani S (2015) Evolutional and developmental understanding of the spinal accessory nerve. *Zool Lett* 1: 4.

Tajbakhsh S, Rocancourt D, Cossu G, Buckingham M (1997) Redefining the genetic hierarchies controlling skeletal myogenesis: *Pax-3* and *Myf-5* act upstream of *MyoD*. *Cell* 89, 127–138.

Takahasi N (1925) On the homology of the cranial muscles of the cypriniform fishes. *J Morphol* 40, 1–109.

Takezaki N, Figueroa F, Zaleska-Rutczynska Z, Klein J (2003) Molecular phylogeny of early vertebrates: Monophyly of the agnathans as revealed by sequences of 35 genes. *Mol Biol Rev* 20, 287–292.

Takio Y, Pasqualetti M, Kuraku S, Hirano S, Rijli FM, Kuratani S (2004) Comment on: Lamprey *Hox* genes and the evolution of jaws. *Nature* 429, 1–2.

Taylor J (1925) An unusual variation of the omo-hyoid muscle. *J Anat* 59, 331–332.

Tchernavin VV (1947a) Six specimens of Lyomeri in the British Museum (with noted on the skeleton of the Lyomeri). *J Linn Soc Lond (Zool)* 41, 287–350.

Tchernavin VV (1947b) Further notes on the structure of the bony fishes of the order Lyomeri (Eurypharynx). *J Linn Soc Lond (Zool)* 41, 387–393.

Tchernavin VV (1953) The feeding mechanisms of a deep-sea fish, *Chauliodus sloani* Schneider. *Br Mus Nat Hist (Zool)* 1953, 1–101.

Terminologia Anatomica (1998) Federative Committee on Anatomical Terminology. Georg Thieme, Stuttgart.

Thacher JK (1877) Median and paired fins, a contribution to the history of vertebrate limbs. *Trans Conn Acad* 3, 281–310.

Theis S, Patel K, Valasek P, Otto A, Pu Q, Harel I, Tzahor E, Tajbakhsh S, Christ B, Huang R (2010) The occipital lateral plate mesoderm is a novel source for vertebrate neck musculature. *Development* 137, 2961–2971.

Thewissen JGM, Babcock SK (1991) Distinctive cranial and cervical innervation of wing muscles: New evidence for bat monophyly. *Science* 251, 934–936.

Thewissen JGM, Babcock SK (1993) The implications of the propatagial muscles of flying and gliding mammals for archontan systematics. In *Primates and their relatives in phylogenetic perspective* (ed. MacPhee RDE), pp. 91–107. Plenum Press, New York.

Thompson P, Hillier WT (1905) The myology of the hind limb of the marsupial mole. *Notoryctes typhlops*. *J Anat Physiol* 39, 308–331.

Thorsen DH, Hale ME (2005) Development of zebrafish (*Danio rerio*) pectoral fin musculature. *J Morphol* 266, 241–255.

Thorsen DH, Westneat MW (2005) Diversity of pectoral fin structure and function in fishes with labriform propulsion. *J Morphol* 263, 133–150.

Tirosh-Finkel L, Elhanany H, Rinon A, Tzahor E (2006) Mesoderm progenitor cells of common origin contribute to the head musculature and the cardiac outflow tract. *Development* 133, 1943–1953.

Tokita M (2004) Morphogenesis of parrot jaw muscles: Understanding the development of an evolutionary novelty. *J Morphol* 259, 69–81.

Tokita M, Nakayama T, Schneider RA, Agata K (2012) Molecular and cellular changes associated with the evolution of novel jaw muscles in parrots. *Proc Biol Sci* 280, 20122319.

Tokita M, Schneider R (2009) Developmental origins of species-specific muscle pattern. *Dev Biol* 331, 311–325.

Tolley K, Herrel A (2013) An Introduction. In *The Biology of Chameleons*. University of California Press, Berkeley, CA.

Tolley KA, Townsend TM, Vences M (2013) Large-scale phylogeny of chameleons suggests African origins and Eocene diversification. *Proc R Soc Lond B Bio Sci* 280, 20130184.

Trainor PA, Melton KR, Manzanares M (2003) Origins and plasticity of neural crest cells and their roles in jaw and craniofacial evolution. *Int J Dev Biol* 47, 541–453.

Tretjakoff D (1926) Das Skelett und die Muskulatur im Kopfe des Flussneunauges. *Z Wiss Zool* 128, 267–304.

Tretjakoff D (1929) Die schleimknorpeligen Bestandteile im Kopfskelett von Ammocoetes. *Zeitschrift Wissen Zool* 133, 470–516.

Trinajstic K, Sanchez S, Dupret V, Tafforeau P, Long J, Young G, Senden T, Boisvert C, Power N, Ahlberg PE (2013) Fossil musculature of the most primitive jawed vertebrates. *Science* 341, 160–164.

Tsuihiji T (2007) Homologies of the longissimus, iliocostalis, and hypaxial muscles in the anterior presacral region of extant diapsida. *J Morphol* 268, 986–1020.

Tsukahara N, Kamata N, Nagasawa M, Sugita S (2009) Bilateral innervation of syringeal muscles by the hypoglossal nucleus in the jungle crow (*Corvus macrorhynchos*). *J Anat* 215, 141–149.

Tucker HM (1993) *The Larynx* (2nd ed). Thieme, New York.

Tulenko FJ, McCauley DW, MacKenzie EL, Mazan S, Kuratani S, Sugahara F, Kusakabe R, Burke AC (2013) Body wall development in lamprey and a new perspective on the origin of vertebrate paired fins. *PNAS* 110, 11899–11904.

Tzahor E (2009) Heart and craniofacial muscle development: A new developmental theme of distinct myogenic fields. *Dev Biol* 327, 273–279.

Vaglia J, Smith KK (2003) Early development of cranial neural crest in the marsupial (*Monodelphis domestica*). *Evol Dev* 5, 121–135.

Valasek P, Theis S, DeLaurier A, Hinits Y, Luke GN, Otto AM, Minchin J, He J, Christ B, Brooks G, Sang G, Evans DJ, Logan M, Huang R, Patel K (2011) Cellular and molecular investigations into the development of the pectoral girdle. *Dev Biol* 357, 108–116.

Valentine JW (2004) On the origin of phyla. University of Chicago Press, Chicago, IL.

Vanden Berge JC (1975) Aves myology. In *The Anatomy of the Domestic Animals (Vol. I)* (ed. Getty R), pp. 1802–1843. W.B. Saunders, Philadelphia, PA.

Vargas AO, Aboitiz F (2005) How ancient is the adult swimming capacity in the lineage leading to euchordates? *Evo Devo* 7, 171–174.

Vargas AO, Fallon JF (2005a) Birds have dinosaur wings, the molecular evidence. *J Exp Zool B Mol Dev Evol* 304, 86–90.

Vargas AO, Fallon JF (2005b) The digits of the wing of birds are 1, 2, and 3, a review. *J Exp Zool B Mol Dev Evol* 304, 206–219.

Vargas AO, Kohlsdorf T, Fallon JF, Vandenbrooks J, Wagner GP (2008) The evolution of HoxD-11 expression in the bird wing: Insights from *Alligator mississippiensis*. *PLoS ONE* 3, e3325.

Vaughn PP (1956) The phylogenetic migrations of the ambiens muscle. *J Elisha Mitchell Sci Soc* 72, 243–262.

Veilleux CC, Kirk EC (2014) Visual acuity in mammals: Effects of eye size and ecology. *Brain Behav Evol* 83, 43–53.

Veldman MB, Lin S (2008) Zebrafish as a developmental model organism for pediatric research. *Pediatr Res* 64, 470–476.

Vermeij GJ (1973) Biological versatility and earth history. *Proc Nat Acad Sci* 70, 1936–1938.

Versluys J (1904) Entwicklung der Columella auris bei den Lacertiliern. *Zool Jahrb Abt Anat Ontog Tiere* 19, 108–188.

Vetter B (1874) Untersuchungen zur vergleichenden Anatomie der Kiemen und Kiefermusculature der Fische. *Jenaische Zeitschriften für Naturwissenshaften* XII.

Vick S-JJ, Waller BM, Parr LA, Smith Pasqualini MC, Bard KA (2007) A cross-species comparison of facial morphology and movement in humans and chimpanzees using the facial action coding system (FACS). *J Nonverbal Behav* 31, 1–20.

Vicq-d'Azyr F (1774) Parallèle des os qui composent les extrémités. *Mém Acad Sci* 1774, 519–557.

Vrba ES (1968) Contributions to the functional morphology of fishes - part V, the feeding mechanism of *Elops saurus* Linnaeus. *Zool Afr* 3, 211–236.

Wagner GP (1994) Homology and the mechanisms of development. In: *Homology: The hierarchical basis of comparative biology* (ed. Hall BK), pp. 273–299. Academic Press, San Diego, CA.

Wagner GP (2014) Homology, genes, and evolutionary innovation. Princeton University Press, Princeton, NJ.

Wagner GP, Altenberg L (1996) Perspective: Complex adaptations and the evolution of evolvability. *Evolution* 50, 967–976.

Wagner GP, Chiu CG (2002) Genetic and epigenetic factors in the origin of tetrapod limbs. In: *Origination of Organismal Form* (eds. Muller GB, Newman S). MIT Press, Cambridge, MA.

Wagner GP, Chiu CH (2001) The tetrapod limb: A hypothesis on its origin. *J Exp Zool Mol Dev Evol* 291, 226–40.

Waitt C, Little AC, Wolfensohn S, Honess P, Brown AP, Buchanan-Smith HM, Perrett DI (2003) Evidence from rhesus macaques suggests that male coloration plays a role in female primate mate choice. *Proc R Soc Lond B Biol Sci* 270, 144–146.

Wakamatsu Y, Nomura T, Osumi N, Suzuki K (2014) Comparative gene expression analyses reveal heterochrony for Sox9 expression in the cranial neural crest during marsupial development. *Evol Dev* 16, 197–206.

Walker J, Westneat M (1997) Labriform propulsion in fishes: Kinematics of flapping aquatic flight in the bird wrasse *Gomphosus varius* (Labridae). *J Exp Biol* 200, 1549–1569.

Walker WF (1947) The development of the shoulder region of the turtle, *Chrysemys picta marginata*, with special reference to the primary musculature. *J Morphol* 80, 195–249.

Walker WF (1954) *Vertebrate dissection*. W. B. Saunders Company, London.

Walker WF (1973) The locomotor apparatus of testudines. In *Biology of the Reptilia, Vol. 14* (eds. Gans C, Parsons TS), pp. 1–100. Academic Press, New York.

Walker WF, Homberger DG (1997) *Anatomy and Dissection of the Rat* (3rd ed). Freeman Company, New York.

Walthall JC, Ashley-Ross MA (2006) Postcranial myology of the California newt, *Taricha torosa*. *Anat Rec A* 288, 46–57.

Walther WG (1922) *Die Neu-Guinea-Schieldkröte Carettochelys insculpta Ramsay*. PhD Thesis, University of Leiden, Leiden.

Warburton NM (2003) *Functional morphology and evolution of marsupial moles (Marsupialia: Notoryctemorphia)*. PhD thesis, University of Western Australia, Perth.

Warburton NM, Marchal CR (2017) Forelimb myology of carnivorous marsupials (Marsupialia: Dasyuridae): Implications for the ancestral body plan of the Australidelphia. *Anat Rec* 300, 1589–1608.

Webb GJW, Gans C (1982) Galloping in *Crocodylus johnstoni*—A reflection of terrestrial activity? *Rec Aust Mus* 34, 607–618.

Webb PW (1973) Kinematics of pectoral fin propulsion in *Cymatogaster aggregata*. *J Exp Biol* 59, 697–710.

Weisz PB (1945a) The development and morphology of the larva of the South African clawed toad, *Xenopus laevis*, I, the third-form tadpole. *J Morphol* 77, 163–192.

Weisz PB (1945b) The development and morphology of the larva of the South African clawed toad, *Xenopus laevis*, II, the hatching and the first- and second- form tadpoles. *J Morphol* 77, 193–217.

Wells LH, Thomas EA (1927) A note on two abnormal laryngeal muscles in a Zulu. *J Anat* 61, 340–343.

Wendling O, Dennefeld C, Chambon P, Mark M (2000) Retinoid signaling is essential for patterning the endoderm of the third and fourth pharyngeal arches. *Development*, 127 1553–1562.

West-Eberhard MJ (2003) *Developmental plasticity and evolution*. Oxford University Press, New York.

Westneat MW, Thorsen DH, Walker JA, Hale ME (2004) Structure, function, and neural control of pectoral fins in fishes. *IEEE J Oceanic Engen* 29, 674–683.

Whidden HP (2000) Comparative myology of moles and the phylogeny of the Talpidae (Mammalia, Lipotyphla). *Am Museum Novitates* 3294, 1–53.

Wible JR (2009) The ear region of the pen-tailed treeshrew, *Ptilocercus lowii* Gray, 1848 (Placentalia, Scandentia, Ptilocercidae). *J Mammal Evol* 16, 199–233.

Wicht H, Northcutt RG (1995) Ontogeny of the head of the Pacific hagfish (*Eptatretus stouti*, Myxinoidea): Development of the lateral line system. *Philos Trans R Soc Lond B Biol Sci* 349, 119–134.

Wiley EO (1979a) Ventral gill arch muscles and the interrelationships of gnathostomes, with a new classification of the Vertebrata. *J Linn Soc Lond (Zool)* 67, 149–179.

Wiley EO (1979b) Ventral gill arch muscles and the phylogenetic interrelationships of *Latimeria*. *Occ Pap Calif Acad Sci* 134, 56–67.

Wilga CD, Wainwright PC, Motta PJ (2000) Evolution of jaw depression mechanisms in aquatic vertebrates: Insights from Chondrichthyes. *Biol J Linn Soc* 71, 165–185.

Wilhelm BC, Du TY, Standen EM, Larsson HCE (2015) Polypterus and the evolution of fish pectoral musculature. *J Anat* 226, 511–522.

Wilkinson M, Nussbaum RA (1997) Comparative morphology and evolution of the lungless caecilian *Atretochoana eiselti* (Taylor) (Amphibia: Gymnophiona: Typhlonectidae). *Biol J Linn Soc* 62, 39–109.

Willey A (1894) *Amphioxus and the ancestry of the vertebrates*. MacMillan and Co, New York.

Willmer P (2003) Convergence and homoplasy in the evolution of organismal form. In: *Origination of organismal form - beyond the gene in developmental and evolutionary biology* (eds. Muller G, Newman S). MIT Press, Cambridge, MA.

Wilson-Pauwels L, Akesson EJ, Stewart PA, Spacey SD (2002) *Cranial Nerves in Health and Disease.* B.C. Decker, Hamilton, ON.

Wind J (1970) On the phylogeny and ontogeny of the human larynx. Wolters-Noordhoff, Copenhagen.

Windle BCA (1889) The flexors of the digits of the hand, I, The muscular masses in the fore-arm. *J Anat Physiol* 24, 72–84.

Windle BCA, Parsons FG (1897) On the myology of the terrestrial carnivora, muscles of the head, neck, and fore-limb. *Proc Zool Soc Lon* 65, 370–409.

Windner SE, Steinbacher P, Obermayer A, Kasiba B, Zweimueller-Mayer J, Stoiber W (2011) Distinct modes of vertebrate hypaxial muscle formation contribute to the teleost body wall musculature. *Dev Genes Evol* 221, 167–178.

Winder SJ, Lipscomb L, Angela Parkin C, Juusola M (2011) The proteasomal inhibitor MG132 prevents muscular dystrophy in zebrafish. *PLoS Curr* 3, RRN1286 10.1371.

Winterbottom R (1974) A descriptive synonymy of the striated muscles of the Teleostei. *Proc Acad Nat Sci (Phil)* 125, 225–317.

Winterbottom R (1993) Myological evidence for the phylogeny of recent genera of surgeonfishes (Percomorpha, Acanthuridae), with comments on Acanthuroidei. *Copeia* 1993, 21–39.

Witmer LM (1995a) The extant phylogenetic bracket and the importance of reconstructing soft tissues in fossils. In *Functional Morphology in Vertebrate Paleontology* (ed. Thomason JJ), pp. 19–33. Cambridge University Press, New York.

Witmer LM (1995b) Homology of facial structures in extant archosaurs (birds and crocodilians), with special reference to paranasal pneumaticity and nasal conchae. *J Morphol* 225, 269–327.

Witmer LM (1997) The evolution of the antorbital cavity of archosaurs: A study in soft-tissue reconstruction in the fossil record with an analysis of the function of pneumaticity. *J Ver Paleontol*, suppl. 1, 17, 1–73.

Woltering JM, Duboule D (2010) The origin of digits: Expression patterns versus regulatory mechanisms. *Dev Cell* 18, 526–32.

Wood J (1866) Variations in human myology observed during the Winter Session of 1865–6 at King's College, London. *Proc R Soc Lond* 15, 229–244.

Wood J (1867a) On human muscular variations and their relation to comparative anatomy. *J Anat Physiol* 1867, 44–59.

Wood J (1867b) Variations in human myology observed during the Winter Session of 1866–7 at King's College London. *Proc R Soc Lond* 15, 518–545.

Wood J (1868) Variations in human myology observed during the winter session of 1867–68 at King's College, London. *R Soc Lond* 16, 483–525.

Wood J (1870) On a group of varieties of the muscles of the human neck, shoulder, and chest, with their transitional forms and homologies in the Mammalia. *PhilosTrans R Soc Lond* 160, 83–116.

Wu X-C (2003) Functional morphology of the temporal region in Rhyncocephalia. *Can J Earth Sci* 40, 589–607.

Wyneken J (2001) The Anatomy of Sea Turtles. NOAA Technical Memorandum NMFS-SEFSC-470. US Department of Commerce, Miami, FL.

Yalden DW (1985) Feeding mechanisms as evidence for cyclostome monophyly. *Zool J Linn Soc* 84, 291–300.

Yamane A (2005) Embryonic and postnatal development of masticatory and tongue muscles. *Cell Tissue Res* 322, 183–189.

Yano T, Abe G, Yokoyama H, Kawakami K, Tamura K (2012) Mechanism of pectoral fin outgrowth in zebrafish development. *Development* 139, 4291–4291.

Yasui K, Kaji T, Morov AR, Yonemura S (2013) Development of oral and branchial muscles in lancelet larvae of *Branchiostoma japonicum*. *J Morphol* 275, 465–477.

Yoshida K, Nakahata A, Treen N, Sakuma T, Yamamoro T, Sakakura Y (2017) Hox-mediated endodermal identity patterns the pharyngeal muscle formation in the chordate pharynx. *Development* 144, 1629–1634.

Young AH (1880) Intrinsic Muscles of the Marsupial Hand. *J Anat Physiol* 14, 149–165.

Young GC, Barwick RE, Campbell KSW (1989) Pelvic girdles of Lungfish (Dipnoi). In: *Pathways in Geology: Essays in Honour of Edwin Sherbon Hills* (ed. LeMaitre RW), pp. 59–75. Blackwell Scientific, Melbourne.

Youson JH (1980) Morphology and physiology of lamprey metamorphosis. *Can J Fisheries Aquatic Sci* 37, 1687–1710.

Youson JH (1997) Is lamprey metamorphosis regulated by thyroid hormones? *Amer Zool* 37, 441–460.

Zhang J, Wagh P, Guay D, Sanchez-Pulido L, Padhi BK, Korzh V, Andrade-Navarro MA, Akimenko M-A (2010) Loss of fish actinotrichia proteins and the fin-to-limb transition. *Nature* 466, 234–237.

Zheng Y, Wiens JJ (2016) Combining phylogenomic and supermatrix approaches, and a time-calibrated phylogeny for squamate reptiles (lizards and snakes) based on 52 genes and 4162 species. *Mol Phylogenet Evol* 94, 537–547.

Zhu M, Yu X, Ahlberg PE, Choo B, Lu J, Qiao T, Qu Q, Zhao W, Jia L, Blom H, Zhu YA (2013) A Silurian placoderm with osteichthyan-like marginal jaw bones. *Nature* 502, 188–193.

Zhu M, Yu X, Choo B, Wang J, Jia L (2012) An antiarch placoderm shows that pelvic girdles arose at the root of jawed vertebrates. *Biol Lett* 23, 453–456.

Ziermann JM (2008) *Evolutionäre Entwicklung larvaler Cranialmuskulatur der Anura und der Einfluss von Sequenzheterochronien.* Unpublished PhD, Friedrich Schiller University, Jena.

Ziermann JM, Diogo R (2013) Cranial muscle development in the model organism *Ambystoma mexicanum*: Implications for tetrapod and vertebrate comparative and evolutionary morphology and notes on ontogeny and phylogeny. *Anat Rec* 296, 1031–1048.

Ziermann JM, Diogo R (2014) Cranial muscle development in frogs with different developmental modes: Direct development vs. biphasic development. *J Morphol* 275, 398–413.

Ziermann JM, Diogo R (in press) Development of head muscles in fishes and notes on phylogeny-ontogeny links: A basis for evo-devo and developmental research on fish muscles. In: *Evolution and Development of Fishes* (eds. Johanson Z, Richter M, Underwood C). Cambridge University Press, Cambridge, MA.

Ziermann JM, Freitas R, Diogo R (2017) Muscle development in the shark *Scyliorhinus canicula*: Implications for the evolution of the gnathostome head and paired appendage musculature. *Frontiers Zool* 14, 31.

Ziermann JM, Infante C, Hanken J, Olsson L (2013) Morphology of the cranial skeleton and musculature in the obligate carnivorous tadpole of *Lepidobatrachus laevis* (Anura: Ceratophryidae). *Acta Zoologica (Stockholm)* 94, 101112.

Ziermann JM, Miyashita T, Diogo R (2014) Cephalic muscles of Cyclostomes (hagfishes and lampreys) and Chondrichthyes (sharks, rays, and holocephalans): Comparative anatomy and early evolution of the vertebrate head muscles. *Zool J Linn Soc* 172, 771–802.

Ziermann JM, Olsson L (2007) Patterns of spatial and temporal cranial muscle development in the African clawed frog, *Xenopus laevis* (Anura: Pipidae). *J Morphol* 268, 791–804.

Zinck JM, Duffield DA, Ormsbee PC (2004) Primers for identification and polymorphism assessment of Vespertilionid bats in the Pacific Northwest. *Mol Ecol Notes* 4, 239–242.

Index

Page numbers followed by f and t indicate figures and tables, respectively.

B

D

E

F

G

H

I

J

K

L

M

N

O

P

Q

R

T

U

V

X

Z